Novel Drug Delivery Systems

DRUGS AND THE PHARMACEUTICAL SCIENCES

A Series of Textbooks and Monographs

edited by

James Swarbrick
School of Pharmacy
University of North Carolina
Chapel Hill, North Carolina

Volume 1. PHARMACOKINETICS, *Milo Gibaldi and Donald Perrier*

Volume 2. GOOD MANUFACTURING PRACTICES FOR PHARMACEUTICALS: A PLAN FOR TOTAL QUALITY CONTROL, *Sidney H. Willig, Murray M. Tuckerman, and William S. Hitchings IV*

Volume 3. MICROENCAPSULATION, *edited by J. R. Nixon*

Volume 4. DRUG METABOLISM: CHEMICAL AND BIOCHEMICAL ASPECTS, *Bernard Testa and Peter Jenner*

Volume 5. NEW DRUGS: DISCOVERY AND DEVELOPMENT, *edited by Alan A. Rubin*

Volume 6. SUSTAINED AND CONTROLLED RELEASE DRUG DELIVERY SYSTEMS, *edited by Joseph R. Robinson*

Volume 7. MODERN PHARMACEUTICS, *edited by Gilbert S. Banker and Christopher T. Rhodes*

Volume 8. PRESCRIPTION DRUGS IN SHORT SUPPLY: CASE HISTORIES, *Michael A. Schwartz*

Volume 9. ACTIVATED CHARCOAL: ANTIDOTAL AND OTHER MEDICAL USES, *David O. Cooney*

Volume 10. CONCEPTS IN DRUG METABOLISM (in two parts), *edited by Peter Jenner and Bernard Testa*

Volume 11. PHARMACEUTICAL ANALYSIS: MODERN METHODS (in two parts), *edited by James W. Munson*

Volume 12. TECHNIQUES OF SOLUBILIZATION OF DRUGS, *edited by Samuel H. Yalkowsky*

Volume 13. ORPHAN DRUGS, *edited by Fred E. Karch*

Volume 14. NOVEL DRUG DELIVERY SYSTEMS: FUNDAMENTALS, DEVELOPMENTAL CONCEPTS, BIOMEDICAL ASSESSMENTS, *Yie W. Chien*

Volume 15. PHARMACOKINETICS, SECOND EDITION, REVISED AND EXPANDED, *Milo Gibaldi and Donald Perrier*

Volume 16. GOOD MANUFACTURING PRACTICES FOR PHARMACEUTICALS: A PLAN FOR TOTAL QUALITY CONTROL, SECOND EDITION, REVISED AND EXPANDED, *Sidney H. Willig, Murray M. Tuckerman, and William S. Hitchings IV*

Volume 17. FORMULATION OF VETERINARY DOSAGE FORMS, *edited by Jack Blodinger*

Volume 18. DERMATOLOGICAL FORMULATIONS: PERCUTANEOUS ABSORPTION, *Brian W. Barry*

Volume 19. THE CLINICAL RESEARCH PROCESS IN THE PHARMACEUTICAL INDUSTRY, *edited by Gary M. Matoren*

Volume 20. MICROENCAPSULATION AND RELATED DRUG PROCESSES, *Patrick B. Deasy*

Volume 21. DRUGS AND NUTRIENTS: THE INTERACTIVE EFFECTS, *edited by Daphne A. Roe and T. Colin Campbell*

Volume 22. BIOTECHNOLOGY OF INDUSTRIAL ANTIBIOTICS, *Erick J. Vandamme*

Volume 23. PHARMACEUTICAL PROCESS VALIDATION, *edited by Bernard T. Loftus and Robert A. Nash*

Volume 24. ANTICANCER AND INTERFERON AGENTS: SYNTHESIS AND PROPERTIES, *edited by Raphael M. Ottenbrite and George B. Butler*

Volume 25. PHARMACEUTICAL STATISTICS: PRACTICAL AND CLINICAL APPLICATIONS, *Sanford Bolton*

Volume 26. DRUG DYNAMICS FOR ANALYTICAL, CLINICAL, AND BIOLOGICAL CHEMISTS, *Benjamin J. Gudzinowicz, Burrows T. Younkin, Jr., and Michael J. Gudzinowicz*

Volume 27. MODERN ANALYSIS OF ANTIBIOTICS, *edited by Adjoran Aszalos*

Volume 28. SOLUBILITY AND RELATED PROPERTIES, *Kenneth C. James*

Volume 29. CONTROLLED DRUG DELIVERY: FUNDAMENTALS AND APPLICATIONS, SECOND EDITION, REVISED AND EXPANDED, *edited by Joseph R. Robinson and Vincent H. Lee*

Volume 30. NEW DRUG APPROVAL PROCESS: CLINICAL AND REGULATORY MANAGEMENT, *edited by Richard A. Guarino*

Volume 31. TRANSDERMAL CONTROLLED SYSTEMIC MEDICATIONS, *edited by Yie W. Chien*

Volume 32. DRUG DELIVERY DEVICES: FUNDAMENTALS AND APPLICATIONS, *edited by Praveen Tyle*

Volume 33. PHARMACOKINETICS: REGULATORY - INDUSTRIAL - ACADEMIC PERSPECTIVES, *edited by Peter G. Welling and Francis L. S. Tse*

Volume 34. CLINICAL DRUG TRIALS AND TRIBULATIONS, *edited by Allen E. Cato*

Volume 35. TRANSDERMAL DRUG DELIVERY: DEVELOPMENTAL ISSUES AND RESEARCH INITIATIVES, *edited by Jonathan Hadgraft and Richard H. Guy*

Volume 36. AQUEOUS POLYMERIC COATINGS FOR PHARMACEUTICAL DOSAGE FORMS, *edited by James W. McGinity*

Volume 37. PHARMACEUTICAL PELLETIZATION TECHNOLOGY, *edited by Isaac Ghebre-Sellassie*

Volume 38. GOOD LABORATORY PRACTICE REGULATIONS, *edited by Allen F. Hirsch*

Volume 39. NASAL SYSTEMIC DRUG DELIVERY, *Yie W. Chien, Kenneth S. E. Su, and Shyi-Feu Chang*

Volume 40. MODERN PHARMACEUTICS, SECOND EDITION, REVISED AND EXPANDED, *edited by Gilbert S. Banker and Christopher T. Rhodes*

Volume 41. SPECIALIZED DRUG DELIVERY SYSTEMS: MANUFACTURING AND PRODUCTION TECHNOLOGY, *edited by Praveen Tyle*

Volume 42. TOPICAL DRUG DELIVERY FORMULATIONS, *edited by David W. Osborne and Anton H. Amann*

Volume 43. DRUG STABILITY: PRINCIPLES AND PRACTICES, *Jens T. Carstensen*

Volume 44. PHARMACEUTICAL STATISTICS: PRACTICAL AND CLINICAL APPLICATIONS, SECOND EDITION, REVISED AND EXPANDED, *Sanford Bolton*

Volume 45. BIODEGRADABLE POLYMERS AS DRUG DELIVERY SYSTEMS, *edited by Mark Chasin and Robert Langer*

Volume 46. PRECLINICAL DRUG DISPOSITION: A LABORATORY HANDBOOK, *Francis L. S. Tse and James J. Jaffe*

Volume 47. HPLC IN THE PHARMACEUTICAL INDUSTRY, *edited by Godwin W. Fong and Stanley K. Lam*

Volume 48. PHARMACEUTICAL BIOEQUIVALENCE, *edited by Peter G. Welling, Francis L. S. Tse, and Shrikant V. Dighe*

Volume 49. PHARMACEUTICAL DISSOLUTION TESTING, *Umesh V. Banakar*

Volume 50. NOVEL DRUG DELIVERY SYSTEMS, SECOND EDITION, REVISED AND EXPANDED, *Yie W. Chien*

Volume 51. MANAGING THE CLINICAL DRUG DEVELOPMENT PROCESS, *David M. Cocchetto and Ronald V. Nardi*

Volume 52. GOOD MANUFACTURING PRACTICES FOR PHARMACEUTICALS: A PLAN FOR TOTAL QUALITY CONTROL, THIRD EDITION, *edited by Sidney H. Willig and James Stoker*

Additional Volumes in Preparation

PRODRUGS: TOPICAL AND OCULAR DRUG DELIVERY, *edited by Kenneth B. Sloan*

Novel Drug Delivery Systems

Second Edition, Revised and Expanded

Yie W. Chien

Controlled Drug-Delivery Research Center
College of Pharmacy
Rutgers University
Piscataway, New Jersey

CRC Press is an imprint of the
Taylor & Francis Group, an **informa** business

CRC Press
Taylor & Francis Group
6000 Broken Sound Parkway NW, Suite 300
Boca Raton, FL 33487-2742

First issued in paperback 2019

ISBN-13: 978-0-8247-8520-8 (hbk)
ISBN-13: 978-0-367-40291-4 (pbk)

Visit the Taylor & Francis Web site at
http://www.taylorandfrancis.com

and the CRC Press Web site at
http://www.crcpress.com

To Margaret, my wife, for
her understanding and encouragement

To Linda, my daughter, for
her efforts and support

Preface

This second edition of *Novel Drug Delivery Systems* has maintained the same objective as that of the first edition: to present a comprehensive, coherent treatment of the science, technology, and regulation of rate-controlled administration of therapeutic agents, with comprehensive coverage of the basic concepts, fundamental principles, biomedical rationales, and potential applications.

Since the successful introduction of the first edition in 1982, much progress has been made in the science and technology of rate-controlled drug administration. It was my intention to rewrite this well-received book with incorporation of all the important scientific discoveries in recent years into the second edition. All the chapters have been extensively rewritten and updated, and new chapters have been added on the rate-controlled delivery of drugs across various mucosae and digestive tract membranes. A special chapter has been added on the issues associated with the systemic delivery of peptide/protein drugs, a new generation of therapeutic agents, via parenteral and nonparenteral routes of administration.

Over the years, the responses we have received from the readers have been very encouraging and have demonstrated to us that the book has been successful in providing a useful source of scientific information for biomedical researchers and pharmaceutical R&D scientists/managers with diverse backgrounds who need to acquire the core knowledge critical to the conceptualization, development, and optimization of rate-controlled drug delivery. In addition, I also received several constructive comments and suggestions that I have implemented in writing this new edition to further enhance the quality of the book. One example of such changes is that the chapter on the fundamental aspects of rate-controlled drug delivery has been moved to the beginning of the book, preceding the chapters on rate-controlled drug delivery through various routes of administration. Also, the chapter on regulatory considerations in controlled drug delivery has been expanded to incorporate new guidelines used in the regulatory approval process.

Overall, the objective set for writing this new edition has been the same as for the first edition, that is, to provide the readers with a broad spectrum of scientific information in a concise, systematic manner, but with new vision and wider scope.

Yie W. Chien

Contents

Novel Drug Delivery Systems

1
Concepts and System Design for Rate-Controlled Drug Delivery

I. INTRODUCTION

For many decades treatment of an acute disease or a chronic illness has been mostly accomplished by delivery of drugs to patients using various pharmaceutical dosage forms, including tablets, capsules, pills, suppositories, creams, ointments, liquids, aerosols, and injectables, as drug carriers. Even today these conventional drug delivery systems are the primary pharmaceutical products commonly seen in the prescription and over-the-counter drug marketplace. This type of drug delivery system is known to provide a prompt release of drug. Therefore, to achieve as well as to maintain the drug concentration within the therapeutically effective range needed for treatment, it is often necessary to take this type of drug delivery system several times a day. This results in a significant fluctuation in drug levels (Figure 1).

Recently, several technical advancements have been made. They have resulted in the development of new techniques for drug delivery. These techniques are capable of controlling the rate of drug delivery, sustaining the duration of therapeutic activity, and/or targeting the delivery of drug to a tissue (1–4). Although these advancements have led to the development of several novel drug delivery systems that could revolutionalize the method of medication and provide a number of therapeutic benefits, they also create some confusion in the terminology between "controlled release" and "sustained release." Unfortunately these terms have been often used interchangeably in the scientific literature and technical presentations over the years.

The term "sustained release" is known to have existed in the medical and pharmaceutical literature for many decades. It has been constantly used to describe a pharmaceutical dosage form formulated to retard the release of a therapeutic agent such that its appearance in the systemic circulation is delayed and/or prolonged and its plasma profile is sustained in duration. The onset of its pharmacologic action is often delayed, and the duration of its therapeutic effect is sustained.

The term "controlled release," on the other hand, has a meaning that goes beyond the scope of sustained drug action. It also implies a predictability and reproducibility in the drug release kinetics, which means that the release of drug ingre-

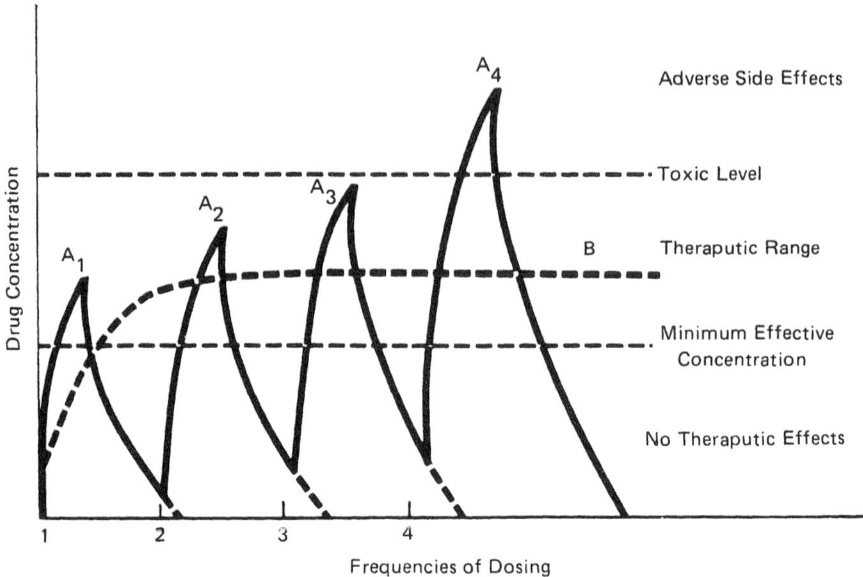

Figure 1 Hypothetical drug concentration profiles in the systemic circulation resulting from the consecutive administration of multiple doses of an immediate-release drug delivery system (A_1, A_2, . . .) compared to the ideal drug concentration profile (B) required for treatment. (Adapted from Reference 1.)

dient(s) from a controlled-release drug delivery system proceeds at a rate profile that is not only predictable kinetically, but also reproducible from one unit to another (1,4).

The difference between controlled-release and sustained-release drug delivery is illustrated in Figure 2, in which the plasma profiles of phenylpropanolamine in humans taking Acutrim, a controlled-release drug delivery system (which is prepared on the basis of osmotic pumping technology), and Dexatrim, a sustained-release drug delivery system (which is prepared on the basis of spansule coating technology), are compared. Dexatrim yields a sustained (but not constant) plasma profile with a duration longer than that achieved by solution formulation; Acutrim produces a constant steady-state plasma level with further prolongation in the duration.

II. CLASSIFICATION OF RATE-CONTROLLED DRUG DELIVERY SYSTEMS

Based on their technical sophistication controlled-release drug delivery systems that have recently been marketed or are under active development can be classified (Figure 3) as

1. Rate-preprogrammed drug delivery systems
2. Activation-modulated drug delivery systems
3. Feedback-regulated drug delivery systems
4. Site-targeting drug delivery systems

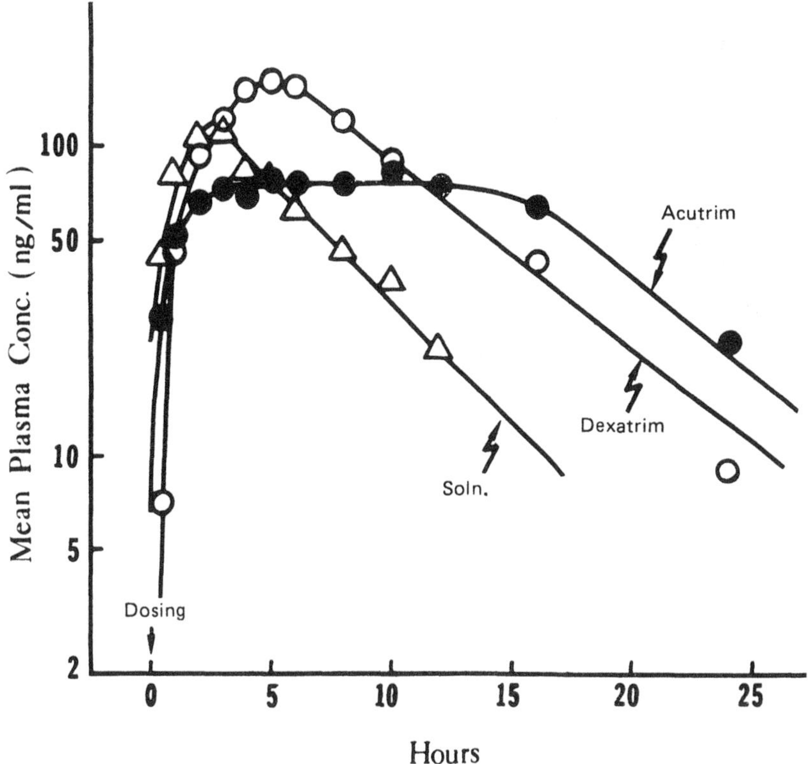

Figure 2 Comparative plasma profiles of phenylpropranolamine (PPA) in 18 healthy human volunteers resulted from oral administration of PPA in solution formulation and delivery by sustained-release Dexatrim or controlled-release Acutrim.

In this chapter, the fundamental concepts and technical principles behind the development of various controlled-release drug delivery systems in each category are outlined and discussed.

III. RATE-PREPROGRAMMED DRUG DELIVERY SYSTEMS

In this group of controlled-release drug delivery systems, the release of drug molecules from the delivery systems has been preprogrammed at specific rate profiles. This was accomplished by system design, which controls the molecular diffusion of drug molecules in and/or across the barrier medium within or surrounding the delivery system. Fick's laws of diffusion are often followed. These systems can be further classified as follows.

A. Polymer Membrane Permeation-Controlled Drug Delivery Systems

In this type of preprogrammed drug delivery systems, a drug formulation is totally or partially encapsulated within a drug reservoir compartment. Its drug release surface is covered by a rate-controlling polymeric membrane having a specific permeability. The drug reservoir may exist in solid, suspension, or solution form. The

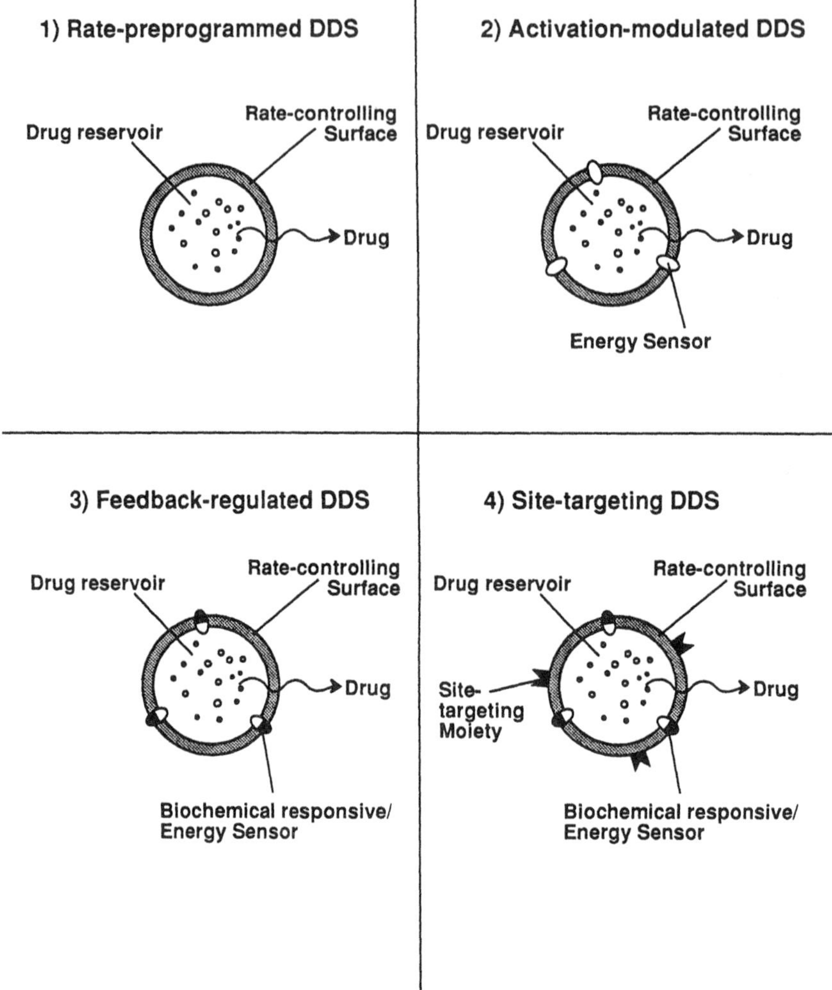

Figure 3 The four major classes of controlled-release drug delivery systems.

polymeric membrane can be fabricated from a nonporous (homogeneous or heterogeneous) polymeric material or a microporous (or semipermeable) membrane. The encapsulation of drug formulation inside the reservoir compartment is accomplished by injection molding, spray coating, capsulation, microencapsulation, or other techniques. Different shapes and sizes of drug delivery systems can be fabricated (Figure 4).

The rate of drug release Q/t from this polymer membrane permeation-controlled drug delivery system should be a constant value and is defined by

$$\frac{Q}{t} = \frac{K_{m/r}K_{a/m}D_dD_m}{K_{m/r}D_mh_d + K_{a/m}D_dh_m} C_R \tag{1}$$

where $K_{m/r}$ and $K_{a/m}$ are the partition coefficients for the interfacial partitioning of

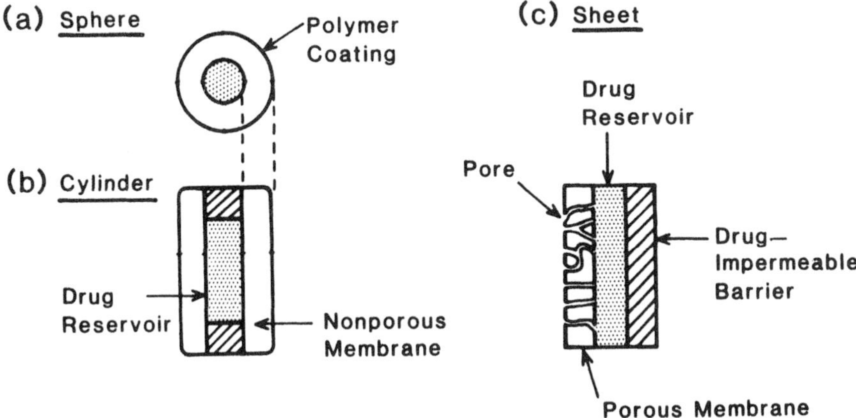

Figure 4 Various types of polymer membrane permeation-controlled drug delivery systems: (a) sphere, (b) cylinder, and (c) sheet. Both polymeric membrane, which is either porous or nonporous, and diffusion layer have a controlled thickness (h_m and h_d, respectively).

drug molecules from the reservoir to the rate-controlling membrane and from the membrane to the surrounding aqueous diffusion layer, respectively; and D_m and D_d are the diffusion coefficients in the rate-controlling membrane (with thickness h_m) and in the aqueous diffusion layer (with thickness h_d). For a microporous or semipermeable membrane, the porosity and tortuosity of the pores in the membrane should be included in the determination of D_m and h_m. C_R is the drug concentration in the reservoir compartment.

The release of drug molecules from this type of rate-controlled drug delivery systems is controlled at a preprogrammed rate by controlling the partition coefficient and diffusivity of the drug molecule and the thickness of the rate-controlling membrane. Representatives of this type of drug delivery system are as follows:

In the *Progestasert IUD*, an intrauterine device, the drug reservoir is a suspension of progesterone crystals in silicone medical fluid and is encapsulated in the vertical limb of a T-shaped device walled by a nonporous membrane of ethylene-vinyl acetate copolymer. It is engineered to deliver natural progesterone continuously in the uterine cavity at a daily dosage rate of at least 65 μg/day to achieve contraception for 1 year (Figure 5).

The *Norplant subdermal implant* is fabricated from nonporous silicone medical-grade tubing (with both ends sealed with silicone medical-grade adhesive) to encapsulate either levonorgestrel crystals alone (generation I) or a solid dispersion of levonorgestrel in silicone elastomer matrix (generation II). It is designed for the continuous subcutaneous release of levonorgestrel at a daily dosage rate of 30 μg, to each subject (6 units of I or 2 units of II) for up to 7 years (Figure 6) (5–8).

In the *Ocusert system* (an ocular insert) the drug reservoir is a thin disk of pilocarpine alginate complex sandwiched between two transparent sheets of microporous ethylene-vinyl acetate copolymer membrane (Figure 7). The microporous membranes permit the tear fluid to penetrate into the drug reservoir compartment to dissolve-pilocarpine from the complex. By system design, pilocarpine molecules are then

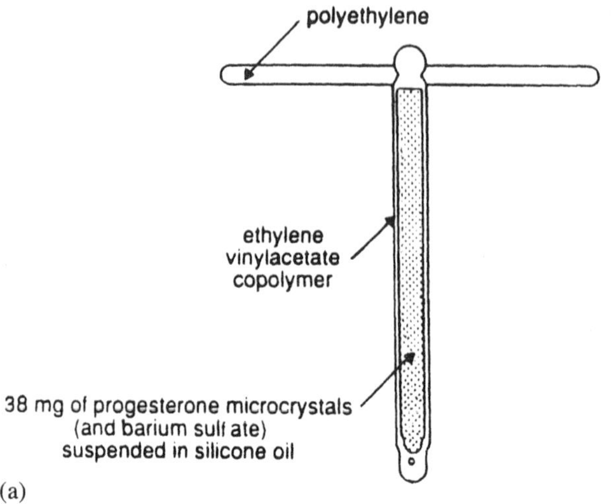

(a)

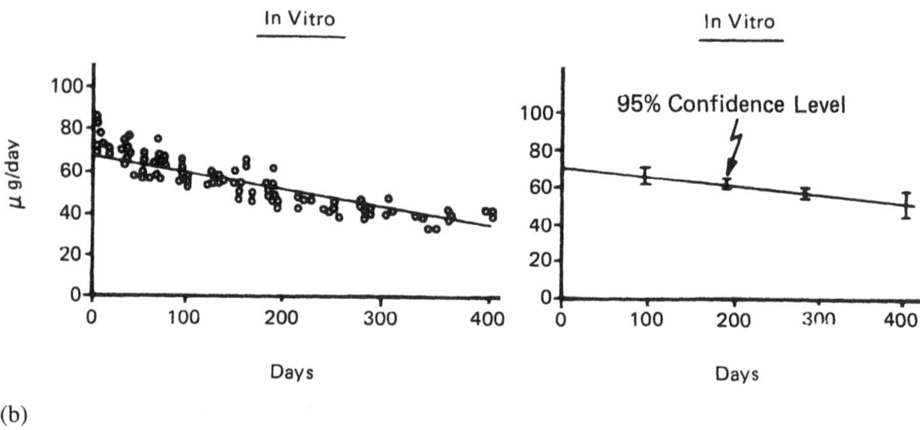

(b)

Figure 5 (a) Various structural components of Progestasert IUD. (b) The in vitro and in vivo release rate profiles of progesterone from the progestasert IUD for up to 400 days.

released at a constant rate of either 20 or 40 μg/hr for the management of glaucoma for up to 7 days (1,4,9).

Transderm-Nitro is a transdermal therapeutic system in which the drug reservoir, a dispersion of nitroglycerin-lactose triturate in the silicone medical fluid, is encapsulated in a thin ellipsoidal patch. The transdermal patch is constructed from a drug-impermeable metallic plastic laminate as the backing membrane and a constant surface of rate-controlling microporous membrane of ethylene-vinyl acetate copolymer as the drug-releasing surface (Figure 8). This device is fabricated by an injection molding process. Additionally, a thin layer of pressure-sensitive silicone adhesive polymer is coated on the surface of the rate-controlling membrane to achieve an

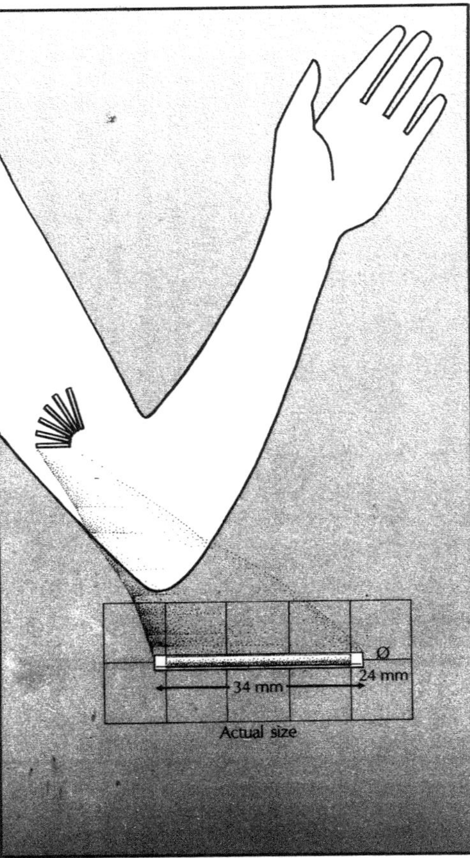

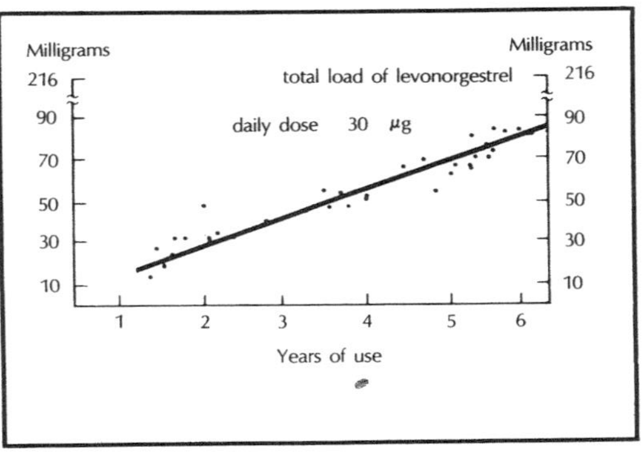

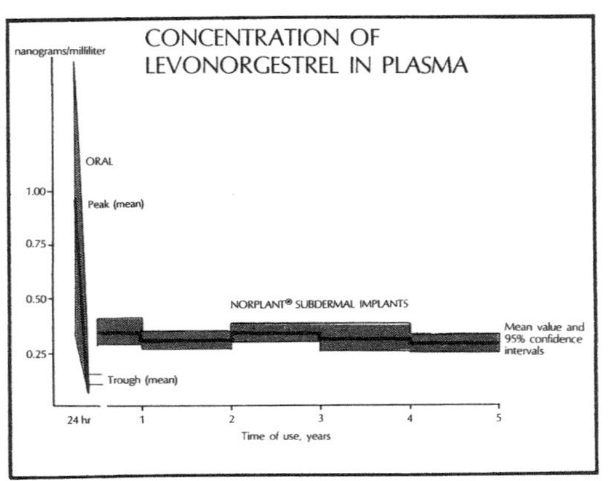

Figure 6 Subcutaneous implantation of Norplant subdermal implants in female volunteers for up to 6 years. The subcutaneous release profile of levonorgestrel and the resultant plasma profile are compared to those obtained by oral administration. (Adapted from References 5–8.)

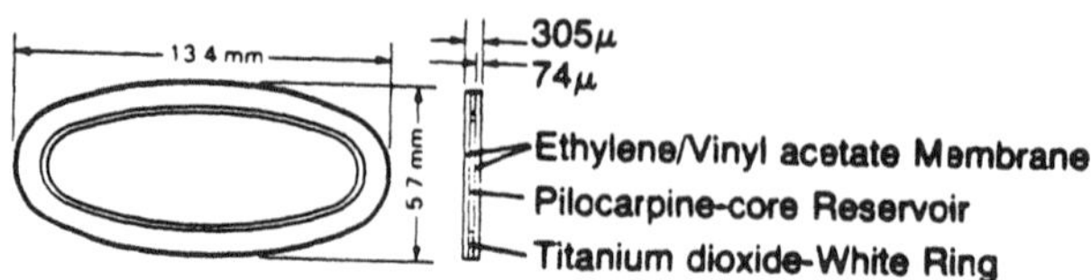

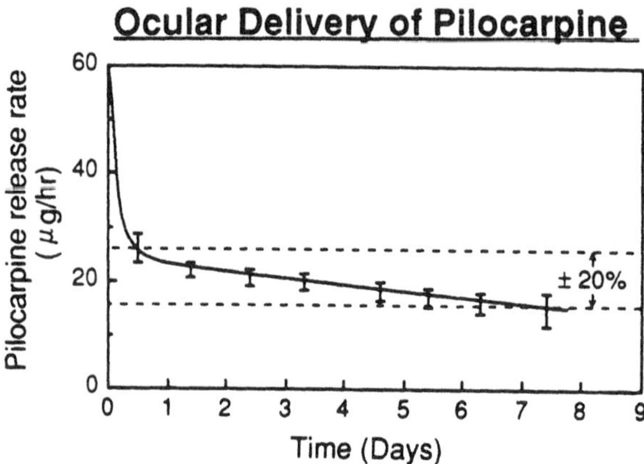

Figure 7 Various structural components of Ocusert and the ocular release rate profile of pilocarpine from Ocusert Pilo-20. (Replotted with modification from Reference 9.)

intimate contact of the drug release surface with the skin. It is engineered to deliver nitroglycerin at dosage rate of 0.5 $mg/cm^2/day$ for transdermal absorption to provide daily relief of anginal attacks (10,11). The same technology has also been recently utilized in the development of (1) Estraderm, which administers a controlled dose of estradiol transdermally over 3–4 days for the relief of postmenopausal syndrome (12,13); (2) Transderm-Scop, which provides a transdermal controlled administration of scopolamine for 72 hr prevention of motion-induced nausea (14,15), and (3) Catapres-TTS, which controls the transdermal permeation of clonidine for 7 days in the treatment of hypertension (16). Some variations have been made in the system design. In the Estraderm system, for example, estradiol is dissolved in ethanolic solution to form a solution drug reservoir, whereas with Transderm-Scop and Catapres-TTS the drug is directly dispersed in the adhesive polymer [e.g., poly(isobutylene) adhesive] to form a solid drug reservoir.

B. Polymer Matrix Diffusion-Controlled Drug Delivery Systems

In this type of preprogrammed drug delivery system the drug reservoir is prepared by homogeneously dispersing drug particles in a rate-controlling polymer matrix fabricated from either a lipophilic or a hydrophilic polymer. The drug dispersion in the polymer matrix is accomplished by either (1) blending a therapeutic dose of finely ground drug particles with a liquid polymer or a highly viscous base polymer, followed by cross-linking of the polymer chains, or (2) mixing drug solids with a rubbery polymer at an elevated temperature. The resultant drug-polymer dispersion is then molded or extruded to form a drug delivery device of various shapes and sizes designed for specific application (Figure 9). It can also be fabricated by dissolving the drug and the polymer in a common solvent, followed by solvent evaporation at an elevated temperature and/or under a vacuum.

The rate of drug release from this polymer matrix diffusion-controlled drug delivery system is time dependent and is defined at steady state by

$$\frac{Q}{t^{1/2}} = (2AC_R Dp)^{1/2} \tag{2}$$

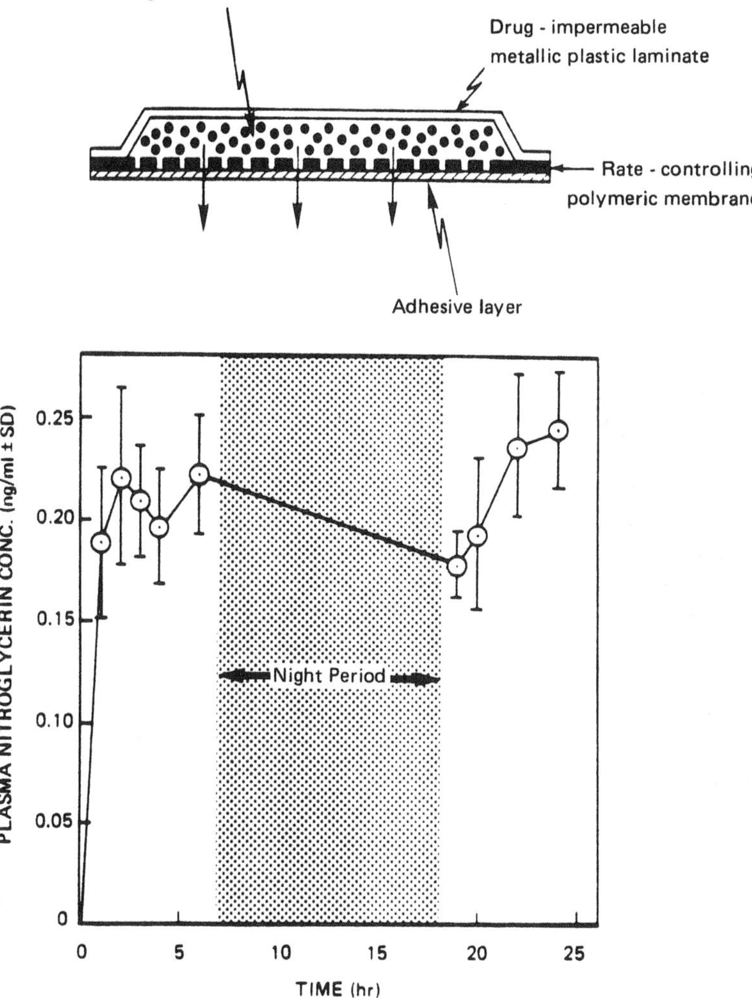

Figure 8 Cross-sectional view of Transderm-Nitro showing various structural components and the 24 hr plasma concentration profiles of nitroglycerin in 14 human volunteers, each receiving one unit of Transderm-Nitro (20 cm^2 with a daily dosage rate of 10 mg/day) for 1 day. (Based on References 9 and 10.)

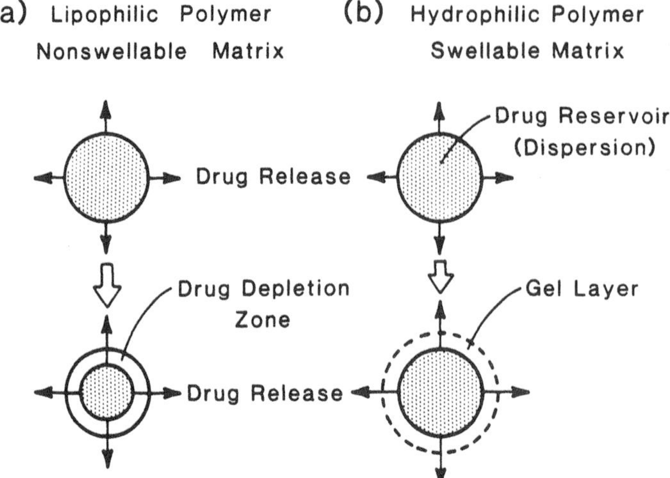

Figure 9 A polymer matrix diffusion-controlled drug delivery systems with a drug reservoir exists as a homogeneous dispersion in (a) a lipophilic, nonswellable polymer matrix with a growing thickness of the drug depletion zone or (b) a hydrophilic, swellable polymer matrix with a growing thickness of the drug-depleted gel layer, which resulted from drug release from the drug-dispersing polymer matrix.

where A is the initial drug loading dose in the polymer matrix; C_R is the drug solubility in the polymer, which is also the drug reservoir concentration in the system, and D_p is the diffusivity of the drug molecules in the polymer matrix.

The release of drug molecules from this type of controlled release drug delivery systems is controlled at a preprogrammed rate by controlling the loading dose, polymer solubility of the drug, and its diffusivity in the polymer matrix. Representatives of this type of drug delivery system are as follows:

Nitro-Dur is a transdermal drug delivery (TDD) system fabricated by first heating an aqueous solution of water-soluble polymer, glycerol, and polyvinyl alcohol. The temperature of the solution is then gradually lowered and nitroglycerin and lactose triturate are dispersed just above the congealing temperature of the solution. The mixture is solidified in a mold at or below room temperature and then sliced to form a medicated polymer disk. After assembly onto a drug-impermeable metallic plastic laminate, a patch-type TDD system is produced with an adhesive rim surrounding the medicated disk (Figure 10). It is designed for application onto intact skin for 24 hr to provide a continuous transdermal infusion of nitroglycerin at a dosage rate of 0.5 $mg/cm^2/day$ for the treatment of angina pectoris (10,17,18).

The drug reservoir can also be formulated by directly dispersing the drug in a pressure-sensitive adhesive polymer, e.g., poly(isobutylene)- or poly(acrylate)-based adhesive polymer, and then spreading the medicated adhesive polymer by solvent casting onto a flat sheet of drug-impermeable backing support to form a single layer or multiple layers of drug reservoir. This type of controlled-release TDD system is best illustrated by the development and marketing of an isosorbide dinitrate-dispersed poly(acrylate)-based TDD system in Japan (Frandol tape by Toaeiyo-Yamanouchi) and of nitroglycerin-dispersed poly(acrylate)-based TDD system in the United States

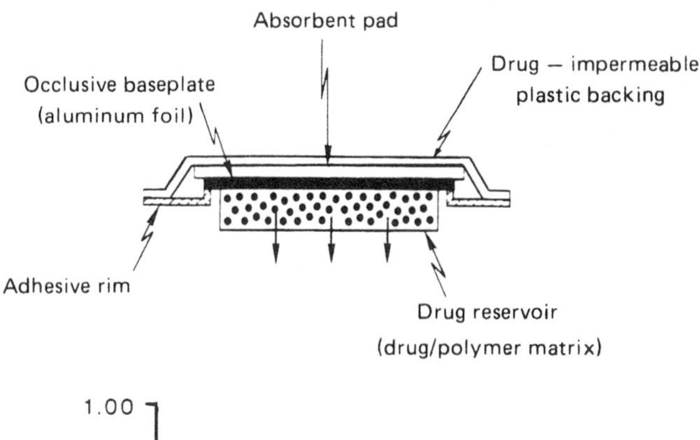

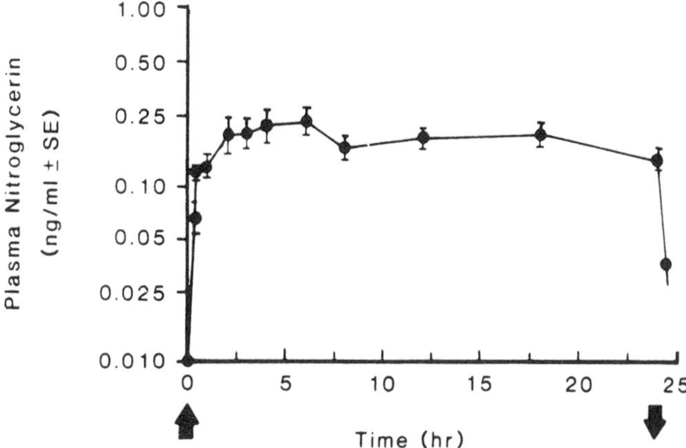

Figure 10 Cross-sectional view of Nitro-Dur showing various structural components and the 24 hr plasma nitroglycerin concentration profiles in 24 healthy male volunteers, each receiving randomly one unit of Nitro-Dur (20 cm^2 with a delivery rate of 10 mg/day) over the chest for 24 h. (Based on References 10 and 18.)

(Nitro-Dur II system by Key) for once-a-day medication of angina pectoris. This second generation of TDD system (NitroDur II) has recently received FDA approval for marketing. Nitro-Dur II compares favorably with Nitro-Dur (Figure 11) and has gradually replaced the first-generation Nitro-Dur in the marketplace.

The *Compudose subdermal implant* is fabricated by dispersing micronized estradiol crystals in a viscous silicone elastomer and then coating the estradiol-dispersing polymer around a rigid (drug-free) silicone rod by extrusion to form a cylindrical implant (Figure 12). This subdermal implant is developed for subcutaneous implantation in steers for growth promotion and to release a controlled dose of estradiol for as long as 200 or 400 days (19).

To improve the Q versus $t^{1/2}$ drug release profiles [Eq. (2)] this polymer matrix diffusion-controlled drug delivery system can be modified to have the drug loading level varied in an incremental manner to form a gradient of drug reservoir to compensate for the increase in diffusional path. The rate of drug release from this drug reservoir gradient-controlled drug delivery systems is defined by

$$\frac{dQ}{dt} = \frac{K_{a/r}D_a}{h_a(t)} C_p(h_a) \tag{3}$$

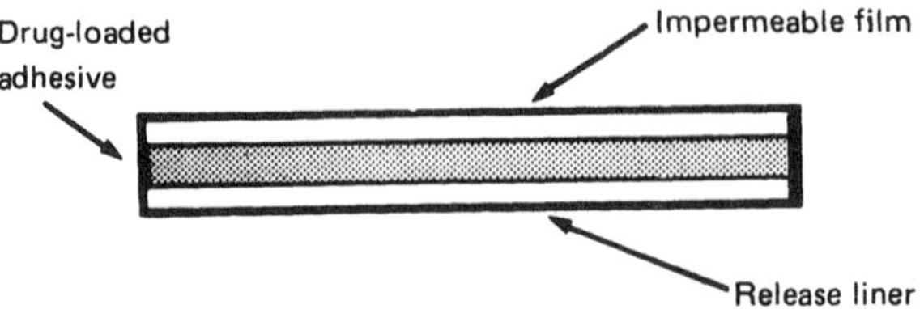

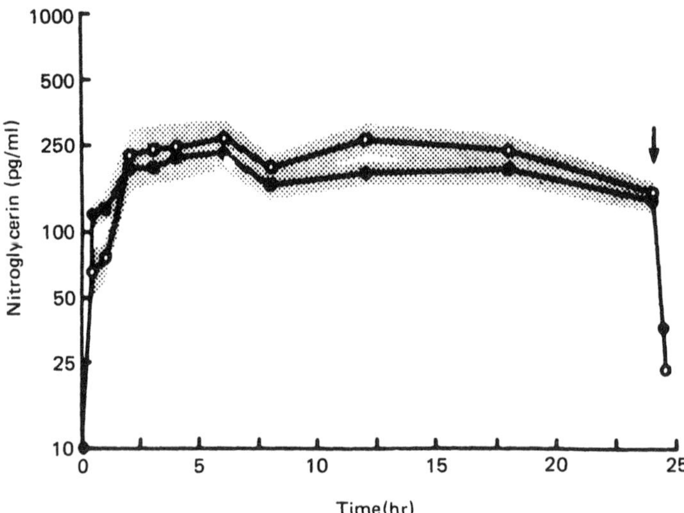

Figure 11 Cross-sectional view of Nitro-Dur II showing various structural components and the comparative 24 hr plasma nitroglycerin concentration profiles in 24 healthy male volunteers, each receiving randomly one unit of Nitro-Dur II (open circle) or Nitro-Dur (closed circle), 20 cm^2 each with a delivery rate of 10 mg/day, over the chest for 24 h (the arrow indicates unit removal). (Based on Reference 18.)

In Equation (3) the thickness of the diffusional path through which drug molecules diffuse increased with time ($h_a(t)$). To compensate, the loading level and/or the polymer solubility of the impregnated drug are increased in proportion ($C_p(h_a)$). A constant drug release profile is thus obtained. This type of controlled-release drug delivery system is best illustrated by the development of the nitroglycerin-releasing Deponit system (Figure 13) by Pharma-Schwartz/Lohmann in Europe (20). Wyeth-Ayerst has received FDA approval for marketing this system in the United States.

Furthermore, it was recently demonstrated that the release of a drug, such as propranolol, from the multilaminate adhesive-based TDD system can be maintained at zero-order kinetics by varying the particle size distribution of drug crystals in the various laminates of adhesive matrix (21).

C. Microreservoir Partition-Controlled Drug Delivery Systems

In this type of preprogrammed drug delivery systems the drug reservoir is fabricated by microdispersion of an aqueous suspension of drug using a high-energy dispersion

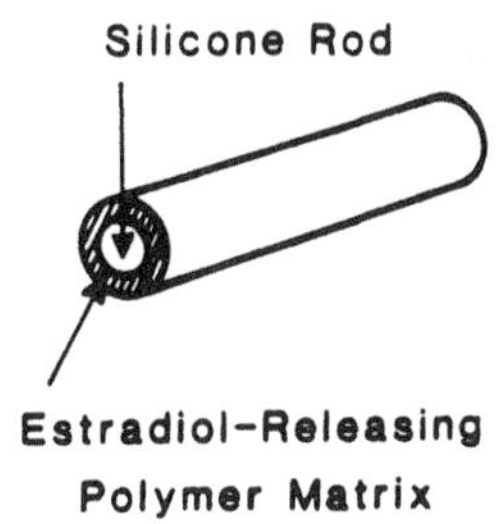

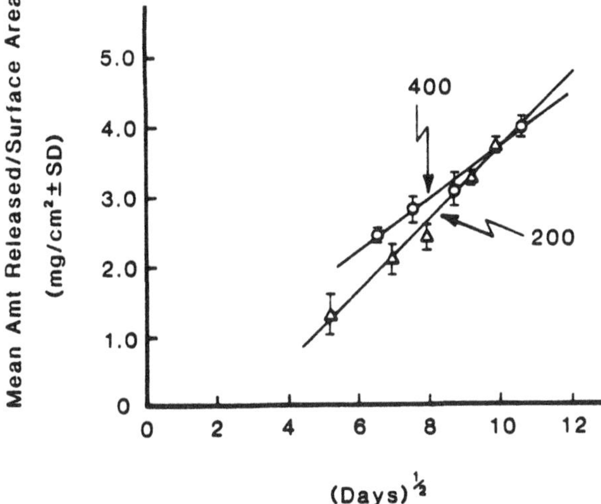

Figure 12 The Compudose subdermal implant and subcutaneous release profiles of estradiol from the implants in rats.

technique (22–24) in a biocompatible polymer, such as silicone elastomers, to form a homogeneous dispersion of many discrete, unleachable, microscopic drug reservoirs. Different shapes and sizes of drug delivery devices can be fabricated from this microreservoir drug delivery system by molding or extrusion. Depending upon the physicochemical properties of drugs and the desired rate of drug release, the device can be further coated with a layer of biocompatible polymer to modify the mechanism and the rate of drug release.

The rate of drug release dQ/dt from this type of drug delivery systems is defined by

$$\frac{dQ}{dt} = \frac{D_p D_d m K_p}{D_p h_d + D_d h_p m K_p} \left[n S_p - \frac{D_l S_l (1 - n)}{h_l} \left(\frac{1}{K_l} + \frac{1}{K_m} \right) \right] \tag{4}$$

where $m = a/b$ and n is the ratio of drug concentration at the inner edge of the interfacial barrier over the drug solubility in the polymer matrix (1). a is the ratio of drug concentration in the bulk of elution solution over drug solubility in the same medium, and b is the ratio of drug concentration at the outer edge of the polymer-coating membrane over drug solubility in the same polymer. K_l, K_m, and K_p are the partition coefficients for the interfacial partitioning of drug from the liquid com-

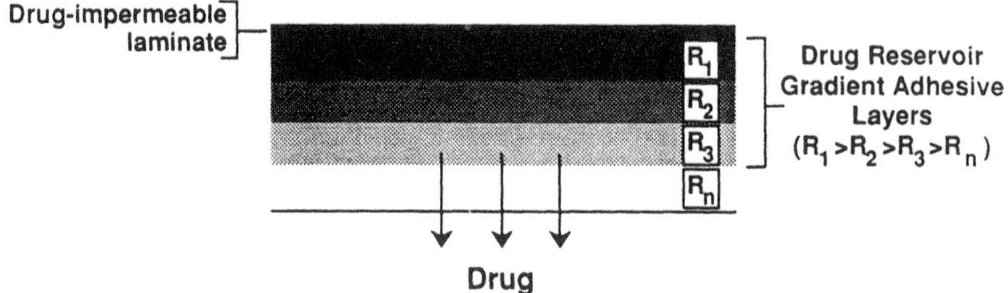

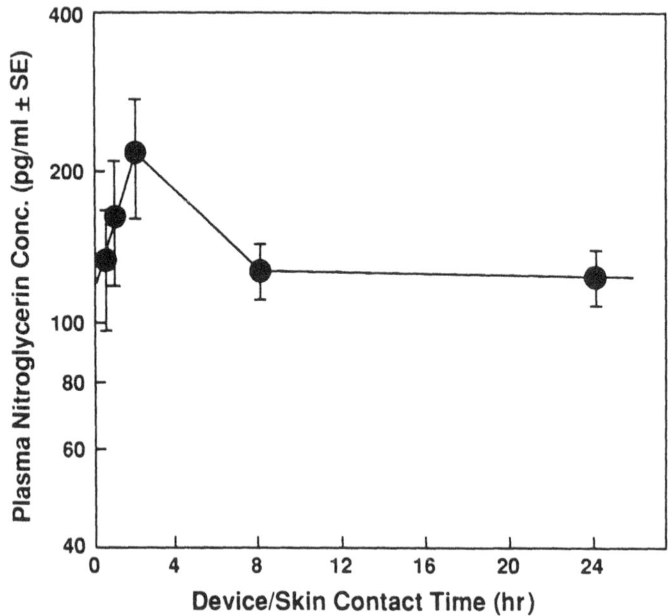

Figure 13 Cross-sectional view of Deponit showing various structural components and the 24 hr plasma nitroglycerin concentration profiles in six human volunteers, each receiving one unit of Deponit (16 cm^2 with a daily dosage rate of 5 mg/day) for 24 h. (Plasma profiles are plotted from data in Reference 20.)

partments to the polymer matrix, from the polymer matrix to the polymer-coating membrane, and from the polymer-coating membrane to the elution solution, respectively. D_l, D_p, and D_d are the diffusivities of the drug in the liquid layer surrounding the drug particles, the polymer-coating membrane enveloping the polymer matrix, and the hydrodynamic diffusion layer surrounding the polymer-coating membrane with thickness h_l, h_p, and h_d. S_l and S_p are the solubilities of the drug in the liquid compartments and in the polymer matrix, respectively.

Release of drug molecules from this type of controlled-release drug delivery system can follow either a dissolution- or a matrix diffusion-controlled process depending upon the relative magnitude of S_l and S_p (25). The rate of release is controlled at a preprogrammed rate by controlling various physicochemical parameters in Equation (4). Representatives of this type of drug delivery systems are as follows:

In the transdermal *Nitrodisc system* (Figure 14) the drug reservoir is formed by

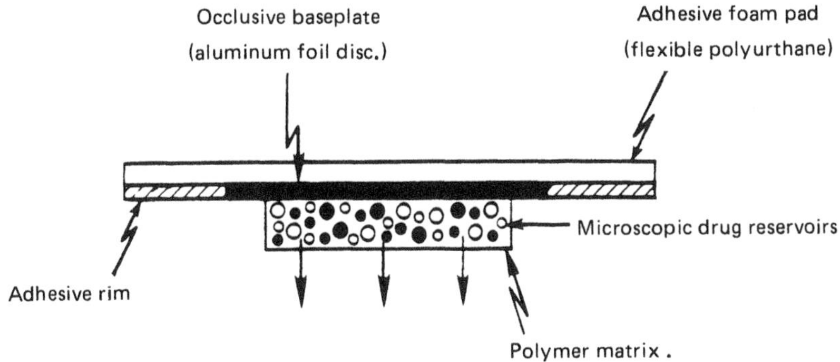

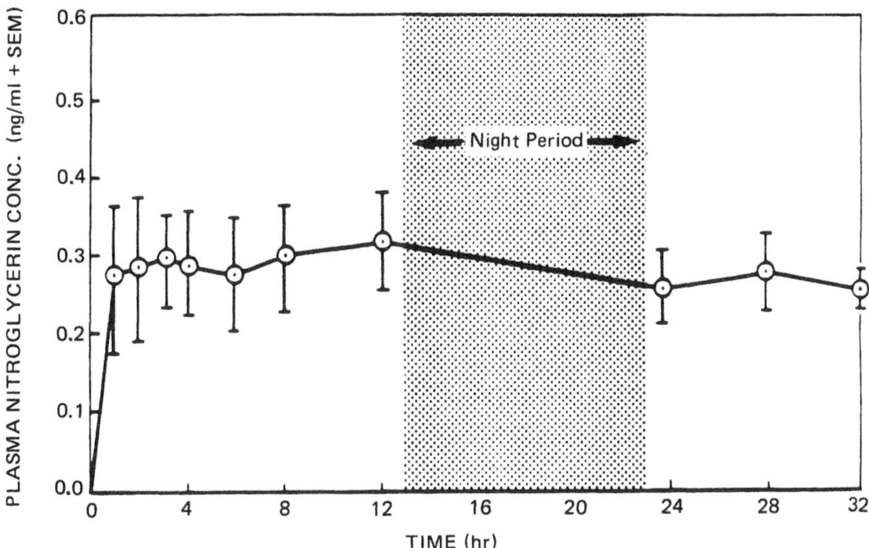

Figure 14 Cross-sectional view of Nitrodisc showing various structural components and the 32 hr plasma nitroglycerin concentration profiles in 12 human volunteers each receiving one unit of Nitrodisc (16 cm^2 with a dialy dosage rate of 10 mg/day) for 32 h. (Plasma profiles are plotted from data in Reference 23.)

first preparing a suspension of nitroglycerin and lactose triturate in an aqueous solution of 40% polyethylene glycol 400 and dispersing it homogeneously with isopropyl palmitate, as dispersing agent, in a mixture of viscous silicone elastomer by high-energy mixing and then cross-linking the polymer chains by catalyst (26). The resultant drug-polymer dispersion is then molded to form a solid medicated disk in situ on a drug-impermeable metallic plastic laminate, with surrounding adhesive rim, by injection molding under instantaneous heating. It is engineered to provide a transdermal administration of nitroglycerin at a daily dosage of 0.5 mg/cm^2 for once-a-day medication of angina pectoris (27,28). A Q versus $t^{1/2}$ (matrix diffusion-controlled) release profile is obtained.

The *transdermal contraceptive device* is based on a patentable micro-drug-reservoir technique (29) to achieve a dual-controlled release of levonorgestrel, a potent

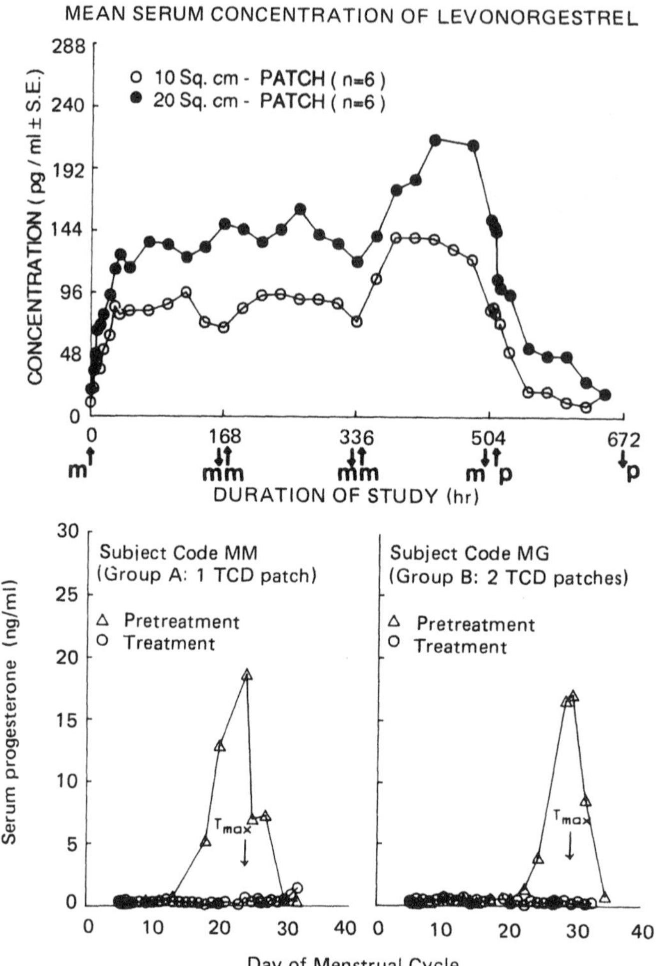

Figure 15 (Above) The 4 week serum levonorgestrel profiles in 12 human volunteers, each receiving one or two units of a transdermal contraceptive system (10 cm^2 with daily dosage of 28.3 μg/day) once a week, consecutively for 3 weeks, and the same size of placebo on week 4. (Below) Comparative serum concentration profiles of progesterone during the pretreatment and treatment cycles in two subjects, each as the representative for group A (receiving 10 cm^2) and group B (receiving 20 cm^2), respectively. The suppression of progesterone peak during the treatment cycle is an indication of effective fertility control.

synthetic progestin, and estradiol, a natural estrogen, at constant and enhanced rates continuously for a period of 7 days (30–35). By applying one unit (10 or 20 cm^2) of transdermal contraceptive device per week, beginning on day 5 of the individual's cycle, for 3 consecutive weeks (3 weeks on and 1 week off), steady-state serum levels of levonorgestrel were obtained and progesterone peak was effectively suppressed (Figure 15).

The subdermal *Syncro-Mate-C implant* is fabricated by dispersing the drug reservoir, which is a suspension of norgestomet in an aqueous solution of PEG 400, in a viscous mixture of silicone elastomers by high-energy dispersion. After the addition of catalyst the suspension is delivered into silicone medical-grade tubing, which serves as the mold as well as the coating membrane, and then polymerized in situ. The polymerized drug-polymer composition is then cut into a cylindrical drug delivery device with open ends (Figure 16). This tiny cylindrical implant is designed to be inserted into the subcutaneous tissue of the livestock's ear flap and to release norgestomet for up to 20 days for the control and synchronization of estrus and ovulation as well as for up to 160 days for growth promotion. A constant Q versus t (dissolution-controlled) release profile is achieved compared to the Q versus $t^{1/2}$ release profile (matrix diffusion-controlled drug release) for Nitrodisc.

IV. ACTIVATION-MODULATED DRUG DELIVERY SYSTEMS

In this group of controlled-release drug delivery systems the release of drug molecules from the delivery systems is activated by some physical, chemical, or biochemical processes and/or facilitated by the energy supplied externally (Figure 3). The rate of drug release is then controlled by regulating the process applied or energy input. Based on the nature of the process applied or the type of energy used, these

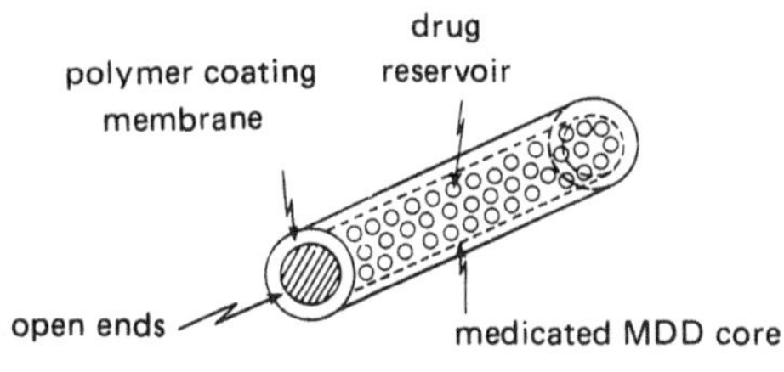

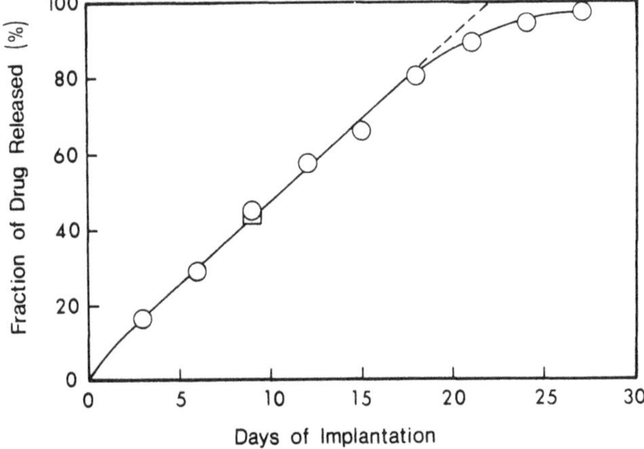

Figure 16 Cross-sectional view of the Syncro-Mate-C subdermal implant fabricated from the microreservoir dissolution-controlled drug delivery system, and the subcutaneous controlled release of Norgestomet, a potent synthetic progestin, at a constant rate for up to 18 days (with 80% loading dose released). The open ends on the implant do not affect the zero-order in vivo drug release profile. (Adapted from Reference 25.)

activation-modulated drug delivery systems (DDS) can be classified into the following categories:

1. *Physical means*
   a. Osmotic pressure-activated DDS
   b. Hydrodynamic pressure-activated DDS
   c. Vapor pressure-activated DDS
   d. Mechanically activated DDS
   e. Magnetically activated DDS
   f. Sonophoresis-activated DDS
   g. Iontophoresis-activated DDS
   h. Hydration-activated DDS
2. *Chemical means*
   a. pH-activated DDS
   b. Ion-activated DDS
   c. Hydrolysis-activated DDS
3. *Biochemical means*
   a. Enzyme-activated DDS
   b. Biochemical-activated DDS

Several systems have been successfully developed and applied clinically to the controlled delivery of pharmaceuticals. These are outlined and discussed in the following sections.

A. Osmotic Pressure-Activated Drug Delivery Systems

This type of activation-controlled drug delivery system depends on osmotic pressure to activate the release of drug. In this system the drug reservoir, which can be either a solution or a solid formulation, is contained within a semipermeable housing with controlled water permeability. The drug is activated to release in solution form at a constant rate through a special delivery orifice. The rate of drug release is modulated by controlling the gradient of osmotic pressure.

For the drug delivery system containing a solution formulation, the intrinsic rate of drug release Q/t is defined by

$$\frac{Q}{t} = \frac{P_w A_m}{h_m}(\pi_s - \pi_e) \tag{5}$$

For the drug delivery system containing a solid formulation, the intrinsic rate of drug release should also be a constant and is defined by

$$\frac{Q}{t} = \frac{P_w A_m}{h_m}(\pi_s - \pi_e)S_d \tag{6}$$

where P_w, A_m, and h_m are the water permeability, the effective surface area, and the thickness of the semipermeable housing, respectively; $\pi_s - \pi_e$ is the differential osmotic pressure between the drug delivery system with osmotic pressure π_s and the environment with osmotic pressure π_e; and S_d is the aqueous solubility of the drug contained in the solid formulation.

The release of drug molecules from this type of controlled-release drug delivery system is activated by osmotic pressure and controlled at a rate determined by the water permeability and the effective surface area of the semipermeable housing as well as the osmotic pressure gradient. Representatives of this type of drug delivery systems are as follows:

In the implantable or insertable *Alzet osmotic pump* the drug reservoir, which is normally a solution formulation, is contained within a collapsible, impermeable polyester bag whose external surface is coated with a layer of osmotically active salt, such as sodium chloride. This reservoir compartment is then completely sealed inside a rigid housing walled with a semipermeable membrane (Figure 17). At the implantation site the water component in the tissue fluid penetrates through the semipermeable housing at a rate determined by P_wA_m/h_m to dissolve the osmotically active salt. This creates an osmotic pressure π_s in the narrow spacing between the flexible reservoir compartment wall and the rigid semipermeable housing. Under the osmotic pressure differential created ($\pi_s - \pi_e$), the reservoir compartment is forced to reduce its volume and the drug solution is delivered at a controlled rate [Eq. (5)] through the flow moderator in the system (36,37). By varying the drug concentration in the solution, different doses of drug can be delivered at a constant rate for a period of 1–4 weeks.

In addition to its application in the subcutaneous controlled administration of drugs, such as vasopressin (Figure 17), for pharmacological studies, this technology has recently been extended to the controlled administration of drugs in the rectum by zero-order kinetics. The hepatic first-pass metabolism of drugs is thus bypassed (38).

Acutrim tablet, an oral rate-controlled drug delivery device, is a solid tablet of water-soluble and osmotically active phenylpropanolamine (PPA) HCl enclosed within a semipermeable membrane made from cellulose triacetate (39,40). The surface of the semipermeable membrane is further coated with a thin layer of PPA dose for immediate release (Figure 18). In the gastrointestinal tract the gastrointestional fluid dissolves the immediately releasable PPA layer, which provides an initial dose of PPA, and its water component then penetrates through the semipermeable membrane at a rate determined by P_wA_m/h_m to dissolve the controlled-release dose of PPA. Under the osmotic pressure differential created ($\pi_s - \pi_e$), the PPA solution is delivered continuously at a controlled rate [Eq. (6)] through an orifice predrilled by a laser beam (36,37). It is designed to provide a controlled delivery of PPA over a duration of 16 hr for appetite suppression in a weight control program (41). The same delivery system has also been utilized for the oral controlled delivery of indomethacin. An extension of this technology is the development of a pushup osmotically-controlled drug delivery system (37) for the oral controlled delivery of nifedipine and metroprolol.

B. Hydrodynamic Pressure-Activated Drug Delivery Systems

In addition to the osmotic pressure already discussed, hydrodynamic pressure has also been explored as the possible source of energy to activate the delivery of therapeutic agents (39,42).

A hydrodynamic pressure-activated drug delivery system can be fabricated by enclosing a collapsible, impermeable container, which contains a liquid drug formulation to form a drug reservoir compartment, inside a rigid shape-retaining hous-

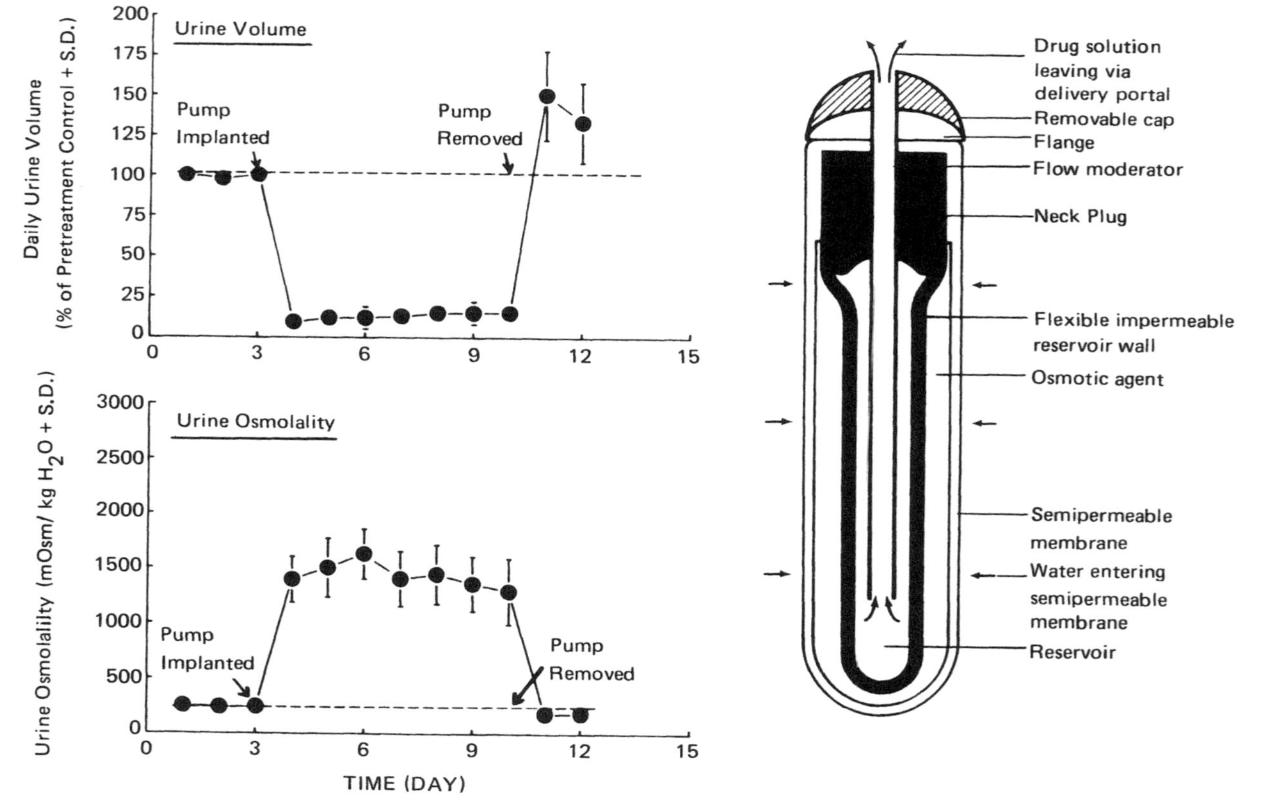

Figure 17 (Left) Cross-sectional view of the Alzet osmotic pump, an osmotic pressure-activated drug delivery system. (Right) The effect of 7 day subcutaneous delivery of antidiuretic hormone (vasopressin) on the daily volume of urinary excretion and urine osmolality in Brattleboro rats with diabetes insipidus.

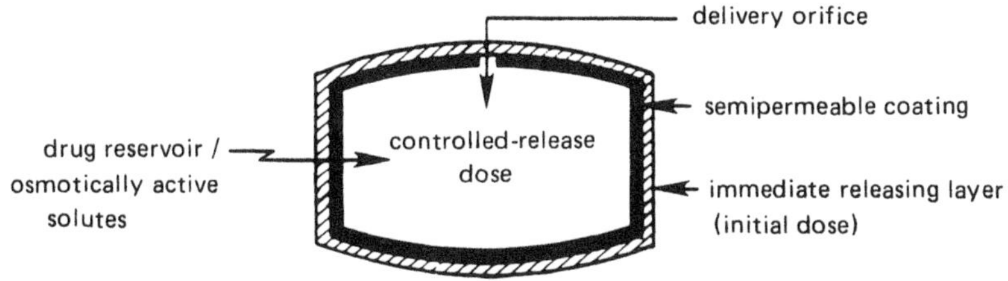

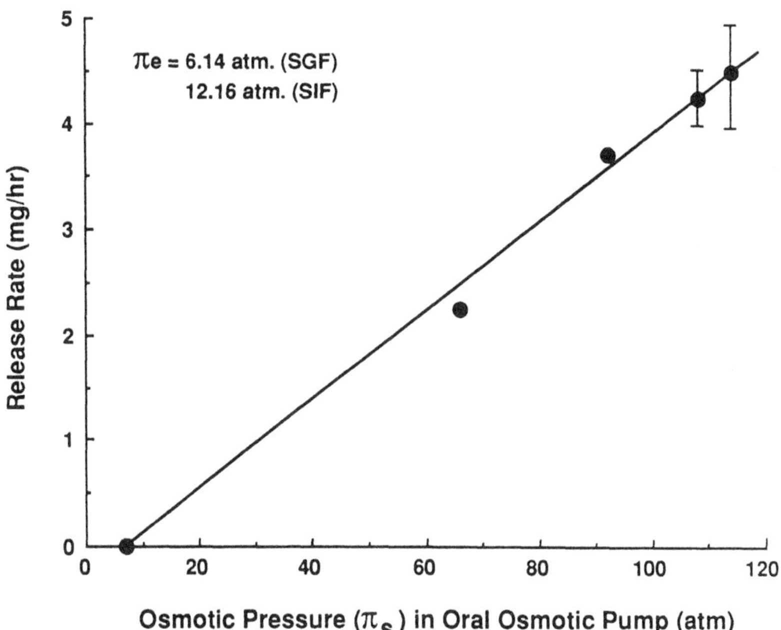

Figure 18 Cross-sectional view of the Acutrim tablet, an osmotic pressure-activated drug delivery system, and the effect of increased osmotic pressure in the delivery system on the release profiles of phenylpropanolamine HCl from the Acutrim tablet at intestinal condition. (Adapted from References 39 and 41.)

ing (Figure 19). A composite laminate of an absorbent layer and a swellable, hydrophilic polymer layer is sandwiched between the drug reservoir compartment and the housing. In the gastrointestinal tract the laminate absorbs the gastrointestinal fluid through the annular openings at the lower end of the housing and becomes increasingly swollen, which generates hydrodynamic pressure in the system. The hydrodynamic pressure thus created forces the drug reservoir compartment to reduce in volume and causes the liquid drug formulation to release through the delivery orifice at a rate defined by

$$\frac{Q}{t} = \frac{P_f A_m}{h_m} (\theta_s - \theta_e) \qquad (7)$$

where P_f, A_m, and h_m are the fluid permeability, the effective surface area, and the

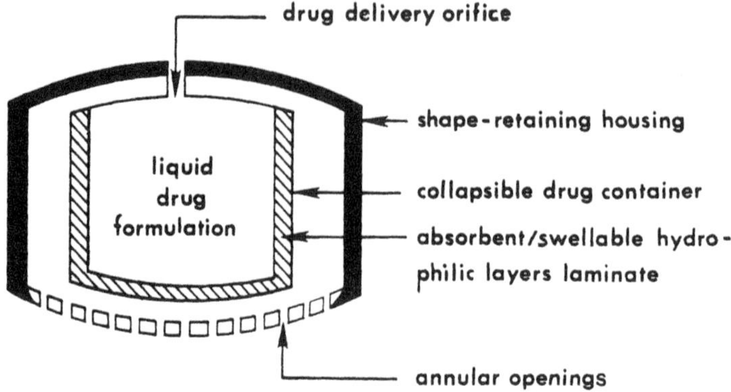

Rate Of Zero-order Drug Release:

$$\frac{Q}{t} = \frac{P_f A_m}{h_m} (\theta_s - \theta_e)$$

Figure 19 Cross-sectional view of a hydrodynamic pressure-activated drug delivery system showing various structural components. (Reproduced from References 39 and 42.)

thickness of the wall with annular openings, respectively; and $\theta_s - \theta_e$ is the difference in hydrodynamic pressure between the drug delivery system (θ_s) and the environment (θ_e).

The release of drug molecules from this type of controlled-release drug delivery system is activated by hydrodynamic pressure and controlled at a rate determined by the fluid permeability and effective surface area of the wall with annular openings as well as by the hydrodynamic pressure gradient.

C. Vapor Pressure-Activated Drug Delivery Systems

Vapor pressure has also been discovered as a potential energy source to activate the delivery of therapeutic agents (43). In this type of drug delivery system the drug reservoir, which also exists as a solution formulation, is contained inside the infusion compartment. It is physically separated from the pumping compartment by a freely movable partition (Figure 20). The pumping compartment contains a fluorocarbon fluid that vaporizes at body temperature at the implantation site and creates a vapor pressure. Under the vapor pressure created the partition moves upward. This forces the drug solution in the infusion compartment to be delivered through a series of flow regulator and delivery cannula into the blood circulation at a constant flow rate (1,43). The rate of delivery Q/t is defined by

$$\frac{Q}{t} = \frac{d^4(P_s - P_e)}{40.74\mu l} \tag{8}$$

where d and l are the inner diameter and the length of the delivery cannula, re-

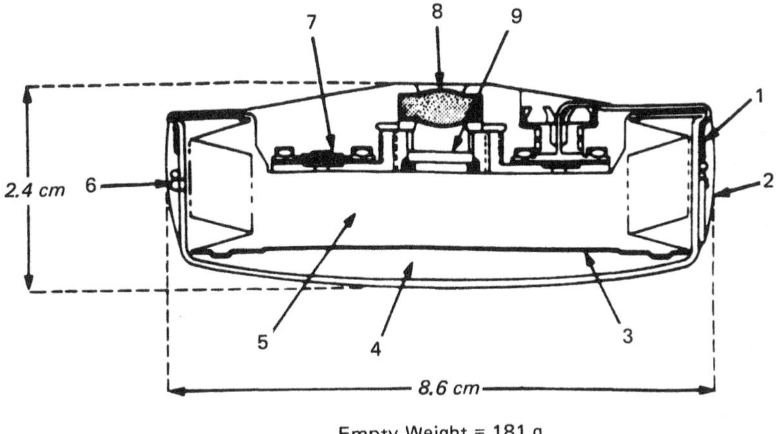

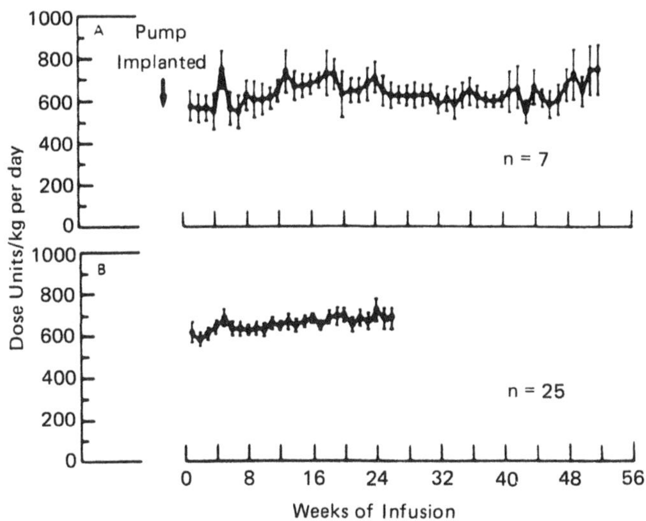

Figure 20 Cross-sectional view of Infusaid, a vapor pressure-activated drug delivery system, and daily heparin dose (mean ± SEM) delivered to 25 dogs for 6 months and to 7 dogs for 12 months. (1) Flow regulator, (2) silicone polymer coating, (3) partition, (4) pumping compartment, (5) infusate compartment, (6) fluorocarbon fluid filling tube (permanently sealed), (7) filter assembly, (8) inlet septum for percutaneous refill of infusate, (9) needle stop. (Adapted from Reference 44.)

spectively; $P_s - P_e$ is the difference between the vapor pressure in the pumping compartment P_s and the pressure at the implantation site P_e; and μ is the viscosity of the drug formulation used.

The release of drug from this type of controlled-release drug delivery system is activated by vapor pressure and controlled at a rate determined by the differential vapor pressure, the formulation viscosity and the size of the delivery cannula.

A typical example is the development of an implantable infusion pump (Infusaid by Metal Bellows) for the constant infusion of heparin in anticoagulation treatment

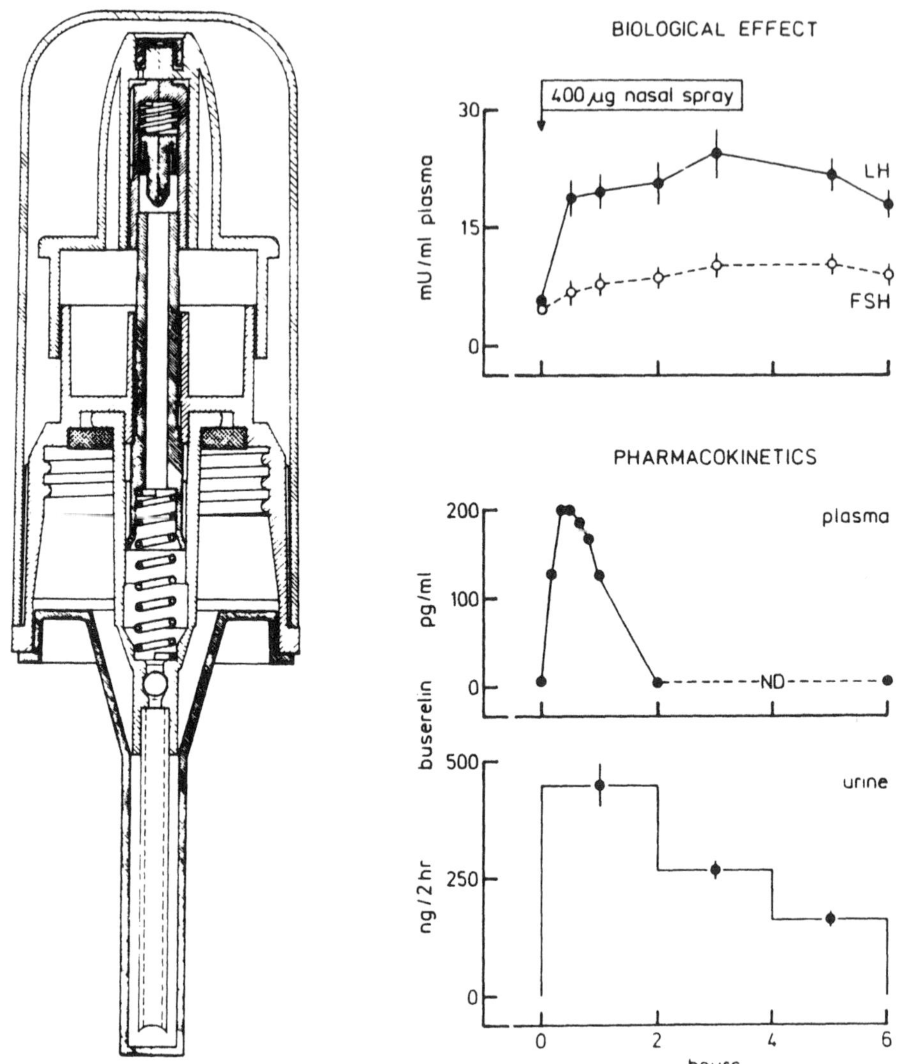

Figure 21 Cross-sectional view of the pumping system for metered-dose nebulizers, mechanically activated drug delivery systems, for the intranasal delivery of a precise dose of peptide solution; The pharmacokinetic profiles of buserelin, a peptide drug, following nasal absorption and its biological effects. (Adapted from Reference 46.)

(44), of insulin in antidiabetic medication (43), and of morphine for patients suffering from the intense pain of terminal cancer (45).

D. Mechanically Activated Drug Delivery Systems

In this type of activation-controlled drug delivery system the drug reservoir is a solution formulation retained in a container equipped with a mechanically activated pumping system. A measured dose of drug formulation is reproducibly delivered into a body cavity, for example the nose, through the spray head upon manual activation of the drug delivery pumping system. The volume of solution delivered is controllable, as small as 10–100 μl, and is independent of the force and duration of activation applied as well as the solution volume in the container.

A typical example of this type of rate-controlled drug delivery system is the development of a metered-dose nebulizer (Figure 21) for the intranasal administration of a precision dose of buserelin, a synthetic analog of luteinizing hormone releasing hormone (LHRH) and insulin. Through nasal absorption the hepatic first-pass elimination of these peptide-based pharmaceuticals is thus avoided (46).

E. Magnetically Activated Drug Delivery Systems

In this type of activation-controlled drug delivery system the drug reservoir is a dispersion of peptide or protein powders in a polymer matrix from which macromolecular drug can be delivered only at a relatively slow rate. This low rate of delivery can be improved by incorporating an electromagnetically triggered vibration mechanism into the polymeric delivery device. Combined with a hemispherical design, a zero-order drug delivery profile is achievd (47). A subdermally implantable, magnetically activated drug delivery device is fabricated by first positioning a tiny magnet ring in the core of a hemispherical drug-dispersing polymer matrix and then coating its external surface with a drug-impermeable polymer, such as ethylene-vinyl acetate copolymer or silicone elastomers, except one cavity at the center of the flat surface. This uncoated cavity is positioned directly above the magnet ring, which permits a peptide drug to be released (Figure 22).

The hemispherical magnetic delivery device produced has been used to deliver protein drugs, such as bovine serum albumin, at a low basal rate, by a simple diffusion process under nontriggering conditions (Figure 22). As the magnet is activated to vibrate by an external electromagnetic field, the drug molecules are delivered at a much higher rate.

F. Sonophoresis-Activated Drug Delivery Systems

This type of activation-controlled drug delivery systems utilizes ultrasonic energy to activate (or trigger) the delivery of drugs from a polymeric drug delivery device (Figure 23). The system can be fabricated from either a nondegradable polymer, such as ethylene-vinyl acetate copolymer, or a bioerodible polymer, such as poly[bis(*p*-carboxyphenoxy)alkane anhydride] (48). The potential application of sonophoresis (or phonophoresis) to regulate the delivery of drugs was recently reviewed (49).

G. Iontophoresis-Activated Drug Delivery Systems

This type of activation-controlled drug delivery systems uses electrical current to activate and to modulate the diffusion of a charged drug molecule across a biological membrane, like the skin, in a manner similar to passive diffusion under a concentration gradient, but at a much facilitated rate. The iontophoresis-facilitated skin permeation rate of a charged molecule *i* consists of three components and is expressed by

$$J_i^{isp} = J^p + J^e + J^c \tag{9}$$

where:

$$J^p = \text{passive skin permeation flux}$$
$$= K_s D_s \frac{dC}{h_s} \tag{10}$$

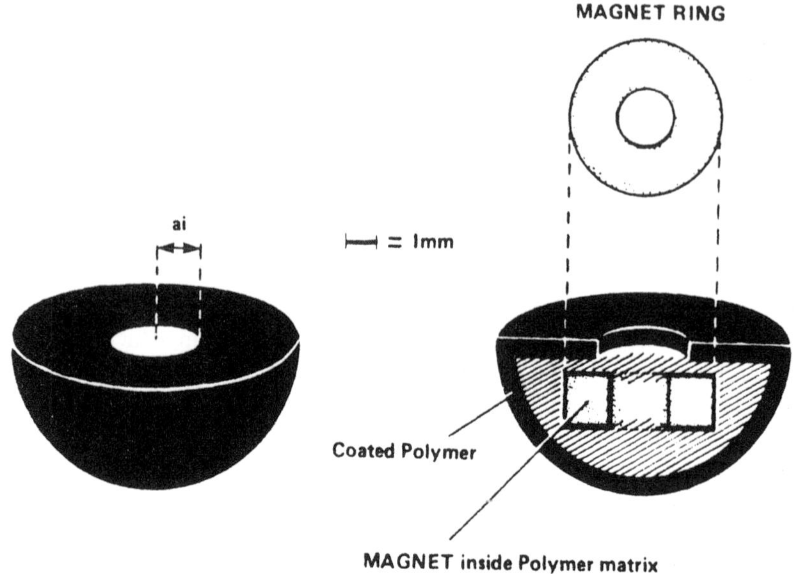

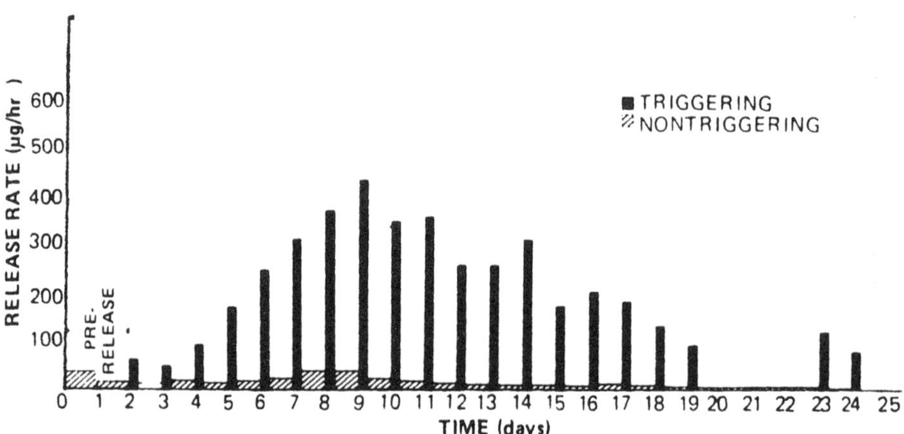

Figure 22 Cross-sectional view of hemispheric magnetically activated drug delivery system showing various structural components, and magnetic modulation of the release rate of bovine serum albumin, a protein molecule of 69,000 daltons. (Reproduced from Reference 47.)

J^e = electrical current-driven permeation flux

$$= \frac{Z_i D_i F}{RT} C_i \frac{dE}{h_s} \tag{11}$$

J^c = Convective flow-driven skin permeation flux

$$= kC_s I_d \tag{12}$$

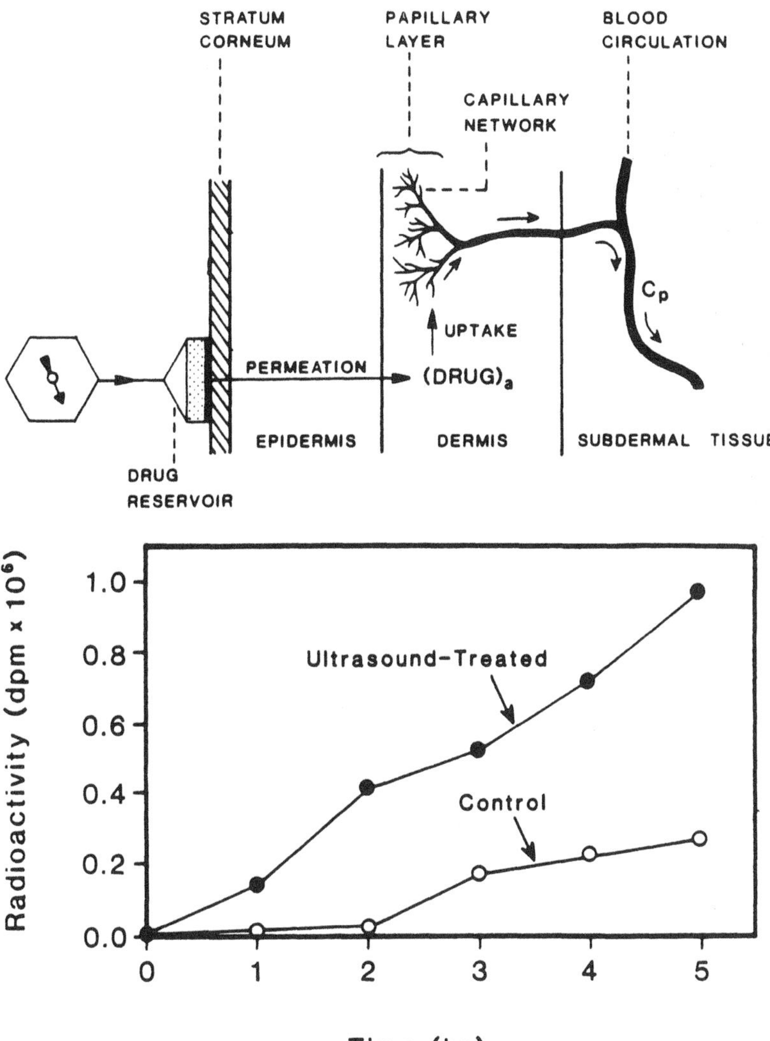

Figure 23 Application of ultrasound to activate the transdermal delivery of a hydrophilic molecule, like mannitol, across the intact rat skin.

and where:

K_s = partition coefficient for interfacial partitioning from donor solution to stratum corneum
D_s = diffusivity across the skin
D_i = diffusivity of ionic species i in the skin
C_i = donor concentration of ionic species i in the skin
C_s = concentration in the skin tissue
$\frac{dE}{h_s}$ = electrical potential gradient across the skin

$\frac{dC}{h_s}$ = concentration gradient across the skin
Z_i = electrical valence of ionic species i
I_d = current density applied
F = faraday constant
k = proportionality constant
R = gas constant
T = absolute temperature

A typical example of this type of activation-controlled drug delivery system is the development of an iontophoretic drug delivery system (Phoresor by Motion Control) to facilitate the percutaneous penetration of anti-inflammatory drugs, such as dexamethasone sodium phosphate (50–52), to surface tissues.

Further development of the iontophoresis-activated drug delivery technique has recently yielded a new design of iontophoretic drug delivery system, the transdermal periodic iontotherapeutic system (TPIS). This new system, which is capable of delivering a physiologically acceptable pulsed direct current in a periodic manner with a special combination of waveform, intensity, frequency, and on/off ratio, for programmed duration. It has significantly improved the efficiency of transdermal delivery of peptide and protein drugs (53). A typical example is the iontophoretic transdermal delivery of insulin, a protein drug, in the control of hyperglycemia in diabetic animals (Figure 24). More extensive discussion will be found in Chapter 11.

H. Hydration-Activated Drug Delivery Systems

This type of activation-controlled drug delivery systems depends on the hydration-induced swelling process to activate the release of drug. In this system the drug reservoir is homogeneously dispersed in a swellable polymer matrix fabricated from a hydrophilic polymer. The release of drug is controlled by the rate of swelling of the polymer matrix. Representatives of this type of hydration-activated drug delivery systems are as follows.

The *Syncro-Mate-B implant* is fabricated by dissolving norgestomet, a potent progestin, in the alcoholic solution of a linear ethylene glycomethacrylate polymer (Hydron S). The polymer chain is then cross-linked with ethylene dimethacrylate, a cross-linking agent, to form a solid cylindrical drug-dispersed Hydron implant (54). This tiny subdermal implant is engineered to release norgestomet upon hydration in the subcutaneous tissue at a rate of 504 $\mu g/cm^2/day^{1/2}$ for up to 16 days for the control and synchronization of estrus in livestock (55).

The *Valrelease tablet* is prepared by a simple pharmaceutical granulation process of homogeneous dispersion of Valium, a tranquilizer, in hydrocolloid and pharmaceutical excipients. The granules are then compressed to form a compressed tablet. After oral intake the hydrocolloid in the tablet absorbs gastric fluid and forms a colloidal gel that starts from the tablet surface and grows inward (Figure 25). The release of Valium molecules is then controlled by matrix diffusion through this gel barrier; the tablet remains buoyant in the stomach as a result of the density difference between the gastric fluid ($d > 1$) and the gelling tablet ($d < 1$) (2,39).

In addition to the hydrophilic polymers, lipophilic polymers, such as silicone elastomer, can also be modified to have swelling properties (56–60). This is achieved

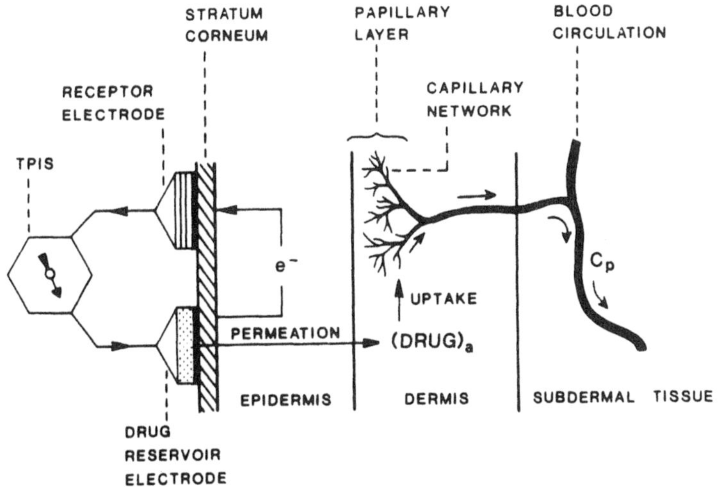

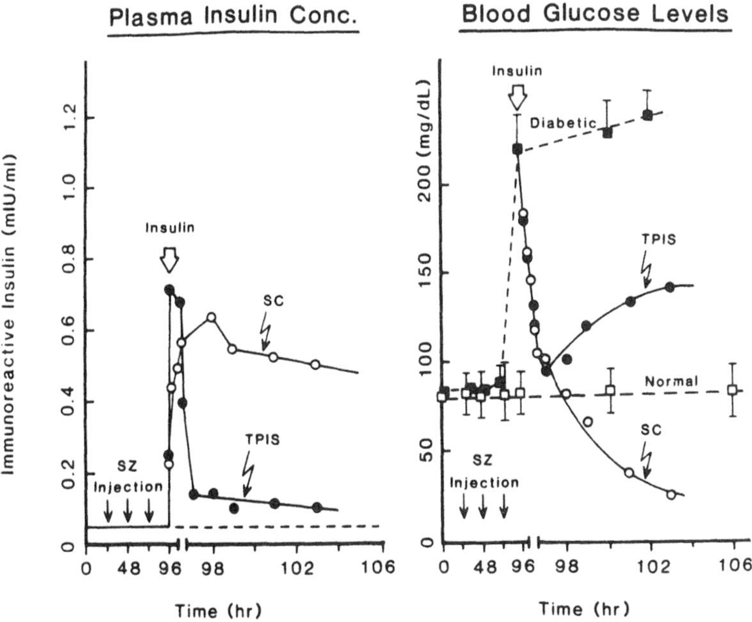

Figure 24 Application of the transdermal periodic iontotherapeutic system (TPIS), an iontophoresis-activated drug delivery system, for the transdermal controlled delivery of insulin to diabetic rabbits and the reduction in blood glucose level from the hyperglycemic state compared to conventional subcutaneous administration. (Reproduced from Reference 53.)

by impregnating a water-miscible liquid, such as glycerol, and/or a water-soluble salt, such as sodium chloride, in the lipophilic polymer matrix. When it contacts with an aqueous medium, the modified lipophilic polymer becomes swollen as a result of continuous absorption of water by the hydrophilic additives in the polymer matrix. This activates the release of drugs, peptide and nonpeptide, at a significantly enhanced rate.

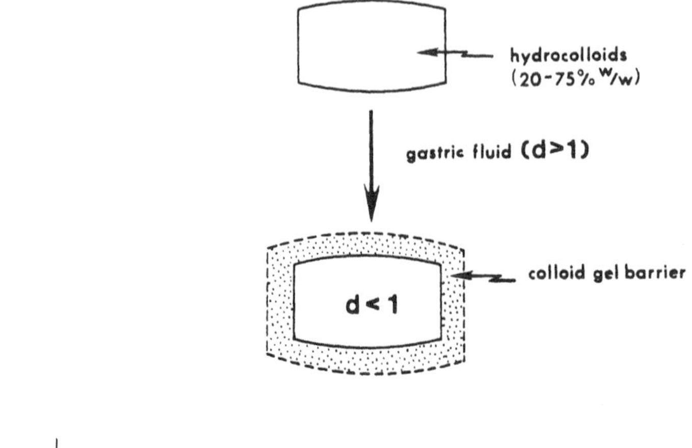

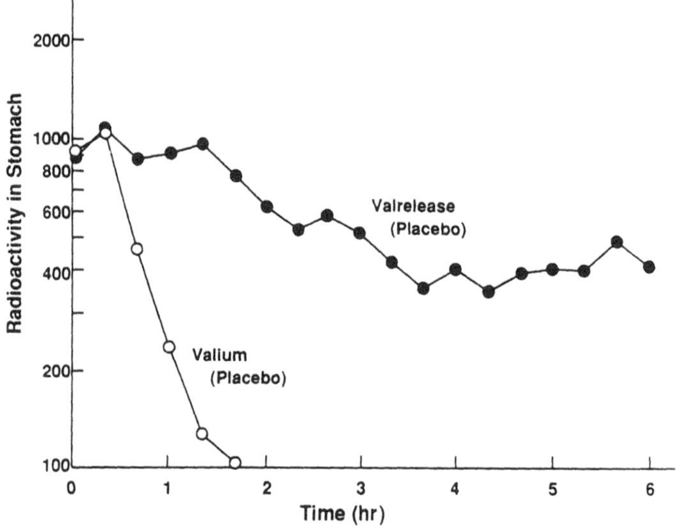

Figure 25 (Top) The Valrelease tablet, a hydration-activated drug delivery system, and the hydration-induced formation of a colloid gel barrier. (Bottom) Comparative gastric retention profiles of the Valrelease and Valium tablets.

I. pH-Activated Drug Delivery Systems

This type of activation-controlled drug delivery system permits targetting the delivery of a drug only in the region with a selected pH range (39). It is fabricated by coating the drug-containing core with a pH-sensitive polymer combination. For instance, a gastric fluid-labile drug is protected by encapsulating it inside a polymer membrane that resists the degradative action of gastric pH, such as the combination of ethylcellulose and hydroxylmethylcellulose phthalate (Figure 26).

In the stomach, the coating membrane resists the action of gastric fluid (pH < 3) and the drug molecules are thus protected from acid degradation. After gastric emptying the drug delivery system travels to the small intestine and the intestinal fluid (pH > 7.5) activates the erosion of the intestinal fluid-soluble hydroxylmethylcellulose phthalate component from the coating membrane. This leaves a microporous membrane constructed from the intestinal fluid-insoluble polymer of ethyl-

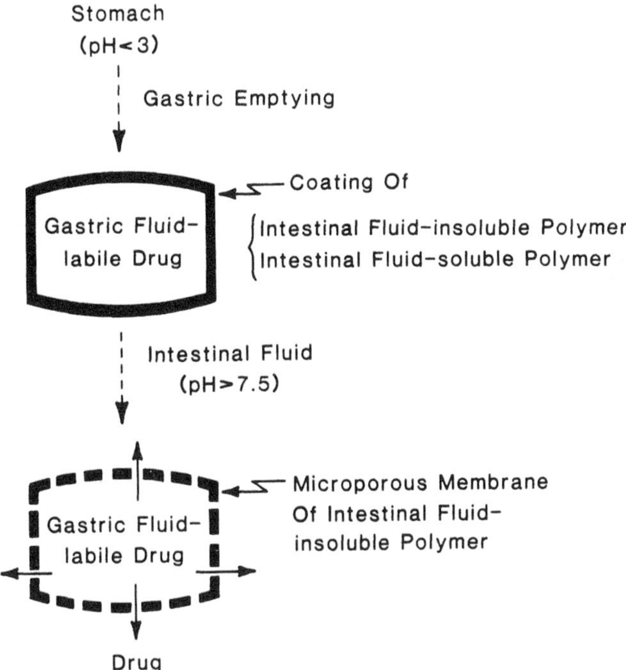

Figure 26 A pH-activated drug delivery system and the pH-dependent formation of microporous membrane in the intestinal tract.

cellulose, which controls the release of drug from the core tablet. The drug solute is thus delivered at a controlled manner in the intestine by a combination of drug dissolution and pore-channel diffusion. By adjusting the ratio of the intestinal fluid-soluble polymer to the intestinal fluid-insoluble polymer, the membrane permeability of a drug can be regulated as desired.

J. Ion-Activated Drug Delivery Systems

In addition to the iontophoresis-activated drug delivery system just discussed, an ionic or a charged drug can be delivered by an ion-activated drug delivery system (39).

Such a system is prepared by first complexing an ionic drug with an ion-exchange resin containing a suitable counterion, for example, by forming a complex between a cationic drug with a resin having a SO_3^- group or between an anionic drug with a resin having a $N(CH_3)_3^+$ group. The granules of drug-resin complex are first treated with an impregnating agent, such as polyethylene glycol 4000, to reduce the rate of swelling in an aqueous environment and then coated, by air suspension coating, with a water-insoluble but water-permeable polymeric membrane, such as ethylcellulose. This membrane serves as a rate-controlling barrier to modulate the influx of ions as well as the release of drug from the system. In an electrolyte medium, such as gastric fluid, ions diffuse into the system, react with the drug-resin complex, and trigger the release of ionic drug (Figure 27).

This system is exemplified by the development of Pennkinetic (by Pennwalt Pharmaceuticals), which permits the formulation of liquid suspension dosage forms

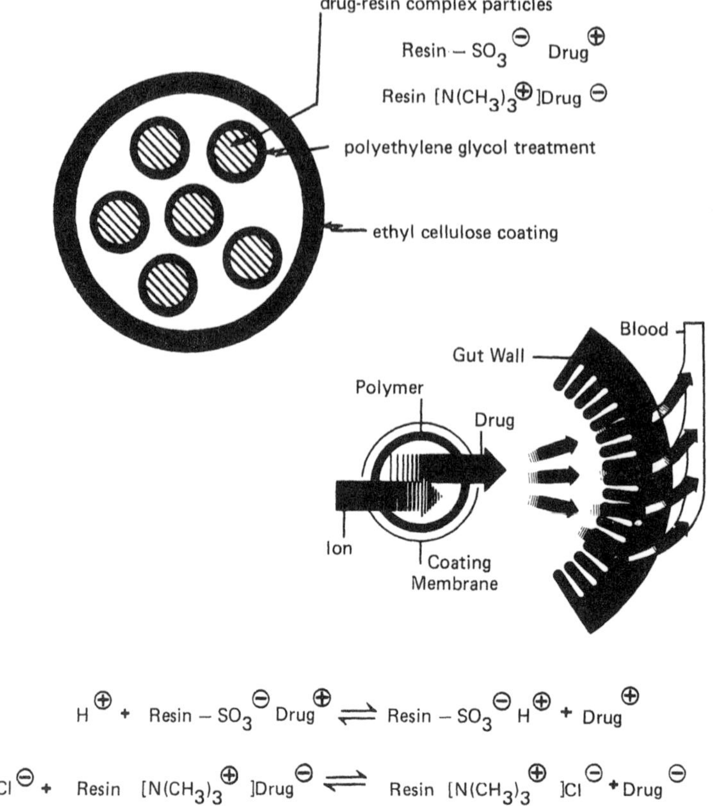

Figure 27 Cross-sectional view of an ion-activated drug delivery system showing various structural components, with a diagram of ion-activated drug release. (Adapted from Reference 39.)

with sustained drug release properties for oral administration (61–63). Since the gastrointestinal fluid regularly maintains a relatively constant level of ions (i.e., constant ionic strength), theoretically the delivery of drug from this ion-activated oral drug delivery system can be maintained at a relatively constant rate.

K. Hydrolysis-Activated Drug Delivery Systems

This type of activation-controlled drug delivery system depends on the hydrolysis process to activate the release of drug molecules. In this system the drug reservoir is either encapsulated in microcapsules or homogeneously dispersed in microspheres or nanoparticles for injection. It can also be fabricated as an implantable device. All these systems are prepared from a bioerodible or biodegradable polymer, such as co(lactic-glycolic)polymer, poly(orthoester), or poly(anhydride). The release of a drug from the polymer matrix is activated by the hydrolysis-induced degradation of polymer chains and controlled by the rate of polymer degradation (64). A typical example of a hydrolysis-activated drug delivery system is the development of LHRH-releasing biodegradable subdermal implant, which is designed to deliver goserelin, a synthetic LHRH analog, for once-a-month treatment of prostate carcinoma (Figure 28).

L. Enzyme-Activated Drug Delivery Systems

This type of activation-controlled drug delivery system depends on the enzymatic process to activate the release of drug. In this system the drug reservoir is either physically entrapped in microspheres or chemically bound to polymer chains from biopolymers, such as albumins or polypeptides. The release of drugs is activated by the enzymatic hydrolysis of the biopolymers by a specific enzyme in the target tissue (65–67). A typical example of this enzyme-activated drug delivery system is the development of albumin microspheres that release 5-fluorouracil in a controlled manner by protease-activated biodegradation.

V. FEEDBACK-REGULATED DRUG DELIVERY SYSTEMS

In this group of controlled-release drug delivery systems the release of drug molecules from the delivery systems is activated by a triggering agent, such as a biochemical substance, in the body and also regulated by its concentration via some feedback mechanisms (Figure 3). The rate of drug release is then controlled by the

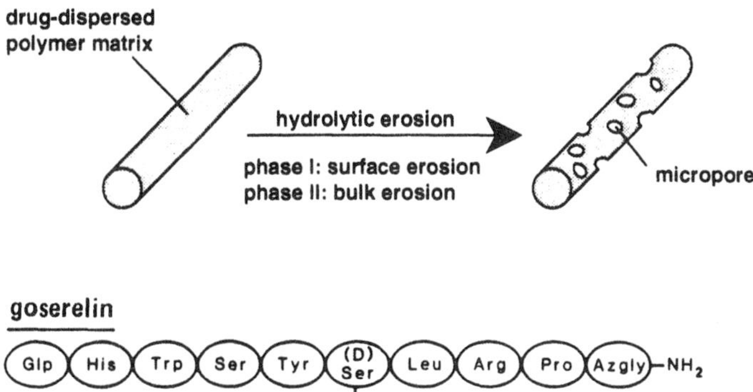

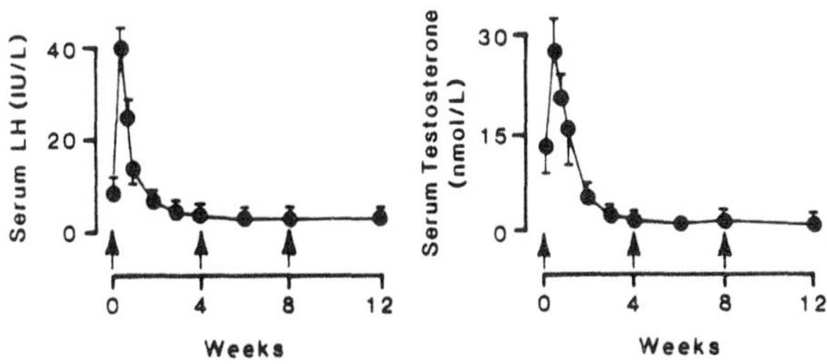

Figure 28 Amino acid sequence of goserelin, a synthetic LHRH, and the effect of subcutaneous controlled release of goserelin from the biodegradable poly(lactide-glycolide) subdermal implant on the serum levels of LH and testosterone.

concentration of triggering agent detected by a sensor in the feedback-regulated mechanism.

A. Bioerosion-Regulated Drug Delivery System

The feedback-regulated drug delivery concept was applied to the development of a bioerosion-regulated drug delivery system by Heller and Trescony (68). The system consisted of drug-dispersed bioerodible matrix fabricated from poly(vinyl methyl ether) half-ester, which was coated with a layer of immobilized urease (Figure 29). In a solution with near neutral pH, the polymer only erodes very slowly. In the presence of urea, urease at the surface of drug delivery system metabolizes urea to form am-

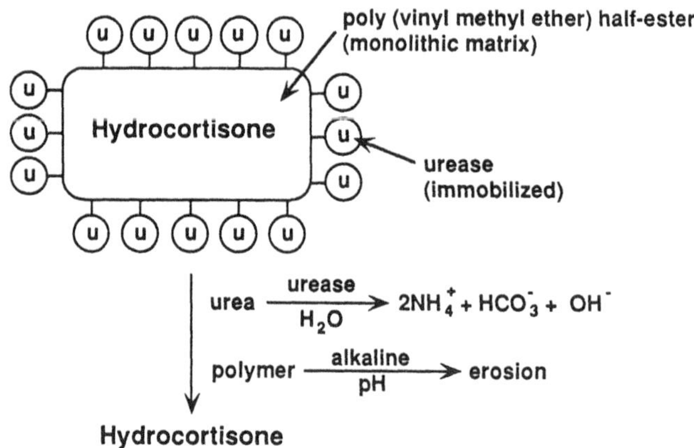

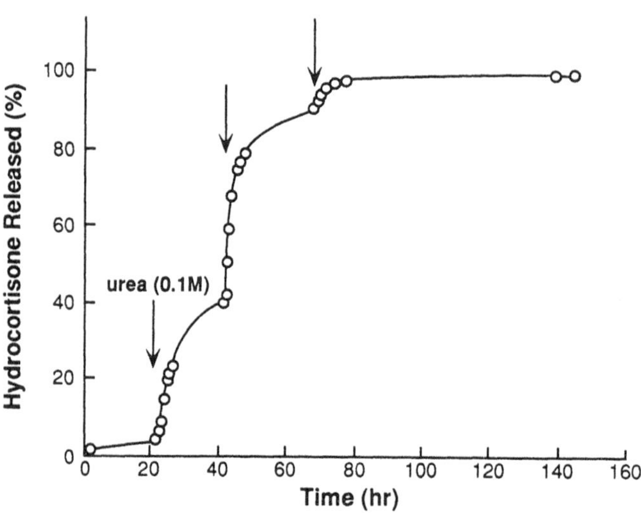

Figure 29 Cross-sectional view of a bioerosion-regulated hydrocortisone delivery system, a feedback-regulated drug delivery system, showing the drug-dispersed monolithic bioerodible polymer matrix with surface-immobilized ureases. The mechanism of release and time course for the urea-activated release of hydrocortisone are also shown. (Based on Reference 68.)

monia. This causes the pH to increase and a rapid degradation of polymer matrix as well as the release of drug molecules.

B. Bioresponsive Drug Delivery Systems

The feedback-regulated drug delivery concept has also been applied to the development of a bioresponsive drug delivery system by Horbett et al. (69). In this system the drug reservoir is contained in a device enclosed by a bioresponsive polymeric membrane whose drug permeability is controlled by the concentration of a biochemical agent in the tissue where the system is located.

A typical example of this bioresponsive drug delivery system is the development of a glucose-triggered insulin delivery system in which the insulin reservoir is encapsulated within a hydrogel membrane having pendant NR_2 groups (Figure 30). In alkaline solution the $-NR_2$ groups are neutral and the membrane is unswollen and impermeable to insulin. As glucose, a triggering agent, penetrates into the membrane, it is oxidized enzymatically by the glucose oxidase entrapped in the membrane

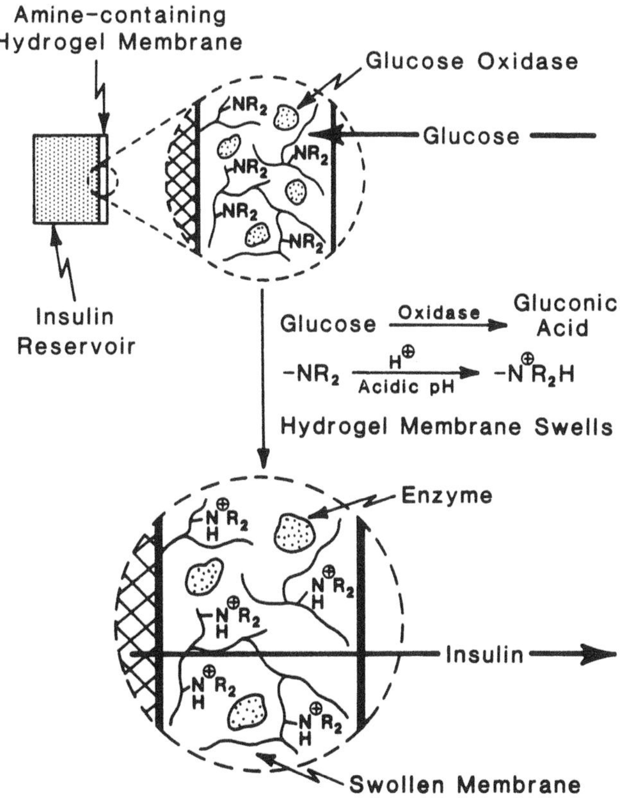

Figure 30 Cross-sectional view of a bioresponsive insulin delivery system, a feedback-regulated drug delivery system, showing the glucose oxidase-entrapped hydrogel membrane constructed from amine-containing hydrophilic polymer. The mechanism of insulin release in response to the influx of glucose is also illustrated. (Based on Reference 69.)

to form gluconic acid. The $-NR_2$ groups are protonated to form $-NR_2H^+$, and the hydrogel membrane then becomes swollen and permeable to insulin molecules (Figure 30). The amount of insulin delivered is thus bioresponsive to the concentration of glucose penetrating the insulin delivery system.

C. Self-Regulating Drug Delivery Systems

This type of feedback-regulated drug delivery systems depends on a reversible and competitive binding mechanism to activate and to regulate the release of drug. In this system the drug reservoir is a drug complex encapsulated within a semipermeable polymeric membrane. The release of drug from the delivery system is activated by the membrane permeation of a biochemical agent from the tissue in which the system is located.

Kim et al. (70) first applied the mechanism of reversible binding of sugar molecules by lectin into the design of self-regulating drug delivery systems. It first involves the preparation of biologically active insulin derivatives in which insulin is coupled with a sugar (e.g., maltose) and this into an insulin-sugar-lectin complex. The complex is then encapsulated within a semipermeable membrane. As blood glucose diffuses into the device and competitively binds at the sugar binding sites in lectin molecules, this activates the release of bound insulin-sugar derivatives. The released insulin-sugar derivatives then diffuse out of the device, and the amount of insulin-sugar derivatives released depends on the glucose concentration. Thus a self-regulating drug delivery is achieved. However, a potential problem exists: that is, the release of insulin is nonlinear in response to the changes in glucose level (71). For instances, a glucose level of 500 mg/dl triggers the release of insulin at only twice the rate of that at 50 mg/dl.

Further development of the self-regulating insulin delivery system utilized the complex of glycosylated insulin-concanavalin A, which is encapsulated inside a polymer membrane (72). As glucose, the triggering agent, penetrates the system, it activates the release of glycosylated insulin from the complex for controlled delivery out of the system (Figure 31). The amount of insulin delivered is thus self-regulated by the concentration of glucose penetrating the insulin delivery system.

VI. SUMMARY

The controlled-release drug delivery systems outlined here have been steadily introduced into the biomedical community since the middle of the 1970s. There is a growing belief that many more of the conventional drug delivery systems we have been using for many decades will be gradually replaced in coming years by these controlled-release drug delivery systems based on high technology.

Delivery of a drug to a target tissue that needs medication is known to be a complex process consisting of multiple steps of diffusion and partitioning. The controlled-release drug delivery systems outlined here generally address only the first step of this complex process. They have been designed to control the rate of drug release from the delivery systems, but the path for the transport of drug molecules from the delivery system to the target tissue remains largely uncontrolled.

Ideally, the path of drug transport should also be under control. Then, the ultimate goal of optimal treatment with maximal safety can be reached. This can be

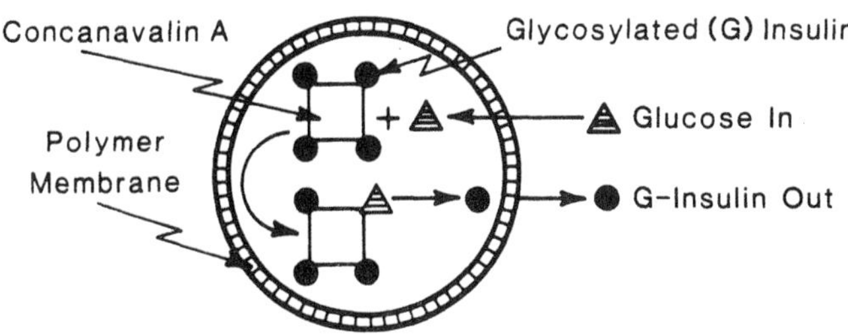

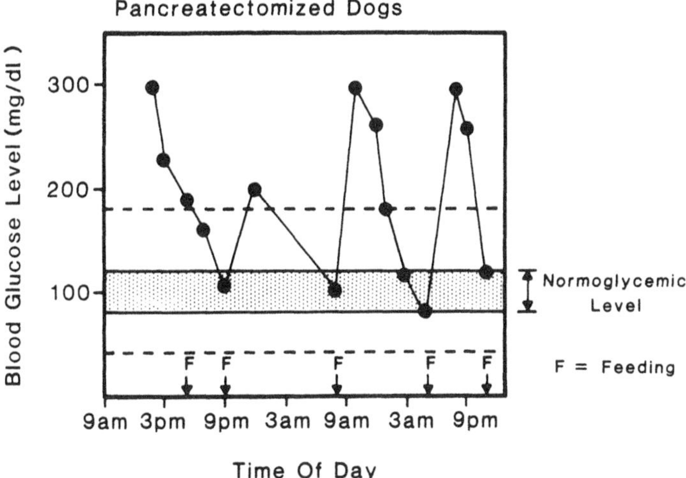

Figure 31 Various components of a self-regulating insulin delivery system, a feedback-regulated drug delivery system, and its control of blood glucose level in pancreatectomized dogs. (Based on Reference 72.)

reasonably accomplished by the development of site-targeting controlled-release drug delivery systems (Figure 3). An ideal site-targeting drug delivery system has been proposed by Ringsdorf (73). A model is shown in Figure 32. It is constructed from a nonimmunogenic and biodegradable polymer backbone having three types of attached functional groups: (1) a site-specific targeting moiety that leads the drug delivery system to the vicinity of a target tissue (or cell); (2) a solubilizer that enables the drug delivery system to be transported to and preferentially taken up by a target tissue, and (3) a drug moiety that is convalently bonded to the polymer backbone through a spacer and contains a cleavalbe group that can be cleaved only by a specific enzyme(s) at the target tissue. Unfortunately, this ideal site-targeting controlled-release drug delivery system is only in the conceptual stage. Its construction remains a largely unresolved, challenging task for the biomedical and pharmaceutical sciences.

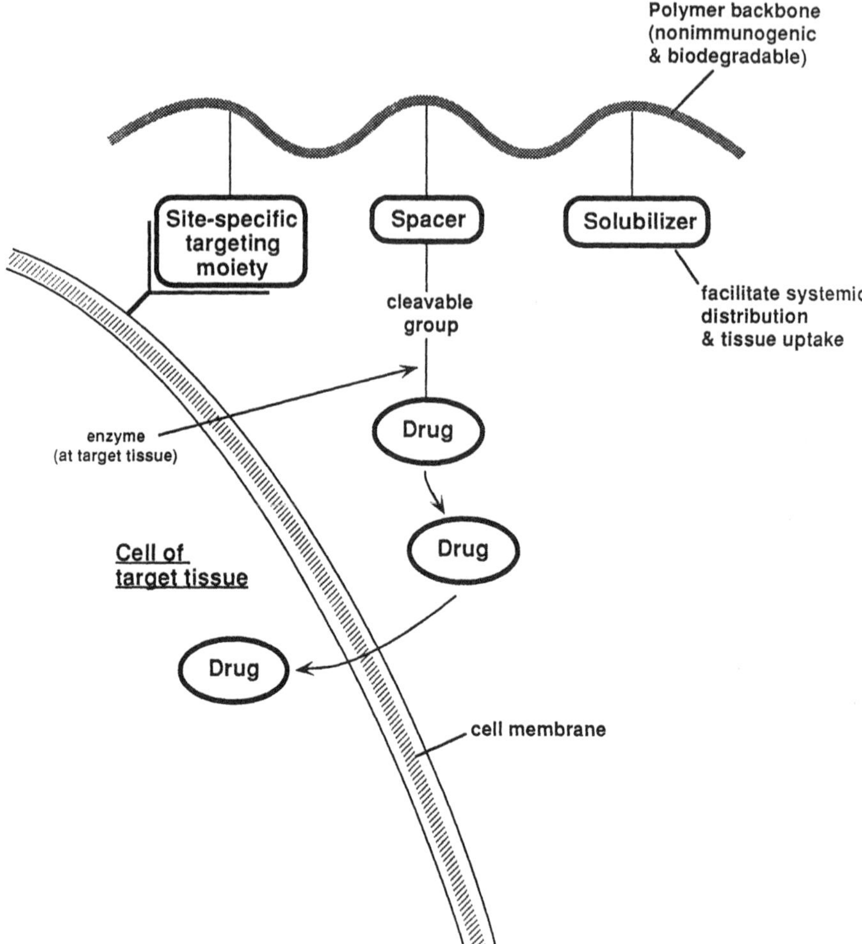

Figure 32 An ideal site-targeting controlled-release drug delivery. (Based on Reference 73.)

REFERENCES

1. Y. W. Chien; *Novel Drug Delivery Systems: Fundamentals, Developmental Concepts and Biomedical Assessments.* Dekker, New York (1982).
2. 1983 Industrial Pharmaceutical R&D Symposium on Oral Controlled Drug Administrations, Rutgers University, College of Pharmacy, Piscataway, New Jersey, January 19 and 20, 1983. Proceedings published in Drug Develop. & Ind. Pharm., 9:1077–1396 (1983).
3. 1985 Internatioinal Pharmaceutical R&D Symposium on Advances in Transdermal Controlled Drug Administration for Systemic Medication, Rutgers University, College of Pharmacy, Piscataway, New Jersey, June 20 and 21, 1985. Proceedings published in Transdermal Controlled Systemic Medications (Yie W. Chien, Ed.), Dekker, New York (1987).
4. Y. W. Chien; Rate-control drug delivery systems: Controlled release vs. sustained release. Medical Progress Through Technology, 15:21–46 (1989).

5. S. J. Segal; The development of Norplant implants. Studies in Family Planning, 14:161 (1983).
6. S. Diaz et al.; A Five-year clinical trial of levonorgestrel silastic implants (Norplant). Contraception, 25:447 (1982).
7. E. Weiner et al.; Plasma Levels of d-norgestrel after oral administration. Contraception, 23:197 (1981).
8. H. B. Croxatto et al.; Plasma levels of levonorgestrel in women during long-term use of Norplant. Contraception, 23:197 (1981).
9. R. W. Baker and H. K. Lonsdale; Controlled delivery—an emerging use for membranes. Chemtech., 5:668 (1975).
10. Y. W. Chien; Logics of transdermal controlled drug administration. Drug Develop. & Ind. Pharm., 9:497–520 (1983).
11. W. R. Good; Tranderm-Nitro: Controlled delivery of nitroglycerin via the transdermal route. Drug Develop. & Ind. Pharm., 9:647–670 (1983).
12. L. Schenkel, J. Balestra, L. Schmitt, and J. Shaw; in: *Second International Conference on Drug Absorption—Rate Control in Drug Therapy,* University of Edinburgh, Edinburgh, Scotland, September 21–23, 1983, p. 41.
13. L. R. Laufer, J. L. De Fazio, J. K. H. Lu, D. R. Meldrum, P. Eggena, M. P. Sambhi, J. M. Hershman, and H. L. Judd; Am. J. Obstet. Gynecol. 146:533 (1983).
14. S. K. Chandrasekaran, A. S. Michaels, P. Campbell, and J. E. Shaw; Scopolamine permeation through human skin in vitro. AIChE. J., 22:828 (1976).
15. J. E. Shaw and S. K. Chandrasekaran; Controlled topical delivery of drugs for systemic action. Drug Metab. Rev., 8:223 (1978).
16. J. E. Shaw; Pharmacokinetics of nitroglycerin and clonidine delivered by the transdermal route. Am. Heart J., 108:217–223 (1984).
17. A. D. Keith; Polymer matrix considerations for transdermal devices. Drug Develop. & Ind. Pharm., 9:605–625 (1983).
18. P. K. Nonoon, M. A. Gonzalez, D. Ruggirello, J. Tomlinson, E. Babcock-Atkinson, M. Ray, A. Golub and A. Cohen; Relative bioavailability of a new transdermal nitroglycerin delivery system. J. Pharm. Sci., 75:688 (1986).
19. D. S. T. Hsieh, N. Smith and Y. W. Chien; Subcutaneous controlled delivery of estradiol by Compudose implants: In vitro and in vivo evaluations, Drug Develop. & Ind. Pharm., 13:2651–2666 (1987).
20. M. Wolff, G. Cordes and V. Luckow; In vitro and in vivo release of nitroglycerin from a new transdermal therapeutic system. Pharm. Res., 1:23–29 (1985).
21. M. Corbo, J.-C. Liu and Y. W. Chien; Transdermal controlled delivery of propanolol from a multilaminate adhesive device. Pharm. Res., 6:753–758 (1989).
22. Y. W. Chien and H. J. Lambert; Microsealed Pharmaceutical Delivery Device, U.S. Patent 3,992,518 (November 16, 1976).
23. Y. W. Chien and H. J. Lambert; Microsealed Pharmaceutical Delivery Device, U.S. Patent 4,053,580 (October 11, 1977).
24. Y. W. Chien; Microsealed drug delivery systems: Fabrication and performance. In: Methods in Enzymology (K. J. Widder and R. Green, Eds.), Volume 112, Academic Press, New York (1985) 461–470.
25. Y. W. Chien; Microsealed drug delivery systems: Theoretical aspects and biomedical assessments. In: Recent Advances in Drug Delivery Systems (J. M. Anderson and S. W. Kim, Eds.), Plenum, New York (1984) 367–387.
26. D. R. Sanvordeker, J. G. Cooney and R. C. Wester; Transdermal Nitroglycerin Pad, U.S. Patent 4,336,243 (June 22, 1982).
27. Y. W. Chien, P. R. Keshary, Y. C. Huang, and P. P. Sarpotdar; Comparative controlled skin permeation of introglycerin from marketed transdermal delivery systems. J. Pharm. Sci., 72:968 (1983).

28. A. Karim; Transdermal absorption: A unique opportunity for constant delivery of nitroglycerin. Drug Develop. & Ind. Pharm., 9:671 (1983).
29. Y. W. Chien, T.-Y. Chien and Y. C. Huang; Transdermal Fertility Control System and Process, U.S. Patent 4,818,540 (April 4, 1989).
30. T.-Y. Chien, Y. C. Huang and Y. W. Chien; In: Abstracts of the Japan-United States Congress of Pharmaceutical Sciences (1987), No. N3-W-30.
31. T.-Y. Chien, Y. C. Huang and Y. W. Chine; In: Abstracts of the Japan-United States Congress of Pharmaceutical Sciences (1987), No. N3-W-31.
32. T.-Y. Chien and Y. W. Chien; In: Abstracts of the Japan-United States Congress of Pharmaceutical Sciences (1987), No. N3-W-32.
33. Y. W. Chien, T.-Y. Chien and Y. C. Huang; Proceed. Int. Symp. Control. Rel. Bioact. Mater., 15:286–287 (1988).
34. Y. W. Chien, T.-Y. Chien, Y. C. Huang and R. E. Bagdon; Proceed. Int. Symp. Control. Rel. Bioact. Mater., 15:288–289 (1988).
35. Y. W. Chien, T.-Y. Chiem, R. E. Bagdon, Y. C. Huang and R. H. Bierman; Transdermal dual-controlled delivery of contraceptive drugs: Formulation development, in vitro and in vivo evaluations, and clinical performance. Pharm. Res., 6:1000–1010 (1989).
36. F. Theeuwes; Elementary osmotic pump. J. Pharm. Sci., 64:1987–1991 (1975).
37. F. Theeuwes and S. I. Yum; Principles of the design and operation of generic osmotic pumps for the delivery of semi-solid or liquid drug formulations. Ann. Biomed. Eng., 4:343–353 (1976).
38. L. G. J. De Leede; Ph.D. Thesis, Rate-controlled and Site-specified Rectal Drug Delivery, State University of Leiden, Leiden, The Netherlands, 1983.
39. Y. W. Chien; Potential developments and new approaches in oral controlled release drug delivery systems. Drug Develop. & Ind. Pharm., 9:1291–1330 (1983).
40. F. Theeuwes; Oros-osmotic system development. Drug Develop. & Ind. Pharm., 9:1331–1357 (1983).
41. J. C. Liu, M. Farber and Y. W. Chien; Comparative release of phenylpropranolamine HCl for long-acting appetite suppressant products: Acutrim vs. Dexatrim. Drug Develop. & Ind. Pharm., 10:1639–1661 (1984).
42. A. S. Michaels; Device for Delivering Drug to Biological Environment, U.S. Patent 4,180,073 (December 25, 1979).
43. P. J. Blackshewar, T. D. Rohde, J. C. Grotling, F. D. Dorman, P. R. Perkins, R. L. Varco and H. Buchwald; Control of blood glucose in experimental diabetes by means of a totally implantable insulin infusion device. Diabetes, 28:634–639 (1979).
44. P. J. Blackshear, T. D. Rohde, R. L. Varco and H. Buchwald; One year of continuous heparinization in the dog using a totally implantable infusion pump. Surg. Gynecol. Obstet., 141:176–186 (1975).
45. American Pharmacy, Implantable pump for morphine. NS24:20 (1984).
46. Y. W. Chien; Transnasal Systemic Medication, Elsevier, Amsterdam (1985) Chapters 6 and 7.
47. D. S. T. Hsieh and R. Langer; Zero-order drug delivery systems with magnetic control. In: (T. J. Roseman S. Z. Mansdorf, Eds.) Controlled Release Delivery Systems, Dekker, New York (1983), Chapter 7.
48. J. Kost, K. W. Leong and R. Langer; Ultrasonic controlled polymeric delivery systems. Proc. Int. Symp. Control. Rel. Bioact. Mater., 13:177–178 (1986).
49. P. Tyle and P. Agrawala; Drug delivery by phonophoresis. Pharm. Res., 6:355–361 (1989).
50. L. E. Bertolucci; Introduction of anti-inflammatory drugs by iontophophoresis: Double blind study. J. Orthopaed. & Sports Phys. Ther., 4:103 (1982).

51. J. M. Glass, R. L. Stephen and S. C. Jacobson; The quantity of radiolabelled dexamethasone delivered to tissue by iontophoresis. Int. J. Dermatol., 19:519 (1980).
52. P. R. Harris; Iontophoresis: Clinical research in musculoskeletal inflammatory conditions. J. Orthopaed. & Sports Phys. Ther., 4:109 (1982).
53. Y. W. Chien, O. Siddiqui, Y. Sun, W. M. Shi and J. C. Liu; Transdermal iontophoretic delivery of therapeutic peptides/proteins: (I) Insulin. Ann. N.Y. Acad. Sci., 507:32–51 (1987).
54. Y. W. Chien; *Novel Drug Delivery Systems: Fundamentals, Developmental Concepts and Biomedical Assessments.* Dekker, New York (1982).
55. Y. W. Chien and E. P. K. Lau; Controlled drug release from polymeric delivery devices (IV): In vitro-in vivo correlation on the subcutaneous release of Norgestomet from hydrophilic implants, J. Pharm. Sci., 65:488–492 (1976).
56. D. S. T. Hsieh, K. Mann and Y. W. Chien; Enhanced release of drugs from silicone elastomers. I. Release kinetics of pineal and steroidal hormones. Drug Develop. & Ind. Pharm., 11:1391 (1985).
57. D. S. T. Hsieh and Y. W. Chien; Enhanced release of drugs from silicone elastomers. II. Induction of swelling and changes in microstructure. Drug Develop. & Ind. Pharm., 11:1411 (1985).
58. D. S. T. Hsieh and Y. W. Chien; Enhanced release of drugs from silicone elastomers. III. Subcutaneous controlled administration of melantonin for early onset of estrus cycles in ewes. Drug Develop. & Ind. Pharm., 11:1433 (1985).
59. D. S. T. Hsieh, P. Mason and Y. W. Chien; Enhanced release of drugs from silicone elstomers. IV. Subcutaneous controlled release of indomethacin and in vivo-in vitro correlations. Drug Develop. & Ind. Pharm., 11:1447 (1985).
60. P. Mason, D. S. T. Hsieh and Y. W. Chien; Enhanced release of sulfonamides from medicated artificial skin. Pharm. Research, 3:59S (1986).
61. L. P. Amsel; Controlled Release Suspensions, *APhA/APS Midwest Reg. Meet.*, Chicago, Illinois, April 2, 1984.
62. Y. Raghunathan; Prolonged Release Pharmaceutical Preparations, U.S. Patent 4,221,778 (September 9, 1980).
63. Y. Raghunathan, L. Amsel, O. Hinsvark and W. Bryant; Sustained-release drug delivery system I: Coated ion-exchange resin system for phenylpropranolamine and other drugs. J. Pharm. Sci., 70:379 (1981).
64. J. Heller; Biodegradable polymers in controlled drug delivery. CRC Critical Reviews in Therapeutic Drug Carrier Systems, 1:39 (1984).
65. Y. Morimoto and S. Fujimoto; Albumin microspheres as drug carriers. CRC Critical Reviews in Therapeutic Drug Carrier Systems, 2:19 (1985).
66. H. Sezaki and M. Hashida; Micromolecular drug conjugates in targeted cancer chemotherapy. CRC Critical Reviews in Therapeutic Drug Carrier Systems, 1:1 (1984).
67. J. Heller and S. H. Pengburn; A triggered bioerodible Naltrexone delivery system Proceed. Int. Symp. Control. Rel. Bioact. Mater., 13:35–36 (1986).
68. J. Heller and P. V. Trescony; Controlled drug release by polymer dissolution. II. Enzyme-mediated delivery device. J. Pharm. Sci., 68:919 (1979).
69. T. A. Horbett, B. D. Ratner, J. Kost and M. Singh; A bioresponsive membrane for insulin delivery, In: *Recent Advances in Drug Delivery Systems* (J. M. Anderson and S. W. Kim, Eds.), Plenum Press, New York (1984), pp. 209–220.
70. S. W. Kim, S. Y. Jeong, S. Sato, J. C. McRea and J. Feijan; Self-regulating insulin delivery system—a chemical approach. In Recent Advances in Drug Delivery Systems (J. M. Anderson and S. W. Kim, Eds.), Plenum, New York (1983), p. 123.
71. R. W. Baker; *Controlled Release of Biologically-active Agents*, J. Wiley & Sons, New York (1987).

72. S. Y. Jeong, S. W. Kim, M. J. D. Eenink and J. Feijen; Self-regulating insulin delivery systems. I. Synthesis and characterization of glycosylated insulin, J. Control. Rel., 1:57–66 (1984).
73. H. Ringsdorf; Synthetic polymeric drugs. In Polymeric Delivery Systems (R. J. Kostelnik, Ed.), Gordon and Brech, New York (1978).

2
Fundamentals of Rate-Controlled Drug Delivery

I. INTRODUCTION

In Chapter 1 the concept of rate-controlled drug delivery was introduced. This laid the foundation for the development of various controlled-release drug delivery systems.

Depending upon the sophistication of the technology in use, controlled-release drug delivery systems can be classified into the following three categories:

1. Rate-preprogrammed drug delivery systems
2. Activation-modulated drug delivery systems
3. Feedback-regulated drug delivery systems

All three categories of controlled-release drug delivery systems (Figure 1) consist of the following common structural features:

1. Drug reservoir compartment
2. Rate-controlling element
3. Energy source

As discussed in Chapter 1, the rate-preprogrammed drug delivery system can be viewed as the first generation of controlled-release drug delivery systems (DDS), providing the technological basis for the later development of activation-modulated drug delivery systems, as the second generation of controlled-release DDS, and of feedback-regulated drug delivery systems, as the third generation of controlled-release DDS. Therefore, analysis of the mechanisms and fundamentals involved in the rate-controlled delivery of drugs from rate-preprogrammed drug delivery systems will assist us in gaining a basic understanding of the scientific principles and technological foundation behind various types of controlled-release DDS and in laying the framework for the development of activation-modulated and feedback-regulated drug delivery systems. The knowledge gained will also aid the rational design of an ideal site-targeting controlled-release drug delivery system in the future.

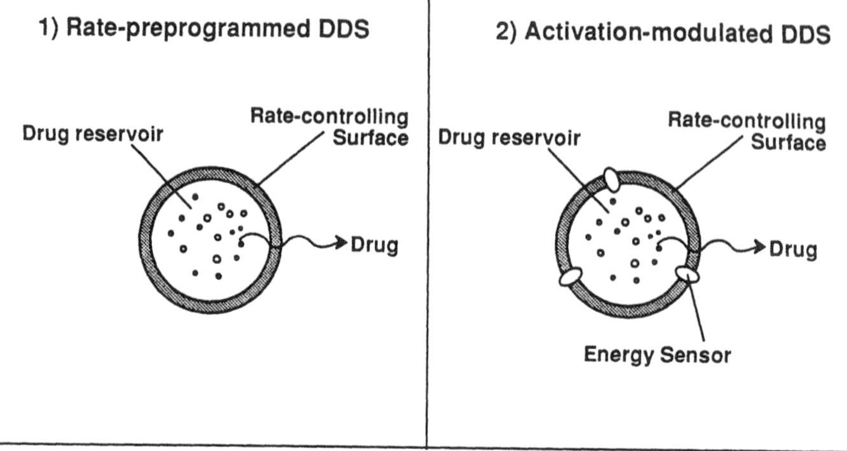

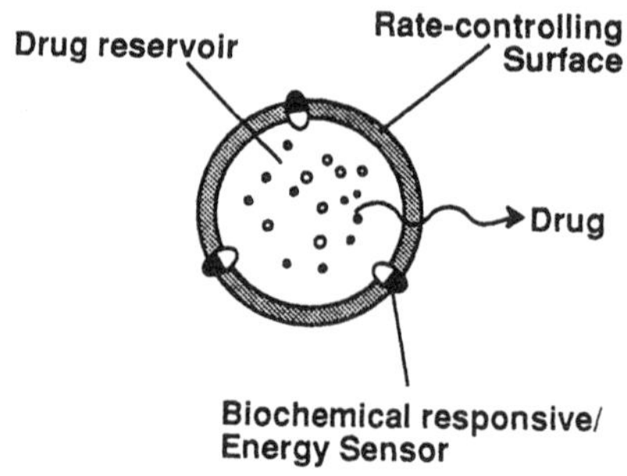

Figure 1 Various classes of controlled-release drug delivery systems. (Modified from Baker, 1987.)

Essentially, rate-preprogrammed drug delivery systems are constructed from the following two basic types of controlled-release drug delivery systems.

A. Polymer Membrane Permeation-Controlled Drug Delivery Systems

Examples are the progesterone-releasing intrauterine device (IUD) for yearly intrauterine contraception, the pilocarpine-releasing ocular insert for 4–7 day continuous glaucoma treatment, the nitroglycerin-releasing Transderm-Nitro therapeutic system for daily prophylaxis or therapy of angina pectoris, and the levonorgestrel-releasing subdermal implant for 3–7 year subcutaneous fertility control. In this type of controlled-release drug delivery system the drug reservoir compartment is encapsulated inside a polymeric membrane, which acts as a rate-controlling element, and the release of drug is controlled by its permeation through the rate-controlling membrane (Figure 2).

B. Polymer Matrix Diffusion-Controlled Drug Delivery Systems

Examples are the nitroglycerin-releasing Nitro-Dur therapeutic systems for the prophylaxis or therapy of angina pectoris, the medroxyprogesterone acetate-releasing

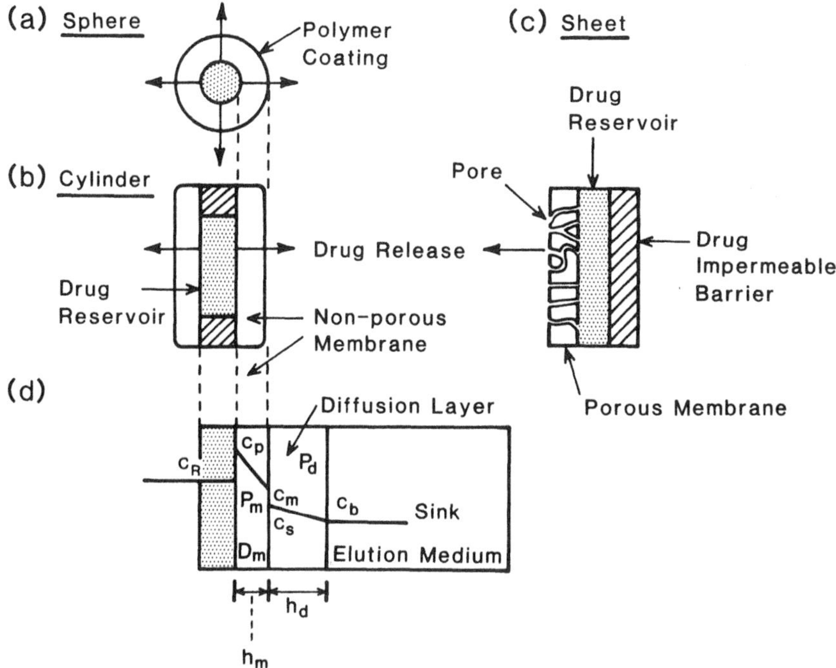

Figure 2 The controlled release of drug molecules from a membrane permeation-controlled reservoir-type drug delivery device of various shapes in which drug is contained in a reservoir compartment enclosed by a polymer membrane. C_R, drug concentration in the drug reservoir compartment; C_p, the solubility of drug in the polymer phase; C_m, the concentration of drug at the polymer/solution interface; C_s, the concentration of drug at the solution/polymer interface; C_b, the concentration of drug in the bulk of elution solution; h_m, thickness of the membrane wall; and h_d, thickness of the hydrodynamic diffusion layer.

vaginal ring for cyclic contraception, and the norgestomet-releasing subdermal hydrophilic implants for controlled estrus synchronization. In this type of controlled-release drug delivery system the drug reservoir is homogeneously dispersed throughout a polymer matrix, which acts as rate-controlling element, and the release of drug is thus controlled by its diffusion through the rate-controlling polymer matrix (Figure 3).

Theoretically, the controlled release of drugs from both membrane permeation- and matrix diffusion-controlled drug delivery systems is governed by Fick's laws of diffusion (1), which define the flux of diffusion J_D across a plane surface of unit area as follows:

$$J_D = -D \frac{dc}{dx} \tag{1}$$

where D is the diffusivity of drug molecule in a medium of solid, solution, or gas; dc/dx is the concentration gradient of the drug molecule across a diffusional path with thickness dx; and a negative sign is used to define the direction of diffusion from a region with high concentration to a region with low concentration. The drug

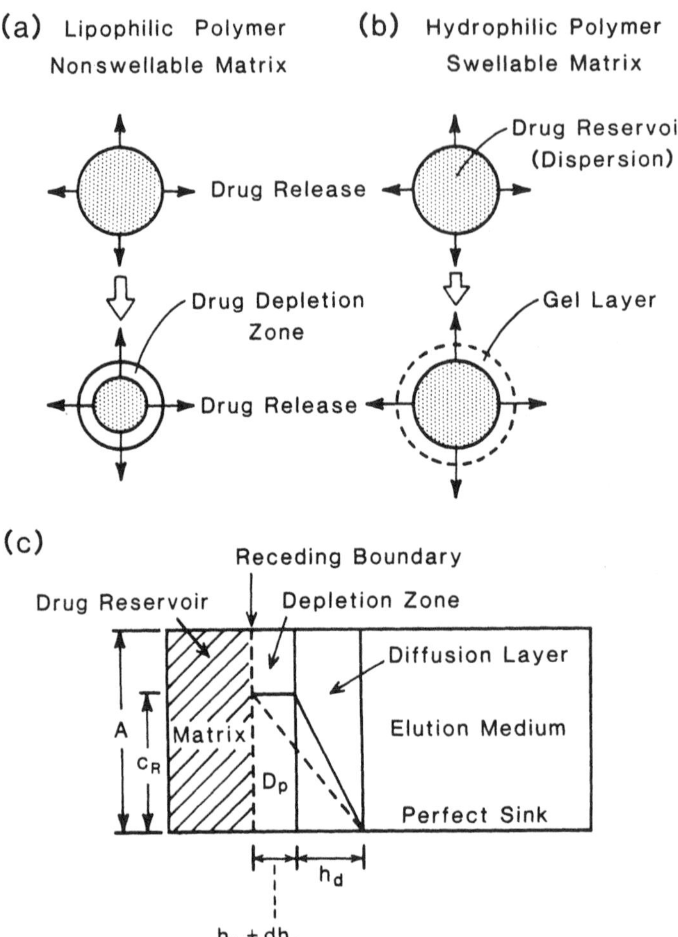

Figure 3 The controlled release of drug molecules from a polymer matrix diffusion-controlled dispersion-type drug delivery device in which solid drug is homogeneously dispersed in the polymer matrix. A, the initial amount of drug solids impregnated in a unit volume of polymer matrix; C_R, the reservoir concentration or saturated concentration of drug in the polymer phase; D_p, the diffusivity of drug in the polymer matrix; h_p and h_d, the thicknesses of the drug depletion zone in the matrix and of the hydrodynamic diffusion layer on the immediate surface of the device, respectively; and $d(h_p)$, the differential thickness of the depletion zone formed following the release of more drug solids.

concentration gradient acts as the energy element for the diffusion of drug molecules.

Kinetically, however, the release of drug molecules from these two types of drug delivery system is essentially controlled by two different mechanistic patterns that result from the time dependency of the diffusional flux J_D [Equation (1)]. In the case of the membrane permeation-controlled drug delivery devices, the drug concentration gradient dc/dx across a constant thickness of polymer membrane is essentially constant and thus is invariable with time (Figure 2). On the other hand, the concentration gradient in matrix diffusion-controlled drug delivery devices is time dependent and decreases progressively in response to the growing increase in the thickness of the diffusional path dx (Figure 3) as time goes on (2).

Therefore, mechanistic analysis should be conducted using different physical models for gaining fundamental understanding of the rate-controlled delivery of drugs from these basic types of controlled-release drug delivery systems.

II. MECHANISTIC ANALYSIS OF CONTROLLED-RELEASE DRUG DELIVERY

For the mechanistic analysis of rate-controlled drug release from both membrane permeation-controlled and matrix diffusion-controlled drug delivery systems and their hybrids, the following assumptions must be established:

1. Dissolution of drug crystals into their surrounding medium is the first step of the drug release process.
2. A pseudo-steady state exists in the process of controlled drug release.
3. The diffusion coefficient of a drug molecule in a given medium is invariable with time and distance.
4. The interfacial partitioning of a drug molecule from polymer toward solution is related to its solubilities in polymer (C_p) and in solution (C_s) as defined by

$$K = \frac{C_s}{C_p} \tag{2}$$

where K is defined as the distribution (or partition) coefficient.

A. Polymer Membrane Permeation-Controlled Drug Delivery Systems

The release of drug molecules from a drug delivery device having a rate-controlling polymer membrane can be visualized schematically as in Figure 2. A reservoir of drug particles is encapsulated in a compartment of various shapes with walls made from rate-controlling polymer membranes with a well-defined thickness h_m. Microscopically the drug molecules on the outermost layer of particles must dissociate themselves from the crystal state and dissolve in the surrounding medium, partition and diffuse through the polymer structure, and finally partition into the elution medium surrounding the device. A thickness h_d of the hydrodynamic diffusion layer, a stagnant solution layer in which drug molecules diffuse by natural convection under a concentration gradient, is also present on the immediate surface of the device (3). If the thickness of the hydrodynamic diffusion layer is maintained constantly small in magnitude than the surface area of the device available for the diffusion of drug molecules, the diffusion of drug molecules to and from the device surface across the hydrodynamic diffusion layer may be treated, for simplicity, as one-dimensional diffusion to a plane surface (4).

The cumulative amount of drug Q released from a unit surface area of a polymer membrane permeation-controlled drug delivery device can be described by the mathematical expression

$$Q = \frac{C_p K D_d D_m}{K D_d h_m + D_m h_d} t - \frac{D_m D_d}{K D_d h_m + D_m h_d} \int_0^t C_{b(t)} dt \tag{3}$$

where C_p is the solubility of drug in the polymer phase, K was defined earlier by Equation (2); D_m and D_d are the diffusivities in the polymer membrane with thickness h_m and in the aqueous solution in the hydrodynamic diffusion layer with thickness h_d, respectively, $C_{b(t)}$ is the concentration of drug in the interface of diffusion layer/bulk solution, t is time, and dt is a differential length of time.

If a sink condition is maintained throughout the course of controlled drug release, i.e., $C_{b(t)} \simeq 0$ or $C_s \gg C_{b(t)}$, Equation (3) can be reduced to

$$Q = \frac{C_p K D_d D_m}{K D_d h_m + D_m h_d} t \tag{4}$$

Equation (4) suggests that the controlled release of drug molecules from polymer membrane permeation-controlled drug delivery devices should yield a constant drug release profile. The rate of drug release is defined by

$$\frac{Q}{t} = \frac{C_p K D_d D_m}{K D_d h_m + D_m h_d} \tag{5}$$

This Q versus t relationship is also followed fairly well in the controlled release of progesterone from a Progestasert IUD, which contains a suspension of progesterone crystals in membrane-impermeable silicone medical fluid as a drug reservoir in which progesterone molecules must be dissolved in the suspending medium before partitioning onto the nonporous membrane of the ethylene-vinyl acetate copolymer (Chapter 10); in the controlled release of nitroglycerin from the Transderm-Nitro system, in which nitroglycerin molecules are also dissolved in the silicone medical fluid partition and diffuse through the liquid-filled microporous membrane of the ethylene-vinyl acetate copolymer and silicone adhesive layer (Chapter 7); and in the controlled release of pilocarpine from the Ocusert ophthalmic insert, in which pilocarpine-alginate, the drug reservoir core, is encapsulated in a semipermeable membrane of an ethylene-vinyl acetate copolymer and pilocarpine molecules must be dissolved in the tear fluid, which penetrates into the device, before being diffused out of the drug delivery device (Chapter 6). From a system design viewpoint, the Progestasert, Transderm-Nitro, and Ocusert controlled-release drug delivery systems are extensions of the polymer membrane permeation-controlled drug delivery system. A typical set of results is illustrated by the controlled release of nitroglycerin from the Transderm-Nitro system (Figure 4), in which the intrinsic rate of drug release is described by

$$\frac{Q}{t} = \frac{K_{m/r} K_{a/m} D_a D_m}{K_{m/r} D_m h_a + K_{a/m} D_a h_m} C_R \tag{6}$$

where $K_{m/r}$ and $K_{a/m}$ are the partition coefficients for the interfacial partitioning of drug molecules from the drug reservoir to the rate-controlling membrane and from the membrane to the surface adhesive layer, respectively; D_m and D_a are the diffusion coefficients in the rate-controlling membrane (with thickness h_m) and in the adhesive layer (with thickness h_a), respectively. Since it is a microporous membrane, the porosity and tortuosity of the pores in the membrane should be included in the determination of D_m and h_m. C_R is the drug concentration in the reservoir compartment.

The constant drug release profile (Q versus t linearity) is maintained as long as

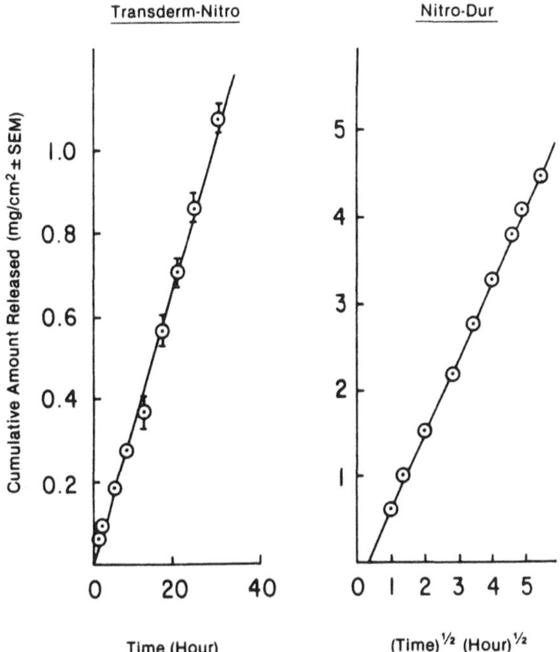

Figure 4 Comparative release profiles of nitroglycerin from the Transderm-Nitro system, a membrane permeation-controlled reservoir-type drug delivery device, and from the Nitro-Dur system, a polymer matrix diffusion-controlled dispersion-type drug delivery device.

the release of drugs is under membrane permeation control and the concentration of drug stays at the saturation level.

Two extreme conditions may exist for Equation (5), depending upon the relative magnitude of the KD_dh_m term compared with the D_mh_d term.

Condition 1. If the KD_dh_m term is significantly greater than the D_mh_d term, as in a thick polymer membrane (h_m) and/or a membrane fabricated from a polymer material with an extremely low drug diffusivity (D_m), Equation (5) is simplified to Equation (7):

$$\frac{Q}{t} = \frac{C_pD_m}{h_m} \tag{7}$$

Under this condition the release of drug molecules is governed by the membrane-modulated permeation process and the rate of drug release is a function of the polymer solubility C_p and membrane diffusivity D_m and is inversely proportional to the thickness of the polymer membrane h_m.

Condition 2. If the KD_dh_m term is substantially smaller than the D_mh_d term, as in a drug species with a very low interfacial partition coefficient (K) or a thick hydrodynamic diffusion layer (h_d), Equation (5) can be reduced to Equation (8):

$$\frac{Q}{t} = \frac{KD_dC_p}{h_d} \tag{8a}$$

or, since $KC_p = C_s$,

$$\frac{Q}{t} = \frac{C_s D_d}{h_d} \tag{8b}$$

Under this condition the release of drug molecules is governed by the diffusion layer-limiting partition-controlled process and the rate of drug release is a linear function of both the solution solubility C_s and solution diffusivity D_d in the diffusion layer and is inversely proportional to the thickness of the hydrodynamic diffusion layer h_d. Close examination of Equations (7) and (8b) indicates that both equations share the same features of those of Fick's expression for diffusional flux [Equation (1)].

B. Polymer Matrix Diffusion-Controlled Drug Delivery Systems

In this type of drug delivery device the drug reservoir compartment is a homogeneous dispersion of drug as discrete crystals (or solid particles) in a matrix environment formed by the cross-linking of linear polymer chains. The dispersing drug crystals (or particles) cannot move away from their individual positions in the polymer matrix. It is believed that the drug molecules can elute out of the matrix only by, first, dissolution in the surrounding polymer and then diffusion through the polymer structure. It is visualized that, microscopically, the drug solids in the layer closer to the surface of the device are the first to elute, and when this layer becomes depleted the drug solids in the next layer then begin to dissolute and elute. This is schematically illustrated in Figure 3C. There exists a drug depletion zone with a thickness h_p. This thickness becomes greater and greater as more drug solids elute out of the device, leading to inward advancement of the interface of the drug dispersion zone/drug depletion zone farther into the core of the device for another thickness $d(h_p)$.

Microscopically there also exists a thin layer of stagnant solution, known as the hydrodynamic diffusion layer, on the immediate surface of the device. The thickness h_d of this stagnant solution layer varies with the hydrodynamics of the solution in which the device is immersed. If this thickness is controlled at a finite and constant value, which can be achieved by rotating the device at a constant angular rotation (6), and is smaller than the surface area of the device available for the diffusion of drug molecules, the diffusion of drug molecules to and from the device surface may also be treated, for simplicity, as one-dimensional diffusion to a plane surface (4).

The cumulative amount of drug Q released from a thickness of drug dispersion zone h_p in the polymer matrix diffusion-controlled drug delivery system is defined by

$$Q = \left(A - \frac{C_p}{2}\right)h_p \tag{9}$$

where A is the initial amount of drug solid impregnated in a unit volume of polymer matrix (mg/cm^3) and h_p is a time-dependent variable. Its magnitude depends on the physicochemical interactions between drug molecules and polymer matrix as expressed mathematically by the complicated relationship

$$h_p^2 + \frac{2(A - C_p)D_p h_d h_p}{(A - C_p/2)D_d \bar{k} K} = \frac{2C_p D_p}{A - C_p/2}t \tag{10}$$

where $\bar{k}$ is a constant that accounts for the relative magnitude of the concentration gradients in both diffusion layer and depletion zone. The other terms have the same meaning as defined in Equations (3) through (9).

The complicated relationship of Equation (10) can be simplified if the following conditions exist.

Condition 1. At the very early stage of drug release process, only a small amount of drug has been released from the surface layers of drug-dispersed polymer matrix so the thickness of depletion zone h_p is so small that the following condition exists:

$$h_p^2 \ll \frac{2(A - C_p)D_p h_d h_p}{(A - C_p/2)D_d \bar{k} K} \tag{11}$$

and Equation (10) is reduced to

$$h_p \simeq \frac{\bar{k} D_d K C_p}{(A - C_p)h_d} t \tag{12}$$

Substituting Equation (12) for the h_p term in Equation (9) gives

$$Q = \frac{\bar{k} D_d K C_p}{h_d} t \tag{13a}$$

or

$$Q = \frac{\bar{k} D_d C_s}{h_d} t \tag{13b}$$

since $A - C_p \simeq A - C_p/2 \simeq A$ and $KC_p = C_s$.

The same drug release pattern (Q versus t profile) is also observed if the magnitude of the partition coefficient K is so small and/or the thickness of the hydrodynamic diffusion layer h_d is so large that the condition defined by Equation (11) exists throughout the whole course of a drug release process (7). This results in a constant drug release rate as described by

$$\frac{Q}{t} = \frac{\bar{k} D_d C_s}{h_d} \tag{14}$$

The similarity of Equation (14) to Equation (8b) suggests that, under the diffusion layer-limiting partition-controlled process, both membrane permeation- and matrix diffusion-controlled drug delivery systems produce essentially a constant drug release profile. The presence of the $\bar{k}$ term in Equation (14) implies that polymer matrix diffusion-controlled drug delivery devices are more sensitive than polymer membrane permeation-controlled drug delivery systems to the relative difference in the concentration gradients between the depletion zone and the diffusion layer.

Condition 2. If the magnitude of the partition coefficient K is very large and/or the thickness of the diffusion layer h_d is rather small, or the thickness of the drug depletion zone h_p, after a finite time has passed, becomes substantially large in its magnitude, the following condition exists:

$$h_p^2 \gg \frac{2(A - C_p)D_p h_d h_p}{(A - C_p/2)D_d \bar{k} K} \tag{15}$$

so Equation (10) is reduced to

$$h_p \simeq \left(\frac{2C_pD_p}{A - C_p/2}\, t\right)^{1/2} \tag{16}$$

Substituting Equation (16) for the h_p term in Equation (9) yields

$$Q = [(2A - C_p)C_pD_pt]^{1/2} \tag{17}$$

Equation (17) indicates that, after a finite period of drug elution, a matrix diffusion-controlled process becomes the predominant step in determining the rate of drug release from the drug-dispersing polymer matrix; The cumulative amount of drug Q released from a unit surface area of a polymer matrix diffusion-controlled drug delivery device becomes directly proportional to the square root of time $t^{1/2}$. This Q versus $t^{1/2}$ relationship has been demonstrated in the intravaginal controlled release of progestins from various types of vaginal rings (Chapter 9), in the transdermal controlled release of nitroglycerin from the Nitro-Dur transdermal infusion system (Chapter 7), and in the subcutaneous controlled release of norgestomet from Hydron subdermal implants (Chapter 8). A typical set of results for the Q versus $t^{1/2}$ drug release profile is shown in Figure 4. The flux of drug release is defined by

$$\frac{Q}{t^{1/2}} = [(2A - C_p)C_pD_p]^{1/2} \tag{18}$$

Kinetically, a remarkable difference in the patterns of drug release profiles is expected between the matrix-controlled process [Equation (17)] and the diffusion layer-limiting partition-controlled process [Equation (13b)], as well as in the rates of drug release at steady state from the polymer matrix diffusion-controlled drug delivery system [Equation (18)] and from the polymer membrane permeation-controlled drug delivery system [Equations (7) and (8)]. This is demonstrated by the difference in the release profiles of nitroglycerin from the Transderm-Nitro and Nitro-Dur systems (Figure 4).

The release of drugs from semisolid ointment (8) and solid wax matrix (9) systems was also reported to follow the Q versus $t^{1/2}$ relationship as defined by Equation (17).

The transition between a partition-controlled process and a matrix-controlled process was investigated in great detail using a homologous series of the esters of *p*-aminobenzoic acid (10). It was observed that the transition time t_{trans}, at which the diffusional resistances across the hydrodynamic diffusion layer h_d and across the drug depletion zone h_p in the polymer matrix are equal in their magnitudes, is defined by

$$t_{\text{trans}} = \frac{3\beta^2}{\alpha}\frac{1}{KC_s} \tag{19}$$

where K and C_s are as defined earlier and α and β are defined by the expressions

$$\alpha = \frac{2AD_p\epsilon}{\theta} \tag{20}$$

$$\beta = \frac{AD_ph_d\epsilon}{D_d\theta} \tag{21}$$

where ϵ and θ are the porosity and tortuosity of the polymer matrix, A, D_p, D_d, and h_d are as defined earlier.

Equation (19) suggests that the transition time t_{trans} is dependent upon a number of system parameters, such as drug loading A; the diffusivities D_p and D_d, solution solubility C_s, and partition coefficient K of the drug; the porosity ϵ and tortousity θ of the polymer matrix; and the thickness of the hydrodynamic diffusion layer h_d.

The controlled release of drug molecules from a matrix-type drug delivery system fabricated from a biodegradable or bioerodible polymer is more complicated and can be viewed as the result of two concurrent kinetic processes: the diffusion of drug molecules in the polymer matrix and the first-order hydrolytic erosion (or degradation) of the polymer (Chapter 8).

C. (Membrane-Matrix) Hybrid Drug Delivery Systems

Recently, several (membrane-matrix) hybrid (or sandwich) drug delivery systems were developed with the objective of combining the constant drug release kinetics of polymer membrane permeation-controlled drug delivery systems with the mechanical superiority of polymer matrix diffusion-controlled drug delivery systems. The hybrid system is exemplified by the development of clonidine-releasing and scopolamine-releasing transdermal therapeutic systems (Catapres-TTS and Transderm-Scop, Chapter 7), levonorgestrel-releasing subdermal implants (Norplant II, Chapter 8), and the *d*-norgestrel-releasing vaginal ring (Chapter 9), in which a rate-controlling nonmedicated polymeric membrane is added to coat the surface of the drug-dispersing polymer matrix, and the release of drug molecules thus becomes controlled by membrane permeation instead of matrix diffusion.

The release profile of drug from a sandwich drug delivery system (Figure 5) is constant, and the instantaneous rate of drug release is defined by

$$\frac{dQ}{dt} = \frac{AC_pD_p}{[D_pK_m(1/P_m + 1/P_d)]^2 + 4AC_pD_pt^{1/2}} \tag{22}$$

where A is the initial amount of drug solid impregnated in a unit volume of polymer matrix with solubility C_p and diffusivity D_p; K_m is the partition coefficient for the interfacial partitioning of drug molecules from polymer matrix toward polymer coating membrane; P_m is the permeability coefficient of the polymer coating membrane with thickness h_m; and P_d is the permeability coefficient of the hydrodynamic diffusion layer with thickness h_d.

At the initial stage of the drug release process, the thickness of the drug depletion zone $h_p(t)$ is very small in magnitude and the rate-controlling step in the release of drug from a sandwich drug delivery system resides in the polymer coating membrane (and the hydrodynamic diffusion layer). In such a circumstance Equation (22) can be reduced to

$$\frac{dQ}{dt} = \frac{C_pP_mP_d}{K_m(P_m + P_d)} \tag{23}$$

A zero-order (Q versus t) drug release profile results, and the rate of drug release is described by

$$\frac{Q}{t} = \frac{C_pP_mP_d}{K_m(P_m + P_d)} \tag{24}$$

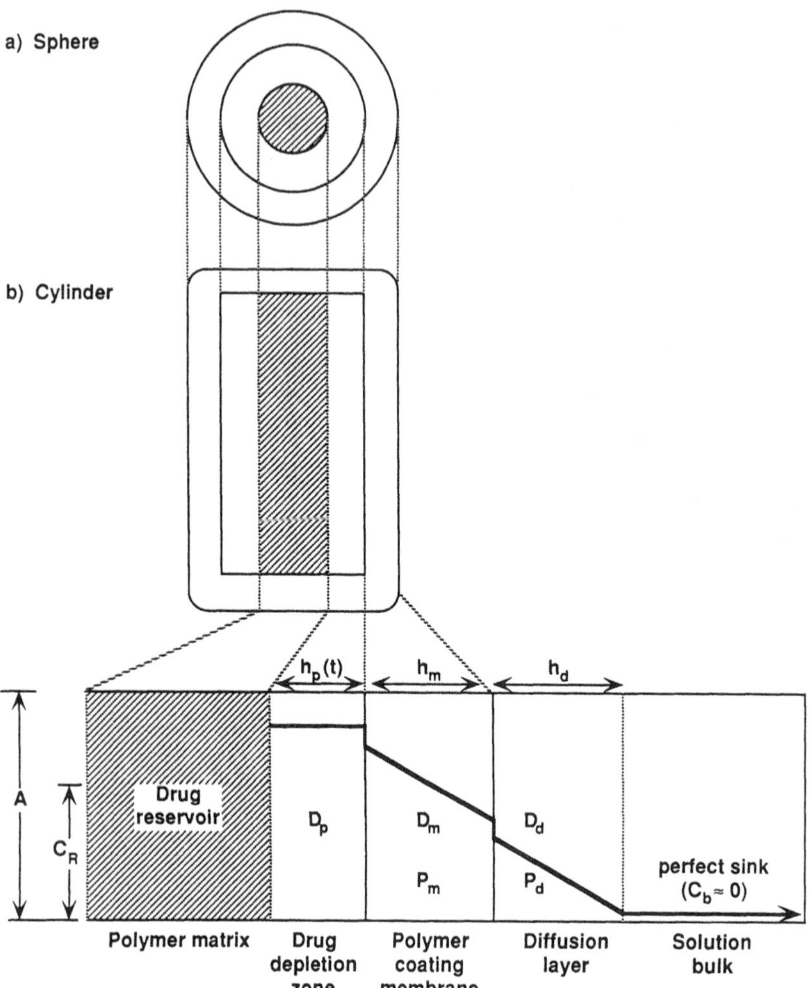

Figure 5 The controlled release of drug molecules from a (membrane-matrix) hybrid (or sandwich)-type drug delivery system in which solid drug is homogeneously dispersed in a polymer matrix, which is then encapsulated inside a polymeric membrane, where D, P, and h are the diffusivity, permeability, and thickness, respectively, and the subscripts p, m, and d denote the drug depletion zone in the polymer matrix, polymer coating membrane, and diffusion layer, respectively.

As the duration of drug release is prolonged, the receding interface of the drug depletion zone/drug dispersion zone progresses to the extent that the diffusion process through the drug depletion zone becomes the predominant step in determining the release of drug molecules from the sandwich drug delivery system. In this situation Equation (22) is simplified to

$$\frac{dQ}{dt} = \frac{1}{2}\left(\frac{AC_pD_p}{t}\right)^{1/2} \tag{25}$$

and a non-zero–order (Q versus $t^{1/2}$) drug release profile results and the flux of drug release is described by

$$\frac{Q}{t^{1/2}} = (2AC_pD_p)^{1/2} \tag{26}$$

D. Microreservoir Dissolution-Controlled Drug Delivery Systems

Another approach to achieve a constant drug release profile from a matrix-type drug delivery system is illustrated by the development of microreservoir dissolution-controlled drug delivery systems (MDD) (11). MDD are composed of a homogeneous dispersion of numerous microscopic spheres (<30 μm) of drug suspension in a solid polymer matrix. These drug-containing spheres exist homogeneously throughout the cross-linked polymer matrix as a discrete, immobilized, unleachable liquid compartment (Figure 6). The drug molecules can elute out of the MDD only by, first, dissolution in the liquid compartment and partitioning into and diffusion through the polymer matrix, and then partitioning into the surrounding elution solution. Mech-

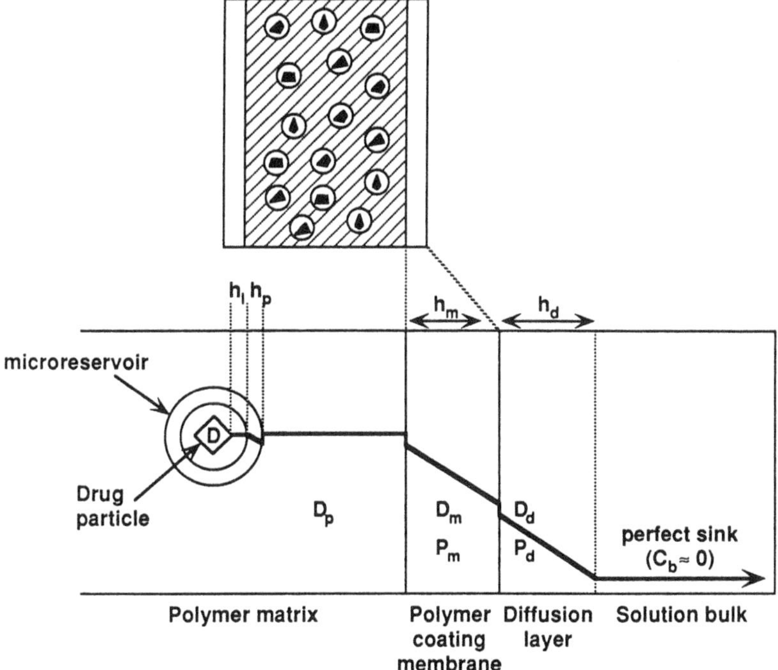

Figure 6 A cross-section of a microreservoir dissolution-controlled drug delivery system. The microscopic liquid compartments, which encapsulate drug particles, are homogeneously dispersed as discrete, immobilized, unleachable spheres (with diameter ≤ 30 μm) in a cross-linked polymer matrix. D, P, and h are the diffusivity, permeability, and thickness, respectively, and the subscripts p, m, and d denote the polymer matrix, polymer coating membrane, and diffusion layer, respectively.

anistic analysis suggested that drug release from MDD can be expressed by the relationship

$$\frac{Q}{t} = \frac{D_m D_d m K_m}{D_m h_d + D_d h_m m K_m}\left[nC_p - \frac{D_l S_l(1-n)}{h_l}\left(\frac{1}{K_l} + \frac{1}{K_p}\right)\right] \tag{27}$$

where $m = a/b$, in which a is the ratio of drug concentration in the bulk of elution solution over drug solubility in the same medium and b is the ratio of drug concentration at the outer edge of the polymer coating membrane over drug solubility in the same polymer. n is the ratio of drug concentration at the inner edge of the interfacial barrier over the drug solubility in the polymer matrix. K_l, K_p, and K_m are the partition coefficients for the interfacial partitioning of drug from the microreservoir compartments to the polymer matrix, from the polymer matrix to the polymer coating membrane, and from the polymer coating membrane to the elution solution, respectively. D_l, D_m, and D_d are the diffusivities of the drug in the liquid layer surrounding the drug particles, polymer coating membrane surrounding the polymer matrix, and the hydrodynamic diffusion layer surrounding the polymer coating membrane, respectively, with thickness h_l, h_m, and h_d, respectively. S_l and C_p are the solubilities of the drug in the microreservoir compartments and in the polymer matrix, respectively.

Equation (27) suggests that the controlled release of drug from MDD is zero order as illustrated by the linear Q versus t relationship. This is exemplified by the subcutaneous controlled release of norgestomet from subdermal MDD implants in heifers and cows (Chapter 8) and the subcutaneous controlled release of desoxycorticosterone acetate in rats for up to 129 days (12).

The mathematical expression of Equation (27) can be simplified if

$$nC_p \gg \left[\frac{D_l S_l(1-n)}{h_l}\left(\frac{1}{K_l} + \frac{1}{K_p}\right)\right] \tag{28}$$

then

$$\frac{Q}{t} = \frac{D_d D_m K_m m n C_p}{D_d h_m K_m m + D_m h_d} \tag{29}$$

This situation can exist if K_l and K_p, the partition coefficient for the interfacial partitioning of drug molecules from microreservoir compartment toward polymer matrix and from polymer matrix toward polymer coating membrane, respectively, are high while S_l, the solubility of drug molecules in the microreservoir compartment, is very low but C_p, the solubility of drug molecules in the polymer matrix, is extremely high.

Equation (29) is very similar to Equation (5) and can also be simplified under the following conditions.

Condition 1. If the $D_d h_m m K_m$ term is substantially greater than the $D_m h_d$ term, as when a thick polymer coating membrane (h_m) exists on the surface of a medicated polymer matrix or the drug diffusivity D_m in the polymer coating membrane is extremely low, Equation (29) can be reduced to

$$\frac{Q}{t} = \frac{nD_m C_p}{h_m} \tag{30a}$$

or, since $C_p = C_m/K_p$,

$$\frac{Q}{t} = \frac{nD_mC_m}{h_mK_p} = \frac{n}{K_p}\frac{D_mC_m}{h_m} \tag{30b}$$

Under this condition the release of drug molecules from the microreservoir dissolution-controlled drug delivery system is governed by the membrane-controlled permeation process and the rate of drug release is a linear function of the diffusivity D_m and solubility C_m of drug in the polymer coating membrane. It is also inversely proportional to the thickness of the polymer coating membrane h_m and the partition coefficient K_p for the interfacial partitioning of drug molecules from polymer matrix toward polymer coating membrane. In the case of MDD, n, the ratio of drug concentration at the inner edge of the interfacial barrier over drug solubility in the polymer matrix, also plays a rate-limiting role in the controlled release of drugs.

Condition 2. If, on the other hand, the $D_dh_mmK_m$ term is significantly smaller than the D_mh_d term, as when a thick hydrodynamic diffusion layer (h_d) exists or the drug molecules have a very low partition coefficient K_m, Equation (29) can be simplified to

$$\frac{Q}{t} = \frac{D_dmnK_mC_p}{h_d} \tag{31a}$$

or, since $K_mC_p = C_s$,

$$\frac{Q}{t} = \frac{nmD_dC_s}{h_d} = nm\frac{D_dC_s}{h_d} \tag{31b}$$

Under this condition the release of drug molecules from a microreservoir dissolution-controlled drug delivery system is governed by the diffusion layer-limiting partition-controlled process and the rate of drug release is linearly proportional to the solution diffusivity D_d and solution solubility C_s and is inversely proportional to the thickness h_d of the hydrodynamic diffusion layer. In the case of MDD, n and m also play rate-limiting roles in the controlled release of drugs.

Equations (30) and (31) are closely related to Equations (7) and (8). They all share features common to Fick's expression for diffusional flux [Equation (1)].

III. EFFECTS OF SYSTEM PARAMETERS ON CONTROLLED-RELEASE DRUG DELIVERY

The mechanistic analysis of controlled-release drug delivery conducted in Section II revealed that the partition coefficient, diffusivity, solubility, diffusional path thickness, and other system parameters play varying degrees of rate-limiting roles in controlling the release of drug molecules from membrane permeation-, matrix diffusion-, (membrane-matrix) hybrid-, or microreservoir-type rate-preprogrammed drug delivery systems. The importance of these system (physicochemical) parameters in controlling drug release was illustrated in the subcutaneous release of progestins from silicone capsules in rats (5,12). A fairly good agreement was established between the rates of subcutaneous drug release and the rates calculated from Equation (5) using the literature values for the partition coefficient, diffusivities, polymer solubility, and polymer wall thickness. In this calculation a physiological diffusion layer (h_d, which is not measurable) of 80 μm is assumed to exist on the surface of subdermal implants (Table 1). A good correlation was also reported to exist between the in vitro release rates for the release of steroids from matrix-type silicone devices and the values calculated from Equation (17) in a matrix diffusion-controlled process (13). A typical set of results is illustrated in Table 2.

In this section these rate-controlling system parameters are singled out, and the effects of each parameter on the overall controlled drug release profiles are analyzed and discussed.

A. Polymer Solubility

In the controlled release of a drug species from either polymer membrane permeation-controlled or polymer matrix diffusion-controlled drug delivery devices or other rate-preprogrammed drug delivery devices, the drug particles are visualized as not being releasable from the device until the drug molecules on the outermost surface layer of a drug particle dissociate from their crystal lattice structure, dissolve or partition into the surrounding polymer (in membrane or matrix form), diffuse through it, and finally partition into the elution medium surrounding the drug delivery device (Figures 2, 3, 5, and 6). This mechanistic analysis suggests that the solubility of a drug species in a rate-controlling polymer membrane or matrix plays a rate-controlling role in its release from a polymeric device. To release at an appropriate rate the drug requires adequate polymer solubility. The importance of polymer solubility in determining the rate of drug release from membrane permeation-, matrix diffusion-, (membrane-matrix) hybrid-, and microreservoir-type drug delivery systems can be appreciated by examining Equations (5), (17), (22), and (27), respectively.

Equation (5) indicates that the rate of drug release Q/t from a polymer membrane permeation-controlled drug delivery system is directly proportional to the magnitude of polymer solubility (C_p or C_m). The linear relationship between Q/t and C_p is illustrated in Figure 7. This linear relationship should also be followed in the controlled release of drug from a (membrane-matrix) hybrid-type drug delivery device [Equation (22)], as well as from a microreservoir-type drug delivery device [C_m in Equation (30b)].

On the other hand, in the controlled release of drugs from a polymer matrix diffusion-controlled drug delivery system the magnitude of $Q/t^{1/2}$ values is a function of the square root of polymer solubility [$C_p^{1/2}$ in Equation (17)]. The linear dependence of $Q/t^{1/2}$ on $C_p^{1/2}$ is illustrated in the controlled release of progesterone derivatives from matrix-type silicone devices (Figure 8).

The difference in polymer solubilities among drugs is very striking. For example, the solubility of steroids in silicone polymer can range from 1.2 μg/ml to as high as 1511.8 μg/ml (Table 3). This dramatic difference in polymer solubility among drugs is very much dependent on the difference in their chemical structures. Variation in functional groups and their stereochemical configurations greatly affects the magnitude of the polymer solubility of drugs. A typical example is demonstrated by the addition of one or more hydroxyl groups to positions 11, 17, and 21 on the progesterone skeleton (Table 3). The polymer solubility of progesterone in the lipophilic silicone polymer is significantly reduced with the addition of one OH group,

Table 1 Comparison Between the Observed In Vivo Release Rate from a Capsule-Type Silicone-Based Subdermal Implant and the Rate Calculated from Various Physicochemical Parameters

	Physicochemical Parameters						Q/t (μg/cm^2/day)	
Progestins	K	D_d (cm^2/day × 10^2)	D_p (cm^2/day × 10^2)	C_p (μg/ml)	h_d (cm)	h_m (cm)	Calculated[a]	Observed
Progesterone	0.022	4.99	14.26	513	0.008	0.080	65.40	64.50
Norgestomet	0.115	3.64	2.31	166	0.008	0.065	35.03	45.72
Chlormadinone acetate	0.122	2.42	2.45	82	0.008	0.080	13.70	18.10
Medroxyprogesterone acetate	0.037	3.60	3.21	87	0.008	0.080	10.24	9.20

[a]Calculated from Equation (5).

Source: Compiled from the data by Chien, Chem. Pharm. Bull., 24:1471 (1976).

Table 2 Comparison Between the Observed In Vitro Release Rates of Ethynodiol Diacetate from a Matrix-Type Silicone Device and the Rates Calculated from Various Physicochemical Parameters

Cosolvent[a] (%v/v)	Q/t ($\mu g/cm^2/day$) Calculated[b]	Q/t ($\mu g/cm^2/day$) Observed	$Q/t^{1/2}$ ($mg/cm^2/day^{1/2}$) Calculated[c]	$Q/t^{1/2}$ ($mg/cm^2/day^{1/2}$) Observed
Partition control				
20	65.5	82.6		
30	114.2	109.8		
40	176.1	141.3		
50	273.7	245.7		
55	344.7	360.0		
Matrix control				
65			3.09	2.99
70			3.09	3.12
75			3.09	3.09
80			3.09	3.04
85			3.09	3.05

[a]Volume fraction of polyethylene glycol 400 as cosolvent system in aqueous elution solution.
[b]Calculated from Equation (14).
[c]Calculated from Equation (18).

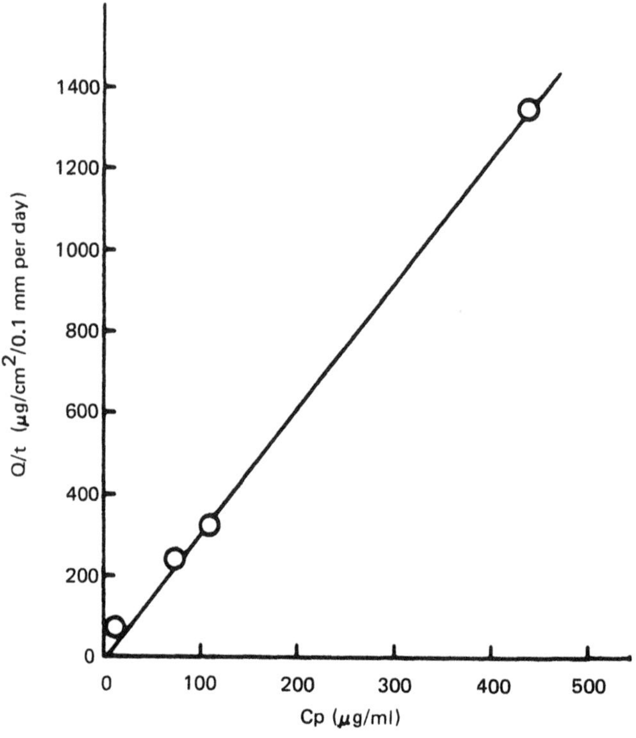

Figure 7 Linear relationship between the subcutaneous release rates of progestins (Q/t) from reservoir-type drug delivery devices and the magnitude of their polymer solubilities C_p.

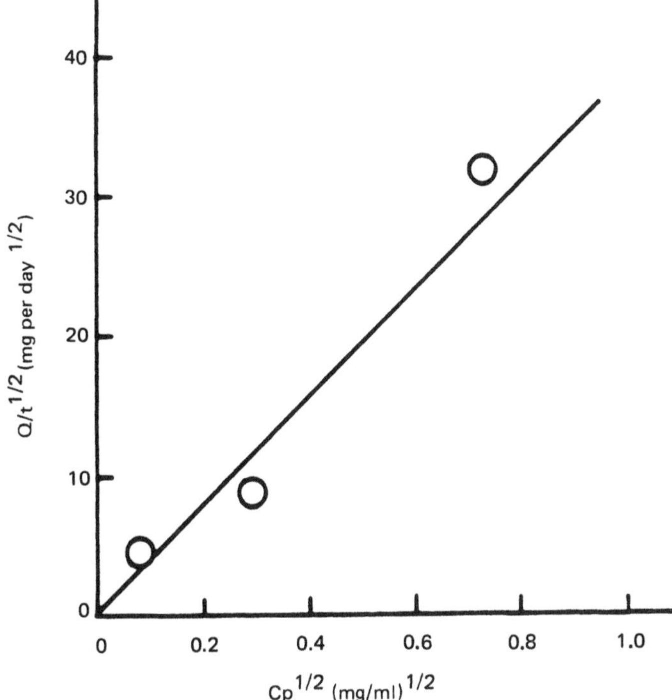

Figure 8 Linear relationship between the release flux $Q/t^{1/2}$ of progestins from matrix silicone devices and the square root of drug solubility in a liquid silicone polymer $C_p^{1/2}$. [Reproduced, with permission, from Roseman; J. Pharm. Sci., 61:46 (1972).]

which varies in the extent of reduction depending upon the position where the OH group is added. The polymer solubility of progesterone derivatives is further reduced with the substitution of two or more hydroxyl groups. On the other hand, the esterification of these hydrophilic OH groups tends to improve the polymer solubility of steroids substantially. This is exemplified by the esterification of β-estradiol to form 17-valerate and 3,17-dipropionate. The solubility of β-estradiol in silicone polymer has been enhanced 9- and 80-fold, respectively (16).

It is common practice to incorporate various quantities of finely ground fillers, such as siliceous earth, into silicone elastomers to enforce the mechanical strength of the elastomers. The presence of fillers was reported to increase the polymer solubility C_p of drugs as a result of the Langmuir adsorption of drug molecules onto the filler particles.

If polymer is treated, microscopically, as a solid solution, then the mole fraction solubility $\bar{C}_p$ of a drug in the polymer composition (17) can be described by

$$\log \bar{C}_p = \log \frac{C_p}{C_p + X_p} = -\log r_p - \frac{\Delta H_{fm}}{2.303RT_s} \frac{T_m - T_s}{T_m} \tag{32}$$

Table 3 Polymer Solubility of Steroids in a Silicone Polymer

Steroids	Polymer solubility[a] (μg/ml)
Androgen family	
Testosterone	134.7
17α-Methyltestosterone	190.5
17α-Ethynyltestosterone	4.8
Testosterone-17-benzoate	106.9
Testosterone-17-acetate	518.8
Testosterone-17-propionate	564.1
19-Nortestosterone	134.4
17α-Ethyl-19-nortestosterone	180.0
17α-Ethynyl-19-nortestosterone	14.7
17α-Ethynyl-19-nortestosterone-3,17-diacetate	1511.8
Estrogen family	
β-Estradiol	16.3
17α-Ethynylestradiol	74.5
Estradiol-3-methyl ether	736.8
17α-Ethynylestradiol-3-methyl ether	179.1
Estradiol-3-benzoate	36.8
Estradiol-17-valerate	150.2
Estradiol-3,17-dipropionate	1282.3
Progestogen family	
Progesterone	594.7
21-Hydroxyprogesterone	205.9
17α-Hydroxyprogesterone	26.5
11α-Hydroxyprogesterone	9.1
17α,21-Dihydroxyprogesterone	1.5
11α,21-Dihydroxyprogesterone	1.2

[a]In silicone medical fluid (DC 360) at 37°C.
Source: Compiled from the data by Chien, *ACS Symposium Series 33: 53 (1976)*, and *Chien et al., J. Pharm. Sci., 68:689 (1979)*.

where X_p is the mole fraction of the polymer composition; r_p is the activity coefficient of the drug solute in the polymer structure; ΔH_{fm} is the molar heat of fusion absorbed when the drug crystals melt into the polymer structure; T_m is the melting point temperature; and T_s is the temperature of the system investigated.

Microscopically, the solution process of a drug crystal in a polymer composition can be visualized as consisting of two consecutive steps: (i) the dissociation of drug molecules from their crystal lattice structure, and (ii) the solvation of these dissociated drug molecules into the polymer structure. The first step requires a dissociation energy and is a T_m-dependent process; the second step requires a solvation energy and thus depends upon T_s. If this is the case, then the energy term in Equation (32) may be split as follows:

$$\log \bar{C}_p = -\log r_p + \frac{\Delta H_d}{2.303R}\frac{1}{T_m} - \frac{\Delta H_s^p}{2.303R}\frac{1}{T_s} \tag{33}$$

where ΔH_d is the energy required in the process of dissociation of the drug molecules from their crystal lattice structure, and ΔH_s^p is the energy required in the process of solvation of the drug molecules into the polymer structure.

Under controlled conditions both the log r_p and $\Delta H_d/2.303RT_m$ terms are constant values for a given drug species, and Equation (33) can be reduced to Equation (34) to define the dependence of the mole fraction polymer solubility $\bar{C}_p$ on the system temperature T_s:

$$\log \bar{C}_p = \text{constant} - \frac{\Delta H_s^p}{2.303R}\frac{1}{T_s} \tag{34}$$

This linear dependence of polymer solubility on T_s^{-1} has frequently been observed experimentally (15,16). A typical example is illustrated by the dissolution of norgestomet in silicone polymer (Figure 9). A ΔH_s^p value of 6.60 kcal/mol was calculated as the energy required in the solvation of norgestomet in the short-chained silicone polymer. The magnitude of the ΔH_s^p value was observed to be sensitive to

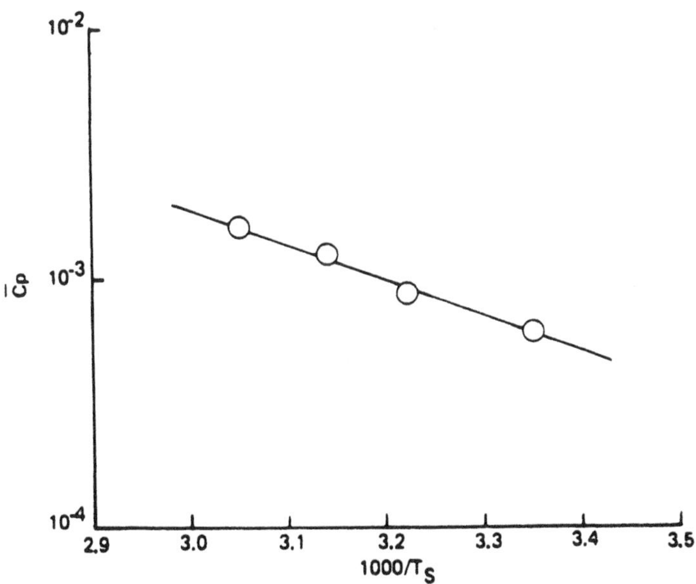

Figure 9 Exponential relationship between the mole fraction solubility $\bar{C}_p$ of norgestomet in silicone medical fluid and the reciprocal of the system temperature T_s. [Reproduced, with permission, from Chien; ACS Symposium Series, 33:53 (1976).]

the addition of hydroxyl groups and their number and location on the progesterone skeleton (Table 4). The physical state of the polymer also affects the magnitude of the ΔH_s^p values.

If the variation in the log r_p and $\Delta H_s^p/2.303RT_s$ values are negligibly small among a homologous series of drug analog, Equation (33) can also be simplified to

$$\log \bar{C}_p = \text{constant} + \frac{\Delta H_d}{2.303R}\frac{1}{T_m} \tag{35}$$

which suggests that the mole fraction solubility $\bar{C}_p$ of a drug species is exponentially dependent upon the reciprocal of its melting point temperature T_m^{-1}. A typical example is illustrated by the dissolution of 27 steroids in silicone polymer (Figure 10). A ΔH_d value of 14.12 kcal/mol was calculated as the energy required for the dissociation of steroid molecules from their crystal lattice structure (16).

B. Solution Solubility

Kincl and his associates observed that the release of progesterone from silicone capsules, a polymer membrane permeation-controlled drug delivery device, into distilled water was decreased by one-half when the volume of distilled water was reduced from 100 to 25 ml (18). It was also reported that the in vitro release rates of norgestrel and megestrel acetate through silicone capsules was greater when the elution medium was plasma than when it was distilled water (19). Schuhman and Taubert (20) carried out a comparative study of the in vitro release of various steroids from silicone capsules and discovered that the release rates in human plasma were 2–15 times greater than those in normal saline (Table 5). This increase in the rate of steroid release was rationalized as the result of enhancement in the solubility of steroids in the elution medium by protein binding. These studies also point to the importance of the solution solubility C_s in determining the rate of drug release from a drug delivery system.

Because in vivo a sink condition is effectively maintained by active hemoper-

Table 4 Energies ΔH_s^p Required for Solvation of a Progesterone Derivative in Silicone Polymer

	ΔH_s^p (kcal/mol)	
Progesterone derivatives	Liquid state[a]	Solid state[b]
Progesterone	5.49	4.86
21-Hydroxyprogesterone	8.80	7.18
17α-Hydroxyprogesterone	8.27	2.11
11α-Hydroxyprogesterone	7.88	8.07
17α,21-Dihydroxyprogesterone	8.72	12.58
11β,21-Dihydroxyprogesterone	12.99	13.18
11β,17α,21-Trihydroxyprogesterone	12.71	3.01

[a]In silicone medical fluid (DC 360, 20 cv) at 37°C.
[b]In cross-linked silicone elastomer 382 at 37°C.
Source: Compiled from the data by Chien et al., J. Pharm. Sci., 68:689 (1979).

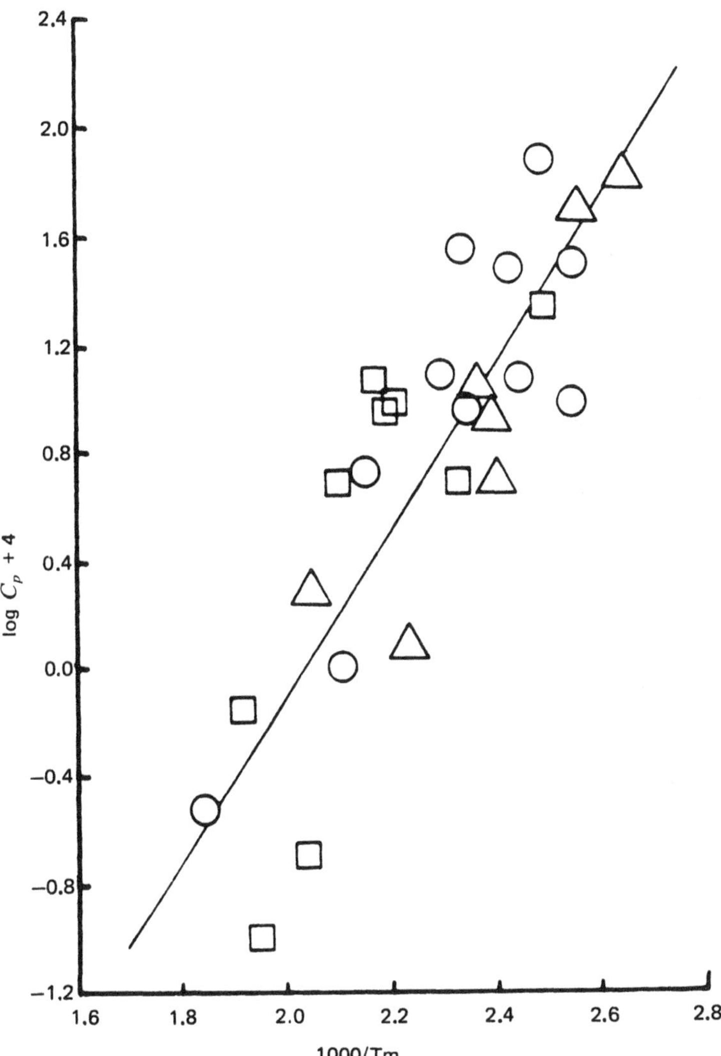

Figure 10 Exponential relationship between the mole fraction solubility $\bar{C}_p$ of androgens (○), progestins (□), and estrogens (△) in silicone medical fluid and the reciprocal of their melting point temperatures T_m. [Reproduced, with permission, from Chien, ACS Symposium Series, 33:53 (1976).]

fusion, it is logical to have all in vitro drug release studies conducted under perfect sink conditions. In this way the in vivo condition is better simulated and correlation of in vitro drug release profiles with in vivo drug administrations can thus be achieved.

To gain a better understanding of the mechanisms of controlled drug release from membrane permeation-, matrix diffusion-, (membrane-matrix) hybrid-, and/or microreservoir-type drug delivery systems, it is necessary to maintain a sink condition so that the release of drug is solely controlled by the delivery system and is not affected or complicated by the solution solubility factor already discussed. This sink

Table 5 Effect of Human Plasma on the In Vitro Release Rate of Various Steroids from a Silicone Capsule

Steroids	$Q/t(\mu g/cm^2/day)$[a] Normal saline	Human plasma	Enhancement[b] (%)
Progesterone	186	276	148
17α-OH-progesterone	63	182	289
17α-OH-progesterone capronate	22	180	818
Gestonorcapronate	14	200	1429
Norgestrel	27	250	926
Medroxyprogesterone acetate	200	236	118
Methylnortestosterone	107	280	1667
17α-Ethynyl-19-nortestosterone	15	250	177
17α-Ethynyl-19-nortestosterone acetate	113	200	

[a]At 37°C.
[b]Enhancement (%) = $(Q/t)_{plasma}/(Q/t)_{saline} \times 100\%$.
Source: Compiled from the data by Schuhman and Taubert, Acta Biol. Med. Germ., 24:897 (1970).

condition may be satisfactorily accomplished either by maintaining the drug concentration in the bulk solution (C_b in Figures 2, 3, 5, and 6) very close to zero or by making solution solubility much greater than the bulk solution concentration ($C_s \gg C_b$) (7). In practice, the first approach has been extensively utilized in many conventional dissolution and drug elution studies and is best illustrated by the use of a large volume (60 L/day) of distilled water as the circulating elution solution in studying the release of medroxyprogesterone acetate (with C_s of 3.25 μg/ml) from matrix-type silicone devices (21).

The second approach was developed by using a water-miscible cosolvent as a solubilizer and addition of the cosolvent into the elution solution to increase the solution solubility of drugs (13). A small volume (100–200 ml) of aqueous solution of polyethylene glycol 400, a water-soluble hydrophilic polymer, was used to enhance the aqueous solubility of ethynodiol diacetate (with C_s of 13.7 μg/ml), which achieved a 3- to 584-fold enhancement in aqueous solubility (22). The solution solubility was enhanced exponentially as a function of the volume fraction of cosolvent added (Figure 11). The same phenomenon was also reported for progestins, estrogens, and androgens (23) using one or more cosolvents.

The aqueous solubility of drugs varies remarkably from one drug species to another. Similar to polymer solubility, as illustrated in Table 3, the difference in aqueous solubility among drugs is also very much dependent upon the difference in their chemical structure, the types and physicochemical nature of the functional groups, and the variations in their stereochemical configurations (Table 6). The aqueous solubility of testosterone and nandrolone was reportedly reduced by esterification of their 17-OH group (24). The aqueous solubility of testosterone esters and nanodrolone esters was found to be correlated with their molar volume in an exponential relationship (Figure 12).

The aqueous solubility of most steroids is very low (Table 6), and this low solubility usually presents a difficult task for investigators in their attempts to maintain a perfect sink condition in conducting in vitro drug release studies. Several phar-

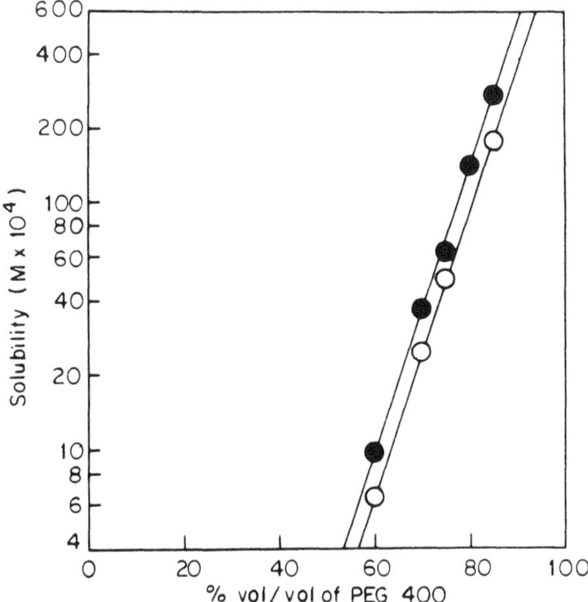

Figure 11 Exponential increase in the solution solubility of ethynodiol diacetate with the addition of varying volume fractions of polyethylene glycol 400 as the cosolvent in the elution solution at 25°C (○) and 37°C (●). [Reproduced, with permission, from Chien et al.; J. Pharm. Sci., 63:365 (1974).]

Table 6 Aqueous Solubility of Steroids

Steroids	Solubility[a] (μg/ml)	Reference[b]
Cortisol	543	1
Progesterone	27	1
Testosterone	25	1
Norgestomet	19.0	2
6α-Methyl-11β-hydroxyprogesterone	16.6	3
Ethynodiol diacetate	13.7	4
19-Norprogesterone	12.0	1
Chlormadinone acetate	10.0	5
Norethindrone	9.0	1
17α-Hydroxyprogesterone	8.1	3
Estradiol	5.0	1
Norgestrel	5.0	1
Medroxyprogesterone acetate	3.3	3
Melengestrol acetate	3.0	1
Megestrol acetate	2.0	1
Mestranol	1.5	1

[a]In distilled water at 37°C

[b]Compiled from the data by (1) Sundaram and Kincl; Steroids, 12:517 (1968); (2) Chien; Chem. Pharm. Bull., 24:1471 (1976); (3) Roseman; J. Pharm. Sci., 61:46 (1972); (4) Chien et al.; J. Pharm. Sci., 63:365 (1974); (5) Haleblian et al.; J. Pharm. Sci., 60:541 (1971).

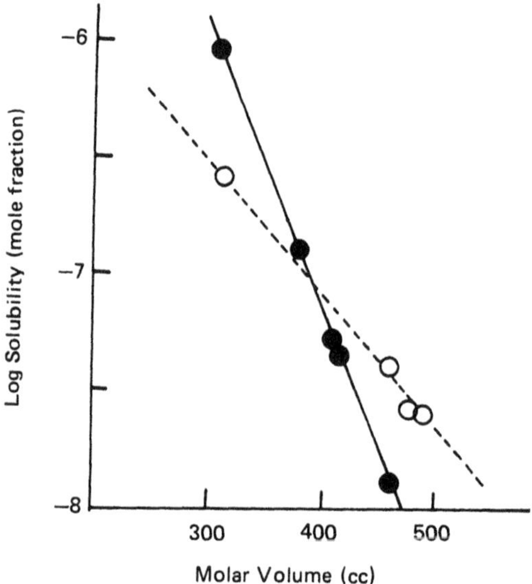

Figure 12 Semilogarithmic relationship between the aqueous solubilities of (●) testosterone esters and (○) nandrolone esters and their molar volumes. [Plotted from the data by Chaudry and James; J. Med. Chem., 17:157 (1974).]

maceutical approaches can be applied to improve the aqueous solubility of poorly soluble drugs, such as micelle formation (25), complexation (26), and cosolvency (23,24,27,28), without chemically modifying the drug molecules.

Solubilization of poorly soluble drugs in aqueous solution can be effectively accomplished by using multiple cosolvent systems. This was demonstrated by the solubilization of steroids (23) and metronidazole (28). It was reported that the aqueous solubility of steroids, which are lipophilic in nature, could be remarkably enhanced by the addition of a combination of more than one miscible cosolvent into distilled water (23). The apparent (gross) solubility of a steroid in various multiple cosolvent systems is described by the following relationships:

1. Binary cosolvent system:

$$\log C_x = \log C_w + e_x f_x \tag{36}$$

2. Ternary cosolvent system:

$$\log C_{a,x} = \log C_w + e_a f_a + e_x f_x \tag{37}$$

3. Quaternary cosolvent system:

$$\log C_{a,b,x} = \log C_w + e_a f_a + e_b f_b + e_x f_x \tag{38}$$

where C's are the apparent solubilities of a steroid in distilled water w and aqueous solutions of various cosolvents a, b, and/or x; and e's are the slopes for the semilogarithmic relationship between solubility and the volume fraction f of a specific cosolvent (Figure 11).

In the case of metronidazole, a polar drug, the exponential linearity between its solution solubilities and the volume fractions of various cosolvent combinations was followed very closely until a maximum solubility was reached (29). This maximum solubility was observed to occur in an aqueous solution with a dielectric constant of 41.5 ± 1.8. Beyond this maximum solubility, any increase in the volume fraction of a cosolvent yielded a reduction in solution solubility (Figure 13), which showed a linear decline (with a negative e_x value). The phenomenon of maximum solubility at optimum dielectric constant was observed in all the cosolvent combinations tested that had a dielectric constant very close to that of metronidazole.

The solution solubility C_s was reported to affect the magnitude of drug release profiles from polymer membrane permeation-controlled drug delivery systems (5). It was observed that the cumulative amount of norgestomet released from silicone capsules, a membrane permeation-controlled drug delivery device, increased with the increase in the solution solubility of the drug in the elution media (Figure 14). This solution solubility dependence of drug release profiles is predictable from the theoretical expression of Equation (8b).

Similarly, the magnitude of drug release rates Q/t from membrane-matrix hybrid and microreservoir-type drug delivery devices should also be a function of drug solubility in an elution solution [Equation (31b)]. Equations (8b) and (31b) can be ex-

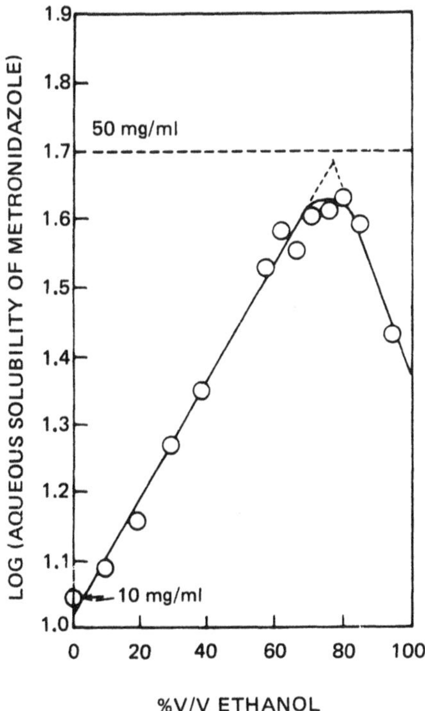

Figure 13 Biphasic semilogarithmic relationship between the aqueous solubility of metronidazole and the volume fraction of ethanol in the ethanol-water cosolvent system.

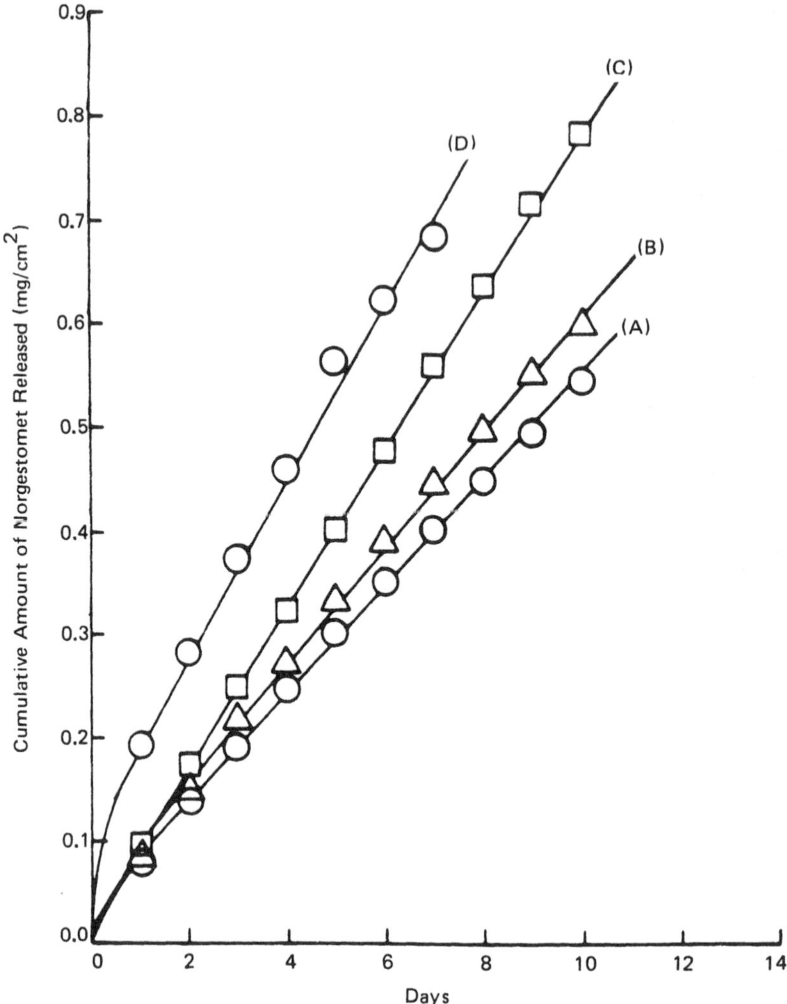

Figure 14 Dependence of the controlled release profiles of norgestomet from silicone capsules on its solution solubility. Solubility (mg/ml): (A) 0.104, (B) 0.239, (C) 1.968, and (D) 2.682. [Reproduced, with permission, from Chien; Chem. Pharm. Bull., 24:1471 (1976).]

pressed alternatively as follows to define the effect of solution solubility C_s on the rates of drug release from both membrane permeation- and microreservoir-type drug delivery systems:

$$\log \frac{Q}{t} = \log \frac{D_d}{h_d} + \log C_s \tag{39}$$

$$\log \frac{Q}{t} = \log\left(nm \frac{D_d}{h_p} \right) + \log C_s \tag{40}$$

The linear relationship between log (Q/t) and log C_s is demonstrated by the controlled release of norgestomet from silicone capsules and from microsealed drug delivery devices (Figure 15).

In matrix-type drug delivery systems, the effect of solution solubility on controlled drug release profiles is more complex in nature. It was reported that solution solubility affects both the mechanisms and the rate profiles of controlled drug release from matrix-type drug delivery systems (7,13,22). When the solution solubility of ethynodiol diacetate was maintained at a level below 200 μg/ml, the release of ethynodiol diacetate was observed to follow a partition-controlled process, and a constant (Q versus t) release pattern resulted as predicted from Equation (14). The rate of drug release Q/t was directly proportional to the magnitude of the C_s values (Table 7). On the other hand, as the solution solubility was further increased to a level beyond 780 μg/ml, a matrix-controlled process became predominant and Q versus $t^{1/2}$ release pattern was observed as expected from Equation (17). The $Q/t^{1/2}$ values were essentially constant and independent of the variation in solution solubility (Table 7).

The controlled release of drugs from matrix-type drug delivery devices prepared from hydrophilic polymers shows a dependence on solution solubility different from that of the devices prepared from lipophilic polymers. Instead of giving biphasic drug release profiles as discussed earlier, the controlled release of drugs from a drug-dispersing hydrophilic polymer matrix shows the Q versus $t^{1/2}$ release pattern from the beginning as a result of the increase in the diffusional path thickness in response to the uptake of elution solution by the polymer. The effect of solution solubility on such a hydrophilic drug delivery system is illustrated by the controlled release of norgestomet from Hydron implants (Syncro-Mate-B implants, Chapter 8). The mag-

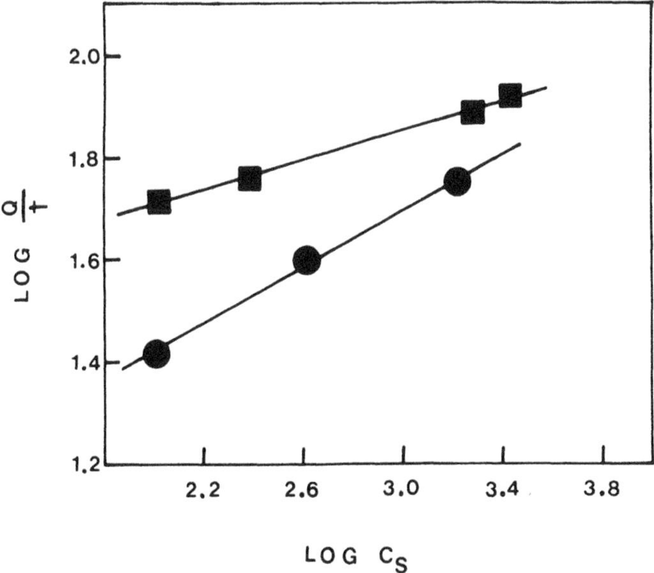

Figure 15 Effect of solution solubility C_s in the elution medium on the release rates Q/t of norgestomet from (■) a silicone capsule and (●) a microsealed drug delivery system.

Table 7 Effect of Solution Solubility C_s on the Mechanisms and Rate Profiles of Ethynodiol Diacetate Release from a Matrix-Type Silicone Device

C_s[a] (μg/ml)	Q/t (μg/cm²/day)	$(Q/t^{1/2})$ (mg/cm²/day$^{1/2}$)
Partition control		
37	82.6	
65	109.8	
100	141.3	
156	245.7	
196	360.0	
Transition phase		
437		
648		
Matrix control		
780		2.99
1390		3.12
2430		3.09
4450		3.04
8000		3.05

[a]The aqueous solubility of ethynodiol diacetate at 37°C is 13.7 μg/ml.

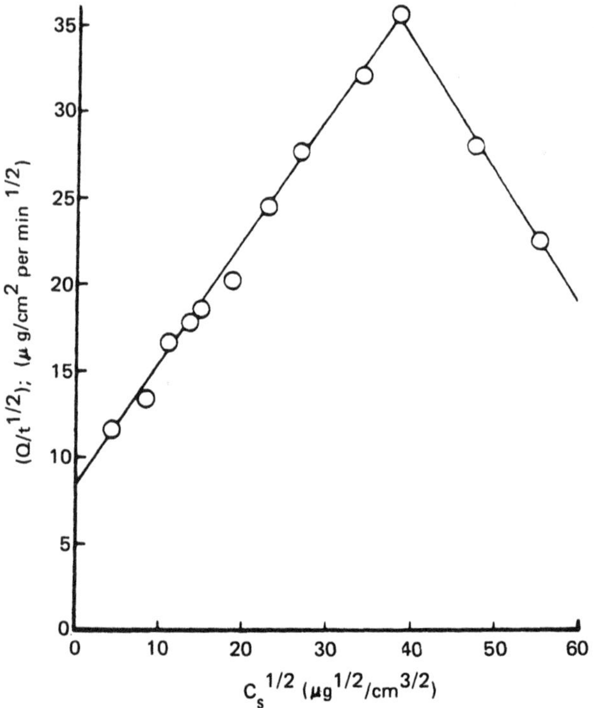

Figure 16 Effect of solution solubility on the initial release profiles $(Q/t^{1/2})_i$ of norgestomet from a matrix-type hydrophilic drug delivery device.

nitude of $Q/t^{1/2}$ values increases linearly with the square root of the solution solubility $C_s^{1/2}$ (Figure 16). After reaching a peak value, the $Q/t^{1/2}$ value decreases in a manner that is directly proportional to the increase in the $C_s^{1/2}$ value.

Thermodynamically, the dissolution of drug crystals in an aqueous solution is also an energy-dependent process. The temperature dependence of the solution solubility, as represented by mole fraction solubility $\bar{C}_s$, is expressed by the relationship

$$\log \bar{C}_s = \text{constant} - \frac{\Delta H_{T,s}}{2.303R} \frac{1}{T_s} \tag{41}$$

The linear relationship between log $\bar{C}_s$ and T_s^{-1} was illustrated by the dissolution of norgestomet crystals in distilled water and various aqueous solutions of polyethylene glycol 400 (16). A typical example is shown in Figure 17. The energy of solvation $\Delta H_{T,s}$ required for the dissolution of norgestomet in distilled water was calculated to be only 5.88 kcal/mol. The incorporation of 10–100% (v/v) of polyethylene glycol 400, a water-miscible polymer, into the distilled water was found to affect the energy of solvation (Figure 18).

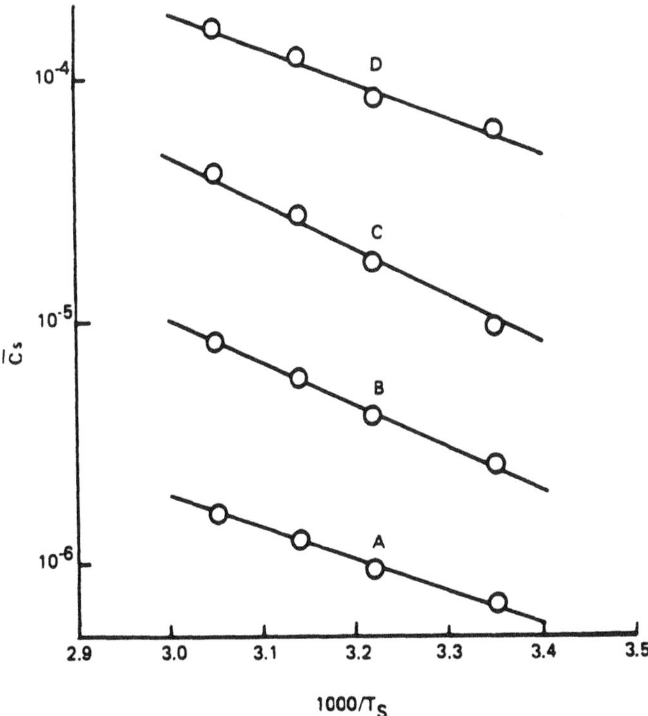

Figure 17 Temperature dependence of the mole fraction solubility $\bar{C}_s$ of norgestomet in various aqueous solutions of polyethylene glycol (PEG) 400. (A) Distilled water alone; (B) 20% (v/v), (C) 40% (v/v), and (D) 60% (v/v) of PEG 400 in aqueous solution. [Reproduced, with permission, from Chien; ACS Symposium Series, 33:53 (1976).]

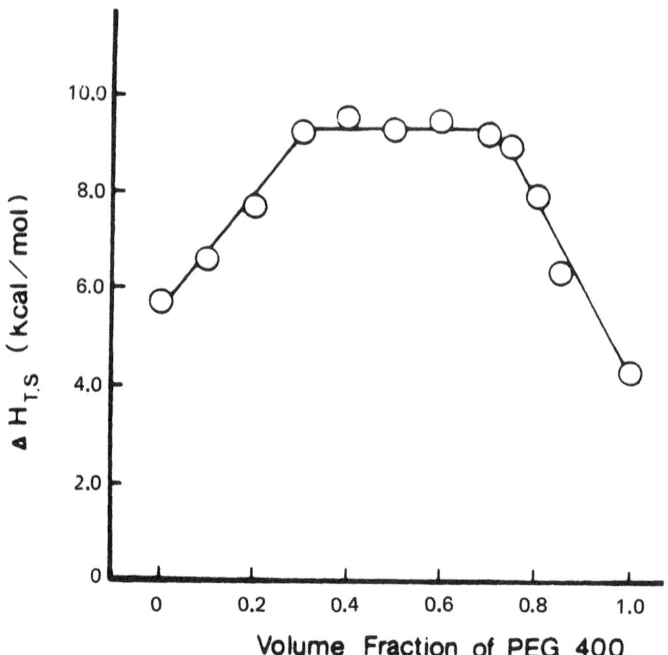

Figure 18 Effect of polyethylene glycol (PEG) 400 on the energy of solvation $\Delta H_{T,s}$ for the dissolution of norgestomet in various aqueous PEG 400 solutions.

C. Partition Coefficient

The partition coefficient K of a drug for its interfacial partitioning from the surface of a drug delivery device toward an elution medium is, as defined in Equation (2), the ratio of its solubility in the elution solution C_s over its solubility in the polymer composition C_p of the device. Any variation in either the C_s or the C_p value results in an increase or decrease in the magnitude of the K value. A typical example is illustrated by the in vitro release of norgestomet from silicone capsules into elution media with varying solution solubilities (Figure 14). By changing the solubility of norgestomet in the elution solution the magnitude of the partition coefficient thus varies, leading to a variation in the drug release rate. As expected from Equation (8a), the magnitude of the Q/t values is a linear function of K (Figure 19).

The effect of the partition coefficient on the controlled release of drugs from a matrix-type drug delivery device was reported to be biphasic: both the mechanism and the rate profile of drug release were dependent upon the variation in the partition coefficient (22). A typical example has been illustrated by the controlled release of ethynodiol diacetate from matrix-type silicone devices. The release profiles in Figure 20 indicate that when the magnitude of the partition coefficient is small, a Q versus t release profile results and the magnitude of the Q/t values increases linearly with the increase in the partition coefficient. This region is defined as governed by a partition-controlled process and is described by Equation (13a). When the partition coefficient is increased beyond a critical point ($K \simeq 0.5$), the matrix-controlled mechanism became predominant and a Q versus $t^{1/2}$ release profile is then observed. As predicted by Equation (17), the magnitude of the $Q/t^{1/2}$ values is virtually in-

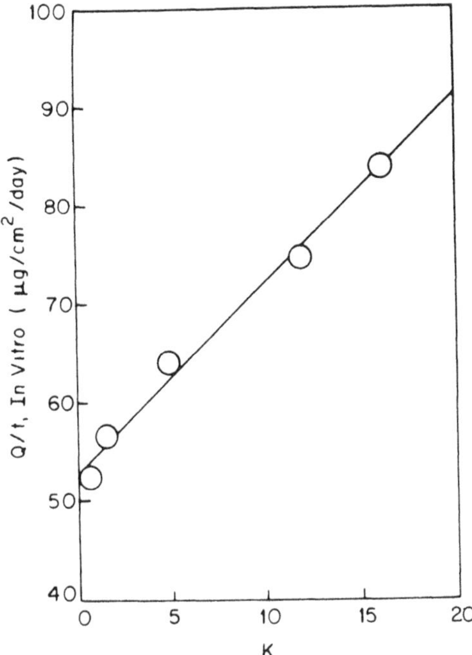

Figure 19 Linear dependence of the in vitro release rate Q/t of norgestomet from silicone capsules on the partition coefficient K. [Reproduced, with permission, from Chien; Chem. Pharm. Bull., 24:1471 (1976).]

dependent of any variation in the partition coefficient. This behavior is obviously quite different from the observation discussed earlier on the controlled release of drug from membrane-encapsulated reservoir-type drug delivery devices (Figures 2 and 19).

Between the partition-control and matrix-control regions there exists a transition phase. As indicated by Equation (19), t_{trans}, the time at which the drug release profile undergoes transition from a partition-controlled process to matrix-controlled process, is inversely proportional to the partition coefficient. This is exemplified by the controlled release of a homologous series of alkyl-*p*-aminobenzoates from matrix-type silicone devices (Table 8). As the alkyl chain length of the ester increases, the partition coefficient of *p*-aminobenzoates from the silicone device toward the elution solution decreases and the time for the transition t_{trans} from the partition-controlled process to the matrix-controlled process becomes longer. The dependence of the transition time on the partition coefficient can be appreciated by converting Equation (19) to Equation (42):

$$\log t_{\text{trans}} = \log \frac{3\beta^2}{\alpha} + \log \frac{1}{KC_s} \tag{42}$$

Equation (42) suggests that a linear relationship should exist between $\log t_{\text{trans}}$ and $\log (1/KC_s)$ with slope equal to unity (Figure 21). Experimentally this linearity is followed irrespective of a variation in drug loading doses in polymeric devices.

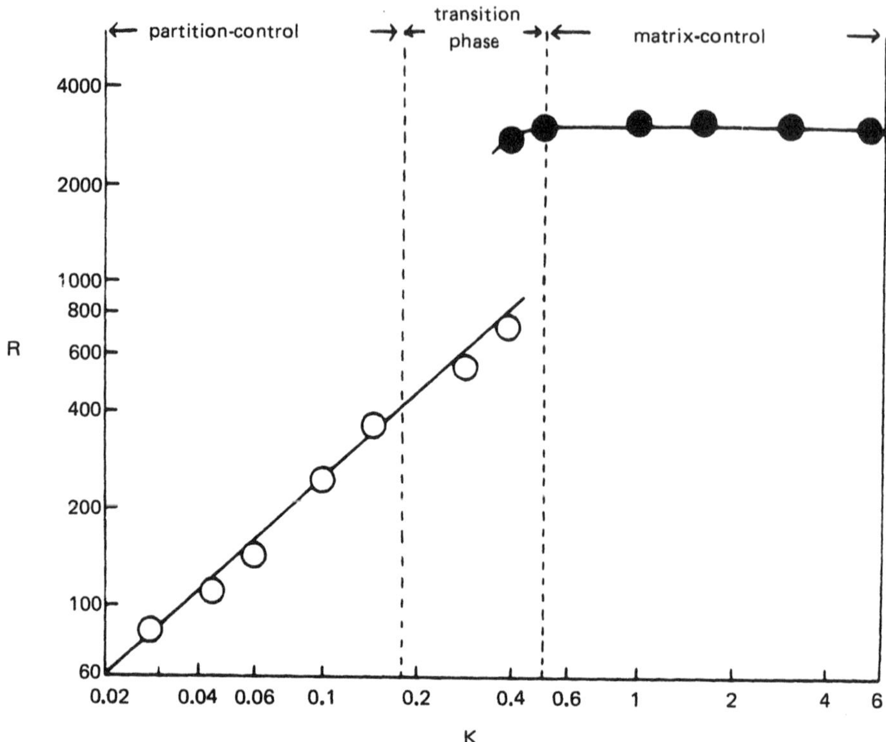

Figure 20 Effect of the partition coefficient K on the mechanism and the rate profile R of the controlled release of ethynodiol diacetate from matrix-type silicone devices. In the region of the partition-controlled process the Q/t value increases linearly as the partition coefficient increases. In the region of the matrix-controlled process the $Q/t^{1/2}$ value is independent of the variation in the partition coefficient. [Reproduced, with permission, from Chien and Lambert; J. Pharm. Sci., 63:515 (1974).]

Table 8 Dependence of Transition Time t_{trans} on Partition Coefficient K

p-Aminobenzoates	t_{trans}[a] (min)	$1/K$
Methyl ester	3.82×10^{-2}	2.08×10^{-1}
Ethyl ester	3.37×10^{-1}	8.17×10^{-1}
Propyl ester	2.28×10^{0}	2.75×10^{0}
Butyl ester	2.15×10^{1}	1.03×10^{1}
Pentyl ester	2.95×10^{2}	3.95×10^{1}
Hexyl ester	3.10×10^{3}	1.05×10^{2}
Heptyl ester	4.75×10^{4}	4.00×10^{2}

[a]Loading dose of alkyl-*p*-aminobenzoates is 5% (w/w).
Source: Compiled from the data by Roseman and Yalkowsky; ACS Symposium Series 33: 33 (1976).

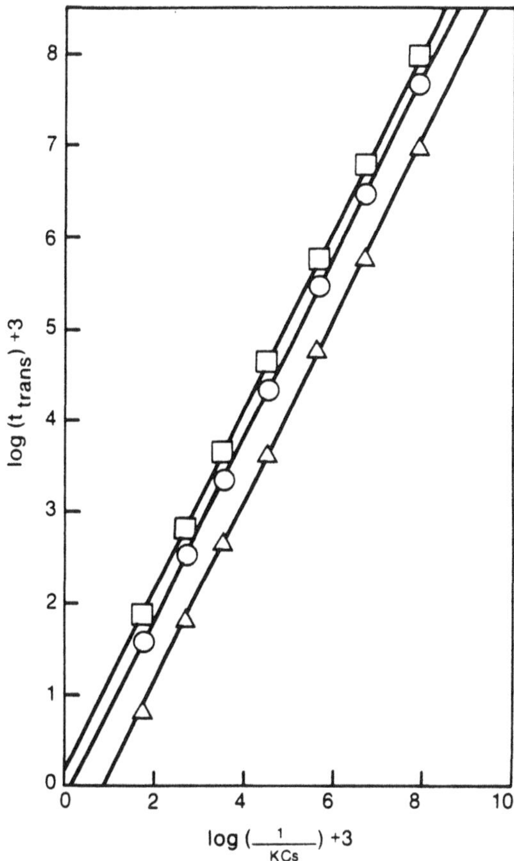

Figure 21 Linear logarithmic relationship between t_{trans} and $1/KC_s$ at various drug loading levels: (△) 1%, (○) 5%, and (□) 10%. [Plotted from the data by Roseman and Yalkowsky; ACS Symposium Series, 33:33 (1976).]

The effect of alkyl chain length on the magnitude of the partition coefficient is exponential, as defined by Equation (43):

$$\log K_n = \log K_0 - n\pi_{CH_2} \tag{43}$$

where K_n is the partition coefficient for the compound with an alkyl chain length of n CH_2 groups; K_0 is the y-axis intercept at zero carbon number; and π_{CH_2} is the slope of the log K_n versus n plots (Figure 22). The attainment of a negative slope results from the fact that as alkyl chain length increases, the polymer solubility C_p of alkyl *p*-aminobenzoates is enhanced at the expense of their solution solubility C_s, leading to a reduction in the partition coefficient K_n.

On the other hand, the addition of hydrophilic functional groups, such as hydroxyl groups, to a drug molecule tends to improve the solution solubility at the expense of the polymer solubility in a lipophilic polymer (15). A typical example is demonstrated by the effect of the addition of hydroxyl groups on the solubility of progesterone in silicone polymer and elution solution (Table 9). This results in a

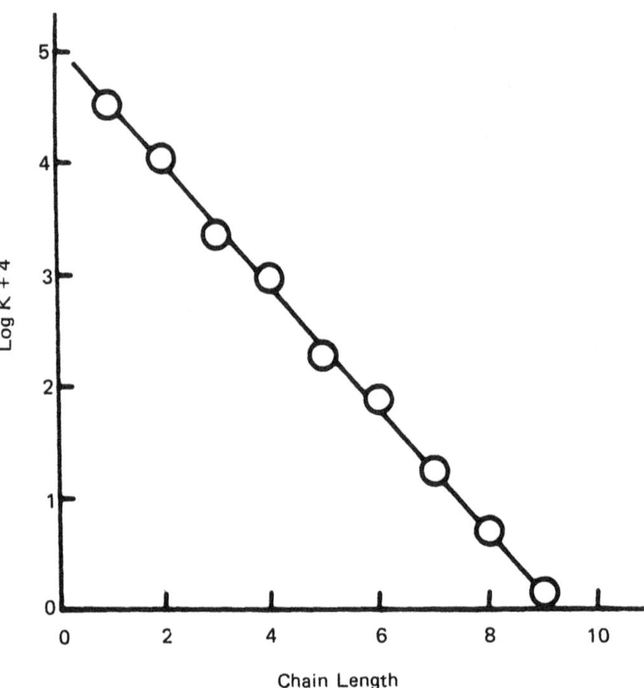

Figure 22 Semilogarithmic relationship between the partition coefficient of alkyl-*p*-aminobenzoates and the length of alkyl chain. [Reproduced, with permission, from Chien; in: *Sustained and Controlled Release Drug Delivery Systems* (Robinson, Ed.), Dekker, New York, 1978, Chapter 4.]

Table 9 Effect of Hydroxy Groups on the Solubility and Partition Coefficient

Progesterone derivatives	Solubilities (μg/ml) C_s[a]	Cp[b]	Partition coefficient[c]
Progesterone	353.3	594.7	0.6
21-Hydroxyprogesterone	1402.1	205.9	6.8
17α-Hydroxyprogesterone	442.0	26.5	16.7
11α-Hydroxyprogesterone	575.0	9.1	63.2
17α,21-Dihydroxyprogesterone	994.1	1.5	671.7
11β,21-Dihydroxyprogesterone	839.0	1.2	682.1
11β,17α,21-Trihydroxyprogesterone	3987.6	3.8	1054.9

[a]Solution solubility C_s in 50% (v/v) polyethylene glycol 400 in water at 37°C.
[b]Polymer solubility C_p in silicone medical fluid 360 at 37°C.
[c]Calculated on the basis of Equation (2).

progressive increase in the partition coefficient in response to the addition of OH groups to the progesterone molecule. This yields an exponential dependence of the partition coefficient K_{OH} on the number n of hydroxyl groups which follows the relationship

$$\log K_{OH} = \log K_p + n\pi_{OH} \tag{44}$$

where K_p and K_{OH} are the partition coefficients of progesterone and its hydroxyl derivatives, respectively; n is the number of OH groups added; and π_{OH} is the slope of the log K_{OH} versus n plots (Figure 23). A positive slope value is obtained because the addition of hydroxy groups improves the hydrophilicity of progesterone and thus enhances the interfacial partitioning behavior of a progesterone molecule from a lipophilic silicone polymer toward the aqueous solution.

The variation in the magnitude of partition coefficients among drugs is striking; a several thousandfold difference in values was demonstrated in steroids with change in functional groups around the steroidal skeleton (Table 10). The partition coefficient was also reported to vary greatly from one type of biomedical polymer to another (Table 11).

Since partitioning is a process involving molecular equilibrium at the interface,

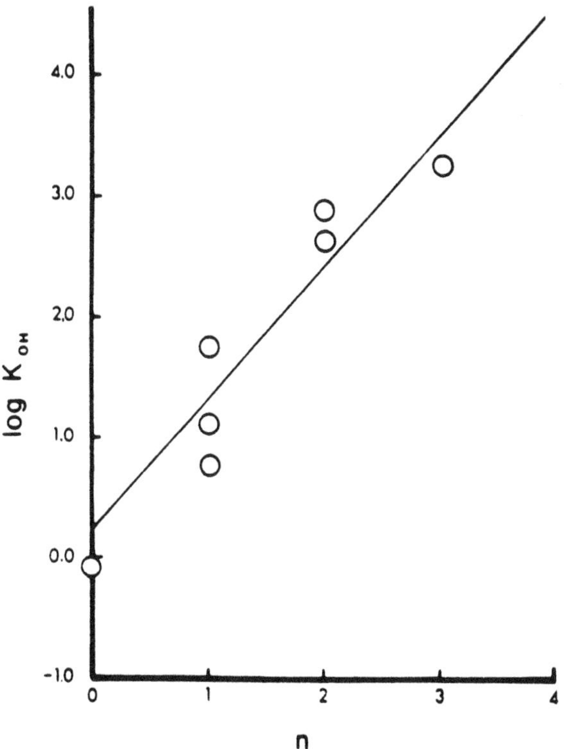

Figure 23 Dependence of log K_{OH}, the (solution /polymer) partition coefficient, on the number of hydroxy groups n on the progesterone molecule. [Reproduced, with permission, from Chien et al.; J. Pharm. Sci., 68:689 (1979).]

Table 10 Partition Coefficient of Steroids from Silicone Polymer Toward Water at 37°C

Steroids	Partition coefficient[a]
Cortisol	181
Estradiol	1.25
Norethindrone	0.818
Norgestrel	0.313
Testosterone	0.232
Melengestrol acetate	0.053
Progesterone	0.044
19-Norprogesterone	0.030
Megestrol acetate	0.028
Mestranol	0.010

[a]Calculated from the solubility at 37°C data by Sundaram and Kincl; Steroids, 12:517 (1968).

a partition coefficient is therefore an equilibrium constant directly related to the standard Gibbs free energy ΔF_d of desorption, i.e., the energy gained by a molecule on desorption from the polymer phase into the elution solution (30):

$$\Delta F_d = -RT \ln K \tag{45}$$

A number of data indicate that when two structurally closely related molecules are partitioning in the same system of two immiscible phases, the difference in the free energies of interfacial transfer ΔF for these molecules is directly related to a specific structural modification (31–34). This generality has been found applicable to compounds that form a regular solution in both immiscible phases, which can be expressed mathematically by

$$\Delta F_{fg} = -2.303RT\,(\log K_{\text{derivative}} - \log K_{\text{parent}}) \tag{46}$$

where ΔF_{fg} is the free energy of partitioning for a particular functional group in the derivative and is theoretically related to the lipophilicity π_{fg} of this functional group, as follows:

$$\pi_{fg} = \frac{\Delta F_{fg}}{2.303RT} \tag{47}$$

Table 11 Partition Coefficient of Progesterone from Various Biomedical Polymers Toward Water

Polymers		Partition coefficient[b]
Regenerated cellulose (Cuprophan)		1.49
Hydroxyethyl methacrylate (Hydron)		7.75×10^{-3}
with ethylene glycol dimethacrylate[a]	0.75%	6.62×10^{-3}
	3.75%	6.21×10^{-3}
	5.25%	4.31×10^{-3}
Polydimethylsiloxane (Silicone)		5.56×10^{-3}
Polyurethane I (Biomer)		4.74×10^{-4}
Polyurethane II (Pellethane)		2.00×10^{-4}

[a]As cross-linking agent at three levels.
[b]Calculated at 25°C from the data by Zentner et al.; J. Pharm. Sci., 67:1347 and 1352 (1978).

The partition coefficient for a derivative, $K_{derivative}$, can be estimated from the partition coefficient of the parent compound, K_{parent}, by Equation (48) and the π_{fg} values for all the functional groups present:

$$\log K_{derivative} = \log K_{parent} + \Sigma\pi_{fg} \tag{48}$$

Application of this concept of summation of constituent properties [Equation (48)] to estimate or to approximate the physicochemical properties of the whole molecule has recently gained some acceptance. It has been successfully applied to correlate the pharmacological activities of a series of drug analogs with their molecular structures (31–37). Sufficient evidence has also been accumulated on the additive-constitutive character of the partition coefficient (31,33,34). A typical example has recently been illustrated by the establishment of a good correlation between the calculated and experimentally determined lipophilicities of 45 corticosteroids (34).

The lipophilicity of a functional group π_{fg} can be calculated using Equation (48) from any two analogs whose structures are different by only one functional group. The magnitude of π_{fg} values varies from one type of functional group to another and also changes substantially as the functional group moves from one position to another (Table 12). For example, at position 16α, the chloro group is more lipophilic than the methyl group, which in turn is more lipophilic than the fluoro group. The methyl group is slightly more lipophilic at position 16β than at 16α, and the fluoro group is more lipophilic at position 6α than at 16α, which in turn is slightly more lipophilic than at position 9α. On replacement of the methyl, fluoro, or chloro group at position 16 with a hydroxy group, the hydrophilicity of corticosteroids increases at the expense of lipophilicity.

Table 12 Lipophilicity of Various Functional Groups on Corticosteroids

Functional groups	Position	Lipophilicity[a]
Methyl	6α	0.456
	16α	0.403
	16β	0.493
Fluoro	6α	0.248
	9α	0.149
	16α	0.176
Chloro	16α	0.696
Acetate	17	0.456
	21	1.286
Desoxy	11	1.061
	17	0.609
	21	0.895
Hydroxy	11α	−0.820
	16α	−0.305
	17α	−0.469
	21	−0.472

[a]Calculated from the partition coefficient data using Equations (43) and (44).

Source: Compiled from the data by Flynn; J. Pharm. Sci., 60:345 (1971).

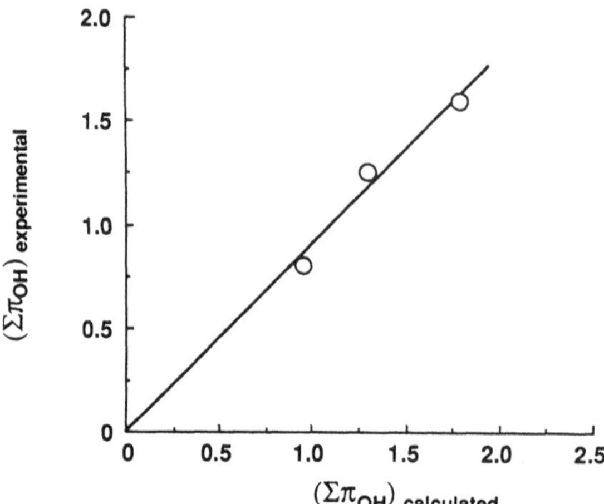

Figure 24 Correlation of observed $\Sigma\pi_{OH}$ values with calculated $\Sigma\pi_{OH}$ values for various hydroxy derivatives of progesterone. The calculated $\Sigma\pi_{OH}$ values were computed from the summation of the π_{fg} data for hydroxy groups at various positions (Table 12), and the observed $\Sigma\pi_{OH}$ values were determined in an *n*-octanol/phosphate buffer (pH 7.4) system at room temperature.

The additive-constitutive character of π_{fg} values is illustrated in Figure 24. A fairly good correlation is achieved between the experimentally observed π_{OH} values for the progesterone derivatives with multiple OH substitutions and the values calculated from the π_{OH} values for each OH group at different positions (Table 12) by the concept of summation.

It was reported earlier (Section IID) that a cosolvent system has been used in microreservoir dissolution-controlled drug delivery systems (MDD) to form a homogeneous dispersion of microscopic drug-saturated liquid compartments in a solid polymer matrix (Figure 6). Variation in the composition of this cosolvent system results in a change in drug solubility [Equations (36) through (38)] and, hence, a increase or decrease in the partition coefficient K_l for the interfacial partitioning of drug molecules from the liquid compartment toward the polymer matrix [Equation (27)]. The final outcome of this sequence of variation is a change in the controlled release rate profile of drug from this type of drug delivery system.

The effect of a cosolvent on the permeation of drugs across a polymer membrane was investigated in depth (38). It was observed that the steady-state permeability of *p*-aminoacetophenone, as the model drug, across silicone membranes from a binary cosolvent system of propylene glycol and water was controlled primarily by the thermodynamic activity of the drug in the applied phase. When the applied phase contains a saturated drug solution, the rate of membrane permeation is independent of the volume fraction of propylene glycol in the solution and, thus, the saturation solubility of the drug at a given propylene glycol concentration (Table 13). On the other hand, if the applied phase contains a nonsaturated solution with a constant concentration of *p*-aminoacetophenone, the rate of membrane permeation of *p*-aminoacetophenone was seen to decrease proportionally as the volume fraction of propylene glycol increased (Figure 25). This decrease in membrane permeability in re-

Table 13 Effect of Propylene Glycol on the Membrane Permeation of *p*-Aminoaceto-phenone from Saturated Solution

Propylene glycol[a] (v/v)	C_s (mg/ml)	$Q/t^b \times 10^5$ (mg/cm²/sec)
0.05	12.71	1.39
0.10	15.60	1.43
0.15	18.99	1.42
0.20	23.47	1.37
0.25	28.80	1.48
0.30	35.58	1.47
0.35	44.14	1.53
0.40	56.08	1.47
0.45	71.16	1.48
0.50	85.64	1.41

[a]Volume fraction of propylene glycol in aqueous solution.
[b]Permeation through a silicone membrane.
Source: Compiled from the data by Flynn and Smith; J. Pharm. Sci., 61:61 (1972).

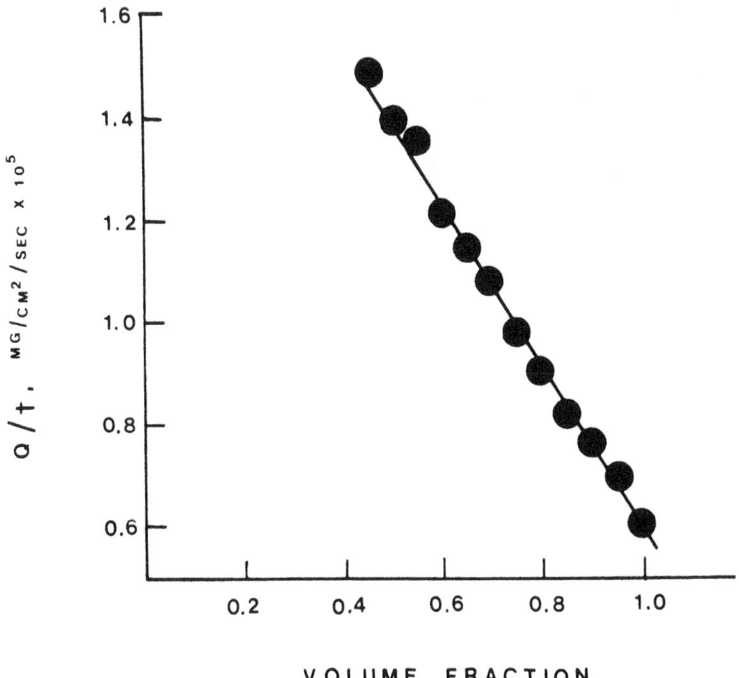

Figure 25 Effect of the volume fraction of propylene glycol in solution on the permeation of *p*-aminoacetophenone across a silicone membrane from a nonsaturated solution with a constant drug concentration. [Plotted from the data by Flynn and Smith; J. Pharm. Sci., 61:61 (1972).]

sponse to the increase in the volume fraction of a cosolvent can be rationalized as due to the decline in the degree of saturation. The addition of propylene glycol was observed to have no effect on the polymer membrane diffusivity D_p.

Membrane permeability P_m is defined as

$$P_m = D_p K_l \tag{49}$$

where K_l, the partition coefficient for the interfacial partitioning of drug molecules from an applied phase toward a polymer membrane, is related to the solution solubility C_s of the drug in the applied phase as follows:

$$\begin{aligned} \log K_l &= \log C_p - \log C_s \\ &= \log C_p - \log C_w - e_x f_x \end{aligned} \tag{50}$$

where C_w, e_x, and f_x are as defined in Equation (36) for a binary cosolvent system; and $\log C_p$ and $\log C_w$ are constants under a given condition.

As defined in Equation (50), the partition coefficient K_l of *p*-aminoacetophenone from the propylene glycol-water system toward the silicone membrane is expected to decrease in an exponential fashion as one increases the volume fraction f_x of propylene glycol. This exponential relationship between K_l and f_x is followed perfectly in both drug-saturated and nonsaturated solutions (Figure 26). The e_x value from the saturated solution (1.834) was found to be approximately two-fold the e_x value obtained in the nonsaturated solution (0.795). The difference in e_x values was attributed thermodynamically to the occurrence of a significant solute-solute interaction in the saturated drug solution, which results in an increase in the entropy and a decrease in the activity coefficient of drug molecules in the solution phase.

Addition of a cosolvent system into a drug solution could also affect the membrane permeability of drug molecules as a result of its effect on the interfacial partitioning of drug molecules from a drug solution toward a polymer membrane (K_l). The effect of a cosolvent system on the membrane permeability P_m of a drug species can be visualized theoretically by substituting Equation (50) for the K_l term in Equation (48) to give

$$\log P_m = (\log D_p C_p - \log C_w) - e_x f_x \tag{51}$$

where $\log D_p C_p$ and $\log C_w$ are constants for a given drug. Equation (51) suggests that the membrane permeability P_m of drug molecules from a cosolvent-H_2O combination decreases in an exponential manner as the volume fraction f_x of the cosolvent in the aqueous solution increases. This exponential relationship between P_m and f_x is followed very well in both drug-saturated and nonsaturated solutions (Figure 27). Again, a different e_x value has been obtained for saturated and nonsaturated drug solutions. As in partitioning, the e_x value from the saturated solution is approximately twice the e_x value found in a nonsaturated solution. The observation indicates that the effect of a cosolvent system on the membrane permeability P_m is due to its effect on interfacial partitioning K_l, not on membrane diffusivity D_p [Equation (49) and (50)]. The same phenomenon was also observed in the release of drugs with low aqueous solubility from the polymer matrix (39).

The linear relationship between membrane permeability P_m and the partition coefficient K_l as defined by Equation (49) is also applicable to the biological envi-

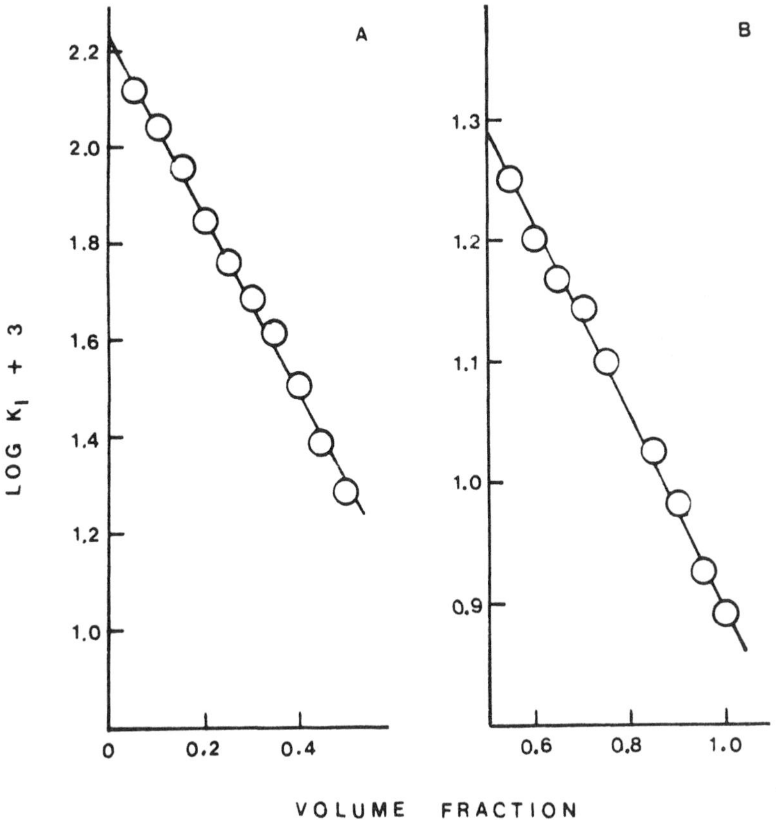

Figure 26 Effect of the volume fraction of propylene glycol in solution on the partition coefficient K_1 of *p*-aminoacetophenone from aqueous solution toward a silicone membrane. (A) Saturated drug solution, and (B) nonsaturated solution with constant drug concentration. [Plotted from the data by Flynn and Smith; J. Pharm. Sci., 61:61 (1972).]

ronment. A typical example is demonstrated later in the discussion of the transdermal permeation of steroids (Chapter 7, Figure 15).

The addition of cosolvent in drug formulations was also reported to affect the absorption and pharmacological responses of drugs. A typical example was illustrated by the effect of propylene glycol in topical formulations on the in vitro percutaneous permeability and in vivo vasoconstrictive responses of fluocinolone acetonide and its acetate (Figure 28) (40). The exponential relationship between permeability and the volume fraction of a cosolvent system was also followed quite well. A positive e_x value was seen before a maximum permeability was reached at a vehicle composition containing the minimum volume fraction of cosolvent to solubilize all the drug dose in the formulation (Figure 28A). Beyond the maximum permeability, a further increase in the volume fraction of cosolvent resulted in the formation of a nonsaturated drug solution, and thus a negative e_x value, as predicted in Equation (51), was obtained. The observation of a positive e_x value before the maximum permeability is reached can be rationalized as follows: an increase in the

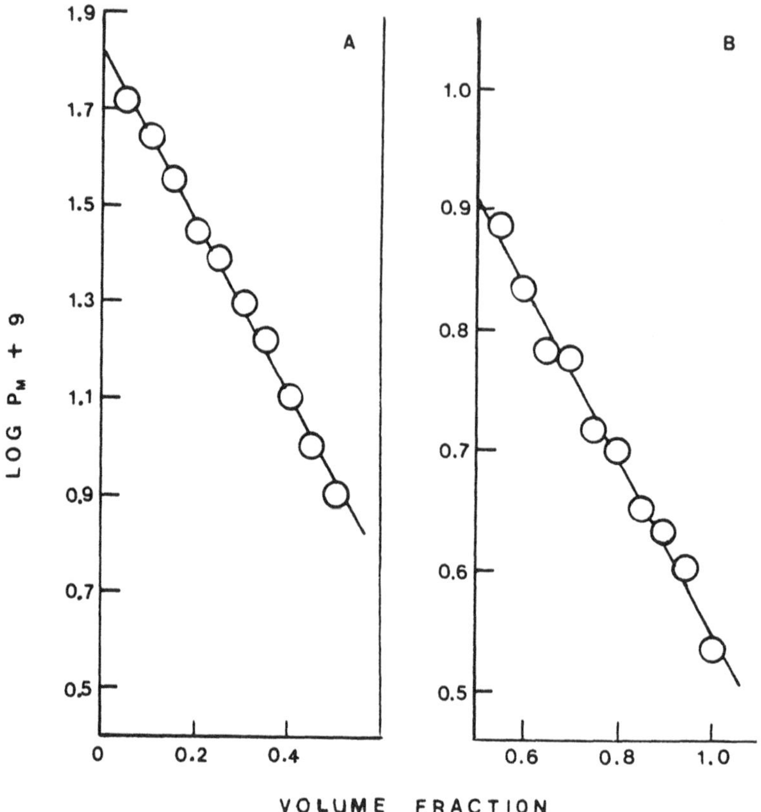

Figure 27 Effect of the volume fraction of propylene glycol in aqueous solution on the membrane permeability P_m of *p*-aminoacetophenone across a silicone membrane. (A) Saturated solution, and (B) nonsaturated solution with constant drug concentration. [Plotted from the data by Flynn and Smith; J. Pharm. Sci., 61:61 (1972).]

volume fraction of cosolvent in a formulation with a constant drug dose results in an increase in the saturation solubility of the drug in the vehicle and, hence, an increase in the drug concentration on the skin surface at equilibrium, which yields an increase in percutaneous absorption.

D. Polymer Diffusivity D_p

The diffusion of small molecules in a polymer structure is an energy-activated process in which the diffusant molecules move to a successive series of equilibrium positions when a sufficient amount of energy, called the energy of activation for diffusion E_d, has been acquired by the diffusant and its surrounding polymer matrix (1). This energy-activated diffusion process is frequently described by the following Arrhenius relationship:

$$D_p = D_0 e^{-(E_d/RT)} \tag{52}$$

where D_0 is a temperature-independent frequency factor; E_d is the energy of acti-

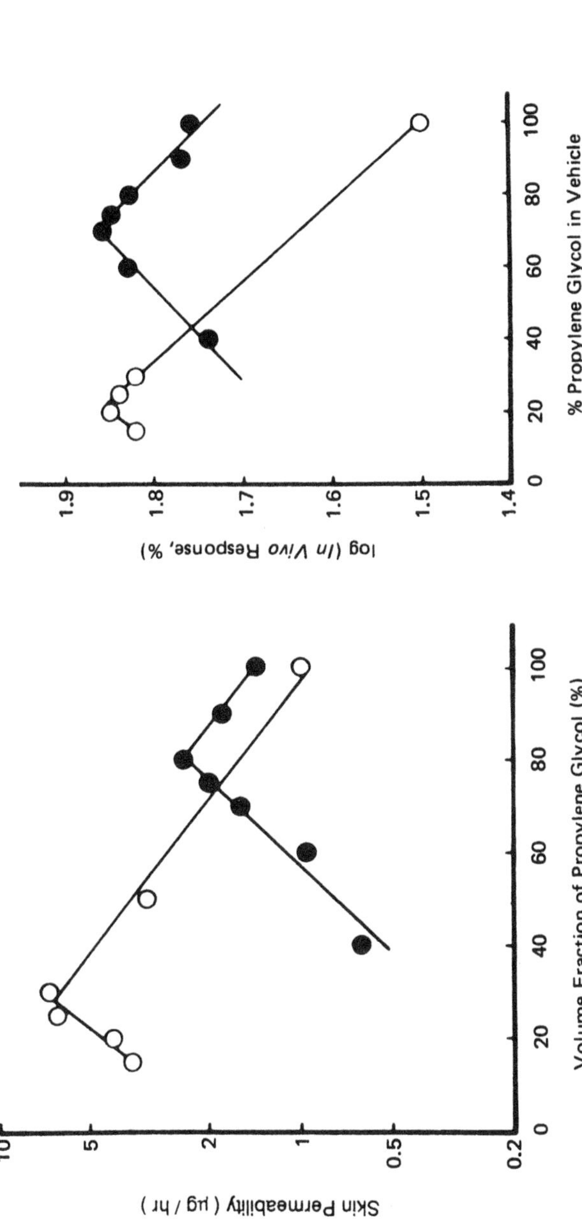

Figure 28 Effect of propylene glycol in topical formulations on the permeability of fluocinolone acetonide (○) and its acetate (●) through the human abdominal skin and their in vivo vasoconstrictive response in humans. Maximal permeability and vasoconstrictive activity were achieved at a volume fraction of propylene glycol needed to solubilize all the drug dose incorporated. [Plotted from the data by Ostrenga et al., J. Pharm. Sci., 60:1175 (1971).]

vation for polymer diffusion, distributed throughout many degrees of freedom in the system; and R and T have their usual thermodynamic meaning.

This energy-activated diffusion process is best described by the *activated state model* of Brandt (Figure 29). In this model the activated state, for the simplicity of model calculation, is visualized as involving only two neighboring polymer chains that have moved apart to permit the passage of a diffusant molecule. The molecular motions that lead to this activated state are assumed to involve: (i) the bending of polymer chains to make room for the diffusing molecule; (ii) the intermolecular repulsion from their neighboring polymer chains and, simultaneously, the intramolecular resistance from the rigid bond distances and bond angles within the molecule, which have made the bending polymer chains to seek alternate routes and produced a partial rotation of chain units out of their equilibrium position against a hindering potential of internal rotation; the resultant total torsional strain is evenly distributed over the entire polymer chain segment; and (iii) the number of degrees of freedom found in a segment of the polymer chain is proportional to the length of the segment (41).

The energy of activation for polymer diffusion E_d is thus the sum of the energy of intramolecular bending E_b and the energy of intermolecular repulsion E_r:

$$E_d = E_b + E_r \tag{53}$$

The results of model calculation indicated that the magnitude of E_b is very high for short segments of polymer chain but decreases as the polymer chain becomes longer. On the other hand, E_r increases steadily as the polymer chain becomes longer; i.e., the degree of freedom becomes larger. A consequence is that the E_d value goes through a distinct minimum at a moderate length of polymer chain segment. It was also suggested that the molecular diameter of the diffusant affects very strongly the magnitude of E_d. On the other hand, certain features of the molecular structure of the polymer, such as the shape of the potential barrier for hindered rotation and the

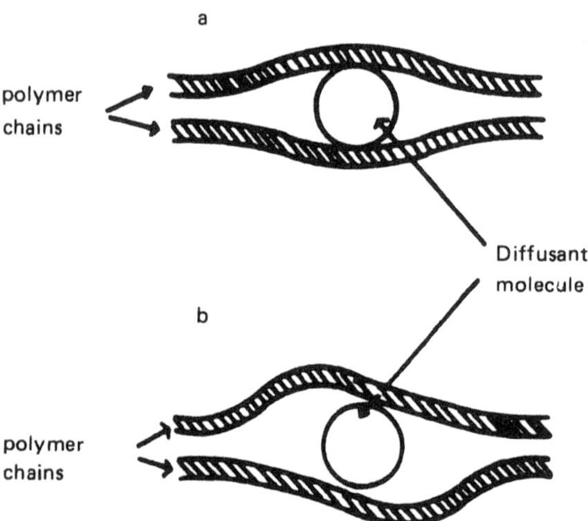

Figure 29 Brandt's activated-state models for the diffusion of a small molecule in a polymer structure. (a) Symmetrical model and (b) unsymmetrical model. [Reproduced with permission from Brandt; J. Phys. Chem., 63:1080 (1959).]

number of degrees of freedom per monomer unit, have only a small effect on the diffusion profiles.

The self-diffusion coefficient of liquid silicone polymers with varying molecular weights can be determined using the nuclear magnetic resonance spin-echo technique (42). It was found that the magnitude of the self-diffusion coefficient of silicone polymers decreases remarkably as the length of the linear polydimethylsiloxane chain is increased (Figure 30).

By comparing the diffusivities of simple gas molecules in a silicone polymer with the self-diffusion coefficient of the linear silicone polymer chain (Figure 30), it was revealed that the diffusion of simple gas molecules, such as O_2, N_2, and CO_2, requires motion of a segment of the silicone polymer chain of the order of three monomer units in length (43). Applying the same approach, it was later estimated that the diffusion of larger molecules, such as ethynodiol diacetate, medroxyprogesterone acetate, and chlormadinone acetate, in a silicone polymer matrix structure requires the molecular motion of a silicone chain segment of the order of 10 monomer units in length to create an opening sufficiently large for the progestin molecules to diffuse (14).

Both the model calculation and diffusion measurements just discussed emphasize the critical importance of the molecular diameter of a diffusant in determining the

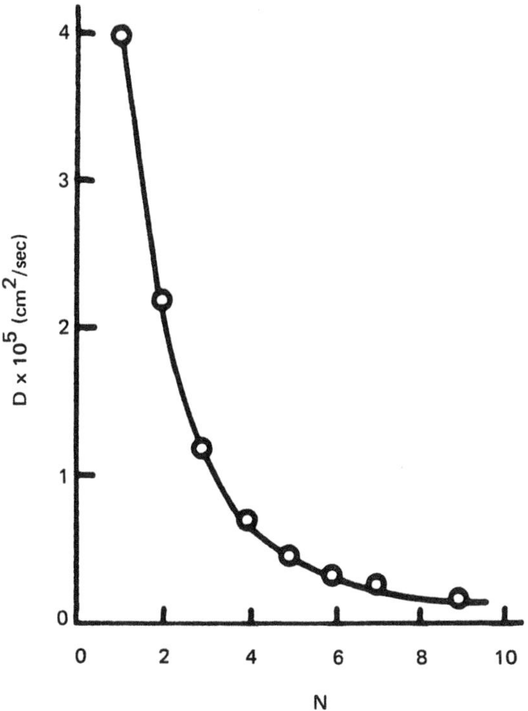

Figure 30 Dependence of the diffusion coefficient D for the self-diffusion of liquid silicone polymer on the number n of dimethylsiloxane monomer $[OSi(CH_3)_2]$ units in $(CH_3)_3Si[OSi(CH_3)_2]_{n-2}OSi(CH_3)_3$ at room temperature. [Reproduced, with permission, from Robb; Ann. N.Y. Acad. Sci., 146:119 (1968).]

magnitude of its polymer diffusivity. In such cases the polymer diffusivity of a diffusant molecule must be inversely proportional to the cube root of its molecular weight (44). This relationship was demonstrated in the diffusion of antitumor drugs across the hydroxyethyl methacrylate polymer membrane (45). The polymer diffusivities of these antitumor drugs were found, as expected, in linear proportion with the reciprocal of the cube root of their molecular weights (Figure 31).

The results in Figure 31 suggest that the diffusion of a diffusant in a polymer structure is also very sensitive to the composition of the polymer. As a fraction of hydroxyethyl methacrylate is replaced by butyl methacrylate, the dependence of polymer diffusivity on molecular weight is still followed, but the magnitude of polymer diffusivity is reduced significantly. As suggested by the theory of Cohen and Turnbull (46), the diffusion process in the polymer structure is governed by the segmental motion of the polymer chain; the bulkier the functional groups attached to the polymer chain, the more difficult the segmental motion is and the lower the polymer diffusivity. This was illustrated by the effect of phenyl replacement in a silicone polymer on the polymer diffusivity of small gas molecules (43). The polymer diffusivity of gas molecules, such as H_2, O_2, N_2, and CO_2, was considerably reduced

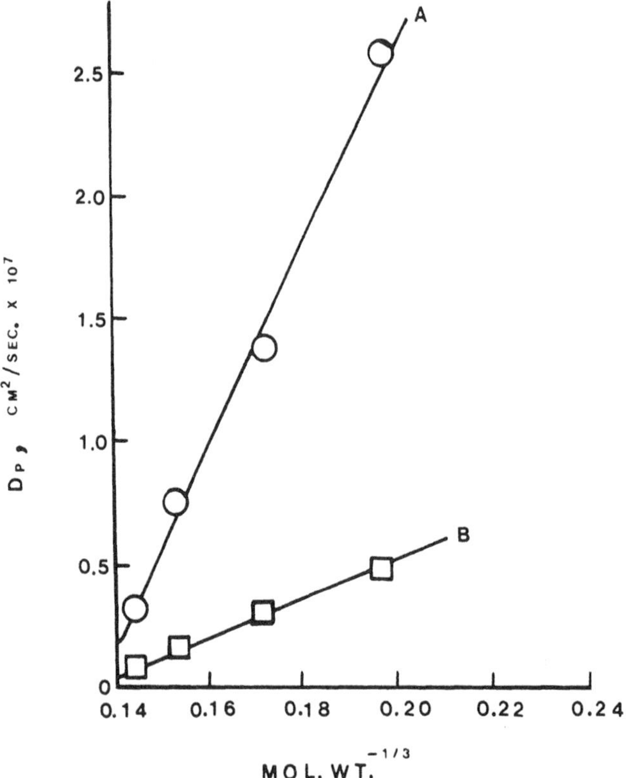

Figure 31 Linear dependence of the polymer diffusivity D_p of antitumor drugs in hydrophilic polymer on the reciprocal of the cubic root of their molecular weight (MW): (A) 100% hydroxyethyl methacrylate membrane; (B) 92.5% hydroxyethyl methacrylate-7.5% butyl methacrylate membrane. [Plotted from the data by Drobnik et al.; J. Biomed. Mater. Res., 8:45 (1974).]

Table 14 Dependence of Polymer Diffusivity on the Chemical Composition of Silicone Polymer

Composition of side-chain groups on silicon atom		$D_p \times 10^6$ (cm^2/sec)			
Methyl	Phenyl	H_2	O_2	N_2	CO_2
100	0	42.5	15.5	14.7	11.1
95	5	—	11.6	10.7	—
80	20	12.5	3.3	2.6	2.5
67	33	5.1	1.2	0.74	1.0
50	50	2.9	0.37	0.25	—
33	67	0.8	0.09	0.04	0.07
20	80	0.8	0.05	0.02	0.04

Source: Compiled from the data by Robb; Ann. N.Y. Acad. Sci., 146:119 (1968).

as the methyl groups attached to the silicone atom were replaced by the bulkier phenyl groups. The extent of reduction in polymer diffusivity was found to be in proportion to the percentage of phenyl replacement (Table 14). The reduction in polymer diffusivity also appeared to be dependent upon the type and molecular diameter of the diffusants. The magnitude of polymer diffusivity varies greatly from one type of polymer to another (Table 15) and is also sensitive to the composition of the copolymer (Tables 16 and 17).

The magnitude of polymer diffusivity D_p is also dependent upon the type of functional group and their stereochemical positions in the diffusant molecule. This is demonstrated by the effect of hydroxyl groups on the polymer diffusivity of progesterone derivatives in a silicone polymer matrix (Table 18).

1. *Effect of Cross-Linking*

During the polymerization of hydrogel, varying amounts of ethylene glycol dimethacrylate, a cross-linking agent, are added to the solution of a linear polymr to produce a three-dimensional ethylene glycomethacrylate gel with various degrees of cross-linkage (47,48). The diffusivity of progestins in the hydrogel matrix was observed to be reduced in response to the addition of a cross-linking agent (49,50). The reduction in polymer diffusivity D_p was found to be a linear function of the reciprocal of the extent of cross-linkage (Figure 32). It was rationalized that the

Table 15 Polymer Diffusivity of Progesterone in Various Biomedical Polymers

Polymers	D_p (cm^2/sec)
Regenerated cellulose (Cuprophan)	5.39×10^{-7}
Polydimethylsiloxane (Silicone)	2.15×10^{-8}
Hydroxyethyl methacrylate (Hydron)	4.38×10^{-9}
Polyurethane I (Biomer)	4.74×10^{-10}
Polyurethane II (Pellethan)	5.75×10^{-12}

Source: Compiled from the 25°C data by Zentner et al.; J. Pharm. Sci., 67:1347 (1978).

Table 16 Polymer Diffusivity of Progesterone in Copolymers of Hydroxyethyl Methacrylates

Copolymers of methacrylate (%)[a]			$D_p \times 10^9$ (cm²/sec)
A	B	C	
100	—	—	4.38
80	20	—	0.98
67	33	—	0.90
34	—	66	7.67
—	—	100	12.70

[a]A, hydroxyethyl methacrylate; B, methoxyethyl methacrylate; C, methoxyethoxyethyl methacrylate.
Source: Compiled from the data by Zentner et al.; J. Pharm. Sci., 67:1352 (1978).

Table 17 Polymer Diffusivity of Tetracycline in Co (Hydroxyethyl Methacrylate-Methyl Methacrylate)Polymers

Copolymer composition (mole ratio)		$D_p \times 10^9$ (cm²/sec ± SD)
Hydroxyethyl methacrylate	Methyl methacrylate	
2	98	8.0 (± 4.7)
14	86	12.0 (± 2.1)
22	78	25.1 (± 11.0)
63	37	43.2 (± 8.9)

Source: Compiled from the data by Olanoff et al.; J. Pharm. Sci., 68:1147 (1979).

addition of a cross-linking agent results in the cross-linkage of some polymer chains, which leads to a reduction in the mobility of the polymer chain and consequently a decrease in porosity ε as well as an increase in tortuosity θ for the diffusioin of drug molecules in the polymer structure (49). The combination of reduced porosity and increased tortuosity resulted in a reduction in polymer diffusivity as expected from the relationship

Table 18 Dependence of Polymer Diffusivity on the Position of Hydroxy Groups in the Progesterone Molecule

	$D_p \times 10^5$ (cm²/sec)		
	Silicone elastomer[a]	Silicone adhesive[b]	
Diffusants		*A*	*B*
Progesterone	1.76	7.06	7.94
21-Hydroxyprogesterone	1.25	9.72	11.21
11-Hydroxyprogesterone	0.82	6.12	5.32
17-Hydroxyprogesterone	0.51	1.10	0.89

[a]Calculated from the data determined in silicone polymer matrix at 45°C [Chien et al.; J. Pharm. Sci., 68:689 (1979)].
[b]Compiled from the data determined in silicone adhesives at 37°C: A = X7-2920 and B = DC-355 [Toddywala and Chien; J. Control. Release, 14: 29–41 (1990)].

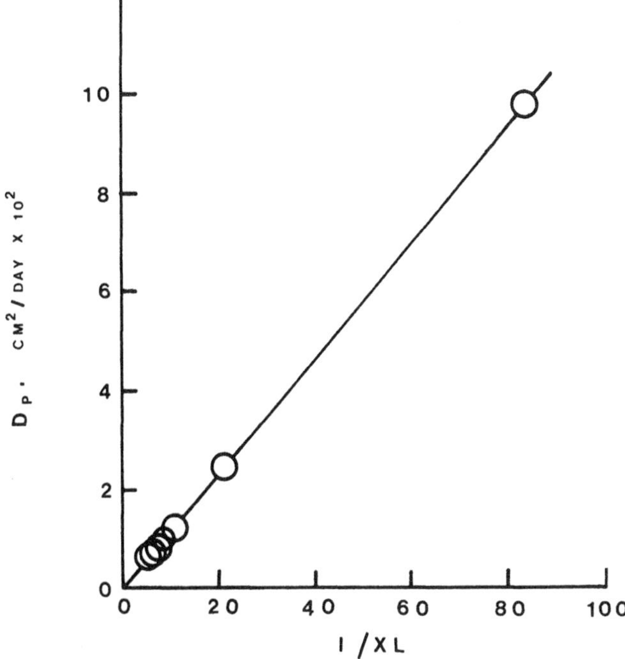

Figure 32 Linear dependence of the polymer diffusivity D_p of norgestomet in a hydrogel matrix on the extent of cross-linkage XL^{-1} of the hydrogel. [Plotted from the data by Chien and Lau; J. Pharm. Sci., 65:488 (1976).]

$$D_p = D \frac{\varepsilon}{\theta} \tag{54}$$

where D is the intrinsic diffusivity of a diffusant molecule (1).

The effect of cross-linking on polymer diffusivity was also observed for a simple gas molecule like N_2 (1). Along with the reduction in polymer diffusivity, both the energy of activation for polymer diffusion E_d and the frequency factor D_0 were found to increase in proportion to the increase in the extent of cross-linkage (Table 19). These findings suggested that the cross-linking process yields a reduction in the mobility of a polymer chain, which in turn requires a higher energy of activation and a greater frequency factor for its motion.

Table 19 Effect of Cross-Linking on the Diffusion of N_2 in Natural Rubber

XL[a] (%)	$D_p \times 10^6$ (cm²/sec)	D_o	E_d (kcal/mol)
1.7	1.08	0.74	8.0
2.9	0.80	1.26	8.5
7.2	0.27	3.24	9.7
11.3	0.11	12.00	11.0

[a]Extent of cross-linking.

Source: Compiled from the data by Barrer and Skirrow; J. Polym. Sci., 3:549 (1948).

Table 20 Dependence of In Vitro and In Vivo Release Profiles of Norgestomet and Polymer Diffusivities on Extent of Cross-Linkage in Hydrogel Implants

XL[a] (%)	$D_p \times 10^3$ (cm^2/day)	Release flux (mg/cm^2/day$^{1/2}$)	
		In Vitro	In Vivo[b]
1.2	97.2	0.605	0.640
4.8	24.2	0.396	0.504
9.6	12.1	0.185	—
12.0	9.7	0.133	—
14.4	8.1	0.101	—
16.8	6.9	0.074	—
19.2	6.1	0.058	0.129

[a]Extent of cross-linkage.
[b]Results from the subcutaneous release studies of norgestomet-releasing Hydron implants in 39 cows for 16 days.
Source: Compiled from the data by Chien and Lau; J. Pharm. Sci., 65:488 (1976).

As the result of the reduction in polymer diffusivity, the release profiles of drugs from hydrogel-based drug delivery systems was observed to decrease as the degree of cross-linking increased (49,50). A typical example is demonstrated in the controlled release of norgestomet, in vitro and in vivo, from hydrogel-based subdermal implants (Table 20). As expected from Equation (17), a linear relationship exists between the drug release flux $Q/t^{1/2}$ and $D_p^{1/2}$ (Figure 33).

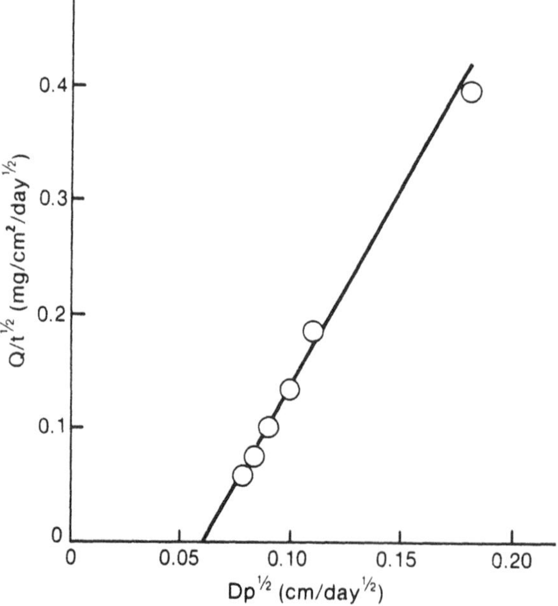

Figure 33 Linear relationship between the in vitro release flux $Q/t^{1/2}$ of norgestomet from hydrogel implants and the square root of polymer diffusivity $D_p^{1/2}$. [Plotted from the data by Chien and Lau; J. Pharm. Sci., 65:488 (1976).]

The effect of cross-linking agents and copolymer compositions on the polymer diffusivity of drugs in hydrophilic polymers was found to be related to the water content in the polymers (50,51). Permeation across a water-swollen polymer membrane can be visualized as primarily by diffusion through the microscopic, water-saturated pore channels within the polymer structure (52,53), and the polymer diffusivity is exponentially dependent upon the reciprocal of the degree of hydration of the hydrophilic copolymer (Figure 34). The addition of a cross-linking agent results in a decrease in porosity and an increase in tortuosity of these pore channels, leading to a reduction in polymer diffusivity [Equation (48)].

2. *Effect of Crystallinity*

It is known that low-density polyethylene (LDPE) has a higher degree of side-chain branching than high-density polyethylene (HDPE). Therefore, LDPE has a lower degree of crystallinity than does HDPE. The crystallites act similarly to the cross-linking agents just discussed. The crystallinity introduces regions of very low diffusion relative to the diffusion in the surrounding amorphous structure, which leads to a significant reduction in gross polymer diffusivity. The effect of density on polymer diffusion is illustrated in Table 21, in which polymer diffusivity decreases as the density of the polyethylene membrane is increased. The effect of density on polymer diffusivity was also demonstrated in simple gas molecules (54). The poly-

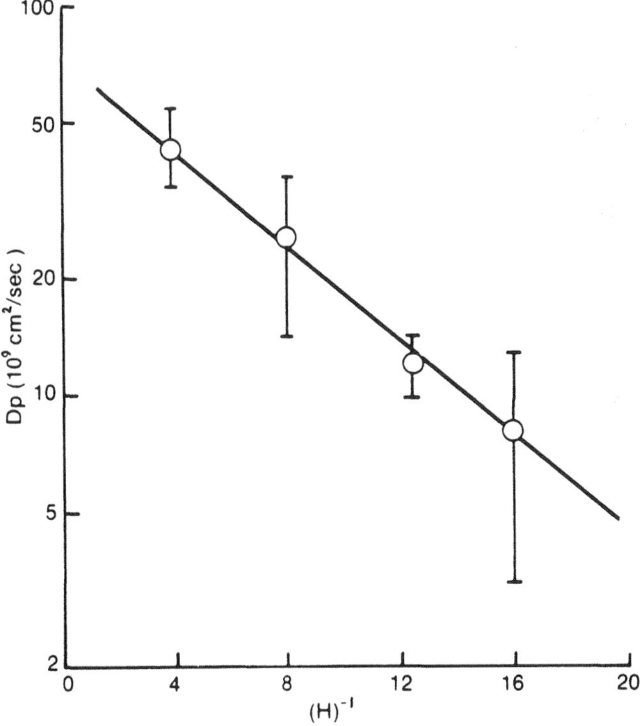

Figure 34 Semilogarithmic relationship between the polymer diffusivity D_p of tetracycline and the reciprocal of the degree of hydration $(H)^{-1}$ of the co(2-hydroxyethyl methacrylate-methyl methacrylate)polymer. [Plotted from the data by Olanoff et al.; J. Pharm. Sci., 68:1147 (1979).]

Table 21 Effect of Density in Polyethylene Membrane on the Polymer Diffusivity of Ethylene Oxide

Density (g/cm^3)	$D_p \times 10^8$ (cm^2/sec)
0.913	6.17
0.917	3.52
0.930	1.83
0.963	1.30

Source: Compiled from the 23°C data by White and Bradley; J. Pharm. Sci., 62:1634 (1973).

mer diffusivity D_p of gas molecules in LDPE is much greater than that in HDPE (Table 22). However, the activation energy for polymer diffusion E_d was observed to be slightly higher for low-density polyethylene than for high-density polyethylene, and the frequency factor D_0 was also found to be much greater for low- than for high-density polyethylene. These observations can be attributed to the higher degree of side-chain branching in LDPE, which requires a high energy of activation for the motion of polymer chains and also a greater frequency factor in response to the increase in the entropy of activation ΔS_d, as expected from the relationship

$$\log D_0 = \text{constant} + \frac{\Delta S_d}{2.303R} \tag{55}$$

Thermodynamically, the energy of activation for diffusion E_d and the entropy of activation ΔS_d are linearly related (1).

3. *Effects of Fillers*

Fillers are often incorporated into a polymer to enhance its mechanical strength. For example, very pure and finely ground silica (or diatomaceous earth) particles are usually added as fillers to silicone elastomers in an amount as high as 20–25% to enforce the mechanical strength of the elastomers. The effect of fillers on polymer diffusion is more complicated than the effect of crystallinity and cross-linking already

Table 22 Dependence of Polymer Diffusion on the Density of Polyethylene Membrane

	$D_p \times 10^6$ (cm^2/sec)		E_d (kcal/mol)		D_o	
Diffusants	LDPE[a]	HDPE[b]	LDPE	HDPE	LDPE	HDPE
H_e	6.8	3.07	5.9	5.6	0.13	0.04
O_2	0.46	0.17	9.6	8.8	4.48	0.43
CO	0.33	0.096	9.5	8.8	2.82	0.25
CO_2	0.37	0.124	9.2	8.5	1.85	0.19
CH_4	0.19	0.057	10.9	10.4	17.2	2.19

[a]Density of low-density polyethylene (LDPE) = 0.910–0.935 g/cm^3.
[b]Density of high-density polyethylene (HDPE) = 0.940–0.965 g/cm^3.
Source: Compiled from the data by Michaels and Bixler; J. Polym. Sci., 50:393 (1961).

Table 23 Effect of Siliceous Earth Filler on the Polymer Diffusivity of Steroids in Silicone Polymer Matrix

Steroids	$D_p \times 10^7$ (cm^2/sec) No filler	With filler
Progesterone	5.78	4.50
17α-Hydroxyprogesterone	5.65	3.88
6α-Methyl-17-acetoxyprogesterone	4.17	3.36

Source: Compiled from the data by Roseman; J. Pharm. Sci., 61:46 (1972).

discussed. The presence of fillers was reported to affect polymer diffusivity as determined by lag time techniques (55).

The effect of fillers was noted in the matrix diffusion of steroids in silicone elastomer containing siliceous earth fillers (56). A reduction in polymer diffusivity by 19–31% was observed with some steroids (Table 23). This reduction can be attributed to the Langmuir adsorption of steroid molecules onto filler particles (57), leading to a prolongation of lag time and, hence, a lower estimated value of diffusion coefficient (56). The reduction in diffusivity is thus a function of filler content in the polymer and can run as high as 15 times in the presence of as little as 25% filler (Table 24).

Diffusion through a homogeneous (i.e., fillerless) polymer may be characterized as Fickian, but diffusion through a heterogeneous (filler-containing) polymer is made complex because of the Langmuir adsorption of diffusant molecules onto the active filler. The role of active filler in the reduction of polymer diffusivity can be quantitated by assessing the effect of Langmuir adsorption on the movement of diffusant molecules in a heterogeneous polymer structure (58,59). The apparent polymer diffusivity $D_{p,f}$ in a filler-containing polymer can be related to the effective polymer diffusivity D_p in a fillerless polymer by the relationship (57)

$$\frac{1}{D_{p,f}} = \frac{1}{D_p} + \frac{K_f}{D_p} V_f \tag{56}$$

Table 24 Dependence of Polymer Diffusivity of Ethyl-*p*-Aminobenzoate on the Content of Silica Filler in a Silicone Membrane

Filler content (% w/w)	$D_p \times 10^6$ (cm^2/sec)
0.0	1.78
12.5	0.24
15.0	0.15
25.0	0.12

Source: Compiled from the data by Most; J. Appl. Polym. Sci., 14:1019 (1970).

Table 25 Adsorption Capacity K_f of Silica Filler

Diffusants	K_f
Ethyl-*p*-aminobenzoate	111.04
17-Hydroxyprogesterone	3.56
Progesterone	2.22
Medroxyprogesterone acetate	1.88

where K_f is the adsorption capacity of the filler and V_f is its volume fraction in the polymer. According to Equation (56), a plot of $1/D_{p,f}$ versus V_f should yield a straight line with an intercept of $1/D_p$ and a slope of K_f/D_p. From the values of intercept and slope, the adsorption capacity K_f of a filler for various drugs can be determined (Table 25). The adsorption capacity of a filler varies greatly from one drug to another.

4. *Determination of Polymer Diffusivity*

The diffusivity of a drug molecule through the rate-controlling membrane of a polymer membrane permeation-controlled drug delivery system can be determined from the relationship

$$D_p = \frac{h_p^2}{6t_l} \tag{57}$$

where t_l is the time-axis intercept of the extrapolation through the steady-state drug release data (Figure 35). This time lag effect is usually observed in a freshly fabricated drug delivery device (1), and it should be treated carefully. The extrapolation should be drawn through those steady-state release data generated after a period of at least two lag times t_l; otherwise, significant errors may result in the determination of extrapolated lag time and in the computation of polymer diffusivity (6,60).

Very often drug delivery devices are stored for varying lengths of time before being used. The release of drug from such shelved devices usually results in a burst effect phenomenon due to the instantaneous release of drug molecules accumulated on the device surface during storage (61). The polymer diffusivity D_p in such an aged device can be determined from the relationship

$$D_p = \frac{h_p^2}{3t_b} \tag{58}$$

where t_b is the negative time-axis intercept of the extrapolation through the steady-state release data (Figure 35).

As discussed earlier, elastomeric polymers often contain very pure and finely ground filler particles to enforce their mechanical strength. In such cases the polymer diffusivity $D_{p,f}$ may be determined by the following relationships [Equations (59) through (61)].

1. If the filler is inert and its presence adds only a tortuosity factor ∂ to the diffusion process,

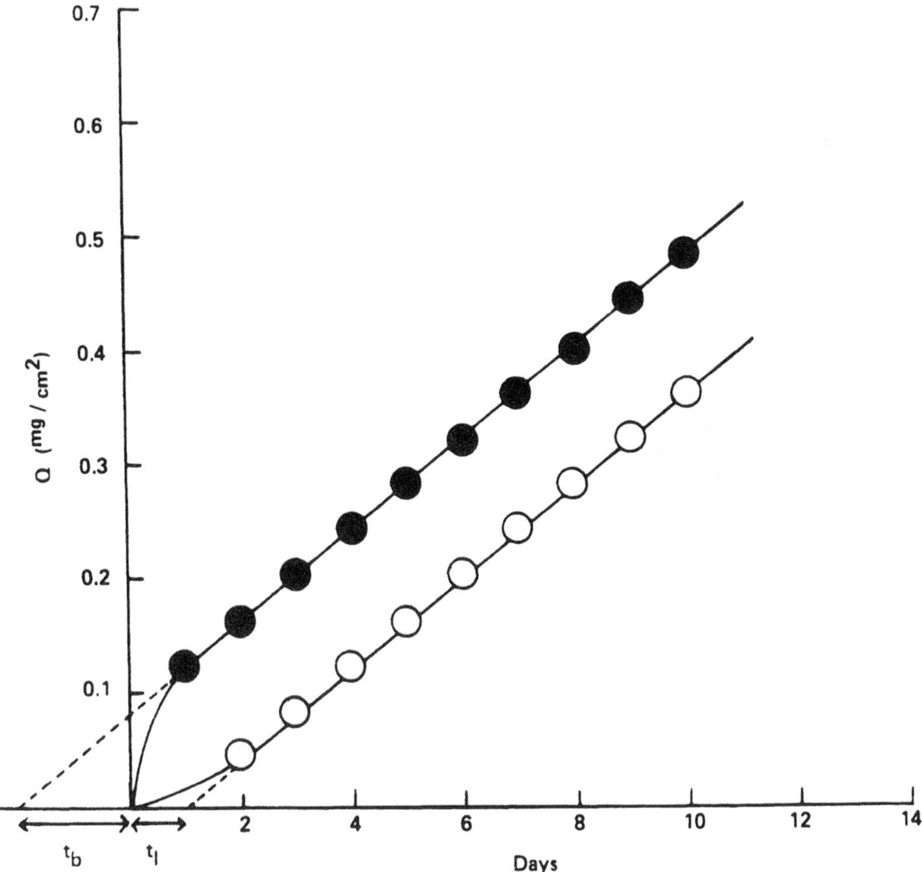

Figure 35 Approaching the steady-state release profile of norgestomet from a silicone capsule subjected to (○) a time lag effect or (●) a burst effect. [Reproduced, with permission, from Chien; in: *Sustained and Controlled Release Drug Delivery Systems* (Robinson, Ed.), Dekker, New York, 1978, Chapter 4.]

$$D_{p,f} = \frac{\partial^2 h_p^2}{6t_l} \tag{59}$$

2. If the filler is active and has a constant adsorption capacity K_f,

$$D_{p,f} = \frac{h_p^2}{t_l}\left(\frac{1}{4} + \frac{K_f V_f}{2K_p C_a D_p}\right) \tag{60}$$

3. If the filler is active and its adsorption capacity K_f is in direct proportion with the local concentration of diffusant molecules,

$$D_{p,f} = \frac{h_p^2}{6t_l}(V_p + K_f V_f) \tag{61}$$

where V_p and V_f are the volume fractions of the polymer and the filler, respectively; K_p is the true partition coefficient for the interfacial partitioning of drug from the solution medium toward the polymer phase; and C_a is the concentration of diffusant in the applied solution medium.

On the other hand, the diffusivity of a drug species through the polymer matrix of a polymer matrix diffusion-controlled drug delivery device cannot be determined from lag time and burst effect techniques. However, it can be conveniently calculated from the linear $Q/t^{1/2}$ versus $[(2A - C_p)C_p]^{1/2}$ plots (Figure 36), and diffusivity of drug molecules in the polymer matrix can be determined by

$$D_p = \frac{(Q/t^{1/2})^2}{[(2A - C_p)C_p]} \tag{62}$$

E. Solution Diffusivity D_s

The diffusion of solute molecules in a solution medium may be considered to result from the random motion of molecules. Under a concentration gradient molecules diffuse spontaneously from a region of higher concentration to a region of lower concentration until the concentration in the whole solution medium becomes uniform throughout.

The solution diffusion process is, microscopically, slightly different from the

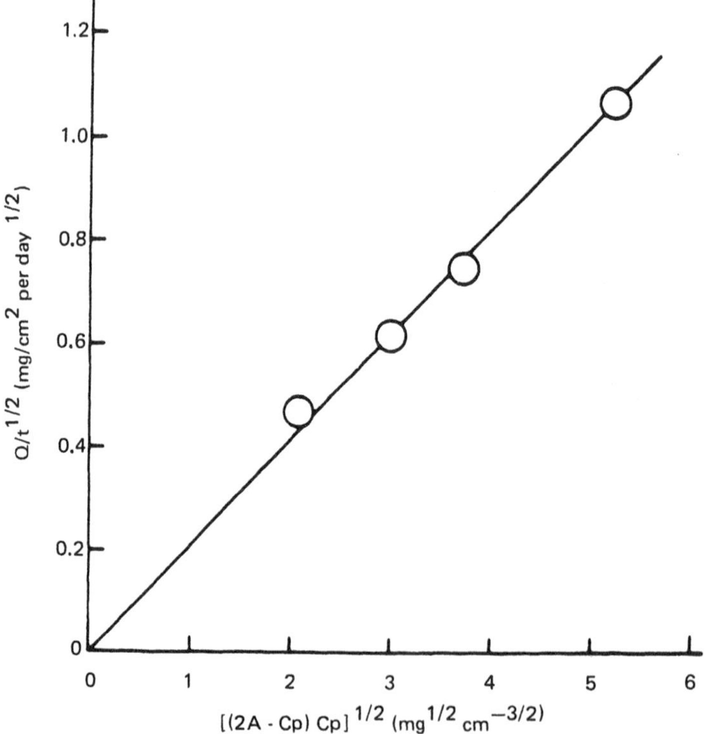

Figure 36 Linear relationship between the release flux $Q/t^{1/2}$ of desoxycorticosterone acetate from a matrix-type silicone device and $[(2A - C_p)C_p]^{1/2}$ values. [Reproduced, with permission, from Chien et al.; ACS Symposium Series, 33:72 (1976).]

Figure 37 Bueche's void occupation model. Drug molecule X jumps from its own void to a new equilibrium position in the adjacent void.

polymer diffusion discussed in the last section. The model of void occupation proposed by Bueche (62) and the theory of free volume by Kumins and Kwei (63) can be applied to the analysis of the solution diffusion process.

Let us consider a simple solution of a drug species. In the solution medium each drug molecule is confined to a cell defined by a group of solvent molecules, like the one shown in Figure 37. Motion of the center of mass of drug molecule X may take place in a number of ways. Suppose one of the neighboring cells becomes vacant for some reason and during the same time interval, if drug molecule X acquires sufficient thermal energy and also moves in the proper direction, drug molecule X can jump from its own cell to a new equilibrium position in the neighboring cell. Such a dislocation gives rise to diffusive motion if another drug molecule jumps into the vacancy left by molecule X before X can return to its original position. The solution diffusivity D_s is thus related to the jump frequency ν and the jump distance λ by

$$D_s = \frac{1}{6} \nu \lambda^2 \tag{63}$$

The variation in solution diffusivity from system to system, often many orders of magnitude, is primarily a result of the difference in jump frequencies ν.

The probability that a diffusional jump takes place is proportional to the probability that a hole of sufficient size is adjacent to molecule X. The probability that the presence of a hole with size V_h can be located is, according to the Boltzmann distribution, proportional to $e^{-(E_h/RT)}$. Therefore, the solution diffusivity may also be defined by the relationship

$$D_s = D_0 e^{-(E_h/RT)} \tag{64}$$

where D_0, the preexponential factor, is assigned to take into account the jump distance and the entropy of activation associated with hole formation (64). A molecule ordinarily vibrates around its equilibrium position at a frequency of 10^{12}–10^{13} vibrations/sec.

For a solute whose molar volume is equal to or greater than the molar volume of water molecules, the diffusivity of the solute molecules in the aqueous solution (at 25°C) is inversely proportional to the cube root of their molar volume. The molar volume of a solute molecule is an additive property of its constituent atoms and functional groups. It is possible to estimate with reasonable accuracy the molar volvume of a drug molecule from its chemical formula by summation of the partial molar volumes of its constituent atoms and functional groups (Table 26).

Table 26 Partial Molar Volume of Some Common Atoms and Groups

Atom	Partial molar volume ($Å^3$/molecule)	Group	Partial molar volume ($Å^3$/molecule)
		CH_2	26.90
H	5.15	CH_3	32.04
H^+	−7.47	NH_2	12.78
N	2.49	$N(CH_3)_3^+$	110.08
N^+	13.95	COO^-	19.09
C	16.44	COOH	31.55
O (=O or -O-)	9.13	C_2H_5	58.61
O (-OH)	3.82	C_3H_7	85.84
O (diol)	0.66	C_4H_9	112.73
S	25.73	C_6H_{13}	166.53
P	28.23	C_8H_{17}	220.32
P^+	47.32	$C_{10}H_{21}$	274.12
Li^+	−8.63	$C_{12}H_{25}$	327.91
Na^+	−9.46	$C_{14}H_{29}$	381.70
K^+	7.47	OCH_2CH_2	62.93
Cl^-	37.02	One ring	−13.45
Br^-	48.48	Two fused rings	−43.83
I^-	67.74	Steroid rings	317.12

Source: Calculated from the data by Flynn et al.; J. Pharm. Sci., 63:479 (1974).

When the solution diffusivities of various chemical classes are compared on the basis of molecular volume, the alkanes are the most rapidly diffusing chemicals, with the relative rates of diffusion: alkanes > alcohols > amides > acids > amino acids > dicarboxylic acids. This order of solution diffusivity is anticipated when one considers the increase in the hydrodynamic volume of traveling diffusant molecules whose −COOH, $-NH_2$, or −OH groups are tightly hydrogen bonded with water molecules (65). The effect of such hydration is proportionally smaller for larger molecules.

The diffusivity of a solute molecule in an aqueous solution usually decreases as its concentration increases. This reduction is frequently related to the increase in solution viscosity that usually accompanies the increase in solute concentration. In general, however, diffusivity determined at a concentration below 0.10 M is very close to the value determined by extrapolation to infinite dilution (65).

The effect of viscosity on the solution diffusivity D_s can be described by the simple relationship

$$D_s = \frac{\omega}{\mu} \tag{65}$$

where ω is a proportionality constant and μ is a viscosity coefficient. Equation (65) indicates that when all the other parameters are constant the solution diffusivity is inversely proportional to the viscosity coefficient of the aqueous solution. The viscosity coefficient may be determined by the Poiseuille equation,

$$\mu = \frac{\pi \Delta P d^4}{128 V l} t \tag{66}$$

where t is the time required for the V ml of a solution to flow through a capillary tube of length l and diameter d under applied pressure ΔP, and π is a constant.

When an aqueous solution contains various volume fractions of a cosolvent, such as polyethylene glycols, to enhance the aqueous solubility of a drug species [Equations (36) through (38) and Figure 11], the intrinsic viscosity μ^* of a cosolvent system can be estimated from the relationship

$$\mu^* = \lim_{f \to 0} \frac{\mu_{sp}}{f} \tag{67}$$

by plotting the values of μ_{sp}/f_x, the ratio of the specific viscosity μ_{sp} over its corresponding volume fraction f_x, against the volume fractions of a cosolvent in a series of aqueous solutions and then extrapolating this linear relation to the zero volume fraction, i.e., pure distilled water. The values of specific viscosity for various cosolvent-water combinations can be determined from the relationship

$$\mu_{sp} = \frac{\mu_x}{\mu_w} - 1 \tag{68}$$

where μ_x and μ_w are the viscosities of an aqueous solution containing x volume fraction of a cosolvent system and pure water, respectively. Pure water has a μ_w value of 8.95×10^{-3} g/cm · sec at 25°C.

The viscosity of most liquids decreases with an increase in temperature. This is expected from the void occupation and of free volume models discussed earlier. This temperature-dependent reduction in viscosity gives an increase in solution diffusivity [Equation (65)]. The temperature dependence of the solution diffusivity D_s can be derived from Equation (64):

$$\log D_s = \log D_0 - \frac{E_h}{2.303R}\frac{1}{T} \tag{69}$$

The energy of activation for solution diffusion E_h calculated from the diffusion data in aqueous solutions is generally between 4.5 and 5.0 kcal/mol. These values are very close to the 4.2 kcal/mol value for the viscous flow of water (65).

F. Thickness of Polymer Diffusional Path h_p

The controlled release of a drug species from both polymer membrane- and polymer matrix-controlled drug delivery systems is essentially governed by the same physicochemical principle as defined by Fick's law of diffusion [Equation (1)]. The observed difference in their drug release patterns [Equations (4) and (17)] is a result of the difference in the time dependence of the thickness of their polymer diffusional paths h_p. For the polymer membrane-controlled reservoir-type drug delivery devices fabricated from nonbiodegradable and nonswollen polymers, such as silicone elastomer, the h_p value is defined by a polymer wall with a constant thickness that is invariable with the time span (Figure 2). On the other hand, in matrix-type drug delivery devices fabricated from nonbiodegradable polymers the thickness of the diffusional path in the polymer matrix as defined by drug depletion zone (Figure 3) grows progressively in proportion to the square root of time [Equation (16) and Figure 38]. The rate of growth in the h_p value can be defined mathematically by Equation (70):

$$\frac{h_p}{t^{1/2}} = \left(\frac{2C_pD_p}{A - C_p/2}\right)^{1/2} \tag{70}$$

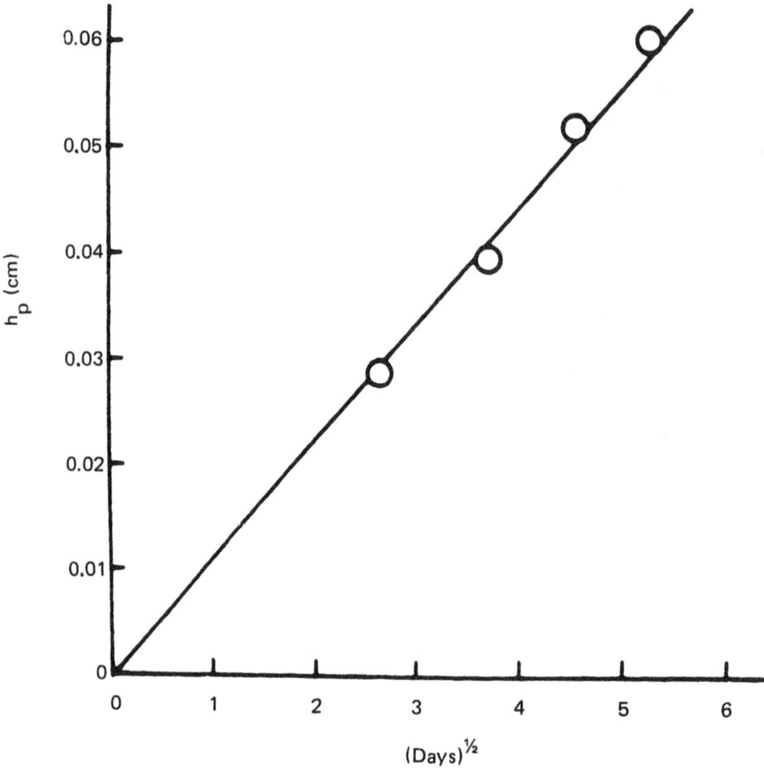

Figure 38 Linear relationship between the thickness of the drug depletion zone h_p in a matrix-type silicone device and the square root of time.

Equation (70) indicates that the rate of growth in the thickness of the drug depletion zone (polymer diffusional path), $h_p/t^{1/2}$, is dependent upon the solubility C_p, diffusivity D_p, and loading dose A of the drug in the polymer. Theoretically, the $h_p/t^{1/2}$ values, estimated from the slope of the linear h_p versus $t^{1/2}$ plots, should be a linear function of the reciprocal of the square root of $A - C_p/2$ (Figure 39).

In the matrix-type drug delivery devices fabricated from a biodegradable polymer, such as co(lactide-glycolide)polymer, the h_p versus $t^{1/2}$ and $h_p/t^{1/2}$ versus $(A - C_p/2)^{-1/2}$ relationships is complicated by the concurrent biodegradation of the polymer matrix (Chapter 8). In matrix-type drug delivery devices fabricated from nonbiodegradable, hydrophilic polymers, such as hydroxyethyl methacrylates, these relationships should also be followed after the equilibrium hydration state is reached (Figure 3).

Effect of the time-invariant polymer diffusional path (with constant membrane thickness) on the controlled release of a drug species from reservoir- and hybrid-type drug delivery systems (Figures 2 and 5) can be appreciated by examining Equations (7) and (30), respectively. The effect of membrane thickness was observed experimentally in the membrane permeation of megestrol acetate (18) and chlormadinone acetate (66) across the lipophilic silicone polymers, antineoplastic 5-fluorouracil across hydrophilic Hydron membranes, and naltrexone from biodegradable

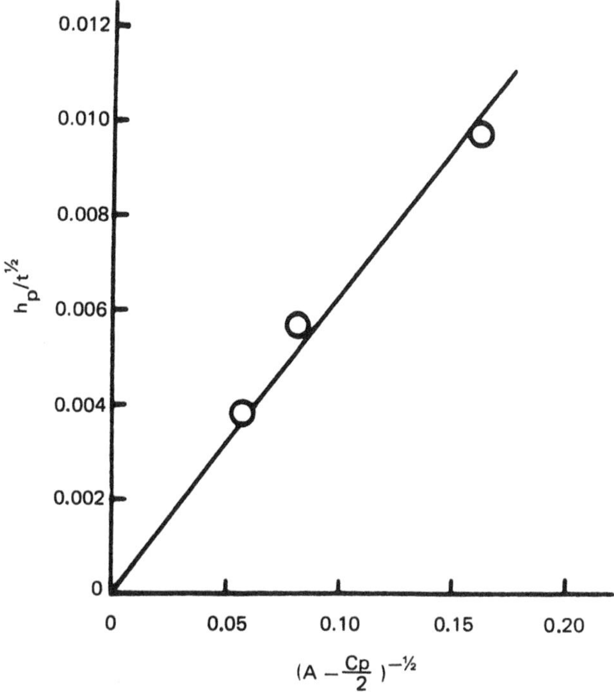

Figure 39 Linear relationship between the rate of growth in the thickness of the drug depletion zone $h_p/t^{1/2}$ in a matrix-type silicone device and the reciprocal of the square root of $A - C_p/2$.

copolymer-coated co(lactide-glycolide)polymer beads (67). The Q versus t linearity is followed for all membranes with varying thickness (Figure 40). As expected from Equation (7), the rate of membrane permeation Q/t is linearly dependent upon the reciprocal of membrane thickness (Figure 41). The thickness of the polymer membrane also affects, as suggested by Equation (57), the lag time t_l for membrane permeation (Figure 42). The linear Q/t versus h_p^{-1} and t_l versus h_p^2 relationships provide the possibility of altering the rate of drug release from the reservoir- and hybrid-type drug delivery systems by varying the thickness of the polymer membrane h_p. This was applied in the subcutaneous controlled release of steroids from silicone capsules in rats (Chapter 8) and in the 90 day intravaginal controlled administration of progesterone from silicone vaginal rings (Figure 43). The rates of in vivo drug release were reduced in proportion to the increase in the thickness of the polymer membrane.

G. Thickness of Hydrodynamic Diffusion Layer h_d

The rate-limiting role of the hydrodynamic diffusion layer h_d in determining drug release profiles can be visualized by considering that as a device is immersed in a stationary position in a solution, a stagnant layer is established on the immediate surface of the device. The effective thickness of this stagnant layer is dependent on the solution diffusivity D_s and varies with the square root of time (3):

$$(h_d)_{nr} = (\pi D_s)^{1/2} t^{1/2} \tag{71}$$

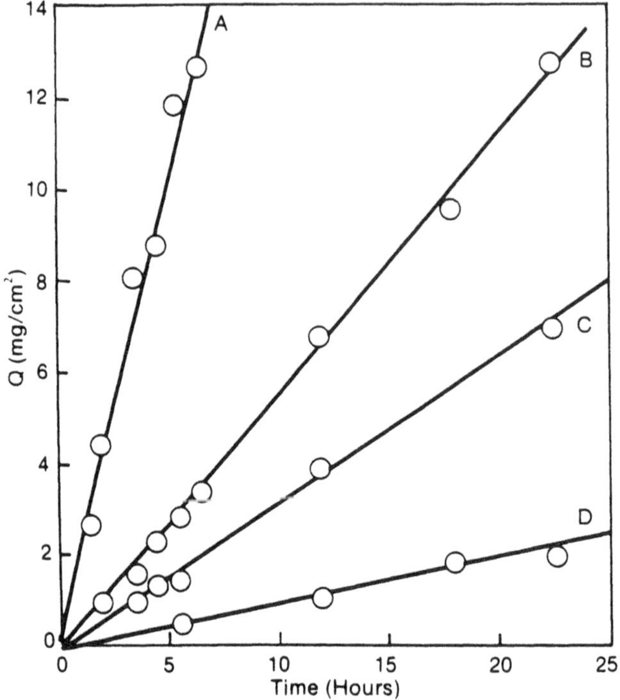

Figure 40 Linear Q versus t relationship for the permeation of 5-fluorouracil across Hydron membranes of varying thickness h_m: (A) 0.15 mm, (B) 0.60 mm, (C) 1.0 mm, and (D) 2.50 mm.

where the subscript *nr* refers to the stationary (or, nonrotationary) state and π is a constant.

If the device is forced to rotate at a constant speed, convection results. The convective diffusion of the drug molecules to and from the surface of the device is much greater than when natural diffusion (under the concentration gradient) operates on a stationary device because the concentration gradient $[C_s - C_{b(t)}]/h_d$ of the drug now extends over a thinner stagnant layer (68). The effective thickness of the hydrodynamic diffusion layer on the rotating device becomes time independent and is now defined by the Levich equation,

$$(h_d)_r = 1.62D_s^{1/3}V^{1/6}W^{-1/2} \qquad (72)$$

where the subscript r stands for the rotation state and V and W represent the kinematic viscosity of the elution medium and the angular rotation speed of the drug delivery device, respectively. Equation (72) allows the calculation of $(h_d)_r$ on a rotating device if D_s, V, and W are known or predetermined.

It was discussed earlier that in the subcutaneous controlled administration of progestins from silicone capsules in rats (5,12) a fairly good agreement can be established between the rates of subcutaneous drug release and the rates calculated from Equation (5) using the literature values for the partition coefficient, diffusivity,

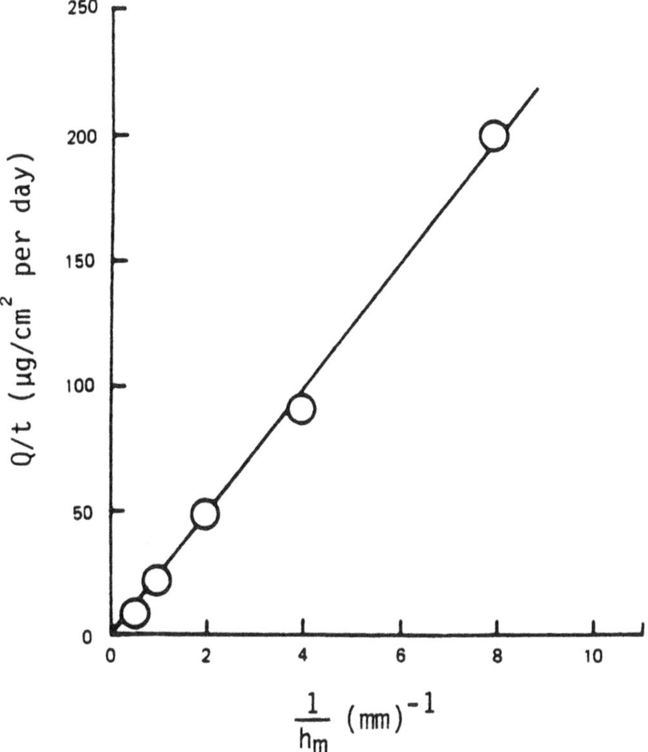

Figure 41 Linear dependence of the in vitro release rate Q/t of megestrol acetate from silicone capsules on the reciprocal of wall thickness h_m. [Plotted from the data by Kincl et al.; Steroids, 11:673 (1968).]

polymer solubility, and polymer wall thickness if a physiological diffusion layer (h_d, which is not measurable) of 80 μm is assumed to exist on the surface of subdermal implants (Table 1). It was also reported that a better in vivo-in vitro correlation can be achieved for the intravaginal controlled release of medroxyprogesterone acetate from a silicone vaginal ring if a physiological diffusion layer of 580 μm is presumed to exist on the vaginal ring in situ (21). The observed difference in the magnitude of the physiological diffusion layers can be attributed to the degree of contact between the drug delivery device inserted and the tissue at the site of insertion (or implantation), the diffusivity of drug molecules in the tissue fluid surrounding the inserted (implanted) device, the kinematic viscosity of the tissue fluid (69), and the motion of the device at the insertion (or implantation) site.

The rate-limiting role of h_d in determining the rates of controlled release of drug from reservoir- and hybrid-type drug delivery systems becomes apparent when one examines Equations (5) and (22). The effect of the hydrodynamic diffusion layer on the drug release profiles $Q/t^{1/2}$ from a matrix-type polymeric device was not considered in the derivation of Equation (18) because of the assumption that the diffusion of drug molecules in the polymer structure is the rate-determining step. Experimentally it was observed, however, that the drug release profile $Q/t^{1/2}$ is a function of

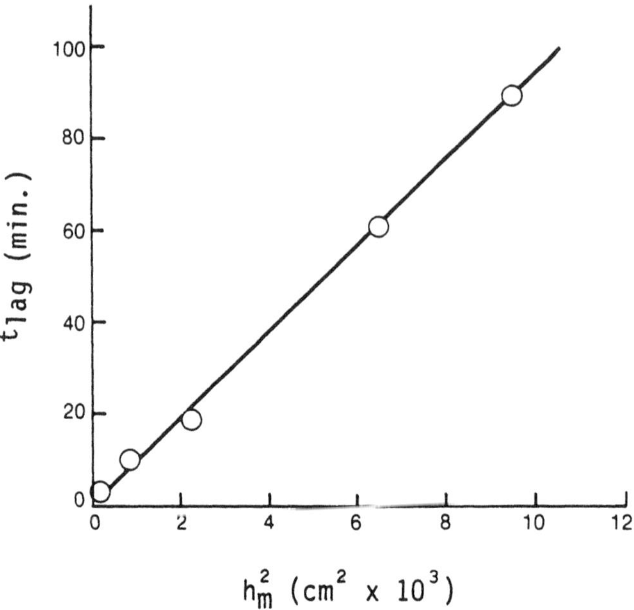

Figure 42 Linear relationship between lag time t_{lag} for the membrane permeation of *p*-aminoacetophenone and the square of silicone membrane thickness h_m^2. [Plotted from the data by Flynn and Roseman; J. Pharm. Sci., 60:1788 (1971).]

the variation in the thickness of the hydrodynamic diffusion layer on the surface of a matrix-type drug delivery device (70). The magnitude of $Q/t^{1/2}$ values decreases as the thickness of hydrodynamic diffusion layer h_d is increased, although the linear Q versus $t^{1/2}$ drug release pattern is followed irrespective of the change in the fluid hydrodynamics around drug delivery devices in the elution solution (Figure 44).

For in vitro release studies of a drug species from a drug delivery device, the thickness of the hydrodynamic diffusion layer h_d on the immediate surface of the device should be controlled at a very small but constant magnitude, via the control of the angular rotation speed W [Equation (72)]; thus the real mechanism and rate profiles of membrane permeation can be better understood by minimizing the complication from the effect of hydrodynamic diffusion properties. This was analyzed and demonstrated by the use of a polymeric membrane-covered varying-speed rotating-disk electrode technique (6).

The physiological hydrodynamics for the diffusion of drug molecule after its release from an inserted (or implanted) drug delivery device are analyzed in depth using the complex multibarrier flow coordinate model of Scheuplein (71). This model has been applied to in vitro membrane transport (72) as well as to the subcutaneous (73) and intravaginal absorption (70) of drugs delivered by matrix-type drug delivery devices.

H. Drug Loading Dose *A*

In the preparation of a drug delivery device varying loading doses of drug are incorporated into the device as required for different lengths of treatment.

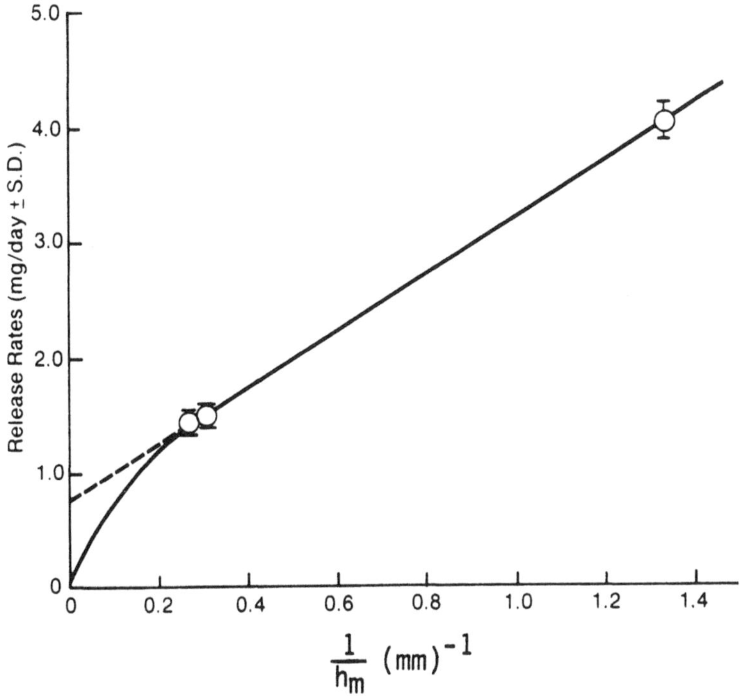

Figure 43 Effect of the thickness h_m of a silicone membrane on the 90 day intravaginal controlled release of progesterone from a silicone vaginal ring. [Plotted from the data by Burton et al.; Contraception, 17:221 (1978).]

Equation (17) indicates that variation in the drug loading dose A in matrix-type drug delivery systems also affects the magnitude of the drug release flux $Q/t^{1/2}$. The effect of drug loading on $Q/t^{1/2}$ values is observed in drug release studies both in vitro (Figure 45) and in vivo (Figure 46). The results suggest that any intention to prolong the duration of medication by incorporating a higher loading dose of a therapeutic agent into a matrix-type drug delivery device also inevitably produces a greater value for the drug release flux $Q/t^{1/2}$.

Variation in drug loading doses was also found to influence the drug release profiles from matrix-type drug delivery devices fabricated from biodegradable co(lactide-glycolide)polymers (67).

On the other hand, the rate of drug release from a membrane permeation-controlled reservoir-type polymeric drug delivery device is independent of the drug loading dose [Equation (5)]. The independence of the drug release rate Q/t from the drug loading dose in reservoir-type drug delivery systems has been observed experimentally (Table 27). Variation in loading doses results only in a change in the duration of constant drug release profiles (Figure 47). The release rate of norgestrienone from silicone capsules with a loading dose of 2.5 mg/cm is essentially the same as that from capsules with a loading level of either 7.5 or 12.5 mg/cm for a duration of 180 days. After this period almost no steroid remains in the capsules. On the other hand, the release rates of norgestrienone from capsules with loading

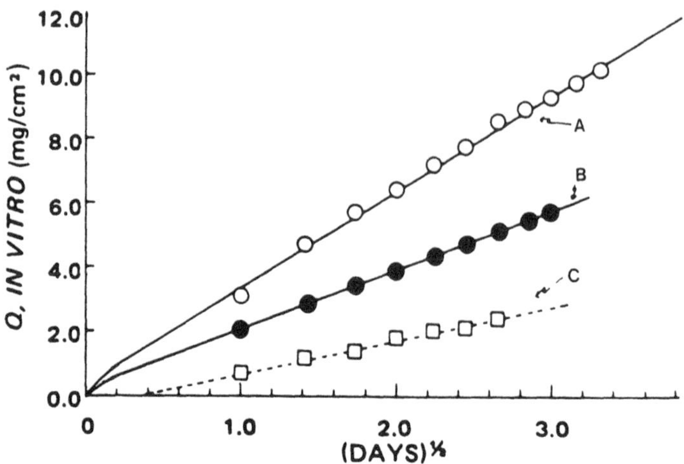

Figure 44 Effect of the hydrodynamic diffusion layer on the linear Q versus $t^{1/2}$ relationship for the in vitro controlled release of ethynodiol diacetate from a silicone vaginal ring. $Q/t^{1/2}(h_d)$: (A) 3.17 mg/cm^2/day$^{1/2}$ (154.7 μm); (B) 1.86 mg/cm^2/day$^{1/2}$ (254 μm); and (C) 1.07 mg/cm^2/day$^{1/2}$ (~500 μm). [Reproduced, with permission, from Chien et al.; J. Pharm. Sci., 64:1776 (1975).]

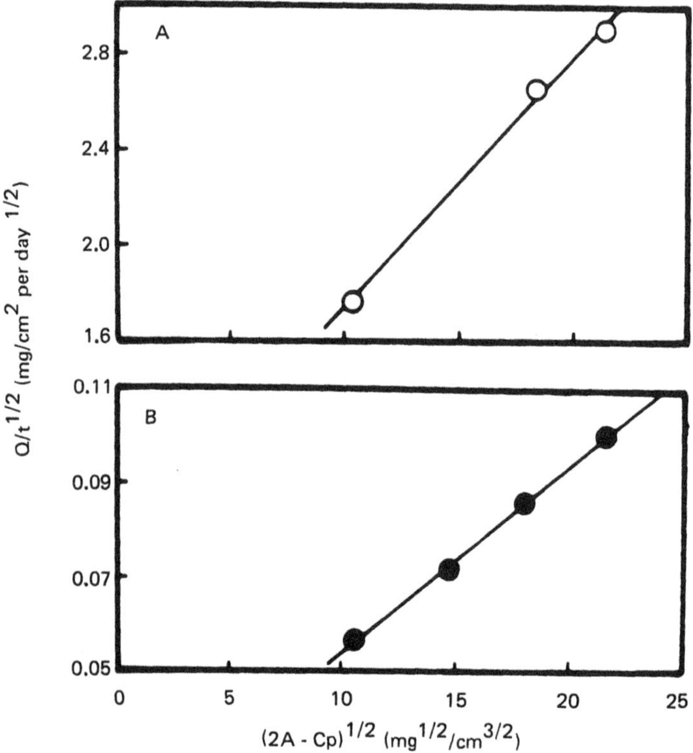

Figure 45 Linear dependence of $Q/t^{1/2}$ on $(2A - C_p)^{1/2}$: (A) progesterone and (B) hydrocortisone. [Reproduced, with permission, from Chien et al.; J. Pharm. Sci., 68:689 (1979).]

doses of 7.5 and 12.5 mg/cm show no difference for a period of more than 1 year (74).

Furthermore, it has been reported that the drug loading level in matrix-type drug delivery systems also affects the time required for the transition of drug release profiles from a partition-controlled process [Equations (13) and (14)] to a matrix-controlled process [Equation (18)]. The transition time t_{trans} was found to be prolonged as the drug loading dose in the devices was increased (Figure 48). The effect of drug loading dose on the transition time can be realized by substituting Equations (20) and (21) for α and β in Equation (19) to form

$$t_{trans/(KC_s)^{-1}} = \frac{3D_p h_d^2 \epsilon}{2D_s^2 \partial} A \tag{73}$$

where K is the partition coefficient for the interfacial partitioning of drug from polymer toward elution solution; C_s is the solubility of drug in the elution solution; D_p and D_s are the diffusivities of drug in polymer and elution solution, respectively; h_d is the thickness of the hydrodynamic diffusion layer; and ϵ and ∂ are the porosity and the tortousity of the polymer structure.

I. Surface Area

The dependence of the rate of drug release on the surface area of a drug delivery device is well known theoretically and experimentally. A typical example is illustrated in Figure 49. Both the in vitro and in vivo rates of drug release are observed to be dependent upon the surface area of drug delivery device.

For the sake of comparison, therefore, the contribution of surface area to the release profiles of a drug from various controlled-release drug delivery systems has been taken into consideration in the mathematical derivation of Equations (5), (17), (22), and others and built into the calculation of the Q value, which by definition is the cumulative amount of drug released from a unit surface area of the device (mg/cm^2).

IV. EVALUATIONS OF CONTROLLED-RELEASE DRUG DELIVERY

A. In Vitro Evaluations

A review of the current literature has revealed that a formidable task facing a researcher in his or her successful development of a viable controlled-release drug delivery device and experimental assessments of the delivery rate profiles of drugs from the device is the proper design of an in vitro drug elution system that permits the accurate evaluation and mechanistic analysis of controlled drug release profiles. Such variables as the specificity and sensitivity of the assay, maintenance of the sink condition throughout the duration of a study, the hydrodynamic characteristics of the solution, and the reproducibility of sampling, as well as the volume and temperature of the system, must be carefully evaluated during the initial stage of design and

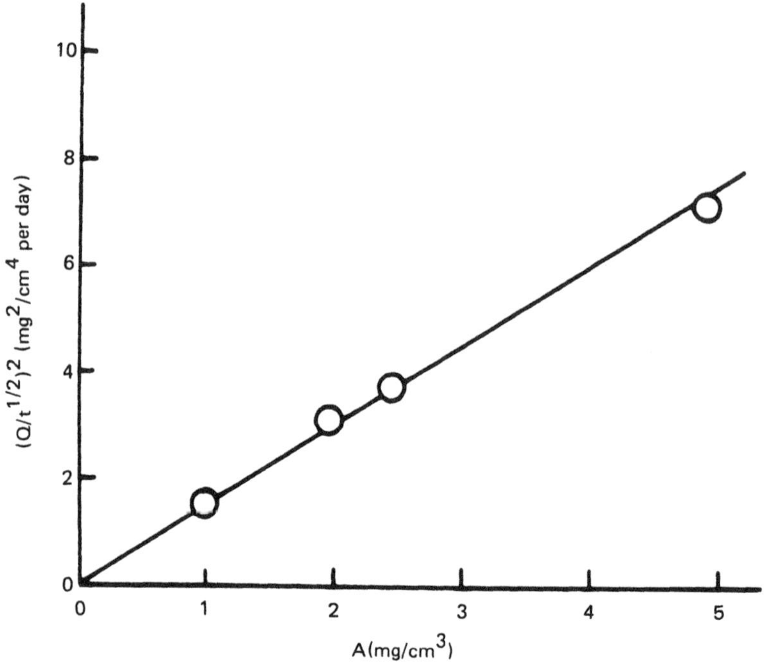

Figure 46 Effect of drug loading dose A on the intravaginal controlled release flux $Q/t^{1/2}$ of norgestrel from matrix-type silicone vaginal rings in humans. [Plotted from the data by Mishell et al.; Contraception, 12:253 (1975).]

Table 27 Independence of the In Vitro Release Rate of Methyl-CCNNU from a Silicone Capsule from the Variation in Drug Loading Dose

Drug loading dose		
Weight[a] (mg/capsule)	Packing extent	Q/t[b] (mg/cm^2/day)
4.0	Very loose	0.598
6.8	Firm	0.518
7.8	Firm	0.598
8.0	Firm	0.558
8.2	Firm	0.598
8.3	Firm	0.677
10.9	Very firm	0.637
		0.598 (±8.6%)

[a]Inside a capsule of 2 mm (ID) × 2.5 mm (length).
[b]In vitro release into normal saline at 37°C.
Source: Compiled from the data by Rosenblum et al.; Cancer Res., 33:906 (1973).

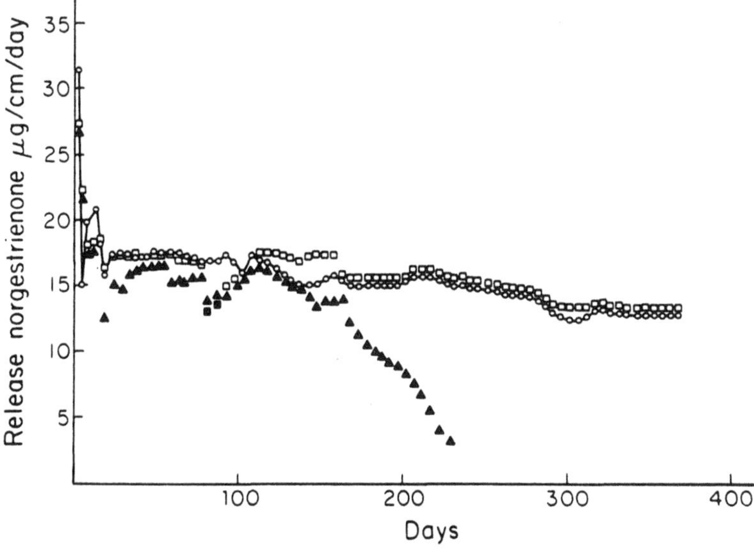

Figure 47 In vitro release profiles of norgestrienone from silicone capsules loaded with different doses of steroid: (○) 12.5 mg/cm, (□) 7.5 mg/cm, and (▲) 2.5 mg/cm. [Reproduced, with permission, from Nash et al.; Contraception, 18:367 (1978).]

precisely controlled during operation. Many designs of in vitro drug elution systems have been devised and used (18,20,21,75). Considering that most of them have been designed for the same purposes, the diversity in their designs is remarkable. In this section only two in vitro drug elution systems, which have been assessed and found to fulfill most of these criteria, are discussed.

1. Drug Elution Systems

Continuous Flow Apparatus. The continuous flow apparatuses designed and used by Kalkwarf et al. (75) and Roseman and Higuchi (21) are very similar in principle. A typical setup is illustrated in Figure 50. The prototype of a drug delivery device under development is positioned in a thermostated drug elution column and exposed to a continuous flow of an elution solution, such as distilled water, isotonic saline, or an isotonic solution of bovine serum albumin (3%). The flow rate of the elution solution can be regulated to maintain solution hydrodynamics that simulate various rates of drug uptake in the body. Samples of the effluent drug solution are then collected and assayed at various time intervals (75,76).

As the flow rate of the elution solution is increased, the rate of drug release from the device also increases and eventually reaches a plateau level (Figure 51). Beyond this level any increase in the rate of solution flow does not produce enhancement of the rate of drug delivery. This value represents the maximum rate of drug delivery from the drug delivery device since the thickness of the hydrodynamic diffusional layer has been reduced to the minimum (Section IIIG) and the sink condition is maintained throughout the study (Sections IIIB). Using this continuous flow apparatus the in vitro release of progesterone from a matrix-type polyethylene-based drug delivery device was investigated and found to be under a matrix diffusion-

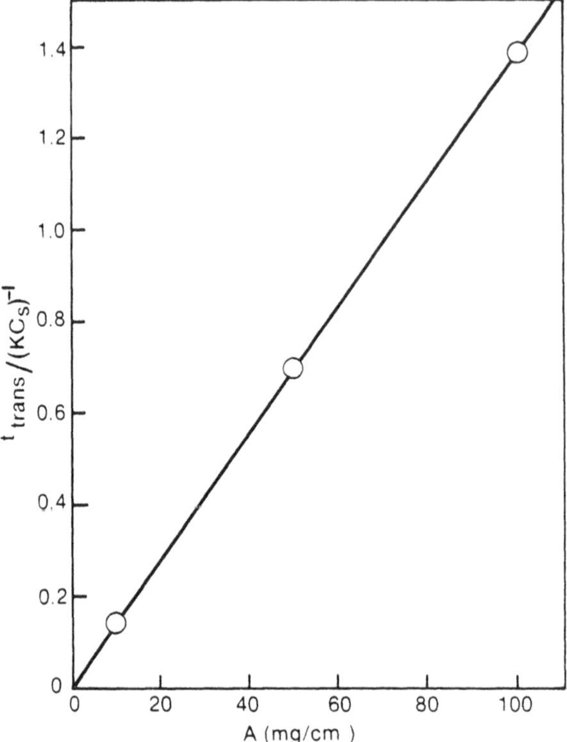

Figure 48 Effect of the loading dose A of p-aminobenzoic acid esters on the transition time t_{trans} as defined in Equation (73). [Plotted from the data by Roseman and Yalkowsky, ACS Symposium Series, 33:33 (1976).]

controlled process, as defined by Equation (17), and a Q versus $t^{1/2}$ drug release profile was obtained irrespective of the type of elution solution used (Figure 52). This release profile was achieved with a very high flow rate of elution solution (300 ml/min or 432 L/day). With this flow rate the use of either a 3% bovine serum albumin (BSA) solution or an isotonic saline solution as the elution medium was observed to yield only a less than 6% difference in drug release profiles (the $Q/t^{1/2}$ value was 0.313 mg/day$^{1/2}$ for BSA solution and 0.296 mg/day$^{1/2}$ for isotonic saline). The results suggest that sink conditions have been maintained. It was reported that the in vitro drug release data determined by this continuous flow apparatus simulate closely, in terms of the mechanism and rate profiles of drug release, the intrauterine and the intraperitoneal controlled release of progesterone from matrix-type polyethylene-based drug delivery systems (75).

Because of the low aqueous solubility commonly observed in many therapeutic agents, particularly the steroidal drugs (Table 6), a large volume of an elution solution must be used in in vitro drug release studies to maintain a sink condition (e.g., 300 ml/min or 432 L/day for progesterone). The direct assay of drug concentrations in the elution solution becomes extremely difficult and labor intensive, although it can be done by special analytic instrumentations, e.g., radioactivity measurement. The release profiles have therefore been alternatively determined by assaying the

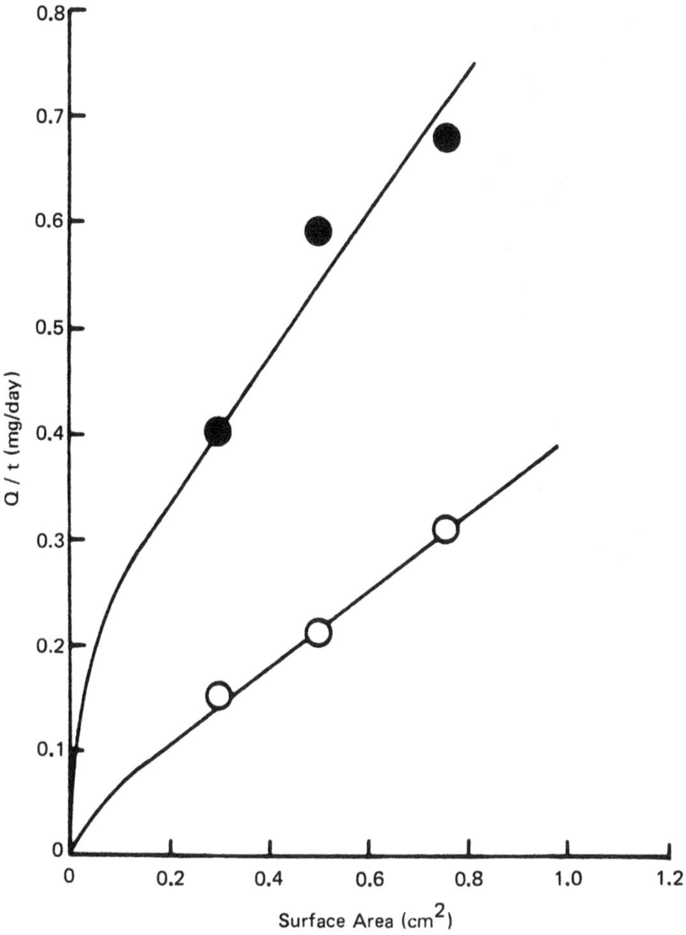

Figure 49 Effect of surface area on the release rate Q/t of 1-(2-chloroethyl)-3-(*trans*-4-methylcyclohexy)-1-nitrosourea from silicone capsules: (●) in vivo release in rats; (○) in vitro release in normal saline at 37°C. [Plotted from the data by Rosenblum et al.; Cancer Res., 33:906 (1975).]

residual drug content in drug delivery devices (75). The precision of analysis of the kinetics of drug release profiles may be jeopardized (6,77,78).

The same mechanical principles were also applied to the construction of a continuous circulation drug elution system to investigate the controlled release of progesterone and its derivatives from matrix-type medicated vaginal rings fabricated from silicone elastomer (21,56). The controlled release of progestins from matrix-type silicone devices was also found to follow the Q versus $t^{1/2}$ relationship when the devices were exposed to the continuous circulation of 60 L/day of distilled water (Figure 53). The difference in the drug release flux $Q/t^{1/2}$ of progestins from drug-dispersing matrix-type polymeric devices can be attributed to the difference in their polymer solubility C_p and polymer diffusivity D_p [Equation (18)] when a constant drug loading dose A is used. Experimentally, the magnitude of $Q/t^{1/2}$ values was

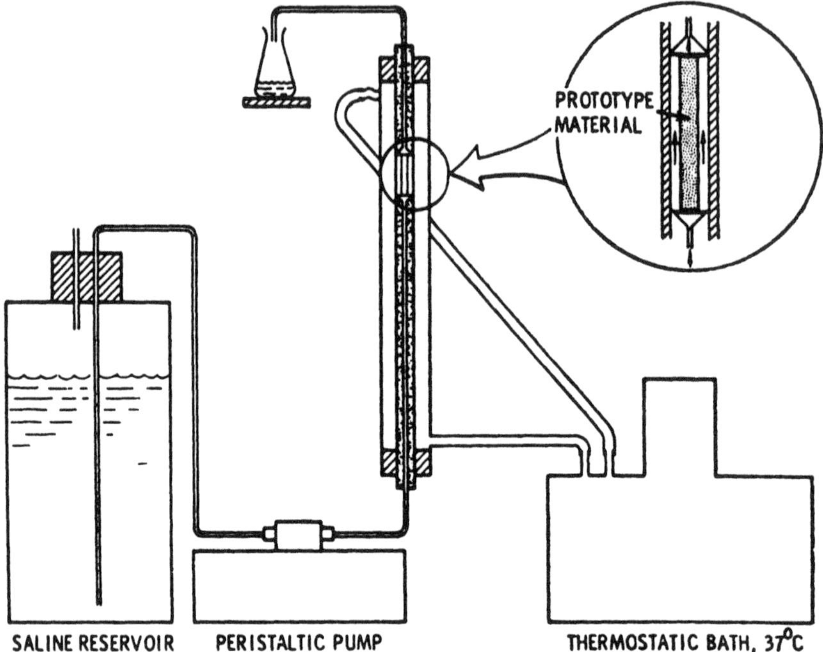

Figure 50 The experimental setup for an in vitro drug release study using a continuous flow apparatus. (Courtesy of Dr. D. R. Kalkwarf, Battelle Pacific Northwest Laboratories.)

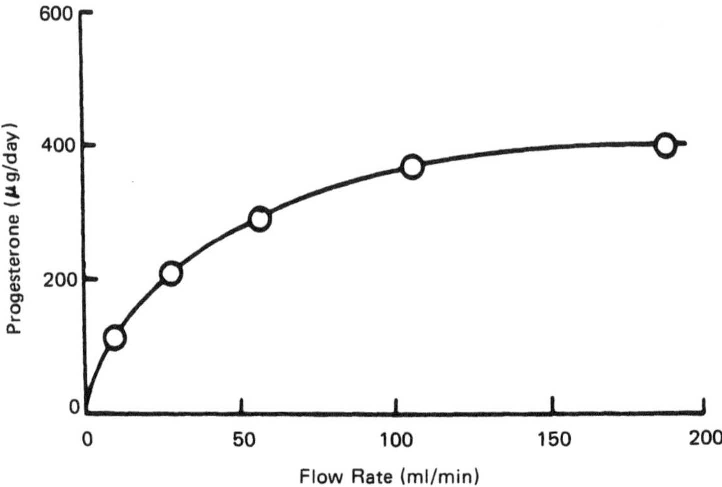

Figure 51 The daily amount of progesterone released from a polyethylene-based drug delivery device as a function of the flow rate (ml/min) of the elution solution in the continuous flow apparatus. (Courtesy of Dr. D. R. Kalkwarf, Battelle Pacific Northwest Laboratories.)

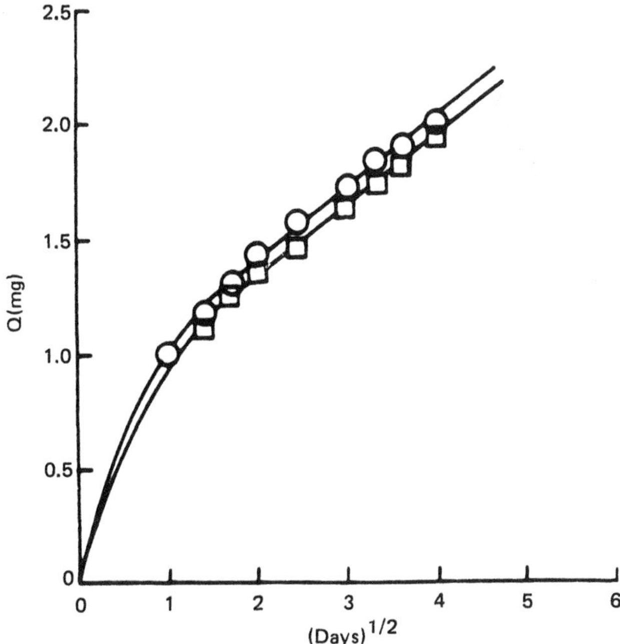

Figure 52 Release profiles of progesterone from a polyethylene-based drug delivery device studied by the continuous flow apparatus with (○) a 3% bovine serum albumin solution or (□) an isotonic normal saline solution (300 ml/min) as the elution solution. [Plotted from the data by Kalkwarf et al.; Contraception, 6:423 (1972).]

found to be linearly dependent upon the square root of the polymer solubilities of these progestins (Figure 8). The effect of polymer solubility and polymer diffusivity on drug delivery rate profiles has been analyzed in Sections IIIA and IIID, respectively.

The continuous flow or circulation apparatus is also applicable to in vitro studies of controlled release rate profiles of drugs from membrane permeation-controlled and (membrane-matrix) hybrid drug delivery systems.

Constant Rotation Apparatus. A primary concern in the in vitro evaluations of a potential drug delivery system before initiation of costly animal testing is the development of an in vitro method that permits determination of the mechanisms and rates of drug delivery reliably and rapidly. The in vitro method developed should also make possible the direct assessment of drug release flux. With these considerations in mind, Chien and his associates (14) developed a constant rotation apparatus that is relatively simple in design and easy to construct (Figure 54). The polymeric drug delivery device is mounted in circular shape in a plexiglass holder (with a spin bar at the center of the holder) and then rotates in the elution medium, maintained under a sink condition, at a constant angular rotation speed to achieve the constant solution hydrodynamics required [Equation (72)]. This in vitro drug elution system permits the maintenance of a constant thickness of the hydrodynamic diffusion layer h_d on the immediate surface of the drug delivery device and a homogeneous drug concentration in the bulk of the elution solution. The effect of h_d

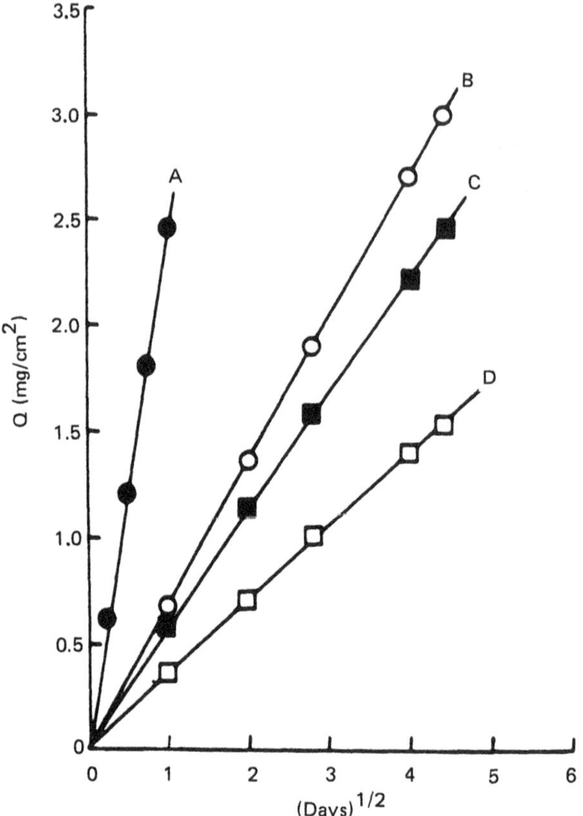

Figure 53 Release profiles of progestins from medicated silicone-based vaginal rings as determined by the continuous circulation apparatus: (A) progesterone, (B) medroxyprogesterone, (C) 6α-methyl-11β-hydroxyprogesterone, and (D) 17α-hydroxyprogesterone. [Plotted from the data by Roseman; J. Pharm. Sci., 61:46 (1972).]

on the magnitude of the drug release flux $Q/t^{1/2}$ was discussed earlier (Section IIIG, Figure 44).

Additionally, the sink conditions required for in vitro drug release studies are maintained in the constant rotation apparatus by the use of a water-miscible cosolvent compatible with the drugs and the polymeric drug delivery devices investigated to enhance the aqueous solubility of relatively insoluble therapeutic agents, such as steroidal drugs (Section IIIB). The effect of the addition of a cosolvent in the water on the aqueous solubility was defined in Equations (36) through (38) and demonstrated in Figure 11 and Table 13. Using cosolvent systems a perfect sink condition can be maintained throughout the in vitro drug elution studies, and only a small volume (e.g., 100–150 ml) of an elution solution is required. The drug concentrations in the solution are sufficiently high that hourly or daily direct assays become feasible and drug release profiles can be closely monitored and rapidly assessed. The combination of constant rotation and cosolvency thus developed provides a rapid methodology for reliable characterization of the mechanisms and rate profiles of controlled-release drug delivery from polymer-based drug delivery devices.

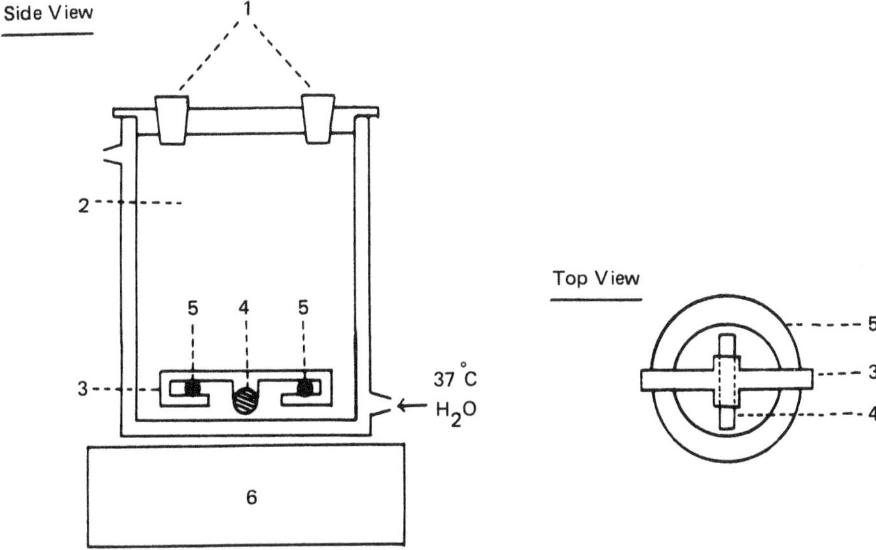

Figure 54 A unit of the drug elution cells in the constant rotation apparatus used to measure in vitro drug release profiles from a polymeric drug delivery device. (1) Sampling holes with Teflon stopcocks; (2) water-jacketed cell; (3) plexiglass holder; (4) Teflon-coated spin bar; (5) a ring-shaped polymeric device; and (6) a magnetic stirrer. Up to six cells can be connected in series and thermostated at 37°C. [Reproduced, with permission, from Chien et al.; J. Pharm. Sci., 63:365 (1974).]

Several drug elution cells can be connected in series and thermostated at the same temperature for in vitro evaluations of several drug delivery devices under the same experimental conditions.

The in vitro evaluations conducted in the constant rotation apparatus indicated that the release of desoxycorticosterone acetate from a matrix-type silicone pellet follows a matrix diffusion-controlled process and, as defined in Equation (17), a Q versus $t^{1/2}$ drug release pattern (Figure 55). On the other hand, the release of the same steroidal drug from a silicone pellet (11,12,79) fabricated from microreservoir dissolution-controlled drug delivery systems (Section IID) shows a Q versus t drug release pattern (Figure 56) as projected from Equation (27). The difference in the controlled drug release patterns between matrix- and microreservoir-type drug delivery systems is remarkable when the daily release data are compared (Figure 57). The daily dose of desoxycorticosterone acetate released from the matrix-type drug delivery device is much higher than from the microreservoir-type drug delivery device initially, resulting from burst release, and then gradually decreases with time (12). On the other hand, the microreservoir-type drug delivery device maintained a constant drug release profile (Figure 57).

Drug elution studies of a membrane permeation-controlled reservoir-type drug delivery system (Section IIA) in the constant rotation apparatus also give a constant drug release profile (Q versus t) as defined by Equation (4) (Figure 58). A similar drug release pattern was also observed when the same drug was released from the membrane-matrix hybrid drug delivery system.

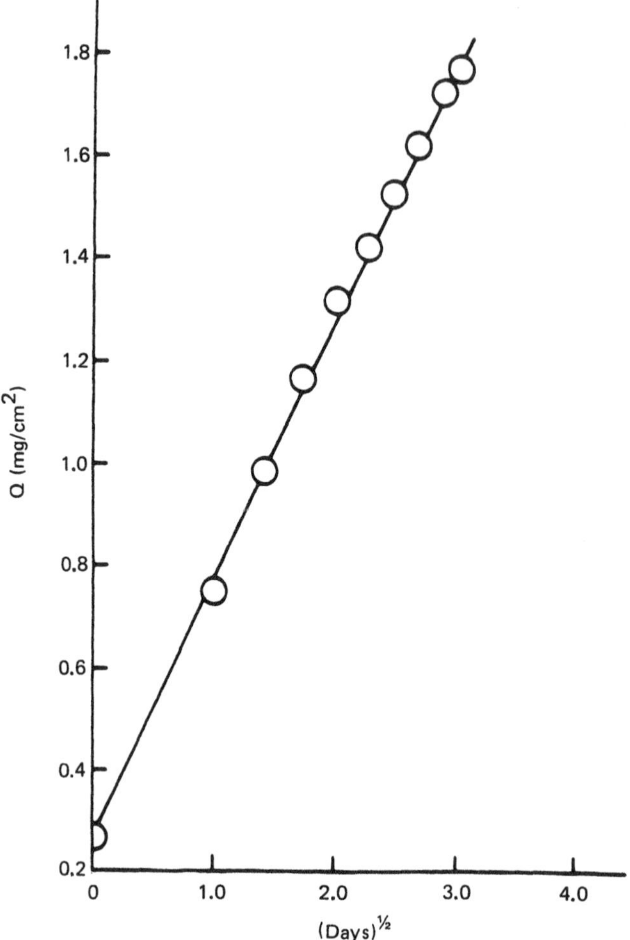

Figure 55 Release profile of desoxycorticosterone acetate from a matrix-type silicone implant determined in the constant rotation apparatus. [Reproduced, with permission, from the data by Chien et al.; ACS Symposium Series, 33:72 (1976).]

The rate profiles of drug release determined in the constant rotation apparatus can be varied to simulate various in vivo hydrodynamic conditions by changing the solution solubility of drugs in the elution medium (Figure 14 and Table 7). The solution solubility of a drug can be easily controlled by varying the volume fraction of cosolvent incorporated into the elution solution [Figure 11 and Equations (36) through (38)], leading to a variation in the mechanism and rate profiles of drug release from the matrix-type drug delivery system (Table 2).

The constant rotation apparatus approach was also applied to the design of other similar apparatuses for the in vitro study of the mechanisms and rate profiles of drug release from other controlled-release drug delivery devices, such as transdermal drug delivery systems (80–83), intravaginal drug delivery systems (84,85), and oral drug delivery systems (86). These in vitro drug release apparatuses were analyzed (87), and their hydrodynamic characteristics were determined theoretically and experi-

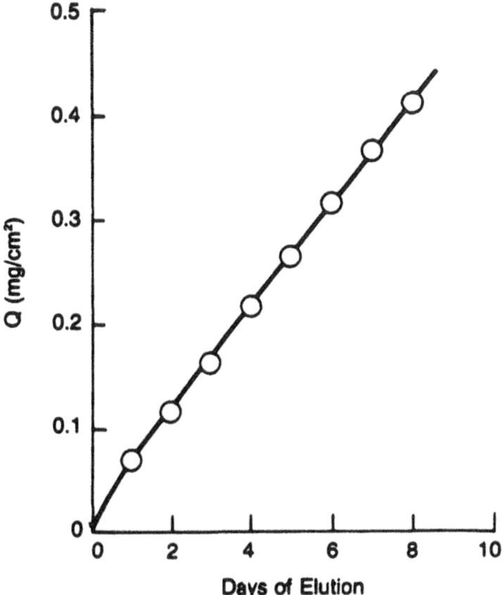

Figure 56 Release profile of desoxycorticosterone acetate from an MDD-type silicone implant determined in the constant rotation apparatus. [Reproduced, with permission, from the data by Chien et al.; J. Pharm. Sci., 67:214 (1978).]

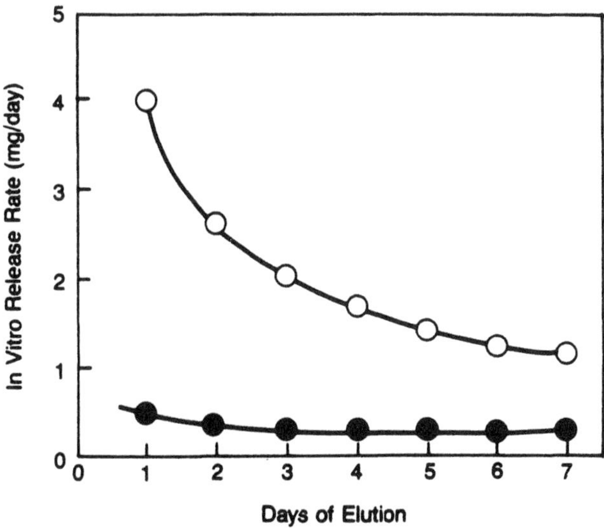

Figure 57 Daily release profiles of desoxycorticosterone acetate from (○) a matrix-type silicone implant and (●) a MDD-type silicone implant. (Both have the same surface area and loading dose.) [Reproduced, with permission, from the data by Chien et al.; J. Pharm. Sci., 67:214 (1978).]

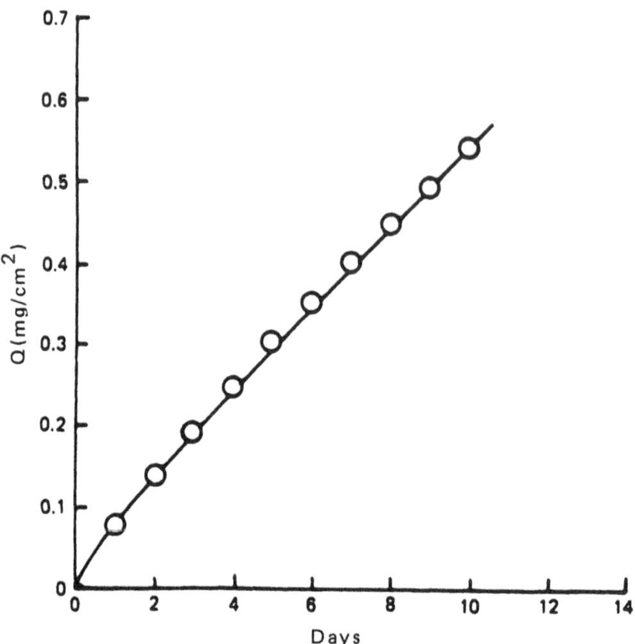

Figure 58 Release profile of norgestomet from a membrane permeation-controlled reservoir-type silicone device determined in the constant rotation apparatus. [Reproduced, with permission, from the data by Chien; in *Sustained and Controlled Release Drug Delivery Systems* (J. Robinson, Ed.), Dekker, New York (1978), Chapter 4.]

mentally (88). These in vitro apparatus are discussed individually in each pertinent chapter.

Continuous Flow Apparatus versus Constant Rotation Apparatus. Using medroxyprogesterone acetate as the model drug, the in vitro drug release profiles from a matrix-type drug delivery system, as determined by a continuous flow apparatus and a constant rotation apparatus, were compared with the data estimated from theoretical calculations using Equation (18) and literature data (14). The result indicated that in vitro drug elution studies conducted in both a continuous flow apparatus and a constant rotation apparatus produce the same drug release pattern (linear Q versus $t^{1/2}$ relationship), and the $Q/t^{1/2}$ data obtained are comparable to the theoretical value (Table 28).

In conclusion, both drug elution systems are ideal for the measurements of in vitro drug release rate profiles, especially the constant rotation apparatus. Both these techniques are capable of maintaining a sink condition to better simulate the biological sink resulting from hemoperfusion. By doing so, a better in vitro-in vivo correlation can be achieved, and the in vitro-in vivo correlation can then be applied to the development and optimization of a controlled-release drug delivery device.

2. *Thermodynamics of Controlled-Release Drug Delivery*

In in vitro drug elution studies in vitro the drug elution system is normally thermostated at 37°C to simulate the body temperature. The rate-limiting role of temperature in the in vivo controlled release of drugs is demonstrated in the 9 day subcutaneous controlled administration of norgestomet, a potent estrus-synchronizing

Table 28 Comparison in the Calculated and Experimental Release Fluxes of Medroxyprogesterone Acetate from Matrix-Type Silicone Devices

Methods	$Q/t^{1/2}$ ($mg/cm^2/day^{1/2}$)
Theoretical calculation	0.241[a]
Experimental determination	
Continuous flow apparatus	0.282[b]
Constant rotation apparatus	0.218[c]

[a]Calculated from Equation (18) using literature data on C_p, D_p, and A.
[b]Determined from the data by Roseman and Higuchi; J. Pharm. Sci., 59:353 (1970).
[c]Determined from the data by Chien et al.; J. Pharm. Sci., 63:365 (1974).

progestin, in heifers (Chapter 8). Because of the subzero temperatures encountered during the winter of 1977, the subcutaneous release rates of norgestomet from three subdermal implant formulations in the heifers were found to be lower than the projected release rates by as much as 26–36% (Table 29). On the other hand, the results obtained from the same formulations conducted in the autumn months with mild temperatures (above 14°C) were in perfect agreement with the projected rates of release. The effect of field temperature on the subcutaneous controlled delivery of norgestomet was also found to produce a reduction in the biological effectiveness of MDD implants in terms of the suppression of estrus during implantation and estrus synchronization after implant removal. A similar reduction in the rates of subcutaneous release and in the biological efficacy, in terms of the percentage of cows in estrus and in pregnancy, was also observed with the animals stabled in hilltop barns compared to those in barns located in the valley. Again, this may have resulted from the reduction in temperature at the top of the hill due to a windchill effect.

To further determine the effect of temperature on the in vivo controlled delivery of drugs, the subcutaneous release of 17β-estradiol from MDD subdermal implants was evaluated in a group of eight heifers that were either stabled in the barn with temperature maintained at approximately 10°C or subjected to the subzero winter temperatures in the open field. The release rate of 17β-estradiol was noted to decrease from 21.96 $\pm$ 5.34 $\mu g/cm^2/day$ when they were stabled inside the barn to 8.70 $\pm$ 3.08 $\mu g/cm^2/day$ when they stayed outside in the field. The temperature dependency of the in vivo drug release rate can be correlated with the in vitro drug release rate profiles by an Arrhenius relationship (Figure 59). By doing so, the temperature at the site of implantation (e.g., the dorsal surface of the ears), which is regulated by hemoperfusion, can be estimated and was found to be 18.74 and 1.27°C, respectively, for animals housed in the barn with room temperature controlled at approximately 10°C and those free in the field at subzero winter temperatures.

The results from these investigations clearly indicate that the controlled release of drugs from a drug delivery device is a temperature-dependent, energy-requiring process. This temperature dependence operates both in vitro and in vivo. From the Arrhenius relationship in Figure 59, the energy requirements for the controlled de-

Table 29 Effect of Field Temperature on Subcutaneous Release Rates of Norgestomet from Subdermal Implants in Heifers and Their Biological Effectiveness

Implant formulations[a]	Q/t (mg/day)			Biological effectiveness[d] (%)			
		Observed[c]		Estrus suppression[e]		Estrus synchronization[f]	
	Projected[b]	Autumn	Winter	Autumn	Winter	Autumn	Winter
A	143.4	139.2	92.4	92.9	65.0	85.7	55.0
B	217.3	218.9	155.7	100.0	72.2	87.5	50.0
C	286.7	291.7	211.9	100.0	100.0	100.0	46.7

[a]Developed from a microsealed drug delivery system (Chien and Lambert, U.S. Patent 4,053,580).
[b]Projected rates based on in vitro drug release studies at 37°C.
[c]Field temperature (mean of dew point temperatures measured every 3 hr by the National Climatic Center): autumn (8/26–9/20)/77: 13.97 to 15.17°C; winter (11/26–12/30)/77: −7.51 to −12.34°C.
[d]Observations made by veterinarians (Dr. A. Peterson, Searle Animal Sciences).
[e]Percentage of heifers with estrus in suppression during the 9 day subcutaneous administration of implants in the ears.
[f]Percentage of heifers with estrus synchronized within 2 days following the removal of implants.

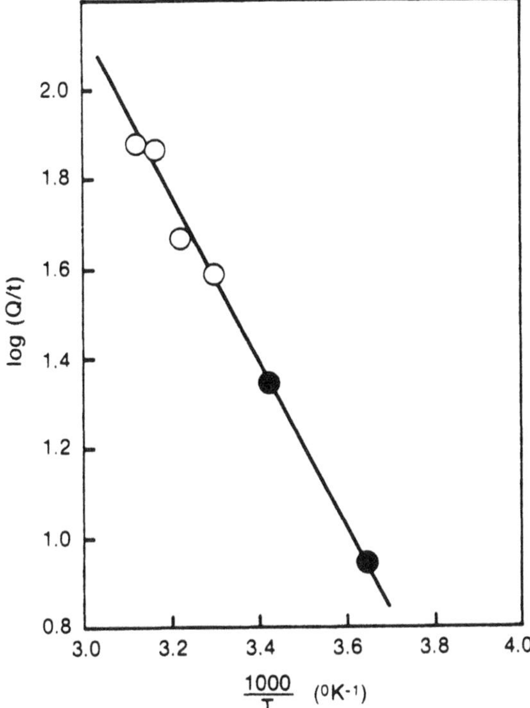

Figure 59 Effect of temperature on the controlled release of 17β-estradiol from MDD subdermal implants. Key: (○) in vitro release and (●) subcutaneous release in heifers.

livery of drugs from drug delivery systems can be calculated. Values of 4.94 and 8.49 kcal/mol, respectively, were found to be required for the controlled release of norgestomet and 17β-estradiol from MDD drug delivery devices.

Using norgestomet as the model drug, the thermodynamics of controlled drug release from a matrix-type silicone device was investigated (16). As described by Equation (14), drug release profiles at the initial state are predominantly governed by the diffusion layer-limiting partition-controlled process, and as expected, a constant (zero-order) drug release pattern was observed (Figure 60). This linear Q versus t relationship was followed at all temperatures studied, and the rate of drug release Q/t was observed to increase approximately four times as the temperature of the elution solution was raised from 30 to 50°C (16). The temperature dependence of the Q/t value is theoretically linked to two energy-activated processes, the solvation and diffusion of drug molecules in the elution solution [Equations (41) and (69)], as defined by the relationship

$$\log \frac{Q}{t} = \text{constant} - \frac{E_h + \Delta H_{T,s}}{2.303R} \frac{1}{T_s} \tag{74}$$

where E_h is the activation energy for solution diffusion, $\Delta H_{T,s}$ is the energy of solvation in the elution solution, and T_s is the temperature of the elution solution.

At steady state the matrix diffusion-controlled process outweighs the partition-controlled process and becomes the rate-limiting step that dictates the whole course

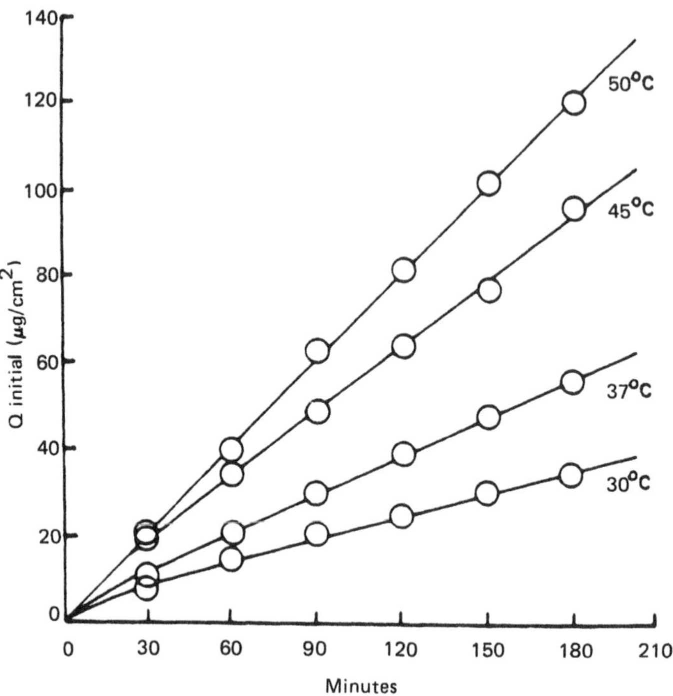

Figure 60 Initial-state release profiles of norgestomet from a matrix-type silicone device at various temperatures. [Reproduced, with permission, from the data by Chien, ACS Symposium Series, 33:53 (1976).]

of controlled drug release from the polymer. So, at the steady state, a Q versus $t^{1/2}$ release pattern is expected [Equation (17)]. This linear Q versus $t^{1/2}$ drug release profile was also observed to be temperature dependent (Figure 61). The magnitude of drug release flux $Q/t^{1/2}$ was seen to increase by approximately twofold as the temperature of the elution solution was raised from 30 to 50°C. This temperature dependence of $Q/t^{1/2}$ values is also theoretically related to two energy-activated processes, the solvation and diffusion of drug molecules in the polymer structure [Equations (34) and (52)], and is defined by the relationship

$$\log \frac{Q}{t^{1/2}} = \text{constant} - \frac{E_d + \Delta H_{s,p}}{4.606R} \frac{1}{T_s} \tag{75}$$

where E_d is the activation energy for matrix diffusion and $\Delta H_{s,p}$ is the energy of solvation in the polymer.

The Arrhenius relationship for the temperature dependence of Q/t and $Q/t^{1/2}$ values, as defined in Equations (74) and (75), is shown in Figure 62. The results suggest that the energy requirements for an interfacial partition-controlled process (at initial state) and a matrix diffusion-controlled process (at steady state) are different in magnitude and vary from one type of polymeric device to another (Table 30). The energies required for various microscopic steps, i.e., diffusion and solvation, in the controlled release of norgestomet under matrix diffusion- and interfacial partition-controlled processes are outlined in Table 31.

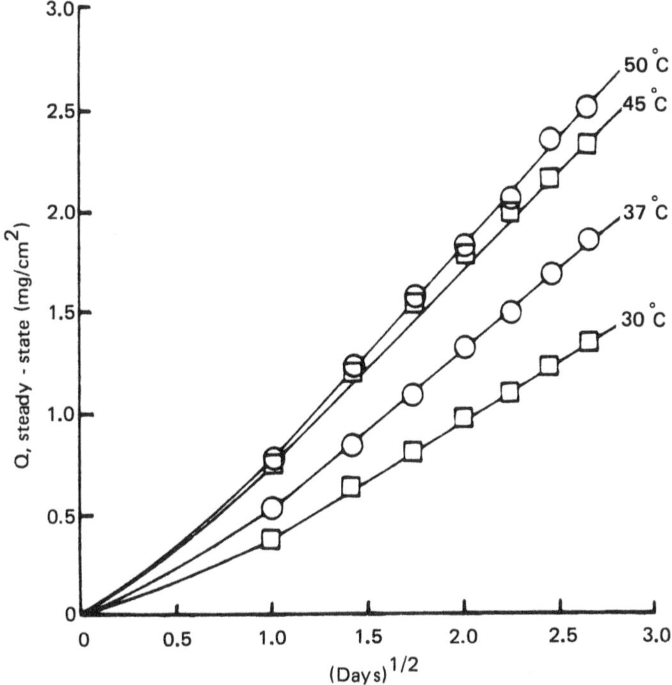

Figure 61 Steady-state release profiles of norgestomet from a matrix-type silicone device at various temperatures. [Reproduced, with permission, from the data by Chien; ACS Symposium Series, 33:53 (1976).]

The controlled release of norgestomet from a hydrophilic polymer matrix system was also studied under in vitro conditions (49). The Q versus $t^{1/2}$ release pattern was observed at both initial and steady states. The magnitude of the $Q/t^{1/2}$ values was also found to be temperature dependent. The energy required for the release of lipophilic drugs, like norgestomet, from hydrophilic hydrogel-based drug delivery devices at steady state was found to be slightly higher than that from lipophilic silicone elastomer-based polymeric devices (Table 30). The controlled release of progesterone and its six hydroxy derivatives from silicone devices was investigated, and the results suggested that it requires an energy in the range of 15.25–23.62 kcal/mol (15).

B. In Vivo Evaluations

In the development of a controlled-release drug delivery system for long-term rate-controlled administration of a therapeutic agent, it becomes necessary to conduct the in vivo evaluations once a satisfactory in vitro drug release profile has been achieved. The in vivo studies of the mechanisms and rate profiles of drug release at the target site(s) should be performed in appropriate animal models to establish the in vitro-in vivo correlation. The established correlation can then be utilized to standardize the in vitro experimental setup and conditions for drug elution studies. The standardized in vitro drug elution system and experimental conditions can then be utilized for

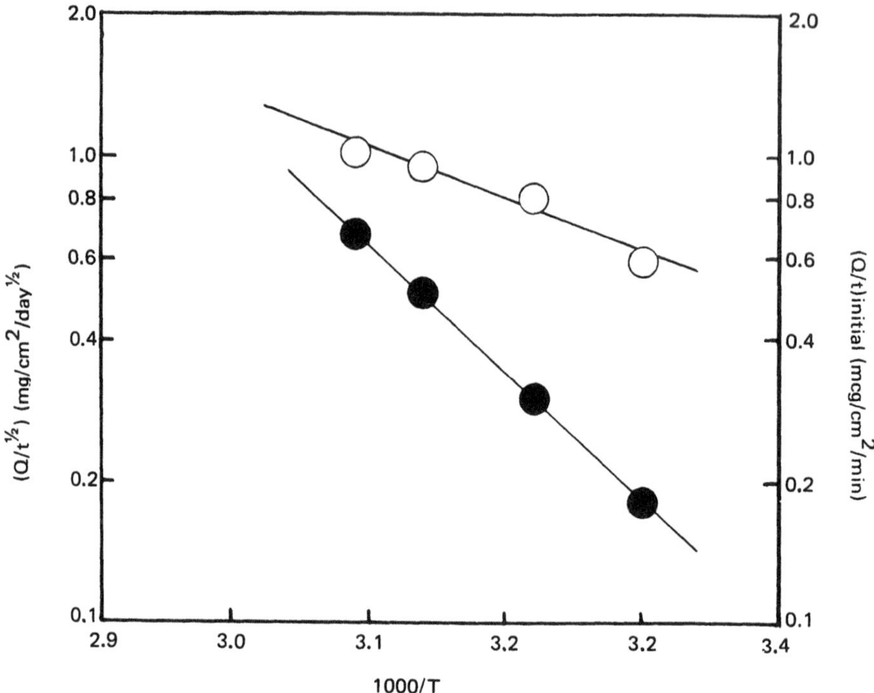

Figure 62 Temperature dependence of the initial-state (●) and steady-state (○) release profiles of norgestomet from a matrix-type silicone device as defined in Equations (74) and (75). [Reproduced, with permission, from the data by Chien; ACS Symposium Series, 33:53 (1976).]

formulation optimization and to refine the controlled release characteristics of the drug delivery system.

The in vivo evaluation of controlled-release drug delivery systems can vary from one type of drug delivery system to another. Depending on the projected biomedical applications, proper animal models and site(s) of administration must be selected. Numerous examples can be found in Chapters 3 through 10 for nasal, ocular, oral, rectal, vaginal, uterine, transdermal, parenteral, and subcutaneous controlled drug administrations. To avoid any redundancy, the in vivo evaluations of a particular drug delivery system for a specific biomedical application should be referred to the chapter related to that biomedical area.

Table 30 Comparison of Energy Requirements for the Controlled Release of Norgestomet from Matrix-Type Drug Delivery Systems

	Energy requirements (kcal/mol)	
Drug delivery systems	Initial state	Steady state
Silicone devices	12.67	10.30
Hydrogel devices	12.18	16.48

Source: Compiled from the data by Chien; ACS Symposium Series, 33:53 (1976); and Chien and Lau; J. Pharm. Sci., 65:488 (1976).

Table 31 Comparison of Energies Required for the Controlled Release of Norgestomet from Silicone Matrix Devices

	Diffusion energies (kcal/mol)		Solution energies (kcal/mol)	
Processes	E_d	E_h	$\Delta H_{s,p}$	$\Delta H_{T,s}$
Matrix diffusion controlled	3.70	—	6.60	—
Interfacial partition controlled	—	3.85	—	8.82

Source: Compiled from the data by Chien; ACS Symposium Series, 33:53 (1976).

C. In Vitro-In Vivo Correlations

In establishing a good in vitro-in vivo correlation, one has to consider not only the pharmaceutics aspects of controlled-release drug delivery systems, but also the biopharmaceutics and pharmacokinetics of the therapeutic agent in the body after its delivery from a drug delivery system, as well as the pharmacodynamics of the therapeutic agent at the site of drug actions (scheme 1). As illustrated in scheme 2, the pathways which the drug molecules take after their release from a drug delivery system in situ consist of a number of intermediate steps in series, and each of these steps may determine to varying degrees the time course and bioavailability of the drug to the target tissue and, hence, the onset, intensity, and duration of its pharmacological activities.

In scheme 2, the drug molecules, which are released from a drug delivery system in situ at a controlled rate, are visualized to dissolve, first, in the tissue fluid at the site of administration before they are absorbed, at a rate constant of k_a and trans-

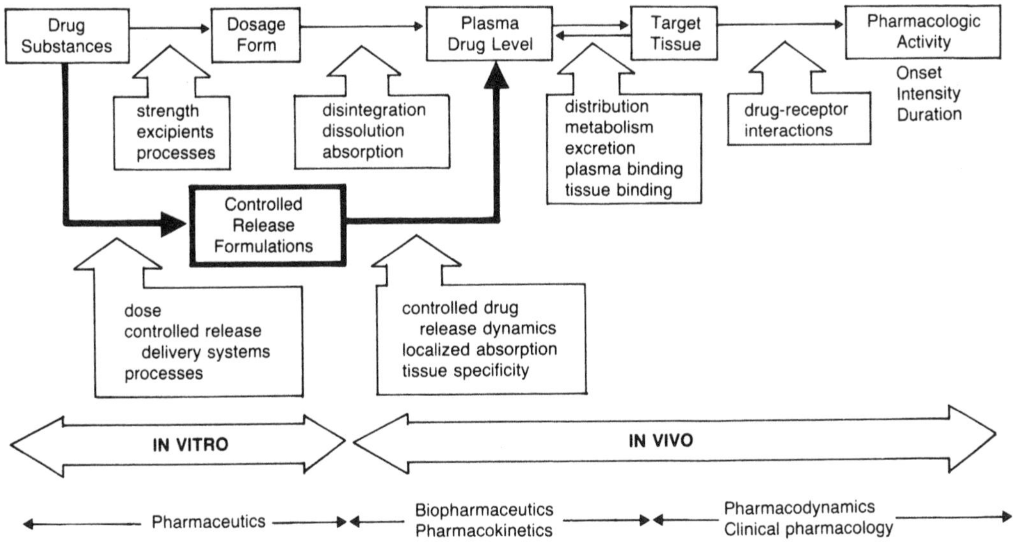

Scheme 1

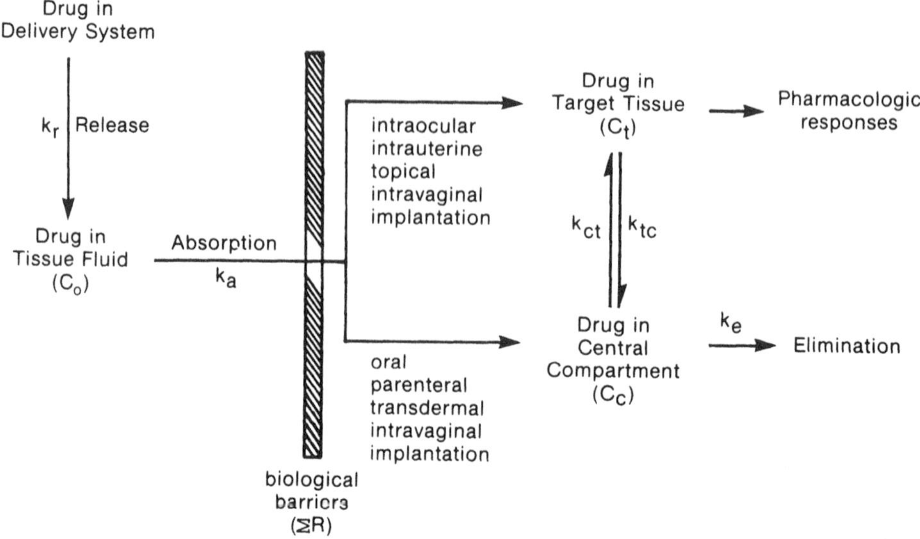

Scheme 2

ported through a series of biological barriers to reach the microcirculation network; through the capillary blood vessel the drug molecules are then transported to the central compartment, a biological sink, by active hemoperfusion. The drug molecules in the systemic circulation are rapidly distributed to targt tissue(s) through reversible diffusion processes at rate constants k_{ct} and k_{tc}. In the target tissue(s) the drug molecules interact with the receptor sites to produce pharmacological responses. This sequence of absorption, distribution, and drug action steps is the primary pathway for therapeutic agents delivered by drug delivery systems administered by a nasal, buccal, oral, parenteral, or transdermal route of administration.

For drug delivery systems administered by an ocular, uterine, or topical route, the drug molecules absorbed are transported directly to the target tissue(s) in the vicinity of the site of administration to execute their pharmacological activities before being distributed to the central compartment.

In vaginal and rectal controlled drug administration, both pathways are possible depending upon the mechanisms of drug action and the location of the target tissues relative to the site of drug delivery. Similarly, for controlled drug administrations by implantation, either pathway is possible depending upon the proximity of the target tissue to the site of implantation.

In any of these situations the rate of drug permeation $(dQ/dt)_a$ across a unit surface area of a biological barrier is directly proportional to the drug concentration in the tissue fluid C_o at the site of administration and inversely proportional to the total diffusional resistance ΣR that the drug molecules must overcome during the course of permeation through the biological barrier:

$$\left(\frac{dQ}{dt}\right)_a = \frac{C_o}{\Sigma R} \tag{76}$$

Based on the complex multibarrier flow coordinate model of Scheuplein (71),

ΣR can be defined as the sum of the diffusional resistances across multiple layers of biological barriers ΣR_{bi} and the aqueous diffusion layers on the absorption side R_a and on the circulation (or target tissue) side R_c of the barriers; that is,

$$\Sigma R = R_a + \sum_{i=1}^{n} R_{bi} + R_c \tag{77}$$

where ΣR_{bi} is the gross diffusional resistance across the ith layers of biological barriers with $i = 1, 2, 3, \ldots, n$.

Alternatively, Equation (77) can be expressed as

$$\Sigma R = \frac{\delta_a}{D_a K_{bi/a}} + \sum_{i=1}^{n} \frac{\delta_{bi}}{D_{bi} K_{i/i-1}} + \frac{\delta_c}{D_c K_{c/bi}} \tag{78}$$

where δ, D, and K represent the thickness, diffusivity, and partition coefficient across two phases in intimate contact, respectively; and the subscripts bi, a, and c are the ith layer of biological barriers and the physiological diffusion layers on the absorption and the circulatory hemoperfusion sides of the barriers, respectively. The δ_a and δ_c terms are hydrodynamic quantities that depend inter alia on the rate of shear near the surface of the biological barrier and the kinematic viscosity of the tissue fluids interfacing with the biological barrier (69). Equation (78) is a simplified form of Scheuplein's model after considering that only the biological barriers per se and the physiological diffusion layers contribute significantly to the total diffusional resistance ΣR.

The total diffusional resistance across cellular membranes under normal physiological conditions was found to range from 1.43 to 33.3×10^2 sec/cm (89). In fact, the magnitude of ΣR is also dependent upon the lipophilicity or hydrophilicity of the drug molecules in permeation and can be varied from 5×10^7 to 2×10^2 sec/cm as the partition coefficient of the drug species increases from 10^{-5} for extremely hydrophilic (polar) molecules to 10^3 for very lipophilic (nonpolar) molecules (71). It was reported that the physiological diffusion layers δ_a and δ_c could become the principle diffusional resistance to the permeation of nonpolar molecules (71), which was illustrated in the vaginal absorption of progesterone and estrone (Chapter 9). On the other hand, for drugs like testosterone or hydrocortisone, vaginal uptake was found to be determined predominantly by molecular transport across the vaginal membrane.

The complex multibarrier flow coordinate model of Scheuplein [Equation (78)] was first applied to mechanistic analyses of the vaginal absorption of steroidal drugs from a vaginal device in rabbits to derive a simple mathematical expression for the in vitro-in vivo correlation (70). Development of a simple in vitro-in vivo relationship is critically important in view of the complex biological processes involved in the absorption, distribution, and pharmacological action of a therapeutic agent following its release from a drug delivery system (scheme 1). These efforts were also extended later to the subcutaneous controlled administration of drugs from subdermal implants in rats (73). The development of such an in vitro-in vivo relationship can be illustrated as follows.

The instantaneous rate of drug release from a matrix-type polymeric device at time t is defined by Equation (79):

$$\frac{dQ}{dt} = \frac{1}{2}\left[\frac{(2A - C_p)D_pC_p}{t}\right]^{1/2} \quad (79)$$

At steady state, the rate of drug permeation [Equation (76)] should be in equilibrium with the rate of drug release at the site of administration [Equation (79)]; this yields

$$2C_0t^{1/2} = \Sigma R[(2A - C_p)D_pC_p]^{1/2} \quad (80)$$

Equation (80) suggests that the drug concentration C_o in the tissue fluid surrounding the drug delivery system should decrease with the square root of time $t^{1/2}$ as the drug molecules are absorbed and actively transported to the biological sink by hemoperfusion; that is, because the release of drug molecules from the drug delivery system is a rate-limiting step in scheme 2, the magnitude of $2C_ot^{1/2}$ value is a constant (since both the ΣR and $[(2A - C_p)D_pC_p]^{1/2}$ terms are constant values under controlled conditions). Equation (80) also suggests that the magnitude of C_o at a given time point may be varied with the magnitude of ΣR when the $[(2A - C_p)D_pC_p]^{1/2}$ term is a constant. In other words, the $2C_ot^{1/2}$ value varies when the same drug delivery system is applied to a different animal or a different tissue of the same animal (a different value of ΣR is expected to exist for different animals or different tissues). Experimentally, the values of ΣR, which can neither be easily nor directly measured, can be estimated alternatively from in vivo drug release profiles by the relationship

$$\Sigma R_{\text{in vivo}} = \frac{2C_ot^{1/2}}{[(2A - C_p)D_pC_p]^{1/2}_{\text{in vivo}}} \quad (81a)$$

or

$$\Sigma R_{\text{in vivo}} = \frac{2C_ot^{1/2}}{(Q/t^{1/2})_{\text{in vivo}}} \quad (81b)$$

provided that the drug concentration C_o in the tissue fluid is known or measurable.

In in vitro drug elution studies, on the other hand, attempts are made to maintain the elution solution at perfect sink conditions throughout the studies (Section IVA1). In this situation there exists only a thickness h_d of hydrodynamic diffusion layer between the surface of the drug delivery system and the bulk of the elution solution maintained under sink conditions. Equation (78) is thus reduced to Equation (82) to define the in vitro diffusional resistance $R_{\text{in vitro}}$ across the hydrodynamic diffusion layer:

$$\Sigma R_{\text{in vitro}} = \frac{h_d}{D_s} \quad (82)$$

if $\delta_a = h_d$, $D_a = D_s$, and $K_{bi/a} = 1$. Furthermore, the $\Sigma R_{\text{in vitro}}$ value can be calculated from in vitro drug release data by the relationship

$$\Sigma R_{\text{in vitro}} = \frac{2C_ot^{1/2}}{(Q/t^{1/2})_{\text{in vitro}}} \quad (83)$$

A simple in vitro-in vivo relationship can be established by conducting simultaneously in vitro and in vivo evaluations of a potential drug delivery system. When the mechanism of in vivo drug release is proven to be in good agreement with that observed in in vitro drug elution studies, an in vitro-in vivo correlation factor γ can be determined from the relationship

$$\gamma = \frac{(C_o t^{1/2})\Sigma R_{\text{in vitro}}}{(C_o t^{1/2})\Sigma R_{\text{in vivo}}} \tag{84a}$$

or

$$\gamma = \frac{(Q/t^{1/2})_{\text{in vivo}}}{(Q/t^{1/2})_{\text{in vitro}}} \tag{84b}$$

Now it becomes possible to estimate the magnitude of the in vivo drug release profile $(Q/t^{1/2})_{\text{in vivo}}$ from an in vitro drug elution study by the equation

$$(Q/t^{1/2})_{\text{in vivo}} = \gamma(Q/t^{1/2})_{\text{in vitro}} \tag{85}$$

Using Equation (84b) the γ values can be calculated from comparative studies of in vitro and in vivo drug release profiles. Some results calculated from the literature data are illustrated in Table 32. The calculated γ values can be used to standardize the in vitro drug elution system. As soon as the drug elution system is standardized, the γ values can be utilized to refine the controlled release characteristics of the drug delivery system under development and also to predict long-term in vivo drug release profiles from a short-term in vitro drug elution study [Equation (85)]. This can be achieved when in vitro drug elution studies are conducted under well-controlled con-

Table 32 In Vitro-In Vivo Correlation Factor γ for Various Matrix-Type Drug Delivery Systems

Drug delivery systems	Drugs	Studies	$Q/t^{1/2}$ ($mg/cm^2/day^{1/2}$)	γ
Silicone devices	Desoxycorticosterone	In vitro	1.072	
	acetate	Subcutaneous[b]	1.025	0.96
	Ethynodiol diacetate	In vitro	3.16	0.66
		Intravaginal[c]	2.100	
Hydron implants[a]	Norgestomet			
1.2%		In vitro	0.605	
		Subcutaneous[d]	0.640	1.08
4.8%		In vitro	0.396	
		Subcutaneous[d]	0.504	1.27
19.2%		In vitro	0.058	
		Subcutaneous[d]	0.129	2.22
Polyethylene devices	Progesterone	In vitro	0.313	
		Intrauterine or intraperitoneal	0.250[e]	0.80

[a]Cross-linked with different amounts of cross-linking agent [Chien and Lau; J. Pharm. Sci., 65:488 (1976)].
[b]A 104 day subcutaneous implantation in rats [Chien et al.; ACS Symposium Series, 33:72 (1976)].
[c]A 52 day intravaginal insertion in rabbits [Chien et al.; J. Pharm. Sci., 64:1776 (1975)].
[d]A 16 day subcutaneous implantation in cows [Chein and Lau; J. Pharm. Sci., 65:488 (1976)].
[e]Calculated from the data by Kalkwarf et al.; Contraception, 6:423 (1972).

Table 33 In Vitro-In Vivo Correlation Factor γ for Various Drug Delivery Systems with Constant Drug Release Profiles

Drug delivery systems	Drugs	Studies	Q/t ($\mu g/cm^2/day$)	γ
Silicone capsules	Norgestomet	In vitro	52.4	0.87
		Subcutaneous[a]	45.7	
MDD devices	Norgestomet	In vitro	100.0	0.83
		Subcutaneous[b]	82.6	
	Desoxycorticosterone acetate	In vitro	49.3	1.09
		Subcutaneous[c]	53.7	
	17^{β}-Estradiol	In vitro	72.0	
		Subcutaneous[d]	22.0	0.31
		Subcutaneous[d]	8.7	0.12
	Testosterone	In vitro	40.25	0.54
		Transdermal[e]	21.82	

[a]A 32 day subcutaneous implantation in heifers [Chien; Chem. Pharm. Bull., 24:1471 (1976)].
[b]A 9 day subcutaneous implantation of four formulations in heifers (Chapter 8).
[c]A 49 day subcutaneous implantation in rats [Chien et al.; J. Pharm. Sci., 67:214 (1978)].
[d]A 56 day subcutaneous implantation in heifers exposed to either 10°C or subzero temperature.
[e]A 32 day topical application in monkeys (determined from urinary radioactivity excretion data).

ditions, under which a constant $\Sigma R_{\text{in vitro}}$ value [Equation (83)] is maintained and reproducible $(Q/t^{1/2})_{\text{in vitro}}$ data are obtained. This was demonstrated in the studies by Chien et al. (70,73), who used Equation (85) as the working equation to predict, from 7 day in vitro drug release data, the 104 day subcutaneous controlled administration of desoxycorticosterone acetate in rats (Chapter 8) and the 56 day vaginal release of ethynodiol diacetate in rabbits (Chapter 9).

For membrane permeation-controlled reservoir-type and membrane-matrix hybrid drug delivery systems a similar relationship can also be derived to correlate in vivo and in vitro drug release data:

$$(Q/t)_{\text{in vivo}} = \gamma(Q/t)_{\text{in vitro}} \tag{86}$$

The γ values calculated from the literature data of $(Q/t)_{\text{in vitro}}$ and $(Q/t)_{\text{in vivo}}$ are summarized in Table 33. The data suggest that γ values are dependent upon the sites of administration and the environmental conditions to which the animals are exposed during treatment.

In any case, Equations (85) and (86) provide a simple but practical relationship between in vivo and in vitro drug release rate profiles. They can be used as the working equations for optimization of the controlled drug release profile of a drug delivery system.

REFERENCES

1. J. Grant and G. S. Park; *Diffusion in Polymers*, Academic Press, New York, 1968, Chapters 2 and 3.
2. Y. W. Chien; in: *Sustained and Controlled Release Drug Delivery Systems* (J. R. Robinson, Ed.), Dekker, New York, 1978, Chapter 4.
3. W. Nernst; Z. Phys. Chem., 47:52 (1904).

4. W. G. Perkins and D. R. Begeal; J. Chem. Phys., 54:1683 (1971).
5. Y. W. Chien; Chem. Pharm. Bull., 24:1471 (1976).
6. Y. W. Chien, C. Olsen and T. Sokoloski; J. Pharm. Sci., 62:435 (1973).
7. Y. W. Chien, H. Lambert and T. Lin; J. Pharm. Sci., 64:1643 (1975).
8. T. Higuchi; J. Pharm. Sci., 52:1145 (1963).
9. J. B. Schwartz, A. P. Simonelli and W. I. Higuchi; J. Pharm. Sci., 57:278 (1968).
10. T. J. Roseman and S. H. Yalkowsky; in: *Controlled Release Polymeric Formulations* (D. R. Paul and F. W. Harris, Eds.), ACS Symposium Series 33, American Chemical Society, Washington, D.C., 1976, Chapter 4.
11. Y. W. Chien and H. J. Lambert; U.S. Patent 3,946,106, March 23, 1976.
12. Y. W. Chien, L. F. Rozek and H. J. Lambert; J. Pharm. Sci., 67:214 (1978).
13. Y. W. Chien and H. J. Lambert; in: *The International Symposium on Bioactivation and Controlled Drug Release*, Stockholm, Sweden, April 21–23, 1976.
14. Y. W. Chien, H. Lambert and D. Grant; J. Pharm. Sci., 63:365 (1974).
15. Y. W. Chien, D. M. Jefferson, J. G. Cooney and H. J. Lambert; J. Pharm. Sci., 68:689 (1979).
16. Y. W. Chien; in: *Controlled Release Polymeric Formulations* (Paul and Harris, Eds.), ACS Symposium Series 33, American Chemical Society, Washington, D.C., 1976, Chapter 5.
17. A. N. Martin, J. Swarbrick and A. Cammarata; *Physical Pharmacy*, 2nd Ed., Lea and Febiger, Philadelphia, 1969, Chapter 12.
18. F. A. Kincl, G. Benagiana and I. Angee; Steroids, 11:673 (1968).
19. K. Sundaram and F. A. Kincl; Steroids, 12:517 (1968).
20. R. Schuhman and H. D. Taubert; Acta Biol. Med. Germ., 24:897 (1970).
21. T. J. Roseman and W. I. Higuchi; J. Pharm. Sci., 59:353 (1970).
22. Y. W. Chien and H. J. Lambert; J. Pharm. Sci., 63:515 (1974).
23. Y. W. Chien and H. J. Lambert; Chem. Pharm. Bull., 23:1085 (1975).
24. M. A. Q. Chaudry and K. C. James; J. Med. Chem., 17:157 (1974).
25. P. H. Elworthy, A. T. Florence and C. B. MacFarlane; *Solubilization by Surface-Active Agents*, Chapman and Hall, London, 1968.
26. T. Higuchi and J. L. Lach; J. Am. Pharm. Assoc., Sci. Ed., 43:349, 525, 527 (1954).
27. S. H. Yalkowsky, G. L. Flynn and G. L. Amidon; J. Pharm. Sci., 61:983 (1972).
28. Y. W. Chien and D. M. Jefferson; U.S. Patent 4,032,645, June 28, 1977.
29. Y. W. Chien; J. Parent. Sci. Technol., 38:32 (1984).
30. J. T. Davies and E. K. Rideal; *Interfacial Phenomena*, Academic Press, New York, 1963, pp. 154–156.
31. R. Collander; Acta Chem. Scand., 4:1085 (1950).
32. J. C. McGowan; J. Appl. Chem., 4:41 (1954).
33. J. Iwasa, T. Fujita and C. Hansch; J. Med. Chem., 8:150 (1965).
34. G. L. Flynn; J. Pharm. Sci., 60:345 (1971).
35. C. Hansch, R. M. Muir, T. Fujita, P. P. Maloney, C. F. Geiger and M. J. Streich; J. Am. Chem. Soc., 85:2817 (1963).
36. C. Hansch and T. Fujita; J. Am. Chem. Soc., 86:1616 (1964).
37. C. Hansch and A. R. Steward; J. Med. Chem., 7:691 (1964).
38. G. L. Flynn and R. W. Smith; J. Pharm. Sci., 61:61 (1972).
39. Y. W. Chien, H. J. Lambert and T. K. Lin; J. Pharm. Sci., 64:1643 (1975).
40. M. Katz and B. J. Poulsen; J. Soc. Cosmet. Chemists, 23:565 (1972).
41. W. W. Brandt; J. Phys. Chem., 63:1080 (1959).
42. D. W. McCall, E. A. Anderson and C. M. Huggins; J. Chem. Phys., 34:804 (1961).
43. W. L. Robb; Ann. N.Y. Acad. Sci., 146:119 (1968).

44. F. Daniels and R. A. Alberty; *Physical Chemistry*, 3rd Ed., J. Wiley & Sons, New York, 1967, Chapter 11.
45. J. Drobnik, P. Spacek and O. Wichterle; J. Biomed., Mater. Res., 8:45–51 (1974).
46. M. H. Cohen and D. Turnbull; J. Chem. Phys., 31:1164 (1959).
47. O. Wichterle and D. Lim; Nature, 185:117 (1960).
48. O. Wichterle and D. Lim; U.S. Patent 2,976,576, March 28, 1961.
49. Y. W. Chien and E. P. K. Lau; J. Pharm. Sci., 65:488 (1976).
50. G. M. Zentner, J. R. Cardinal and S. W. Kim; J. Pharm. Sci., 67:1352 (1878).
51. L. Olanoff, T. Koinis and J. M. Anderson; J. Pharm. Sci., 68:1147 (1979).
52. H. Yasuda, C. E. Lamaze and L. D. Ikenberry; Die MaKromol. Chem., 118: (1968).
53. H. Yasuda, C. E. Lamaze and A. Peterlin; J. Polym. Sci. (A-2), 9:1117 (1971).
54. A. S. Michaels and H. J. Bixler; J. Polym. Sci., 50:393 (1961).
55. C. F. Mosot, Jr.; J. Appl. Polym. Sci., 14:1019 (1970).
56. T. J. Roseman; J. Pharm. Sci., 61:46 (1972).
57. G. L. Flynn and T. J. Roseman; J. Pharm. Sci., 60:1788 (1971).
58. W. I. Higuchi and T. Higuchi; J. Am. Pharm. Ass., Sci. Ed., 49:598 (1960).
59. K. F. Finger, A. P. Lemberger, T. Higuchi, W. Busse and D. E. Wurster; J. Am. Pharm. Ass., Sci. Ed., 49:569 (1960).
60. R. D. Siegel and R. W. Coughlin; J. Appl. Polym. Sci., 14:3145 (1970).
61. R. W. Baker and H. K. Lonsdale; in: *Controlled Release of Biologically Active Agents* (A. C. Tanquary and R. E. Lacey, Eds.), Plenum Press, New York, pp. 15–71.
62. F. Bueche; *Physical Properties of Polymers*, Interscience, New York, 1962, Chapters 3 & 4.
63. C. A. Kumins and T. K. Kwei; in: *Diffusion in Polymers* (J. Crank and G. S. Park, Eds.), Academic Press, New York, 1968, Chapter 4.
64. V. Stannett; in: *Diffusion in Polymers* (J. Crank and G. S. Park, Eds.), Academic Press, New York, Chapter 2.
65. G. L. Flynn, S. G. Yalkowsky and T. J. Roseman; J. Pharm. Sci., 63:479 (1974).
66. J. Halebliam, R. Runkel, N. Mueller, J. Christopherson and K. Ng; J. Pharm. Sci., 60:541 (1971).
67. A. D. Schwope, D. L. Wise and J. F. Howes; Life Sci., 17:1877 (1975).
68. V. G. Levich; *Physicochemical Hydrodynamics*, Prentice-Hall, Englewood Cliffs, N.J., 1962.
69. H. Schlichting; *Boundary Layer Theory* (translated by J. Kestin), McGraw-Hill, New York, 1960.
70. Y. W. Chien, S. E. Mares, J. Berg, S. Huber, H. J. Lambert and K. F. King; J. Pharm. Sci., 64:1776 (1975).
71. R. J. Scheuplein; J. Theor. Biol., 18:72 (1968).
72. G. L. Flynn and S. H. Yalkowsky; J. Pharm. Sci., 61:838 (1972).
73. Y. W. Chien, H. J. Lambert and L. F. Rozek; in: Controlled Release Polymeric Formulations (D. Paul and F. Harris, Eds.), ACS Symposium Series 33, American Chemical Society, Washington, D.C., 1976, pp. 72–86.
74. H. A. Nash, D. N. Robertson, A. J. M. Young and L. E. Atkinson; Contraception, 18:367 (1978).
75. D. R. Kalkwarf, M. R. Sikov, L. Smith and R. Gordon; Contraception, 6:423 (1972).
76. G. W. Duncan and D. R. Kalkwarf; in: *Human Reproduction: Conception and Contraception* (Hafez and Evans, Eds.), Harper and Row, Hagerstown, Md., 1973, Chapter 22.
77. C. L. Olson, T. D. Sokolowski, S. N. Pagay and D. Michaels; Anal. Chem., 41:865 (1969).
78. G. L. Flynn and E. W. Smith; J. Pharm. Sci., 60:1713 (1971).
79. Y. W. Chien and H. J. Lambert; U.S. Patent 3,992,518, Nov. 16, 1976.

80. Y. W. Chien; J. Pharm. Sci., 73:1064 (1984).
81. P. R. Keshary and Y. W. Chien; Drug Dev. Ind. Pharm., 10:883–913 (1984).
82. P. R. Keshary, Y. C. Huang and Y. W. Chien; Drug Dev. Ind. Pharm., 11:1213–1253 (1985).
83. Y. W. Chien, K. H. Valia and E. B. Doshi; Drug Dev. Ind. Pharm., 11:1195–1212 (1985).
84. M. Kabadi and Y. W. Chien; Drug Dev. Ind. Pharm., 11:1271–1312 (1985).
85. M. Kabadi and Y. W. Chien; Drug Dev. Ind. Pharm., 11:1313–1361 (1985).
86. E. L. Tan, J. C. Liu and Y. W. Chien; Int. J. Pharm., 42:161–169 (1988).
87. K. Tojo, M. Ghannam, Y. Sun and Y. W. Chien; J. Control. Release, 1:197–203 (1985).
88. K. Tojo, J. A. Masi and Y. W. Chien; I&EC Fundamentals, 24:368–373 (1985).
89. D. A. T. Dick; J. Theor. Biol., 7:504 (1964).

3
Oral Drug Delivery and Delivery Systems

I. INTRODUCTION

Oral drug delivery has been known for decades as the most widely utilized route of administration among all the routes that have been explored for the systemic delivery of drugs via various pharmaceutical products of different dosage forms. The reasons that the oral route achieved such popularity may be in part attributed to its ease of administration as well as the traditional belief that by oral administration the drug is as well absorbed as the foodstuffs that are ingested daily. In fact, the development of a pharmaceutical product for oral delivery, irrespective of its physical form (solid, semisolid, or liquid dosage form), involves varying extents of optimization of dosage form characteristics within the inherent constraints of gastrointestinal (GI) physiology.

Pharmaceutical products designed for oral delivery and currently available on the prescription and over-the-counter markets are mostly the *immediate-release* type, which are designed for immediate release of drug for rapid absorption. Because of their clinical advantages over immediate-release pharmaceutical products containing the same drugs, *sustained-release* pharmaceutical products, such as those formulated on the basis of spansule coating technology, have over the past decade gradually gained medical acceptance and popularity since their introduction into the marketplace. Recently, a new generation of pharmaceutical products, called *controlled-release* drug delivery systems, such as those developed from the osmotic pressure-activated drug delivery system, have recently received regulatory approval for marketing, and their pharmaceutical superiority and clinical benefits over the sustained-release and immediate-release pharmaceutical products have been increasingly recognized (1).

All the pharmaceutical products formulated for systemic delivery via the oral route of administration, irrespective of the mode of delivery (immediate, sustained, or controlled release) and the design of dosage forms (either solid, dispersion, or liquid), must be developed within the intrinsic characteristics of GI physiology. Therefore, a fundamental understanding of various disciplines, including GI physiology, pharmacokinetics, pharmacodynamics, and formulation design, is essential to achieve a systematic approach to the successful development of an oral pharmaceutical dosage form (or drug delivery system). The more sophisticated a delivery system, the greater is the complexity of these various disciplines involved in the design and optimization of the system. In any case, the scientific framework required for the successful development of an oral drug delivery system consists of a basic understanding of the following three aspects: (i) physicochemical, pharmacokinetic, and pharmacodynamic characteristics of the drug, (ii) the anatomic and physiologic characteristics of the gastrointestinal tract (Table 1), and (iii) physicomechanical characteristics and the drug delivery mode of the dosage form to be designed.

Although it is often impractical to alter the physicochemical, pharmacokinetic, and/or pharmacodynamic characteristics of a drug to be delivered by a chemical approach, such as the synthesis of an analog, or medically undesirable to modify the anatomic and physiologic characteristics of the gastrointestinal tract, the design of a controlled-release oral dosage form by optimization of dosage form characteristics with GI anatomy and physiology taken into consideration could provide some opportunity to rationalize the systemic delivery of drugs and maximize their therapeutic benefits.

The term "controlled-release oral dosage form" is not new to most people working in various fields of pharmaceutical research and development. In fact, approximately 30 years ago, the U.S. Food and Drug Administration (FDA) published regulatory requirements for controlled-release products. Unfortunately, there has been a proliferation of controlled-release dosage forms on the marketplace that may have little rationale and provide no advantages over the same drugs in conventional dosage forms. Over the last decades there has also been an increase in the (possibly unwarranted) use of controlled-release labeling claims (2).

As defined in Chapter 1, controlled-release drug administration means not only prolongation of the duration of drug delivery, similar to the objective in sustained release and prolonged release, but the term also implies the predictability and reproducibility of drug release kinetics (2). Oral controlled-release drug delivery is thus a drug delivery system that provides the continuous oral delivery of drugs at

Table 1 Gastrointestinal Tract: Physical Dimensions and Dynamics

Region	Surface area (m^2)	Length (m)	Transit time	
			Fluid[a]	Digestible Solid
Gastrointestinal tract	200	—	—	—
Stomach[b]	0.1–0.2	—	50 min	8 hr[d]
Small intestine	100 4500[c]	3.0	2–6 hr	4–9 hr[e]
Large intestine	0.5–1.0	1.5	2–6 hr	3 hr to 3 days

[a]Isotonic saline solution, 500 ml, was ingested.
[b]Residual volume = 50 ml.
[c]Taking intestinal microvilli area into account.
[d]Solid food, 50 g, was ingested.
[e]Food first appeared at the cecum after 4 hr and all indigestible material entered the large intestine within 9 hr.

predictable and reproducible kinetics for a predetermined period throughout the course of GI transit. Also included are systems that target the delivery of a drug to a specific region within the GI tract for either a local or a systemic action.

For the oral controlled administration of drugs, several research and development activities have shown encouraging signs of progress in the development of programmable controlled-release dosage forms as well as in the search for new approaches to overcome the potential problems associated with oral drug administration. These potential developments and new approaches are discussed in this chapter along with an overview of GI physiology. Since an understanding of the basic concepts involved in the development of various controlled-release drug delivery systems is vital for the future development of a new generation or a more sophisticated novel drug delivery system, particular emphasis is placed on the rationale of system design and the mechanism of drug delivery for such controlled-release oral drug delivery systems developed in relationship to the GI environment and dynamics.

In the exploration of oral controlled-release drug administration, one encounters three areas of potential challenge (2):

1. Development of a drug delivery system: to develop a viable oral controlled-release drug delivery system capable of delivering a drug at a therapeutically effective rate to a desirable site for a duration required for optimal treatment.
2. Modulation of gastrointestinal transit time: to modulate the GI transit time so that the drug delivery system developed can be transported to a target site or to the vicinity of an absorption site and reside there for a prolonged period of time to maximize the delivery of a drug dose.
3. Minimization of hepatic first-pass elimination: if the drug to be delivered is subjected to extensive hepatic first-pass elimination, preventive measures should be devised to either bypass or minimize the extent of hepatic metabolic effect.

It is the intention of this chapter to review various potential developments and new approaches that have been recently explored to meet these challenges.

II. DEVELOPMENT OF NOVEL DRUG DELIVERY SYSTEMS FOR ORAL CONTROLLED-RELEASE DRUG ADMINISTRATION

A review of the literature has revealed the recent development of several novel drug delivery system that can be utilized for the controlled delivery of drugs in the alimentary canal. These potential developments are outlined and discussed in the following sections.

A. Osmotic Pressure-Controlled Gastrointestinal Delivery Systems

These are systems fabricated by encapsulating an osmotic drug core containing an osmotically active drug (or a combination of an osmotically inactive drug with an osmotically active salt, e.g., NaCl) within a semipermeable membrane made from biocompatible polymer, e.g., cellulose acetate (Figure 1). A delivery orifice with a controlled diameter is drilled, using a laser beam, through the coating membrane for controlling the release of drug solutes (3–5).

This polymer membrane is not only semipermeable in nature but is also rigid

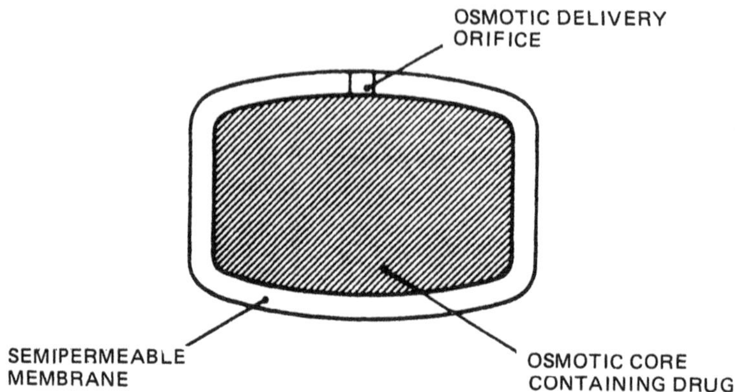

Figure 1 Osmotic pressure-controlled drug delivery system. (Adapted from Reference 3.)

and capable of maintaining the structural integrity of the gastrointestinal delivery system during the course of drug release. Because of its semipermeable characteristics, it is permeable to the influx of water in the gastrointestinal tract; on the other hand, it is impermeable to drug solutes. When in use, water is continuously absorbed into the drug reservoir compartment through the semipermeable membrane to dissolve the osmotically active drug and/or salt. A gradient of osmotic pressure is thus created, under which the drug solutes are continuously pumped out over a prolonged period of time through the delivery orifice at a rate defined by following relationship (3,4)

$$\left(\frac{Q}{t}\right)_z = \frac{P_w A_m}{h_m}(\pi_s - \pi_e)S_D \qquad (1)$$

where P_w, A_m, and h_m are the water permeability, the effective surface area, and the thickness of the semipermeable membrane, respectively; π_s is the osmotic pressure of the saturated solution of osmotically active drug or salt in the system; π_e is the osmotic pressure of the gastrointestinal fluid; and S_D is the solubility of the drug (4).

In principle this type of drug delivery system dispenses drug solutes continuously at a zero-order rate until the concentration of the osmotically active ingredient in the system drops to a level below the saturation solubility. A non-zero–order release pattern then results at a rate described by

$$\frac{dQ}{dt} = \frac{(Q/t)_z}{\{[1 + (Q/t)_z/S_D V_t](t_r - t_z)\}^2} \qquad (2)$$

Where $(Q/t)_z$ is the rate of zero-order drug release; V_t is the total volume of the drug reservoir compartment; t_z is the total length of time in which the system delivers the drug at a zero-order rate; and t_r is the duration of residence time.

Equation (1) indicates that the rate for the zero-order release of drug from osmotic pressure-controlled drug delivery systems can be programmed by varying the thickness of the coating membrane h_m, thereby controlling the duration of drug delivery (Figure 2). This equation also suggests that the rate of drug release is dependent upon the differential osmotic pressure $\pi_s - \pi_e$. The greater the difference in osmotic pressure, the higher is the rate of drug release (Figure 3).

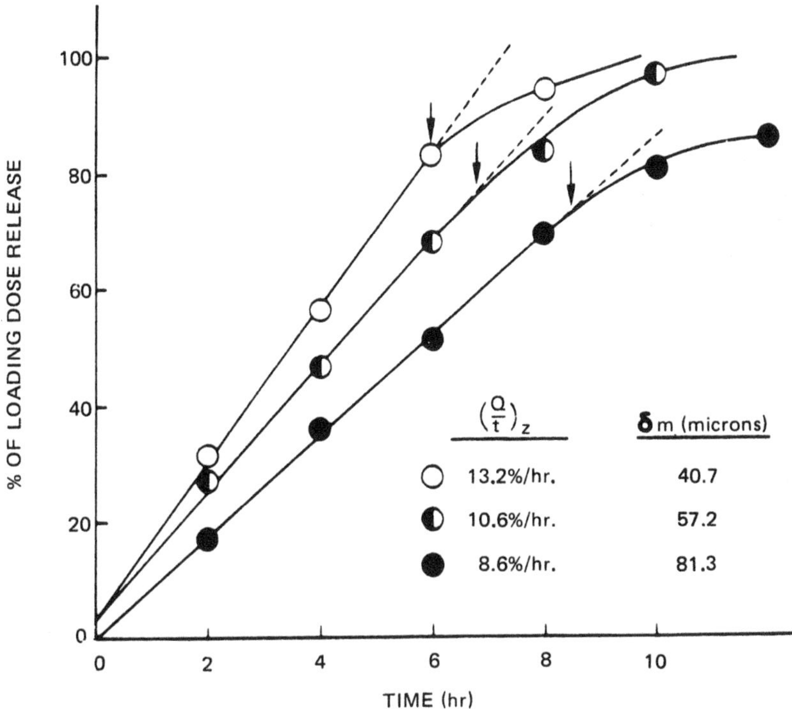

Figure 2 Effect of coating membrane thickness on the rate and duration of zero-order release of indomethacin from osmotic pressure-controlled gastrointestinal delivery system.

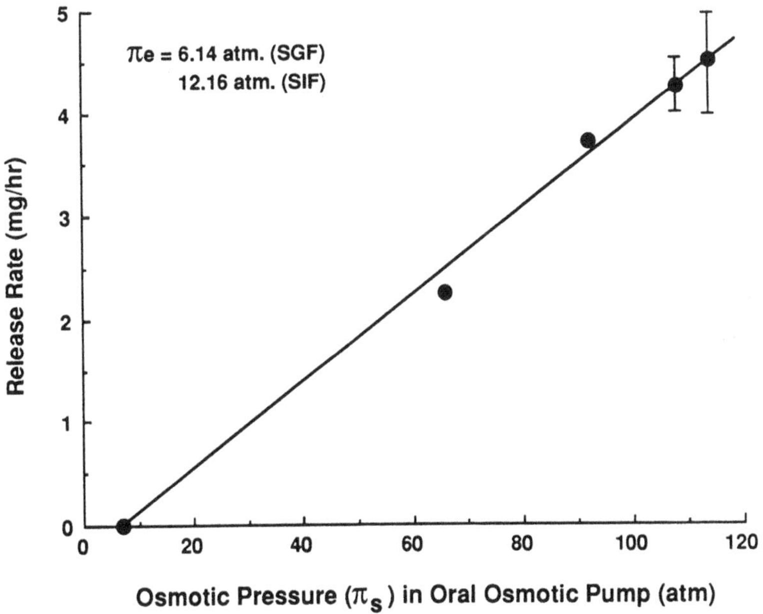

Figure 3 Effect of osmotic pressure π_s in the osmotic pressure-controlled gastrointestinal delivery system on the rate of release of phenylpropanolamine.

This osmotic pressure-controlled gastrointestinal delivery system has been applied to regulate the gastrointestinal delivery of indomethacin (3). The delivery profiles of indomethacin in the gastrointestinal tract from the gastrointestinal delivery system (GIDS) with two delivery rates are evaluated clinically with indomethacin capsules (6). The results compared in Figure 4, indicate that a prolonged delivery of indomethacin is achieved by GIDS with a drug delivery profile bioequivalent to that of a 12 hr indomethacin dose in capsules taken three times a day at 0, 4, and 8 hr. The data in Figure 4 also demonstrate that the GIDS also avoids the surge in the systemic concentration of indomethacin observed with the oral delivery of a 12 hr indomethacin dose taken at once, which minimizes the potential adverse effect of indomethacin.

The observed slower onset of drug delivery from this GIDS, which resulted from the need for activation of the system, can be overcome by incorporating an immediate-release dose to the GIDS. This is accomplished by dividing a therapeutic dose into two fractions. One-third of the therapeutic dose is designated as the "immediate-release" fraction and the remaining two-thirds as the "controlled-release" fraction. The controlled-release fraction is encapsulated inside the semipermeable membrane coating; the immediate-release fraction is used to coat the external surface of the semipermeable membrane (Figure 5) to provide the initial dose upon oral administration (1,7). This approach has been successfully applied to the development of the

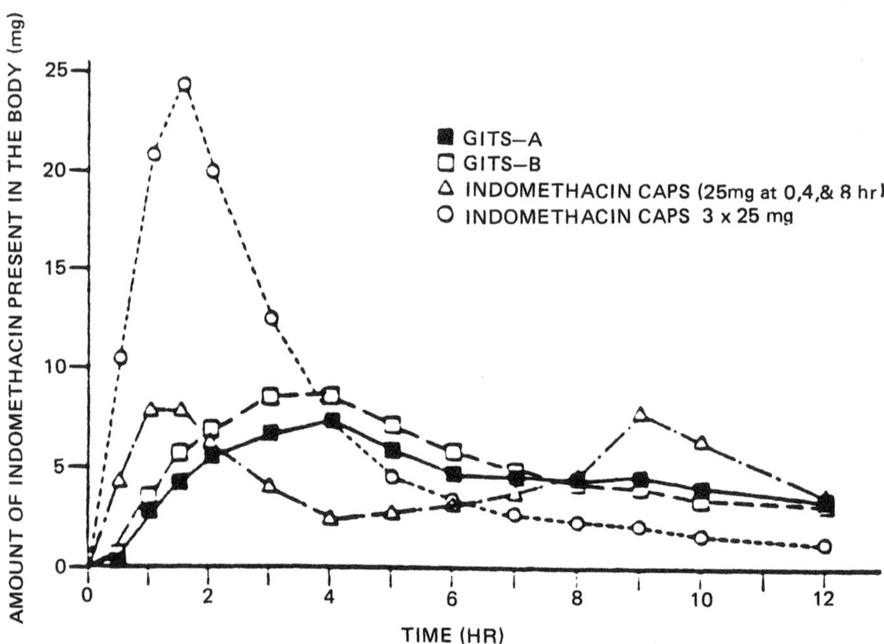

Figure 4 Comparison of the time course for the amount of indomethacin delivered to humans using various dosage forms. Indomethacin capsules: (○) 3 units taken at time 0 and (△) 1 unit each taken at time 0, 4, and 8 hrs. Indomethacin-gastrointestinal therapeutic systems: (■) 1 unit of system A (7 mg/hr) taken at time 0, and (□) 1 unit of system B (9 mg/hr) taken at time 0.

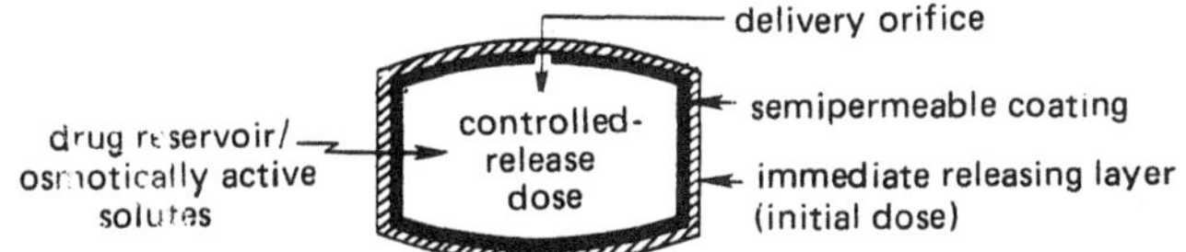

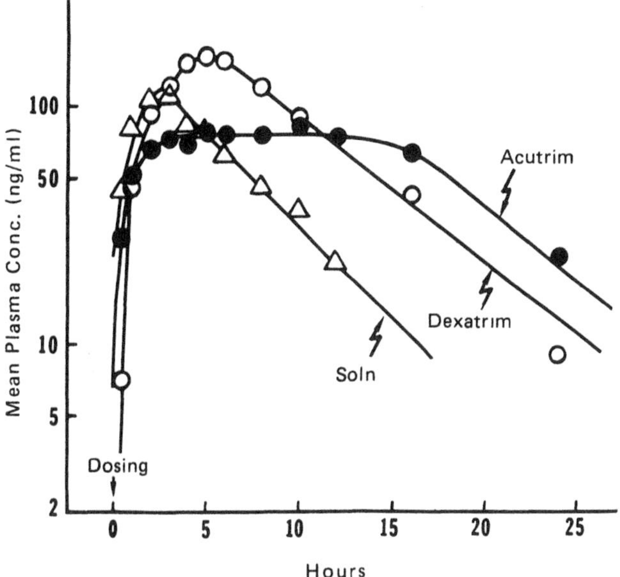

Figure 5 Comparison of the plasma profiles of phenylpropanolamine (PPA) in humans delivered from various dosage forms: (△) PPA in solution formulation, (○) PPA in Dexatrim capsules, and (●) PPA in Acutrim tablets.

Acutrim tablet to achieve a 16 hr oral controlled delivery of phenylpropanolamine (PPA) for the daily suppression of appetite. The plasma profile of PPA resulting from the controlled-release delivery of PPA via once-a-day administration is compared with that of the sustained-release delivery of PPA from the Dexatrim capsule and that of the immediate-release delivery of PPA from the solution formulation in Figure 5. The data clearly demonstrate that a steady-state plasma level of PPA is attained by the controlled-release Acutrim tablet, not by the sustained-release Dexatrim capsule or the immediate-release solution formulation. The difference in clinical performance in terms of the pharmacokinetic profiles of PPA between the Acutrim tablet, an osmotic pressure-based controlled-release drug delivery system, and the Dexatrim capsule, a spansule technology-based sustained-release drug delivery system, can be attributed to the difference in drug release profiles (7). Dissolution studies indicated that, following the rapid release of the immediate-release dose, the PPA dose in the controlled-release core in the Acutrim tablet is delivered continuously in a controlled manner for a prolonged duration; the PPA dose from the Dexatrim tablet is released with only a slight prolongation in the drug release profile (Figure 6).

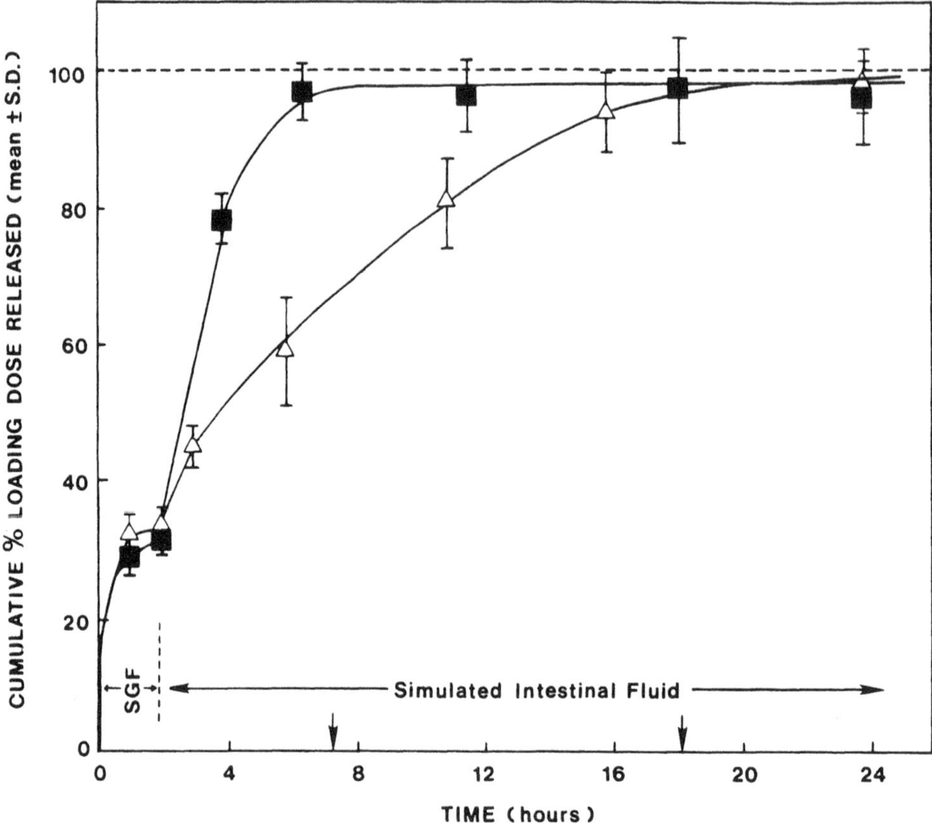

Figure 6 Comparison of the release profiles of phenylpropanolamine (PPA) from (■) Dexatrim capsule and (△) Acutim tablet in dissolution medium (simulated gastric fluid for the first 2 hr and simulated intestinal fluid for the remaining 22 hr).

On the other hand, the external surface of the semipermeable membrane can also be coated with a layer of bioerodible polymer, e.g., enteric coating, to regulate the penetration of gastrointestinal fluid through the semipermeable membrane (8) and target the delivery of a drug to the lower region of the gastrointestinal tract.

Furthermore, the coating membrane of the delivery system can also be constructed from a laminate of two or more semipermeable membranes with differential permeabilities (9) or a laminate of a semipermeable membrane and a microporous membrane (Great Britain Patent 1,556,149) to modulate the rate of water influx and so program the rate of drug delivery.

The osmotic pressure-controlled gastrointestinal delivery system can be further modified to constitute two compartments separated by a movable partition (Figure 7). The osmotically active compartment absorbs water from the gastrointestinal fluid to create an osmotic pressure that acts on the partition and forces it to move upward and to reduce the volume of the drug reservoir compartment and to release the drug formulation through the delivery orifice (2). It should be pointed out that there is a specific range of diameters in which the zero-order release rate is independent of the

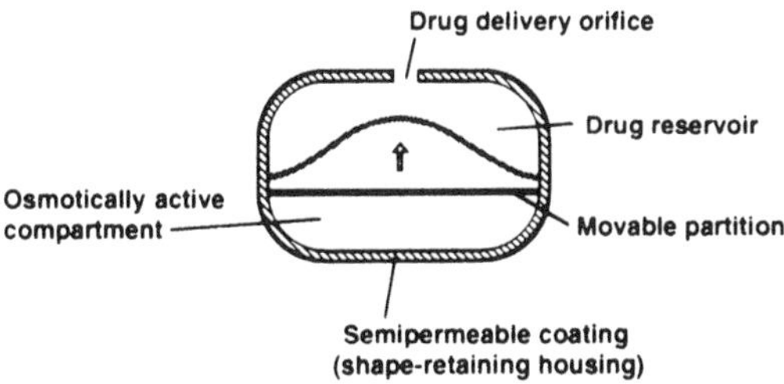

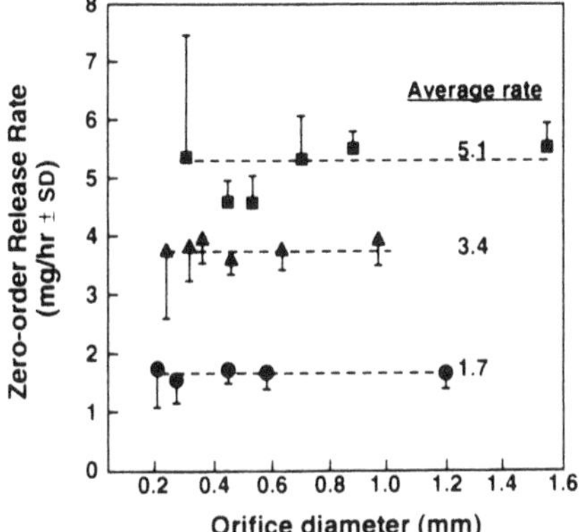

Figure 7 (Top) Second generation of an osmotic pressure-controlled drug delivery system having a drug reservoir compartment and an osmotically active compartment separated by a movable partition (9).
(Bottom) Range of the diameter of a delivery orifice in which a constant rate of zero-order release of nifedipine is maintained. Release rate (mg/hr)/loading dose (mg/system): (●) 1.7/30, (▲) 3.4/60, (■) 5.1/90. I = one half of the total standard deviation, that is, the statistical sum of standard deviations between and within systems.

diameter of the delivery orifice. The system has been applied to the development of a gastrointestinal delivery system for the oral controlled delivery of nifedipine (10,11). A zero-order release profile of nifedipine was attained, which maintained a fairly constant plasma level of nifedipine throughout the course of 24 hr (Figure 8). An excellent in vitro-in vivo correlation was also achieved for the release profiles of nifedipine from a GIDS.

The clinical benefits of drugs delivered in a controlled-release manner are demonstrated by the improvement in therapeutic efficacy as well as the reduction in side effects. For example, by controlled delivery through a GIDS night pain and weight-bearing pain have been relieved and functional activity has been significantly improved (Figure 9), and the GI side effects have been substantially reduced (Figure 10), compared to conventional capsule dosage form.

The osmotic pressure-controlled gastrointestinal delivery system has also been utilized for the gastrointestinal controlled administration of metoprolol and oxprenolol (12,13). Furthermore, a two-compartment GIDS (Figure 7) has been applied to

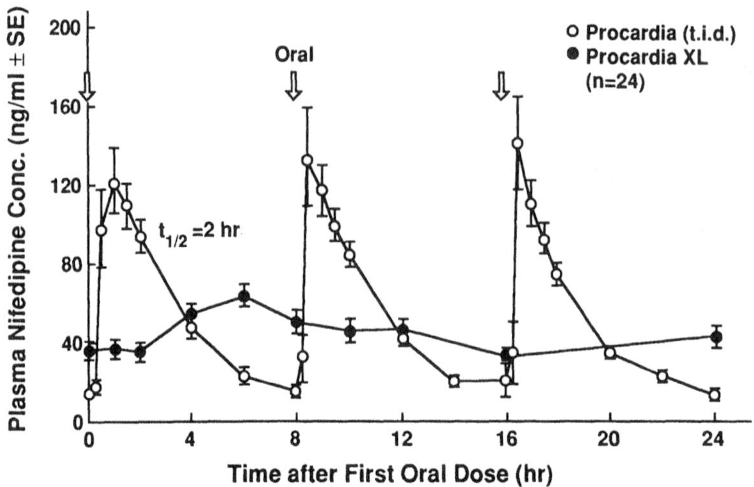

Figure 8 Comparative pharmacokinetic profiles of nifedipine delivered from procardia XL, an osmotic pressure-controlled drug delivery system, once-a-day versus that from procardia, an immediate-release dosage form, taken on time 0, 8 and 16 in human volunteers. (Modified from Zaffaroni, 1991.)

the simultaneous gastrointestinal controlled delivery of two drugs, such as oxprenolol sebacinate and hydralazine HCl, from separate compartments, simultaneously and independently at different delivery rates (14).

B. Hydrodynamic Pressure-Controlled Gastrointestinal Delivery System

In addition to osmotic pressure, hydrodynamic pressure is also a potential energy source for controlling the release of therapeutic agents (15). A hydrodynamic pressure-controlled gastrointestinal drug delivery system can be fabricated by enclosing a collapsible drug compartment inside a rigid shape-retaining housing (Figure 11). The space between the drug compartment and the external housing contains a laminate of swellable, hydrophilic cross-linked polymer, e.g., polyhydroxyalkyl methacrylate, which absorbs the gastrointestinal fluid through the annular openings in the bottom surface of the housing. This absorption causes the laminate to swell and expand, which generates hydrodynamic pressure in the system and forces the drug compartment to reduce in volume and induce the delivery of a liquid drug formulation through the delivery orifice.

At steady state a zero-order drug release profile can be attained, with the rate of drug release $(Q/t)_h$ defined by

$$\left(\frac{Q}{t}\right)_h = \frac{P_f A_m}{h_m}(\theta_s - \theta_e) \tag{3}$$

where P_f, A_m, and h_m are the fluid permeability, the effective surface area, and the thickness of the annular openings, respectively; and θ_s and θ_e are the hydrodynamic pressure in the system and in the gastrointestinal tract, respectively.

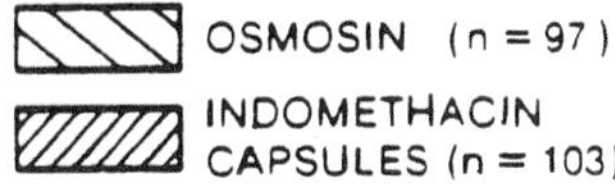

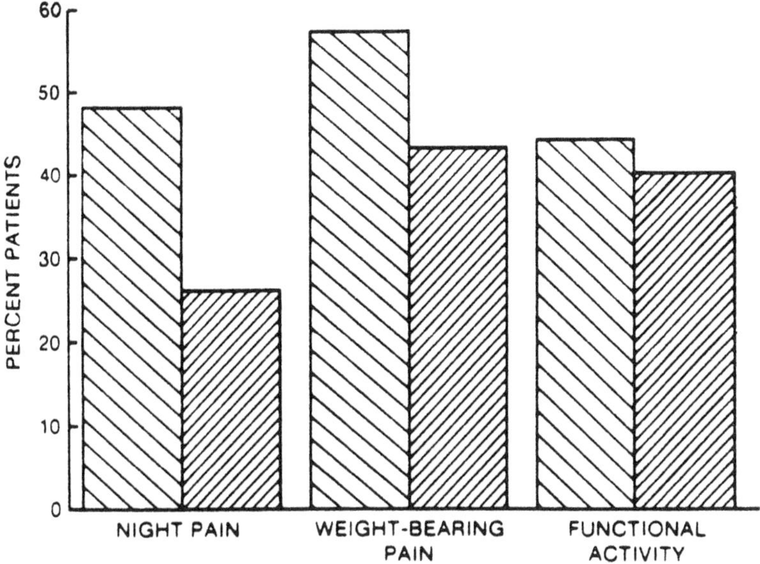

Figure 9 Comparison of the efficiency of clinical improvement in night pain, weight-bearing pain, and functional activity in patients during treatment with indomethacin delivered by osmotic pressure-controlled Osmosin tablets (with delivery rate of 7 mg/hr) or by conventional capsule formulation (25 mg dose). (Adapted from Reference 3.)

C. Membrane Permeation-Controlled Gastrointestinal Delivery Systems

The membrane permeation process has been successfully applied to the development of controlled-release drug delivery systems for the transdermal controlled delivery of nitroglycerin (Transderm-Nitro system), estradiol (Estraderm system), scopolamine (Transderm-Scop system), and clonidine (Catapres-TTS) through the intact skin for systemic medication for 1–7 days (Chapter 7). It has also been utilized for the controlled delivery of drugs for site-specific administration, such as ocular delivery of pilocarpine for up to 7 days treatment of glaucoma (Chapter 6) and intrauterine administration of progesterone for once-a-year contraception (Chapter 10). These polymer membrane permeation-controlled drug delivery systems are known to use a prefabricated microporous or nonporous membrane to meter the release of therapeutic agents.

The membrane permeation process has also been utilized in the development of oral controlled-release drug delivery systems in which the microporous membranes are produced, during the course of transit in the gastrointestinal tract, directly from a nonporous polymer coating. Several potential developments that have proven feasible are outlined as follows.

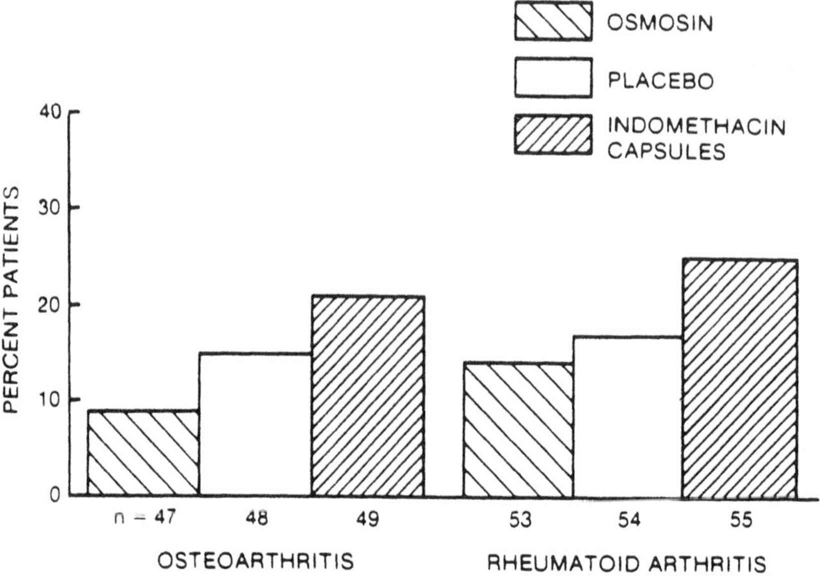

Figure 10 Comparison of the extent of gastrointestinal side effects experienced by osteoarthritis and rheumatoid arthritis patients during treatment with indomethacin delivered by osmotic pressure-controlled Osmosin tablets (with delivery rate of 7 mg/hr), conventional capsule formulation (25 mg dose), or placebo formulation. (Adapted from Reference 3.)

1. Microporous Membrane Permeation-Controlled Gastrointestinal Delivery Device

This is prepared by first compressing the crystals (or particles) of a water-soluble drug, in combination with appropriate pharmaceutical excipients, into a core tablet and then coating the tablet with a layer of non-GI–erodible polymer, e.g., a copolymer of vinyl chloride and vinyl acetate. The polymer coating contains a small amount of water-solube pore-forming inorganic agents, e.g., magnesium lauryl sulfate, which create porosity when the tablet comes into contact with gastrointestinal fluid (Figure 12). The porosity of the polymer coating can be varied by controlling the loading

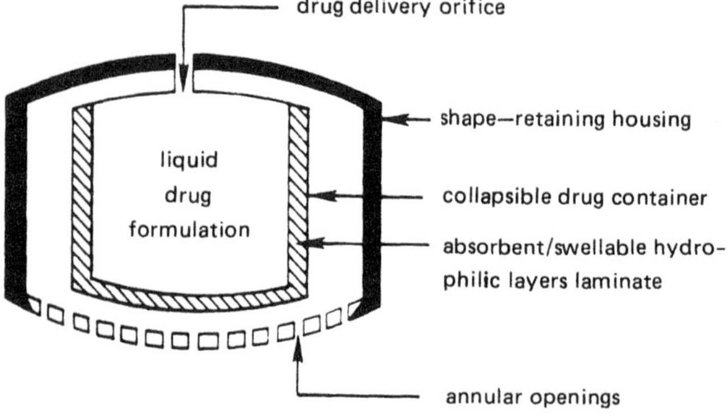

Figure 11 Hydrodynamic pressure-controlled gastrointestinal drug delivery system.

level of inorganic agent to give a slow or fast release at constant rates (Netherlands Patent 7,313,696).

Alternatively, the core tablet may be coated with a layer of non-GI–erodible thermoplastic polymer, e.g., polyvinyl chloride, which contains a high loading of plasticizer, e.g., dioctyl ophthalate. During the course of GI transit, this plasticizer is dissolved away by gastrointestinal fluid, in a manner similar to that of the pore-forming inorganic agent, to form a microporous membrane. The rate of drug release can be controlled and predetermined by regulating the concentration of plasticizer in the polymer coating (Belgium Patent 814,491). An example is shown in Figure 13.

2. *Gastric Fluid-Resistant Intestine-Targeted Controlled-Release Gastrointestinal Delivery Device*

This device, which is designed to release a gastric fluid-labile drug only in the intestinal region at a controlled rate, is prepared by coating a core tablet of the drug with a combination of an intestinal fluid-insoluble polymer, e.g., ethylcellulose, and an intestinal fluid-soluble polymer, e.g., methylcellulose (or hydroxymethylcellulose phthalate). Since both polymer components resist the attack of gastric fluid, this gastrointestinal delivery system thus stays intact in the stomach to protect the drug.

When the device arrives in the intestinal tract, the intestinal fluid-soluble polymer component is dissolved away by the intestinal fluid, leaving a microporous membrane of intestinal fluid-insoluble polymer (Figure 14). This microporous membrane renders a controlled release of drug in the intestine (Japanese Patent 8-0,018,694). A typical example is illustrated by the controlled release of potassium chloride (Figure 15).

D. Gel Diffusion-Controlled Gastrointestinal Delivery Systems

This type of gastrointestinal delivery system is fabricated from gel-forming polymers. It can be prepared by first dispersing the therapeutic dose of a drug in layers

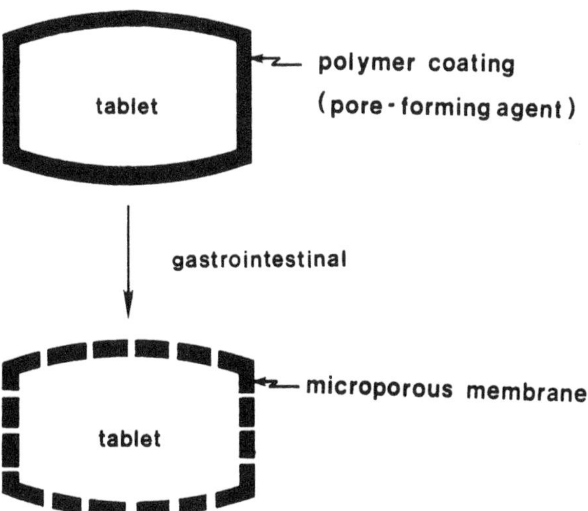

Figure 12 Microporous membrane permeation-controlled gastrointestinal drug delivery system (11).

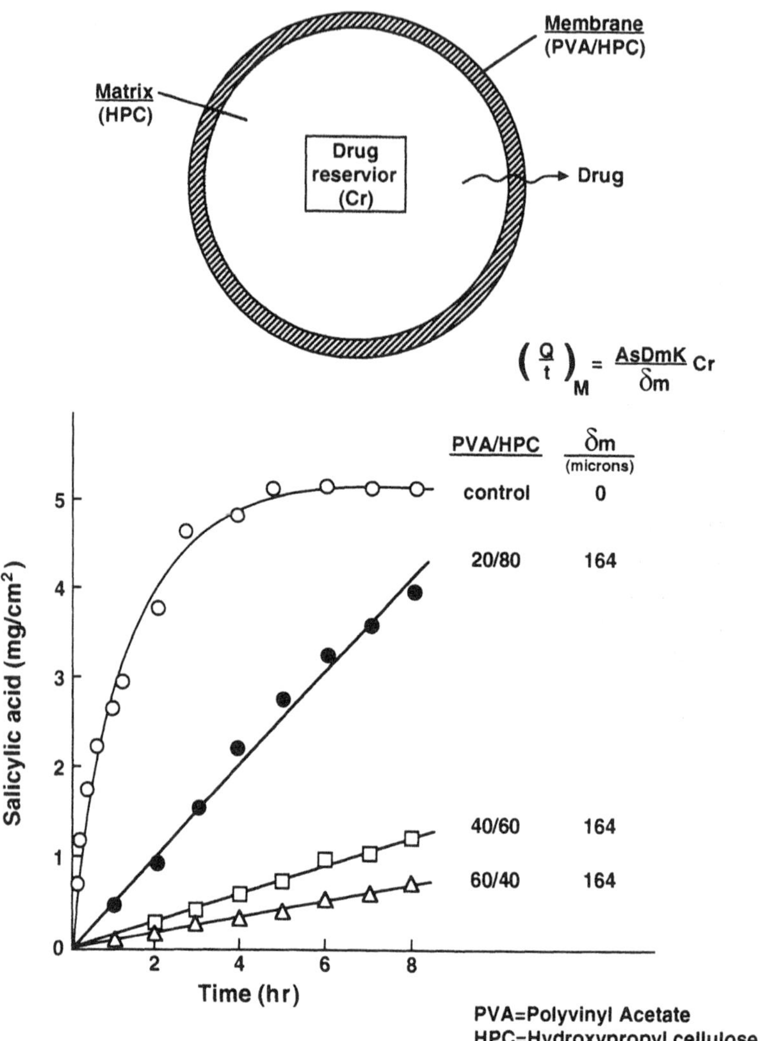

Figure 13 Release profile of drug from the reservoir layer of hydroxypropylcellulose containing 20% salicylic acid and the effect of hydroxypropylcellulose-polyvinyl acetate membrane.

of water-soluble carboxymethylcellulose (CMC), sandwiching the drug-loaded CMC layers between layers of cross-linked carboxymethylcellulose (which is water insoluble but water swellable) and then compressing these layers to form a multilaminated device. This device can be further coated with a polymer coating material to form a gastrointestinal delivery device (Figure 16). Furthermore, with incorporation of a sealing layer two incompatible drugs can be formulated in the same device.

In the gastrointestinal tract the cross-linked CMC layers become swollen and gelatinized to create a colloidal gel barrier, which controls the release of drug from the CMC layers (16).

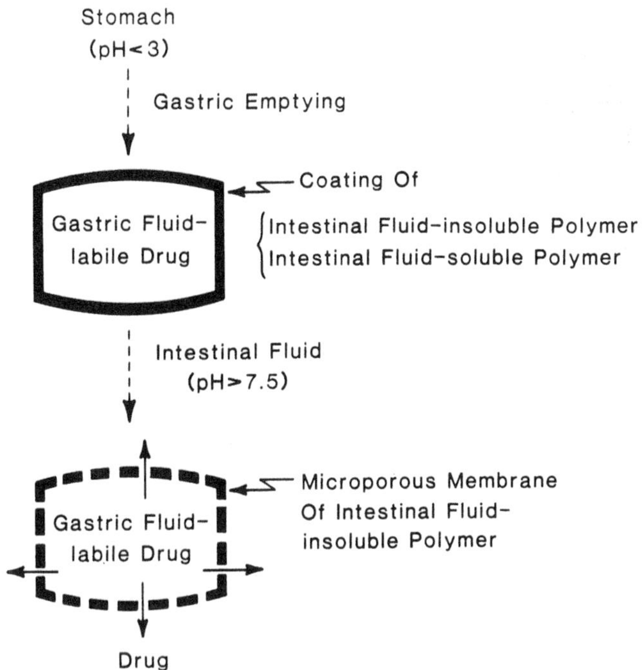

Figure 14 A gastric fluid-resistant intestinal controlled-release delivery system before and after the dissolution of intestinal fluid-soluble polymer in the intestinal tract.

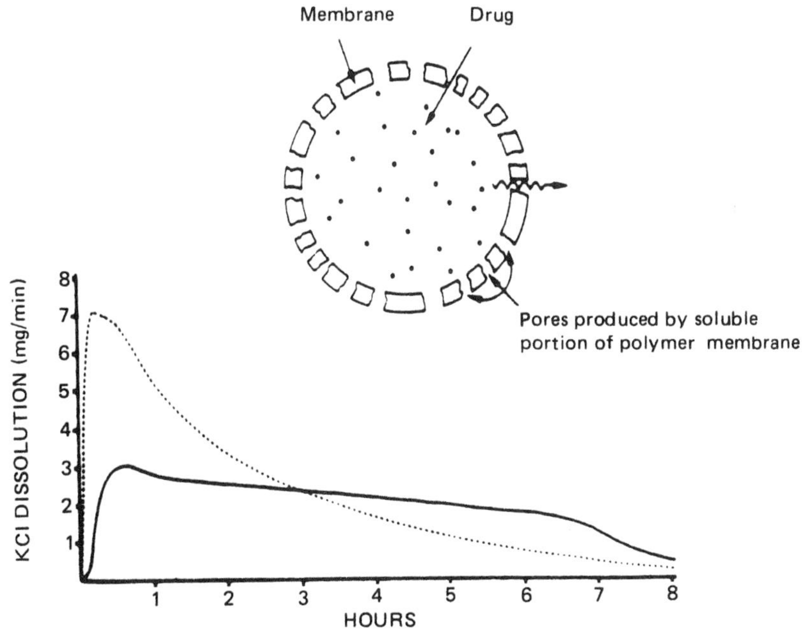

Figure 15 In vitro dissolution profile of KCl from a conventional formulation (····) and its release profile from the gastric fluid-resistant intestinal controlled-release delivery system (———).

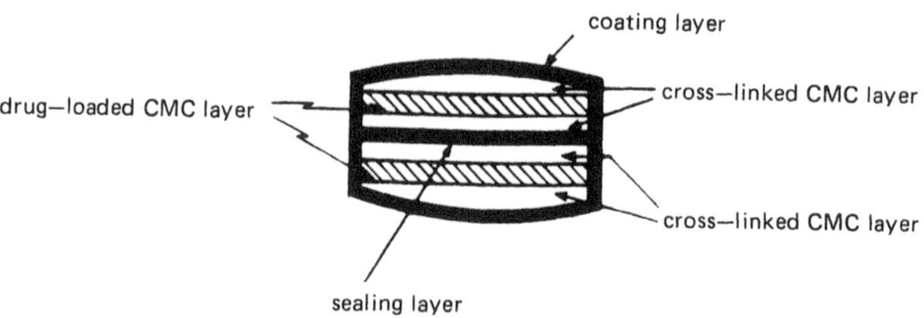

Figure 16 A gastrointestinal drug delivery device consists of a gel-forming multilaminate structure to control the release of drug.

E. pH-Controlled Gastrointestinal Delivery Systems

This type of gastrointestinal delivery system is designed for the controlled release of acidic (or basic) drugs in the gastrointestinal tract at a rate independent of the variation in gastrointestinal pH (Table 2). It is prepared by first blending an acidic (or basic) drug with one or more buffering agents, e.g., a primary, secondary, or tertiary salt of citric acid, granulating with appropriate pharmaceutical excipients to form small granules, and then coating the granules with a gastrointestinal fluid-permeable film-forming polymer, e.g., cellulose derivatives (Figure 17).

The polymer coating controls the permeation of gastrointestinal fluid. The gastrointestinal fluid permeating into the device is adjusted by the buffering agents to an appropriate constant pH, at which the drug dissolves and is delivered through the membrane at a constant rate regardless of the location of the device in the alimentary canal (Germany Patent 2,414,868).

Table 2 Gastrointestinal Tract: Physiologic Characteristics

Region	pH	Lumen constituents	Osmolarity[a]	Dilution	Sensitivity to absorption promotor
Gastrointestine	Varying	—	—	—	—
Stomach[b]	1.2–3.5	HCl	Nonisoosmotic	Large	Low
Fasted	1–3	Mucus	(variable)		
Fed	3–5	Pepsin			
		Rennin			
		Cathepsin			
		Lipase			
		Intrinsic factor			
Small intestine	4.7–6.5	Amylase	Isoosmotic	Large	Medium
Duodenum	4.6–6.0	Bile acids	(330 mOsmol)		
Jejunum	8	Mucus			
		Glucohydrolase			
		Galactohydrolase			
		Lipase			
		Trypsin			
		Chymotrypsin			
Large intestine	7.5–8.0	Mucus	Nonisoosmotic	Small	High
Colon	5–7	Flora			

[a]Relative to the osmolarity of the blood.
[b]Residual volume = 50 ml.

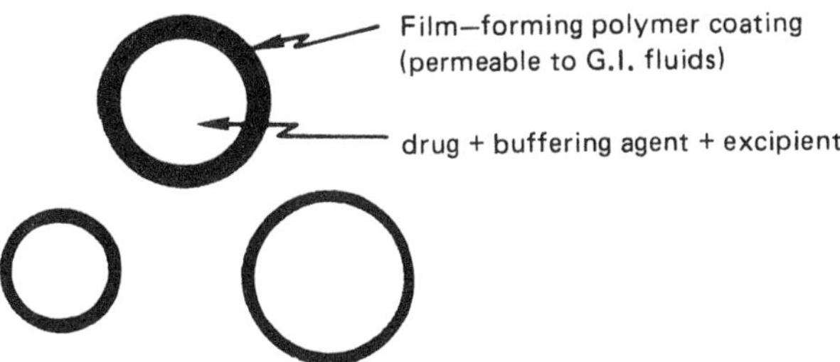

Figure 17 pH-controlled gastrointestinal drug delivery device.

F. Ion-Exchange–Controlled Gastrointestinal Delivery Systems

This type of gastrointestinal delivery system is designed to provide the controlled release of an ionic (or ionizable) drug for intragastric delivery. It is prepared by first absorbing an ionized drug onto the ion-exchange resin granules, such as codeine base with Amberlite (RTM) IRP-69, and then, after filtration from the alcoholic medium, coating the drug-resin complex granules with a water-permeable polymer, e.g., a modified copolymer of polyacrylic and methacrylic ester, and then spray drying the coated granules to produce the polymer-coated drug-resin preparation (Figure 18, top). This approach has been utilized in several patentable developments (17–19).

Further improvement of this ion-exchange drug delivery system has resulted in the development of the Pennkinetic system by Pennwalt Corporation (Figure 18, middle). In this system the drug-resin complex granules are further treated with an impregnating agent, e.g., polyethylene glycol 4000, to retard the rate of swelling in the water and are then coated by an air suspension technique with a water-permeable polymer membrane, e.g., ethylcellulose, to act as a rate-controlling barrier to regulate the release of drug from the system.

It is known that the ionic strength of the gastrointestinal fluid is normally maintained at a relatively constant level. In the GI tract ions diffuse through the ethylcellulose membrane and react with the drug-resin complex to activate the release of drug ions (Figure 18, bottom).

1. Cationic Drugs

A cationic drug forms a complex with an anionic ion-exchange resin, e.g., a resin with a SO_3^- group. In the GI tract hydronium ion (H^+) in the gastrointestinal fluid penetrates the system and activates the release of cationic drug from the drug-resin complex:

$$H^+ + \text{resin} - SO_3^-\text{drug}^+ \rightleftharpoons \text{resin} - SO_3^-H^+ + \text{drug}^+$$

2. Anionic Drugs

An anionic drug forms a complex with a cationic ion-exchange resin, e.g., a resin with a $[N(CH_3)_3^+]$ group. In the GI tract the chloride ion (Cl^{-1}) in the gastrointestinal fluid penetrates the system and activates the release of anionic drug from the drug-resin complex:

$$Cl^- + \text{resin} - [N(CH_3)_3^+]\text{drug}^- \rightleftharpoons \text{resin} - [N(CH_3)_3^+]Cl^- + \text{drug}^-$$

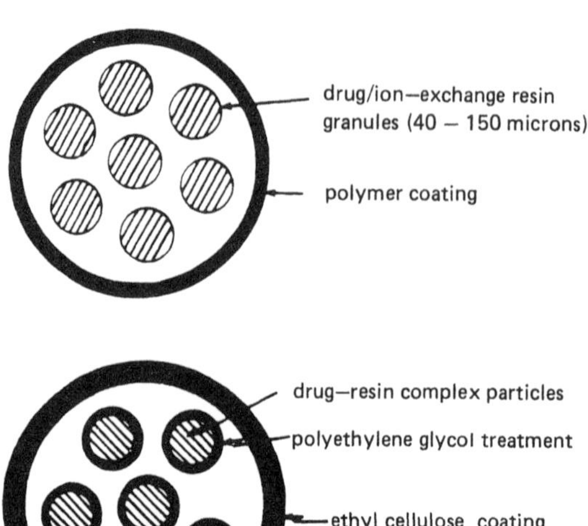

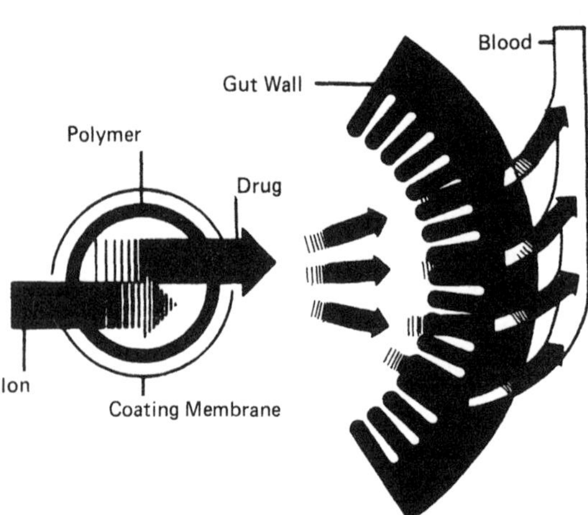

Figure 18 Ion-exchange–controlled gastrointestinal drug delivery systems.
(Top) First generation.
(Middle) Second generation.
(Bottom) Process involved in the ion-exchange–controlled release and GI absorption of ionic drugs from the second generation of ion-exchange–controlled gastrointestinal drug delivery systems (Pennkinetic system).

The free charged drug molecules then diffuse into the gastrointestinal fluid, through the coating membrane, for GI absorption.

The advantages of this type of gastrointestinal drug delivery system are that (i) the rate of drug release is not dependent upon the pH conditions, enzyme activities, temperature, or volume of the GI tract; (ii) the system is administered in the form of a large number of particles, which may eliminate the effect of gastric emptying; and (iii) it can be formulated as a stable liquid suspension-type pharmaceutical dosage form.

By combining different ratios of polymer-coated and uncoated granules in the formulation, a range of dissolution profiles and blood levels can be achieved (Figure 19). Uncoated granules serve as the immediate-release component to provide the initial dose for rapid absorption, and coated granules serve as the sustained-release component to provide the maintenance dose for maintaining a prolonged systemic drug level.

III. MODULATION OF GASTROINTESTINAL TRANSIT TIME

All the oral controlled-release drug delivery systems already discussed have only limited utilization in the gastrointestinal controlled administration of drugs if the

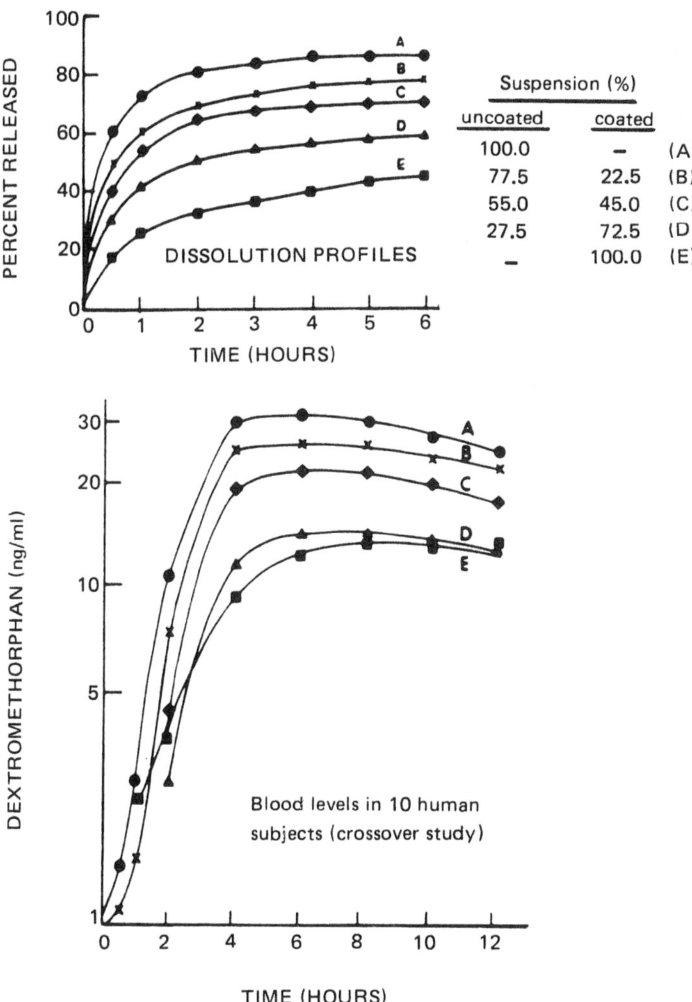

Figure 19 In vitro dissolution profiles and blood concentration profiles in 10 human subjects of dextromethorphan delivered by the Pennkinetic system containing various combinations of immediate-release and sustained-release components (%/%): (A) 100/0, (B) 77.5/22.5, (C) 55/45, (D) 27.5/72.5, and (E) 0/100.

systems cannot remain in the vicinity of the absorption site for the lifetime of the drug delivery.

The transit time for the mouth to the anus varies from one person to another. It also depends upon the physical properties of the object ingested and the physiological conditions of the alimentary canal (Table 1). A study conducted by Hinton and his associates (20) suggested that the alimentary canal transit time for an indigestible object can vary by as much as 8–62 hr, which shows a pyramidal curve with a peak excretion rate of approximately 40% at around 27 hr (Figure 20). Analysis of the data indicates that 50% of human subjects excrete the object within 24 hr ($t_{50\%}$). Therefore, the majority of the oral controlled-release or sustained-release drug delivery systems designed for drug delivery in the gastrointestinal tract are subjected to the restriction of this physiological residence time.

Furthermore, there also exists an absorption surface in the upper small intestine region that is known to have a transit time of only 2–3 hr (21). A concept called anatomic reserve length was recently introduced by Ho et al. (22) to define the length of the small intestine still available for absorption of drugs.

To successfully modulate the GI transit time of a drug delivery system for maximal gastrointestinal absorption of drugs, one needs to have a good fundamental understanding of the anatomic and physiological characteristics of the human gastrointestinal tract (Figure 21). To date, the design of oral drug delivery systems has

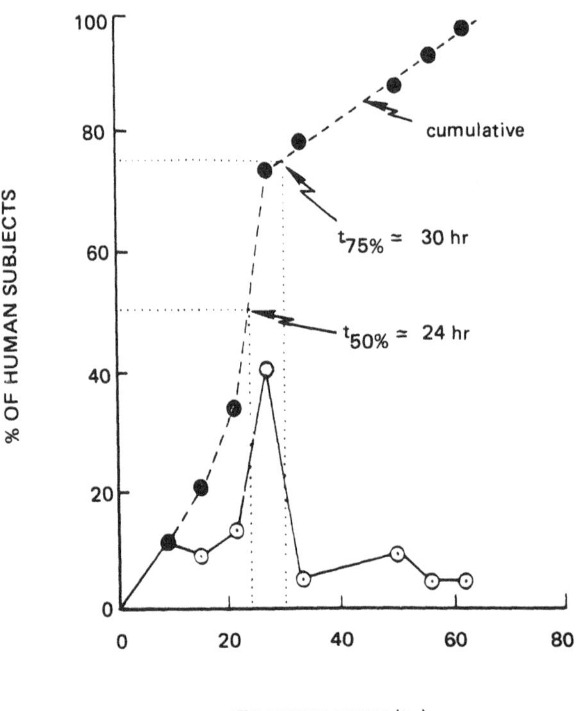

Figure 20 Cumulative excretion profile of an indigestible marker following oral administration in humans and the time course for transit of the marker in the alimentary canal. (Plotted from the data in Reference 25.)

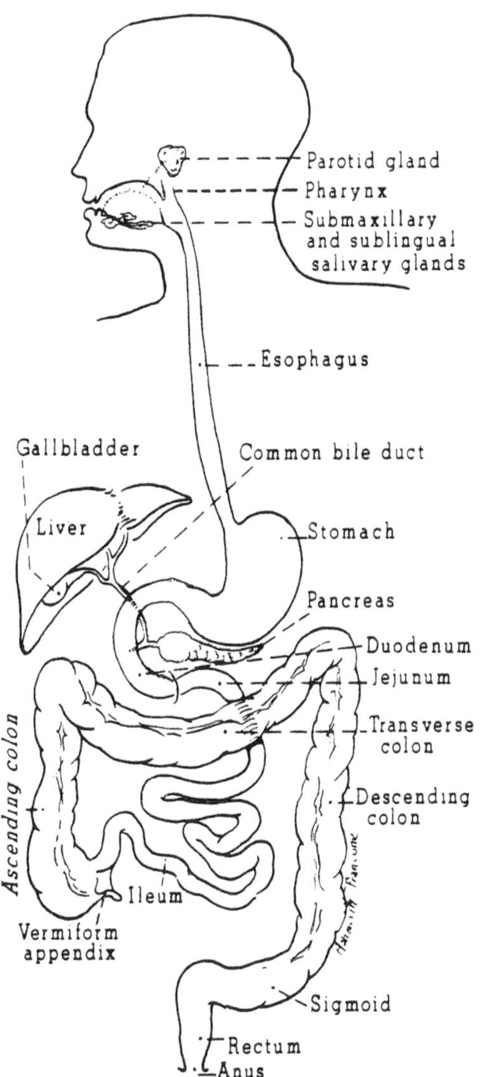

Figure 21 The human digestive system and associated organs. (Reproduced from Jacob and Francone, 1970.)

largely been based on an empirical understanding of GI anatomy and physiology. However, for successful development of a long-acting oral pharmaceutical dosage form, especially a controlled-release gastrointestinal drug delivery system, it is critically important to gain some basic understanding of those physiological factors in the GI tract that may affect the performance of a gastrointestinal drug delivery system. These are outlined and briefly discussed in the following sections.

A. Gastrointestinal Anatomy and Dynamics

A schematic illustration of the human gastrointestinal tract is shown in Figure 21.

1. Anatomy

The stomach is an organ with a capacity for storage and mixing. Its fundus and body regions are capable of displaying a large expansion to accommodate food without much increase in the intragastric pressure. The stomach lining is devoid of villi but consists of a considerable number of gastric pits that contribute to the storage capacity of the stomach. The antrum region is known to be responsible for the mixing and grinding of gastric contents. Under fasting conditions the stomach is a collapsed bag with a residual volume of 50 ml and contains a small amount of gastric fluid (pH 1–3) and air (Tables 1 and 2). There are two main secretions, mucus and acid, produced by specializied cells in the stomach lining. Mucus is secreted by goblet cells and gastric acid by oxyntic (parietal) cells. The mucus spreads and covers the mucosal surface of the stomach as well as the rest of the GI tract. The thickness of this mucus coating varies from one region of the GI tract to another.

Under physiological conditions, the gastric absorption of most drugs is insignificant as a result of its limited surface area (0.1–0.2 m^2) covered by a thick layer of mucous coating, the lack of villi on the mucosal surface, and the short residence time of most drugs in the stomach.

The contents of the stomach are emptied through the pylorus into the proximal duodenal region of the small intestine. In humans the gastroduodenal junction controls the unidirectional passage from the stomach to the duodenum, even though a duodenogastric reflux occurs in some animals. In the proximal duodenum the contents of the gallbladder (e.g., bile) and pancreas as well as some duodenal secretions, including bicarbonate, are emptied. The pancreatic juice contributes lipolytic, proteolytic, and carbohydrate splitting enzymes (Table 2). Several enzymes in the intestinal secretion, including leucine aminopeptidase, were actually found to reside on the striated border of the intestinal absorptive cells as integral parts of the microvilli of the brush border. Following transit through the small intestine, which has a length of approximately 3 m, the contents are passed through the terminal ileum into the colon via a junction known as the ileocecal valve.

Unlike the stomach, the small intestine is a tubular viscous organ and has enormous numbers of villi on its mucosal surface that create a huge surface area (4500 m^2 compared to only 0.1–0.2 m^2 for the stomach). These villi are minute fingerlike projections of the mucosa and have a length of 0.5–1.5 mm, depending upon the degree of distention the intestinal wall and the state of contraction of smooth muscle fibers in their own interiors. They cover the entire surface of the mucosa, with numbers varying from 10 to 40 mm^{-2}. They are most numerous in the duodenum and proximal jejunum. It should be noted that there is a progressive decrease in surface area from the proximal to the distal region of the small intestine and colon (Table 1). As a result the proximal small intestine is the region with the most efficient absorption. For maximal systemic bioavailability the drug should be targeted for delivery in the vicinity of this region.

The colon lacks villi, and its primary function is to store indigestible food residue. It also contains a variety of flora that are normal residents of the GI tract and may degrade the contents of the colon.

2. Dynamics

It should be recognized that the gastrointestinal tract is always in a state of continuous motility. There are two modes of motility pattern: the digestive mode and the interdigestive mode (23) involved in the digestion of food. The interdigestive GI motility is characterized by a cyclic pattern that originates in the foregut and propagates to the terminal ileum and consists of four distinct phases:

Phase I. Period of no contraction
Phase II. Period of intermittent contractions
Phase III. Period of regular contractions at the maximal frequency that migrate distally
Phase IV. Period of transition between phase III and phase I

A complete cycle of these four phases has an average duration of 90–120 min in both human and dog. Certain disease states, such as bacterial overgrowth, mental stress, and diurnal variation, or their combinations, can influence the duration of each individual phase as well as the total cycle (24,25). Phase III has a housekeeping role and serves to clear all indigestible materials from the stomach and the small intestine. Consequently, any controlled-release gastrointestinal drug delivery system designed to stay during the fasted state should be capable of resisting the housekeeping action of phase III if one intends to prolong the GI retention time. The bioadhesive properties added to the gastrointestinal drug delivery system must be capable of adhering to the mucosal membrane strongly enough to withstand the shear forces produced in this phase.

The cyclic motor activity of interdigestive GI motility is also associated with the gastric, pancreatic, and biliary secretory activities of GI tract; both the migratory and secretory activities constitute two aspects of the same periodicity (26). Under fasting conditions, both the migratory and secretory activities of the stomach, small and large intestines, pancreas, and liver change periodically to provide the mechanical and chemical means required for GI housekeeping.

Feeding has been reported to yield an interruption of the interdigestive motility cycle and the appearance of a continuous pattern of contraction, called postprandial contraction, which can be induced in dogs by a gastric content with a volume as small as 150 ml water (27). A normal meal can change the motility pattern from a fasted state to a fed state for a duration of up to 8 hr, depending upon the caloric content of the food ingested (28).

3. GI Transit

The transit time of a gastrointestinal drug delivery system along the GI tract is the most limiting physiological factor in the development of a controlled-release gastrointestinal drug delivery system that is targeted to once-a-day medication. Like the motility pattern, the patterns of GI transit depend on whether the person is in a fasted or fed state. In addition, the physical state of the drug delivery system, either a solid or a liquid, also influences the transit time through the GI tract (Table 1).

Fasted State. The gastric emptying of liquids in the fasted state is a function of the volume administered (27). For a small volume (<100 ml), this is controlled by the existing phasic activity and liquids are emptied at the onset of phase II; most of them are gone before the arrival of phase III. For volumes larger than 150 ml, liquids are emptied by characteristic discharge kinetics irrespective of phasic activity. The half-life of discharge in dogs is 40–50 min for a small volume of liquids but only 8–12 min for a larger volume. The observed difference in transit behavior could be due to the fact that a small volume does not affect the existing motility pattern in the stomach, but a large volume converts the fasted state to a fed state, which in turn creates the fed-state motility pattern. The fasted-state emptying pattern of liquids is independent of the presence of any indigestible solids in the stomach (29). Based on these observations, for prolonged gastric emptying the gastrointestinal drug delivery system should be administered with a small volume of liquid.

Indigestible solids are emptied from the stomach as a function of their physical size. Solids of small particle size (<1 mm) can be emptied with the liquid; solids of 2 mm or greater do not empty until the arrival of phase III activity, at which time they are emptied as a bolus (29). In a fasted dog the gastric emptying of solids is independent of size, density, and surface characteristics (30). Depending upon proximity of the time of ingestion to the next phase III activity, a solid dosage form can therefore stay in a fasted stomach for a duration anywhere in the range of 0–120 min.

During phase I, when contractions are at a minimum, there is little or no movement of liquid or solids through the intestine. In phases II and III, on the other hand, the flow of materials in the intestinal lumen becomes progressively faster. The segregation of liquids and solids also occurs, and liquids tend to migrate during phase II and solids during phase III. The motor activity of the small intestine during the fasted state may not be sufficiently strong to move the solids.

During the fasted state there is relative motion between the dosage form in the small intestine and the luminal fluid content. Shear forces and constant fluid movement around the dosage form may attribute to the difference between the in vivo bioavailability and the in vitro release of drug from the dosage form. For multiunit dosage forms, once the particles have been emptied from the stomach as a bolus there is little, if any, further spreading of particles in the intestine (31). However, once in the colon particles show some tendency to disperse, perhaps as a result of the high viscosity of the luminal contents in this region.

Fed State. Following feeding the fundus of the stomach expands to accommodate food without an appreciable increase in the intragastric pressure. Once in the stomach food begin emptying almost immediately. Liquids are emptied at a rate faster than that of solids, and the rate is controlled by feedback mechanisms from the duodenum and ileum. Solids are not emptied in the fed state unless they have been ground to a particular size of 2 mm or less. There is a sieving mechanism in the fed stomach, which is influenced by the viscosity of the meal. Since grinding and mixing take place in the antral area, dosage forms tend to reside in this area if they are large. Multiunit dosage forms, however, disperse and empty with food and thus achieve a considerable degree of distribution (32).

Gastric secretion also starts following the ingestion of food, and its volume depends upon the nature and volume of the ingested food. The volume emptied is replaced by gastric secretion, and thus the gastric volume may actually remain constant during the first hour of gastric emptying. The total time for gastric emptying varies in the range of 2–6 hr.

In the small intestine, the contents move faster in the fed state than the phase III transit in the fasted state, which helps the transit of smaller particles but not larger particles. The intestinal transit time for both liquids and solids, regardless of their nature, is around 3–4 hr in both the fasted and the fed states. This constancy in intestinal transit can be important in colon-targeted drug delivery.

Studies of the GI transit of dosage forms, such as tablets, capsules, and particles, have demonstrated a transit pattern similar to that of nutrients. Most dosage forms taken orally in the fasted state empty within 90 min. In the fed state nondisintegrating tablets and capsules stay in the stomach for 2–6 hr and are discharged only at the onset of the fasted state. However, disintegrating dosage forms and small particles are emptied together with the food. In all instances the transit time for the small intestine is 3–4 hr (32). Recent studies conducted in five healthy subjects taking a standard breakfast discovered that coadministration of the ammonium salt of myristic acid (a saturated fatty acid and a constituent of fat-rich food) led to an average increase of 23.8% in drug absorption from a conventional capsule-shaped dosage form, which could be due to the delaying effect of ammonium myristate on GI passage (33).

In summary, the total transit time of foods and dosage forms in humans from stomach to the ileocecal junction is approximately 3–6 hr in the fasted state and 6–10 hr in the fed state. This sets an approximately 10 hr limit for the delivery of drugs absorbed solely from the small intestine region.

4. *Ileocecal Junction*

This serves mainly to ensure the unidirectional flow of the luminal contents from the small to the large intestine.

5. *Colon and Gut Flora*

Because of the high water absorption capacity of the colon, the colonic contents are considerably viscous and their mixing is not efficient. Thus the availability of most drugs to the absorptive membrane is low.

The human colon has over 400 distinct species of bacteria as resident flora, a possible population of up to 10^{10} bacteria per gram of colonic content (34). Among the reactions carried out by these gut flora are azoreduction and enzymatic cleavage, i.e., glycosidases (35). These metabolic processes may be responsible for the metabolism of many drugs and may also be applied to colon-targeted drug delivery. For example, the azoreduction process has been utilized in the colon-targeted delivery of peptide-based macromolecules, like insulin, by oral administration (Chapter 11). Enzymatic cleavage has been applied to the colon site-specific delivery of drugs by forming ester-type macromolecular prodrugs of carboxylic acid-containing drugs, like naprozen, with dextran (36).

6. *GI Mucus*

Mucus is continuously secreted by specialized goblet cells located throughout the GI tract. Fresh mucus on the mucosal surface is very thick and becomes diluted and less viscous as it nears the lumen. Its thickness varies depending upon the region of the GI tract.

The primary function of mucus appears to be protection of the surface mucosal cells from gastric acid and peptidase as well as a barrier to antigens, bacteria, and virus. It also acts as a lubricant to assist the passage of solids.

Chemically, mucus is a glycoprotein network that consists of oligosaccharide chains with sialic acid ($pK_a = 2.6$) and holds a varying amount of bound water. The presence of mucus on the GI tract has presented an opportunity to prolong transit time by application of bio(muco)adhesive polymer technology (32,37).

B. Prolongation of GI Retention

Several approaches have recently been developed to extend gastrointestinal transit time by prolonging the residence time of drug delivery systems in the stomach.

1. Hydrodynamically Balanced Intragastric Delivery System

The hydrodynamically balanced gastrointestinal drug delivery system, in either capsule or tablet form, is designed to prolong GI residence time in an area of the GI tract to maximize drug reaching its absorption site in the solution state and, hence, ready for absorption. It is prepared by incorporating a high level (20–75% w/w) of one or more gel-forming hydrocolloids, e.g., hydroxyethylcellulose, hydroxypropylcellulose, hydroxypropylmethylcellulose, and sodium carboxymethylcellulose, into the formulation and then compressing these granules into a tablet (or encapsulating into capsules) (38). Formulation of this device must comply with the following criteria:

1. It must have sufficient structure to form a cohesive gel barrier.
2. It must maintain an overall specific gravity lower than that of the gastric contents (1.004–1.010).
3. It should dissolve slowly enough to serve s a drug reservoir.

On contact with gastric fluid the hydrocolloid in this intragastric floating device starts to become hydrated and forms a colloid gel barrier around its surface with thickness growing with time (Figure 22). This gel barrier controls the rate of solvent penetration into the device and the rate of drug release from the device. It maintains a bulk density of less than 1 and thus remains buoyant in the gastric fluid inside the stomach for up to 6 hr (Figure 23); conventional dosage forms disintegrate completely within 60 min and are emptied totally from the stomach shortly afterward

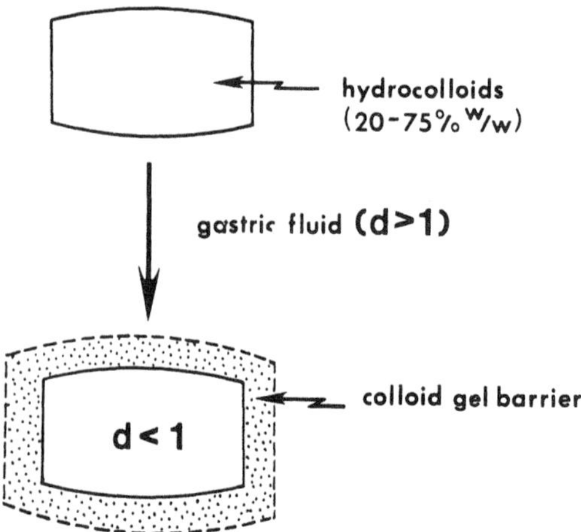

Figure 22 Hydrodynamically balanced intragastric drug delivery device before and after contact with gastric fluid, which activates the formation of the colloid gel barrier.

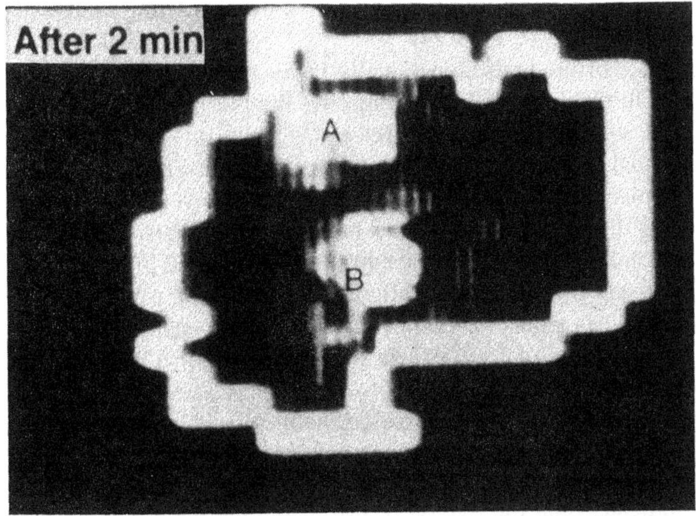

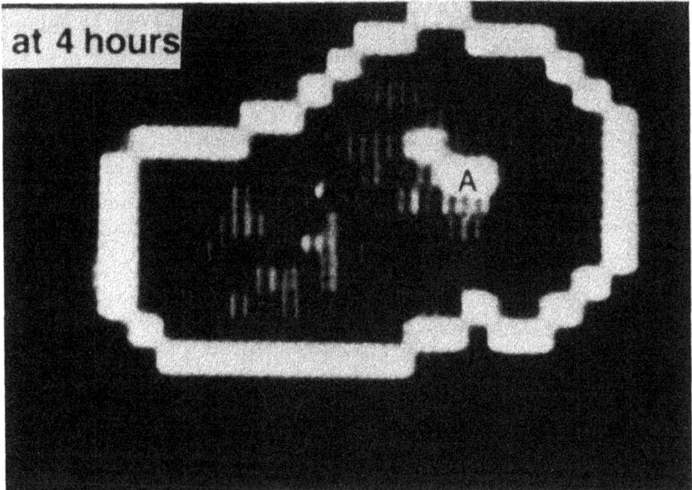

Figure 23 Computerized output of radioactivity from a ^{99m}Tc-labeled hydrodynamically balanced intragastric drug delivery device (A) and a conventional tablet (B) swallowed simultaneously with 200 ml of ^{111}In-labeled water in a human volunteer at 2 min and 4 hr postingestion. The stomach outline is illustrated by the ^{111}In-labeled water. the disproportion between the size of drug delivery systems A and B and the size of the stomach resulted from the intensity of the ^{99m}Tc label, which created a greater scatter of radioactivity. (Reproduced from Reference 45.)

(39,40). Radioactivity measurement by scintigraphy also showed that the gastric retention was substantially prolonged (Figure 24).

The in vivo performance of the hydrodynamically balanced intragastric floating drug delivery system was further assessed by comparing the plasma profiles of diazepam following the oral administration of Valrelease capsule, the hydrodynamically balanced gastrointestinal delivery system containing 15 mg diazepam as a single dose in fasted and fed subjects, and of Valium tablets in the conventional dosage form, each containing 5 mg diazepam, three times a day at 5 hr intervals. The results shown in Figure 25 demonstrate that a Valrelease capsule administered in both fasted and fed states attained a steady plasma level; Valium tablets produce a saw-toothed pattern plasma profile that shows peaks and valleys in response to each dosing. The data also suggest that the ingestion of food delays the attainment of a peak plasma level of diazepam for the Valrelease capsule (40).

The mechanism of drug release from the hydrodynamically balanced gastrointestinal delivery system (GIDS) was investigated using chlordiazepoxide HCl, which shows a 4000-fold difference in aqueous solubility as pH is varied from 3 to 6 (40). In vitro release studies indicated that the release of drug from the hydrodynamically balanced GIDS follows the matrix diffusion-controlled release process in the first 4–6 hr and a Q versus $t^{1/2}$ linearity is attained (Figure 26A). The observed Q versus $t^{1/2}$ relationship could be attributed to the time-dependent growth in the thickness of the colloid gel barrier (Figure 22).

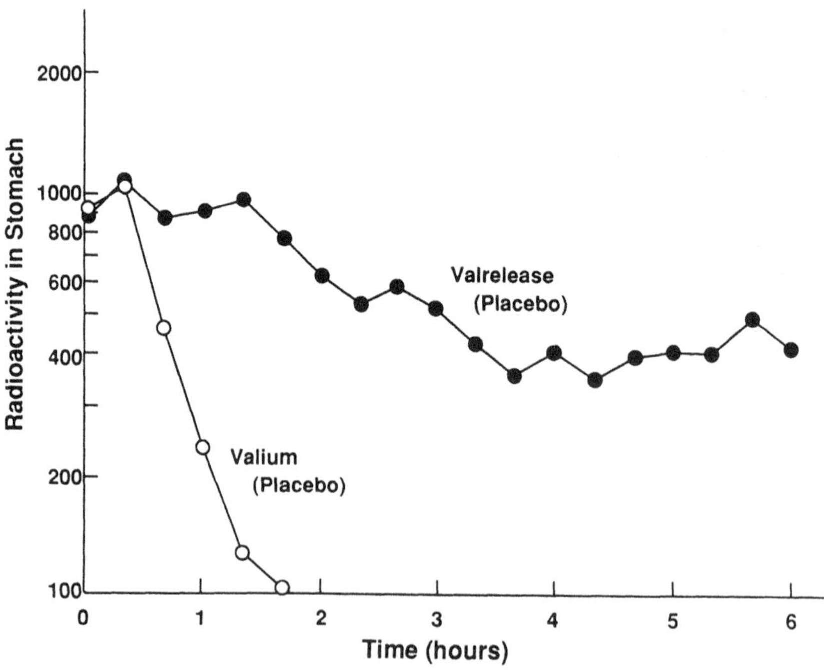

Figure 24 comparison of radioactivity retained in the stomach of a human volunteer between the hydrodynamically balanced intragastric drug delivery device and a conventional tablet; both were ^{99m}Tc-labeled and measured by external scintigraphy. (Replotted from the data in Reference 45.)

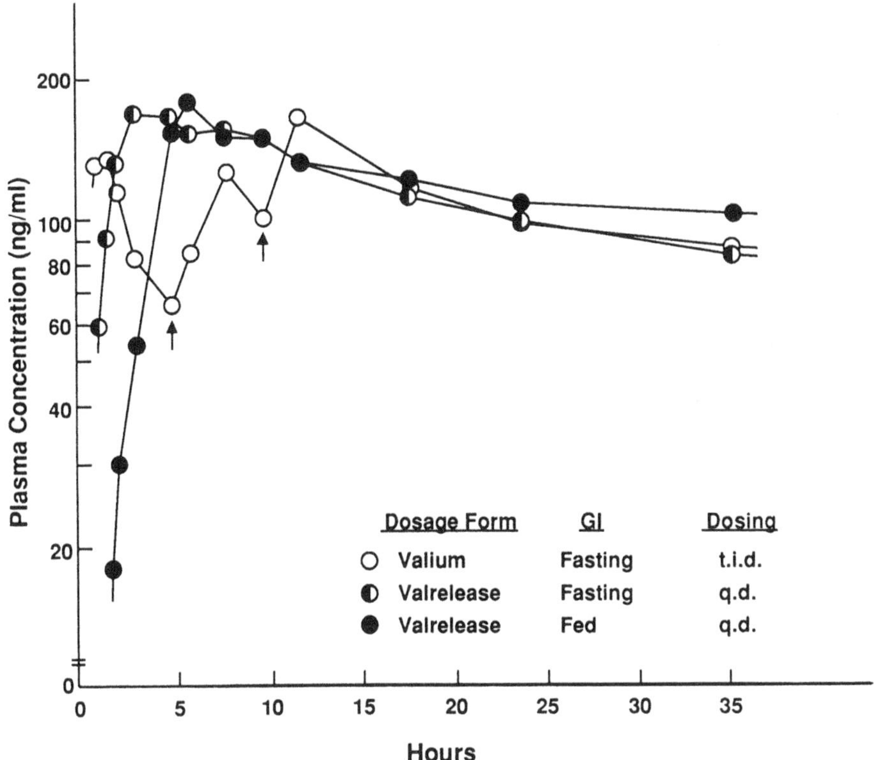

Figure 25 Comparison of plasma concentration profiles of diazepam following single-dose administration of a Valerase capsule, a hydrodynamically balanced intragastric drug delivery device, in fasted and fed subjects, a multidose administration of a Valium tablet, a conventional dosage form, three times a day at 0, 5, and 10 hr. (Replotted from the data in Reference 45.)

In vivo studies of the hydrodynamically balanced GIDS demonstrated that the percentage of the maximum plasma concentration of chlordiazepoxide also shows a linear dependence on the square root of time (Figure 26B), with a slope value in good agreement with that of the in vitro drug release profile (33.63 versus 35.02%/$hr^{1/2}$). This in vitro-in vivo agreement suggests that the plasma drug concentration in the initial perioid of oral administration is dependent primarily upn drug release characteristics (matrix diffusion-controlled release of drug through the colloidal gel barrier). The in vitro-in vivo correlation is further substantiated by the linear relationship observed between the percentage of the maximal plasma concentration attained and the percentage of drug release, which yields a slope value of 1.01 compared to the theoretical value of unity (40).

A bilayer tablet can also be prepared to contain one immediate-release layer and one sustained-release layer. After the initial dose is delivered by the immediate-release layer, the sustained-release layer absorbs the gastric fluid and forms a colloidal gel barrier on its surface. This produces a bulk density less than that of the gastric fluid and remains buoyant in the stomach for an extended period of time (41).

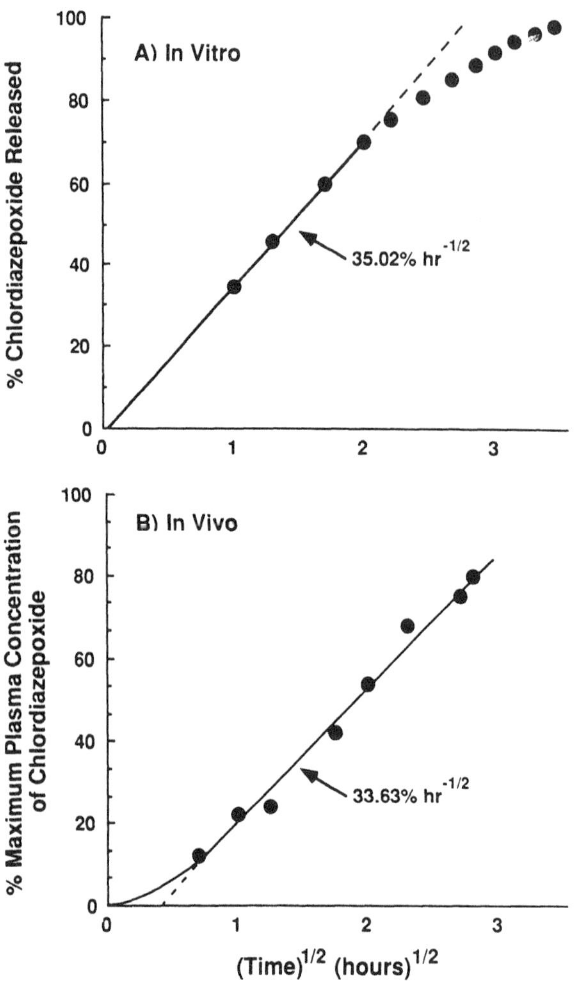

Figure 26 (A) In vitro release profile of chlordiazepoxide from the hydrodynamically balanced intragastric delivery device. (B) Time course for the percentage maximum plasma concentration of chlordiazepoxide delivered by the hydrodynamically balanced intragastric delivery device. There is linear dependence of the percentage maximum plasma concentration on the square root of time. (Replotted from the data in Reference 45.)

2. *Intragastric Floating Gastrointestinal Drug Delivery System*

A gastrointestinal drug delivery system (GIDS) can be made to float in the stomach by incorporating a floatation chamber, which may be a vacum or filled with air or a harmless gas (Figure 27).

A drug reservoir is encapsulated inside a microporous compartment with apertures along its top and bottom walls. The peripheral walls of the drug reservoir compartment are completely sealed to prevent any direct contact of the stomach mucosal surface with the undissolved drug.

In the stomach the floatation chamber causes the GIDS to float in the gastric

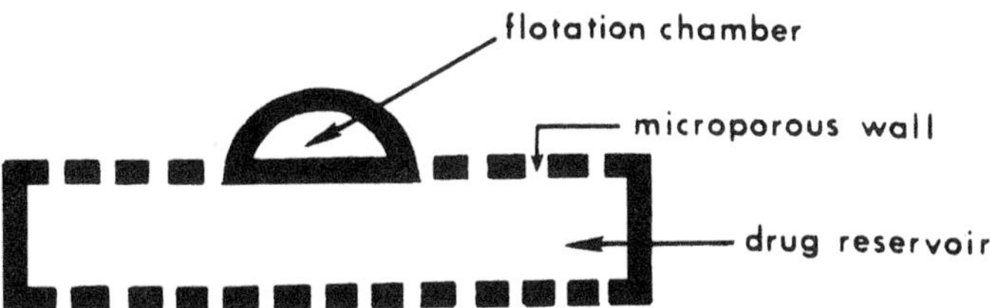

Figure 27 Intragastric floating drug delivery device.

fluids. Fluids enter through the apertures, dissolve the drug, and carry the drug solutes out of the drug delivery system for continuous transport to the intestine for absorption (42).

3. *Inflatable Gastrointestinal Drug Delivery System*

The residence time of the drug delivery device in the stomach can also be sustained by incorporation of an inflatable chamber, which contains a liquid, e.g., ether, that gasifies at body temperaure to cause the chamber to inflate in the stomach (Figure 28).

The inflatable gastrointestinal drug delivery system is fabricated by loading the inflatable chamber with a drug reservoir, which can be a drug-impregnated polymeric matrix, and then encapsulating the unit in a gelatin capsule. After oral ingestion the capsule dissolves to release the drug reservoir compartment together with the inflatable chamber. The inflatable chamber automatically inflates and retains the drug reservoir compartment in the stomach. The drug solutes are continuously released from the reservoir into the gastric fluid.

The inflatable chamber also contains a bioerodible polymer filament, e.g., a copolymer of polyvinyl alcohol and polyethylene, that gradually dissolves in the gastric fluid and finally causes the inflatable chamber to release the gas and become collapsed after a predetermined time period to permit the spontaneous ejection of the inflatable GIDS from the stomach (43).

4. *Intragastric Osmotically Controlled Drug Delivery System*

The osmotic pressure-controlled drug release mechanism discussed earlier can also be incorporated in the inflatable GIDS to control the release of drug in the stomach (Figure 29).

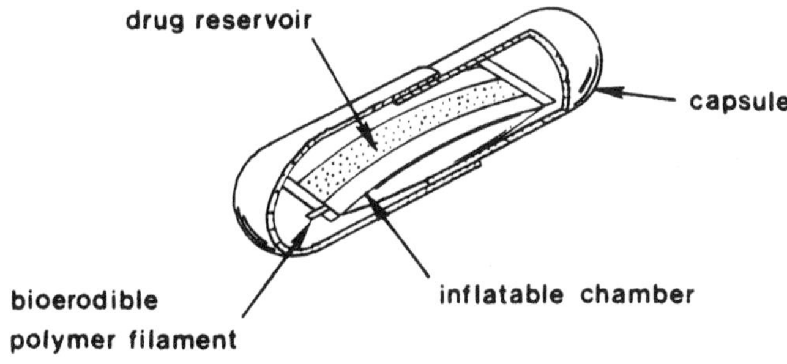

Figure 28 Inflatable gastrointestinal drug delivery device.

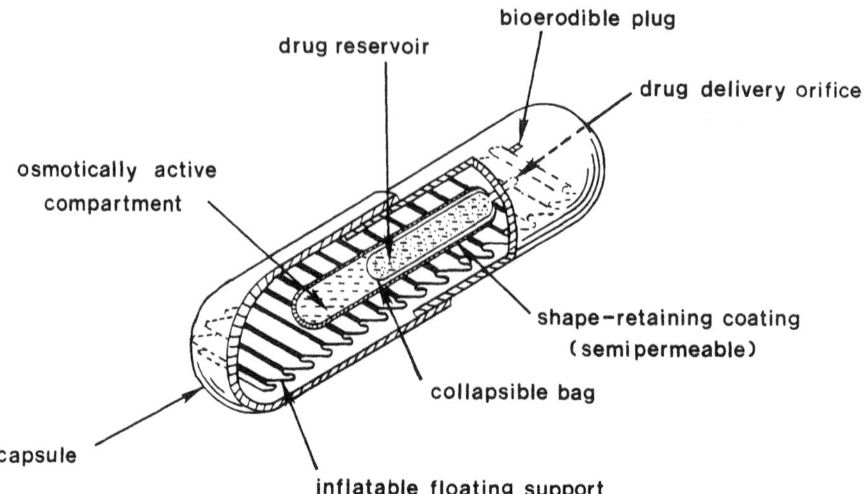

Figure 29 Intragastric osmotically controlled drug delivery system. (Replotted from Reference 49.)

It is comprised of an osmotic pressure-controlled drug delivery device and an inflatable floating support in a bioerodible capsule. When the drug delivery device reaches the site of drug administration, e.g., the stomach, the capsule quickly disintegrates to release the intragastric osmotically-controlled drug delivery device.

The inflatable floating support is made from a deformable holow polymeric bag that contains a liquid that gasifies at body temperature to inflate the bag.

The osmotic pressure-controlled drug delivery device consists of two compartments: (i) a drug reservoir compartment, and (ii) an osmotically active compartment. The drug reservoir compartment is enclosed by a pressure-responsive collapsible bag, which is impermeable to vapor and liquid and has a drug delivery orifice. The osmotically active compartment contains an osmoticially active salt and is enclosed within a semipermeable housing. In the stomach the water in the gastric fluid is continuously absorbed through the semipermeable membrane into the osmotically active compartment to dissolve the osmotically active salt. An osmotic pressure is thus created, which acts on the collapsible bag and, in turn, forces the drug reservoir compartment to reduce its volume and activate the release of a drug solution formulation through the delivery orifice (44).

The floating support is also made to contain a bioerodible plug that erodes after a predetermined time to deflate the support. The deflated drug delivery system is then excreted from the stomach.

5. *Intrarumen Controlled-Release Drug Delivery Device*

This is designed for intraruminal controlled delivery of veterinary drugs in a ruminant animal. It is prepared by compressing a layer of medicated polymer matrix between two layers of water-insoluble polymer film to form a sandwich composition. The laminate is then rolled into a configuration by a gelatin band for easy oral administration. In the rumen the band is dissolved to regenerate the original configuration of the drug delivery device to prolong retention in the rumen (Figure 30).

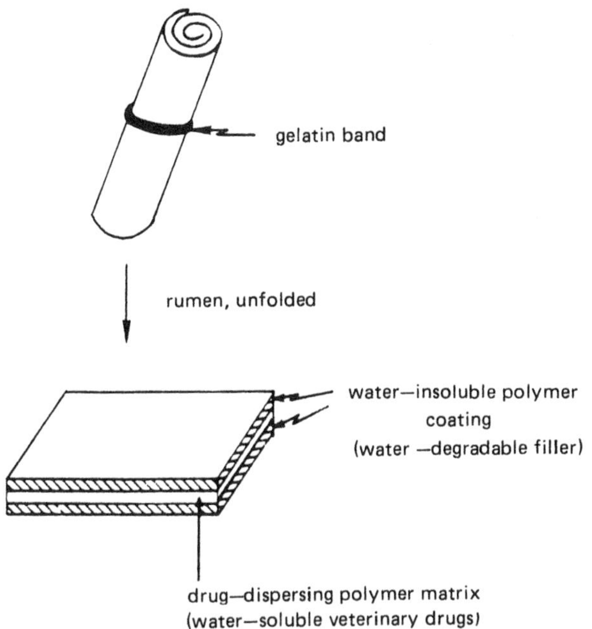

Figure 30 Intrarumen controlled-release drug delivery system.

The medicated polymer matrix is prepared by blending a homogeneous mixture of a water-soluble drug and a water-insoluble polymer, e.g., ethylene-vinyl acetate copolymer, at 100°C and then compressing the mixture into a sheet-shaped drug-dispersed matrix-type polymeric device.

The nonmedicated coating polymer film is prepared by blending a homogeneous mixture of a water-insoluble ethylene-vinyl acetate copolymer and a water-degradable filler, e.g., lactose, also at 100°C, and then compressing the mixture to form the film-shaped polymer coating. after lamination with the medicated polymer matrix to form a sandwich-type drug delivery device, the film coatings control the rate of drug release depending upon the loading levels of the water-degradable filler (Belgium Patent 867,692).

6. *Bio(muco)adhesive Gastrointestinal Drug Delivery Systems*

Another potential approach to extend the gastrointestinal residence time is the development of a bio(muco)adhesive polymer-based drug delivery system, which has been conceptualized on the basis of a GI self-protective mechanism.

It is known that the surface epithelium of the stomach and intestine retains its integrity throughout the course of its lifetime, even though it is constantly exposed to a high concentration of hydrochloric acid (as high as 0.16 N) and powerful protein-splitting enzymes, like pepsin. This self-protective mechanism is due to the fact that the specialized goblet cells located in the stomach, duodenum, and transverse colon continuously secrete a large amount of mucus that remains closely applied to the surface epithelium. The mucus contains mucin, an oligosaccharide chain with terminal sialic acid (pKa = 2.6), which is capable of neutralizing the hydrochloric acid and withstanding the action of pepsin and thus protects the epithelial cell membrane.

The surface epithelium adhesive properties of mucin have been recognized and recently applied to the development of gastrointestinal drug delivery devices based on bio(muco)adhesive polymers.

The concept of using mucoadhesive polymer to extend the GI transit time is shown in Figure 31. The drug delivery system coated with mucoadhesive polymer binds to the mucin molecules in the mucus lining and is therefore retained on the surface epithelium for extended periods of time. The drug molecules contained in the drug delivery device coated with mucoadhesive polymer are constantly released for absorption.

A bio(muco)adhesive polymer is a natural or a synthetic polymer capable of producing an adhesive interaction with a biological membrane, which is then called a bioadhesive polymer, or with the mucus lining on the GI mucosal membrane, which is thus called a mucoadhesive polymer. A bio(muco)adhesive polymer is known to have the following molecular characteristics:

1. It has molecular flexibility.
2. It contains hydrophilic functional groups.
3. It poses a specific molecular weight, chain length, and conformation.

A number of commonly used macromolecular pharmaceutical excipients have been evaluated and found to have bio(muco)adhesive properties (Table 3). From the rank order of bioadhesion (45–47), it appears that polyanions with a high charge density are highly active. Among the various polyanions evaluated, it was found that the polymers containing carboxylic groups, such as polyacrylic polymer, show a high level of bioadhesion (48). Using a tensiometer and fresh pig intestine, Dittgen and Oestereich (49) measured the bioadhesive properties of polyacrylic polymer and co-

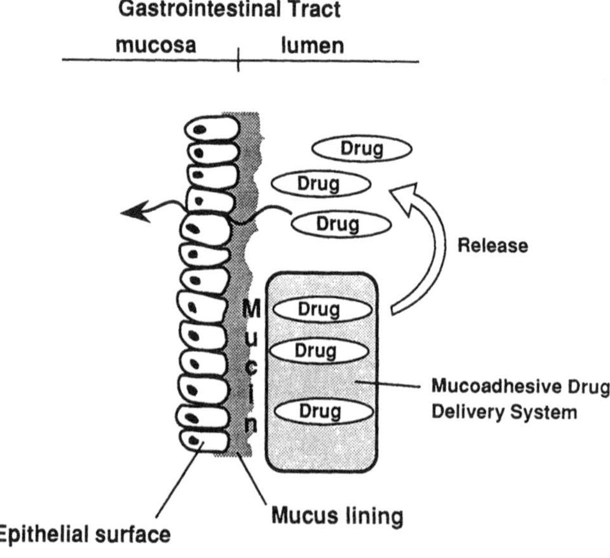

Figure 31 Interaction of a mucoadhesive drug delivery system (ma-DDS) with the mucus layer on the gastrointestinal surface epithelium.

Table 3 Relative Mucoadhesive Performance of Some Potential Bio(Muco)adhesive Pharmaceutical Polymers

Polymers	Relative mucoadhesive force[a]	Qualitative bioadhesion property[b]
Carboxymethylcellulose	193	Excellent
Carbopol	185	Excellent
Polycarbophil	—	Excellent
Tragacanth	154	Excellent
Na alginate	126	Excellent
HPMC	125	Excellent
Gelatin	116	Fair
Pectin	100	Poor
Acacia	98	Poor
Providone	98	Poor

[a]Percentage of a standard, tested in vitro [Smart et al.; (1984). J. Pharm. Sci., 36:295.]
[b]Assessed in vivo [Chen and Cyr; in: *Adhesion in Biological Systems* (R. S. Manly, Ed.), Academic Press, New York, 1970, pp. 161–181.]

polymers. They observed that the bioadhesion A_b of polyacrylate to pig intestine shows a bell shaped dependent on polymer concentration C_{bp}, which can be described by the quadratic equation

$$A_b = a_0 + a_1 C_{bp} + a_2 C_{bp}^{\ 2} \tag{4}$$

where a_0, a_1 and a_2 are constants.

From the data summarized in Figure 32, the maxima of bioadhesion and the minimal polymer concentration needed to attain the maximal bioadhesion are determined and compared in Table 4. The results indicate that the most hydrophilic polyacrylic polymer, carbopol 934, is the most active bioadhesive polymer with the maxima of bioadhesion ($399P_a$) attained at a polymer concentration of only 0.15%. The copolymerization of polyacrylic acid reduces the bioadhesive capacity of polyacrylic polymer, and copolymers thus require a higher polymer concentration to achieve the maxima of bioadhesion. The most hydrophobic polyacrylate ternary copolymer, scopacryl D340, appears to be more bioadhesive than the medium hydrophilic binary copolymer, scopacryl D339. Both are, however, active bioadhesives at a polymer concentration much higher than that of carbopol (Figure 32, top). It was also observed that the addition of pharmaceutical excipients into the bioadhesive polyacrylate polymer tends to reduce the bioadhesion, which is dependent upon the type and concentration of excipients added (Figure 32, bottom).

The systemic bioavailability of drugs taken orally may be limited by the gastrointestinal transit time of the drug delivery system used (21). Especially for drugs absorbed only from the small intestine, their systemic bioavailability is limited by the residence time of the drug in or upstream of the small intestine. Scintigraphic studies have shown that pharmaceutical dosage forms have a gastric emptying time of under 2 hr in the fasted state and up to 4 hr in the fed state; the small intestinal transit time is generally in the range of 2–6 hr irrespective of whether the subject

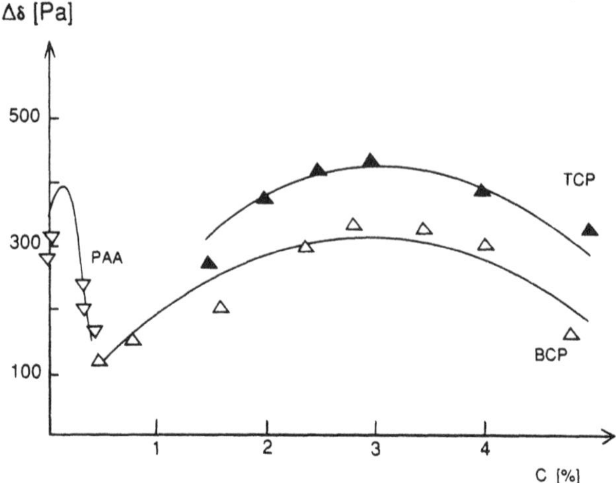

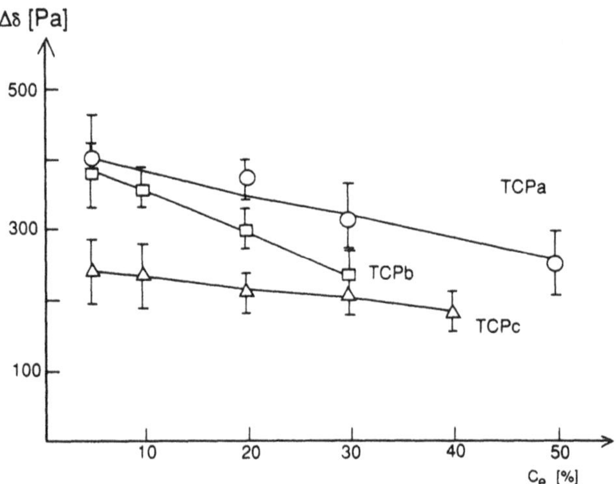

Figure 32 (Top) Relationship between the measured bioadhesion of an aqueous solution of polyacrylic acid (PAA), a binary (BCP), and a ternary (TCP) polyacrylic copolymer and the concentration of polymer (or copolymer). (Adapted from Reference 55.)
(Bottom) Effect of excipients on the measured bioadhesion of an aqueous solution of a ternary polyacrylic copolymer (TCP) and its dependence on excipient concentration. TCP_a = TCP + lactose; TCP_b = TCP + potato starch; and TCP_c = TCP + microcrystalline cellulose.

is fasted or fed (50). For drugs that are absorbed only from the small intestine, a release time of longer than 4–8 hr, as in the case of oral controlled-release dosage forms, is likely to prove ineffective (51).

Recently, Harris et al. (51) investigated the feasibility of using bio(muco)adhesive polymers to extend the GI transit time in rats and humans. The results obtained in rats indicated that among the polymers investigated, polyacrylic polymers, such as carbopol and polycarbophil, are most likely to be of use in delaying gastrointestinal transit; however, the major delay is due to a decrease in the gastric emptying time

Table 4 Bioadhesion of Polyacrylate Polymer and Copolymers to Pig Intestine

Polyacrylate	Neutralizing capacity[a] (ml/g)	Maxima of bioadhesion[b] Removing force (P_a)	Polymer Concentration (%w/v)
Carbopol 934 (homopolymer)	14.1	399	0.15
Scopacryl D339 (binary copolymer)	3.75	301	2.9
Scopacryl D340 (ternary copolymer)	0	421	3.3

[a]Volume of ammonia solution required to neutralize the carboxylic acid groups in 1 g of the polyacrylate solution (or suspension).
[b]The maxima of bioadhesion is expressed as the maximal removing force achieved with the minimal polymer concentration indicated.
Source: From Dittgen and Oestereich (1989).

(Table 5). On the other hand, transit through the upper small intestine is much more rapid than transit through the lower small intestine. Additionally, the concentration and physical state of a bioadhesive polymer tends to affect the GI transit time. However, the results generated in humans were not conclusive (Table 6).

Several approaches have been applied to incorporate drug into bio(muco)adhesive polymers for the preparation of oral drug delivery systems. For water-soluble polymers it is possible to use polymers to coat the surface, totally or partially, of a sheet or macro- or micro-size capsule-shaped drug delivery device. In this case the duration of retention on the mucosal tissues is generally controlled by the dissolution rate of the bio(muco)adhesive polymer. A cross-linked bio(muco)adhesive polymer must first be hydrated to become an effective bio(muco)adhesive, and when it is, it often separates from the rate-controlling drug delivery system or causes a premature

Table 5 Effect of Some Adhesive Polymers on the Gastrointestinal Transit of ^{51}Cr-Labeled Microspheres in Rat

			Emptying half-life (hr)		
				Small intestine	
Polymer	Dosage Form	Concentration (%)	Stomach	Upper	Lower
Carbopol 934	Solution	0.2	1.01	0.83	3.32
	Solution	4.0	3.21	0.24	3.05
	Solution	5.0	3.21	0.24	3.05
	Capsule	5.0	1.49	0.86	1.94
Polycarbophil	Solution	7.5	3.21	0.24	2.23
Hyaluronic acid	Solution	1.0	1.49	0.35	2.45
Hydroxyethyl-cellulose	Solution	1.5	1.49	0.86	2.81
	Capsule	1.5	1.01	0.83	3.32
	Capsule	7.5	1.99	0.36	2.81

Source: From Harris et al. (1989).

Table 6 Effect of Polyacrylate on the Gastrointestinal Transit of ^{99m}Tc-labeled Amberlite Resin in Fasting Male Subjects

Formulation	Amberlite particle size	Emptying half-life (min ± SEM) Stomach	Intestine	Colon arrival time (min ± SEM)
Control				
Lactose	5–50	31 (14)	111 (13)	142 (22)
	750–1000	25 (11)	137 (11)	162 (16)
Polyacrylate				
Polycarbophil	5–50	25 (11)	165 (20)	191 (25)
	750–1000	36 (11)	147 (20)	183 (24)
Carbopol 934	5–50	45 (16)	145 (29)	190 (21)
	750–1000	82 (50)	137 (11)	170 (19)

Source: From Harris et al. (1989).

release of drug, especially water-soluble drugs. A drug that is only sparingly soluble in water can simply be trapped within the gelling polymer for slow release as a result of reduced diffusivity, so the problem is substantially reduced (52). On the other hand, the drug can be directly dispersed in the bio(muco)adhesive polymer to form a bio(muco)adhesive matrix-type drug delivery system, or the matrix-type drug delivery system can be further coated with a layer of bio(muco)adhesive polymer with a similar or different structure (53).

A mucoadhesive gastrointestinal drug delivery system was recently developed (54). This system is a mucoadhesive microsphere (100–500 μm) that consists of a drug-dispersed hydrogel matrix fabricated from polyhydroxy methacrylate (*p*-HEMA) by suspension polymerization and a mucoadhesive coating, a Carbopol 943/Eudragit RL-100 (9:1) blend by an air-suspension process. Using an in situ perfused rat ileal loop model, the intestinal transit of this bioadhesive microcapsule system was investigated. The results, compared in Figure 33, demonstrate that the intestinal transit time of the microspheres was substantially extended by microadhesive coating, with the mean residence time improved by twofold when compared to uncoated microspheres (48.4 ± 5.8 versus 23.6 ± 7.3 min). The adhesive interaction of the microsphere with the intestinal mucosa was further studied using an in vitro vertically perfused intestinal loop model, and the results showed that the mucoadhesive polymer-coated microspheres clearly adhered to the mucosal tissue, whereas the noncoated microspheres circulated freely in the solution. However, the mucoadhesive effect was observed to last for about 2 hr, even though the coating remained intact for more than 6 hr during perfusion. The incorporation of Eudragit RL-100, a water-insoluble polyacrylic copolymer, by up to 10% increased the mechanical stability of the coating without adversely affecting the mucoadhesive strength of Carbopol 934. In aqueous media the mucoadhesive coating swells immediately up to a thickness of about 50 μm.

A similar concept was utilized in the development of bioadhesive polycarbophil-coated albumin beads for affecting the gastrointestinal transit time (55). The results, compared in Figure 34, indicate that the gastric retention time was substantially pro-

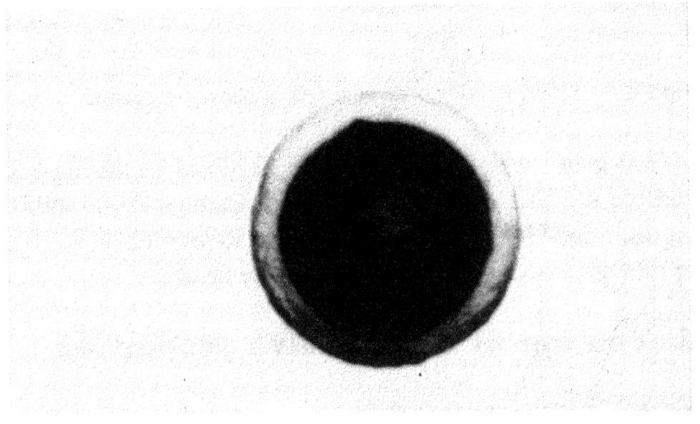

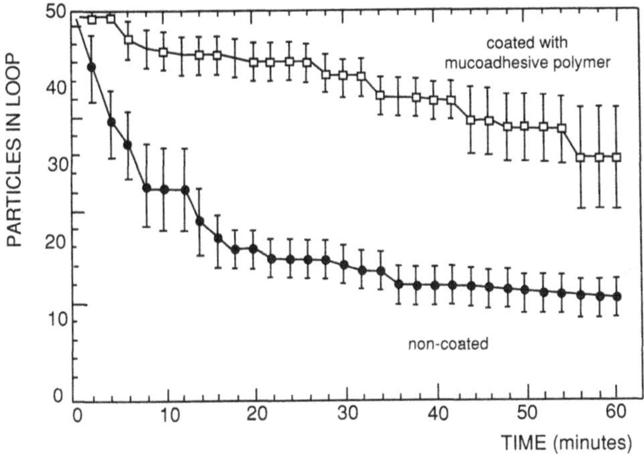

Figure 33 (Top) Micrograph of a mucoadhesive-coated microsphere in the swollen state. The microsphere was prepared from p-HEMA hydrogel and coated with mucoadhesive polymer (90% Carbopol 934 + 10% Eudragit RL 100).
(Bottom) Residence profiles of mucoadhesive-coated microspheres (n = 6) and noncoated microspheres (n = 7) in an in situ perfused rat intestinal loop (Adapted from Reference 58.)

longed by the bioadhesive coating and that the systemic bioavailability of drug was significantly improved with a greater enhancement in the plasma drug concentration.

7. *Coadministration with GI Motility-Reducing Drugs*

A number of antimuscarinic drugs are known to reduce gastrointestinal motility and gastric secretion and induce the drying of mucous membranes (56). Several reports have demonstrated that coadministration with the antimuscarinic drugs, such as propantheline, could slow gastric emptying and intestinal transit. The enhanced oral bioavailability of several drugs, such as hydrochlorothiazide, was achieved (57,58).

IV. OVERCOMING HEPATIC FIRST-PASS ELIMINATION

A gastrointestinal transit-physiological dynamic model is presented in Figure 35 as a physical model for the GI segment of the digestive system (Figure 21). It outlines

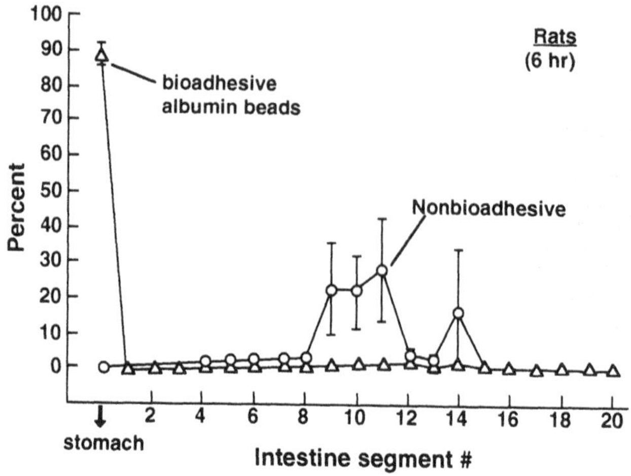

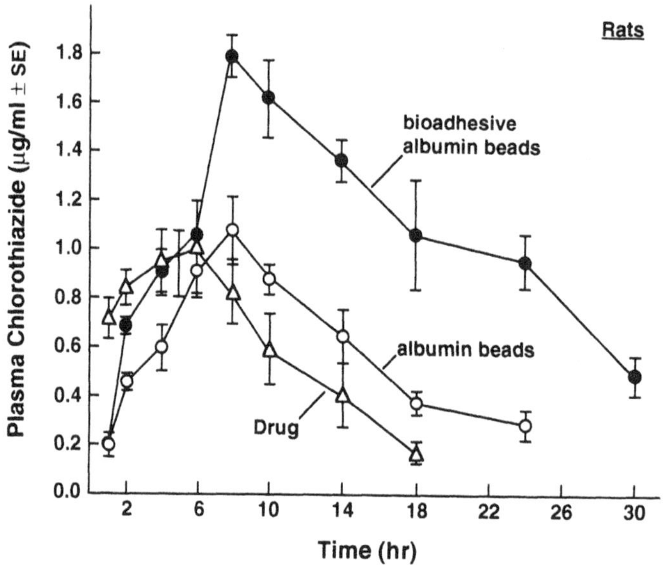

Figure 34 (Top) Distribution of albumin beads in the rat gastrointestinal tract at 6 hr postadministration: (△) bioadhesive polycarbophil-coated albumin beads; (○) noncoated albumin beads.
(Bottom) Plasma profiles of chlorothiazide in rats following oral administration of (△) drug in powder; (○) drug in albumin beads; (●) drug in bioadhesive polycarbophil-coated albumin beads.

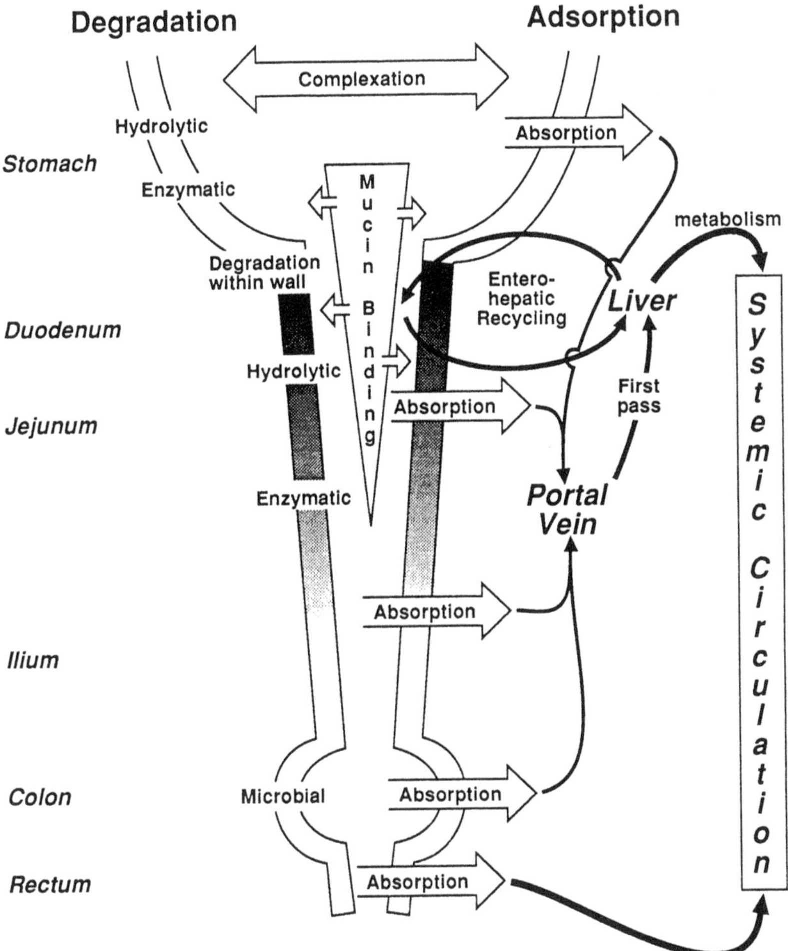

Figure 35 Physical model used to illustrate various physiologic processes encountered by an orally administered drug during the course of GI transit. (Modified from Robinson, 1985.)

various GI physiological processes that a drug molecule encounters during its GI transit, which include complexation, binding to mucin in the mucous lining, and hydrolytic, enzymatic, and/or microbial degradation, as well as absorption.

Following absorption through the mucosal membrane in the various segments of the GI tract, with the exception of the rectum (lower and middle surfaces), all the drug molecules absorbed are pooled in the portal circulatory system and liver, where the drug molecules are subjected to a hepatic first-pass elimination process before they are transported to the heart via the hepatic vein and inferior vena cava for systemic circulation throughout the whole body (Figure 36).

If a drug is subjected to extensive hepatic first-pass elimination, oral administration of the drug delivered by a controlled- or sustained-release process produces no advantage over administration via a conventional immediate-release dosage form. It should be pointed out that by controlled-release or sustained-release drug admin-

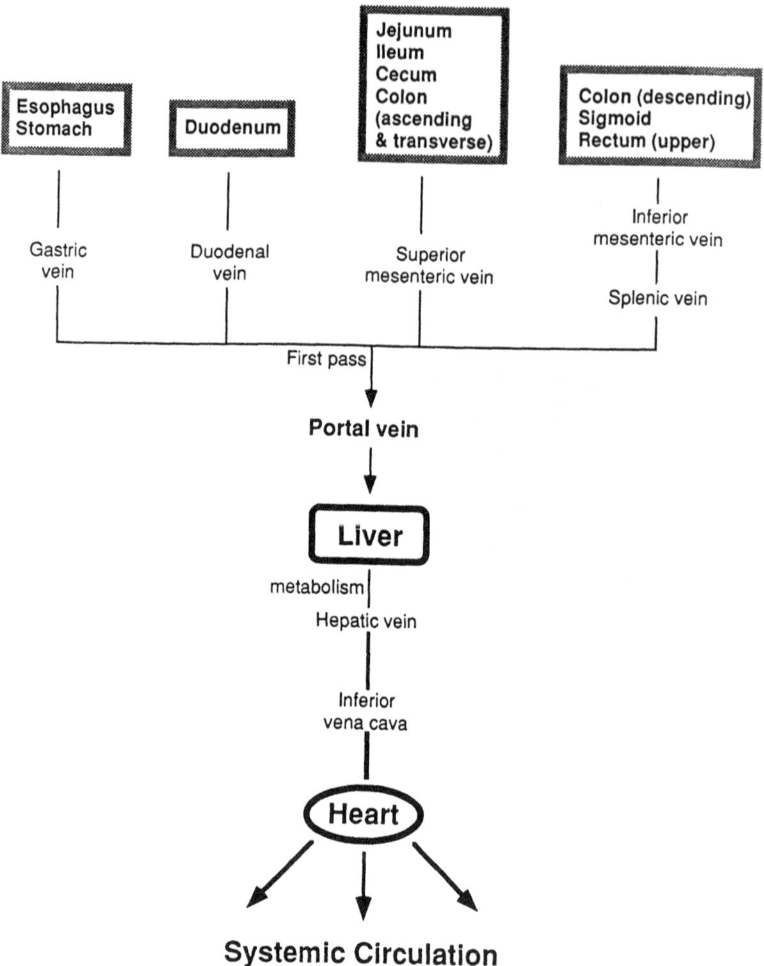

Figure 36 Various circulatory pathways that a drug molecule takes following its absorption from the various segments of gastrointestinal tract.

istration drug molecules may be subjected to a greater extent of hepatic first-pass elimination (59).

A typical example is the metabolism of acetaminophen by the microsomal enzymes in the liver into glucuronide (major metabolite) and sulfate (minor) for elimination through the urinary excretion (Figure 37). This hepatic first-pass metabolism accounts for 90–100% of the elimination of an oral dose of acetaminophen, which means that less than 10% of the acetaminophen remains intact and pharmacologically active (60).

A survey of literature has suggested that very few peptide-based drugs are orally active and numerous organic-based drugs have been reported to have low oral bioavailability as a result of low GI absorption (Table 7) or extensive metabolism (Table 8). Peptide-based pharmaceuticals are mostly difficult to maintain in a bioactive form when taken orally, but the organic-based pharmaceuticals can be made orally active

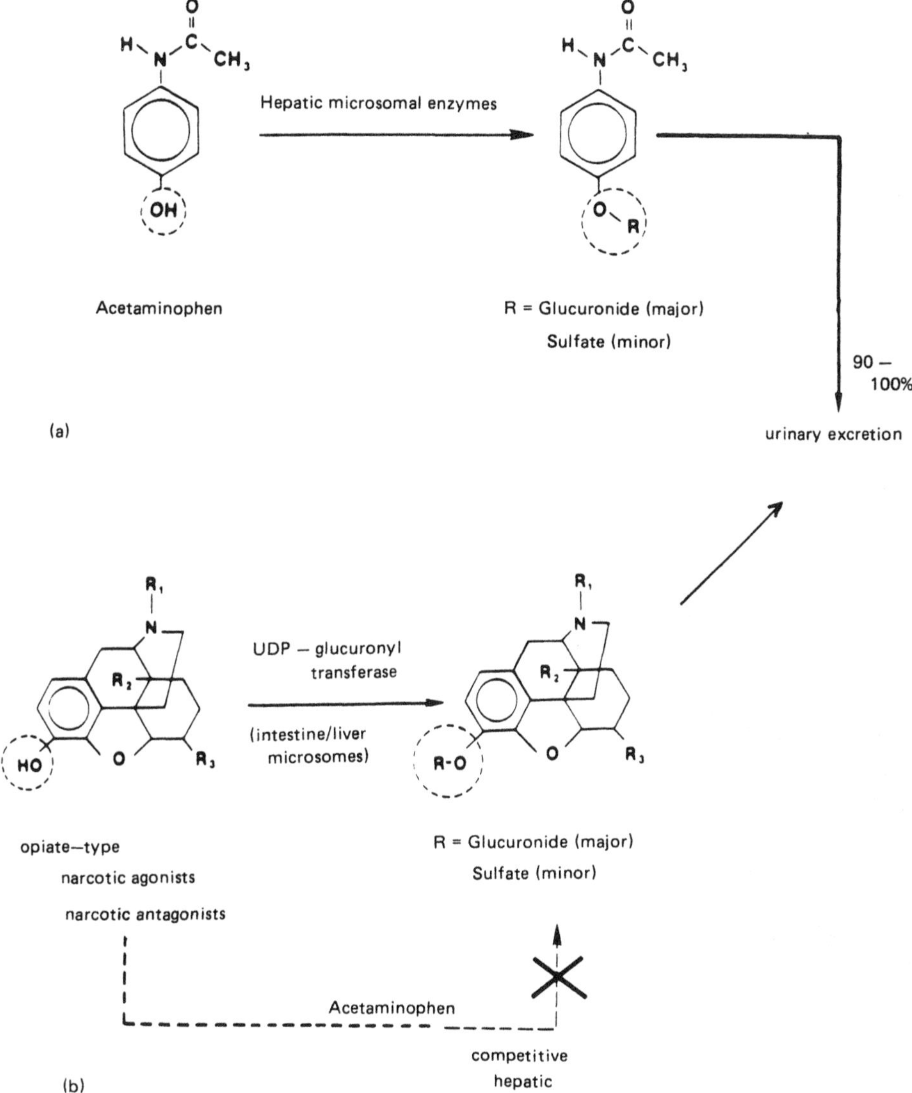

Figure 37 (a) Hepatic first-pass metabolism of drugs, such as the phenolic hydroxy group in the acetaminophen molecule, which under the action of microsomal enzymes, conjugates with glucuronic acid to form glucoronide and is excreted through the urine.
(b) Coadministration with the drug having the phenolic hydroxy group, such as the opiate narcotic agonists or antagonists; acetaminophen is preferentially metabolized because of its lower oxidation-reduction potential and thus protects narcotic drugs from hepatic metabolism.

Table 7 Organic-based Pharmaceuticals Known to Have Low Oral Absorption[a]

Drugs	Fraction of dose absorbed
Acebutolol	0.4
Ampicillin	0.5
Hydralazine	
Fast acetylator	0.22
Slow acetylator	0.38
Kanamycin	0.0
Meperidine	0.5
Mezlocillin	0.0
Morphine	0.4
Nadolol	0.4
Penicillin G	0.3
Penicillin V	0.4
Propoxyphene	0.2
Propranolol	0.35
Streptomycin	0.0
Verapamil	0.15

[a]Drugs with oral absorption less than 0.5.

by administering an excess dose to compensate for the drug loss due to hepato-gastrointestinal first-pass metabolism, by parenteral administration, or by delivery through other nonparenteral routes (e.g., nitroglycerin by transdermal delivery and progesterone by intrauterine administration). The following approaches have been found useful in improving the oral bioavailability of organic-based pharmaceuticals.

A. Biochemical Approaches

The hepatic first-pass metabolism of acetaminophen discussed earlier can be utilized productively to protect the metabolism of some drugs having a liable phenolic OH group (61). An example is the protection of opiate narcotic drugs from the metabolism by the UDP-glucuronyltransferase in the intestinal and hepatic microsomes using the mechanism of competitive metabolism (Figure 37). Because of the preferential metabolism of acetaminophen, the metabolic elimination of narcotic drugs is significantly reduced, thereby greatly enhancing their oral bioavailability and pharmacological activity.

B. Chemical Approaches

On the other hand, the hepatic first-pass elimination of narcotic drugs can also be minimized via the formation of prodrug aspirin derivatives by chemical synthesis. Following gastrointestinal absorption the prodrug narcotic aspirinate is converted via hepatic first-pass elimination (Figure 38) back to narcotic drugs and aspirin to achieve a synergistic analgesic activity.

C. Physiological Approaches

As discussed in Chapter 4, delivery of drugs via the mucosal membranes, like the oral and rectal mucosae, has the advantage of bypassing the hepato-gastrointestinal first-pass elimination associated with oral administration.

Table 8 Organic-based Pharmaceuticals Known to Be Eliminated Extensively in Metabolized Forms

Drugs[a]	Fraction of dose metabolized[b] (%)
Amitriptyline	99.5
Carbamazepine	99
Chlordiazepoxide	~100
Desipramine	90
Dicumarol	99
Digitoxin	92
Glutethimide	99.8
Hexobarbital	~100
Hydralazine (with fast acetylator)	98
Imipramine	~100
Lorazepam	99
Meperidine	90
Meprobamate	90
Mexiletine	90
Morphine	90
Nitrazepam	99
Oxazepam	98
Pentobarbital	99
Phenylbutazone	~100
Phenytoin	95
Primidone	90
Probenecid	95
Propoxyphene	98.5
Propranolol	99
Theophylline	92
Timolol	90

[a]Drugs with extent of metabolism greater than 90%.
[b]In urinary excretion.

1. *Oral Mucosal Drug Delivery*

It has been known for centuries that after buccal and sublingual administration drug solutes are rapidly absorbed into the reticulated vein, which lies underneath the oral mucosa, and transported through the facial veins, internal jugular vein, and braciocephalic vein and are then drained into the systemic circulation. Therefore, the buccal and sublingual routes of administration can be utilized to bypass the hepatic first-pass elimination of drugs.

Oral mucosal delivery has been practiced for many years as evidenced by the development of such pharmaceutical dosage forms as sublingual tablets and lozenges. These oral products have been available for several decades and are commonly used for delivering organic-based pharmaceuticals to the oral mucosa for either local or systemic medication.

The human oral mucosa is a lining tissue that serves to protect the underlying structures. It consists of two parts: the epithelium and the underlying connective tissues. The oral epithelium shows several distinct patterns of maturation that may

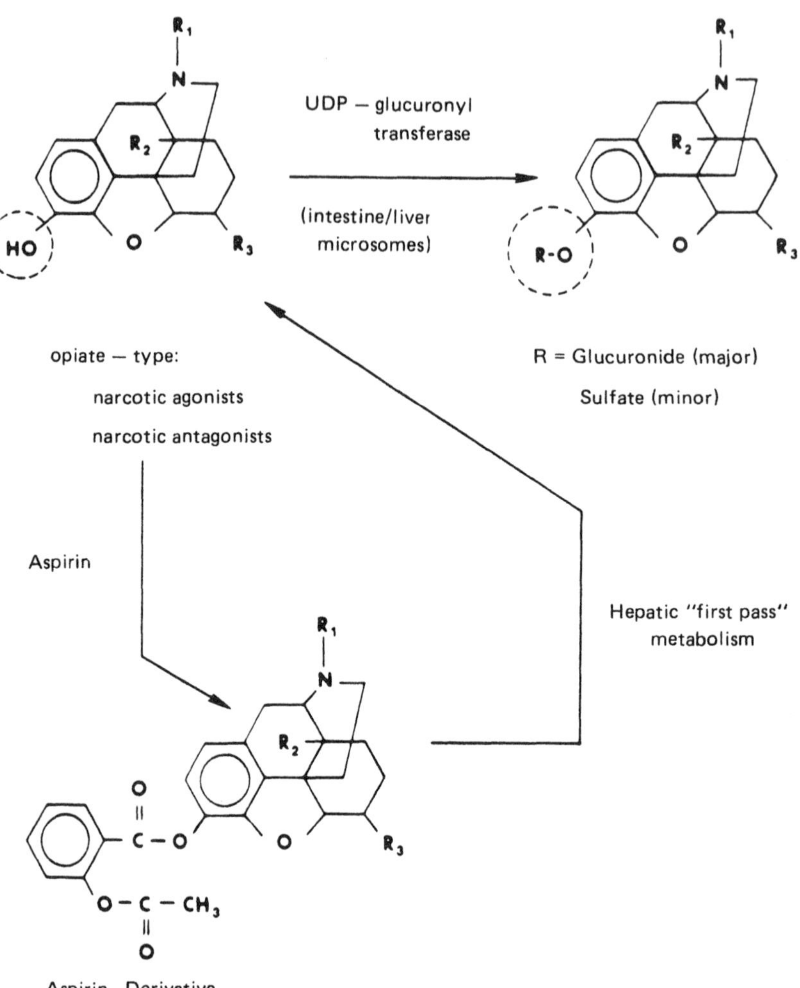

Figure 38 The hepatic first-pass metabolism of opiate narcotic agonists or antagonists can be reduced by protecting the enzyme-labile phenolic hydroxy group by forming a prodrug aspirin derivative that regenerates the parent drugs during transit through the liver in the hepatic first-pass process.

be related to the different functions of the mucosa in the various regions of the oral cavity. The thickness of the epithelium and its extent of keratinization vary from one region of oral mucosa to another, and in some regions keratinized epithelium may coexist with, or be replaced by, parakeratinized epithelium.

Sublingual absorption is mostly rapid in action, but also short-acting in duration. Nitroglycerin, for example, is an effective antianginal drug but is extensively metabolized when taking orally (>90%). It is rapidly absorbed through the sublingual mucosa, and its peak plasma level is reached within 1–2 min. Because of its short biological half-life (3–5 min), however, the blood concentration of nitroglyerin declines rapidly to the level below the therapeutic concentration within 10–15 min (Figure 39). Hence a short duration of hemodynamic activities results. Recently a

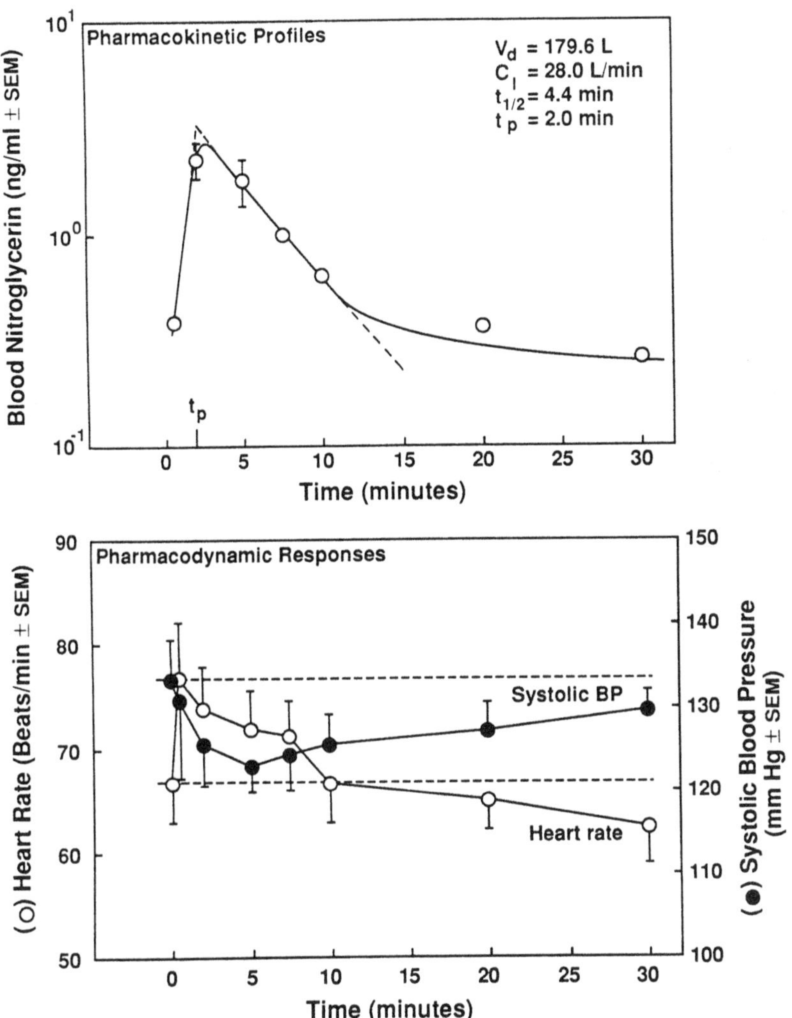

Figure 39 Blood concentration profiles of nitroglycerin following sublingual administration and the time course for its hemodynamic responses. (Plotted from data by Armstrong et al., 1979.)

nitroglycerin-delivering lingual aerosol formulation (nitroglycerin in propellants) in a metered-dose spraying pump, Nitrolingual Spray, was developed. It delivers nitroglycerin by spraying onto or under the tongue in the form of spray droplets. A pharmacokinetic study conducted in 13 healthy men showed that the peak plasma concentration occurs within 4 min and the apparent plasma half-life is approximately 5 min, which is not statistically different from that attained by sublingual nitroglycerin tablets at equal doses (0.8 mg). In a randomized, double-blind study in patients with exertional angina pectoris, a dose-related increase in exercise tolerance was observed following doses of 0.2–0.8 mg nitroglycerin delivered by metered spray.

On the other hand, the buccal route has all the advantages of the sublingual route. In addition, the buccal mucosa permits a prolonged retention of a dosage form. Several oral mucosa bioadhesive drug delivery systems have been developed to take advantage of the unique properties of buccal mucosa to sustain the duration of transmucosal drug delivery.

The oral mucosa bioadhesive drug delivery system is designed to be applied to an oral mucosa site. It consists of a polymeric adhesive composition that, in contact with saliva, becomes adhesive and renders the system attached to the oral mucosa. The system can be designed to stay in a fixed position on the oral mucosa for a duration of up to 12 hr, during which the drug solutes are continuously released into the oral cavity for transmucosal absorption into the systemic circulation.

Several polymeric adhesive compositions have been developed: (i) hydroxypropylcellulose (≥80%) and ethycellulose (≤20%) combinations (United Kingdom (Great Britian) patent 1,279,214; Zambia patent 7,805,528); (ii) hydroxypropylcellulose (0.02–2%) and polyacrylic acid-sodium salt (0.2%) combinations (Japanese Patent 5-4,041,320); (iii) sodium polyacrylate (10–60%) compositions (J. Patent 7-9,038,168); and (iv) alkyl acrylate (<30%) and acrylamide-vinyl pyrrolidone (>70%) combinations (G.B. patent 2,021,610) (2). A typical example of the application of these polymeric adhesive compositions in the formulation of saliva-activated oral mucosa bioadhesive drug delivery systems is demonstrated by the development of the Susadrin transmucosal tablet.

The polymeric adhesive composition of the Susadrin transmucosal tablet is prepared by blending hydroxypropylcellulose and ethylcellulose together in a dry state and humidifying the blend to increase the moisture content to ≥85% and then dehumidifying it to a moisture content of less than 10% to form the saliva-activated polymeric adhesive composition. The resulting composition has been named the Synchron controlled-release base (62).

External gamma scintigraphy, which was demonstrated useful in assessing the in vivo performance of hydrodynamically balanced intragastric floating drug delivery systems (Section IIIB1, Figures 23 and 24), was also utilized to investigate the in vivo release characteristics of a transmucosal tablet prepared from the Synchron controlled-release base in healthy male volunteers. Each subject acted as his own control and was given a Synchron transmucosal tablet or a sublingual immediate-release tablet. Both the transmucosal and sublingual tablets were prepared to contain a gamma-emitting radionuclide technetium 99m (^{99m}Tc). The results of scintigraphic measurement indicated that the release profile of ^{99m}Tc from the Synchron transmucosal tablet applied to the buccal mucosa is very much linear and the duration of radioactivity remaining in the tablet is substantially prolonged over that from the sublingual tablet (Figure 40). Apparently, the Synchron transmucosal tablet can be utilized to achieve the controlled release of drugs for prolonged transmucosal delivery.

Based on the informatioin gained from these in vivo studies, a nitroglycerin-releasing transmucosal tablet, called the Susadrin buccal tablet, was developed from the Synchron controlled-release base (62). It is quite small in size and is designed to be adhered to the buccal mucosa between the lip and gum above the upper incisors or between the cheek and gum for a duration of up to 6 hr. The clinical performance of Susadrin buccal tablets was evaluated in angina patients using a double-blind, crossover exercise tolerance protocol (63,64). The results indicate that the exercise tolerance time is improved by the Susadrin buccal tablet in more than 50% of the

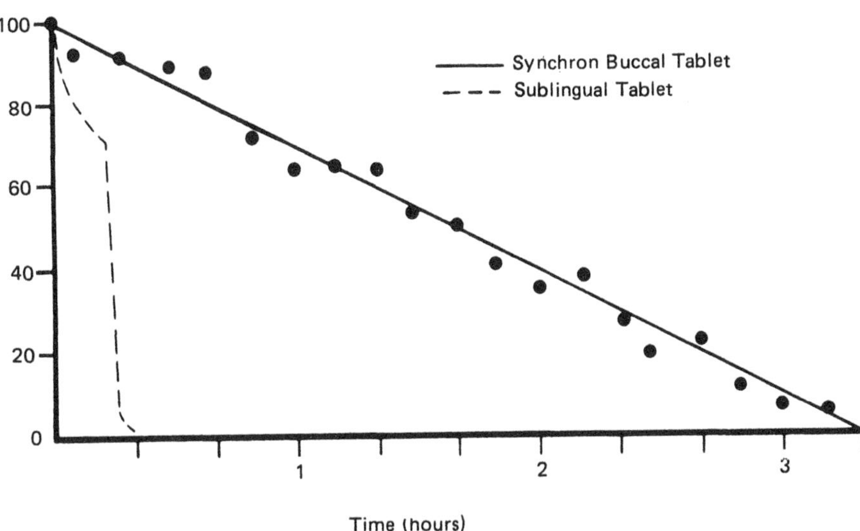

Figure 40 Comparison of the release profiles of ^{99m}Tc, measured as the percentage of radioactivity remaining in the tablet, from a Synchron controlled-release buccal tablet and a sublingual immediate-release tablet. (Data from Schor et al., 1983.)

patients for a duration of over 5–6 hr. The improvement in the clinical performance of exercise tolerance following Susadrin administration was found to be comparable to the peak antianginal effect of nitroglycerin by sublingual administration. This maximal antianginal activity was maintained over a full 5 hr period (Figure 41), which is substantially longer than the 5 min attained by the sublingual tablet. Further clinical studies in anginal patients by chronic dosing have confirmed the results obtained from the exercise tolerance studies and also demonstrated that approximately 50% of the patients become totally angina free when they take Susadrin buccal tablets on a three-times-a-day dosing schedule; the remainder were substantially improved (62). A buccal delivery system, Nitrogard, has recently been approved by the regulatory agency for marketing in the United States.

A new generation of saliva-activated oral mucosa bioadhesive drug delivery system was recently developed (65), which consists of one fast-release layer and one sustained-release layer. In contact with saliva, only the surface of the sustained-release layer becomes adhesive and adheres to the oral mucosa in the gingival area (Figure 42). The advantages of transmucosal drug delivery by this bioadhesive drug delivery system were investigated using nifedipin, an orally metabolizable antihypertensive drug. The results in beagle dogs indicated that some improvements have been made in the pharmacokinetic profiles (Figure 43) and pharmacodynamic responses (Figure 44) of nifedipin over the oral sustained-release tablet (65).

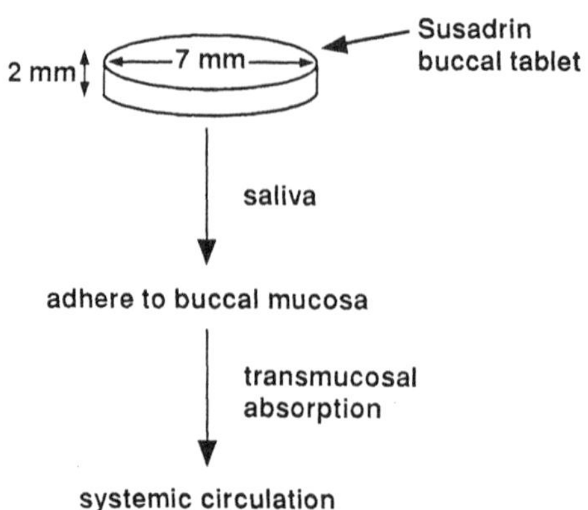

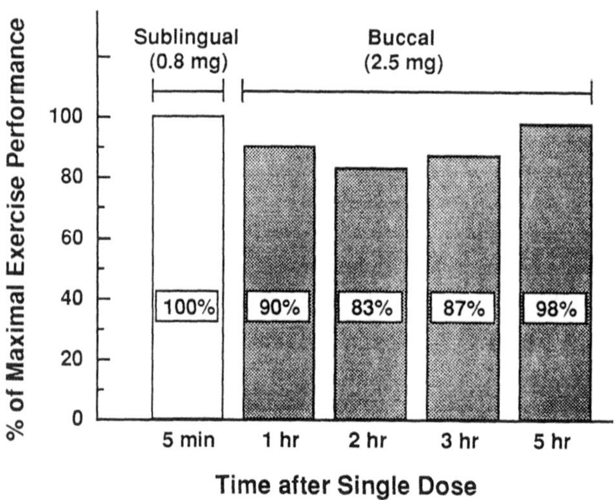

Figure 41 Process involved in the transmucosal systemic delivery of nitroglycerin via a Susadrin buccal tablet and the clinical performance (reported as a percentage of the maximal exercise performance achieved by sublingual nitroglycerin measured at 5 min) over a 5 hr period. (Data from Schor et al., 1983.)

2. *Rectal Mucosal Drug Delivery*

The use of the rectum for the systemic delivery of drugs is a relatively recent idea, even though the administration of drugs in a rectal suppository dosage form for local medication is a very old practice. In contrast to the gastrointestinal route of administration, the rectal route may provide the advantage of bypassing the hepato-gastrointestinal first-pass elimination of orally metabolizable drugs and reducing the proteolytic degradation of peptide-based pharmaceuticals, thus achieving greater systemic bioavailability.

The human rectal mucosa is composed of the epithelium, the lamina propria, and the double-layered muscularis mucosae. The epithelial surface consists of closely packed columnar cells, with some areas interrupted by crypt regions. Within the crypt regions there are mucus-producing goblet cells. Throughout the rectal mucosa

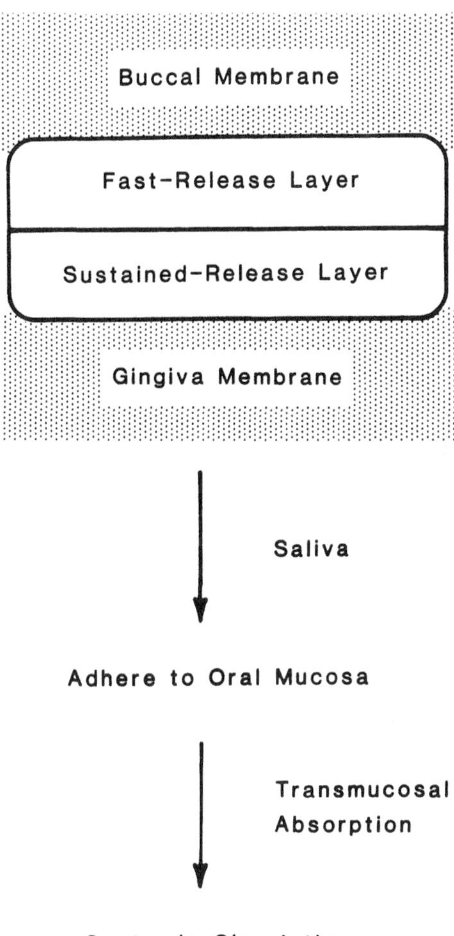

Figure 42 The transmucosal delivery of drugs from a Toyobo saliva-activated oral mucosa bioadhesive drug delivery system.

there are lymphoid nodules covered by columnar surfce epithelium. The lamina propria consists of two layers: a dense acellular collagen layer and a loose connective tissue layer. Within the lamina propria there exist superficial blood vessels and inflammatory cells, including macrophages, eosinophils, and lymphocytes. The muscularis mucosae contains smooth muscle cells and larger blood vessels (66–69).

In the rectum (70) the lower venous drainage system (inferior and middle hemorrhoidal veins) is connected directly to the systemic circulation by the iliac veins and the vena cava (Figure 45), even though the upper venous drainage system (superior hemorrhoidal vein) is connected to the portal system as in the other regions of GI tract (Figure 36). Thus there exists an opportunity to reduce the extent of hepatic first-pass elimination by rectal delivery, especially when the drug is administered in the low region of the rectum (71,72). Additionally, the rectum has a large number of lymphatic vessels that could offer an opportunity to target drug delivery to the lymphatic circulation through intrarectal administration (71).

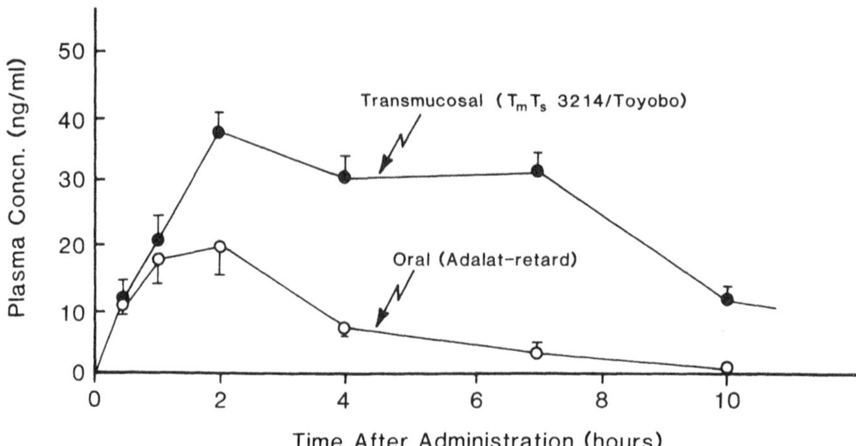

Figure 43 Comparative plasma profiles of nifedipin delivered either transmucosally by a Toyobo oral mucosa bioadhesive drug delivery system or orally by a sustained-release oral tablet (Adalat-retard) in beagle dogs (n = 8). (Courtesy of Yukimatsu, 1989.)

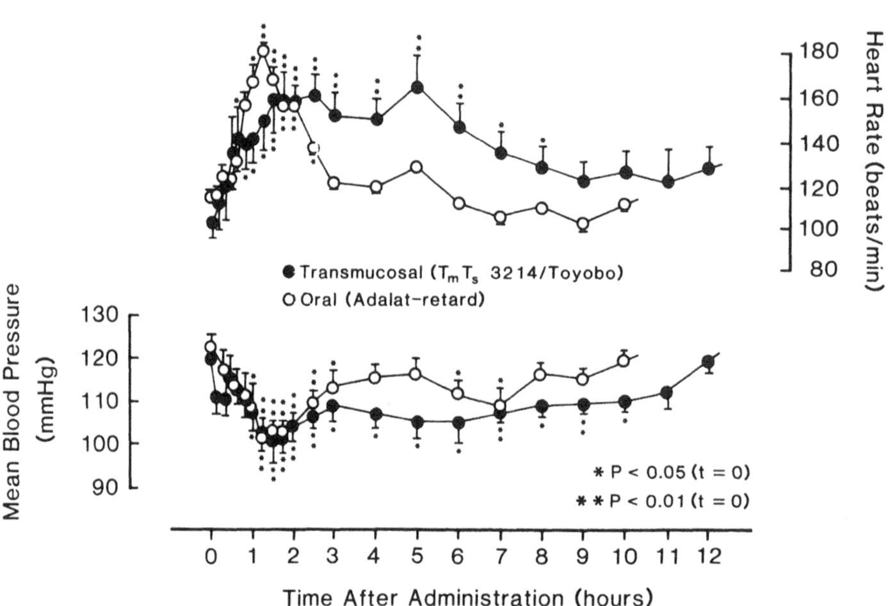

Figure 44 Comparative pharmacodynamic responses, measured as the changes in blood pressure and heart rate, in conscious dogs (n = 6) to nifedipin delivered either transmucosally by a Toyobo oral mucosa bioadhesive drug delivery system or orally by a sustained-release oral tablet (Adalat-retard). (Courtesy of Yukimatsu, 1989.)

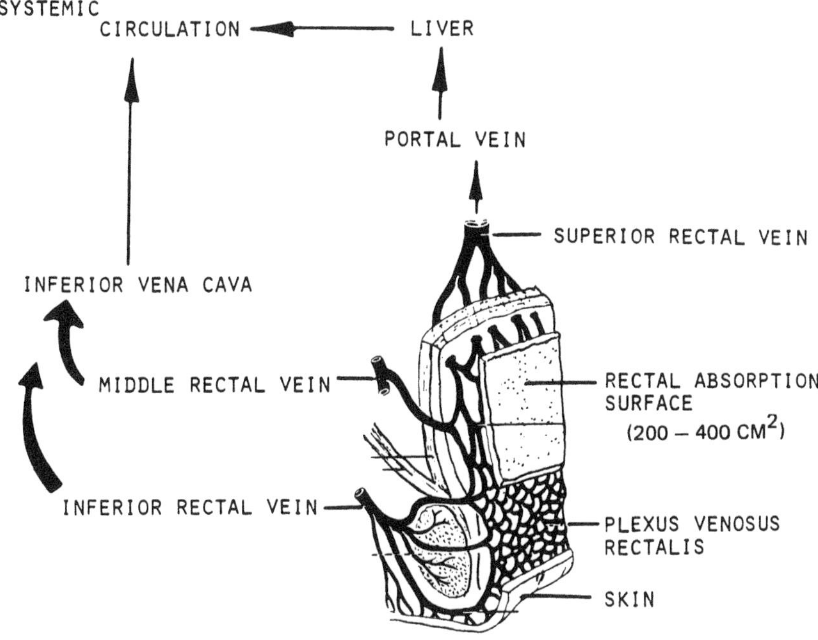

Figure 45 The Venous drainage network in human rectum. (Chien, 1983.)

It was reported that when drug is delivered to the lower region of the rectum it is absorbed into the inferior and middle rectal veins, then passes into the inferior vena cava, and finally drains into the systemic circulation, thereby bypassing the portal vein and the hepatic first-pass metabolism (73). On the other hand, if the drug is delivered to the upper region of the rectum, it is probably absorbed and then transported via the superior rectal veins into the portal vein and subjected to hepatic first-pass elimination before entering the systemic circulation. However, this anatomic situation is complicated by anastomoses among the rectal veins underneath the rectal mucosa (absorption surface; Figure 45).

The effect of rectal administration on the hepatic first-pass metabolism of drugs was demonstrated by studying the systemic bioavailability of a high-clearance drug, e.g., lidocaine (74,75). Studies conducted in six healthy volunteers indicated that the rectal absorption of lidocaine in aqueous solution results in a systemic bioavailability as high as 69%, which compares favorably to the 30.5% by oral administration. It is estimated that the rectal administration of lidocaine has resulted in 55% of the dose bypassing the hepatic first-pass metabolism (75).

The effect of sites of rectal infusion on the systemic bioavailability of lidocaine was also investigated (76). Results suggested that an infusion site located at only 2 cm from the anus produces a greater systemic bioavailability of lidocaine than the infusion site located 4 cm away. This implies that the rectal absorption via inferior rectal veins has a greater chance to bypass the hepatic first-pass metabolism (77).

Using a specially-designed rectal osmotic system (Section IIA), the blood level of theophylline was significantly prolonged by applying 1 unit every 36 hr (Figure 46). A fairly constant blood level was also maintained. On the other hand, the oral

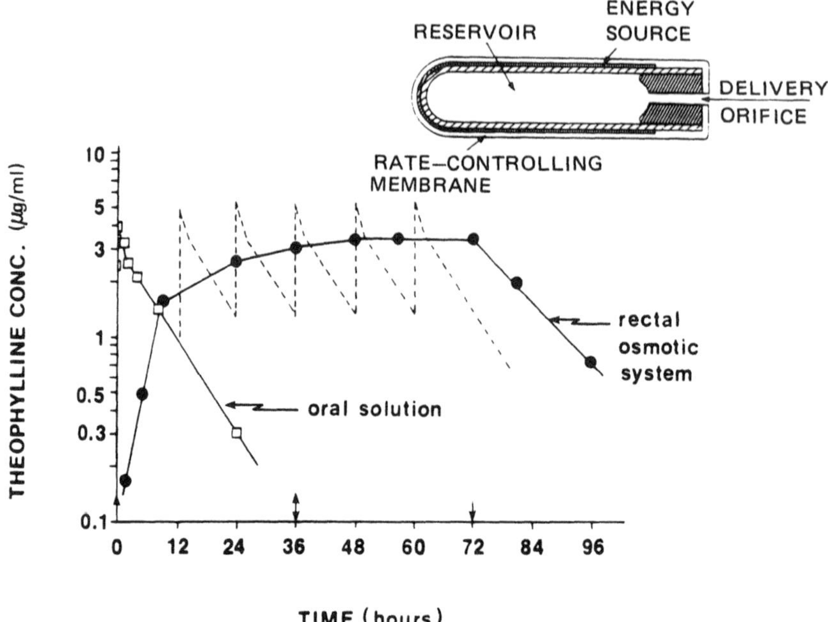

Figure 46 Comparative plasma profile in humans of theophylline delivered either rectally by a rectal osmotic pump or orally by a solution formulation. (Replotted from the data by the Leede et al., 1981.)

administration of theophylline in solution formulation, one dose every 12 hr, produced a fluctuating blood level of theophylline (78).

REFERENCES

1. Y. W. Chien; Rate-control drug delivery systems: Controlled release vs. sustained release. Med. Prog. Technol., 15:21–46 (1989).
2. Y. W. Chien; Potential developments and new approaches in oral controlled-release drug delivery systems. Drug dev. Ind. Pharm., 9:1291–1330 (1983).
3. F. Theeuwes; Osmotic system development. Drug Dev. Ind. Pharm., 9:1331–1357 (1983).
4. Y. W. Chien; Drug delivery systems of tomorrow. Drugs of Today (Spain), 23:31 (1987).
5. R. L. Jerzewski and Y. W. Chien; Osmotic drug delivery, in: *Treatise on Controlled Drug Delivery* (A. Kydonieus, Ed.), Marcel Dekker, Inc., New York, 1991, Chapter 5.
6. J. D. Rogers, R. B. Lee, P. R. Souder, R. O. Davies, K. C. Kwan, F. Theeuwes and R. K. Ferguson; Pharmacokinetic evaluation of osmotically-controlled indomethacin delivery systems in man after single doses. Int. J. Pharm., 16:191 (1983).
7. J. C. Liu, M. Farber and Y. W. Chien; Comparative release of phenylpropanolamine HCl from long-acting appetite suppressant products: Acutrim vs. Dexatrim. Drug Dev. Ind. Pharm., 10:1639–1661 (1984).
8. A. Zaffaroni, A. S. Michaels and F. Theeuwes; Method for administering drugs to the gastrointestinal tract. U.S. Patent 4,096,238, June 20, 1978.
9. F. Theeuwes and A. D. Ayer; Osmotic system with laminated wall formed of different materials. U.S. Patent 4,058,122, Nov. 15, 1977.

10. D. R. Swanson, B. L. Barclay, P. S. L. Wong and F. Theeuwes; Nifidipine gastrointestinal therapeutic system. Am. J. Med., 83(S6B):3–9 (1987).
11. M. Chung, D. P. Reitberg, M. Gaffney and W. Singleton: Clnical pharmacokinetics of nifedipine gastrointestinal therapeutic system. Am. J. Med., 83(S6B):10–14 (1987).
12. F. Theeuwes, D. R. Swanson, G. Guittard, A. Ayer and S. Khanna; Osmotic delivery for the β-adrenoceptor antagonists metoprolol and oxprenolol: design and evaluation of systems for once-daily administration. Br. J. Clin. Pharmacol., 19:69S–76S (1985).
13. F. W. Fara, R. E. Myrback and D. R. Swanson; Evaluation of oxprenolol and metoprolol Oros systems in the dog: Comparions of in vivo and in vivo drug release, and of drug absorption from duodenal and colonic infusion sites. Br. J. Clin. Pharmacol., 19:91S–95S (1985).
14. F. Theeuwes; Osmotic device for dispensing two different medications. U.S. Patent 4,455,143, March, 1982.
15. A. S. Micahaels; Device for delivering drug to biological environment. U.S. Patent 4,180,073, Dec. 25, 1979.
16. A. H. Goldberg and M. L. Franklin; Unit dosage forms. U.S. Patent 4,180,558, Dec. 25, 1979.
17. A. Koff; Castor wax-amprotropine-resin composition. U.S. Patent 3,138,525, June 16, 1961.
18. T. J. Macek, C. E. Shoop and D. R. Stauffer; Palatable coated particles of an ion exchange resin. U.S. Patent 3,499,960, March 10, 1970.
19. S. Borodkin and P. Sundberg; Chewable tablets including coated particles of pseudoephedrine-weak cation exchange resin. U.S. Patent 3,594,470, July 20, 1971.
20. J. M. Hinton, J. E. Lennard-Jones and A. C. Young; Alimentary canal transit time for an inert object in humans. Gut, 10:842 (1969).
21. A. F. Hofmann, J. H. Pressman, C. F. Code and K. F. Witztum; Controlled entry of orally-administered drugs: Physiological considerations. Drug Dev. Ind. Pharm., 9(7):1077 (1983).
22. N. F. H. Ho, H. P. Merkle and W. I. Higuchi; Quantitative mechanistic and physiologically-realistic approaches to the biopharmaceutical design of oral drug delivery systems. Drug Dev. Ind. Pharm., 9(7):111 (1983).
23. E. M. M. Quigly, S. F. Phillips and J. Dent; Distinctive patterns of interdigestive motility at the canine ilio-colonic junction. Gastroenterology, 87:836 (1984).
24. J. E. Kellow, T. J. Borody, S. F. Phillips, R. L. Tucker and A. C. Hadda; Human interdigestive motility: Variations in patterns from esophagus to colon. Gastroenterology, 91:386 (1986).
25. S. McRae, K. Younger, D. G. Thompson and D. L. Wingate; Sustained mental stress alters human jejunal motor activity. Gut. 23:404 (1982).
26. S. J. Konturek, P. J. Thor, J. Bilski, W. Bielanski and J. Laskiewicz; relationships between duodenal motility and pancreatic secretions in fasted and fed dogs. Am. J. Physiol., 250:G570 (1986).
27. P. K. Gupta and J. R. Robinson; Gastric emptying of liquids in the fasted dog. Int. J. Pharm., 43:45 (1988).
28. I. DeWever, C. Eeckhout, G. Vantrappen and J. Hellemans; disruptive effect of test meals on interdigestive motor complex in dogs. Am. J. Physiol., 235:E661 (1978).
29. R. A. Hinder and K. A. Kelly; Canine gastric emptying of solids and liquids. Am. J. Physiol., 233:E335 (1977).
30. P. Gruber, A. Rubinstein, V. H. L. Li, P. Bass and J. R. Robinson; Gastric emptying of non-digestive solids in the fasted dog. J. Pharm. Sci., 76:117 (1986).
31. G. A. Digenis; Gamma scintigraphy in development of controlled release oral delivery systems. Proc. Int. Symp. Bio. Mat., 13:115 (1986).
32. P. K. Gupta and J. R. Robinson; Oral controlled-release delivery. In *Textbook on Con-*

trolled Release Technologies (A. Kydoneius et al., Eds.), Marcel Dekker, New York, 1991, Chapter VI.
33. R. Groning and G. Heun; Dosage forms with controlled gastrointestinal passage-studies on the absorption of nitrofurantoin. Int. J. Pharm., 56:111–116 (1989).
34. R. R. Scheline; Toxicological implications of drug metabolism by intestinal bacteria. Eur. Soc. Study Drug Toxicol. Proc., 13:25 (1972).
35. R. R. Scheline; Drug metabolism by intestinal microorganisms. J. Pharm. Sci., 57:2021 (1968).
36. E. Harboe, C. Larsen, M. Johansen and H. P. Olesen; Macromolecular prodrugs. XV. Colon-targeted delivery—bioavailability of naproxen from orally administered dextran-naproxen ester prodrugs varying in molecular size in the pig. Pharm. Res., 6(11):919–923 (1989).
37. C. M. Lehr, J. A. Bouwstra, J. J. Tukker and H. E. Junginger; Design and testing of a bioadhesive drug delivery system for oral application. S. T. P. Pharmacol., 5 (12):857 (1989).
38. P. R. Sheth and J. L. Tossouniam; Sustained release tablet formulations. U.S. Patent 4,126,672, Nov. 21, 1978.
39. A. H. Goldberg; Hydrodynamically-balanced system. Presented at 1983 Industrial Pharmaceutical R & D Symposium on Oral Controlled Drug Administration: Impact of Science, Technology and Regulation, Rutgers University, College of Pharmacy, New Brunswick, New Jersey, January 19 and 20, 1983.
40. P. R. Sheth and J. L. Tossounian; The hydrodynamically-balanced system: A novel drug delivery system for oral use. Drug Dev. Ind. Pharm., 10(2):313–339 (1984).
41. P. R. Sheth and J. L. Tossounian; Sustained release tablet formulations. U.S. Patent 4,140,755, Feb. 20, 1979.
42. R. M. Harrigan; Drug delivery device for preventing contact of undissolved drug with the stomach lining. U.S. Patent 4,055,178, Oct. 25, 1977.
43. A. S. Michaels, J. D. Bashwa and A. Zaffaroni; Integrated device for administering beneficial drug at programmed rate. U.S. Patent 3,901,232, Aug. 26, 1975.
44. A. S. Michaels; Drug delivery device with self-activated mechanism for retaining device in selected area. U.S. Patent 3,786,813, 1974.
45. K. Park, L. Cooper and J. R. Robinson; Bioadhesive hydrogels, in: *Hydrogels in Medicine and Pharmacy* (N. A. Peppas, Ed.), Vol. III, CRC Press, Boca Raton, Florida, 1987, Chapter 5.
46. M. Dittgen, S. Oestereich and F. Dittrich; Influence of the concentration of the mucous excipient on the bioadhesion ex vivo. Pharmazie, 44:460–462 (1989).
47. J. D. Smart and I. W. Kellaway; In vitro techniques for measurement mucoadhesion. J. Pharm. Pharmacol., 34:70P (1982).
48. K. Park, H. S. Ch'ng and J. R. Robinson; Alternative approches to oral controlled drug delivery: Bioadhesives and in situ systems, in: *Recent Advances in Drug Delivery Systems* (J. M. Anderson and S. W. Kim, Eds.), Plenum Press, New York, 1984, pp. 163–183.
49. M. Dittgen and S. Oestereich; Development of a bioadhesive oral drug delivery system. I. Basic investigation. S.T.P. Pharmacol., 5:867–870 (1989).
50. S. S. Davis, J. G. Hardy and J. W. Fara; Transit of pharmaceutical dosage forms through the small intestine. Gut. 27:886–892 (1986).
51. D. Harris, J. T. Fell, H. Sharma, D. C. Taylor and J. Linch; Studies on potential bioadhesive systems for oral drug delivery. S.T.P. Pharmacol., 5:852–856 (1989).
52. J. R. Robinson; Ocular drug delivery: Mechanism(s) of corneal drug transport and mucoadhesive delivery systems. S.T.P. Pharmacol., 5(12):839–846 (1989).
53. J. D. Smart, I. W. Kellaway and H. E. C. Worthington; An in-vitro investigation of

mucosa-adhesive materials for use in controlled drug delivery. J. Pharm. Pharmacol., 36:295 (1984).
54. C. M. Lehr, J. A. Bouwstra, J. J. Tukker and H. E. Junginger; Design and testing of a bioadhesive drug delivery system for oral application. S.T.P. Pharmacol., 5:857–862 (1989).
55. M. A. Longer, H. S. Ch'ng and J. R. Robinson; Bioadhesive polymers as platforms for oral-controlled drug delivery. III. Oral delivery of chlorthiazide using a bioadhesive polymer. J. Pharm. Sci., 74:406 (1985).
56. A. Osol et al.; *Remington's Pharmaceutical Sciences*, 15th Ed., Mack, Easton, PA, 1975, Chapter 47.
57. V. Mannienen, A. Apajalahti, J. Merlin and M. Karesoja; Altered absorption of digoxin in patients given propantheline and metoclopramide. Lancet, 1:398–400 (1973).
58. B. Beerman and M. Groschinsky-Grind; Enhancement of the gastrointestinal absorption of hydrochlorothiazide by propantheline. Eur. J. Clin. Pharmacol., 13:385–387 (1978).
59. P. G. Welling; Oral controlled drug administration: Pharmacokinetic considerations. Drug Dev. Ind. Pharm., 9:1185–1225 (1983).
60. R. J. Flower, S. Moncada and J. R. Vane; in: *The Pharmacological Basis of Therapeutics* (A. G. Gilman, L. S. Goodman and A. Gilman, Eds.), 6th Ed., Macmillan, New York, 1980, Chapter 29.
61. S. H. Weinstein, M. Pfeiffer, J. M. Schor, L. Franklin, M. Mintz, and E. R. Tutko; J. Pharm. Sci., 62:1416 (1973).
62. J. M. Schor, S. S. Davis, A. Nigalaye and S. Bolton; Susadrin transdermal tablets (nitroglycerin in Synchron controlled-release base). Drug Dev. Ind. Pharm., 9:1359–1377 (1983).
63. N. Reicheck; in: *Controlled Release Nitroglycerin in Buccal and Oral Form*, Vol. 1 of *Advances in Pharmacotherapy* (W. D. Bussmann et al., Eds.), Karger, Basel, 1982, pp. 143, 155.
64. N. Lichstein; in: *Controlled Release Nitroglycerin in Buccal and Oral Form*, Vol. 1 of *Advances in Pharmacotherapy* (W. D. Bussmann et al., Eds.), Karger, Basel, 1982, pp. 166, 174.
65. K. Yukimatsu; personal communication (1989).
66. H. Goldman and D. A. Antonioli; Mucosal biopsy of the rectum, colon, and distal ileum. Hum. Pathol., 13:981 (1982).
67. V. Lorenzsonn and J. S. Trier; The fine structure of human rectal mucosa: The epithelia, lining of the base of the crypt. Gastroenterology, 55:88 (1968).
68. G. I. Kaye, C. M. Fenoglio and R. R. Pascal; Comparative electron microscopic features of normal, hyperplastic, and adenomatous colonic epithelium. Gastroenterology, 64:926 (1973).
69. S. Eidelman and D. Lagunoff; The morphology of the normal human rectal Biopsy. Hum. Pathol., 3:389 (1972).
70. G. Tondury; in: *Angewandte und Topografische Anatomie*, G. Thieme Verlag, Stuttgart, 1959.
71. A. G. deBoer, D. D. Breimer, J. Pronk and J. M. Gubbens-Stibbe; Rectal bioavailability of lidocaine in rats: Absence of significant first-pass elimination. J. Pharm. Sci., 69:804–807 (1980).
72. L. C. J. deLeede, A. G. deBoer, C. P. J. M. Roozen and D. D. Breimer; Avoidance of "first-pass" elimination of rectally administered lidocaine in relation to the site of absorption in rats. J. Pharmacol. Exp. Ther., 225:181–185 (1983).
73. L. Caldwell, T. Nishihata, J. Fix, S. Salk, R. Cargill, C. R. Gardner and T. Higuchi; *Rectal Therapy*, Prous Publishers, Barcelona, Spain 1984, pp. 57–61.
74. A. G. de Boer and D. D. Breimer; in: *Proceedings of International Conference on Drug*

Absorption (L. F. Prescott and W. S. Nimmo, Eds.), Adis Press, Sydney, 1980, pp. 61–72.

75. A. G. de Boer, D. D. Breimer, H. Mattie, J. Pronk and J. M. Gubbens-Stribbe; Clin. Pharmacol. Ther., 26:701 (1979).
76. A. G. de Boer, L. C. G. de Leede and D. D. Breimer; Pharm. Int., 3:267 (1982).
77. A. G. de Boer, F. Moolenaar, L. G. J. de Leede and D. D. Breimer; Clin. Pharmacokinet., 7:285 (1982).
78. L. G. J. de Leede, A. G. de Boer, S. L. van Velzen and D. D. Breimer; J. Pharmacokinet. Biopharm., 10:525 (1982).

4
Mucosal Drug Delivery: Potential Routes for Noninvasive Systemic Administration

I. INTRODUCTION

For systemic delivery, the oral route has been the preferred route of administration for many systemically active drugs. When administered by the oral route, however, many therapeutic agents have been reportedly subjected to extensive presystemic elimination by gastrointestinal degradation and/or hepatic metabolism. Results of low systemic bioavailability, short duration of therapeutic activity, and/or formation of inactive or toxic metabolites have been often reported.

Delivery of drugs via the absorptive mucosa in various easily accessible body cavities, like the ocular, nasal, buccal, rectal and vaginal mucosae, has the advantage of bypassing the hepato-gastrointestinal first-pass elimination (Figure 1) associated with oral administration (1,2). Furthermore, because of the dual biophysical and biochemical nature of these mucosal membranes drugs with hydrophilic and/or lipophilic characteristics can be readily absorbed.

Mucosal membranes, particularly the nasal mucosa, also offer the potential for a rapid absorption of drugs with a plasma profile closely duplicating that from an intravenous (IV) bolus injection (2). This is especially useful in emergency situations. In addition, mucosal membranes may also be useful sites with good accessibility for easy application of drug delivery systems, especially for those with bio(muco)adhesive properties. With the development of mucosal delivery systems having controlled drug release characteristics, the mucosal routes can be exploited for the noninvasive systemic delivery of organic- and peptide-based drugs, with rapid absorption as well as sustained drug action.

II. HUMAN MUCOSAE

A. Physiological Characteristics

The anatomy and histology of nasal, rectal, and vaginal mucosae have been fairly well characterized in the human. The surface of human nasal mucosa is lined with both ciliated columnar epithelium, which covers the nasal septum and turbinates,

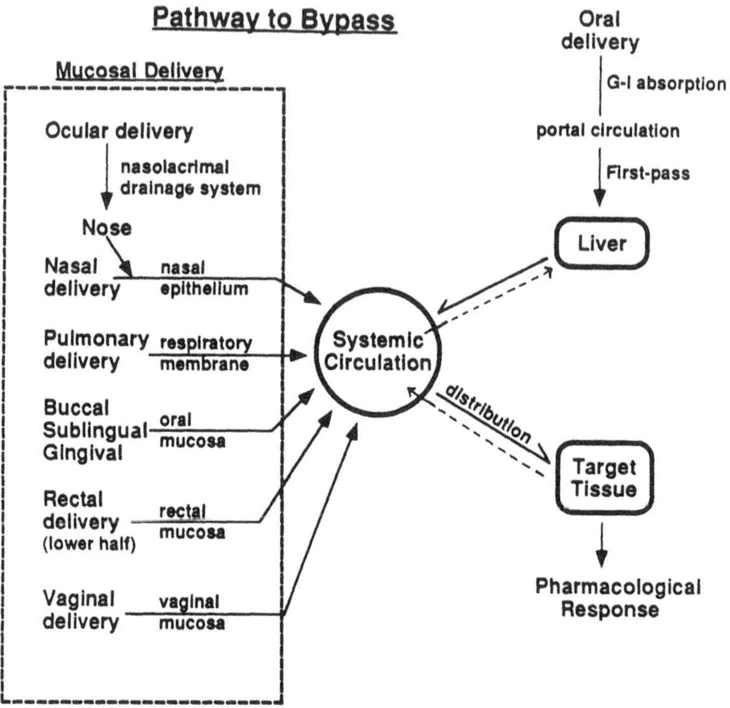

Figure 1 Various mucosal routes as potential pathways to bypass the hepato-gastrointestinal first-pass elimination associated with oral administration.

and squamous cutaneous epithelium. A individual cilium on the columnar epithelial surface is approximately 5 μm in length and 0.2 μm in width. Within the mucosa there exist goblet cells, which produce nasal secretions with a pH of 7.4. On the surface of each goblet cell there are hundreds of clublike microvilli. Beneath the ciliated epithelium and goblet cells several layers of flat polygonal basal cells, which have microvilli-like processes, exist. In the intercellular spacings there exists a composition of homogeneous mucoprotein-like substances. The basement membrane is composed of parallel and transverse reticulum, fibers, and connective tissues (3–7).

Human oral mucosa is a lining tissue that serves to protect the underlying structures. It consists of two parts: the underlying epithelium and the connective tissues. The human oral mucosal epithelium shows several distinct patterns of maturation that may be related to the different functions of the mucosa at the various regions of the oral cavity. The epithelium is of the stratified squamous type and varies in its thickness and the extent of keratinization from one region of oral mucosa to another, and in some regions keratinized epithelium may coexist with, or be replaced by, parakeratinized epithelium.

The human rectal mucosa is composed of the epithelium, the lamina propria, and the double-layer muscularis mucosae. The epithelial surface consists of closely packed columnar cells with some areas interrupted by crypt regions. Within the crypt regions there are mucus-producing goblet cells. Throughout the rectal mucosa there are lymphoid nodules covered by columnar surface epithelium. The lamina propria consists of two layers: a dense acellular collagen layer and a loose connective tissue layer. Within the lamina propria there exist superficial blood vessels and inflammatory cells, including macrophages, eosinophils, and lymphocytes. The muscularis mucosae contains smooth muscle cells and larger blood vessels (8–11).

The human vaginal mucosa is arranged in a pleated design to permit considerable expansion of the vaginal walls. The vaginal mucosa is composed of the epithelium, the lamina propria, the tunica propria, the muscularis mucosae, and the outer fibrous layer. The vaginal epithelium may be further subdivided into a superficial layer, an intermediate layer, and a basal layer. The superficial layer, which undergoes continuous desquamation, is composed of stratified squamous cells in various states of cornification. The intermediate layer consists of 10–30 layers of cells that are polyhedral and are connected together by intercellular bridges. The basal layer consists of cuboidal or columnar cells with distinct intercellular bridges. Beneath the vaginal epithelium is the lamina propria, a connective tissue layer containing small blood vessels and inflammatory cells. The underlying tunica propria contains more extensive vascular channels and elastic tissue. The muscularis mucosae are composed of an inner circular and an outer longitudinal muscular layer. The outer fibrous layer contains quilted bundles of connective tissues and elastic fibers (12–14).

During a menstrual cycle there are subtle changes in the human vaginal mucosa. The increased estrogen levels in the preovulatory and ovulatory phases cause an increase in the cornification of the superficial epithelium (15–17). There are also some changes in the types of intercellular junctions and bridges between the epithelial cells during the menstrual cycle (18,19).

B. Biochemical Characteristics

The surface environment of the nasal, rectal and vaginal mucosae is influenced, to a great extent, by the presence of mucus secreted by the goblet cells. These secretions, which contain proteolytic enzymes and immunoglobulins, may add both enzymatic and diffusional barriers to the mucosal absorption of drugs. Vaginal secretions also contain both glucose and glycogen; both are converted by enzymes and bacteria to lactic acid, resulting in a mucus with pH of 4–5 (20–23).

Although the biochemical composition of the nasal, rectal, and vaginal epithelium has not been well characterized, the lipid composition of other mucosal epithelia, including the buccal mucosa and alimentary tract, have been analyzed (24).

C. Mucosal Metabolism

Drugs may be subject to metabolism during the course of transmucosal permeation, either in the mucosal surface microenvironment or in the mucosal membrane. Peptidases have been noted to be present in the nonoral mucosae, including nasal, rectal, and vaginal musocal homogenates (25). The ability of the aminopeptidases to hydrolyze enkephalins was greater in the rectal mucosa than in the nasal and vaginal mucosae.

Rat nasal mucosa slices have been shown to contain reductase and hydroxysteroid oxidoreductase enzymes capable of metabolizing progesterone (26,27). Porcine nasal epithelium has been shown to possess hydroxysteroid dehydrogenases to metabolize testosterone but lack reductases to metabolize progesterone (28).

The metabolism in the rectal mucosa has not been well characterized, but gut wall metabolism has been demonstrated following oral administration (28,29). The presence of decarboxylases in the rectal mucosa has been documented (30). In addition, drugs administered rectally may be susceptible to hydrolysis or reduction by microorganisms in the rectal lumen (31–33).

Metabolism by the vaginal mucosa has also not been well characterized. The presence of peroxidases and phosphatases has been noted in humans and monkeys (34,35). It is expected that the microorganisms residing in the vaginal lumen may also be capable of metabolizing drugs administered intravaginally.

D. Mucosal Membrane Models

It has been generally accepted that the biological membrane can be represented by the fluid mosaic model originally proposed by Singer and Nicolson (36,37). Figure 2 is a two-dimensional representation of this model, which depicts a biological membrane composed of a fluid-state lipid bilayer embedded with globular integral proteins. In Figure 3 the integral proteins are shown to be either embedded in a portion of the lipoidal membrane or spanning throughout its entire thickness. The amphipathic protein molecules have been hypothesized to minimize the free energy required for transmembrane permeation by maximizing both hydrophilic and lipophilic interactions in the membrane. It is visualized that the ionic and polar portions of the protein molecule remain in contact with the aqueous environment on the membrane surface and its relatively nonpolar portions interact with the alkyl chains in the lipid bilayer (Figure 3A). The integral membrane proteins may also exist as subunit ag-

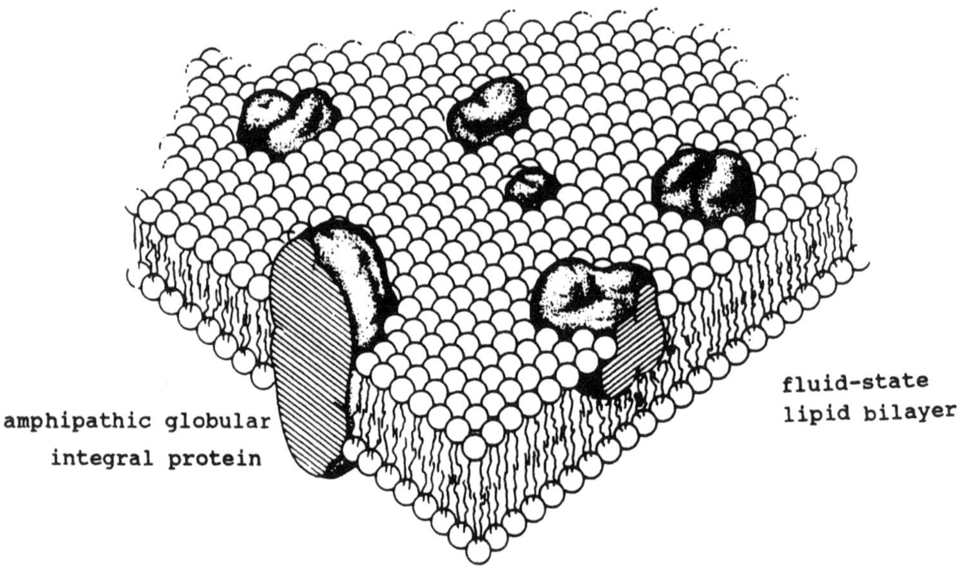

Figure 2 The fluid mosaic model proposed by Singer and Nicolson for the structure of epithelial membrane, which consists of amphipathic globular integral protein molecules embedded in a fluid-state lipid bilayer. (From Singer and Nicolson, 1972.)

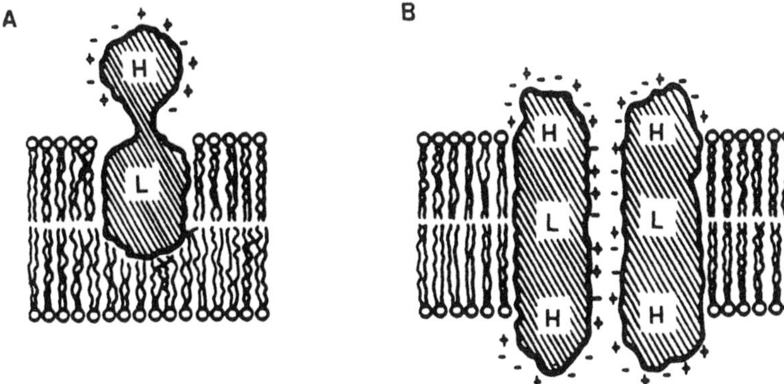

Figure 3 Two thermodynamically favored structures of the globular integral protein molecule. (A) Monodisperse structure in which the hydrophilic component (H) is exposed to the aqueous environment on the surface of epithelial membrane and the lipophilic component (L) is embedded in the lipid bilayer. (B) Subunit aggregate structure in which a subunit aggregate of two protein molecules spanning through the entire thickness of the lipid bilayer to form an aqueous solution-filled pore channel. (Modified from Singer and Nicolson, 1972.)

gregates, which span through the entire thickness of the lipid bilayer to form a continuous water-filled channel (Figure 3B). Thus the mucosa, as a biological membrane, may be considered composed of lipid-rich regions interrupted by aqueous channel pores formed by subunit aggregates of membrane proteins (36–43).

III. TRANSMUCOSAL SYSTEMIC DELIVERY OF DRUGS

As illustrated in Figure 1, mucosal routes provide the potential pathways to bypass hepatogastrointestinal first-pass elimination following oral administration. Transmucosal drug delivery has the potential to achieve greater systemic bioavailability for orally metabolized drugs, including organic- and peptide-based pharmaceuticals.

For organic-based pharmaceuticals, such as progesterone, a series of investigations was recently carried out systematically in rabbits (44) that demonstrated that transmucosal delivery has attained a relatively high systemic bioavailability for orally metabolized progesterone, especially the nasal route, which has bioavailability 5–10 times greater than that by oral administration (Table 1).

However, the transmucosal delivery of peptide-based pharmaceuticals, such as insulin, has achieved a much lower systemic bioavailability than parenteral administration (45). The lower extent of transmucosal absorption of insulin and many other peptide-based pharmaceuticals is probably due to a combined effect of poor mucosal permeability and extensive metabolism at the absorption site. The data in Table 2, which compare the hypoglycemic efficacy of insulin delivered through various mucosal routes, indicate that by rectal delivery insulin is more efficacious than that by the nasal, buccal, and sublingual routes. When administered without an absorption promotor, however, the hypoglycemic effect of insulin delivered by all mucosal routes is substantially low compared to intramuscular administration. On the other hand, the coadministration with an absorption promotor, such as sodium glycocholate, the transmucosal permeation of insulin is greatly improved as demonstrated by a sub-

Table 1 Systemic Bioavailability of Progesterone Following Mucosal Administration

Route	Bioavailability[a] (% ± SEM)
Nasal	88.4 ± 10.1
Rectal	58.8 ± 6.7
Vaginal	46.6 ± 5.4
Oral	9.5 ± 6.2
	7.9 ± 1.6[b]

[a]Obtained in ovariectomized rabbits by radioimmunoassay (mean ± standard error of the mean of four animals) with IV data as control.
[b]From Corbo, Huang, and Chien (1988).
Source: From Corbo, Liu, and Chien (1989).

stantial enhancement in its hypoglycemic efficacy following delivery by all mucosal routes, by as much as 120-fold, with the rank order of nasal > rectal > buccal > sublingual. The nasal and rectal delivery of insulin achieved a therapeutic efficacy almost one-half that of intramuscular insulin administration. It should be pointed out that protease activities in the homogenates of the nasal, buccal, rectal, and vaginal mucosae of the albino rabbit have been found to be substantial and comparable to those in the ileal homogenate. This may partly explain the low transmucosal bioavailability of peptide-based pharmaceuticals. The systemic delivery of peptide-based pharmaceuticals is discussed systematically in Chapter 11.

A. Mechanisms of Transmucosal Permeation

There are two routes potentially involved in drug permeation across epithelial membranes: the transcellular route and the paracellular route (46). In studies dealing with the mechanisms of transmembrane permeation, the structure of the epithelial mem-

Table 2 Relative Hypoglycemic Efficacy of Insulin[a] Delivered Through Various Mucosal Routes

Route[b]	Relative efficacy[c] (% ± SD)	
	No promotor	With promotor[d]
Nasal	0.4 ± 0.2	46.8 ± 4.7
Sublingual	0.3 ± 0.2	11.9 ± 7.6
Buccal	3.6 ± 2.8	25.5 ± 7.5
Rectal	17.0 ± 5.6	40.1 ± 8.2

[a]Bovine insulin solution (pH 7.4; 10 IU/kg).
[b]Group of six male Lewis rats.
[c]Relative to IM administration.
[d]Na glycocholate (5%) as absorption promotor.
Source: Modified from Aungst et al. (1988).

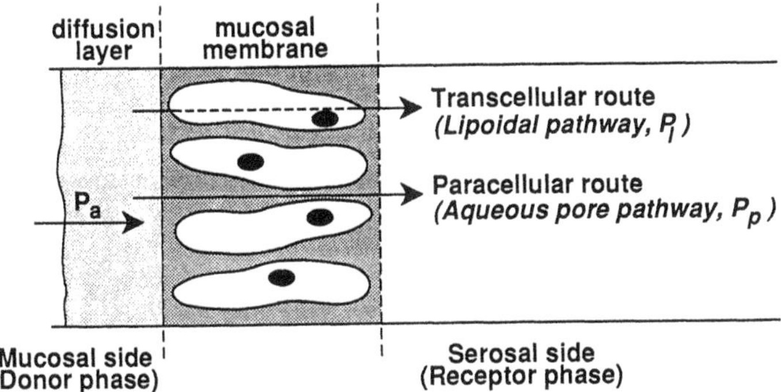

Figure 4 Diagram showing the physical model for transmembrane permeation across a mucosal membrane consisting of lipoidal and aqueous pore pathways in parallel, which is in series with an aqueous diffusion layer on the mucosal surface. P's are the permeabilities across the aqueous diffusion layer P_a, the lipoidal pathway P_l, and the aqueous pore pathway P_p.

brane is frequently simplified to consist of a lipoidal pathway and an aqueous pore pathway (Figure 4). Traditionally, the skin and gastrointestinal mucosa have been viewed as primarily lipoidal barriers, in which the absorption of a drug is determined by the magnitude of its partition coefficient and molecular size until the diffusion through the aqueous diffusion layer (P_a) becomes a rate-limiting step in the course of transmembrane permeation (47–50). However, several investigators have also noted the lack of a linear correlation between penetrant lipophilicity, as indicated by the magnitude of the partition coefficient, and permeability (51–55), which implies that these membranes cannot be regarded as simple lipoidal barriers.

For nasal mucosal membrane there is experimental evidence supporting the existence of both a lipoidal and an aqueous pore pathway. The rate and extent of nasal absorption of β-adrenoreceptor-blocking drugs in humans were reportedly related to the lipophilicity of the drugs, with the most lipophilic propranolol achieving the highest absorption (56). The rate constant for the in situ absorption of progesterone, testosterone, and hydrocortisone in rats was found to be proportional to the lipophilicity of the steroids (57). However, several studies have indicated that although the lipophilicity of a permeant may influence its absorption, it is not the sole determinant of the rate and extent of nasal absorption (58,59). It has been suggested that for lipophilic permeants the nasal mucosa acts as a modified lipoidal barrier (60). For numerous hydrophilic drugs it has been observed that the pH-partition theory, which states that the absorption is dependent upon the concentration of the undissociated species of a permeant molecule, is not applicable (61,62). It has also been reported that a linear correlation between the log of nasal absorption and the log of drug molecular weights exists for many water-soluble compounds (63,64). This implies the presence of aqueous pore channels in the nasal mucosa for the absorption of hydrophilic drugs. Hayashi et al. (65) compared water influx and sieving coefficients between the nasal and gastrointestinal (GI) mucosae in the rat. It was determined that the nasal mucosa has a richer distribution of small aqueous pores than the jejunal membrane.

Transmucosal permeation of polar molecules, such as peptide-based pharmaceuticals, may be by way of paracellular route; however, several barriers exist during the course of paracellular permeation: (i) basal lamina, whose barrier function is dependent upon the molecular weight of the permeant molecule and its reactivity with the barrier as well as the structural and functional factors of the barrier (66,67); (ii) membrane-coating granules, which extrude into the intercellular region of both keratinized and nonkeratinized oral epithelium (68) and prevent the transmucosal penetration of water-soluble peptide or protein, such as horseradish peroxidase (69); and (iii) the keratin layer, whose barrier function in oral mucosa is not as well defined as in the skin. Although the rate of permeation of water was shown to be greater in nonkeratinized than in keratinized oral epithelium (70), the permeation of horseradish peroxidase was found similar in both mucosae (69). The permeability coefficients of 19 solutes across the rabbit lingual frenulum and attached gingiva, determined under in vitro conditions, were found highly correlated (71), which suggests that the epithelia from different regions of the oral mucosa may be rather similar in mechanisms and rates of transmucosal permeation.

In contrast to the complexity of the nasal and oral mucosae, drug absorption through the rectal mucosa is generally in good agreement with the pH-partition theory (72,73). It has been well established that ionic and lipid-insoluble drugs are poorly absorbed when administered rectally, and lipid-soluble drugs are rapidly absorbed (72–76). The rectal mucosa of the rat was reportedly to have a poorer distribution of aqueous pore channels than other regions of the gastrointestinal tract, including the jejunum (71). Coupled with the observations of the correlation between the rectal absorption of drugs and their partition coefficients, this implies that the rectal mucosa may be considered a simple lipoidal barrier (77–79).

The vaginal mucosal membrane, like the nasal, has been proposed to consist of both a lipoidal pathway and an aqueous pore pathway (80,81). In the rabbit vaginal mucosa, it was found that the permeability coefficient of aliphatic alcohols increases with the increase in their lipophilicity (80), indicating the importance of the lipoidal pathway in vaginal permeation. The aqueous diffusion layer on the rabbit vaginal mucosal surface was estimated to have a thickness of approximately 310 μm. Hwang et al. (81) studied the vaginal absorption of a homologous series of straight-chain alkanoic acids in rabbits, and the results indicated the importance of the aqueous pore pathway for the transmucosal permeation of alkanoic acids with a short alkyl chain. It was found that the permeability coefficient of the acids increases as the length of alkyl chain increases but does not correlate linearly with their partition coefficient. This indicates that in the vaginal membrane, like the nasal membrane, there may exist semipolar regions of neither a purely lipophilic nor a purely hydrophilic nature (20,81). Few studies conducted in the rhesus monkey have indicated the presence of intercellular lipids that restrict the vaginal absorption of water-soluble compounds with large molecular weight (82). In the estrous (cycling) animals the absorption of hydrophilic compounds increases in the proestrus and diestrus stages of the estrous cycle as a result of a loosening of the epithelial cell junctions (83,84).

B. Systematic Studies of Transmucosal Permeation in Rabbits

A review of the literature indicates that very few systematic investigations have been designed and performed to compare the mechanisms of transmucosal delivery through various mucosal membranes or the systemic bioavailability and pharmacokinetics of organic- and peptide-based pharmaceuticals following transmucosal absorption, even though many classes of drug entities have been reported to be delivered successfully via several mucosal membranes (Chapters 3, 5, 9, and 11).

As discussed earlier in this chapter, compounds of either a hydrophilic or a lipophilic nature may be readily absorbed through mucosal membranes because of their dual biophysical and biochemical characteristics (Figures 2 and 4), and several studies have demonstrated that nasal (65,85,86), rectal (72,75,79), and vaginal (80,81) membranes may be characterized as consisting of a lipoidal pathway and an aqueous pore pathway. The relative influences of these lipoidal and aqueous pathways on the transmucosal permeation of organic- and peptide-based pharmaceuticals through these mucosal membranes have not been well characterized and thoroughly compared.

To gain a fundamental understanding of the mechanisms involved in transmucosal permeation and the effect of mucosal routes on the systemic delivery of organic-based pharmaceuticals, a series of in vitro and in vivo investigations have recently been systematically conducted in rabbits using progesterone, a lipophilic molecule known to be subject to extensive hepatogastrointestinal first-pass metabolism when taken orally, as the model penetrant; its hydroxy derivatives, with increasing hydrophilicity, have also been investigated to study the effect of the physicochemical properties of the penetrant molecule (87). Progesterone has been reported to have a higher absorption and bioavailability following nasal (88–90), rectal (91,92), and vaginal (93–95) delivery than oral administration. Recently, the extent of absorption of these progestins in ovariectomized rabbits was found to depend on the hydrophilicity of the penetrant molecule and the mode of nasal delivery via either an immediate-release spray solution formulation or a controlled-release device (96).

1. Rabbit Mucosae

The surface of the rabbit nasal mucosa is lined with both ciliated columnar epithelium and olfactory epithelium. As with human nasal mucosa, the columnar epithelium possess both cilia and microvilli. Olfactory epithelium, which consists of sensory, Sertoli's (sustentacular), and basal cells, may also possess a few cilia. Mucus glands and goblet cells, which produce nasal secretions, are associated with both ciliated and olfactory epithelia (97–100).

Since rabbit lacks an estrous cycle, its vaginal epithelium is not subject to a time-dependent cyclic variation in histology. Since the histological changes induced by the estrous cycle have been shown to affect the transmucosal permeability of vaginal mucosa in cyclic animals, like rhesus monkey, rabbit has been used as the animal model to investigate the vaginal absorption and intravaginal delivery of drugs without the complication of an estrous cycle (Chapter 9).

2. Biophysicochemical Characteristics of Rabbit Mucosae

A systematic investigation of rabbit mucosae was recently conducted (101), which suggested that the structures of rabbit nasal, rectal, and vaginal mucosae are relatively similar to those of the human (99,102–104). Like the human, the rabbit nasal mucosa possesses a single layer of ciliated columnar epithelium that covers the goblet and basal cells. The rabbit rectal mucosa is characterized by many small folds and is covered by a single layer of closely packed columnar epithelium. The rabbit vaginal mucosa consists of many layers of stratified squamous cells, beneath which are basal cells and the mucosal and muscularis mucosae layers. Since the rabbit is a noncyclic animal, its vaginal epithelium is not subject to the time-dependent histological changes seen in humans (104,105), and this implies that the membrane permeability of vaginal mucosa is expected to be fairly constant (106,107). Therefore, the results of histological studies suggest that the rabbit could be a good animal model for studying the transmucosal delivery of drugs across the nasal, rectal, and vaginal mucosae.

However, histological studies showed some basic differences in the epithelial membrane barrier of the nasal, rectal, and vaginal mucosae (101). The difference in the physiological functions of these mucosal membranes are reflected in the differences in their "real" surface areas available for transmucosal permeation. The epithelia of the nasal and rectal mucosae are characterized by many small folds; the vaginal epithelium is characterized by several large folds (to permit expansion).

The biophysicochemical characteristics of the rabbit mucosal membranes are compared in Table 3. There are no significant differences in composition, such as the percentage of unbound water, proteins, or lipids, but a substantial difference has been seen in the thickness between nasal mucosa and rectal and vaginal mucosae.

Using differential scanning calorimetry (DSC) the qualitative differences in the microscopic interactions in the lipid and protein domains of these mucosae can be examined. The DSC scans of these mucosal membranes (Figure 5) indicate that the nasal and rectal mucosae have similar lipid interactions, and the thermal transitions in the lipid region (70–100°C) for the vaginal mucosa differ from those in the nasal and rectal mucosae, indicating that different lipid types or interactions could occur in the vaginal mucosa. In the protein region (150–170°C) the rectal mucosa does not possess the same interactions seen in the nasal and vaginal mucosae.

3. *Mechanisms and Kinetics of Transmucosal Permeation*

To investigate the mechanism of transmucosal permeation and to assess the relative contribution of the lipoidal and aqueous pore pathways to the transmucosal permeation of drugs, the permeation kinetics of progesterone, as the model lipophilic molecule, and mannitol, as the model hydrophilic molecule, was studied using mucosal membranes excised freshly from female rabbits and mounted in a Valia-Chien permeation cell (101). The results in Figure 6 indicate that both the lipophilic progesterone molecule and the hydrophilic mannitol molecule are capable of permeating through all three mucosal membranes at zero-order kinetics, with the nasal mucosa showing a significantly higher rate of permeation and a shorter lag time than the

Table 3 Biophysicochemical Characteristics of Rabbit Mucosae

	Biophysical and biochemical properties			
		Composition (% ± SD)		
Mucosal membrane	Thickness (μm)	Water (unbound)	Proteins	Lipids
Oral mucosa[a]				
Buccal	594	—	—	—
Palatal	158–224	—	—	—
Sublingual	111	—	—	—
Nonoral mucosae[b]				
Nasal	53.5	75.7	12.3	12.0
Rectal	175.3	78.1	14.2	7.7
Vaginal	165.1	78.3	7.5	14.2

[a]Data from Gigoux (1962).
[b]Mean Data from 6 animals (Corbo, Liu, and Chien, 1990).

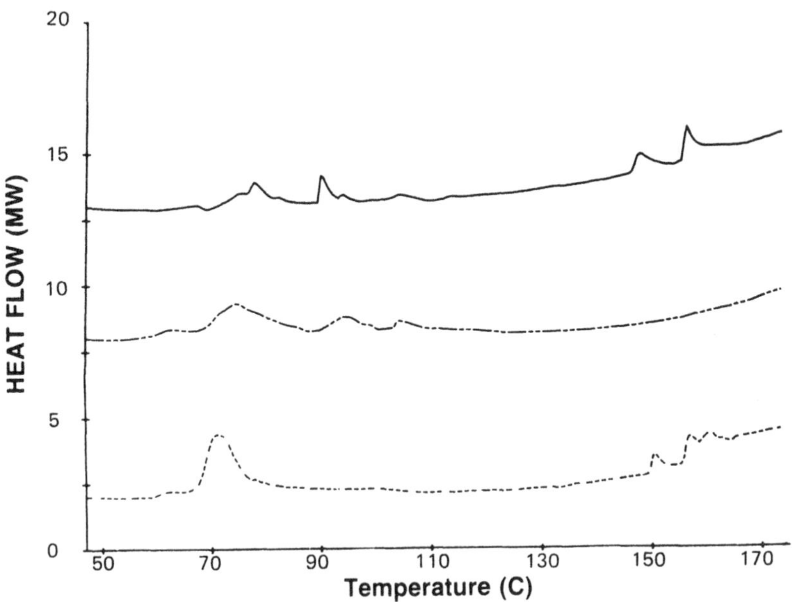

Figure 5 The DSC thermogram of the rabbit nasal (———), rectal (— ––), and vaginal (----) mucosae. (From Corbo et al., 1990.)

rectal and vaginal mucosae. The permeation characteristics of these mucosae to lipophilic and hydrophilic molecules are compared in Table 4, which indicates that the mucosal permeability of progesterone is substantially greater than that of mannitol. This suggests that these mucosae are characterized as primarily lipophilic membranes. However, the nasal mucosa appears to have a much more hydrophilic character than the vaginal mucosa, and the rectal mucosa is the least hydrophilic in character.

For further characterization of the mechanisms and kinetics of transmucosal permeation, mucosal permeation studies of a series of progesterone derivatives with systematic variation in their hydrophilicity were conducted (44). A representative set of transmucosal permeation profiles illustrating the effect of hydrophilicity is shown in Figure 7. The results indicate that, similar to the transmucosal permeation of progesterone (Figure 6), the hydroxy derivatives of progesterone also permeated through the nasal, rectal, and vaginal mucosae in a zero-order fashion throughout the course of the 24-hr study. The duration of the lag time was noted to increase in the order nasal < rectal < vaginal mucosa. The longer lag time observed for vaginal permeation may be attributed to the multiple-cell layer nature of its epithelial barrier, as opposed to the single-cell layer character of the nasal and rectal epithelia (44). The results of the comparative mucosal permeation studies shown in Figure 7 suggest that with the exception of tri-OH-progesterone, the more hydrophilic the progestin molecule, the lower is the rate of transmucosal permeation. The higher rate of transmucosal permeation observed with trihydroxyprogesterone could be attributed to its high concentration gradient. The effect of penetrant hydrophilicity on the apparent permeability through various mucosal membranes is illustrated in Table 5, which indicates that the nasal mucosa has a significantly higher permeability than the rectal and vaginal mucosae for progesterone as well as its hydroxy derivatives. It is inter-

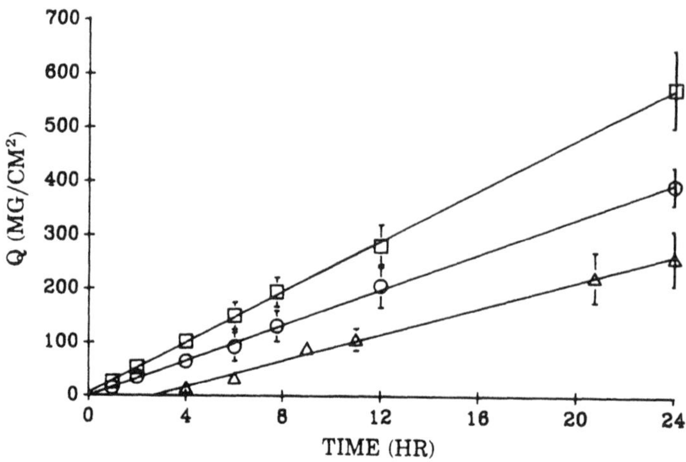

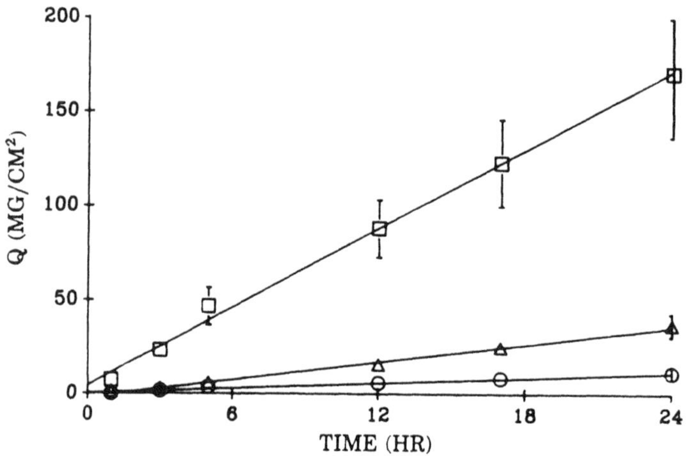

Figure 6 (Top) In vitro permeation profiles of progesterone across nasal (□), rectal (○), and vaginal (△) mucosae. (Bottom) In vitro permeation profiles of mannitol across nasal (□), rectal (○), and vaginal (△) mucosae. (From Corbo et al., 1990.)

esting to note that mucosal permeability decreases as the number of hydroxy groups on the progesterone molecule increases. This trend is followed for all the mucosal membranes investigated. This finding suggests that the lipophilicity of a penetrant molecule is an important factor in determining its mucosal permeability.

On the other hand, the variation in penetrant hydrophilicity has little influence on the diffusivity of progestins through the mucosal membranes. The finding is not surprising when one considers that addition of hydroxy groups only slightly increases the molecular weight or size of progesterone.

The mucosal permeability of progestins was found to be dependent upon their

Table 4 Comparative Permeation Characteristics Among Various Rabbit Mucosae

Biophysical parameters	Mucosa	Mean value (± SD) Mannitol	Progesterone	P/M[a]
Permeability, (cm/hr)[b]	Nasal	2.24 (0.41)	4.6 (0.8)	205
	Rectal	0.22 (0.05)	3.1 (0.7)	1409
	Vaginal	0.48 (0.09)	2.6 (0.7)	542
Diffusivity, $(cm^2/sec) \times 10^8$	Nasal	1.89 (0.56)	3.95 (1.23)	2.09
	Rectal	0.50 (0.12)	1.22 (0.38)	2.44
	Vaginal	0.57 (0.10)	9.88 (2.23)	17.33

[a]The ratio of permeability of progesterone over mannitol.
[b]Unit: mannitol, cm/hr $\times 10^3$; progesterone, cm/hr $\times 10^1$.
Source: From Corbo, Liu, and Chien (1990).

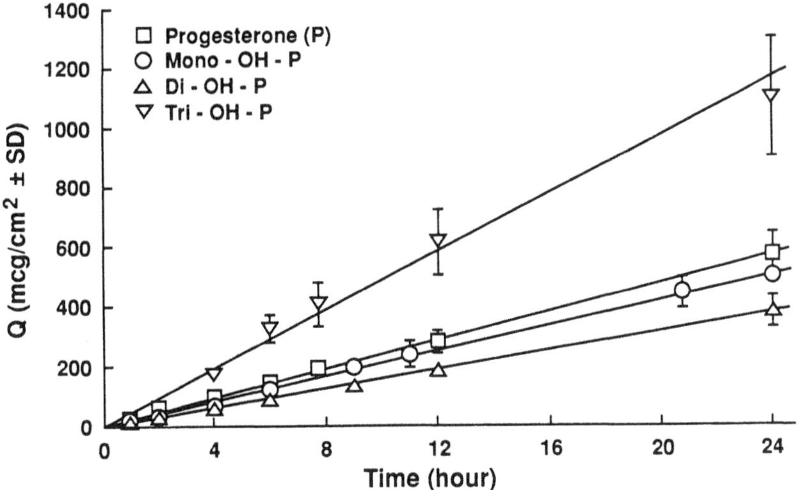

Figure 7 In vitro permeation profiles of progesterone (□) and its monohydroxy (○), dihydroxy (△), and trihydroxy (▽) derivatives across rabbit nasal mucosa (n = 3). (Plotted from the data by Corbo, 1989.)

Table 5 Effect of Penetrant Hydrophilicity on Apparent Permeability

	Apparent permeability (cm/hr ± SD)			
		Hydroxyprogesterone		
Mucosa	Progesterone	Monohydroxy	Dihydroxy	Trihydroxy
Nasal	0.48 (0.09)	0.34 (0.05)	0.12 (0.02)	0.05 (0.01)
Rectal	0.27 (0.05)	0.06 (0.02)	0.05 (0.01)	0.02 (0.01)
Vaginal	0.21 (0.04)	0.07 (0.02)	0.04 (0.01)	0.01 (0.01)

Source: Compiled from the data by Corbo et al. (1990).

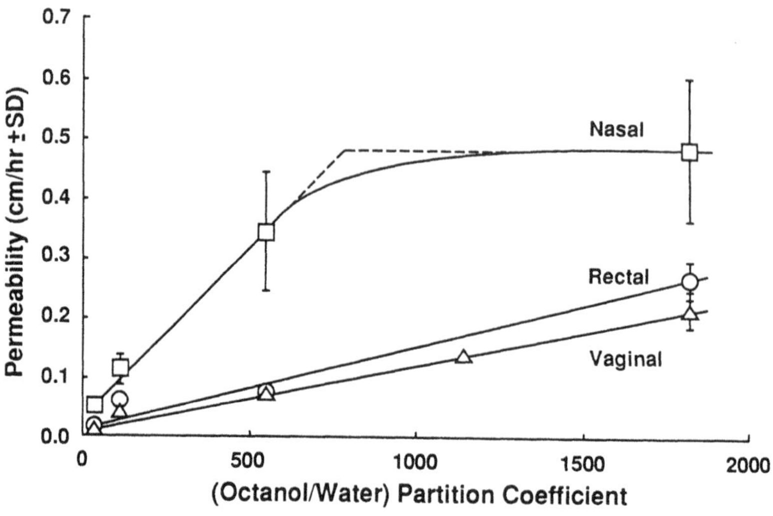

Figure 8 Relationship between the apparent permeability of progestins across the nasal (□), rectal (○), and vaginal (△) mucosae and their (octanol/water) partition coefficients. (Plotted from the data by Corbo, 1989).

partition coefficients determined in the octanol-water system (Figure 8). The results show that for all the mucosal membranes investigated, the apparent permeability increases proportionally as the partition coefficient (or lipophilicity) of the penetrant molecule is increased. However, the linear relationship exists over a wider range of partition coefficient for rectal and vaginal mucosae than for nasal mucosa. Nasal mucosa shows a greater dependency on the lipophilicity of the penetrant than the rectal mucosa, and vaginal mucosa is least dependent upon lipophilicity. The data are in good agreement with the effect of penetrant hydrophilicity shown in Table 5. The results in Table 5 and Figure 8 suggest that nasal mucosa may have a rather different nature of interaction (or binding) with the penetrant molecule, and the rectal and vaginal membranes are very similar in their interactions.

To investigate the possible role of mucosal lipids in the interaction of progestins with the mucosa, the effect of delipidization on the mucosal partition coefficient was examined. The results summarized in Table 6 indicate that with the exception of the monohydroxy derivative, the mucosal partition coefficient of progestins decreases substantially after the removal of mucosal lipids. This finding suggests that progestin, as a lipophilic molecule itself, should interact primarily with the lipid domains in the mucosa during the course of transmucosal permeation. For monohydroxyprogesterone, however, delipidization has resulted in enhancement of the partition coefficient for all the mucosae. This implies that the addition of a hydroxy group at position 17 affects the nature of the interaction of the progesterone molecule with the mucosal membrane.

On the other hand, the removal of lipid domains by delipidization was found to have no effect on the zero-order transmucosal permeation kinetics, but it did reduce the transmucosal permeation rates of progesterone across all the mucosal membranes investigated (Figure 9). Delipidization reduced, to a different extent, the rate of permeation across the various mucosae. The rate was reduced by as much as 35 times for rectal, 7 times for vaginal, and 4 times for nasal mucosae. The results suggest

Table 6 Effect of Delipidization on Mucosal Partition Coefficients of Progestins

Progestins	Delipidization	Mucosal partition coefficient[a] Nasal	Rectal	Vaginal
Progesterone (P)	Before	180.2 (40.1)	76.7 (18.8)	95.6 (19.9)
	After	74.7 (19.6)	34.8 (6.9)	86.4 (16.5)
Mono-OH-P	Before	441.7 (113.6)	41.3 (8.9)	51.5 (10.2)
	After	1484.5 (360.7)	340.2 (80.2)	294.6 (62.3)
Di-OH-P	Before	99.6 (19.7)	37.6 (6.5)	50.2 (9.7)
	After	18.4 (3.7)	7.8 (2.9)	9.4 (2.6)
Tri-OH-P	Before	26.3 (6.5)	27.9 (6.8)	45.2 (8.2)
	After	1.3 (0.3)	2.9 (0.5)	4.5 (0.9)

[a]Compiled from the data by Corbo (1989), expressed as mean (± SD) of three determinations.

that among the mucosae examined, rectal mucosa is most sensitive to the removal of lipid domains. It was also reported (101) that the mucosal permeabilities of progesterone and its monohydroxy derivative decrease and the trihydroxy derivative increase, but the permeability of the dihydroxy derivative decreases or is not changed. These mixed results are very different from the skin permeability of these progestins, which was observed to increase after delipidization (108). Furthermore, it was noted that, following delipidization, the dependence of mucosal permeability on the hydrophilicity of progestins observed for intact mucosae (Table 5) disappears.

Based on the transmucosal permeation model presented in Figure 4, the following equation has been developed (44) to describe the apparent permeability coefficient P_{app} of progestins:

$$P_{app} = \frac{1}{(1/P_a) + [1/(P_p + P_l)]} \tag{1}$$

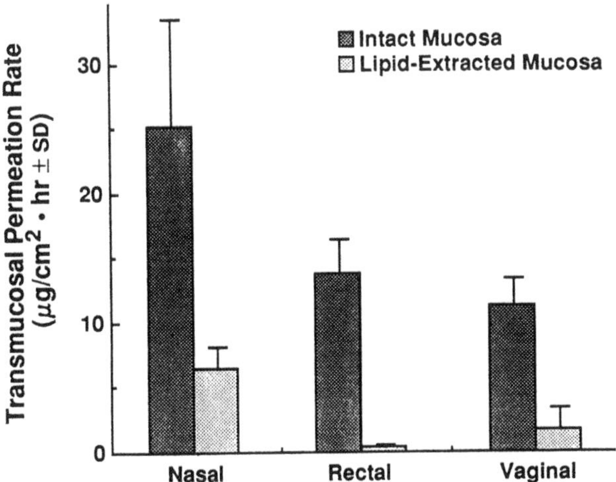

Figure 9 Effect of delipidization on the transmucosal permeation rate of progesterone across the various mucosae. (Plotted from the data by Corbo, 1989.)

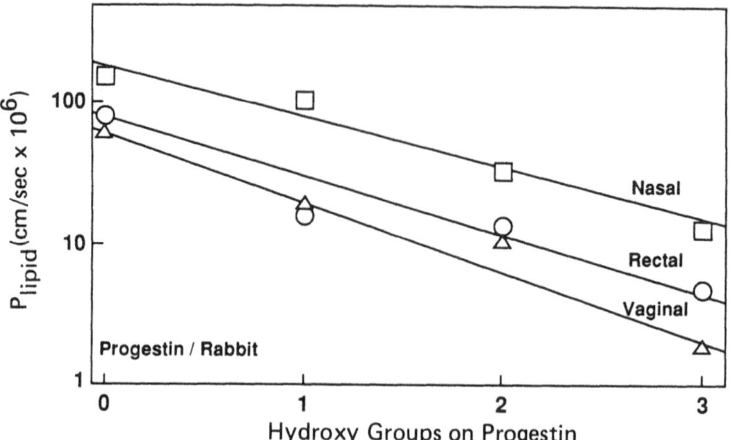

Figure 10 Semilogarithmic relationship between permeability coefficient aross the lipoidal pathway P_l and the number of hydroxy groups on the progestin molecule. (Plotted from the data by Corbo, 1989.)

where P_a, P_p, and P_l are the permeability coefficient across the aqueous diffusion layer, the aqueous pore pathway, and the lipoidal pathway, respectively. The permeability coefficient of the lipoidal pathway P_l is further defined by

$$\log P_l = \log P_l^0 - n\pi_{OH} \tag{2}$$

where P_l^0 is the permeability coefficient of the lipoidal pathway for a progestin molecule with zero hydroxy groups ($n = 0$); n is the number of hydroxy groups on the progestin molecule; and π_{OH} is the hydrophilicity incremental constant.

A linear semilogarithmic relationship should exist between P_l, the permeability coefficient, and n, the number of hydroxy groups. The results in Figure 10 demonstrate that the P_l values decrease exponentially with the increase in the hydrophilicity of progestin molecule predicted from Equation (2). The values of P_l^0 and π_{OH} can be calculated from the y-intercept and slope, respectively. The P_l^0 and π_{OH} values for various rabbit mucosae studied are summarized in Table 7 with P_p values

Table 7 Contribution of Lipoidal and Polar Pathways to Permeation of Progestins Through Rabbit Mucosae

	Specific permeability[a]		
Mucosa	$P_l \times 10^5$ (cm/sec)	$P_p^0 \times 10^7$ (cm/sec)	π_{OH}[b]
Nasal	18.6	6.22	0.92
Rectal	6.3	0.61	1.14
Vaginal	6.2	1.32	1.38

[a]P_l = lipoidal pathway permeability coefficient; P_p = pore pathway permeability coefficient.
[b]Calculated from the slope value of log (P_l) versus n.
Source: From Corbo (1989).

calculated from the mannitol permeation data on the assumption that progestins and mannitol pass through pores of a similar size (44). The results in Table 7 suggest that the permeability coefficient for the lipoidal pathway P_l is 3–10 times greater than that for the aqueous pore pathway P_p, which means that the lipoidal pathway has a greater contribution to the transmucosal permeation of progestin molecule than does the aqueous pore pathway. Among the mucosal membranes tested, the P_l value for nasal mucosa is 3-fold that of both rectal and vaginal mucosae. On the other hand, the P_p value for nasal mucosa is approximately 10 times higher than that for rectal mucosa and 5 times greater than that for vaginal mucosa. The π_{OH} values suggest that the effect of hydrophilicity on the P_l value varies slightly from one mucosa to another, with increasing order as nasal < rectal < vaginal mucosa.

4. *Mucosal Absorption and Systemic Bioavailability of Progestins*

The pharmacokinetics and systemic bioavailability of progesterone following nasal, rectal, and vaginal delivery were compared with oral administration in ovariectomized rabbits using radioimmunoassay (87). The plasma profiles in Figure 11 suggest that, following mucosal delivery using a metered-dose sprayer, progesterone is also rapidly absorbed, as does oral delivery by intragastric administration, with peak plasma concentrations reached within 30 min. By transmucosal absorption progesterone achieved a higher peak plasma concentration and greater systemic bioavailability than by GI absorption. The results of pharmacokinetic analysis (Table 8) indicate that the rate constant K_a for nasal absorption is significantly greater than that for oral absorption; the rate constants for absorption through rectal and vaginal mucosae are substantially lower than that by nasal absorption. The data suggest that a 3- to 10-fold higher C_{max} value and a 5–9 times greater systemic bioavailability were achieved by mucosal delivery than by oral administration. It is interesting to note that nasal delivery produced a C_{max} value and a systemic bioavailability almost equivalent to those attained by IV administration.

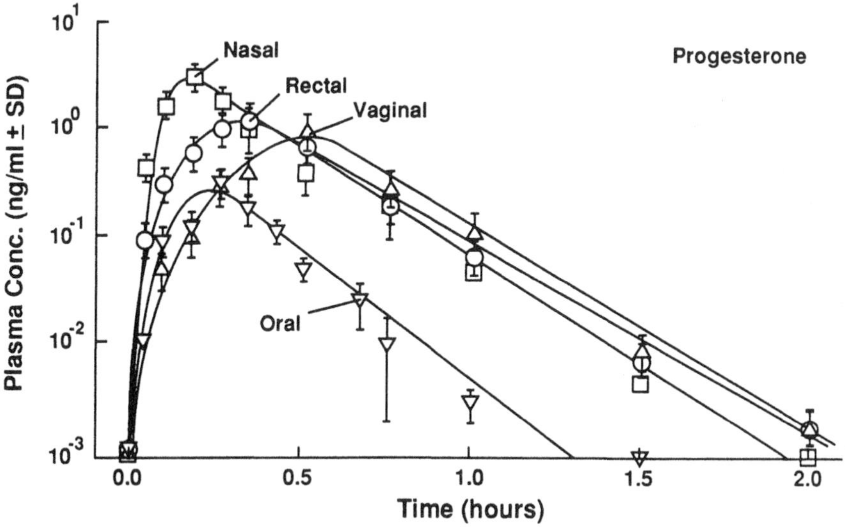

Figure 11 Comparative plasma profiles of progesterone in ovariectomized rabbits ($n = 4$) following oral and mucosal delivery. (Plotted from the data by Corbo et al., 1989.)

Table 8 Pharmacokinetic Parameters Following Various Routes of Administration[a]

Route	Rate constant k_a (hr^{-1})	Rate constant k_e (hr^{-1})	C_{max}[b] (ng/ml)	Systemic bioavailability[c] (%)
IV	—	6.4 (1.7)	2.7 (0.7)	100.0
Oral	7.3 (0.9)	6.1 (1.2)	0.3 (0.1)	9.5 (6.2)
Nasal	11.4 (1.1)	5.4 (1.3)	3.1 (0.8)	88.4 (10.1)
Rectal	4.5 (0.6)	5.3 (1.0)	1.2 (0.4)	58.8 (6.7)
Vaginal	4.8 (0.5)	5.0 (0.9)	1.0 (0.4)	46.6 (5.4)

[a]Data are mean (± standard deviation) of four ovariectomized rabbits.
[b]Normalized for administered dose.
[c]Relative to intravenous administration.
Source: Compiled from the data by Corbo et al. (1989).

The hydroxy derivatives of progesterone were also observed to be rapidly absorbed following mucosal delivery, reaching peak plasma concentrations within 30 min. A typical set of plasma profiles is shown in Figure 12, which indicates that the time to reach the peak plasma concentration t_{max} appears to be delayed as the hydrophilicity of penetrant molecule increases. In addition, penetrant hydrophilicity also appears to affect the elimination kinetics of progestins; the slope of the elimination curves decreases as the hydrophilicity of progestins increases. There are no significant differences in the elimination rate constants of each progestin following nasal, rectal, or vaginal delivery (Table 9).

The rate constants for absorption K_a of progestins following various mucosal routes of delivery are compared in Table 9. Analysis of the data indicate that, fol-

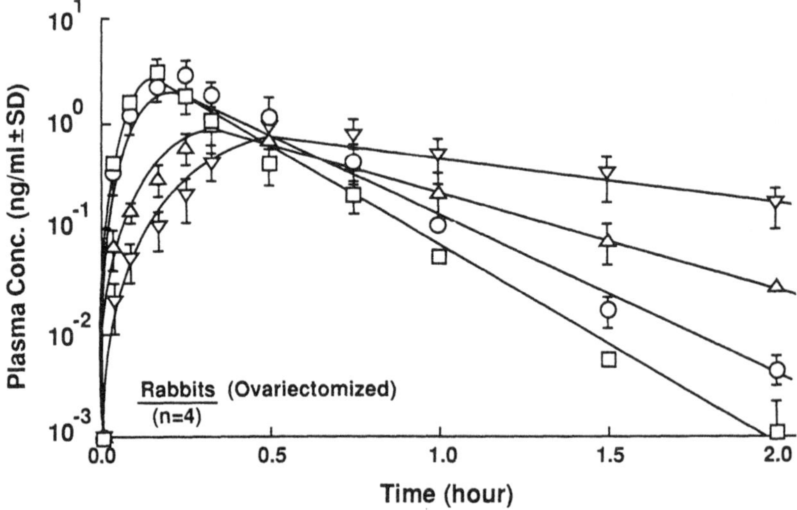

Figure 12 Comparative plasma profiles in ovariectmized rabbits (n = 4) following nasal delivery of progesterone (□) and its monohydroxy (○), dihydroxy (△), and trihydroxy (▽) derivatives in solution formulation. (Plotted from the data by Corbo et al., 1989.)

Table 9 Rate Constants of Absorption and Elimination Following Mucosal Delivery of Progestins[a]

Progestin	First-order rate constant ($hr^{-1} \pm$ SEM)[b]						
	Absorption			Elimination			
	Nasal	Rectal	Vaginal	Nasal	Rectal	Vaginal	Intravenous
Progesterone (P)	11.4 (1.1)	4.5 (0.6)	4.8 (0.5)	5.4 (1.3)	5.3 (1.0)	5.0 (0.9)	6.4 (1.7)
Mono-OH-P	9.0 (0.7)	4.9 (0.7)	5.4 (0.7)	4.2 (0.9)	4.2 (1.2)	3.7 (0.8)	4.5 (1.2)
Di-OH-P	6.0 (0.6)	3.2 (0.4)	2.0 (0.3)	2.2 (0.4)	2.8 (0.6)	2.1 (0.5)	2.3 (0.7)
Tri-OH-P	3.6 (0.2)	1.4 (0.3)	1.8 (0.3)	1.6 (0.2)	1.5 (0.3)	1.4 (0.3)	1.7 (0.3)

[a]In ovariectomized rabbits.
[b]Mean (± standard error of the mean) of four rabbits.
Source: Compiled from the data by Corbo (1989).

lowing nasal delivery, there is a substantial reduction in the K_a value as the number of hydroxy groups on the progesterone molecule increases. A similar trend is also followed for the rectal and vaginal routes, although the rate constants for the rectal and vaginal absorption of progesterone and its monohydroxy derivative are not statistically different. As observed for progesterone, the K_a values for all the hydroxyprogesterones are consistently higher following nasal delivery than rectal and vaginal delivery. The rate constants for absorption via various mucosal membranes were found to be linearly dependent upon the lipophilicity of the penetrant molecule (Figure 13). This linear dependence suggests that lipophilicity is an important factor in determining the rate constant of mucosal absorption. These findings are very similar to the results observed earlier, under in vitro conditions, for the relationship between the permeability coefficient and the partition coefficient (Figure 8). Both in vitro and in vivo studies have demonstrated that nasal mucosa is more sensitive to the effect of penetrant hydrophilicity than rectal and vaginal mucosae.

As reported earlier, following the mucosal delivery of progesterone in a soluton formulation the highest systemic bioavailability was achieved by the nasal route followed by the rectal route and then by the vaginal route (Table 8). In all cases, the noninvasive delivery of progesterone through the mucosal membranes of these easily accessible body cavities has attained a systemic bioavailability that is substantially greater than the bioavailability (9.5%) by intragastric administration. This finding demonstates that the mucosal delivery of progesterone can minimize the extent of presystemic elimination associated with oral administration, which has resulted in a 5- to 9-fold increase in systemic bioavailability.

Further investigations indicated that the systemic bioavailability of progestins by mucosal delivery is dependent upon their hydrophilicity (87). The results summarized in Table 10 suggest that the systemic bioavailability of progestins by mucosal

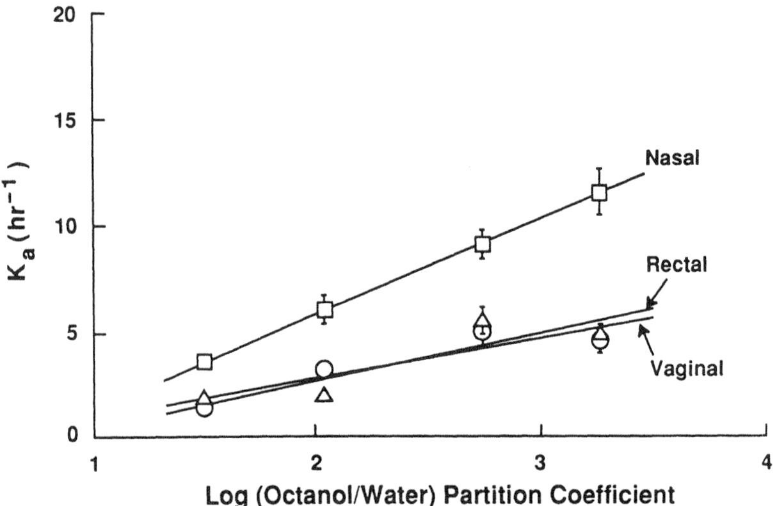

Figure 13 Linear relationship between the rate constants of mucosal absorption following nasal (□), rectal (○), and vaginal (△) delivery and the logarithmic (octanol/water) partition coefficients of the progestins. (Replotted from the data by Corbo et al., 1989.)

Table 10 Systemic Bioavailability of Progestins Following Mucosal Delivery via Immediate-Release Spray Solution Formulation and Effect of Penetrant Hydrophilicity

	Systemic bioavailability (% ± SEM)[b]			
		Hydroxyprogesterone		
Route[a]	Progesterone	Mono-OH	Di-OH	Tri-OH
Mucosal				
Nasal	88.4 (10.1)	90.7 (9.4)	74.1 (7.6)	55.3 (6.2)
Rectal	58.8 (6.7)	52.8 (6.2)	43.6 (5.1)	37.9 (4.2)
Vaginal	46.6 (5.4)	43.8 (6.7)	34.0 (4.3)	32.1 (3.9)
Oral	9.5 (6.2)	6.8 (4.6)	5.6 (3.8)	2.7 (1.8)

[a]Ovariectomized rabbits.
[b]Mean (± standard error of the mean) of four rabbits.
Source: Compiled from the data by Corbo (1989) and Corbo et al. (1989).

delivery decreases as the number of hydroxy groups on the progesterone molecules increases, with the exception of the nasal delivery of monohydroxyprogesterone. However, in all cases mucosal delivery substantially improved the systemic bioavailability of progestins. The extent of bypassing the hepatic first-pass metabolism is dependent upon the route of transmucosal absorption and the hydrophilicity of progestins.

All the in vivo results have shown that nasal delivery is characterized by a faster absorption and a greater bioavailability than rectal and vaginal administration (Tables 9 and 10). These differences may be attributed to the variations in drug absorption and/or metabolism in these mucosae. The results of in vitro permeation studies reported earlier also indicated that nasal mucosa has greater mucosal permeability than rectal and vaginal mucosae (Figure 6 and Table 4).

The results of in vivo studies outlined in Table 10 suggest that the systemic bioavailability, an indication of the extent of absorption, following mucosal delivery decreases as the number of hydroxy groups on the progesterone molecule increases (which leads to an increase in the hydrophilicity of the penetrant molecule). This trend may be related to the effect of penetrant hydrophilicity on the rate constant of mucosal absorption (Figure 13). As expected, the systemic bioavailability of a progestin is dependent upon its rate constant of mucosal absorption (Figure 14). The results indicate that the systemic bioavailabilities of progesterone and its hydroxy derivatives increase linearly at first, with the increase in the rate constant of mucosal absorption K_a, and then plateau at K_a values above 5 hr^{-1}. This relationship is followed for all the mucosal membranes studied, but the slope of the bioavailability versus K_a linearity varies from one mucosa to another, with a rank order of nasal > rectal ≥ vaginal mucosa. The findings are logical since the systemic bioavailability of a drug molecule is dependent upon the relative magnitude of the rates between absorption and elimination. A penetrant with a faster rate of absorption through an administration site and/or a slower rate of elimination would have yielded a higher level of systemic bioavailability.

From the results in Figure 14 and Table 9 it is also apparent that the nasal route is characterized by a greater systemic bioavailability and a faster rate of absorption than the rectal and vaginal routes. The observed differences may be attributed to the

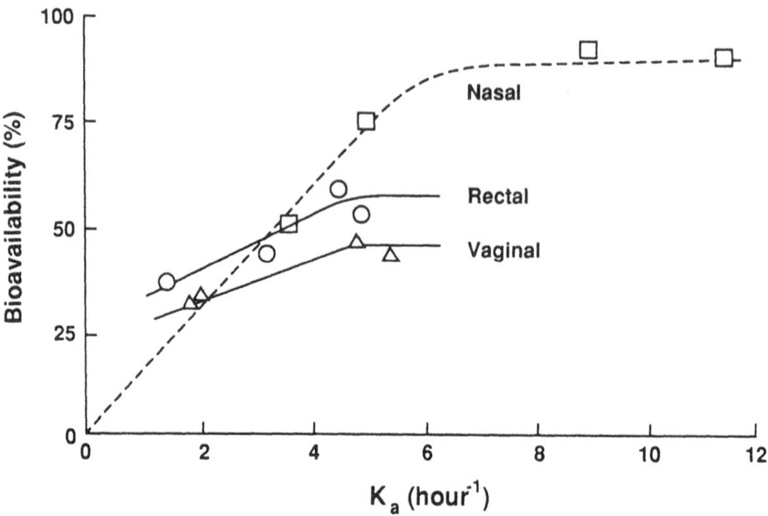

Figure 14 Relationship between the systemic bioavailability of progestins following nasal (□), rectal (○), and vaginal (△) absorption and their rate constants of mucosal absorption (K_a). (Replotted from the data by Corbo et al., 1989.)

differences in the apparent permeabilities of these mucosae to progesterone and its hydroxy derivatives (Table 5).

C. Rate-Controlled Transmucosal Delivery of Progestins

The effect of the mode of drug delivery on the pharmacokinetics of transmucosal drug delivery was first studied using the nasal route. The plasma profile of progesterone following nasal delivery from microporous membrane-controlled reservoir-type nasal delivery device was compared with a nasal solution formulation delivered by a metered-dose sprayer in ovariectomized rabbits (88). The results in Figure 15 indicate that the nasal delivery of progesterone by the immediate-release spray formulation led to a rapid attainment of the peak plasma level of progesterone ($t_{max} <$ 2 min), indicating a rapid absorption of progesterone by the nasal mucosa. On the other hand, the intranasal administration of progesterone by the controlled-release nasal delivery device produced a gradual increase in the plasma concentration of progesterone, which reaches a plateau level within 15–30 min and remains at the steady-state level for 3–4 hr during the course of a 6-hr study. Apparently, a significant prolongation of the plasma progesterone level was achieved by the controlled delivery of progesterone from the microporous membrane-controlled nasal delivery device. A systemic bioavailability of 72.4 ± 25.7% was obtained by the controlled-release nasal device, which is not significantly different from the 82.5 ± 13.5% achieved by the immediate-release nasal spray ($p < 0.05$).

Using a cylindrical matrix-type silicone polymer insert (106,107), the effect of the drug delivery mode on the pharmacokinetics of transmucosal drug delivery was further studied, including nasal as well as the rectal and vaginal routes. The plasma profile of progesterone delivered from this controlled-release silicone device was compared again with the immediate-release spray formulation in ovariectomized rab-

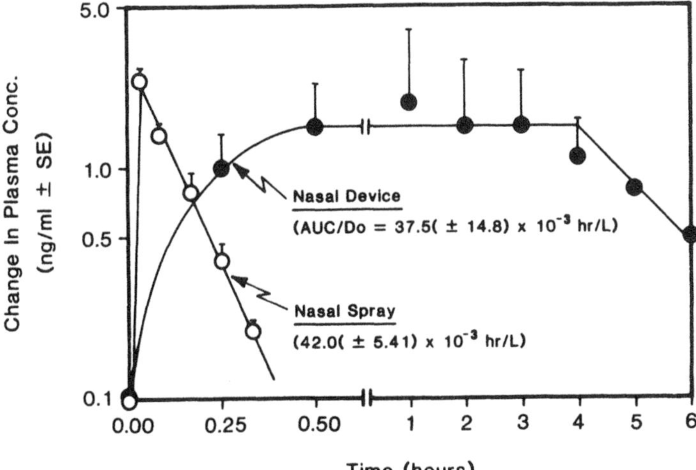

Figure 15 Comparative plasma profiles of progesterone in ovariectomized rabbits following nasal delivery by an immediate-release spray formulation and a controlled-release nasal device. (Plotted from the data by Corbo et al., 1988.)

bits (44). A representative set of results is shown in Figure 16, which indicates that, similar to the nasal route of administration, the rectal delivery of progesterone by the immediate-release spray formulation also led to a rapid attainment of the peak plasma level of progesterone ($t_{max} \leq 20$ min). In comparison, the controlled-release silicone device yielded a gradual increase in the plasma concentration of progester-

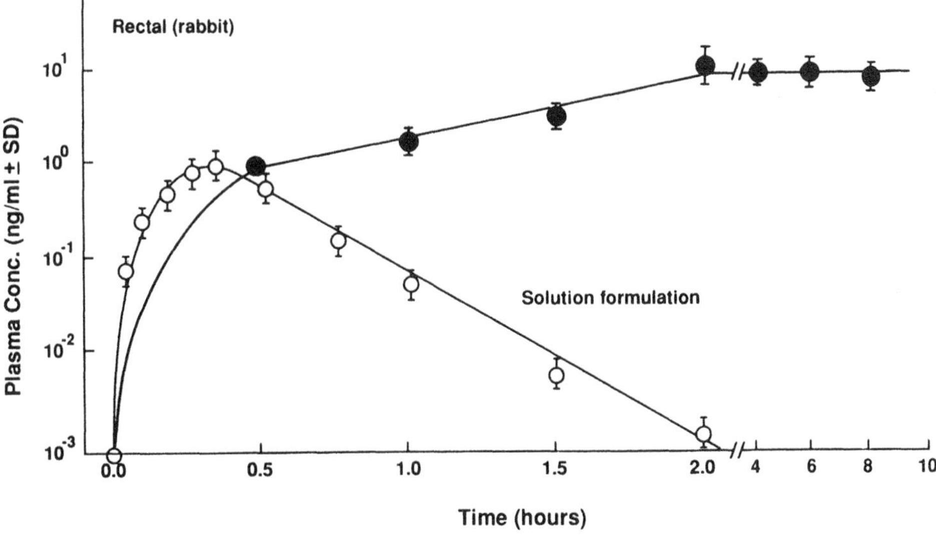

Figure 16 Comparative plasma profiles of progesterone in ovariectomized rabbits following rectal delivery by an immediate-release spray formulation and a controlled-release silicone polymer insert. (Plotted from the data by Corbo, 1989.)

one, and it takes approximately 2 hr to reach a plateau level. This steady-state plateau level, which is almost 10-fold higher than that attained by the spray formulation, was maintained throughout the course of the 8-hr study. Similar results were also obtained with the transmucosal delivery via the nasal and vaginal routes (Figure 17). It is interesting to note that the controlled delivery of progesterone via the controlled-release silicone device achieved essentially the same extent of systemic bioavailability as that attained by the immediate-release spray formulation for each route of transmucosal delivery (Figure 18). Both types of mucosal drug delivery systems yielded a systemic bioavailability 5–10 times greater than that via gastrointestinal administration.

The investigation was extended to evaluate the effect of the variation in hydrophilicity on the mucosal controlled delivery of the progestin molecule. The plasma profiles in Figure 19 indicate that by controlled delivery via a controlled-release silicone polymer insert, progesterone and its hydroxy derivatives permeate through the vaginal mucosa at zero-order kinetics and reach peak plasma concentrations within 3–4 hr. These peak plasma concentrations were maintained at a steady-state level throughout the course of the 8-hr study. The rate of vaginal permeation, as indicated by the slope of the absorption phase (Figure 19) and the magnitude of the steady-state plasma levels, appears to decrease as the number of hydroxy groups on the progestin molecule, or the hydrophilicity, increase. The same trend was also observed for mucosal controlled permeation through the nasal and rectal mucosae (Table 11). The results are in good agreement with that observed earlier on the effect of hydrophilicity on the apparent permeability of various mucosae determined by in vitro studies (Table 5). The role of hydrophilicity in determining the mucosal permeation of progestins was also reflected in the systemic bioavailability (Table 12).

In vitro mucosal permeation studies were also performed to investigate the effect

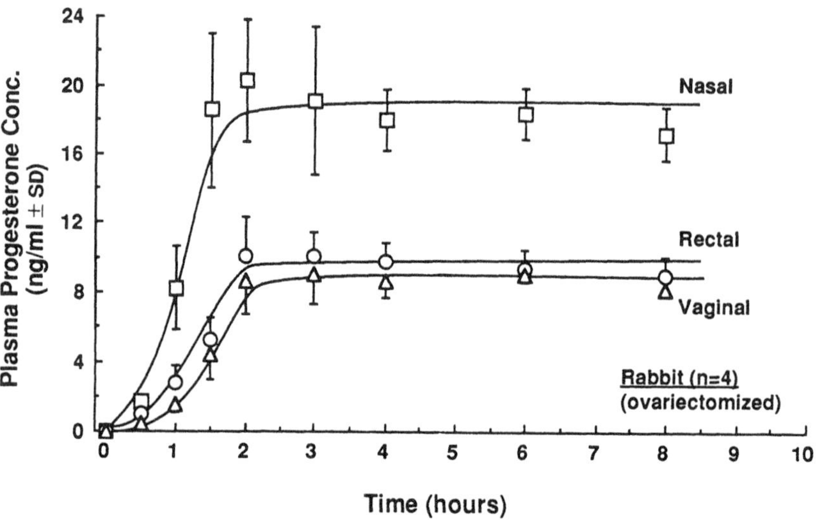

Figure 17 Comparative plasma profiles of progesterone in ovariectomized rabbits following mucosal delivery by a controlled-release silicone polymer insert: (□) nasal (○) rectal, and (△) vaginal mucosae. (Plotted from the data by Corbo, 1989.)

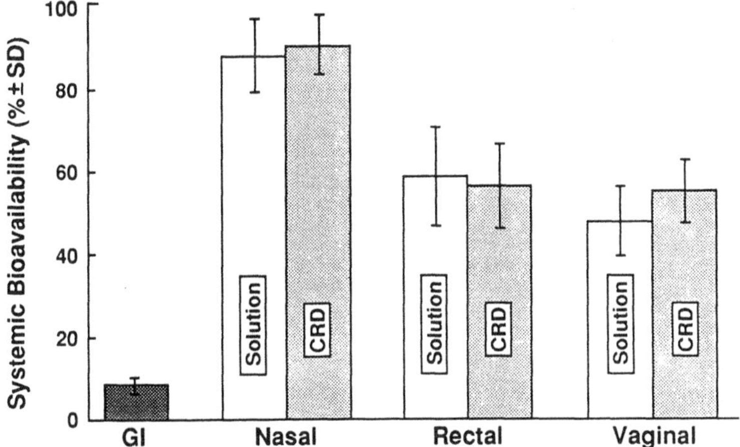

Figure 18 Comparison of the systemic bioavailability of progesterone following mucosal delivery by an immediate-release spray formulation and a controlled-release silicone polymer insert. (Plotted from the data by Corbo, 1989.)

of drug delivery by a controlled-release mode on the mucosal permeation kinetics of progestins (44). A representative set of permeation profiles is shown in Figure 20. The results indicate that the controlled delivery of progesterone achieved zero-order permeation kinetics, similar to that from a solution formulation (Figure 6). The same trend was also followed for the hydroxy derivatives of progesterone. The data in Table 13 demonstrate that the transmucosal permeation rates of progestins decrease as the number of hydroxy groups on the progestin molecule (or hydrophilicity)

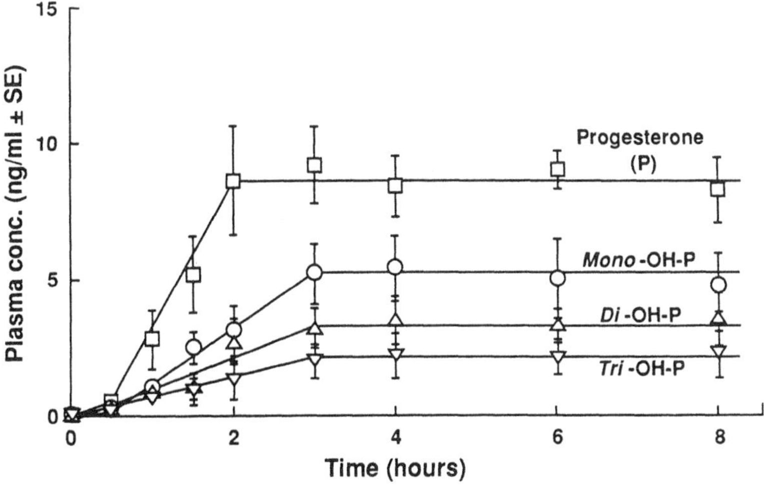

Figure 19 Comparative plasma profiles of progestins in ovariectomized rabbits following vaginal delivery by a controlled-release silicone polymer insert containing progesterone (□) and its monohydroxy (○), dihydroxy (△), and trihydroxy (▽) derivatives. (Plotted from the data by Corbo, 1989.)

Table 11 Steady-State Plasma Concentration of Progestins Following Controlled Mucosal Delivery[a] and Effect of Penetrant Hydrophilicity

	Steady-state plasma concentration (ng/ml ± SEM)[c]			
		Hydroxyprogesterone		
Route[b]	Progesterone	Mono-OH	Di-OH	Tri-OH
Nasal	19.0 (2.9)	9.6 (1.4)	5.3 (0.8)	5.0 (0.7)
Rectal	10.0 (1.5)	6.2 (0.9)	3.7 (0.5)	2.3 (0.3)
Vaginal	8.8 (1.3)	5.4 (0.8)	2.3 (0.3)	2.1 (0.3)

[a]Progestin-releasing silicone polymer insert.
[b]Ovariectomized rabbits.
[c]Mean (± standard error of the mean) of four rabbits.

Table 12 Systemic Bioavailability of Progestins Following Controlled Mucosal Delivery[a] and Effect of Penetrant Hydrophilicity

	Systemic bioavailability (% ± SEM)[c]			
		Hydroxyprogesterone		
Route[b]	Progesterone	Mono-OH	Di-OH	Tri-OH
Nasal	89.6 (10.1)	85.2 (11.6)	82.1 (10.6)	69.9 (8.2)
Rectal	55.1 (9.7)	51.3 (8.1)	55.9 (6.4)	41.2 (5.0)
Vaginal	53.5 (7.8)	45.4 (6.3)	48.5 (6.7)	40.0 (4.8)

[a]Progestin-releasing silicone polymer insert.
[b]Ovariectomized rabbits.
[c]Mean (± standard error of the mean) of four rabbits.
Source: Compiled from the data by Corbo (1989).

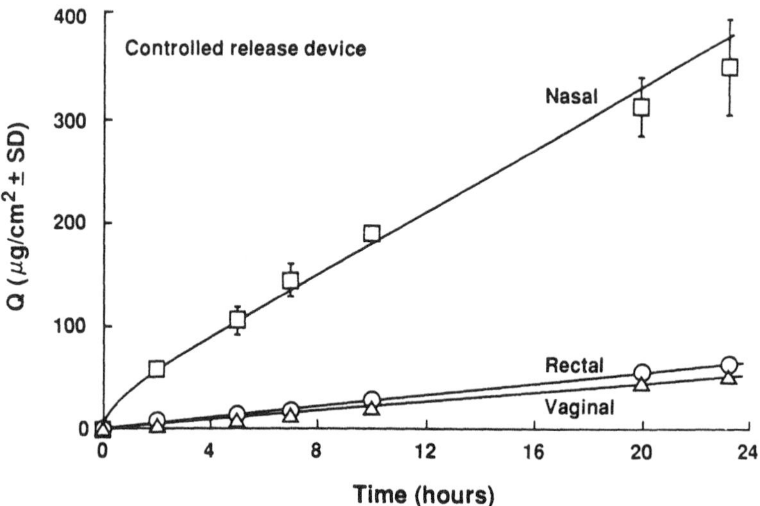

Figure 20 In vitro mucosal permeation profiles of progesterone delivered by controlled-release silicone polymer device: (□) nasal, (○) rectal, and (△) vaginal mucosa. (Plotted from the data by Corbo, 1989.)

Table 13 Transmucosal Permeation Rates of Progestins Delivered from a Controlled-Release Silicone Insert and Effect of Penetrant Hydrophilicity

	Rate of permeation ($\mu g/cm^2 \cdot hr \pm SD$)[b]			
		Hydroxyprogesterone		
Route[a]	Progesterone	Mono-OH	Di-OH	Tri-OH
Nasal	12.6 (2.6)	3.2 (1.2)	3.4 (0.9)	0.7 (0.3)
Rectal	2.6 (0.6)	1.3 (0.5)	0.9 (0.3)	0.6 (0.1)
Vaginal	2.2 (0.5)	1.1 (0.3)	0.9 (0.2)	0.5 (0.1)

[a]Mucosa was freshly excised from rabbits.
[b]Mean (± standard deviation) of three determinations.
Source: Compiled from the data by Corbo (1989).

increases. Similar to the results on apparent permeability determined from the in vitro permeation studies of progestins in a solution formulation (Table 5), the rates of transmucosal permeation achieved by the controlled delivery of progestins also have the same rank order of mucosal delivery: nasal > rectal ≥ vaginal mucosa (Table 13).

REFERENCES

1. A. K. Banga and Y. W. Chien; Systemic delivery of therapeutic peptides and proteins. Int. J. Pharm., 48:15–50 (1988).
2. Y. W. Chien, K. S. E. Su and S. F. Chang; *Nasal Systemic Drug Delivery,* Dekker, New York, 1989.
3. H. Lenz; Three-dimensional surface representation of the cilia free nasal mucosa of man. Acta Otolaryngol. (Stockh.), 76:47 (1973).
4. N. Mygind and P. Bretlau; Scanning electron microscopic studies of the human nasal mucosa in normal persons and in patients with perennial rhinitis. Acta Allergol., 28:9 (1973).
5. J. L. Carson, A. M. Collier, M. R. Knowles, R. C. Boucher and J. C. Rose; Morphometric aspects of ciliary distribution and ciliogenesis in human nasal epithelium. Proc. Natl. Acid. Sci. USA, 78:6996 (1981).
6. M. Okuda and T. Kanda; Scanning electron microscopy of the nasal mucous membrane. Acta Otolaryngol., (Stockh.) 76:283 (1973).
7. P. P. C. Graziadei; The mucous membranes of the nose. Ann. Otol., 79:433 (1970).
8. H. Goldman and D. A. Antonioli; Mucosal biopsy of the rectum, colon and distal ileum. Hum. Pathol., 13:981 (1982).
9. V. Lorenzsonn and J. S. Trier; The fine structure of human rectal mucosa: The epithelial lining of the base of the crypt. Gastroenterology, 55:88 (1968).
10. G. I. Kaye, C. M. Fenoglio and R. R. Pascal; Comparative electron microscopic features of normal, hyperplastic and adenomatous colonic epithelium. Gastroenterology, 64:926 (1973).
11. S. Eidelman and D. Lagunoff; The morphology of the normal human rectal biopsy. Hum. Pathol., 3:389 (1972).
12. J. B. Bernstine and A. E. Rakoff; *Vaginal Infections, Infestations and Discharges,* The Blakiston Co., New York, 1953.
13. R. L. Dickinson; *Human Sex Anatomy,* Williams & Wilkins, Baltimore, 1949.

14. D. H. Nichols and C. L. Randall; *Vaginal Surgery,* Williams & Wilkins, Baltimore, 1976.
15. R. Cruickshank and A. Sharman; The biology of the vagina in the human subject. Br. J. Obstet. Gynecol., 41:190 (1934).
16. A. E. Rackoff, L. G. Feo and L. Goldstein; The biologic characteristics of the normal vagina. Am. J. Obstet. Gynecol., 47:467 (1944).
17. H. F. Traut, P. W. Bloch and A. Kuder; Cyclical changes in the human vaginal mucosa. Surg. Gynecol. Obstet., 63:17 (1936).
18. M. H. Burgos and C. E. R. Vargas-Linares; Cell junctions in the human vaginal epithelium. J. Anat., 108:571 (1972).
19. H. E. Karrer; Cell connections in normal human cervical epithelium. J. Biophys. Biochem. Cytol., 7:181 (1960).
20. S. Hwang, E. Owada, L. Suhardj., N. F. H. Ho, G. L. Flynn and W. I. Higuchi; Systems approach to vaginal delivery of drugs. IV. Methodology for determination of membrane surface pH. J. Pharm. Sci., 66:778 (1977).
21. J. S. Beilly; Determination of pH of vaginal secretion as an index of ovarian activity in hypoovarian states. Endocrinology, 26:959 (1940).
22. R. Cruickshank; Conversion of glycogen of vagina into lactic acid. J. Pathol. Bacteriol., 39:213 (1934).
23. F. W. Oberst and E. D. Plass; pH of human discharge. Am. J. Obstet. Gynecol., 32:22 (1936).
24. W. Curatolo; The lipoidal permeability barriers of the skin and alimentary tract. Pharm. Res., 4:271 (1987).
25. S. D. Kashi and V. H. L. Lee; Enkephalin hydrolysis in homogenates of various absorptive mucosae of the albino rabbit: Similarities in rates and involvement of aminopeptidases. Life Sci., 38:2019 (1986).
26. E. B. Brittebo; Metabolism of progesterone by the nasal mucosa in mice and rats. Acta Pharmacol. Toxicol. (Copenh.) 51:441 (1982).
27. E. B. Brittebo and J. J. Rafter; Steroid metabolism by rat nasal mucosa: Studies on progesterone and testosterone. J. Steroid Biochem., 20:1147 (1984).
28. M. R. Hancock and D. B. Gower; Properties of 3-alpha and 3-beta-hydroxysteroid dehydrogenases of porcine nasal epithelium: Some effects of steroid hormones. Biochem. Soc. Trans., 14:1033 (1986).
29. C. T. Dollery and D. S. Davies; Differences in the metabolism of drugs upon their routes of administration. Ann. N.Y. Acad. Sci., 172:108 (1972).
30. R. J. Moorehead, M. Hoper and S. T. D. McKelvey; Assessment of ornithine decarboxylase activity in rectal mucosa as a marker for colorectal adenomas and carcinomas. Br. J. Surg., 74:364 (1987).
31. A. G. de Boer, F. Moolenaar, L. G. J. de Leede and D. D. Breimer; Rectal drug administration: Clinical pharmacokinetic considerations. Clin. Pharmacokinet., 7:285 (1982).
32. H. G. Boxenbaum, I. Bekersky, M. L. Jack and S. A. Kaplan; Influence of gut microflora on Bioavailability. Drug Metab. Rev., 9:259 (1979).
33. R. R. Scheline; Metabolism of foreign compounds by gastrointestinal microorganisms. Pharmacol. Rev., 25:451 (1973).
34. J. C. M. Taibris, P. W. Langenberg, F. S. Khan-Dawood, W. N. Spellacy; Cervicovaginal peroxidases: Sex hormone control and potential clinical uses. Fertil. Steril., 44:236 (1985).
35. B. F. King; Ultrastructural localization of acid phosphatase in human primate vaginal epithelium. Cell Tissue Res., 239:249 (1985).
36. S. J. Singer and G. L. Nicolson; The fluid mosaic model of the structure of cell membranes. Science, 175:720 (1972).

37. S. J. Singer; The molecular organization of membranes. Annu. Rev. Biochem., 43:805 (1974).
38. S. Abrahamsson and I. Pascher; *Structure of Biological Membranes,* Plenum Press, New York, 1977, pp. 443–461.
39. E. E. Bittar; *Membrane Structure and Function,* John Wiley and Sons, New York, 1980, pp. 4–41.
40. R. C. Aloia; *Membrane Fluidity in Biology: General Principles,* Vol. 2, Academic Press, New York, 1979, pp. 141–143.
41. G. Benga; *Structure and Properties of Cell Membranes,* Vol. 1, CRC Press, Boca Raton, Florida, 1985, pp. 159–168.
42. R. A. Capaldi; *Membrane Proteins and Their Interactions With Lipids,* Marcel Dekker, New York, 1977, pp. 11–16.
43. T. E. Andreoli, J. F. Hoffman, D. D. Fanestil and S. G. Schultz; *Membrane Physiology*, Plenum Medical Book Company, New York, 1986, pp. 3–16.
44. D. C. Corbo; Transmucosal drug delivery and absorption: Kinetics and mechanisms of progestin permeation, Ph.D. Dissertation, Rutgers University, New Brunswick, N.J.
45. B. J. August, N. J. Rogers and E. Shefter; Comparison of nasal, rectal, buccal, sublingual and intramuscular insulin efficacy and the effects of a bile salt absorption promotor. J. Pharmacol. Exp. Ther., 224:23–27 (1988).
46. D. W. Powell; Barrier function of epithelia. Am. J. Physiol., 241:G275 (1981).
47. R. J. Scheuplein, I. H. Blank, G. J. Brauner and D. J. MacFarlane; Percutaneous absorption of steroids. J. Invest. Dermatol., 52:63 (1969).
48. R. J. Scheuplein; Mechanism of perceutaneous absorption. I. Routes of penetration and the influence of solubility. J. Invest. Dermatol., 45:334 (1966).
49. T. W. Schultz and H. Y. Ando; A relationship between diffusional transport in lipid membranes and a lipophilic-eutectic parameter. J. Theor. Biol., 123:367 (1986).
50. T. Nook, E. Doelker and P. Buri; Intestinal absorption kinetics of various model drugs in relation to partition coefficients. Int. J. Pharm., 43:119 (1988).
51. C. Akermann, G. L. Flynn and W. M. Smith; Ether-water partitioning and permeability through nude mouse skin in vitro. II. Hydrocortisone 21-n-alkyl esters, alkanols and hydrophilic compounds. Int. J. Pharm., 36:67 (1987).
52. J. E. Treherne; The permeability of skin to some non-electrolytes. J. Physiol. (Lond.) 133:171 (1956).
53. R. J. Scheuplein and I. H. Blank; Mechanism of percutaneous absorption. IV. Penetration of non-electrolytes (alcohols) from aqueous solutions and from pure liquids. J. Invest. Dermatol., 60:286 (1973).
54. K. Tojo, C. C. Chiang and Y. W. Chien; Drug permeation across the skin: Effect of penetrant hydrophilicity. J. Pharm. Sci., 76:123 (1987).
55. W. E. Jetzer, A. S. Hou, N. F. H. Ho, G. L. Flynn, N. Duraiswamy and L. Condie; Permeation of mouse skin and silicone rubber membranes by phenols: Relationship to in-vitro partitioning. J. Pharm. Sci., 75:1098 (1986).
56. G. S. M. J. E. Duchateau, J. Zuidema, W. M. Albers and F. W. H. M. Merkus; Nasal absorption of alprenolol and metoprolol. Int. J. Pharm., 34:131 (1986).
57. R. E. Gibson and L. S. Olanoff; Physicochemical determinants of nasal drug absorption. J. Control. Rel., 6:361 (1987).
58. C. H. Huang, R. Kimura, R. Bawarshi-Nassar and A. Hussain; Mechanism of nasal absorption of drugs. I. Physicochemical parameters influencing the rate of in situ nasal absorption of drugs in rats. J. Pharm. Sci., 74:608 (1985).
59. C. H. Huang, R. Kimura, R. Bawarshi-Nassar and A. Hussain; Mechanism of nasal absorption of drugs. II. Absorption of L-tyrosine and the effect of structural modification on its absorption. J. Pharm. Sci., 74:1298 (1985).

60. S. Hirai, T. Yashiki, T. Matsuzawa and H. Mima; Absorption of drugs from the nasal mucosa of rat. Int. J. Pharm., 7:317 (1981).
61. Y. Kaneo; Absorption from the nasal mucous membrane. Acta Pharm. Suec., 20:379 (1983).
62. T. Ohwaki, H. Ando, F. Kakimoto, K. Uesugi, S. Watanabe, K. Miyake and M. Kayano; Effect of dose, pH and osmolarity on nasal absorption of secretin in rats. II. Histological aspects of the nasal mucosa in relation to the absorption variation due to the effects of pH and osmolarity. J. Pharm. Sci., 76:695 (1987).
63. A. N. Fisher, K. Brown, S. S. Davis, G. D. Parr and D. A. Smith; The effect of molecular size on the nasal absorption of water-soluble compounds in the albino rat. J. Pharm. Pharmacol., 39:357 (1987).
64. C. McMartin, L. E. F. Hutchinson, R. Hyde and G. E. Peters; Analysis of the structural requirements for the absorption of drugs and macromolecules from the nasal cavity. J. Pharm. Sci., 76:535 (1987).
65. M. Hayashi, T. Hirasawa, T. Muraoka, M. Shiga and S. Awazu; Comparison of water influx and sieving coefficient in rat jejunal, rectal and nasal absorptions of antipyrine. Chem. Pharm. Bull., 33:2149–2152 (1985).
66. M. C. Alfono, A. I. Chasens and C. W. Masi; Autoradiographic study of the penetration of radiolabeled dextrans and inulin through non-keratinized oral mucosa in vitro. J. Periodont. Res., 12:368–377 (1977).
67. I. A. Siegel; Permeability of the oral mucosa, in: *The Structure and Function of Oral Mucosa* (J. Meyer et al., Eds.), Pergamon Press, New York, 1984, pp. 95–108.
68. C. A. Squier; Membrane-coating granules in non-keratinizing oral epithelium. J. Ultrastruct. Res., 60:212–220 (1977).
69. C. A. Squier; The permeability of keratinized and non-keratinized oral epithelium to horseradish peroxidase. J. Ultrastruct. Res., 43:160–177 (1973).
70. S. Kaaber; The permeability and barrier functions of the oral mucosa with respect to water and electrolytes. Acta Odontal. Scand., 32(Suppl.):66 (1974).
71. I. A. Siegel and K. T. Izutsu; Permeability of oral mucosa to organic compounds. J. Dent. Res., 59:1604–1605 (1980).
72. K. Morimoto, T. Iwamoto and K. Morisaka; Possible mechanisms for the enhancement of rectal absorption of hydrophilic drugs and polypeptides by aqueous polyacrylic acid gel. J. Pharmacobiodyn., 10:85–91 (1987).
73. H. Yamada and R. Yamamoto; Bipharmaceutical studies on factors affecting rate of absorption of drugs. Chem. Pharm. Bull., 13:1279 (1965).
74. L. S. Schanker; Absorption of drugs from the rat colon. J. Pharmacol. Exp. Ther., 126:283 (1959).
75. K. Kakemi, T. Arita, R. Hori, R. Konishi, K. Nishimura, H. Matsui and T. Nishimura; Absorption and excretion of drugs. XXXIV. An aspect of the mechanism of drug absorption from the intestinal tract in rats. Chem. Pharm. Bull., 17:255 (1969).
76. P. A. Shore, B. B. Brodie and C. A. M. Hogben; The gastric secretion of drugs: A pH-partition hypothesis. J. Pharmacol. Exp. Ther., 119:361 (1957).
77. S. Muranishi; Characteristics of drug absorption via the rectal route. Methods Find. Exp. Clin. Pharmacol., 6:763–772 (1984).
78. C. A. M. Hogben, D. J. Tocco, B. B. Brodie and L. S. Schander; On the mechanism of intestinal absorption of drugs. J. Pharmacol. Exp. Ther., 125:321 (1959).
79. K. Kakemi, T. Arita and S. Muranishi; Absorption and excretion of drugs. XXV. On the mechanism of rectal absorption of sulfonamides. Chem. Pharm. Bull., 13:861–869 (1965).
80. S. Hwang, E.Owada, T. Yotsuyanagi L. Suhardja, N. F. H. Ho, G. L. Flynn and W.

I. Higuchi; Systems approach to vaginal delivery of drugs. II. In situ vaginal absorption of unbranched aliphatic alcohols. J. Pharm. Sci., 65:1574–1578 (1976).
81. S. Hwang, E. Owada, L. Suhardja, N. F. H. Ho, G. L. Flynn and W. I. Higuchi; Systems approach to vaginal delivery of drugs. III. In situ vaginal absorption of 1-alkanoic acids. J. Pharm. Sci., 66:781–784 (1977).
82. B. F. King; The permeability of nonhuman primate vaginal epithelium: A freeze-fracture and tracer-perfusion study. J. Ultrastruct. Res., 83:99 (1983).
83. H. Okada, T. Yashiki and H. Mima; Vaginal absorption of a potent luteinizing hormone-releasing hormone analogue (Leuprolide) in rats. III. Effects of estrous cycle on vaginal absorption of hydrophilic model compounds. J. Pharm. Sci., 72:173 (1983).
84. E. Winterhager and W. Kuhnel; Diffusion barriers in the vaginal epithelium during the estrous cycle in guinea pigs. Cell Tissue Res., 241:325 (1985).
85. S. Hirai, T. Yashiki, T. Matsuzawa and H. Mima; Int. J. Pharm., 7:317–325 (1981).
86. C. McMartin, L. E. F. Hutchinson, R. Hyde and G. E. Peters; J. Pharm. Sci., 76:535–540 (1987).
87. D. C. Corbo, J. C. Liu and Y. W. Chien; Drug absorption through mucosal membranes: Effect of mucosal route and penetrant hydrophilicity, Pharm. Res., 6:846–850 (1989).
88. D. C. Corbo, Y. C. Huang and Y. W. Chien; Nasal delivery of progestational steroids in ovariectomized rabbits. I. Progesterone—comparison of pharmacokinetics with intravenous and oral administration. Int. J. Pharm., 46:133–140 (1988).
89. G. F. X. David, C. P. Puri, T. C. A. Kumar and C. Anand; Experientia, 37:533–534 (1981).
90. A. A. Hussain, S. Hirai and R. Bawarsh; J. Pharm. Sci., 70:466–467 (1981).
91. S. Askel and G. S. Jones; Am. J. Obstet. Gynecol., 118:466–472 (1974).
92. S. J. Nillus and E. D. B. Johansson; Am. J. Obstet. Gynecol., 110:470–477 (1971).
93. B. Villaneuva, R. F. Casper and S. S. C. Yen; Fertil. Steril., 35:433–437 (1981).
94. E. R. Meyers, S. J. Sondheimer, E. W. Freeman, J. F. Strauss and K. Richels; Fertil. Steril., 47:71–75 (1987).
95. C. M. A. Glazener, I. Bailey and M. G. R. Hull; Br. J. Obstet. Gynecol., 92:364–368 (1985).
96. D. C. Corbo, Y. C. Huang and Y. W. Chien; Nasal delivery of progestational steroids in ovariectomized rabbits. II. Effect of penetrant hydrophilicity. Int. J. Pharm. 50:253–260 (1989).
97. T. Kanda and D. Hilding; Development of respiratory tract cilia in fetal rabbits. Acta Otolaryngol., (Stockh.) 65:611 (1968).
98. B. D. Mulvaney and H. E. Heist; Mapping of rabbit olfactory cells. J. Anat., 107:19 (1970).
99. B. D. Mulvaney and H. E. Heist; Regeneration of rabbit olfactory epithelium. Am. J. Anat., 131:241–251 (1971).
100. A. C. Allison and R. T. Turner Warwick; Quantitative observations on the olfactory system of the rabbit. Brain, 72:186 (1948).
101. D. C. Corbo, J. C. Liu and Y. W. Chien; Characterization of the barrier properties of mucosal membranes. J. Pharm. Sci., 79:202–206 (1990).
102. M. Okuda and T. Kanda; Acta Otalaryngol. (Stockh.), 76:283–294 (1973).
103. S. Eidelman and D. Lagunoff; Hum. Pathol., 3:390–401 (1972).
104. R. L. Dickinson; *Human Sex Anatomy,* Williams & Wilkins, Baltimore, 1949, Chapters 2 and 3.
105. B. F. King; J. Ultrastruct. Res., 82:1–18 (1983).
106. Y. W. Chien; *Novel Drug Delivery Systems,* Marcel Dekker, New York, 1982, Chapter 3.

107. Y. W. Chien, S. E. Mares, J. Berg, S. Huber, H. J. Lambert and K. F. King; Controlled drug release from polymeric delivery devices. III. In vitro/in vivo correlation for intravaginal release of ethynodiol diacetate in rabbits. J. Pharm. Sci., 64:1776 (1975).
108. U. B. Doshi, C. C. Chiang, K. Tojo and Y. W. Chien; Mode of skin permeation. I. Effect of delipidization. Proc. Int. Symp. Control. Relat. Bioact. Mater., 13:138 (1986).

5
Nasal Drug Delivery and Delivery Systems

I. INTRODUCTION

Historically, nasal drug delivery has received intensive interest since ancient times (1–4). Therapy through intranasal administration has been an accepted form of treatment in the Ayuredic system of Indian medicine. Psychotropic drugs and hallucinogens have been used in the form of snuff by Indians in South America. In more recent years many drugs have been shown to achieve a better systemic bioavailability by self-medication through the nasal route than by oral administration. The systemic bioavailability by nasal delivery of some peptide and protein drugs with low nasal absorption has been improved by coadministering them with absorption promotors, enzyme inhibitors, and/or microspheres fabricated from bioadhesive and bioerodible polymers.

Intranasal administration appears to be an ideal alternative to the parenterals for systemic drug delivery (1–4). The advantages of nasal drug delivery include (i) avoidance of hepatic first-pass elimination, gut wall metabolism, and/or destruction in the gastrointestinal tracts; (ii) the rate and extent of absorption and the plasma concentration versus time profiles are relatively comparable to that obtained by intravenous medication; (iii) the existence of a rich vasculature and a highly permeable structure in the nasal mucosa for systemic absorption; and (iv) the ease and convenience of intranasal drug administration.

II. PHYSIOLOGICAL ASPECTS OF THE NOSE

The nasal passage, which runs from the nasal vestibule, i.e., nasal valve, to the nasopharynx, has a depth of approximately 12–14 cm (Figure 1) (5). The lining is ciliated, highly vascular, and rich in mucus glands and goblet cells. The blanket of nasal mucus is transported in a posterior direction by the synchronized beat of the cilia. An individual cilium is approximately 5 μm in length and 0.2 μm in diameter and moves at a frequency of about 20 beats/sec (6,7).

The rate of diffusion of a nasal preparation through the mucus blanket and its rate of clearance from the nasal cavity may be influenced by the physicochemical

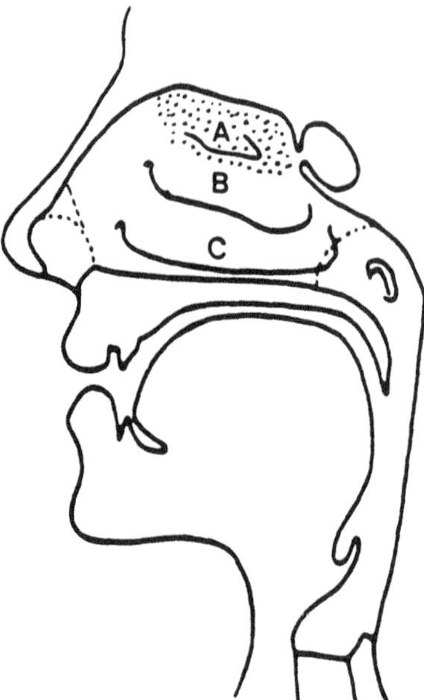

Figure 1 The upper airways seen from the midline. The dashed line just behind the nostril marks the beginning of the nasal valve, whereas the dashed line near the nasopharynx indicates the posterior termination of the nasal septum. A, superior turbinate; B, middle turbinate; C, inferior turbinate. The dotted areas indicate the olfactory region. (Modified from Reference 5.)

properties of the formulation vehicle, the particle size and surface charge of a drug, and any additives incorporated (which may add or reduce the diffusional resistance of the mucus blanket).

Nasal secretions in the adults have a normal pH in the range of 5.5–6.5 and often contain a variety of enzymes (1,8). Insulin (zinc-free) was reportedly hydrolyzed slowly by leucine aminopeptidase in nasal secretions (9), and prostaglandin E, progesterone, and testosterone were also found to be inactivated by other nasal enzymes (1).

An important consideration in nasal drug delivery is the effect of drugs and additives administered intranasally on nasal ciliary functions, and many drugs and additives were noted to have a negative effect, e.g., cocaine, atropine, antihistamines, propranolol, bile salts, xylometazoline, preservatives, and hair spray (1).

Nasal polyposis, atrophic rhinitis, and severe vasomotor rhinitis can reduce the capacity of nasal absorption of such drugs as cerulein (10). The common cold or any pathological conditions involving mucociliary dysfunctions can greatly affect the rate of nasal clearance and subsequently the therapeutic efficacy of drugs administered intranasally.

The vehicle for nasal formulations and mode of application can be optimized to deliver drugs to the absorptive turbinate region. Several factors should be considered in the optimization of nasal drug delivery: (i) methods and techniques of administration, (ii) site of deposition, (iii) rate of clearance, and (iv) minimization of any pathological conditions.

Absorption promotors have long been used to achieve a better systemic bioavailability of nasally-administered drugs. However, the long-term use of an absorption promotor and the chronic use of a promotor-containing nasal formulation could affect the biochemical and biophysical characteristics and the functions of nasal mucosa and thus the efficiency of transnasal permeation. Therefore, the local toxic effects, the possibility of antibody formation, and the tolerance potential of nasal formulations all need to be evaluated.

III. FUNDAMENTALS OF NASAL ABSORPTION

A. Physical and Chemical Parameters

The physical and chemical properties of a drug candidate should be evaluated before the development of nasal drug delivery system.

1. Effect of Molecular Size

It has been reported (11) that nasal absorption falls off sharply for a drug molecule with a molecular weight of greater than 1000 daltons; oral absorption declines even more steeply when the molecular weight goes beyond 400 daltons (Figure 2).

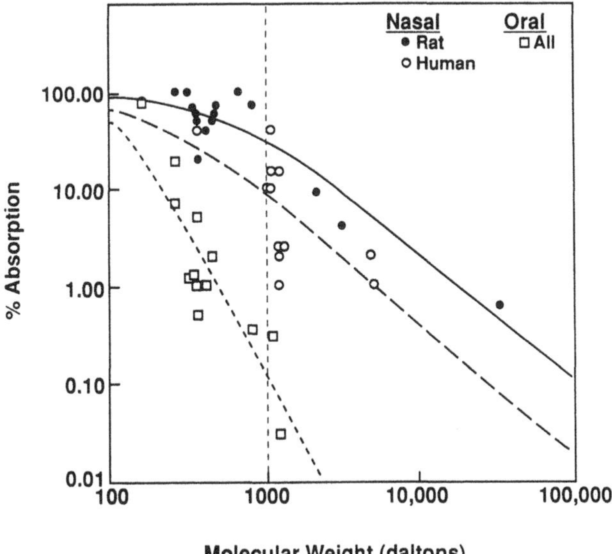

Figure 2 Log-log relationship between the nasal and oral absorption of various molecules and their molecular weight: (●) rat, nasal; (○) human, nasal; (□) all species, oral. The curves are the results of least-squares fitting of the function: % absorption = $100[1 + a(\mathrm{MW})^b]$. Curves: Solid line, rat, nasal ($a = 0.003$, $b = 1.3$); dashed line, human, nasal ($a = 0.001$, $b = 1.35$); dotted line, all species, oral ($a = 8.4 \times 10^{-7}$, $b = 3.01$). (Modified from McMartin et al., 1987.)

The nasal absorption of a wide range of water-soluble compounds with different molecular weights, like inulin, dextran, and *p*-aminohippuric acid, was studied in Wistar rats (12). The results indicated that a good linear correlation exists between the log (percentage of drug absorbed nasally) and the log (molecular weight), suggesting the participation of aqueous channels in the nasal absorption of water-soluble molecules.

2. Effect of Perfusion Rate

Using the ex vivo nasal perfusion technique (Section VB), the nasal administration of phenobarbital was evaluated. The results showed that as the perfusion rate increases the nasal absorption is first increased and then reaches a plateau level that is independent of the rate of perfusion (>2 ml/min) (13).

3. Effect of Perfusate Volume

As the volume of the perfusate solution increases, the first-order disappearance rate of phenobarbital from the perfusion solution has been observed to decrease (Figure 3) (13). Results from studies using drugs with different molecular structures suggested that the intrinsic rate constant varies from one drug to another.

4. Effect of Solution pH

The effect of the pH of a perfusion solution on nasal absorption was examined using a water-soluble ionizable compound, such as benzoic acid (pK_a = 4.2) in the pH range 2.0–7.1. It was found that the extent of absorption is pH dependent, which is higher at a pH lower than the pK_a and decreases as the pH increases beyond the pK_a (Figure 4) (13). The rate of nasal absorption decreased as the pH increased owing to the ionization of the penetrant molecule. A good linear relationship was found to exist between the absorption rate constant of hydralazine and the fraction of its undissociated species calculated from the pK_a value (Figure 5) (14). The nasal absorption rate of decanoic, octanoic, and hexanoic acids was also found to be pH dependent and reached a maximum at pH 4.5, beyond which it decreased steadily as the solution became more acidic or basic (15).

The variation in solution pH was also observed to affect the nasal absorption of peptide-based drugs, such as insulin. For example, the reduction in plasma glucose levels in dogs was noted to depend upon the pH of an insulin solution administered intranasally (Figure 6) (16,17). At pH 6.1 only a slight hypoglycemic effect was attained, whereas at pH 3.1 a reduction of about 55% in the glucose level was achieved. The insulin molecule is known to have an isoelectric point at pH 5.4 and becomes positively charged at a pH lower than its isoelectric pH and is negatively charged at a pH higher than its isoelectric pH (Figure 7).

5. Effect of Drug Lipophilicity

The effect of lipophilicity on the extent of nasal absorption was studied using a series of barbiturates at pH 6.0, at which the barbiturates (pK_a = 7.6) exist entirely in the undissociated form (13). Only a fourfold change in the nasal absorption was noted between pentobarbital and barbital, even though the magnitude of their partition coefficients was different by as much as 40 times. Furthermore, it was also observed that the great difference in the partition coefficient between propranolol and 1-tyrosine resulted in only a very small variation in the rate constant of nasal absorption (13). Results from nasal delivery studies of a series of progestational steroids, which

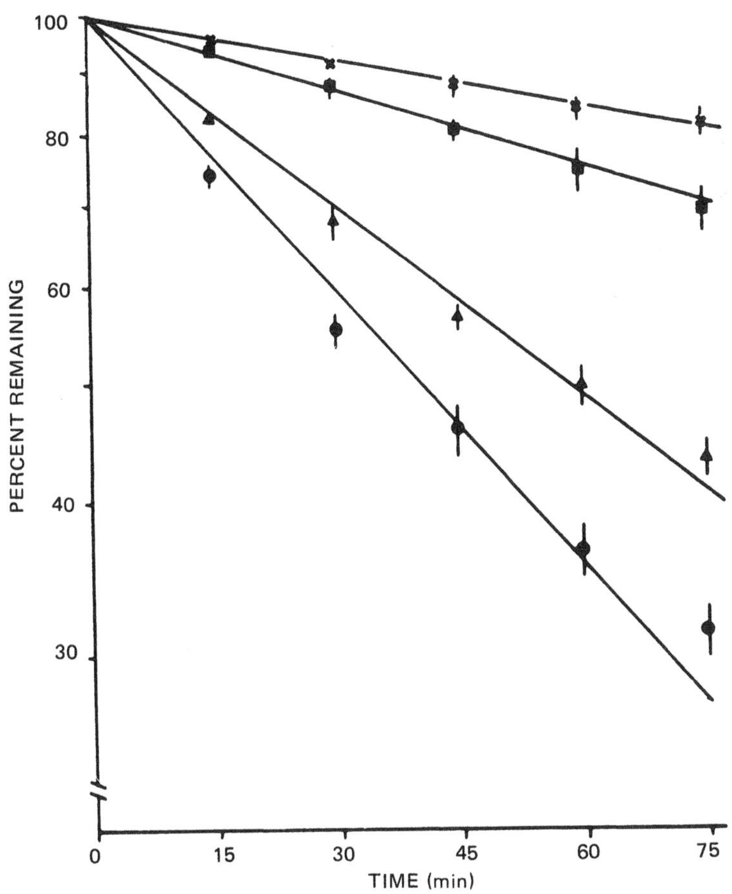

Figure 3 Time course for the percentage (mean ± SEM) of phenobarbital remaining in the perfusion solution (at pH 6 and 37°C) as a function of perfusion volume used in *ex vivo* nasal perfusion studies in rats (n = 3–5): (x) 20 ml; (■) 10 ml; (▲) 5 ml; (●) 3 ml. (From Reference 13; reproduced with permission.)

varied in their hydrophilicity, in ovariectomized rabbits, demonstrated that the partition coefficient determined in an octanol-water system does not predict well the permeation behavior of these progestational steroids across the nasal mucosa (18). The results, however, showed that the systemic bioavailability of progesterone and its hydroxy derivatives correlates well with the partition coefficient determined in nasal mucosa-buffer system (Figure 8). This observation indicates that transnasal permeation behavior cannot be predicted by the lipophilicity measured in a simple octanol-water system.

6. *Effect of Drug Concentration*

The effect of a variation in the drug concentration in the perfusion solution for nasal absorption was studied by monitoring the disappearance of 1-tyrosyl-1-tyrosine and the formation of 1-tyrosine using the ex vivo nasal perfusion technique in the rat (Section VB). It was found that the nasal absorption of 1-tyrosine depends upon its

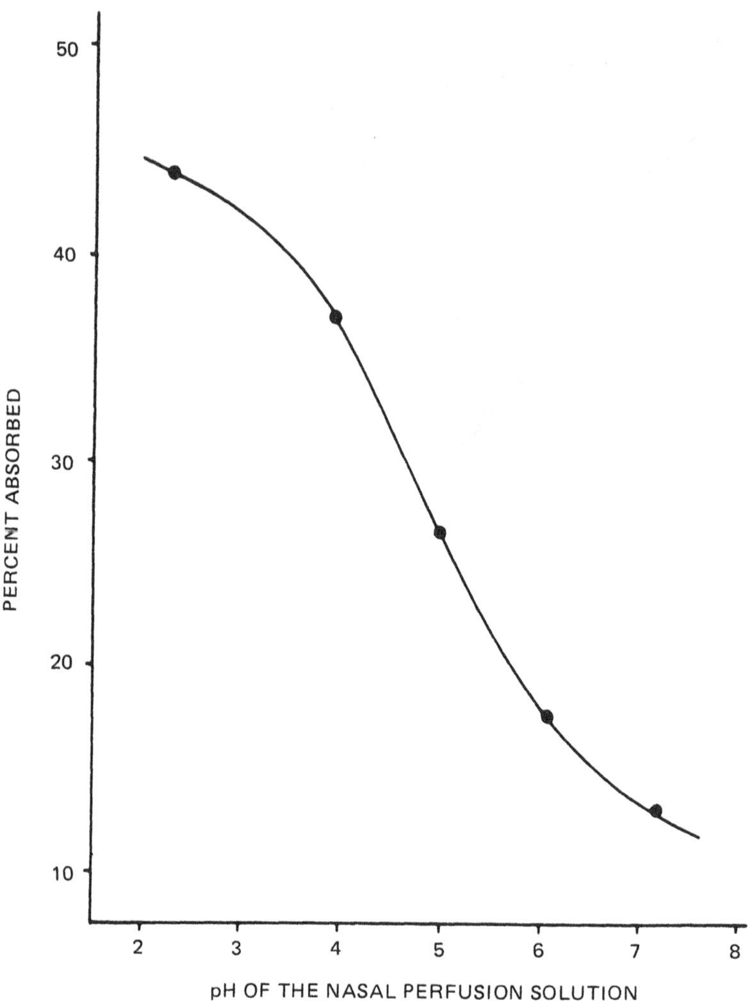

Figure 4 The extent of nasal absorption (in 60 min) of benzoic acid as a function of pH of the perfusion solution. (From Reference 13; reproduced with permission.)

concentration since the formation of 1-tyrosine is dependent upon the initial concentration of 1-tyrosyl-1-tyrosine (Figure 9) (13).

B. Mechanisms and Pathways

1. *Mechanisms*

The mechanisms of nasal delivery was investigated in rats (11) using SS-6, an octapeptide, and horseradish peroxidase, a protein molecule; two mechanisms of transport are involved, a fast rate, which is lipophilicity dependent, and a slower rate, which is sensitive to the variation in molecular weight. The results of nasal absorption are inconsistent with the nonspecific diffusion of penetrant molecule through the aqueous channels between the nasal mucosa cells, which impose a molecular

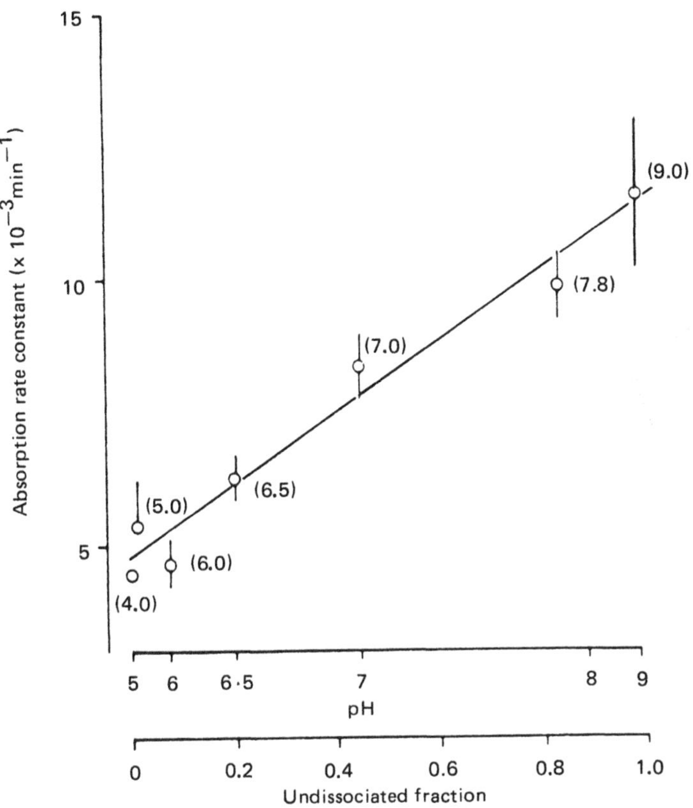

Figure 5 Linear relationship between the rate constant for the nasal absorption of hydralazine (1 mM) in rats (n = 3) and the fraction of its undissociated species at various pH values. The numbers in parentheses indicate the pH value. The data are expressed as mean ± SEM. (From Reference 16; reproduced with permission.)

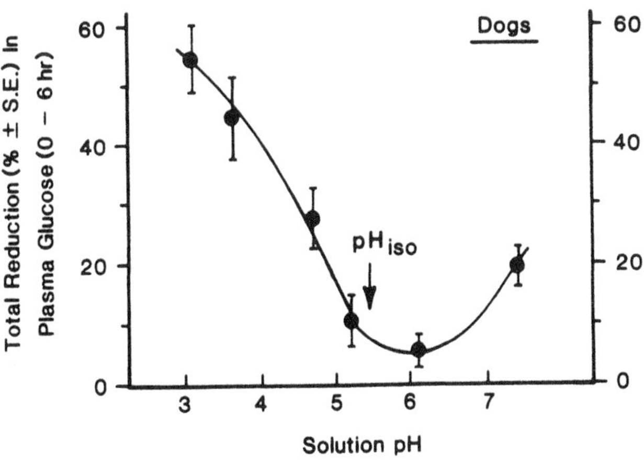

Figure 6 Effect of pH in the nasal insulin solution on the reduction of plasma glucose levels in dogs (dose 50 IU/dog; n = 4). (From Reference 14; reproduced with permission.)

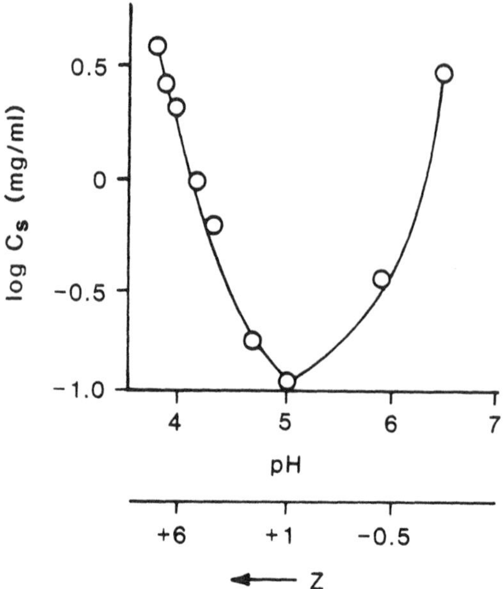

Figure 7 Aqueous solubility-pH profile of insulin as well as the type and density of charge on the insulin molecule as a function of solution pH. (Modified from Klostermeyer and Humbel, 1966.)

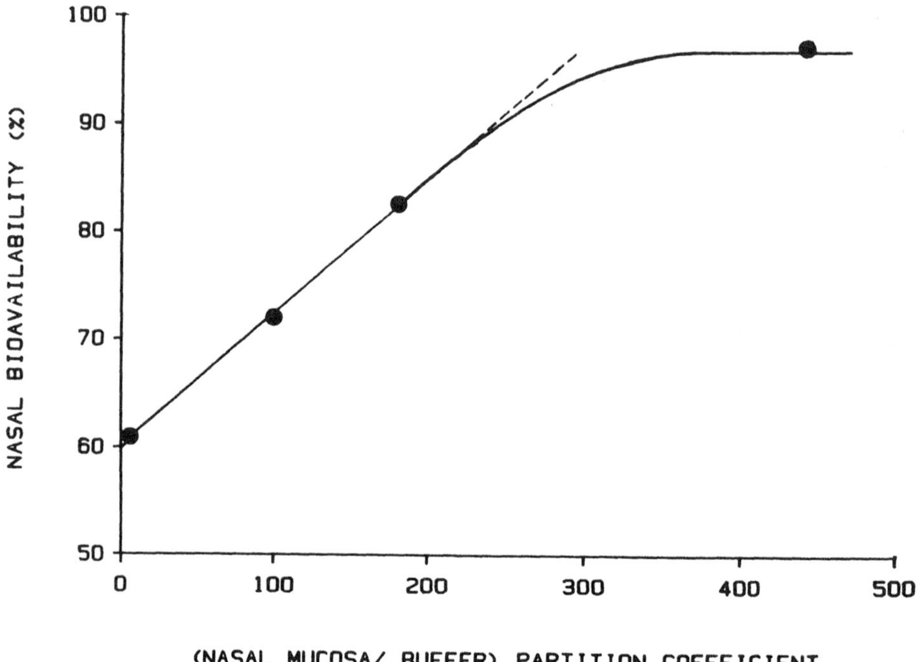

Figure 8 Relationship between the systemic bioavailability of progesterone and its hydroxy derivatives in ovariectomized rabbits following nasal spray administration and (nasal mucosa/buffer) partition coefficient. (From Reference 18; reproduced with permission.)

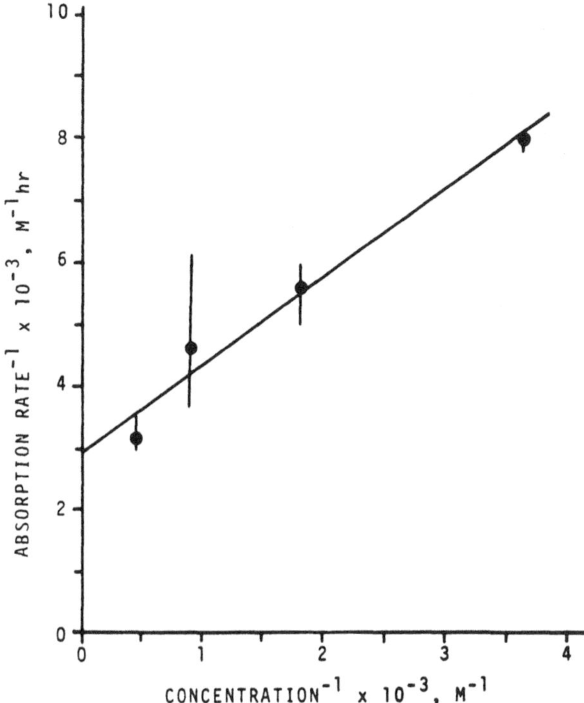

Figure 9 Linear relationship between the nasal absorption rate of 1-tyrosine and 1-tyrosine concentration (at pH 7.4 and 37°C). Values are Mean ± SEM (n = 4–5). (From Reference 13; reproduced with permission.)

size-dependent nasal permeability. These data, combined with other literature results, indicated that a good systemic bioavailability can be achieved for molecules with a molecular weight of up to 1000 daltons when no enhancer is used; with the assistance of enhancers, a good bioavailability can be extended to a molecular weight of at least 6000 daltons.

Water-soluble compounds, such as sodium cromoglycate, were found well absorbed. Their nasal absorption is likely to be dependent upon diffusion through aqueous channels (pores) (19). The molecular size of such a compound is a determinant for the rate of nasal absorption.

Wheatly et al. in 1988 used Ussing chambers to study the mechanism of transport across nasal mucosal tissue (20). They noted that the nasal delivery of insulin, mannitol, or propranolol follows a transport mechanism involving passive diffusion. The addition of deoxycholate (0.1%) to the mucosal bathing solution was found to reversibly increase transepithelial conductance across the nasal membrane, and the nasal delivery of mannitol and insulin was enhanced by 10 to 20-fold.

The transport of amino acids, like tyrosine (Tyr) and phenylalanine (Phe) across the rat nasal mucosa was studied by Tengamnuay and Mitra (21), using the ex vivo nasal perfusion technique (Section VB). They observed that both amino acids are absorbed by an active but saturable transport process and that the transport appears to be Na^+ dependent and may require metabolic energy as a driving force, since the uptake of *l*-Phe is inhibited by quabain and 2,4-dinitrophenol. When d-Tyr and d-Phe were used as the substrates the extent of their nasal absorption was significantly lower, indicating a specific affinity of the carrier for the *l*-amino acids. When a mixture of *l*-Tyr and *l*-Phe were administered, both *l*-amino acids were found to be concomitantly absorbed, but in a competitive manner.

2. *Pathways*

The olfactory epithelium (Figure 1) is known to be a portal for a substance to enter the central nervous system (CNS) and the peripheral circulation following nasal absorption (1). In addition, the nasopharynx has been shown to act as the portal of entry for viruses that induce some common viral diseases, i.e., measles, the common cold, smallpox, chickenpox, and poliomyelitis. There appears to be communication between the nasal cavity and the subarachnoid space, between the lymphatic plexus in the nasal mucosa and the subarachnoid space, as well as between the perineural sheaths in the olfactory nerve filaments and the subarachnoid space. The process of drug transport across the nasal membrane involves either the diffusion of drug molecules through the pore channels in the nasal mucosa or participation of some nonpassive pathways before they reach the bloodstream (15).

It is important to investigate the pathways involved in the nasal absorption of a drug species before studying how to improve and/or control its nasal delivery rate and transnasal bioavailability. The potential pathways involved in the nasal absorption of drugs reported in the literature are summarized in Table 1.

C. Distribution and Deposition

The distribution of drug (or formulation) delivered intranasally in the nasal cavity is one of the important factors that could affect the efficiency of nasal absorption (1). The mode of administration could influence the distribution of a drug in the nasal cavity, which in turn determines the efficiency of its absorption. Using a cast of a human nose, the distribution of drug following intranasal administration by different types of nasal delivery systems, including nose drops, a plastic bottle nebulizer, an atomized pump, and a metered-dose pressurized aerosol, was evaluated and a significant difference was demonstrated. Among the systems evaluated the atomized pump was found to be the best nasal delivery system because it delivered a constant dose and achieved a very uniform distribution on the nasal mucosa. The results also suggested that the use of a large volume of a dilute solution is preferable to a small volume of a concentrated solution. A simulated nasal cavity made of acryl resin was also developed for studying the distribution of beclomethasone dipropionate (BD) aerosol particles in the nasal cavity (22). No significant difference was noted among the gas-, liquid-, and powder-type preparations. The highest concentration of BD was usually detected in the anterior portion of the middle turbinate.

The nasal deposition of particles is also related to an individual's nasal resistance to airflow (23). With nasal breathing nearly all particles having an aerodynamic size of 10–20 μm are often found to be deposited on nasal mucosa (5). One should avoid deposition in both the poorly absorptive stratified epithelium of the anterior atrium and in the posterior nasopharyngeal region, which leads to drug loss to the stomach by swallowing. Insoluble particles, if deposited in the main nasal passage, are likely to be transported posteriorly by ciliary movement and dispatched to the stomach. If

Table 1 Potential Pathways for Nasal Absorption

Substances	Possible pathways
Albumin	
Albumin (labeled with Evans blue and horseradish peroxidase)	Nasal mucosa → sensory nerve cells of olfactory epithelium → subarachnoid space → bloodstream
Egg albumin	Nasal mucosa → lymphatic stream
Serum albumin	Nasal mucosa → lymphatic stream
Amino acids (arginine, glutamic acid, glycine, γ-aminobutyric acid, proline, serine, tritiated leucine)	Nasal mucosa → blood vessel (active transport) Nasal mucosa → olfactory nerve fiber → CNS
Bacteria	
Rabbit virulent type III pneumococci	Nasopharyngeal epithelium → lymphatics → cervical lymphatic vessel → blood vessel
E. coli and *Staphylococus*	Nasal mucosa → lymphatics → blood
Clofilium tosylate	Nasal mucosa → epithelial cells → systemic circulation
Dopamine	Nasal mucosa → cerebrospinal fluid (CSF) and serum
Hormones	
Estradiol	1). Nasal membrane → CSF (<1 min) 2). Nasal membrane → olfactory neurons → brain and CSF
Norethisterone	1). Nasal membrane → olfactory dendrites → nervous system → supporting cells in the olfactory mucosa → submucosal blood vascular system 2). Nasal membrane → peripheral circulation (high levels) and CSF (low levels) → CNS
Progesterone	1). Nasal membrane → olfactory dendrites → nervous system → supporting cells in the olfactory mucosa → submucosal blood vascular system → CSF 2). Nasal membrane → CSF (<1 min)
Penicillins	Nasal membrane → bloodstream
Virus	
Herpesvirus encephalitis	Nasal mucosa → peripheral and cranial nerves → CNS
Herpesvirus simplex	Nasal mucosa → cranial nerve → CNS
Mouse passage strains of herpesvirus	Nasal mucosa → trigeminal and olfactory pathways → CNS
Neutropic virus and poliomyelitis virus	Nasal mucosa → olfactory nerve → CNS
Vaccina virus	Nasal mucosa → submucous lymphatics → cervical lymphatic pathways → CNS
Water (distilled water)	Nasopharynx → cervical lymph

the drug is introduced as a vapor (or a soluble particle) it may readily diffuse into the lining secretions and then be absorbed from there into the microcirculation.

The deposition of aerosols in the respiratory tract is also a function of particle size as well as respiratory patterns (24). The density, shape, and hydroscopicity of the particles and pathological conditions in the nasal passage influence the deposition of particles, whereas particle size distribution determines the site of deposition and affects the subsequent biological responses. A uniform distribution throughout the nasal mucosa could be achieved by delivering the drug particles from a new nasal spray using a pressurized gas propellant. A metered-dose delivery system developed for the nasal delivery of flunisolide, a synthetic fluorinated corticosteroid, was assessed and found to provide a consistent dose delivery and spray pattern that affects the deposition of droplets in the nasal cavity (25).

The particle (or droplet) size of an aerosol is important in determination of both efficacy and toxicity. For example, the metered-dose flunisolide aerosol formulation reported earlier requires that the majority of particles to be delivered intranasally should have a diameter of greater than 10 μm to achieve a localized delivery in the nasal cavity and to avoid any potential undesired effects resulting from any deposition of flunisolide particles in the lung (26).

On the other hand, the nasal delivery of beclomethasone dipropionate (BD) particles was accomplished by spraying with a Freon propellant. The results indicated that the shape of the nasal cavity produces a greater effect on the deposition of BD from the gas spray than from the powder spray. This difference is probably due to the spray angle and to the size and speed of the aerosol particles. The powder spray is preferable with regard to the deposition and distribution of drug particles in the nasal cavity.

Among the three mechanisms usually taken into consideration when one assesses the deposition of particles in the respiratory tract, i.e., inertia, sedimentation, and diffusion, inertial deposition was found to be a dominant mechanism in nasal deposition (27). Particles with an aerodynamic diameter of 50 μm or greater do not enter the nasal passage. It was demonstrated that 60% of aerosolized particles with an aerodynamic diameter of 2–20 μm are deposited in the anterior region of the nostrils (28). The site of drug deposition within the nasal cavity depends upon the type of delivery system used and the technique of administration applied (29). It was found that, following administration by nose drops, greater coverage of the nasal walls is achieved, which is independent of the volume administered (over the range 0.1–0.75 ml) (30). The particles, once deposited at the anterior region of the nasal cavity may be again conveyed posteriorly by inhaled air, ciliary movement, and/or diffusion in the mucous layer.

The pattern of nasal deposition and the rate of clearance were studied in normal subjects using ^{99m}Tc-labeled human serum albumin (HSA) delivered by either nasal spray or nose drops (31). Nasal spray was observed to deposit HSA anteriorly in the nasal cavity, with little of the dose reaching the turbinates. In contrast, nose drops dispersed the dose throughout the length of the nasal cavity, from the atrium to the nasopharynx, and the dosing with three drops resulted in greater coverage of the nasal walls compared to that of a single drop. The solution deposited anteriorly in the nasal cavity was noted to be slow to clear, especially with spray administration. The dose administered by nose drops cleared more rapidly than that administered by nasal spray.

The regional deposition of drug discharged from a pressurized aerosol product and a metered-pump spray product was compared in a model nose (32). The results indicated that these two products produce no significant difference in regional deposition. In addition, most of the drug in each case was observed to deposit in the anterior region of the nose by inertia impaction, with little nasal penetration of the drug. The anteriorly deposited drug can be spread backward by mucocilial flow and general surface flow. The initial distribution and subsequent clearance of aerosol discharged from a nasal pump spray was recently studied (33). These results also showed that aerosol is concentrated mainly in the anterior region of the nose, but the area of deposition varies from one subject to another. On the average, 56% of the dose was retained at the initial site of deposition 30 min after administration; the remaining 44% of the dose was cleared to the nasopharynx. Whaley et al. in 1988 developed a method to expose the nasal cavity of a beagle dog to a radiolabeled aerosol without exposure of the remainder of the respiratory tract (34). The results of these studies suggested that the efficiency of deposition is at 15 ± 2% of the inhaled activity; the maximum deposition is noted to occur in the anterior third of the nasal cavity (which contains 78 ± 4% of the total radioactivity deposited). On the other hand, the middle third of the nasal cavity receives 13 ± 3% and the posterior third 9 ± 2% of the deposited radioactivity.

A mathematical model was recently developed to describe the rate processes involved in the deposition of drug delivered into the human nasal cavity by a delivery system (35). The model contains a series of parallel first-order rate processes consisting of the convective transport of drug and carrier by fluid flow, mucocilial clearance and peristalsis, and drug decomposition, as well as a series of sequential irreversible first-order rate processes consisting of the release and absorption of the drug before its appearance in the systemic circulation. Simulation using this model showed that the use of a bioadhesion technique could improve bioavailability and, in the meantime, reduce the variability in absorption that could result from removal of the drug from the nasal cavity by sniffing, blowing, or wiping the nose.

D. Enhancement in Absorption

Several methods have been used to facilitate the nasal absorption of drugs (1):

Structural modification: The chemical modification of the molecular structure of a drug has been often used to modify the physicochemical properties of a drug, and hence it could also be utilized to enhance the nasal absorption of a drug.

Salt or ester formation: The drug could be converted to form a salt or an ester for achieving better transnasal permeability, such as formation of a salt with increased solubility or of an ester with better nasal membrane permeability.

Formulation design: Proper selection of formulation excipients could improve the stability and/or enhance the nasal absorption of drugs.

Surfactants: Incorporation of surfactants into nasal formulations could modify the permeability of nasal mucosa, which may facilitate the nasal absorption of drugs.

Several approaches have been developed and successfully applied to enhance the efficiency of nasal drug delivery (36).

The first approach involves the use of various surfactants or bile salts to promote nasal absorption (1,21,36–49). Some of the surfactants or bile salts that have been used to improve the nasal absorption of drugs are shown in Table 2. A number of surfactants have been reported to enhance the absorption of drugs through the nasal mucosa to a level sufficient to achieve their systemic effects. Mild surfactants at low concentrations may only alter membrane structure and permeability, whereas certain surfactants at high concentrations may disrupt and even dissolve nasal membranes. The mechanisms of action for some surfactants or bile salts are shown in Table 3.

The second approach to enhance the transnasal delivery of drugs applied various

Table 2 Representative Surfactants and Bile Salts Used as Absorption Promotors for Enhancement of Nasal Drug Permeation

Drugs	Promotor used
Atropine	Sodium lauryl sulfate
Buserelin	Bacitracin
[$Asu^{1,7}$]eel calcitonin	Polyacrylic acid
Calcitonin	Polyacrylic acid
	Sodium glycocholate
Cholecystokinin	Sodium deoxycholate
Enkephalin analogs	Sodium glycocholate
Gentamycin	Sodium glycocholate
	Lysophosphatidylcholine
Glucagon	Sodium glycocholate
Met-hGH (γ-hGH)	Sodium glycocholate
	Laureth-9 (BL-9)
Hydralazine	Sodium glycocholate
	BL-9
Insulin	BL-9
	1α-Lysophosphatidylcholine
	Disodium carbenoxolone
	Dipotassium glycyrrhizinate
	Saponin
	Sodium caprate
	Sodium caprylate
	Sodium deoxycholate
	Sodium glycocholate
	Sodium glycyrrhizinate
	Sodium laurate
	Sodium taurodihydrofusidate (STDHF)
Interferon	Azone
	Sodium cholate
	Sodium glycocholate
[D-Arg^2]kyotorphin	Mixed micelles of sodium glycocholate and oleic acid (or linoleic acid)
LHRH	STDHF
Phenol red	BL-9
	Sodium deoxycholate
	STDHF
Progesterone	Polysorbate 80
Scarlet fever toxin	Sodium taurocholate
Testosterone	Polysorbate 80

Table 3 Mechanisms of Action for Some Nasal Absorption Promotors

Absorption promotors	Mechanisms of action
Bile salts (i.e., sodium glycocholate, sodium deoxycholate, others)	1). Inhibit aminopeptidase activity in nasal mucosa 2). Form transient hydrophilic pores in the membrane bilayer 3). Reduce the viscosity of mucus 4). Remove the epithelial cells, which constitute a major permeability barrier 5). Solubilize drug in bile salt micelles, thus creating a transmembrane concentration gradient; form reverse micellar structures within the mucosal membranes, which act as temporary aqueous pores
Bile salts and unsaturated fatty acid mixed micelle	Bile salts can solubilize fatty acid in the mixed micelles, thus making it more available at the mucosal surface for absorption
BL-9	Irreversibly removes membrane proteins or lipids
EDTA	Increases paracellular transport by removal of luminal calcium, thus affecting the permeability of the tight junctions
Fatty acid salts (i.e., sodium caprate, sodium laurate, others)	1). Inhibit leucine aminopeptidase activity 2). Act to create intercellular space by temporarily extracting calcium ions from nasal mucosa
Glycyrrhetic acid derivatives	Inhibit leucine aminopeptidase in nasal mucosa
Lysophosphatidylcholine	Affects enzyme activity and induces morphological changes in nasal membranes
Sodium taurodihydrofusidate	1). Inhibits leucine aminopeptidase activity 2). Forms peptide-adjuvant complex 3). Facilitates the paracellular transport from nasal cavity to the capillary bed

bioadhesive polymers, such as methylcellulose, carboxymethylcellulose, hydroxypropylcellulose, or polyacrylic acid (36,50,51). The enhancement of nasal drug absorption by bioadhesives presumably results from the increase in the residence time of drug in the nasal cavity and a higher local drug concentration in the mucus lining on the nasal mucosal surface. One of the properties of a bioadhesive polymer is its ability to swell by absorbing water from the mucous layer in the nasal cavity and thereby forming a gel-like layer in which the polymer forms a bond with the glycoprotein chains of the mucin. Methylcellulose and polyacrylic acid gel have been shown to enhance the transnasal bioavailability of propranolol in the rats and cause some irritation of nasal mucosa with certain morphological changes (52). A nasal formulation of meclizine (50 mg/ml) was prepared in 85% propylene glycol and 10% glycerol (53), and by nasal delivery it yielded an absorption of about 50%, which was as effective as the intravenous injection but was about six times more effective than oral administration (which was only 8%). The mean time to peak plasma levels was found to be about 8.5 min for the nasal delivery compared to 49.0 min after oral administration. Enhancement of the nasal absorption of insulin was demonstrated in dogs by using a powder formulation prepared from hydroxypropyl-cellulose and neutralized polyacrylic acid (Carbopol 934p), which forms a gel when in contact with the nasal mucosa (50). Similarly, a polyacrylic acid (0.1%) gel base was utilized to enhance the absorption of insulin and calcitonin in rats by intranasal administration (51). Formulations with pH in the range 4.5–7.5 appeared to have no influence on the nasal absorption of insulin from the gel formulation, which showed a behavior different from that of the solution formulation (shown earlier in Figure 6); however, the lower viscosity gels resulted in a more rapid onset of hypoglycemic response than did the higher viscosity gels.

The third approach that can be used to improve the efficiency of nasal absorption of drugs is to deliver the drugs in microspheres that have good bioadhesive characteristics and that swell easily when in contact with the nasal mucosa (36,54). This microsphere drug delivery system has the ability to control the rate of drug clearance from the nasal cavity as well as protect the drug from enzymatic degradation in nasal secretions, thereby provide a potential for increasing the systemic bioavailability of drugs. The microspheres prepared from bioadhesive polymers, such as starch, albumin, gelatin, and dextran, were retained in the nasal cavity with half-life of clearance increased to 3 hr or longer (55). This bioadhesive microsphere drug delivery system has now been investigated for its feasibility of altering the nasal bioavailability of several drugs, such as gentamycin, insulin, rose bengal, and cromoglycate disodium (55–59).

The fourth approach to enhance nasal drug absorption involves the use of "physiological modifying agents" (36). These agents have vasoactive properties and exert their action by increasing nasal blood flow, which include histamine, leukotriene D_4, prostaglandin E_1, and the β-adrenergic agents (isoprenaline and terbutaline). These physiological modifying agents may represent a new class of nasal absorption promotors distinct from bioadhesive or permeability-ehancing surfactants (or bile salts).

E. Effect of Delivery Systems

Several types of drug delivery systems have been used for delivering drugs to the nasal cavity, such as the nasal spray, nose drops, the aerosol spray, the saturated cotton pledget, and the insufflator (1).

The metered-dose nebulizer has recently been introduced as a potential nasal drug delivery device that operates by mechanical actuation and delivers a predetermined volume with precision into the nasal cavity. The dose of active ingredient administered intranasally depends upon the volume of drug solution delivered at each actuation and the concentration of drug in the formulation. The metered-dose nebulizer has already been successfully utilized as the nasal delivery system for several topical drugs, such as corticosteroids (Extracort), beclomethasone diproprionate (Aldecin, Beconase, and Becotide), flunisolide (Nasalide), tramazoline (Tobispray), and nasal decongestant (Rhinospray). In addition, the metered-dose nebulizer has also been explored as the nasal drug delivery system for the systemically-active drugs, such as DDAVP, enviroxime, insulin, and nitroglycerin.

An inflatable nasal device with its wall constructed from a microporous membrane was developed to provide the long-acting, controlled delivery of drugs from a drug suspension formulation. Following insertion into the rabbit nasal passage the nasal device is inflated by filling with drug suspension (Figure 10) (60). It is inflated and conforms to the contour of the nasal passage, and it is left in place for a prolonged duration. Using progesterone as the model drug and ovariectomized rabbits as the animal model, the plasma profile of progesterone delivered by this controlled-release device was compared to that attained by an immediate-release nasal spray (Figure 11). The results indicated that the nasal spray produces a peak plasma level of progesterone within 2 min, suggesting the rapid absorption of progesterone by the nasal mucosa (60). On the other hand, the nasal delivery of progesterone by the experimental controlled-release nasal device led to a gradual increase in the plasma progesterone concentration, which reaches a plateau level within 20–30 min and remains at an elevated level throughout the course of a 6-hr insertion. An equivalent AUC/D_0 value was attained for progesterone by both methods of nasal delivery (Figure 11). The systemic bioavailability of progesterone following intranasal administration by an immediate-release spray formulation (82.5 ± 13.5%) and a controlled-release nasal device (72.4 ± 25.7%) were not statistically different, but they were both significantly greater than that resulting from oral administration (7.9 ± 1.6%).

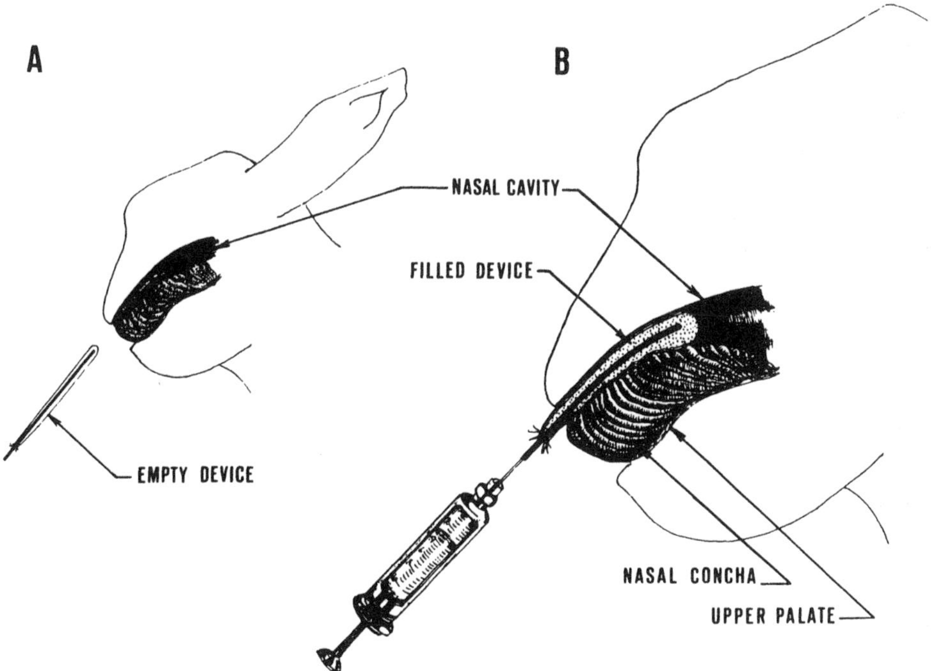

Figure 10 The controlled-release nasal delivery device before insertion into the rabbit nasal passage (A), and the in situ filling of the nasal delivery device with the progesterone formulation in the nasal cavity (B). (From Reference 59; reproduced with permission.)

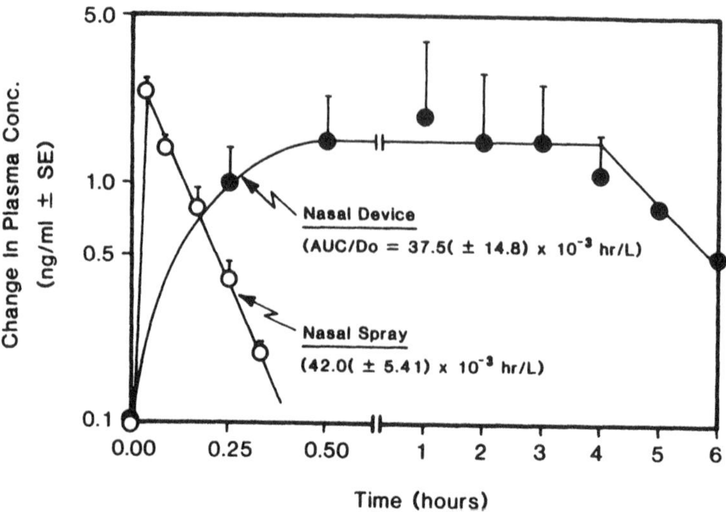

Figure 11 Time course for the change in the plasma concentration of progesterone in ovariectomized rabbits ($n = 3$) after intranasal administration of progesterone by immediate-release nasal spray (2 μg/kg) and a controlled-release nasal device (60 μg/kg). Equivalent bioavailability was accomplished with both methods of nasal delivery. (From Reference 59; reproduced with permission.)

The results suggest that this controlled-release nasal drug delivery technique achieved a 9- to 10-fold reduction in the extent of presystemic metabolism for progesterone while maintaining a prolonged and steady plasma level (Figure 11).

Sustained-release formulations have been developed using methylcellulose for propranolol, and studies carried out in rats achieved a low initial but prolonged blood level and bioavailability equal to that following intravenous administration (61). Hydrogel, such as Carbopol 934p, may be useful for formulation of a sustained-release nasal delivery system since they can achieve a long-term intimate contact with the nasal mucosa and thus provide a prolonged therapeutic effect without suppressing mucociliary functions. Microspheres fabricated from various bioerodible materials have potential to become controlled-release drug delivery systems for nasal administration (55–59). Microspheres prepared in suitable sizes can be loaded with a variety of drugs, which can then be released in a sustained manner at the desired site in the nasal cavity. The efficient trapping of a microsphere-type drug delivery system in the nasal cavity is very much dependent upon the size and the surface characteristics of the microspheres. Microspheres with a particle size larger than 10–15 μm should be suitable for targeting its deposition in the nasal cavity.

As discussed earlier, Nagai et al. reported the use of a bioadhesive powder dosage form and a mucoadhesive powder spray formulation for the nasal delivery of insulin (50,62). Studies conducted in rabbits and dogs indicated that the bioadhesive powder dosage form produces a drug absorption that is more effective and less irritating than liquid forms. The results generated from the mucoadhesive powder spray formulation suggest that it gives a large reduction in plasma glucose level to almost the same extent, without great pH dependence (23,50). The final powder dosage form successfully produced a hypoglycemia one-third of that attained by intravenous injection. The nasal absorption of insulin was fast and sustained when delivered in the formulations prepared from crystalline cellulose, starch, and neutralized CMC-Na. For the mucoadhesive powder spray, the powder mixture was prepared with hydroxypropylcellulose (HPC) as the mucoadhesive base and was administered by a special applicator called a Pulizer. The advantages of using HPC were found to be (i) the drug dose may be reduced, (ii) the side effect may be lowered, and (iii) a longer duration of effect is expected. In the nasal cavity the powder absorbs nasal fluid and becomes swollen and adheres to the mucosal membrane for as long as 6 hr. The results of drug disposition studies showed that beclomethasone remains very much longer in the body than do regular nasal drops.

Another means of nasal drug delivery is the application of a dosage form with better adhesion but less irritating to the nasal mucosa by using a small amount of absorption enhancer. Illum et al. in 1987 investigated a nasal delivery system in the form of bioadhesive microspheres, which could not be cleared easily from the nasal cavity because of their intimate contact with the mucosa (57). For the development of bioadhesive microsphere systems, albumin, starch, and DEAE-dextran were used as the formulation base. Using rose bengal and sodium cromoglycate as the model drugs, these studies demonstrated that it is possible to control the release of drugs from the bioadhesive microsphere system. Studies have been conducted in human volunteers using a variety of microsphere systems with good bioadhesive properties. The microspheres were labeled with a technetium complex, using powder and solution forms as the controls, and the nasal clearance was monitored using gamma scintigraphic technique. The results showed that the microspheres prepared from starch and DEAE-Sephadex as the formulation bases are effective in delaying nasal mucociliary clearance. The half-life of clearance for starch-based microspheres was found to be of the order of 240 min compared to 15 min for control (liquid and powder) formulations. The albumin-based microsphere formulation was less effective than these two but was still more effective than the powder and solution dosage forms. Nasal absorption promotors could also be delivered with the drug in this system. The retention of drug and promotor on the nasal mucosa for extended periods of time could lead to an improvement in the systemic bioavailability of drugs, including peptides and proteins.

The possibility of improving the bioavailability of gentamycin administered intranasally by means of a gelling microsphere delivery system was investigated in rats and sheep (55). It was found that by using the microsphere delivery system the uptake of gentamycin across the nasal membrane was increased. The uptake was furthered enhanced by incorporation of an absorption promotor, called lysophosphatidylcholine (LPC), into the microspheres. The results indicated that administration of gentamycin solution alone yields a poor bioavailability, whereas addition of LPC as the absorption promotor gives rise to a five fold increase in peak level (Figure 12) (55). If gentamycin was administered in combination with starch-based microspheres, a significant increase in bioavailability was obtained (Figure 13) (55). An even more dramatic effect was observed when the gentamycin and the absorption promotor were coadministered in the starch-based microsphere formulation; the blood level of gentamycin peaked at 6.3 μg/ml compared to 0.4 μg/ml when administered in solution formulation. In fact, the data in Figure 13 demonstrate that the nasal delivery of gentamycin by microspheres containing LPC produces a plasma concentration versus time profile very similar to that obtained by intravenous

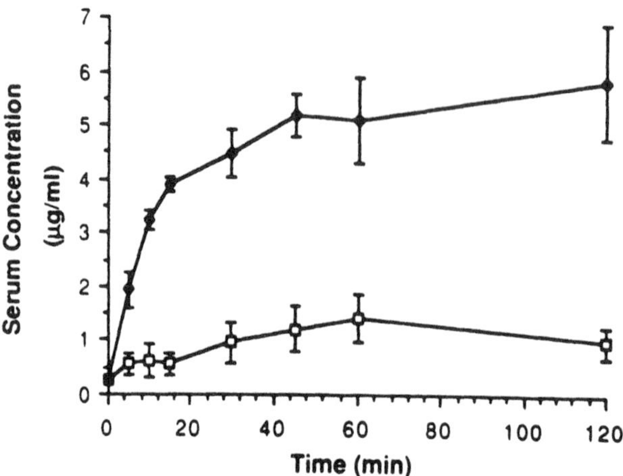

Figure 12 Serum concentration (mean ± SEM) profiles of gentamycin following intranasal administration to rats (5 mg/kg) (in situ model) from (□) a phosphate buffered solution (20 mg/ml) and (◆) with addition of lysophosphatidylcholine (2 mg/ml). (Modified from Reference 55.)

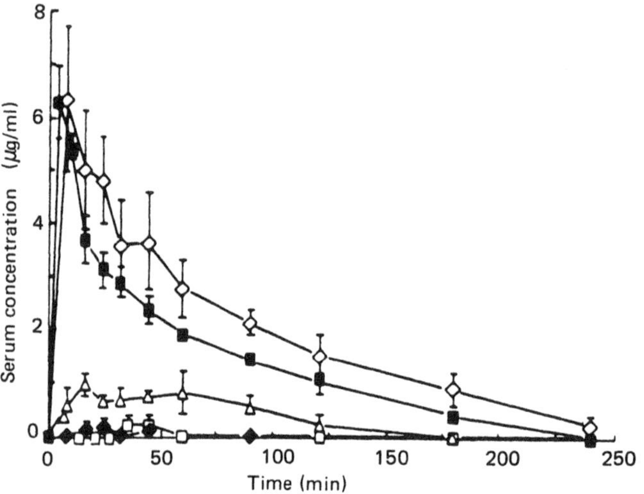

Figure 13 Serum concentration profiles of gentamycin in sheep (mean ± SEM) following intranasal administration of gentamycin (5 mg/kg) from (□) solution formulation (n = 2); (◆) solution formulation containing 2 mg/ml of lysophosphatidylcholine (0.026 mg/kg), (n = 3); (△) starch microspheres (n = 3); (◇) starch microspheres containing lysophosphatidylcholine (0.2 mg/kg), (n = 3); (■) intravenous administration of gentamycin (2 mg/kg) in solution formulation (n = 3). (Modified from Reference 55.)

administration. In comparison, a bioavailability of less than 1% was achieved for a simple nasal gentamycin solution.

The feasibility of using degradable starch microspheres (DSM) for the nasal delivery of insulin was also evaluated in rats (56), and the results indicated that insulin-DSM preparations administered nasally as a dry powder resulted in a dose-dependent reduction in blood glucose levels and a concomitant increase in serum insulin concentrations. The blood glucose was reduced by 40 and 64% at 30 and 40 min, respectively. The insulin peak was reached 8 min after dosing with a bioavailability of approximately 30%. DSM or insulin alone showed no effect on hypoglycemic activity. The results indicated that DSM offer a system for improving the nasal absorption of drugs. Recently, Farraj et al. also reported that when using the starch-based microspheres as a bioadhesive delivery system, the half-time for the clearance of insulin from the nasal cavity is prolonged to 240 min from the 15 min for the nasal solution. Insulin administered in the starch-based microsphere formulation in sheep produced a fivefold increase in the AUC for plasma insulin compared to the insulin solution. When insulin (16.7 IU/kg) was coadministered with lysophosphatidylcholine (0.5%) the plasma glucose level was reduced by 65%. When insulin (2.0 IU/kg) was given to conscious sheep with LPC in starch-based microspheres (2.5 mg/kg), the area under the plasma concentration versus time curve (AUC) for insulin increased by almost 17 times compared to a simple insulin solution. Using the combined system, the nasal delivery of insulin achieved a systemic bioavailability of 31.5% compared to that obtained by subcutaneous administration. The relationship between the pharmacodynamic responses, as demonstrated by the reduction in plasma glucose levels, and the pharmacokinetic profiles, as shown by the increase in plasma insulin concentrations, is shown in Figure 14. The pharmacokinetic parameters and systemic bioavailability of insulin as a function of the route of administration, as well as the effect of formulation on the improvement in the nasal delivery of insulin, are summarized and compared in Table 4. Both pharmacodynamic and pharmacokinetic results clearly demonstrate the potential of this system for the nasal delivery of insulin (59).

F. Pharmacokinetics and Bioavailability

Factors that have been reported to affect the pharmacokinetics and bioavailability of drugs following intranasal administration (63) include the following:

1. Physiological factors:
    a. Speed of mucus flow
    b. Change in physiological state
    c. Atmospheric conditions in the nasal cavity
2. Dosage form factors:
    a. Physicochemical properties of the active drug
    b. Concentration of the active drug
    c. Physicochemical properties of the pharmaceutical excipients used
    d. Density, viscosity, and pH characteristics of the formulation
    e. Toxicity of the dosage form
3. Administration factors:
    a. Size of dose
    b. Site of deposition

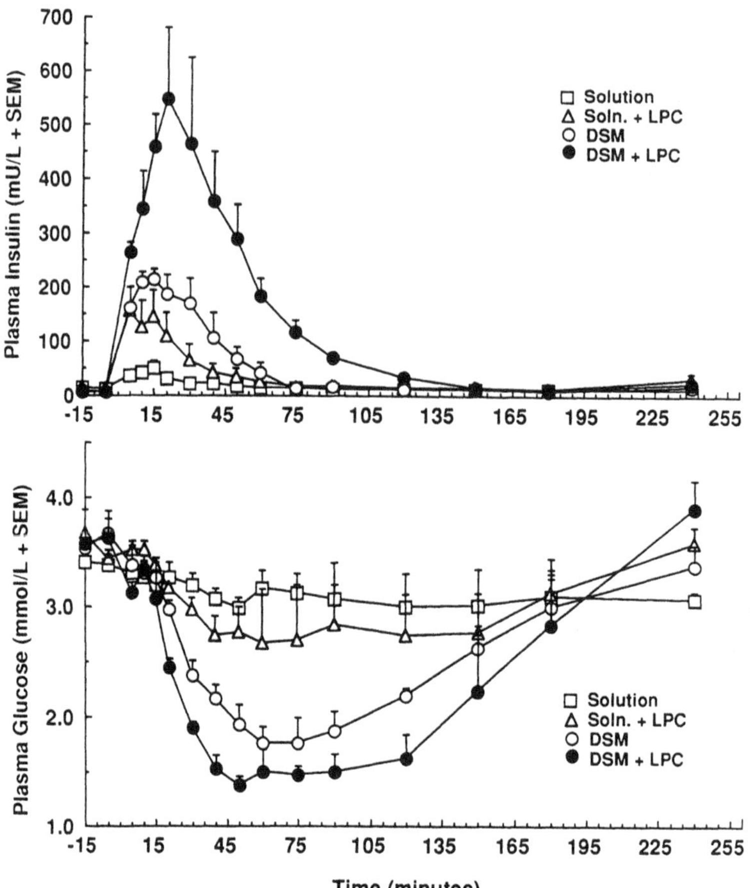

Figure 14 (Top) Change in plasma insulin concentrations in sheep following intranasal administration of insulin (2.0 IU/kg) in (□) phosphate buffer solution (pH 7.3), (△) solution containing 2 mg/ml of lysophosphatidylcholine (LPC), (○) degradable starch microspheres (DSM), and (●) DSM containing LPC. (Bottom) Corresponding change in plasma glucose levels in sheep following intranasal administration of insulin (2.0 IU/kg) via the formulation compositions outlined in the top panel. (Plotted from the data in Reference 59.)

c. Mechanical loss posteriorly into the esophagus
d. Mechanical loss to other regions in the nose
e. Mechanical loss anteriorly from the nose

The bioavailability of a drug after intranasal administration may be expressed in terms of the absolute nasal absorption B_n determined from the area under the plasma concentration versus time curve following the intravenous (IV) and intranasal (IN) dose:

$$B_n = \frac{(\mathrm{AUC})_{\mathrm{IN}}(\mathrm{dose})_{\mathrm{IV}}}{(\mathrm{AUC})_{\mathrm{IV}}(\mathrm{dose})_{\mathrm{IN}}} \tag{1}$$

Table 4 Comparative Systemic Bioavailability and Pharmacokinetics of Insulin and Effect of Formulation on Nasal Delivery

		Insulin[c]	Pharmacokinetic parameters[d]			Bioavailability	
Route[a]	Formulation[b]	dose (IU/kg)	t_{max} (min)	C_{max} (mU/L)	AUC (mU/L · min)	Relative[e] (%)	Absolute[f] (%)
Intravenous	Solution	0.1	5 (0)	671 (34)	10,348 (748)	240.2	100.0
Subcutaneous	Solution	0.2	70 (20)	95 (14)	8,615 (1297)	100.0	41.6
Nasal	Solution	2.0	13 (2)	52 (13)	1,546 (417)	1.8	0.8
	Solution + LPC	2.0	8 (3)	172 (37)	5,053 (1823)	5.9	2.4
	DSM	2.0	12 (3)	232 (8)	9,227 (2201)	10.5	4.5
	DSM + LPC	2.0	22 (4)	595 (104)	27,160 (5556)	31.5	13.1

[a]In sheep.
[b]DSM = degradable starch microsphere formulation;
LPC = lysophosphatidylcholine.
[c]Semisynthetic Zn insulin.
[d]Mean (±SEM).
[e]Relative to subcutaneous administration.
[f]Relative to intravenous administration.
Source: Modified from the data by Farraj et al. (1990).

where the AUC was extrapolated to an infinite time following the administration of a single intravenous or intranasal dose. Two types of kinetic profile may be involved in the transnasal permeation of drugs.

1. Zero-Order Transnasal Permeation Kinetics

When the absorption of drugs from the nasal site of administration follows zero-order kinetics, e.g., a controlled delivery of drug at a constant rate of absorption, the plasma profile of the drug may be described by

$$\frac{dX_B}{dt} = K_0 - K_e X_B \tag{2}$$

where K_0 is the zero-order absorption rate constant, K_e is the overall rate constant for plasma elimination, and X_B is the amount of drug absorbed into the body or in the blood circulation, i.e., the central compartment. Then, the plasma concentration C_p of drug may be expressed as

$$C_p = \frac{K_0}{CL} 1 - e^{-K_e t} \tag{3}$$

where CL is the total body clearance, and t represents any specified time interval following the intranasal drug administration. Following zero-order transnasal permeation of the drug, the plasma drug level increases to a steady-state plateau level $(C_p)_{ss}$ and then begins to decline exponentially after time t_p, which is the time when there is no more absorption of the drug from the nasal cavity.

2. First-Order Transnasal Permeation Kinetics

When the absorption of drugs from the nasal site of administration follows first-order kinetics, the plasma profile of the drug can be described by

$$\frac{dX_B}{dt} = F_a X_{\mathrm{IN}} K_a - K_e X_B \tag{4}$$

where K_a is the first-order absorption rate constant, F_a is the fraction of applied dose absorbed, and X_{IN} is the amount of drug administered intranasally to the absorption site. The plasma concentration C_p of drug can then be expressed as

$$C_p = \frac{F_a X^0_{\mathrm{IN}} k_a}{V_d(K_a - K_e)} (e^{-k_e t} - e^{-k_a t}) \tag{5}$$

where X^0_{IN} is the initial drug dose delivered intranasally to the site of absorption at time zero, and V_d is the volume of distribution.

The plasma concentration of drugs following nasal delivery shows a concentration profile more similar to that of intravenous bolus injection than to those following other nonparenteral routes of administration. Drugs with poor oral absorption (e.g., sulbenicillin, cephacetrile, cefazoline, phenol red, and disodium cromoglycate) and drugs subjected to extensive hepatic first-pass metabolism (e.g., progesterone, estradiol, testosterone, insulin, hydralazine, propranolol, cocaine, buprenorphine, naloxone, and nitroglycerin) can be rapidly absorbed through the nasal mucosa with a systemic bioavailability of near 100% (1). The biopharmaceutical data, including relative bioavailability, for the nasal absorption of organic- and peptide-based pharmaceuticals are summarized in Tables 5 and 6, respectively.

Several factors have been known to affect the systemic bioavailability of drugs, such as the deposition and clearance of drug in the nasal cavity and the transnasal permeation rate of drug into the microcirculation. Some compounds have very satisfactory absorption characteristics if sufficient time is allowed for maintaining contact with the absorptive regions of the nasal mucosa; therefore, the poor nasal absorption of some drugs may not necessarily be attributed to the problem of low nasal permeability. The rapid clearance of drugs by the mucociliary clearance mechanism in the nasal cavity may also be the factor affecting the extent of nasal bioavailability.

The relationship between the systemic bioavailability of hydrophilic compounds by nasal delivery and their molecular weight was recently examined (12). It was observed that the nasal uptake of drugs is decreased with the increase in molecular weight; however, polar molecules with a relative high molecular weight could still be taken up to a significant extent. On the other hand, the studies conducted at the same time by another group of investigators (11) demonstrated that the rate of permeation across the nasal membrane falls off sharply at a molecular weight higher than 1000 (Figure 2). Low-molecular-weight drugs, such as propranolol, naloxone, or steroids, were found to be well absorbed following intranasal administration from a nasal spray formulation; the resultant plasma level versus time profiles are almost indistinguishable from those obtained by intravenous injection (Figures 15 through 17).

IV. Biomedical Applications of Nasal Drug Delivery

A. Nasal Delivery of Organic-Based Pharmaceuticals

Organic-based pharmaceuticals that have been investigated for the feasibility of nasal delivery are tabulated in Table 7, and their biopharmaceutical data are shown in Table 5. Drugs with extensive presystemic metabolism, such as progesterone, estradiol, testosterone, hydralazine, propranolol, cocaine, naloxone, and nitroglycerin, can be rapidly absorbed through the nasal mucosa (Figures 15 through 17) with a systemic bioavailability of approximately 100% (52,64–66). Water-soluble organic-based compounds, such as sodium cromoglycate, were also found to be well absorbed (Table 5), and their nasal absorption is likely to be dependent upon aqueous channel diffusion.

B. Nasal Delivery of Peptide-Based Pharmaceuticals

Because of their physicochemical instability and susceptibility to hepatogastrointestinal first-pass elimination, peptide and protein pharmaceuticals have a generally low oral bioavailability and are normally administered by parenteral routes. Most nasal formulations of peptide or protein pharmaceuticals have been simply prepared in simple aqueous (or saline) solution with preservatives. Recently, more research and development work has been done on the development of delivery systems for the nasal delivery of peptides or proteins.

In the United States, currently only four nasal pharmaceutical products are in-

Table 5 Biopharmaceutical Data for Some Organic-Based Pharmaceuticals

Pharmaceuticals	Animal model	t_{max}	Relative bioavailability: Nasal (%)	Relative bioavailability: Other[a] (%)
Buprenorphine	Rat	2–5 min	95	9.7 (ID)
	Human	5 min	48	
Clofilium tosylate	Rat	<10 min	69.6	1.3 (PO)
Cocaine	Human	15–60 min	—	—
	Human	58 min (solution)	—	—
		35 min (crystals)	—	—
Cromoglycate disodium	Rat	20 min	60 (plasma)	—
		15–30 min	53 (bile)	—
Dopamine	Rhesus monkey	15 min	—	—
Diazepam	Humans	60 min	72–84	—
Ergotamine tartrate	Rat	20 min	62	12.7 (ID)
Ergotamine tartrate with caffeine	Rat	20 min	65.4	5.1 (ID)
Gentamycin	Human	30 min		
Hydralazine				
pH 3.0	Rat	30 min	127	—
pH 6.5	Rat	<10 min	83	—
pH 3.0 (+BL-9, 0.5%)	Rat	<10 min	113	—
Lorazepam	Humans	0.5–4 hr	51	—
Meclizine	Rat	8.5 min	51	8 (PO)
	Dog	12 min	89	22 (PO)
Naloxone	Rat	20 min	101	1.5 (ID)
Nitroglycerin	Human	1–2 min		
Norethisterone	Rhesus monkey	5–30 min		
Phenol red				
With Na deoxycholate	Rabbit	—	98	—
With BL-9	Rabbit	—	87	—
With STDHF	Rabbit	—	83	—
Progesterone				
	Rat	6 min	100	1.2 (ID)
	Rabbit	15 min	88	10 (PO)
	Rabbit	5 min (spray)	82.5	7.9 (ID)
	Rhesus monkey	5.5 min	91	—
Propranolol	Rat	5–6.3 min	100	15–19 (PO)
	Dog	5 min	103	7 (PO)
	Human		109	—
Testosterone	Rat	<2 min	90–99	1 (ID)
Verapamil	Dog	5 min	37	13 (PO)

[a]ID, intraduodenal administration; PO, oral administration.

Table 6 Biopharmaceutical Data for Nasal Delivery of Some Peptide-Based Pharmaceuticals

Pharmaceuticals	Animal model	t_{max}	Relative bioavailability (%)
Alsactide (ACTH-17)	Rat	1 hr	12
α(1–18)-ACTH	Human	<4 hr	12
Buserelin	Human	1–6 hr (LH)	—
		2 hr (FSH)	—
Calcitonin (with sodium glycocholate)	Human	15 min	—
[Asu^{1,7}]eel calcitonin (with polyacrylic acid, pH 6.5)	Rat	30 min	—
Cerulein	Human	—	1
DDAVP	Human	—	10–20
β-Endorphin	Bonnet monkey	30 min	—
Enkephalin analogs			
Leucine enkephalin	Rat	—	<10
DADLE (in saline)	Rat	—	59
DADLE (with 1% Na glycocholate)	Rat	—	94
Mekephamid	Rat	10 min	102
GnRh	Human	<30–90 min (LH)	—
		<90–120 min (FSH)	—
Glucagon (with Na glycocholate)	Human	10 min	50
Growth hormone (hpGRF-40)	Human	<30 min	1–2
Met-hGH			
No enhancer	Rat	30 min	<1
With enhancer			
BL-9	Rat	10–15 min	57–79
Na glycocholate	Rat	10–15 min	7–8
Horseradish peroxidase	Rat	5–10 min	0.6
Insulin			
No promotor	Rat	—	5
With promotor	Rat	—	10
Saponin	Dog	—	30
Na glycocholate	Dog	—	25–33
	Human	13.5 min	12.5
Na deoxycholate	Human	10 min	10–20
BL-9	Human	<15 min	7–10
Na_2 carbenoxolone	Rat	15 min	15
Na caprate	Rat	5 min	98
Na caprylate	Rat	5 min	27
K_2 glycyrrhizinate	Rat	15 min	13
Na glycyrrhizinate	Rat	5 min	27
Na laurate	Rat	5 min	55
Lysophosphatidylcholine	Rat	8 min	30
	Sheep	8 ± 3 min	2.4
STDHF	Rat	10 min	18
	Rabbit	10 min	5
	Sheep	10 min	16

Table 6 Continued

Pharmaceuticals	Animal model	t_{max}	Relative bioavailability (%)
Freezed-dried powder with Carbopol 934p	Dog	30 min	33
β-Interferon	Rabbit	15–60 min	22
[D-Arg[2]]kyotorphin	Rat	—	47
LHRH	Human	60 min	1
Lypressin	Human	—	14
Nafarelin acetate	Rhesus monkey	<15 min	2
		30 min	5
Oxytocin	Human	<10–20 min	—
	Human	—	<1–2
	Rabbit	—	1–10
Pentagastrin	Human	—	20–33
Secretin	Rat	—	10
SS-6	Rat	5–10 min	73
Substance *P*	Rat	2–5 min	—
Thyroxine-releasing hormone	Rat	15 min	20

tended for systemic delivery, i.e., desmopressin (DDAVP), lypressin (Diapid), oxytocin (Syntocinon), and nafarelin acetate (Synarel). If the therapeutic index of a peptide- or protein-based pharmaceutical is broad and its cost is low, nasal delivery is a viable alternative to parenteral administration. For peptide or protein drugs with a narrow therapeutic index and a high bulk material cost, nasal delivery is not an acceptable route of administration because of its low bioavailability (Table 6).

The efficacy of a peptide or protein pharmaceutical delivered intranasally is highly dependent upon its molecular structure and molecular size. Respiratory epithelial cells are capable of absorbing peptide or protein molecules by a vesicular transport mechanism followed by transfer to the extracellular spaces and subsequent uptake by the submucosal vascular network (8,67).

The extent of systemic delivery of peptides or proteins by transnasal permeation may depend on (i) the structure and size of the molecules, (ii) the partition coefficient, (iii) the susceptibility to proteolysis by nasal enzymes, (iv) nasal residence time, and (v) formulation variables (pH, viscosity, and osmolarity).

High-molecular-weight peptides, such as calcitonin, and proteins, such as insulin and growth hormone-releasing factor, all have poor stability in the nasal cavity and low systemic bioavailability. More investigations have been initiated to concentrate on the enhancement of the nasal absorption of peptides or proteins using the following agents:

1. Viscosity-enhancing agents, i.e., methylcellulose, hydroxymethylcellulose, hydroxypropylcellulose, polyethylene glycol, propylene glycol, or polyacrylic acid
2. Bile salts, i.e., sodium salt of deoxycholic or glycocholic acid

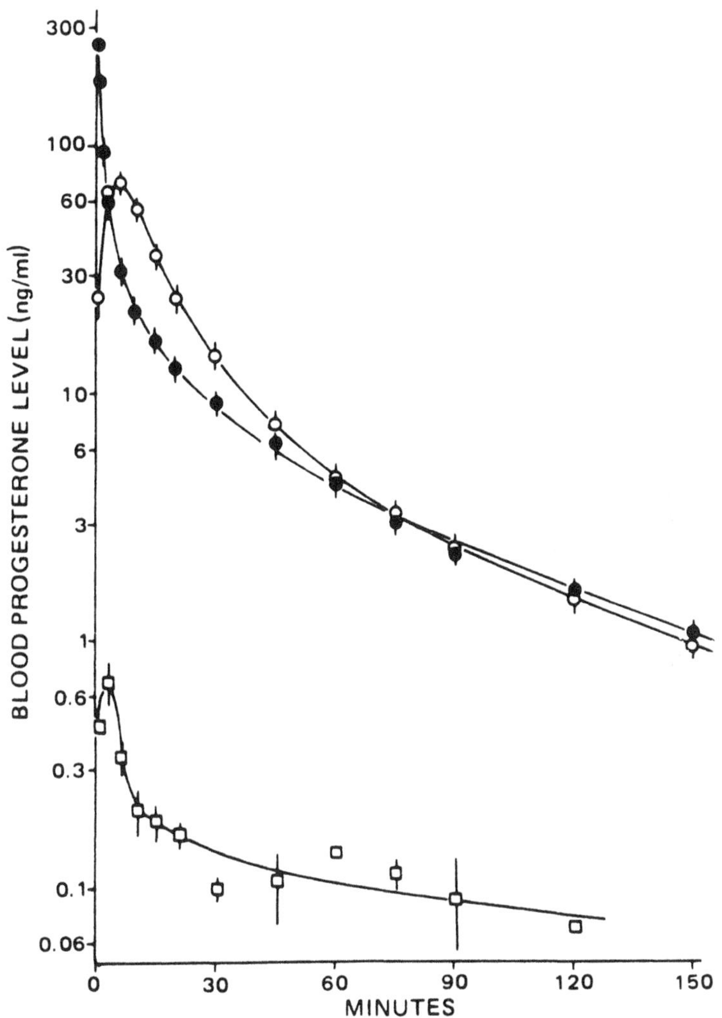

Figure 15 Plasma concentration profiles of progesterone in rats after (○) nasal, (●) intravenous, and (□) intraduodenal administration of progesterone (50 μg per rat). The lines were drawn through the points, and the vertical bars represent standard error of the mean. (From Hussain et al., 1981; reproduced with permission.)

3. Surfactants, i.e., polyoxyethylene-9-lauryl ether
4. Enzyme inhibitors, i.e., aprotinin or amastatin
5. Mucoadhesive or bioadhesive polymers, i.e., starch, albumin, or gelatin

The peptides or proteins that have been studied for feasibility of nasal delivery are listed in Table 8 together with amino acids and bilogical products, and their pharmacokinetic parameters and bioavailabilities are shown in Table 6 (1,36–49,68–72). Several examples that illustrate the increase in the nasal systemic bioavailability of peptides or proteins by absorption promotors are shown in Figures 18 and 19 (68–70).

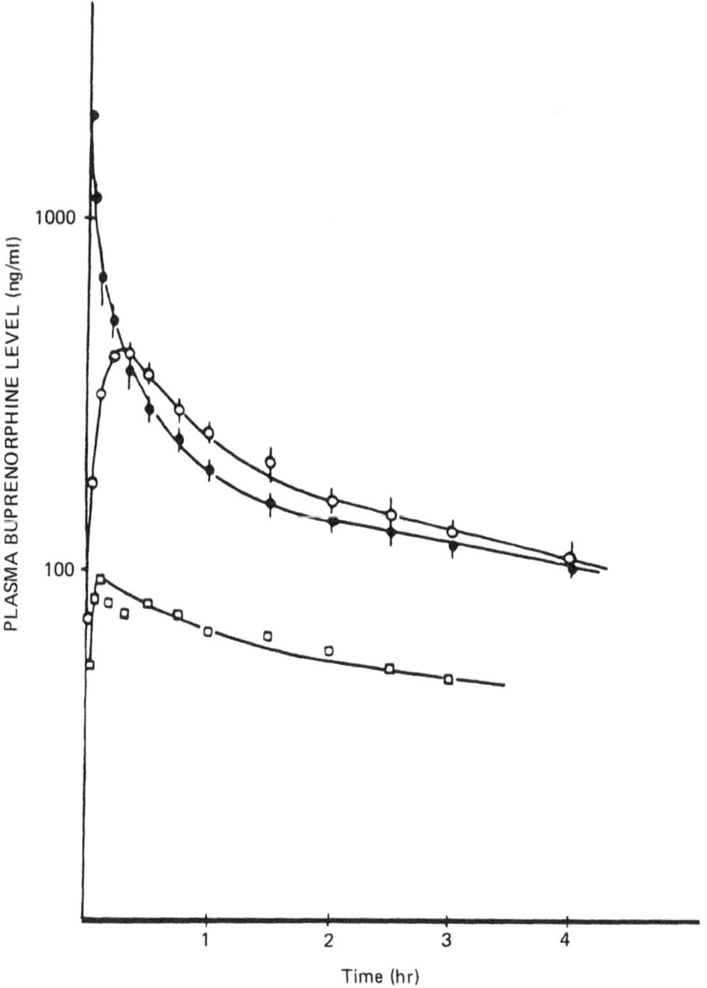

Figure 16 Plasma concentration profiles of buprenorphine in rats after (○) nasal, (●) intravenous, and (□) intraduodenal administration of buprenorphine (135 μg per rat). Points represent mean (± SEM) of three animals. (From Hussain, 1984; reproduced with permission.)

V. ANIMAL MODELS

A. In Vivo Nasal Absorption Models

1. *Rat Model*

The rat model used for studying the nasal absorption of drugs was first presented in the late 1970s (1,61). The surgical preparation for in vivo nasal absorption is described as follows. The rat is anesthetized by intraperitoneal injection of sodium pentobarbital. After an incision is made in the neck, the trachea is cannulated with a polyethylene tube. Another tube is inserted through the esophagus toward the posterior part of the nasal cavity. The passage of the nasopalatine tract is sealed sur-

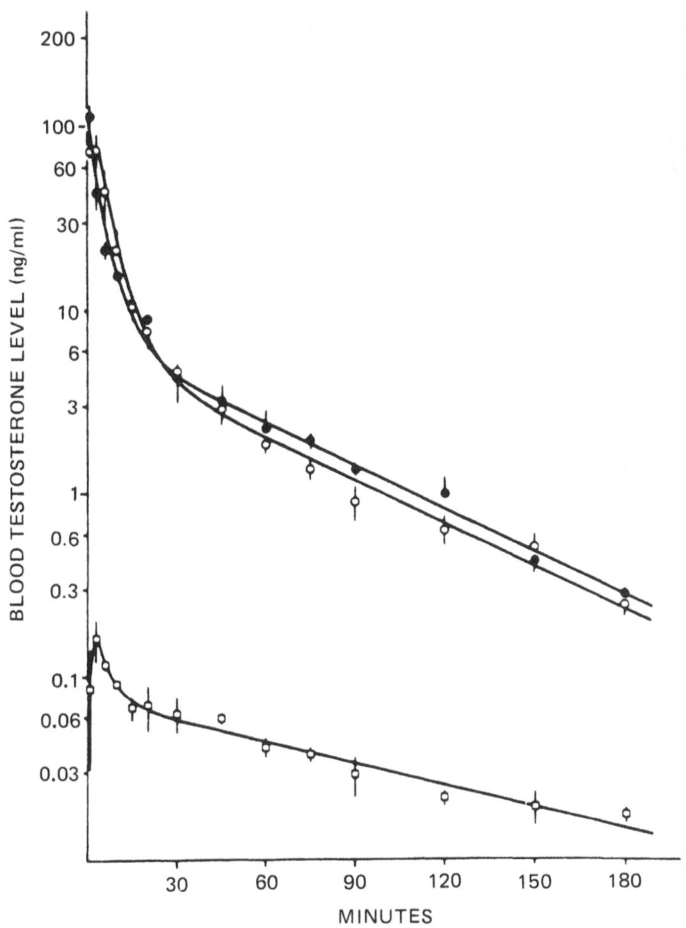

Figure 17 Blood concentration profiles of testosterone in rats (n = 3) after (○) nasal, (●) intravenous, and (□) intraduodenal administration of testosterone (25 μg per rat). Points represent mean (± SEM) values (From Hussain et al., 1984; reproduced with permission.)

gically to prevent the drainage of drug solution from the nasal cavity into the mouth. The drug solution is delivered to the nasal cavity through either the nostril or the esophageal tubing. The blood samples are then collected from the femoral vein. Since all the possible outlets in this rat model are blocked after surgical preparation, the only possible way for the drug to be absorbed and transported into the systemic circulation is penetration (or permeation) through the nasal mucosa.

2. *Rabbit Model*

The rabbit model for nasal drug delivery studies can be prepared as follows. A rabbit weighing approximately 3 kg is either anesthetized or maintained in the conscious state, depending upon the purpose of experiment. In the anesthetized model the rabbit is anesthetized by an intramuscular injection of a combination of ketamine and xylazine. The drug solution is delivered by nasal spray into each nostril while the rabbit's head is held in an upright position. During the study the rabbit is permitted

Table 7 Organic-Based Pharmaceuticals Being Studied for Nasal Delivery

1. Adrenal corticosteroids
2. Antibiotics
   a. Aminoglycosides
      (1) Gentamycin
      (2) Streptomycin
   b. Cephalosporins
   c. Penicillins
   d. Tyrothricin
3. Antimigraine drugs
   a. Diergotamine
   b. Ergotamine tartrate
4. Antiviral agents
   a. Enviroxime
   b. Phenyl-*p*-guanidinobenzoate (PGB)
5. Autonomic nervous system drugs
   a. Sympathomimetics
      (1) Dobutamine
      (2) Dopamine
      (3) Ephedrine
      (4) Epinephrine
      (5) Phenylephrine
      (6) Tramazoline
      (7) Xylometazoline
   b. Parasympathomimetics
      (1) Methacholine
      (2) Nicotine
   c. Parasympatholytics
      (1) Atropine
      (2) Ipratropium
      (3) Prostaglandins
      (4) Scopolamine
6. Cardiovascular drugs
   a. Angiotensin II antagonist
   b. Clofilium tosylate
   c. Hydralazine
   d. Isosorbide dinitrate
   e. Nitroglycerin
   f. Propranolol
   g. Verapamil
7. Cental nervous system drugs
   a. Stimulants
      (1) Cocaine
      (2) Lidocaine
   b. Depressants
      (1) Diazepam
      (2) Lorazepam
8. Histamines and antihistamines
   a. Histamines
   b. Antihistamines
      (1) Disodium cromoglycate
      (2) Meclizine
9. Narcotics and antagonists
   a. Buprenorphine
   b. Naloxone
10. Sex hormones
   a. Estradiol
   b. Norethindrone
   c. Progesterone
   d. Testosterone
11. Vitamins

to breath normally through the nostrils, and the body temperature is maintained at 37°C by a heating pad. The blood samples are collected via an indwelling catheter in the marginal ear vein (1,60).

The rabbit is a relatively inexpensive and readily available animal and can be easily maintained in a laboratory setting. The blood volume of the rabbit is sufficiently large (approximately 300 ml) to permit multiple blood samplings (1–2 ml each) at a frequency that permits full characterization of the pharmacokinetic profile of a drug candidate following nasal delivery. The rabbit model described here has been used to study the nasal absorption and controlled delivery of progesterone and its hydroxy derivatives (18,60).

3. Dog Model

The procedure for preparing a dog model for nasal absorption studies is briefly outlined as follows. The dog is either anesthetized or maintained in the conscious state depending upon the purpose of the study and the characteristics of the drug. In the anesthetized model the dog is anesthetized by intravenous injection of sodium thio-

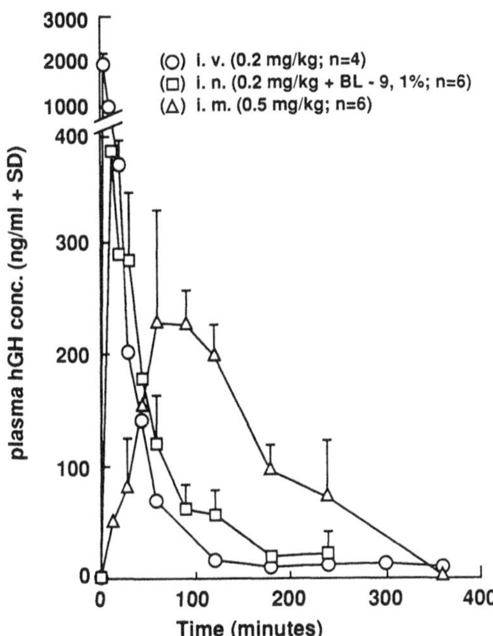

Figure 18 Plasma concentration profiles of hGH after (□) intranasal (0.2 mg/kg with 1% BL-9; n = 6), (○) intravenous (0.2 mg/kg; n = 4), and (△) intramuscular (0.5 mg/kg; n = 6) administration. Each point is expressed as the mean ± SD. (Replotted from Daugherty et al., 1988.)

pental and maintained in an anesthetized state with sodium phenobarbital. A positive-pressure pump provides ventilation through a cuffed endotracheal tube, and a heating pad keeps the body temperature at 37–38°C. Blood samples are collected from the jugular vein (1).

4. Sheep Model

The sheep model for studying nasal drug delivery is prepared using basically the same procedure as that described for the dog model. A male inhouse-bred sheep is used because it lacks nasal infectious diseases.

Because of their larger nostril and body size compared to the rat model, rabbit, and dog, sheep are suitable and practical animal models for the evaluation of pharmacokinetic and pharmaceutical parameters involved in the nasal delivery of drugs from a more sophisticated formulation.

5. Monkey Model

The monkey model for nasal absorption studies is prepared using the following procedure. A monkey weighing about 8 kg is tranquilized, anesthetized, or maintained in the conscious state depending upon the purpose of the study. The monkey is tranquilized by intramuscular injection of ketamine HCl solution or is anesthetized by intravenous injection of sodium phenobarbital. While holding the head in an upright position, the drug solution is delivered into each nostril. The monkey is then placed in a supine position in a metabolism chair for 5–10 min following intranasal

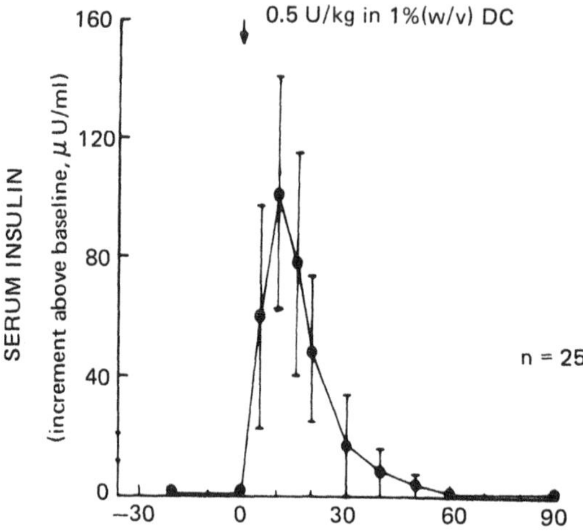

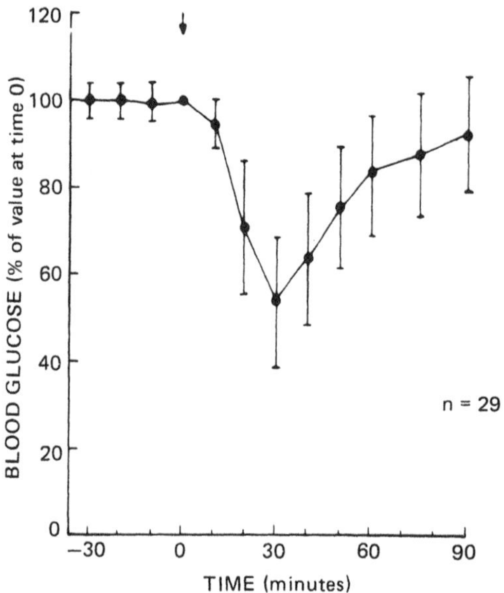

Figure 19 (Top) Serum insulin concentration (mean ± 1 SD) profile for 25 normal subjects who each received a nasal spray of 0.5 IU/kg of insulin in 1% (*w*/*v*) deoxycholate at time 0. (Bottom) Blood glucose level as percentage of the value at time 0 (mean ±1 SD) versus time for 29 normal subjects who each received 0.5 IU/kg of insulin in 1% (*w*/*v*) deoxycholate aerosol at time 0. (From Moses et al., 1983; reproduced with permission.)

administration. During the entire course of the study, the monkey breathes normally through the nostrils. Blood samples are collected via an indwelling catheter in the vein (1).

In summary, the rat is small, easy to handle, and low in cost, as well as inexpensive to maintain. Unfortunately its applications are limited by its small body

Table 8 Amino Acids, Peptides, Proteins, and Biological Products Being Studied for Nasal Delivery

1. Amino acids
2. Peptides
   a. Calcitonin
   b. Cerulein
   c. Cholecystokinin
   d. Enkephalins
   e. Kyotorphin
   f. Pentagastrin
   g. Secretin
   h. SS-6
   i. Substance *P*
   j. Thyrotropin-releasing hormone
3. Polypeptides and proteins
   a. Albumins
   b. Anterior pituitary hormones
      (1) Adrenal corticotropic hormone
      (2) Gonadotropin-releasing hormone
      (3) Growth hormones
   c. Horseradish peroxidase
   d. Pancreatic hormones
      (1) Insulin
      (2) Glucagon
   e. Posterior pituitary hormones
      (1) Oxytocin
      (2) Vasopressin
4. Biological products
   a. Interferons
   b. Vaccines

size, and hence it is useful only for preliminary studies of nasal drug absorption. The primate model continues to be a very useful one, even though it is expensive and has come under increasing pressure from animal rights groups. Alternatively, rabbit, dog, and sheep models are excellent models that are particularly useful for formulation and pharmacokinetic studies. However, the constraint of an animal model still plays an important role in the assessment of the nasal delivery of drugs.

B. Ex Vivo Nasal Perfusion Model

The experimental setup for the ex vivo (in situ) nasal perfusion studies is shown in Figure 20. The same surgical preparation as that described earlier for the in vivo rat model may be followed (1). During the perfusion studies a funnel is provided underneath the nose to lead the drug solution, which is flowing out of the nasal cavity, into the drug reservoir. The reservoir solution of a drug candidate to be evaluated is placed in the container, which is maintained at 37°C, and is circulated through the nasal cavity of the rat by means of a peristaltic pump. The perfusion solution passes out from the nostril and through the funnel and flows into the drug reservoir solution again. The reservoir is stirred constantly, and the amount of drug absorbed

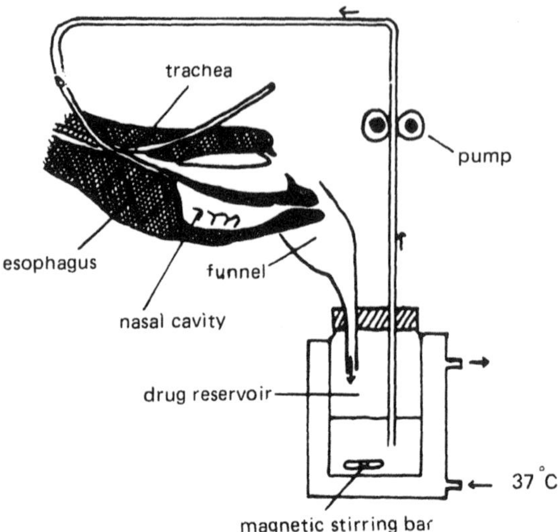

Figure 20 The experimental setup for ex vivo nasal perfusion studies. (From Chien, 1985; reproduced with permission.)

is then determined by measuring the drug concentration remaining in the solution after a period of perfusion. Because of the experimental conditions, the possible loss of drug activity due to physicochemical stability, such as the loss of peptides and proteins by proteolysis, aggregation, and other factors, must be considered.

Using the rabbit as the animal model, the ex vivo nasal perfusion model can also be used for studying the pharmacokinetics of drugs following nasal absorption (73). The kinetics of nasal absorption is monitored by simultaneous measurements of the drug in the perfusion solution as well as in the systemic circulation (Figure 21).

VI. IRRITATION AND TOXICOLOGICAL EVALUATIONS

Assessment of the irritation and toxicity of drugs or formulation to the nasal mucosa is difficult. There are great anatomic differences in nasal tissues between small laboratory animals and humans, and therefore different results in disposition and concentration of drugs in the nasal cavity are expected following the intranasal administration of an applied dose. Nevertheless, the nasal mucosa of experimental animals should be examined histologically to ensure that intranasal administration does not result in any damage to the absorptive cells of the nasal mucosa (74).

Toxicological evaluations of intranasal administration present a number of unique challenges. In addition to evaluation of the adverse local and systemic effects of a drug candidate for suitability of intranasal administration, many other issues arise and deserve special considerations, such as determination of an appropriate dose for clinical applications and selection of concentration and the volume to be administered for a test drug or its formulation in animal models, which may differ greatly in nasal anatomy from humans, as well as the method of administration (1,75).

Both the drug candidate and the final nasal dosage form for clinical use must first be submitted to both acute and subchronic toxicity studies. The pharmaceutical

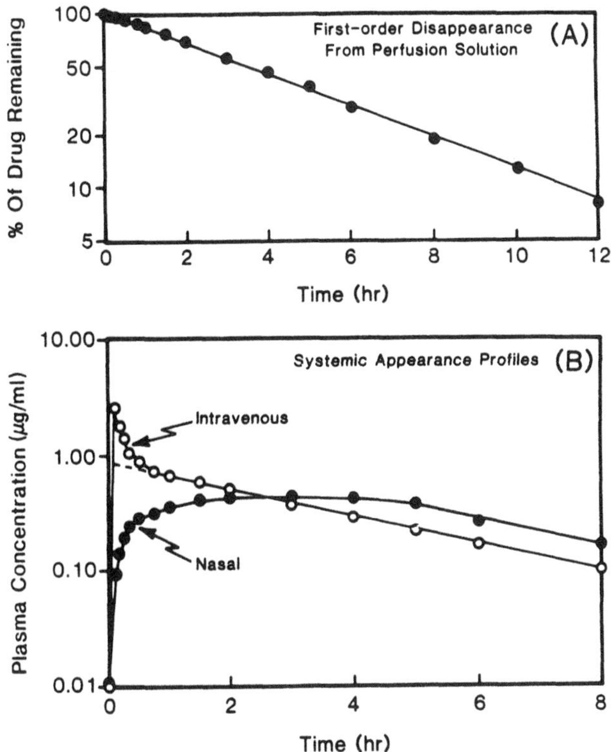

Figure 21 The ex vivo nasal absorption study of hydromorphone in rabbits (n = 3). (A) First-order disappearance of drug from the perfusion solution. (B) Comparative plasma profiles of drug following nasal absorption and intravenous administration at the same dose (5 mg/kg). (From Reference 14; reproduced with permission.)

excipients and additives used in the nasal formulation are preferably those of GRAS materials (generally regarded as safe), so no major toxicity issues need be addressed. If the toxicity of an excipient or an absorption promotor added to the formulation is unknown, an extensive program designed to define the absorption, distribution, metabolism, excretion, and toxicity or carcinogenicity of the agent may be in order (1).

Subchronic toxicity studies usually include a range of doses (or concentrations) and require testing in two different animal species. Other studies, such as fertility, teratology, and carcinogenicity, should also be considered. In general, a 30-day toxicity study on the finished product is sufficient to support a single-dose clinical trial, whereas a 90-day toxicity evaluation is needed to justify multiple-dose clinical trials. The nasal mucosa of experimental animals should be examined histologically to ensure the integrity of the nasal mucosa exposed. Local effects, including irritation, cell damage, and mucociliary clearance, must also be evaluated. During a clinical study gross examination of the nasal cavity of the subjects should be conducted by an ear, nose, and throat specialist (1).

Long-term local toxic effects are even more important for formulations to which promotor is added to increase the permeability of nasal mucosa. Chronic erosion of

Table 9 Toxic Effects of Some Nasal Absorption Promotors

Promotors	Toxic effects
Bile salts	Irritation and congestion effects to nasal mucosa; decrease ciliary function; cause ultrastructural abnormalities
BL-9	Hemolytic and protein-releasing actions on nasal mucosa.
EDTA	Ciliostatic action on nasal mucosa
Fatty acid salts	Hemolytic action on nasal mucosa
STDHF	Ciliostatic action on nasal mucosa

the mucous membrane by certain surfactants may result in inflammation, hyperplasia, metaplasia, and deterioration of normal nasal functions (49). Certain bile salts are known to break down the structure of mucus membrane, accelerate the release of phospholipid and protein from membranes, and damage interstitial mucosa. Some examples of the potential toxic effects of several nasal absorption promotors are outlined in Table 9.

ACKNOWLEDGMENTS

This chapter was rewritten, with updated literature data added, from Chapter 9, Nasal Drug Delivery, authored by S. F. Chang and Y. W. Chien, for the *Treatise on Controlled Drug Delivery*, edited by A. Kydonieus, as well as from the book *Nasal Systemic Drug Delivery*, by Y. W. Chien, K. S. E. Su, and S. F. Chang.

REFERENCES

1. Y. W. Chien, K. S. E. Su, and S.- F. Chang; *Nasal Systemic Drug Delivery*, Dekker, New York, 1989.
2. Y. W. Chien and S.- F. Chang; Crit. Rev. Ther. Drug Carrier Syst., 4:67–194 (1987).
3. Y. W. Chien and S.- F. Chang; in *Transnasal Systemic Medications: Fundamental Concepts and Biomedical Assessments* (Y. W. Chien, Ed.), Elsevier Science Publishers, Amsterdam, 1985, pp. 1–99.
4. S.- F. Chang and Y. W. Chien; Pharm. Int., 5:287 (1984).
5. D. F. Proctor; Am. Rev. Respir. Dis., 115:97 (1977).
6. G. Ewert; Acta Otolaryngol. (Stockh.), 220(Suppl.):1 (1965).
7. G. A. Laurenzi; J. Occup. Med., 15:174 (1973).
8. R. E. Stratford, Jr. and V. H. L. Lee; Int. J. Pharm., 30:73 (1986).
9. E. L. Smith, R. L. Hill and A. Borman; Biochim. Biophys. Acta, 29:207 (1958).
10. D. F. Proctor; in: *Transnasal Systemic Medications* (Y. W. Chien, Ed.), Elsevier, Amsterdam, 1985, pp. 101–106.
11. C. McMartin, L. E. F. Hutchinson, R. Hyde and G. E. Peters; J. Pharm. Sci., 76:535–540 (1987).
12. A. N. Fisher, K. Brown, S. S. Davis, G. D. Parr and D. F. Smith; J. Pharm. Pharmacol., 39:357–362 (1987).
13. A. A. Hussain, R. Bawarshi-Nassar and C. H. Huang; in *Transnasal Systemic Medications* (Y. W. Chien, Ed.), Elsevier, Amsterdam, 1985, pp. 121–137.

14. Y. Kaneo; Acta Pharm. Suec., 20:379 (1983).
15. R. E. Gibson and L. S. Olanoff; J. Control. Release, 6:361–366 (1987).
16. S. Hirai, R. Ikenaga and T. Matsuzawa; Diabetes, 27:296 (1978).
17. S. Harai, R. Yashiki, T. Matsuzawa and H. Mima; Int. J. Pharm., 7:317 (1981).
18. D. C. Corbo, Y. C. Huang and Y. W. Chien; Int. J. Pharm., 50:253 (1989).
19. A. N. Fisher, K. Brown, S. S. Davis, G. D. Parr and D. A. Smith; J. Pharm. Pharmacol., 37:38–41 (1985).
20. M. A. Wheatly, J. Dent, E. B. Wheeldon and P. L. Smith; J. Control. Release, 8:167 (1988).
21. P. Tengamnuay and A. K. Mitra; Life Sci., 43:585 (1988).
22. T. Unno, Y. Okude, O. Yanai and S. Onodera; Jpn. J. Otol. (Tokyo), 85:277 (1982).
23. R. F. Hounan, A. Black and M. Walsh; Aerosol Sci., 2:47 (1971).
24. B. O. Stuart; Arch. Intern. Med., 131:60 (1973).
25. C. D. Yu, R. E. Jones, J. Wright and M. Henesian; Drug Dev. Ind. Pharm., 9:473 (1983).
26. C. D. Yu, R. E. Jones and M. Henesian; J. Pharm. Sci., 73:344 (1984).
27. J. Wolfsdorf, D. L. Swift and M. E. Avery; Pediatrics, 43:799 (1969).
28. F. A. Fry and A. Black; J. Aerosol Sci., 4:113 (1973).
29. N. Mygind; *Nasal Allergy*, 2nd Ed., Blackwell, Oxford, 1979, pp. 260–262.
30. J. G. Hardy, S. W. Lee and C. G. Wilson; J. Pharm. Pharmacol., 37:294 (1985).
31. F. Y. Aoki and J. C. W. Crawley; Br. J. Clin. Pharmacol., 3:869 (1976).
32. G. W. Hallworth and J. M. Padfield; J. Allergy Clin. Immunol., 77:348–353 (1986).
33. S. P. Newman, F. Moren and S. W. Clarke; Rhinology, 25:77–82 (1987).
34. S. L. Whaley, S. Renken, B. A. Muggenburg and R. K. Wolff; J. Toxicol. Environ. Health, 23:519–525 (1988).
35. I. Gonda and E. Gipps; Pharm. Res., 7:69–75 (1990).
36. L. S. Olanoff and R. E. Gibson; in *Controlled Release Technology—Pharmaceutical Applications* (P. I. Lee and W. R. Good, Eds.), ACS Symposium Series 348, American Chemical Society, Washington, D.C., 1987, pp. 301–308.
37. M. Mishma, Y. Wakita and M. Nakano; J. Pharmacobiodyn., 10:624–631 (1987).
38. J. P. Longnecker, A. C. Moses, J. S. Flier, R. D. Silver, M. C. Carey and E. J. Dubov; J. Pharm. Sci., 76:351–355 (1987).
39. M. J. M. Deurloo, W. A. J. J. Hermens, S. G. Romeyn, J. C. Verhoef and F. W. H. M. Merkus; Pharm. Res., 6:853–856 (1989).
40. B. J. Aungst and N. J. Rogers; Pharm. Res., 5:305–308 (1988).
41. L. Illum, N. F. Farraj, H. Critchley, B. R. Johanson and S. S. Davis; Int. J. Pharm., 46:261 (1988).
42. M. Mishma, S. Okada, Y. Wakita and M. Nakano; J. Pharmacobiodyn., 12:31–36 (1989).
43. B. J. Aungst, N. J. Rogers and E. Shefter; J. Pharmaco. Exp. Ther., 244:23–27 (1988).
44. E. Hayakawa, A. Yamamoto, Y. Shoji and V. H. L. Lee; Life Sci., 45:167–174 (1989).
45. G. S. M. J. E. Duchateau, J. Zuidema and S. W. J. Basseleur; Int. J. Pharm. 39:87–92 (1987).
46. S. C. Raehs, J. Snadow, K. Wirth and H. P. Merkle; Pharm. Res., 5:689–693 (1988).
47. S. J. Hersey and R. T. Jackson; J. Pharm. Sci., 76:876–879 (1987).
48. W. A. J. J. Hermens, P. M. Hooymans, J. C. Verhoef and F. W. H. M. Merkus; Pharm. Res., 7:144–146 (1990).
49. W. A. Lee and J. P. Longnecker; Biopharmaceutics, 1:30–37 (1988).
50. T. Nagai, Y. Nishimoto, N. Nambu, Y. Suzuki and K. Sekine; J. Control. Release, 1:15 (1984).
51. K. Morimoto, K. Morisaka and A. Kamada; J. Pharm. Pharmacol., 37:134 (1985).
52. A. A. Hussain, S. Harai and R. Bawarshi; J. Pharm. Sci., 68:1196 (1979).

53. Y. Kaneo; Acta Pharm. Suec., 20:379 (1983).
54. S. S. Davis, L. Illum, D. Burges, J. Ratcliffe and S. N. Mills; in *Controlled Release Technology, Pharmaceutical Applications*, (P. I. Lee and W. R. Good, Eds.), American Chemical Society, Washington, D.C., 1987, pp. 201–213.
55. L. Illum, N. Farraj, H. Critchley and S. S. Davis, Int. J. Pharm., 46:261–265 (1988).
56. E. Bjok and P. Edman; Int. J. Pharm., 47:233–238 (1988).
57. L. Illum, H. Jorgensen, H. Bisgaard, O. Krogsgaard and N. Rossing; Int. J. Pharm., 39:189–199 (1987).
58. N. F. Farraj, L. Illum, S. S. Davis and B. R. Johansen; Diabetologia, 32:486A (1989).
59. N. F. Farraj, B. R. Johansen, S. S. Davis and L. Illum; J. Control. Release, 13:253–261 (1990).
60. D. C. Corbo, Y. C. Huang and Y. W. Chien; Int. J. Pharm., 46:133 (1988).
61. A. Hussain and R. Bawarshi; J. Pharm. Sci., 69:1411 (1980).
62. T. Nagai and Y. Machida; in *CRC Bioadhesive Drug Delivery Systems* (V. Lenaerts and R. Gurney, Eds.), CRC Press, Boca Raton, Florida, 1990, pp. 179–178.
63. J. L. Colaizzi; in *Transnasal Systemic Medications* (Y. W. Chien, Ed.), Elsevier, Amsterdam, 1985, pp. 107–119.
64. A. A. Hussain, R. Kimura, C. H. Huang and T. Kashihara; Int. J. Pharm., 21:233 (1984).
65. A. A. Hussain, S. Hirai and R. Bawarshi; J. Pharm. Sci., 70:466 (1981).
66. A. A. Hussain, R. Kimura and C. H. Huang; J. Pharm. Sci., 73:1300 (1984).
67. J. Richardson, T. Bouchard and C. C. Ferguson; Lab. Invest., 35:307–314 (1976).
68. A. L. Daugherty, H. D. Liggitt, J. G. McCabe, J. A. Moore and J. S. Patton; Int. J. Pharm., 45:197 (1988).
69. A. C. Moses, G. S. Gordon, M. C. Carey and J. S. Flier; Diabetes, 32:1040 (1983).
70. K. S. E. Su, K. M. Campanale, L. G. Mendelsohn, G. A. Kerchner and C. L. Gries; J. Pharm. Sci., 74:394 (1985).
71. B. Tarquini, V. Cavallini, A. Cariddi, M. Checchi, V. Sorice and M. Cecchettin; Chronobiol. Int., 5:149–152 (1988).
72. Y. Maitani, T. Igawa, Y. Machida and T. Nagai; Drug Design Delivery, 1:65–70 (1986).
73. S.-F. Chang, L. C. Moore and Y. W. Chien; Pharm. Res., 5:718 (1988).
74. K. S. E. Su, K. M. Campanale and C. L. Gries; J. Pharm. Sci., 73:1251 (1984).
75. M. A. Dorato; in *Nasal Delivery of Peptides and Proteins: Toxicologic Considerations*, presented at a symposium on nasal administration of peptide and protein drugs, Princeton, New Jersey, October 1987.

6
Ocular Drug Delivery and Delivery Systems

I. INTRODUCTION

Most ocular treatments call for the topical administration of ophthalmically active drugs to the tissues around the ocular cavity. Several types of dosage forms can be applied as the delivery systems for the ocular delivery of drugs. The most prescribed dosage form is the eye drop solution, for example, ocular decongestant eye drops and aqueous antiglaucoma pilocarpine solutions.

The eye drop dosage form is easy to instill but suffers from the inherent drawback that the majority of the medication it contains is immediately diluted in the tear film as soon as the eye drop solution is instilled into the cul-de-sac and is rapidly drained away from the precorneal cavity by constant tear flow, a process that proceeds more intensively in inflamed than in the normal eyes, and lacrimal-nasal drainage. Therefore, only a very small fraction of the instilled dose is absorbed into the target tissues (e.g., 1.2% is available to the aqueous humor), and relatively concentrated solution is required for instillation to achieve an adequate level of therapeutic effect. The frequent periodic instillation of eye drops becomes necessary to maintain a continuous sustained level of medication. This gives the eye a massive and unpredictable dose of medication, and unfortunately, the higher the drug concentration in the eye drop solution, the greater the amount of drug lost through lacrimal-nasal drainage system. Subsequent absorption of this drained drug, if it is high enough, may result in undesirable systemic side effects (1).

Furthermore, the intraocular concentration of medication surges to a peak everytime eye drops are instilled; the drug level then declines rapidly at an exponential pattern as time passes. A plot of intraocular drug concentration versus time yields a series of peaks of drug level, which may surpass the toxic threshold of the drug, separated by extended valleys of drug level below the critical level required to achieve the desired therapeutic efficacy.

Suspension-type pharmaceutical dosage forms have also been widely used for ocular medication; hydrocortisone acetate and prednisolone acetate are typical of drugs currently marketed as suspensions. Suspension formulations also have some inherent drawbacks. For example, they are generally formulated with relatively water-insoluble drugs to avoid the intolerably high toxicity created by saturated solutions of water-soluble drugs. However, the rate of drug release from the suspension is dependent upon the rate of dissolution of the drug particles in the medium, which varies constantly in its composition with the constant inflow and outflow of lacrimal fluid.

A basic concept shared by most scientists in ophthalmic research and development is that the therapeutic efficacy of an ophthalmic drug can be greatly improved by prolonging its contact with the corneal surface. For achieving this purpose, viscosity-enhancing agents, such as methylcellulose, are added to eye drop preparations, or the ophthalmic drug is formulated in a water-insoluble ointment formulation to sustain the duration of intimate drug-eye contact. Unfortunately, these dosage forms give only marginally more sustained drug-eye contact than eye drop solutions and do not yield a constant drug bioavailability as originally hoped. Repeated medications are still required throughout the day.

Recently, drug-presoaked hydrogel contact lenses and pledgets have gained some popularity in an attempt to bypass the need for repetitive drug dosing and to avoid the peak-and-valley activity-time curves resulted from periodic applications of eye drops and ointments (2–5). A micropump-type delivery system has also been developed for the continuous administration of fluid to dry eyes or medications to infected eyes (6). These drug delivery systems have succeeded in significantly reducing the frequency of dosing and also in remarkably improving the therapeutic efficacy of ophthalmic drugs.

However, the prime objective of developing a truly continuous, controlled delivery of ophthalmically active drugs to the eye was not satisfactorily accomplished until the development of the Ocusert system (Figure 1) for the ocular delivery of pilocarpine at predetermined rates to glaucoma patients (7).

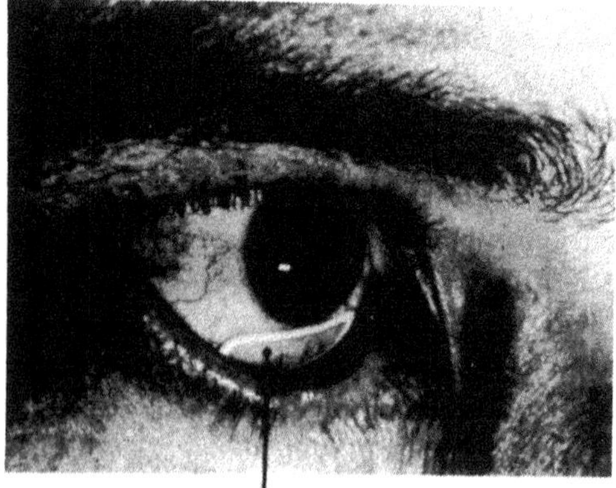

Figure 1 A unit of Ocusert system, which releases pilocarpine continuously at a constant rate when inserted in the upper or lower cul-de-sac of the eye for the relief of intraocular pressure for 7 days. [Reproduced, with permission, from Baker and Lonsdale; Chemtech., 5:668 (1975).]

II. PHYSIOLOGY OF THE EYE

The internal structures of the eye and blood supply are illustrated in Figure 2. The cornea, lens, and vitreous body are all transparent media with no blood vessels; oxygen and nutrients are transported to these nonvascular tissues by the aqueous humor. The aqueous humor has a high oxygen tension and about the same osmotic pressure as blood.

The cornea also derives part of its oxygen need from the atmosphere, and if oxygen is excluded the anaerobic metabolism results in an increase in the intracorneal lactic acid concentration. This can produce edema sufficient to lead to the loss of corneal transparency and a temporary interference with vision. This may occur if a contact lens applied to the cornea does not permit the exchange of atmospheric oxygen or interferes with the capillary blood supply at the limbus.

The cornea is covered by a thin epithelial layer continuous with the conjunctiva

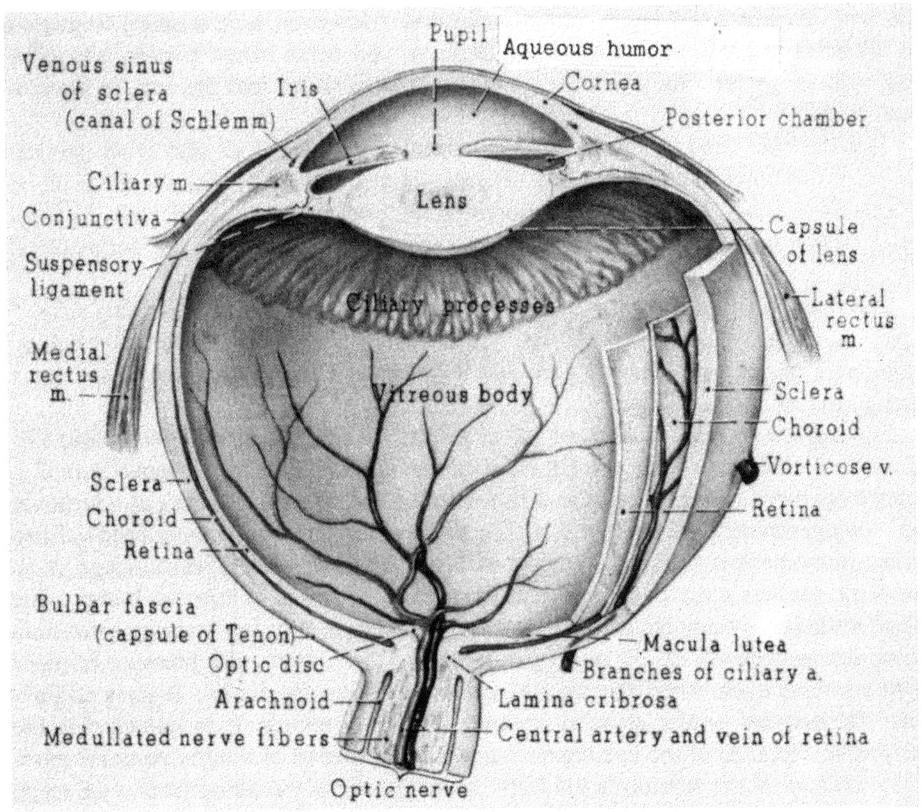

Figure 2 Midsagittal section through the eyeball showing the internal structures of the eye and blood supply. [Reproduced, with permission, from Jacob and Francone; *Structure and Function in Man*, W. B. Saunders, Philadelphia, 1970, p. 257.]

at the cornea-sclerotic junction. The main bulk of the cornea is formed of criss-crossing layers of collagen and is bounded by elastic laminae on both front and back. Its posterior surface is covered by a layer of endothelium. The cornea is richly supplied with free nerve endings. The transparent cornea is continued posteriorly into the opaque white sclera, which consists of tough fibrous tissue. Both cornea and sclera withstand the intraocular tension constantly maintained in the eye.

The eye is constantly cleansed and lubricated by the lacrimal apparatus (8), which consists of four structures: (i) lacrimal glands, (ii) lacrimal canals, (iii) lacrimal sac, and (iv) nasolacrimal duct. The lacrimal fluid secreted by the lacrimal glands is emptied on the surface of the conjunctiva of the upper eyelid at a turnover rate of 16%/min (9). It washes over the eyeball and is swept up by the blinking action of the eyelids. Muscles associated with the blinking reflex compress the lacrimal sac. When these muscles relax, the sac expands, pulling the lacrimal fluid from the edges of the lids, along the lacrimal canals, into the lacrimal sacs. Gravitational force, in turn, moves the fluid down the nasolacrimal duct into the inferior meatus of the nose. Thus the eyeball is continually irrigated by a gentle stream of lacrimal fluid that prevents it from becoming dry and inflamed. The amount of lacrimal fluid renewed by frequent involuntary blinking movements normally is just sufficient to keep pace with its disappearance from the conjuctiva. However, an excessive formation and secretion of lacrimal fluid, or lacrimation, can occur when foreign bodies or other irritants get into the eye, when a bright light is shone into the eye, or in emotional stress.

The lacrimal fluid in humans has a normal volume of 7 μl and is an isotonic aqueous solution of bicarbonate and sodium chloride (pH 7.4) that serves to dilute irritants or to wash the foreign bodies out of the conjunctival sac. It contains lysozyme, whose bacteriocidal activity reduces the bacterial count in the conjunctival sac.

The rate of blinking varies widely from one person to another (ranging from 5 to 50 blinking movements per minute), with an average of approximately 20 blinking movements per minute. During each blink movement the eyelids are closed for a short period of about 0.3 sec.

The aqueous humor in humans has a volume of approximately 300 μl that fills the anterior chamber of the eye (in front of the lens; Figure 2). Aqueous humor is secreted by the ciliary processes and flows out of the anterior chamber at a turnover rate of approximately 1%/min (9,10). The drainage system has recently been defined at the sinus venosus sclerae, a region of low blood pressure. This drainage is an unspecific mechanical process different from the production of aqueous humor. The rate of drainage is comparable to the rate of production, thus maintaining a constant intraocular tension of 25–30 mmHg in humans. This intraocular pressure remains fairly constant even when the arterial pressure widely fluctuates. It rises slightly when the external ocular muscles contract and on winking. It is known that the focusing mechanism of the eye depends upon the existence of a fairly constant intraocular tension. If the tension is too high, as in glaucoma (resulting from a defect in the outflow of fluid), the ciliaris muscles may not be able to bring about accommodation. The high intraocular pressure may also cause the restriction of the retinal circulation with resultant damage to the retina. On the other hand, an excessive reduction in the intraocular tension may slacken the suspensory ligaments of the lens and allows the latter to bulge.

III. GLAUCOMA AND ITS MANAGEMENT

Increased intraocular pressure has been detected in over 2% of the U.S. population over the age of 40. It is also the symptom of glaucoma.

Glaucoma is a disease characterized mainly by an increase in intraocular tension that, if sufficiently high and persistent, leads to irreversible blindness (12). It can be classified into three types: primary, secondary, and congenital.

Primary glaucoma is further subdivided into narrow-angle (acute, congestive) and wide-angle (chronic, simple) types, based on the configuration of the angle of the anterior chamber where reabsorption of the aqueous humor occurs. Narrow-angle glaucoma is nearly always a medical emergency that usually develops in an eye having a shallow anterior chamber with increased curvature of the anterior lens surface (13,14). It has long been known as a painful cause of blindness (15,16) and drugs are essential to control its acute attack, but long-range management is usually based predominantly on surgery. Wide-angle glaucoma, on the other hand, has a gradual, insidious onset, is not generally amenable to surgical improvement, and requires drug therapy on a permanent basis (12).

Progressive loss of the visual field by infarction of nerve fiber bundles at the disk can happen as long as the intraocular pressure continues to remain high. In an attack of narrow-angle glaucoma, it is a matter of ophthalmological urgency to reduce the intraocular tension to physiological levels (25–30 mmHg) and maintain it there for the duration of the attack. In general, an anticholinesterase agent is instilled into the conjunctival sac in combination with a parasympathomimetic agent for greater effectiveness. One such combination that is frequently employed is physostigmine salicylate (1% solution) plus pilocarpine nitrate (4% solution). This combination should be instilled six times at 10 min intervals, then three times at 30 min intervals, and thereafter as required. The secretion of aqueous humor can be reduced by the intravenous injection of acetazolamide as adjunctive therapy.

For the treatment of wide-angle glaucoma (or secondary) glaucoma, for which therapy must be continued indefinitely, a wide choice of parasympathomimetic drugs and anticholinesterase agents is available. In general it is desirable to use the lowest concentration of drug at the longest interval between instillations that will maintain a satisfactory reduction in intraocular pressure to minimize ocular and systemic side effects (12).

IV. OCULAR DELIVERY OF PILOCARPINE

For nearly a century pilocarpine has remained the preferred ocular hypotensive agent for the treatment of increased intraocular pressure. It is the chief alkaloid extracted from the leaflets of South American shrubs known as *Pilocarpus jaborandi* and *Pilocarpus microphyllus*. Pilocarpine produces its cholinomimetic effects mainly by a direct action on autonomic effector cells.

Pilocarpine is traditionally employed at a concentration of 4% in the intensive pilocarpine treatment regimen involving the administration of one or two drops every minute for 5 min, every 5 min for half an hour, and then every 15 min for an hour and a half. These 17 doses deliver approximately 40–80 mg pilocarpine to the eye. When the reduction in intraocular pressure fails to reach normal physiological levels, this intensive treatment is often continued or repeated (3). The total dose of pilocarpine administered is well over the maximum advised dose of 12 mg for systemic administration (17). This intensive pilocarpine treatment is required since pilocarpine is an effective but short-acting hypotensive drug and only 2–3% of the instilled pilocarpine doses is absorbed into the aqueous humor (18,19). Most of the pilocarpine doses are lost through the nasolacrimal drainage system.

Additionally, studies have shown that administration of pilocarpine in eye drop dosage forms has serious patient noncompliance problems. Approximately 30–60% of the patients fail to take pilocarpine eye drops as directed (20,21), thereby making the management of intraocular pressure levels erratic. The noncompliance issues, the low intraocular drug bioavailability, and the potential systemic side effects associated with pilocarpine eye drops were significantly improved with the development of the pilocarpine-releasing Ocusert system (22,23).

The Ocusert system (Alza Corporation, Palo Alto, California) consists of a pilocarpine-aliginate core sandwiched between two transparent, rate-controlling enthylene-vinyl acetate copolymer membranes (Figure 3). When this is placed under the upper or lower eyelid (Figure 1), the pilocarpine molecules dissolved in the lacrimal fluid are released through the rate-controlling membranes in a zero-order fashion at preprogrammed rates and diffuse to the corneal surface for absorption (Section VI). Two Ocusert products, Ocusert Pilo-20 and Ocusert Pilo-40, which deliver pilocarpine continuously at a constant rate of 20 and 40 μg/hr, respectively, for 7 days were approved for marketing in late 1974 by the U.S. Food and Drug Administration.

V. BIOPHARMACEUTICS OF OCULAR PILOCARPINE ADMINISTRATION

The topical application of opthalmically active drugs to the eye is the most prescribed route of administration for the treatment of various ocular disorders. It is generally agreed that the intraocular bioavailability of topically applied drugs is extremely poor; for example, in the treatment of glaucoma with pilocarpine eye drops only a very small fraction (i.e., 2–3%) of the instilled pilocarpine doses is actually absorbed into the aqueous humor, the target tissue for the reduction of intraocular tension in glaucoma patients (18,19).

Many factors can affect the intraocular bioavailability of topically applied ophthalmically active drugs: (i) the presence of lacrimal fluid in the cul-de-sac dilutes

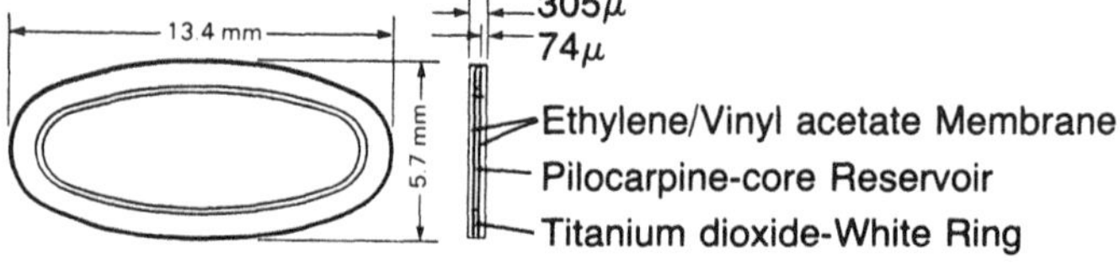

Figure 3 The Ocusert therapeutic system, a pilocarpine-releasing ocular insert. Each Pilo-20 unit measures 5.7 × 13.4 mm on its axes and 0.3 mm in thickness and contains 5 mg pilocarpine. Each Pilo-40 unit measures 5.5 × 13 mm on its axes and 0.5 mm in thickness and contains 11 mg pilocarpine. [Reproduced, with permission, from Baker and Lonsdale; Chemtech., 5:668 (1975).]

the drug solution instilled into the precorneal area of the eye, and the continual inflow and outflow of lacrimal fluid can also cause a significant loss of applied drug; (ii) the efficient nasolacrimal drainage, used for the drainage of lacrimal fluid, also acts as a conduit through which an instilled drug solution may be drained away from the precorneal area; (iii) the substances normally present in the lacrimal fluid, such as protein, can interact and/or degrade the drugs introduced into the ocular cavity; and (iv) productive and nonproductive absorption of topically applied drugs into various ocular tissues, most notably the cornea and conjuctiva. The effects of these factors on intraocular drug bioavailability using pilocarpine as the model drug are analyzed in this section. Pilocarpine has the molecular structure.

A. Precorneal Disposition of Pilocarpine

The drug solution instilled as eye drops into the ocular cavity may disappear from the precorneal area of the eye by any or a composite of the following routes: (i) nasolacrimal drainage, (ii) tear turnover, (ii) productive corneal absorption, and (iv) nonproductive conjunctival uptake (Scheme I).

Following the instillation of a pilocarpine eye drop dose (50–70 μl) into the precorneal area of the eye, a major portion of the drug solution is drained away within 5 min by the nasolacrimal drainage system until the solution volume returns

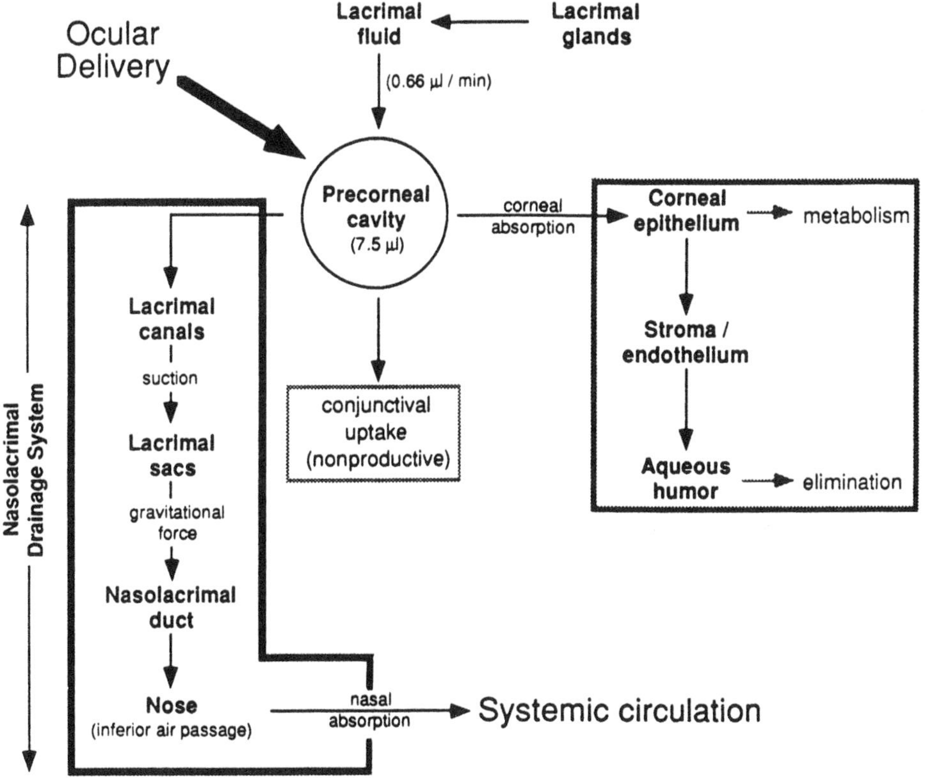

Scheme I

to the normal resident tear volume of 7.5 μl (24). This drainage results in a loss of 80% of the drug dose administered, but it does not affect the concentration of drug in the solution to any significant extent (25). After the precorneal solution returns to its normal resident volume, the solution volume stays constant but its drug concentration declines as a result of continuing dilution by tear turnover (0.66 μl/min) and active uptake by corneal and conjunctival tissues. This results in a biphasic decline of both the dose and the concentration of pilocarpine in the precorneal area: initially the drug dose decreases very rapidly as a result of nasolacrimal drainage; the drug concentration declines very slowly as a result of tear turnover and corneal and conjunctival absorption. As the solution volume in the precorneal area returns to its normal resident level the residual drug dose declines slowly since drainage is no longer operative; on the other hand, the drug concentration declines rapidly by the dilution effect of incoming tear fluid and absorption by the cornea and conjunctiva.

The precorneal disposition of a pilocarpine solution by various routes was observed to follow a first-order kinetic pattern. The first-order rate constants for various routes of precorneal disposition of instilled pilocarpine nitrate are compared in Table 1. Apparently the nasolacrimal drainage is the major route of precorneal disposition for drug applied to the eye. As much as 75% of the drug dose introduced into the precorneal area is lost to the nasolacrimal drainage system within 5 min following each eye drop instillation. Because of this nasolacrimal drainage, the intraocular bioavailability of pilocarpine nitrate is reduced by as much as 71%, and overall only a very small fraction (i.e., 1.2%) of the instilled philocarpine nitrate dose is absorbed and actually reaches the target tissue, the aqueous humor (Table 2).

The effect of the solution volume of a pilocarpine eye drop instilled into the precorneal area on the intraocular drug bioavailability was investigated in rabbits using either a constant drug dose (24) or a varying drug dose (26). Both studies demonstrated that the rate of nasolacrimal drainage is dependent upon the volume of eye drop instilled: the smaller the instillation volume, the higher the intraocular bioavailability of pilocarpine (Table 3). This observation can be attributed to the effect of instillation volume on the first-order rate constant for nasolacrimal drainage (k_{nl}). The k_{nl} value decreases as the volume of the instilled eye drop is reduced (Table 3). The reduction in the k_{nl} value was observed as a linear relationship with the volume of eye drop instilled (24).

It was suggested that the rate of nasolacrimal drainage could be greater in humans than that measured in rabbits; therefore, to maximize intraocular bioavailability

Table 1 Precorneal Disposition of Pilocarpine Nitrate Applied Topically to Rabbit Eyes

Routes of precorneal disposition	First-order rate constant (min^{-1}) $\times 10^2$
Overall precorneal disposition	7.80
Specific precorneal disposition	
Nasolacrimal drainage	4.97
Dilution by tear turnover	1.16
Corneal and conjunctival absorption	2.17

Source: Compiled from the data by Patton and Robinson; J. Pharm. Sci., 65:1295 (1976).

Table 2 Transcorneal Bioavailability of Pilocarpine to Various Ocular Tissues Following Eye Drop Instillation

Ocular tissues[a]	Tissue size[b] (mg)	Bioavailability[c] (% instilled dose)
Cornea	53.1	1.1
Aqueous humor	300	1.2
Iris	19.7	0.1
Ciliary body	20.7	0.1
Lens	257	0.28
Vitreous humor	1077	0.52

[a]Of albino rabbit eye.
[b]The mean data of 100 determinations.
[c]Following a single instillation of 25 μl pilocarpine nitrate eye drop (0.01 M).
Source: Compiled from the data by Makoid and Robinson; J. Pharm. Sci., 68:435 (1979).

and the biological activity of ophthalmically active drugs in humans, the instillation volume of the eye drop solution should be substantially reduced from its present eye drop size of 50–70 μl to, at most, 5–10 μl (24). The more the instillation volume is reduced, the greater the intraocular bioavailability, and the pharmacological activity as well, is achieved.

B. Transcorneal Permeation of Pilocarpine

The prevailing theories of the transcorneal permeation of pilocarpine often include, all or in part, the following hypotheses: (i) the existence of a permeation barrier in the lipophilic corneal epithelium; (ii) the uptake and permeation of pilocarpine through

Table 3 Effects of Eye Drop Instillation Volume on the Bioavailability of Pilocarpine Nitrate

Instillation volume[a] (μl)	Intraocular bioavailability: Varying drug dose[b]: AUC/dose[d] (min/ml)	% Absorbed	Constant drug dose[c]: AUP/dose[e] (cm/min · μg)	Peak height[f] (mm)	k_{nl}[g] (sec^{-1})
75	—	—	1.27	2.05	—
50	—	—	1.45	2.40	0.82
25	0.90	1.16	1.65	2.53	0.55
15	1.57	1.92	—	—	—
10	1.89	2.23	2.62	3.53	0.37
5	2.79	3.01	4.60	4.17	0.31

[a]The volume of eye drop solution instilled into the rabbit eye.
[b]Data compiled from Patton; J. Pharm. Sci., 66:1058 (1977).
[c]Data compiled from Chrai et al.; J. Pharm. Sci., 62:1112 (1973).
[d]The area under the aqueous humor drug concentration-time profile (AUC) achieved by a unit drug dose.
[e]The area under the pupillary diameter-time profile (AUP) achieved by a unit drug dose.
[f]The mean of the peak height of the change in pupillary diameter-time profiles.
[g]The first-order rate constant for nasolacrimal drainage.

the cornea are rapid processes; (iii) transport of pilocarpine from the cornea to the anterior chamber is controlled by the corneal endothelium; and (iv) the presence of a pilocarpine depot somewhere in the cornea.

The existence of the transcorneal permeation barrier was recently characterized, located at the outermost cell layer of the corneal epithelium (27,28). This barrier varies in the magnitude of its permeability in relation to the solubility characteristics of a penetrant molecule. A very water-soluble drug does not easily penetrate this barrier; a very lipid-soluble drug may penetrate it easily, but it may not be able to diffuse into the deep ocular tissue from this barrier. Therefore, some degree of solubility in both the aqueous and lipid media is necessary for optimal transcorneal permeability (29,30).

The mechanisms of transcorneal pilocarpine permeation were recently investigated in albino rabbits (31). The results indicated that following the instillation of pilocarpine eye drops the drug concentrations reached a peak level within 5 min in the cornea and within 20 min in the aqueous humor (Figure 4). Similar results were also obtained by other investigators (32–34). First-order kinetic analyses indicated

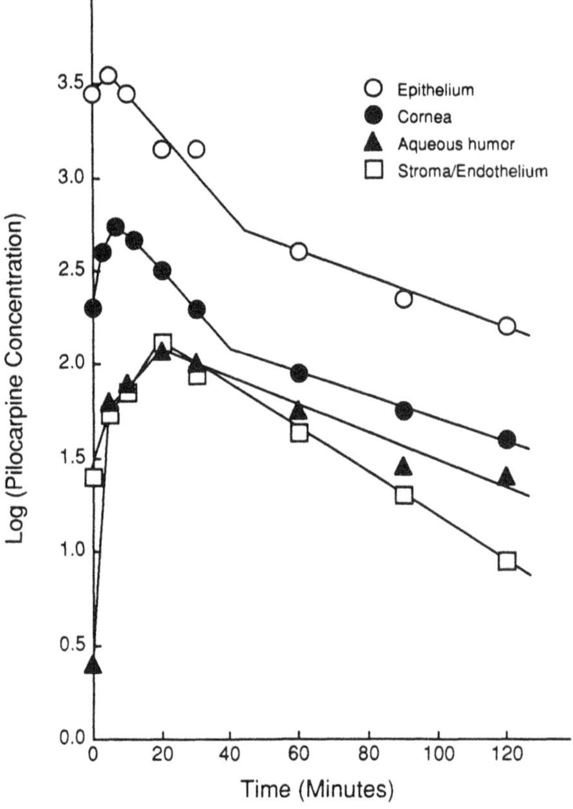

Figure 4 Concentration-time profiles of pilocarpine in (▲) aqueous humor, (●) cornea, (○) epithelium, and (□) stroma-endothelium. [Plotted from the data by Sieg and Robinson; J. Pharm. Sci., 65:1816 (1976).]

that pilocarpine is absorbed by the cornea at a rate constant of 0.57 min^{-1} (Table 4) and penetrates the cornea at a slower rate (with an apparent transcorneal rate constant of 0.08 min^{-1}). The concentration profiles in Figure 4 indicate that immediately after the peak drug concentrations are reached within 5 min pilocarpine concentrations in the epithelium and in the cornea begin to decline in a biexponential manner: a fast distribution within the first 30 min (α-phase) and a slower elimination thereafter (β-phase). On the other hand, the pilocarpine levels in the aqueous humor and in the stroma-endothelium declined after reaching the peak drug concentration within 20 min in a monoexponential pattern. The pilocarpine levels in both cornea and aqueous humor were eliminated at approximately the same magnitude of elimination rate constant (Figure 4), which is very much the same as the elimination rate constant (β-phase) for pilocarpine from the epithelium (Table 4). Removal of the corneal epithelium before medication was found to produce a seven- to eightfold increase in the aqueous humor pilocarpine concentration, and the peak drug level was achieved within a much shorter time, i.e., 5 min after eye drop instillation. These observations confirmed the belief that the corneal epithelium is the major barrier to the transcorneal permeation of pilocarpine.

The results illustrated in Figure 4 also clearly indicate that pilocarpine in the cornea mainly concentrates in the epithelial layer. The pharmacokinetic profile of epithelial pilocarpine concentration showed a relationship parallel to that of the cornea. Further analyses of the corneal pilocarpine levels suggested that at the very early stage of transcorneal penetration of pilocarpine, up to 80% of the total drug content taken up by the cornea has been found to reside in the corneal epithelium. This percentage decreases with time as more pilocarpine molecules diffuse into the stroma. Even up to 2 hr, the epithelium was observed to retain the major fraction of the corneal pilocarpine content and no buildup of drug could be detected in the corneal endothelium layer. This observation confirms that the corneal epithelium layer plays a dual role in the transcorneal permeation of pilocarpine, both as the permeability barrier to transcorneal permeation and as a reservoir for the prolonged release of pilocarpine to the underlying layers of cornea, i.e., the stromal-endothelial layer.

Table 4 Pharmacokinetic Parameters for Pilocarpine Disposition in Corneal Tissues and Aqueous Humor Following Topical Dosing

Ocular tissues[a]	Peak time (min)	k_a[b] (min^{-1})	k_e[c] (min^{-1}) Phase α	Phase β
Intact cornea	5	0.57	—	—
Corneal tissues				
Epithelium	4	0.50	0.06	0.016
Stroma-endothelium	20	0.09	0.021	—
Aqueous humor	20	0.08	0.017	—

[a]Of albino cornea.
[b]First-order rate constant for absorption.
[c]First-order rate constant for elimination.
Source: Compiled from the data by Sieg and Robinson; J. Pharm. Sci., 65:1816 (1976).

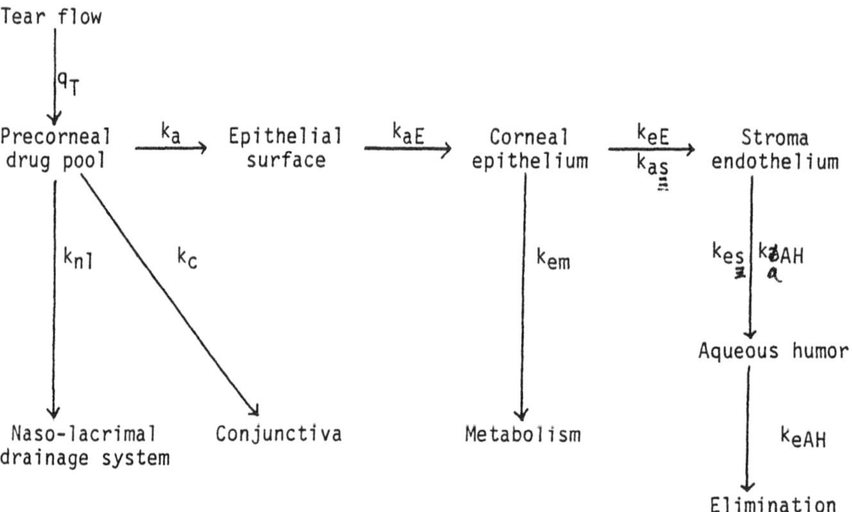

Scheme II

C. Pharmacokinetics of Transcorneal Pilocarpine Permeation

The transcorneal permeation of pilocarpine following its instillation into the cul-de-sac is shown in scheme II

where:
q_T = normal production rate of tear fluid (0.66 μl/min)
k_{nl} = composite first-order elimination rate constant of nasolacrimal drainage
k_c = apparent rate constant for conjuctival uptake of pilocarpine
k_a = apparent rate constant for epithelial uptake of pilocarpine
k_{aE} = apparent rate constant for absorption into epithelium
k_{eE} = apparent rate constant for elimination from epithelium
k_{aS} = apparent rate constant for absorption into stroma-endothelium
k_{eS} = apparent rate constant for elimination from stroma-endothelium
k_{aAH} = apparent rate constant for absorption into aqueous humor
k_{eAH} = apparent rate constant for elimination from aqueous humor
k_{em} = apparent rate constant for metabolism from epithelium

By considering the eye as consisting of two major compartments, the precorneal area and the aqueous humor (35), a simplified pharmacokinetic model can be established to analyze the transcorneal permeation of pilocarpine. This pharmacokinetic model mathematically defines the rate profile for the disappearance of pilocarpine from the precorneal compartment as

$$\frac{dC_T}{dt} = \frac{-q_T C_T - k_p S_c / h_c (C_T - C_{AH})}{V_D e^{-k_{nl} t} + V_0} \tag{1}$$

and the rate profile for the appearance of pilocarpine in the aqueous humor compartment is defined by

$$\frac{dC_{AH}}{dt} = \frac{k_p S_c}{V_{AH} hc}(C_T - C_{AH}) - k_{eAH} - \frac{C_{AH}}{V_{AH}} \tag{2}$$

where

C_T = drug concentration in the tear fluid
C_{AH} = drug concentration in the aqueous humor
k_p = specific transcorneal permeation rate (3.675×10^{-4} μl/cm·min)
k_{nl} = $(0.25 + 0.0113\ V_D)\text{min}^{-1}$
S_c = surface area of the cornea (2 cm^2)
h_c = thickness of the cornea (0.035 cm)
V_0 = normal resident tear volume (7.5 μl)
V_D = drop size of the drug solution instilled
V_{pc} = volume of drug pool in precorneal area after instillation of eye drops
V_{AH} = volume of the aqueous humor

Pharmacokinetic analysis of the concentration-time profile of pilocarpine in epithelium (Figure 4) indicates that the elimination of pilocarpine from the epithelium following epithelial uptake is biphasic. The first elimination (α) phase, which covers the period after reaching the peak drug concentration (at 4 min) to approximately 40 min following dosing, has an apparent absorption rate constant of 0.50 min^{-1} and an apparent elimination rate constant of 0.06 min^{-1} (Table 4). The second elimination (β) phase, which carries through the remainder of the 2-hr observation period, has an apparent elimination rate constant of 0.016 min^{-1}. On the other hand, the drug concentration-time profile for the stroma-endothelium consists of an absorption phase and a single elimination phase, which is similar to the pharmacokinetic pattern in the aqueous humor. Pharmacokinetic parameters for the transcorneal permeation of pilocarpine through various ocular tissues are compared in Table 4.

The most striking feature of the results illustrated in Figure 4 and Table 4 is the similarity of the pharmacokinetic profiles between the stroma-endothelium and the aqueous humor. This suggests a free diffusion of pilocarpine molecules from the stroma-endothelium to the aqueous humor (scheme II); that is, $k_{aAH} = k_{eS}$. In other words, the endothelial layer does not act as a barrier for the movement of pilocarpine molecules from the cornea to the aqueous humor (31,33). No buildup of pilocarpine could be detected in the stromal-endothelial structure.

The pharmacokinetic data in Table 4 also suggest that the apparent rate constant for the epithelial uptake of pilocarpine ($k_a = 0.50\ \text{min}^{-1}$) agrees fairly well with the apparent rate constant for the uptake of pilocarpine by the intact cornea (0.57 min^{-1}). Additionally, the α-phase elimination rate constant of pilocarpine from the epithelium ($k_{eE} = 0.06\ \text{min}^{-1}$) correlates quite nicely with the apparent rate constant for absorption into the stroma-endothelium ($k_{aS} = 0.09\ \text{min}^{-1}$) and into the aqueous humor ($k_{aAH} = 0.08\ \text{min}^{-1}$). On the other hand, the β-phase elimination rate constant of pilocarpine from the epithelium ($k_{eE} = 0.016\ \text{min}^{-1}$) equates well with the elimination rate constants from the stroma-endothelium ($k_{eS} = 0.021\ \text{min}^{-1}$) and from the aqueous humor ($k_{eAH} = 0.017\ \text{min}^{-1}$). All these pharmacokinetic analyses confirmed the earlier conclusion that epithelium is the rate-controlling barrier in the transcorneal permeation of pilocarpine (35).

Since the cornea is not vascularized, lateral distribution of drug does not occur in the corneal tissues. A lateral movement of drug can occur in the precorneal area, however, where a composite of several parallel processes, including nasolacrimal drainage, tear fluid turnover, and conjunctival uptake (scheme I), removes a major fraction of the drug dose instilled (24,25) and in the anterior chamber, where multiple tissue equilibria take place (34). Therefore, the elimination profile of pilocarpine from the aqueous humor becomes multiphasic at times beyond 2 hr after medication (Figure 5).

The pilocarpine concentration in the aqueous humor can be increased by multiple instillations (36). By multiple instillation of one eye drop dose every 30 min for five dosings, the pilocarpine concentration profile in the aqueous humor compartment was found to be significantly increased and prolonged compared to single instillation (Figure 6).

D. Effects of Protein Binding on Intraocular Pilocarpine Bioavailability

1. *Protein Composition in Ocular Tissues*

The lacrimal fluid in humans contains a total protein content of approximately 0.7% and an albumin content of 0.4% (37), whereas in rabbits the total protein content is 0.5% and the albumin content is 0.3% (38). Emotional stress, irritation, and other factors have reportedly affected the protein content in the lacrimal fluid significantly (39).

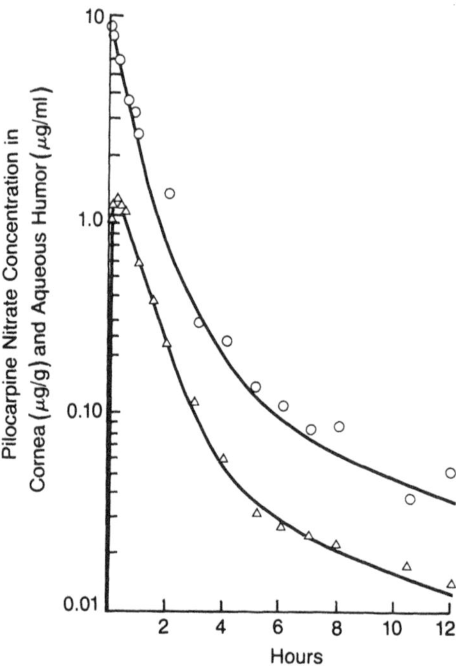

Figure 5 Long-term concentration profiles of pilocarpine in cornea (○) and in aqueous humor (△) following instillation of 25 μl pilocarpine nitrate eye drop (0.01 M). [Reproduced, with permission, from Makoid and Robinson; J. Pharm. Sci., 68:435 (1979).]

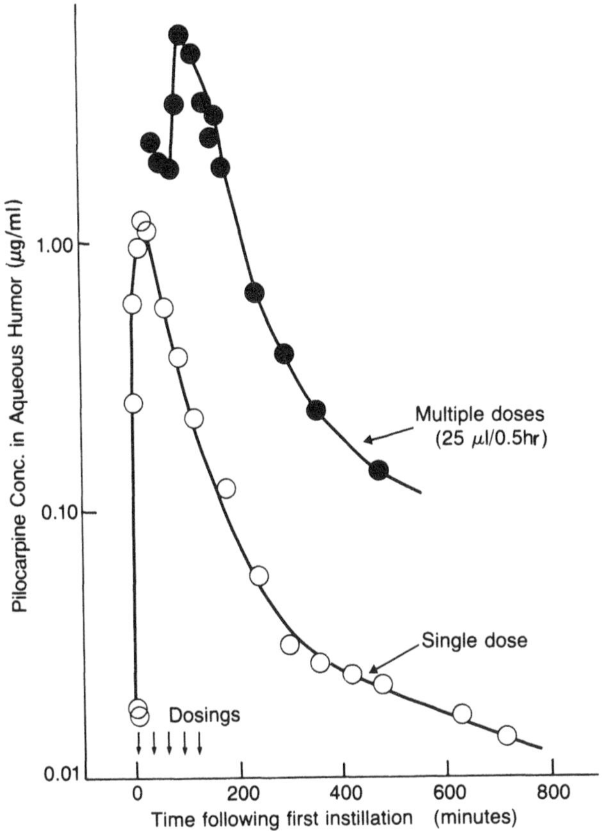

Figure 6 Comparative pilocarpine concentration profiles in aqueous humor following single dosing (○) and during multiple dosings (●, 0.5 hr × 5) of 25 μl pilocarpine nitrate eye drops (0.01 M). [Plotted from the data by Makoid and Robinson; J. Pharm. Sci., 68:435 (1979).]

The cornea consists of 18.4% collagen along with 0.15% albumin and globulin (37). Albumin diffuses from the peripheral capillaries toward the center of the cornea and then diffuses at the corneal surface into the lacrimal fluid and aqueous humor. In humans albumin turns over at a rate of 10–20%/day (40).

The human aqueous humor contains a total protein content of only 0.01–0.02%, whereas in rabbits it consists of a total protein content of 0.05%, of which 0.025% is albumin (10). The protein content in the aqueous humor is dependent upon the integrity of the ocular blood vessels, which are the blood-aqueous humor barrier, and may result in partial or complete breakdown in response to drugs or certain disease states. This breakdown yields a concomitant increase in protein concentration in the aqueous humor, leading to the development of plasmoid aqueous humor. It can, at its extreme, possess a similar protein composition to that of plasma (7% in humans or 5.6% in rabbits).

2. *Multiple Drug-Protein Interactions*

Upon medication the drug concentration is immediately diluted in the lacrimal fluid and enters into rapid equilibrium with the protein composition. Only the free, unbound species of the drug is available for uptake by the corneal epithelium. The nasolacrimal drainage and the turnover of lacrimal fluid remove a major fraction of both free and protein-bound species of the drug in the precorneal area.

When drug molecules enter the cornea they immediately participate in another rapid equilibrium with the corneal proteins. This drug-protein interaction could produce a lag time for the penetration of drug molecules into the anterior chamber. The length of this lag time appears to be proportional to the equilibrium constant for binding, corneal drug concentration, and transcorneal permeability of the free drug species, as well as the thickness of the cornea.

After entering the anterior chamber the drug molecules are subjected to another drug-protein interaction. The extent of binding of drug molecules to the protein composition in the aqueous humor depends on the physiological condition of the eye, which varies the content and composition of aqueous humor proteins. Additionally, the turnover of aqueous humor results in some loss of both free and bound species of drug molecules.

3. *Effects on Intraocular Drug Bioavailability*

The effects of drug-protein interactions on the intraocular bioavailability of ophthalmically active drugs have been demonstrated even for drug species with a low binding affinity for albumin, such as pilocarpine nitrate (39). The miotic activity of pilocarpine nitrate was reported to be significantly reduced with the addition of rabbit serum albumin to the instilled dose. The peak height in the pupillary diameter-time profile, which was reached at approximately 26 min after dosing, was found to be reduced as much as 25.2% with the addition of serum albumin at 1% and 47.9% at 3%. A similar reduction in the miotic activity was also observed when the protein content in the aqueous humor of several animal species was increased by induced paracentesis. This reduction in the ophthalmological activity of pilocarpine nitrate can be attributed to the decrease in its transcorneal bioavailability in response to drug-protein interactions in various ocular tissues, including the aqueous humor.

The observed depression in the miotic activity of pilocarpine nitrate can be fully reversed with the addition of a competitive inhibitor that competively occupies the binding sites on the protein for pilocarpine, leading to an increase in the concentration of free pilocarpine (41). With the presence of cetylpyridinium chloride in the precorneal area, for instance, pilocarpine nitrate was reported to show a 10-fold increase in its pharmacological activity.

VI. CONTROLLED OCULAR PILOCARPINE DELIVERY BY OCUSERT SYSTEMS

The Ocusert system is an oval flexible ocular insert that consists of a core reservoir made from complexation of pilocarpine with alginic acid sandwiched between two sheets of a transparent, lipophilic rate-controlling membrane of ethylene-vinyl acetate copolymer (Figure 3). When it is inserted in the cul-de-sac, the pilocarpine molecules are dissolved in the lacrimal fluid penetrating into the system and released through the rate-controlling membranes according to a zero-order kinetics process as defined by the mathematical expression (22)

$$\left(\frac{dQ}{dt}\right)_r = \frac{D_p K_m (C_R - C_T)}{h_m} \tag{3}$$

where $(dQ/dt)_r$ is the release rate of pilocarpine from a unit surface area of the Ocusert system; D_p is the diffusivity of pilocarpine in the ethylene-vinyl acetate copolymer membranes with thickness h_m; K_m is the partition coefficient of pilocarpine toward the membrane; and $C_R - C_T$ is the difference in pilocarpine concentrations between the pilocarpine reservoir in the medicated core (C_R) and the tear fluid (C_T).

If an infinite sink condition is maintained in the ocular cavity, that is $C_R \gg C_T$, then Equation (3) can be simplified to

$$\left(\frac{dQ}{dt}\right)_r = \frac{D_p K_m C_s}{h_m} \tag{4}$$

by maintaining a saturated pilocarpine solution in the Ocusert system (with a saturation solubility of C_s). The release rate of pilocarpine from the Ocusert system should remain constant until the drug concentration in the reservoir C_R drops off below the pilocarpine saturation level C_s. The in vitro release of pilocarpine from various Ocusert systems was investigated (7) and found to follow the zero-order rate profile as projected from Equation (4) (Figure 7). The release rates can be varied by changing the composition and thickness of ethylene-vinyl acetate copolymer membranes and, hence, the magnitude of membrane permeability constant $D_p K_m / h_m$ of pilocarpine.

A typical in vivo release rate profile of pilocarpine from the Ocusert Pilo-20 system is illustrated in Figure 8. During the first few hours the system releases pilocarpine at a rate three times higher than the programmed rate, i.e., 20 μg/hr. The programmed rate is then achieved within approximately 6 hr. A total of 0.3 mg pilocarpine is released during this initial 6-hr period. During the remainder of the 7-day treatment period the release rate is maintained within ±20% of the programmed rate. The system administers a total of less than 70% of the pilocarpine loading dose to the eye at the end of 7-day medication (42).

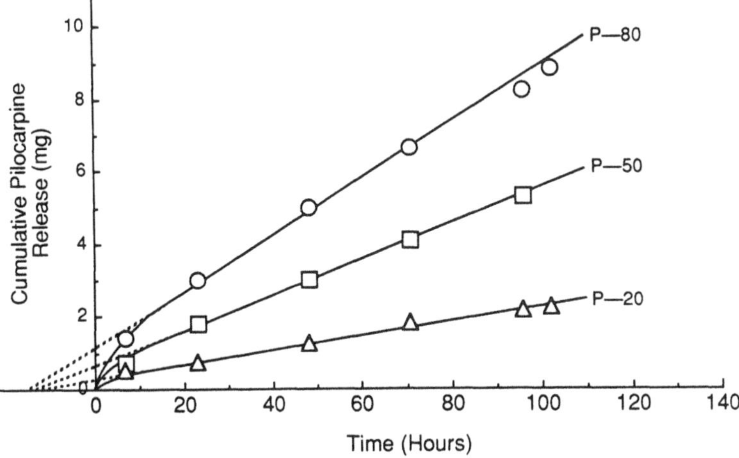

Figure 7 In vitro release pattern of pilocarpine from various Ocusert systems. The rates of pilocarpine release are Ocusert Pilo-20 (16.9 μg/hr); Ocusert Pilo-50 (47.9 μg/hr); Ocusert Pilo-80 (71.9 μg/hr). [Plotted from the data by Armaly and Rao; Invest. Ophthalmol., 12:491 (1973).]

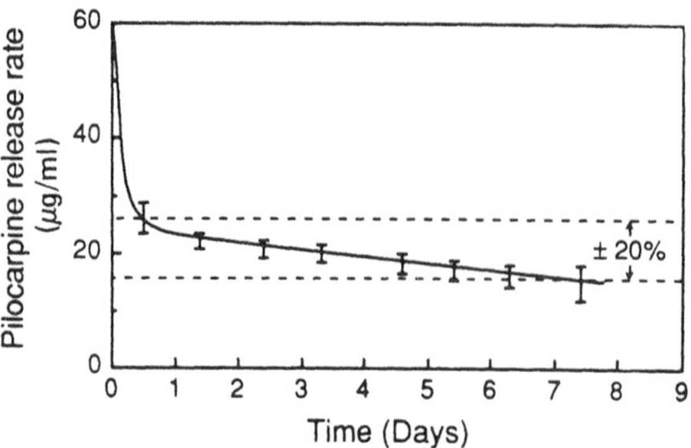

Figure 8 In vivo release rate profile of pilocarpine from Ocusert Pilo-20. [Modified from Baker and Lonsdale; Chemtech., 5:668 (1975).]

The membrane permeability of pilocarpine across the rate-controlling membrane of a ethylene-vinyl acetate copolymer can also be modified by incorporating additives during fabrication of the membrane, for example the addition of di-(2-ethylhexyl)phthalate into the membrane of the Ocusert Pilo-40 system to achieve a higher rate of membrane permeation for pilocarpine (42).

Ocusert systems can be used concomitantly with various ophthalmic medications. The release rate of pilocarpine was found not affected by coadministration of carbonic anhydrase inhibitors, epinephrine, or antiinflammatory steroid ophthalmic solutions (42).

VII. CLINICAL EFFICACY OF PILOCARPINE-RELEASING OCUSERT SYSTEMS

A. Comparative Pharmacological Activities

The hypotensive effect of the pilocarpine-releasing Ocusert system was compared with the conventional antiglaucoma formulation (i.e., 2% pilocarpine eye drops) in patients with open-angle glaucoma or ocular hypertension (43).

On the pretreatment days the mean intraocular pressures for both treatment groups were essentially equal and exhibited similar variations with the time of the day (Figure 9). The patients in the Ocusert-treated group had a pretreatment intraocular pressure of 24.60 ± 1.31 mmHg, which is statistically no different from the 24.34 ± 1.02 mmHg in the eye drop-treated group. Following the insertion of one Ocusert Pilo-20, which releases pilocarpine at the programmed rate of 20 μg/hr (Figure 8), the intraocular pressure immediately showed a noticeable reduction at the first measurement taken at 4 hr after medication. This intraocular pressure continued to drop to a level of 20.89 ± 1.80 mmHg, which was maintained throughout the period with one Ocusert Pilo-20 system inserted in the morning every day. The group of patients treated with 2% pilocarpine eye drops (two drops per dose, four doses a day) also showed a similar pattern of reduction in the intraocular pressure. Both treatments produced a hypotensive effect that was statistically significant at the 99% confidence

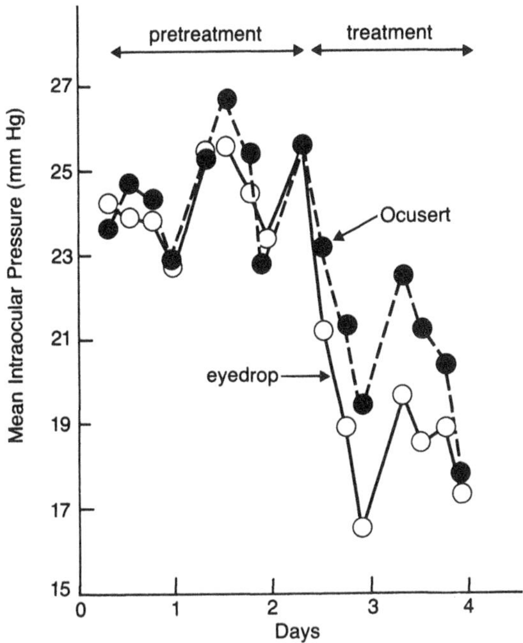

Figure 9 Comparative hypotensive activity of Ocusert Pilo-20 and 2% pilocarpine eye drops on the reduction of intraocular pressure in 20 patients. Treatment schedule is instillation of two drops per dose at 8 a.m., 1 p.m., 7 p.m., and 11 p.m. for 2 days; or insertion of one new Ocusert Pilo-20 at 8 a.m. every day for 2 days. [Plotted from the data by Armaly and Rao; in: *Symposium on Ocular Therapy* (I. H. Leopold, Ed.), C. V. Mosby, St. Louis, Vol. 6, 1974.]

level compared to the pretreatment levels and the control eyes administered with a corresponding placebo formulation.

The magnitude of reduction in intraocular pressure response to 2% pilocarpine eye drops treatment appeared to be greater than that in response to the Ocusert Pilo-20 system (Figure 9), even though the difference between these two treatment groups was not at all statistically significant (18.9 ± 1.5 mmHg for the eye drop group compared to 20.9 ± 1.8 mmHg for the Ocusert group). These results suggest that the hypotensive activity of the Ocusert Pilo-20 system is comparable to that achieved by the instillation of two drops of pilocarpine (2%) solution four times a day. It was also reported that with daily insertion of one Ocusert Pilo-20 system the intraocular pressure in hypertensive and/or glaucoma patients can be controlled within physiological levels (43).

The miotic effect of the pilocarpine-releasing Ocusert system on pupil diameter was also compared with that of the 2% pilocarpine eye drops in the same group of patients (43). The results indicated that immediately following the treatment with the Ocusert Pilo-20 system the pupil diameter showed a significant reduction from 3.28 ± 0.07 to 2.45 ± 0.01 mm and remained constricted during the course of Ocusert treatment (Figure 10). Treatment with 2% pilocarpine eye drops also produced a similar pattern in miotic activity, with the pupil diameter reduced from the

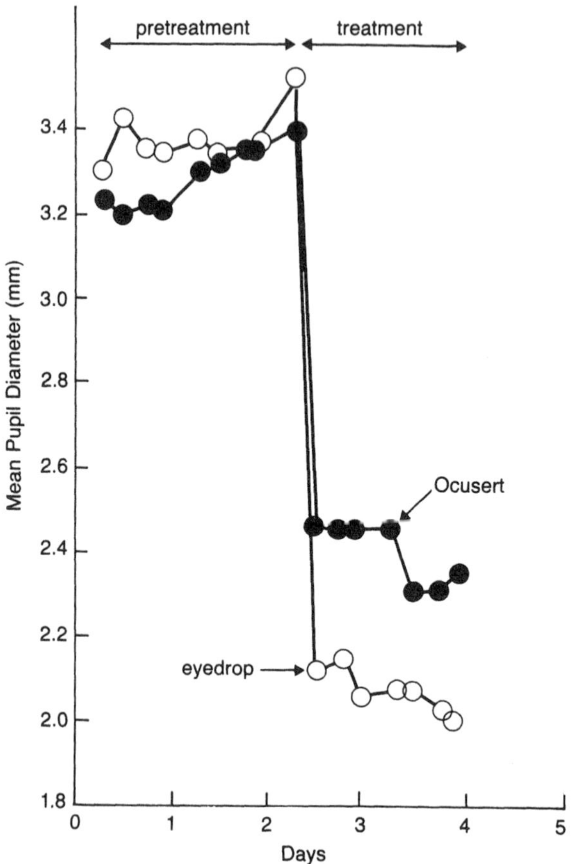

Figure 10 Comparative miotic activity of Ocusert Pilo-20 and 2% pilocarpine eye drops on the reduction in pupil diameter in 20 patients. Same treatment schedule as in Figure 9. [Plotted from the data by Armaly and Rao; in: *Symposium on Ocular Therapy* (I. H. Leopold, Ed.), C. V. Mosby, St. Louis, Vol. 6, 1974.]

pretreatment level of 3.37 ± 0.06 to 2.07 ± 0.05 mm; the miotic activity after the eye drop treatment (2.07 mm) was substantially lower than that achieved by the Ocusert system (2.45 mm).

The results of clinical evaluations clearly demonstrated that the pilocarpine-releasing Ocusert system produced a significant, prolonged reduction in the intraocular pressure and pupil diameter throughout the 24-hr insertion of one unit of the Ocusert Pilo-20 system. The administration of one Ocusert Pilo-20 system delivers to the eye a daily pilocarpine dose of only 400–500 μg. This daily dose is only one-eighth to one-tenth of the pilocarpine dose (4 mg/day) administered by the instillation of two drops of 2% pilocarpine solution four times a day. Apparently, application of the controlled drug administration mechanism, the Ocusert system significantly minimizes the dose requirements of pilocarpine for the effective management of intraocular pressure; in other words, the therapeutic efficacy of pilocarpine in glaucoma treatment has been improved 8–10 times with the use of the Ocusert system to control the ocular delivery of pilocarpine.

This comparison of the dose-activity relationship is further magnified when single instillations are considered. Every time two drops of pilocarpine (2%) solution are instilled, 1 mg pilocarpine HCl is delivered to the conjunctival sac in only a few seconds. This dose is immediately diluted by tear fluid and drained away by nasolacrimal drainage as well as tear turnover, so only a small fraction of the instilled dose remains available for productive transcorneal permeation. Within the same time span, on the other hand, the Ocusert Pilo-20 system releases only a microdose of pilocarpine (11 μg initially and 2–3 μg at steady state) into the conjunctival sac. This microdose of pilocarpine is certainly also subjected to the same precorneal disposition as the eye drop instillation. However, continuous delivery of this microdose of pilocarpine by the Ocusert system with no increase in the resident tear volume in the precorneal area is expected to reduce various precorneal dispositions of pilocarpine but also greatly minimize the chances of ocular and/or systemic side effects which could result from the uptake of large dose of pilocarpine during the process of nasolacrimal drainage (43).

The magnitude of reduction in the intraocular pressure was found in proportion to the pilocarpine dose delivered by Ocusert system (Table 5). As discussed earlier, the Ocusert Pilo-20 system, which gives an in vitro release rate of 16.9 μg/hr of pilocarpine, produced a smaller magnitude of hypotensive effect (although the difference is not statistically significant) than that attained by the instillation of 2% pilocarpine eye drops. The increase in the release rate of pilocarpine from the Ocusert systems, such as Pilo-50 and Pilo-80, was observed to achieve a greater reduction in the outflow pressure (Table 5). For example, the Ocusert Pilo-50 system, which releases pilocarpine at a rate 2.8-fold greater than that from Ocusert Pilo-20 system, produces a hypotensive activity 2.7 times greater than that by the Ocusert Pilo-20 system. The magnitude of reduction in the outflow pressure (45.4 ± 4.0 and 48.0 ± 3.2%) produced by Ocusert Pilo-50 and Pilo-80 systems is statistically significantly higher than that produced by Ocusert Pilo-20 and 2% pilocarpine eye drops (Table 5).

Table 5 Comparative Hypotensive Activity of Various Ocusert Systems and Pilocarpine Eye Drops

Treatments	Pilocarpine doses	% Reduction in outflow pressure[a] (mean ± SD)
Pilocarpine	2 drops/dose	26.1 (± 3.6)
eye drops (2%)	4 doses/day	28.4 (± 3.6)
Ocusert systems		
Pilo-20	16.9 μg/hr[b]	17.0 (± 3.6)
Pilo-50	47.9 μg/hr[b]	45.4 (± 3.9)
Pilo-80	71.9 μg/hr[b]	48.0 (± 3.2)

[a]Percentage change in applanation pressure (mean ± standard deviation) at 4 hr after medication compared to pretreatment level. Compiled from the data by Armaly and Rao (1973, 1974).

[b]Data calculated from the slope of the Q verus t plots (Figure 7).

B. Continuous Treatment with Ocusert Systems

The hypotensive and miotic activities illustrated in Figures 9 and 10 were achieved by daily insertion of one unit of Ocusert Pilo-20 system into the hypertensive eye. In vitro and in vivo studies, however, have demonstrated that the Ocusert system can deliver pilocarpine at controlled rates continuously for a duration of up to 7 days (Figures 7 and 8).

The prolonged, continuous release of pilocarpine from the Ocusert system and its sustained clinical efficacy in controlling the intraocular pressure were investigated in patients with glaucoma or ocular hypertension (44). The results indicated that continued good management of intraocular pressure can be achieved with the continuous wearing of one unit of Ocusert Pilo-20 system for 4 days (Figure 11). This clinical testing further substantiated the results reported earlier that the pilocarpine-releasing Ocusert Pilo-20 system is safe, well tolerated and as effective as 2% pilocarpine eye drops in reducing the ocular hypertension in glaucoma patients with a much lower dose, which can be as low as only 1/40 the total dose administered by

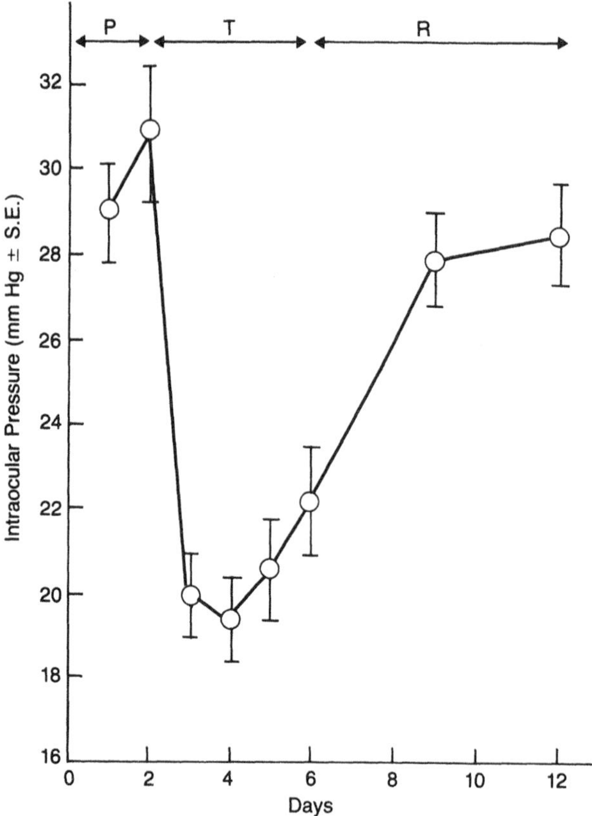

Figure 11 Long-acting hypotensive action of Ocusert Pilo-20 in the management of intraocular pressure of 14 outpatients. *P*, pretreatment phase; *T*, treatment phase (4 days) with one Ocusert Pilo-20 system in situ; *R*, recovery phase. [Plotted from the data by Friederich; Ann. Ophthalmol., 6:1279 (1974).

eye drop instillation. Furthermore, this controlled pilocarpine-releasing therapeutic system has several advantages over the conventional eye drop medication: it provides better patient compliance, less frequent dosing, around-the-clock protection for at least 4 days, fewer ocular and systemic side effects, and a possible delay in the refractory state.

The duration of clinical efficacy of Ocusert systems in the management of ocular hypertension is varied, depending upon the individual and the unit used. Two typical cases are illustrated in Figure 12. The patient in the first case was able to control his intraocular pressure with the use of one unit of Ocusert Pilo-20 system in each eye every 4 days for a period of 7 weeks. On the other hand, the patient in the second case was able to maintain his pressure level below 20 mmHg for only 3 days with the same strength of Ocusert system (Pilo-20). By shifting to a higher strength of Ocusert system, like Ocusert Pilo-50 system, the intraocular pressure of this patient can be controlled well below the 20 mmHg level for 4 days. The patient in the second case successfully controlled his glaucoma condition for $2^{1}/_{2}$ months by wearing a new pair of Ocusert Pilo-50 systems every 4 days.

A patient whose intraocular pressure has been controlled by 1 or 2% pilocarpine eye drop solution has a higher probability of pressure control with the Pilo-20 system

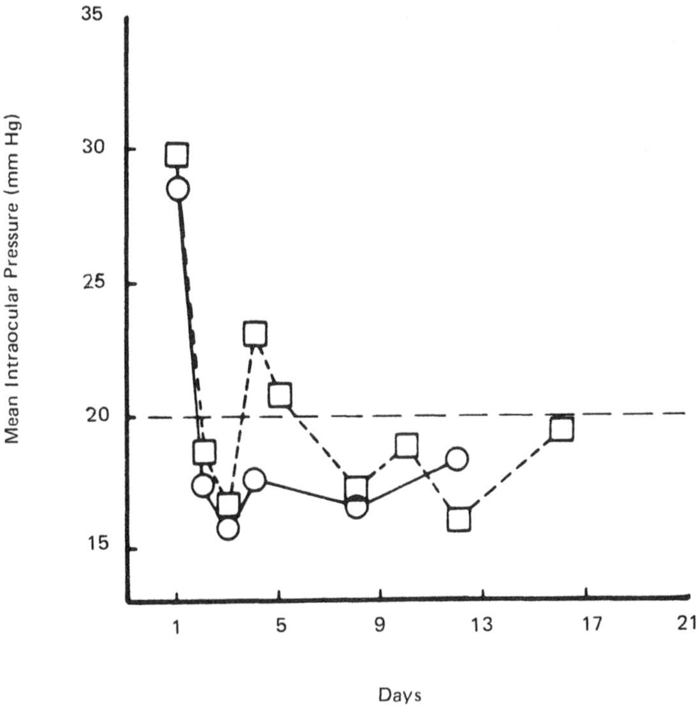

Figure 12 Duration of intraocular pressure control with consecutive treatment with various pilocarpine-releasing Ocusert systems: (○) insertion of one Ocusert Pilo-20 system on days 1, 4, 8, and 12; (□) insertion of one Ocusert Pilo-20 system on day 1 and then one Ocusert Pilo-50 system on days 5, 8, and 12. [Plotted from the data by Armaly and Rao; Invest. Ophthalmol., 12:491 (1973).]

than a patient who has used a higher strength pilocarpine solution, who might require treatment with the Pilo-40 system. However, there is no direct correlation between the strengths of the Ocusert system and the strengths of pilocarpine eye drop solutions required to achieve a given level of pressure reduction (42).

The concurrent use of the Ocusert systems was reported to increase the rate of corneal absorption for an autonomic drug, such as epinephrine; however, the Ocusert system does not induce the occurrence of mild bulbar conjunctival edema, which frequently results from the administration of epinephrine ophthalmic solution (42).

Of the 302 patients who used the Ocusert systems in clinical studies for more than 2 weeks, 75% showed a preference over the conventional eye drop formulations (42). This percentage was found to increase with improved wearing experience.

C. Comparative Tissue Distribution

The distribution of pilocarpine in various ocular tissues during treatment with the Ocusert systems was compared in rabbits with conventional eye drop administration using radiolabeled pilocarpine. Constant low levels of pilocarpine in the ciliary body and iris were achieved with Ocusert treatment.

On the other hand, following the instillation of the conventional eye drop formulation the initial levels of pilocarpine in the cornea, aqueous humor, ciliary body, and iris were three to five times higher than the corresponding levels achieved by the Ocusert systems; it then declined over the next six hr to approximately the tissue drug concentrations maintained by the Ocusert system. However, the pilocarpine concentration in the conjunctiva, lens, and vitreous humor remained consistently high from eye drops and did not return to the constant low levels observed with the Ocusert system. Additionally, pilocarpine was observed not to accumulate in ocular tissues during Ocusert treatment.

D. Comparative Physiological Effects

Continuous delivery of pilocarpine from the Ocusert systems at the rates of 20–80 μg/hr was found to produce little or no effect on tear fluid pH in rabbit eyes (45). On the other hand, administration of pilocarpine in salt form by eye drop instillation or by a spray formulation was reported to acidify immediately the tear film by as much as 1.1–1.6 pH units. The pH remained below the pretreatment level of pH 7.47 ± 0.03 for as long as 45–60 min after medication (Figure 13). The observed difference in tear film pH following the administration of pilocarpine may partially explain the four- to eightfold difference in the total pilocarpine doses required by treatment with an Ocusert system compared to the conventional eye drop and spray formulations. At a physiological pH of 7.47, more pilocarpine molecules, which have a pK_a value of 6.8, are deprotonated than at pH 5.87–6.37 resulting from eye drop instillation. It is believed that the cornea is more permeable to the un-ionized, neutral species of the pilocarpine molecule than to the protonated species existing in an acid solution. The same phenomenon was also observed with epinephrine bitartrate (46).

It was reported that some 30–60% of patients fail to take pilocarpine eye drops as directed (21). This patient noncompliance may be attributed to the fact that the instillation of pilocarpine eye drops often creates some visual disturbances (miosis and myopia), discomfort, and sometimes disability. The myopia induced by pilo-

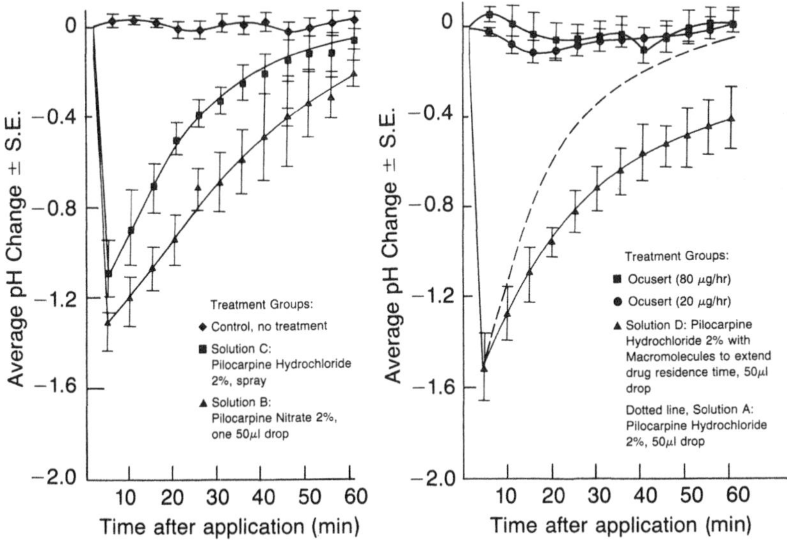

Figure 13 Comparison of the effect of pilocarpine salts, which were delivered in eye drop solution, spray formulation, or from Ocusert systems, on rabbit tear film pH. [Reproduced, with permission, from Longwell et al.; J. Pharm. Sci., 65:1654 (1976).]

carpine eye drop instillations has a diopters-time profile very similar to the tear film pH change-time profile illustrated in Figure 13. These adverse effects may be sufficient to discourage patient compliance with a given treatment regimen.

The incidence of visual disturbances was significantly reduced by treatment with the Ocusert systems. During the first several hours after insertion of an Ocusert system into the conjunctival cul-de-sac, induced myopia may also occur. In contrast to the fluctuating, high levels of induced myopia often observed with the administration of pilocarpine eye drop formulations, the amount of induced myopia following the use of Ocusert systems decreases, after the initial several hours, to a low baseline level of approximately 0.5 diopters or less, which persists throughout the therapeutic life of the Ocusert system (42).

VIII. POTENTIAL DEVELOPMENT OF OCULAR CONTROLLED DRUG DELIVERY

A. Development of Epinephrine-Releasing Ocular Therapeutic Systems

Like pilocarpine, epinephrine is also recommended for the management of intraocular pressure. While pilocarpine is a parasympathomimetic drug that reduces intraocular pressure by increasing the facility of outflow of aqueous humor, epinephrine is a sympathomimetic drug that diminishes intraocular pressure by the dual actions of reducing the secretion of aqueous humor and increasing the facility of outflow. These two drugs are commonly used in separate eye drop solutions for controlling glaucoma, and sometimes they are used in combination for the same therapeutic purpose.

Recently an epinephrine-releasing ocular therapeutic system was also developed, and its intraocular hypotensive efficacy was compared to that of eye drop instillation in both normotensive and hypertensive rabbit eyes (46). The results indicated that in the normotensive eyes the epinephrine-releasing ocular therapeutic systems, with release rates of 2 and 4 μg/hr, are able to achieve a hypotensive efficacy equivalent to eye drop solutions containing 0.5% epinephrine hydrochloride and 2% epinephrine bitartrate, respectively. These two eye drop formulations administered 500 and 1100 μg of epinephrine, which are more than 20 times the epinephrine dose delivered by the ocular therapeutic systems during the same length of treatment.

On the other hand, the intraocular hypertension induced by an intragastric water load in rabbits was found to be significantly reduced by the epinephrine-releasing ocular systems, which administer epinephrine continuously at a rate of 3 or 6 μg/hr. The intraocular hypotensive effectiveness achieved by the epinephrine-releasing ocular systems is also equivalent to that accomplished by 2% epinephrine bitartrate eye drop solutions, which administered as much as 15–40 times more epinephrine to the eyes than the ocular system.

Another new ocular therapeutic system was also recently developed to deliver both pilocarpine and epinephrine at controlled rates for the treatment of glaucoma (47). The simultaneous administration of epinephrine was observed to affect the bioavailability of pilocarpine into the target tissue, aqueous humor (Table 6). A maximum enhancement of the bioavailability of pilocarpine was achieved with a combination of 20 μg/hr of pilocarpine and 3 μg/hr of epinephrine. A threefold increase in the aqueous humor concentration of pilocarpine was achieved. This is equivalent to that achieved by the pilocarpine-releasing ocular system with a release rate of 40 μg/hr.

The observed enhancement in the intraocular bioavailability of pilocarpine may be attributed to the vasoconstrictive action of epinephrine. As reported earlier, predosing with vasoactive agents could affect the first-order rate constant for the disappearance of the precorneal pilocarpine concentration (48). Epinephrine bitartrate, a vasoconstrictor, was found to reduce the rate constant by a factor of 1.5; on the other hand, histamine dihydrochloride, a vasodilator, increased it by the same magnitude. Alternatively, the concurrent use of pilocarpine-releasing Ocusert systems increases the rate of the corneal absorption of epinephrine (42).

Table 6 Effect of Epinephrine on Steady-State Pilocarpine Concentrations in Rabbit Aqueous Humor

Release rates (μg/hr)[a]		Aqueous humor concentration[b]
Pilocarpine	Epinephrine	(μg/ml ± SD)
20	0	0.601 (± 0.175)
20	1	0.401 (± 0.027)
20	3	1.688 (± 0.255)
20	6	1.351 (± 0.127)
40	0	1.466 (± 0.255)

[a]Epinephrine was delivered simultaneously with pilocarpine from the same ocular system.

[b]Calculated from the data by Gale et al.; U.S. Patent 4,190,642, Feb. 26, 1980.

B. Hydrophilic Contact Lenses as Ophthalmic Drug Delivery Systems

As discussed in earlier sections, the therapeutic efficacy of an ophthalmically active drug can be greatly improved by systaining the duration of intimate drug-eye contact. The drug-eye contact time can be substantially prolonged by the use of hydrophilic contact lenses. In humans, the Bionite lens, which is manufactured from the hydrophilic polymer of 2-hydroxyethyl methacrylate, has been shown to produce a greater penetration of fluorescein (2) and to yield a more profound and prolonged ocular response to pilocarpine or phenylephrine (3,49–51). The lens is inserted into the eye after being presoaked in the drug solution. Sauflon hydrophilic contact lenses, which are fabricated from a vinylpyrrolidone-acrylic copolymer with a high water content, has also been successfully applied in the treatment of acute close-angle glaucoma with pilocarpine.

The dynamics of the uptake and release of drug by various types of hydrophilic contact lenses was studied (52). The results demonstrated that the drug action can be prolonged for up to several hours with drug-saturated contact lenses compared to eye drops, which are capable of remaining in contact with the cornea and conjunctiva for less than 2 min; the addition of hydroxypropylmethylcellulose as the viscosity-enhancing agent in the eye drop formulation can prolong the contact time to 10 min.

The possibility of using soft contact lenses as the ophthalmic drug delivery system was further investigated recently (4). It was reported that drug-presoaked Soflen lenses markedly increase the intensity of the drug-induced peak miotic response, the duration of miosis, and the areas under the temporal miotic response versus intensity curve for pilocarpine, carbachol, and echothiophate iodide. The greatest enhancement was achieved for echothiophate iodide, a quaternary ammonium cation antiglaucoma drug (Figure 14). It appears that echothiophate iodide is the drug that most benefited from administration by the Soflens contact lens. The maximum miotic response intensity (I_{max}) was enhanced approximately fivefold (from 0.13 to 0.59), and the duration of the miotic response was prolonged by nearly 17 times (from 2.5 to 42 hr) by Soflens medication. On the other hand, Soflens administration enhanced the I_{max} of pilocarpine by less than three times (from 0.088 to 0.203) and prolonged its duration by only four times (from 1.5 to 6 hr).

The observed difference between the quaternary ammonium cationic echothiophate iodide and pilocarpine, in terms of the enhancing effect of Soflens administration, may be related to the difference in their transport from the lense to the cornea and the interaction of cationic echothiophate with the anionic corneal surface (53). On the other hand, the miotic activity of pilocarpine can be improved by using a more permeable Bioplex lens (54,55).

The potential biomedical applications of the hydrophilic contact lenses for ocular drug delivery have also been extended to the ocular controlled administration of antibiotics, such as tetracycline HCl and chloramphenicol (5).

C. Ocular Controlled Delivery of Antiinflammatory Steroids

Topical steroids, such as prednisolone acetate and hydrocortisone acetate, were reported to be extremely effective in the treatment of conjunctival inflammation as well as in some cases of anterior uveitis of either nonspecific or antigenic origin (56,57).

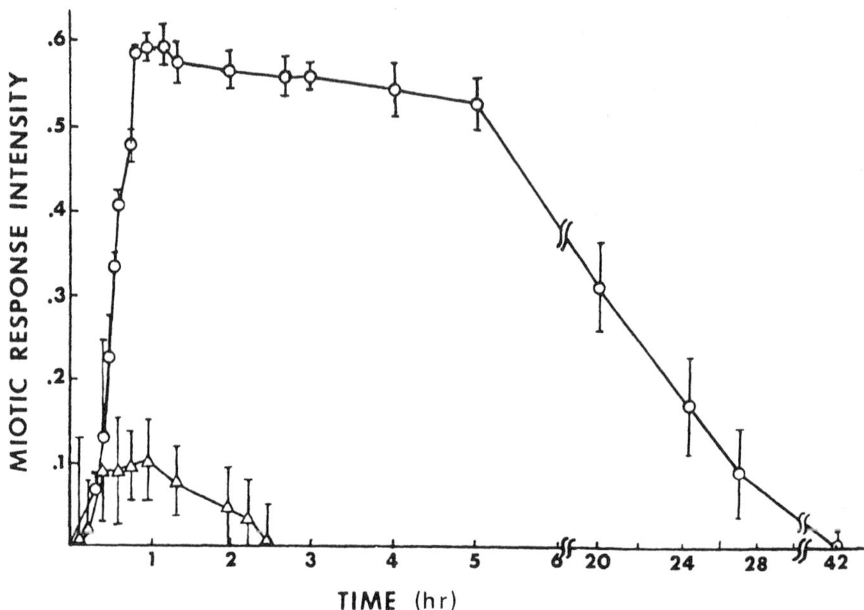

Figure 14 Temporal variation in miotic response intensity following the dosing of rabbits with Soflenses presoaked with 0.03% (*w*/*v*) echothiophate iodide (○) compared to 0.03% (*w*/*v*) echothiophate iodide ophthalmic solution (△). Each point is an average (± standard error of the mean) of three experiments. [Reproduced, with permission, from Smolen et al.; Drug Dev. Commun., 1:479 (1974).]

An ocular insert, which is an elliptical system with prednisolone acetate (or hydrocortisone acetate) uniformly dispersed in a polypeptide matrix, was recently developed for delivering the antiinflammatory steroid at a controlled rate to the inflamed eye (58). Its antiinflammatory activity was found to be significantly greater than that achieved by the same steroid given at equivalent daily doses intermittently in eye drops (Figure 15). After continuous administration of prednisolone delivered from the ocular insert, which releases prednisolone acetate continuously at a rate of 10 μg/hr, inflammation was completely controlled after 4 days of treatment (the inflammation score was reduced by 93.4% compared to the control group); on the other hand, after treatment with a prednisolone acetate eye drop solution at a higher dose (50 μg) with a schedule of five doses a day, the inflammation was also significantly reduced but to a lesser extent (the inflammation score was reduced by 77.5% after treatment for 4 days). The results clearly demonstrated that the continuous delivery of antiinflammatory prednisolone acetate from the controlled-release ocular insert is more efficacious than the same steroid given intermittently by eye drops. Similar results and conclusions have also been reached for hydrocortisone acetate-releasing ocular inserts. However, prednisolone acetate was observed to be two to six times more effective than hydrocortisone acetate when delivered in the same ocular delivery system (Figure 16).

The distribution pattern of the topical antiinflammatory steroids among various ocular tissues of an inflamed rabbit eye was reported to vary with the method of

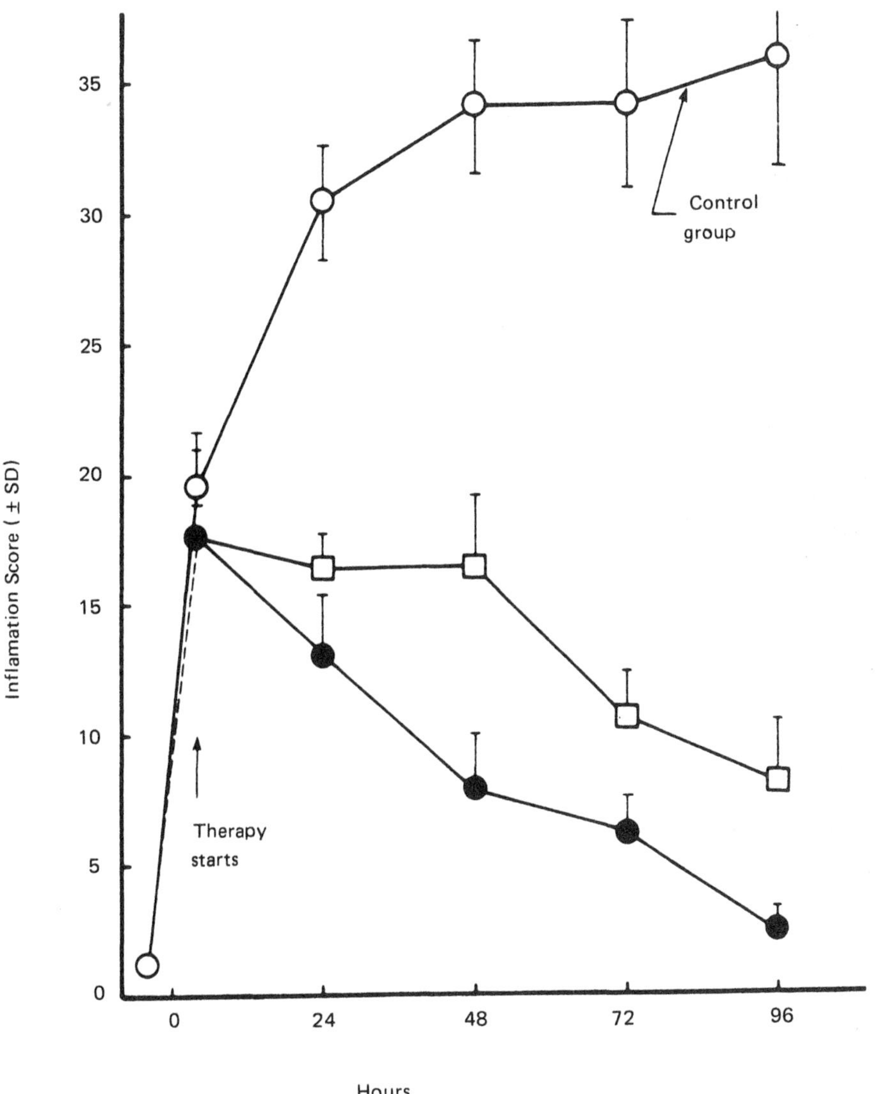

Figure 15 Effect of prednisolone acetate administered by ocular insert (●) and eye drops (□) to rabbit eye on the inflammation score compared to controls treated with placebo (○). [Plotted from the data by Place and Benson; J. Steroid Biochem., 6:717 (1975).]

drug administration (59). With continuous administration of hydrocortisone acetate from the ocular insert, the percentage of total drug in the ocular tissues declined steadily from the external tissues to the internal tissues of the eye. The conjunctiva and sclera, the target tissues, accounted for 92.4% of the drug present in the whole eye. With intermittent eye drop instillations, on the other hand, the conjunctiva and sclera accounted for 65.8 and 79.5%, respectively, in high-frequency eye drop treatment (once an hour for 8 hr) and low-frequency eye drop treatment (twice a day for 4 5 days). Another 17.7–29% of the drug administered to the eyes by eye drops was

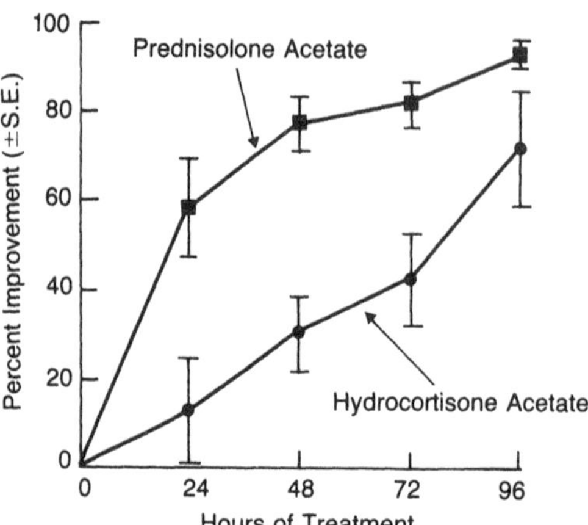

Figure 16 Comparison of the antiinflammatory efficacy between prednisolone acetate- and hydrocortisone acetate-releasing ocular inserts. Data expressed as percentage improvement in group mean Draize scores. [Reproduced, with permission, from the data by Keller et al.; Arch. Ophthalmol., 94:644 (1976).]

found in the nontarget tissues, cornea and aqueous humor, compared to only 5.5% detected in these tissues during continuous drug delivery by ocular insert (1,60,61). It was also found that the relative amount of drug in the lens during eye drop treatment was two to six times that observed during ocular insert medication. This is noteworthy because of the potential danger of cateract, which can develop after prolonged steroid therapy (62).

Recently, a bioerodible ocular insert was developed for the controlled ocular delivery of hydrocortisone acetate (63). This hydrocortisone acetate-releasing bioerodible ocular insert is a small, pliable flat disk, oval or round in shape. It is made of either cross-linked polypeptides or polysaccharides with hydrocortisone acetate dispersed homogeneously throughout the polymer matrix. These polymer materials dissolve slowly in aqueous fluids, such as the lacrimal fluid in the precorneal area, thereby releasing the antiinflammatory hydrocortisone acetate continuously at controlled rates.

The antiinflammation efficacy of hydrocortisone acetate-releasing bioerodible ocular inserts was evaluated in rabbits with corneal inflammation, and the results demonstrated that the bioerodible ocular inserts, which release hydrocortisone acetate at a rate ranging from 0.1 to 7.0 μg/hr, have a marked effect in suppressing the development of corneal and uveal inflammation. A steady antiinflammatory effect was maintained from the first through the second week of treatment. The antiinflammatory activity resulting from the insertion of the hydrocortisone acetate-releasing bioerodible ocular inserts was found comparable to that produced by the instillation of hydrocortisone acetate ophthalmic suspension administered four times a day (64). The bioerodible ocular inserts were found to be well tolerated by all the rabbits tested. No or only traces of reactive hyperemia or chemosis were evident. No corneal abrasion could be detected. A retention rate of 77.8–89.1% was achieved.

REFERENCES

1. R. D. Schoenwald and V. F. Smolen; J. Pharm. Sci., 60:1039 (1971).
2. S. R. Waltman and H. E. Kaufman; Invest. Ophthalmol., 9:250 (1970).
3. J. S. Hillman; Br. J. Ophthalmol., 58:674 (1974).
4. V.F. Smolen, R. Vermuri, T. S. Miya and E. J. Williams; Drug Dev. Commun., 1:479 (1974–75).
5. R. Praus, I. Brettschneider, L. Krejci and D. Kalvodova; Ophthalmologica, 165:62 (1972).
6. C. H. Dohlman, M. G. Doane and C. S. Reshmi; Ann. Ophthalmol., 3:126 (1971).
7. M. F. Armaly and K. R. Rao; Invest. Ophthalmol., 12:491 (1973).
8. S. W. Jacob and C. A. Francone; *Structure and Function in Man*, 2nd Ed., W. B. Saunders, Philadelphia, 1970, Chapter 17.
9. S. Mishima, A. Gasset, S. D. Klyce and J. L. Baum; Invest. Ophthalmol., 5:264 (1966).
10. H. Davson; *The Eye*, Vol. 1, Academic, New York, 1969.
11. L. S. Harris, T. W. Mittag and M. A. Galin; Arch. Ophthalmol., 86:1 (1971).
12. A. G. Gilman, L. S. Goodman and A. Gilman (Eds.); *The Pharmacological Basis of Therapeutics*, 6th ed., Macmillan, New York, 1980.
13. R. F. Lowe; *Am. J. Ophthalmol.*, 67:87 (1969).
14. R. F. Lowe; *Br. J. Ophthalmol.*, 56:409 (1972).
15. R. Banister; *A Treatise of One Hundred and Thirteen Diseases of The Eyes and Eyelids*, 2nd Ed., T. Man, London, 1622.
16. C. De Saint-Yves; *A New Treatise of the Diseases of the Eyes*, 2nd Ed., T. Osborne, London, 1744.
17. Martindale; *Martindale—The Extra Pharmacopoeia*, 26th Ed., (N. W. Blacow, Ed.), Pharmaceutical Press, London, 1972, p. 1158.
18. J. E. Harris; in: *Symposium on Ocular Therapy*, Vol. 3 (I. H. Leopold, Ed.), C. V. Mosby, St. Louis, 1968, p. 99.
19. C. F. Asseff, P. L. Weisman, S. M. Podos and B. Becker; *Am. J. Ophthalmol.*, 75:212 (1973).
20. P. Vincent; Sight Sav. Rev., Winter:213 (1972).
21. G. Spaeth; Invest. Ophthalmol., 9:73 (1970).
22. J. Shell and R. Baker; Ann. Ophthalmol., 6:1037 (1974).
23. J. Shell; Ophthalmic Surg., 5:73 (1974).
24. S. S. Chrai, T. F. Patton, A. Mehta and J. R. Robinson; J. Pharm. Sci., 62:1112 (1973).
25. T. F. Patton and J. R. Robinson; J. Pharm. Sci., 65:1295 (1976).
26. T. F. Patton; J. Pharm. Sci., 66:1058 (1977).
27. A. M. Tonjum; Acta Ophthalmol. (Copenh.) 52:650 (1974).
28. T. Iwata, M. Uyama and K. Ohkawa; Jpn. J. Ophthalmol., 19:139 (1975).
29. K. C. Swan and N. G. White; Am. J. Ophthalmol., 25:1043 (1942).
30. A. Kupferman, M. V. Pratt, K. Suckewer and H. M. Leibowitz; Arch. Ophthalmol., 91:373 (1974).
31. J. W. Sieg and J. R. Robinson; J. Pharm. Sci., 65:1816 (1976).
32. D. L. Krohn and J. M. Breitfeller; Invest. Ophthalmol., 13:312 (1974).
33. M. C. Van Hoose and F. E. Leaders; Invest. Ophthalmol., 13:377 (1974).
34. R. Lazare and M. Horlington; Exp. Eye Res., 21:281 (1975).
35. K. J. Himmelstein, I. Guvenir and T. F. Patton; J. Pharm. Sci., 67:603 (1978).
36. M. C. Makoid and J. R. Robinson; J. Pharm. Sci., 68:435 (1979).
37. F. H. Adler; Physiology of the Eye, C. V. Mosby, St. Louis, Mo. 1965, pp. 38, 45.
38. S. Duke-Elder; *System of Ophthalmology*, Vol. 4, C. V. Mosby, St. Louis, Mo., 1968.

39. T. J. Mikkelson, S. S. Chrai and J. R. Robinson; J. Pharm. Sci., 62:1648 (1973).
40. D. M. Maurice and D. G. Watson; Exp. Eye Res., 4:355 (1965).
41. T. J. Mikkelson, S. S. Chrai and J. R. Robinson; J. Pharm. Sci., 62:1942 (1973).
42. Alza product information 71-4064-1-3, March 1978.
43. M. F. Armaly and K. R. Rao; in: *Symposium on Ocular Therapy*, Vol. 6 (I. H. Leopold, Ed.), C. V. Mosby, St. Louis, Mo., 1974.
44. R. L. Friederich; Ann. Ophthalmol., 6:1279 (1974).
45. A. Longwell, S. Birss, N. Keller and D. Moore; J. Pharm. Sci., 65:1654 (1976).
46. S. A. Birss, A. Longwell, S. Heckbert and N. Keller; Ann. Ophthalmol., 10:8 (1978).
47. R. M. Gale, M. Ben-Dor and N. Keller; U.S. Patent 4,190,642, Feb. 26, 1980.
48. V. H. Lee and J. R. Robinson; J. Pharm. Sci., 68:673 (1979).
49. H. E. Kaufman, M. H. Uotila, A. R. Gasset, T. O. Wood and E. D. Ellison; Trans. Am. Acad. Ophthalmol. Otolaryngol., 75:361 (1971).
50. S. M. Podos, B. Becker, C. Asseff and J. Hartstein; Am. J. Ophthalmol., 73:336 (1972).
51. S. M. Podos; Invest. Ophthalmol., 12:3 (1973).
52. L. Krejci, I. Brettschneider and R. Praus; Cs. Oftal., 27:285 (1971).
53. V. F. Smolen, C. S. Park and E. J. Williams; J. Pharm. Sci., 64:520 (1975).
54. F. E. Leaders, G. Hecht, M. Van Hoose and M. Kellogg; presented at The Association for Research in Ophthalmology meeting, April 1971.
55. Y. T. Maddox and H. N. Bernstein; Ann. Ophthalmol., 4:789 (1972).
56. S. B. Aronson, E. K. Goodner, E. Yamamoto and M. Foreman; Arch. Ophthalmol., 73:402 (1965).
57. N. Keller, A. M. Longwell and S. A. Birss; *The Association for Research in Vision and Ophthalmology*, Sarasota, Florida, May 6, 1973.
58. V. Place and H. Benson; J. Steroid Biochem., 6:717 (1975).
59. S. A. Birss, N. Keller and A. M. Longwell; *The Association for Research in Vision and Ophthalmology*, Sarasota, Florida, April 25, 1974.
60. P. G. Burch and C. J. Migeon; Arch. Ophthalmol., 79:174 (1968).
61. J. R. Nursall; Am. J. Ophthalmol., 59:29 (1965).
62. A. M. Longwell, S. A. Birss and N. Keller; Ann. Ophthalmol. 8:600 (1976).
63. A. S. Michaels; U.S. Patent 3,867,519, Feb. 18, 1975.
64. C. H. Dohlman, D. Pavan-Langston and J. Rose; Ann. Ophthalmol., 4:823 (1972).

7
Transdermal Drug Delivery and Delivery Systems

I. INTRODUCTION

Continuous intravenous infusion at a programmed rate has been recognized as a superior mode of drug delivery not only to bypass the hepatic first-pass elimination but also to maintain a constant, prolonged, and therapeutically-effective drug level in the body. A closely monitored intravenous infusion can provide both the advantages of direct entry of drug into the systemic circulation and control of circulating drug levels. However, such a mode of drug delivery entails certain risks and therefore necessitates hospitalization of patients and close medical supervision of the medication. Recently there has been an increasing awareness that the benefits of intravenous drug infusion can be closely duplicated, without its potential hazards, by continuous transdermal drug administration through intact skin (1,2).

In response to this new idea several transdermal drug delivery (TDD) systems have recently been developed, aiming to achieve the objective of systemic medication through topical application to the intact skin surface (3). They were exemplified first with the development of a scopolamine-releasing TDD system (Transderm-Scop) for 72 hr prophylaxis or treatment of motion-induced nausea (4), then by the successful marketing of nitroglycerin-releasing TDD systems (Deponit, Nitrodisc, Nitro-Dur, Transderm-Nitro and others) and an isosorbide dinitrate-releasing TDD system (Frandol tape) for once-a-day medication of angina pectoris (3–6), as well as a clonidine-releasing TDD system (Catapres-TTS) for the weekly therapy of hypertension (3,6) and of an estradiol-releasing TDD system (Estraderm) for the twice-a-week treatment of postmenopausal syndromes (7). Most recently another fentanyl-releasing TDD system (Duragesic) received FDA approval for the twice-a-week analgesic in cancer patients.

II. THE SKIN SITE FOR TRANSDERMAL DRUG ADMINISTRATION

The skin is one of the most extensive and readily accessible organs of the human body. The skin of an average adult body covers a surface area of approximately 2 m^2 (or 3000 $inch^2$) and receives about one-third of the blood circulating through the body (8). It is elastic, rugged, and, under normal physiological conditions, self-regenerating. With a thickness of only a few millimeters (2.97 ± 0.28 mm), the skin separates the underlying blood circulation network and viable organs from the outside environment. It serves as a barrier against physical and chemical attacks and shields the body from invasion by microorganisms.

A. Anatomy of the Skin

Microscopically the skin is a multilayered organ composed of, anatomically, many histological layers, but it is generally described in terms of three tissue layers: the epidermis, the dermis, and the subcutaneous fat tissue (Figure 1).

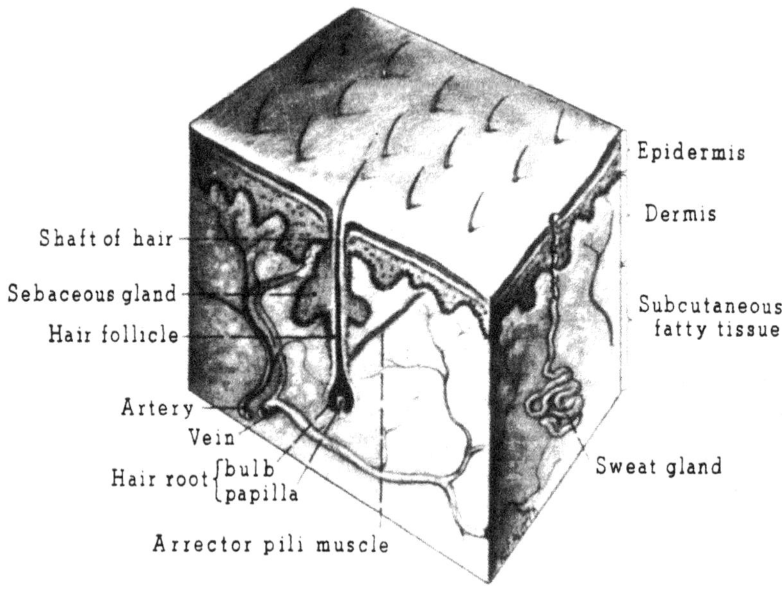

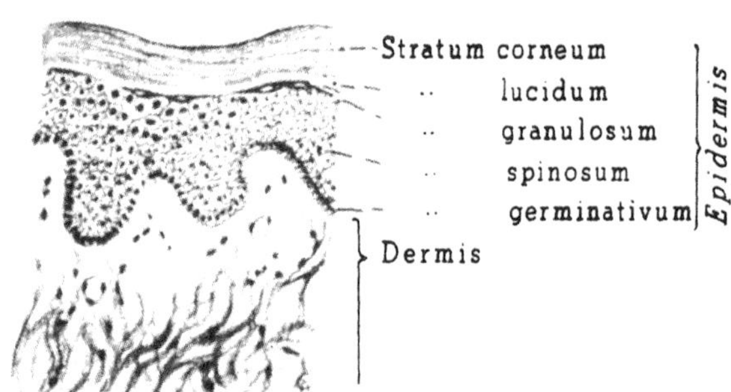

Figure 1 Three-dimensional view of the skin showing various skin tissue layers and appendages and cross-sectional view of various epidermal layers and the dermis. (From Jacob and Francone, 1970.)

1. Epidermis

The outer (epidermal) layer of the skin is composed of stratified squamous epithelial cells. The epithelial cells are held together mainly by highly convoluted interlocking bridges, which are responsible for the unique integrity of the skin. The epidermis is thickest in the areas of the palms and soles and becomes thinner over the ventral surface of the trunk (8).

Microscopic sections of the epidermis show two main parts: the stratum corneum and the stratum germinativum. The stratum corneum forms the outermost layer of the epidermis and consists of many layers of compacted, flattened, dehydrated, keratinized cells in stratified layers. These horny cells have lost their nuclei and are physiologically rather inactive. They are formed and continuously replenished by the slow upward migration of cells produced by the basal cell layer of the stratum germinativum, which is the regenerative layer of the epidermis. Studies have shown that the stratum corneum is replenished about every 2 weeks in a mature adult (9,10). In normal stratum corneum the cells have a water content of only approximately 20% compared to the normal physiological level of 70% in the physiologically active stratum germinativum. The stratum corneum requires a minimum moisture content of 10% (*w/w*) to maintain flexibility and softness. It becomes rough and brittle, resulting in so-called dry skin, when its moisture content decreases at a rate faster than can be resupplied from the underlying tissues (11).

The stratum corneum is responsible for the barrier function of the skin. It also behaves as the primary barrier to percutaneous absorption (12). The thickness of this layer is mainly determined by the extent of stimulation of the skin surface by abrasion and bearing of weight; hence thick palms and soles are developed (8). In the thicker parts of the skin, the transition from the living cells of the germinativum zone to the dead, cornified cells of the stratum corneum is made prominent by three layers, the stratum spinosum (prickly layer), stratum granulosum (granular layer), and stratum lucidum (clear layer), (Figure 1). In the process of degeneration, which occurs in this transition zone, granules of keratohyalin appear in the cells. When these granules have completely changed into keratin, the cells assume a homogeneous appearance to form the stratum lucidum. Like stratum corneum, the stratum granulosum and stratum lucidum are also physiologically important. Removal of these three upper epidermal layers results in water loss and an enhancement of skin permeability (13).

The normal epidermis is renewed as quickly as it is worn off. Examination of skin scrapings show that there are free amino acids among the keratin; these may act as buffers and protect the skin from the action of acids and alkalis (9).

2. Dermis

Electron microscopic examination shows that the dermis is made up of a network of robust collagen fibers of fairly uniform thickness with regularly spaced cross-striations. This network may, however, be an artifact of histological fixation since examination of unfixed dermis by fluorescence microscopy suggests that it is a gel containing oriented tropocollagen (polypeptide) macromolecules. The network or gel structure is responsible for the elastic properties of the skin. Beneath the dermis, the fibrous tissue opens out and merges with the fat-containing subcutaneous tissue. On the other hand, the upper portion of the dermis is formed into ridges (or papillae) projecting into the epidermis, which contains blood vessels, lymphatics, and nerve endings. Only the nerve fibers reach into the germinative zone of the epidermis (9).

3. Subcutaneous Tissue

This is a sheet of fat-containing areolar tissue, known as the superficial fascia, attaching the dermis to the underlying structures.

B. Biochemistry of the Skin

1. Keratinization

Histochemical tests indicate that in the region of the granular layer of normal human skin there is a high-energy system responsible for the synthesis of keratin from polypeptides in the cytoplasm of epidermal cells. Sulfhydryl groups, phosopholipids, and glycogen are concentrated in the granular layer, and all are absent in the immediately overlying keratin. At this site the polypeptide chains are presumably unfolded and broken down and resynthesized into keratin molecules (9).

The keratin was reported to be present in the cornified envelope beneath the plasma membrane of horny cells (14). It is not soluble even in a boiling solution of sodium dodecylsulfate and a reducing agent, and approximately 18% of its lysine residues could be found in the form of γ-glutamyl-ε-lysine, a cross-linking dipeptide.

The appearance of particular proteins at different developmental stages has been correlated to the morphology of mouse epidermis (15). A histidine-rich protein appears simultaneously with the production of mature keratohyalin granules in 18-day fetal mouse epidermis and then decreases as the number of keratohyalin granules diminishes with increasing age. This protein is localized exclusively in the keratohyalin granules.

A recent study in fetal rats indicated that as keratinization proceeds, the lipids in the skin decrease their phospholipid content with an increase in triglycerides and sterol esters (16).

2. Skin Surface Lipids

Sebum, the product of the sebaceous glands, is reportedly a mixture of triglycerides, free fatty acids, waxes, sterols, squalene, and paraffins. The free fatty acids give sebum bacteriocidal and fungicidal activities. Sebum is produced most abundantly on the forehead, less on the trunk, and none, or very little, on the extremities (for instance, the palms have no sebaceous glands). Sebaceous activity is stimulated by androgens and increases at puberty. It is greater in the second half of the menstrual cycle than the first half, but is very low during pregnancy (9).

Significant changes in season and climate affect the rate of production and contents of skin surface lipids (17).

In addition to sebum, the skin surface lipids also contain a noncollagenous, fucosylated glycoprotein, which is an *s-s* linked aggregate released by human skin fibroblasts (18). The proliferation of human fibroblasts was increased 10- to 20-fold with the addition of human epidermal growth factor (19). This growth stimulation depended on the presence of serum and was enhanced by the addition of ascorbic acid.

3. Skin Fatty Acids

Skin contains two essential unsaturated fatty acids: linoleic acid and arachidonic acid. Linoleic acid has recently been identified as playing an important role in regulating the barrier functions of the skin. On the other hand, arachidonic acid may act to furnish prostaglandins (20).

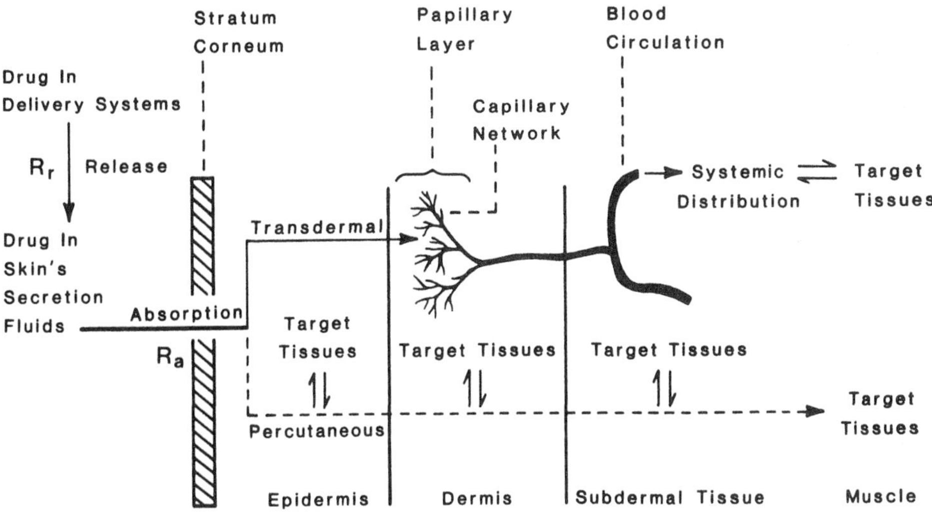

Figure 2 Drug release and absorption across skin tissues for localized therapeutic action in tissues directly underneath the site of drug administration or for systemic medication in tissue remote from the site of topical drug application.

4. *Proteinase in Topical Inflammatory Responses*

A proteinase, active at physiological pH, was recently isolated from human skin and is believed to participate in modulating the inflammatory response to cellular injury (21).

For many decades, the skin has been commonly used as the site for the topical administration of dermatological drugs to achieve a localized pharmacological action in skin tissues. In this case the drug molecule is considered to diffuse to a target tissue in the proximity of drug application to produce its therapeutic effect before it is distributed to the systemic circulation for elimination (Figure 2). The use of hydrocortisone for dermititis, benzoyl peroxide for acne, and neomycin for superficial infection (22) are typical examples of the application.

When the skin serves as the port of administration for systemically active drugs, the drug applied topically is distributed, following absorption, first into the systemic circulation and then transported to target tissues, which can be relatively remote from the site of drug application, to achieve its therapeutic action (Figure 3). This new

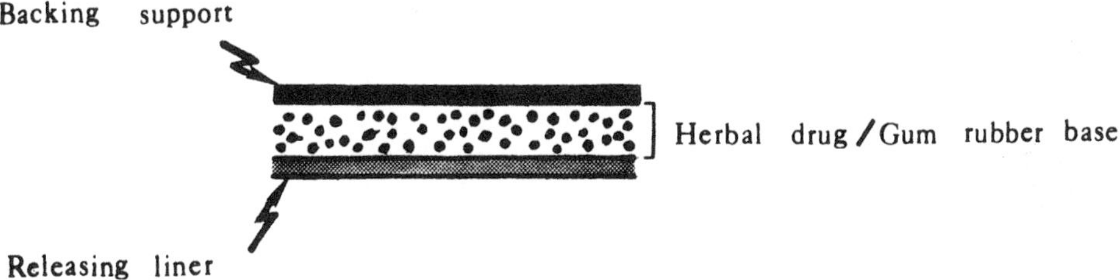

Figure 3 Cross-sectional view of a medicated plaster showing various major structural components.

application is exemplified by the transdermal controlled delivery of nitroglycerin or isosorbide dinitrate to myocardium for the prevention of anginal attacks, of scopolamine to the vomiting center for the prevention of motion-induced sickness, and of estradiol to various estradiol receptor sites for the relief of postmenopausal syndromes (23–25).

III. CONCEPTUAL ORIGIN OF TRANSDERMAL DRUG DELIVERY

The potential of using the intact skin as the port of drug administration has been recognized for several decades, as evidenced by the development of medicated plaster (Figure 3). By definition, the plaster is also a drug delivery system designed for external applications. Historically, the medicated plaster can be viewed as the first development of transdermal drug delivery. It is designed to bring medication into close contact with the skin, so drug can be delivered transdermally (26).

Figure 4 Representative Oriental medicated plasters. The composition for one of the medicated plasters, Yang-Cheng plaster, is outlined in Table 1.

Table 1 Chinese Medicated Plaster

Main ingredients, %	
Fossilia oassis mastodi,	10.42
Eupolyphagasinensis Walker,	10.42
Sanguis draconis,	4.17
Catechu,	6.25
Myrrha,	6.25
Rhizoma drynariae,	4.17
Radix dipsaci,	4.17
Flos carthami,	9.17
Rhizoma rhei,	8.33
Herba taraxaci,	8.33
Mentholum,	20.00
Methylis salicylas,	8.32

Description and Action: This plaster is prepared on the basis of the dialectic therapeutics of traditional Chinese medicine. The elements of various drugs and herbs, when applied to the skin, penetrate the subcutaneous tissues to stimulate circulation and produce a local analgesic effect. This plaster helps to curb inflammation in the muscles and to promote the healing of bone fractures.

Indications: Bruises, fractures, sprains, swelling and pains, bad blood circulation, injuries and wounds, rheumatic arthritis, neuralgia, limb languor, others.

Directions: Cut a piece of desired size from the roll, remove the cellophane, and apply it to the affected part. Medicinal effect lasts 24 hr.

To date, the historical development of medicated plasters has not been well documented. However, the use of medicated plasters can be traced several hundred years back to ancient China. Two representatives of Chinese medicated plasters, which are still available for medical practice, are shown in Figure 4. As shown in Table 1, this early generation of medicated plasters tend to contain multiple ingredients of herbal drugs and are indicated for localized action in the tissues directly underneath the site of application.

Medicated plasters are also very popular in Japan as over-the-counter pharmaceutical dosage forms. They are also commonly called cataplasms (27). Salonpas, which is also shown in Figure 4, is a typical example. It is still been formulated to contain multiple ingredients, including six purified therapeutically active agents.

Medicated plasters have also been available in Western medicine for several decades. For example, Allcock's porous plasters of England and the ABC (arnica/belladonna/capsicum) pflaster of Germany. In the United States, three medicated plasters have been listed in the official compendia since as early as 40 years ago (28,29): belladonna plaster, mustard plaster, and salicylic acid plaster. Like the Oriental plasters, the Western medicated plasters are rather simple in formulation and were developed mainly for local medication.

IV. RECENT DEVELOPMENTS IN TRANSDERMAL DRUG DELIVERY

The potential of using intact skin as the site for continuous transdermal infusion of drug has been recently recognized beyond the boundary of local medication. The development of female syndromes in male operators working in the production of estrogen-containing pharmaceutical dosage forms challenged the old theory that the skin is a perfectly impermeable barrier and also triggered the research curiosity of several biomedical scientists to evaluate the feasibility of transdermal delivery of systemically active drugs. The findings, accumulated over the years, have practically revolutionized the old concept of an impermeable skin barrier and also motivated a number of pharmaceutical scientists to develop patch-type drug delivery systems for the rate-controlled transdermal administration of drugs to achieve systemic medication (1–4,30).

Over a decade of intensive research and development efforts, several rate-controlled transdermal drug delivery systems have been successfully developed and commercialized (Figure 5). They can be classified according to the technological basis of their approach into the following four categories:

Figure 5 Representative transdermal drug delivery (TDD) systems. (1) Transderm-Nitro system, (2) Catapres-TTS system, (3) Transderm-Scop system, (4) Frandol tape, (5) Deponit system, (6) Nitro-Dur system, (7) Nitrodisc system.

1. Polymer membrane permeation-controlled TDD systems: Transderm-Scop system (scopolamine-releasing TDD system); Transderm-Nitro system (nitroglycerin-releasing TDD system); Catapres-TTS system (clonidine-releasing TDD system); Estraderm system (estradiol-releasing TDD system)
2. Adhesive polymer dispersion TDD systems: Deponit system (nitroglycerin-releasing TDD system); Frandol tape (isosorbide dinitrate-releasing TDD system); Minitran system (nitroglycerin-releasing TDD system); Nitro-Dur II system (nitroglycerin-releasing TDD system)
3. Nonadhesive polymer dispersion TDD systems: Nitro-Dur system (nitroglycerin-releasing TDD system); NTS system (nitroglycerin-releasing TDD system)
4. Microreservoir dissolution-controlled TDD systems: Nitrodisc system (nitroglycerin-releasing TDD system); transdermal contraceptive system (progestin-estrogen–releasing TDD system)

By now at least 11 TDD systems have been launched on the worldwide prescription drug market: Transderm-Scop system (Ciba), Transderm-Nitro system (Ciba), Catapres-TTS system (Boehringer-Ingelheim), Estraderm system (Ciba), Minitran system (3M Riker), Nitro-Dur system (Key), Nitrodisc system (Searle), NTS system (Bolar), Deponit system (Pharma-Schwarz), and Frandol tape (Toaeiyo-Yamanouchi) (Figure 5). Furthermore, several TDD systems have been submitted for regulatory review and approval.

V. FUNDAMENTALS OF SKIN PERMEATION

Until the turn of the century, the skin was generally regarded in the scientific literature as an impermeable barrier, except possibly to gases. By the beginning of this century, however, enough systematic investigations had been carried out to reach some general conclusions: lipid-soluble substances, such as nonelectrolytes, have a comparatively greater skin permeability than water-soluble substances, like electrolytes.

As early as 1853 it was recognized that the various layers of skin are not equally permeable (31). It was later noted that the epidermis is much less permeable than the dermis (32). The concept of the stratum corneum as the permeation barrier of the skin was not immediately accepted, and for the following 50 years this concept was the focus of much scientific debate. The basis for the controversy seems to relate to a misbelief that the stratum corneum is a grossly porous membrane through which not only ions but also large molecules move easily (33). Water permeation studies using isolated epidermis and stratum corneum conducted in 1951 removed all reasonable doubts that stratum corneum is indeed the barrier to skin permeation, at least for water molecules (34). The best direct evidence that the stratum corneum is essentially a uniformly good permeation barrier came from studies using isotopic tracers. The largest amount of isotope was always detected in the outer layers, decreasing proportionally toward the dermis. This observation suggests that the outer layers of the skin greatly impede permeation (35–37).

A. Stratum Corneum as the Skin Permeation Barrier

The composite structure of the skin permeation barrier in humans is represented by three distinct layers: the stratum corneum (15 μm thick), the viable epidermis (150

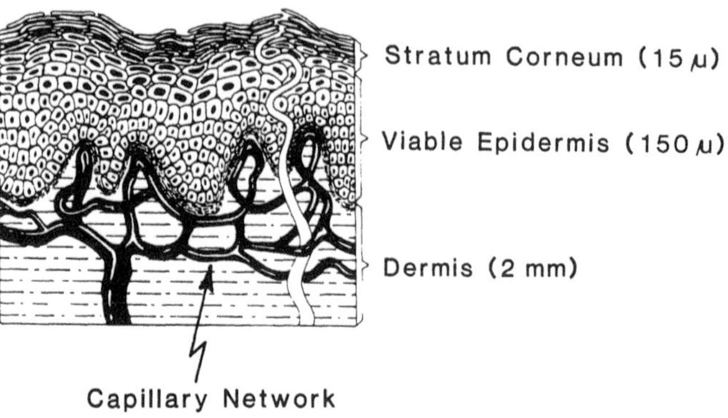

Figure 6 The structural relationship among stratum corneum, viable epidermis, and the capillary network at the dermoepidermal junction.

μm thick), and the papillary layer of the dermis (100–200 μm in thickness; Figure 6). The papillary arterial plexus forms the lower boundary of this composite slab. This composite structure is pierced in various places by two types of potential diffusion shunts: hair follicles and sweat glands (Figure 1).

The average human skin surface is known to contain on average 40–70 hair follicles and 200–250 sweat ducts on every square centimeter of skin area. These skin appendages, however, actually occupy, grossly, only one-tenth of 1% (0.1%) of the total human skin surface. Even though foreign agents, especially water-soluble substances, may be able to penetrate into the skin via these skin appendages at a rate faster than that through the intact area of the stratum corneum, this transappendageal route of percutaneous absorption has provided a very limited contribution to the overall kinetic profile of skin permeation. Therefore, the skin permeation of most neutral molecules at steady state can thus be considered primarily a process of passive diffusion through the intact stratum corneum in the interfollicular region. Thus for studying the fundamentals of skin permeation and the mechanisms of transdermal drug delivery (38), the organization of the skin can be represented by a simplified four-layer model as shown in Figure 7.

The phenomenon of percutaneous absorption (or skin permeation) can be visualized as consisting of a series of steps in sequence: sorption of a penetrant molecule onto the surface layers of stratum corneum, diffusion through it and the viable epidermis, and finally, at the papillary layer of the dermis, the molecule is taken up into the microcirculation for subsequent systemic distribution. The viable tissue layers and the capillaries are relatively permeable, and the peripheral circulation is sufficiently rapid (39), so that for the great majority of penetrants, diffusion through the stratum corneum is often the rate-limiting step. The stratum corneum acts as a passive, but not an inert, diffusion medium (40). No active transport process has been shown to be involved in skin permeation (12,34,41–51). As discussed earlier (Section IIA1), the rate-limiting stratum corneum is composed of dead, keratinized, metabolically inactive horny cells.

Scanning electron photomicrography illustrated that these keratinized cells are closely packed together and the bulk of the stratum corneum is mechanically coherent (with the exception of a few desquamating layers (52–54). The intercellular spaces

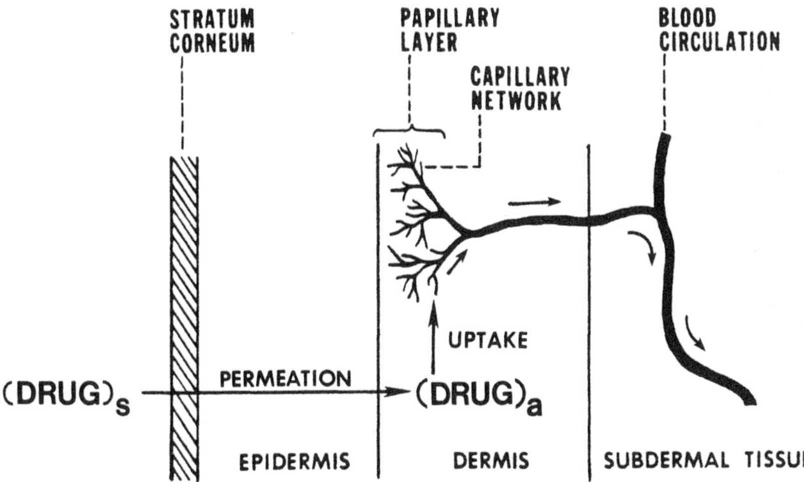

Figure 7 A multilayer skin model showing the sequence of transdermal permeation of drug for systemic delivery: sorption by stratum corneum, permeation across the viable epidermis, and then uptake by the capillary network in the dermoepidermal junction for systemic distribution.

are not only narrow but also filled with the overlapping interdigitation of adjacent cells and cell walls (Figure 8).

A typical horny cell is made up of an amorphous matrix of mainly lipid and nonfibrous protein, within which keratin filaments (60 ~ 80 Å in diameter) are distributed (55). A rough composition of the stratum corneum is compiled in Table 2. The conversion of aqueous epidermal cells (with a water content of 70%) into dried, compact, keratin-containing horny cells (containing only 20% water) constitutes a crucial event in the continuously developing epidermis. It leads to a transformation from an aqueous fluid medium, which is characterized approximately by solution-state diffusion, to dry, semi-solid "keratinized cells" characterized by a much lower fiber-like diffusion (56). The ultrastructure of the intercellular keratin may also play a role in the process of skin permeation, particularly in accounting for the selective permeability of polar and nonpolar molecules.

The horny cells of the stratum corneum were recently reported to comprise a trilayered membrane with rather thin outer and middle layers and a thick inner layer (70 ~ 150 Å in thickness) (57). The inner layer was isolated and found to be elastic and highly resistant to chemical reagents and solvents. The membrane of horny cells has apparently doubled in thickness during keratinization as a result of the inclusion or deposition of chemically resistant materials (58,59).

As discussed earlier, the stratum corneum is partially hydrated (with a water content of approximately 20%). In vitro studies of the hydration of isolated stratum corneum demonstrated that it is not immediately hydrated upon immersion in water. Although maceration may be detected within a few minutes, swelling can continue for 3 days. When fully hydrated stratum corneum was found to absorb water up to five to six times its weight, and the water absorbed was found strongly bound within the intercellular keratin (60). The hydrated intercellular keratin appears to comprise

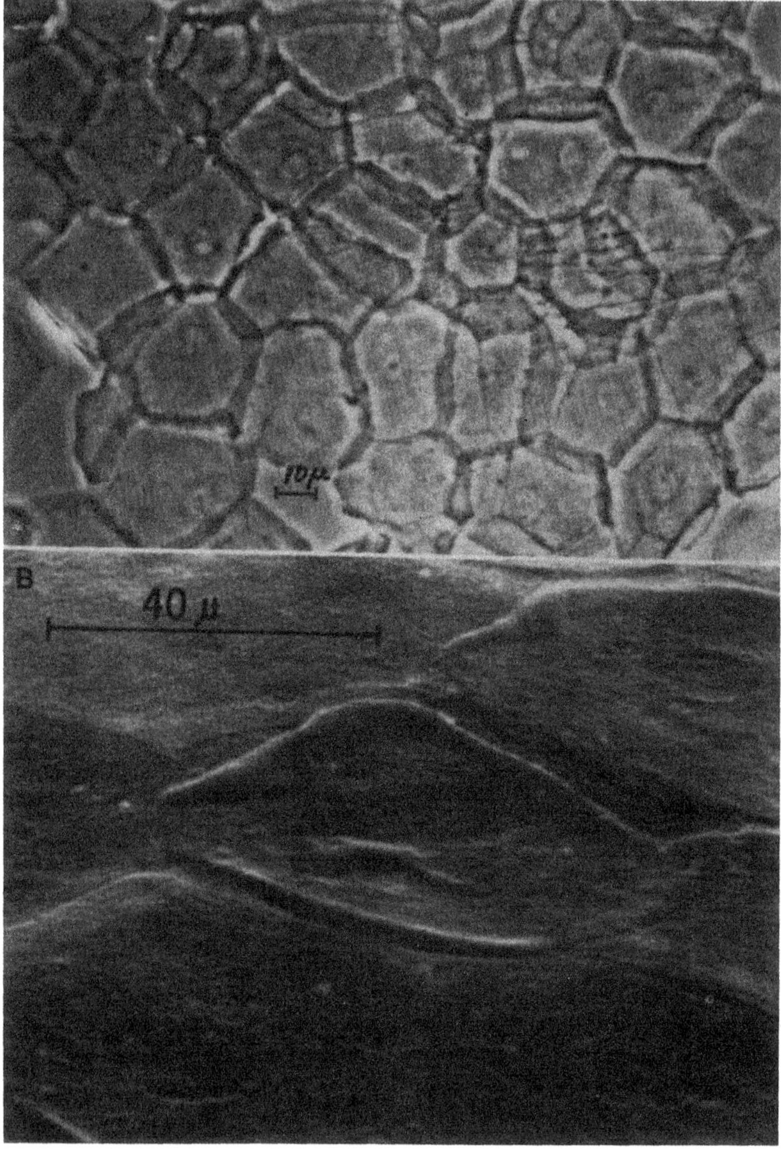

Figure 8 (A) A "monocellular layer" of stratum corneum (phase contrast). Dark bands with bordering cells are regions of cell overlap. (B) Scanning electron photomicrograph of stratum corneum surface. (Reproduced with permission from Scheuplein and Blank, 1971.)

a stable two-phase system at the macromolecular level: a continuous, water-rich polar region intermingled with a network of nonpolar lipids. Hydration apparently increases the thickness of stratum corneum by several fold.

There are significant differences in the structure and chemistry of human stratum corneum from one region of the body to another that are reflected in the skin's permeability. For instance, plantar and palmar callus can be as thick as 400–600

Table 2 Composition of Human Stratum Corneum

Components	%	Gross biochemical compositions
Cell membranes	5	Lipids and nonfibrous proteins
Cell contents	85	Lipids (20%)
		α-Protein (50%)
		β-Protein (20%)
		Nonfibrous proteins (10%)
Intercellular materials	10	Lipids and nonfibrous proteins

Source: Compiled from the data by Scheuplein and Blank (1971).

μm compared to only 10–20 μm for the back, arms, legs, and abdomen (Table 3). Despite its greater thickness, however, the callus is not a good diffusional barrier (61). The callus can even be dissolved by a weekly alkaline solution, whereas the thinner stratum corneum from the other sites remains relatively unaffected (62). However, the thin stratum corneum may also show some minor regional variations, which measurably affect skin permeability (Table 3). The skin permeability of a simple molecule, like cortisol, was observed to follow broadly the same order of permeability as that of a water molecule (63).

The data accumulated to date have demonstrated that skin permeation is primarily controlled by diffusion across the bulk of stratum corneum. The rate of diffusion of penetrant molecules across the hydrodynamic diffusion layer on the surface of the stratum corneum and the uptake rate of penetrant molecules by stratum corneum from the applied vehicle are both very rapid compared to diffusion through the rate-limiting stratum corneum (56). As an illustration, the diffusivities of water molecules across the stratum corneum and other biological membranes are compared in Table 4. It appears that the diffusion coefficient of water molecules across the stratum corneum is as small or slightly smaller than that for the physiologically important biological membranes. However, the diffusivity is three to four orders of magnitude lower than that in the dermis, which is fivefold thicker, or that in a stagnant water layer of the same thickness (2×10^{-5} cm^2/sec) (64,65).

On the other hand, even though the mobility of water molecules in the stratum corneum (4.2×10^{-10} cm^2/sec) is almost the same as in the erythrocyte membrane (6.5×10^{-9} cm^2/sec), the total diffusional resistance of the stratum corneum toward

Table 3 Regional Variation in Water Permeability of Stratum Corneum

Skin region	Thickness (μm)	Permeation rate (mg/cm^2/hr)	Diffusivity (cm^2/sec × 10^{10})
Abdomen	15.0	0.34	6.0
Volar forearm	16.0	0.31	5.9
Back	10.5	0.29	3.5
Forehead	13.0	0.85	12.9
Scrotum	5.0	1.70	7.4
Back of hand	49.0	0.56	32.3
Palm	400.0	1.14	535.0
Plantar	600.0	3.90	930.0

Source: Compiled from the data by Scheuplain and Blank (1971).

Table 4 Water Diffusivity through Stratum Corneum and Other Biological Membranes

Biological barriers[a]	h_m (cm × 10^7)	D_m^b (cm^2/sec × 10^{10})	Reference
Stratum corneum	40,000	4.2	Scheuplein (1965)
Dermis	200,000	20,000	Blank and Scheuplein (1969)
Erythrocyte membrane	5	65	Harris (1960)
Leukocyte membrane	10	28	Harris (1960)

[a]Of human origin.

[b]Diffusivity of water molecule in a stagnant water layer with thickness of 4 × 10^{-3} cm is 2 × 10^{-5} cm^2/sec. (Davies and Rideal, 1963.)

the permeation of water molecules is approximately five orders of magnitude greater than that for erythrocyte membrane (9 × 10^6 versus 80 sec/cm), resulting from the greater thickness of stratum corneum (Table 4).

The great diffusional resistance of stratum corneum was further demonstrated in the comparative absorption of drugs, like hydrocortisone (66). Mucous membranes in the rectal and vaginal regions permitted the absorption of 26–29% of the steroid applied, but only 2% of the applied dose was absorbed through the skin. Absorption of drugs through intestinal membranes has also been recognized as much faster than the skin permeation of drugs. The lower skin permeation was rationalized as due to the high keratin content in the stratum corneum and its low water level (67).

Heparin, despite its high molecular weight (~6000–20,000) was recently reported to penetrate the intact stratum corneum to reach biologically significant concentrations and to exert its activity (68).

B. Mechanistic Analysis of Skin Permeation

Before a topically applied drug can act either locally or systemically, it must penetrate the stratum corneum, the skin permeation barrier. Drug molecules may diffuse through the skin by three different routes: the intact stratum corneum, the hair follicle region, and the sweat gland ducts.

Despite the many investigations already carried out on various areas of skin permeation, there is no total agreement on the mechanism responsible for permeation through the intact skin. Many researchers believe that essentially all penetration occurs through the transfollicular route; other equally recognized scientists believe that the major pathway is diffusion through the intact, cornified horny cells of the stratum corneum. Recently, a growing number of investigators are inclined to accept both routes, with the relative importance depending upon the characteristics of the penetrating molecules. In the initial transient diffusion stage, the drug molecules may penetrate the skin along the hair follicles or sweat ducts and are then absorbed through the follicular epithelium and the sebaceous glands. When a steady diffusion state has been reached, diffusion through the stratum corneum becomes the dominant pathway (69).

Despite the similarity between these two routes of drug movement through the skin structure, there are relationships that are valid irrespective of the correct absolute mechanism. Skin permeation by both pathways can be analyzed mathematically by employing the diffusion process through a passive membrane as a model system (69).

The stratum corneum is not simply an inert structural material, but one with an affinity for a solute species applied (or deposited) on its surface (56). Thus the concentration of the solute species applied to the surface layers is not usually equal to but is related to its concentration in the applied vehicle in accordance with the stratum corneum-vehicle sorption isotherm. The linear isotherm can be defined in terms of the distribution coefficient between vehicle and stratum corneum. The steady-state permeation flux of the penetrant molecules J_s through the skin barrier is defined by the relationship

$$J_s = P_s \Delta C_s \tag{1}$$

where ΔC_s is the concentration difference across the skin barrier, and P_s is the permeability coefficient of the skin barrier with a thickness of h_{sb} and is defined by

$$P_s = \frac{K_s D_{sb}}{h_{sb}} \tag{2}$$

where D_{sb} is the average diffusion coefficient across the skin barrier, and K_s is the stratum corneum/vehicle partition coefficient.

According to Equation (1), the steady-state skin permeation flux J_s should be directly proportional to ΔC_s, the concentration difference across the skin barrier. The results of transdermal permeation studies with aqueous solutions of butanol across an isolated epidermis indicated that the linear J_s versus ΔC_s is followed for solutions with low donor concentrations (Figure 9); a positive deviation is seen at higher concentrations ($\geq$0.12 mol/L). The observed deviation was found to arise from an expansion of the epidermis as a result of increased sorption of the nonpolar solutes at

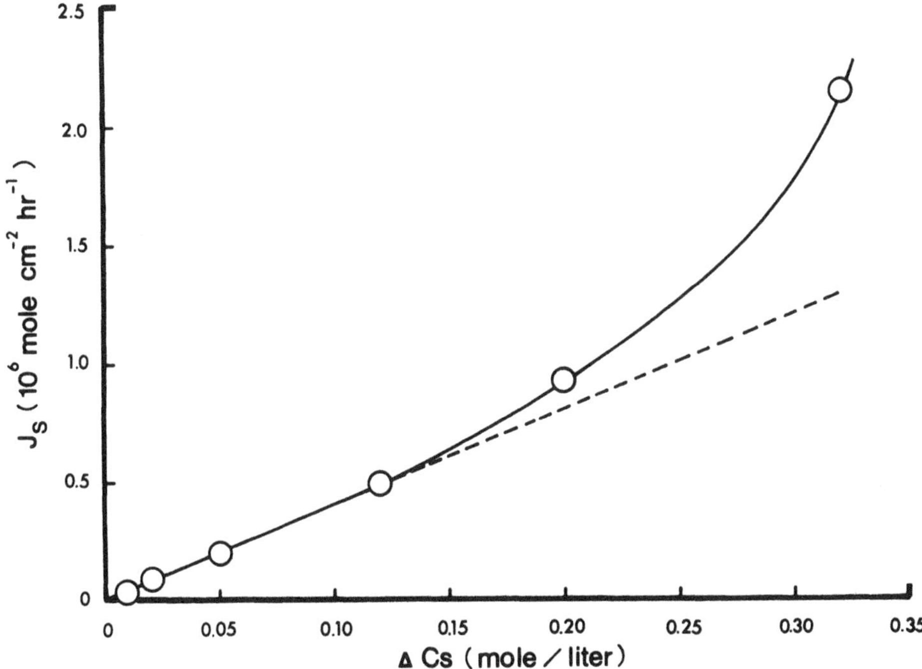

Figure 9 Dependence of the steady-state skin permeation flux J_s across an isolated epidermis on the concentration of butanol ΔC_s in aqueous solution. (Plotted from the data by Scheuplein and Blank, 1971.)

higher concentrations (70). Highly lipophilic molecules appear to produce a great extent of effect, and when sorbed by the stratum corneum, they also tend to increase the skin permeability of other chemical species administered with them.

The skin permeability of alkanols has been thoroughly investigated (12,49,50,70). The effect of selective introduction of polar and nonpolar groups in the alkanol molecule on skin permeability was examined. Skin permeability was observed to vary with the alkyl chain length of the alkanol (Figure 10). As the alkyl chain length increased, the steady-state skin permeation flux J_s increased to a maximum value at $n = 6$ (hexanol) and then decreased when the number of methylene (CH_2) groups was higher than C_6. A bell-shaped J_s versus n relationship thus resulted. The difference in J_s values can be as high as 14 times (compare hexanol with methanol). It is interesting to note that the skin permeability coefficient P_s shows an exponential dependence on the alkyl chain length of alkanols (Figure 11). This linear relationship is defined by

$$\log (P_s)_n = \log (P_s)_0 + (\pi)_{CH_2}(n)_{CH_2} \qquad (3)$$

where $(P_s)_n$ is the skin permeability coefficient of an alkanol with n number of (CH_2)

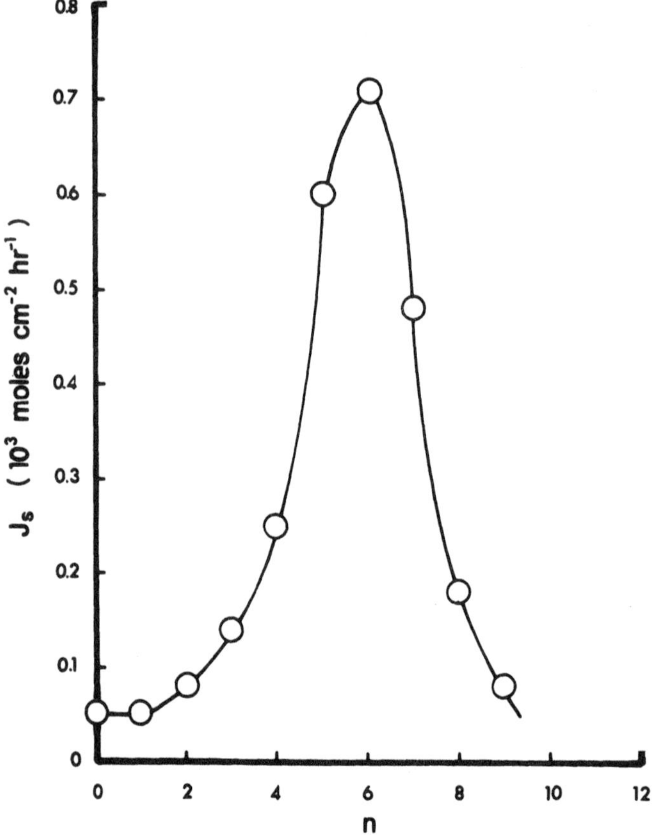

Figure 10 Dependence of the steady-state skin permeation flux J_s across an isolated epidermis on the alkyl chain length n of alkanols in aqueous solution. (Plotted from the data by Scheuplein and Blank, 1971.)

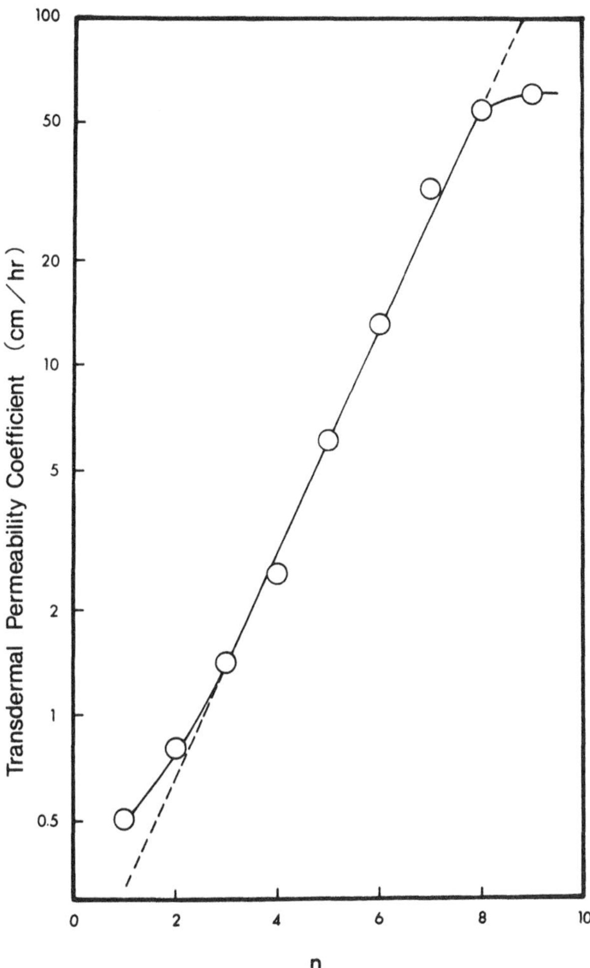

Figure 11 Dependence of skin permeability coefficients on the alkyl chain length n of alkanols in aqueous solution. (Plotted from the data by Scheuplein and Blank, 1971.)

groups; $(P_s)_0$ is the skin permeability coefficient of a hypothetical alkanol with zero carbon atom; $(\pi)_{CH_2}$ is the incremental constant for the methylene group; and $(n)_{CH_2}$ is the number of CH_2 groups in the alkanol. Equation (3) is followed fairly well when the number of methylene groups is between 3 and 8. Beyond this range a positive deviation is observed with lower alkyl chain lengths ($\leq C_3$), and a negative deviation is seen with higher alkanols ($\geq C_8$).

Based on Equation (2), the variation in the skin permeability coefficient P_s should be reflected in the magnitude of the K_s value, the distribution coefficient between stratum corneum and applied vehicle. When the value of D_{sb}/h_{sb} is constant, a linear relationship should exist between P_s and K_s (Figure 12). K_s values should also show an exponential dependence on the length of the alkyl chain (Figure 13). This linearity is mathematically expressed by

$$\log (K_s)_n = \log (K_s)_0 + (\pi)_{CH_2}(n)_{CH_2} \tag{4}$$

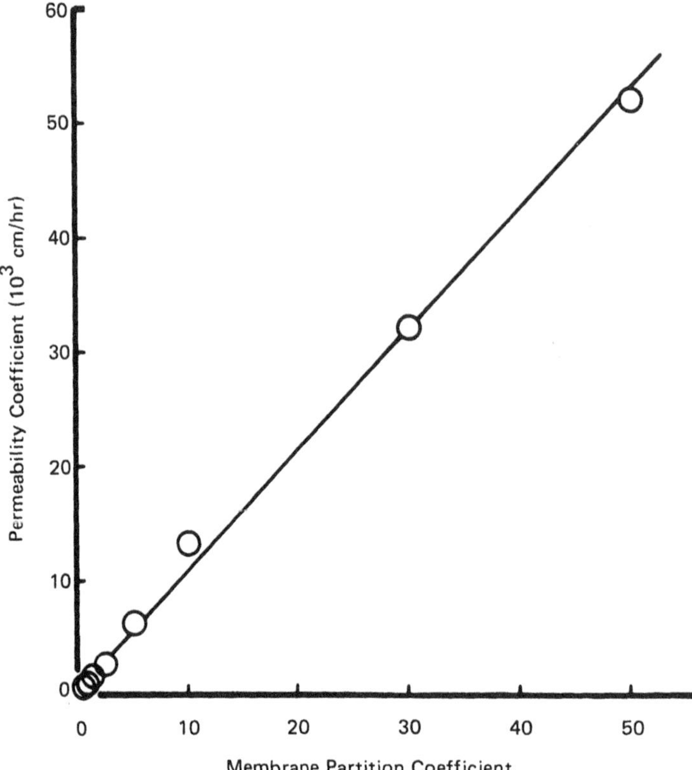

Figure 12 Linear relationship between the skin permeability coefficient and the stratum corneum/aqueous vehicle partition coefficient (or membrane partition coefficient). (Plotted from the data by Scheuplein and Blank, 1971.)

where $(K_s)_n$ is the stratum corneum/vehicle partition coefficient for an alkanol with n number of methylene groups; $(K_s)_0$ is the partition coefficient of a hypothetical alkanol with zero carbon atom; and $(\pi)_{CH_2}$ has the same meaning as defined in Equation (3). A positive deviation from linearity is also observed with lower alkyl chain length ($\leq C_3$). This explains the same deviation noted in the P_s versus n profile (Figure 11).

Basically, the hydrated stratum corneum has an affinity for both water-soluble and lipid-soluble nonelectrolytes. This bifunctional solubility arises from its inherently mosaic, filament-matrix ultrastructure, which allows hydrophilic and lipophilic regions to exist separately. This bifunctional structure and its unique, encompassing solubility characteristics cannot be adequately duplicated over the full range of polarity by any lipoidal solvents, such as olive oil or ether (42,46,71,72).

The major pathway for the skin permeation of water-soluble molecules is primarily transcellular, i.e., through cells and cell membranes alike without discrimination. The keratin structure inside the cells has been pinpointed as the site responsible for the skin's major diffusional resistance for water-soluble molecules. The water molecules within this structure are in a "bound" state, and the diffusion of the penetrant molecules dissolved within this bound water occurs very slowly.

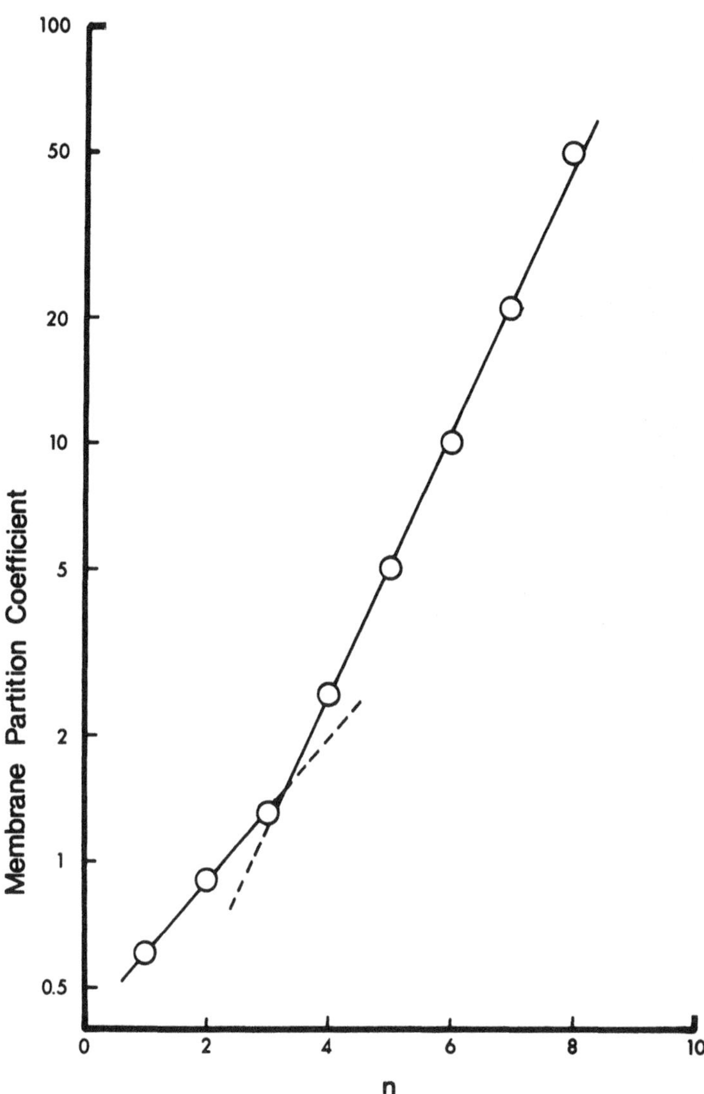

Figure 13 Effect of the alkyl chain length *n* of alkanols in aqueous solution on the partition coefficient for the interfacial partitioning of alkanols from aqueous phase to stratum corneum. (Plotted from the data by Scheuplein and Blank, 1971.)

On the other hand, the pathway for lipid-soluble molecules is not well understood and presumably follows the endogenous lipids within the stratum corneum. These lipids are located both intracellularly and between the keratin filaments within horny cells (Table 2) (52).

C. Biphasic Mathematical Model for Skin Permeation

The transdermal permeation of drugs and other small molecules through human skin can be considered as consisting of a process of dissolution, partitioning, and molecular diffusion through a composite of multilayered skin cells with principal barrier resides within the stratum corneum. A simplistic two-phase model was proposed for stratum corneum, which describes it as a dispersion of hydrophilic protein gel in a continuous lipid matrix through which penetrant molecules migrate by dissolution and Fickian diffusion (Figure 14). A mathematical expression was derived (73) to define the permeability coefficient P_{sc} across the biphasic stratum corneum, as follows:

$$P_{sc} \approx \frac{K_{pa}D_{pg}}{h_{sc}} \left(\frac{1.16}{0.16(D_{pg}/K_{pl}D_{lm}) + 1} + 0.0017 \frac{K_{pl}D_{lm}}{D_{pg}} \right) \tag{5}$$

where K_{pa} and K_{pl} are the partition (or distribution) coefficients of the penetrant molecules between the protein gel and its surrounding aqueous solution and between the lipid and the protein phases, respectively; D_{pg} and D_{lm} are the diffusion coefficients of the penetrant molecules in the protein gel and in the lipid matrix, respectively; and h_{sc} is the thickness of stratum corneum (Figure 14). Equation (5) suggests that the permeability coefficient of stratum corneum to any penetrant molecule should be determined by two physicochemical terms, $K_{pa}D_{pg}/h_{sc}$ and $K_{pl}D_{lm}/D_{pg}$.

The permeability of highly hydrated protein gels to relatively small molecules (with molecular weight 1000 or less) can be approximated, with fair accuracy, if the

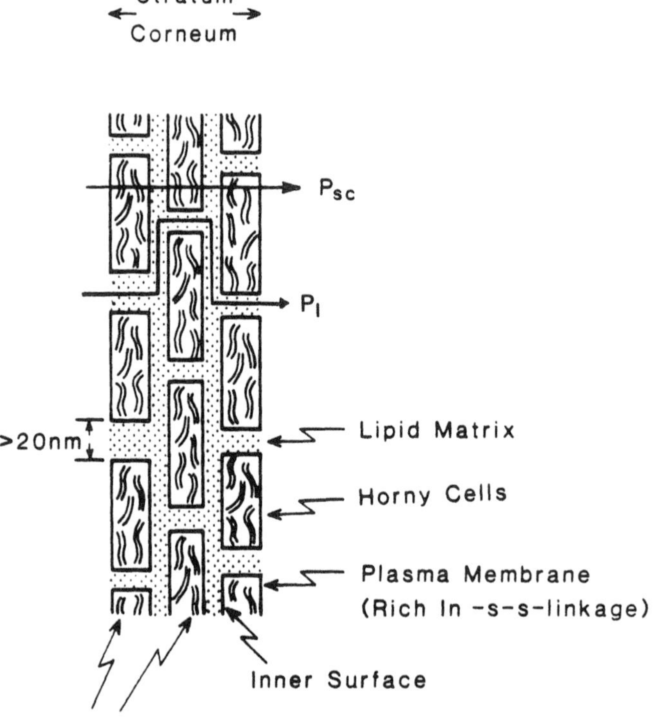

Figure 14 The microstructure of stratum corneum in which horny cells are dispersed like layers of bricks in a lipid matrix.

penetrant molecules do not specifically bind to or associate with the protein molecules. Under such circumstances the distribution coefficient K_{pa} of the penetrant molecules between the protein gel and its equilibrating aqueous solution may be approximated to be equal to the volume fraction of water content in the protein gel, i.e., about 75–90%. The diffusion coefficient of the penetrant molecules in the protein gel D_{pg} can be approximated to be equal to one-tenth of its solution diffusivity in the water. For penetrant molecules with molecular weight in the range 300–500, a value of $D_{pg} \approx 2 \times 10^{-7}$ cm^2/sec is a reasonable approximation (44). If the stratum corneum has an average thickness h_{sc} of 40 μm, Equation (5) can be reduced to Equation (6) to define a normalized stratum corneum permeability coefficient $\bar{P}_{sc}$:

$$\bar{P}_{sc} = 0.135 \frac{K_{pl}D_{lm}}{D_{pg}} \left(\frac{1.16 + 0.0017K_{pl}D_{lm}/D_{pg}}{0.16 + (K_{pl}D_{lm}/D_{pg})} \right) \tag{6}$$

Two situations may exist. If $K_{pl}D_{lm}/D_{pg}$ is very small, Equation (6) can be further simplified to

$$\bar{P}_{sc} = 0.98 \frac{K_{pl}D_{lm}}{D_{pg}} \tag{7a}$$

If $K_{pl}D_{lm}/D_{pg}$ is extremely large, Equation (6) can also be reduced to

$$\bar{P}_{sc} = 2.3 \times 10^{-4} \frac{K_{pl}D_{lm}}{D_{pg}} \tag{7b}$$

Experimentally it was observed that the value of K_{pl} for most skin penetrants can be approximated by the oil/water partition coefficient determined experimentally, as long as the oil phase used has a cohesive energy density similar to that of lipid matrix in the stratum corneum (73).

Equation (6) suggests that a penetrant molecule that is either exceedingly water insoluble and/or has a very low partition coefficient displays a low rate of skin permeation. Conversely, a penetrant that is both highly water soluble and has a strong tendency to partition into the oil phase gives a relatively high permeation rate through the skin.

This analysis leads to the prediction that the maximum attainable rate of permeation through intact skin depends primarily upon: (i) the aqueous solubility of the drug, which determines the magnitude of the concentration gradient across the skin tissues (i.e., the driving force for skin permeation); and (ii) the oil/water partition coefficient of the drug, which governs the specific skin permeability to that drug (Table 5). For nonionogenic drugs one should therefore expect an increase in the permeability coefficient P_s as increasing the lipophilicity of drugs. The results in Figure 15 indicate that the skin permeability coefficient increases linearly with lipophilicity until a plateau value is reached with partition coefficient values greater than 10,000.

D. Skin Permeation of Polar Nonelectrolytes

Introduction of polar groups, such as hydroxy groups, was reported to reduce the skin permeability of steroids (46). The magnitude of the reduction showed a first-

Table 5 Steady-State Skin Permeation Fluxes J_s and Skin Permeability P_s, of Several Therapeutically Active Agents and Their Physicochemical Characteristics

Drugs	J_s ($\mu g/cm^2/hr$)	P_s[a] (cm/hr × 10^3)	C_s[b] (mg/ml)	PC[c]
Ephedrine	300	6.0	50	1.0
Diethylcarbamazine	100	0.13	800	0.064
Nitroglycerin	13	11.0	1.3	10.0
Scopolamine	3.8	0.05	75	0.026
Chlorpheniramine	3.5	2.2	1.6	0.46
Fentanyl	2.0	10.0	0.2	200
Atropine	0.02	0.0086	2.4	0.006
Estradiol	0.016	5.2	0.003	12
Quabain	0.008	0.00078	10.0	0.00026
Digitoxin	0.00013	0.013	0.01	0.014

[a]Calculated from $P_s = J_s/C_s$.
[b]Maximum aqueous solubility at 30°C.
[c]Partition coefficient determined in a mineral oil/water system.
Source: Compiled from the data (at 30°C) by Michaels et al., AIChE. J., 21:985 (1975).

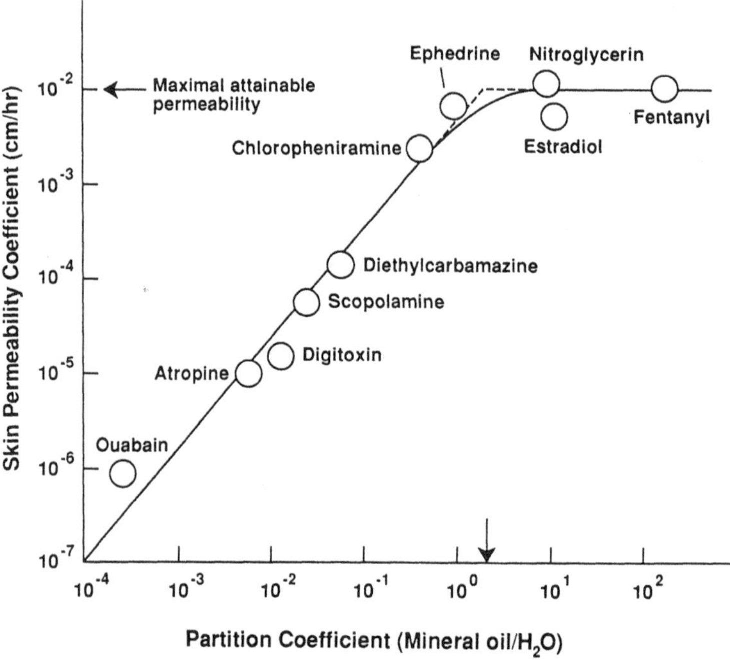

Figure 15 Dependence of the skin permeability coefficients of a series of drugs on their partition coefficients determined in a mineral oil-water system. (Plotted from the data by Michaels et al., 1975.)

order dependence on the number of OH groups added to the progesterone skeleton (Figure 16). The relationship can be expressed as

$$\log (P_s)_{OH} = \log (P_s)_p - (\pi)_{OH}(n)_{OH} \tag{8}$$

where $(P_s)_p$ and $(P_s)_{OH}$ are the permeability coefficients for progesterone and its hydroxyl derivatives, respectively; $(\pi)_{OH}$ is the incremental constant for the hydroxyl group; and $(n)_{OH}$ is the number of OH groups.

As indicated in Equation (2), the reduction in the skin permeability coefficient of progesterone as a result of the addition of hydroxy groups is thus linearly dependent upon the variation in its partition coefficient (Figure 17).

A group contribution factor for the effect of hydroxy groups at various positions on the skin permeability coefficients of various hydroxylprogesterone derivatives is illustrated in Table 6. The results indicate that the more OH groups added, the greater

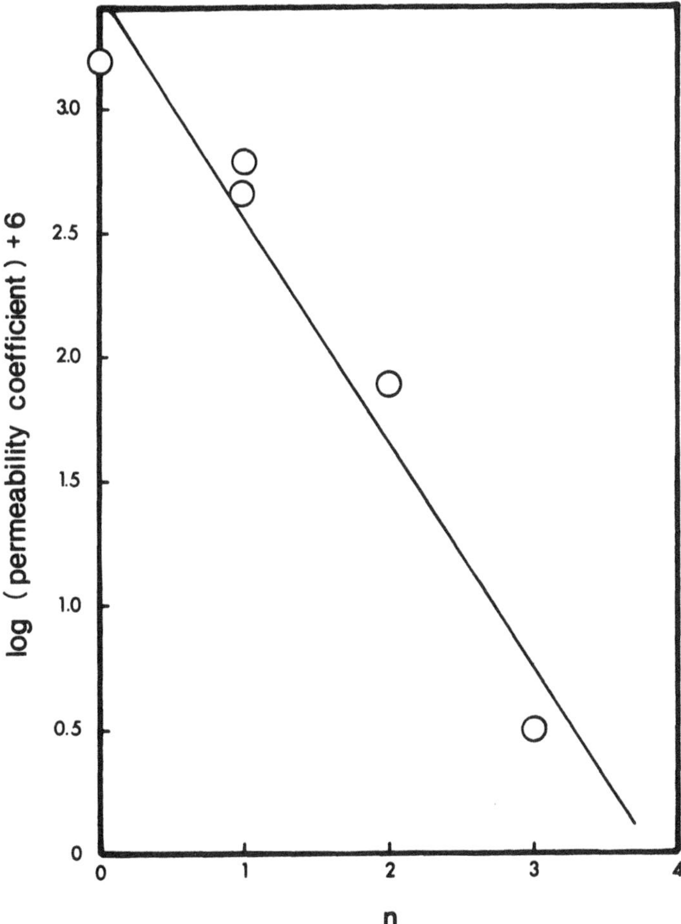

Figure 16 First-order dependence of the transdermal permeability coefficients of progesterone and its derivatives on the number n of hydroxy groups. (Plotted from the data by Scheuplein et al., 1969.)

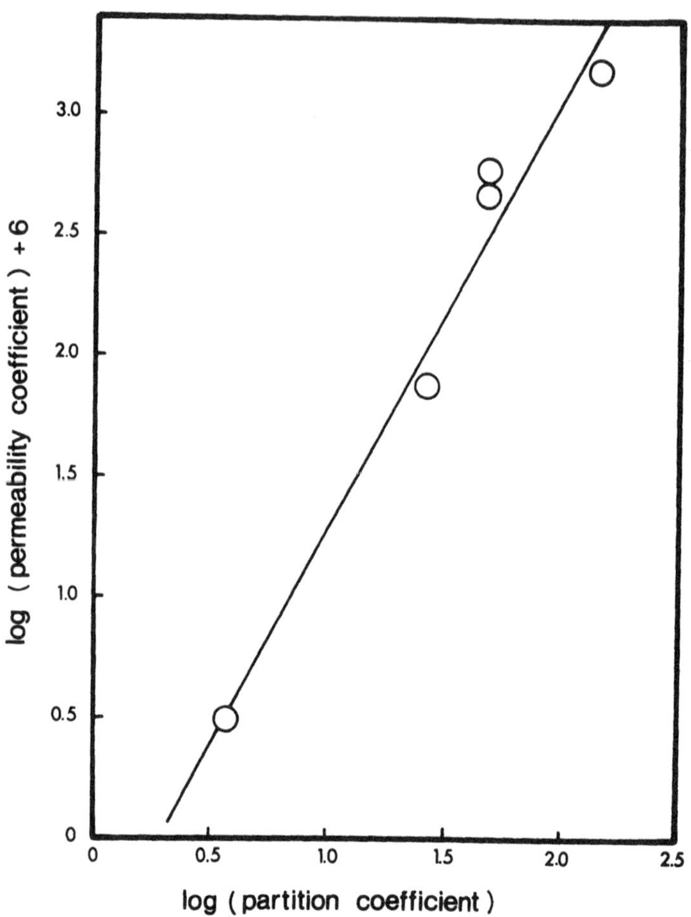

Figure 17 Linear relationship between the transdermal permeability coefficients of progesterone and its hydroxy derivatives and their partition coefficients measured in an *n*-octanol/buffer system at pH 7.4.

Table 6 Group Contribution Factor of Various Hydroxy Groupings in Hydrocortisone

Steroids	OH groupings	Group contribution factor[a]
17-OH-progesterone	17α-OH	−0.40
Desoxycorticosterone	21-OH	−0.53
17-OH-desoxycorticosterone	17α-OH 21-OH	−1.30
Hydrocortisone	11β-OH 17α-OH 21-OH	−2.70

[a]Calculated from the skin permeability coefficient data by Scheuplein and Blank (1971) by the equation: group contribution factor = $\log (P_s)_{OH} - \log (P_s)_p$.

is the magnitude of the group contribution factor and the lower the skin permeability coefficient, as observed experimentally in Figure 16.

The addition of hydroxyl groups was also found to modify the pharmacokinetic profile of steroids following topical administration (Figure 18). The addition of a OH group to the steroidal skeleton of progesterone slowed the rate of elimination, and the introduction of two or more OH groups reduced both the rates of absorption and elimination. Additionally, the total percentage of dose absorbed percutaneously showed a biphasic dependence on the number of hydroxyl groups added (Figure 19). With the addition of only one OH group, percutaneous absorption was enhanced slightly, which resulted from the increased solubility of steroid in the hydrated region

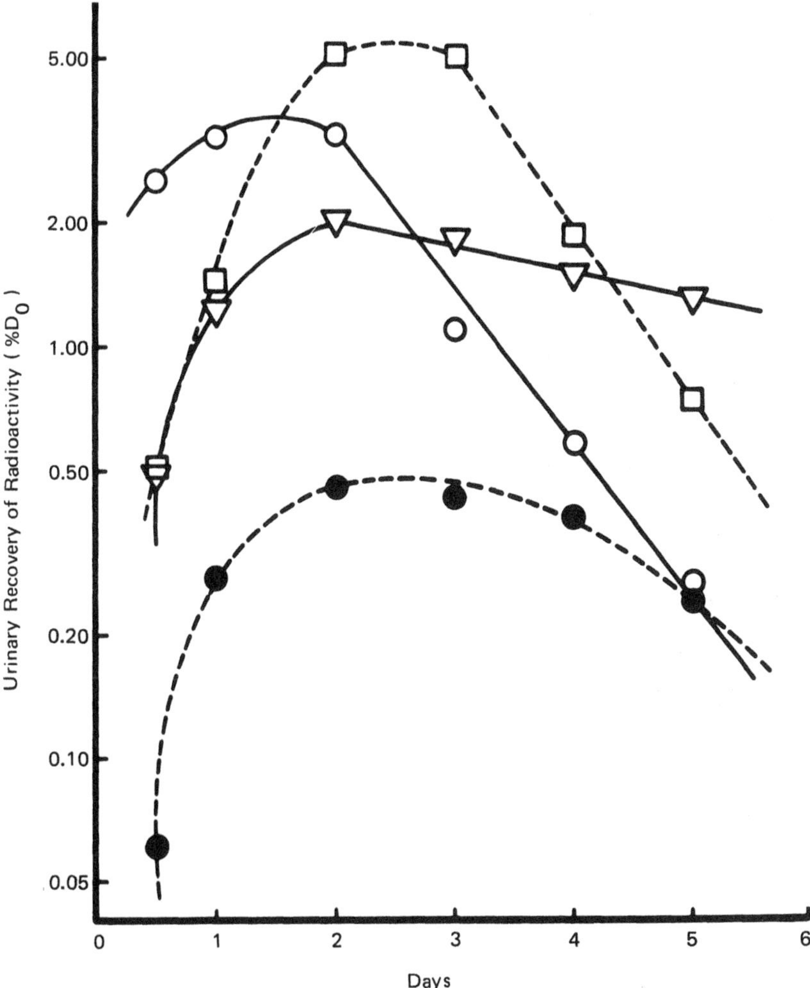

Figure 18 Urinary excretion profiles of progesterone and its hydroxy derivatives following transdermal absorption in humans: (○) progesterone, (□) 17-OH-progesterone, (▽) 17,21-OH-progesterone, and (●) hydrocortisone. (Plotted from the data by Feldman and Maibach, 1969.)

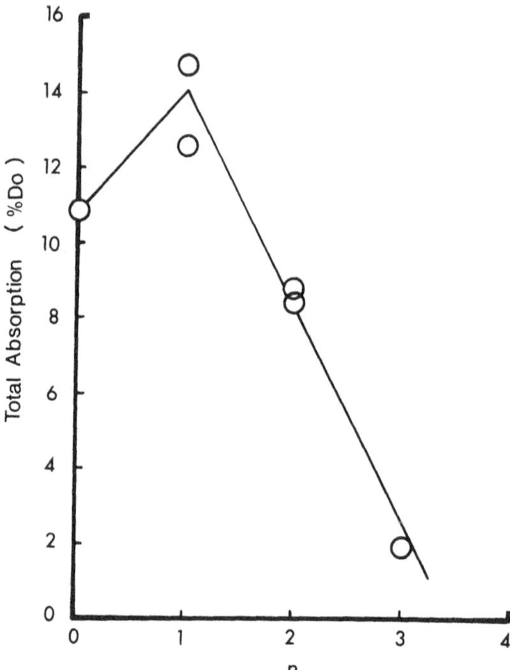

Figure 19 Effect of hydroxy groups on the total fraction of the applied dose absorbed following topical application of progesterone and its hydroxy derivatives (where n is the number of OH groups). (Plotted from the data by Feldman and Maibach, 1969.)

of skin tissue. The introduction of more OH groups markedly decreased the percutaneous absorption as a result of reduction in the lipophilicity of the hydroxylprogesterone derivatives (Figure 17).

E. Skin Permeation of Ionogenic Compounds

For ionogenic compounds the process of skin permeation is likely to be complicated by the simultaneous presence of both ionized (or protonated) and nonionized (or neutral) species in the solution, each of which permeates the skin at different rates. However, if it can be assumed that each species transports through skin tissue by simple diffusion at a rate governed solely by its own concentration gradient and unaltered by the presence of other species, the overall skin permeability of an ionogenic drug can be expressed in terms of the sum of the specific skin permeabilities for all the species present and their respective concentrations and concentration gradients (73). For a weakly basic drug, for example, the overall skin permeability $P_{B(0)}$ is thus defined by

$$P_{B(0)} = \frac{P_B + P_{BH} + (H^+/K_a)}{1 + (H^+/K_a)} \times \frac{C_{B(0)}}{h_{sc}} \tag{9}$$

where $C_{B(0)}$ is the total concentration of the weakly basic drug, in all forms, in the solution; P_B and P_{BH^+} are the specific skin permeabilities of the free, unprotonated base species B and of the protonated species BH^+ of the drug, respectively; H^+ is the concentration of hydronium ion; and K_a is the acid ionization constant of the drug.

Three situations could exist. If $(H^+)/K_a \ll 1$, i.e., the pH of the drug solution is much higher than the pK_a of the drug, then

$$P_{B(0)} \simeq P_B \frac{C_{B(0)}}{h_{sc}} \tag{10a}$$

If $(H^+)/K_a \gg 1$, i.e., the pH of the drug solution is much lower than the pK_a of the drug, then

$$P_{B(0)} \simeq P_{BH^+} \frac{C_{B(0)}}{h_{sc}} \tag{10b}$$

If $(H^+)/K_a = 1$, i.e., the pH of drug solution is in the neighborhood of the pK_a for the drug, then

$$P_{B(0)} = \frac{P_B + P_{BH^+}}{2} \frac{C_{B(0)}}{h_{sc}} \tag{10c}$$

Equation (9) suggests that in a solution with a constant pH, i.e., constant H^+ concentration, the overall rate of skin permeation should be proportional to the total drug concentration $C_{B(0)}$ in the solution in intimate contact with the stratum corneum.

The specific skin permeabilities of the free base species P_B and of the protonated species P_{BH^+} are defined by

$$P_B = K_{pl(B)} \frac{D_{lm(B)}}{D_{pg(B)}} h_{sc} \tag{11a}$$

and

$$P_{BH^+} = K_{pl(BH^+)} \frac{D_{lm(BH^+)}}{D_{pg(BH^+)}} h_{sc} \tag{11b}$$

respectively.

If it is assumed that

$$D_{lm(B)} \simeq D_{pg(B)} \tag{12a}$$

and

$$D_{lm(BH^+)} \simeq D_{pg(BH^+)} \tag{12b}$$

then the specific permeabilities of the stratum corneum to the un-ionized (free base) species and the ionized (protonated) species of the drug are in proportion to their oil/water partition coefficients:

$$\frac{P_B}{P_{BH^+}} = \frac{K_{pl(B)}}{K_{pl(BH^+)}} \tag{13}$$

In fact, the free energy change associated with the interfacial transfer of a pair of ions from an aqueous phase (with high dielectric constant) to oil phase (with low dielectric constant) is invariably greater than that for the transfer of its neutral species, and one should expect to observe that

$$K_{pl(B)} \gg K_{pl(BH^+)} \tag{14}$$

and therefore,

$$\frac{P_B}{P_{BH^+}} \gg 1 \tag{15}$$

Hence one should find that the specific permeability of stratum corneum to the unionized (or unprotonated) form of an ionogenic drug is substantially greater than that of the ionized (or protonated) species (Table 7).

However, the specific steady-state skin permeation flux of each species is also a function of its maximum achievable concentration in the solution in intimate contact with the stratum corneum:

$$\frac{J_B}{J_{BH^+}} = \frac{P_B C_B}{P_{BH^+} C_{BH^+}} \approx \frac{K_{pl(B)} C_B}{K_{pl(BH^+)} C_{BH^+}} \tag{16}$$

If the aqueous solubility of the free base of the drug C_B is much less than that of its protonated species C_{BH^+}, the specific steady-state skin permeation flux J_B of the free base species may be lower than that of its protonated form J_{BH^+}, even though the intrinsic permeability of the skin to the free base species P_B may be much greater than the intrinsic permeability P_{BH^+} of the protonated species.

F. Thermodynamics of Skin Permeation

The difference in the mechanisms of skin permeation between water-soluble and lipid-soluble nonelectrolytes is demonstrated by very different values of activation energy E_a, preexponential factor $\bar{A}$, enthalpy ΔH, and entropy ΔS (Table 8).

1. Activation Energy for Skin Permeation

The skin permeation of nonelectrolytes, like alkanols, was reported to be an energy-requiring process and to follow the Arrhenius relationship

$$J_s = \bar{A} e^{-E_a/RT} \tag{17}$$

where $\bar{A}$ is the preexponential factor and E_a is the activation energy.

For hydrophilic alkanols ($\leq C_5$) the activation energy is constant and 16.5 ± 0.1

Table 7 Specific Skin Permeabilities of the Free Base P_B and Protonated Species P_{BH^+}

Drugs	pK_a	Specific skin permeability (cm/hr × 10^5)	
		P_B/h_{sc}	P_{BH^+}/h_{sc}
Ephedrine	9.65	600	33
Chlorpheniramine	9.10	220	0.8
Scopolamine	7.35	5	0.3

Source: Compiled from the data by Michaels et al.; AIChE. J., 21:985 (1975).

Table 8 Thermodynamic Quantities Associated with Skin Permeation of *n*-Alkanols

	Thermodynamic quantities				
Alkanols	$\bar{A}$[a] (cm^2/sec)	E_a (kcal/mol)	ΔF[b] (kcal/mol)	ΔH (kcal/mol)	ΔS[b] (cal/mol·deg)
Ethanol	119	16.4	10.2	15.8	18.7
Propanol	502	16.5	10.2	15.9	19.2
Butanol	1760	16.7	10.4	16.1	21.2
Pentanol	3060	16.5	9.7	15.9	20.9
Hexanol	1.71	10.9	8.9	10.3	4.67
Heptanol	0.41	9.9	8.4	9.3	−0.36
Octanol	0.076	8.7	9.4	8.1	−4.17

[a]This includes the membrane partition coefficient.
[b]The membrane partition coefficient was excluded from determinations of the ΔF and ΔS values (Glasstone et al., 1941).
Source: Compiled from the data by Scheuplein and Blank (1971).

kcal/mol is required for their permeation through hydrated stratum corneum (Table 8). On the other hand, for the skin permeation of lipophilic alkanols ($>C_5$) a lower activation energy (10.9 kcal/mol or less) is needed at temperatures higher than 25–30°C (56). The observed lower energy requirement for lipophilic alkanols may be attributed to the possibility that these lipid-soluble aliphatic alcohols penetrate the skin via a lipoidal pathway; at temperatures higher than 25–30°C the lipid components in the stratum corneum become less viscous. Thus a lower energy of activation is needed for their skin permeation. A lower value of entropy ΔS also results. The longer is the alkyl chain length, the more lipophilic the alkanols and thus, the lower the activation energy and also the entropy (Table 8).

However, both hydrophilic and lipophilic alkanols share a common feature: that is, a substantially higher free energy ΔF value is needed for diffusion in the stratum corneum (9.60 ± 0.75 kcal/mol) than in the solution. This may be the underlying reason that skin permeation occurs so slowly (56).

2. *Free Energy of Desorption for Stratum Corneum/Vehicle Partitioning*

The stratum corneum/vehicle partition coefficient K_s is theoretically related to the standard Gibbs free energy of desorption F_d, i.e., the energy gained by a solute molecule upon desorption from the stratum corneum into the vehicle applied (65):

$$\Delta F_d = \text{RT} \ln K_s \tag{18}$$

A large ΔF_d value implies a greater solubility in the stratum corneum, a greater K_s, and consequently greater skin permeability [Equation (2)].

If it is assumed that ΔF_d can be expressed additively in terms of the individual contribution of the polar groups (ΔF_{OH}) and the lipophilic group (ΔF_{CH2}), then

$$\Delta F_d = \Delta F_{\text{OH}} + n\, \Delta F_{\text{CH2}} \tag{19}$$

then,

$$\log K_s = \text{constant} + \frac{n\, \Delta F_{\text{CH2}}}{2.303\text{RT}} \tag{20}$$

Following Equation (20), a plot of log K_s versus n for a homologous series of alkanols should permit calculation of ΔF_{CH_2}, the free energy of desorption for a single methylene group. The results suggest that the addition of a methylene group gives a ΔF_{CH_2} of 241 cal/mol for $n < 3$ and 448 cal/mol for $n > 3$. The lower ΔF_{CH_2} value observed for lower alkanols ($<C_3$) can be explained by the miscibility of these alkanols with water. Since the stratum corneum has a similar, well-hydrated environment, a lower value of ΔF_{CH_2} results. The lower ΔF_{CH_2} values also explain the positive deviation observed in Figures 11 and 13.

Comparison of Equation (20) with Equation (4) suggests that the methylene group incremental constant $(\pi)_{CH_2}$ is thermodynamically defined by

$$(\pi)_{CH_2} = \frac{\Delta F_{CH_2}}{2.303RT} \tag{21}$$

Equations (18) and (19) also imply that if a vehicle that has physicochemical characteristics identical to the stratum corneum is used, ΔF_d equals zero and the K_s value is unity; skin permeability then becomes independent of chain length. On the other hand, if a vehicle that is more lipophilic than the stratum corneum is used, a negative value of ΔF_d results and the K_s value becomes smaller than unity; skin permeability thus decreases with the increase in alkyl chain length. This behavior is demonstrated by the use of isopropyl palmitate as the vehicle for straight-chain aliphatic alcohols (Figure 20). A ΔF_{CH_2} value of -326.5 cal/mol is estimated. With the replacement of aqueous solution by isopropyl palmitate as the vehicle, the free energy of desorption of straight-chain alkanols changes from an increase of ~241–448 cal/mol to a decrease of 326.5 cal/mol for each additional CH_2 group.

3. *Effect of Polar Groups on the Free Energy of Desorption*

It was reported earlier that the introduction of polar groups, such as hydroxy groups, reduces the skin permeability coefficient of steroids (Figure 16). The first-order dependence of the skin permeability coefficient $(P_s)_{OH}$ on the number n of OH groups [Equation (8)] can be defined alternatively by

$$\log (P_s)_{OH} = \log (P_s)_p - \frac{n\Delta F_{OH}}{2.303RT} \tag{22}$$

where $(\pi)_{OH}$ is thermodynamically equivalent to $(\Delta F_{OH})/2.303RT$, and ΔF_{OH}, the Gibbs free energy of desorption contributed by OH groups, is related to the standard Gibbs free energy of desorption as follows:

$$\Delta F_d = \Delta F_p + n\,\Delta F_{OH} \tag{23}$$

where ΔF_p is the Gibbs free energy of desorption for the interfacial partitioning of progesterone molecules. A ΔF_{OH} value of -1.291 kcal/mol was estimated for the addition of every OH group into the progesterone molecule.

G. Diffusional Resistance for Skin Permeation

The total diffusional resistance ΣR_s, which is the reciprocal of the overall skin permeability across the composite skin diffusion barrier, is the sum of the individual diffusional resistance (45,74) and can be expressed mathematically as

$$\Sigma R_s = R_{sc} + R_e + R_{pd} \tag{24a}$$

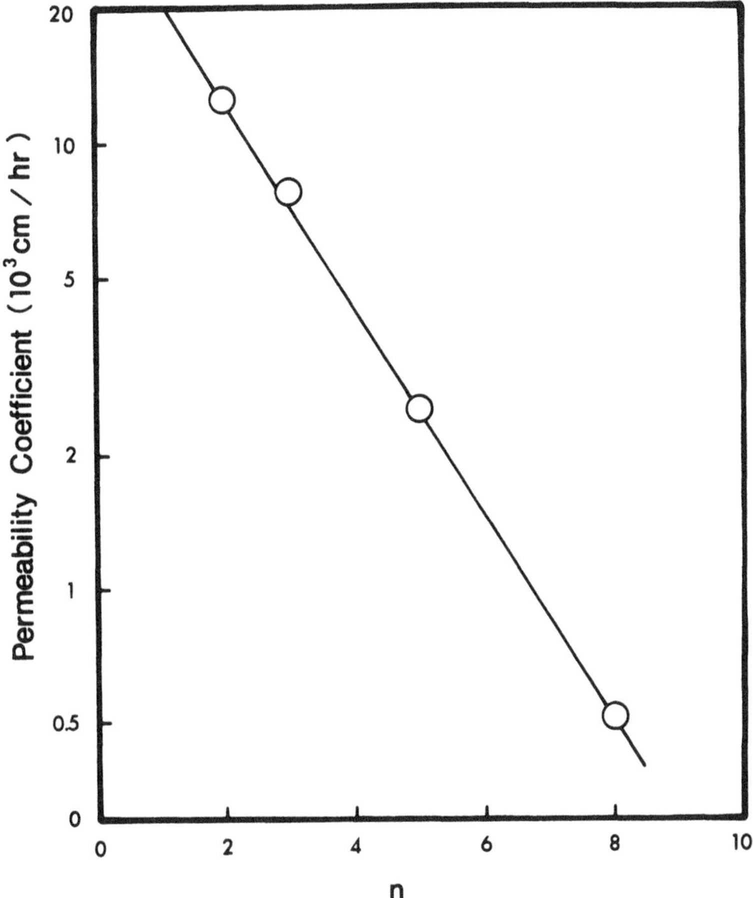

Figure 20 Effect of alkyl chain length n on the skin permeability coefficients of alkanols from a lipophilic vehicle of isopropyl palmitate. (Plotted from the data by Scheuplein and Blank, 1971.)

or

$$\Sigma R_s = \frac{h_{sc}}{D_{sc}K_{sc}} + \frac{h_e}{D_eK_e} + \frac{h_{pd}}{D_{pd}K_{pd}} \tag{24b}$$

where h, D, and K are the thickness, diffusivity, and partition coefficient, respectively; and the subscripts s, sc, e, and pd are skin, stratum corneum, viable epidermis, and papillary layer of the dermis.

Permeability studies conducted with isolated dermis, epidermis, and stratum corneum allow the quantification of their relative contributions to overall skin permeability through whole skin tissue. The diffusional resistance profiles in Figure 21 suggest that the diffusional resistance from the dermis is negligibly small compared to that from the stratum corneum and intact epidermis. Trypsin digestion of the epidermis to remove all the viable cells from stratum corneum was observed to yield no measurable effect on the diffusional resistance and permeability of the epidermis,

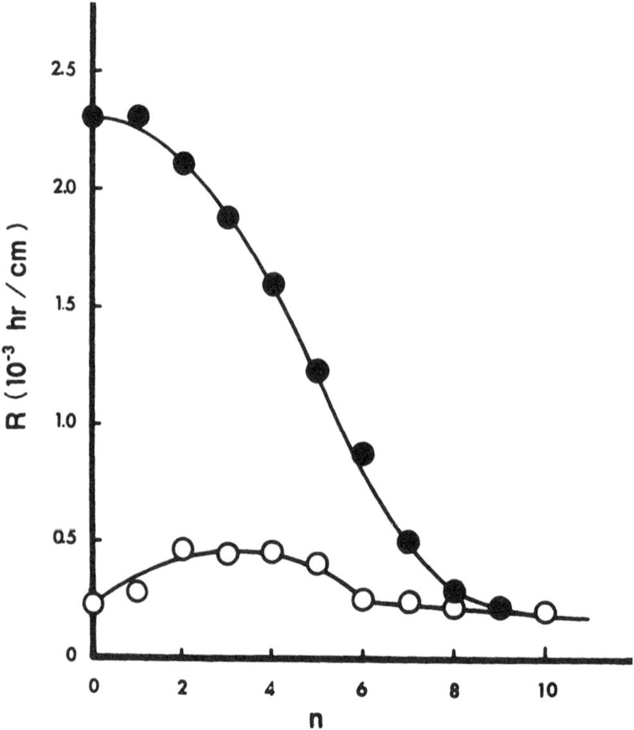

Figure 21 Effect of alkyl chain length n on the diffusional resistance R of isolated epidermis (●) and dermis (○) to alkanols. (Plotted from the data by Scheuplein and Blank, 1971.)

indicating that the contribution to the diffusional resistance of epidermis is solely that of the stratum corneum and that the diffusional resistance of the viable epidermis is very small (56). The R versus n curve for dermis is the contribution of the full thickness of the dermal tissue, not just the papillary layer; it thus tends to exaggerate the contribution of the dermal papillae to the total diffusional resistance by a factor of at least 10 (56). The diffusional resistance of the stratum corneum to a molecule like water is approximately 1000 times the resistance of the viable epidermis or the papillary region of the dermis. Equation (24) can thus be simplified to

$$\Sigma R_s \simeq R_{sc} = \frac{h_{sc}}{D_{sc}K_{sc}} \tag{25}$$

The reciprocal relationship between diffusional resistance and skin permeability coefficient thus becomes apparent. [Equations (25) and (2)].

The data in Figure 21 also suggest that as the alkyl chain length of the alkanols increases, the diffusional resistance through the epidermal tissue (or stratum corneum) decreases; this leads to an increase in the skin permeability coefficients (Figure 11) due to the enhancement in the membrane partition coefficient (Figure 13).

For larger molecules or molecules with several polar groups, such as hydrocortisone, the relative contribution of the stratum corneum in the diffusional resistance of the skin increases even further as result of their smaller diffusion coefficients in the stratum corneum and/or lower stratum corneum/vehicle partition coefficients K_{sc}. The group contribution of polar OH groups to the skin permeability coefficient has already been discussed (Table 6).

H. Intercellular Versus Transcellular Diffusion

Electron photomicroscopy shows that intercellular regions in the stratum corneum are filled with lipid-rich amorphous material (53,54,59), as illustrated in Figure 14. During cornification the lipid composition shifts from polar to neutral constituents (75,76). In the dry stratum corneum the intercellular volume may be as high as 5% and at least 1% in the fully hydrated stratum corneum. This intercellular volume is at least an order of magnitude larger than that (~0.1%) estimated for the transappendageal pathway; thus intercellular diffusion could be rather significant.

For intercellular diffusion to occur as the predominant mechanism for skin permeation, the intercellular boundaries (the cell membranes) must confine the movement of the majority of the penetrating molecules to the intercellular regions. The diffusivity of the cell membranes must be near 10^{-11} cm^2/sec or many times smaller to agree with the observed skin permeability of 10^{-7} cm/sec (12,64). Various diffusion data suggest that diffusion cannot be primarily intercellular, but transcellular (with a diffusivity of 5–10 $\times$ 10^{-10} cm^2/sec); that is, diffusion occurs through intercellular regions and across cell membranes without discrimination (12,49).

I. Intra- and Transappendageal Diffusion

Hair follicles and sweat ducts can act as diffusion shunts, i.e., relatively easy pathways for diffusion through the rate-limiting stratum corneum. The effect of this appendageal diffusion is usually at a minimum since their total fractional area is relatively small, e.g., approximately 0.1% on abdominal skin. However, for ions, polyfunctional polar compounds (e.g., cortisol), and extremely large molecules, these transappendageal diffusion pathways can be quite significant and must be considered (56,77).

The diffusion coefficients for low-molecular-weight molecules moving through the appendages (~5–20 $\times$ 10^{-8} cm^2/sec for hair follicles and ~1–20 $\times$ 10^{-6} cm^2/sec for sweat ducts) are much greater than those for molecules permeating through the stratum corneum (4.2 $\times$ 10^{-10} cm^2/sec; Table 4). However, the data accumulated to date have shown that the intraappendageal network provides a limited contribution to the steady-state skin permeation of the low-molecular-weight nonelectrolytes, and all other compounds that diffuse through this diffusion shunt rapidly, since the higher diffusivity of the appendages is offset by their extremely small effective fractional area (0.1%) of the skin surface (42,49,50,78).

Diffusion through hair follicles and sweat ducts can be very significant during the initial phase before steady-state skin permeation is established; that is, intraappendageal diffusion can be very significant in the initial state of nonlinear, time-dependent transient diffusion (right after medication). The intraappendageal pathway has a shorter lag time as well as a greater diffusivity. It is known that the rate of transient diffusion depends exponentially, rather than linearly, on diffusivity. There is substantial evidence in the literature demonstrating that diffusion through the sweat ducts and hair follicles could occur as early as within 5 min (79,80).

The precise contribution of intraappendageal diffusion to total skin permeation obviously cannot be assessed for a system as complicated and variable as the skin. It can be concluded, however, that intraappendageal diffusion may predominate, for some period of time, in the initial stage right after the topical application of a penetrant, and that initial concentration levels in the viable epidermis cells that surround the appendages can be vastly greater than corresponding values in the bulk of the stratum corneum (56).

J. Significance of Peripheral Blood Circulation

The blood supply of the human skin is from a microcirculating bed (81,82). The arterioles and venules form three important plexuses in the dermis: a horizontal network in the papillary dermis, from which the capillary loops of the dermal papillae arise, and individual plexuses around hair follicles and eccrine sweat glands (83).

Electron microscopy of the ultrastructure of the capillary loops in the dermal papillae suggested that the capillary loops can be structurally divided into two segments: intrapapillary and extrapapillary portions (84). The intrapapillary portion has the ultrastructural characteristics of an arterial capillary—a homogeneous-appearing basement membrane without bridged fenestrations. On the other hand, the extrapapillary portion has both the ultrastructural characteristics of an arterial capillary in its ascending limb as well as the venous characteristics—a multilayered basement membrane in the descending limb in the same portion.

The vascular surface available for the exchange of penetrant molecules between tissue and blood in the papillary layer of the dermis is far less than that in the muscle, between 1 and 2 cm^2/cm^2 of skin (33). At ambient temperatures the average flow rate of peripheral blood circulation through the skin is approximately 0.05 $ml/cm^3 \cdot min$ for the trunk, arms, and legs; on the other hand, the rate of blood flow is higher in the cheek, forehead, palms, soles, and digits (85,86). If the rate at which molecules accumulate by skin permeation is faster than the blood perfusion rate, then the rate of skin permeation could be controlled by the transfer of molecules into the capillaries, not by permeation through the stratum corneum.

A transfer resistance term R_t, which is the reciprocal of the product of peripheral blood flow rate θ and the thickness of the diffusional path for skin permeation h_s, should be added to the total diffusional resistance ΣR_s in Equation (24) to form

$$\Sigma R_s = R_{sc} + R_e + R_{pd} + R_t \tag{26a}$$

or

$$\Sigma R_s = \frac{h_{sc}}{D_{sc}K_{sc}} + \frac{h_e}{D_eK_e} + \frac{h_{pd}}{D_{pd}K_{pd}} + \frac{1}{\theta h_s} \tag{26b}$$

The transfer resistance R_t provided by blood perfusion is approximately 6×10^3 sec/cm. This value is small in comparison to the diffusional resistance of stratum corneum, except possibly for small, highly lipid-soluble molecules, such as *n*-octanol ($R_{sc} = 2 \times 10^4$ sec/cm), and for permanent gasses ($R_{sc} = \sim 3–9 \times 10^3$ sec/cm). The skin permeation of compounds that penetrate more rapidly than octanol or damage the skin barrier is also limited by blood perfusion. For the majority of innocuous substances, however, skin permeation is predominantly controlled by transport across the stratum corneum. In other words, R_{sc} is much larger than R_t, R_e, and R_{pd} (56).

VI. MECHANISMS OF RATE-CONTROLLED TRANSDERMAL DRUG DELIVERY

For a systemically active drug to reach a target tissue remote from the site of drug administration on the skin surface, it must possess physicochemical properties that facilitate the sorption of drug by the stratum corneum, the penetration of drug through the viable epidermis, and also the uptake of drug by microcirculation in the dermal papillary layer (Figure 7). The rate of permeation dQ/dt across various layers of skin tissues can be expressed mathematically (87) as

$$\frac{dQ}{dt} = P_s(C_d - C_r) \tag{27}$$

where C_d and C_r are, respectively, the concentrations of a skin penetrant in the donor phase, e.g., the concentration of drug on the stratum corneum surface as delivered from a TDD system, and in the receptor phase, e.g., systemic circulation; and P_s is the overall permeability coefficient of the skin tissues to the penetrant and is defined by

$$P_s = \frac{K_{s/d}D_{ss}}{h_s} \tag{28}$$

where $K_{s/d}$ is the partition coefficient for the interfacial partitioning of the penetrant molecule from a transdermal drug delivery system onto the stratum corneum; D_{ss} is the apparent diffusivity for the steady-state diffusion of the penetrant molecule through the skin tissues; and h_s is the overall thickness of the skin tissues for penetration. The permeability coefficient P_s for a skin penetrant can be considered a constant value if the $K_{s/d}$, D_{ss}, and h_s terms in Equation (28) are essentially constant under a given set of conditions.

Analysis of Equation (27) suggests that to achieve a constant rate of drug permeation one needs to maintain a condition in which the drug concentration on the surface of stratum corneum C_d is consistently and substantially greater than the drug concentration in the body C_r; i.e., $C_d \gg C_r$. Under such conditions Equation (27) can be reduced to

$$\frac{dQ}{dt} = P_sC_d \tag{29}$$

and the rate of skin permeation dQ/dt should be a constant, if the magnitude of C_d value remains fairly constant throughout the course of skin permeation. To maintain the C_d at a constant value, it is necessary to deliver the drug at a rate R_d that is either constant or always greater than the rate of skin absorption R_a; i.e., $R_d \gg R_a$ (Figure 22). By making R_d greater than R_a the drug concentration on the skin surface C_d is maintained at a level equal to or greater than the equilibrium (or saturation) solubility of the drug in the stratum corneum C_s^e; i.e., $C_d \geq C_s^e$. A maximum rate of skin permeation $(dQ/dt)_m$, as expressed by Equation (30), is thus achieved:

$$\left(\frac{dQ}{dt}\right)_m = P_sC_s^e \tag{30}$$

Apparently, the magnitude of $(dQ/dt)_m$ is determined by the permeability coef-

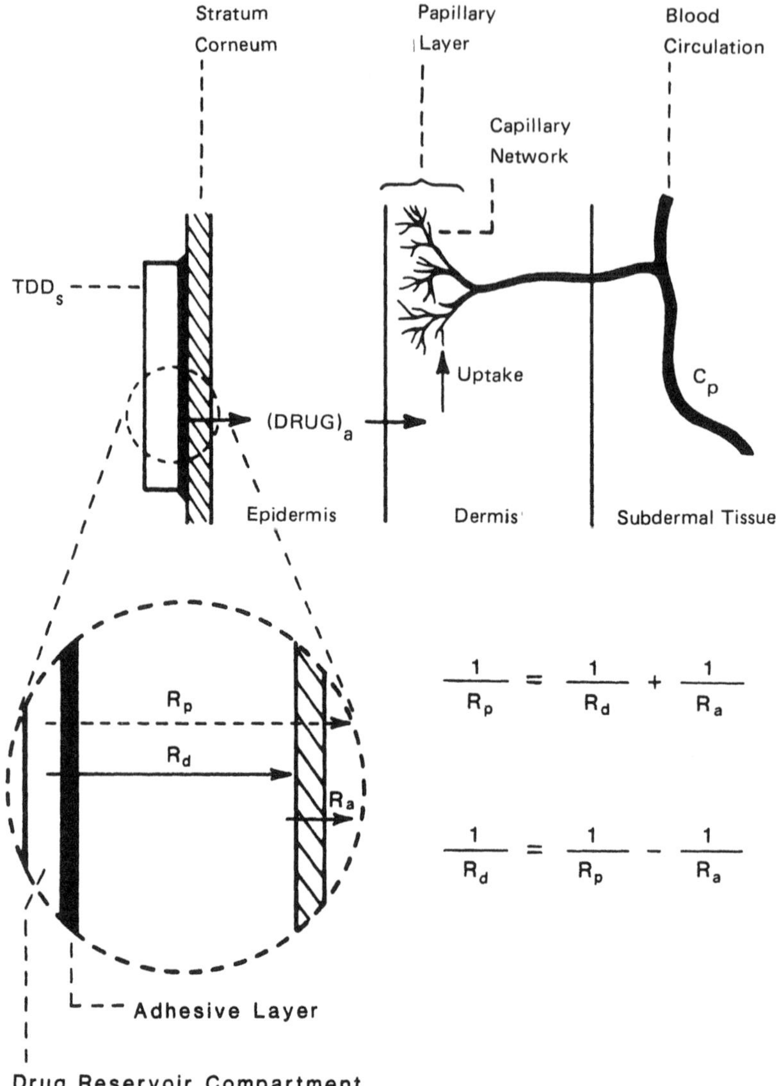

Figure 22 Relationship among the rate of skin permeation R_p of a drug, the rate of drug delivery R_d from a TDD system, and the rate of drug absorption R_a by the skin.

ficient P_s of the skin to the drug and the equilibrium solubility of the drug in the stratum corneum C_s^e. This concept of stratum corneum-limited skin permeation was investigated by depositing various doses of pure radiolabeled nitroglycerin, dissolved in a volatile organic solvent, onto rhesus monkey skin with a controlled surface area (88). Analysis of the urinary recovery data indicated that the rate of skin permeation dQ/dt increases with the increase in nitroglycerin dose C_d applied on a unit surface area of the skin (Figure 23). It appears that a maximum rate of skin permeation (1.585 mg/cm^2/day) is achieved when the applied dose of nitroglycerin reaches a level of 4.786 mg/cm^2 or greater.

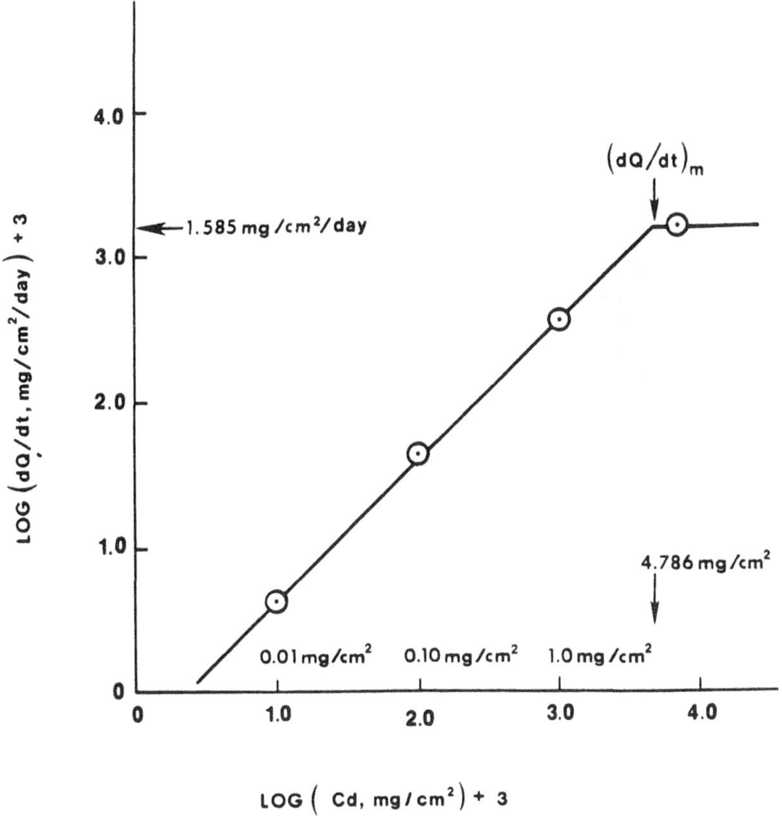

Figure 23 Linear relationship between the skin permeation rate of nitroglycerin dQ/dt determined from the daily urinary recovery data and the nitroglycerin dose applied to the rhesus monkey skin (C_d). (Plotted from the data by Sanvordeker et al., 1982.)

The kinetics of skin permeation can be more precisely analyzed by studying the permeation profiles of drug across a freshly excised skin specimen mounted on a diffusion cell, such as the Franz diffusion cell (Figure 24). A typical skin permeation profile is shown in Figure 25 for nitroglycerin. The results indicated that nitroglycerin penetrates through the freshly excised abdominal skin of hairless mouse at a zero-order rate of 19.85 ± 1.71 $\mu g/cm^2/hr$, as expected from Equation (30), when the pure nitroglycerin in oily liquid form is directly deposited on the surface of stratum corneum (in this case the skin permeation of drug is under *no* influence from either an organic solvent or a rate-controlled drug delivery system) (89). Using a hydrodynamically well-calibrated horizontal skin permeation cell (90) the same observations were also made in a series of long-term skin permeation kinetic studies for estradiol (91), which provides a critical analysis of the relationships among skin permeation rate, permeability coefficient, partition coefficient, diffusivity, and solubility. The effects of skin uptake, binding, and metabolism kinetics of estradiol on its skin permeation profiles were also evaluated and illustrated (92).

To gain a fundamental understanding of the skin permeation kinetics of drugs and to assist the formulation development of transdermal drug delivery systems, in

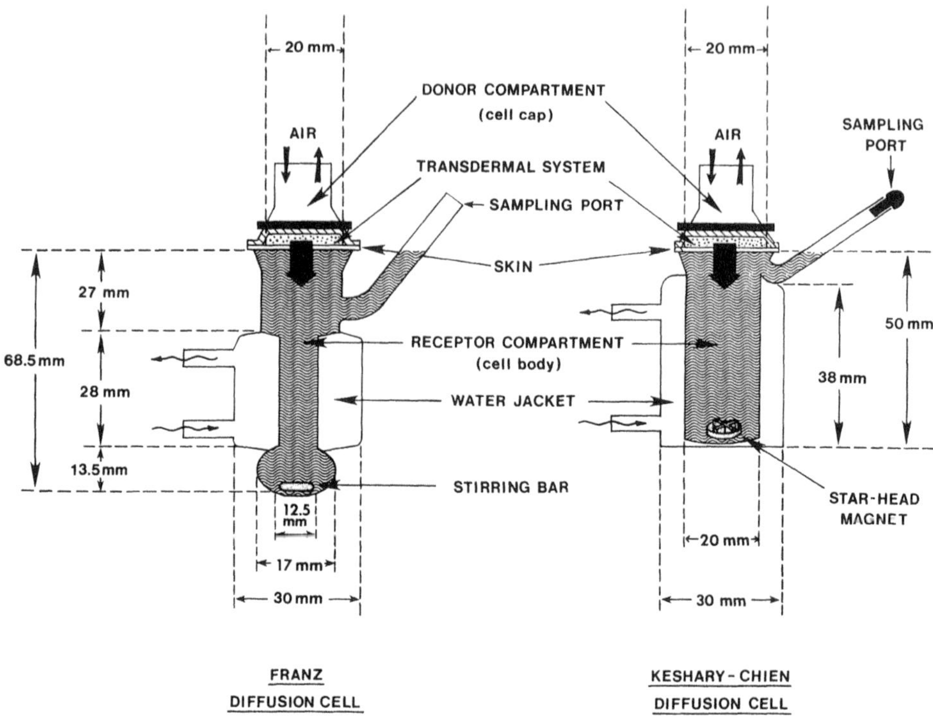

Figure 24 The vertical-type in vitro skin permeation system; the Franz diffusion cell is shown along with the Keshary-Chien skin permeation cell. (Reproduced from Keshary and Chien, 1984.)

vitro skin permeation studies using a freshly excised skin sample mounted in a hydrodynamically well-calibrated skin permeation cell are considered a cost- and time-saving approach and a *must* before initiation of costly and time-consuming in vivo evaluations in human volunteers.

VII. TECHNOLOGIES FOR DEVELOPING TRANSDERMAL DRUG DELIVERY SYSTEMS

Several technologies have been successfully developed to provide rate control over the release and skin permeation of drugs (93). These technologies can be classified into four basic approaches and are discussed in this section.

A. Polymer Membrane Permeation-Controlled TDD Systems

In this system the drug reservoir is sandwiched between a drug-impermeable backing laminate and a rate-controlling polymeric membrane (Figure 26). The drug molecules are permitted to release only through the rate-controlling polymeric membrane. In the drug reservoir compartment the drug solids are dispersed homogeneously in a solid polymer matrix (e.g., polyisobutylene), suspended in a unleachable, viscous liquid medium (e.g., silicone fluid) to form a pastelike suspension, or dissolved in a releasable solvent (e.g., alkyl alcohol) to form a clear drug solution. The rate-

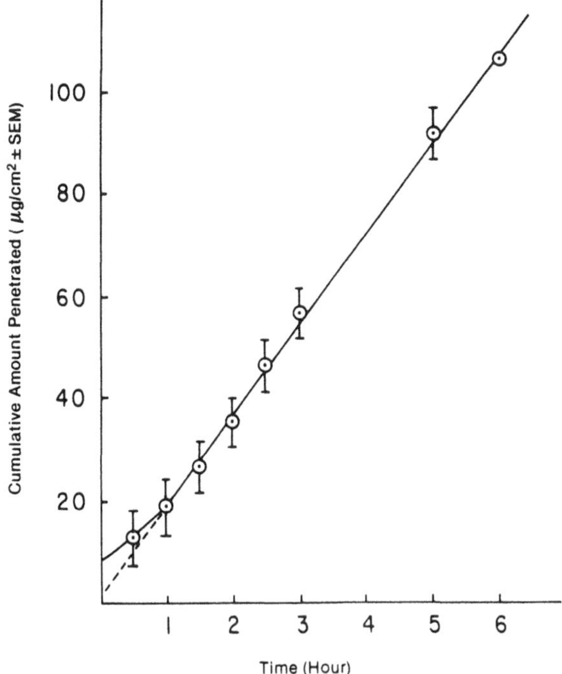

Figure 25 Permeation profile of pure nitroglycerin across the abdominal skin of hairless mouse mounted on a Franz diffusion cell at 37°C. A constant skin permeation profile was obtained with a permeation rate of 19.85 ± 1.71 $\mu g/cm^2/hr$. (From Keshary et al., 1984.)

controlling membrane can be either a microporous or a nonporous polymeric membrane, e.g., ethylene-vinyl acetate copolymer, with a specific drug permeability. On the external surface of the polymeric membrane a thin layer of drug-compatible, hypoallergenic pressure-sensitive adhesive polymer, e.g., silicone adhesive, may be applied to provide intimate contact of the TDD system with the skin surface. The rate of drug release from this TDD system can be tailored by varying the composition of the drug reservoir formulation and the permeability coefficient and/or thickness of the rate-controlling membrane. Several TDD systems have been successfully developed from this technology and approved by the FDA for marketing, such as the Transderm-Nitro system for once-a-day medication of angina pectoris (94,95), the Transderm-Scop system for 3-day protection of motion sickness (4), the Catapres-TTS system for weekly therapy of hypertension (96–98), the Estraderm system for twice-a-week treatment of postmenopausal syndromes, and the Duragesic system for 72-hr management of chronic pain (7,99,100).

The intrinsic rate of drug release from this type of TDD system is defined by

$$\frac{dQ}{dt} = \frac{K_{m/r}K_{a/m}D_aD_m}{K_{m/r}D_mh_a + K_{a/m}D_ah_m} C_R \tag{31}$$

where C_R is the drug concentration in the reservoir compartment; $K_{m/r}$ and $K_{a/m}$ are the partition coefficients for the interfacial partitioning of drug from the reservoir to

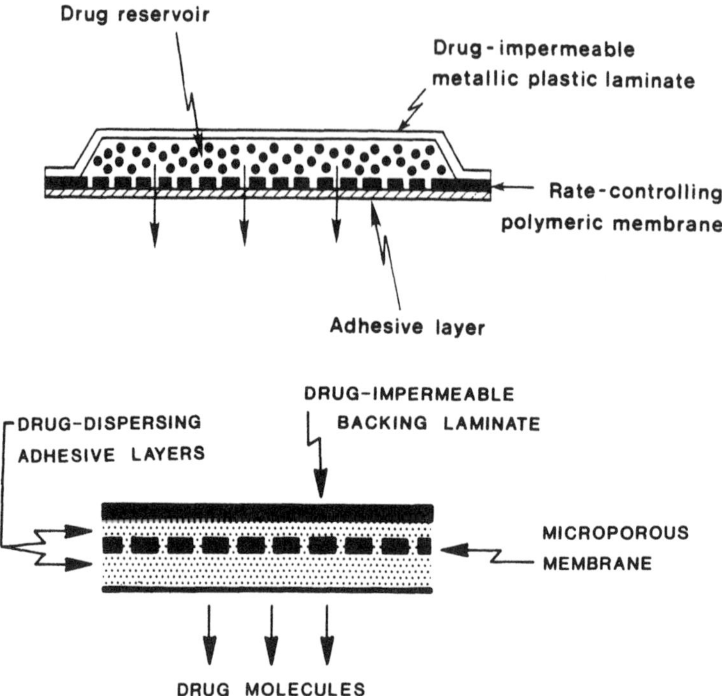

Figure 26 Cross-sectional view of a polymer membrane permeation-controlled TDD system showing various major structural components, with a liquid drug reservoir (top) or a solid drug reservoir (bottom). (Reproduced from Y. W. Chien, 1987.)

the membrane and from the membrane to the adhesive, respectively; D_m and D_a are the diffusion coefficients in the rate-controlling membrane and in the adhesive layer, respectively; and h_m and h_a are the thickness of the rate-controlling membrane and the adhesive layer, respectively. For a microporous membrane the porosity and tortuosity of the membrane should also be taken into account in calculation of the D_m and h_m values.

The polymer membrane permeation-controlled transdermal drug delivery technology has also been applied to the development of TDD systems for the rate-controlled percutaneous absorption of prostaglandin derivative (101).

B. Polymer Matrix Diffusion-Controlled TDD Systems

In this approach the drug reservoir is formed by homogeneously dispersing the drug solids in a hydrophilic or lipophilic polymer matrix, and the medicated polymer formed is then molded into medicated disks with a defined surface area and controlled thickness. This drug reservoir-containing polymer disk is then mounted onto an occlusive baseplate in a compartment fabricated from a drug-impermeable plastic backing (Figure 27). Instead of coating the adhesive polymer directly on the surface of the medicated disk, as shown earlier in the first type of TDD systems (Figure 26), in this system the adhesive polymer is applied along the circumference of the patch to form a strip of adhesive rim surrounding the medicated disk. The rate of drug release from this polymer matrix drug dispersion-type TDD system is defined as

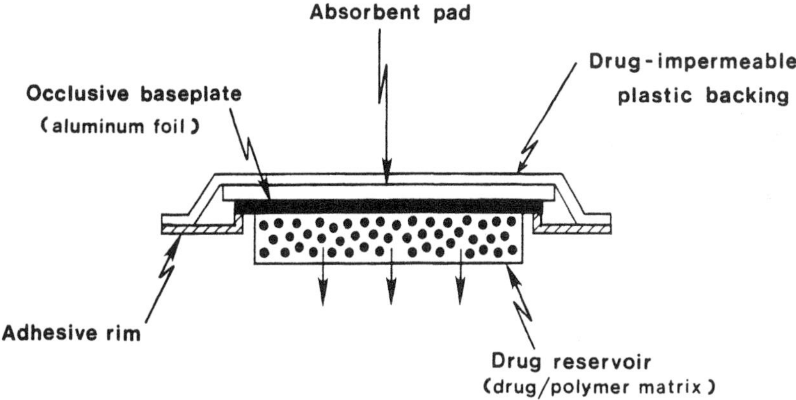

Figure 27 Cross-sectional view of a polymer matrix diffusion-controlled TDD systems showing various major structural components. (Reproduced from Y. W. Chien, 1985.)

$$\frac{dQ}{dt} = \left(\frac{L_d C_p D_p}{2t}\right)^{1/2} \tag{32}$$

where L_d is the drug loading dose initially dispersed in the polymer matrix; and C_p and D_p are the solubility and diffusivity of the drug in the polymer matrix, respectively. Because only the drug species dissolved in the polymer can release, C_p is practically equal to C_R.

At steady state a Q versus $t^{1/2}$ drug release profile is obtained (102) as defined by

$$\frac{Q}{t^{1/2}} = [(2L_d - C_p)C_p D_p]^{1/2} \tag{33}$$

This type of TDD system is exemplified by the development and marketing of the Nitro-Dur system (103) and the NTS system (104), which have been approved by the FDA for the once-a-day medication of angina pectoris.

Alternatively, the polymer matrix drug dispersion-type TDD system can be fabricated by directly dispersing the drug in a pressure-sensitive adhesive polymer, e.g., polyacrylate, and then coating the drug-dispersed adhesive polymer by solvent casting or hot melt onto a flat sheet of a drug-impermeable backing laminate to form a single layer of drug reservoir (Figure 28). This yields a thinner and/or smaller TDD patch. The release profiles of drug from this type of TDD system also follow a Q

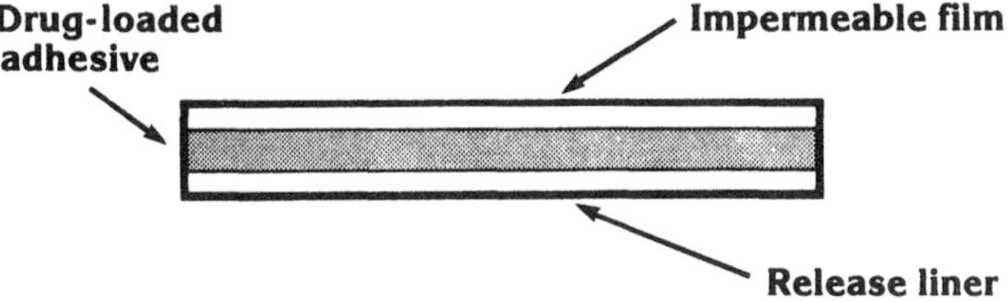

Figure 28 Cross-sectional view of an adhesive polymer drug dispersion-type TDD system showing various major structural components.

versus $t^{1/2}$ pattern, as expected from the matrix diffusion process [Equation (32)]. This type of TDD system is best illustrated by the development and marketing of the nitroglycerin-releasing TDD system, the Minitran system and the Nitro-Dur II system, and the isosorbide dinitrate-releasing TDD system (Frandol tape) for the once-a-day medication of angina pectoris.

C. Drug Reservoir Gradient-Controlled TDD Systems

To overcome the non-zero–order (Q versus $t^{1/2}$) drug release profiles [Equation (33)], polymer matrix drug dispersion-type TDD systems can be modified to have the drug loading level varied in an incremental manner, forming a gradient of drug reservoir along the diffusional path across the multilaminate adhesive layers (Figure 29). The rate of drug release from this type of drug reservoir gradient-controlled TDD system can be expressed by

$$\frac{dQ}{dt} = \frac{K_{a/r} D_a}{h_a(t)} L_d(h_a) \tag{34}$$

In this system the thickness of diffusional path through which drug molecules diffuse increases with time, i.e., $h_a(t)$ [Equation (34)]. To compensate for this time-dependent increase in diffusional path as a result of drug depletion due to release, the drug loading level in the multilaminate adhesive layers is also designed to increase proportionally, i.e., $L_d(h_a)$. This, in theory, should yield, a more constant drug release profile. This type of TDD system is best illustrated by the development of a nitroglycerin-releasing TDD system, the Deponit system (105).

D. Microreservoir Dissolution-Controlled TDD Systems

This type of drug delivery system can be considered a hybrid of the reservoir- and matrix dispersion-type drug delivery systems. In this approach the drug reservoir is formed by first suspending the drug solids in an aqueous solution of a water-miscible drug solubilizer, e.g., polyethylene glycol, and then homogeneously dispersing the drug suspension, with controlled aqueous solubility, in a lipophilic polymer, by high-shear mechanical force, to form thousands of unleachable microscopic drug reservoirs (Figure 30). This thermodynamically unstable dispersion is quickly stabilized by immediately cross-linking the polymer chains in situ, which produces a medicated polymer disk with a constant surface area and a fixed thickness. A TDD system is

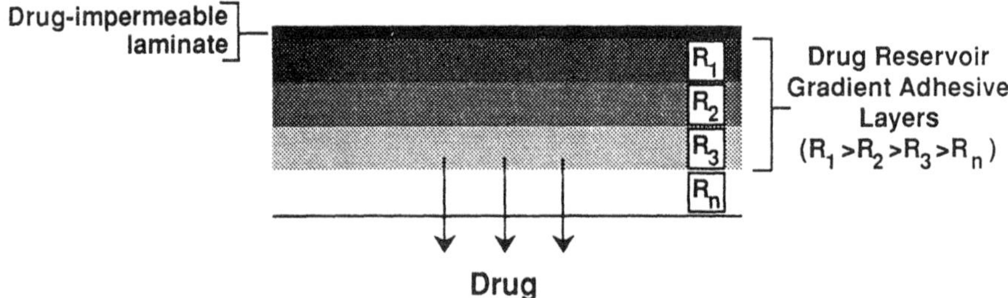

Figure 29 Cross-sectional view of a drug reservoir gradient-controlled TDD system showing various major structural components.

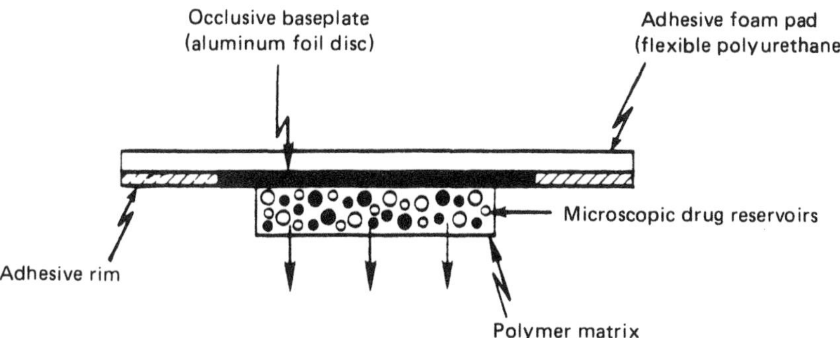

Figure 30 Cross-sectional view of a microreservoir dissolution-controlled TDD system showing various major structural components. (Reproduced from Y. W. Chien, 1985.)

then produced by mounting the medicated disk at the center of an adhesive pad. This technology has been successfully utilized in the development and marketing of the Nitrodisc system, which has been approved by the FDA for the once-a-day treatment of angina pectoris (88,106–111).

The rate of drug release from a microreservoir drug delivery system is defined (102,110) by

$$\frac{dQ}{dt} = \frac{D_p D_s A K_p}{D_p h_d + D_s h_p A K_p} \left[B S_p - \frac{D_1 S_1 (1 - B)}{h_1} \left(\frac{1}{K_1} + \frac{1}{K_m} \right) \right] \tag{35}$$

where $A = a/b$. a is the ratio of the drug concentration in the bulk of elution solution over the drug solubility in the same medium, and b is the ratio of the drug concentration at the outer edge of the polymer coating membrane over the drug solubility in the same polymer composition; B is the ratio of the drug concentration at the inner edge of the interfacial barrier over the drug solubility in the polymer matrix. K_l, K_m, and K_p are the partition coefficients for the interfacial partitioning of drug from the liquid compartment to the polymer matrix, from the polymer matrix to the polymer coating membrane, and from the polymer coating membrane to the elution solution (or skin), respectively; D_l, D_p, and D_s are the drug diffusivities in the liquid compartment, polymer coating membrane, and elution solution (or skin), respectively; S_l and S_p are the solubilities of the drug in the liquid compartment and in the polymer matrix, respectively; and h_l, h_p, and h_d are the thickness of the liquid layer surrounding the drug particles, the polymer coating membrane around the polymer matrix, and the hydrodynamic diffusion layer surrounding the polymer coating membrane, respectively.

The release of drugs from a microreservoir-type drug delivery system can follow either a partition-control or matrix diffusion-control process depending upon the relative magnitude of S_l and S_p (110). Thus a Q versus t or Q versus $t^{1/2}$ release profile results (112,113).

Further research of microreservoir dissolution-controlled drug delivery techniques has resulted in the development of a transdermal contraceptive device capable of delivering, simultaneously, a combination of a potent progestin and a natural estrogen at different daily dosage rates for weekly fertility regulation in females (114–116).

Development of other types of drug delivery systems is also underway for possible application in the transdermal controlled delivery of drugs, exemplified by the disposition of drugs in a poroplastic membrane (117) and formation of a hydrophilic polymeric reservoir (118). Both may be viewed as a drug solution-saturated porous polymer matrix.

VIII. EVALUATIONS OF TRANSDERMAL DRUG DELIVERY KINETICS

The release and skin permeation kinetics of drug from these technologically different TDD systems can be evaluated using a two-compartment diffusion cell assembly under identical conditions. This is carried out by individually mounting a skin specimen excised from either a human cadaver or a live animal (119) on a vertical diffusion cell, such as the Franz diffusion cell and its modifications (Figure 24), or a horizontal diffusion cell, such as the Valia-Chien skin permeation cell (Figure 31). Each unit of the TDD system is then applied with its drug-releasing surface in intimate contact with the stratum corneum surface of the skin (113,114). The skin permeation profile of the drug is followed by sampling the receptor solution at predetermined intervals until the steady-state flux is established and assaying drug concentrations in the samples by a sensitive analytic method, such as high-performance liquid chromatography (HPLC). The release profiles of drug from these TDD systems can also be investigated in the same diffusion cell assembly without a skin specimen.

In the actual determination of drug release and skin permeation kinetics studies, the rate profiles obtained may be well below the intrinsic rates calculated from Equations (31), (33), (34), and (35) owing to the effect of mass transfer across the hydrodynamic diffusion layer on the surface of the drug delivery system or dermis (120). The magnitude of the reduction is related to the thickness of the hydrodynamic diffusion layer and the physicochemical properties of the drugs (121,122). It is rather important to take these effects into consideration for accurate determination of drug release and skin permeation rate profiles.

A. In Vitro Drug Release Kinetics

Using a Franz diffusion cell assembly, the mechanisms and the rates of drug release from these technologically different TDD systems were evaluated and compared (123). The results indicated that nitroglycerin is released at a constant rate profile (Q versus t) from the TDD systems like the Transderm-Nitro system [a polymer membrane permeation-controlled TDD system; Equation (31)] and the Deponit system [a drug reservoir gradient-controlled TDD system; Equation (34)]; (Figure 32). The release rate of nitroglycerin from the Transderm-Nitro system (0.843 ± 0.035 mg/cm^2/day) is almost three times greater than that from the Deponit system (0.324 ± 0.011 mg/cm^2/day). This suggests that diffusion through the rate-controlling adhesive polymer matrix in the Deponit system plays a greater rate-limiting role over the release of nitroglycerin than does permeation through the rate-controlling polymer membrane in the Transderm-Nitro system.

On the other hand, the release profiles of nitroglycerin from the Nitrodisc and Nitro-Dur systems are not constant but are observed to follow a linear Q versus $t^{1/2}$

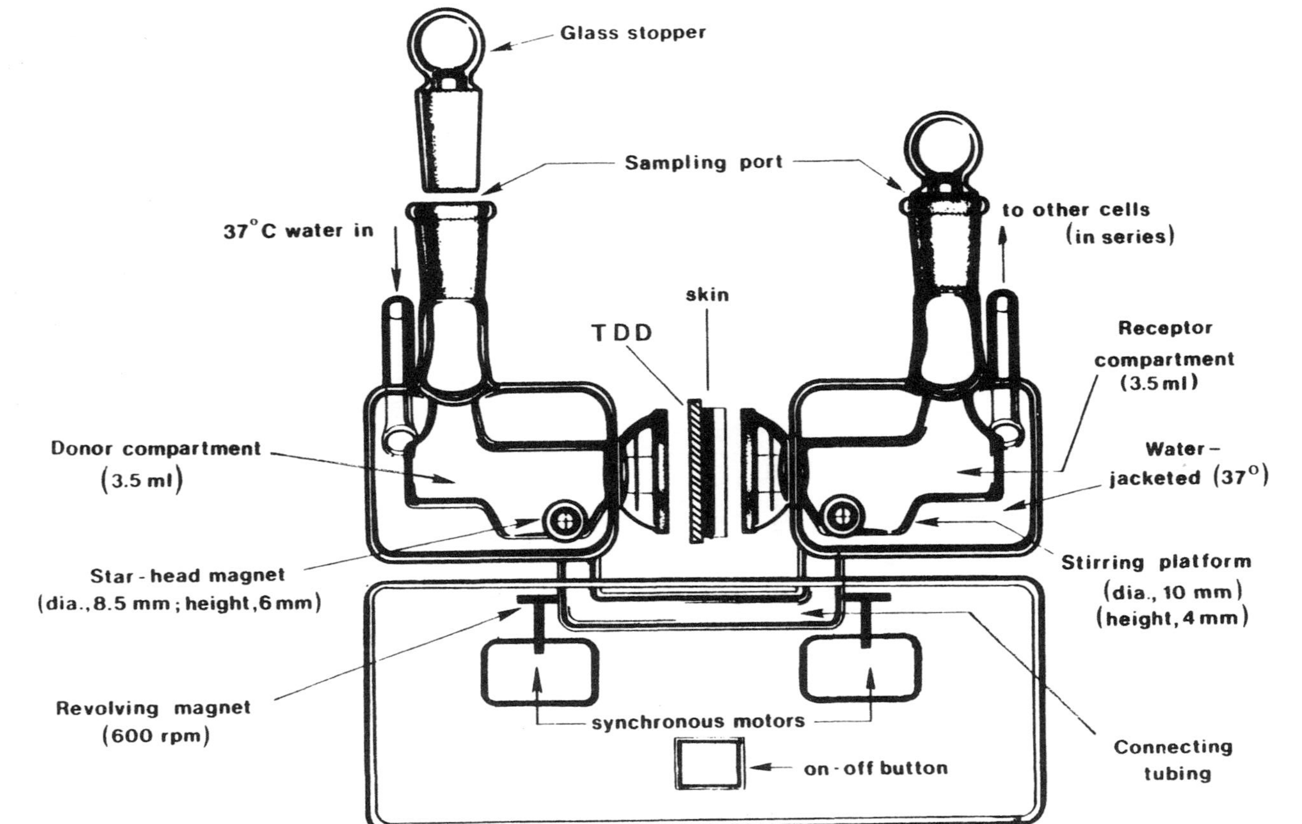

Figure 31 The hydrodynamically well-calibrated horizontal skin permeation system used for studying the controlled release and skin permeation of drugs from transdermal drug delivery (TDD) systems. (Reproduced from Chien et al., 1985.)

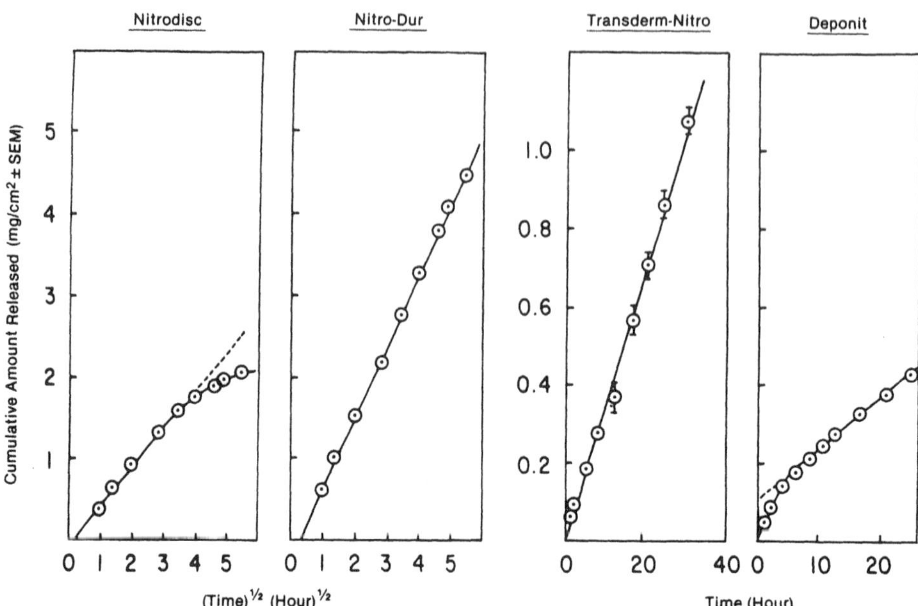

Figure 32 Comparative release profiles of nitroglycerin from various TDD systems into saline solution under sink conditions at 37°C. The release flux of nitroglycerin was: Nitrodisc system (2.443 ± 0.136 $mg/cm^2/day^{1/2}$), Nitro-Dur system (4.124 ± 0.047 $mg/cm^2/day^{1/2}$), Transderm-Nitro system (0.843 ± 0.035 $mg/cm^2/day$), and Deponit system (0.324 ± 0.011 $mg/cm^2/day$). (From Keshary et al., 1984.)

pattern as expected from matrix diffusion-controlled drug release kinetics [Equation (33)]. The release flux of nitroglycerin from the Nitro-Dur system (a polymer matrix diffusion-controlled TDD system) is almost twice greater than that from the Nitrodisc system (a microreservoir dissolution-controlled TDD system; 4.124 ± 0.047 versus 2.443 ± 0.136 $mg/cm^2/day^{1/2}$).

Apparently the mechanisms and/or rates of nitroglycerin release from these four TDD systems are quite different from one another, as expected from Equations (31), (33), (34), and (35).

B. In Vitro Skin Permeation Kinetics

1. Animal Skin Model

The skin permeation studies of these TDD systems suggested that all four systems give a constant rate of skin permeation (Figure 33), as expected from Equation (29). The highest rate of skin permeation was observed with the Nitrodisc system (0.426 ± 0.024 $mg/cm^2/day$), which is, however, statistically no different from the rate of skin permeation for pure nitroglycerin (0.476 ± 0.041 $mg/cm^2/day$; Figure 25). For the Nitro-Dur system practically the same rate of skin permeation (0.408 ± 0.024 $mg/cm^2/day$) was obtained initially and 12 hr later; however, the rate slowed to 0.248 ± 0.018 $mg/cm^2/day$. On the other hand, the rate of skin permeation of nitroglycerin from the Transderm-Nitro system (0.388 ± 0.017 $mg/cm^2/day$) was

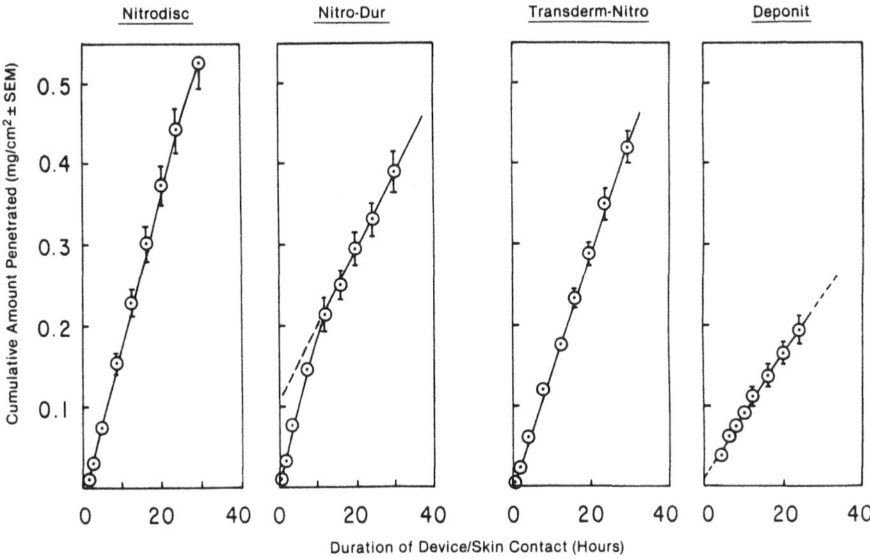

Figure 33 Comparative permeation profiles of nitroglycerin from various TDD systems through hairless mouse abdominal skin at 37°C. The rate of skin permeation was: Nitrodisc system (0.426 ± 0.024 $mg/cm^2/day$), Nitro-Dur system [0.408 ± 0.024 $mg/cm^2/day$ (<12 hr); 0.248 ± 0.018 $mg/cm^2/day$ (>12 hr)], Transderm-Nitro system (0.338 ± 0.017 $mg/cm^2/day$), and Deponit system (0.175 ± 0.016 $mg/cm^2/day$). (From Keshary et al., 1984.)

found to be one-third lower than the rate achieved by pure nitroglycerin or one-quarter slower than that of the Nitrodisc system. The lowest rate of skin permeation was obtained by the Deponit system (0.175 ± 0.016 $mg/cm^2/day$), which achieved only one-third the skin permeation rate for pure nitroglycerin.

A comparison made between the rate of skin permeation (Figure 33) and the rate of release (Figure 32) suggests that under sink conditions all TDD systems deliver nitroglycerin at a rate greater than its rate of permeation across the skin ($R_d > R_a$; Figure 22). For example, nitroglycerin was delivered by the Transderm-Nitro system, which is a polymer membrane permeation-controlled drug delivery system, at a rate (0.843 $mg/cm^2/day$) 2.5 times greater than its rate of permeation across the skin (0.338 $mg/cm^2/day$). Likewise, the rate of delivery by the Deponit system, which is a multilaminate adhesive dispersion-type drug delivery system with the slowest rate of nitroglycerin delivery (0.324 $mg/cm^2/day$), was found to be almost twofold faster than the rate of skin permeation (0.175 $mg/cm^2/day$). The same observations were also true for the Nitrodisc and Nitro-Dur systems. This phenomenon is an indication that the stratum corneum plays a rate-limiting role in the transdermal delivery of drugs, including the relatively skin-permeable nitroglycerin, as a result of its extremely low permeability coefficient [Equation (27)]. The difference in skin permeation rates among the various TDD systems could be attributed to the variation in formulation design that affects the magnitude of the partition coefficient [$k_{s/d}$ in Equation (28)].

2. Human Cadaver Skin Model

The permeation of nitroglycerin across the skin of human cadaver was also investigated for various nitroglycerin-releasing TDD systems using the Valia-Chien skin permeation cell assembly (Figure 31). The results (124) indicated that the skin permeation of nitroglycerin through human cadaver skin following the delivery from all the TDD systems evaluated also follows the same zero-order kinetic profile (Figure 34) as observed with hairless mouse abdominal skin (Figure 33).

It has been found that differences in the type and thickness of a skin specimen and variation in the hydrodynamics of in vitro skin permeation cells could affect

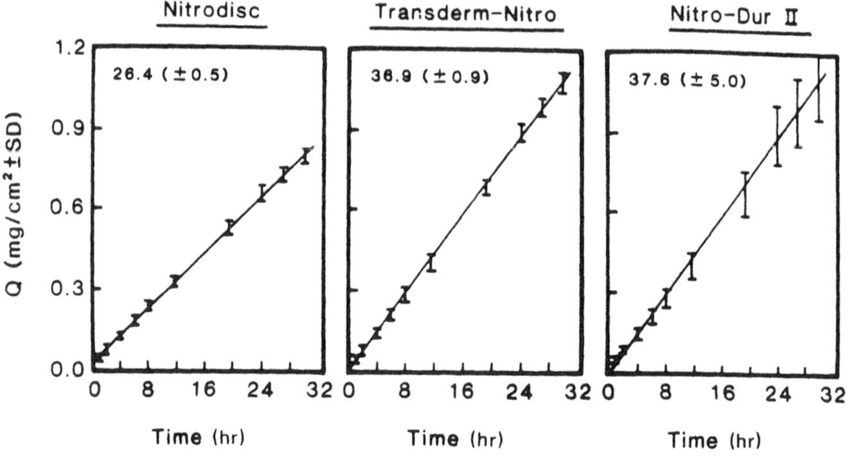

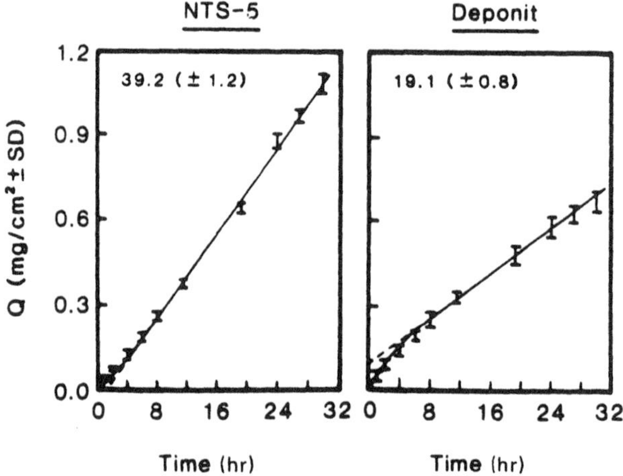

Figure 34 Comparative permeation profiles of nitroglycerin from various TDD systems across human cadaver dermatomed skin. The rate of skin permeation was: Nitrodisc system (26.4 ± 0.5 $\mu g/cm^2/hr$), Transderm-Nitro system (36.9 ± 0.9 $\mu g/cm^2/hr$), Nitro-Dur II system (37.6 ± 5.0 $\mu g/cm^2/hr$), NTS-5 system (39.2 ± 1.2 $\mu g/cm^2/hr$), and Deponit system (19.1 ± 0.8 $\mu g/cm^2/hr$).

Table 9 Interspecies Correlation in the In Vitro Transdermal Controlled Delivery of Drugs from TDD Systems

Drugs	TDD system	Permeation rate (μg/cm^2/hr)	
		Human cadaver[a]	Hairless mouse[b]
Nitroglycerin	Transderm-Nitro	19.21	14.08
	Nitro-Dur	20.29	17.01
	Nitrodisc	—	17.75
	Deponit	—	7.29
Estradiol	Estraderm	0.27	0.40
Clonidine	Catapres-TTS	2.05	3.62

[a]Nitroglycerin data were from Chien (1987); estradiol and clonidine data were determined in Valia-Chien skin permeation cells (Figure 31) at 37°C (h_{aq} = 0.0054 cm).
[b]Nitroglycerin data were determined in Franz diffusion cells (Figure 24) at 37°C (h_{aq} = 0.0338 cm); estradiol and clonidine data were determined in Valia-Chien skin permeation cells (Figure 31) at 37°C (h_{aq} = 0.0054 cm).

interspecies correlation in skin permeation rates (Table 9). For a better correlation, a skin model with a controlled source, such as hairless mouse, and a skin permeation cell with well-calibrated hydrodynamics, like Valia-Chien skin permeation cells, should be used in skin permeation kinetics studies (126). By doing so, the transdermal permeation rates of nitroglycerin across human cadaver skin are correlated with those across hairless mouse skin (Table 10). Similarly, the skin permeation rates of progesterone and its hydroxy derivatives across human cadaver skin are also correlated with that across hairless mouse skin (Figure 35).

C. In Vivo Transdermal Bioavailability in Humans

The transdermal bioavailability of nitroglycerin resulting from the 24–32 hr topical applications of various TDD systems in human volunteers is shown in Figures 36 through 39. The results suggest that a prolonged, steady-state plasma level of nitroglycerin was achieved and maintained throughout the duration of TDD system applications for at least 24 hr as a result of continuous transdermal infusion of drug at a controlled rate from the TDD systems.

Table 10 Interspecies Correlation in Transdermal Permeation Rates of Nitroglycerin

TDD systems	Permeation rate[a] (μg/cm^2/hr ± SD)	
	Human cadaver[b]	Hairless mouse
Transderm-Nitro	36.9 (± 0.9)	34.7 (± 0.8)
Nitrodisc	26.4 (± 0.5)	44.3 (± 0.8)
Nitro-Dur II	37.6 (± 5.0)	46.1 (± 6.1)
NTS-5	39.2 (± 1.2)	55.9 (± 1.7)
Deponit	19.1 (± 0.8)	16.7 (± 0.7)

[a]Measured in Valia-Chien skin permeation cells (Figure 31) at 37°C.
[b]Calculated from the data in Figure 34.
Source: From Chien, unpublished data (1989).

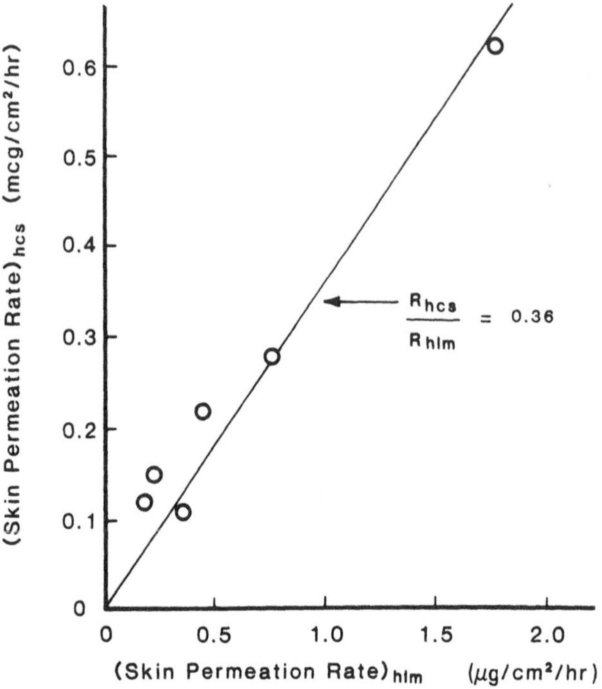

Figure 35 Correlation of the permeation rates of progesterone and its hydroxyl derivatives across the skin of human cadaver (*hcs*) and of hairless mouse (*hlm*) of approximately the same thickness.

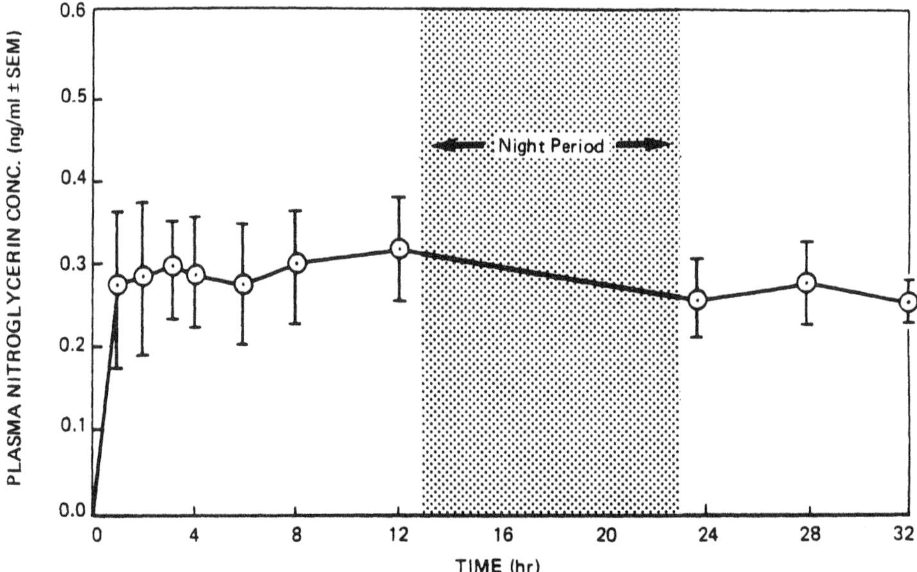

Figure 36 Plasma profiles of nitroglycerin in 12 healthy male volunteers; each received 1 unit of Nitrodisc system (16 cm^2) on the chest for 32 hr. A mean steady-state plasma level $(C_p)_{ss}$ of 280.6 ± 18.7 pg/ml was obtained. (Plotted from the data by Karim, 1983.)

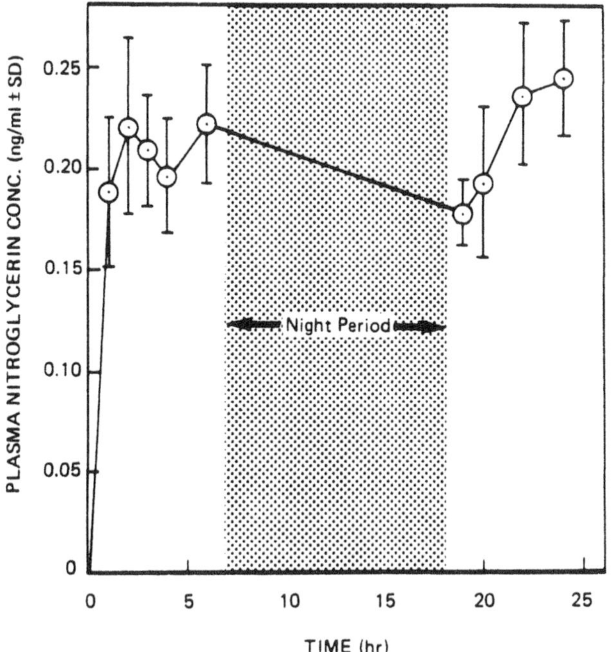

Figure 37 Plasma profiles of nitroglycerin in 14 healthy human subjects; each received 1 unit of Transderm-Nitro system (20 cm^2) for 24 hr. A $(C_p)_{ss}$ value of 209.8 ± 22.8 pg/ml was attained. (Plotted from the data by Gerardin et al., 1981.)

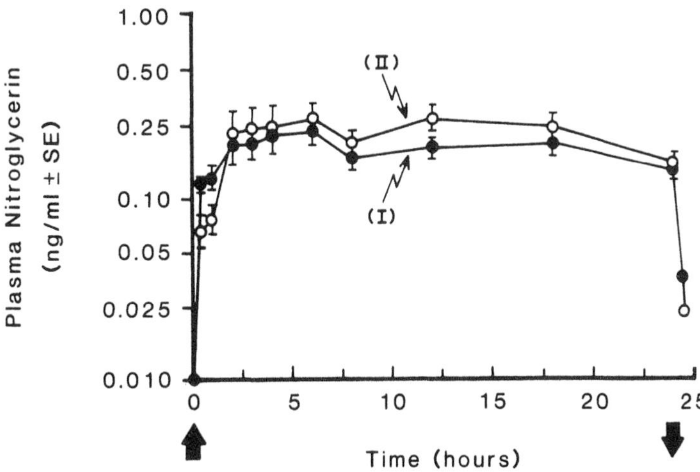

Figure 38 Comparative plasma profiles of nitroglycerin in 24 healthy male volunteers; each received randomly 1 unit of Nitro-Dur system I or II (20 cm^2) over the chest for 24 hr. A $(C_p)_{ss}$ value of 182 ± 114 (I) and 224 ± 172 (II) pg/ml was achieved. (Plotted from the data by Noonan et al., 1986.)

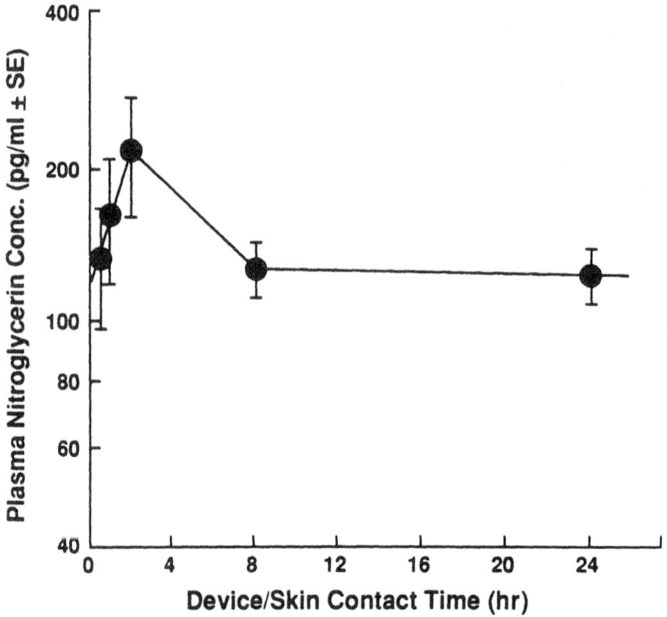

Figure 39 Plasma profiles of nitroglycerin in six healthy male volunteers; each received 1 unit of the Deponit system (16 cm^2) over the chest for 24 hr. A $(C_p)_{ss}$ value of 125 ± 50 pg/ml was obtained. (Plotted from the data by Wolff et al., 1985.)

A comparative systemic bioavailability study was initiated, and the results demonstrated that there is no statistically significant difference among the plasma profiles of nitroglycerin achieved by transdermal delivery from the Transderm-Nitro, Nitrodisc, and Nitro-Dur systems (96).

The plasma level was found to be linearly proportional to the area of the drug-releasing surface of TDD systems in intimate contact with the skin (94). The plasma drug level can thus be easily tailored to reach a target therapeutic concentration by simply controlling the drug-releasing surface area of TDD device applied to the skin.

Further investigations demonstrated that the transdermal bioavailability of nitroglycerin and estradiol delivered by TDD systems is independent of the site of application (94,99). The results of repeated daily applications also showed an excellent day-to-day reproducibility, whereas no drug accumulation was detected (94).

Similar to sublingual administration, transdermal drug delivery is also capable of bypassing hepatogastrointestinal first-pass elimination. Comparative clinical evaluations were initiated to study the antianginal activity of a transdermal nitroglycerin patch using sublingual nitroglycerin administration as a positive control and placebo patch as the negative control (127,128). A typical set of results shown in Figure 40 demonstrates that by transdermal delivery nitroglycerin is capable of improving the exercise performance of anginal patients for as long as 10 hr with exercise tolerability equivalent to the maximal performance achieved by sublingual nitroglycerin (measured at 5 min after administration). However, the duration of

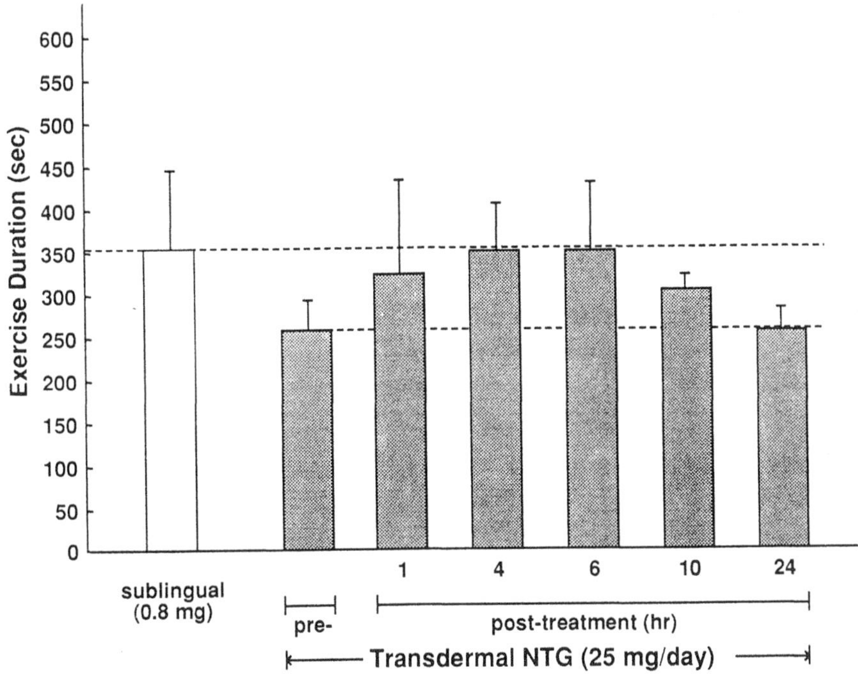

Figure 40 Antianginal effect of nitroglycerin delivered by a transdermal patch using the maximal exercise performance achieved by sublingual nitroglycerin (measured at 5 min) as the positive control and the time course for the improvement in exercise performance. (Plotted from the data in Reicht/Chien, 1987.)

the antianginal effect has not been sustained for 24 hr, even though the plasma concentration of nitroglycerin was maintained at a steady-state level throughout the course of the 24-hr application (Figures 36 through 39). This observation was attributed to the development of tolerance following long-term organic nitrate administration (127). Implementation of a washout phase between treatments has been recommended.

The clinical performance of the Estraderm system was compared with that of oral products: Premarin, which administers a combination of conjugated estrogens, and Estrace, which administers estradiol (E_2). The results in Figure 41 indicate that the transdermal delivery of E_2 at daily dose of only 0.10 mg/day is capable of achieving a sustained serum level of E_2 comparable to that attained by an oral E_2 dose 20 times higher (2 mg/day) or oral conjugated estrogens at 1.25 mg/day. Additionally, the physiological estradiol-estrone (E_2/E_1) ratio was recovered by transdermal estradiol, not by oral administration of estradiol or conjugated estrogens. The physiological levels of estradiol in the premenopause (at follicular phase) can be achieved and maintained by the transdermal controlled delivery of estradiol via the Estraderm system, not by oral conjugated estrogens (Figure 42). While the bone metabolism and vaginal cytology in postmenopausal women can be recovered to premenopausal levels (Figures 43 and 44) by both treatments, the hepatic protein activity appears to be affected by oral conjugated estrogen treatment (Figure 45).

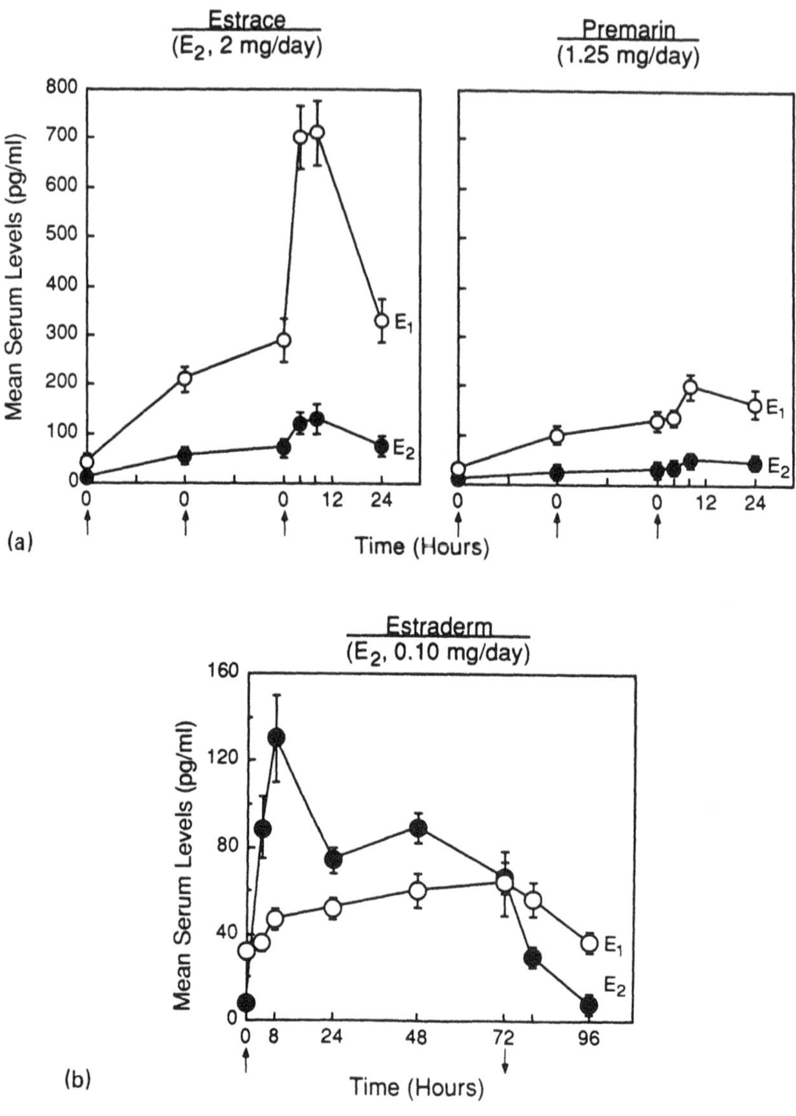

Figure 41 Clinical pharmacokinetic profiles of estradiol (E_2) delivered transdermally by the Estraderm system (E_2, 0.10 mg/day) for a 72 hr application compared to oral administration of Estrace tablets (E_2, 2 mg/day), once a day, and of Premarin tablets (conjugated estrogens, 1.25 mg/day), once a day. (Plotted from the data by Shaw et al., 1987.)

D. Correlation of In Vivo with In Vitro Transdermal Permeation Rates

The in vivo rate of transdermal permeation $(Q/t)_{\text{i.v.}}$ can be calculated from the steady-state plasma level $(C_p)_{ss}$ data (Figures 36 through 39) using the relationship (129)

$$\left(\frac{Q}{t}\right)_{\text{i.v.}} = (C_p)_{ss} K_e \frac{V_d}{A_s} \tag{36}$$

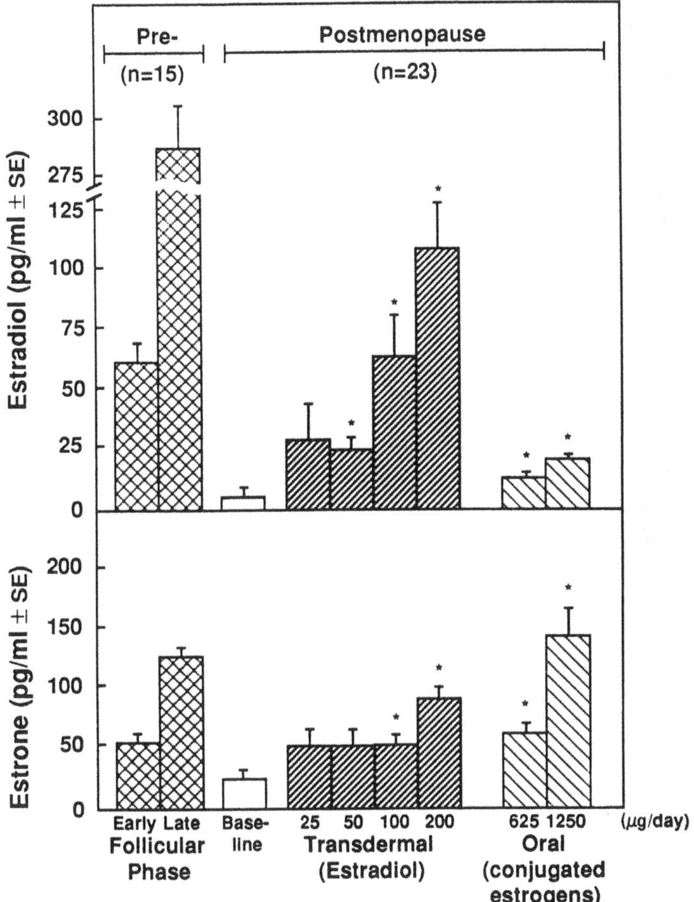

Figure 42 Systemic bioavailability of estradiol and serum levels of estrone in postmenopausal women (n = 23) following the transdermal delivery of estradiol from the Estraderm system with various daily dosage rates and oral administration of various daily doses of conjugated estrogens from Premarin tablets. The serum levels of estradiol and estrone in the premenopausal women (n = 15) are also shown for comparison. (Plotted from the data by Chetkowski et al., 1986.)

where K_e and V_d are the intrinsic first-order rate constant for elimination and the apparent volume of distribution for the drug, respectively, and A_s is the drug-releasing surface area of the TDD device in contact with the skin.

The results in Table 11 indicate that the in vivo rates of transdermal permeation calculated on the basis of Equation (36) show a good agreement with the in vitro data determined from either the epidermis or the dermatomed skin of human cadaver.

In view of the uncertainty involved in the availability of human cadaver skin and the variability in the source of its supply (Table 12) (130), the in vivo-in vitro agreement achieved, as shown in Table 11, suggests that hairless mouse skin could be an acceptable skin model as an alternative to human cadaver skin in studying the

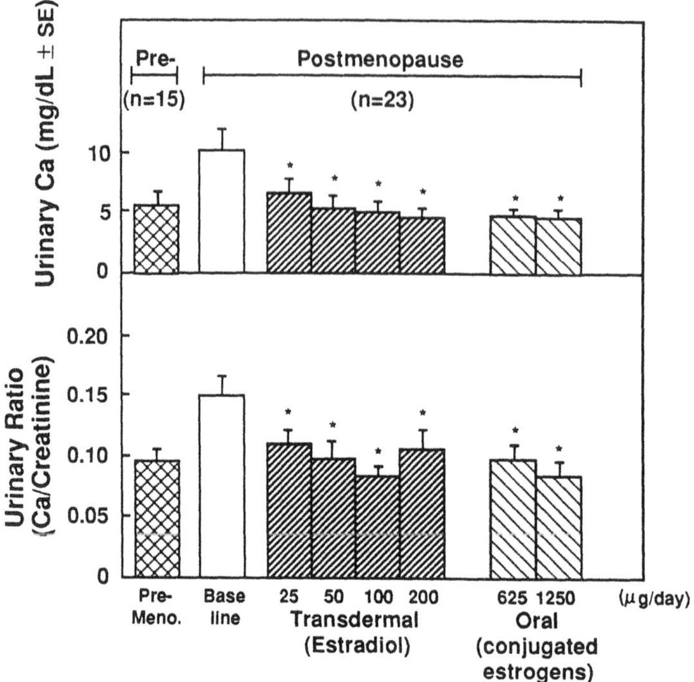

Figure 43 Comparative effect on bone metabolism in postmenopausal women (n = 23) following the transdermal delivery of estradiol from the Estraderm system with various daily dosage rates and oral administration of various daily doses of conjugated estrogens from Premarin tablets. The calcium levels in the urine and the urinary calcium-creatinine ratios in premenopausal women (n = 15) are also shown for comparison. (Plotted from the data by Chetkowski et al., 1986.)

transdermal permeation kinetics of systemically active drugs and for prediction of the in vivo rate of transdermal drug delivery in humans.

IX. OPTIMIZATION OF TRANSDERMAL CONTROLLED DRUG DELIVERY

To formulate a TDD system one should take into consideration the relationship between the rate of drug delivery R_d to the skin surface and the maximum achievable rate of drug absorption R_a by skin tissue (Figure 22). This is particularly important because the stratum corneum is known to be highly impermeable to most drugs. A TDD system should ideally be designed to have a skin permeation rate determined by the rate of drug delivery from the TDD system, not by the skin permeability. In such a case the transdermal bioavailability of a drug becomes less dependent upon any possible intra- and/or interpatient variabilities in skin permeability.

The rate of skin permeation of a drug at steady state $(R_p)_{ss}$ is mathematically related to the actual rate of drug delivery from a TDD system $(R_d)_a$ to the skin surface

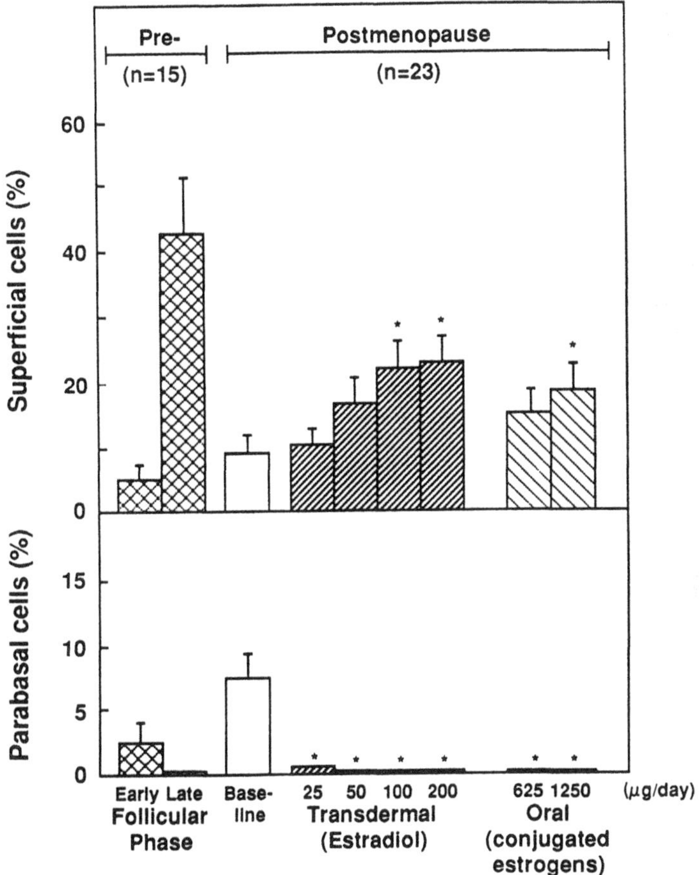

Figure 44 Comparative effect on vaginal cytology in postmenopausal women ($n = 23$) following the transdermal delivery of estradiol from the Estraderm system with various daily dosage rates and oral administration of various daily doses of conjugated estrogens from Premarin tablets. The percentage of superficial cells and parabasal cells in the premenopausal women ($n = 15$) is also shown for comparison. (Plotted from the data by Chetkowski et al., 1986.)

and the maximum achievable rate of skin absorption $(R_a)_m$ by the relationship (131)

$$\frac{1}{(R_p)_{ss}} = \frac{1}{(R_d)_a} + \frac{1}{(R_a)_m} \tag{37}$$

The actual rate of drug delivery from a TDD system to the skin surface, which acts as the receptor medium in the clinical applications, can thus be determined from

$$\frac{1}{(R_d)_a} = \frac{1}{(R_p)_{ss}} - \frac{1}{(R_a)_m} \tag{38}$$

If we consider the rate of skin permeation of pure nitroglycerin, which is free of any influence by the formulation or vehicle, as the value for $(R_a)_m$, the actual

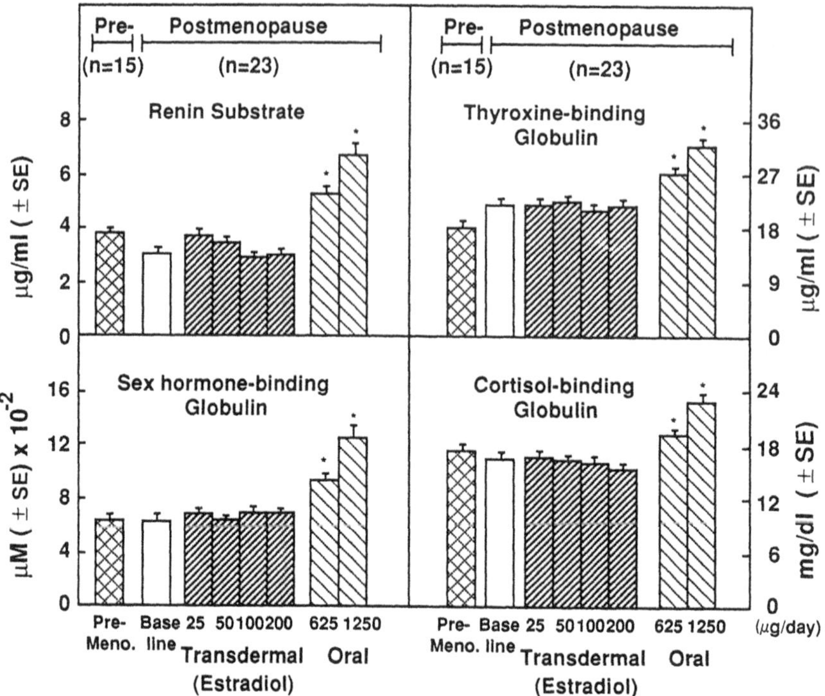

Figure 45 Comparative effect on hepatic protein activity in postmenopausal women (n = 23) following the transdermal delivery of estradiol from the Estraderm system with various daily dosage rates and oral administration of various daily doses of conjugated estrogens from Premarin tablets. The serum levels of various hepatic proteins in the premenopausal women (n = 15) are also shown for comparison. (Plotted from the data by Chetkowski et al., 1986.)

delivery rate of nitroglycerin from various TDD systems can be determined. The results in Table 13 indicate that the delivery rates of nitroglycerin from the Nitrodisc, Nitro-Dur, and Transderm-Nitro systems are three- to ninefold greater than the maximum achievable rate of skin permeation for nitroglycerin (0.476 mg/cm^2/day). On the other hand, the rate of delivery from the Deponit system is only 58% of the maximum rate of skin permeation. The data suggest that the delivery rate of nitroglycerin from all four TDD systems has not been adequately optimized.

Using a matrix diffusion-controlled TDD system the relationship between the rate of skin permeation and the rate of drug delivery from the TDD system can be established. The skin permeation rate of a drug at steady state, $(R_p)_{ss}$, is related to the drug delivery rate from a matrix-type TDD system, $(Q/t^{1/2})$, as follows (132):

$$(R_p)_{ss} = \frac{m(Q/t^{1/2})^2}{1 + n(Q/t^{1/2})^2} \tag{39}$$

where m and n are composite constants and defined as

$$m = \frac{1}{2k} \tag{40}$$

Table 11 Comparison in In Vitro and In Vivo Transdermal Permeation Rates

		Permeation rates ($\mu g/cm^2/day$)		
		In vitro		In vivo[a]
Drugs	Delivery systems	Hairless mouse	Human cadaver	
Nitroglycerin	Nitrodisc	435.6[b]	—	713.0
	Nitro-Dur	400.1[b]	487.9[c]	411.6
	Transderm-Nitro	349.2[b]	461.5[c]	427.9
	Deponit	269.5[d]	—	282.5
Estradiol	Estraderm	9.6[e]	6.5[e,f]	5.0
Clonidine	Catapres-TTS	86.9[e]	49.2[e,g]	38.9

[a]Calculated from steady-state plasma profiles using Equation (36).
[b]Determined in Franz diffusion cells (Figure 24) at 37°C (h_{aq} = 0.0304 cm). (From Chien et al., 1983.)
[c]Determined from skin permeation studies at 37°C using epidermis isolated from human cadaver abdominal skin (D. E. Magnuson, 1983).
[d]Determined in Keshary-Chien skin permeation cells (Figure 24) at 37°C (h_{aq} = 0.0108 cm). (From Keshary and Chien, 1984.)
[e]Determined in Valia-Chien skin permeation cells (Figure 31) at 37°C (h_{aq} = 0.0054 cm). (From Chien et al., 1986.)
[f]Determined from skin permeation studies at 37°C using human cadaver skin (male, 381 μm).
[g]Determined from skin permeation studies at 37°C using human cadaver skin (black female, left anterior leg, 620 μm).

Table 12 Intersubject Variability in Human Cadaver Skin Permeability

Human cadaver skin[a]		Skin permeation rate ($\mu g/cm^2/hr \pm SD$)	
Age (years)	Location	Estraderm 50[b]	Transderm-Nitro[c]
40	Anterior trunk	0.37 (± 0.04)	—
40	Left posterior leg	0.46 (± 0.06)	47.40 (± 7.43)
63	Left posterior leg	0.19 (± 0.01)	13.18 (± 1.39)
67	Posterior trunk	0.61 (± 0.06)	40.69 (± 8.47)

[a]White male dermatomed skin obtained from a skin bank.
[b]Lot 1F107735.
[c]Lot 100047.
Source: C. C. Chiang and C. S. Lee, unpublished data, (1990).

Table 13 Delivery Rate of Nitroglycerin from Various TDD Systems

TDD systems	Delivery rate[a] ($mg/cm^2/day$)
Nitrodisc	4.058
Nitro-Dur	2.857
Transderm-Nitro	1.166
Deponit	0.277

[a]Calculated from the hairless mouse data in Table 9 using Equation (38) and $(R_a)_m$ = 0.476 $mg/cm^2/day$.

$$n = \frac{R_{sc}}{2K_1kC_p}\left(1 + \frac{R_{vs}K_3 + R_{aq}}{K_2K_3R_{sc}}\right) \tag{41}$$

where k is a constant; K_1, K_2, and K_3 are the partition coefficients for the interfacial partitioning between the stratum corneum and polymer matrix, between viable skin and stratum corneum, and between receptor solution and viable skin, respectively; C_p is the drug solubility in the polymer matrix; and R_{sc}, R_{vs}, and R_{aq} are the diffusional resistances for stratum corneum, viable skin, and receptor solution on the dermis side, respectively.

Equation (39) indicates that a hyperbolic relationship should exist between $(R_p)_{ss}$ and $(Q/t^{1/2})^2$. Experimentally it was found that when the rate of drug delivery from the TDD system is low [i.e., $n(Q/t^{1/2})^2 < 1$], skin permeation is controlled by the delivery rate from the TDD system (Figure 46). As the rate of drug delivery is increased, the rate of skin permeation increases in a hyperbolic manner and then reaches a plateau level [i.e., $n(Q/t^{1/2})^2 > 1$] at which the rate of skin permeation becomes rate limited by the inherent permeability of stratum corneum to the drug species to

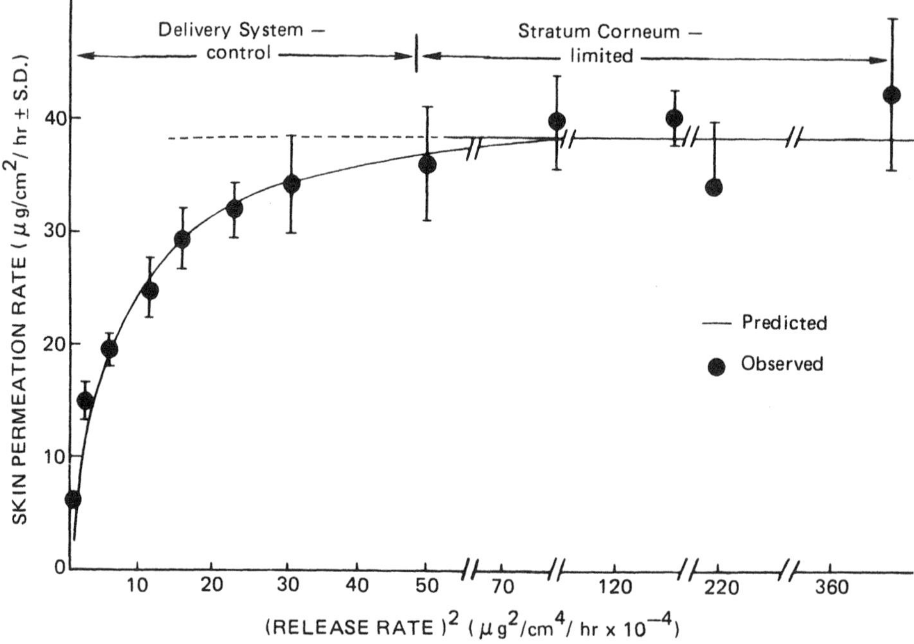

Figure 46 The hyperbolic relationship between the skin permeation rate and the square of the release flux of nitroglycerin delivered by the matrix diffusion-controlled TDD system as predicted from Equation (13). It was observed that when the $(Q/t^{1/2})^2$ value is equal to or less than 48 $\mu g^2/cm^4/hr$, the skin permeation rate of nitroglycerin is controlled by the delivery system; when the $(Q/t^{1/2})^2$ value is greater than 48 $\mu g^2/cm^4/hr$, the skin permeation rate becomes limited by the stratum corneum permeability. (Reproduced, with modification, from Keshary et al., 1985.)

be delivered transdermally. The magnitude of this plateau skin permeation rate is thus determined by the ratio m/n [Equations (39) through (41)].

Using Equation (39) one can optimize the formulation and the design of a TDD system with the rate of skin permeation controlled by the rate of drug delivery from the TDD system.

X. ADVANCES IN TRANSDERMAL CONTROLLED DRUG DELIVERY RESEARCH

It has been recognized that transdermal rate-controlled drug delivery offers one or more of the following potential biomedical benefits:

1. Avoid the risks and inconveniences of intravenous therapy
2. Bypass the variation in the absorption and metabolism associated with oral administration
3. Permit continuous drug administration and the use of drugs with a short biological half-life
4. Increase the bioavailability and efficacy of drugs through the bypass of hepatic first-pass elimination
5. Reduce the chance of over- or underdosing through the prolonged, preprogrammed delivery of drug at the required therapeutic rate
6. Provide a simplified therapeutic regime leading to better patient compliance
7. Permit a rapid termination of the medication, if needed, by simply removing the TDD system from the skin surface

The intensity of interest in the potential biomedical applications of rate-controlled transdermal drug administration has been demonstrated by a substantial increase in the research and development activities in many health care institutions aiming to develop viable TDD systems for the prolonged continuous transdermal infusion of therapeutic agents (5,133,134). The drug candidates evaluated include antihypertensive, antianginal, antihistamine, antiinflammatory, analgesic, antiarthritic, steroidal, and contraceptive drugs. It has been estimated by marketing research experts that within the next 5 years over 10% of drug products will be marketed in TDD systems.

On the other hand, it has been increasingly recognized that not every drug can be delivered transdermally at a rate high enough to achieve a blood level that is therapeutically beneficial for systemic medication. An increasing number of biomedical researchers working in the fields of transdermal drug delivery have cautioned the potential limitations of transdermal systemic drug delivery (133,135).

Because of difference in the technologies applied to the development and formulation of various TDD systems, different patch sizes are needed to deliver the same daily dose of nitroglycerin. For example, for transdermal systemic delivery of a daily dose of 5 mg nitroglycerin, the Nitrodisc system with a drug-releasing surface of 8 cm^2 is sufficient as compared to a patch size of 16 cm^2 for the Deponit system and 10 cm^2 for the Transderm-Nitro, Nitro-Dur I and II, and NTS systems. Similarly, for an ointment formulation an application area of 10 cm^2 over a duration of 4–8 hr also provides the equivalent dose. These TDD systems have been found to be capable of establishing a mean steady-state plasma level of less than 0.18 ng/ml (129), which is substantially lower than the plasma level (0.29–0.41 ng/ml) achieved by IV infusion of nitroglycerin at a rate of 3.4 μg/min (i.e., a daily dose of 4.9 mg, Figure 47). To achieve the same plasma level as the IV infusion, it is estimated that a transdermal patch with a nitroglycerin-releasing surface of 20–30 cm^2 is needed, depending upon which type of TDD system is used. The same requirement is also applied to the 2% ointment formulation, but with a more frequent dosing schedule (three times a day) (136).

To achieve and to maintain a plasma drug concentration above the minimum therapeutic level, the barrier properties of the skin must be overcome before the effective transdermal controlled delivery of drugs can be successfully accomplished. The following approaches have been shown to be potentially promising for accomplishing the goals of reducing skin's barrier properties and enhancing the transdermal permeation of drugs (137):

1. Physical approach
   a. Stripping of stratum corneum
   b. Hydration of stratum corneum
   c. Iontophoresis

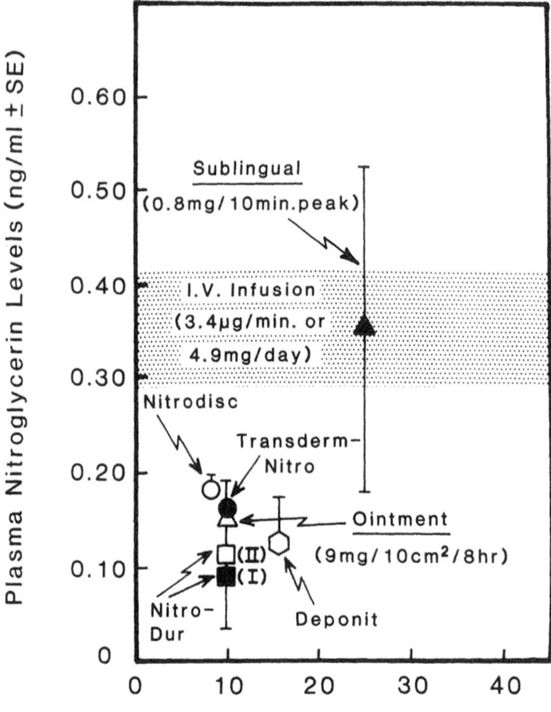

Figure 47 Comparison of the steady-state plasma levels of nitroglycerin achieved by various TDD systems in relation to those by ointment application, sublingual administration, and IV infusion. (Reproduced from Chien, 1987.)

   d. Phonophoresis
   e. Thermal energy
2. Chemical approach
   a. Synthesis of lipophilic analogs
   b. Delipidization of stratum corneum
   c. Coadministration of skin permeation enhancer
3. Biochemical approach
   a. Synthesis of bioconvertible prodrugs
   b. Coadministration of skin metabolism inhibitors

Some examples of the successful approaches are discussed in the following sections.

A. Improved Transdermal Permeation by Bioconvertible Prodrugs

Prodrugs can be viewed as the therapeutically inactive derivatives of a therapeutically active drug. In a biological environment they undergo bioconversion, either by hydrolytic or enzymatic transformation, to regenerate the therapeutically active parent drug before reaching the target tissues to exhibit their pharmacological activities (138).

The objective of applying bioconvertible prodrug concept to transdermal controlled drug delivery is to modify the skin permeability of a drug by altering its physicochemical properties such that its rate of transdermal permeation is greatly enhanced (139).

A drug with poor skin permeability may be chemically modified to form a prodrug with improved skin permeation characteristics. During the course of skin permeation the prodrug is transformed by the metabolic processes within the skin tissues to regenerate the active parent drug (Figure 48). In other words, if an active drug has a rather low affinity to the stratum corneum it therefore does not easily partition into it to any significant extent for permeation. The partition behavior of this drug can be improved by simple chemical modification to form a lipophilic prodrug; and

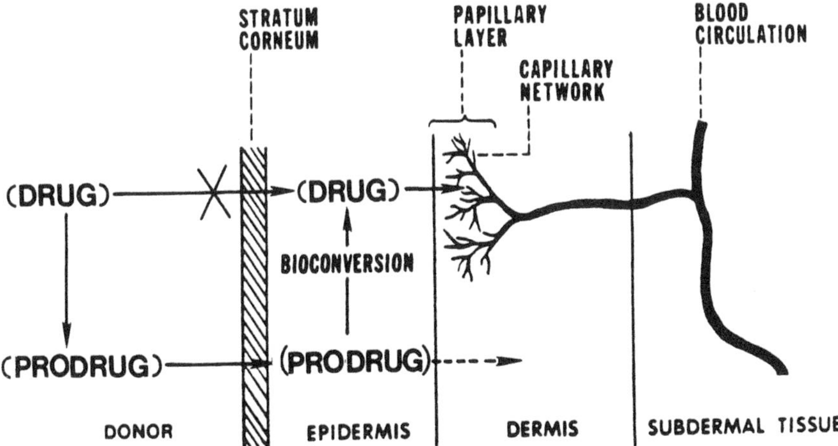

Figure 48 Multilayered skin model showing enhancement of the skin permeation of a drug by formation of a prodrug that has greater skin permeability. In the viable epidermal layers the prodrug is metabolized and bioconverted to its parent drug.

Table 14 Chemical Structure of Estradiol and Its Ester-type Prodrugs

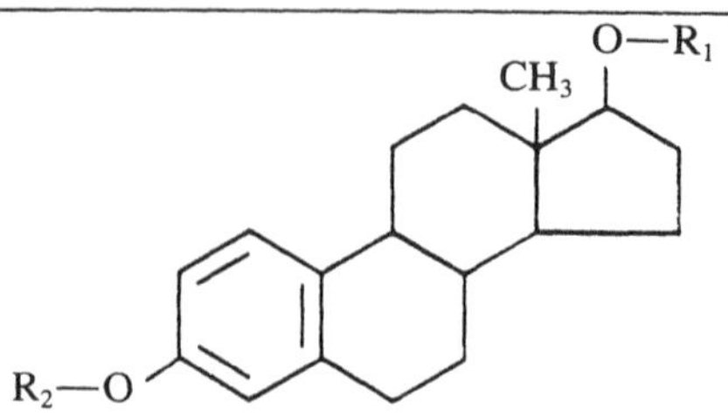

$O-R_1$, CH_3, R_2-O

	R_1	R_2
17β-Estradiol	$-H$	$-H$
Estradiol-17-Acetate	$-\underset{\overset{\|}{O}}{C}-CH_3$	$-H$
Estradiol-3,17-Diacetate	$-\underset{\overset{\|}{O}}{C}-CH_3$	$-\underset{\overset{\|}{O}}{C}-CH_3$
Estradiol-17-Valerate	$-\underset{\overset{\|}{O}}{C}-(CH_2)_3-CH_3$	$-H$
Estradiol-17-Heptanoate	$-\underset{\overset{\|}{O}}{C}-(CH_2)_5-CH_3$	$-H$
Estradiol-17-Cypionate	$-\underset{\overset{\|}{O}}{C}-CH_2-CH_2$	$-H$

the transport of the drug across the stratum corneum is substantially enhanced. Upon absorption and penetration through the skin tissues the prodrug is rapidly metabolized to regenerate the active parent drug. A typical example of such an approach is the esterification of less-skin-permeable estradiol to form lipophilic estradiol esters (Table 14).

In vitro skin permeation studies (139) demonstrated that the ester-type prodrugs of estradiol are extensively metabolized during the course of skin permeation by the esterase in skin tissues to regenerate estradiol (Figure 49), and the rate of regeneration of estradiol from its esters by bioconversion during the course of transdermal permeation was observed to follow the order diacetate > valerate > heptanoate > cypionate > acetate. The metabolism of estradiol esters was found to follow first-order kinetics, and the rate constant for the enzymatic hydrolysis of the ester group at position 3 was observed to be substantially greater than that at position 17 (139,140).

Based on the results of simultaneous skin permeation and bioconversion profiles, a transdermal bioactivated hormone delivery (TBHD) device was developed using the microreservoir dissolution-controlled drug delivery system (Figure 30) for the transdermal controlled delivery of estradiol esters (141). Using the hydrodynamically well-calibrated Valia-Chien skin permeation cell (Figure 31), the in vitro drug release and skin permeation profiles of estradiol and its esters were investigated simultaneously. The results demonstrated that estradiol and its esters are released from the

O—R_1

CH_3

R_2—O

Estradiol-3,17-Diester

k_1 Esterase

O-R_1

CH_3

HO

Estradiol-17-Ester

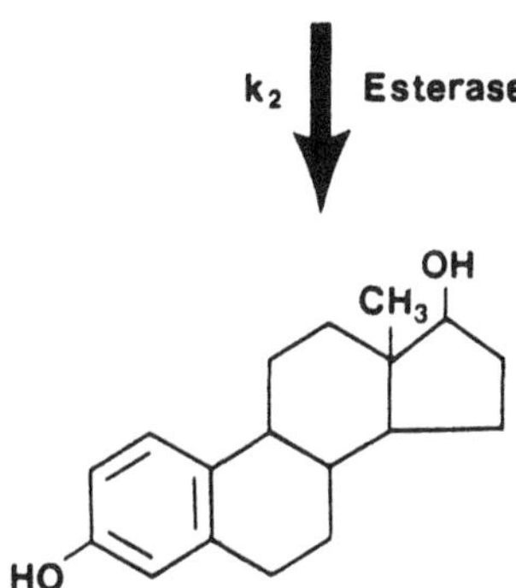

Estradiol

Figure 49 The bioconversion of estradiol-3,17-diester by the esterase in the cutaneous tissue to estradiol-17-ester as the intermediate and then to the biologically active estradiol. (Reproduced from Valia et al., 1985.)

TBHD device at a constant, zero-order rate profile, and the release rate of estradiol increases after the hydrophilic OH groups at positions 3 and/or 17 have been esterified to form a lipophilic ester (Figure 50). The release rate of estradiol-17-esters from the lipophilic TBHD device was found to be dependent upon the alkyl chain length of the ester at position 17.

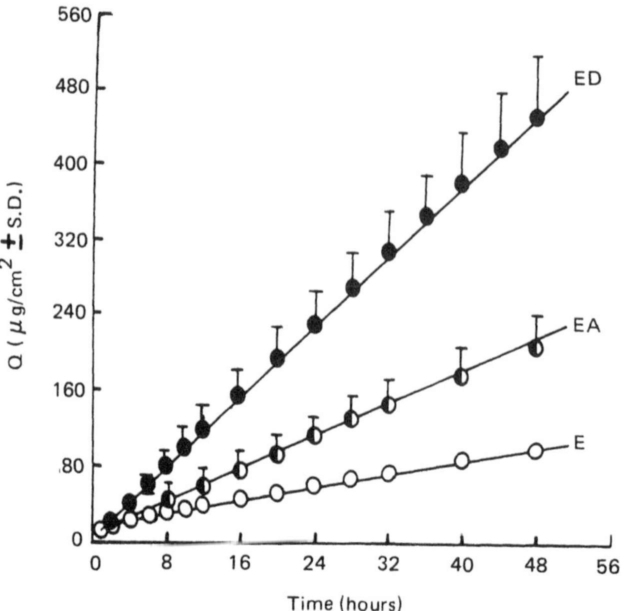

Figure 50 The cumulative release profiles of estradiol and its esters from a TBHD system: *E* (estradiol); *EA* (estradiol-17-acetate); *ED* (estradiol-3,17-diacetate). (Reproduced from Chien et al., 1985.)

The skin permeation studies suggested that all the esters of estradiol, except extradiol-3-acetate and 3,17-diacetate, are totally metabolized by the esterase in the viable skin during the course of skin permeation to regenerate the biologically active estradiol (Figure 51). The rate of regeneration of estradiol from the esters was also found to be dependent upon the alkyl chain length (Table 15). The appearance of estradiol from diacetate and valerate achieved a rate that was four- and two-fold greater, respectively, than the rate of estradiol appearance by skin permeation alone.

B. Enhanced Transdermal Permeation by Skin Permeability Promotors

The skin permeability of drugs can also be greatly improved by treating the stratum corneum with an appropriate skin permeation promotor. Representative classes of the potential skin permeation promotors are shown in Figure 52.

The concept of promoting the skin permeability of drugs by skin permeation enhancers has recently been applied to the practice of transdermal controlled drug delivery by developing a skin permeation enhancing (SPE) transdermal drug delivery (TDD) system (Figure 53) (137). This new generation of transdermal drug delivery system is capable of releasing one or a combination of two or more skin permeation enhancers to the surface of the stratum corneum to modify the skin's barrier properties and render the skin more permeable to the drug, before the controlled delivery of active drug (Figure 54). In vitro skin permeation studies in the hydrodynamically well-calibrated Keshary-Chien skin permeation cell (Figure 24) demonstrated that the addition of simple pharmaceutical excipients, such as the straight-chain saturated alkanoic acids, can substantially enhance the transdermal permeation rate of pro-

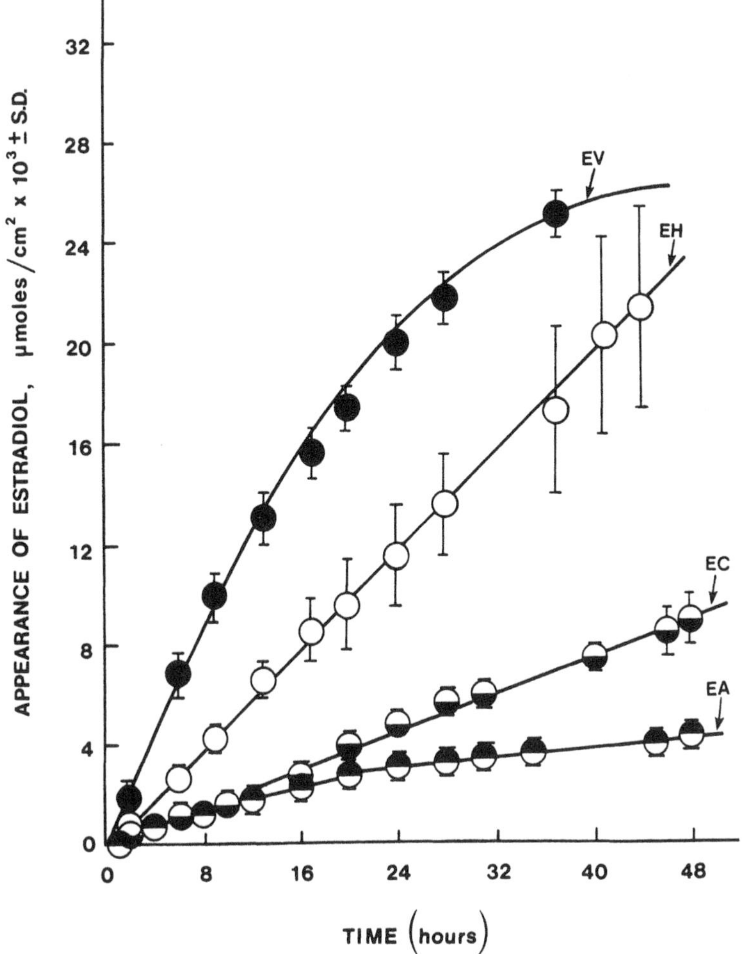

Figure 51 Time course for the regeneration of estradiol from various estradiol-17-esters over the course of skin permeation by metabolism following the transdermally controlled administration of various estradiol ester-releasing TBHD systems: *EV* (estradiol-17-valerate); *EH* (estradiol-17-heptanoate); *EC* (estradiol-17-cypionate); *EA* (estradiol-17-acetate). No unmetabolized esters, except the mono- and diacetates, were detectable in the dermal solution. (Reproduced from Chien et al., 1985.)

gesterone and also significantly reduce the duration of the time lag while the zero-order skin permeation rate profile is maintained (Figure 55). The extent of enhancement in skin permeability appears to be dependent upon the alkyl chain length and the terminal carboxylic group of the straight-chain alkanoic acid (Figure 56). The alkanoic acids appear to be more effective in enhancing skin permeability than their propyl ester.

The efficiency of skin permeability enhancement also shows dependence on the molecular structure of the penetrants. Alkanoic acids are capable of enhancing the skin permeation rate of progesterone to a greater extent than that of nitroglycerin.

Table 15 Appearance Rate Profiles of Estradiol Following the Transdermal Controlled Administration of Various Estradiol Esters[a]

Species	Rate of appearance ($\mu g/cm^2/hr \pm SD$)	Enhancement Factor[b]
Estradiol	0.117 (± 0.027)	1.0
Estradiol (*E*) esters		
E-Diacetate	0.490 (± 0.250)	4.2
E-Acetate	0.057 (± 0.013)	0.5
E-Valerate	0.227 (± 0.042)	1.9
E-Heptanoate	0.061 (± 0.013)	0.5
E-Cypionate	0.016 (± 0.002)	0.1

[a]Delivered by transdermal bioactivated hormone delivery system.
[b]Enhancement factor = $(\text{rate of appearance})_{ester}/(\text{rate of appearance})_{estradiol}$.

The results in Figures 55 and 56 also demonstrate that maximum enhancement has been achieved by alkanoic acid with alkyl chain lengths of $n = 6$ for progesterone and of $n = 8$ for nitroglycerin.

In addition to progesterone, which is a relatively skin-permeable steroidal drug, the propyl esters of myristic acid (a long-chain saturated fatty acid) and of oleic acid (a long-chain unsaturated fatty acid; Figure 52), are also capable of promoting the rate of skin permeation for the less permeable steroidal antiinflammatory agents (e.g.,

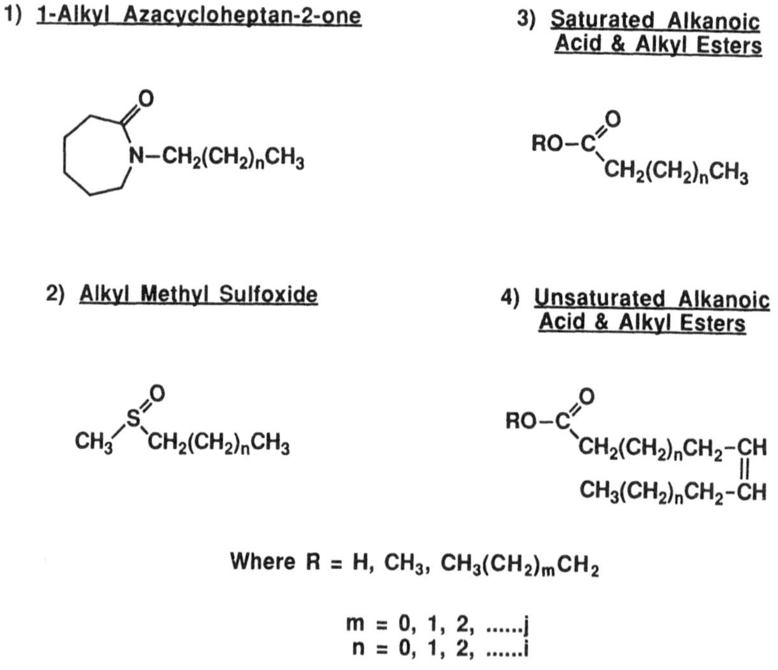

Figure 52 Representative classes of potential skin permeation promotors. (Reproduced from Chien, 1988.)

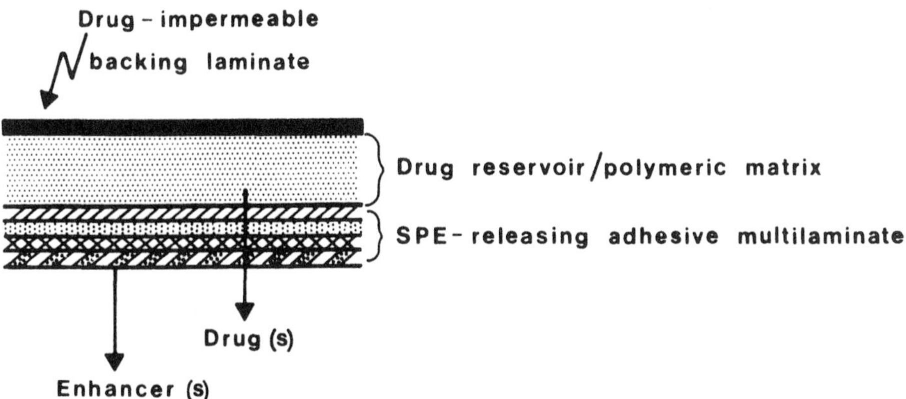

Figure 53 Cross-sectional view of a skin permeability-enhancing TDD system showing various major structural components. (Reproduced from Chien and Lee, 1985.)

hydrocortisone), nonsteroidal antiinflammatory drugs (e.g., indomethacin), and estrogenic steroids (e.g., estradiol; Table 16).

In addition to the esters of saturated and unsaturated fatty acids, azone and decylmethylsulfoxide are also shown to be very effective in improving the skin permeability of drugs (Table 16). The results appear to suggest the possible existence of a relationship between the skin permeability enhancement of a drug and its molecular structure, as well as the type and concentration of promotor used (Figure 57). A synergistic effect in skin permeability enhancement could be achieved by incorporating a combination of two or more enhancers in the adhesive layers (Figure 53).

The mechanisms of action of various skin permeability promotors may be attributed to their activity on lipophilic lipid matrix and/or hydrophilic protein gel in the stratum corneum (137,142).

The concept of improving transdermal permeation by skin permeability promotors has been recently practiced in the development of transdermal contraceptive devices that achieve dual-controlled transdermal delivery of a potent progestin and a natural estrogen for weekly fertility regulation in females (114–116).

C. Facilitated Transdermal Permeation by Iontophoresis

Iontophoresis is a process that facilitates the transport of ionic species by the application of a physiologically acceptable electrical current (143). It was first proposed by Pivati in 1747 (144) and has been used with success for the local delivery of drugs (145–148). More recently it has been explored for the transdermal systemic delivery of systemically active drugs (148–152). The mechanism of iontophoresis-facilitated transdermal delivery of systemically active drugs is illustrated in Figure 58.

The results of various investigations (149–152) have demonstrated that the rate of transdermal delivery of ionic drugs can be substantially enhanced with the application of iontophoresis. A typical example is shown in Figure 59, in which the skin permeation of oxycodone cation (pK_a = 8.5), a systemically active narcotic anal-

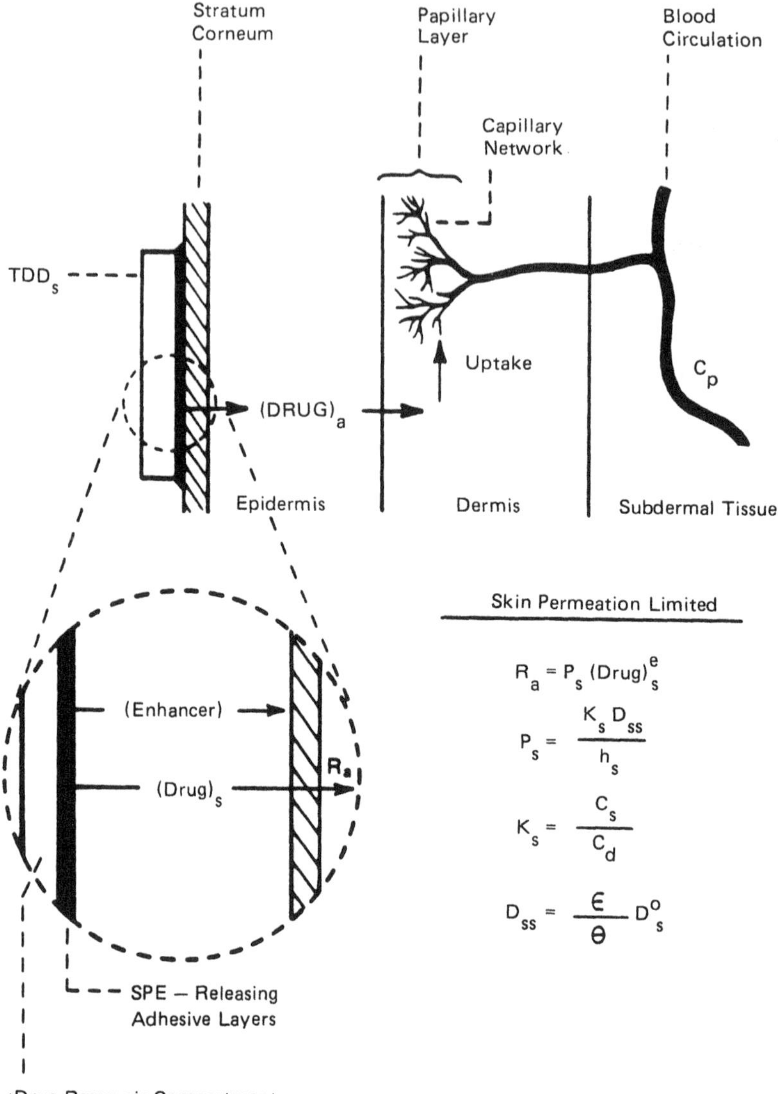

Figure 54 The concept of enhancing the skin permeation of drugs by first releasing one or more enhancers to the skin surface to modify the permeability characteristics of the stratum corneum before the controlled delivery of a systemically active drug.

gesic, has been significantly improved, with a substantial reduction in lag time, by the application of a pulse current delivered from a transdermal periodic iontotherapeutic system (TPIS). The enhancement in the rate of transdermal permeation was found to be dependent upon the magnitude of current intensity applied (Figure 60) as well as the duration of iontophoretic treatment (Figure 61).

The transdermal permeation of peptide drugs received limited attention in the past, possibly because of the old belief that for a drug with a molecular mass as large as peptides or proteins, in combination with their polar properties and chemical

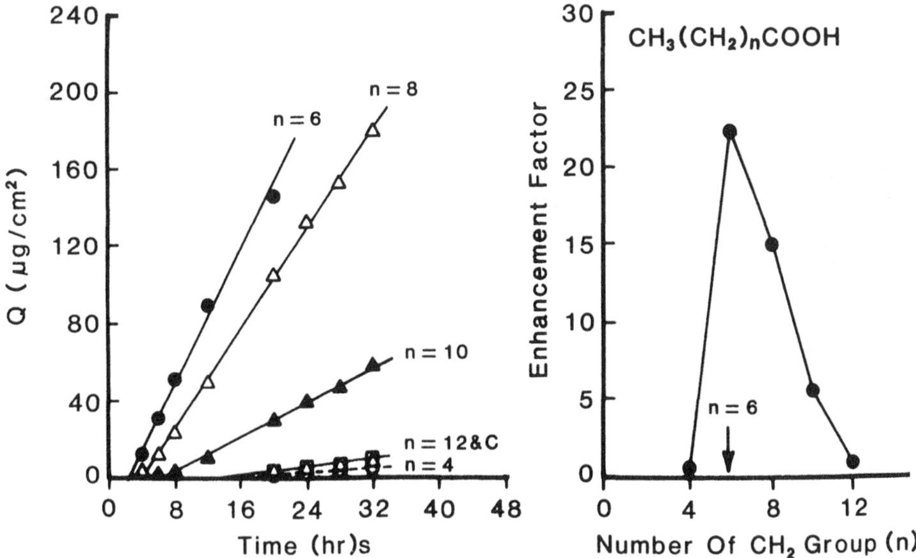

Figure 55 The skin permeation profiles of progesterone by straight-chain saturated alkanoic acids released from the adhesive coating layer and the dependence of skin permeability enhancement on the alkyl chain length *n* of alkanoic acids. The enhancement factor is calculated from the relationship: enhancement factor = (normalized skin permeation rate)$_{enhancer}$/(normalized skin permeation rate)$_{control}$. (From Chien and Lee, unpublished data, 1990.)

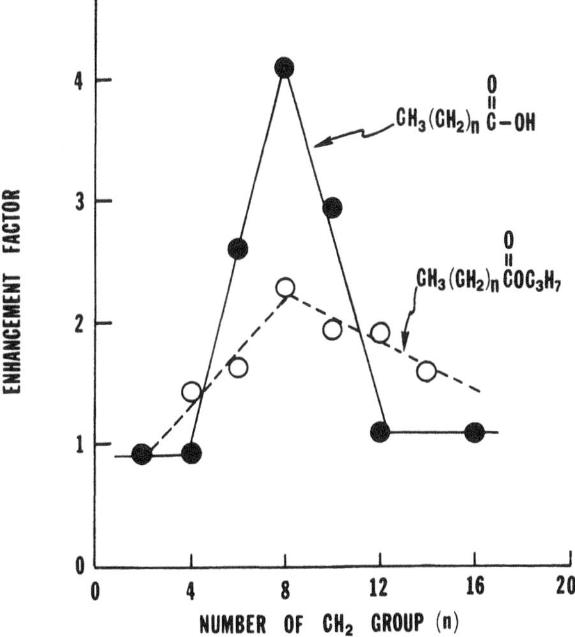

Figure 56 Dependence of the enhancement factor for the skin permeation of nitroglycerin on the alkyl chain length *n* of the straight-chain saturated alkanoic acid and its propyl ester. (Chien and Lee, unpublished data, 1990.)

Table 16 Enhancement of Skin Permeability by Different Types of Promotors

Drugs	Skin Permeation rate[a] ($\mu g/cm^2/day \pm SD$)	Enhancement factor[b]			
		Propyl myristate	Propyl oleate	Azone	Decylmethyl sulfoxide
Progesterone	36.72 (± 10.32)	4.6	5.4	6.0	11.0
Estradiol	29.29 (± 24.48)	9.3	14.6	20.2	12.6
Indomethacin	9.36 (± 0.08)	3.8	4.7	14.5	15.7
Hydrocortisone	1.10 (± 0.12)	4.6	5.0	61.3	25.2

[a]Skin permeation rate of the drug from the TDD system containing no promotor.
[b]Unit concentration of promotor in the adhesive layer = 3.2 mg/cm^2.
Source: Compiled from the data by Chien and Lee (1985).

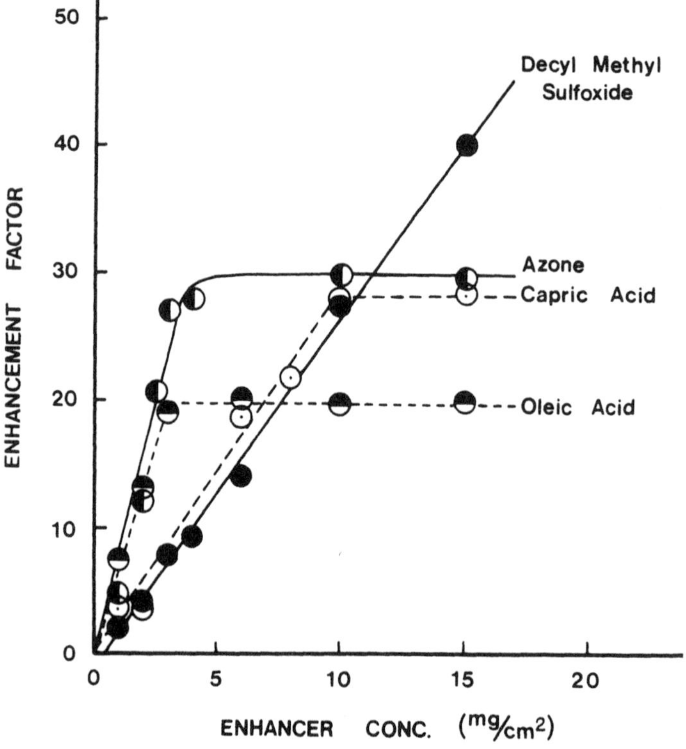

Figure 57 Dependence of the enhancement factor for the skin permeation of progesterone on the concentration of various skin permeation enhancers. (Reproduced from Chien et al., 1986.)

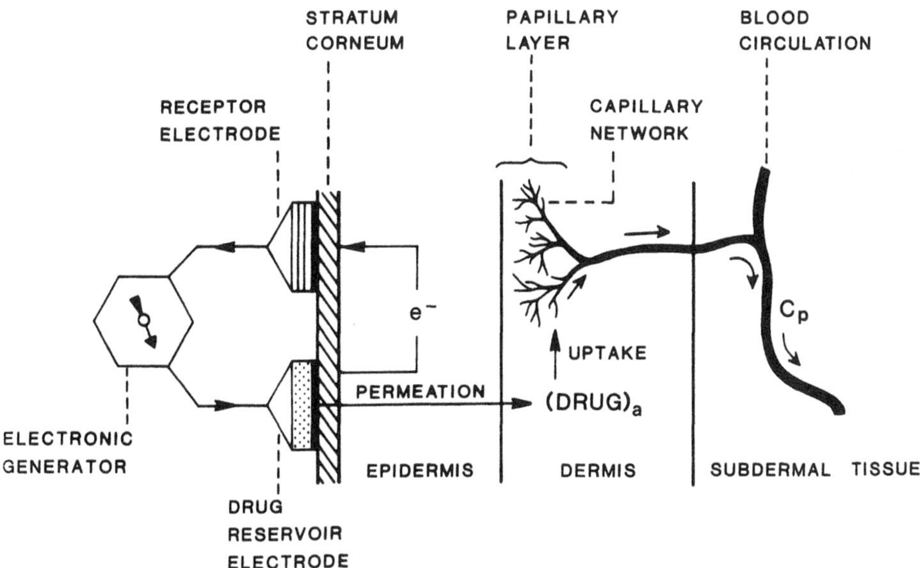

Figure 58 The mechanism involved in the iontophoresis-facilitated transdermal delivery of polar or ionic drugs.

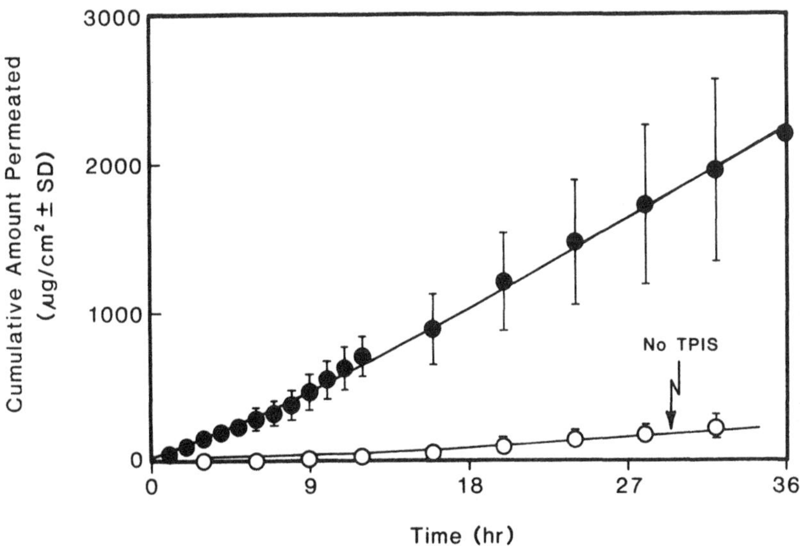

Figure 59 Comparative skin permeation profiles of oxycodone cation from buffer solution (pH 4.0) with (●) and without (○) the application of pulse current from the transdermal periodic iontotherapeutic system (TPIS). (Plotted from the data by Kuo, 1987.)

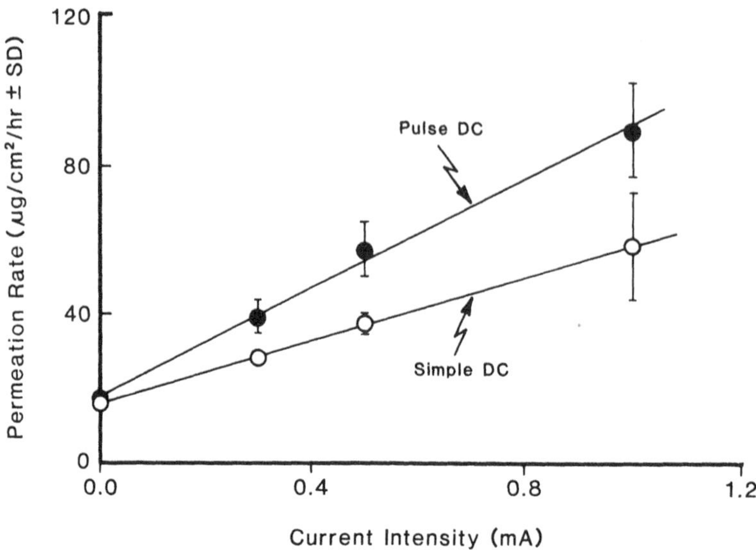

Figure 60 Linear dependence of the iontophoresis-facilitated skin permeation rate of oxycodone cation on the intensity of pulse or simple direct current. (Plotted from the data by Kuo, 1987.)

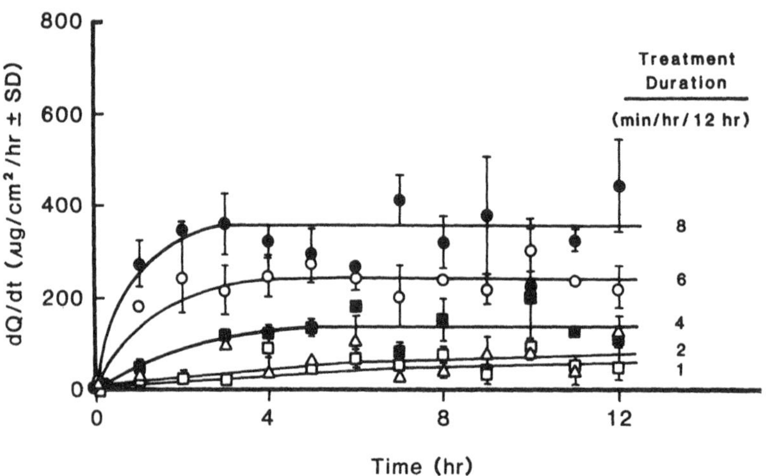

Figure 61 Skin permeation rate profiles of oxycodone cation following the periodic application of a pulse current for varying durations. The plateau rate of iontophoresis-facilitated skin permeation is increased with the increase in the duration of iontophoresis treatment. (Plotted from the data by Kuo, 1987.)

instability, skin permeability would be insignificant (153). Most recently the transdermal delivery of peptide-based pharmaceuticals, such as insulin, has become feasible by the application of iontophoresis (154–157). The results accumulated to date have demonstrated that the barrier properties of stratum corneum can be overcome by iontophoresis and some of the polar, charged peptide-based pharmaceuticals can be successfully delivered transdermally to produce a systemic therapeutic effect (156–158). The iontophoresis-facilitated transdermal systemic delivery of therapeutic peptides and proteins will be discussed in more detail in Chapter 11.

XI. CONCLUSION

The future of transdermal rate-controlled drug delivery in medicine is undoubtedly adminisbright. The scope of the biomedical application of this new form of drug delivery will be increasingly expanded for many years to come, especially with the successful development of new approaches capable of enhancing the skin permeability of drugs (137,139–158).

REFERENCES

1. Y. W. Chien; *Transdermal Controlled Systemic Medications*, Dekker, New York, 1987, Chapter 1.
2. J. E. Shaw, S. K. Chandrasekaran and P. Campbell; J. Invest. Dermatol., 67:677 (1976).
3. Y. W. Chien; *Transdermal Controlled Systemic Medications*, Dekker, New York, 1987, Chapters 2, 9, 10, 11, and 12.
4. J. E. Shaw and S. K. Chandreasekaran; Drug Metab. Rev., 8:223 (1978).
5. 1982 Industrial Pharmaceutical R & D Symposium, *Transdermal Controlled Release Medication*, Rutgers College of Pharmacy, Piscataway, New Jersey, January 14 and 15, 1982. Proceedings published in Drug Dev. Ind. Pharm. 9(4):497–744 (1983).
6. World Congress of Clinical Pharmacology Symposium, *Transdermal Delivery of Cardiovascular Drugs*, Washington, D.C., August 5, 1983. Proceedings published in Am. Heart J., 108(1):195–236 (1984).
7. W. R. Good, M. S. Powers, P. Campbell and L. Schenkel; J. Control Release, 2:89–97 (1985).
8. S. W. Jacob and C. A. Francone; *Structure and Function of Man*, 2nd Ed., W. B. Saunders, Philadelphia, 1970, Chapter 4.
9. G. H. Bell, J. N. Davidson and H. Scarbrough; *Textbook of Physiology and Biochemistry*, 5th Ed., E. & S. Livingston, Edinburgh, 1963, Chapter 37.
10. W. Montagna; *The Structure and Function of Skin*, 2nd Ed., Academic, New York, 1961, p. 454.
11. D. M. Pillsbury, W. B. Shelley and A. M. Kligman: *Dermatology*, W. B. Saunders, Philadelphia, 1956.
12. R. J. Scheuplein; J. Invest. Dermatol., 45:334 (1965).
13. F. Reiss; Am. J. Med. Sci., 252:588 (1966).
14. R. H. Rice and H. Green; Cell, 11:417 (1977).
15. A. Balmain, D. Loehren, J. Fisher and A. Alonso; Dev. Biol., 60:442 (1977).
16. R. K. Freinkel and T. N. Traczyl; J. Invest. Dermatol., 69:413 (1977).

17. M. Gloor, J. Handke, C. Baumann and H. C. Freidrich; Dermatol. Monatsschr., 161:996 (1975).
18. G. H. J. Sears, M. E. Grant and D. S. Jackson; Biochem. Biophys. Res. Commun., 71:379 (1976).
19. G. Carpenter and S. Cohen; J. Cell Physiol., 88:227 (1976).
20. P. J. Hartop and C. Prottey; Br. J. Dermatol., 95:255 (1976).
21. V. B. Hatcher, G. S. Lazarus, N. Levine, P. G. Burk and F. J. Yost; Biochim. Biophys. Acta, 483:160 (1977).
22. E. K. Kastrup and J. R. Boyd; *Drug: Facts and Comparison*, J. B. Lippincott, Philadelphia, 1983, pp. 1634–1708.
23. J. E. Shaw, W. Bayne and L. Schmidt; Clin. Pharmacol. Ther., 19:115 (1976).
24. P. W. Armstrong, J. A. Armstrong and G. S. Marks; Am. J. Cardiol., 46:670 (1980).
25. R. Sitruk-Ware, B. de Lignieres, A. Basdevant and P. Mauvais-Jarvis; Maturitas, 2:207 (1980).
26. A. Osol; *Remington's Pharmaceutical Sciences*, 16th Ed., Mack, Easton, Pennsylvania, 1980, p. 1534.
27. First Transdermal Therapeutic System Symposium on Evaluation of External Adhesive System, Tokyo, Japan, July 26, 1985.
28. *National Formulary (NF)*, VIII Ed., 1946.
29. *United States Pharmacopeia (USP)*, XIV Ed., 1950.
30. A. S. Michaels, S. K. Chandrasekaran and J. E. Shaw; AIChE J. 21:985 (1975).
31. A. Homalle; Union Med., 7:462 (1953).
32. F. Duriau; Arch. Gen. Med. T., 7:161 (1856).
33. S. Rothman; *Physiology and Biochemistry of the Skin*, University of Chicago Press, Chicago, 1954.
34. G. S. Berenson and G. E. Burch; Am. J. Trop. Med. Hyg., 31:842 (1951).
35. I. H. Blank and E. Gould; J. Invest. Dermatol., 37:311 (1962).
36. T. Fredriksson; Acta Dermato-Venereol., 41:353 (1962).
37. A. G. Matoltsy, M. Matoltsy and A. Schrogger; J. Invest. Dermatol., 38:251 (1962).
38. Y. W. Chien; Drug Dev. Ind. Pharm., 9:497 (1983).
39. W. Perl; Ann. N.Y. Acad. Sci., 108:92 (1963).
40. R. T. Tregear; in: *Monographs in Theoretical and Experimental Biology*, Vol. 5, Academic, New York, 1966.
41. R. T. Tregear; J. Invest. Dermatol., 46:16 (1966).
42. J. E. Treherne; J. Physiol. (Lond.), 133:171 (1956).
43. R. J. Scheuplein; Biophys. J., 6:1 (1966).
44. R. J. Scheuplein; J. Invest. Dermatol., 48:79 (1967).
45. R. J. Scheuplein; J. Theor. Biol., 18:72 (1968).
46. R. J. Scheuplein, I. H. Blank, G. J. Brauner and D. J. MacFarlane; J. Invest. Dermatol., 52:63 (1969).
47. G. E. Burch and T. Winsor; Arch. Dermatol., 53:39 (1944).
48. I. H. Blank and R. J. Scheuplein; Br. J. Dermatol., 81(Suppl. 4):4 (1969).
49. I. H. Blank, R. J. Scheuplein and D. J. MacFarlane; J. Invest. Dermatol., 49:582 (1967).
50. I. H. Blank; J. Invest. Dermatol., 43:415 (1964).
51. M. J. Ainsworth; J. Soc. Cosmet. Chem., 11:69 (1960).
52. I. Brody; J. Ultrastruct. Res., 4:264 (1960).
53. I. Brody; J. Invest. Dermatol., 42:27 (1964).
54. I. Brody; Nature (Lond.), 209:472 (1966).
55. B. K. Filskie and G. E. Rogers; J. Mol. Biol., 3:784 (1961).
56. R. J. Scheuplein and I. H. Blank; Physiol. Rev., 51:702 (1971).

57. A. G. Matoltsy; *Biochem. Cutaneous Epidermal Differ. Proc. Jpn.-U.S. Semin.*, 1976, pp. 93–109 (published in 1977).
58. A. G. Matoltsy and C. A. Balsamo; J. Biophys. Biochem. Cytol., 1:339 (1955).
59. A. G. Matoltsy and P. F. Parakkal; J. Cell. Biol., 24:297 (1965).
60. R. J. Scheuplein and L. J. Morgan; Nature (Lond.), 214:456 (1967).
61. R. F. Rushmer, K. J. K. Buettner, J. M. Short and G. F. Odland; Science, 154:343 (1966).
62. A. M. Kligman; in: *The Epidermis* (W. Montagna and W. C. Lobitz, Eds.), Academic, New York, 1964, pp. 387–433.
63. R. J. Feldman and H. I. Maibach; J. Invest. Dermatol., 48:181 (1967).
64. E. J. Harris; in: *Transport and Accumulation in Biological Systems*, Butterworths, London, 1960, p. 39.
65. J. T. Davies and E. K. Rideal; *Interfacial Phenomena*, Academic, New York, 1963, pp. 154–156.
66. G. W. Liddle; J. Clin. Endocrinol. Metab., 16:557 (1956).
67. H. D. Onken and C. A. Meyer; Arch. Dermatol., 87:584 (1963).
68. A. Zesch and H. Schaefer; Arzneimittalforsch., 26:1365 (1976).
69. T. Higuchi; J. Soc. Cosmet. Chem., 11:85 (1960).
70. R. J. Scheuplein and L. Ross; J. Soc. Cosmet. Chem., 21:853 (1970).
71. W. E. Clendenning and R. B. Stoughton; J. Invest. Dermatol., 39:47 (1962).
72. E. Cronin and R. B. Stoughton; Arch. Dermatol., 87:445 (1963).
73. A. S. Michaels, S. K. Chandrasekaran and J. E. Shaw; AIChE J., 21:985 (1975).
74. J. Crank; *Mathematics of Diffusion*, Clarendon Press, Oxford, 1967, Chapter IV.
75. P. M. Elias, J. Goerke and D. S. Friend; J. Invest. Dermatol., 69:535 (1977).
76. M. L. Williams and P. M. Elias; CRC Crit. Rev. Ther. Drug Carrier Syst., 3:95 (1987).
77. R. J. Scheuplein; J. Invest. Dermatol., 67:672 (1976).
78. R. T. Tregear; J. Physiol. (Lond.), 156:303 (1961).
79. H. A. Abramson and M. H. Gorin; J. Phys. Chem., 44:1094 (1940).
80. G. M. Mackee, M. B. Sulzberger, F. Herrmann and R. L. Baer; J. Invest. Dermatol., 6:43 (1945).
81. G. Moretti; in: *Handbuch der Haut- and Geschlechtskrank-heiten* (O. Gans and G. K. Steigleder, Eds.), Springer-Verlag, Berlin, 1968, Vol. 1, pp. 567–609.
82. R. K. Winkelmann, S. R. Scheen, Jr., R. A. Pyka and M. B. Coventry; in: *Advances in Biology of Skin* (W. Montagna and R. A. Ellis, Eds.), Pergamon, New York, 1961, Vol. 2, pp. 1–19.
83. A. Yen and I. M. Braverman; J. Invest. Dermatol., 66:131 (1976).
84. I. M. Braverman and A. Yen; J. Invest. Dermatol., 68:44 (1977).
85. A. B. Hertzmann; Am. J. Phys. Med., 32:233 (1953).
86. A. B. Hertzmann; in: *Advances in Biology of the Skin* (W. Montagna and R. A. Ellis, Eds.), Pergamon, New York, 1961, Vol. 2, pp. 98–116.
87. Y. W. Chien; *Novel Drug Delivery Systems: Fundamentals, Developmental Concepts and Biomedical Assessments*, Dekker, New York, 1982, Chapter 5.
88. D. R. Sanvordeker, J. G. Cooney and R. C. Wester; U.S. Patent 4,336,243, June 22, 1982.
89. P. R. Keshary, Y. C. Huang and Y. W. Chien; unpublished data, personal communication, 1984.
90. Y. W. Chien and K. H. Valia; Drug Dev. Ind. Pharm., 10:575 (1984).
91. K. H. Valia and Y. W. Chien; Drug Dev. Ind. Pharm., 10:951 (1984).
92. K. H. Valia and Y. W. Chien; Drug Dev. Ind. Pharm., 10:991 (1984).
93. Y. W. Chien; Pharm. Technol., 9:50–66 (1985).

94. W. R. Good; Drug Dev. Ind. Pharm., 9:647 (1983).
95. A. Gerardin, J. Hirtz, P. Fankhauser and J. Moppert; in: APhA/APS 31st National Meeting Abstracts, 11(2):84 (1981).
96. J. E. Shaw; Am. Heart J., 108:217 (1984).
97. M. A. Weber and J. I. M. Drayer; Am. Heart J., 108:231 (1984).
98. D. Arndts and K. Arndts; Eur. J. Clin. Pharmacol., 26:79 (1984).
99. L. Schenkel, J. Balestra, L. Schmitt and J. Shaw; in: *Second International Conference on Drug Absorption—Rate Control in Drug Therapy*, Edinburgh, Scotland, September 21–23, 1983, p. 41.
100. L. R. Laufer, J. L. De Fazio, J. K. H. Lu, D. R. Meldrum, P. Eggena, M. P. Sambhi, J. M. Hershmann and H. L. Judd; Am. J. Obstet. Gyncol., 146:533 (1983).
101. T. J. Roseman, R. M. Bennett, J. J. Biermacher, M. E. Tuttle and C. H. Spilman; in: *Proceedings of 11th International Symposium on Controlled Release Bioactive Materials* (W. E. Meyers and R. L. Dunn, Eds.), Fort Lauderdale, Florida, 1984, p. 50.
102. Y. W. Chien; *Novel Drug Delivery Systems*, Dekker, New York, 1982, Chapter 9.
103. A. D. Keith; Drug Dev. Ind. Pharm., 9:605 (1983).
104. A. F. Kydonieus and B. Berner; *Transdermal Drug Delivery Systems*, CRC Press, Boca Raton, Florida, 1987, Volume I, Chapter 11.
105. M. Wolff, G. Cordes and V. Luckow; Pharm. Res., 1:23 (1985).
106. Y. W. Chien and H. J. Lambert; U.S. Patent 3,946,106, March 23, 1976.
107. Y. W. Chien and H. J. Lambert; U.S. Patent 3,992,518, November 16, 1976.
108. Y. W. Chien and H. J. Lambert; U.S. Patent 4,053,580, October 11, 1977.
109. A. Karim; Drug Dev. Ind. Pharm., 9:671 (1988).
110. Y. W. Chien; in: *Recent Advances in Drug Delivery Systems* (J. M. Anderson and S. W. Kim, Eds.), Plenum, New York, 1984, p. 367.
111. Y. W. Chien; Drug and enzyme targeting. Methods Enzymol., 112:461–470 (1985).
112. Y. W. Chien; J. Pharm. Sci., 73:1064 (1984).
113. Y. W. Chien, P. R. Keshary, Y. C. Huang and P. P. Sarpotdar; J. Pharm. Sci., 72:968 (1983).
114. Y. W. Chien, T. Y. Chien and Y. C. Huang; Proc. Int. Symp. Control. Release Biol. Mater., 15:286 (1988).
115. Y. W. Chien, T. Y. Chien, Y. C. Huang and R. E. Bagdon; Proc. Int. Symp. Control. Release Biol. Mater., 15:288 (1988).
116. Y. W. Chien, T. Y. Chien and Y. C. Huang; U.S. Patent 4,818,540, April 4, 1989.
117. S. J. Davidson, L. D. Nichols, A. S. Obermeyer, M. B. Allen, E. J. Murphy and R. N. Hurd; Proc. Int. Symp. Control. Release Biol. Mater., 11:58 (1984).
118. A. C. Hymes; in: APhA/APS Midwest Regional Meeting Abstract, April 2, 1984, Chicago, Illinois, p. 4.
119. H. Durrheim, G. L. Flynn, W. I. Higuchi and C. R. Behl; J. Pharm. Sci., 69:781 (1980).
120. K. Tojo; in: *Transdermal Controlled Systemic Medications* (Y. W. Chien, Ed.), Dekker, New York, 1987, Chapter 6.
121. K. Tojo, J. A. Masi and Y. W. Chien; I&EC Fundament., 24:368–373 (1985).
122. K. Tojo, M. Ghannam, Y. Sun and Y. W. Chien; J. Control. Release, 1:197–203 (1985).
123. Y. C. Huang; in: *Transdermal Controlled Systemic Medications* (Y. W. Chien, Ed.), Dekker, New York, 1987, Chapter 7.
124. Y. W. Chien; unpublished data, personal communication (1988).
125. Y. W. Chien; *Drugs of Today*, 23:625 (1987).
126. Y. W. Chien; Alternative membranes for in-vitro skin permeation studies, in: *American Association of Pharmaceutical Scientists Pre-meeting Workshop on Transdermal*, Washington, D.C., October 31–November 1, 1986.

127. N. Reichek; in: *Transdermal Controlled Systemic Medications*, (Y. W. Chien, Ed.), Dekker, New York, 1987, Chapter 9.
128. D. E. Rezakovic; in: *Transdermal Controlled Systemic Medications* (Yie W. Chien, Ed.), Dekker, New York, 1987, Chapter 10.
129. Y. W. Chien; Am. Heart J., 108:207 (1984).
130. C. C. Chiang and C. S. Lee; unpublished data, personal communication (1986).
131. J. E. Shaw, S. K. Chandrasekaran, A. S. Michaels and L. Taskovich; in: *Animal Models in Dermatology* (H. Maibach, Ed.), Churchill–Livingstone, Edinburgh, 1975, Chapter 14.
132. P. R. Keshary, Y. C. Huang, and Y. W. Chien; Drug Dev. Ind. Pharm., 11:1213–1254 (1985).
133. 1985 International Pharmaceutical R & D Symposium on Advances in Transdermal Controlled Drug Administration for Systemic Medications, Rutgers University, College of Pharmacy, June 20 and 21, 1985.
134. 1986 Neu-Ulm Conference on Transdermal Drug Delivery System, University of Ulm, West Germany, December 1–3, 1986.
135. Symposium on Problems and Possibilities for Transdermal Drug Delivery, Schools of Medicine and Pharmacy, University of California, San Francisco, California, February 2–3, 1985.
136. P. W. Armstrong, J. A. Armstrong and G. S. Marks; Am. J. Cardiol., 46:670 (1980).
137. Y. W. Chien and C. S. Lee; in: *Controlled Release Technology: Pharmaceutical Applications* (P. I. Lee and W. R. Good, Eds.), American Chemical Society, Washington, D.C., 1987, Chapter 21.
138. T. Higuchi and V. Stella; *Prodrugs as Novel Drug Delivery Systems*, American Chemical Society, Washington, D.C., 1975.
139. K. H. Valia, K. Tojo and Y. W. Chien; Drug Dev. Ind. Pharm., 11:1133–1173 (1985).
140. K. Tojo, K. H. Valia, G. Chotani and Y. W. Chien; Drug Dev. Ind. Pharm., 11:1175–1193 (1985).
141. Y. W. Chien, K. H. Valia and U. B. Doshi; Drug Dev. Ind. Pharm., 11:1195–1212 (1985).
142. E. R. Cooper; Abstracts of 1985 International Pharmaceutical R & D Symposium on Advances in Transdermal Controlled Drug Administration for Systemic Medications, Rutgers University, College of Pharmacy, June 20 and 21, 1985, p. 7.
143. A. K. Banga and Y. W. Chien; J. Control. Release., 7:1–24 (1988).
144. K. Stillwell; *Therapeutic Electricity and Ultraviolet Radiation*, 3rd Ed., Williams & Wilkins, Baltimore, 1983, p. 33.
145. S. Licht; *Therapeutic Electricity and Ultraviolet Radiation*, 2nd Ed., Waverley Press, Baltimore, 1967, pp. 167–171.
146. J. Russo, A. Lipman, T. Constock, B. Page and R. Stephen; Am. J. Hosp. Pharm., 37:843–847 (1980).
147. P. Harris; J. Orthopaed. Sports Phys. Ther., 4:109–112 (1982).
148. P. Tyle; Pharm. Res., 3:318–326 (1986).
149. K. Okabe, H. Yamaguchi and Y. Kawai; J. Control Release, 4:79–85 (1986).
150. L. Wearley, J. C. Liu and Y. W. Chien; J. Control. Release, 8:237–250 (1989).
151. L. Wearley, J. C. Liu and Y. W. Chien; Iontophoresis-facilitated transdermal delivery of verapamil. II. Factors affecting the reversibility of skin permeability. J. Control. Release, 9:231–242 (1989).
152. P. C. Kuo; Kinetics of transdermal permeation of oxycodone HCL and the enhancement by iontophoresis, M.Sc. Thesis, Rutgers University, New Brunswick, N.J.
153. O. Siddiqui and Y. W. Chien; Crit. Rev. Ther. Drug Carrier Syst., 3:195–208 (1987).
154. B. Kari; Diabetes, 35:217 (1986).
155. R. Stephen, T. Petelenz and S. Jacobsen; Biomed. Biochim. Acta, 43:553–558 (1984).

156. O. Siddiqui, Y. Sun, J. C. Liu and Y. W. Chien; J. Pharm. Sci., 76:341–345 (1987).
157. Y. W. Chien, O. Siddiqui, Y. Sun, W. M. Shi and J. C. Liu: Ann. N.Y. Acad. Sci., 507:32–51 (1987).
158. Y. W. Chien, O. Siddiqui, W. M. Shi, P. Lelawongs and J. C. Liu; J. Pharm. Sci., 78:376 (1989).

8
Parenteral Drug Delivery and Delivery Systems

I. INTRODUCTION

Unlike mucosal and transdermal drug delivery, in which the systemic bioavailability of a drug is always limited by its permeability across a permeation barrier (epithelial membrane or stratum corneum) and oral drug delivery, in which the systemic bioavailability of a drug is often subjected to variations in gastrointestinal transit and biotransformation in the liver by "first-pass" metabolism, parenteral drug delivery, especially intravenous injection, can gain easy access to the systemic circulation with complete drug absorption and therefore reach the site of drug action rapidly. Parenteral drug delivery via intramuscular or subcutaneous administration, although not as fast as intravenous injection, still achieves therapeutically effective drug levels rapidly if the drugs are administered in aqueous solution. This rapid drug absorption is unfortunately also accompanied by a rapid decline in drug levels in the systemic circulation. The usual outcome is the production of a fairly rapid onset but relatively short-acting therapeutic action. For the sake of effective treatment it is often desirable to maintain systemic drug levels within the therapeutically effective concentration range for as long as a treatment calls for (Chapter 1, Figure 1).

Continuous intravenous infusion has been recognized as a superior mode of systemic drug delivery that can be tailored to maintain a constant and sustained drug level within a therapeutic concentration range for as long as required for effective treatment. It also provides a means of direct entry into the systemic circulation for drugs that are subjected to hepatic first-pass metabolism and/or suspected of producing gastrointestinal incompatibility. Unfortunately, such a mode of drug administration entails certain health hazards and therefore necessitates continuous hospitalization during treatment and requires close medical supervision.

To duplicate the benefits of intravenous drug infusion without its potential hazards, much effort has been invested in the development of depot-type parenteral controlled-release formulations. These developmental efforts have generated a number of injectable depot formulations, such as penicillin G-procaine suspensions (Duracillin, Lilly; Crysticillin, Squibb; and Wycillin, Wyeth), cyanocobalamin-Zn-tannate suspensions (Depinar, Armour), medroxyprogesterone acetate suspensions

(Depo-Provera, Upjohn), fluphenazine enanthate- and decanoate-in-oil solutions (Prolixin enanthate and Prolixin decanoate, Squibb), Adenocorticotropin–Zn-tannate gelatin preparations (H. P. Acthar, Armour), microcrystalline desoxycorticosterone pivalate in oleaginous suspension (Percorten pivalate, Ciba), testosterone enanthate (Delatestryl, Squibb), testosterone enanthate–estradiol valerate in ethyl oleate BP repository vehicle (Ditate-DS, Savage), nandrolone decanoate injection (Deca-Durabolin, Organon), and insulin-zinc suspensions (Ultralente, Lente, and Semilente, Novo), levonorgestrel-releasing subdermal implant (Norplant, Wyeth) and goserelin acetate-releasing biodegradable implant (Zoladex implant, ICI), to name just a few.

Parenteral administration of a drug in depot formulation, which is an aqueous (or oleaginous) suspension or an oleaginous solution, into subcutaneous or muscular tissue results in the formation of a depot at the site of injection. This depot acts as a drug reservoir that releases the drug molecules continuously at a rate determined to a large extent by the characteristics of the formulation, leading to the prolonged absorption of drug molecules from the formulation. The nature of the vehicle, either aqueous or oleaginous, used in the formulation and the physicochemical characteristics of the drug (or its derivatives), as well as the interactions of drug with vehicle and tissue fluid, determine the rate of drug absorption and hence the duration of therapeutic activity.

The sustained or controlled release of drugs from parenteral depot formulations in many cases reduces the inherent disadvantages of conventional "immediate-release" parenteral dosage forms. Benefits derived from the parenteral controlled-release formulations are primarily the achievement of a relatively constant and substantially sustained therapeutic drug level with a reduction in the frequency of injection. In other words, the injectable depot formulation was developed with primary objective of simulating the continuous drug administration of intravenous (IV) infusion on a more practical basis. It often results in such additional benefits as reduced drug dose, decreased side effects, enhanced patient compliance, and improved drug utilization.

II. INJECTABLE DRUG DELIVERY

A. Approaches

Several pharmaceutical formulation approaches may be applied to the development of parenteral controlled-release or sustained-release formulations. The most commonly used techniques are as follows.

1. Use of viscous, water-miscible vehicles, such as an aqueous solution of gelatin or polyvinylpyrrolidone
2. Utilization of water-immiscible vehicles, such as vegetable oils, plus water-repelling agent, such as aluminum monostearate
3. Formation of thixotropic suspensions
4. Preparation of water-insoluble drug derivatives, such as salts, complexes, and esters
5. Dispersion in polymeric microspheres or microcapsules, such as lactide-glycolide homopolymers or copolymers
6. Coadministration of vasoconstrictors

These techniques may be used alone, for example formation of aqueous insulin zinc suspensions, or in combination, for example penicillin G-procaine suspension in vegetable oil gelled with aluminum monostearate. Application of these techniques has produced a variety of depot formulations. They may be classified on the basis of the process used for controlled drug release as follows.

1. Dissolution-Controlled Depot Formulations

In this depot formulation the rate of drug absorption is controlled by the slow dissolution of drug particles in the formulation or in the tissue fluid surrounding the formulation. The rate of dissolution $(Q/t)_d$ under sink conditions (1) is defined by

$$\left(\frac{Q}{t}\right)_d = \frac{S_a D_s C_s}{h_d} \tag{1}$$

where S_a is the surface area of drug particles in contact with the medium; D_s is the diffusion coefficient of drug molecules in the medium; C_s is the saturation solubility of the drug in the medium; and h_d is the thickness of the hydrodynamic diffusion layer surrounding each drug particle.

Basically, two approaches can be utilized to control the dissolution of drug particle to prolong the absorption and hence the therapeutic activity of the drug.

Formation of Salt or Complexes with Low Aqueous Solubility. A water-soluble basic (or acid) drug can be rendered effective as a depot by transforming it into a salt with an extremely low aqueous solubility. Typical examples are preparations of penicillin G procaine (C_s = 4 mg/ml) and penicillin G benzathine (C_s = 0.2 mg/ml) from the highly water-soluble alkali salts of penicillin G and preparations of naloxone pamoate (2) and naltrexone-Zn-tannate (3,4) from the water-soluble hydrochloride salts of naloxone and naltrexone, respectively. Aqueous suspensions of benzathine penicillin G (Bicillin L-A, Wyeth), procaine penicillin G (Crysticillin A. S., Squibb), and penicillin benzathine and penicillin G procaine combination (Bicillin C-R, Wyeth) as well as the oleaginous suspensions of procaine penicillin G, naloxone pamoate, and naltrexone-Zn-tannate in vegetable oil (gelled with aluminum monostearate), all produce prolonged therapeutic activities.

Suspension of Macrocrystals. Large crystals are known to dissolve more slowly than small crystals. This is called the macrocrystal principle and can be applied to control the rate of drug dissolution. Typical examples are the aqueous suspension of testosterone isobutyrate for intramuscular administration (Figure 1), and of diethylstilbestrol monocrystals for subcutaneous injection.

An exception to the macrocrystal principle was observed with penicillin G procaine suspension in gelled peanut oil for intramuscular injection (5,6). With large particles ($>150\ \mu m$) the serum levels of penicillin rise rapidly, reach a peak level, and then fall relatively fast. With micronized particles ($\leq 5\ \mu m$), on the other hand, the peak serum concentration of penicillin is reduced in magnitude and the therapeutic effective dose level is prolonged substantially. In contrast, with the suspension in plain peanut oil, that is, without gelation with aluminum monostearate, or with the aqueous suspension this macrocrystal principle was followed fairly well. The observed exception was thought to be due to the possibility that the micronized particles of penicillin G procaine may reach the interface of tissue fluid and gelled suspension at a rate significantly slower than the large particles. A lower rate of

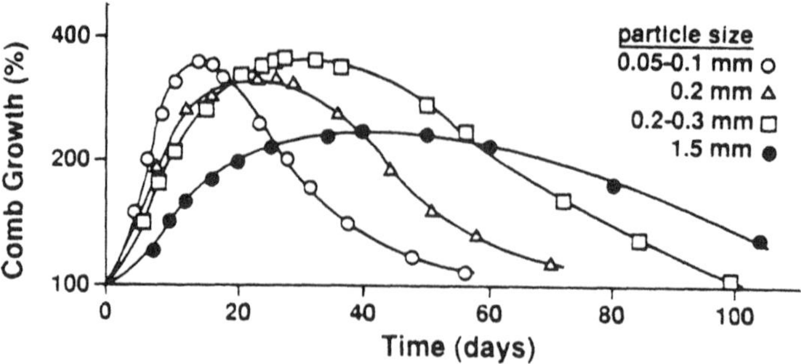

Figure 1 Effect of particle size of testosterone isobutyrate on the growth of the comb of capons (%). [Reproduced with permission from Lippold; Pharmacy International, 1:60 (1980).]

interfacial dissolution is thus achieved for micronized particles than for large particles (1).

The major drawback of these two injectable depot formulations is that the release of drug molecules is not of zero-order kinetics as expected from the theoretical model defined by Equation (1). Two reasons account for this deviation: (i) the surface area S_a of the drug particles diminishes with time because of increased drug dissolution; and (ii) the saturation solubility C_s of the drug at the injection site cannot be easily maintained because of rapid absorption.

2. *Adsorption-type Depot Preparations*

This depot preparation is formed by the binding of drug molecules to adsorbents. In this case only the unbound, free species of the drug is available for absorption. As soon as the unbound drug molecules are absorbed a fraction of the bound drug molecules is released to maintain equilibrium. The equilibrium concentration of free, unbound drug species $(C)_f$ is determined by the Langmuir relationship,

$$\frac{(C)_f}{(C)_b} = \frac{1}{a(C)_{b,m}} + \frac{(C)_f}{(C)_{b,m}} \tag{2}$$

where $(C)_b$ is the amount of drug (mg) adsorbed by 1 g adsorbent; $(C)_{b,m}$ is the maximum amount of drug (mg) adsorbed by 1 g adsorbent and can be estimated from the slope value of the linear plots of $(C)_f/(C)_b$ versus $(C)_f$; and a is a constant and can be determined from the intercept and $(C)_{b,m}$.

This depot preparation is exemplified by vaccine preparations in which the antigens are bound to highly dispersed aluminum hydroxide gel to sustain their release and hence prolong the duration of stimulation of antibody formation.

3. *Encapsulation-type Depot Preparations*

This depot preparation is prepared by encapsulating drug solids within a permeation barrier or dispersing drug particles in a diffusion matrix. Both permeation barrier and diffusion matrix are fabricated from biodegradable or bioabsorbable macromolecules, such as gelatin, dextran, polylactate, lactide-glycolide copolymers, phospholipids, and long-chain fatty acids and glycerides. Typical examples are naltrexone pamoate-releasing biodegradable microcapsules (7), liposomes (8), and norethindrone-releasing biodegradable lactide-glycolide copolymer beads. The release of drug molecules is controlled by the rate of permeation across the permeation barrier and the rate of biodegradation of the barrier macromolecules.

4. *Esterification-type Depot Preparations*

This depot preparation is produced by esterifying a drug to form a bioconvertible prodrug-type ester and then formulating it in an injectable formulation. This formulation forms a drug reservoir at the site of injection. The rate of drug absorption is controlled by the interfacial partitioning of drug esters from the reservoir to the tissue fluid and the rate of bioconversion of drug esters to regenerate active drug molecules. It is exemplified by the fluphenazine enanthate (Prolixin enanthate, Squibb), nandrolone decanoate (Deca-Durabolin, Organon), and testosterone 17β-cypionate (depo-testosterone cypionate, Upjohn) in oleaginous solution.

B. Development of Injectable Controlled-Release Formulations

1. *Long-acting Penicillin Preparations*

Penicillin in the form of a water-soluble sodium or potassium salt is rapidly absorbed from subcutaneous and intramuscular sites of parenteral administration. The intramuscular route is preferred. Following intramuscular administration of the aqueous solution of the sodium or potassium salt of penicillin G, rapid absorption and high peak serum levels of penicillin are obtained. The high serum penicillin concentrations then decline rapidly as a result of the rapid urinary excretion of penicillin following its absorption from the site of injection.

It is known that 80% of urinary excretion is by tubular secretion and the remaining 20% is by glomerular filtration. The renal tubular secretion of penicillin can be interfered with or blocked by coadministration of such drugs as phenylbutazone, aspirin, and indomethacin, leading to enhancement and prolongation of the effective blood levels of penicillin (9).

A number of pharmaceutical techniques have been utilized to extend the duration of therapeutic activity of penicillin by preparing long-acting formulations of penicillin for intramuscular administration. The earliest approach was to reduce the aqueous solubility of penicillin by converting the water-soluble sodium or potassium salt into salts with an extremely low aqueous solubility, such as penicillin G procaine (with an aqueous solubility of 4 mg/ml). Intramuscular administration of penicillin G procaine in a vegetable oil produces a depot effect that sustains the therapeutic blood level of penicillin for 24–48 hr. Gelation of this oil suspension with 2% aluminum monostearate further prolongs the therapeutic blood level of penicillin to 96 hr.

Another approach is to prepare an aqueous suspension of relatively water-insoluble salts. This type of depot preparation was found most acceptable to medical professionals, but it has achieved only limited success in sustaining the therapeutic blood levels of penicillin during the initial stage of development. The aqueous suspension of penicillin G procaine, for example, was able to maintain the therapeutic blood level of penicillin for only 12–24 hr. It was later discovered that the absorption and hence the therapeutic blood levels of penicillin can be significantly prolonged by maintaining a high solid-water ratio in the aqueous suspension of penicillin G procaine (10). Further development has resulted in the preparation of long-acting thixotropic suspensions of penicillin G procaine (11), such as Duracillin (Lilly), Crysticillin (Squibb), and Wycillin (Wyeth).

1) Oleaginous Suspensions. An injectable depot formulation of penicillin was developed by dispersing the micronized crystals of penicillin G procaine in vegetable oil, such as peanut or sesame oil. This dispersion was then gelled with aluminum monostearate (5). Intramuscular injection of this formulation was reported to produce therapeutic blood levels of penicillin in both animals and humans for sustained duration (5,6,12). The duration of the depot effect of this parenteral sustained-release drug formulation depends on the following formulation variables:

a) Type and size of penicillin G procaine crystals. The crystal type of penicillin G procaine, even after micronization, was found to affect the magnitude of the peak serum concentration of penicillin and also the duration of therapeutic penicillin levels (Table 1). The differences in bioavailability observed among the crystals precipitated from different solvent systems may be due to the variation in polymorphic form or the change in solvate formation of the penicillin crystals. The duration of therapeutically effective serum levels of penicillin was sustained markedly, but the peak serum penicillin concentration was suppressed when a small particle size of penicillin G procaine was administered (Figure 2). The result is an exception to the macrocrystal principle demonstrated earlier by testosterone isobutyrate (Figure 1). However, the macrocrystal principle is followed by penicillin G procaine crystals when they are formulated in aqueous suspensions (Table 2) or in vegetable oil suspensions without the addition of aluminum monostearate.
b) Type and amount of aluminum stearate. Without gelling with aluminum monostearate the oleaginous suspension of micronized penicillin G procaine crystals in either peanut oil or sesame oil shows no advantages over the aqueous suspension in maintaining the sustained therapeutic blood levels of penicillin. In either aqueous or oleaginous suspension the therapeutic effective levels could be maintained for at most 24 hr.

With the addition of aluminum monostearate to the vegetable oils to form a gel, the intramuscular bioavailability of penicillin from a penicillin G procaine suspension

Table 1 Effect of Crystal Type of Penicillin G Procaine on the Intramuscular Bioavailability of Penicillin

	Serum penicillin concentration	
Crystallization medium and crystal types[a]	Peak level[b] (units/ml)	Duration[c] (hr)
Water	1.18	162
Acetone/water	0.59	146
Propanol		
Large crystals	2.19	93
Small crystals	0.76	152

[a]Suspensions were prepared to contain micronized penicillin G procaine (300,000 units/ml) of one crystal type in peanut oil gelled with 2% aluminum monostearate.
[b]Peak serum penicillin levels in rabbits were reached within 1 hr.
[c]Duration with serum penicillin concentration higher than the minimum therapeutic concentration (0.03 units/ml).
Source: Compiled from the data by Buckwalter and Dickison; J. Am. Pharm., 47:661 (1958).

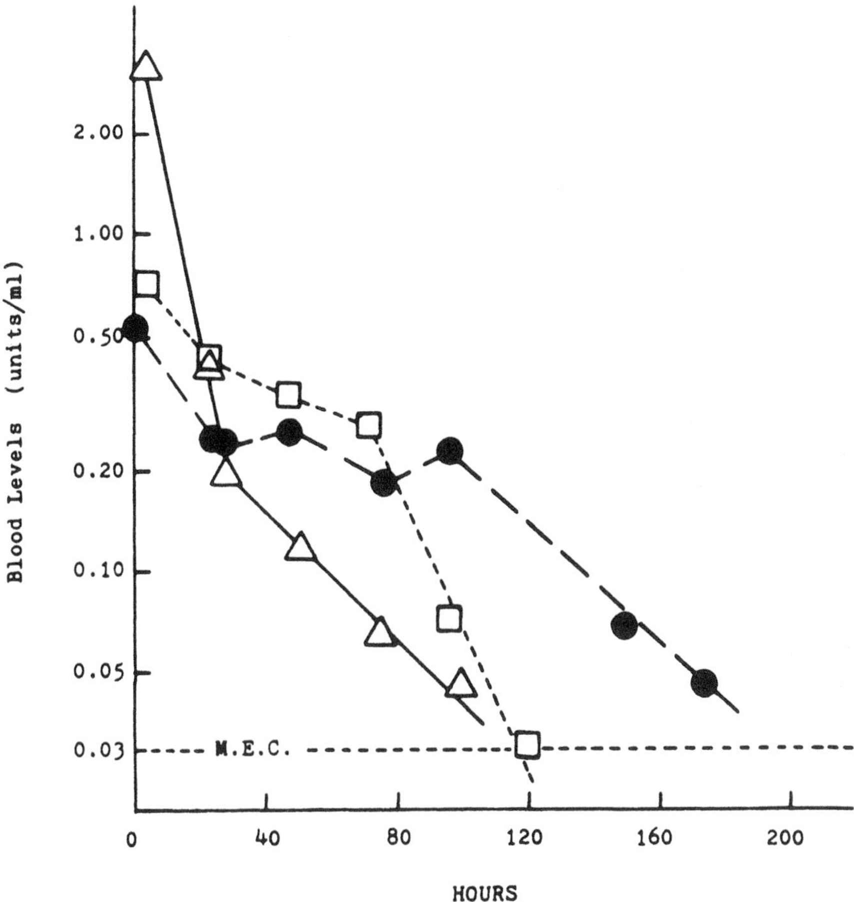

Figure 2 Effect of particle size of penicillin G procaine on the blood profiles of penicillin following intramuscular administration (50,000 units/kg) in rabbits. Each suspension contains 300,000 units/ml of penicillin G procaine with a particle size of (△) 150–175 μm, (□) 45–60 μm, or (●) <5 μm (in peanut oil gelled with 2% aluminum monostearate). (Plotted from the data by Buckwalter; U.S. Patent 2,507,193; May 9, 1950.)

is significantly prolonged and the peak serum level of penicillin is suppressed (Figure 3). The effect appears to depend on the concentration of aluminum monostearate. The therapeutic effective concentration of penicillin is prolonged approximately five times and the peak plasma level is suppressed by as much as sevenfold when the suspension is gelled with 1% aluminum monostearate. Serum penicillin concentrations are maintained over the minimum therapeutic level for longer than 168 hr when the aluminum monostearate content is doubled from 1 to 2%. However, incorporation of more than 2% aluminum monostearate into the vegetable oils appears to add only limited benefits to the prolongation of the duration of effective penicillin levels. On the contrary, suspensions that contain more than 2% aluminum monostearate are too viscous for practical use. Aluminum monostearate produces a relatively similar effect in sesame and peanut oils.

Table 2 Effect of Particle Size of Penicillin G Procaine on Serum Penicillin Levels

Particle size[a] (μm)	Average serum levels[b] (units/ml)					
	1 hr	4 hr	24 hr	28 hr	48 hr	72 hr
150–250	1.37	1.29	0.82	0.86	0.31	0.12
105–150	1.24	1.50	0.76	0.28	0.16	c
58–105	1.54	1.44	0.47	0.25	0.12	c
35–38	1.64	1.51	0.62	0.33	0.15	c
<35	2.40	2.36	0.33	0.16	0.07	c
1–2	2.14	2.22	0.06	c	c	c

[a]Each aqueous suspension contains penicillin G procaine (300,000 units/ml) with a specific particle size range indicated.
[b]Average serum levels of penicillin in rabbits (units/ml).
[c]Concentration is lower than the therapeutic effective levels of penicillin.
Source: Compiled from the data by Buckwalter and Dickison; J. Am. Pharm. Assoc., 47:661 (1958).

It is interesting to note that aluminum distearate and tristearate appear to have an effect very similar to that of monostearate on the intramuscular bioavailability of penicillin. On the other hand, prolongation of the intramuscular bioavailability of penicillin by aluminum stearate cannot be duplicated by beeswax. Addition of beeswax (up to 5%) produced only a slight effect, if any, on the duration of therapeutic penicillin levels (6).

The depot effect of penicillin G procaine suspension in gelled vegetable oil appears to be related to the combined effects of (i) a reduction in the aqueous solubility of penicillin G by the formation of a procaine salt with low aqueous solubility; and (ii) retardation of intramuscular drug absorption by the formation of compact cohesive depots within the muscular tissue as soon as the suspension is injected.

The clinical bioavailability of penicillin G procaine suspension in gelled peanut oil was compared with that of other formulations in Figure 4. The results indicate that penicillin G procaine (50% or more with particle size over 50 μm) in peanut oil gelled with 2% aluminum monostearate produces a longer depot action than the ungelled peanut oil. On the other hand, the water-soluble sodium salt and water-insoluble aluminum salt of penicillin G procaine salt in gelled peanut oil produced a duration similar to that in ungelled oil. In complete contrast to the macrocrystal principle, micronized penicillin G procaine (95% with particle size less than 5 μm) in gelled peanut oil produced a much longer repository action than large penicillin G procaine crystals in the same formulation.

Further evaluation in 1008 patients demonstrated that depot penicillin formulations with micronized penicillin G procaine crystals suspended in gelled peanut oil are capable of maintaining a sustained therapeutic effective concentration (≥0.03 units/ml) in 90% of patients for 80 hr (Figure 5). In comparison, the same formulation with nonmicronized penicillin G procaine crystals was able to maintain the $ED_{90\%}$ for only 30 hr. Without gelation with 2% aluminum monostearate, the micronized penicillin G procaine suspension in plain peanut oil maintained the $ED_{90\%}$ for approximately 10 hr.

The concepts and techniques of depot penicillin oleaginous suspensions have also been satisfactorily applied to the development of a parenteral controlled-release

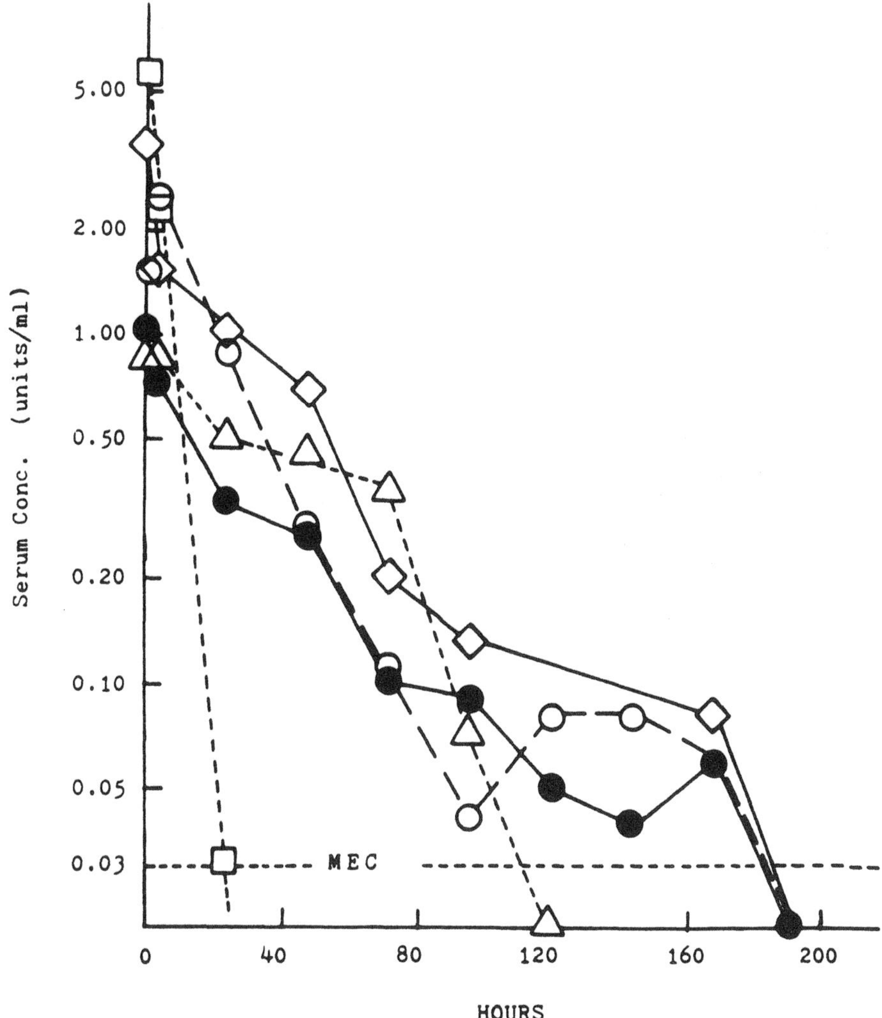

Figure 3 Effect of aluminum monostearate content in penicillin G procaine suspension on the intramuscular bioavailability of penicillin in rabbits. Each suspension contains 300,000 units/ml of micronized penicillin G procaine in sesame oil gelled with aluminum monostearate: (□) 0%, (△) 1%, (●) 2%, (◇) 3%, and (○) 4%. [Plotted from the data by Buckwalter and Dickison; J. Am. Pharm. Assoc., 47:661 (1958).]

formulation for relaxin (13), naloxone and naltrexone palmoate (2), naltrexone-Al-tannate and naltrexone-Zn-tannate (3), and aurothioglucose (Solganal, Schering).

2. Aqueous Suspensions. Several years following the development of depot penicillin oleaginous suspensions, the scientists at Abbott Laboratories also discovered that the therapeutic serum concentration of penicillin can be substantially prolonged by formulating penicillin G procaine in an aqueous thixotropic suspension (10,11). This was accomplished by maintaining a high solid-vehicle ratio (40–70% *w/v* of milled and micronized penicillin G procaine particles). Its prolonged

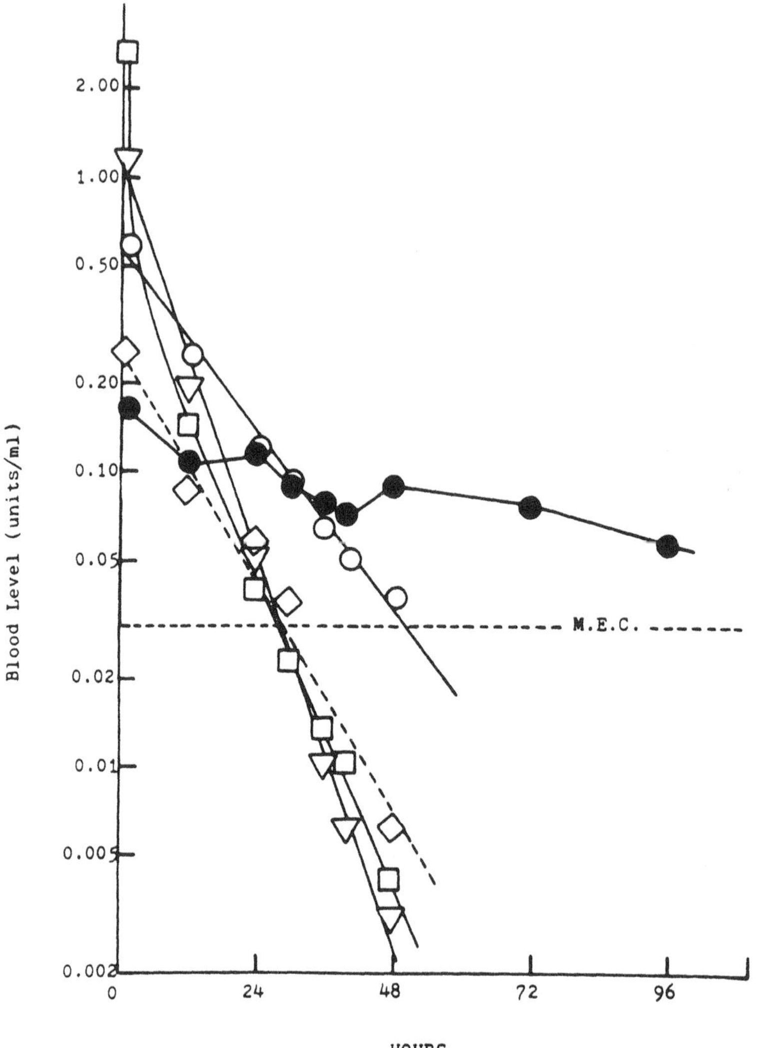

Figure 4 Comparative clinical bioavailability of penicillin from intramuscular administration of various penicillin G salts in peanut oil suspension gelled with 2% aluminum monostearate. (●) Penicillin G procaine (95% ≤ 5 μm); (○) penicillin G procaine (≥50% over 50 μm); (□) penicillin G sodium (≥50% over 50 μm); and (◇) penicillin G aluminum (≥50% over 50 μm). (▽) Penicillin G procaine (≥50% over 50 μm) in ungelled peanut oil suspension. [Plotted from the data by Thomas et al.; J. Am. Med. Assoc., 135:1517 (1948).]

action is partly because these thixotropic suspensions tend to form compact and cohesive depots at the site of intramuscular injection, leading to the slow release of penicillin G procaine and partly because of the low aqueous solubility of the procaine salt of penicillin G, which renders the intramuscular absorption of penicillin under the control of dissolution of penicillin G procaine in the tissue fluid.

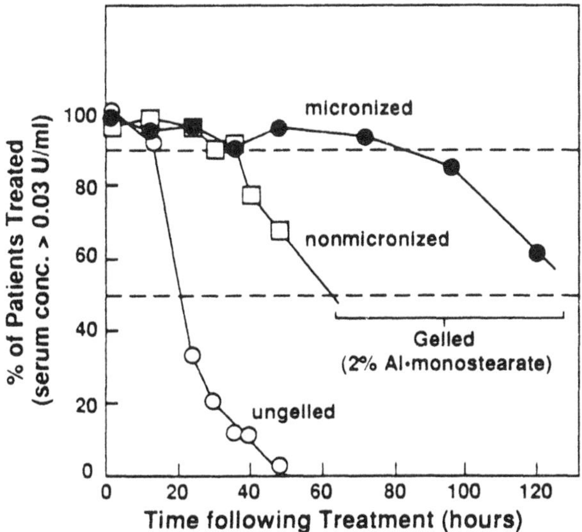

Figure 5 Comparative clinical performance of penicillin G procaine suspension. The perentage of patients treated has serum penicillin concentrations greater than the minimum therapeutic level of 0.03 units/ml (which is used as the indicator for clinical effectiveness): (○) penicillin G procaine suspension in peanut oil; (□) large particles of penicillin G procaine in peanut oil suspensions gelled with 2% aluminum monostearate; (●) micronized particles of penicillin G procaine in peanut oil suspension gelled with 2% aluminum monostearate. [Plotted from the data by Buckwalter and Dickison; J. Am. Pharm. Assoc., 47:661 (1958).]

Thixotropy is a desirable rheological behavior that gives a stable suspension in storage and also confers the necessary flow properties required for manufacturing, processing, and injection. Soon after intramuscular injection a suspension regains its original structure and forms a compact depot at the site of injection. The most important feature of the rheogram for a thixotropic suspension is the existence of a structural breakdown point, a measure of the structure of a suspension before the application of a shear force.

The ability of a suspension to form a depot structure can be evaluated in vivo or in vitro (11). It was concluded that to form a spherical depot aqueous suspensions of penicillin G procaine must possess a structural breakdown point of at least 10^5 dyn-cm (Table 3). Suspensions that possess no structural breakdown point, even though they were pastelike and contained a high penicillin solid content per unit suspension, produced no depot at all.

The structural breakdown point of aqueous penicillin G procaine suspensions was reported to depend upon the solids content in the suspension and their specific surface area. The structural breakdown point was observed to increase in proportion to the increase in both the specific surface area and the solids content. A series of combinations of specific surface area and solids content for powders with broad particle size distributions can be selected to give suspensions of any desired structural breakdown point.

The clinical bioavailability of several aqueous suspensions of penicillin G procaine was compared in Figure 6, which indicates that those suspensions that have a

Table 3 Effect of Structural Breakdown Point and Other Physical Properties of Aqueous Penicillin G Procaine Suspensions on Depot Formation

Suspensions[a]	Particle size (μm)	Specific surface (cm^2/g)	Structural breakdown point (dyn-cm)	Depot shape[b]
A	12.–105	>5,000	0	Flat and fanned
B	8.5–89	7,620	1.02×10^5	Oval
C	2.5–90	20,200	4.11×10^5	Spherical
D	3.2–95	25,000	4.14×10^5	Spherical
E	2.3–79	24,000	5.78×10^5	Spherical

[a]Each suspension contains 55% of penicillin G procaine solids.
[b]Determined by injecting, using a 20 gauge needle, the trial suspensions into a 2% gelatin gel. The formation of depot and its shape can be observed and photographed.
Source: Compiled from the data by Ober et al.; J. Am. Pharm. Assoc., 47:667 (1958).

structural breakdown point greater than 10^5 dyn-cm and are capable of forming a spherical depot structure tend to provide a longer therapeutic penicillin serum level.

The question of possible plugging of hypodermic needles by suspensions is also important from a practical point of view. It was found that suspensions that are made up of penicillin G procaine crystals with a specific surface area exceeding 10^4 cm^2/g and structural breakdown point values below 10^6 dyn-cm have good injectability through a 20 gauge hypodermic needle.

2. *Long-acting Insulin Preparations*

There are more than 50 million people in the world today who suffer from diabetes (World Health Organization, May–June, 1991). Clinically they need regular administration of exogenous insulin for the control of their hyperglycemia.

Insulin is a protein macromolecule. The possibility of delivering insulin orally is attractive, but has been little success in achieving acceptable bioavailability by the oral route because of inactivation of insulin by gastrointestinal enzymes, such as pancreatic enzymes (14,15). Several attempts were made to facilitate the oral absorption of intact insulin by coadministration of trypan red to inhibit gastric digestion, of malachite green to prevent tryptic digestion, and of saponin to aid insulin absorption (16). One of the most recent attempts involved the incorporation of insulin in liposomes (17) and an azopolymer coating (Chapter 11). All these efforts produced a net normoglycemic activity significantly lower than that of parenteral insulin administration, due to extensive hepatic first-pass metabolism following oral absorption (Chapter 3).

Other routes for the systemic delivery of insulin that have been investigated include the ocular, nasal, buccal or sublingual, rectal, and transdermal, which aim to bypass gastrointestinal and hepatic metabolism (Chapter 11). All these routes have yielded incomplete and/or uncertain absorption of this macromolecular drug because of its large molecular size, hydrophilicity, and low permeability across the mucosa membranes and skin.

Because of these failures insulin has been given to diabetic patients exclusively by parenteral administration via the subcutaneous route. The therapeutic activity of insulin is evident within about 1 hr following subcutaneous injection. The duration

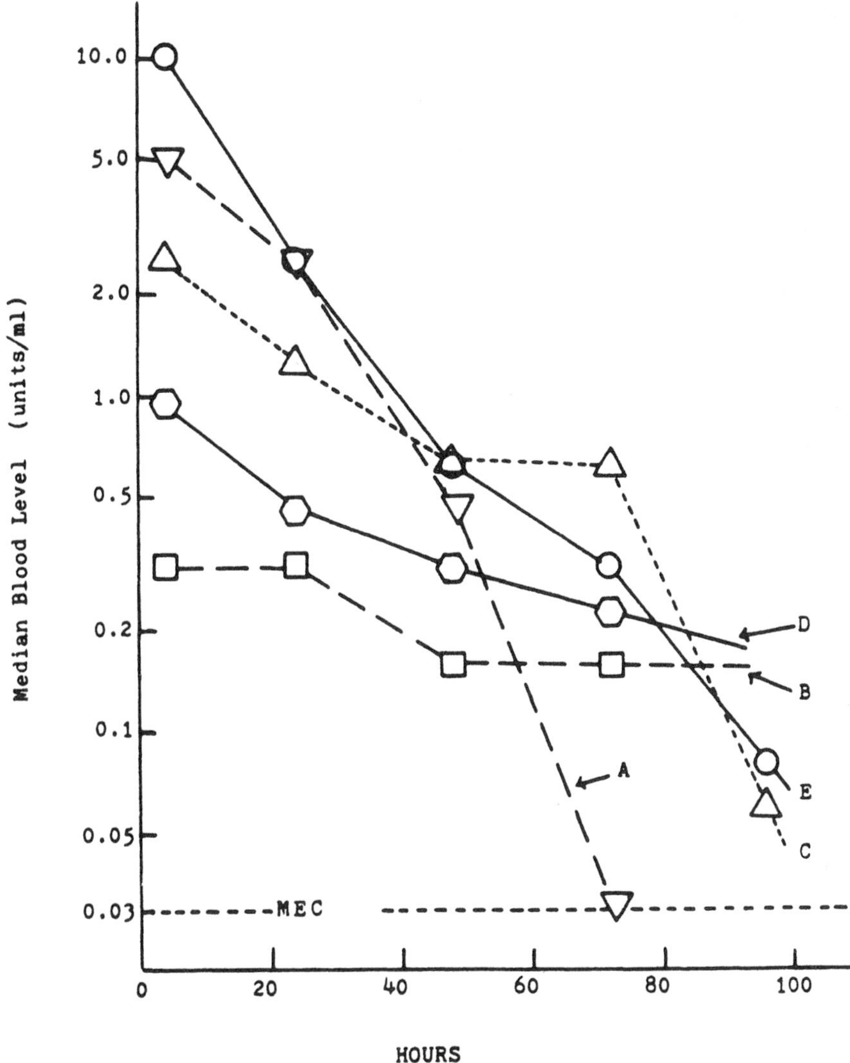

Figure 6 Comparative clinical bioavailability of penicillin from the intramuscular administration of various aqueous penicillin G procaine suspensions listed in Table 3.

of action is relatively short, however, with the plasma half-life approximately 40 min for Insulin ^{131}I and less than 9 min for unlabeled insulin. It was also found that the duration of action is not linearly proportional to the dosage of insulin injected but is a simple function of the logarithm of the dose. This dose dependence is because insulin is inactivated by the liver at a rate proportional to its concentration in the systemic circulation. The usual duration of a regular insulin USP injection is 4–8 hr (Table 4); two to four injections daily are therefore required for the proper control of severe diabetes. For practical reasons diabetic patients are trained to inject themselves at home (18).

Improper injection techniques have been found to be responsible for the poor

Table 4 Normoglycemic Activity and Duration of Some Commercial Insulin Products

Insulin preparations	Normoglycemic activity[a] Onset (hr)	Peak (hr)	Duration (hr)
Insulin injection, USP	0.5–1.0	2–3	4–8
Insulin-zinc complex[b]			
Semilente insulin	0.5–1.0	5–7	12–16
Lente insulin	1.0–1.5	8–12	24
Ultralente insulin	4.0–8.0	16–18	>36
Insulin-Zn-protein complex			
Globin-Zn-insulin injection, USP	2.0	8–16	24
Isophane insulin suspension, USP	1.0–1.5	8–12	24
Protamine-Zn-insulin suspension, USP	4.0–8.0	14–20	36

[a]The time at which activity is evident.
[b]Developed by Novo Terapeutisk Laboratorium A/S.

parenteral bioavailability of insulin. A diabetic patient can easily inject the correct dosage too deeply into the muscular tissue or not deeply enough (intradermally or into the subcutaneous fat tissues). Insulin that is not properly injected into the subcutaneous region, where the normal absorption of insulin can take place, may become temporarily trapped, resulting in areas of lumpiness or swelling. The lumpiness also tends to occur when injections are given in the same region for several days. If the patient traumatizes or exercises the lumpy area, however, insulin can be suddenly absorbed into the blood circulation, resulting in adverse insulin reactions.

Considering these problems associated with the parenteral administration of insulin, development of a long-acting injectable insulin preparation becomes a critical need. It is important for good control of a patient's blood glucose and for obviating the need for multiple daily injections.

An important advance in achieving the prolongation of insulin's activity was made by complexing insulin with protamine, a water-soluble, strongly basic simple protein isolated from the sperm or the mature testes of fish (19). This protamine-insulin complex has an isoelectric point at pH 7.3 and is therefore relatively insoluble in tissue fluids at physiological pH. When injected subcutaneously it slowly releases the insulin for a prolonged period of time (up to 24 hr). The protamine-insulin complex was found to be unstable, however. Its stability was later improved by adding zinc chloride to the preparation to form a protamine-Zn-insulin complex (20). When injected into the loose subcutaneous tissue the protamine-zinc-insulin suspension provides a prolonged normoglycemic activity (over 36 hr). Unfortunately, the protamine-Zn-insulin preparation has a slow onset of action and the normoglycemic action of insulin is not usually evident until 6–8 hr after administration and takes 14–20 hr to reach its peak level (Table 4). Isophane Insulin suspension USP is a preparation similar to protamine-Zn-insulin but with fairly rapid onset (1–1.5 hr) and moderate duration of activity (24 hr). Globin-Zn-insulin injection USP is another typical example of a protein-Zn-insulin complex prepared from zinc chloride and globin, a conjugated protein isolated from beef blood. It provides an insulin release pattern, in terms of the onset and duration of action, similar to that of Isophane Insulin suspension USP.

It was later discovered that the duration of normoglycemic activity of insulin can be sustained substantially without adding proteins, such as protamine and globin, to the preparation. This was achieved by controlling the crystallinity of insulin in the presence of zinc chloride (21).

The insulin molecule reacts with zinc ion and precipitates as a water-insoluble Zn-insulin complex. Depending upon the pH of the solution, it may precipitate as an amorphous or a crystalline solid. Insulin crystals (10–40 μm with a high zinc content) can be precipitated from acetate buffer at pH 5–6. This crystalline insulin-zinc complex is absorbed very slowly and has a prolonged normoglycemic activity. After subcutaneous injection of these Zn-insulin crystals as a suspension in buffer (pH 7.3), the insulin is slowly released and absorbed (4–8 hr to the onset of action) and retains its activity for more than 36 hr (22,23). This preparation is called Ultralente Insulin (Table 4).

On the other hand, the amorphous Zn-insulin complex precipitated at a higher pH (pH 6–8). This amorphous insulin has a low zinc content and is absorbed more readily and achieves a duration of action shorter than that of Ultralente. When administered subcutaneously the insulin in the amorphous Zn-insulin suspension (with a particle size <2 μm) is quickly released and absorbed (with onset of action within 1 hr) and has a shorter duration of normoglycemic activity (12–16 hr). This preparation is called Semilente Insulin (Table 4). Both Ultralente and Semilente insulin, however, produce a longer duration of normoglycemic activity than the regular insulin injection USP (Figure 7). The clinical efficacy of the long-acting Ultralente Insulin is illustrated in Figure 8.

A single injection a day, which is desired by both diabetic patients and physicians, requires adequate timing of the insulin preparation. There is no simple relationship between the blood glucose response of a diabetic patient and the activity range of the insulin preparation.

Patients can be classified according to the promptness of their response to an insulin injection as fast, medium, and slow reaction types. These three types of diabetic patients respond very differently to the same kind of insulin preparation. The blood glucose response of a diabetic patient also depends upon the activity range of the insulin preparation used. Clinical studies have concluded that fast-reaction patients need a relatively slow-acting insulin preparation to maintain the blood glucose level within the range 100–200 mg/dl; slow-reaction patients require a relatively fast-acting preparation (24).

One of the advantages is that the Ultralente and Semilente insulin preparations are mutually miscible, and because of this a range of activity can be formulated. A typical combination is Lente Insulin (Table 4), which consists of seven parts of crystalline and three parts of amorphous insulin-zinc complexes. It provides an intermediate-acting form of insulin. Its effect is evident in approximately 1–1.5 hr, reaches its peak level within 8–12 hr, and has a duration of action of 24 hr.

These three long-acting insulin preparations, Ultralente, Lente, and Semilente, with activity ranging from 12 to 36 hr (Table 4), provide the three reaction types of diabetic patients with a good selection of insulin preparations for single daily injection depending upon their blood glucose response. Clinical evaluation of these in-

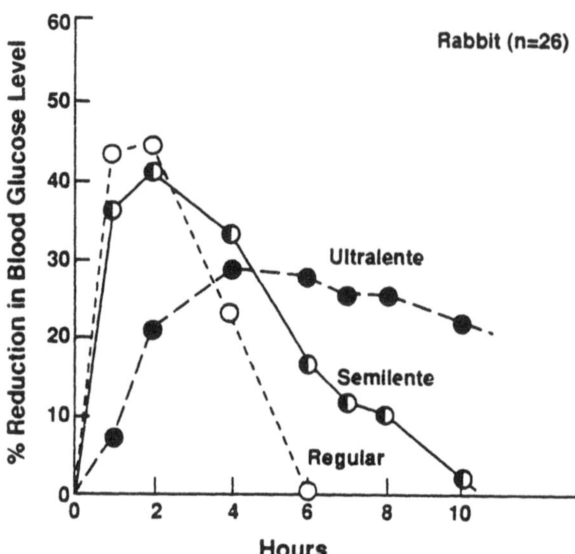

Figure 7 Comparative normoglycemic activity of Ultralente, Semilente, and regular insulin preparations in 26 rabbits by crossover test. Each rabbit received 3 units insulin: (○) regular insulin; (◐) Semilente insulin; and (●) Ultralente insulin. [Plotted from the data by Hallas-Moller; Diabetes, 5:7 (1956).]

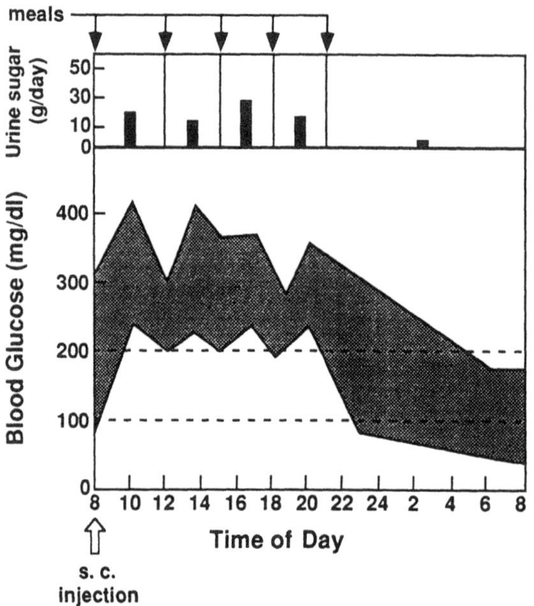

Figure 8 The 24-hr blood glucose profiles in nine diabetic patients after one subcutaneous injection of Ultralente insulin suspension (with average dose of 55 units) at 8 am. [Reproduced with permission from Hallas-Moller et al.; Science, 16:394 (1952).]

sulin-zinc suspensions in 65 patients with severe diabetes indicated that the satisfactory control of blood glucose levels can be obtained in these patients with only single daily injection of the appropriate preparation (24). These long-acting insulin preparations have been marketed in the United States by both Lilly and Squibb under the license of Novo Terapeutisk Laboratorium A/S. The same sustained-release technologies have also been applied to the development of long-acting human insulin preparations, such as Lente human insulin and NPH human insulin isophane suspension (Chapter 11, Table 3). With the recent development of the Novopen insulin delivery device (designed to deliver insulin by a push-button mechanism) and of the Novolinpen, a dial-a-dose insulin delivery device (Figure 9), the subcutaneous administration of insulin has become a convenient routine.

Several controlled-release delivery devices have been developed for the systemic delivery of insulin and are discussed in Chapter 11. These novel approaches for maintaining long-term normoglycemia in diabetes include invasive and noninvasive means of insulin delivery that control blood glucose concentrations by means of preprogrammed, activation-controlled, or feedback-regulated drug delivery mechanisms (25,26). These approaches are discussed in detail in Chapter 11.

3. *Long-acting Vitamin B_{12} Preparations*

By oral administration the systemic bioavailability of vitamin B_{12} is limited and variable because its absorption from the gastrointestinal tract is mediated by two separate

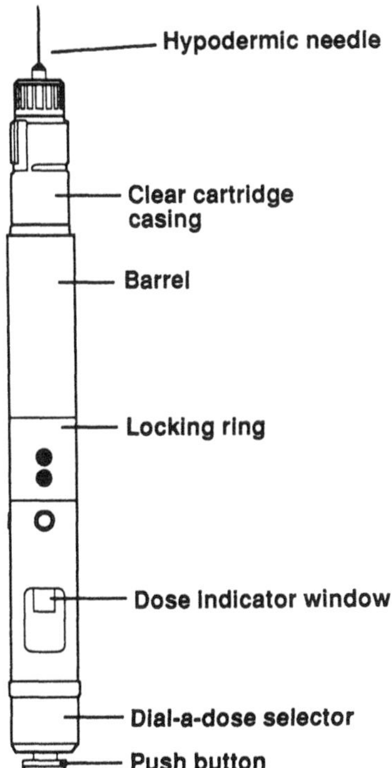

Figure 9 Novolinpen, a mechanically activated drug delivery system, which is designed as a dial-a-dose insulin delivery device to inject human insulin in 2 unit increments (with a dosage accuracy of 99.5%).

and distinct mechanisms. The more important one is mediated by the gastric intrinsic factor of Castle, a glycoprotein secreted by the gastric parietal cells. This gastric intrinsic factor is easily saturated by vitamin B_{12} in an amount as small as only 1.5–3 μg; additionally, attachment of the vitamin B_{12}-intrinsic factor complex to receptors on the ileal surface requires calcium and a pH >6. After a delay of several hours at and within the ileal mucosa, the vitamin is transported by the bloodstream to the liver and other organs. It was reported that interference with this mechanism is responsible for the overwhelming majority of the megaloblastic anemias seen in the United States. Another mechanism of vitamin B_{12} absorption is independent of Castle's gastric intrinsic factor and appears to be a mass action effect, possibly diffusion, and accounts for the absorption of approximately 1% of any dose of free vitamin B_{12} along the entire length of the small intestine. These two mechanisms for the oral absorption of vitamin B_{12} overlap to a variable degree depending upon the daily intake of dietary vitamin B_{12} and the quantity of the vitamin released from its bound state.

On the other hand, vitamin B_{12} is rapidly and quantitatively absorbed from intramuscular and subcutaneous sites of injection. The plasma level of vitamin B_{12} reaches its peak concentration within 1 hr of intramuscular injection. The parenteral administration of vitamin B_{12} given by intramuscular or deep subcutaneous injection is therefore the medication of choice for the treatment of pernicious anemia and other vitamin B_{12} deficiency states. The usual maintenance treatment of patients with pernicious anemia is by monthly injections of vitamin B_{12} (0.03–1.0 mg). Unfortunately, much of the injected doses is lost in the urine. For these reasons, plus the convenience of injection at intervals greater than a month (because pernicious anemia is a chronic disease), it becomes highly desirable to develop a parenteral controlled-release formulation for vitamin B_{12}.

The first approach to the development of a parenteral controlled vitamin B_{12} release formulation was to formulate the vitamin B_{12} in a concentrated partially hydrolyzed gelatin (32%) solution. However, no sustained-release behavior was observed in human testing (27).

In the second approach crystalline vitamin B_{12} was suspended in sesame oil gelled with aluminum monostearate (2%). This approach achieved a significant prolongation of the absorption of vitamin B_{12} compared to an aqueous solution of vitamin B_{12}. Unfortunately, the urinary excretion of this vitamin was still unacceptably excessive.

The third approach was to synthesize an insoluble derivative of vitamin B_{12}, the vitamin B_{12}–zinc-tannate complex, and then suspend it in sesame oil gelled with aluminum monostearate (2%). This preparation achieved a significant prolongation of the absorption of vitamin B_{12} and urinary loss was significantly reduced (Figure 10). On the other hand, the simple salts of vitamin B_{12}-zinc and vitamin B_{12}-tannate did not produce prolongation of vitamin B_{12} absorption. These results are in strong contrast with the sustained release accomplished by aqueous insulin zinc suspensions (Ultralente Insulin and Semilente Insulin, Novo) for diabetes and vasopressin tannate (Pitressin tannate in oil, Parke-Davis) for antidiuresis, in which the prolongation of biological activity is achieved by the formation of a simple salt.

In clinical testing cyanocobalamin-Zn-tannate preparations also produce a more steady and prolonged serum level of vitamin B_{12} (Figure 11). This developmental effort resulted in the marketing of a parenteral long-acting vitamin B_{12} product, De-

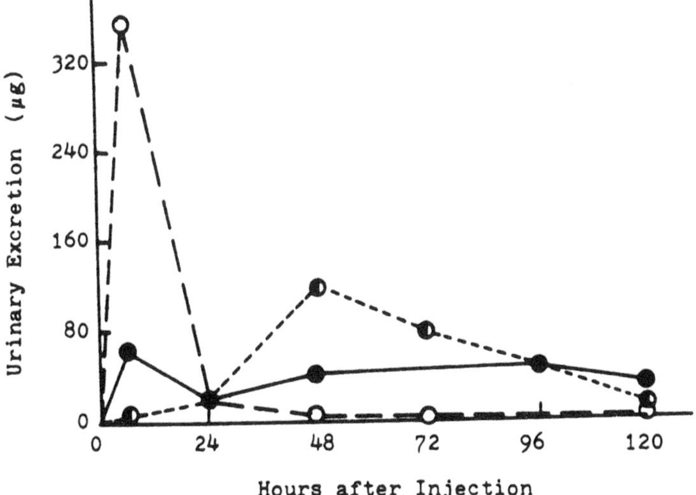

Figure 10 Comparative subcutaneous bioavailability of vitamin B_{12} in rats from various vitamin B_{12} preparations: (○) B_{12} in saline; (◐) B_{12}-Zn-tannate complex in gelled sesame oil; and (●) Depinar (mean of three tests on three commercial lots). [Plotted from the data by Thompson; Bull. Parent. Drug Assoc., 14:6 (1960).]

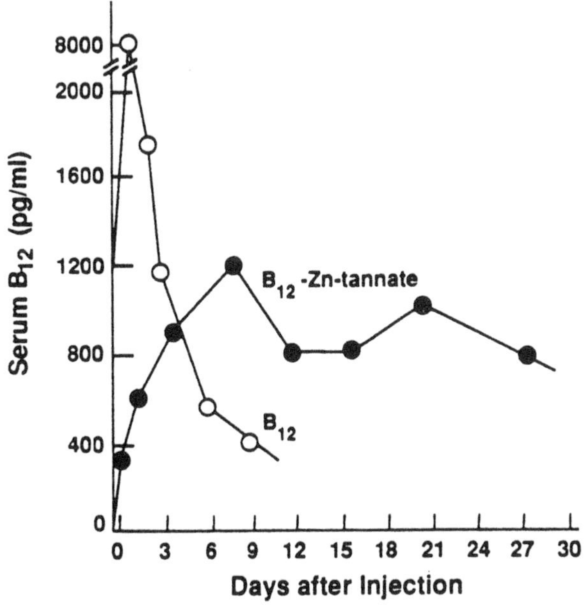

Figure 11 Comparative serum levels of vitamin B_{12} in humans after intramuscular injection of 500 μg vitamin B_{12} (○) and a molar equivalent amount of vitamin B_{12}-Zn-tannate complex (●). [Plotted from the data by Thompson and Hecht; Am. J. Clin. Nutr., 7:311 (1959).]

pinar (Armour Pharmaceutical Co.), which consists of a combination of a readily absorbable vitamin B_{12} and a sustained-release vitamin B_{12}–Zn-tannate. This preparation produces a very small urinary loss (Figure 10), and much of the vitamin B_{12} dose is gradually released from the injection site and deposited in the liver. Adequate maintenance therapy has been provided when given at a dose of 1 mg/ml and at intervals of 8–12 weeks.

4. *Long-acting Adrenocorticotropic Hormone Preparations*

Adrenocorticotropic hormone (ACTH) is a polypeptide hormone that stimulates and regulates the secretion of adrenal steroids, mainly the corticosteroids, from the adrenal cortex. The adrenocorticotropic activity of ACTH is easily destroyed by proteolytic enzymes in the gastrointestinal tract; exogenous ACTH is therefore ineffective when given orally.

On the other hand, ACTH is readily absorbed from parenteral sites of administration and is usually administered by intramuscular injection and occasionally by intravenous infusion. Following intravenous administration ACTH rapidly disappears from the systemic circulation with a plasma half-life of only 15 min in the human. In the rat it was observed that only 0.2% of the hormonal activity is detected in the target organ, the adrenal, at 5–15 min after a rapid intravenous injection. On the other hand, 10–20% of the activity can be found in the nontarget organ, the kidneys. A proteolytic system, probably fibrinolysin, which may be responsible for the inactivation of ACTH, has been detected in the blood (28).

The maximum effect of ACTH on the adrenals can be achieved when an optimum amount of the hormone is delivered continuously. In studies with continuous infusion of a fixed dose of ACTH for varying periods it was observed that the excretion of ketosteroid increases with the duration of infusion. This means that when the ACTH is given by slow intravenous infusion for a prolonged duration much of the injected dose can act on the adrenal cortex. This also occurs, but to a lesser extent, when ACTH is administered intramuscularly in aqueous solution (28). Development of an injectable sustained-released ACTH formulation thus becomes critical.

ACTH is commercially prepared from bovine, porcine, ovine, and cetacean pituitaries. This hormone shows a strong affinity to tissue proteins. The adsorption onto tissue proteins was reportedly responsible for the low effectiveness of ACTH following a single subcutaneous or intramuscular injection of ACTH in aqueous solution. Only a small fraction of the administered dose is actually absorbed into the bloodstream (29). Gelatin was found to inhibit the protein binding of ACTH. Addition of a partially hydrolyzed gelatin into the injectable ACTH solution was noted to enhance adrenal ascorbic acid responses in hypophysectomized rats (Figure 12). This activity was observed to increase as the gelatin concentration in the ACTH injection was increased. The results of this investigation provided the foundation for the development of a repository corticotropin injection (H. P. Acthar gel, Armour), a long-acting injectable formulation that contains a highly purified preparation of ACTH in 16% gelatin solution. It gives a rapid onset and prolonged action to stimulate the functioning adrenal cortex to produce and secrete adrenal steroids. This preparation is active for 24 hr, a significant improvement over the regular corticotropin injection (Acthar, Armour), which has a duration of action of only 8 hr.

The subcutaneous absorption and biological activity of ACTH can also be sus-

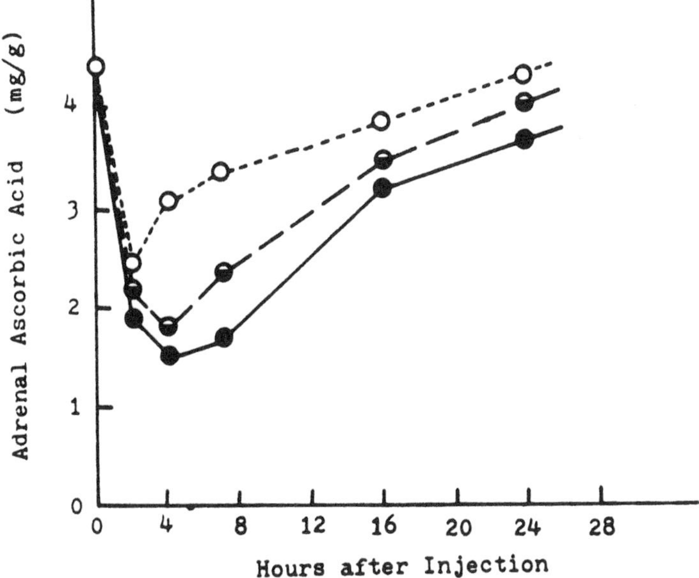

Figure 12 Effect of gelatin on the adrenal ascorbic acid response of hypophysectomized rats to ACTH injections: ACTH in saline (○) with 16% gelatin (◒) and 32% gelatin (●). [Plotted from the data by Thompson; Bull. Parent. Drug Assoc., 14:6 (1960).]

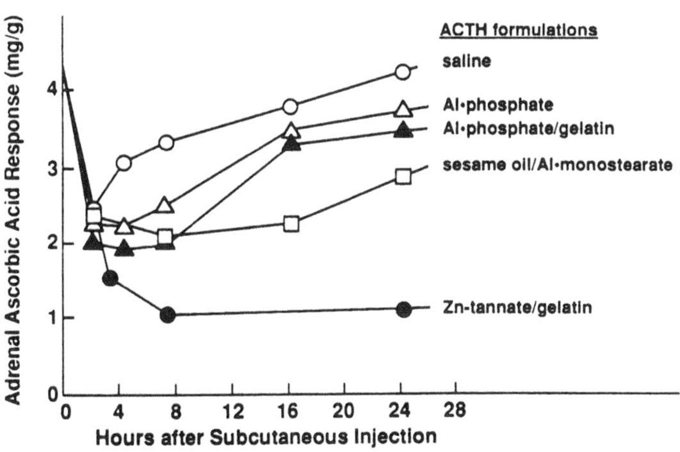

Figure 13 Comparative adrenal ascorbic acid responses in hypophysectomized rats to various ACTH preparations: (○) ACTH in saline; (△) ACTH with aluminum phosphate; (▲) ACTH with aluminum phosphate and 16% gelatin; (□) ACTH suspension in sesame oil gelled with 2% aluminum monostearate; and (○) ACTH-Zn-tannate complex with gelatin. [Plotted from the data by Thompson; Bull. Parent. Drug Assoc., 14:6 (1960).]

tained to varying degrees by adsorption onto aluminum phosphate, by suspension in sesame oil gelled with aluminum monostearate, or by formation of ACTH-Zn-tannate complex (Figure 13). Addition of hydrolyzed gelatin was found to further enhance the activity of various ACTH preparations. The ACTH-Zn-tannate complex in gelatin solution was found to produce the most efficient and long-acting adrenocorticotropic effect.

With prolonged absorption ACTH-Zn-tannate–gelatin preparation achieved a relatively more intense response on adrenal ascorbic acid depletion, adrenal weight, and thymus weight in hypophysectomized rats than an ACTH-saline or an ACTH-gelatin preparation at the same dose by a single injection. A significantly longer duration of activity was produced by the ACTH-Zn-tannate–gelatin preparation.

The activity-time profile of the ACTH-Zn-tannate–gelatin preparation was further compared in humans, using the cumulative urinary excretion of 17-hydroxysteroid as the indicator, with the ACTH-gelatin preparation (H. P. Acthar gel) and the ACTH-Zn-tannate complex (Figure 14). Apparently the ACTH-Zn tannate–gelatin preparation produced a vastly greater effect than the complex alone or the ACTH-gelatin preparation.

5. *Long-acting Steroid Preparations*

Thus far the discussion has been focused on the use of insoluble salts, such as penicillin G procaine, or insoluble complexes, such as vitamin B_{12}–Zn tannate to achieve prolongation of drug action. Alternatively, sustained drug activity can also be ac-

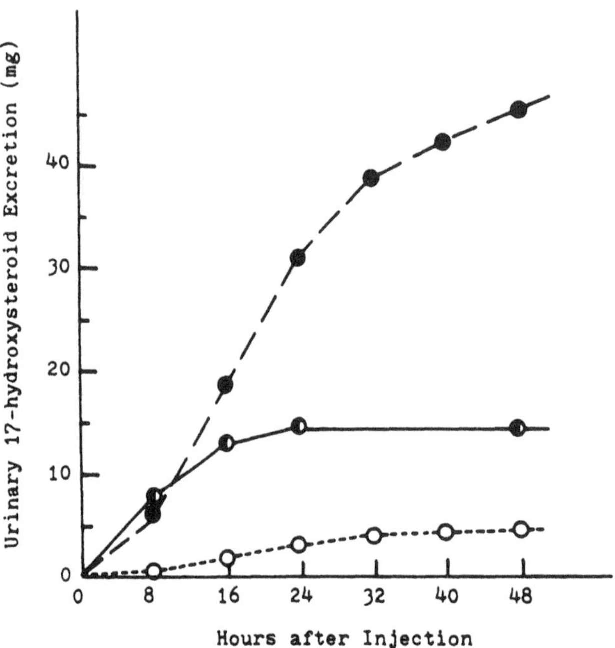

Figure 14 Time course for the cumulative urinary excretion of 17-hydroxy-steroid in humans in response to the subcutaneous administration of various ACTH preparations: (○) ACTH-Zn-tannate-gelatin; (◐) ACTH-gelatin; and (○) ACTH-Zn-tannate (each contains 80 units ACTH). [Plotted from the data by Thompson; Bull. Parent. Drug Assoc., 14:6 (1960).]

complished by esterification of drugs to form water-insoluble but oil-soluble prodrugs. The area in which the prodrug approach has been utilized most extensively is probably in the development of long-acting injectable steroid preparations.

Androgenic Steroids. Testosterone is readily absorbed orally, but such administration is almost completely ineffective because the hormone is subject to pre-systemic metabolism by the liver before reaching the systemic circulation. Testosterone is also quickly metabolized and excreted after parenteral administration in an oleaginous solution, and hence the androgenic effect is small. The androgenic activity of testosterone was reportedly enhanced and prolonged by esterification (30). The biological half-life and the time of maximum effect for a series of testosterone esters administered in an oleaginous intramuscular injectable formulation were found to be closely related to the oil/water partition coefficient of these fatty acid esters (31,32). Further investigations revealed that the difference in the oil/water partition coefficients of the homologous series of testosterone esters is largely determined by the difference in their aqueous solubilities, since their solubilities in an oil like ethyl oleate are approximately equal.

It was further demonstrated by several investigators that the duration of the androgenic activity of testosterone by intramuscular administration can be further sustained by acylation of the 17β-hydroxy group in the testosterone molecule (33–42). The duration of activity of testosterone was effectively prolonged as the chain length of the acyl group increased. It was thought that the longer the chain length of acyl group, the slower was the release of testosterone esters from the oleaginous vehicle. This is a result of the reduction in the water/oil partition coefficient, leading to a decrease in the rate of regeneration of the active testosterone moiety in the body to exhibit its androgenic activity. The effect of chain length and partition coefficient on androgenic activity is discussed subsequently. Miescher et al. (30) were the first to promote the use of long-chain esters of testosterone as an injectable depot form of testosterone.

Studies of a series of saturated and unsaturated esters of testosterone later revealed that the β-cyclopentylpropionate ester of testosterone injected intramuscularly in cottonseed oil is a superior depot formulation of testosterone, as demonstrated by results on the relative growth of seminal vesicles in castrated rats. These investigations provided the foundation for the marketing of testosterone 17β-cypionate injection USP (depo-testosterone cypionate, Upjohn). The enanthate ester of testosterone (Delatestryl, Squibb) is another testosterone ester that provides excellent sustained androgenic activity. Following a single intramuscular injection both formulations produce sustained androgenic activity over a period of about 4 weeks. The continuous release of testosterone from the esters is thought to resemble closely the endogenous production of testosterone.

Testosterone enanthate has also been formulated with estradiol valerate, a long-acting estrogen, in ethyl oleate BP repository vehicle (Ditate-DS, Savage) for the prevention of postpartum breast engorgement.

The esterification approach was also applied to 19-nortestosterone to prolong its anabolic activity by intramuscular (or subcutaneous) injection (43–47). The effect of esterification on the anabolic activity of 19-nortestosterone is illustrated by the formation of a nonanoate, nandrolone nonanoate, in Figure 15, which indicates that following the injection of nandrolone nonanoate the weight of levator ani muscle in

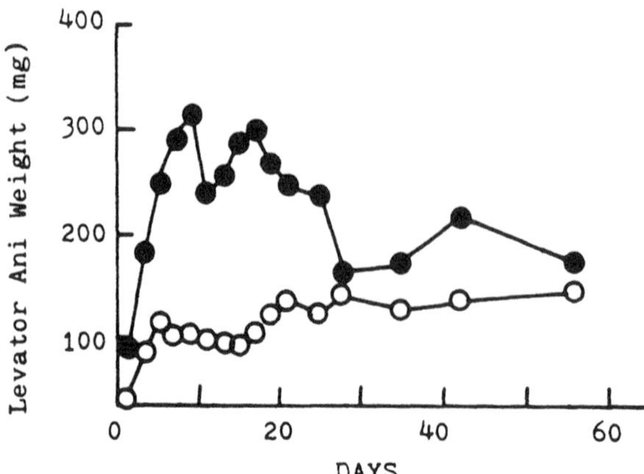

Figure 15 Effect of nandrolone (19-nortestosterone) nonanoate on the growth of levator ani in castrated rats: (●) injection of 1 mg nandrolone nonanoate; (○) controls. [Plotted from the data by Chaudry and James; J. Med. Chem., 17:157 (1974).]

castrated rats was substantially increased compared to control rats for a prolonged period of time.

It is interesting to observe the existence of two peaks of activity for all the nandrolone esters at all dose levels. Multiple regression analysis of these maxima suggested that the times for the second peak of activity are semilogarithmically related to the ethyl oleate/water partition coefficient, but not the first maximum. Similar results were also observed with testosterone esters (31). The biological half-life of testosterone esters in rats was found to be linearly related to the ethyl oleate/water partition coefficient, but the half-life for their disappearance from the injection site was not. These observations were attributed to the distribution to and the accumulation of esters in body fat tissues and their subsequent release (scheme 1).

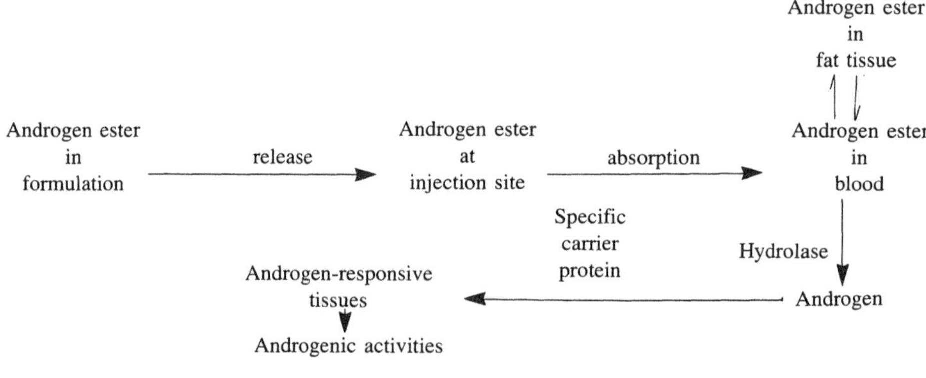

Scheme 1

Based on scheme 1, the first peak activity noted in the levator ani weight-time profiles (Figure 15) is the result of the androgen released from the androgen ester

Table 5 Anabolic Activities and Times to Maximum Anabolic Effect of Nandrolone Esters

	Time to maximum effect (days)		Anabolic activity (g/day)[a]	
Esters	First maximum	Second maximum	1 mg Dose	1 mM Dose[b]
Butyrate	6	11	1.49	0.61
Hexanoate	11	15	3.73	1.46
Heptanoate	9	17	6.56	1.94
Octanoate	13	19	5.56	1.91
Nonanoate	9	19	5.08	1.84
Decanoate	14	21	7.74	2.56
Undecanoate	14	21	6.58	1.56

[a]Corrected for control weights. Each data point was computed from at least 32 determinations.
[b]Molar concentration of nandrolone base.
Source: Compiled from the data by Chaudry and James; J. Med. Chem., 17:157 (1974).

dose absorbed into the systemic circulation. The second maximum is produced by the androgen released from the androgen ester accumulated earlier in body fat tissues.

The anabolic activities of various nandrolone esters, which can be calculated from the area under the levator ani weight-time curves (Figure 15), are compared in Table 5. The results show that on a weight basis the butyrate has the lowest anabolic activity and the decanoate the highest in the homologous series of nandrolone esters. A similar pattern also develops when the anabolic activity of nandrolone esters is compared on the basis of nandrolone alone (mM).

The data in Table 5 were found to be related to the ethyl oleate/water partition coefficient in the following ways:

1. The time to the second peak of anabolic activity $t_{p,2}$ is highly correlated with the partition coefficient (PC) as follows:

$$\log (t_{p,2}) = 0.219 \log \mathrm{PC} - 0.015 \qquad r = 0.96 \tag{3}$$

2. The anabolic activity (AA) shows a binomial relationship with the partition coefficient (PC) as follows:

$$\log (\mathrm{AA}) = 7.33 \log \mathrm{PC} - 0.636 \log (\mathrm{PC})^2 - 17.8 \qquad r = 0.97 \tag{4}$$

The parabolic relationship demonstrated by the anabolic activity of nandrolone esters has been frequently observed with many other drugs and systems (48). The maximum in the parabola represents an optimum combination of hydrophilicity and lipophilicity in a drug molecule. The good correlation obtained in Equations (3) and (4) suggests that the anabolic activity of nandrolone esters and its duration are highly dependent upon the lipophilic nature of the esters. This analysis concludes that nandrolone decanoate produces a maximum anabolic activity in this series of esters (43).

These investigations provide a fundamental understanding of the mechanism underlying the prolonged anabolic activity of nandrolone decanoate (Deca-Durabolin, Organon). The decanoate ester is longer acting than another commercially available long-acting anabolic product, nandrolone phenylpropionate (Durabolin, Organon). The nandrolone decanoate can be administered to the gluteal muscle on a monthly basis, whereas nandrolone phenylpropionate must be injected weekly.

Estrogenic Steroids. By oral administration the natural estrogens, such as estradiol, are subjected to extensive hepatic first-pass metabolism. They are mostly converted by metabolism to less active metabolites, such as estriol, by oxidation to unknown nonestrogenic substances and by conjugation with sulfuric and glucuronic acids (49). By parenteral administration estradiol is rapidly absorbed and quickly metabolized, with a plasma half-life of only about 1 hr.

Long-acting estrogenic steroids are useful in the treatment of estrogen deficiency states in women, such as menopause, thus become desirable and have been of interest to a number of scientists. A prodrug approach by esterification of the 3- and/or 17-hydroxyl groups of estradiol has been shown to produce a prolonged estrogenic action when administered by intramuscular injection (50–55). It was found that the aryl and alkyl esters of estradiol become less polar as the size of the substituents increases; correspondingly, the rate of parenteral absorption is progressively slowed and the duration of estrogenic activity prolonged. Several long-acting estrogenic steroids have been made commercially available.

Estradiol benzoate (Progynon Benzoate, Schering). This is prepared by benzoylation of the 3-hydroxyl group of estradiol. An oil-soluble ester from which the active β-estradiol is slowly released from the oleaginous formulation at the site of intramuscular injection, providing a sustained therapeutic level of estrogen for several days. Estradiol benzoate has an intrinsic biological activity about half that of natural estradiol. Owing to its sustained action, however, it is more efficacious than estradiol. It is used in the treatment of the same condition as estrone but is several times more active.

Estradiol 17-valerate (Delestrogen, Squibb; Duratrad, Ascher). This is prepared by esterifying estradiol at both the 3 and 17 positions by treatment with valeryl chloride (in pyridine) and then removing the 3-valerate by treatment with potassium carbonate (in aqueous methanol). It is very slowly released and absorbed from an oil suspension injected intramuscularly and provides a therapeutic level of estradiol for a duration of about 3 weeks.

It has also been formulated with the long-acting testosterone enanthate in ethyl oleate BP repository vehicle (Ditate-DS, Savage), which has been used for the prevention of postpartum breast engorgement and for the postmenopausal syndrome in those patients not improved by estrogen alone.

Estradiol 17-valerate may also be administered along with a progestational drug in the management of primary or secondary amenorrhea and functional uterine bleeding.

Estradiol dipropionate (Ovocylin Dipropionate, Ciba). This is prepared by esterifying estradiol at both the 3 and 17 positions by treatment with propionyl chloride (in pyridine). It is about half as potent as estradiol benzoate, but because of its more sustained depot action the dipropionate is more potent than benzoate with respect to the cumulative maintenance dosage. It is also longer acting than estradiol 17-valerate when injected as a solution in vegetable oil.

Estradiol 17β-cypionate (Depo-estradiol, Upjohn). This is prepared by the same method described for the valerate but using cyclopentylpropionyl chloride as the esterificant. Similar to that demonstrated earlier by testosterone cypionate (depo-testosterone cypionate, Upjohn), the 17-cyclopentylpropionate ester of estradiol also produces sustained hormonal activity. By intramuscular injection in vegetable oil, estradiol 17β-cypionate has shown a more prolonged estrogenic activity than benzoate, valerate, or dipropionate. Its average duration of action is 3–8 weeks.

Progestational Steroids. The rate of turnover for progesterone is extremely rapid, with a plasma half-life of only about 5 min. Inactivation takes place largely in the liver. The characteristic metabolite is pregnane-3α,20α-diol, which is excreted in the urine in conjugated form with glucuronic acid (49).

By parenteral administration in oleaginous solution progesterone is readily absorbed, but it is degraded at a rate that is too rapid for optimal therapeutic effect. In fact, it is extremely difficult to achieve effective blood levels with any convenient dosing schedules. In animal tests, administration of several doses per day was found to be more effective than the same dose given once daily. It was also noted that less frequent dosing is quite ineffective (49,56).

The first parenterally active, long-acting progestational steroid formulation was Depo-Provera (Upjohn), an aqueous suspension of micronized medroxyprogesterone acetate crystals. By intramuscular administration, this formulation has produced a sustained progestational activity for 2–4 weeks.

The prodrug concept has also been applied to the development of long-acting injectable progestational preparations, but primary interest has been centered on the development of an injectable depot formulation for contraception, such as norgestrel 17β-fatty acid esters. This development is discussed later.

The incorporation of medroxyprogesterone acetate (or norgestrel) in controlled-release vaginal devices for long-term (monthly) intravaginal contraception is discussed in length in Chapter 9. Natural progesterone has also been incorporated in medicated intrauterine devices (IUDs) for yearly intrauterine fertility control (Chapter 10). The feasibility of transdermal and mucosal delivery of progesterone for bypassing hepatic first-pass elimination and thus improvement in systemic bioavailability has also been evaluated (Chapters 4, 5, and 7).

Adrenal Steroids. As discussed earlier, the adrenal cortex synthesizes two classes of adrenal steroids under the regulation of ACTH: the corticosteroids and the adrenal androgens from cholesterol (28).

The corticosteroids have numerous and diverse physiological functions and pharmacological actions. They affect the metabolism of carbohydrate, protein, fat, and purines; electrolyte and water balance; and the functional capacity of the cardiovascular system, the kidney, skeletal muscle, the nervous system, and other organs and tissues. Furthermore, the corticosteroids endow the human with the capacity to resist all types of noxious stimuli and environmental change (28). Traditionally they have been classified into mineralocorticoids and glucocorticoids. Desoxycorticosterone, the prototype of the mineralocorticoids, is highly potent in sodium retention but has practically no activity in liver glycogen deposition. On the other hand, cortisol, the prototype of the glucocorticoids, is highly potent in liver glycogen deposition but weak in sodium retention activity.

The corticosteroids are required for replacement therapy in adrenal insufficiency, such as Addison's disease. Both mineralocorticoids and glucocorticoids may be needed to approximate the equivalent of their physiological concentrations. Glucocorticoids are additionally used to treat rheumatic, inflammatory, allergic, neoplastic, and other disorders (56).

The treatment of Addison's disease has been greatly advanced by the use of desoxycorticosterone. Although the defects in carbohydrate and protein metabolism are not corrected by this mineralocorticoid alone, life can be maintained by its intelligent administration.

All the natural corticosteroids are poorly effective when taken orally because of their rapid and extensive hepatic first-pass metabolism. They must be given parenterally for systemic effect. Desoxycorticosterone has a serum half-life of about 70 min; thus daily intramuscular injection becomes necessary for adequate maintenance treatment of Addison's disease. Since medication is needed for life, development of a parenteral long-acting desoxycorticosterone preparation is therefore most desired.

The corticosteroid activity of desoxycorticosterone was remarkably prolonged by converting its acetate to pivalate. Intramuscular injection of microcrystalline desoxycorticosterone pivalate in oleaginous suspension (Percorten pivalate, Ciba) produced a very long duration of action. It is administered once every 4 weeks compared to the daily injection required for desoxycorticosterone acetate (Percorten acetate in oil, Ciba).

Other commercially available long-acting injectable corticosteroid preparations with activity prolonged by esterification are betamethasone acetate (Celestone Soluspan suspension, Schering), 6α-methylprednisolone-21-acetate (Depo-Medrol, Upjohn), and triamcinolone hexacetonide suspension (Aristospan, Lederle), to name just a few.

6. *Long-acting Antipsychotic Preparations*

Use of antipsychotic drugs for the treatment of psychotic disorders has become widespread only since the mid-1950s. Today the phenothiazines, the most prescribed antipsychotic agents, as a class are among the most widely used drugs in the practice of medicine. They have been utilized on a grand scale to modify the attitude and emotion of psychiatric patients by a tranquilizing effect.

Fluphenazine, the trifluoromethylphenothiazine, is the most potent derivative of phenothiazine available for the management of manifestations of psychotic disorders. Both laboratory and clinical studies have demonstrated that this drug exhibits several important features that are different from those of other phenothiazines. It is more potent, exhibits a more prolonged duration of action, is less likely to induce hypotension, is less sedative, and does not potentiate central nervous system (CNS) depressants and anesthetics to the same degree as other phenothiazines.

In the management of psychotic disorders, patient compliance has always presented a major challenge to the medical profession. Development of a long-acting injectable antipsychotic preparation will certainly provide a practical solution to minimize this noncompliance problem.

Prolongation of the antipsychotic activity of fluphenazine can be accomplished by esterification (57–60), exemplified by the development of fluphenazine enanthate (Prolixin enanthate, Squibb) and fluphenazine decanoate (Prolixin decanoate, Squibb). The esterification of fluphenazine with the enanthate moiety markedly prolongs its duration of antipsychotic action without unduly attenuating its beneficial effects (61).

Parenteral (IM or SC) administration of either ester in sesame oil produces antipsychotic action for an average duration of 2 weeks. The onset of action generally appears within 24–72 hr following injection, and its effect on psychotic symptoms becomes significant within 48–96 hr. Amelioration of symptoms continues for 1–3 weeks or longer. These preparations have been prescribed for the management of schizophrenia and other psychotic disorders and are particularly effective in modifying psychotic behavior patterns and ameliorating such symptoms as agitation, delusion, and hallucination.

The esterification approach has also been utilized to render a longer duration of action for other neuroleptic drugs (62–67). Typical examples are the development of α-fluphenthixol. The neuroleptic activity has been substantially prolonged with conversion from the dihydrochloride salt (with an oral daily dose of 5 mg/kg) to the decanoate ester (which requires an IM injection of 10 mg/kg for a 10-day period). A single intramuscular injection of α-fluphenthixol decanoate in oil produces a fairly steady neuroleptic activity, in terms of the inhibition of a conditional avoidance response, for a duration much longer than that achieved by multiple oral daily intake of the dihydrochloride salt (Figure 16). The neuroleptic activity that results from the oral administration of α-fluphenthixol dihydrochloride fluctuates. The metabolic pattern of the esters was found to be identical to that of the parent compound (62).

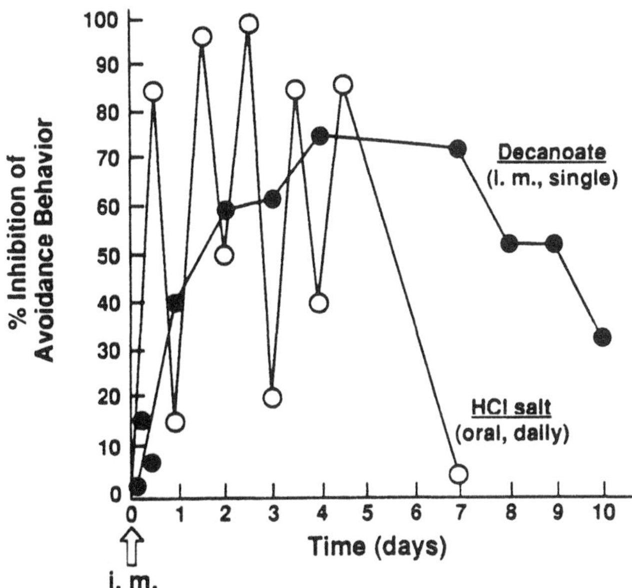

Figure 16 Comparison in the neuroleptic activity and duration in terms of the inhibition of a conditional avoidance behavior between a single intramuscular injection of α-fluphenthixol decanoate in oil (●) and multiple oral daily administrations of α-fluphenthixol dihydrochloride (○). [Plotted from the data by Nymark et al.; Acta Pharmacol. Toxicol., 33:363 (1973).]

7. *Long-acting Antimalarial Preparations*

Antimalarial drugs may be classified rather broadly into two quite distinct groups. The drugs in the first group are rapid in their schizontocidal action and nonspecific in their mechanism of action and make it difficult for originally sensitive strains to develop resistance. They interact with and alter the properties of the DNA in both parasite and host without discrimination. The selectivity in their toxicity depends upon selective accumulation in the intracellular milieu of the parasite. The second group is characterized by a schizontocidal effect that is slow in onset and dependent upon the stage of multiplication of the parasite. The mechanism of action appears to be much more specific than that of the first group. Drugs in the second group either interfere with the incorporation of *p*-aminobenzoic acid into folic acid, a process that does not occur in mammals, or bind to plasmodial dihydrofolate reductases (68). These specific mechanisms of action are exemplified by acedapsone and chloroguanide, respectively.

Elslager reviewed the chemotherapy of malaria and discussed at great length various means by which the duration of antimalarial action could be extended (69). The acylation of acedapsone and the formation of sparingly water-soluble cycloguanil pamoate are the two most interesting examples.

Cycloguanil Pamoate. Cycloguanil is the active metabolite of chloroguanide, whose antimalarial action is related to the binding and inhibition of plasmodial dihydrofolate reductase. It has a short duration of action because of rapid excretion. This stimulated Thompson and his coworkers to synthesize compounds that would act for a much longer period of time (70). Their objective was to develop a single-dose, parenteral depot preparation that would, in a well-tolerated dose, protect humans from malarial infection for a prolonged duration and preferably have immediate action as well.

Various less water-soluble salts of cycloguanil were prepared with the objective of sustaining the antimalarial activity of cycloguanil. It became apparent later that the ability of cycloguanil to be released from an intramuscular depot formulation is highly dependent upon its solubility in the tissue fluid at the injection site relative to its solubility in the formulation vehicle. If the drug has low solubility in physiological fluid, following intramuscular injection the depot formulation forms a drug reservoir in muscular tissues and gradually releases the active drug moiety for absorption into the systemic circulation. It is known that the aqueous solubility of any amine salt is dependent upon the type of counteranion used. The resulting solubility is equal to the square root of the solubility product of the amine and its counteranion and is highly pH dependent. Investigations revealed that the duration of the antimalarial activity of various cycloguanil salts is linearly proportional to their aqueous solubility (Figure 17). This linear relationship can be expressed mathematically by

$$\log(\text{duration of ED}_{50}) = -0.71 \log C_s + 0.077 \tag{5}$$

Equation (5) suggests that the lower the aqueous solubility of the cycloguanil salts, the longer the duration of antimalarial action. Results indicated that the pamoate salt of cycloguanil, with an aqueous solubility of 0.03 mg/ml (at pH 7), possesses the desired properties.

Laboratory tests with *Plasmodium berghei*-infected mice and *Plasmodium cynomolgi*-infected monkeys showed that when given subcutaneously well-tolerated doses cycloguanil pamoate has protected mice for up to 8.5 weeks (70,71). The

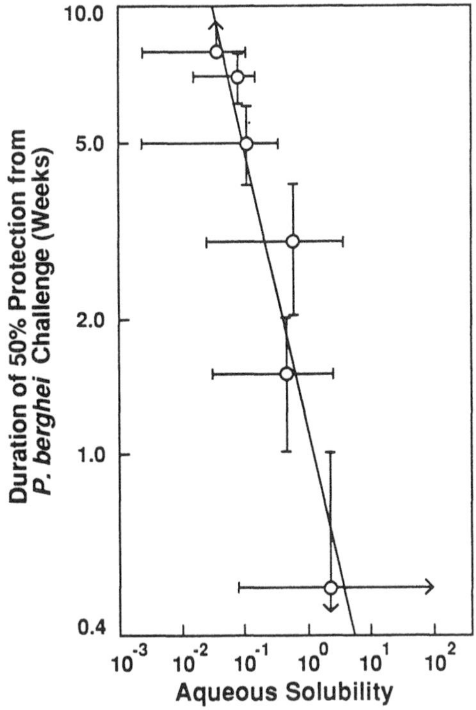

Figure 17 Linear relationship between the log (antimalarial activity duration), indicated by 50% of mice protected from *P. berghei* challenge, and log C_s, the aqueous solubility of cycloguanil salts. [Plotted from the data by Higuchi and Stella; ACS Symposium Series 14:51 (1975).]

prolonged antimalarial activity was confirmed to be the result of slow release of the active moiety from a depot formed at the injection site, not to the formation of a systemic reservoir (72).

Subsequent studies in humans showed that a single intramuscular injection of cycloguanil pamoate (5 mg/kg) provides protection against *Plasmodium falciparum* and *Plasmodium vivax* for several months. It acts to provide prolonged infusion of a therapeutically effective concentration of soluble dihydrotriazine into the blood, which appears to prevent the growth or even the survival of both the erythrocytic and the pre-erythrocytic parasites. Up to 80% of the injected dose was found to remain at the injection site 2 weeks after administration, and small amounts could be detected for as long as 56 weeks later (70). Absorption of the drug from the intramuscular site was determined to follow a first-order rate process that was influenced greatly by the particle size of the cycloguanil pamoate (73), as expected from the kinetics of dissolution.

Clinically it was observed that with the coadministration of acedapsone the duration of antimalarial protection of cycloguanil pamoate is further extended. This combination (Dapolar) also delayed the emergence of resistant strains and even provided some protection against strains already less sensitive to either drug alone (68).

Acedapsone. Acedapsone (Hansolar), is produced by acylation from 4,4′-diaminodiphenyl sulfone. It was reported to produce prolonged antimalarial action and is useful in the treatment of *P. faliciparum* in humans for up to 42 days after a single intramuscular injection. Following intramuscular injection it is hydrolyzed slowly in the body to regenerate the active antimalarial 4,4′-diaminodiphenyl sulfone (74). The antimalarial activity of 4,4′-diaminodiphenyl sulfone has been attributed to its inhibition of folic acid synthesis in these parasites.

8. Long-acting Antinarcotic Preparations

Narcotic addiction affects the lives of hundreds of thousands of individuals and creates a great deal of socioeconomic hardship to the community. These narcotic addicts become psychologically dependent upon the drug and believe that the effects produced by self-administration of the narcotics are necessary for maintaining an optimal state of well-being. In extreme forms the abuser exhibits the characteristics of a chronic relapsing disease (75).

Recently attention has focused in the use of narcotic antagonists for the treatment of narcotic addiction (76–80). The rationale is that the antagonist blocks the euphoric effects of a narcotic drug taken during an addict's rehabilitation and makes it pleasureless, thus extinguishing the drug-seeking behavior and removing the addict's incentive for continued use.

For the treatment of narcotic addition and dependence, methadone was found useful and used extensively. Unfortunately it introduced some complications of its own, since methadone is basically a substitute for heroin and is itself also an addicting drug subject to potential abuse (81,82).

An ideal antagonist should be free of agonistic (morphinelike) properties and effective for at least 1 week. Naloxone, for example, is known to be a pure narcotic antagonist that does not possess any agonistic properties (e.g., respiratory depression) (83,84). Long-term administration of naloxone is not associated with dysphoria, nor is there an abstinence syndrome on abrupt withdrawal. Unfortunately its duration of action is short, and because of extensive hepatic first-pass metabolism its oral bioavailability is poor. Therefore oral doses, as large as 3 g/day, are required for a 24 hr effective blockade to heroin challenge (82,85). Furthermore, naloxone, unlike methadone, provides no incentive to addicts to return for frequent maintenance therapy, and patient compliance thus becomes a critical issue. The development of long-acting antinarcotic preparations using naloxone or other narcotic antagonists is thus urgent for maximum patient compliance and successful addict rehabilitation.

Oral Bioavailability and Hepatic First-Pass Metabolism. The therapeutic effectiveness of both opiate-type narcotic agonists, such as morphine, and antagonists, such as naloxone, by oral administration was reported to be extremely poor (75). The poor therapeutic effectiveness was found not to be a result of low gastrointestinal absorption but was attributed to an extensive hepatic first-pass metabolism (86).

Naloxone is rapidly absorbed, with a peak plasma level reached within 5 min after oral administration, and absorption is virtually complete at 90 min. However, the maximum amount of intact naloxone molecules detected in rat plasma is only 0.19% of the naloxone dose administered orally. This extensive elimination was attributed to the enzymatic metabolism of naloxone by liver. The major pathway for the hepatic metabolism of naloxone in most species appears to be the conjugation of the 3-phenolic OH group with glucuronic acid to form naloxone-3-glucuronide, the major metabolite. Several minor metabolites, such as 3-sulfate and free and conjugated naloxol can also be detected in the urine (87,88). These results lead to the conclusion that the low oral therapeutic effectiveness of naloxone is the result of rapid presystemic elimination (86). The reduction of the 6-keto group and the N-dealkylation of naloxone were reported also to occur in humans (89,90).

Pharmacokinetic analyses indicated that naloxone is rapidly absorbed in humans, with an absorption half-life of 0.72 min, and is also rapidly eliminated, with an elimination half-life of 16 min (86). This elimination half-life value agrees well with the 20 min determined in rats (91) and the 19 min obtained in guinea pigs (92).

Increasing knowledge of the pharmacokinetic behavior of narcotic agonists has opened a new possibility for a more rational approach to antinarcotic treatment. Together with studies on other drug-induced biochemical changes, pharmacokinetic investigations can provide better insight into the mechanisms of narcotic antagonism.

Comparative Pharmacokinetics of Narcotic Antagonism. The concentration profiles of naloxone and of morphine in the serum and brain tissues of rats following the subcutaneous administration of an equivalent dose are compared in Figure 18. It appears that the peak serum concentrations, time to peak serum concentrations, and serum half-life for naloxone and morphine are comparable (93,94).

The brain entry and egress of these two compounds differ markedly, however the peak brain level of naloxone occurs within 30 min and declines by 50% within 1 hr, whereas the peak brain level of morphine is reached within 1 hr and is sustained for up to 2 hr. At peak concentration the brain-serum ratio is only 0.1 for morphine whereas for naloxone it is 15 times greater. As serum levels decline the ratio becomes greater for morphine (approaches 0.5), but for naloxone it remains within a narrow range of 1.5–2.0. At equivalent serum levels one therefore finds three to four times more naloxone in the brain than morphine. These observations suggest that naloxone enters the brain more easily than morphine. The maximum levels of naloxone in brain tissues achieved by subcutaneous administration were found to be approxi-

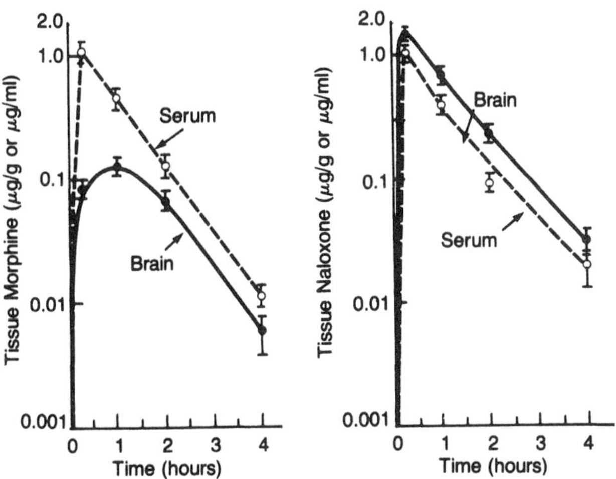

Figure 18 Serum and brain concentration profiles of morphine and naloxone in rats following subcutaneous administration (5 mg/kg). Each data point represents the mean for 3–5 animals. Vertical bars are standard error of the mean. [Reproduced with permission from Berkowitz et al.; J. Pharmacol. Exp. Ther., 195:499 (1975).]

mately 11-fold greater than those of morphine with an equivalent dose (88). The rapid uptake by brain tissues and high ratio of brain to serum concentrations for naloxone may well explain the potent antinarcotic activity of naloxone. Unfortunately, the rate of disappearance of naloxone from the CNS is much faster than that of morphine (88,93). Rapid hepatic metabolism and a high elimination rate appear to account for the short duration of action of naloxone.

The effect of long-term subcutaneous morphine administration on the ratio of brain to serum morphine concentrations was also studied. It was found that this ratio is maintained within a narrow range of 0.45–0.54 as the subcutaneous morphine dose is increased by fivefold and the duration of treatment is prolonged from 3 to 10 days (95).

If both narcotic agonist and antagonist are eliminated from the receptor site compartment according to a first-order rate process, the time course of the narcotic action can be expressed by

$$\log \frac{D_N - D_N^\circ}{D_N^\circ} = (\log f_{NA} D_{NA} + \log K_{NA}) - 0.43 k_{eNA} t \tag{6}$$

where D_N and D_N° are the doses of a narcotic agonist administered in the presence and absence of a given dose of antagonist (D_{NA}) to produce an equivalent pharmacological response; k_{eNA} is the first-order rate constant for the elimination of narcotic antagonist from the receptor site compartment; f_{NA} is the fraction of the narcotic antagonist dose absorbed into the receptor site compartment; and K_{NA} is the affinity constant of the antagonist-receptor complex (91).

As expected from Equation (6), plots of $\log [(D_N - D_N^\circ)/D_N^\circ]$ against time should yield a series of parallel straight lines (Figure 19), with

$$\text{Slope} = -0.43 k_{eNA} \tag{7}$$

and

$$\text{Intercept} = \log f_{NA} D_{NA} + \log K_{NA} \tag{8}$$

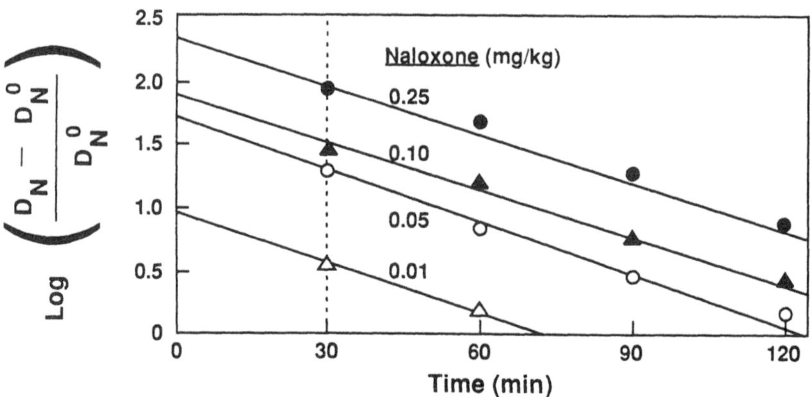

Figure 19 Linear relationship of $\log [(D_N - D_N^\circ)/D_n^\circ]$ with time for morphine, using the tail compression test in rats, in the present of naloxone at (△) 0.01, (○) 0.05, (▲) 0.10, and (●) 0.25 mg/kg. [Plotted from the data by Tallarida; in: *Factors Affecting the Action of Narcotics* (Adler et al., Eds.), Raven Press, New York (1978).]

The relationship between the narcotic antagonistic activity P_{NA} and the doses of narcotic antagonist can be established from Equation (8):

$$P_{NA} = \log K_{NA} + \log f_{NA} \tag{9a}$$

$$P_{NA} = \text{intercept} - \log D_{NA} \tag{9b}$$

A P_{NA} value of 2.57 ± 0.03 was obtained for the narcotic antagonism between naloxone and morphine (91). This P_{NA} value, determined from animal studies, is a useful and reproducible pharmacological constant in assessing the selectivity of antagonists and also in characterizing pharmacological receptors in different isolated tissues.

The competitive antagonism of naloxone was demonstrated in an in vitro comparative binding study in rat brain tissue with an opiate agonist, dihydromorphine (96). The binding of dihydromorphine to rat brain tissues may be separated into two components: one is a saturable component and the other is nonsaturable. A marked regional difference was observed in the distribution of the saturable component in the brain as a result of the variation in the concentration of saturable binding sites within various brain regions. However, the saturable binding sites from various brain regions appeared to have similar affinities for dihydromorphine, except the binding sites from cerebral cortex, which have a higher affinity. In contrast, the saturable binding sites for naloxone in various brain regions showed varying degrees of affinity for naloxone.

Naloxone was found to contain at least two types of saturable binding sites, one of which is not available for the binding of dihydromorphine. The saturable binding sites in cerebellum are predominantly naloxone specific, whereas those in striatum are capable of binding both naloxone and dihydromorphine.

Development of Long-acting Antinarcotic Formulations. As discussed earlier, a most desirable component of addiction treatment is that the antinarcotic activity of a narcotic antagonist formulation be long-acting so that frequent dosing is not necessary, helping addicts to dissociate from their opiate-taking desire.

There seems to be some difference of opinion about how long-acting an ideal antinarcotic formulation should be. Some researchers believe that 1 or 2 days is long enough, whereas others believe that a duration of 20 months or longer would be better (97). The current consensus is that a duration of at least 1–3 months is desirable. To achieve an antinarcotic activity with a desired duration of 1–3 months, a logical approach is to develop long-acting drug delivery systems for the controlled administration of an acceptable but short-acting antagonist.

Efforts to develop long-acting drug delivery systems were launched in the early 1970s by the New York City Public Health Department and by the National Institute of Drug Abuse (98). Two approaches have been successfully utilized for the development of parenteral controlled-release antinarcotic formulations.

Oleaginous suspensions of narcotic antagonist-salt complexes. One approach to achieve a long-acting antinarcotic activity is development of an insoluble salt and ionic complex of potential narcotic antagonists. This approach has already been successfully applied in several therapeutic areas, as discussed earlier (Section II.B), such as penicillin G procaine, vitamin B_{12}–Zn tannate, and cycloguanil pamoate. For example, the systemic level of penicillin G was markedly prolonged by the use of intramuscular depot preparations consisting of relatively insoluble salts of penicillin

G in a gelled oleaginous suspension. It was reported that the therapeutic blood level of penicillin G persists for 48–96 hr if penicillin G is administered as procaine salt, an ion-pairing salt, in an oil suspension gelled with aluminum monostearate. By changing to a different type of ion-pairing salt, such as benzathine penicillin G, which has an aqueous solubility 20 times lower than that of the procaine salt, the therapeutic blood levels of penicillin G were further sustained to 1 week or longer (99).

Approximately 100 monobasic and polybasic organic acids, such as salicylic and tannic acids, were evaluated for their ability to form water-insoluble salts with narcotic antagonists, such as naloxone and naltrexone, or water-insoluble complexes with polyvalent metallic ions, such as Zn^{2+}, Mg^{2+}, Ca^{2+}, and Al^{3+} (100).

Using the mouse tail-flick test the duration of antinarcotic activity of representative salts and zinc complexes of naloxone was evaluated (3,101,102). Results indicated that the antinarcotic activity of naloxone following intramuscular administration is significantly prolonged by the formation of insoluble salts and Zn complexes (Table 6). Naloxone HCl, a water-soluble salt itself, showed significant antagonistic activity for only 4 hr even at a very high dose (4 mg/kg of naloxone base equivalent). This activity was sustained to at least 8 hr with the formation of pamoate and tannate salts and up to 16 hr with 5-*tert*-octylsalicylate. The narcotic antagonistic activity was further prolonged to 24 hr with the formation of a Zn-tannate complex. On the other hand, incorporation of zinc had no prolonging effect on the antagonistic activity of pamoate and 5-*tert*-octylsalicylate. It appeared that the incorporation of zinc has the most significant sustaining action on the duration of the antagonistic activity of naloxone tannate preparations. Similar results were also observed with other narcotic antagonists, such as cyclazocine and naltrexone. The proposed structure of narcotic antagonist-Zn-tannate is shown in Figure 20.

The duration of antinarcotic activity of the salts and Zn complexes of naloxone was observed to be roughly correlated with their degree of dissociation: hydrochloride > pamoate > 5-*tert*-octylsalicylate > tannate. The incorporation of zinc

Table 6 Duration of Antinarcotic Activity of Naloxone Salts and Zinc Complexes in Mice

	% Antagonism[b]					
Naloxone derivatives[a]	40 min	4 hr	8 hr	16 hr	24 hr	48 hr
Salts						
Hydrochloride	99	79	16	3	10	—
Pamoate	100	100	94	11	1	—
Tannate	99	87	66	6	3	—
5-*t*-Octylsalicylate	100	78	58	43	18	17
Complexes						
Zn-pamoate	100	100	86	9	0	—
Zn-tannate	95	99	90	41	42	23
Zn-5-*t*-octylsalicylate	99	96	29	57	25	14

[a]In peanut oil (4 mg/kg of naloxone base equivalent) by intramuscular injection.
[b]Percentage antagonism to a standard morphine sulfate dose (20 mg/kg) administered intraperitoneally 30 min before mouse tail-flick tests (12 mice each).
Source: Compiled from the data by Gray and Robinson; in: *Narcotic Antagonists* (Braude et al., Eds.), Raven Press, N.Y. (1974), p. 555.

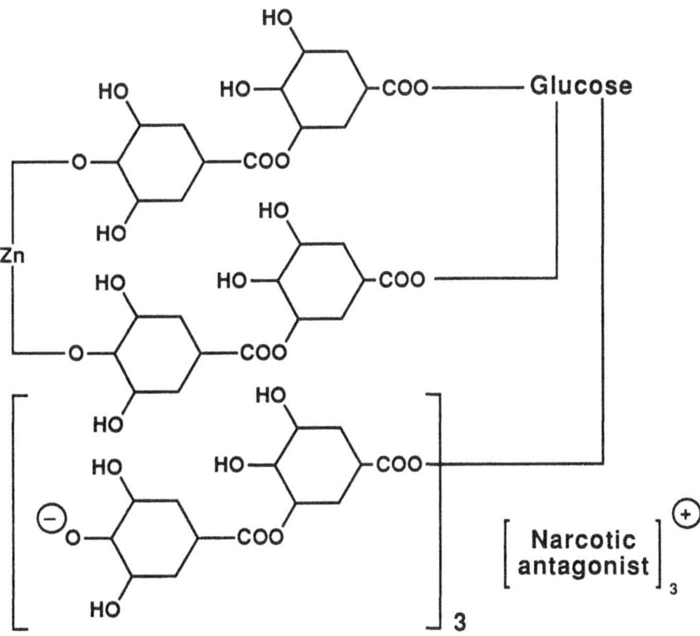

Figure 20 The proposed chemical structure of narcotic antagonist-Zn-tannate.

was found to substantially reduce the dissociation of tannate from 53.3 to 13.9%, but not pamoate (3).

A long-acting injectable naloxone pamoate formulation was also developed and patented (2). This consisted of a naloxone pamoate suspension in peanut or sesame oil gelled with 2% aluminum monostearate. The duration of narcotic antagonistic activity of this long-acting formulation was evaluated in mice by the Straub tail test and in rats by the narcosis test (103). The results demonstrated that with naloxone HCl (50 mg/kg) the Straub tail reaction of oxymorphine HCl (2 mg/kg) is blocked at 15 min but not at 24 hr. With the naloxone pamoate suspension complete protection (100%) was achieved for a duration of up to 24 hr and 77.8% of the mice are still protected at 48 hr.

Comparative antinarcotic studies in rats also illustrated that the aqueous naloxone HCl solution produces a complete narcotic blockade in rats only at 15 min but not at 24 hr. On the other hand, naloxone pamoate suspension showed a significantly longer narcotic blockade: 75% of the treated rats that were protected from oxymorphone-produced narcosis were still protected at 24 hr, 62.5% at 48 hr, and 37.5% at 72 hr (101).

The duration of narcotic antagonistic activity of this naloxone pamoate suspension was also evaluated in 12 dogs. Intramuscular injection of naloxone pamoate suspension (10 mg/kg) was found to protect against serious narcotic depression in all the treated dogs for at least 7 days, in most of the animals for at least 14 days, and in a few dogs for 21 days. In comparison, the same dosage of naloxone HCl in aqueous solution gave adequate protection for less than 1 day. In additional investigations in nonnarcotized animals the intramuscular administration of naloxone pamoate suspension at a dosage equivalent to 10 mg/kg of naloxone HCl produced no significant narcotic depression as that produced by naloxone HCl. At higher doses naloxone pamoate showed substantially less convulsant action and toxicity than naloxone HCl, presumably resulting from the sustained release of naloxone from the relatively insoluble pamoate salt in oleaginous suspension (naloxone pamoate has an oil solubility that is 12 times lower than that of naloxone itself).

The effect of formation of the insoluble salts of narcotic antagonists with polybasic organic acids or of insoluble complexes with polyvalent metallic ions on the duration of their antinarcotic activity can be illustrated by naltrexone, a close analog of naloxone. Naltrexone is essentially as pure an antagonist as naloxone but is two to three times as potent as naloxone. When administered orally for narcotic blockade in rats naltrexone HCl is about eightfold as active and about three times as long-acting as naloxone HCl on a mg-dose basis (104). The difference between naloxone and naltrexone in the duration of oral activity may be related to the difference in the pharmacokinetic processes involved in their permeation from the blood circulation to the target tissue containing the receptor sites.

The intramuscular antinarcotic activity of naltrexone was slightly prolonged by the conversion of its hydrochloride to tannate salt. Furthermore, the activity of naltrexone showed at least threefold increase in its duration with the formation of a Zn-tannate complex (Figure 21). Prolongation of the antinarcotic activity was further

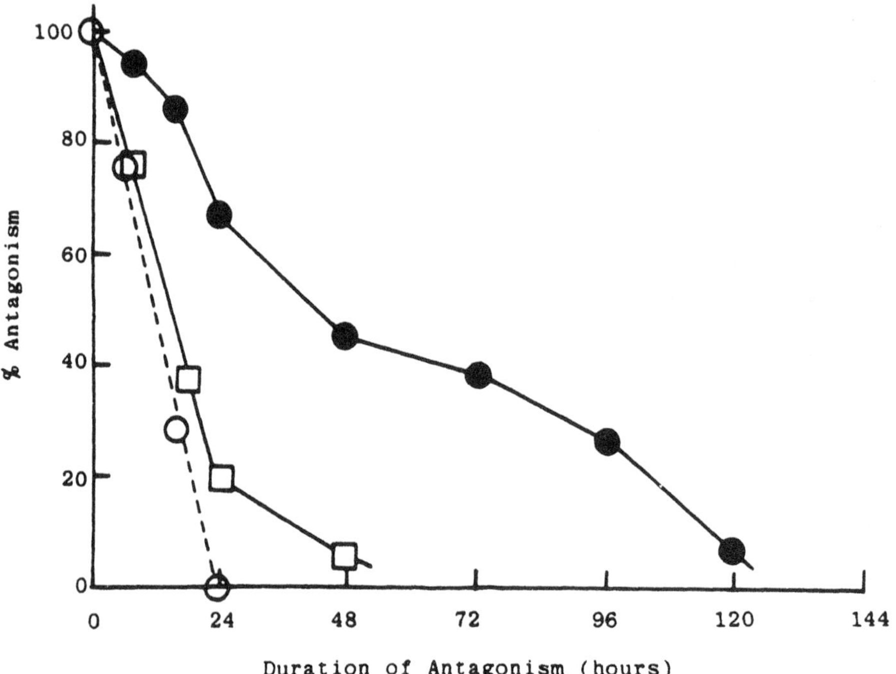

Figure 21 Comparative antagonistic activity of naltrexone preparations in peanut oil against the analgesic effect of morphine sulfate in mice: (○) Naltrexone HCl; (□) Naltrexone tannate; (●) Naltrexone-Zn-tannate (4 mg/kg of Naltrexone base, each). [Plotted from the data by Gray and Robinson; J. Pharm. Sci., 63:159 (1974).]

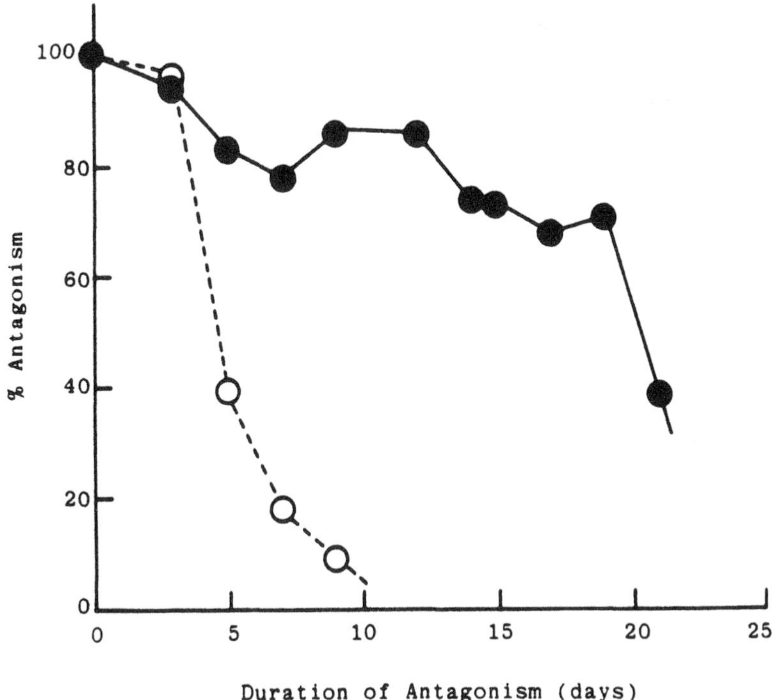

Figure 22 Comparative antagonistic activity of naltrexone-Zn-tannate (40 mg/kg of Naltrexone base) against the analgesic effect of morphine sulfate (20 mg/kg) in mice: (○) in peanut oil and (●) in peanut oil gelled with aluminum monostearate. [Plotted from the data by Gray and Robinson; J. Pharm. Sci., 63:159 (1974).]

prolonged by roughly threefold when the peanut oil suspension of naltrexone-Zn-tannate was gelled with aluminum monostearate (Figure 22). This naltrexone-Zn-tannate suspension still produced 70% antagonism against the analgesic effect of morphine sulfate in mice 19 days after intramuscular administration (4). The effect of aluminum monostearate on the intramuscular activity of naltrexone-Zn-tannate may be explained by its water repellency, which resulted in a reduction in the interfacial dissolution rate of naltrexone-Zn-tannate from the oil suspension (99).

As illustrated by naloxone, the duration of the antinarcotic activity of naltrexone salt and its Zn-tannate complex has also been related to the extent of dissociation (Table 7), although all of them produced the same ED_{80} value for narcotic antagonism as the naltrexone base (at a dose of 0.02 mg/kg).

Biodegradable antinarcotic beads and microcapsules in aqueous suspensions. An implantable or injectable drug delivery system can be fabricated from biodegradable polymers, such as the homopolymer or copolymer of lactic acid and glycolic acid (105). In a biological environment these biodegradable lactide-glycolide copolymers are degraded by hydrolysis into innocuous metabolites, such as carbon dioxide and water, and the narcotic antagonist impregnated within the polymer matrix or encapsulated in the microcapsules is thus released into the tissue fluid at the site of implantation or injection (Figure 23). The biodegradable polymers have been

Table 7 Extent of Dissociation of Naltrexone Salts and Zn-Complex

Narcotic antagonist preparation	% Dissociation[a]
Naloxone	
HCl	100
Pamoate	74.1
5-*t*-Octylsalicylate	62.5
Tannate	53.3
Zn-tannate	13.9
Naltrexone	
HCl	100
Tannate	38.2
Zn-tannate	11.5

[a]In an isotonic phosphate buffer at pH 7.3 and 37°C.
Source: Compiled from the data by Gray and Robinson; J. Pharm. Sci., 63:159 (1974).

extensively applied to the development of long-acting delivery systems for narcotic antagonists (106–110).

A considerable advantage would be achieved in the long-term administration of narcotic antagonists if the drug-containing polymeric drug delivery system could be delivered into the body in the form of very fine particles, such as microcapsules, microspheres, or nanoparticles, by parenteral administration using a hypodermic needle. The long-term bioavailability of narcotic antagonists, such as naltrexone, from the lactide-glycolide copolymer after parenteral administration in mice is illustrated in Figure 24. The cumulative urinary excretion profile suggests that the in vivo release of naltrexone from lactide-glycolide copolymer beads is fairly constant for up to 53 days, and 0.83% of the dose was released daily. The antinarcotic activity of this long-acting naltrexone formulation was evaluated and found to achieve a significant antinarcotic activity against morphine challenge for 2 months (Figure 25). With polylactide beads as the delivery system, naltrexone was able to achieve narcotic antagonism for 3–4 weeks in several animal species (Table 8).

Under the sponsorship of NIDA, an injectable microcapsule formulation has also been developed from the biodegradable poly(lactic acid) polymer as a long-acting delivery system for naltrexone and its pamoate (111).

9. *Long-acting Contraceptive Preparations*

It is known that progestational steroids, such as natural progesterone, in high doses suppress the pituitary release of luteinizing hormone (LH) and the hypothalmic release of LH-releasing factor (LRF), thus preventing ovulation (56).

In addition to the development of progestational steroid-releasing vaginal rings for monthly intravaginal contraception (Chapter 9) and progesterone-releasing IUDs for yearly intrauterine contraception (Chapter 10), several successful attempts have also been made to develop an injectable, long-acting contraceptive formulation to achieve long-term ferility control without interruption.

The first attempt was to inject natural progesterone in oleaginous solution, and progesterone was found to be readily absorbed from the site of injection. However,

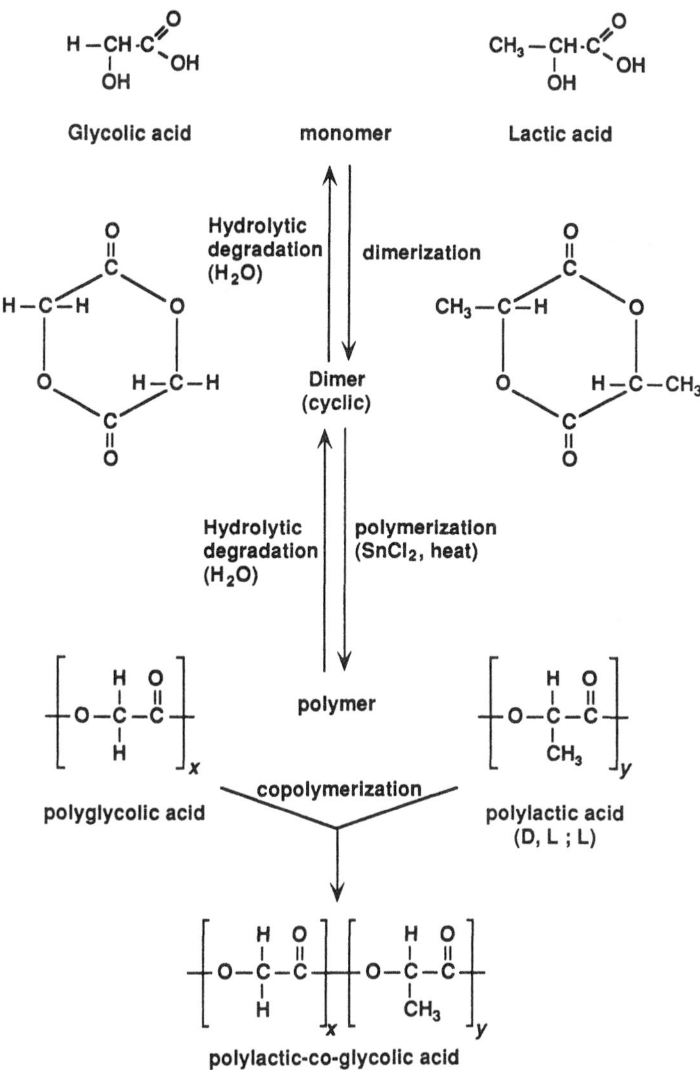

Figure 23 Pathways for the synthesis of the biodegradable homopolymer and copolymer of lactide-glycolide polymers and for their biodegradation.

it is also degraded at a rate that is too rapid for optimal therapeutic effectiveness. In fact, the results suggested that it is extremely difficult to achieve effective blood levels for progesterone with any convenient dosing schedules. In animal tests the administration of several doses per day is more efficacious than the same dose administered once daily, and less frequent dosing is always ineffective (49,56).

The development of an injectable, long-acting contraceptive formulation was made possible with the use of long-acting derivatives of progesterone. Relatively successful results have been obtained in preclinical and/or clinical studies with the following preparations:

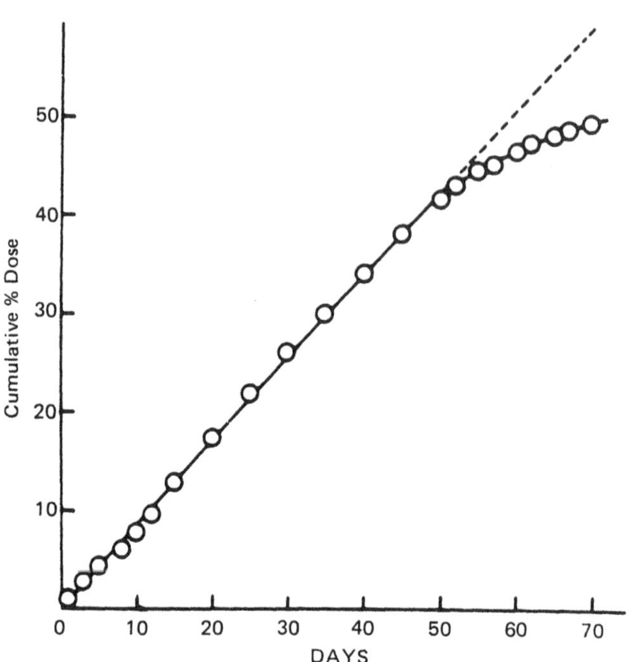

Figure 24 Cumulative urinary excretion of [^{3}H]-Naltrexone (% of administered dose) from poly(lactic-glycolic) beads (90:10) injected into 10 mice. A release rate of 0.831 (± 0.023)%/day was estimated from the slope through Day 53. [Plotted from the data by Schwope et al.; NIDA Research Monograph Series 4 (Willette, Ed.), 1976, p. 13.]

1. Medroxyprogesterone acetate in aqueous suspension (Depo-Provera, Upjohn)
2. Dihydroxyprogesterone acetophenide and estradiol enanthate in oleaginous solution (Deladroxate, Squibb)
3. Norethindrone in a biodegradable polymer beads
4. Norethindrone enanthate in oleaginous solution (Norigest, Schering AG)
5. Norgestrel 17β-fatty acid esters in oleaginous solution

These long-acting injectable contraceptive preparations contain either progestin alone or in combination with an estrogen (112–116). Their development is discussed individually.

Depo-Provera C-150. Depo-Provera C-150 (Upjohn) is an aqueous suspension of microcrystalline medroxyprogesterone acetate (150 mg). It is recommended for intramuscular injection deep into the gluteal muscle, one dose every 3 months.

Depo-Provera has been made commercially available as a long-acting injectable preparation for various gynecological and obstetric indications since 1960 and for the treatment of inoperative, advanced endometrial carcinoma since 1972. However, its potential as an injectable long-acting contraceptive formulation was not officially accepted until late 1974. It has been used as an injectable contraceptive preparation in the British Commonwealth.

The long-acting contraceptive activity of medroxyprogesterone acetate administered parenterally in aqueous suspension formulation is believed to be the result of one or more of the following actions (117–119):

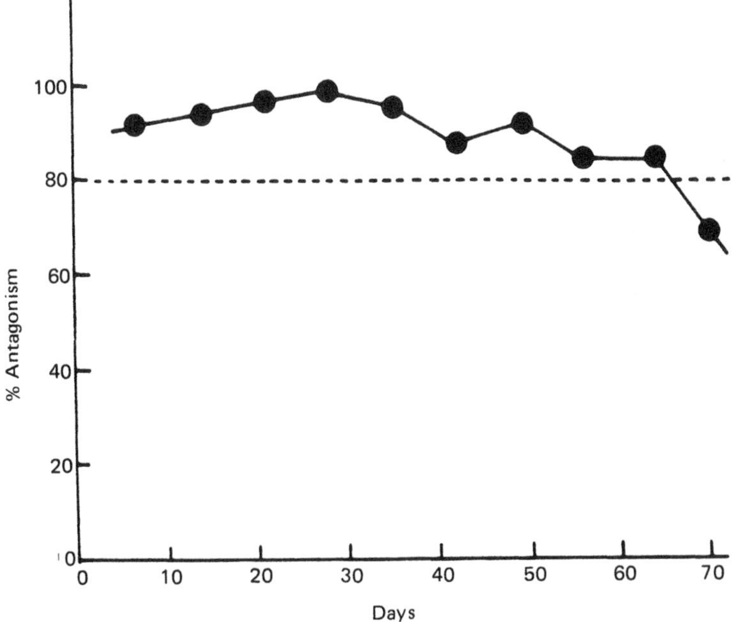

Figure 25 Antinarcotic activity of naltrexone-dispersed beads of poly(lactic-glycolic) copolymer in mice against morphine challenge. [Plotted from the data by Schwope et al.; NIDA Research Monograph Series 4 (Willette, Ed.), 1976, p. 13.]

Table 8 Duration of Antinarcotic Activity of Naltrexone-Dispersed Poly(lactic Acid) Beads in Animals

Animal species	Dose[a] (mg/kg)	Duration[b] (days)
Dogs	17	29
Monkeys	30	20
Mice	39	21
Rats	240	24

[a]Injectable suspension of 35% naltrexone-poly(lactic acid) beads (500–710 μm) in 7% CMC gel.
[b]Duration of antinarcotic activity was determined in rats and mice by the tail pinch test; in dogs by measuring the flexor reflex, the skin twitch reflex, the pulse rate, and the pupillary diameter; and in monkeys by determining the changes in morphine-induced prolongation of interblinking time.
Source: Compiled from the data by Yolles et al.; J. Pharm. Sci., 64:348 (1975).

1. Suppressing ovulation by inhibition of the preovulatory surge of luteinizing hormone (LH) and follicle stimulating hormone (FSH)
2. Impeding nourishment of the blastocyst within the endometrial cavity by alteration of the secretory transformation of endometrium
3. Reducing the penetration of spermatozoa into the uterus by increasing the viscosity of cervical mucus

The first action was substantiated clinically by gonadotropin bioassay in patients receiving a single injection of Depo-Provera C-150 (118) and by double-antibody radioimmunoassay in 84 patients who had been on Depo-Provera for 3–24 months (119). Both measurements demonstrated that LH and FSH levels in the treated patients do not differ significantly from control values, except for the observation that in treated patients the preovulatory peaks of both gonadotropic hormones are eliminated.

Between July 1963 and March 1973 a total of 11,500 women were treated with various dosage regimens of Depo-Provera for a total contraceptive experience of 208,894 woman-months (120). Ovarian biopsies from the subjects receiving Depo-Provera revealed the existence of follicles in all stages of development with small follicular cysts and the absence of corpora lutea. Histomorphological and histochemical studies of the ovaries from the Depo-Provera–treated patients were found to be comparable with those of a normal ovary, except for the absence of a corpus luteum (121).

The serum concentration profile of medroxyprogesterone acetate following the intramuscular administration of a single 150 mg dose of Depo-Provera was monitored using radioimmunoassay for a duration much longer than the prescribed treatment period of 90 days (118,120–122). Concentrations of medroxyprogesterone acetate in the systemic circulation were found in the range of 10–25 ng/ml during the first 20 days after injection and then dropped to a steady-state level of 4–8 ng/ml for a period of up to 140 days (Figure 26) and to <3 ng/ml during the remaining period from 150 to 200 days.

The serum levels of medroxyprogesterone acetate in patients receiving multiple doses of Depo-Provera, one injection every 90 days, were also monitored at various time intervals. A median serum level of 3.5 ng/ml was obtained. A comparison of the median values from various injections revealed that the serum medroxyprogesterone acetate levels at the end of injection period 7 are no different from the serum concentrations detected at the end of injection period 1. This observation suggested that there is no accumulation of medroxyprogesterone acetate in the body over the treatment period.

There is a wide variation in the rates of continuation at the end of 1 year of use reported by various clinical studies. Life table analysis indicated that approximately 58–67% of the patients who have chosen this method of contraception continue to use it after 1 year of medication (Table 9) (123,124).

The clinical effectiveness of Depo-Provera C-150 was evaluated in 1123 subjects for a total of over 14,000 months (125). The results indicated that the intramuscular administration of Depo-Provera C-150 at 3 month intervals is a suitable and effective means of contraception. Only one method-related pregnancy and three patient-failure pregnancies occurred in over 14,000 months of use. This can be translated to a pregnancy rate of 0.3 per 100 woman-years. This result is comparatively better than

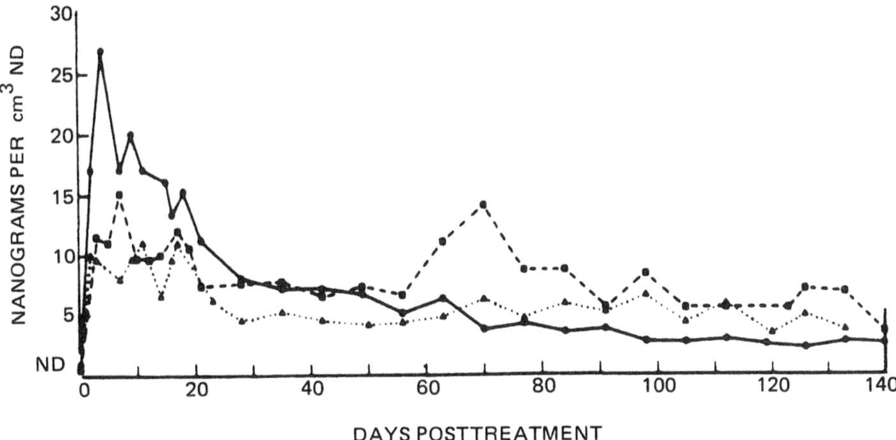

Figure 26 Long-term serum concentration profiles of medroxyprogesterone acetate in three women following intramuscular injection of a single dose of Depo-provera C-150. [Reproduced with permission from Cornette et al.; J. Clin. Endocrinol. Metab., 33:459 (1971).]

that with other methods of fertility control (Table 10). In contrast to the metabolic effects commonly noted with the use of oral contraceptives containing a progestin-estrogen combination, no apparent change in liver functions, lipid metabolism, or blood pressure has been detected during Depo-Provera treatment (126).

After discontinuing Depo-Provera treatment conception was attained in 12 months in >60% of the users (125). In another investigation the results also indicated that resumption of ovulation and fertility occurs in the majority of women within 1 year after stopping treatment (123); however, a delayed resumption of regular menses and ovulation for longer than 1 year was observed in 20–25% of users.

Table 9 Life Table Analysis of the Clinical Effectiveness of Depo-Provera

	Clinical effectiveness[a]	
Events	C-150	C-300
Pregnancies	0.3	2.3
Termination		
Medical	19.1	10.3
Personal	22.5	20.7
Continuation rate	58.1	66.7

[a]The statistic is based on more than 25,000 woman-months of use with Depo-Provera C-150, 150 mg dose every 3 months, and more than 10,000 woman-months of use with C-300, 300 mg dose every 6 months. Expressed as events per 100 women through 12 months of use.

Source: Compiled from Schwallie and Assenzo; Fertil. Steril., 24:331 (1973); Scutchfield et al.; Contraception, 3:21 (1971).

Table 10 Comparison in Pregnancy Rate between Depo-Provera and Other Methods of Birth Control

Methods	Rate of pregnancy[a]	
	Method failure	Combined failure
Depo-Provera C-150	0.085	0.3
Oral contraceptives		
Combined regimen	0.1	0.7
Sequential regimen	0.5	1.4
Intrauterine devices[b]		
Large Lippes loop	1.9	2.7
Saf-T-Coil	1.9	2.8
Condom or diaphragm	2.6	—
U.S. clinics		
Diaphragm and spermicidal jelly (or cream)	—	17.9
Vaginal foam	—	28.3
Vaginal jelly (or cream) alone	—	36.8

[a]Expressed as the number of pregnancy per 100 woman-years.
[b]Nonmedicated, conventional IUDs.
Source: Compiled from the data by Powell and Seymour; Am. J. Obstet. Gynecol., 110:36 (1971).

Elevated progesterone concentrations, which are consistent with the levels found at postovulation, were detected 200–245 days after the treatment. The elevation in progesterone levels signaled the return of the ovaries to normal ovulatory cycles at the end of medication.

Vaginal cytology and endogenous estrogen secretion were also determined in a group of women who had been on continuous Depo-Provera treatment for 21–42 months. The results showed that the mean estrogen values in the treatment group were suppressed by approximately 50% below mean values in normal menstruating women on Days 7–8. Mean estradiol values were still higher than the levels measured in postmenopausal women, however. No atrophy pattern and no absence of superficial cells were noted.

Urinary estrogen levels measured in 20 patients were found to be moderately depressed after a single injection of Depo-Provera but returned to pretreatment levels within 7–20 weeks after treatment. However, the results of another study revealed no statistically significant difference from the controls in the urinary estrogen levels of 12 women treated with Depo-Provera for over 1 year.

Serum estradiol levels were also analyzed in 121 women who received Depo-Provera for 1–5 years. Mean serum estradiol levels were in the range normally found in the early follicular phase of the ovulatory cycle. Nearly all serum estradiol measurements revealed levels higher than those found in postmenopausal women (127). Separate investigations in 20 women who were treated with Depo-Provera for at least 18 months indicated that the plasma ratio of estradiol to estrone is 2:1, which is the same as in untreated women (128). They also observed a variation in estrogen levels in the 12-week treatment period with a significant rise at the end of treatment, indicating no possible accumulation of medroxyprogesterone acetate in the body at the end of medication.

The IND on Depo-Provera for injectable contraceptive uses was filed in July 1963. The NDA on Depo-Provera C-150 received U.S. Food and Drug Administration (FDA) approval in September 1974 as an injectable, long-acting contraceptive for limited use in patients who are unwilling or unable to use other contraceptives. The dose recommended for contraception is one injection of Depo-Provera C-150 every 3 months.

Several undesirable side effects, primarily the frequent amenorrhea and the erratic, unpredictable interval between the injection and the occurrence of menstruation, were reported. These adverse side effects probably accounted for the high patient dropout rate reported, which reached an average of 42% at the end of 9 months of treatment. These side effects were reportedly mitigated later by coadministration of long-acting estradiol 17β-cypionate with Depo-Provera treatment (129). The results of intramuscular injection of the combination, 25 mg medroxyprogesterone acetate and 5 mg estradiol 17β-cypionate, in 104 patients, one dose every 28–32 days, for 4–15 months indicated that 100% fertility control was achieved. Aside from a minor change in menstrual patterns, no significant side effects were noted. This monthly injectable contraceptive formulation caused no discomfort to users, and the dropout rate was found to be less than 5% at the end of 9 months compared to the 42% reported with Depo-Provera C-150.

Endometrial biopsies performed in the first month of treatment revealed distinct phases of proliferation and secretion with varying degrees of hypoplasia, which became the dominating feature after 6–7 months of medication. No apparent change in breast morphology and function was seen, and neither mastalgia nor breast discomfort was reported.

Deladroxate. Deladroxate (Squibb) is a once-a-month intramuscular contraceptive preparation. It is composed of dihydroxyprogesterone acetophenide (150 mg) and estradiol enanthate (10 mg) and provides, when administered parenterally, simultaneous prolongation of progestational and estrogenic activities for approximately 3 weeks. It was reported that it gives a steroid ratio that is adequate for the control of ovulation (130). This combination produced a menstrual cycle that simulates most closely the normal menstrual pattern in humans in terms of length and quantity of menstrual flow. When given on Day 8 of the cycle it gave a mean cycle length of 27.8 days.

Extensive clinical testing was conducted in 60 centers worldwide with over 70,000 cycles in 8000 women. Only an occasional pregnancy was reported to occur during the first treatment cycle (131). The 22 patients who have completed 24 consecutive monthly injections of Deladroxate every month on Day 7, 8, or 9 following the onset of menstruation for 1255 cycles also reported no pregnancies during a 24-month period (132).

When used alone dihydroxyprogesterone acetophenide (Deladroxone, Squibb) was observed to cause a delay in the appearance of the LH peak, which shows up approximately 4 days after the expected midcycle surge, and inhibition of ovulation. The use of dihydroxyprogesterone acetophenide and estradiol enanthate in combination (Deladroxate), on the other hand, markedly depressed the level of LH in addition to delaying its appearance. The results of both immunoassay of LH and radioimmunoassay of serum gonadotropin levels further confirmed that the suppression of the midcycle LH peak is the mechanism of contraceptive action for Deladroxate. Both LH and FSH levels are depressed and ovulation is inhibited during its use.

Norethindrone-Releasing Biodegradable Polymer Bead Suspension. A long-acting, injectable contraceptive norethindrone formulation was developed by first preparing biodegradable norethindrone-dispersing polymeric beads (90–180 μm in particle size) of lactide-glycolide copolymer (90:10) and then suspending them (1% *w/v*) in aqueous methylcellulose solution (133).

Intramuscular administration of this formulation to four female baboons indicated that norethindrone is released at a fairly constant rate with a daily dose of approximately 1.075 mg (or 1.92% of the administered dose) for the first 3 weeks. The burst effect is very minimal (Figure 27). The release rate in Weeks 3–7 is also fairly constant, but it decreases by threefold to a daily dose of 364 μg. Beyond Week 7 the amount of norethindrone released daily becomes negligibly small. Overall, 58.4% of the administered dose was excreted during the 7-week observation.

The intramuscular bioavailability of norethindrone from this injectable contraceptive preparation was also evaluated in small animals (rats) and larger animals (dogs). It was found that the in vivo release rate of norethindrone, represented by the total excretion rate in both urine and feces, is linearly proportional to the initial dose of norethindrone impregnated in the polymer matrix (Figure 28). It appears that the intramuscular release rate of norethindrone from biodegradable polymer beads is

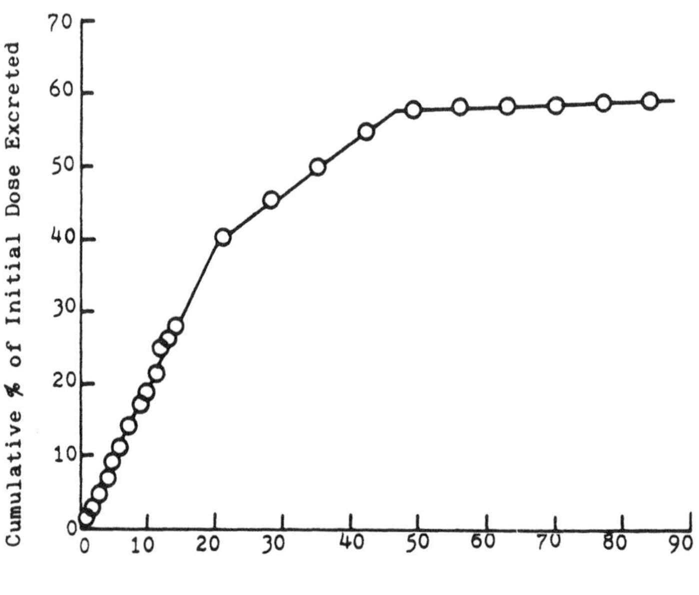

Figure 27 Cumulative excretion of radioactivity (expressed as a percentage of initial dose of norethindrone) from four baboons injected intramuscularly with biodegradable polymer beads of lactide-glycolide copolymer (90:10) containing norethindrone (20% *w/w*). [Plotted from the data by Gresser et al.; Contraception, 17:253 (1978).]

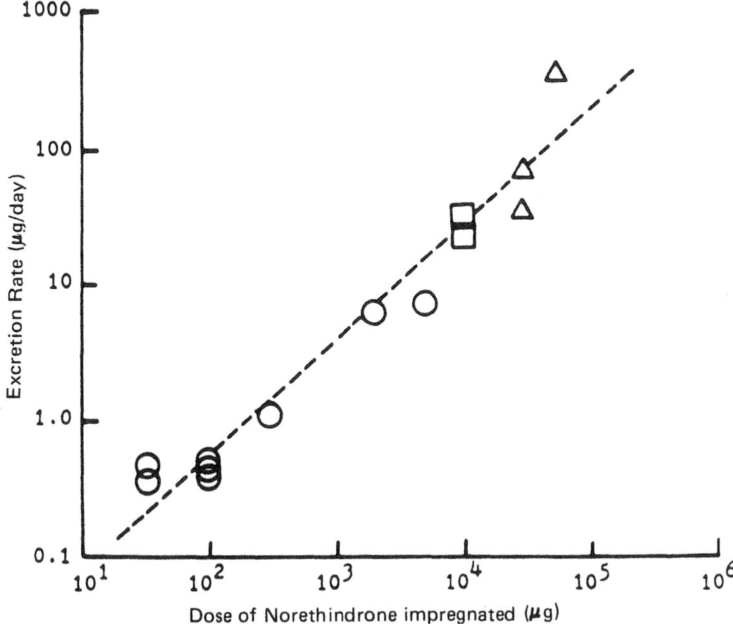

Figure 28 Dependence of the excretion rate of norethindrone in three animal species on the initial dose of norethindrone impregnated in the polymer beads. [Plotted from the data by Gresser et al.; Contraception, 17:253 (1978).]

independent of the animal species. The linear relationship between the rate of excretion of norethindrone and its dosage in polymer beads is defined by

$$\log \left(\frac{Q}{t}\right)_e = 0.82 \log (\text{dose}) - 1.85 \qquad (10)$$

where $(Q/t)_e$ is the total excretion rate of norethindrone measured in urine and feces. Equation (10) suggests that the intramuscular release rate of norethindrone can be controlled to provide a therapeutic dose of norethindrone by impregnating adequate amounts of norethindrone into the biodegradable polymer beads.

Norethindrone Enanthate in Oleaginous Solution. Another injectable, long-acting norethindrone formulation was also developed by synthesizing the water-insoluble but oil-soluble enanthate ester of norethindrone. Intramuscular administration of norethindrone enanthate (200 mg) in oleaginous solution (Norigest, Schering AG) to 130 young fertile women, one dose every 3 months, for a total of 2300 months of observation produced only three pregnancies, which occurred during the second or third month after the last injection (134). This result translates into a method-related failure rate of 2.3%. A total of 25.3% of the patients dropped out of the trial because of side effects, such as disturbance of the menstrual cycle, or for other reasons. They became pregnant between months 4 and 7 after the last medication.

Additional clinical trials of this contraceptive formulation in 160 patients, with a dosing schedule of one injection every 10 weeks, suggested that this preparation causes fewer complaints of bleeding and weight gain than Depo-Provera. It also yielded a more rapid return of ovulation following termination of the medication. More clinical evaluations have been initiated in several countries. Clinical trials in the United States were also recommended by the FDA Fertility and Maternal Health Drug Advisory Committee (FDC Reports, 10/1/79).

The mechanism of action for this long-acting contraceptive preparation appears to be attributed to the changes norethindrone induces in the cervical mucus after its delivery, leading to interference with sperm survival and ascent. The modifications observed in the endometrium and endosalpinx may also contribute to some extent to the contraception. Gynecological laparotomy of six women revealed no signs of ovulation or the formation of corpus luteum during the medication (134).

Norgestrel 17β-Fatty Acid Esters. Daily oral administration of *d*-norgetrel at low doses (30 μg) has been recognized as a useful method of hormonal contraception (135,136). Recently, several efforts have been devoted to the development of long-acting *d*-norgestrel formulations for sustained, continuous fertility regulation in females. The systemic bioavailability and contraceptive action of *d*-norgestrel have been effectively prolonged by impregnating the steroid in ring-shaped vaginal silicone devices (Chapter 9), by encapsulating the drug in subdermal implants (Chapter 8), or by esterifying the *d*-norgestrel with long-chain fatty acids to yield an injectable long-acting contraceptive formulation (137,138).

Five *d*-norgestrel 17β-fatty acid esters with alkyl chain length ranging from 6 to 16 carbon atoms ($n = 4$–14) were synthesized and intramuscularly administered in a castor oil-benzyl benzoate formulation to dogs (137). The systemic bioavailability of *d*-norgestrel from these esters was found to be dependent upon the chain length of the fatty acids: the longer the chain length, the slower the release rate of *d*-norgestrel from the depot formulation and, hence, the longer the duration of contraceptive activity (Figure 29). The total percentage dose excreted within the 8-week period following intramuscular injection was found to decrease exponentially as the alkyl chain length of the fatty acid esters increased (Figure 30).

The clinical effectiveness of this injectable *d*-norgestrel depot formulation for long-acting fertility regulation was evaluated in eight women using a single intramuscular injection of 100 mg *d*-norgestrel 17β-undecylate (138). A peak serum level of *d*-norgestrel in the range of 400–700 pg/ml was achieved in six subjects within 1–2 days after administration. This serum *d*-norgestrel concentration dropped thereafter to a level varying between 50 and 150 pg/ml. This drug level was maintained for a period of at least 130 days. A higher serum *d*-norgestrel concentration was seen in the other two patients. Of the eight subjects two had normal first treatment cycles followed by anovulatory cycles. All other subjects had anovulatory first treatment cycles followed by ovulatory cycles. Excessive bleeding was seen in all patients. The observed variation in the systemic bioavailability of *d*-norgestrel among the patients may be related to the variability in the extent of cleavage of *d*-norgestrel from fatty acid esters after release from the depot formulation.

C. Biopharmaceutics

For a drug administered intramuscularly or subcutaneously to reach the site of action to execute its therapeutic activity, it must first be released from its formulation, transported from the injection site into the systemic circulation, and then delivered to the target tissue. For a drug administered parenterally in the form of a suspension

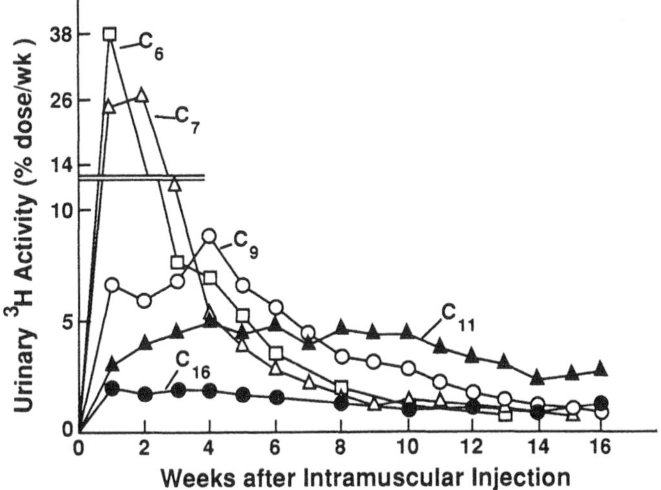

Figure 29 Mean weekly excretion of ^{3}H activity after intramuscular injection of 50 mg *d*-norgestrel esters (800 μCi ^{3}H) in 1 ml castor oil-benzyl benzoate (6:4, v/v) into each of two beagles. Measurements were made for up to 8 weeks in two animals and up to 16 weeks in one animal: (□) hexanoate (C_6); (△) heptanoate (C_7); (○) nonanoate (C_9); (▲) undecylate (C_{11}); and (●) hexadecanoate (C_{16}). [Replotted from the data by Humpel et al.; Contraception, 15:401 (1977).]

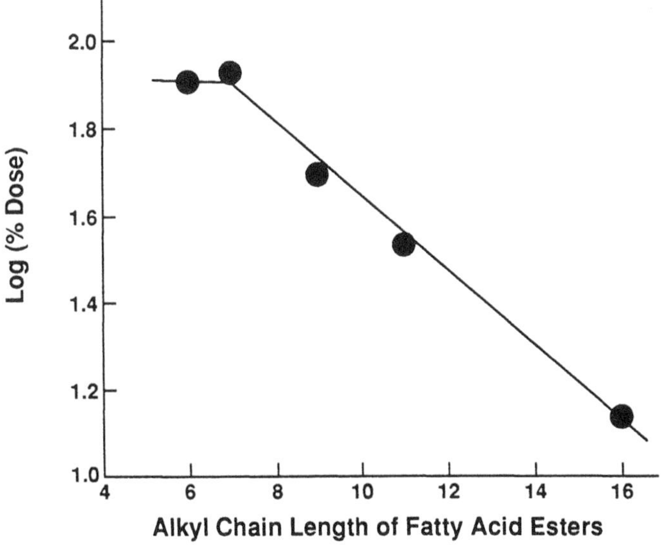

Figure 30 Effect of alkyl chain length of fatty acid esters on the 8-week urinary excretion of *d*-norgestrel from intramuscular *d*-norgestrel depot formulations (50 mg) in beagles. [Plotted from the data by Humpel et al.; Contraception, 15:401 (1977).]

in either aqueous or oleaginous vehicle the rate of delivery and the extent of availability of the drug to the site of drug action is frequently found to be controlled by the slowest (rate-limiting) step in the pharmacokinetic sequence in scheme 2.

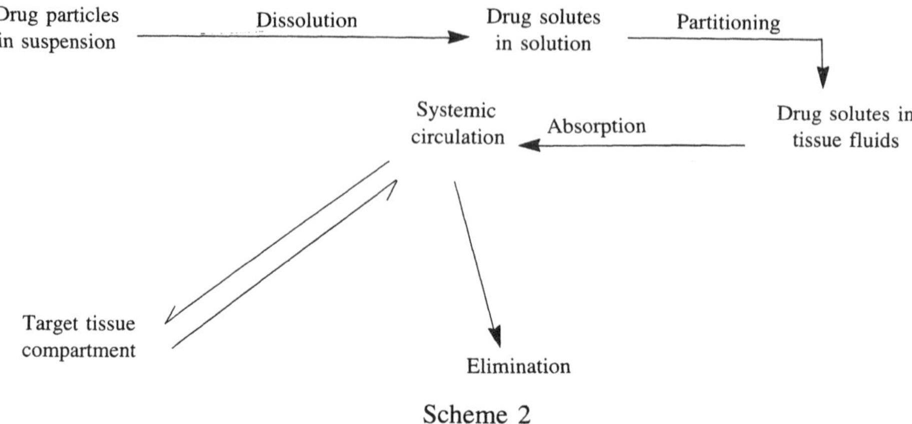

Scheme 2

1. *Effect of Physicochemical Properties*

Parenteral absorption is greatly dependent upon the composite effect of the following physicochemical parameters of the formulation:

1. Rate of dissolution of drug solids in the formulation vehicle
2. Particle size and crystalline habit of drug solids
3. pH value of the formulation
4. pK_a value of the drug
5. Lipophilicity of the drug
6. Tissue fluid/vehicle partition coefficient of the drug
7. Solubility of drug in biological fluids at the injection site
8. Presence of other ingredients in the formulation and their interaction with the drug molecule

All these parameters may play important roles in determining the time of onset, the intensity, and the duration of a therapeutic response of a drug delivered by an injectable controlled-release or sustained-release formulation. In addition, the physiological conditions, such as blood flow around the injection site, can also affect the onset and intensity of therapeutic activity.

In many instances the slowest step, that is, the rate-determining step, in scheme 2 is the dissolution of drug solids in the formulation vehicle and/or the interfacial partitioning of drug molecules from the vehicle to the surrounding tissue fluid. Any factor that affects the rate of dissolution ultimately produces effects on the parenteral absorption of drug. For example, reduction in the particle size of drug solids in the suspension leads to an increase in the total surface area of drug particles available for dissolution. This generally results in an enhancement in the rate of drug dissolution. The effect of particle size was demonstrated by the dependence of the intramuscular bioavailability of phenobarbital suspension on its particle size (139). The blood levels of phenobarbital were reportedly increased by reducing the particle size

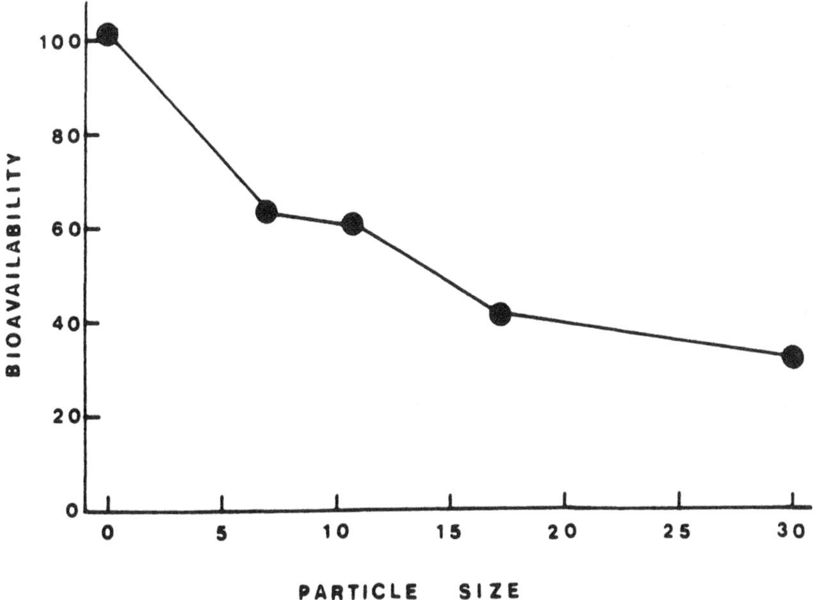

Figure 31 Dependence of intramuscular bioavailability (%) of phenobarbital in beagle dogs on the particle size (μm) of phenobarbital in suspension. [Plotted from the data by Miller and Fincher; J. Pharm. Sci., 60:1733 (1971).]

of phenobarbital (Figure 31). The intramuscular absorption of penicillin G procaine suspension was also reported to be affected by the particle size (140). The average peak levels of penicillin G were found to increase by twofold, that is, from 1.24 to 2.40 units/ml, when the particle size was reduced from 105–150 μm to smaller than 35 μm (Table 2).

Alternatively, the effect of particle size on the dissolution of drug solids can also be applied to prolong the release of a drug. This can be achieved by increasing the particle size of drug solids in the formulation (macrocrystal principle), as illustrated earlier by the effect of testosterone isobutyrate particle size on the growth of comb (Figure 1). Results suggest that the larger the particle size, the more sustained is the serum drug level and the longer the duration of biological activity. One exception to the macrocrystal principle was demonstrated by suspensions of micronized penicillin G procaine in vegetable oil gelled with aluminum monostearate (Figure 2).

Altering the solubility of drug in the formulation can either increase or decrease the rate of dissolution. A number of parameters can alter drug solubility. For example, polymorphism was reportedly responsible for a number of clinically significant differences in drug activity. It is known that approximately one in every three organic compounds exhibits polymorphic behavior. In general there is only one stable crystalline form. Other polymorphic forms are less stable and usually have greater solubility. The polymorph with greater solubility has a faster rate of dissolution and therefore a more rapid absorption. This was demonstrated by chloramphenicol and novobiocin, in which only the amorphous form exhibits biological activity (140).

For weakly acid or weakly basic drugs, drug molecules can exist in either an un-ionized or an ionized state, and the degree of ionization depends upon the dissociation constant K_a of the drug and the pH of the medium as defined by the Henderson-Hasselbalch equation: for a weakly acid drug,

$$\log \frac{[A^-]}{[HA]} = \text{pH} - pK_a \tag{11}$$

for a weakly basic drug,

$$\log \frac{[B]}{[BH^+]} = \text{pH} - pK_a \tag{12}$$

For acid drugs in medium with pH values below the pK_a the drug molecule would exist predominantly in the un-ionized form (HA). On the other hand, the basic drug molecule would exist mainly in the protonated form (BH^+). The un-ionized (HA or B) and ionized (A^- or BH^+) species exhibit different solubility behaviors and lipophilicities. The solubility of these drugs can be altered by changing the pH in the vehicle, leading to an increase (or decrease) in the rate of dissolution in the hydrodynamic diffusion layer and the extent of partitioning of drug molecules from the formulation to the tissue fluids at the injection site.

The viscosity of the suspension can also affect the rate of dissolution by altering the solution diffusivity of drug molecule in the vehicle. This was demonstrated by observations that use of an aqueous solution having 35% (v/v) glycerin as the vehicle decreased the acute subcutaneous toxicity of isoniazid and streptomycin sulfate.

Certain suspensions show thixotropic behavior: when stirred very gently or left standing for some time, they show a nearly infinite viscosity, but they become more fluidy in their consistency and flow more readily when shaken or stirred vigorously. A thixotropic suspension has three advantages: (i) in storage the suspension is stabilized by its structure and high viscosity; (ii) when the suspension is shaken before injection it becomes fluid enough to pass through a hypodermic needle; and (iii) once the suspension reaches injection site in the muscle tissue the suspension structure regenerates and a compact depot results. This thixotropic behavior was demonstrated in the development of penicillin G procaine suspension (11). It has been noted that those suspensions having a structural breakdown point value in excess of 10^5 dyn-cm give spherical depots (Table 3). The structural breakdown point is the point at which the viscosity of the suspension begins to break in response to an increase in the stirring rate. It was also observed that the penicillin G procaine suspension, which has a structural breakdown point value greater than 10^6 dyn-cm, tends to yield excessive needle plugging. This behavior may be due to a high specific surface area or a high solids content, or both. It was concluded that an acceptable intramuscular suspension formulation should contain drug solids with a specific surface of 1×10^4 to 3×10^4 cm^2/g and a structural breakdown point of 10^5–10^6 dyn-cm.

As early as 1936, the influence of the vehicle used in the formulation on the biological activity of drugs was recognized. It was reported that the biological activity of androgens greatly depends upon the types of vehicle used (Table 11). It is interesting to note that androsterone is more than twice as effective in stimulating organ growth when administered in a water-miscible propylene glycol vehicle than in a water-immiscible olive oil vehicle. The difference may result from the precip-

Table 11 Effect of Vehicles on the Biological Activity of Androgens

Androgens	Vehicles[a]	Biological activities[b] Prostate glands (mg)	Biological activities[b] Seminal vesicles (mg)
Testosterone[c]	Control	41	14
	Glycerol (50%)[d]	36	12
	Sesame oil	70	41
	Wool fat	128	98
	Mineral oil	44	15
	Mineral oil + palmitic acid[e]	145	126
Androsterone[f]	Olive oil	78	18
	Arachis oil	133	21
	Castor oil	149	19
	Propylene glycol	199	37

[a]Subcutaneous injection of 50 μg testosterone in 0.5 ml vehicle or 10 mg androsterone in 2 ml vehicle.
[b]The weight of organs 1 day after the last injection.
[c]Compiled from the data by Ballard; in: *Sustained and Controlled Release Drug Delivery Systems* (Robinson, Ed.), Dekker, New York, 1978, Chapter 1.
[d]Glycerol 50% in water.
[e]Dose of 50 mg/day.
[f]Compiled from the data by Ballard and Nelson; in: *Remington's* Pharmaceutical Sciences (Oslo and Hoover, Eds.), 15th ed., Mack, Easton, Pennsylvania, 1975, Chapter 91.

itation of steroid from propylene glycol vehicle after injection because of dilution of the vehicle by tissue fluid. This yields a crystalline mush at the injection site (141). However, the use of 50% glycerol for testosterone did not achieve the same effect (Table 11).

The data in Table 11 suggest that the androgenic activity of testosterone in an oleaginous solution is very much dependent upon the type of oil vehicle used. For example, the maximum activity is attained for testosterone when it is administered in wool fat, but no increase in activity is observed (compared to control) when it is administered in mineral oil. However, the addition of palmitic acid into mineral oil significantly improves the biological activity of testosterone. Furthermore, the subcutaneous activity of testosterone has also been found to depend on the volume of oil vehicle injected (Table 12). The androgenic activity of testosterone in arachis oil solution was observed to increase as the volume of vehicle used to dissolve the same dosage of testosterone increased. However, a reverse trend was observed when testosterone was administered in sesame oil (Table 13). On the other hand, administration of testosterone propionate in 0.8 ml sesame oil gave a substantially greater androgenic activity than in 0.2 ml oil. In both volumes testosterone propionate produces a greater biological activity than testosterone itself at the same dose level. This result suggests that the androgenic activity of testosterone can be greatly enhanced by formation of a long-acting ester, such as propionate, and this improvement in activity is almost three times greater by increasing the volume of vehicle from 0.2 to 0.8 ml.

These studies have illustrated how important the formulation can be in affecting biological responses to a drug delivered parenterally by an injectable controlled-re-

Table 12 Effect of the Volume of Arachis Oil on the Biological Activity of Testosterone in Rats

Volume[a] (ml)	Biological activities[b] Prostate glands (mg)	Seminal vesicles (mg)
2	61	27
5	183	118
10	219	163

[a]Subcutaneous administration to two groups of rats in 10 equal daily injections. It contains 2 mg testosterone.
[b]The weight of organs 1 day after the last injection.
Source: Compiled from the data by Ballard and Nelson; in: *Remington's Pharmaceutical Sciences* (Oslo Hoover et al., Eds.), 15th ed., Mack, Easton, Pennsylvania, 1975, Chapter 91.

lease or sustained-release formulation. The ideal oil utilized as a vehicle for a long-acting, injectable formulation, as either solution or suspension, should meet the following requirements (142):

1. Chemically it should be a stable oil and neutral in composition, that is, containing minimal free fatty acids. It should not react with the drug to produce toxic degradation products.
2. Physically it should not be too viscous to pass readily through a hypodermic needle. It should have good thermal stability at both high and low temperatures.
3. Biologically it should be inert and nonirritating. The oil should be free of antigenic properties, be rapidly absorbed from the injection site soon after the end of medication, and leave no residue.

Before a drug molecule can reach the systemic circulation for delivery to its target tissues it must first penetrate a series of biological membranes. Biological membranes are highly complex in composition and structure but can be described as composed of lipids and proteins. The lipid layer, which is made of phospholipids

Table 13 Comparative Androgenic Activities of Testosterone and Its Propionate and Effect of Solution Volume

Androgens[a]	Seminal vesicle weight[b] 0.2 ml	0.8 ml
Testosterone	28	10
Testosterone propionate	58	140

[a]Testosterone or its propionate, 5 mg, dissolved in sesame oil and injected as a single dose subcutaneously into each of castrated male rats.
[b]The change in the weight (in mg per 100 g rat) of the seminal vesicle at the end of 10 days.
Source: Compiled from the data by Honrath et al.; Steroids, 2:425 (1963).

and cholesterol sandwiched between two protein layers, is the backbone of each biological membrane. The phospholipid molecules are arranged in a bilayer structure with the lipophilic portion of each molecule directed inward and the hydrophilic portion facing toward the outer protein layers, which lends mechanical strength to the membrane. From the structure and composition it is logical to assume that the biological membrane presents a lipoidal barrier to the permeation of drug molecules. This means that drug molecules require a proper lipophilicity to penetrate the membrane. This lipophilicity can be correlated with the partitioning of drug molecules across an oil/water interface. Permeation of drugs through the cell membranes of the oral mucosa, gastrointestinal epithelium, skin, bile, central nervous system, and kidney has been reportedly related to the oil/water partition coefficient of the drug molecules. Even with intravenous administration, in which absorption of drug into the blood circulation is not required, distribution of the drug from the circulation to the site of action (that is, target tissues), still depends greatly on the lipophilic characteristics of the drug molecules.

For intramuscular or subcutaneous drug administration, an additional absorption step is required before drug molecules can gain access to the blood circulation for tissue distribution. Thus any factors that influence this absorption step also affect the rate at which active drug enters the systemic circulation. In addition to the physicochemical properties of the drug molecule itself, as discussed, physiological conditions, such as blood flow from the injection site, can also be relatively important in determining the onset and intensity of drug activity. It was demonstrated that epinephrine delays the subcutaneous absorption of a number of drugs as a result of its constricting action on the vascular bed around the absorption site by reducing blood flow. On the other hand, drug absorption was reportedly enhanced by the incorporation of hyaluronidase as a result of its enhancing effect on the spreading of injected drug solution over a larger area of connective tissue, which leads to the exposure of drug molecules to a greater surface area for absorption (140).

2. *Effect of Physiological Conditions*

An increase in muscular activity, which produces an increase in blood flow to the muscles, may yield an enhancement in the rate of drug absorption from the injection site. The degree of body movement, for instance, was reported to affect the duration of effective penicillin blood levels following the intramuscular administration of penicillin G procaine suspension in oil. The plasma penicillin concentration was maintained above 0.039 units/ml for a mean duration of 33 hr in pneumonia patients compared to only 12 hr in ambulatory patients.

The effect of body movement on intramuscular bioavailability was also observed in patients administered aqueous suspension of penicillin G procaine injected into the gluteal region. The mean peak serum levels attained at 1–2 hr in outpatients was more than twice as high as that for hospitalized patients (0.72 versus 0.29 units/ml). Particularly high serum concentrations were detected in those outpatients who performed active sports during medication. The higher plasma drug levels observed in the outpatients can be attributed to muscular movement during walking and running, which promotes the intramuscular absorption of penicillin from the gluteal region possibly as a result of a decrease in the physiological diffusion layer surrounding the depot formulation, formed by tissue fluid, and an increase in the regional blood flow during and after exercise.

The extent of the effect on drug bioavailability is also dependent upon the site of intramuscular injection. It was found that the intramuscular absorption of penicillin G benzathine from a depot formulation produced a longer penicillinemia in the active training group of Navy recruits than in the less active, hospitalized group. The difference was observed when the traditional upper, outer quadrant of the gluteal region was used for intramuscular administration. On the other hand, no difference was noticed when the injection was made on the anterior gluteal region on the lateral thigh.

The criteria used to determine the route and the site of parenteral drug administrations are as follows (143):

1. Desired rate and extent of systemic absorption
2. Total volume of the formulation to be administered
3. Dosing frequency
4. Inherent irritation, acidity or basicity, and/or concentration
5. Extent of local tissue irritation, nerve damage, and inadvertent blood vessel entry
6. Age and physical condition of the patient

The two major routes of administration for injectable, long-acting depot formulations are the subcutaneous and the intramuscular. The subcutaneous route is generally limited to a nonirritating, water-soluble drug that is well absorbed from the adipose and connective tissue sites, which are poorly perfused with blood compared to muscular tissues (144,145). The subcutaneous administration of insulin preparations is a well-known example. It is extremely important that the sites used for repeated subcutaneous injections, as in the self-administration of insulin in diabetic individuals, be rotated frequently to prevent local tissue damage as well as the accumulation of a depot of unabsorbed drug. The volume for a single subcutaneous injection is usually small, normally in the range of 0.5–1.5 ml (Table 14).

The ideal site for intramuscular injection is deep in the muscle and away from major nerves and arteries. The best sites are the gluteal, the deltoid, and the vastus lateralis (146).

The gluteal muscle has become the most common site for intramuscular injection since it has a greater muscle mass, permitting the injection of larger volumes of fluid (Table 14). The best suited area is the upper outer quadrant, which has a minimum risk of the needle piercing the sciatic nerve or the superior gluteal artery. Careful localization of the injection site is of primary importance. Injections should be made lateral and superior to a line drawn from the posterior superior iliac spine to the greater trochanter. In infants the anterior or lateral thigh is recommended (147).

The deltoid muscle is thick and also has a superior blood supply, which provides the fastest absorption and systemic action of all the intramuscular sites (147,148). This site is found 2 cm below the acromion. Owing to the nonyielding tendinous septa in the upper and lower regions of this muscle, only a small area in the center provides a satisfactory site for drug administration, and the volume of a single injection is also smaller than for the gluteal muscle (Table 14).

The subcutaneous and intramuscular bioavailability of drugs was compared using butorphanol tartrate in dogs (149). The results suggested that there is basically no significant difference between these two routes of administration in terms of ab-

Table 14 Routes, Sites, and Volumes Commonly Used for Parenteral Controlled Drug Administrations

Routes	Sites of injection: Tissue	Sites of injection: Location	Volume[a] Usual (ml)	Volume[a] Range (ml)
Subcutaneous	Subcutaneous fatty layer	Abdomen (at naval level)		
		Buttocks (lateral upper hips)		
		Thighs		
		Upper arm (back middle)	0.5	0.5–1.5
Intramuscular	Gluteus medius	Upper outer quadrant of buttocks	2–4	1–6
	Ventrogluteal	Central upper hip	1–4	1–6
	Quadriceps femoris	Central midthigh	1–4	1–6
	Vastrus lateralis	Outer midthigh	1–4	1–6
	Deltoid	Outer triangular muscle (extreme upper arm at shoulder)	0.5	0.5–2

[a]Volume for single injection.
Source: Compiled from the data by Newton and Newton; J. Am. Pharm. Assoc., NS17:685 (1977).

sorption lag times, peak serum drug concentrations, pharmacokinetics, or the areas under the serum concentration versus time curves. These routes of drug administration can be considered bioequivalent, and both are very sensitive to the extent of regional blood and lymph flow.

3. *Pharmacokinetic Basis*

To maintain drug concentrations within the therapeutic range over the prescribed period of treatment, it is important that the amount of drug delivered from an injectable rate-controlled drug delivery system be sufficiently large to compensate for the quantity of drug eliminated from the body during the same period of time. Two pharmacokinetic models should be discussed.

In Model A, if the rate of drug release from an injectable long-acting drug delivery system follows pseudo–zero-order kinetics with a rate constant for zero-order drug delivery k_0, then

$$(D)_D \xrightarrow{k_0} (D)_B \xrightarrow{k_e} (D)_E$$

Scheme 3

where $(D)_D$, $(D)_B$, and $(D)_E$ are the loading dose of drug in the drug delivery system injected intramuscularly or subcutaneously, the amount of drug absorbed into the systemic circulation from the injection site, and the amount of drug eliminated out of the body, respectively; k_0 is the pseudo–zero-order rate constant for drug delivery; and k_e is the first-order rate constant for elimination.

The rate of change in the total amount of drug in the body $(D)_B$, is thus defined by

$$\frac{d(D)_B}{dt} = k_0 - k_e(D)_B \tag{13}$$

or

$$\frac{d(D)_B}{dt} = k_0 - k_e C_B V_D \tag{14}$$

where C_B is the drug concentration in the sampling compartment, which is often the blood circulation, and V_D is the volume of distribution for the drug delivered.

Integration of Equation (14) gives

$$C_B = \frac{k_0}{k_e V_D}(1 - e^{-k_e t}) \tag{15}$$

That Equations (13) and (14) do not contain the $(D)_D$ term suggests that the concentration of drug in the sampling compartment C_B and thus $(D)_B$, the total amount of drug in the body, are independent of the loading dose of drug in the parenteral drug delivery system $(D)_D$ when the drug molecules are programmed to release at a psuedo–zero-order rate profile.

k_e, the first-order rate constant for the elimination of drug from the body, appears in both the denominator and the exponential term of Equation (15). This suggests that if the biological half-life of the drug is short, that is, a short-acting drug, such as naloxone and ACTH, then the k_e term is large and the magnitude of k_0 value must be large to maintain its therapeutic blood level C_B. In other words, the rate of release of a short-acting drug from the parenteral controlled-release drug delivery system must be sufficiently high to maintain a constant, therapeutically effective drug level in the body.

At some time after drug administration when the $e^{-k_e t}$ term is approaching zero, a steady-state drug concentration is reached, which is defined by

$$(C_B)_{ss} = \frac{k_0}{k_e V_D} \tag{16}$$

This steady-state drug concentration is maintained for as long as the drug is being released at the rate constant k_0 from the parenteral drug delivery system. Equation (16) indicates that the steady-state systemic concentration of a drug $(C_B)_{ss}$ can be maintained at various levels by controlling the rate of drug delivery k_0 from the long-acting drug delivery system.

In model B, if the rate of drug release from an injectable long-acting drug delivery system follows first-order kinetics with a first-order rate constant for drug release, K_1 then

$$(D)_D \xrightarrow{k_1} (D)_B \xrightarrow{k_e} (D)_E$$

Scheme 4

The rate of change in the total amount of drug in the body $(D)_B$ is thus defined by

$$\frac{d(D)_B}{dt} = k_1(D)_D - k_e(D)_B \tag{17}$$

The drug concentration in the sampling compartment can be described by

$$C_B = \frac{k_1(D)_D(e^{-k_e t} - e^{-k_1 t})}{V_D(k_1 - k_e)} \tag{18}$$

Equation (18) suggests that C_B is dependent upon the loading dose of drug in the parenteral drug delivery system $(D)_D$. This means that as $(D)_D$ becomes smaller as a result of the continuous release of drug, C_B becomes smaller proportionally. Furthermore, both the $e^{-k_e t}$ and $e^{-k_1 t}$ terms also decrease with time. In other words, this pharmacokinetic model indicates that if the release of drug from the long-acting drug delivery system is first-order in nature, the concentration of drug in the blood circulation C_B and the amount of drug in the body $(D)_B$, which are in direct proportion to drug loading in the drug delivery system $(D)_D$, thus decrease with time.

From these pharmacokinetic analyses we can conclude that a parenteral drug delivery system that contains a drug with a long biological half-life and releases it according to a zero-order rate process is preferable to a system that does not have these features (150).

III. IMPLANTABLE DRUG DELIVERY

A. Historical Development

In 1861 Lafarge pioneered the concept of implantable therapeutic systems for long-term continuous drug administration with development of a subcutaneously implantable drug pellet. The technique was then rediscovered in 1936 by Deanesly and Parkes (151,152), who administered crystalline hormones in the form of solid steroid pellets to mimic the steady, continuous secretion of hormones from an active gland for hormone substitution therapy (153).

The subcutaneous release rate of steroids from the pellets was found to be slowed and hormonal activities prolonged by dispersing the steroids in a cholesterol matrix during fabrication of the pellets (154). Unfortunately it was observed that the subcutaneous absorption of steroids from the cholesterol pellets varies greatly from one condition to another. Subcutaneous drug administration by pellet implantation was then subjected to modification by several investigators (155,156).

The clinical use of implantable pellets for human health care has declined in recent years. For example, there were three steroid pellets still commercially available in 1979 for medication: (i) a testosterone pellet (Oreton, Schering), (ii) a desoxycorticosterone acetate pellet (Percorten, Ciba), and (iii) an estradiol pellet (Progynon, Schering) (157). These products were no longer available in 1989 (158). On the other hand, the laboratory use of implantable pellets for experimental purposes is still popular (159–168).

Subcutaneous drug administration by pellet implantation is known to have several undesirable drawbacks. The primary one is that the release profile of drugs from the pellet is not constant and cannot be readily controlled in terms of the precision of the release rate and the duration of action. This has triggered the research and development of novel implantable therapeutic systems with rate-controlled drug release kinetics to replace pellets for long-term, continuous subcutaneous administration of drugs.

Application of biocompatible polymers to the construction of implantable therapeutic systems for achieving a better control of drug release, in terms of the precision of drug delivery rate and the duration of therapeutic action, was not realized until the accidental discovery of the rate-controlled drug delivery characteristics of silicone elastomers.

The story began approximately 30 years ago at the U.S. Naval Research Center, where the paths of two unrelated experiments unexpectedly converged (169). At the time, Folkman was trying to treat an experimental heart block with thyroid hormone pellets that would release a minute dose of the hormone locally at a steady rate. At the same time Long was studying how well silicone-based heart valves performed under stress in turbulent water. For photographic purposes he attempted to stain the translucent valves with various types of dye. He observed that certain oil-soluble dyes, such as rhodamine and Sudan III, penetrated the lipophilic silicone valves and stain them easily, whereas most water-soluble dyes, like methylene blue and chlorazol black, did not. This observation triggered him to apply the reversible sorption of lipophilic dyes by silicone elastomers as the means to control the release of drugs for long-term medication.

To evaluate this potential biomedical application a very small capsule-shaped implant was fabricated from silicone polymer tubing to encapsulate thyroid hormone. In vitro elution studies demonstrated that the silicone capsule releases thyroid hormone at a steady rate, day after day. When it was implanted in dog myocardium with heart block, the controlled release of thyroid hormone was found to produce a localized hyperthyroid myocardium with fast pacemaker activity. The same controlled drug release behavior was also observed for isoproterenol, digitoxin, and EDTA when encapsulated in silicone (170,171). Atropine and histamine were also reportedly delivered by silicone capsules for a prolonged period of time (172).

Powers reported 1 year after this pioneering finding that silicone elastomer can be used for the preparation of pyrimethamine-releasing implants to protect chicks from malaria as well as piperazine HCl- and antimony dimercaptosuccinate-releasing implants to provide effective protection in mice against schistosomal infection (173).

Then, in 1966, Dziuk and Cook demonstrated that many steroidal hormones and their synthetic derivatives penetrate readily through the silicone polymer. They also observed that when cattle are implanted with progesterone-releasing silicone capsules their fertility is inhibited for more than 1 year (174). These results were later confirmed by Kincl and his associates (175). Further investigations indicated that steroids penetrate silicone membrane at rates that are substantially greater than that across the membrane fabricated from other synthetic polymers, such as polyethylene. The rate of membrane permeation was found to be controlled by the thickness and surface area of the membrane as well as the polarity of the penetrants. These results stimulated several clinical studies to evaluate the feasibility of using a progestin-releasing silicone capsule to provide a long-term subcutaneous contraception in humans.

All these historic developments have explored the potential of using a silicone capsule as the implantable therapeutic system to achieve the precision of long-term, continual dosing and to control the duration of therapeutic effect. The silicone capsule was later extended to achieve the controlled administration of insecticides (176), methoxyflurane for chronic analgesia (177), nicotine for studying its chronic effect on tobacco smokers (178), anesthetic agents, like ether, for general anesthesia (179,180), and L-dopa to various focal brain targets (181). Continuous research and development efforts over the years have resulted in the regulatory approval of several silicone elastomer-based implantable therapeutic systems for biomedical use, such as Norplant subdermal implants designed for the subcutaneous controlled release of levonorgestrel, a synthetic progestin, for female contraception for up to 7 years (182–186), and Compudose implants designed for the subcutaneous controlled delivery of estradiol, a natural estrogen, for growth promotion in steers for up to 200–400 days (187).

Toward the end of the 1960s a concentrated effort was made to expand the silicone elastomer-based implantable therapeutic system technology to other biocompatible polymers for the controlled release of water-soluble molecules. Some of the implantable therapeutic systems that have evolved from this effort include the non-biodegradable sandwich-type silicone implant for levonorgestrel (184,186) and the ethylene-vinyl acetate implant for 3-ketodesogestrel (188), the biodegradable (lactic-glycolic) copolymer for the subcutaneous and intramuscular controlled administration of narcotic antagonists and synthetic LHRH analogs, the bioerodible polysaccharide polymers for the ocular controlled administration of antiinflammatory steroids, and the hydrophilic Hydron implant for the subcutaneous controlled administration of estrus synchronizing agents (189), as well as implantable therapeutic systems activated by such energy sources as osmotic pressure, vapor pressure, magnetism, and ultrasound (Section III.B).

B. Approaches to the Development of Implantable Drug Delivery Systems

Historically, the subcutaneuous implantation of drug pellets is known to be the first biomedical approach aiming to achieve the prolonged and continuous administration of drugs. This first generation of implantable drug delivery systems was produced by compressing drug crystals, with or without a small fraction of pharmaceutical excipients, into tiny, cylindrical solid pellets that could be readily implanted into subcutaneous tissue by means of a Kearns pellet injector or by making a small skin incision. Subcutaneous tissue is essentially a sheet of areolar tissue lying directly underneath the skin. It is rich in fat but poor in hemoperfusion and nerves. Because of its ready access to implantation, slow drug absorption, and low reactivity to the insertion of foreign materials, it has become a route of choice for prolonged drug administration.

Over the years a number of approaches have been developed to achieve the controlled administration of biologically active agents via implantation (or insertion) in tissues. These approaches are outlined as follows:

I. Controlled drug delivery by diffusion process
   A. Polymer membrane permeation-controlled drug delivery using

1. Nonporous membranes
2. Microporous membranes
3. Semipermeable membranes

B. Matrix diffusion-controlled drug delivery using
1. Lipophilic polymers
2. Hydrophilic (swellable) polymers
3. Porous polymers

C. Microreservoir partition-controlled drug delivery using
1. Hydrophilic reservoir in lipophilic matrix
2. Lipophilic reservoir in hydrophilic matrix

D. Membrane-matrix hybrid-type drug delivery using
1. Lipophilic membrane with hydrophilic matrix
2. Hydrophilic membrane with lipophilic matrix

II. Controlled drug delivery by activation process
A. Osmotic pressure-activated drug delivery
B. Vapor pressure-activated drug delivery
C. Magnetically activated drug delivery
D. Phonophoresis-activated drug delivery
E. Hydration-activated drug delivery
F. Hydrolysis-activated drug delivery

III. Controlled drug delivery by feedback-regulated process
A. Bioerosion-regulated drug delivery
B. Bioresponsive drug delivery

These approaches and the various implantable drug delivery devices developed are briefly described as follows.

1. Polymer Membrane Permeation-controlled Drug Delivery Devices

In this implantable controlled-release drug delivery device the drug reservoir is encapsulated within a capsule-shaped or spherical compartment that is totally enclosed by a rate-controlling polymeric membrane (Chapter 1, Figure 4). The drug reservoir can be either drug solid particles or a dispersion of drug solid particles in a liquid or a solid dispersing medium. The polymeric membrane can be fabricated from a homogeneous or heterogeneous nonporous polymeric material or a microporous (and/or semipermeable) membrane. The encapsulation of drug reservoir inside the polymeric membrane can be accomplished by encapsulation, microencapsulation, extrusion, molding, or other techniques. Different shapes and sizes of implantable drug delivery devices can be fabricated. An example of this type of implantable drug delivery device is the Norplant subdermal implant (Figure 32, top). The mechanisms and kinetics of drug release from this polymer membrane permeation-controlled drug delivery device are analyzed in Chapter 2 (Section II.A). As expected from the physical model that describes the membrane permeation-controlled drug delivery, the subcutaneous release of levonorgestrel from Norplant subdermal implants in human females is constant over a duration of more than 6 years (182–187) with a daily dosage rate of 29.9 μg/day (Figure 32, bottom).

2. Polymer Matrix Diffusion-controlled Drug Delivery Devices

In this implantable controlled-release drug delivery device the drug reservoir is formed by homogeneous dispersion of drug solid particles throughout a lipophilic or hydro-

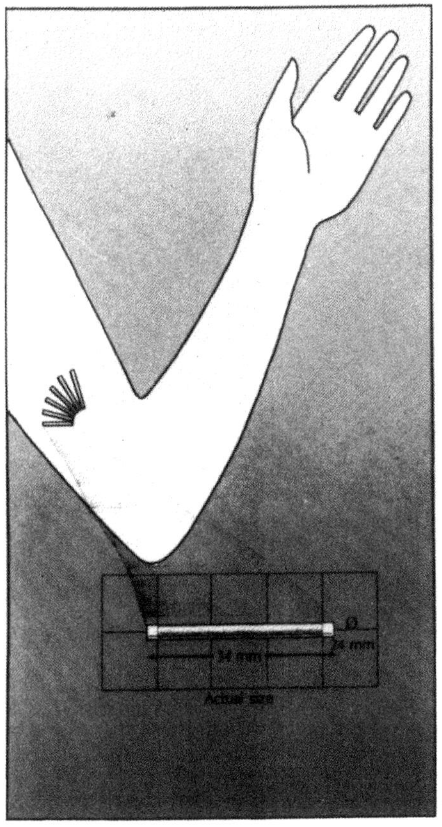

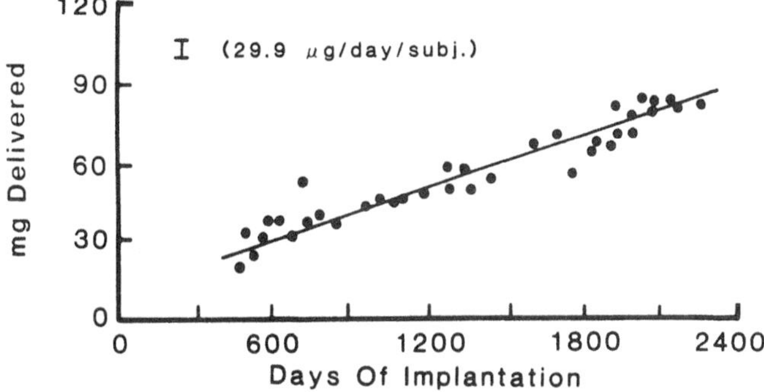

Figure 32 Implantation of 6 units Norplant subdermal implants in the subcutaneous tissue of a human subject's arm and long-term subcutaneous release profile of levonorgestrel from the implants for a duration of up to 6 years.

philic polymer matrix (Chapter 1, Figure 9). The dispersion of drug solid particles in the polymer matrix can be accomplished by blending drug solids with a viscous liquid polymer or a semisolid polymer at room temperature followed by cross-linking of polymer chains or by mixing drug solids with a melted polymer at an elevated temperature. These drug-polymer dispersions are then molded or extruded to form a drug delivery device of various shapes and sizes. An example of this type of im-

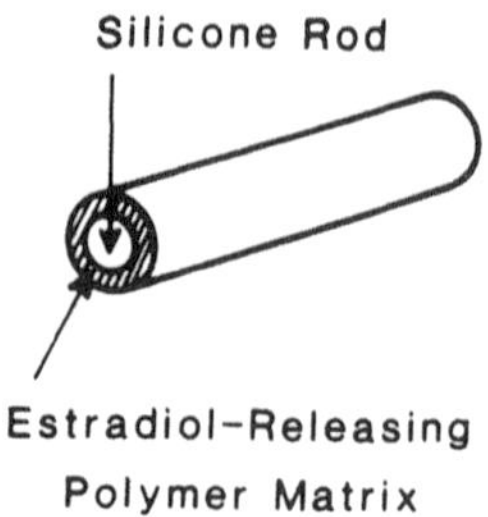

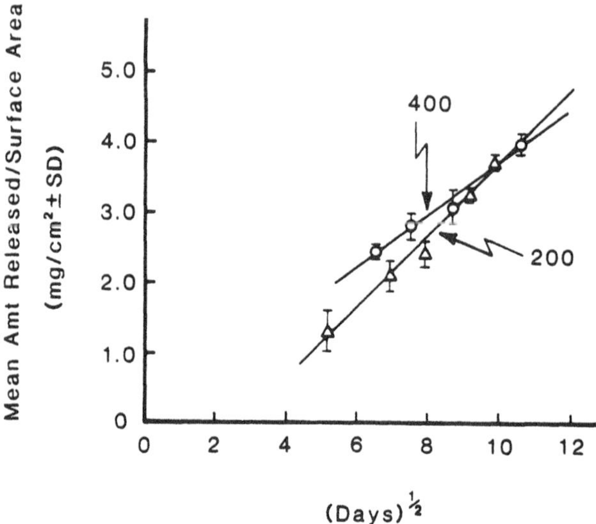

Figure 33 The Compudose implant and the long-term subcutaneous release profiles of estradiol from the implants, Compudose-200 and Compudose-400, in rats for up to 120 days.

plantable drug delivery device is the Compudose implant (Figure 33). It can also be fabricated by dissolving the drug solid and/or the polymer in a common organic solvent followed by cocaevation and solvent evaporation at an elevated temperature and/or under a vacuum to form microspheres. The mechanisms and kinetics of drug release from this polymer matrix diffusion-controlled drug delivery device are analyzed in Chapter 2 (Section II.B). As expected from the physical model describing matrix diffusion-controlled drug delivery, the subcutaneous release of estradiol from Compudose implants in rats follows the Q versus $t^{1/2}$ relationship (Figure 33) (188).

3. *Membrane-Matrix Hybrid-type Drug Delivery Devices*

This type of implantable controlled-release drug delivery device is a hybrid of the polymer membrane permeation-controlled drug delivery system and the polymer matrix diffusion-controlled drug delivery system. It aims to take advantage of the constant drug release kinetics maintained by the membrane permeation-controlled drug delivery system while minimizing the risk of dose damping from the reservoir compartment of this type of drug delivery system. Basically the drug reservoir is also formed by homogeneous dispersion of drug solid particles throughout a polymer matrix, as in the matrix-type drug delivery devices discussed above, but the drug reservoir is further encapsulated by a rate-controlling polymeric membrane as in reservoir-type drug delivery (Section III.B) to form a sandwich-type drug delivery device. An example of this type of implantable drug delivery device is the Norplant II subdermal implant. The mechanisms and kinetics of drug release from this sandwich-type drug delivery device are discussed in Chapter 2 (Section II.C). As expected from the physical model describing membrane permeation-controlled drug delivery, the subcutaneous release of levonorgestrel from Norplant II subdermal implants in human females is constant over a time span of over 4 years (184,186) with a dosage rate of 17.6 μg/day (Figure 34).

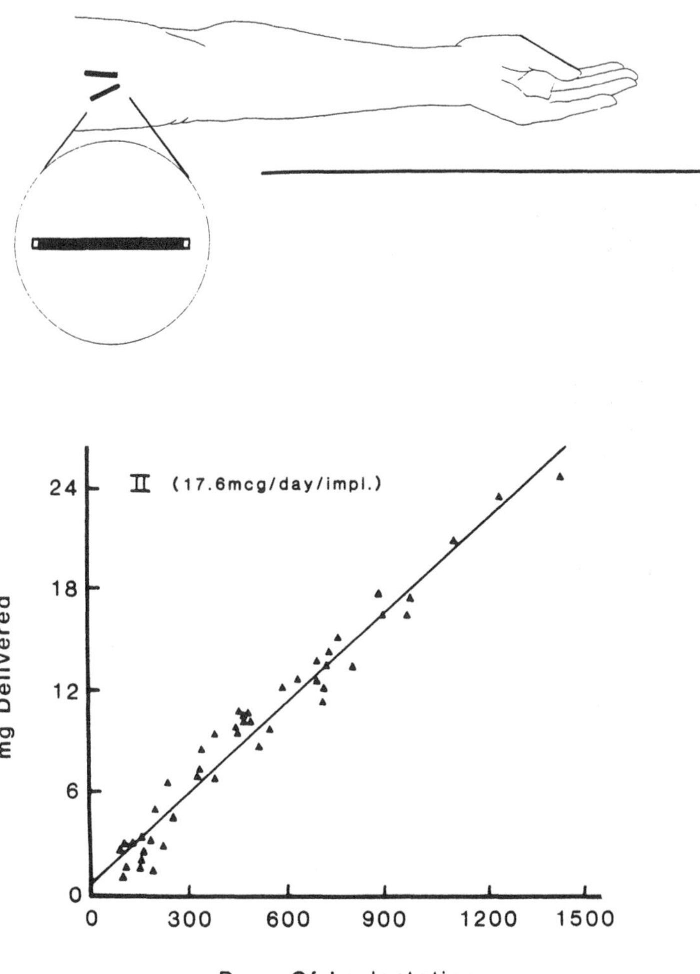

Figure 34 The Norplant-II subdermal implant and the long-term subcutaneous release profiles of levonorgestrel from the implants for a duration of over 4 years.

4. *Microreservoir Partition-controlled Drug Delivery Devices*

In this implantable controlled-release drug delivery device the drug reservoir, which is a suspension of drug crystals in an aqueous solution of a water-miscible polymer, forms a homogeneous dispersion of millions of discrete, unleachable, microscopic drug reservoirs in a polymer matrix (Figure 35). Microdispersion is accomplished by a high-energy dispersion technique (190–193). Different shapes and sizes of drug delivery devices can then be fabricated from this microreservoir-type drug delivery system by molding or extrusion. Depending upon the physicochemical properties of drugs and the desired rate of drug release, the device can be further coated with a layer of biocompatible polymer to modify the mechanism and the rate of drug release. An example of this type of implantable drug delivery device is the Syncro-Mate implant (Figure 36). The mechanisms and kinetics of drug release from this microreservoir dissolution-controlled drug delivery device are analyzed in Chapter 2 (Section II.D). As expected from the physical model describing microreservoir dis-

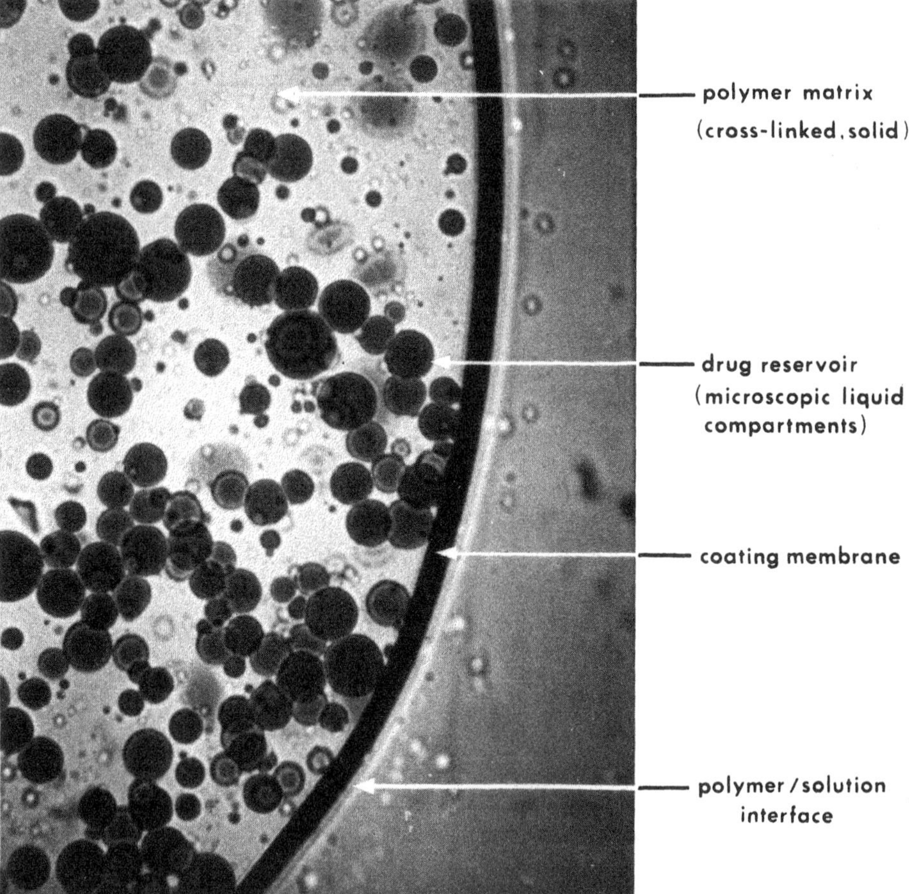

Figure 35 Photomicrograph of a microreservoir-type drug delivery system that shows the microscopic structure of various components.

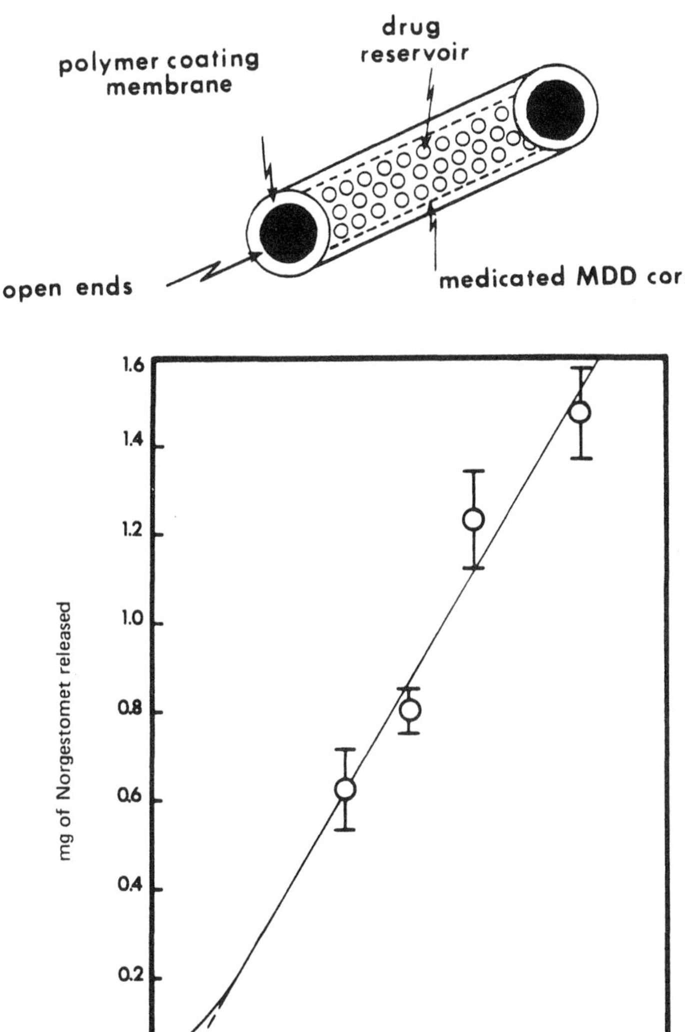

Figure 36 The Syncro-Mate implant, a subdermal implant fabricated as a microreservoir drug delivery system to release norgestomet at a constant rate. The open ends on the implant did not affect the zero-order subcutaneous release profile of norgestomet from the subdermal implants in 20 U.S. cows for up to 20 days. The cumulative amount of norgestomet (Q) released is linearly dependent upon the length of implantation.

solution-controlled drug delivery, the subcutaneous release of norgestomet from Syncro-Mate implants in heifers follows the Q versus t relationship (Figure 36) (193).

5. Osmotic Pressure-activated Drug Delivery Devices

In this implantable controlled-release drug delivery device osmotic pressure is used as the energy source to activate and modulate the delivery of drugs. The drug res-

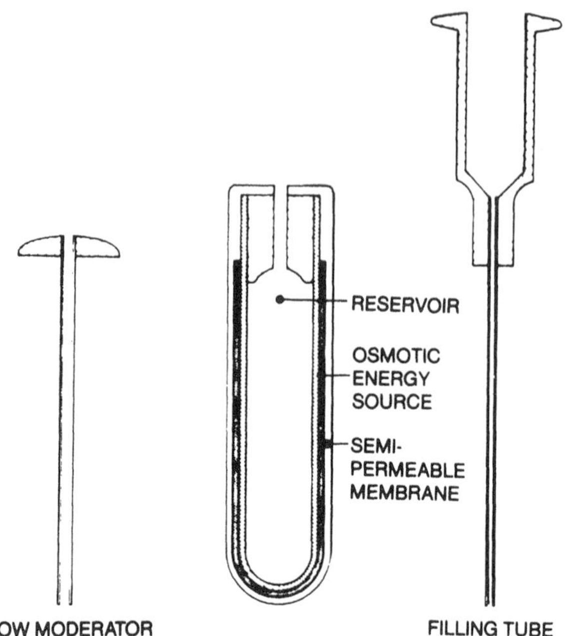

Figure 37 The Alzet osmotic pump with major components. (Reproduced from Reference 197.)

ervoir, which is either a solution or a semisolid formulation, is contained within a semipermeable housing with controlled water permeability. The drug solution is activated to release at a controlled, constant volume as determined by the equation

$$\frac{dv}{dt} = \frac{P_w A_m}{h_m} [\sigma(\pi_s - \pi_e) - (P_s - P_e)] \tag{19}$$

where P_w, A_m, and h_m are the water permeability, the effective surface area, and the thickness of the semipermeable housing, respectively; σ is the reflection coefficient of the membrane; $(\pi_s - \pi_e)$ is the differential osmotic pressure between the drug delivery system with an osmotic pressure of π_s and the environment with an osmotic pressure of π_e; and $(P_s - P_e)$ is the differential hydrostatic pressure between the drug delivery system with a hydrostatic pressure of P_s and the environment with a hydrostatic pressure of P_e. An example of this type of implantable drug delivery device is the Alzet osmotic pump (Figure 37).

An Alzet osmotic pump consists of an external component, which is a rigid, semipermeable housing made of substituted cellulosic polymers, and an internal component, which is a drug reservoir compartment enclosed by a flexible, water- and osmotic agent-impermeable polyester bag. Sandwiched between the two compartments is an osmotic energy source that is a layer of osmotically-active salt coating the external surface of the polyester bag. A rigid polymeric plug is used to form a leakproof seal between the drug reservoir compartment and the osmotic energy source. Following the filling of a drug solution using the specially-designed filling tube, the osmotic pump is assembled with the stainless steel flow modulator having

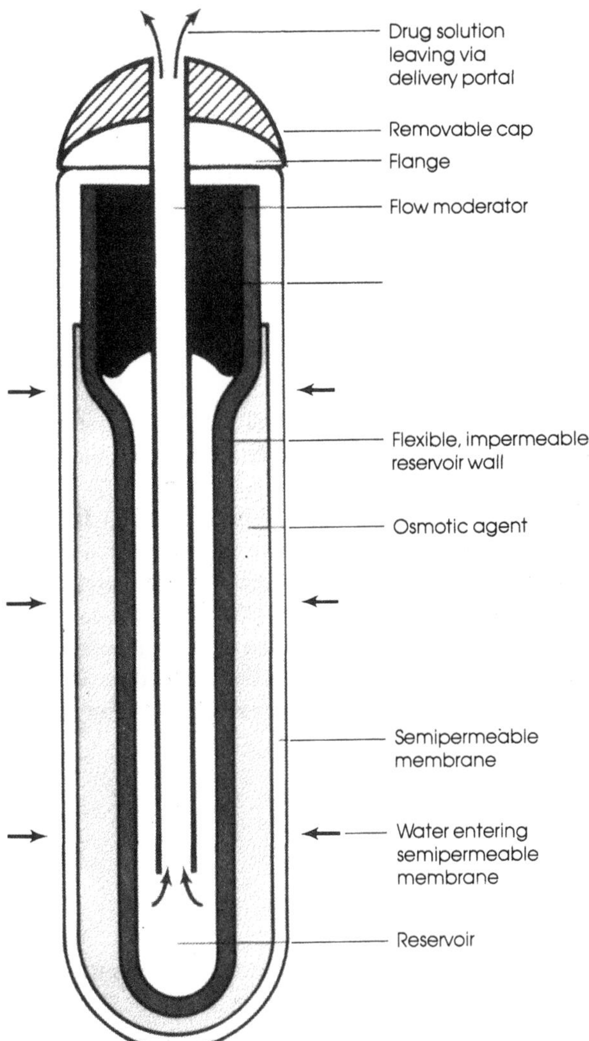

Figure 38 The osmotic pump in operation.

a specific flow rate and then is ready for implantation. At the implantation site the water in the tissue fluid penetrates the semipermeable membrane to dissolve the osmotically active salt, creating osmotic pressure in the narrow space between the flexible reservoir wall and the rigid semipermeable housing. Under the osmotic pressure created the reservoir compartment is reduced in volume and the drug solution is forced to release at a controlled rate through the flow moderator (Figure 38). By varying the drug concentration in the solution different amounts of drug can be delivered at a constant rate, following Equation (26), for a duration of 1–4 weeks (194,195).

This pump design provides some versatility in the volume of drug reservoir, the rate and duration of steady-state drug solution delivery, and the size of the device

Table 15 Implantable Osmotic Pressure-Activated Drug Delivery Devices

Delivery device	Drug reservoir volume (ml/unit)	Steady-state delivery Rate (μl/hr)	Steady-state delivery Duration (hr)	Dimension (cm × cm)
Miniosmotic pump	0.2	1.0	168	0.7 × 2.5
		0.5	336	
Osmotic pump	2.0	10.0	168	1.3 × 4.5
		5.0	336	
		2.5	672	

Source: Alza Corporation product information on Alzet osmotic minipump.

(Table 15). The implantable osmotic pump is especially well suited for assessment of the pharmacokinetic and pharmacodynamic profiles of a drug, which could assist the rational design of an optimized drug delivery system.

The implantable osmotic pump has been applied to examine the effect of a bleomycin dosage regimen on its pharmacologic activity in mice with Lewis lung carcinoma (196). In this investigation the continuous administration of bleomycin from a miniosmotic pump was compared with two intermittent injection frequencies, twice weekly and 10 injections per week, over a 7-day period using a similar dosage regimen in the range of 0–80 mg/kg. The results indicated that the continuous infusion of bleomycin is superior to intermittent infusion in reducing tumor size (Figure 39). The zero-order rate of drug delivery $(Q/t)_z$ is defined by

$$\left(\frac{Q}{t}\right)_z = \frac{P_m A_m}{h_m}[\sigma(\Delta\pi) - (\Delta P)]C_r \tag{20}$$

$$\frac{dQ}{dt} = \frac{(Q/t)_z}{\{1 + [(Q/t)_z/S_D V_t](t - t_z)\}^2} \tag{21}$$

where V_t is the total volume inside the system, C_r and S_D are the concentration and aqueous solubility of drug to be delivered, respectively, and t_z is the total duration at which the system delivers the drug at zero-order kinetics.

6. *Vapor Pressure-activated Drug Delivery Devices*

In this implantable controlled-release drug delivery device vapor pressure is used as the power source to activate the controlled delivery of drugs. The drug reservoir, which is a solution formulation, is contained inside an infusate chamber. By a freely-movable bellows the infusate chamber is physically separated from the vapor pressure chamber, which contains a vaporizable fluid, such as a fluorocarbon (Figure 40). The fluorocarbon vaporizes at body temperature and creates a vapor pressure that pushes the bellows to move upward and forces the drug solution in the infusate chamber to deliver, through a series of flow regulators and delivery cannulas, into the blood circulation at a constant flow rate (199,200) defined by

$$\frac{dV}{dt} = \frac{3.14 d^4\, \Delta p_v}{128 \mu l} \tag{22}$$

where d and l are the inner diameter and the length of the delivery cannula, Δp_v is

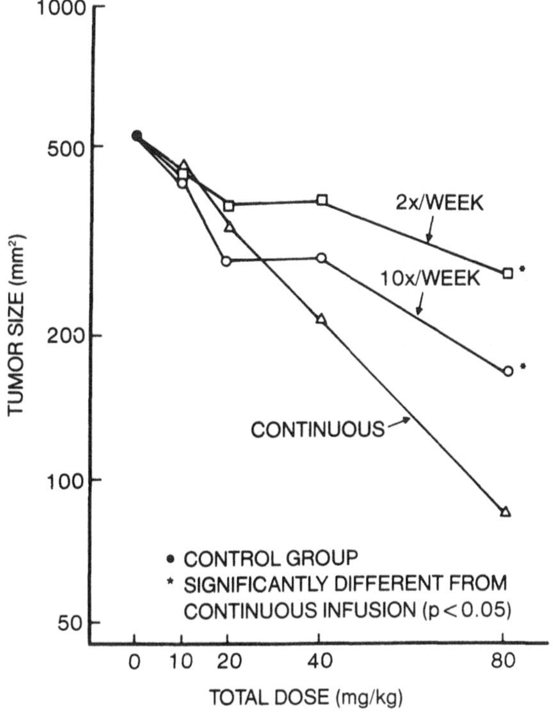

Figure 39 Effect of drug delivery mode on the antitumor effect of bleomycin against Lewis lung carcinoma with dosage ranging from 0 to 80 mg/kg: (△) continuous bleomycin infusion through miniosmotic pump; (○) intermittent SC injection at a frequency of 10 times per week; and (□) intermittent SC injections at a frequency of twice a week. The tumor size was measured on Day 15 after each treatment. The treatment was conducted for 1 week. [Reproduced from Reference 198.)

the differential vapor pressure between the vapor pressure chamber and the implantation site, and μ is the solution viscosity of the drug reservoir.

A typical example is the development of Infusaid, an implantable infusion pump, for the controlled infusion of morphine for the patients suffering from the intensive pain of terminal cancer (201), of heparin for anticoagulation treatment (202), and of insulin for the treatment of diabetes (202). For illustration, the daily dosage rate profile of heparin delivered by the Infusaid pump to dogs for up to 1 year is shown in Figure 41. A fairly constant delivery rate profile has been attained.

7. *Magnetically Activated Drug Delivery Devices*

In this implantable controlled-release drug delivery device electromagnetic energy is used as the power source to activate the delivery of drugs and to control the rate of drug delivery. A magnetic wave-triggering mechanism is incorporated into the drug delivery device, and drug can be triggered to release at varying rates depending upon the magnitude and the duration of electromagnetic energy applied (203). Coupled with a hemispherical design (Figure 42), a zero-order drug release profile was achieved. This subdermally implantable, magnetically modulated hemispherical drug delivery device was fabricated by positioning a tiny donut-shaped magnet at the center of a

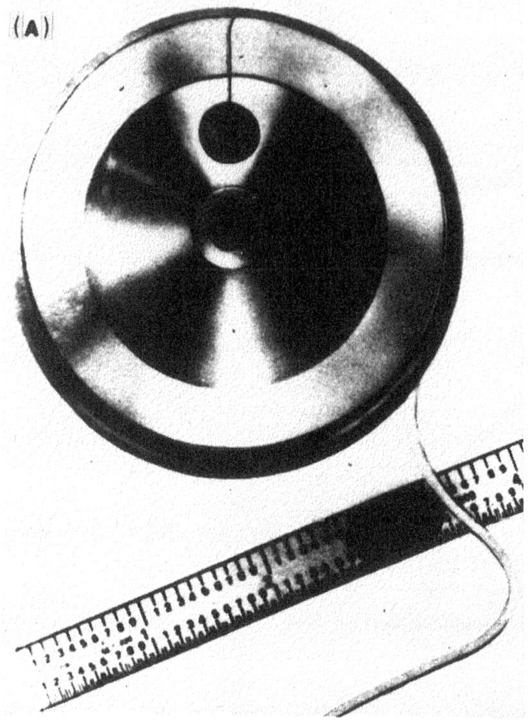

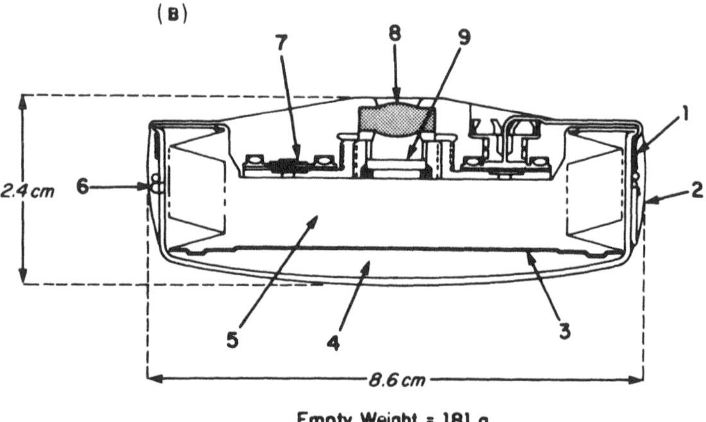

Figure 40 Cross-sectional views of a vapor pressure-activated drug delivery device for example, Infusaid: (1) flow regulator, (2) silicone polymer coating, (3) bellows, (4) fluorocarbon chamber, (5) infusate chamber, (6) fluorocarbon fluid-filling tube (permanently sealed), (7) filter assembly, (8) inlet septum for percutaneous refill of infusate, and (9) needle stop. (From Reference 202.)

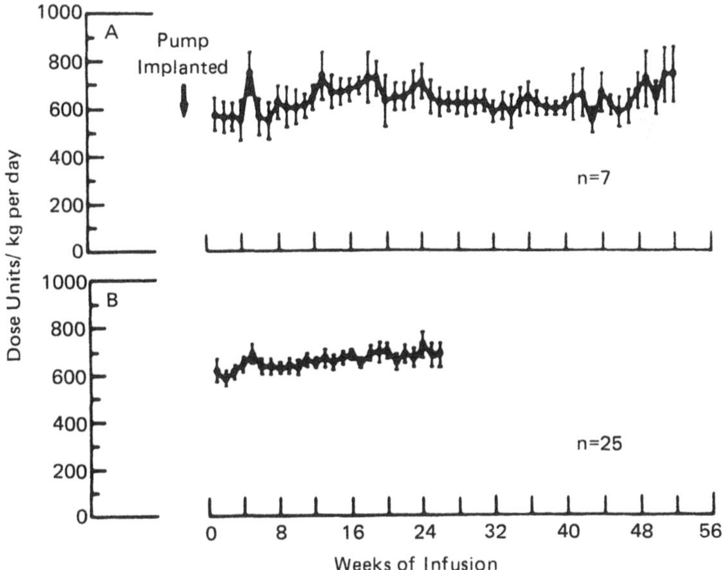

Figure 41 Calculated daily heparin dose (mean ± SEM) delivered by Infusaid pumps to 25 dogs for 6 months (*B*) and to 7 for 12 months (*A*). (From Reference 202.)

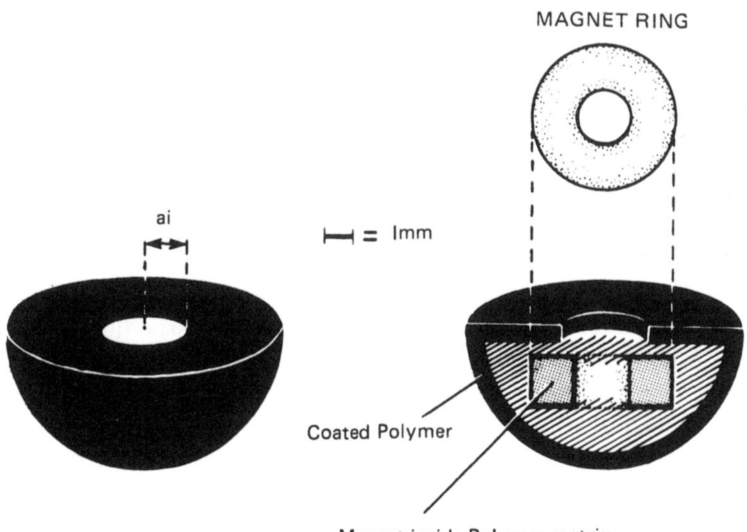

Figure 42 An implantable, hemispheric-shaped magnetically-activated drug delivery device. (From Reference 203.)

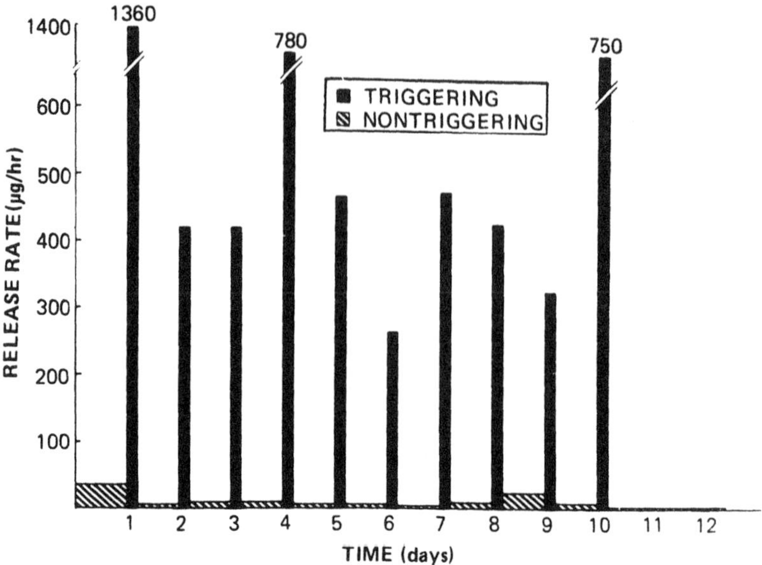

Figure 43 Magnetic modulation of the release of bovine serum albumin (BSA). Before triggering the pellets were prereleased for 1 day. The pellets were triggered for 5 hr followed by nontriggering for 19 hr. The cycle of triggering and nontriggering was repeated daily. (From Reference 203.)

medicated polymer matrix that contains a homogeneous dispersion of a drug with low polymer permeability at a rather high drug-polymer ratio to form a hemispherical pellet. The external surface of the hemispherical pellet is further coated with a pure polymer, such as ethylene-vinyl acetate copolymer or silicone elastomers, on all sides, except one cavity at the center of the flat surface, which is left uncoated to permit the drug molecules to be delivered through the cavity.

Under nontriggering conditions this implantable hemispherical magnetic pellet releases drugs at a controlled basal rate by diffusion alone. By applying an external magnetic field the drugs are activated by the electromagnetic energy to release from the pellet at a much higher rate of delivery. A typical release rate profile is shown in Figure 43 for bovine serum albumin under triggering and nontriggering conditions.

8. *Hydration-activated Drug Delivery Devices*

This type of implantable controlled-release drug delivery device releases drug molecules upon activation by hydration of the drug delivery device by tissue fluid at the implantation site. To achieve this the drug delivery device is often fabricated from a hydrophilic polymer that becomes swollen upon hydration. Drug molecules are released by diffusing through the microscopic water-saturated pore channels in the swollen polymer matrix. The hydration-activated implantable drug delivery device is exemplified by the development of the norgestomet-releasing Hydron implant for estrus synchronization in heifers (204). This was fabricated by polymerizing ethylene glycol methacrylate (Hydron S) in an alcoholic solution that contains norgestomet, a cross-linking agent (such as ethylene dimethacrylate), and an oxidizing catalyst to form a cylindrical water-swellable (but insoluble) Hydron implant. This tiny sub-

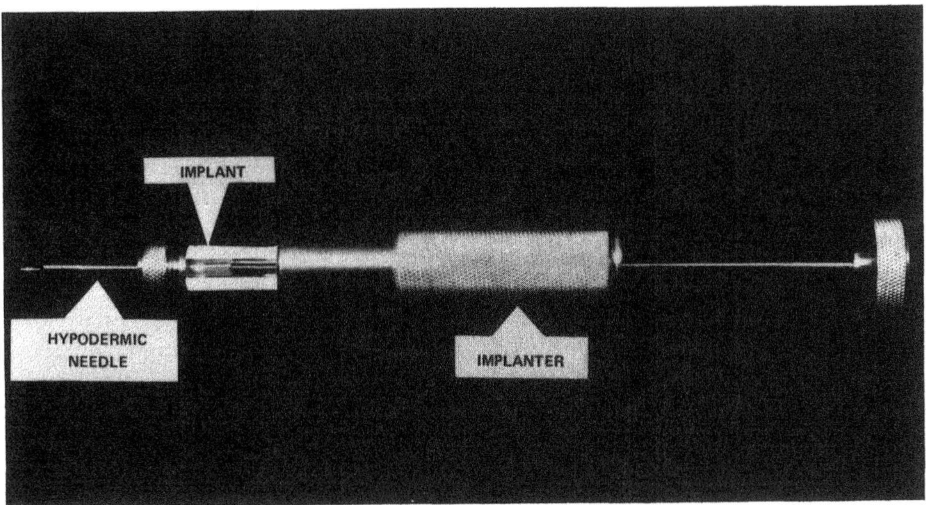

Figure 44 Implanter, with a unit of implant in position, designed specifically for the subcutaneous implantation of Syncro-Mate-B. (Courtesy of Dr. S. E. Mares.)

dermal implant is small (2 × 21 mm) and can be easily implanted subcutaneously into the animal's ear flap (dorsal side) by a specially-designed implanter (Figure 44). The release of norgestomet to the subcutaneous tissue was observed to follow a matrix diffusion process resulting from the time-dependent increase in the thickness of the diffusional path in the swelling polymer matrix triggered by hydration. As expected from the matrix diffusion-controlled process (Chapter 2, Section II.B), a Q versus $t^{1/2}$ drug release profile was observed (Figure 45). The rate of drug release can be modulated by controlling the degree of cross-linking (205).

9. *Hydrolysis-activated Drug Delivery Devices*

This type of implantable controlled-release drug delivery device is activated to release drug molecules upon the hydrolysis of the polymer base by tissue fluid at the implantation site. To achieve this the drug delivery device is fabricated by dispersing a loading dose of solid drug, in micronized form, homogeneously throughout a polymer matrix made from bioerodible or biodegradable polymer, which is then molded into a pellet- or bead-shaped implant. The controlled release of the embedded drug particles is made possible by the combination of polymer erosion by hydrolysis and diffusion through the polymer matrix. The rate of drug release is determined by the rate of biodegradation, polymer composition and molecular weight, drug loading, and drug-polymer interactions.

The rate of drug release from this type of drug delivery system is not constant and is highly dependent upon the erosion process of the polymer matrix (Figure 46). It is exemplified by the development of biodegradable naltrexone pellets fabricated from poly(lactide-glycolide) copolymer for the antinarcotic treatment of opioid-dependent addicts (Figure 47) (206,207).

In addition to poly(lactide-glycolide) copolymer, several other biodegradable or bioerodible polymers, such as polysaccharide, polypeptide, and homopolymer of polylactide or polyglycolide, polyanhydride, and polycapolactone, can also be used

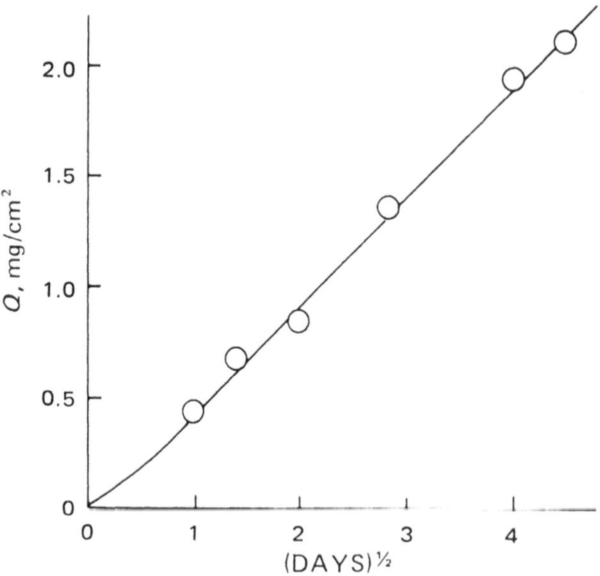

Figure 45 Linear relationship between the cumulative amount of norgestomet (Q) released from 1 unit Syncro-Mate-B implants inserted subcutaneously in 13 cows and the square root of implantation time ($t^{1/2}$). (Reproduced from Chien and Lau, 1976.)

to prepare biodegradable, implantable controlled-release drug delivery devices. Recently a new biodegradable polymer, poly(ortho esters), was developed and used for the controlled release of contraceptive steroids, such as levonorgestrel (208). This polymer contains linkages in the polymer backbones that are relatively stable at the physiological pH of 7.4 but become progressively unstable as the pH is lowered. The erosion rate of such polymers is controlled by using a buffering agent, such as calcium lactate, which is physically impregnated in the polymer matrix and produces a pH that activates the polymer to hydrolyze at a desirable rate in the presence of water. If the polymer is maintained at a rather high hydrophobicity then only the buffering agent in the surface layers is exposed to water and the hydrolysis of polymer occurs only in the surface layers. A constant (zero-order) rate of drug release thus results.

C. Biomedical Applications

Unlike transdermal drug delivery (Chapter 7), by which the percutaneous absorption of most drugs is largely limited by the highly impermeable stratum corneum, oral drug administration (Chapter 4), by which the bioavailability of drugs is often subjected to the variability in gastrointestinal absorption and biotransformation by hepatic first-pass metabolism, and intravenous drug administration, by which the duration of drug action is often short for the majority of therapeutic agents and frequent injections become necessary, the parenteral controlled administration of drugs via subcutaneously or intramuscularly implantable drug delivery systems can gain easy access to the systemic circulation to achieve a total bioavailabililty of drugs as well

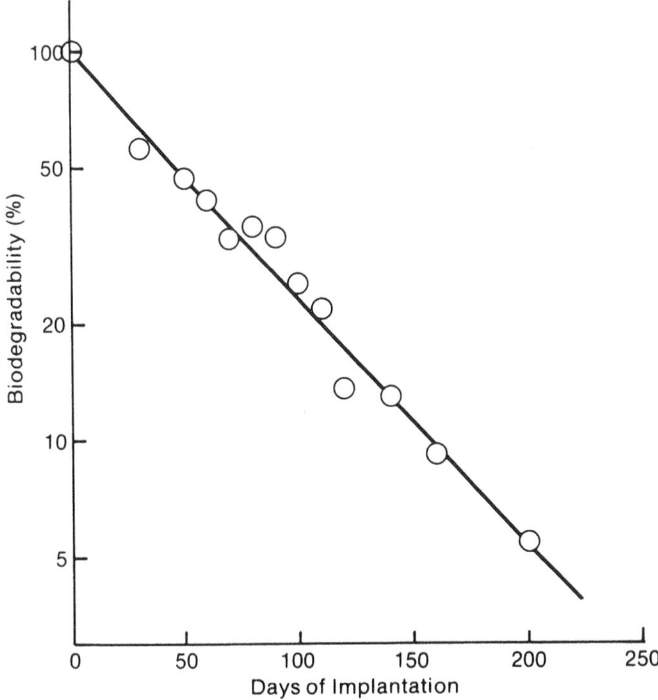

Figure 46 First-order biodegradation profile of poly(lactide-glycolide) copolymer pellets in mice during long-term subcutaneous implantation. The copolymer was prepared from 90% lactic acid and 10% glycolic acid and had a molecular weight of 90,400. (From Reference 199.)

as a continuous delivery of drugs at a controlled rate without the limitations of the transdermal, oral, or intravenous routes of administration. In addition, the implantable drug delivery systems offer a unique advantage over the injectable controlled-release formulations discussed earlier in this chapter (Section II.B), a retrievable mechanism. This feature permits a readily reversible termination of the medication and drug delivery into the body whenever medical and/or personal reasons dictate such a need. This retrievable mechanism is especially important for biomedical applications in humans.

An implantable controlled-release drug delivery system must fulfill one or more of the following requirements:

1. It should facilitate and maximize patient compliance with a therapeutic regimen over an extended time period required for effective treatment.
2. It should be capable of delivering a drug at a controlled rate throughout the period of treatment and of maximizing the dose-response-duration relationship and, in the meantime, reducing the adverse side effects.
3. It can be readily implanted or inserted without requiring a major surgical procedure and repeated motivation on the part of patients.
4. It should be free of potential medical complications such that close supervision

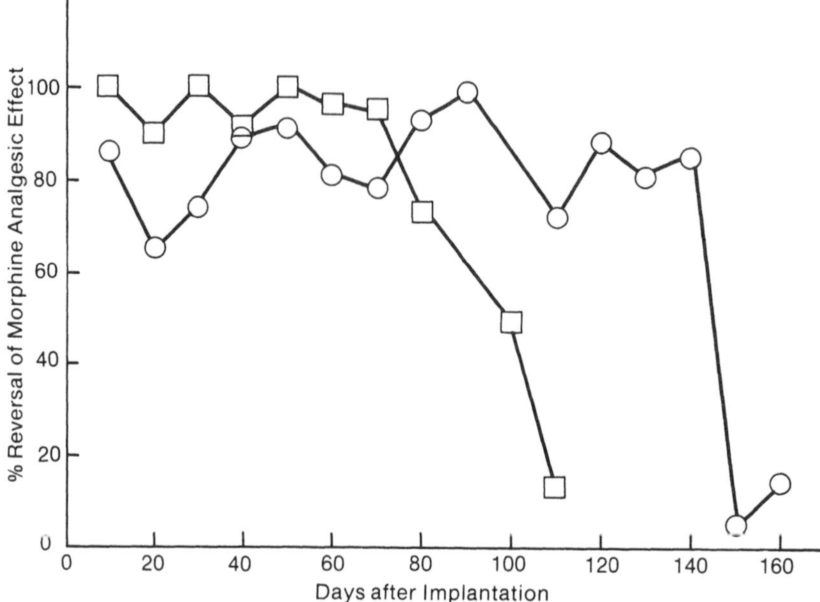

Figure 47 Narcotic antagonistic activity of naltrexone in mice after subcutaneous implantation of naltrexone pellets fabricated from a copolymer of 90% L(+)-lactic acid and 10% glycolic acid (○) and a homopolymer of 100% L(+)-lactic acid (□). Both pellets were coated with polylactide polymer. (From Reference 199.)

by medical personnel is not necessary and infrequent subsequent medical follow-up does not contraindicate its use.

5. It should have a minimum risk of misuse or unauthorized termination of medication by nonmedical personnel.
6. It should be readily retrievable by medical personnel whenever medical reasons dictate.
7. It should be simple to produce and relatively inexpensive.

A number of implantable controlled-release drug delivery devices have recently been developed for human or veterinary uses with one or more of these requirements in mind. These drug delivery devices are discussed in details here.

1. *Human Applications*

Several types of implantable drug delivery systems have been developed that aim to achieve the continuous administration of systemically-active drugs for the long-term regulation of a physiological process, such as a contraceptive steroid for fertility regulation, or uninterrupted treatment of a chronic illness, such as heparin for refractory thromboembolic disease. Such long-term medication can be beneficially accomplished by a specially-designed implantable drug delivery device, such as a progestin-releasing subdermal implant for contraception, an infusion pump for the intravenous infusion of heparin for the prevention of blood clotting, and a biodegradable gonadotropin subdermal implant for the treatment of prostate carcinoma.

Contraceptive Subdermal Implants. Several progestin-releasing subdermal implants have been evaluated for long-term uninterrupted contraception via the subcutaneous continuous delivery of progestogen at a contraceptively-effective dosage rate. One of them, Norplant (Figure 32), has recently received FDA approval for subcutaneous contraception.

Norplant. The Norplant subcutaneous contraceptive system consists of a set of six silicone-based capsule-shaped subdermal implants (each measures 3.4 cm in length and 2.4 mm in outside diameter and contains 36 mg levonorgestrel crystals). The six implants are inserted by minor surgery into the subcutaneous tissue via a trocar and positioned in a fan-shaped arrangement on the inside of the female's upper arm (Figure 32). They are designed to provide the continuous delivery of levonorgestrel at a dosage rate of 29.9 μg/day for a period of up to 5 years. The subcutaneous release of levonorgestrel from these six subdermal implants was reported to have an initial rate of approximately 75 μg/day, which is in good agreement with the in vitro release rate of 72–84 μg/day (209). It then declined gradually during the first year of implantation. By the end of the first year the subcutaneous rate of release leveled to approximately 30 μg/day throughout the remaining 4 years (187,210). This daily dosage rate produced a plasma level of 350 pg/ml of levonorgestrel during the first year, with no significant change in plasma level thereafter until year 5, when the level was approximately 290 pg/ml (Figure 48).

The mechanism of action for the Norplant subcutaneous contraceptive system is via ovulation suppression during the first 6–12 months of use due to the inhibition of LH release from the pituitary and also to its effects on cervix and endometrium; thereafter its contraceptive action depends on densification of cervical mucus, which makes sperm penetration difficult, and on the suppression of endometrium, which prevents the implantation of a blastocyst (211).

The contraceptive efficacy of Norplant subdermal implants was found compa-

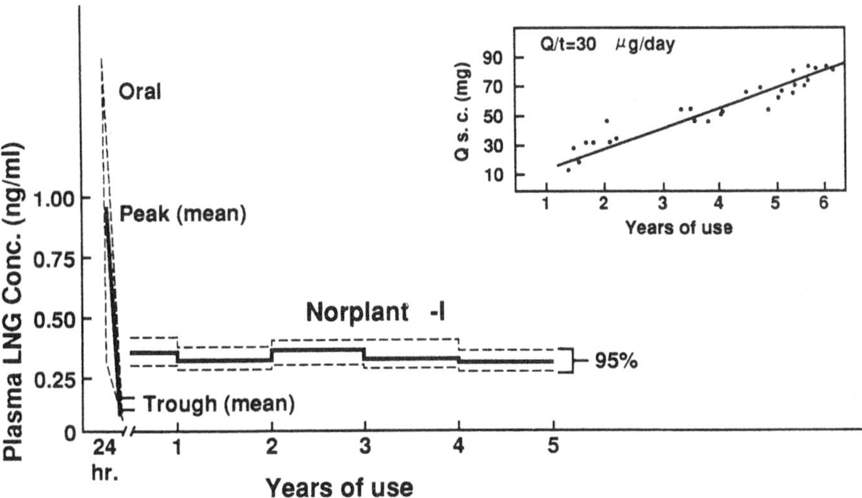

Figure 48 The plasma concentration profile of levonorgestrel throughout the course of a 5-year subcutaneous implantation of Norplant subdermal implants in females compared to the oral administration of levonorgestrel. The subcutaneous release profile of levonorgestrel is shown in the insert. (Plotted from the data by Weiner et al., 1976, and Croxatto et al., 1981.)

Table 16 Annual and Cumulative Gross and Pearl Pregnancy Rates During 5 Year Clinical Studies of Norplant Subdermal Implants

		Number of woman-years		Gross rate (per 100 users)		Pearl rate (per 100 woman-years)	
Year	Subjects[a]	Annual	Cumulative	Annual	Cumulative	Annual	Cumulative
1	778	893.5	893.5	0.3	0.3	0.3	0.3
2	467	565.9	1459.4	0.2	0.5	0.2	0.3
3	383	419.0	1878.4	0.5	1.0	0.5	0.3
4	268	308.7	2187.1	1.3	2.3	1.3	0.5
5	129	193.5	2380.6	0.4	2.7	0.5	0.5

[a]Number of women completing the study in that year. (There were a total of 992 initial acceptors.)
Source: Modified from Sivin et al. (1983).

rable to that of tubal sterilization. The cumulative 5-year pregnancy rate in 992 initial acceptors of the implants was determined to be 2.7 per 100 continuing users (212); the annual pregnancy rate was found to be generally below 0.5 per 100 (Table 16). Further studies conducted in six countries have also confirmed these findings (213–216).

The most common side effect of the Norplant are irregularity of menstrual bleeding, an effect that is particularly marked in the first few months of use. Intermenstrual spotting or amenorrhea may also occur. Users are also found likely to experience an increase in the average number of monthly bleeding and/or spotting days (187). These irregularities usually diminish after 3–6 months of use.

Subcutaneous contraception by Norplant subdermal implants was found to be reversible. There is no evidence that prolonged use of Norplant implants impairs subsequent pregnancy rates (212). Life table analysis of pregnancy rates provides a summary of fecundity after implant removal (Table 17). The 12-month pregnancy rate was 76 per 100 women, and the 24-month rate was estimated as 90. The World Health Organization (WHO) concluded that Norplant is a suitable contraceptive method for use in family-planning programs.

Based on the clinical results generated in some 55,000 women for 20 years, the

Table 17 Life Table Pregnancy Rates After Removal of Norplant Implants for Planned Pregnancy[a]

Month	Cumulative pregnancy rate (per 100 women)
1	18 (± 6)
3	40 (± 8)
6	63 (± 8)
12	76 (± 8)
24	90 (± 6)

[a]45 women had implants removed and 6 were lost to follow-up.
Source: Modified from Sivin et al. (1983).

Table 18 Comparison of Contraceptive Efficacy between Norplant and the Major Existing Contraceptive Methods

Method	User[a] (%)	Efficacy (%) Effectiveness[b]	Pregnancy[c]	Failure rate[d]
Subdermal implant,				
Norplant	—[e]	99.8	0.2	0.2
Oral Contraceptive	18.5	94–97	—	—
Combo	—		0.1	3–6
Minipill	—		0.5	3–9
Intrauterine device	1.2	94	—	—
Cu-T 380A	—		0.8	6
Progestasert	—		2.0	6
Barrier method				
Condom	8.8	86	2	14
Diaphragm	3.5	84	6	16
Cervical cap	1.0	73–92	6	8–27
Sponge	0.7	72–82	6–9	18–28
Spermicide	0.6	79	3	21
Sterilization				
Male	7.0	99.8	0.1	0.15
Female	16.6	99.6	0.2	0.4
Rhythm	1.8	80	1–9	20

[a]Percentage of couples who use a particular method of contraception.
[b]Percentage of users who avoid pregnancy by using a specific method in a given year.
[c]Percentage of women who expect to conceive in a year assuming they always use the method correctly.
[d]Percentage of women who actually conceive in the first year of using a particular method.
[e]Clinical studies were conducted in some 55,000 women for 20 years. Roughly 500,000 women in 15 countries have used Norplant.
Source: S. Findlay (U.S. News & World Report, 12/24/90).

FDA recently approved the marketing of Norplant in the United States (*U.S. News & World Report*, 12/24/90). The contraceptive efficacy of Norplant compares favorably with that of the reversible contraceptive methods currently available (Table 18). In addition to the United States, Norplant has also been approved for 5-year subcutaneous contraception in Germany, France, India, Finland, Pakistan, the Soviet Union, Thailand, Indonesia, and China. This contraceptive system has been marketed by Huhtamaki Oy Leiras Pharmaceuticals of Finland.

Norplant-II. As discussed earlier, WHO-sponsored clinical studies with Norplant subcutaneous contraceptive system have achieved satisfactory results that led to regulatory approval of marketing for 5-year subcutaneous fertility regulation in a number of countries. Even though the results also demonstrated that subcutaneous implantation of Norplant implants requires only a minor surgical procedure (with a 3 mm incision and insertion time of only 5–10 min for experienced health care workers) (187), and expulsion of one or more of the implants has been found to be very rare, efforts to develop a better designed subdermal implant system that is easier to manufacture and produces a more constant drug release profile than Norplant have continued with the development of a second generation of subcutaneous contraception, Norplant-II (Figure 34). This system consists of a set of two silicone-based sandwich-type subdermal implants (each measures 4.4 cm in length and 2.4 mm in outside diameter and contains 70 mg levonorgestrel crystals dispersed homogeneously in the silicone-based polymeric matrix). This new generation of subdermal implant was fabricated by mixing levonorgestrel crystals with silicone elastomer (without silica filler) in a 1:1 ratio and then delivering the homogeneous mixture into a thin-walled silicone tubing (0.4 mm thick) by injection molding (120). After curing the rod is sealed on both ends with silicone medical adhesive to form a sandwich-type rod-shaped subdermal implant. This system design permits the maintenance of a better interfacial contact between the medicated polymer matrix with the internal surface of the silicone tubing and a more uniform concentration gradient, which yields a more constant drug release profile. In vitro drug release studies indicated that Norplant-II implants provide dosage rate of 70–80 μg/day (210), which is equivalent to that achieved by 6 units of Norplant-I. Clinical studies conducted over a period of more than 3 years demonstrated that levonorgestrel is delivered subcutaneously at a dosage rate of 35.2 μg/day from 2 units of Norplant-II (Figure 34), which is again equivalent to the rate achieved by Norplant throughout the period of 2–4 years (Figure 32).

A clinical study was recently performed in 250 women of child-bearing age (18–40 years old) to compare the efficacy and side effect patterns of Norplant-II with Norplant-I for a duration of up to 3 years (217). The results showed that only 1 pregnancy occurred in the 75 Norplant-I users, which happened during Month 27 of use, but no pregnancy has been detected in the 175 Norplant-II users. During the 36-month treatment period, 21.6% of patients discontinued the study (13.6% were side effect related and 8.0% were not side effect related). Both subcutaneous contraceptive systems yielded very much the same degree of side effect-related discontinuation. The most commonly observed side effect leading to removal of the implants was irregular uterine bleeding (7.5% for Norplant-II and 4.0% for Norplant-I) manifested as either prolonged bleeding or too frequent episodes of bleeding. The average duration of use for those women requesting removal of the implants due to side effects was 10.1 months for irregular uterine bleeding (10.4 months for Norplant-II versus 9.0 months for Norplant-I), and 13.3 months for other medical reasons (12.1 months for Norplant-II versus 15.6 months for Norplant-I). A total of 8 subjects discontinued the study to become pregnant. The return to fertility was found to be rapid. The investigators concluded that the ease of insertion and removal of Norplant-II has made it a preferred subcutaneous contraceptive system over Norplant-I system, even though the rate of discontinuation due to bleeding irregularity is slightly higher with Norplant-II than with Norplant-I.

New developments of interest. A new generation of subcutaneous contraceptive implant, Implanon, was recently developed. Similar to Norplant-II, it is also a sandwich-type subdermal implant (each measures 4.0 cm in length and 2.0 mm in outside diameter and contains 67 mg 3-ketodesogestrel). Instead of using silicone elastomer to form the polymer matrix as in Norplant-II, Implanon is fabricated from ethylene-vinyl acetate copolymer (Figure 49). The type of ethylene-vinyl acetate copolymer used for fabrication of the coating membrane (with a thickness of 60 μm) has a permeability for 3-ketodesogestrel that is approximately 20 times lower than that for the type of ethylene-vinyl acetate copolymer used as the core matrix, so the coating membrane becomes the rate-limiting membrane for the release of 3-ketodesogestrel.

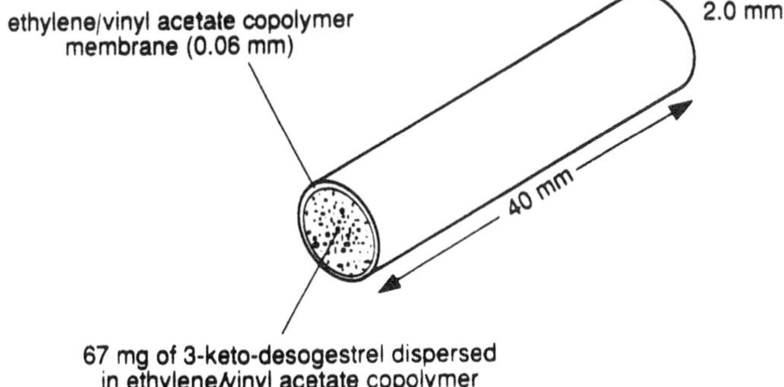

Figure 49 The Implanon subcutaneous contraceptive implant showing various structural components and dimensions. (Courtesy of A. P. Sam.)

This implant was fabricated using fiber spinning technology (218). The main processing step is the melt extrusion of a coaxial fiber, allowing the production of fiber by the mile (187). In the coextrusion process two extruders and two spin pumps supply a spin head with a 3-ketodesogestrel-dispersed ethylene-vinyl acetate copolymer core material and with drug-free ethylene-vinyl acetate copolymer coating material (Figure 50). In the spin head, the core material flows along the central axis and the coating material comes from the outside. The two flows come together near the opening of the spin head, and the coextrudate leaves the spin head as a thick coaxial fiber that is subsequently cooled in a water bath and wound. The fiber is then cut into 4 cm pieces, placed into an inserter, and sterilized by gamma irradiation (187).

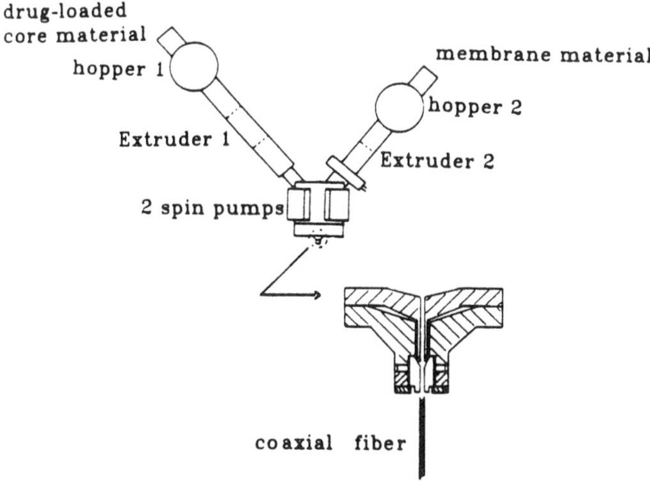

Figure 50 The coextrusion process used in the manufacture of Implanon, in which the melt extrusion of a coaxial fiber using the combination of two extruders and two spin pumps to supply the core and coating materials to a spin head, is shown as the main processing step. (Courtesy of A. P. Sam.)

The implant is designed for simple subcutaneous implantation from a sterile disposable inserter with minimal risk of insertion-related infection and to release 3-ketodesogestrel for a period of 2 years at a dosage rate >30 μg/day needed for complete inhibition of ovulation.

In vivo studies conducted in dogs indicated that a fairly constant release profile of 3-ketodesogestrel has been obtained at a rate comparable to the in vitro rate of release (Figure 51).

Implantable Infusion Pumps. The chronic intravenous infusion of drug solutions at a constant rate has been long recognized as a desirable therapeutic mode for ambulatory patients in a number of diseases. These include (i) continuous heparin infusion in the treatment of refractory thromboembolic disease; (ii) chronic insulin administration in the treatment of diabetes mellitus; and (iii) localized delivery of antineoplastic drugs in the treatment of solid tumors (219).

However, several problems are associated with chronic IV infusion therapy, which may account for its relative lack of popularity compared with oral and periodic parenteral medications. These problems include (i) it requires constant immobilization of the patient; (ii) it needs fairly close supervision from medical staff; (iii) it requires hospitalization of patients; and (iv) it may produce a number of complications as the result of long-term intravenous catheterization; such as local inflammation and sepsis (220).

A number of research efforts have been made to alleviate these problems through the use of implantable infusion pumps; however, the most recent models of electrolytic, spring-powered, and electrically-powered infusion pumps either require external power sources or necessitate periodic battery changes or spring windings, thus preventing the unit from being totally implantable or self-sufficient.

An ideal implantable infusion pump for chronic drug delivery should have the following features:

1. It should be capable of maintaining a constant low rate of infusion, which should be adjustable even when the pump is in situ.
2. Its power source should be virtually inexhaustible, thus making the pump permanently implantable.

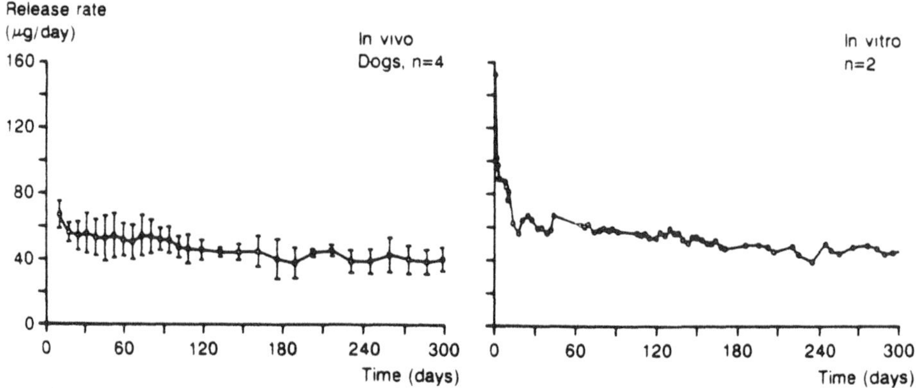

Figure 51 The in vitro and in vivo release rate profiles of 3-keto-desogestrel from the Implanon contraceptive implant. (Courtesy of A. P. Sam.)

3. Its drug reservoir compartment should be sufficiently large to minimize the frequency of refilling, but the overall size of the pump should be kept to a reasonable minimum for ease of implantation.
4. It should permit the refilling of drug reservoir through a simple percutaneous injection.
5. It should, as a whole, have long-term tissue biocompatibility.

An implantable infusion pump (Infusaid, Metal Bellows Corp.) that was recently developed on the basis of the physical concept that a vapor in equilibrium with its liquid phase exerts a constant vapor pressure at a given temperature, regardless of volume, seems to fulfill these requirements. A typical model consists of a hollow titanium disk divided into two chambers by a freely movable titanium bellows (Figure 40). The upper chamber contains the infusate solution, and the lower chamber contains a fluorocarbon fluid (FC-88, which is a stable, chemically inert, nonflammable and nontoxic mixture of perfluoropentane isomers). At body temperature the vapor pressure produced from the vaporization of this fluorocarbon fluid is approximately 300 mmHg greater than the atmospheric pressure. This vapor pressure provides the power needed to exert on the bellows and forces the infusate to flow at a constant low rate through a series of bacterial filters, a flow-regulating resistance element, a silicone polymer delivery cannula, and finally into a vein.

This totally implantable infusion pump can be sterilized by autoclaving and implanted beneath the skin and refilled simply by percutaneous injection through a self-sealing Silastic/Teflon septum. The pressure exerted by the injection fills the upper chamber with a new supply of infusate and simultaneously forces the bellows to move downward and reduce the volume of the lower chamber, which in turn exerts pressure sufficient to condense the fluorocarbon vapor, thus recycling the pump. The rate of flow (dV/dt) through the regulator is governed by the Poisseuille relationship, as defined earlier by Equation (29), and the rate of drug delivery is expressed by

$$\frac{Q}{t} = \frac{dV}{dt} C_r = \frac{3.14 d^4 \, \Delta P_v C_r}{128 \mu l} \tag{23}$$

where C_r is the drug concentration in the infusate (reservoir solution) compartment; d and l are the inner diameter and the length of the narrow-bore stainless steel capillary tubing or Teflon delivery catheter (cm); ΔP_v is the pressure difference (dyn/cm^2) between the vapor pressure in the pump and the atmospheric pressure; μ is the viscosity of the infusate solution (poise). Equation (30) suggests that the flow rate of the infusate solution can be adjusted readily by varying the diameter and length of the capillary tubing and/or altering the viscosity of the infusate by incorporating a small amount of water-soluble high-molecular-weight polymer, such as dextran (220). Addition of dextran in a concentration range of 0–4% was found to produce a wide latitude of variation in flow rate with the pump in situ (1.95–0.06 ml/hr; Table 19). On the other hand, the amount of drug delivered can be controlled by adjusting the drug concentration in the infusate solution (C_r).

The FC-88 in the pump has an effective vapor pressure of approximately 8 pounds/inch^2 (at 37°C). An increase in temperature was observed to produce an additive effect that results in an increase in vapor pressure and a simultaneous decrease in infusate viscosity. This leads to an approximately 5.6% increase in Celsius temperature (221).

Table 19 Effect of Dextran on the Viscosity and Flow Rate of Infusate

Dextran concentration (%)	Viscosity (poise)	Flow rate[a] (ml/hr ± SEM)
0	0.7	1.95 ± 0.08
1	3.1	0.73 ± 0.02
2	9.0	0.26 ± 0.02
4	36.5	0.06 ± 0.001

[a]Determined at 37°C from in vitro studies with duration ranging from 90 to 755 hr.
Source: Compiled from the data by Blackshear et al., Surg. Gynecol. Obstet., 134:51 (1972).

The potential biomedical applications of Infusaid can be illustrated by the continuous subcutaneous controlled infusion of heparin for anticoagulation treatment and of insulin for antidiabetic medication.

1-Year continuous heparinization in anticoagulation treatment. Chronic long-term anti-coagulation is a therapeutic objective in the management of a variety of clinical disorders, including recurrent venous thromboembolism, systemic arterial thrombi, transient ischemic attacks, and cardiac diseases resulting from the placement of valve prostheses and heart transplantation.

Heparin, a mucopolysaccharide that inhibits the clotting of blood, is recognized as the most effective anticoagulant clinically available today (222). The comparative safety and efficacy of heparin administration by continuous intravenous infusion were recently demonstrated in two clinical trials. The use of continuous intravenous administration of heparin has been advocated as a more effective means of preventing thromboembolic phenomena in humans than heparinization by more conventional methods of administration (223).

Despite its desirability as a therapeutic alternative, long-term anticoagulation treatment by continuous intravenous infusion of heparin has not been practiced in ambulatory patients because of the following reasons: (i) no acceptable means was available to accomplish the continuous intravenous administration of heparin on a long-term basis, and (ii) fears of some anticipated complications, such as hemorrhage, thrombocytopenia, and osteoporosis. Requirements for administering heparin by frequent bolus injections or using a cumbersome external infusion apparatus have thus limited heparinization to short-term therapy only (224).

The development of Infusaid pumps has enabled the continuous intravenous administration of heparin to ambulatory patients.

Theoretically, a constant flow rate can be achieved with this infusion pump system when the ambient temperature and pressure are constant [Equation (30)]. In practice, however, flow variability is introduced by the springlike characteristics of the welded titanium bellows. Experimentally it was found that in the proposed operating range of infusate volumes (5–45 ml) the bellows spring constant causes a pressure difference as much as 45 mmHg (or a 15% change in the chamber pressure) (223). This 15% pressure change can be translated into a variability of approximately 11% in flow rate during a single pump cycle.

The in vivo performance of Infusaid pumps was evaluated for continuous heparin infusion in 25 dogs for 6 months and in 7 of these dogs for a period of 1 year. The properly calibrated and sterilized pumps were implanted under aseptic conditions into a subcutaneous pocket in the left iliac fossa region. The silicone infusion cannula was threaded through the deep circumflex iliac vein into the inferior vena cava. After wound healing was complete the bacteriostatic water in the pump was replaced with sodium heparin solution at a concentration designed to provide a heaprin dose of approximately 650 IU/kg/day. This dose was adjusted in the next 1–4 weeks until evidence of an acceptable anticoagulation level was achieved. After this adjustment period the pump was refilled at 1–2 week intervals by percutaneous injection, and blood parameters, such as the Lee-White clotting time, activated partial thromboplastin time, and thrombin time, were monitored for the next 26–52 weeks (223,224).

The results indicated that no marked or statistically significant changes ($p < 0.05$) are detected in the mean body weight, hematocrit value, or platelet count in the treated animals throughout the period of treatment. Both the heparinized and control dogs appeared to be normally active and were free of manifest disease. The daily heparin dose delivered by the Infusaid pumps as required to achieve the desired levels of anticoagulation was not significantly varied during the course of the 6–12 months of study (Figure 41), indicating that heparin tolerance did not develop during long-term heparinization treatment. The overall mean daily heparin dose for the 25 dogs under treatment in the period of weeks 5–26 was calculated as 666 IU/kg/day. All infusion cannulas remained patent during the 52-week application.

The anticoagulation test demonstrated that twice the normal mean values of the Lee-White whole-blood clotting time (>18 min) are achieved within week 4 of heparin infusion (Figure 52) and the clotting time was maintained above 18 min in 80% of the individual values determined after four weeks of treatment. On the other hand, the infusion of bacteriostatic water from the same pump in control animals did not produce changes in the Lee-White clotting time.

The dogs that received a continuous administration of heparin from the implantable infusion pump had a mean activated partial thromboplastin time greater than 17 sec on or before week 5 and thereafter to the completion of 52 weeks of treatment (Figure 53). The time was maintained greater than 17 sec in 82% of the individual test values measured after week 4. There was no marked change in the same anticoagulation measurement in the control group.

In the same period of heparinization treatment by Infusaid pumps 88% of plasma thrombin time values were detected as greater than 120 sec for dogs receiving heparin compared to less than 20 sec in dogs receiving only bacteriostatic water.

In summary, after the initial 1–4 weeks of dosage adjustment period, all three measurements of anticoagulation demonstrated that an adequate degree of heparinization was achieved and maintained for 6–12 months (223). There was no evidence of anemia or thrombocytopenia. Serum alkaline phosphate values did not change significantly throughout the study. The mean temperature at the pump implantation site was observed to drop by only 0.8°C during the course of exposure in a refrigerated room (at 4.6°C) for 40 min.

Hemorrhage was suspected as the most frequent complication at this level of anticoagulation. Of the 17 bleeding episodes experienced by 10 dogs during the 6–12 months of study, 13 were found to occur at percutaneous puncture sites during pump refill or venipuncture. The other four episodes consisted of hematomas on the

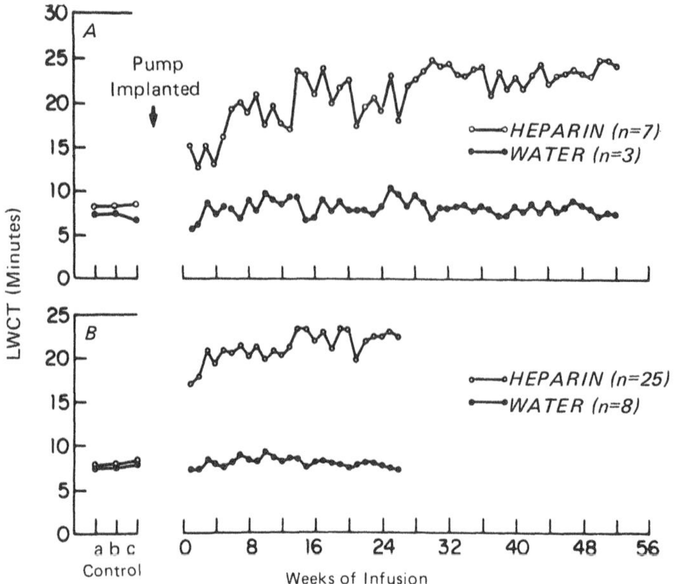

Figure 52 Mean Lee-White whole-blood clotting time (LWCT) for 25 dogs receiving heparin (○) and 8 receiving water (●) from Infusaid pumps for 6 months (*B*) and for 7 receiving heparin (○) and 3 receiving water (●) for 12 months (*A*). [Reproduced from Blackshear et al., Surg. Gynecol. Obstet., 141:176 (1975).]

shoulder or thigh. The bleeding was quickly controlled in the majority of cases by stopping heparin administration for several days.

Only one bacteriological contamination was reported in the 30 Infusaid pumps sampled. The organisms cultured from the pump were identified as two strains of coagulase-positive staphylococcus.

Heparin was the first parenteral drug selected for the long-term clinical evaluation of the Infusaid pumps because of its therapeutic potential, as the anticoagulant of choice in the treatment of acute thromboembolic disorders and also for pragmatic reasons, since it is stable at physiological temperature, easily monitored in vivo by anticoagulation tests, and tends to promote catheter patency.

Following sterilization by autoclaving Infusaid pumps were implanted subcutaneously in the subclavicular fossa of 11 patients with a history of severe or recurrent thromboembolic disorders. The catheter was threaded through the axillary vein into the superior vena cava under direct fluoroscopic observation. Polypropylene tabs were sutured to the underlying fascia to hold the pump in place. Each pump was filled with bacteriostatic water just before implantation and was replaced with heparin solution 2–3 days after surgery. A heparin dose was selected, based on calculation, to maintain a plasma level of 0.2 IU/ml of heparin. Patients were released from the hospital 5–10 days after implantation. Hematological and coagulation parameters were determined weekly, and bone parameters were monitored monthly. Adjustments in the heparin dose delivered were made if necessary by filling the pump with a different concentration of heparin (219,222).

The blood heparin levels of 0.1–0.3 IU/ml plasma were observed to be main-

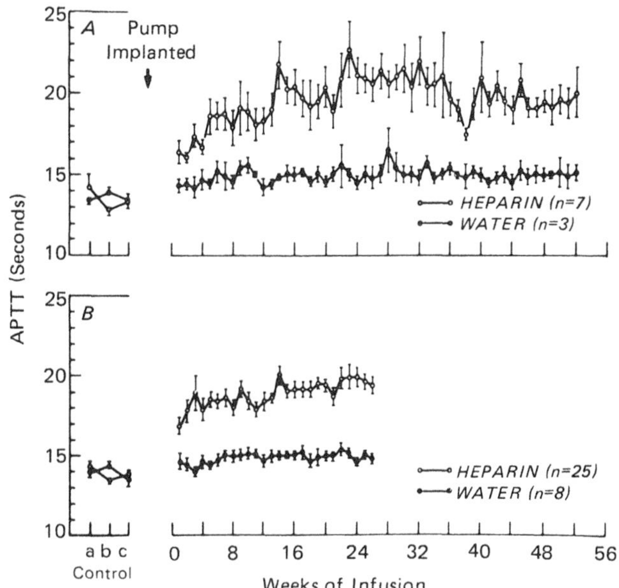

Figure 53 Activated partial thromboplastin time (APTT, mean ± SEM) for 25 dogs receiving heparin (○) and 8 receiving water (●) for 6 months (*B*) and for 7 receiving heparin (○) and 3 receiving water (●) for 12 months (*A*). [Reproduced from Blackshear et al., Surg. Gynecol. Obstet., 141:176 (1975).]

tained in these patients (Figure 54C) for up to 1 year with daily infusion of 15–30 $\times 10^3$ IU. Plasma heparin levels remained essentially constant, but activated partial thromboplastin time (APTT) values were found to increase from 32 sec before treatment to approximately 40 sec during treatment (Figure 54B). On the other hand, no change in platelet counts was observed (Figure 54A), nor did any noteworthy changes occur in bone mineral density or in the serum levels of Ca^{2+}, inorganic phosphate, and alkaline phosphatase. Apparently bone demineralization (i.e., osteoporosis), a reported side effect of prolonged heparin anticoagulation treatment, did not occur in this study.

Transient thrombocytopenia was reported in one patient, and in another an idiosyncratic heaprin-related immune thrombocytopenia also necessitated removal of the pump. Two other pumps were removed because of a wound-healing complication in one patient and dislodging of the catheter in another after 8 months of satisfactory infusion.

No recurrences of thromboembolic phenomena were observed in any of the 11 pump recipients, except for a single brief episode of phlebitis in 1 patient, which responded to a small increase in heparin dose.

Long-term control of blood glucose in diabetes. The conventional diabetic treatment by daily subcutaneous injections of insulin preparations (Section II.B) cannot simulate the physiological postprandial fluctuations of plasma insulin levels because of the slow and variable absorption of insulin from subcutaneous injection sites. This dilemma in diabetic treatment cannot be overcome even with the development of NovolinPen (Figure 9), which has substantially improved the subcuta-

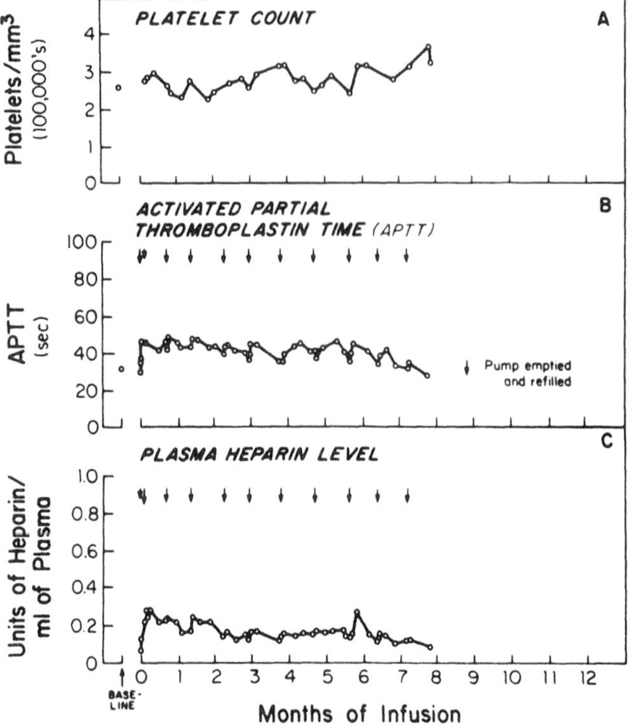

Figure 54 Plasma heparin levels (*C*), activated partial thromboplastin time (*B*), and platelet counts (*A*) in a patient receiving long-term continuous administration of heparin from the Infusaid pump. [Reproduced from Rohde et al., Trans. Am. Soc. Artif. Intern. Organs, 23:13 (1977).]

neous injection of insulin. To maintain the blood levels of glucose, metabolites, and hormones in the normal physiological range, patients with insulin-dependent diabetes mellitus must follow a rigid regimen, including meals, snacks, physical activities, and frequent testing of urine and/or blood for glucose and ketone body levels. Furthermore, there is a possibility of unpredicted attacks of hypoglycemia from vigorous exercise (226).

The standard insulin therapy of diabetes can also produce vascular complications as a result of long-term parenteral injection. In addition, juvenile-onset diabetic patients with brittle diabetes are often difficult to control by standard methods of insulin administration. To resolve these problems investigators have been looking at various means of providing near physiological insulin administration, such as pancreas transplants or Langerhans islet or β cell transplants, as well as a totally implantable artificial pancreas that consists of a glucose sensor, a computer, and a pumping device.

Biological approaches, such as pancreas transplantation, are limited by the costly medical expense, although they are expected to fulfill all the endocrine functions of a normal pancreas. Mechanical approaches, like the artificial pancreas, offer the economical advantages of mass production but fall short of achieving total endocrine

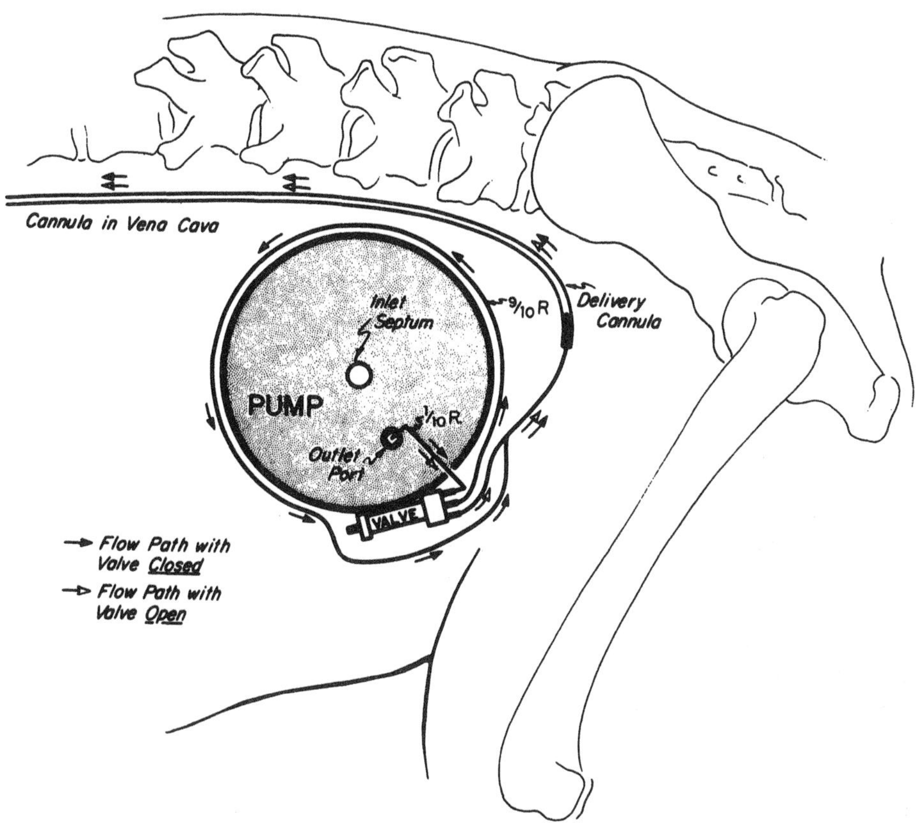

Figure 55 The pump, valve, and flow regulator arrangement showing the approximate anatomic location of the subcutaneous implantation site in the dog. The valve is activated transcutaneously with a small, permanent magnet. [Reproduced from Blackshear et al., Diabetes, 28:634 (1979).]

replacement. Each approach has its inherent drawbacks, and none is currently available for widespread clinical use.

The satisfactory performance of Infusaid pumps in the long-term intravenous infusion of heaprin for the refractory thromboembolic disease in dogs (223,224) and humans (219,222) has encouraged investigators to evaluate the feasibility of applying Infusaid pumps to the intravenous controlled infusion of insulin for the continuous treatment of diabetes (227). This treatment has been made possible with the development of a soluble insulin preparation that is stable at physiological temperatures (228). The basal flow rate can be adjusted at any time by altering the concentration of insulin in the infusate through a percutaneous refill.

The pump unit can be modified to use two flow-regulating resistance elements connected in series (Figure 55). When these resistance elements are used in series a constant basal flow rate is provided, as in unmodified pumps. However, when the valve that connects the two resistance elements is activated transdermally by means of a permanent magnet, the flow rate of the infusate solution can be boosted approximately 15 times by bypassing the high-resistance flow regulator (221). The basal flow rates from the unmodified pumps are typically 1–4 ml/day; the flow rate of the valved pump is 1 ml/day before activation (227).

The evaluation of Infusaid pumps for the long-term intravenous infusion of insulin was first carried out in dogs made diabetic by total pancreatectomy (227). After sterilization by autoclaving the pump was implanted in the iliac fossa and the delivery cannula was threaded into the inferior vena cava through the deep circumflex iliac vein or into the hepatic portal vein via the splenic or a mesenteric vein. The wound was healed and experiments were performed 2–3 weeks after pump implantation.

The results demonstrated that with the continuous intravenous infusion of insulin, such as Lilly's U-100 neutral regular insulin, by the Infusaid pump at the basal rate into the systemic peripheral circulation, a nearly normal blood glucose level was maintained in diabetic dogs in response to intravenous glucose tolerance tests and oral protein ingestion (Figure 56). On the other hand, an elevated blood glucose level was observed in the diabetic animal that received no infusion of insulin.

Further studies were conducted in another diabetic dog to determine the blood glucose level during a 16-hr profile of standard activity compared to a normal, nondiabetic dog on the same study protocol. The results indicated that the initial serum glucose values obtained after 12 hr of fasting were similar between the diabetic dog, which received a continuous insulin infusion at 12 mIU/kg/hr from the Infusaid pump, and the nondiabetic animal. The glucose levels remained very similar during the course of the study, except after a vigorous exercise period, when the diabetic animal developed moderate hypoglycemia, and after the evening meal, when a serum glucose excursion to 183 mg per 100 ml was noted (Figure 57A). After emptying the insulin from the implanted pump at the end of study the blood glucose level was detected to climb to the hyperglycemic level of 398 mg per 100 ml 36 hr later.

The control of blood glucose levels by the implantable insulin infusion pump can be improved by the addition of a second, superimposed higher insulin delivery rate onto the basal rate. This was demonstrated in a preliminary study in which a diabetic dog received a continuous insulin infusion at the basal rate of 20 mIU/kg/hr during the 16 hr profile of activity. By transdermal valve activation insulin was delivered at a higher rate of infusion (300 mIU/kg/hr) into the systemic circulation for several 7-min periods, three after the morning meal and two after the evening meal. A nearly normal blood glucose level was achieved and maintained using this approach (Figure 57B). The moderate hypoglycemia observed before the morning meal and after vigorous exercise and the evening meal suggested the possibility of reducing the insulin doses administered during these periods.

The studies outlined here concluded that the Infusaid infusion pump can be used to deliver a continuous intravenous infusion of insulin at variable rates for maintaining a nearly normal blood glucose level throughout the day in a diabetic dog (227).

Several potential developments of implantable infusion pumps, some of them incorporating a biofeedback-regulated mechanism into the bioresponsive drug delivery process, have recently been reported. They are discussed at some length in Chapter 1 (Section V).

Antitumor Biodegradable Subdermal Implants. A number of major advances have been made in recent years in genetic engineering, and consequently many therapeu-

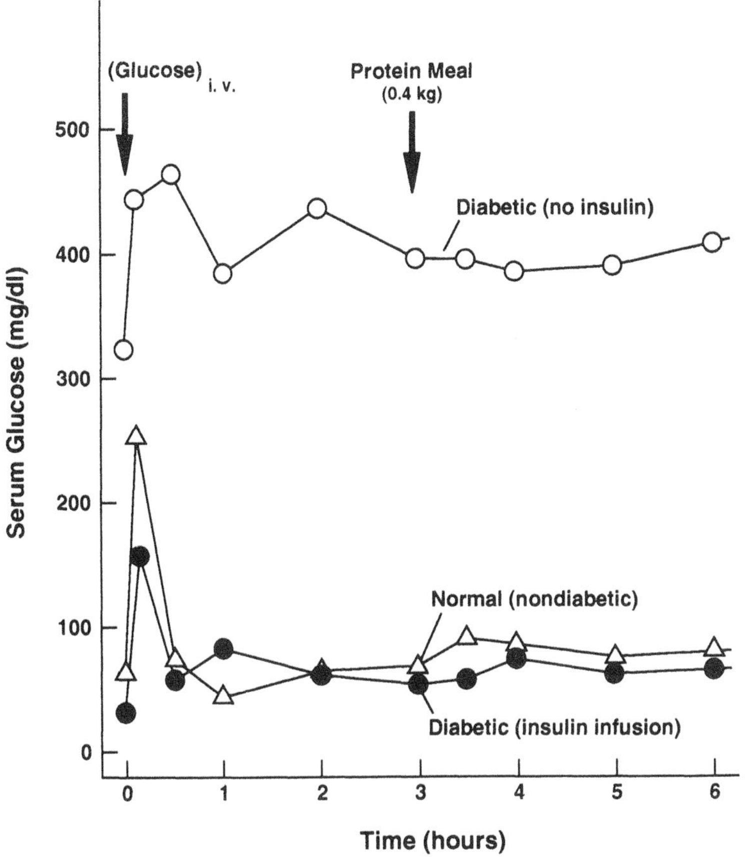

Figure 56 Serum glucose levels in response to intravenous glucose tolerance tests $(\text{Glucose})_{iv}$ three times a day and oral ingestion of protein meal (400 g) at 3-hr intervals: (△) normal dog before pancreatectomy; (○) dog made diabetic 1 week after pancreatectomy, received no insulin infusion; (●) diabetic dog received 2 weeks of pump-infused insulin. [Modified from Blackshear et al., Diabetes, 28:634 (1979).]

tically-active peptides and proteins are produced in commercial quantities. However, the therapeutic potential and commercial success of these peptide and protein drugs will only be fully realized if these advances are also accompanied by improvement in dosage form design or development of controlled-release drug delivery systems that consider the special requirements for the mode, rate, and duration of delivery of these therapeutic peptide and protein molecules to maximize their therapeutic potential.

A typical example is the development of a subdermally-implantable controlled-release drug delivery system from biodegradable and biocompatible polymers for Goseralin (Zoladex, ICI), a highly potent synthetic analog of luteinizing hormone-releasing hormone (LHRH) by total chemical synthesis, for antitumor treatment (229). Its relationship to LHRH and other synthetic analogs is shown in Table 20. As discussed in Chapter 11, LHRH and its synthetic analogs are usually ineffective by oral administration because they are rapidly degraded and deactivated by proteolytic en-

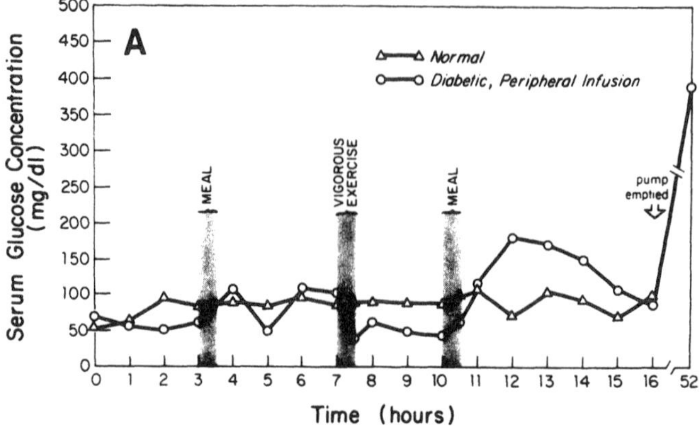

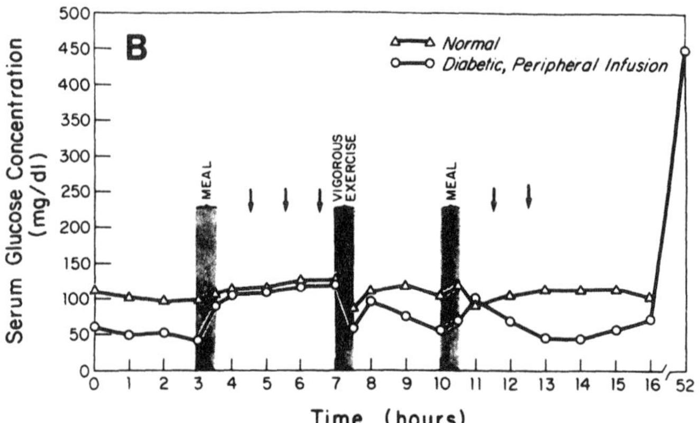

Figure 57 Two dogs, one depancreatized and one normal, were followed for 16 hr during which both animals were given meals consisting of 300 g dog chow and 100 g meat. (A) The depancreatized animal received pump-infused insulin at 12 mU/kg/hr and ate 400 g food during the first meal and 255 g during the second meal, and the control animal ate 370 and 40 g, respectively. (B) The depancreatized animal received a basal insulin infusion at a rate of 20 mU/kg/hr and several 7-min periods of higher insulin infusion rate at 300 mU/kg/hr (as indicated by the vertical arrows). Both the diabetic and nondiabetic dogs ate 400 g food at the first feeding and 200 g at the second. [Reproduced from Blackshear et al., Diabetes, 28:634 (1979).]

zymes in the alimentary tract. Even though the synthetic analogs have substantially improved the biochemical and biophysical stability of LHRH and are more resistant to enzymatic degradation, their molecular weight is often too large for effective gastrointestinal absorption. The feasibility of systemic delivery through the easily assessible mucosal routes (230,231), including the nasal (232–234), buccal (235), vaginal (236–239), and rectal, have been investigated and found to be associated with a low and variable systemic bioavailability. Consequently these peptide drugs

Table 20 Comparison of the Sites of Modification and the Effect on the Biological Activities of Luteinizing Hormone-Releasing Hormone (LHRH) and Its Synthetic Analogs

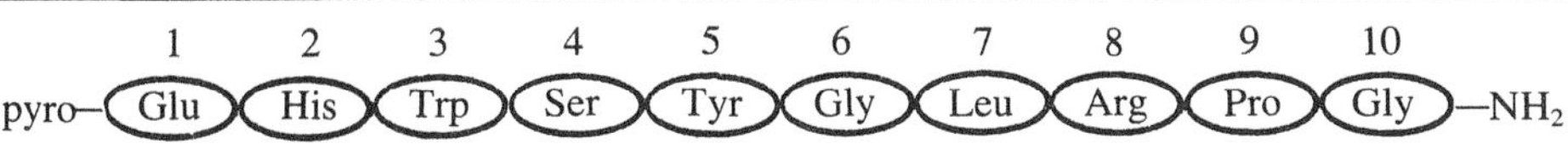

	Sites of modification			Biological activities		
LHRH	6	9	10	Potency	$t_{1/2}$ (min)	Binding
Natural LHRH	Gly	Pro	Gly	1	8	?
Synthetic LHRH						
Leuprorelin	D-Leu	Pro N-Et	0	5–15	–	+
Buserelin	D-Ser t-Bu	Pro N-Et	0	25	80	+
Histrelin	D-His Bzl	Pro N-Et	0	100	–	+
Goserelin	D-Ser t-Bu	Pro	Aza Gly-NH_2	75	–	+
Nafarelin	D-2-napht. -Alanine	Pro	Gly	200	144	?

are normally administered parenterally (subcutaneous, intramuscular, and intravenous injection). Since these naturally-occurring macromolecules or synthetic analogs are very short-acting (because of their short biological half-life), frequent injections are required to produce effective therapy. The development of a long-acting subcutaneously implantable delivery system capable of delivering these peptide drugs continuously at a controlled rate over a period of weeks or even months would be beneficial in optimizing the therapeutic efficacy of these macromolecular drugs, which are bioavailable and pharmacologically active only by parenteral administration. Furthermore, it is preferable that the implantable drug delivery devices developed be biodegradable in nature so that they ultimately disappear from the implantation site at the completion of treatment.

Development of biodegradable subdermal implants. Several biodegradable polymers have been developed, and they have been extensively reviewed by Heller (240). The biodegradable polymers that have been investigated for controlled drug delivery are outlined in Table 21. The biodegradation profile for some of these biodegradable polymers are compared in Figure 58. The homopolymers and copolymers of lactic and glycolic acids have been well investigated for their potential application in controlled drug delivery (241). Long experience with these polymers has shown that poly(lactide) and poly(glycolide) and their copolymers, poly(lactide-glycolide), are physiologically inert and biocompatible with biological tissues and degrade in the physiological environment to toxicologically acceptable metabolites, for example, the naturally-occurring lactic acid. They are therefore the polymers of choice

Table 21 Biodegradable Polymers Investigated for Drug Delivery

Polyglycolide or poly(glycolic acid)
Polylactide or poly(lactic acid)
Poly(lactide-glycolide) copolymer
Poly(ε-caprolactone)
Poly(hydroxybutyric acid)
Poly(ortho esters)
Polyacetals
Polyalkylcyanoacrylate
Polyanhydride
Polydihydropyrans
Proteins (cross-linked)
Synthetic polypeptides

Source: Modified from Hutchinson and Furr (1990).

for the development of implantable delivery devices for antitumor peptide drugs, such as Goserelin (Zoladex, ICI) (242).

Different from the controlled release of organic-based drug molecules, in which the mechanisms most commonly used to achieve controlled drug delivery are membrane permeation, matrix diffusion, and interfacial partitioning (Chapter 2, Section II), peptide and protein molecules, in contrast, are either insoluble in or incompatible with the polymers with a totally dissimilar structure because of entropic and enthalpic factors (243). Consequently, the low or negligible solubility of peptide and protein drugs in a polymer and the log-log relationship between diffusion coefficient and

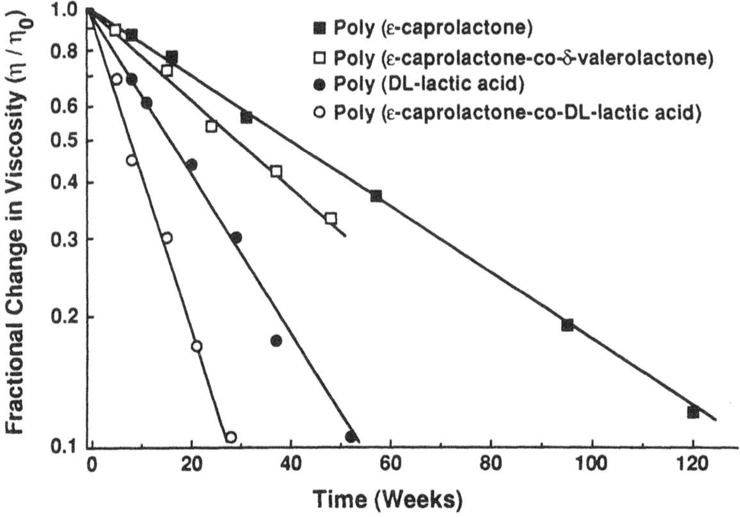

Figure 58 Biodegradation profile of several biodegradable aliphatic polyester-type polymers and copolymers. (Replotted from the data by Heller, 1984.)

molecular weight have prevented molecular diffusion in the polymer phase. Furthermore, peptide and protein molecules are biologically labile and readily degraded by tissue enzymes, and they must therefore be effectively protected at the implantation site if the peptide and protein drug is to be released continuously in an active form. With these considerations in mind it is logical to use biodegradable polymers in the development of implantable delivery devices for the controlled delivery of peptide and protein drugs; they are released at a controlled rate as the polymer is degraded continuously at the implantation site.

The homopolymers and copolymers of lactic and glycolic acids were prepared by ring-opening polymerization or polycondensation (Figure 23). Subdermal implants (each measures 10 mm in length and 1 mm in diameter and contains 3.6 mg Goserelin) were fabricated using conventional polymer extrusion or compression molding (242) and supplied in a ready-to-use syringe applicator. The results of in vitro degradation studies of poly(*d*,*l*-lactide-glycolide) copolymer (at a molar ratio of 50:50) indicated that these polymers have a normal polydispersity ($p \simeq 2$) and that the molecular weight of these polymers decreases exponentially, as a result of the degradation of the polymer chains, immediately following incubation in aqueous medium (Figure 59). The duration of the induction period before weight loss was found to be dependent upon the molecular weight (intrinsic viscosity) of the polymers as well as the rate of water uptake (229). This degradation is not enzyme mediated and must occur by simple hydrolytic cleavage of ester linkages (Figure 23). Electron microscopic examination demonstrated the generation of microporosity. Initiation of pore formation, the size of the micropores generated, and the contiguity or continuity of pores are also functions of the molecular weight, polydispersity, composition, and

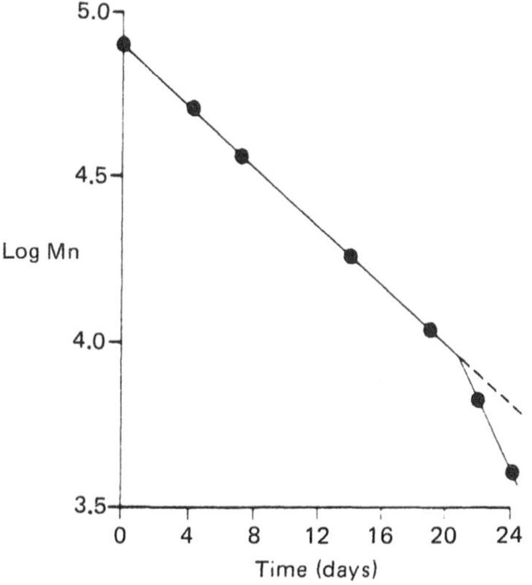

Figure 59 In vitro degradation profile of poly(*d*,*l*-lactide-glycolide) copolymer (molar ratio 50:50; film 0.2 mm thick) in buffer (pH 7.4) at 37°C, in which the number-average molecular weight, M_n decreases exponentially with the time of incubation. (Modified from Hutchinson and Furr, 1990.)

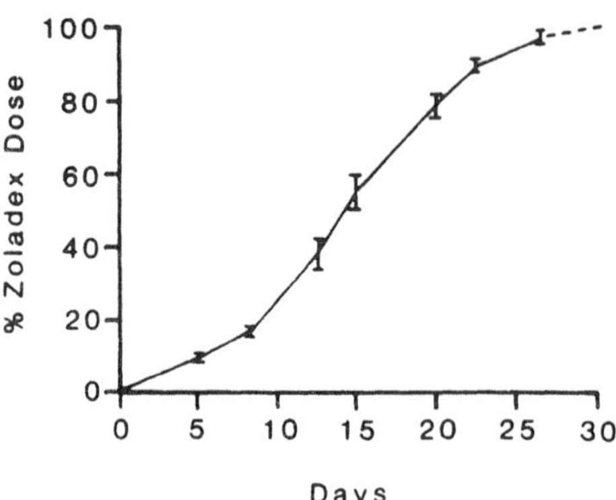

Figure 60 In vitro release profile of goserelin (Zoladex, ICI) from the biodegradable poly(*d,l*-lactide-glycolide) copolymer subdermal implant. (Reproduced from the data in the ICI product bulletin.)

structure of the biodegradable polymers used. The in vitro release profile of goserelin from subdermal implants is shown in Figure 60, which indicates that goserelin is released continuously from implants, with total release by 4 weeks.

In vivo drug release studies and optimization. The subcutaneous controlled release of goserelin from subdermal implants was studied in adult female rats with a regular estrous cycle (cycle length of 4 days). The effective delivery of goserelin can be measured qualitatively by its biological effect elicited in these cyclic animals as indicated by the absence of cornified cells in vaginal smears; female rats therefore show an extended period of diestrus.

The results indicated that the release of goserelin at a rate that is effective to maintain the female rats in a diestrus state is a function of the composition (Figure 61), the molecular weight (Figure 62), and the polydispersity (or even the type of molecular weight distribution) of poly (*d,l*-lactide-glycolide) subdermal implants as well as the loading dose of goserelin (Figure 63). The structure of the copolymers (the degree of heterogeneity and the segmental length of the co-monomer units) could also play an important role in implant degradation and drug release kinetics (229). By proper control of all these variables a subdermal implant that provides the continuous release of goserelin at a therapeutically-effective dosage rate for a duration of over 28 days was developed (Figure 64).

Preclinical efficacy studies. Although the primary objective for the development of goserelin-releasing biodegradable subdermal implants is to produce a drug delivery device that is more convenient to administer and would secure better compliance of prostate cancer patient (244,245), these implants, which release goserelin over a period of at least 28 days, were also observed to have improved the therapeutic efficacy of goserelin in terms of suppression of serum luteinizing hormone levels (Figure 65).

The therapeutic efficacy of goserelin-releasing biodegradable subdermal implants was evaluated in two sex hormone-responsive tumor models: (i) the dimethylbenzanthracene (DMBA)-induced rat mammary carcinoma, which is known to be dependent upon both estrogen and prolactin, and (ii) the Dunning R 3327H trans-

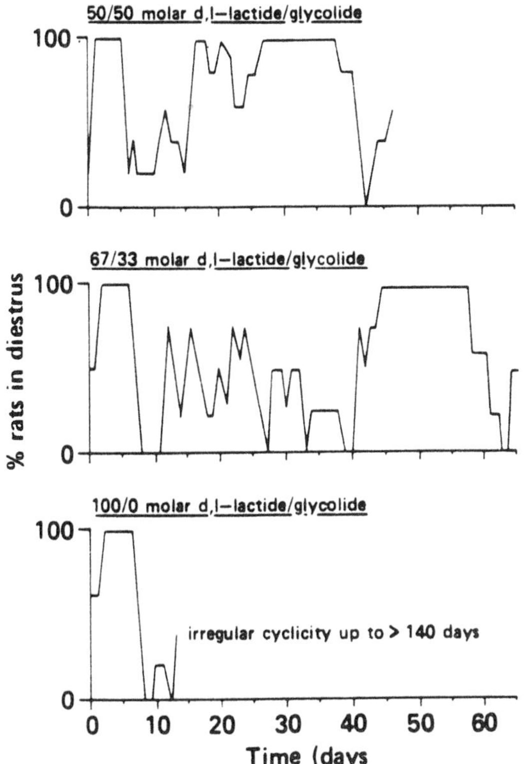

Figure 61 Effect of the copolymer composition of the poly(*d,l*-lactide-glycolide) subdermal implant on the estrous cycle of female rats. Each implant contains 3% (*w/w*) of goserelin. (Reproduced from Hutchinson and Furr, 1990.)

plantable rat prostate adenocarcinoma, which is androgen responsive and has been used extensively as a model for the human tumor.

The results from the studies in tumor model I indicated that the subcutaneous implantation of a single goserelin-releasing implant caused an inhibition of estrogen excretion and the disappearance of cornified cells from vaginal smears, as well as the regression of DMBA-induced mammary tumors (Figure 66). Half the tumors present at the beginning of the treatment were not palpable at 28 days; however, all of them reappeared within 40–60 days as goserelin became exhausted. On the other hand, mammary tumors in the control animals treated with placebo implants were noted to increase in size by more than 50%. Further studies with subcutaneous implantation of the same implants in the same animal model at the treatment regimen of one implant every 4 weeks, for example at weeks 0, 4, and 8, showed a greater regression of the tumors (Figure 67). No tumor was found palpable by week 11. However, regrowth of the tumors occurred by week 16 because the treatment was stopped at week 8, and by week 20 the tumors reattained their pretreatment size. Furthermore, the subcutaneous implantation of one goserelin-releasing implant consecutively at 28-day intervals was observed to produce a more profound inhibition of tumor appearance: only 9 out of 21 rats treated with 15 implants had mammary

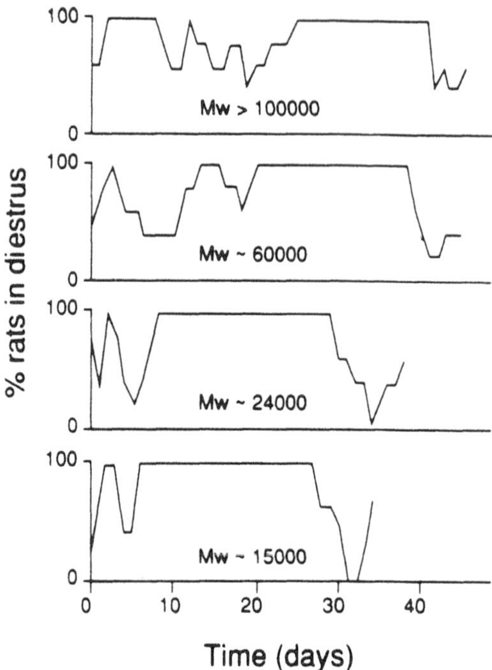

Figure 62 Effect of the molecular weight M_w of poly(*d*,*l*-lactide-glycolide) copolymer (with molar ratio 50:50) on the estrous cycle of female rats. Each implant contains 10% (*w*/*w*) of goserelin. (Reproduced from Hutchinson and Furr, 1990.)

tumors compared to 19 rats treated with 3 implants at the end of the study on Day 450 and to all 21 rats in the control group treated with placebo implant (Figure 68). The tumors found in the 9 rats (Treatment group B) did not regress even after ovariectomy and were classified as nonhormone responsive (229).

Subcutaneous implantation of a single goserelin-releasing implant at 28-day intervals to tumor model II (rats bearing prostate tumors) was found to cause a marked inhibition of tumor growth indistinguishable from that in surgically-castrated rats (Figure 69). The testes in these rats at 21 days after consecutive treatment with eight implants were found to weigh about 10% of those of control rats and showed atrophic histological changes. The weights of ventral prostate gland and seminal vesicle were identical to those in the castrated group (Table 22) and histologically were also completely atrophic.

Serum levels of luteinizing hormone (LH) and testosterone were undetectable in the group of rats treated with goserelin-releasing biodegradable subdermal implants, but the serum concentrations of follicle stimulating hormone (FSH) were decreased by 60–70%. On the other hand, serum levels of prolactin doubled in the goserelin implant-treated rats, as they did in castrated rats. This is probably a result of androgen withdrawal, since there was a significant reduction in serum prolactin levels in female rats following estrogen withdrawal when treated with goserelin implants (246).

The therapeutic efficacy of goserelin-releasing subdermal implants in the prostate tumor-bearing rat model by repeated subcutaneous implantation at 28-day intervals

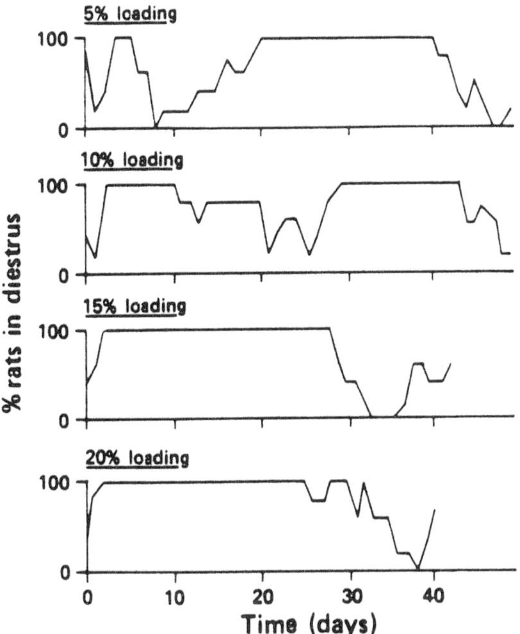

Figure 63 Effect of the loading level of goserelin in poly(*d*,*l*-lactide-glycolide) copolymer (with molar ratio 50:50) on the estrous cycle of female rats. (Reproduced from Hutchinson and Furr, 1990.)

was reassessed in male pigtailed monkeys (247,248). The results demonstrated that the subcutaneous administration of a single goserelin-releasing subdermal implant at Days 0, 28, and 56 suppressed plasma testosterone levels to values close to or within the castrate range and prevented pulsatile testosterone secretion. Daily subcutaneous injections of goserelin for 28 days caused only a slight reduction in plasma testosterone levels with regular pulses still observed. In addition, suppression of the testosterone level was found to be rapidly reversed when goserelin-releasing subdermal implant was removed.

The safety of goserelin-releasing biodegradable subdermal implants was evaluated in both acute (single dose) and chronic (6 month) animal toxicology studies. Of a group of 20 male rats treated for 6 months 2 developed benign pituitary gland microadenomas. However, male rats have also been shown to develop benign pituitary tumors as a result of surgical castration or treatment with another synthetic analog, leuprorelin (Table 20). On the other hand, other species have not shown any increase in pituitary lesions following surgical castration or LHRH analog treatment. It was thus concluded that the effect observed could be species specific in the rat and may not be relevant to humans.

Clinical evaluations of antitumor effect. The clinical trial program began with daily subcutaneous injections of goserelin (Zoladex, ICI) in solution followed by subcutaneous administration of the goserelin-releasing subdermal implant at three dose levels into the anterior abdominal wall. Following the subcutaneous administration of goserelin implants serum concentrations of goserelin rose continuously and

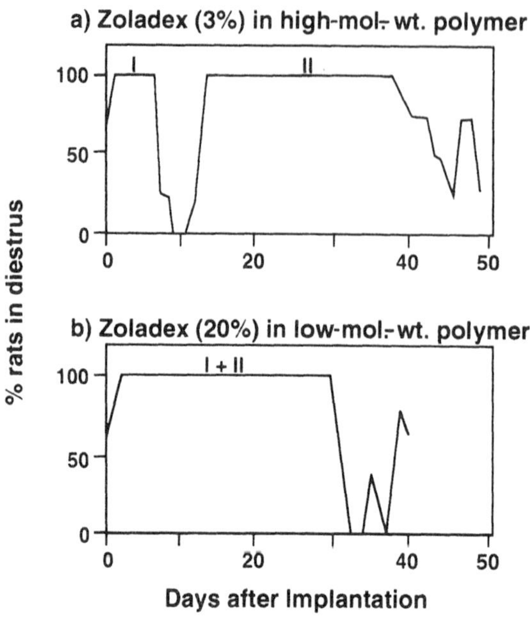

Figure 64 Effect of Zoladex (goserelin)-releasing subdermal implants on the estrous cycle of female rats: I, initial drug release by surface leaching; II, bulk drug release induced by polymer degradation. (a) Subdermal implants contain 3% (w/w) of Zoladex in high-molecular-weight polymer. (b) Subdermal implants contain 20% (w/w) of Zoladex in low-molecular-weight polymer. (Replotted from Hutchinson and Furr, 1990.)

reached the peak level in the middle of the 4-wk implantation and then declined (Figure 70). A similar serum goserelin concentration profile was reproduced in subsequent implantations at 4-week intervals. It was concluded that the 3.6-mg implant is most effective in suppressing testosterone levels (249,250). Normal values of circulating testosterone are around 4.4 ng/ml; castrate values, achieved either surgically or medically, are generally below 0.6 ng/ml.

The initial dose of goserelin-releasing subdermal implant administered to prostate cancer patients resulted in a prompt rise in serum LH concentration, which reached a peak level after approximately 48 hr and then declined gradually to pretreatment values or lower by Day 15. Subsequent monthly administration of the implants was found to produce no increase in these levels (Figure 71), suggesting complete desensitization of pituitary glands (249). Since the production of testosterone is LH dependent, a similar pattern was also observed in testosterone secretion. Serum testosterone concentrations rose on Days 1–4 and then declined gradually. Testosterone was suppressed to castrate values by the end of Week 3 in the majority of patients and by the end of Week 4 in all patients (Figure 71). Goserelin is excreted mainly by the kidneys and has a half-life of approximately 4 hr in prostate cancer patients with normal renal function. Clearance of goserelin from the body is slower in patients with impaired renal function, but no dosage adjustment appears necessary since the

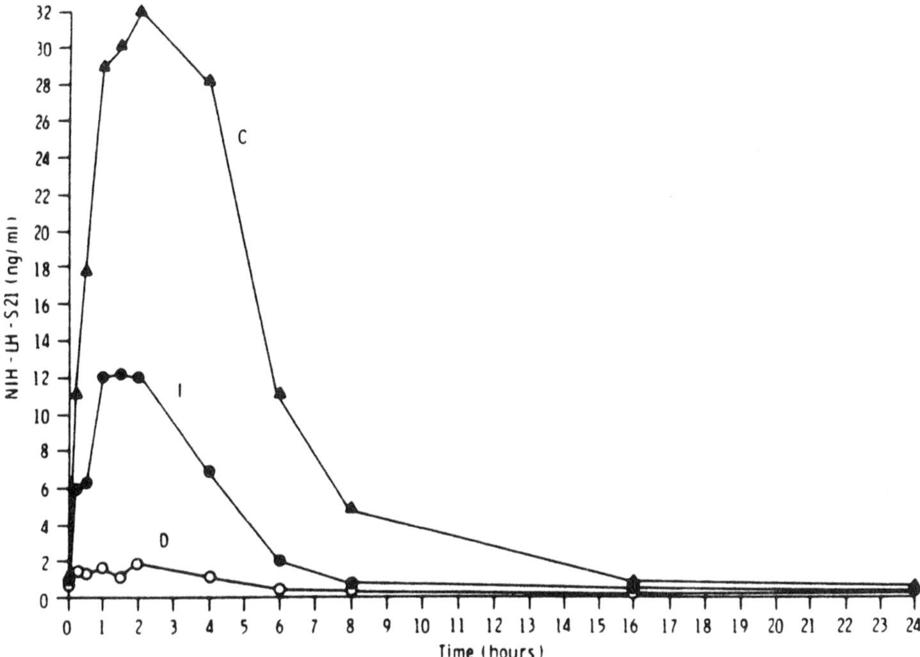

Figure 65 Serum concentration profiles of LH in response to a bolus dose of goserelin (50 μg) in rats pretreated with saline (*C*) or 50 μg goserelin daily for 6 weeks (*I*) or treated with subcutaneous administration of goserelin implants at weeks 0 and 4. (Reproduced from Hutchinson and Furr, 1990.)

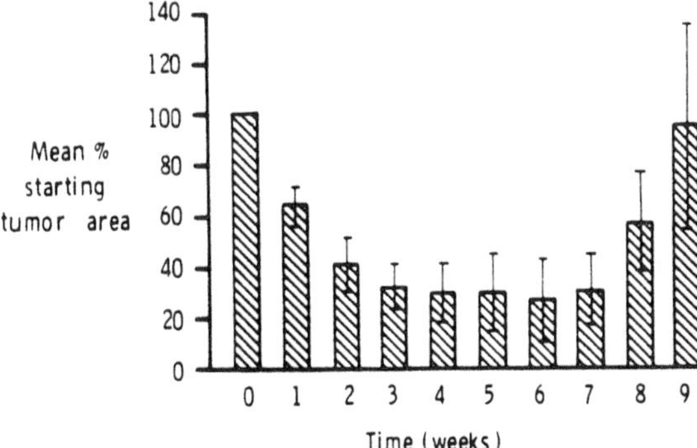

Figure 66 Effect of single subcutaneous administration of goserelin implant at time zero on the growth of DMBA-induced rat mammary tumors. The values shown are mean ±SEM of 10 rats. (Reproduced from Hutchinson and Furr, 1990.)

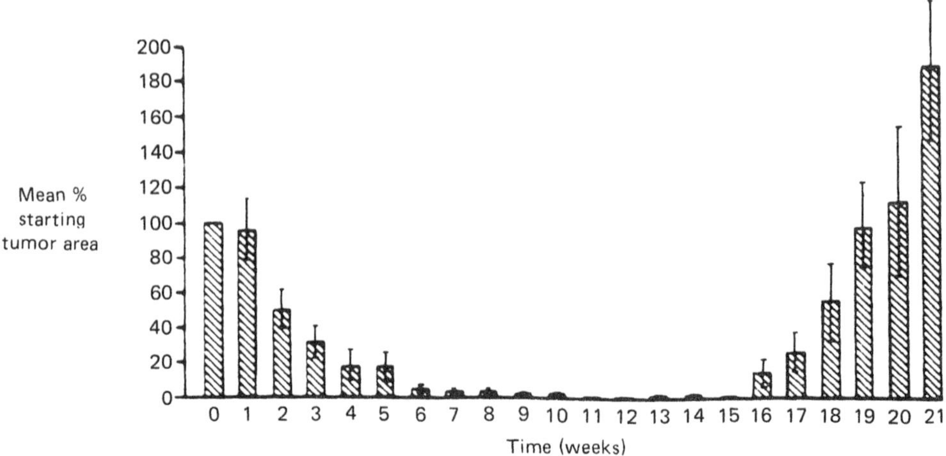

Figure 67 Effect of consecutive subcutaneous administration of one goserelin implant at weeks 0, 4, and 8 on the growth of DMBA-induced rat mammary tumors. The values shown are mean ± SEM for 10 rats. (Reproduced from Hutchinson and Furr, 1990.)

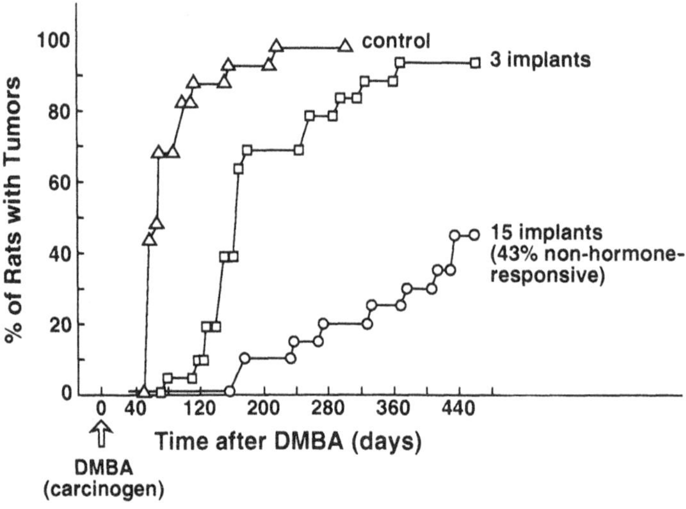

Figure 68 Effect of consecutive subcutaneous administration of one goserelin implant at 28-day intervals (starting 30 days after the induction of DMBA) on the appearance of mammary tumors in rats (21 rats in each group): (△) control group given placebo implant; (□) treatment group A given single goserelin implant on Days 30, 58, and 86; (○) treatment group B given single goserelin implant on Day 30 and every 28 days thereafter for 15 implantations. (Replotted from Hutchinson and Furr, 1990.)

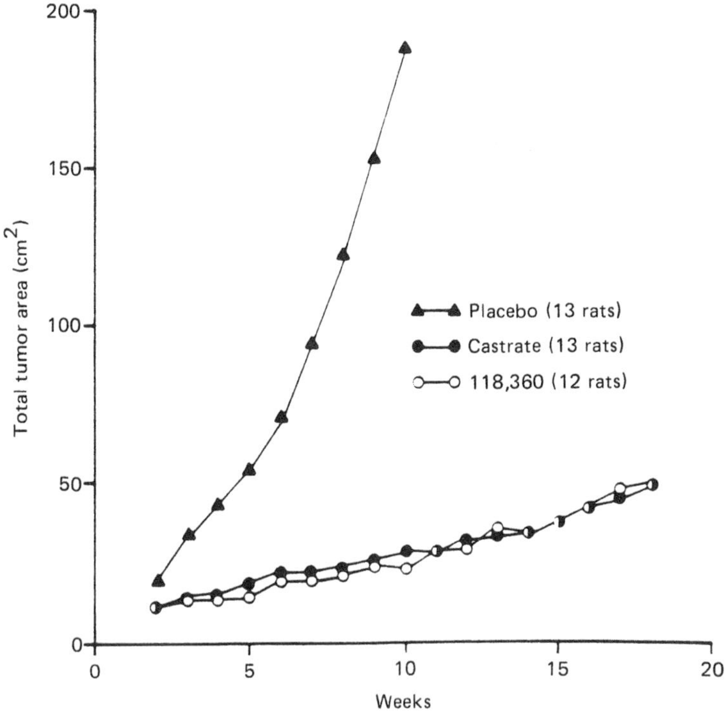

Figure 69 Comparison of the growth of Dunning R3327H transplantable rat prostate tumors among various treatments: (△) control group (n = 13) given single placebo implant every 4 weeks; (○) treatment group (n = 12) given single goserelin (118,360) implant every 4 weeks; (●) castrated group (n = 13) treated by surgical castration. (Reproduced from Hutchinson and Furr, 1990.)

Table 22 Sex Organ Weights and Serum Hormone Concentrations in Goserelin Implant-Treated and Surgically Castrated Rats Bearing Dunning R3327H Prostate Tumors

	Rats with prostate tumors		
Parameter	Goserelin-treated	Castrated	Control rats[a]
Organ weight, mg			
Testes	366.5 (10.2)	—	3500
Ventral prostate	21.3 (1.1)	19.9 (0.7)	250
Seminal vesicle	54.3 (1.0)	53.8 (0.8)	350
Serum concentration, (ng/ml)			
LH	<0.2	12.3 (1.0)	1.5
FSH	174 (6)	1413 (24)	400
Prolactin	63.2 (4.5)	60.9 (7.5)	30
Testosterone	<0.25	<0.26	3

[a]Mean values.
Source: Modified from Hutchinson and Furr (1990).

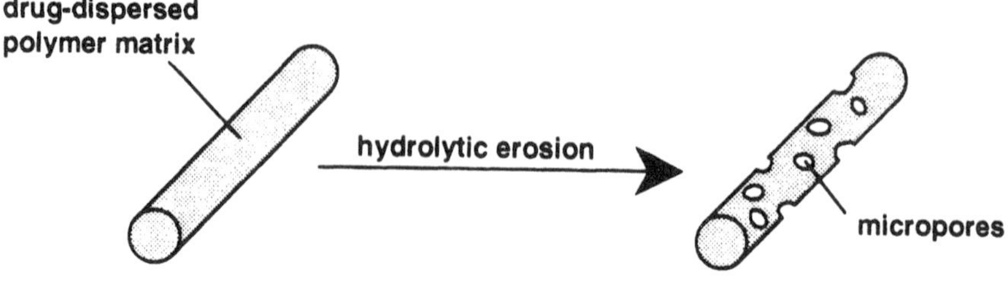

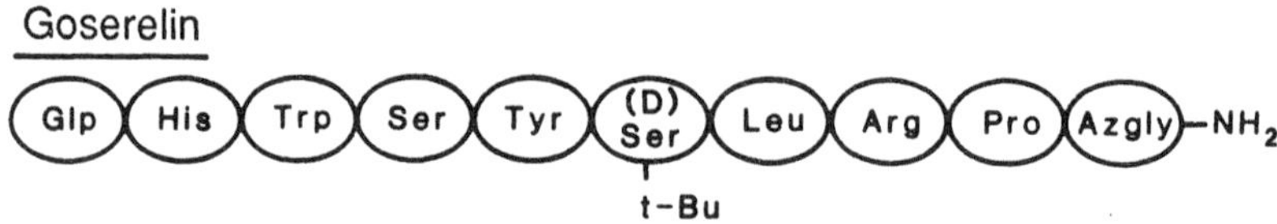

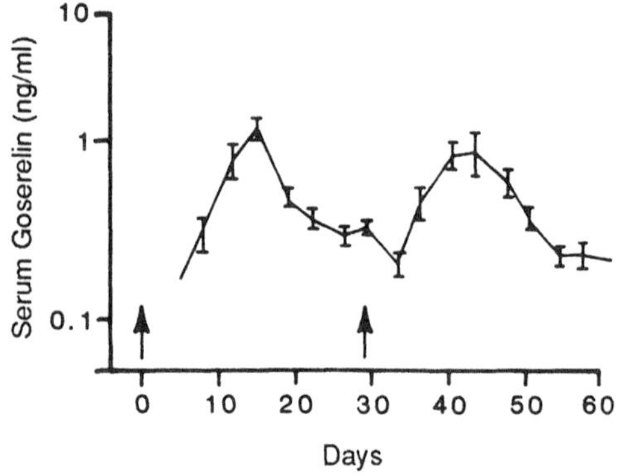

Figure 70 The goserelin-releasing subdermal implant [biodegradable poly(lactide-glycolide)], the amino acid composition of goserelin, and serum profiles of goserelin in prostate cancer patients (n = 10) treated with goserelin-releasing subdermal implants once every 4 weeks for 2 months. (Replotted from the data in the ICI product bulletin.)

overall clearance is relatively rapid even in patients with severe renal impairment (with a half-life of 12 hr).

It was encouraging to note that the continued administration of goserelin implants achieved the long-term suppression of serum testosterone levels with no fluctuations, indicating that pharmacological tolerance does not develop on long-term therapy even after almost 3 years (251).

A multicenter randomized trial was conducted to compare the clinical efficacy of goserelin implants with surgical castration in the management of advanced prostate cancer (250). The results generated from the first 240 patients, at a minimum follow-up of 3 months, indicated that goserelin subdermal implant and orchidectomy are equally effective in terms of the suppression and maintenance of serum testosterone at castrate levels and reduction in the mean serum concentrations of total acid phos-

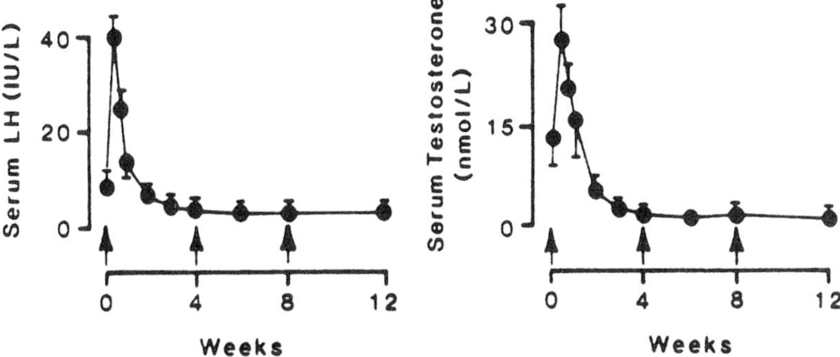

Figure 71 Serum profiles of luteinizing hormone (LH) and testosterone in prostate cancer patients ($n = 8$) following the subcutaneous administration of goserelin subdermal implants once every 4 weeks consecutively for 12 weeks. (Plotted from the data in Allen et al., 1983.)

phatase, and they also showed similar effects on urine flow, activity score, and pain-analgesia scores. The pharmacological effects are similar in both groups of patients (Table 23).

2. *Veterinary Applications*

Several implantable drug delivery devices have been developed from biocompatible (lipophilic or hydrophilic) polymers for veterinary applications. A typical example is the development of norgestomet-releasing subdermal implants for estrus synchronization. Its development and improvement are discussed here.

Norgestomet subdermal implants. Norgestomet, a potent progestin that is completely free of estrogenic effects, was reported to suppress estrus and ovulation by intramuscular injection at a daily dose of 140 μg (252).

Similar to the 18 to 21 day treatment with other progestins (Chapter 9, Section V.B), norgestomet treatment also yielded a reduction in fertility when artificial insemination was carried out at the synchronized estrus (253). This subfertility following treatment with progestins for 18–21 days was later shown as due to a delay in embryo cleavage (254). The level of fertility was found to be improved by a shorter (9-day) treatment period with a norgestomet-releasing subdermal implant in conjunction with an intramuscular estradiol valerate injection at the time of implantation to induce early luteal regression (254,256). Alternatively, a 5-day treatment with this implantable controlled-release norgestomet delivery device in combination with prostaglandin $F_{2\alpha}$ injection at the time of implant removal also demonstrated an improvement in conception rate. Further development established a treatment regimen, Syncro-Mate-B (G. D. Searle & Co.), which consists of subcutaneous implantation of a cylindrical norgestomet-releasing subdermal implant into the earflap between the dorsal skin and the conchal cartilage for 9 days and a supplementary intramuscular injection of norgestomet (3 mg) and estradiol valerate (5 mg) in sesame oil (2 ml) at the time of implantation. The antiluteotropic properties of the norgestomet-estradiol valerate combination inhibit the development of corpus luteum in recently ovulated cattle. This treatment regimen induced a majority of treated animals to exhibit synchronized estrus within 5 days following implant removal and to yield a normal level of fertility when artificial insemination was carried out in the syn-

Table 23 Comparison in Patient Characteristics, Responses, and Pharmacological Effects between Two Groups of Castrations

Parameters	Methods of castration	
	Goserelin implants	Orchidectomy
Patient characteristics		
No. of patients	120	106
Mean age (range), years	72 (49–86)	73 (55–90)
Histological grade		
G1	19	20
G2	50	42
G3	43	34
Metastases		
Bone	111	103
Soft tissue	14	13
Both	7	11
Responses (subjective), %		
By protocol	80	72
Clinician	93	90
Pharmacological effects, %		
Decreased libido	67	78
Impaired erections	76	81
Hot flushes	60	53
Breast swelling	6	3
Breast tenderness	1	2
Complications	0	7.5

Source: Compiled from the data by Kaisary et al. (1987) and Turkes et al. (1987).

chronized estrus period (257). Furthermore, the stage of the cycle at the initiation of treatment was found to have no effect on the proportion of animals in estrus or their fertility (254).

The subdermal implant, which was fabricated by impregnating 6 mg norgestomet in a hydrophilic cross-linked polymer of poly(ethylene glycomethacrylate), also called Hydron, is cylindrical and small (2 × 21 mm) and can be easily inserted into an animal's ear by a specially designed implanter (Figure 44).

Mechanisms of estrus synchronization. As discussed earlier, norgestomet is a potent progestational agent that is completely free of estrogenic effects. By subcutaneous administration it shows 100 times more potency than progesterone.

Like all pure progestational agents, norgestomet has no effect on the uterus of a castrated or prepubertal animal. However, in the presence of estrogens, like estradiol or its long-acting valerate, norgestomet has been reported to induce some very significant proliferative cytological responses in the uterus. A considerable proliferation of the endometrial mucosa is produced, creating an ideal condition for nidation of the ovum on the endometrium (258). Additionally, norgestomet produces no inhibition of gonadotropic hormone production in the pituitary gland but a blockade of the process of its secretion to the systemic circulation.

The continuous administration of norgestomet from subdermal implants apparently acts to prepare the uterine mucosa for nidation of the ovum on the one hand

and to temporarily block the secretion of gonadotropins, such as FSH and LH, on the other hand. The supplementary intramuscular injection of the norgestomet-estradiol valerate combination also apparently yields an antiluteotropic activity against the formation of corpus luteum in cycling animals.

In summary, the mechanism of action of estrus synchronization with Syncro-Mate-B treatment can be visualized such that the injection of norgestomet-estradiol valerate shortens the estrous cycle by inhibiting luteal formation and function (257,258) and the norgestomet-releasing subdermal implant lengthens the cycle by preventing ovulation (259). Without this treatment cows that have just undergone estrus would not be expected to be in estrus again for about 3 weeks. Treatment with Syncro-Mate-B in these animals shortens the cycle because the injection "resets" the ovary and the norgestomet implants keep the cows out of estrus until the implants are removed. On the other hand, a cow that is late in the estrous cycle when treatment is conducted would be kept out of estrous by the norgestomet implant until the implant is removed; thus all cows would come into estrus at approximately the same time (260).

Termination of the estrus control treatment by removal of the norgestomet subdermal implant yields a sudden "withdrawal of medication," which leads to the release of LH and the synchronization of estrus and ovulation. For noncyclic animals studies conducted in France indicated that additional injection of pregnant mare serum gonadotropin (PMSG) at the time of implant removal enhances follicular proliferation and the appearance of synchronized estrus.

Administration of norgestomet by either daily injection for 21 days or continuous administration by controlled release from a subdermal implant for 9 days showed no significant effect on the rates of ovum recovery and fertilization (254). However, treatment with daily injections of norgestomet was associated with the early cleavage of fertilized ovum, an effect absent in those heifers treated with a norgestomet-releasing subdermal implant for 9 days and in the untreated animals.

The effect of intramuscular injection of various combinations of norgestomet and estradiol valerate on the effectiveness of estrus synchronization with a norgestomet-releasing subdermal implant was studied. The administration of 3 mg norgestomet and 5 mg estradiol valerate at the time of implantation was found to maximize the percentage of heifers in estrus (90% compared to 40% without this supplementary dose) (261).

Pharmacokinetics. The plasma profiles of norgestomet in heifers treated with the subcutaneous administration of one norgestomet-releasing implant and concurrently the intramuscular injection of the norgestomet-estradiol valerate combination under field conditions are shown in Figure 72. The results indicated that injection immediately yields a high plasma level of norgestomet, with a peak concentration of around 1.5 ng/ml reached within 1 hr of treatment; 2 days after medication the plasma drug level decreases to a very steady state. This low but quite constant plasma norgestomet level (0.1–0.2 ng/ml) in the last 7 day period of the treatment resulted primarily from the subcutaneous controlled administration of norgestomet from the subdermal implant. As soon as the implant is removed at the end of the 9-day treatment the plasma levels fell off rapidly, with an elimination half-life of around 3 hr, to a concentration below baseline values. It appears that residual norgestomet in the body is eliminated within 48 hr after implant removal. The coadministration of es-

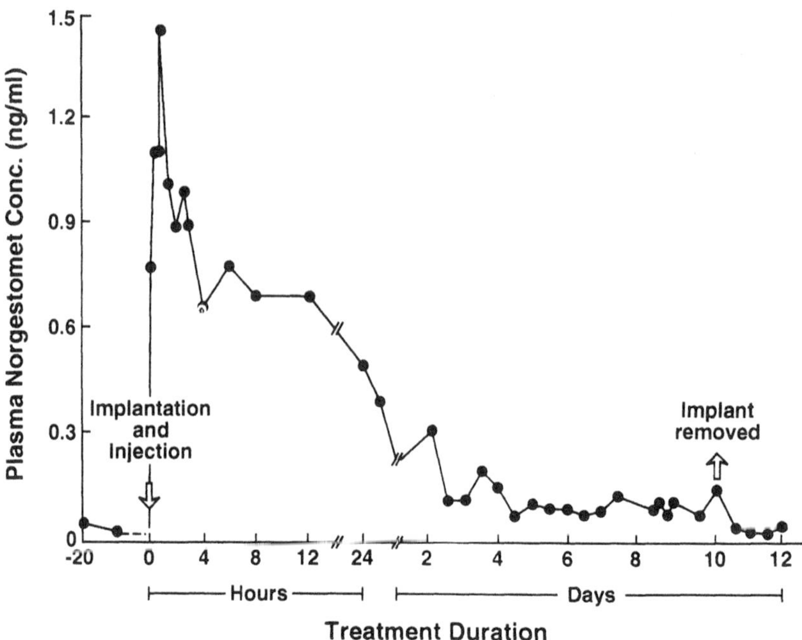

Figure 72 Plasma profile of norgestomet following the 9-day subcutaneous implantation of one Syncro-Mate-B implant plus the intramuscular injection of norgestomet (3 mg) and estradiol valerate (5 mg) in sesame oil (2 ml) on the first day of implantation. (Replotted from the data by S. E. Mares, 1981.)

tradiol valerate at the beginning of the treatment does not modify the course of this plasma profile (262).

The subcutaneous release profile of norgestomet from the Hydron implant is illustrated in Figure 73. The data suggest that the cumulative amount of norgestomet released from the implant in the subcutaneous tissue is directly but not linearly proportional to the duration of implantation (205). This release profile is the result of a matrix diffusion process (Chapter 2, Section II.B). As expected from matrix diffusion-controlled delivery theory, the release profile of norgestomet from the hydrophilic polymer matrix also follows the Q versus $t^{1/2}$ relationship (Figure 45).

Norgestomet in the body is hydrolyzed enzymatically first to the 17β-hydroxy derivative, then reduced to a diol metabolite, and finally oxidized to a 17-keto metabolite. These metabolites have been detected primarily in the bile (which amount to almost 80% of the administered dose) and in the urine (approximately the remaining 20%). These metabolites either exist in free form or form conjugates with glucuronic acid. They have only an extremely weak progestational activity and are eliminated in the feces without evidence of intestinal reabsorption (262).

On the other hand, the plasma concentrations of estradiol resulting from the supplementary intramuscular injection of estradiol valerate (5 mg) at the beginning of treatment was reported to rise gradually to reach a peak level of 40 pg/ml within 12 hr and then decrease to a level lower than 15 pg/ml in the next 8–12 hr (262).

Clinical evaluations in cattle. For a better understanding of the mechanism of

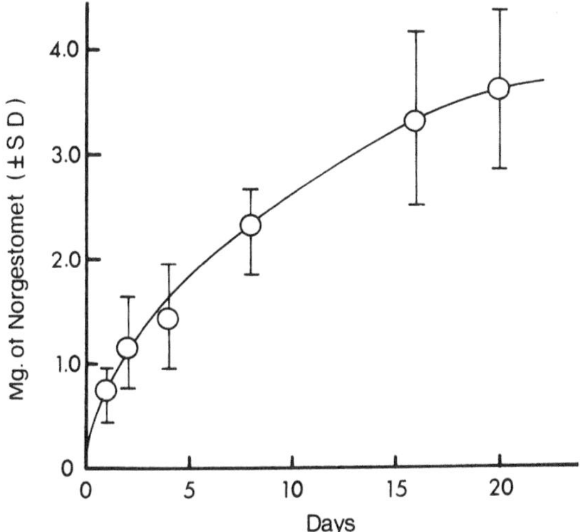

Figure 73 Subcutaneous release profile of norgestomet from Hydron implants in 13 cows.

estrus control treatment in cattle, one should recognize that animals can exhibit either one of two estrus states:

Cyclic state. Dairy cows and heifers show cyclic estrus during their optimum reproductive period. The proportion of cycling females increases as an adequate feeding program is implemented and after a good pasturing.

Acyclic state. Anestrus is mainly observed in the first several weeks following calving (the postpartum period). It can also appear in cows and heifers at the end of a winter season. The length of the postpartum period varies and is known to depend upon a number of parameters, such as suckling, nutrition, and milk production level.

In those cows in a cyclic state the estrous cycle can be effectively controlled and then synchronized by Syncro-Mate-B treatment. It has been established that the stage of the cycle at which the treatment begins does not affect the efficacy of the treatment. If treatment starts at the beginning of an estrous cycle, it produces an antiluteotropic effect; that is, plasma progesterone levels are suppressed, manifested by an arrest of corpus luteum development. After removal of the implant estrus is regenerated and synchronized. On the other hand, if treatment is initiated at the midcycle period the corpus luteum continues to develop normally, but the progesterone level drops on days 16–17. A natural reappearance of estrus occurs following termination of treatment. In another situation, if treatment starts at the end of an estrous cycle the corpus luteum develops normally, but estrus does not appear until termination of suppression by the continuous subcutaneous administration of norgestomet from the Syncro-Mate-B implant. In all three situations cyclic cows and heifers can therefore be effectively controlled and synchronized (263).

In cows in an acyclic state, on the other hand, studies conducted in France indicated that an additional injection of a dose of PMSG, which has an activity similar to that of FSH, on the last day of estrus control treatment is beneficial for achieving the effective synchronization of these anestrus animals. In the United States separation of cows from nursing calves from the time of implant removal until artificial insemination is complete was found to increase the proportion of cows in estrus over a shorter period of time (264,265).

Clinical efficacy studies. Several field trials were conducted in herds representing the major beef-producing areas of the United States and Canada to evaluate the clinical efficacy of Syncro-Mate-B treatment in heifers (260). The herds were first subjected to Syncro-Mate-B treatment 9 days before a planned breeding season, and artificial insemination of each animal was then done according to two schedules: one group of treated animals at approximately 12 hr after the detection of estrus (group A) and another within 48–54 hr after implant removal without estrus detection (group B). The results indicated that the heifers treated with Syncro-Mate-B are in estrus within a time period much earlier than untreated controls (Table 24). Within 5 days after the removal of implants 98% of the treated heifers were found in estrus compared to only 23% of control animals. The percentage of control animals in estrus increased to 91% by day 25.

The rates of pregnancy were found to be about the same for the two treatment groups. This observation suggested that the extra effort of estrus detection before artificial insemination (group A) provides no advantage over insemination at a fixed time (group B) when the heifers are synchronized with Syncro-Mate-B treatment. On the other hand, the data also clearly demonstrated that Syncro-Mate-B–treated animals yield a significantly higher rate of pregnancy in a shorter time than control animals. Pregnancy rates of 53 and 48% were obtained on day 5 for treatment groups A and B, respectively, compared to only 14% for the control group. The feasibility of carrying out early detection of pregnancy provided by this treatment gave the treated animals the opportunity to return to estrus for a second insemination by day 25 or a third insemination by day 45 of the breeding season if pregnancy did not occur.

Similar results were also attained in prepubertal heifers (258). A synchronized estrus was successfully induced in prepubertal heifers by Syncro-Mate-B treatment.

Results from 13 field trials involving 980 treated heifers and 426 untreated controls concluded that (i) 96% of the treated animals were detected in estrus during the first 4–5 days following removal of implants compared to only 23% in controls;

Table 24 Field Efficacy of Syncro-Mate-B Treatment in Heifers Located in U.S. and Canadian Herds

		Efficacy of treatment (%)[a]				
	Number of	Estrus rate		Pregnancy rate		
Groups	heifers	Day 5	Day 25	Day 5	Day 25	Day 45
Control group	184	23	91	14	64	76
Treatment groups[b]						
A	170	98	100	53	71	77
B	339	—	—	48	71	81

[a]Percentage of the total number of animals in each group.

[b]Artificial insemination schedule: group A, inseminated at approximately 12 hr after the detection of estrus; group B, inseminated at approximately 48–54 hr after implant removal without estrus detection.

Source: Compiled from data from Reference 260.

Table 25 Time Relationships of Estrus and Ovulation in Heifers Treated with Norgestomet-Releasing Hydron Implants

Observations	Time (mean ± SD), hr
Duration of estrus	17.8 ± 6.4
Intervals	
Implant removal to estrus	36.0 ± 8.9
End of estrus to ovulation[a]	14.6 ± 2.6
Implant removal to ovulation[a]	68.5 ± 9.7

[a]Ovulation was detected by endoscopy every 4 hr starting 8 hr after the end of estrus.
Source: Compiled from data from Reference 254.

(ii) 46% of treated animals became pregnant when artificial insemination was carried out during the 4–5 day period compared to only 14% in the control group; and (iii) this treatment was also observed to induce earlier puberty in heifers.

Effect on time relationships between estrus and ovulation. The effect of Syncro-Mate-B treatment on the time relationship of estrus and ovulation is illustrated in Table 25. The data appear to suggest that the treatment results in prolongation of the average duration of estrus. Estrus was observed to last for 17.8 ± 6.4 hr compared to 13.5 ± 6.2 hr obtained in heifers treated with daily injections of norgestomet for 21 days and 14.0 ± 4.7 hr in untreated animals. This difference is not statistically significant, however. A total of 86.7% of treated heifers were observed in estrus during the 4-day period following the removal of the implant, and none of the treated heifers ovulated earlier than 10 hr after the end of estrus (with an overall time period of 14.6 ± 2.6 hr). This observation agrees well with endoscopic findings in a large field trial in which animals were treated with daily intramuscular injections of norgestomet (254). Furthermore, it is interesting to note that the percentage of heifers in ovulation relative to those in estrus increases linearly from zero at 48 hr to 96% at 84 hr after removal of the implant (Figure 74).

The data in Tables 24 and 25 as well as Figure 74 suggest that it may be feasible to inseminate the Syncro-Mate-B–treated heifers once or twice on a predetermined schedule after completion of treatment and achieve the same level of fertility as when the insemination is done following the detection of estrus in each animal (254). This possibility was evaluated on three British farms (Table 26). The results suggested that the precision of estrus response is striking, with approximately 80% of the heifers going into estrus in a narrow time interval of 25–48 hr following removal of the implant. Overall, 88% of the 150 treated animals have their estrus synchronized by treatment within 72 hr. The 18 heifers in which estrus was not observed were later confirmed by endoscopic examination to also have ovulated, and their ovulation was also sufficiently controlled by treatment to permit conception by artificial insemination at a scheduled time, even though no estrus was observed.

Effect of artificial insemination schedules. The effect of the timing of artificial insemination on the rate of pregnancy in heifers treated with Syncro-Mate-B was also studied using five insemination schedules. The field trials conducted on three British farms yielded overall pregnancy rates of 51.0, 41.2, and 65.2%, respectively, for insemination schedules I, II, and III (Table 27). The results appear to suggest

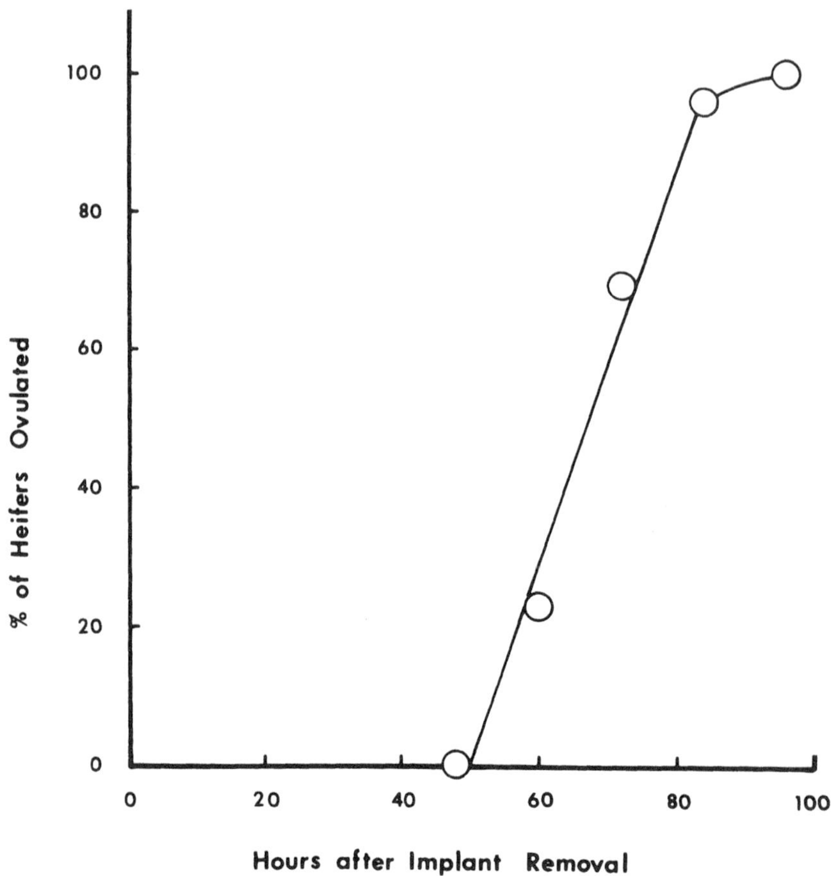

Figure 74 Relationship between the cumulative percentage of ovulating heifers relative to those in estrus and the time after implant removal. [Plotted from the data by Wishart and Young, Vet. Rec., 95:503 (1974).]

Table 26 Distribution of Estrus in Heifers Treated with Norgestomet-Releasing Hydron Implants

Herds		% Animals treated				
		In estrus[a]				
Farm	Number treated	0–24	25–48	49–72	Total	Not in estrus
B	36	0.0	86.1	2.8	88.9	11.1
L	42	2.4	78.6	11.9	92.9	7.1
P	72	0.0	77.8	6.9	84.7	15.3

[a]Hours after implant removal.
Source: Calculated and compiled from data from Reference 254.

Table 27 Effect of the Timing of Artificial Insemination on the Rate of Pregnancy[a]

	Time schedules of artificial insemination[b]				
Farms	I	II[c]	III[c]	IV[c]	V[c]
B	33.3[d]	38.5	72.7	—	—
L	33.3[d]	14.3	53.8	—	—
P	72.7	58.3	68.2	—	—
Overall	51.0	41.2	65.2		
C	—	—	76.0	53.9	70.8
D	—	—	54.3	58.8	53.3
E	—	—	70.6	70.6	70.6
F	—	—	85.7	71.4	71.4
G	—	—	63.2	63.0	67.3
$\bar{x} \pm$ SD			70.0 ± 12.0	63.5 ± 7.5	66.7 ± 7.7

[a]The percentage of those animals inseminated found to be pregnant at 90 days.
[b]Schedule I, 49 heifers were inseminated at approximately 9:30 am on the day after estrus was first detected. Schedule II, 51 heifers were inseminated at 48 hr after implant removal. Schedule III, 46 heifers (B, L, and P) or 148 heifers (C–G) were inseminated twice, at 48 and 60 hr after implant removal. Schedule IV, 145 heifers were inseminated twice, at 48 and 72 hr after implant removal. Schedule V, 135 heifers were inseminated once at 54 hr after implant removal.
[c]In Schedules II–V all the heifers were inseminated whether or not they had been observed in estrus.
[d]All the heifers inseminated in this group were seen in estrus.
Source: Data on farms B, L, and P were from Reference 254. Data on farms C, D, E, F, and G were from Reference 265.

that schedules I and II are very sensitive to the variation in herds from one farm to another, and the effect of herd-to-herd variation on the rate of pregnancy seems to be minimized by applying schedule III, which consists of double insemination at 48 and 60 hr following implant removal. The additional insemination at 60 hr significantly improved the rate of pregnancy from the single insemination either at 48 hr after implant removal (schedule II) or the next morning after the detection of estrus (schedule I). This improvement in conception rate was particularly striking in herds *B* and *L*, in which fertility levels from schedules I and II were extremely low.

The results in Table 27 demonstrate that it is now possible to eliminate the time-consuming and labor-intensive process of estrus detection and to inseminate all treated animals on a predetermined schedule, that is, double insemination at 48 and 60 hr after removal of the implant. Obviously, artificial insemination on a fixed time schedule will greatly relieve farmers of the burden of watching for the appearance of estrus, reduce the frequency of trips made by an inseminator and the cost incurred, and also give nonestrus, but synchronously ovulating, animals a chance to become pregnant (254).

Further field trials on another five British farms (265) confirmed that normal levels of fertility are achievable with the double fixed time insemination schedule at 48 and 60 hr (schedule III) as well as at 48 and 72 hr (schedule IV) after the removal of norgestomet-releasing implants. A similar observation was also reported with prostaglandin treatment (266). In addition, the provision of adequate calorie and protein levels in the diet as well as the choice of bull semen were also important in determining the levels of fertility (254).

Table 28 Effect of the Timing of Artificial Insemination on Rate of Pregnancy in Heifers

Group	Number of animals	Efficacy of treatment (%)[a]				
		Estrus rate		Pregnancy rate		
		Day 5	Day 25	Day 5	Day 25	Day 45
Control group	67	21	90	16	66	70
Treatment group[b]						
A	56	100	—	57	79	79
B	65	—	—	55	82	83
C	65	—	—	68	83	88
D	61	—	—	64	88	88

[a]Percentage of the total animals in each group.
[b]Timing of artificial insemination for group A, inseminated at approximately 12 hr after the detection of estrus; group B, inseminated at 45 hr after implant removal; group C, inseminated at 50 hr after implant removal; group D, inseminated at 55 hr after implant removal.
Source: Compiled from data from Reference 260.

Endoscopic examination suggested that the interval from the end of treatment to ovulation has a mean ± SD value of 68.5 ± 9.7 hr (Table 25). Based on this timing of ovulation artificial insemination at 48 hr after implant removal is considered too early to achieve optimal fertility. This may explain the low pregnancy rate (41.2%) achieved by the single insemination at 48 hr (schedule II, Table 27). This low level of fertility was improved to 66.7 ± 7.7% by delaying the insemination time from 48 to 54 hr (schedule V, Table 27) (267).

Studies conducted on five farms with a total of 430 heifers led to the conclusion that a single insemination at 54 hr instead of 48 hr could yield a normal level of fertility (66.7 ± 7.7%) that is not statistically different from double insemination at either 48 and 60 hr (70.0 ± 12.0%) or at 48 and 72 hr (63.5 ± 7.5%) following the removal of norgestomet implants (Table 27). Similarly, field trials carried out in the United States also indicated that one-time insemination at either 50 or 55 hr produces a higher rate of pregnancy than at 45 hr when detection is made on Day 5 (Table 28). However, the difference became smaller when detection of fertility was made on Day 25 and Day 45. These results demonstrated that artificial insemination can be done over the period from 50 to 55 hr after implant removal without adversely affecting fertility (260).

Factors affecting pregnancy rate. The minimum pregnancy rate in the synchronized animals desired by the French breeders is 55% (268). However, the pregnancy rates obtained in French beef cattle following various estrus synchronization treatments have been found variable and at times even disappointing. Mauleon and his associates (269) recently analyzed these problems and outlined the following factors that may have affected the rate of pregnancy.

Low calving rates have been observed in beef cattle following treatment with prostaglandin injections (Estrumate, ICI), progesterone-releasing vaginal spirals (PRID, Abbott), or norgestomet-releasing Hydron implants (Syncro-Mate-B, Searle) if the treatments were initiated during the end of a winter season (270,271).

Results of a 5-year field trial in France on Salers nursing cows concluded that calving rates after treatment with norgestomet-releasing Hydron implants could be

Table 29 Effect of Norgestomet Dose on the Calving Rate in Salers Nursing Cows

Dose (mg)[a]	Number treated	Calving rates[b] (%)
6	414	44.4
9	81	55.5
6 + 3[c]	258	55.4
12	672	60.1

[a]Dose of Norgestomet in the Hydron implants.
[b]Resulted from the treatment with Norgestomet-releasing Hydron implant alone (no intramuscular injection of Norgestomet-estradiol valerate combination) and double insemination at fixed time schedule.
[c]Intramuscular injection of 3 mg Norgestomet on the first day of treatment.
Source: Compiled from data from Reference 269.

enhanced by increasing the dose of norgestomet in the subdermal implant (Table 29) and concomitantly decreasing the duration of implantation (Table 30). Without the supplementary intramuscular injection of the norgestomet-estradiol valerate combination, a desirable calving rate (≃55%) can be achieved in Salers nursing cows with 7–9 day subcutaneous implantation of two norgestomet implants, each containing 6 mg norgestomet (Table 30).

Calving rates following treatment with norgestomet-releasing Hydron implants were also compared to treatment with progesterone-releasing vaginal spirals under the same physiological conditions. Both treatments produced essentially the same level of fertility (Table 31).

The benefit of estrogen injection at the beginning of estrus control treatment may not be evident in cycling females depending upon the stage of the cycle (257). In nursing cows, however, parenteral administration of either estradiol valerate or estradiol benzoate on the first day of treatment was noted to improve calving rates

Table 30 Effect of the Duration of Implantation on the Calving Rate in Salers Nursing Cows

Duration[a] (days)	Number treated	Calving rates[b] (%)
7	219	58.8
9	1092	54.4
11	114	45.6
13–16	166	28.9

[a]Duration of the subcutaneous implantation of 2 Hydron implants (containing 12 mg Norgestomet).
[b]Results from the double artificial insemination at fixed time following the removal of implants (no Norgestomet-estradiol valerate injection at the time of subcutaneous implantation).
Source: Compiled from data from Reference 269.

Table 31 Comparison in Calving Rates Between Norgestomet-Releasing Subdermal Implants and Progesterone-Releasing Vaginal Spirals

	Calving rates[a] (%)	
Nursing cows	Norgestomet implants[b]	Progesterone spirals[c]
Salers	55.2 (n = 295)	53.2 (n = 94)
Charolais	45.5 (n = 321)	45.9 (n = 74)

[a]n in the parentheses is the number of cows treated.
[b]Double inseminations at 48 and 72 hr after implant removal.
[c]Double inseminations at 56 and 74 hr after spiral removal.
Source: Compiled from data from Reference 269.

after treatment with either norgestomet-releasing subdermal implants or progesterone-releasing vaginal spirals by 6 and 12%, respectively (Table 32).

The intramuscular injection of PMSG on the day of implant removal was also found to substantially affect calving rates. For example, the administration of PMSG (800 IU) in Charolais heifers and nursing cows at the completion of intramuscular norethandrolone treatment improved calving rates from 20.4 to 45.6% and from 28.7 to 41.1%, respectively (269).

The data collected in France led to the conclusion that in beef cattle a high pregnancy rate can be achieved by short-term treatment with either 2 units of a norgestomet-releasing subdermal implant or 1 unit of a progesterone-releasing vaginal spiral plus the injection of estrogen on the first day of treatment or of PMSG on the day of completion of treatment. Artificial insemination is then carried out according to a fixed time schedule (269).

Calving rates were found to be varied from one breed to another even when the same estrus control treatment was used. Such a breed-dependent variation is related to differences in ovarian activity among the breeds (Table 33).

The results in Table 33 also suggest that variation in the calving rate after estrus control treatment also occurs in the same breed. For example, in the Charolais breed the level of fertility is lower in primiparous cows than in multiparous cows. This difference can be explained by the observation that primiparous cows frequently show a lower ovarian activity than multiparous cows during a similar postpartum interval.

On the other hand, the efficiency of treatment with the norgestomet-releasing

Table 32 Effect of Estrogen Injection on the Calving Rates from Progestin-Releasing Drug Delivery Device

	Estrogen injection[a]	
Drug delivery devices	With	Without
Norgestomet-releasing subdermal implants (9-day treatment)	54.1% (n = 37)	48.3% (n = 31)
Progesterone-releasing vaginal spirals (12-day treatment)	53.2% (n = 94)	41.2% (n = 102)

[a]Administration of 5 mg estradiol valerate (with subdermal implantation) or 10 mg estradiol benzoate (with vaginal spirals) on the first day of treatment.
Source: Compiled from data from Reference 269.

Table 33 Relationship Between the Ovarian Activity Before Treatment and the Calving Rate After Treatment with Norgestomet-Releasing Subdermal Implants

Breed	Parity	Cycling cows (%)	Calving rate (%)
Salers	—	27.8	60.1
Charolais	—	18.2	45.8
Charolais	Primiparous	13.3	37.2
Charolais	Multiparous	27.7	53.5

Source: Compiled from data from Reference 269.

Hydron implant in synchronizing noncycling cows (in terms of the percentage ovulation after treatment) is lower in Charolais cows (68.2%) than in Salers cows (95.4%; Table 34). This observation suggests that the difference between breeds in the degree of anestrus is expressed not only by the difference in their ovarian activity but also by the variation in hypophyseal-ovarian refractoriness to synchronization. The relationship between ovarian activity and hypophyseal-ovarian refractoriness exists not only between breeds but also within herds. The induction of ovulation in anestrus cows is more effective in herds in which more cows are cycling. The percentage of active ovarian activity (cycling) within a herd appears to be a critical factor in determining how easy it is to induce ovulation in cycling cows. The fertility of these induced ovulations is apparently related to the efficiency of ovulation induction.

It appears that fertility in beef cattle after estrus control treatment with norgestomet-releasing subdermal implants can be improved by either one of the following two approaches: (i) increase the efficiency of ovulation induction and (ii) increase the number of cows regaining ovarian activity.

The efficiency of ovulation induction can be improved by increasing the norgestomet dose in the subdermal implant (Table 29) with 7-day implantation (Table 30). The second approach can be accomplished by increasing the dose of PMSG injection (Table 35). Unfortunately, the data in Table 35 also suggest that an increase in the PMSG dose also tends to yield an increase in multiple pregnancies, especially

Table 34 Rates of Ovulation and Pregnancy After Norgestomet Implant Treatment and Relation to Ovarian Activity Before Treatment[a]

	Ovarian activity (%)			
Breeds	Cycling	Noncycling	Ovulation after treatment (%)	Rate of pregnancy[b]
Charolais				
Heifers	38.9	—	100.0	73.9
	—	61.1	72.2	69.1
Cows	31.8	—	85.7	73.3
	—	68.2	68.2	58.8
Salers				
cows	45.0	—	100.0	55.5
	—	55.0	95.4	76.1

[a]By progesterone assay (Reference 269).
[b]Every 100 ovulations.

Table 35 Effect of PMSG Doses on the Induction of Ovulation and Multiple Pregnancies in Charolais Cows

Dose (IU)	Ovulation (%)[a]	Number calving[b]	Multiple pregnancy (%)	
			Twins	Triplets and quadruplets
400	67.5	—	—	—
500	83.1	—	—	—
600	92.1	—	—	—
700	—	425	7.3	1.2
800	—	649	9.6	5.0
1000	—	80	20.2	6.3

[a]Total percentage of ovulation in both cycling and noncycling heifers.
[b]From nursing cows.
Source: Compiled from data from Reference 269.

the undesirable triplets and quadruplets. The 600 IU PMSG appear to be the optimum dose.

The effect of cycling activity in nursing cows on the clinical efficacy of Syncro-Mate-B treatment was also evaluated in the United States in three herds with a low cycling activity (18–40%) and five herds with a high cycling activity (average of 77%) (260). The results indicated that Syncro-Mate-B treatment improved both herds with either a low or a high cycling activity (Table 36), especially in treatment group B. It appears that Syncro-Mate-B treatment is more effective and able to achieve a higher rate of pregnancy in nursing cows with a high cycling activity than those with a low cycling activity. In either case, however, Syncro-Mate-B treatment yielded a higher rate of pregnancy in nursing cows than that seen in untreated animals.

Table 36 Effect of Cycling Activity in Nursing Cows on the Efficacy of Syncro-Mate-B Treatment

Cycling activity	Group[a]	Number of animals	Efficacy of treatment (%)[b]				
			Estrus rate		Pregnancy rate		
			Day 5	Day 25	Day 5	Day 25	Day 45
Low (18–40%)	Control	218	5–8	18–40	3	17	54
	Treatment						
	A	262	—	—	16	32	61
	B	254	—	—	26	41	64
High (average of 77%)	Control	406	19	77	12	56	78
	Treatment						
	A	421	67	83	39	59	78
	B	400	—	—	51	66	83

[a]Artificial insemination schedule for treatment groups A, inseminated at approximately 12 hr after the detection of estrus, and B, inseminated at approximately 48–54 hr after implant removal without estrus detection.
[b]Percentage of the total number of animals in each group.
Source: Compiled from data from Reference 260.

Table 37 Effect of Postpartum Interval on the Pregnancy Rate in Registered Simmental Cows in Response to Syncro-Mate-B Treatment

Group	Number of animals	Pregnancy rate (%)[a]		
		Day 5	Day 25	Day 45
Control group	60	12	35	58
Treatment groups[b,c]				
A	61	25	28	51
B	38	47	79	87
C	93	47	73	82

[a]Percentage of the total animals in each group.
[b]Artificial insemination at approximately 48–54 hr after implant removal.
[c]Length of postpartum interval: A, first calf heifers with 60 days postpartum; B, first calf heifers with 90 days postpartum; C, mature cows with 50–90 days postpartum.

The effect of postpartum interval on the clinical efficacy of Syncro-Mate-B treatment was also investigated in a group of first-calf heifers and mature cows (260). The results indicated that heifers with a postpartum interval of 90 days respond as well to treatment as mature cows, but heifers with a shorter postpartum interval (60 days) do not (Table 37). Other studies carried out in Angus, Simmental, and Maine-Anjou cattle failed to demonstrate any effect of postpartum interval on the response to Syncro-Mate-B treatment, even though the pregnancy rate in untreated cows was found to be definitely better in the longer (>60 days) postpartum group than in the 30–60 day group. This observation, however, pointed out the common problem of lower cycling activity with a postpartum interval shorter than 60 days. Since a cow must become pregnant by approximately 80 days after calving to maintain yearly calf production, Syncro-Mate-B treatment certainly provides a useful means to achieve such a goal.

In beef cattle most female animals have been found to be anestrus at the time of breeding, that is, at the end of a winter season (272). This may well be related to the fact that cows are usually maintained at low nutritional levels during the winter with hay of medium or low quality. As a result very few cows cycle during the winter, irrespective of the postpartum interval. After the winter season they go out to pasture and their reproductive activity is activated again and the mean interval from calving is thus shortened.

Ovarian activity is also dependent upon the stabling system used. A higher ovarian activity (35.9%) is detected in those cows that are free than in those that are tied and kept inside (23.2%). This may be because free cows may have a higher winter nutritional level (269).

Nutritional levels also affect the rate of ovulation induced by estrus synchronization treatment. The percentage of females ovulating following treatment can be improved by distributing extra food at the time of breeding (269).

Progestin treatment is known to yield unfavorable conditions in the genital tract for spermatozoal penetration. This emphasizes the importance of the quality as well as the number of spermatozoa used to achieve a better calving rate. The differences in fertility rate among bulls were found to be greater for artificial inseminations carried out at synchronized estrus than at normal estrus. On the other hand, a difference in the fertility rate also occurs between primiparous and multiparous cows synchronized by norgestomet-releasing Hydron implants (Table 33). A higher fertility rate is usually produced in multiparous cows than in primiparous cows inseminated with semen from the same bull (Table 38). The difference observed may be attributed to the difference in nutrition requirements since a primiparous cow is still growing and therefore requires more feed (268).

The nursing relationship between the calf and the cow and stimulation of the udder can prolong the postpartum anestrus period in nursing cows. The number of cows ovulating can be increased by temporary weaning (24–48 hr) alone at the end of each estrus synchronization treatment (268).

In conclusion, the induction of ovulation in noncycling female cows following treatment with an estrus synchronizer can be much improved by shortening the postpartum anestrus period in the winter season through management practice: that is, by raising nutritional levels at the time of breeding, by housing systems, and by temporary weaning before or after treatment.

Potential Developments of Interest. As discussed earlier, the subcutaneous release profile of norgestomet from Hydron implants is not constant but follows a nonlinear Q versus t relationship (Figure 73). The daily amount of norgestomet released from this polymer matrix diffusion-controlled subdermal implant is high initially and gradually decreases with time (205). As expected from the theoretical model describing the matrix-diffusion process (Chapter 2, Section II.B), the cumulative amount of norgestomet released is linearly proportional to the square root of the implantation time (Figure 45).

A new generation of controlled-release drug delivery system, microsealed drug delivery (MDD) system, has recently been developed with the objective of overcoming this non–zero-order drug release kinetics (190–193). MDD is a microreservoir dissolution-controlled drug delivery system, and the mechanisms of controlled drug delivery for this type of controlled-release drug delivery system are analyzed and discussed in Chapter 2 (Section II.D). As expected from the physical model, the subcutaneous release of drug from an MDD system was observed experimentally at a constant rate for an observation period of up to 129 days (273). The feasibility of developing an estrus-synchronizing subdermal implant to administer a constant daily dose of norgestomet has also been explored. The results in Figure 75 indicate

Table 38 Effect of Bull Semen on the Fertility of Charolais Cows after Estrus Control Treatment

	Calving rate (%)	
Bull[a]	Primiparous	Multiparous
I	16.7	38.0
II	12.5	42.9
III	16.7	45.4
IV	54.5	56.0
V	55.6	63.9

[a]Charolais bulls.
Source: Compiled from data from Reference 269.

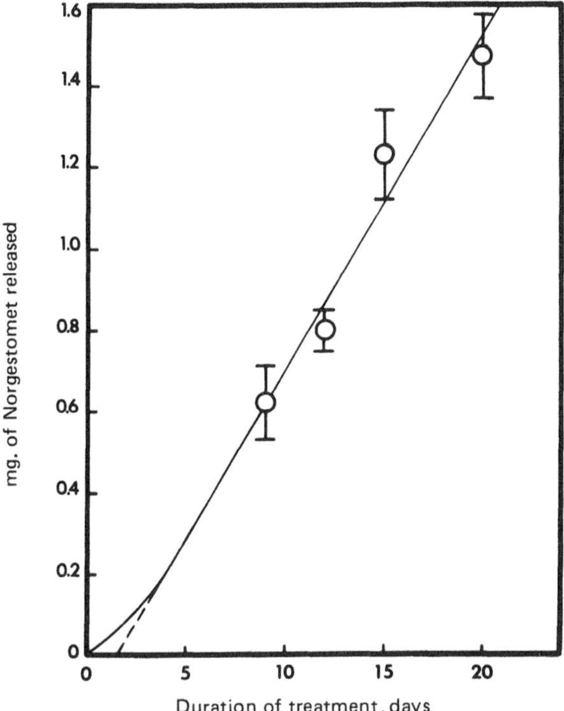

Figure 75 Subcutaneous release profile of norgestomet from MDD-type subdermal implants (MDD 434) in 20 U.S. cows for up to 20 days. The cumulative amount of norgestomet (Q) released is linearly dependent upon the length of implantation.

that subcutaneous implantation of this MDD-type estrus synchronizer has yielded a constant release profile of norgestomet in French heifers as well as in American cows. A good in vitro-in vivo correlation was also achieved with the four MDD subdermal implants with varying release rates of norgestomet (Figure 76).

The effectiveness of norgestomet-releasing MDD subdermal implants in the suppression and subsequent synchronization of estrus is illustrated in Table 39. The results suggest that the biological efficacy of an MDD-type estrus synchronizer depends upon the daily dose of norgestomet delivered subcutaneously from the implant. With a dose of 44.4 μg/day only 10% of treated heifers are effectively suppressed (9 of 10 treated heifers were detected in estrus during treatment). The efficiency of estrus suppression is enhanced by as much as eight times to 80% by doubling the dose of norgestomet to 72.5–87.7 μg/day. Following completion of treatment 70–80% of treated animals show synchronized estrus within 48 hr after removal of the implants. A further increase in the daily norgestomet dose delivered, however, does not improve the efficiency of estrus synchronization. This effective dose (72.5–87.7 μg/day) is significantly lower than the 140 μg/day required for effective estrus synchronization established earlier by daily intramuscular injection of norgestomet (252).

The effectiveness of an MDD-type estrus synchronizer in the suppression and synchronization of estrus was also observed to be further enhanced with the intra-

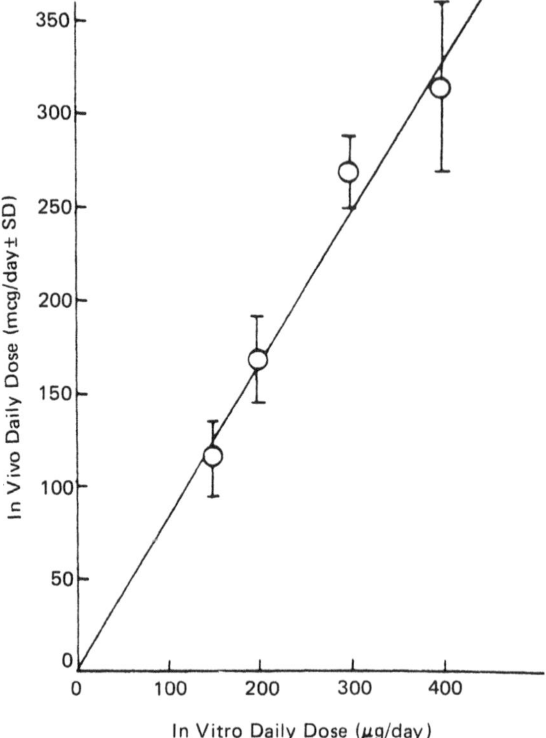

Figure 76 Correlation between the in vitro and in vivo daily doses of norgestomet released during the 9-day subcutaneous implantation of four MDD implant formulations in French heifers.

muscular injection of a supplementary dose of norgestomet (3 mg) and estradiol valerate (5 mg) as demonstrated earlier for the Hydron-type estrus synchronizer. This supplementary injection does not result in any deviation in the correlation between the in vitro and in vivo subcutaneous release rates of norgestomet from three MDD subdermal implants.

The subcutaneous release rates of norgestomet from MDD implants tested on various farms have been found to be quite reproducible and very close to the release rate projected from in vitro drug release studies (Table 40). The frequency of calving, that is, the state of parity, had no significant effect on the subcutaneous release rate of norgestomet (Table 41).

It was reported earlier that it is possible to inseminate animals treated with norgestomet-releasing Hydron implants once or twice according to a predetermined time schedule after completion of treatment (Table 27) and hence the time-consuming, difficult process of estrus detection can be omitted. Normal levels of fertility have been achieved with this fixed-time artificial insemination (254,265). This feasibility was also evaluated in MDD implant-treated heifers. The results, outlined in Table 42, indicate that fertility was maximized with double insemination at 48 and 60 hr for animals treated with an MDD implant releasing norgestomet at a dose of 94.7 μg/day (MDD-1) and at 48 and 72 hr for those treated with the 159.3 μg/day

Table 39 Effectiveness of MDD Implants in Estrus Synchronization in Heifers and Effect of Supplementary Norgestomet-Estradiol Valerate Injection

	Without supplementary dose					With supplementary dose[a]			
		MDD treatment[b]					MDD treatment[b]		
	Control	A	B	C	D	Control	B′	C′	D′
Heifers treated	10	10	10	10	10	10	10	10	10
Implant lost during treatment	—	0	0	0	1	—	0	0	1
Heifers in estrus									
During treatment	10	9	2	2	1	7	0	0	1
48 hr after treatment	—	1	8	7	7	0	9	10	7
Heifers in silent estrus after treatment	0	0	0	1	1	3	1	0	2

[a]Intramuscular injection of 3 mg Norgestomet and 5 mg estradiol valerate in 2 ml sesame oil.
[b]Daily dose (μg/day ± SD) of Norgestomet released from the 9-day subcutaneous implantation of MDD implants: A, 44.4 ± 11.2; B, 72.5 ± 8.7; C, 87.7 ± 15.4; D, 143.3 ± 42.1; B′, 75.8 ± 5.5; C′, 88.1 ± 8.8; D′, 114.8 ± 4.4.
Source: The studies were conducted by Dr. D. Wishart in the United Kingdom.

implant (MDD-2). The levels of fertility achieved by the MDD-type estrus synchronizer (58.8 and 68.8%) are satisfactory and comparable to those (62.5%) attained in heifers treated with the Hydron-type estrus synchronizer (Table 27).

The efficiency of fixed-time artificial insemination in achieving conception seems to be dependent upon the daily norgestomet dose delivered subcutaneously. With the daily dose of 94.7 μg a normal level of fertility was obtained with double inseminations at 48 and 60 hr or with a single insemination at 54 hr, but not with double inseminations at 48 and 72 hr (Table 42). On the other hand, with a dose 159.3 μg/day, the fertility level was maximized with double inseminations at 48 and 72 hr. This difference may be explained by the fact that the timing of estrus synchronization is dose dependent and the synchronized estrus from the higher daily dosage of nor-

Table 40 Variation in Subcutaneous Release Rates of Norgestomet from MDD Implants[a] Among Farms

	Release rates (μg/day ± SD)	
Farm	A	B
1	82.2 ± 14.2	138.7 ± 15.0
2	93.2 ± 8.3	143.1 ± 16.5
3	92.3 ± 17.3	140.5 ± 13.0
4	97.7 ± 12.2	147.5 ± 15.2
5	94.7 ± 8.6	159.3 ± 8.4

[a]MDD 461-S implants with projected daily dose: A, 100 μg/day; B, 150 μg/day.
Source: The studies were conducted by Dr. D. Wishart in the United Kingdom.

Table 41 Effect of Calving Frequency on the Subcutaneous Release Rate of Norgestomet from MDD Implants in French Nursing Cows[a]

Calving frequency	Number of cows	Release rate (μg/day ± SD)
1	17	149.0 ± 12.6
2	5	130.8 ± 18.5
3	4	141.9 ± 19.2
4	3	139.5 ± 8.6
Multiple[b]	6	137.6 ± 6.8

[a]A 10-day subcutaneous implantation of MDD 461-S (conducted by Mr. D. Aguer in France).
[b]With a calving frequency of 5–10.

gestomet appears several hours later than that from the lower daily dose (Figure 77).

The levels of fertility achieved with fixed-time artificial insemination compare favorably with the rate of pregnancy resulting from insemination after estrus detection (compare the data in Table 42 with those in Table 43). It is interesting to note that a (6–30%) higher level of fertility was obtained with artificial insemination carried out according to a predetermined time schedule (without prior estrus detection).

The plasma profiles of norgestomet resulting from the subcutaneous controlled delivery of norgestomet by the MDD implants (with zero-order drug release) and the Hydron implants (with non–zero-order drug release) were compared in Spanish ewes. The results, shown in Figure 78, demonstrate that the constant release of norgestomet from MDD implants (with a release rate of 55.1 ± 2.3 μg/day) produced a fairly constant steady-state plasma level of norgestomet. At the end of 12 days of treatment the norgestomet-releasing MDD implant was removed and the plasma drug level decreased rapidly to a concentration below 20 pg/ml (the pretreatment level). On the other hand, treatment with the norgestomet-releasing Hydron implant yielded a spike plasma concentration of 350 pg/ml within 24 hr, which is more than twice the plateau drug concentration achieved by MDD implant. This plasma drug level

Table 42 Effect of the Time Schedule of Artificial Insemination on the Rate of Pregnancy in Heifers Treated with MDD Implants[a]

	Rate of pregnancy (%)[b]		
Time schedule[c]	MDD-1	MDD-2	Hydron[d]
Single insemination at 54 hr	58.8	29.4	—
Double inseminations at 48 and 60 hr	68.8	43.8	—
Double inseminations at 48 and 72 hr	41.2	58.8	62.5

[a]A 9-day subcutaneous implantation.
[b]Subcutaneous release rates: MDD-1, 94.7 ± 0.9 μg/day (n = 51); MDD-2, 159.3 ± 1.9 μg/day (n = 51).
[c]Time elapsed from the removal of implants.
[d]The Hydron implant was tested as a reference.
Source: The studies were conducted by Dr. D. Wishart in the United Kingdom.

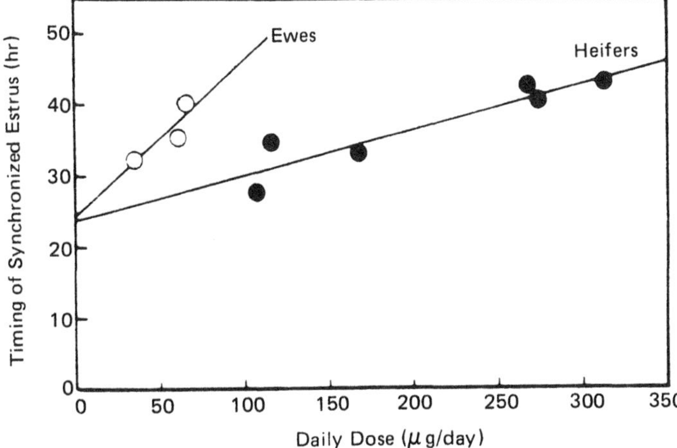

Figure 77 Effect of the daily dose of norgestomet released from the MDD subdermal implants on the timing of synchronized estrus detected in (○) ewes (n = 56) and (●) heifers (n = 77).

decreased continuously during the course of treatment to a concentration of around 50 pg/ml at the time of implant removal (Day 12 of treatment), which was less than one-half the plateau level (120–150 pg/ml) maintained by the MDD implant.

The difference in the plasma profiles of norgestomet resulting from treatment with norgestomet-releasing MDD or Hydron estrus synchronizer implants is also reflected in the biological effectiveness (Table 44). The MDD implant achieves a total

Table 43 Comparative Biological Effectiveness between Hydron Implant Treatment and MDD Implant Treatment with Estrus Detection[a]

	Hydron implant	MDD implant (μg/day ± SD)[b]	
		98.8 ± 17.2	176.8 ± 16.1
Number of heifers treated	29	29	29
Percentage of treated heifers in estrus after implant removal			
24–48 hr	69	72.4	58.6
49–72 hr	17.2	13.8	20.7
Total	86.2	86.2	79.3
Number of heifers inseminated[c]	25	25	23
Percentage of inseminated heifers pregnant[d]	56.0	40.0	52.2

[a]A 9-day subcutaneous implantation with a supplementary dose of 3 mg Norgestomet and 5 mg estradiol valerate in sesame oil injected intramuscularly on the first day of implantation.
[b]Actual subcutaneous release rates.
[c]Only those heifers in estrus were inseminated the next morning following the day of estrus detection.
[d]Rectal examination for pregnancy was conducted 42 days after insemination.
Source: The studies were conducted by Dr. D. Wishart in the United Kingdom.

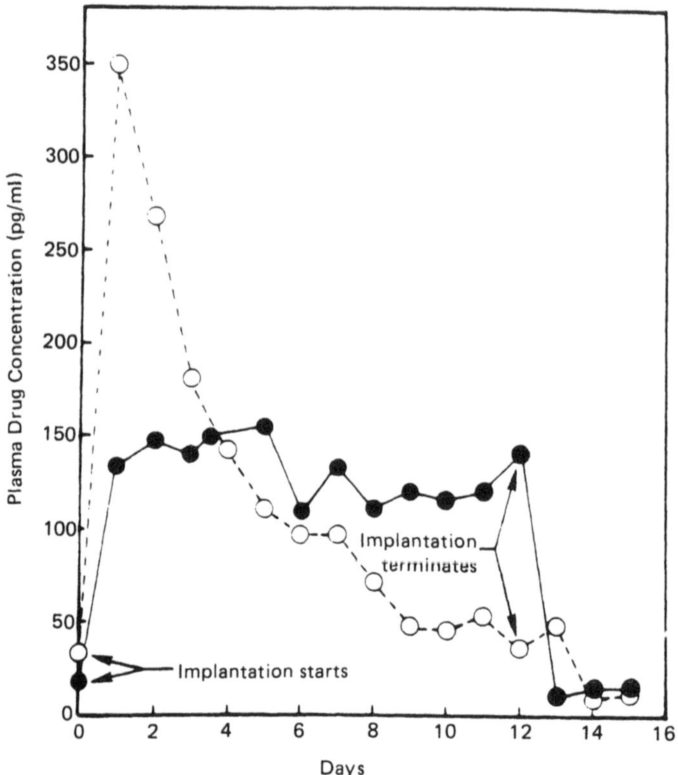

Figure 78 Plasma profiles of norgestomet in Spanish ewes treated with (●) MDD 445A implants or (○) Hydron implants for 12 days. Each data point is the mean of eight determinations conducted in four ewes.

Table 44 Comparison of Biological Effectiveness in Spanish Ewes Treated with MDD or Hydron Subdermal Implants[a]

Implants	Number of ewes	Cumulative percentage of animals in estrus at 22 hr	24 hr	28 hr	32 hr	36 hr	46 hr	Ovulation (%)[b]
MDD 445A[c]	13	84.6	92.3	92.3	100.0	—	—	76.9
Hydron[d]	16	43.8	68.8	75.0	93.8	93.8	100.0	50.0

[a]A 12-day subcutaneous implantation without the intramuscular injection of 3 mg Norgestomet and 5 mg estradiol valerate.
[b]The percentage of ewes ovulated or ovulating 50–55 hr after implant removal.
[c]Subcutaneous release rate = 55.1 ± 2.3 μg/day.
[d]Subcutaneous release rate = 316.4 $\mu g/day^{1/2}$ (estimated).
Source: The studies were conducted by Dr. D. Aguer in France.

estrus synchronization within a shorter duration (32 hr) and with a higher rate of ovulation (76.9%) than those attained by the Hydron implant (within 46 hr and 50%, respectively).

The norgestomet-releasing MDD implant has been licensed to Intervet International B.V. and recently marketed by Intervet in Europe as the Syncro-Mate-B implant. This veterinary product has the same indications as the Syncro-Mate vaginal pessary (Chapter 9).

D. Benefits

The therapeutic benefits of controlled drug administration via implantation can be illustrated by comparing the biological activity and duration resulting from the subcutaneous administration of drug via an implantable controlled-release drug delivery system with those produced by conventional drug administration through subcutaneous injection.

The subcutaneous controlled administration of megestrol acetate, a very potent antiovulation progestin without estrogenic or androgenic activity of its own, from silicone capsule implants was found to be the most efficient way of delivering this progestin (225,274). The biological potency of megestrol acetate in terms of delayed implantation, antifertility, and antiovulation activity was found to be enhanced 7–13 times when it was administered at a controlled rate from silicone capsule implants compared to conventional subcutaneous injection in suspension preparations (Table 45).

The data outlined in Table 45 also indicate that progesterone, when administered by silicone capsule implants, is 13 times more potent than with subcutaneous injection in the delayed implantation test and 25 times more potent in deciduoma formation test (275). On the other hand, 19-nor-progesterone delivered from silicone capsule implants was found to be 25 times more effective than subcutaneous injection

Table 45 Relative Biological Potency of Various Steroid-Releasing Silicone Capsules

Steroid	Biological activity	Relative potency[a]
Magestrol acetate	Delayed implantation	13.4
	Inhibition of fertility	6.7
Progesterone	Delayed implantation	12.5
	Decidouma formation	25.0
19-Nor-progesterone	Antiandrogenicity in immature rats	25.0
Testosterone	Parabiotic rat assay	
	Gonadotropin inhibition	30.0
	Androgenic activity	30.0
	Androgenicity in adult males	15.0
Norgestrel	Delayed implantation	32.0
	Compensatory ovarian hypertrophy	41.0
Estrone	Uterotropic	≥2.0

[a]The biological potency from subcutaneous injection is assumed to be unity.

Source: Compiled from data in Chang and Kincl (1968, 1970) and Zbuzkova and Kincl (1970).

in an oil solution formulation in inhibiting the growth of seminal vesicles in immature male rats.

Furthermore, the androgenic activity of testosterone-releasing silicone capsules was evaluated by two biological assays. In the parabiotic rat assay the silicone capsule implant, which releases 5 μg/day of testosterone, was found to be sufficient to inhibit the secretion and/or the release of pituitary gonadotropins and to stimulate the growth of ventral prostate and seminal vesicles in the castrated male rat. If administered by subcutaneous injection a daily dose of 150 μg testosterone was required to produce the same levels of suppression of ovarian growth and stimulation of the growth of ventral prostate and seminal vesicles (276,277). Apparently, both the gonadotropin inhibition activity and the androgenic activity of testosterone were enhanced 30 times via subcutaneous continuous delivery at a controlled rate by silicone capsule implants (Table 45).

In the castrated adult male rat the silicone capsule implant, which releases 40 μg/day of testosterone, was able to maintain the growth of seminal vesicles, ventral prostate, and levator ani muscle during a 90-day observation. On the other hand, by subcutaneous injection a dose of 600–700 μg/day of testosterone was needed to achieve comparable biological activity (278). The results suggested that the androgenic activity of testosterone is enhanced approximately 15 times via the subcutaneous controlled administration by silicone capsule implants (Table 45).

The biological potency of norgestrel delivered at a controlled rate by silicone capsule implants was also compared with a subcutaneous injectable preparation of norgestrel using the delayed implantation test and the steroid-induced block of ovarian compensatory hypertrophy in hemicastrated rats (275).

The results of the delayed implantation test in the rat indicated that blastocyst viability is fully preserved in all the tested animals receiving a silicone capsule implant that releases a daily norgestrel dose of 8 μg/day. The average number of implantation sites (6.3 ± 0.3), however, was slightly less than the 9.0 ± 0.7 detected in rats treated with daily injections of 160 μg progesterone. By daily subcutaneous injection of 250 μg norgestrel the full activity of norgestrel was achieved, and all the animals treated had an average number of implantation sites (9.8 ± 0.6) similar to those in the progesterone-treated group. It was calculated that the norgestrel-releasing silicone capsule has a relative potency, by the delayed implantation test, 32 times that of the subcutaneous injection of norgestrel (Table 45).

The results of the compensatory ovarian hypertrophy test suggested that the hypertrophy observed in the remaining ovary as the result of removing one of the two ovaries could be blocked by a single daily injection of norgestrel (200 μg/day). When norgestrel was administered continuously at a rate of 2 μg/day from a silicone capsule implant, inhibition equivalent to that attained by a subcutaneous injection of norgestrel at 100 μg/day was achieved. By statistical calculation the norgestrel-releasing silicone capsule was found to be 41 times more effective than daily subcutaneous injection (Table 45). A similar response was also observed when norgestrel-releasing silicone capsules were implanted in the intraperitoneal cavity. It appears that the norgestrel-releasing silicone capsule implant acts as an "artificial gland" at both sites of implantation.

The data summarized in Table 45 show that for a specific biological response a substantially smaller dose of steroids is needed if it is administered subcutaneously through the continuous release mechanism from the silicone capsule compared to the

Table 46 Comparative Biological Activity and Toxicity of Atropine Base Administered by Silicone Capsules and Subcutaneous Injection

Dose (mg/rat)	Mydriasis duration (days)[a]		Mortality rate (%)	
	Capsule	Injection	Capsule	Injection[b]
0	1	1	0	SN
6.25	4	2	0	SN
12.50	4	—	0	—
25.00	11	4	0	SN
50.00	17	—	0	—
100.00	26	6	0	50
200.00	37	—	0	—
400.00	48	—	0	100

[a]The number of days each treatment group had a mean mydriatic response equal to or greater than +1 mydriasis (25% or more dilations over the control group).
[b]Severe necrosis (SN) was observed at the injection site.
Source: Compiled from Bass et al. (1965).

same steroid given by conventional daily subcutaneous injection. The increase in biological potency for the steroids investigated can range from a modest 2-fold increase in the uterotropic activity of estrone to the 41-fold increase in the steroid-induced blockade of ovarian compensatory hypertrophy achieved by norgestrel. The variation in the degrees of enhancement of the effectiveness achieved by controlled drug delivery for different steroids is attributable not only to the difference in the biological tests used but also to the difference in biological half-life from one steroid to another.

Controlled-release silicone capsules were also reported to extend the duration of mydriatic response produced by atropine base as well as to reduce its toxicity in the rat (172). When administered subcutaneously through controlled-release silicone capsules atropine base was found to produce a dose-dependent mydriasis in rats: the higher the loading dose of atropine in the capsules, the longer the duration of myriatic response produced (Table 46). It is interesting to note that the duration of mydriasis produced by the atropine-releasing silicone capsule is significantly longer than that produced by subcutaneous injection of the same atropine dose suspended in an aqueous solution of 1% methylcellulose. This prolongation in mydriatic response was observed to be dose dependent: the higher the atropine dose, the more prolongation the duration. Additionally, the mortality rate and toxicity of the atropine base were markedly reduced by administering it in silicone capsules. This could be due to a more uniform tissue distribution of drug (and its metabolites) when delivered by silicone capsules, which prevent any abnormal accumulation of the drug or its metabolites in any organs (279).

E. Medical Aspects

1. Living Tissues Environment

Animal tissue contains approximately 70% body fluid. This fluid consists of two major compartments: intracellular fluid and extracellular fluid (280). Extracellular fluid, which bathes all tissues, is further subdivided into interstitial fluid and intra-

Table 47 Chemical Composition of Interstitial Fluid[a]

Cations		Anions	
Species	Concentration (mEq/L)	Species	Concentration (mEq/L)
Na^+	140	Cl^-	105
K^+	4	HCO_3^-	30
Ca^{2+}	5	HPO_4^{2-}	5
Mg^{2+}	3	SO_4^{2-}	3
		Organic acids	6
		Proteins	3
Total	152		152

[a]Normal pH value of 7.4.
Source: From Bell et al. (1963).

vascular fluid (which includes plasma and lymph). Interstitial fluid, which the implants mostly encounter at the site of implantation, has the chemical composition shown in Table 47.

Oxygen is freely available and is readily replaced by complex biochemical processes and hemostatic controls (281). Carbon monoxide also exists. However, in certain tissues that have necrosed or are affected by various types of bacterial infections, anaerobic conditions may result. At the cellular level the pH value may be lower than the pH 7.4 measured in extracellular fluid.

Many enzymes, which are capable of oxidation, reduction, or hydrolysis, are present in living tissues. Since many enzymatic reactions are dependent upon various trace metals, thorough investigation is needed to search for the early or long-term effects of chemicals or additives that could leach out or result from the degradation of implanted polymeric materials.

2. *Reactions of Host to Implant*

Inflammation is one of the defensive reactions of the living body to any irritant, whether physical (e.g., burns), chemical (e.g., toxins), or biological (e.g., bacteria) (282). Practically any foreign agent can act as an irritant. The acute phase of the inflammatory reaction leads to the formation of an exudate and fibrinous network at the affected site. Vascular and lymphatic systems are activated. Leukocytes and mast cells permeate out of the capillaries accumulate at the affected site with red blood cells.

The presence of a surgically implanted device calls for major adaptation by the host tissue unless the device is absorbable and replaceable, like catgut ligature. The polymeric device may remain as an incompletely covered foreign body with a barrier of connective tissue forming between it and the surface epithelium. The absorption and permeation of drugs may be effectively blocked (283–285). Successful implantation may also be jeopardized by immune responses from the lymphatic system. Despite the availability of modern techniques to suppress these, natural tissues in the process of disruption and liquefication may become antigenic, resulting in rejection of the implants. In addition, the leaching out of any additives from the implants may also cause toxic reactions.

To minimize reactions any polymeric device should have the minimum surface area by design and a very smooth surface finish. Ideally such an implant should possess the same structural characteristics as the tissue in which it is embedded. A rigid plastic material inserted into soft tissues often becomes infected and rejected.

Little is known about the problems that arise when blood or other tissue fluids make contact with foreign surfaces. Electrical phenomena occurring at the interface could in certain situations lead to thrombosis formation (286). It was hoped that materials with a high negative zeta potential might resist this tendency. However, this negative potential was found to be immediately neutralized on contact with blood, plasma, albumin, fibrinogen, and gammaglobulin (287). To date our knowledge suggests that to prevent thrombosis the surface of vascular implants should be smooth and in contact only with an area of high velocity (288).

Folkman and Long (289) reported that any incision made in the myocardium healed by formation of a smooth envelope of fibrous tissue, which appeared to present a barrier to the permeation of drug from the site of implantation. However, the amount of fibrosis in the myocardium was the same in areas in which only an incision and tunnel were made as it was in areas near a silicone capsule containing either air or a drug (290). On the other hand, subcutaneous and intraperitoneal implantation of silicone capsules resulted in very little foreign body reaction, and no formation of fibrous tissue could be noted (289). The 11-month subcutaneous implantation of silicone capsules in rats resulted in no evidence of local tumor formation, foreign body reaction, or signs of inflammation (291). The implants were merely enclosed by a thin film of connective tissue. As previously mentioned (291), the inflammatory reaction, which was observed occasionally at the subcutaneous implantation site, was attributed to animal hair that was brought inadvertently into the incision at the time of implantation.

The chemistry of polymer degradation in the human body is another area in which knowledge is scarce. Late degradation may lead to the release of potential toxins or antigens and then to failure and rejection of implants. For example, nylon can lose approximately 80% of its tensile strength within 3 years of implantation (282).

As an example of the effects on tissues after incorporation of additives, plasticizers, or catalysts used in the fabrication of a plastic, high-density polyethylene may be cited. A 2% phenolic antioxidant, if used, causes a marked reaction of guinea pig tissues to polymer. A 2% amine antioxidant causes no such reaction. A high concentration of catalyst residues in a Ziegler-type polyethylene can be similarly irritating (282).

More recent investigations revealed that the particle size of foreign substances is most important in determining the extent of tissue reaction: the smaller the particle size, the greater the tissue reaction (292). No direct correlation can be made between the tissue reaction and the surface area (293). The particle size of fillers incorporated in polymers may also play a similar role should they move to the surface or even leach out of the polymeric device during implantation (294).

It becomes obvious that the form in which the polymeric device is implanted, the animal species used for studies, and the duration of implantation all play an important part in deciding the degree of tissue reaction to a foreign material. These factors, as well as the particle size, should all be considered when assessing the healing pattern of tissues surrounding an alloplastic material (295).

Any synthetic polymeric device from which potential toxins leach out, as a result of either faulty fabrication or breakdown during implantation, could lead at any time to immunological or allergic reaction, inflammation, and then rejection.

It was reported that approximately 20% of users of conventional IUDs expel the device within 5 months following the first insertion, although only 5% of these women will not ultimately retain some sort of IUD (296). Expulsion is the result of uterine contraction, and the expulsion rate may be reduced by continuously delivering a low dose of a progestational agent to decrease uterine sensitivity and myometrial contractility (297,298).

Since the early 1920s considerable controversy has existed about the role that the cervicovaginal appendage (or tail) on the IUD plays in the development of bacterial infection of the upper genital tract (299–302). The investigation conducted by Tatum et al. (303) revealed that the tail of the Dalkon Shield is structurally and functionally different from the tails of the four other IUDs tested. The tail of the Dalkon Shield consists of a bundle of monofilaments enclosed freely within a thin plastic sheath. The unique construction of the Dalkon tail theoretically could provide a mechanism whereby pathogenic bacteria from the vagina enter the uterine cavity and cause sepsis.

3. *Reactions of Implant to Host*

After a period of time no polymer is totally impermeable to body fluids. Folkman et al. (304) reported that fat-soluble substances in the blood, such as cholesterol, testosterone, estradiol, and hydrocortisone, were absorbed by the lipophilic silicone elastomer. Most important in the clinical uses of polymeric devices are the effects of tissue enzymes and free radicals as well as the hydrolysis caused by the absorption of body fluids.

Environmental stress cracking can occur in some implantable polymeric materials (305). Changes in physical properties need not be related to the chemical breakdown of polymeric devices. Chemical breakdown itself influences the mechanical strength of the implanted polymeric materials. Therefore, carbon-carbon bond cleavage may account for the loss of tensile strength in hydrophobic polymers like polyethylene. Heat, light, oxidation, moisture, and ionizing radiation can all be responsible for the degradation of polymeric materials.

The use of seemingly innocuous materials can be dangerous. The silicone oil used to coat tubing or oxygenator pumps in open heart surgery was reported to cause fatalities when subsequently released into the bloodstream (306). Certain plastics may be affected by the drugs used in chemotherapy (307). Blood coagulation was observed with silicone shunts when anesthetic liquid was added to the chamber (304). A tightly bonded heparin layer applied to the lumen of these shunts was reported to prevent clot formation yet it did not cause systemic anticoagulation or interfere with the diffusion of anesthetics.

In conclusion, an ideal implantable drug delivery system should be biostable, biocompatible with minimal tissue-implant interactions, nontoxic, noncarcinogenic, and removable if required and should release the drug at a constant, programmed rate for a predetermined duration of medication.

ACKNOWLEDGMENTS

The author expresses his appreciation to Dr. D. Wishart of the United Kingdom, Dr. D. Auger of France, and Drs. S. Mares, L. Peterson, and E. Henderson of the

United States for their cooperation in various animal testings of norgestomet-releasing MDD implants; to Mr. J. Cooney and Ms. D. Jefferson for their preparation and assay of MDD implants; to Dr. M. Woulfe for providing technical information on norgestomet-releasing Hydron implants; to Ms. S. J. Kelly and Dr. S. A. Pasquale for some of their clinical data on Norplant-II implants; and to Dr. A. P. Sam for his sharing of technical data on Implanon.

REFERENCES

1. B. C. Lippold; Pharmacy International, 1:60 (1980).
2. L. Lachman, R. H. Reiner, E. Shami and W. Spector; U.S. Patent 3,676,557, July 11, 1972.
3. A. P. Gray and D. S. Robinson; in: *Narcotic Antagonists* (M. C. Braude, L. S. Harris, E. L. May, J. P. Smith and J. E. Villareal, Eds.), Raven Press, New York, 1974, p. 555.
4. A. P. Gray and D. S. Robinson; J. Pharm. Sci., 63:159 (1974).
5. F. H. Buckwalter; U.S. Patent 2,507,193, May 9, 1950.
6. F. H. Buckwalter and H. L. Dickison; *J. Am. Pharm. Assoc.* (Sci. Ed.), 47:661 (1958).
7. C. Thies; in: NIDA Research, Monograph Series 4 (R. Willette, Ed.), 1976, p. 19.
8. G. Gregoriadis, A. C. Allison, I. P. Braidman, G. Dapergolas and D. E. Neerunjun; in: *Conference on Liposomes and Their Uses in Biology and Medicine*, New York Academy of Sciences, September 14–16, 1977, New York.
9. S. C. Harvey; in *Remington's Pharmaceutical Sciences* (Oslo et al., Eds.), Mack, Easton, Pennsylvania, 1975, Chapter 63.
10. F. J. Kirchmey and H. C. Vincent; U.S. Patent 2,741,573, April 10, 1956.
11. S. S. Ober, H. C. Vincent, D. E. Simon and K. J. Frederick; J. Am. Pharm. Assoc., 47:667 (1958).
12. E. W. Thomas, R. H. Lyons, M. J. Romansky, C. R. Rein and D. K. Kitchen; J. Am. Med. Assoc., 135:1517 (1948).
13. J. Anschel; U.S. Patent 2,964,448, Dec. 13, 1960.
14. J. J. Lewis; Physiol. Rev., 29:75 (1949).
15. J. Koch-Weser; New Engl. J. Med., 291:233 (1974).
16. F. Lasch and E. Schonbrunner; Klin. Wochenschr., 17:1177 (1938).
17. H. M. Patel and B. E. Ryman; FEBS Lett., 62:60 (1976).
18. S. C. Harvey; in: *Remington's Pharmaceutical Sciences* (Oslo et al., Eds.) Mack, Easton, Pennsylvania, 1975, Chapter 51.
19. H. C. Hagedorn, B. N. Jensen, N. B. Krarup and I. Wodstrup; J. Am. Med. Assoc., 106:177 (1936).
20. D. A. Scott and A. M. Fisher; J. Pharm. Exp. Ther., 58:78 (1936).
21. J. Haleblian and W. McCrone; J. Pharm. Sci., 58:911 (1969).
22. K. Petersen, J. Schlichtkrull and K. Hallas-Moller; U.S. Patent 2,882,203, April 14, 1959.
23. K. Hallas-Moller, K. Petersen and J. Schlichtkrull; Science, 116:394 (1952).
24. K. Hallas-Moller, M. Tersild, K. Petersen and J. Schlichtkrull; J. Am. Med. Assoc., 150:1667 (1952).
25. J. V. Santiago, A. H. Clemens, W. L. Clarke and D. M. Kipnis; Diabetes, 28:71 (1979).
26. H. M. Creque, R. Langer and J. Folkman; Diabetes, 29:37 (1980).
27. R. E. Thompson and R. A. Hecht; Am. J. Clin. Nutrition, 7:311 (1959).
28. G. Sayers and R. H. Travis; in: *The Pharmacological Basis of Therapeutics* (Goodman and Gilman, Eds.), 4th ed., MacMillan, London, 1970, Chapter 72.

29. E. A. Lazo-Wasem and S. W. Hier; Proc. Soc. Exper. Biol. & Med., 90:380 (1955).
30. K. Miescher, A. Wettstein and E. Tschopp; Biochem. J., 30:1977 (1936).
31. K. C. James, P. J. Nicholls and M. Roberts; J. Pharm. Pharmacol., 21:24 (1969).
32. K. C. James; Experientia, 28:479 (1972).
33. R. Deanesly and A. S. Parkes; Biochem. J., 30:291 (1936).
34. C. W. Emmens; Endocrinology, 28:633 (1941).
35. C. W. Lloyd and J. Federicks; J. Clin. Endocrinol., 11:724 (1951).
36. A. C. Ott, M. H. Kuizenga, S. C. Lyster and B. A. Johnson; J. Clin. Endocrinol., 12:15 (1952).
37. H. S. Kupperman, S. G. Aronson, J. Gagliani, M. Parsonnet, M. Roberts, B. Silver and R. Postiglioni; Acta Endocrinol. (Copenh.), 16:101 (1954).
38. B. Baggett, L. L. Engel, K. Savard and R. I. Dorfman; J. Biol. Chem., 221:931 (1956).
39. B. Camerino and G. Sala; Prog. Drug Res., 2:71 (1960).
40. E. Diczfalusy; Acta Endocrinol. (Copenh.), 35:59 (1960).
41. E. Diczfalusy, O. Ferno, H. Fex and B. Hogberg; Acta Chem. Scand., 17:2536 (1963).
42. M. Kupchan, A. F. Casy and J. W. Swintosky; J. Pharm. Sci., 54:514 (1965).
43. J. De Visser and G. A. Overbeek; Acta Endocrinol. (Copenh.), 35:405 (1960).
44. Van der Vies; Acta Endocrinol. (Copenh.), 49:271 (1965).
45. R. T. Rapala, R. J. Kraay and K. Gerzon; J. Med. Chem., 8:580 (1965).
46. G. Pala, S. Casadio, A. Mantegani, G. Bonardi and G. Coppi; J. Med. Chem., 15:995 (1972).
47. M. A. Q. Chaudry and K. C. James; J. Med. Chem., 17:157 (1974).
48. C. Hansch, R. M. Muir, T. Fujita, P. O. Maloney, F. Geiger and M. Streitch; J. Am. Chem. Soc., 85:2817 (1963).
49. E. B. Astwood; in: *The Pharmacological Basis of Therapeutics*, (L. S. Goodman and A. Gilman, Eds.), 4th ed., MacMillan, London (1970), Chapter 69.
50. A. S. Parkes; Biochem. J., 31:579 (1937).
51. A. S. Parkes; J. Endocrinol., 3:288 (1943).
52. J. Ferrin; J. Clin. Endocrinol., 12:28 (1952).
53. K. G. Tillinger and A. Westman; Acta Endocrinol. (Copenh.), 25:113 (1957).
54. O. Ferno, H. Fex, B. Hogberg, T. Linderot, S. Veige and E. Diczfalusy; Acta Chem. Scand., 12:1675 (1958).
55. R. Gardi, R. Vitali, G. Falconi and A. Erloli; J. Med. Chem., 16:123 (1973).
56. S. C. Harvey; in: *Remington's Pharmaceutical Sciences* (Oslo et al., Eds.), 15th ed., Mack, Easton, Pennsylvania, 1975, Chapter 51.
57. A. G. Ebert and S. M. Hess; J. Pharmacol. Exp. Ther., 148:412 (1965).
58. J. Krincross and K. D. Charalampous; J. Neuropsychiatr., 1:66 (1965).
59. J. Mischinsky, K. Khazen and F. G. Sulman; Neuroendocrinology, 4:321 (1969).
60. J. Dreyfus, J. J. Ross, Jr. and E. C. Schreiber; J. Pharm. Sci., 60:829 (1971).
61. E. A. Swinyard; in: *Remington's Pharmaceutical Sciences* (Oslo et al., Eds.), 15th ed., Mack, Easton, Pennsylvania, 1975, Chapter 58.
62. A. Jorgensen, K. F. Overø and V. Hansen; *Acta Pharmacol. Toxicol.*, 29:339 (1971).
63. A. Villeneuve and P. Simon; J. Ther., 2:3 (1971).
64. A. Villeneuve, A. Pires, A. Jus, R. Lachance and A. Drolet; Curr. Ther. Res., 14:696 (1972).
65. L. Julou, G. Bourat, R. Ducrot, J. Fournel and C. Garrett; Acta Psychiatr. Scand., 241:9 (1973).
66. M. Nymark, K. F. Franck, V. Pedersen, V. Boeck and I. M. Nielsen; Acta Pharmacol. Toxicol., 33:363 (1973).
67. J. B. Thomsen and D. Birkerod; Acta Psychiatr. Scand., 49:119 (1973).

68. I. M. Rollo; in: *The Pharmacological Basis of Therapeutics* (Goodman and Gilman, Eds.), 4th ed., MacMillan, London, 1970, Chapter 52.
69. E. F. Elslager, D. B. Capps and D. F. Worth; J. Med. Chem., 12:597 (1969).
70. P. E. Thompson, B. J. Olszewski, E. F. Elslager and D. F. Worth; Am. J. Trop. Med. Hyg., 12:481 (1963).
71. L. H. Schmidt, R. N. Rossan and K. F. Fisher; Am. J. Trop. Med. Hyg., 12:494 (1963).
72. J. A. Waitz, B. J. Olszewski and P. E. Thompson; Science, 141:723 (1963).
73. A. J. Glazko, W. A. Dill, T. Chang, R. E. Ober, D. H. Kaump and A. Z. Lane; Proc. Fed. Am. Soc. Exp. Biol., 23:491 (1964).
74. W. Peters; in: *Recent Advances in Pharmacology* (Robinson and Stacey, Eds.), J. & A. Churchill, London, 1968, pp. 503–537.
75. J. H. Jaffe; in: *The Pharmacological Basis of Therapeutics* (L. S. Goodman and A. Gilman, Eds.) 4th ed., MacMillan, London, 1970, Chapter 15.
76. A. M. Freedman, M. Fink, R. Sharoff and A. Zaks; J. Am. Med. Assoc., 202:191 (1967).
77. A. M. Fre, M. Fink, R. Sharoff and A. Zaks; *Am. J. Psychiatr.*, 124:1499 (1968).
78. M. Fink; Science, 169:1005 (1970).
79. R. Resnick, M. Fink and A. M. Freedman; Am J. Psychiatr., 126:1256 (1970).
80. R. Resnick, M. Fink and A. M. Freeman; Compr. Psychiatry, 12:491 (1971).
81. T. M. Maugh, II; Science, 177:249 (1972).
82. A. Zaks, T. Jones, M. Fink and A. M. Freedman; J. Am. Med. Assoc., 215:2108 (1971).
83. F. Foldes, J. Lunn, J. Moore and I. Brown; Am. J. Med. Sci., 245:23 (1963).
84. M. S. Sadove, R. C. Balagot, S. Hatano and E. A. Jobgen; J. Am. Med. Assoc., 183:666 (1963).
85. A. Zaks, A. Bruner, F. Fink and A. Freedman; Dis. Nerv. Syst. (Suppl.), 30:89 (1969).
86. S. H. Weinstein, M. Pfeffer, J. M. Schor, L. Franklin, M. Mintz and E. R. Tutko; J. Pharm. Sci., 62:1416 (1973).
87. K. Oguri, S. Ida, H. Yoshimura and H. Tsukamoto; Chem. Pharm. Bull., 18:2414 (1970).
88. A. L. Misra, C. L. Mitchell and L. A. Woods; Nature (Lond.), 232:48 (1971).
89. J. M. Fujimoto; Proc. Soc. Exp. Biol. Med., 133:317 (1970).
90. S. H. Weinstein, M. Pfeffer, J. M. Schor, L. Indindoli and M. Mintz; J. Pharm. Sci., 60:1567 (1971).
91. R. J. Tallarida; in: *Factors Affecting the Action of Narcotics* (Adler et al., Eds.), Raven Press, New York, 1978.
92. H. W. Kosterlitz and A. J. Watt; Br. J. Pharmacol. Chemother., 33:266 (1968).
93. B. Berkowitz, S. Ngai, J. Hempstead and S. Spector; J. Pharmacol. Exp. Ther., 195:499 (1975).
94. S. Spector, S. H. Ngai, J. Hempstead and B. A. Berkowitz; in: *Factors Affecting the Action of Narcotics* (Adler et al., Eds.), Raven Press, New York, 1978.
95. H. N. Bhargava; J. Pharm. Pharmacol., 30:133 (1978).
96. C. Y. Lee, T. Akera, S. Stolman and T. M. Brody; J. Pharmacol. Exp. Ther., 194:583 (1975).
97. A. Archer; in: *Narcotic Antagonists* (Braude et al., Eds.), Raven Press, New York, 1974, p. 549.
98. R. Willette (Ed.); NIDA Research Monograph Series 4, 1976.
99. A. Osol and J. Hoover; in: *Remington's Pharmaceutical Sciences* (Oslo et al., Eds.), 15th ed., Mack, Easton, Pennsylvania, 1975, Chapter 63.
100. A. P. Gray and W. J. Guardina; in: NIDA Research Monograph 4 (R. Willette, Ed.), 1976, p. 21.

101. L. S. Harris and A. K. Pierson; J. Pharmacol. Exp. Ther., 143:141 (1964).
102. L. S. Harris, W. L. Dewey, J. F. Howes, J. S. Kennedy and H. Pars; J. Pharmacol. Exp. Ther., 169:17 (1969).
103. L. Lachman et al.; U.S. Patent 3,676,557, July 11, 1192.
104. H. Blumberg and H. B. Dayton; in: *Narcotic Antagonists* (Braude et al., Eds.), Raven Press, New York, 1974, p. 33.
105. S. Yolles; U.S. Patent 3,887,699, Dec. 29, 1970.
106. H. R. Woodland, S. Yolles, D. A. Blake, M. Helrich and F. J. Meyer; J. Med. Chem., 16:897 (1973).
107. S. Yolles, J. Eldridge, T. Leafe, J. H. R. Woodland, D. R. Blake and F. Meyer; in: *Controlled Release of Biologically Active Agents* (Tranquary and Lacey, Eds.), Plenum Press, New York, 1974.
108. T. D. Leafe, S. F. Sarner, J. H. R. Woodland, S. Yolles, D. A. Blake and F. J. Meyer; in: *Narcotic Antagonists* (Braude et al., Eds.), Raven Press, New York, 1974, p. 569.
109. S. Yolles, T. D. Leafe, J. H. R. Woodland and F. J. Meyer; J. Pharm. Sci. 64:348 (1975).
110. A. D. Schwope, D. L. Wise and J. F. Howes; in: NIDA Research Monograph Series 4 (R. Willett, Ed.), 1976, p. 13.
111. C. Theis; in: *Controlled Release Polymeric Formulations* (D. R. Paul and F. W. Harris, Eds.), ACS Symposium Series, 33:190 (1976).
112. I. Siegel; Obstet. Gynecol., 21:666 (1963).
113. H. T. Felton, E. W. Hoelscher and D. P. Swartz; Fertil. Steril., 16:665 (1965).
114. E. M. Coutinho, J. C. DeSouza and A. I. Csapo; Fertil. Steril., 17:261 (1966).
115. T. H. Rizkallah and M. L. Taymor; Am. J. Obstet. Gynecol., 94:161 (1966).
116. J. Zanartu, E. Rice-Wray and J. W. Goldzieher; Obstet. Gynecol., 28:513 (1966).
117. M. Kau; S.A. Nurs. J., 41:31 (1974).
118. D. Mischell, A. Parlow, M. Tolas and M. El-Hobasky; Proc. World Cong. Fertil. Steril., 6th Congress, Tel Aviv, Israel, 1970, p. 203.
119. J. Goldzieher, J. Kleber and L. Moses; Contraception, 2:225 (1970).
120. P. C. Schwallie; J. Reprod. Med., 13:113 (1974).
121. J. Zanartu; Fertil. Steril., 21:525 (1970).
122. J. C. Cornette, K. T. Kirton and G. W. Duncan; J. Clin. Endocrinol. Metab., 33:459 (1971).
123. F. D. Scutchfield, W. N. Long, B. Corey and C. W. Tyler, Jr.: Contraception, 3:21 (1971).
124. P. C. Schwallie and J. R. Assenzo; Fertil. Steril., 24:331 (1973).
125. L. C. Powell, Jr. and R. J. Seymour; Am. J. Obstet. Gynecol., 110:36 (1971).
126. E. V. Mackay, S. K. Khoo and R. R. Adam; Aust. N.Z. J. Obstet. Gynecol., 11:148 (1971).
127. D. Mishell, I. Thorneycroft, K. Kharina and R. Nakamura; Am. J. Obstet. Gynecol., 113:372 (1972).
128. S. Jeppsson, E. D. B. Johansson and N. Sjoberg; Contraception, 8:165 (1973).
129. E. M. Coutinho and J. C. De Souza; J. Reprod. Fertil., 15:209 (1968).
130. T. H. Rizkallah and M. L. Taymor; Am. J. Obstet. Gynecol., 95:249 (1966).
131. M. L. Taymor and M. A. Yussman; in: *Advances in Planned Parenthood*, Proceedings of the 6th Annual Meeting of the American Assoc. of Planned Parenthood Physicians (A. J. Sobrero and S. Lewit, Eds.), Excerpta Medica, Amsterdam, 1969, p. 132.
132. G. M. Herzog and S. D. Soule; Obstet. Gynecol., 32:111 (1968).
133. J. D. Gresser, D. L. Wise, L. R. Beck and J. F. Howes; Contraception, 17:253 (1978).
134. J. Zanartu and C. Navarro; Obstet. Gynecol., 31:627 (1968).

135. G. L. Escobar, H. Willomitze, J. R. Sanchez and J. D. Castillo; Rev. Soc. Colomb. Endocrinol., 8:1 (1970).
136. E. Kesseru, A. Larranage, H. Hurtudo and G. Benarides; Int. J. Fertil., 17:17 (1972).
137. M. Humpel, G. Kuhne, P. E. Schulze and U. Speck; Contraception, 15:401 (1977).
138. J. Spona, E. Weiner, B. Nieuweboer, M. Humpel, W. H. F. Schneider and E. D. B. Johansson; Contraception, 15:413 (1977).
139. L. G. Miller and J. H. Fincher; J. Pharm. Sci., 60:1733 (1971).
140. S. Feldman; Bull. Parent. Drug Assoc., 28:53 (1974).
141. B. E. Ballard and E. Nelson; in: *Remington's Pharmaceutical Sciences* (Oslo and Hoover et al., Eds.), 15th ed., Mack, Easton, Pennsylvania, 1975, Chapter 91.
142. W. E. Brown, V. M. Wilder and P. Schwartz; J. Lab. Clin. Med., 29:259 (1944).
143. D. W. Newton and M. Newton; J. Am. Pharm. Assoc., NS 17:685 (1977).
144. M. Pitel; Am. J. Nurs., 71:76 (1971).
145. D. H. Geolot and N. P. McKinney; Am. J. Nurs., 75:788 (1975).
146. S. W. Jacob and C. A. Francone; *Structure and Function in Man*, 2nd ed., W. B. Saunders, Philadelphia, 1970, Chapter 17.
147. D. J. Hanson; G.P., 27:109 (1963).
148. E. W. Martin, S. F. Alexander, W. E. Hassan, Jr. and B. S. Sherman; *Techniques of Medication*, J. B. Lippincott, Philadelphia, 1969, p. 108.
149. F. J. Ayd, Jr.; Int. Drug. Ther. Newslett., 12:1 (1977).
150. M. Pfeffer, R. D. Smyth, K. A. Pittman and P. A. Nardella; J. Pharm. Sci., 69:801 (1980).
151. R. Deanesly and A. S. Parkes; Comparative activities of compounds of the androsterone-testosterone series. Biochem. J., 30:291 (1936).
152. R. Deanesly and A. S. Parkes; Factors influencing the effectiveness of administered hormones. Proc. R. Soc. Lond. [Biol.], 124:279 (1937).
153. B. E. Ballard and E. Nelson; Prolonged-action pharmaceuticals. in: *Remington's Pharmaceutical Sciences* (A. Oslo et al., Eds.), 15th ed., Mack, Easton, Pennsylvania, 1975, Chapter 91.
154. M. B. Shimkin and J. White; Absorption rate of hormone-cholesterol pellets. Endocrinology, 29:1020 (1941).
155. F. Fuenzalida; Absorption of steroids from subcutaneously implanted tablets of the pure hormone and of the hormone mixed with cholesterol. J. Clin. Endocrinol., 10:1511 (1950).
156. A. F. Pi, A. Oriol, L. H. Lasso, M. Maqueo, R. I. Dorfman and F. A. Kincl; Inhibition of fertility in mice by steroid implants. Acta Endocrinol. (Copenh.), 48:602 (1965).
157. *Physicians' Desk Reference*, 33rd ed., Medical Economics Co., Oradell, New Jersey, 1979.
158. *Physician' Desk Reference*, 43rd ed., Medical Economics Co., Oradell, New Jersey, 1989.
159. E. L. Way, H. H. Loh and F. H. Shen; Simultaneous quantitative assessment of morphine tolerance and physical dependence. J. Pharmacol. Exp. Ther., 167:1 (1969).
160. F. H. Shen, H. H. Loh and E. L. Way; Brain serotonin turnover in morphine-tolerant and dependent mice. J. Pharmacol. Exp. Ther., 175:427–434 (1970).
161. S. Algeri and E. Costa; Physical dependence on morphine fails to increase serotonin turnover rate in rat brain. Biochem. Pharmacol., 20:877 (1971).
162. I. Marchall and D. G. Grahame-Smith; Evidence against a role of brain 5-hydroxytryptamine in the development of physical dependence upon morphine in mice. J. Pharmacol. Exp. Ther., 179:634–641 (1971).
163. J. Blasig, A. Herz, K. Reinhold and S. Zieglgansberger; Development of physical

dependence of morphine in respect to time and dosage and quantification of the precipitated withdrawal syndrome in rats. Psychopharmacologia, 33:19–38 (1973).
164. T. J. Cicero and E. R. Meyer; Morphine pellet implantation in rats: Quantitative assessment of tolerance & dependence. J. Pharmacol. Exp. Ther., 184:404–408 (1973).
165. E. Wei, H. H. Loh and E. L. Way; Quantitative aspects of precipitated abstinence in morphine dependent rats. J. Pharmacol. Exp. Ther., 184:398–403 (1973).
166. K. S. Hui and M. B. Roberts; An improved implantation pellet for rapid induction of morphine dependence in mice. J. Pharm. Pharmacol., 27:569 (1975).
167. H. N. Bhargava; Rapid induction and quantitation of morphine dependence in the rat by pellet implantation. Psychopharmacology, 52:55–62 (1977).
168. H. N. Bhargava; Quantitation of morphine tolerance induced by pellet implantation in the rat. J. Pharm. Pharmacol., 30:133 (1978).
169. J. Folkman; Controlled drug release from polymers. Hosp. Pract., 13:127–133 (1978).
170. J. Folkman and D. M. Long, Jr.; The use of silicone rubber as a carrier for prolonged drug therapy. J. Surg. Res., 4:139–142 (1964).
171. J. Folkman and D. M. Long, Jr.; Drug pacemakers in the treatment of heart block. Ann. N.Y. Acad. Sci., 8:857 (1964).
172. P. Bass, R. A. Purdon and J. N. Wiley; Prolonged administration of atropine or histamine in a silicone rubber implant. Nature, 208:591 (1965).
173. K. G. Powers; The use of silicone rubber implants for the sustained release of antimalarial and antischistosonal agents. J. Parasitol., 51:53 (1965).
174. P. J. Dziuk and B. Cook; Passage of steroids through silicone rubber. Endocrinology, 78:208–211 (1966).
175. F. A. Kincl, G. Benagiano and I. Anges; Sustained release hormonal preparation. I. Diffusion of various steroids through polymer membranes. Steroids, 11:673 (1968).
176. C. M. Clifford, C. E. Yunker and M. D. Corwin; Control of the louse *Polyplax serrata* with systemic insecticides administered in silastic rubber implants. J. Econ. Entomol., 60:1210 (1967).
177. J. Folkman, W. Reiling and G. Williams; Chronic analgesia by silicone rubber diffusion. Surgery, 66:194 (1969).
178. T. S. Gaginella and P. Pass; Nicotine: Release from silicone rubber implants in vivo. Res. Commun. Chem. Pathol. Pharmacol., 7:213 (1974).
179. J. Folkman, D. M. Long, Jr. and R. Rosembaum; A new diffusion property useful for general anesthesia. Science, 154:148 (1966).
180. J. Folkman and V. H. Mark; Diffusion of anesthetics and other drugs through silicone rubber: Therapeutic implications. Trans. N.Y. Acad. Sci., 30:1187 (1968).
181. P. Siegel and J. R. Atkinson; In vivo chemode diffusion of L-dopa. J. Appl. Physiol., 30:900 (1971).
182. J. Posti; Design and clinical properties of the contraceptive subdermal implant-Norplant. Sci. Technol. Pract. Pharm., 34:309–312 (1987).
183. D. N. Robertson; in: *Long-Acting Steroid Contraception* (D. R. Mishell, Jr., Ed.), Raven Press, New York, 1983, pp. 127–147.
184. D. N. Robertson; Implantable levonorgestrel rod systems: In vivo release rates and clinical effects, in: *Long-Acting Contraceptive Delivery Systems* (G. L. Zatuchni et al., Eds.), Harper & Row, Philadelphia, 1984, pp. 1–19.
185. C. W. Bardin; Long-acting steroidal contraception: An update. Int. J. Fertil., 34:88–95 (1989).
186. I. Sivin; International experience with Norplant and Norplant-2 contraceptives. Studies Fam. Planning, 19:81–94 (1988).
187. A. P. Sam; Controlled-release contraceptive devices, in: Minutes of 5th Int. Pharm. Techn. Symp., *New Approaches to the Controlled Drug Delivery* (A. Hincal et al., Eds.), Edition de Santé, Paris (1990).

188. D. S. T. Hsieh, N. Smith and Y. W. Chien; Subcutaneous controlled delivery of estradiol by Compudose implants: In vitro and in vivo evaluations. Drug Develop. & Ind. Pharm., 13:2651–2666 (1987).
189. Y. W. Chien and E. P. K. Lau; Controlled drug release from polymeric delivery devices (IV): In vitro-in vivo correlation on the subcutaneous release of Norgestomet from hydrophilic implants. J. Pharm. Sci., 65:488 (1976).
190. Y. W. Chien and H. J. Lambert; Method for making a microsealed delivery device. U.S. Patent 3,992,518, Nov. 16, 1976.
191. Y. W. Chien and H. J. Lambert; Microsealed pharmaceutical delivery devices. U.S. Patent 4,053,580, Oct. 11, 1977.
192. Y. W. Chien; Microsealed drug delivery systems: Theoretical aspects and biomedical assessments, in: *Recent Advances in Drug Delivery Systems* (J. M. Anderson & S. W. Kim, Eds.), Plenum, New York, 1984, pp. 367–387.
193. Y. W. Chien; Microsealed drug delivery systems: Methods of fabrication. *Methods Enzymol.*, 112:Chapter 34, (1985).
194. F. Theeuwes; Elementary osmotic pump. J. Pharm. Sci., 64:1987 (1975).
195. F. Theeuwes and S. I. Yum; Principles of the design and operation of generic osmotic pumps for the delivery of semisolid or liquid drug formulations. Ann. Biomed. Eng., 4:343 (1976).
196. B. Sikic, J. Collins, E. Mimnaugh and T. Gram; Cancer Treat. Rep., 62:2011 (1978).
197. F. Theeuwes; in: *Controlled Release Technologies*, (A. F. Kydonieus, Ed.), CRC Press, Boca Raton, FL, 1980, Chapter 10.
198. S. Yum and R. Wright; in: *Controlled Drug Delivery* (S. D. Bruck, Ed.), Vol. II, CRC Press, Boca Raton, FL, 1983, Chapter 3.
199. Y. W. Chien; Implantable controlled-release drug delivery systems. *Novel Drug Delivery Systems: Fundamentals, Developmental Concepts and Biomedical Assessments*, Dekker, New York, 1982, Chapter 7.
200. P. J. Blackshear, T. D. Rohde, J. C. Grotling, F. D. Dorman, P. R. Perkins, R. L. Varco and H. Buchwald; Control of blood glucose in experimental diabetes by means of a totally implantable insulin infusion device. Diabetes, 28:634 (1979).
201. American Pharmacy; Implantable pump for morphine. NS24, (8):20 (1984).
202. P. J. Blackshear, T. D. Rohde, R. L. Varco and H. Buchwald; One year of continuous heparinization in the dog using a totally implantable infusion pump. Surg. Gynecol. Obstet., 141:176 (1975).
203. D. S. T. Hsieh and R. Langer; Zero-order drug delivery systems with magnetic control, in: *Controlled Release Delivery Systems* (T. J. Roseman and S. Z. Mansdorf, Eds.), Dekker, New York, 1983, Chapter 7.
204. Y. W. Chien; *Novel Drug Delivery Systems*, Dekker, New York, 1982, pp. 429–450.
205. Y. W. Chien and E. P. K. Lau; Controlled release from polymeric delivery devices (IV): In vitro-in vivo correlation on the subcutaneous release of Norgestomet from Hydrophilic Implants, J. Pharm. Sci., 68:689 (1979).
206. A. D. Schwope and D. L. Wise; Development of drug delivery systems for use in treatment of narcotic addiction, in: Dynatech Quarterly Report (No. 1361 to NIDA), January 16, 1976.
207. D. L. Wise, T. D. Fellmann, J. E. Sanderson and R. L. Wentworth; Lactic/glycolic and polymers, in: *Drug Carriers in Biology and Medicine* (G. Gregoriadis, Ed.), Academic Press, New York, 1979, Chapter 12.
208. J. Heller; Controlled drug release from poly(ortho esters), in: *Proceedings of the 11th International Symposium on Controlled Release Bioactive Materials* (W. E. Meyers and R. L. Dunn, Eds.), The Controlled Release Society, Inc., Lincolnshire, IL, 1984, p. 128.

209. J. Posti; Design and clinical properties of the contraceptive subdermal implant, Norplant. Sci. Technol. Pract. Pharm., 34:309–312 (1987).
210. D. N. Robertson; in: *Long-Acting Steroid Contraception* (D. R. Mishell, Jr., Ed.), Raven Press, New York, 1983, pp. 127–147.
211. C. W. Bardin; Long-acting steroid contraception: An update. Int. J. Fertil., 34:88–95 (1989).
212. I. Sivin et al.; A four-year clinical study of Norplant implants. Studies Fam. Planning, 14:184 (1983).
213. M. M. Shaaban et al.; A prospective study of Norplant implants and the TCu 380 Ag IUD in Assint, Egypt. Studies Fam. Planning, 14:163 (1983).
214. S. Satayapan et al.; Perceptions and acceptability of Norplant implants in Thailand. Studies Fam. Planning, 14:170 (1983).
215. P. Marangoni et al.; Norplant implants and the TCu 200 IUD: A comparative study in Ecuador. Studies Fam. Planning, 14:177 (1983).
216. F. Lubis et al.; One-year experience with Norplant implants in Indonesia. Studies Fam. Planning, 14:181 (1983).
217. S. A. Pasquale and S. J. Kelly; personal communication, 1986.
218. M. J. D. Eenink, G. C. T. Maassen, A. P. Sam, J. A. A. Geelen, J. B. J. M. Lieshout, J. Olijslager, H. de Nijs and E. de Jager; Development of a new long-acting contraceptive subdermal implant releasing 3-keto-desogestrel. Proc. Int. Symp. Control. Rel. Bioact. Mater., 15:402–403 (1988).
219. T. D. Rohde, P. J. Blackshear, R. L. Varco and H. Bachwald; Trans. Am. Soc. Artif. Intern. Organs, 23:13–16 (1977).
220. P. J. Blackshear, F. D. Dorman, P. L. Blackshear, Jr., R. L. Varco and H. Buchwald; Surg. Gynecol. Obstet., 134:51 (1972).
221. P. R. Perkins, F. D. Dorman, T. D. Rohde, P. J. Blackshear, P. L. Blackshear, Jr., R. L. Varco and H. Buchwald; Trans. Am. Soc. Artif. Intern. Organs, 24:229–31 (1978).
222. T. D. Rohde, P. J. Blackshear, R. L. Varco and H. Buchwald; Minn. Med., 60:719–722 (1977).
223. P. J. Blackshear, T. D. Rohde, R. L. Varco and H. Buchwald; Surg. Gynecol. Obstet., 141:176–186 (1975).
224. T. D. Rohde, P. J. Blackshear, R. L. Varco and H. Buchwald; Trans. Am. Soc. Artif. Intern. Organs, 21:510–514 (1975).
225. C. C. Chang and F. A. Kincl; Sustained release hormonal preparations. 3. Biological effectiveness of 6-methyl-17α-acetoxypregna-4,6-diene-3,20-dione. Steroids, 12:689–696 (1968).
226. J. Bojsen, T. Deckert, K. Kolendorf and B. Lorup; Diabetes, 28:974 (1979).
227. P. J. Blackshear, T. D. Rohde, J. C. Grotling, F. D. Dorman, P. R. Perkins, R. L. Varco and H. Buchwald; Diabetes, 28:634 (1979).
228. R. L. Jackson, W. O. Storvick, C. S. Hollinder, L. E. Stroeh and J. G. Stilz; Diabetes, 21:235 (1972).
229. F. G. Hutchinson and B. J. A. Furr; Biodegradable polymer systems for the sustained release of polypeptides. J. Control. Rel., 13:279 (1990).
230. O. Siddiqui and Y. W. Chien; Non-parenteral administration of peptides and protein drugs. CRC Crit. Rev. Ther. Drug Carrier Systems, 3:195–208 (1987).
231. A. K. Banga and Y. W. Chien; Systemic delivery of therapeutic peptides and proteins. Int. J. Pharm., 48:15 (1988).
232. Y. W. Chien, K. S. E. Su and S. F. Chang; *Nasal Systemic Drug Delivery*, Dekker, New York, 1989, Chapter 4.
233. S. T. Anik, L. M. Sanders, M. D. Chaplin, S. Kushinsky and C. Nerenbeg; Delivery

systems of LHRH and analogues, in: *LHRH and Its Analogues: Contraceptive and Therapeutic Applications* (B. H. Vickery, J. J. Nestor, Jr. and E. S. E. Hafez, Eds.), MTP Press, Boston, 1984, pp. 421–435.

234. W. Petri, R. Seidel and J. Sandow; Pharmaceutical approaches to long-term therapy with peptides. Int. Cong. Ser.-Excerpta Med., 665:63 (1984).

235. R. Anders, H. P. Merkel, W. Schurr and R. Ziegler; Buccal absorption of protirelin: An effective way to stimulate thyrotropin and prolactin. J. Pharm. Sci., 72:1481 (1983).

236. H. Okada, I. Yamazaki, Y. Ogawa, S. Hirai, T. Yashiki and H. Mima; Vaginal absorption of a potent luteinizing hormone-releasing hormone analog (leuprolide) in rats. I. Absorption by various routes and absorption enhancement. J. Pharm. Sci., 71:1367 (1982).

237. H. Okada, I. Yamazaki, T. Yashiki and H. Mima; Vaginal absorption of a potent luteinizing hormone-releasing hormone analog (leuprolide) in rats. II. Mechanism of absorption enhancement with organic acids. J. Pharm. Sci., 72:75 (1983).

238. H. Okada, T. Yashiki and H. Mima; Vaginal absorption of a potent luteinizing hormone-releasing hormone analog (leuprolide) in rats. III. Effect of estrous cycle on vaginal absorption of hydrophilic model compounds. J. Pharm. Sci., 72:173 (1983).

239. H. Oskada, I. Yamazaki, T. Yashiki, T. Shimamoto and H. Mima; Vaginal absorption of a potent luteinizing hormone-releasing hormone analog (leuprolide) in rats. IV. Evaluation of the vaginal absorption and gonadotropin response by radioimmunoassay. J. Pharm. Sci., 73:298 (1982).

240. J. Heller; Biodegradable polymers in controlled drug delivery. CRC Crit. Rev. Ther. Drug Carrier Systems, 1:39–90 (1984).

241. D. L. Wise, T. D. Fellman, J. E. Sanderson and R. L. Wentworth; Lactic/glycolic acid polymers, in: *Drug Carriers in Biology and Medicine* (G. Gregoriadis, Ed.), Academic Press, London, 1979, pp. 237–270.

242. J. R. Churchill and F. G. Hutchinson; Continuous release formulations. U.S. Patent 4,526,938, July 2, 1985.

243. L. Bohn; Compatible polymers, in: *Polymer Handbook*, 2nd ed., Vol. 3 (J. Brandrup and E. H. Immergut, Eds.), Wiley, New York, 1975, p. 211.

244. G. Williams, D. J. Kerle, S. M. Roe, T. Yeo and S. R. Bloom; Results obtained in the treatment of prostate cancer patients with "Zoladex" in: EORTC Genitourinary Group Monograph 2, Part A, *Therapeutic Principles in Metastatic Prostatic Cancer* (F. H. Schroeder and B. Richards, Eds.), Alan R. Liss, New York, 1985, pp. 287–295.

245. J. B. F. Grant, S. R. Ahmed, S. M. Shalet, C. B. Costello, A. Howell and N. J. Blacklock; Testosterone and gonadotrophin profiles in patients on daily or monthly LHRH analogue ICI 118, 630 (Zoladex) compared with orchiectomy. Br. J. Urol., 58:539 (1986).

246. B. J. A. Furr and R. I. Nicholson; Use of analogues of LHRH for treatment of cancer. J. Reprod. Fertil., 64:529 (1982).

247. F. G. Hutchinson and B. J. A. Furr; Biochem. Soc. Trans., 13:520 (1985).

248. B. J. A. Furr and F. G. Hutchinson; in: *EORTC Genitourinary Group Monograph*, 2, Part A (F. H. Schroder and B. Richards, Eds.), Alan R. Liss, New York, 1985, p. 143.

249. K. J. Walker et al.; J. Endocrinol., 103:R1 (1984).

250. A. O. Turkes, W. B. Peeling and K. Griffiths; J. Steroid Biochem., (1987).

251. R. J. Donnelly and R. A. V. Milsted; in: *LHRH and Its Analogues: Contraception and Therapeutic Applications* (B. H. Vickery and J. J. Nestor, Eds.), Part 2, MTP Press, Lancaster (1991).

252. D. F. Wishart; Vet. Rec., 90:595 (1972).

253. D. F. Wishart and J. B. Snowball; Vet. Rec., 92:139 (1973).
254. D. F. Wishart and I. M. Young; Vet. Rec., 95:503 (1974).
255. R. W. Whitman, J. N. Wiltbank, D. G. Lefeuer and A. H. Denham; Proc. West. Sec. Am. Soc. Anim. Sci., 23:280 (1972).
256. C. Burrell, J. N. Wiltbank, G. Lefever and G. Rodeffer; Proc. West. Sec. Am. Soc. Anim. Sci., 23:547 (1972).
257. M. Lemon; Ann. Biol. Anim. Biochem. Biophys., 15:243 (1975).
258. J. N. Wiltbank and E. Gonzales-Padilla; Ann. Biol. Anim. Biochem. Biophys., 15:255 (1975).
259. D. F. Wishart; Ann. Biol. Anim. Biochem. Biophys., 15:215 (1975).
260. S. E. Mares; in: *Proc. of 5th Int. Symp. on Controlled Release of Bioactive Materials*, National Bureau of Standards, Gaithersburg, MD, August 14–16, 1978.
261. Courtesy of S. E. Mares; Searle Veterinary Research & Development (1981).
262. P. Brunaud; in: *Maitrise des Cycles Sexuels Chez les Bovins*, Janvier 12-13, 1976, Sersia, Paris, p. 51.
263. D. Chupin and D. Aguer: *Maitrise des Cycles Sexuels Chez les Bovins*, Janvier 12-13, 1976, Sersia, Paris, p. 69.
264. J. N. Wiltbank and S. E. Mares; *Proc. 11th Conf. Artif. Insem. Beef Cattle*, Denver, 1977.
265. D. F. Wishart, I. M. Young and S. B. Drew; Vet. Rec., 100:417 (1977).
266. J. F. Roche; World Rev. Anim. Prod., XII:79 (1976).
267. D. F. Wishart, I. M. Young and S. B. Drew; Vet. Rec., 101:230 (1977).
268. J. N. Wiltbank; *Proc. 11th Conf. Artif. Insem. Beef Cattle*, Denver, 1977, p. 57.
269. P. Mauleon, D. Chupin, J. Pelot and D. Aguer; Reunion C.E.E., Galway, September 1977.
270. D. Chupin, F. Deletang, M. Petit, J. Pelot, F. LeProvost, R. Ortavant and P. Mauleon; Ann. Biol. Anim. Biochem. Biophys., 14:27 (1974).
271. D. Chupin, J. Pelot and M. Petit; Bull. Tech. Insem. Artif., 5:2 (1977).
272. D. Chupin, J. Pelot, M. Alonso de Miguel and J. Thimonier: *VIIIth Int. Cong. Anim. Reprod. Artif. Insem.*, Krakow, 1976, p. 346.
273. Y. W. Chien, L. F. Rozek and H. J. Lambert; J. Pharm. Sci., 67:214 (1978).
274. A. Cuadros, A. Brinson and K. Sundaram; Progestational activity of megestrol acetate-polydimethylsiloxane (PDS) capsules in rhesus monkeys. Contraception, 2:29 (1970).
275. C. C. Chang and F. A. Kincl; Sustained release hormonal preparations. 4. Biological effectiveness of steroid hormones. Fertil. Steril., 21:134 (1970).
276. V. Zbuzkova and F. A. Kincl; Sustained release hormonal preparations. 11. Biological effectiveness. Endocrinol. Exp., 4:215 (1970).
277. F. A. Kincl, H. J. Ringold and R. J. Dorfman; Pituitary gonadotrophin inhibition by subcutaneously administered steroids. Acta Endocrinol. (Copenh.), 36:83 (1961).
278. F. A. Kincl and R. I. Dorfman; Pituitary gonadotrophin in inhibitory action of neutral steroids. Acta Endocrinol. (Copenh.), 46:300 (1964).
279. C. R. Moore and D. Price; Some effects of testosterone and testosterone propionate in the rat. Anat. Rec., 71:59 (1938).
280. F. A. Kincl, I. Angee, C. C. Chang and H. W. Rudel; Sustained release hormonal preparations. 9. Plasma levels and accumulation into various tissues of 6-methyl-17α-acetoxy-4,6-pregnadiene-3,20-dione after oral administration or absorption from polydimethylsiloxane implants. Acta Endocrinol. (Copenh.), 64:508 (1970).
281. G. H. Bell, J. N. Davison and H. Scarborough; *Textbook of Physiology and Biochemistry*, 5th ed., Livingston, London, 1963, Chapter 35.
282. B. Bloch and G. W. Hastings; *Plastics Materials in Surgery*, 2nd ed., Charles C. Thomas, Springfield, IL, 1972, Chapter VI.

283. V. Menken; *Biochemical Mechanisms in Inflammation*, Charles C. Thomas, Springfield, IL, 1956.
284. E. M. Coutinho, D. A. M. Ferreira, H. Prates and F. A. Kincl; Excretion of [6-^{14}C]megestrel acetate (6-methyl-17-acetoxypregna-4,6-diene-3,20-dione) released from subcutaneous silastic implants in women. J. Reprod. Fertil., 23:345 (1970).
285. G. Benagiano, M. Ermini, L. Carenza and G. Rolfini; Studies on sustained contraceptive effects with subcutaneous polydimethylsiloxone implants. 1. Diffusion of megestrol acetate in humans. Acta Endocrinol (Copenh.), 73:335 (1973).
286. G. Benagiano and M. Ermini; Continuous steroid treatment by subdermal polysiloxane implants. Acta Eur. Fertil., 3:119 (1972).
287. V. Mirkovitch; Bioelectric phenomena, thrombosis and plastics: A review of current knowledge. Cleveland Clin. Q., 30:241 (1963).
288. V. Mirkovitch, R. E. Beck and P. G. Andrus; The zeta potentials and blood compatibility characteristics of some selected solids. J. Surg. Res., 4:395 (1964).
289. I. Folkman and D. M. Long, Jr.; The use of silicone rubber as a carrier for prolonged drug therapy. J. Surg. Res., 4:139 (1964).
290. J. Folkman and D. M. Long, Jr.; Drug pacemakers in the treatment of heart block. Ann. N.Y. Acad. Sci., 111:857 (1963–1964).
291. R. Schuhman and H. D. Taubert; Long-term application of steroids enclosed in dimethyl-polysilioxane (Silastic): In vitro and in vivo experiments. Acta Biol. Med. Germ., 24:897 (1970).
292. R. H. Hall; Acute toxicity of inhaled beryllium: Observations correlating toxicity with physicochemical properties of beryllium oxide dust. Arch. Ind. Hyg., 2:25 (1950).
293. H. LeVeen and J. R. Barberio; Tissue reaction to plastics and in surgery with special reference to Teflon. Ann. Surg., 12:974 (1949).
294. F. C. Usher and S. A. Wallace; Tissue reaction to plastics: A comparison of nylon, orlon, dacron, Teflon and Marlex. Am. Med. Assoc. Arch. Surg., 76:997 (1958).
295. N. E. Stinson; Tissue reaction induced in guinea pigs by particulate polymethylacrylate, polyethylene and nylon of the same size range. Br. J. Exp. Pathol., 46:135 (1965).
296. C. Tietze; Evaluation of intrauterine devices, 9th Progress Report of the Cooperative Statistical Program. Studies Fam. Planning, 55:1 (July, 1970).
297. L. L. Doyle; Hormone-releasing silicone-rubber intrauterine contraceptive devices: Effect of incorporation of various compounds on intrauterine devices in rats. Am. J. Obstet. Gynecol., 121:405 (1975).
298. L. L. Doyle and T. H. Clewe; Preliminary studies on the effect of hormone-releasing intrauterine devices. Am. J. Obstet. Gynecol., 101:564 (1968).
299. C. C. Bollinger; Bacterial flora of the nonpregnant uterus: A new culture technique. Obstet. Gynecol., 23:251 (1964).
300. M. I. Buchman; A study of the intrauterine contraceptive device with and without an extracervical appendage of tail. Fertil. Steril., 21:348 (1970).
301. H. Lehfeldt; Intrauterine contraception: Tailed vs. tailless devices. Extrail du Livre Jubilaire Offert Au Dr. Jean Dalsace, Masson et Cie, Paris, 1966, pp. 199–213.
302. J. R. Wilson, C. C. Bollinger and W. J. Ledger; The effect of an intrauterine contraceptive device on the bacterial flora of the endometrial cavity. Am. J. Obstet. Gynecol., 90:726 (1964).
303. H. J. Tatum, F. H. Schmidt, D. Phillips, M. McCarty and W. M. O'Leary; The Dalkon Shield controversy: Structural and bacteriological studies of IUD tails. J. Am. Med. Assoc., 231:711 (1975).
304. J. Folkman, W. Reiling and G. Williams; Chronic analgesia by silicone rubber diffusion. Surgery, 66:194 (1969).

305. B. S. Oppenheimer; Further studies of polymers as carcinogenic agents in animals. Cancer Res., 15:333 (1955).
306. Bull. Dow Corning Centre Med. Res., 5:7 (1963).
307. P. J. Dzuik and B. Cook; Passage of steroids through silicone rubber. Endocrinology, 78:208 (1966).

9
Vaginal Drug Delivery and Delivery Systems

I. HISTORICAL ASPECTS OF DEVELOPMENT

It has been known for several decades that various pharmacologically active agents, such as steroids, may be effectively absorbed through the vaginal mucosa (1). The feasibility of vaginal absorption was first demonstrated experimentally by the intravaginal administration of progesterone via a drug-impregnated suppository formulation (2). The orally-inactive progesterone was found active when administered intravaginally.

Using the vagina as the route of administration for contraceptive steroids has several advantages. Among these the most practical is that a drug-releasing vaginal device allows insertion and removal by the user and provides continuous administration of an effective dose level, thus, ensuring better patient compliance. This route of administration materialized in 1970 with the development of a medicated, resilient vaginal ring (Figure 1). This was fabricated from a biocompatible silicone elastomer to contain medroxyprogesterone acetate for intravaginal contraception (3). A clinical study was initiated to determine whether an effective contraceptive dose of medroxyprogesterone acetate could be safely and continuously administrated to normally ovulating women from this vaginal ring, which was positioned in the vagina around the cervix for weeks (4). The results showed that serum drug levels resulting from the absorption of medroxyprogesterone acetate through the vaginal mucosa are sufficiently high to inhibit ovulation as well as to raise basal temperature; the histological pattern of the endometrium is also altered. Upon removal of the vaginal ring absorption of the progestin ceased, withdrawal bleeding occurred, and ovulation resumed promptly.

In subsequent years a series of clinical trials with different designs and sizes of vaginal rings containing various doses of medroxyprogesterone acetate (5,6), chlormadinone acetate (7), norethindrone (8), gestrinone (9,10), and norgestrel (8,11–14) were undertaken in an attempt to develop a contraceptive vaginal ring that would consistently inhibit ovulation and breakthrough bleeding during its 3-week intravaginal insertion and allow withdrawal bleeding during the week it was not in use. These investigations repeatedly demonstrated that the progestin-releasing vaginal ring

Figure 1 Medicated, resilient vaginal rings fabricated from silicone elastomer, like Silastic 382 medical-grade elastomer.

did not cause any significant adverse local effects or discomfort to the wearers or their sex partners. However, irregular breakthrough bleeding often occurred during treatment with vaginal rings containing either medroxyprogesterone acetate or norgestrel. Withdrawal bleeding also failed to occur on occasion. These problems were alleviated with the development of a sandwich-type vaginal ring that administered both progestin and estrogen simultaneously (15).

The concept of controlled drug delivery has also been successfully applied to the intravaginal administration of a synthetic progestin for estrus synchronization in sheep (16–18) and a synthetic prostaglandin derivative for abortion induction (19–21).

II. BENEFITS OF INTRAVAGINAL CONTROLLED DRUG ADMINISTRATION

As discussed in Section I, an intravaginal controlled-release drug delivery system is an effective means for achieving a continuous delivery of therapeutic agents, not only the systemically active drugs, such as contraceptive steroids (6–9,13), but also the locally active drugs, such as metronidazole (22). This continuous "infusion" of drugs through the vaginal mucosa can prevent the possibility of hepatogastrointestinal first-pass metabolism and inefficient therapeutic activity resulting from the alternatively surging and ebbing plasma drug levels that occur with the intermittent use of oral dosage forms, which are ingested as discrete, discontinuous doses (23).

The advantage of intravaginal controlled drug administration over conventional oral administration is best illustrated by the comparison shown in Figure 2. After the oral ingestion of medroxyprogesterone acetate (MPA) in tablet formulation, a peak plasma drug concentration is rapidly reached within 2 hr. This peak plasma level, varying in the range of 1.15 to 5.15 ng/ml, declines in the next 22 hr in a biphasic exponential pattern, with a rapid α-phase and a slower β-phase, to a drug concentration lower than 0.15 ng/ml (12). On the other hand, following the intravaginal controlled delivery of MPA from a vaginal ring in the same group of human volunteers, the plasma drug level resulting from the vaginal absorption of medroxyprogesterone acetate was also found to increase relatively rapidly to attain a rather steady plasma plateau, ranging between 0.37 and 0.63 ng/ml, within 4 hr. This plateau plasma level was maintained throughout the course of the treatment until removal of the vaginal ring.

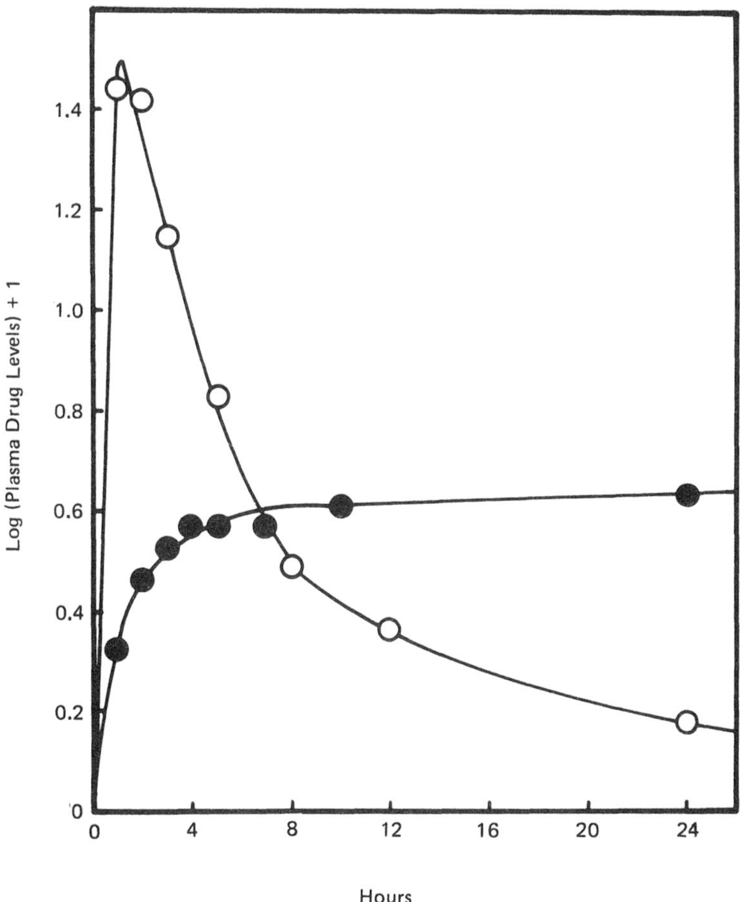

Figure 2 Comparative plasma profiles of medroxyprogesterone acetate (MPA) in five women following the oral administration of MPA (10 mg in tablets) (○) and the intravaginal administration of MPA (100 mg in a vaginal ring), (●). [Plotted from the data by Victor and Johansson, Contraception, 14:319 (1976).]

Moreover, the presystemic elimination associated with oral drug administration can be avoided when drug is administered intravaginally. The perineum venous plexus, which drains the vaginal tissue and rectum, flows into the pudendal vein and ultimately into the vena cava, which circumnavigates the liver on first pass. This is in marked contrast to gastrointestinal blood circulation, which drains into the portal vein and passes directly through the liver before entering the general circulation. Thus the vaginal route may be of great value for drugs like progesterone and estradiol, which are poorly bioavailable when taken orally because they are extensively inactivated by the liver (23). In addition, the intravaginal route of administration can also be beneficial with drugs, such as prostaglandins, that cause adverse gastrointestinal irritation.

III. PHYSIOLOGY AND DYNAMICS OF THE VAGINA

The human vagina is a tubular canal with a length of 4–6 inches, directed upward and backward, extending from the vestibule to the uterus (Figure 3). The vaginal wall consists of a membranous lining and a muscular layer, capable of constriction as well as enormous dilation, separated by a layer of erectile tissue (24). The mucous membrane forms thick transverse folds and is always maintained moist by cervical

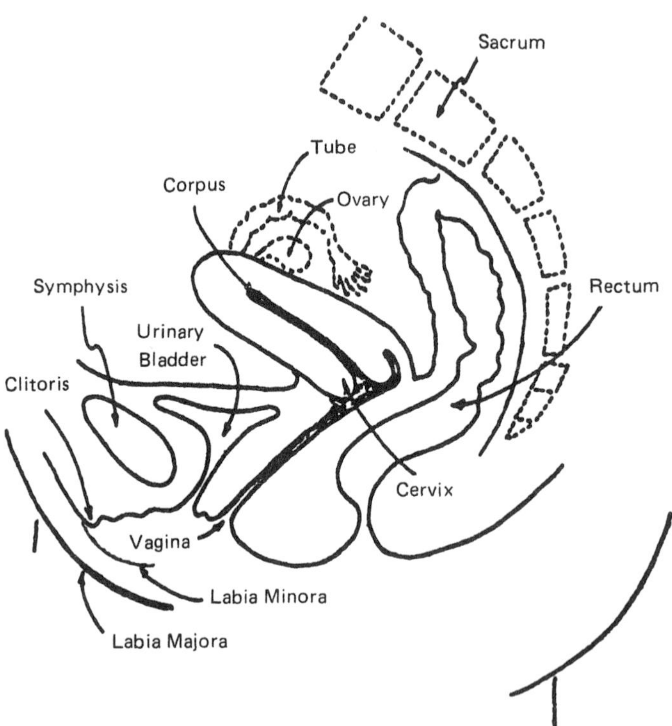

Figure 3 Lateral view of the female pelvis showing the vagina in an unstimulated state. (Reproduced with permission from Masters and Johnson, *Human Sexual Response,* Little, Brown, 1966.)

secretions. The vaginal walls are normally folded in close apposition to each other, forming a collapsed tube.

It is generally accepted that the human vaginal lumen has a normal acidity of pH 4–5. This vaginal pH is a secondary reflection of the active production of ovarian hormones and is also constantly maintained by the sloughing of mature cells in the superficial layers of the vaginal mucosa. Under the influence of an estrogen these cells contain a high content of glycogen, which is metabolized to lactic acid (with pKa of 3.8) in the vaginal canal to maintain the vaginal pH on the acid side. On the other hand, a cyclic luteal influence on the vaginal mucosa significantly elevates the pH value (25).

The effect of vaginal secretions on the pH level is dependent upon the amount and the duration of production. Both vaginal secretions and vaginal pH could affect the controlled-release profile of those vaginal solubility-dependent and/or pH-sensitive drugs from a vaginal drug delivery system (26).

It has also been reported that sexual activity declines during the luteal phase of the mestrual cycle (when circulating progesterone levels are at maximum) in humans (27) and monkeys (28–30). The investigations conducted in rhesus monkeys suggested that this declining sexual activity may be related to the action of progesterone on the vagina (31). On the other hand, estradiol is known to enhance the sexual attractiveness of a female monkey by affecting its vagina (32). This may be a result of stimulation of the emission of olfactory attractant or alteration of the tactile qualities of this tissue by estradiol.

In addition, one should recognize that vaginal lumen is a nonsterile area and is inhabited by a variety of microorganisms, mainly the species *Lactobacillus, Bacteroides,* and *Staphylococcus epidermidis,* as well as potentially pathogenic aerobes (33). The existence of these microbes and their possible metabolites may have some kind of effect on the intravaginal stability of a vaginal drug delivery device and its controlled drug release profiles.

As mentioned, the unstimulated vagina anatomically consists of a luminal space that exists potentially rather than actually. However, in response to sexual excitement some tension-induced anatomic variations occur. These anatomic variations may have effects on the long-term intravaginal residence and controlled drug release profiles of medicated vaginal rings. These variations are reflected in the existence of four phases in a sexual response cycle as outlined by Masters and Johnson (25).

A. Excitement Phase

The first sign of physiological response to a stimulation is the production of a vaginal lubricating fluid. This appears on the vaginal mucosal surface within 10–30 sec after an effective stimulation. As sexual tension progresses individual droplets of transudation-like mucoid material appear scattered throughout the rugal folds of the vaginal lumen and then coalesce to form a smooth, glistening coating over the entire vaginal mucosa surface. This transudative mucoid material results from the activation of a massive localized vasocongestive reaction and marked dilation of the venous plexus that encircles the entire vaginal lumen. This sweating phenomenon provides complete lubrication of the vagina.

Meanwhile, there occurs, initially, a lengthening and distention of the inner two-thirds of the vaginal lumen. As sexual tension mounts toward the plateau phase, the vaginal wall in this area expands involuntarily and then partially relaxes in an irregular, tensionless manner. The demand to expand gradually overcomes the tendency to relax. In addition to the expansive effect in the vaginal fornices, the cervix and corpus pull slowly backward and upward into the false pelvis position. This cervical elevation creates a "tenting effect" at the transcervical depth in the midvaginal plane. This phenomenon always occurs in a normal anteriorly positioned uterus (Figure 4).

In the nulliparous woman, the sexually unstimulated vaginal lumen has a diameter of approximately 2 cm in its transcervical plane (just anterior to the resting cervix of an anteriorly positioned uterus). In response to sexual tension this transcervical diameter expands three times to a range of 5.75–6.25 cm. In the meantime the length of the vagina increases from 7–8 to 9.5–10.5 cm. These expansion and lengthening effects are completely independent of any previously established states of vaginal distention; in other words, the vagina of either nulliparous or multiparous women, regardless of prior degree of vaginal expansion or lengthening, further increases substantially in length and transcervical width with sexual stimulation.

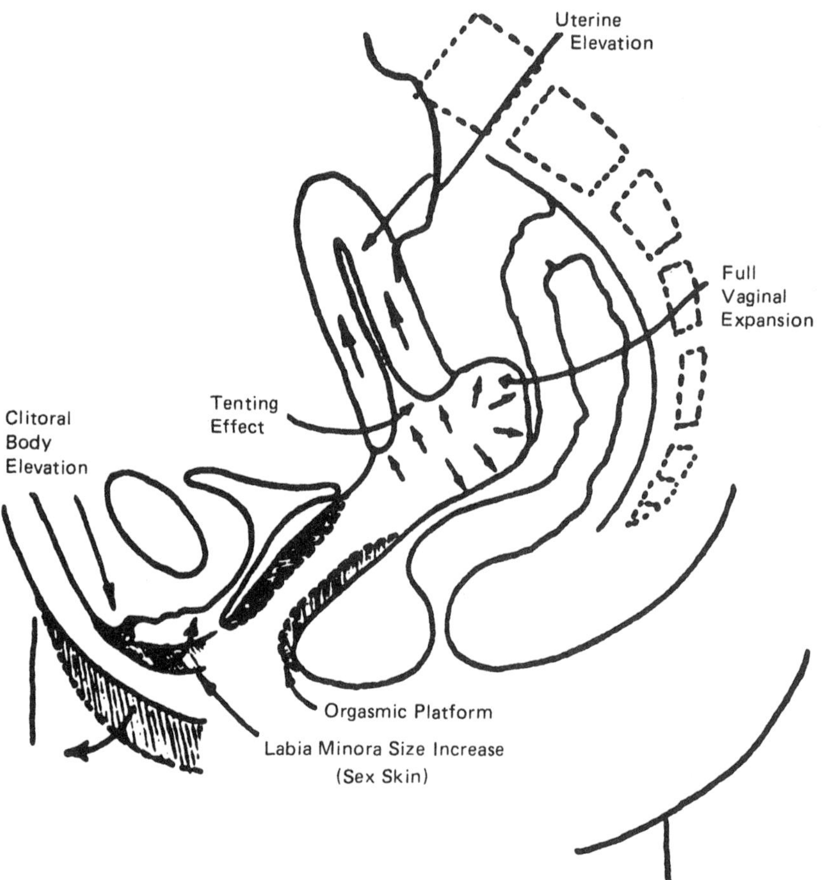

Figure 4 Lateral view of the female pelvis showing the stimulated vagina. (Reproduced with premission from Masters and Johnson, *Human Sexual Response,* Little, Brown, 1966.)

As the excitement-phase responses progress toward the plateau phase, there occurs a flattening of the rugal pattern of the well-stimulated vaginal wall resulting from involuntary expansion of the inner two-thirds of the vaginal lumen.

The sweating phenomenon and vaginal expansion in response to sexual tension in the excitement phase may thus produce a dislocation of a medicated vaginal device in the vagina and variation in its controlled drug release profiles.

B. Plateau Phase

With attainment of the plateau-phase level of sexual tension, a marked localized vasocongestive reaction develops in the outer one-third of the vaginal lumen. The entire area becomes grossly distended with venous blood, and its central lumen is reduced by at least a third from the distention previously established in the excitement phase.

Meanwhile, only a minimal further increase in the width and depth of the vaginal lumen occurs. The production rate of vaginal lubricating fluid also gradually slows, particularly if this level of sexual tension has been experienced for an extended period of time.

C. Orgasmic Phase

During the orgasmic phase the basic response of the inner vaginal lumen is essentially expansive rather than constrictive in character. On the other hand, the bulbar vasoconstriction at the orgasmic platform in the outer one-third of the vaginal lumen contracts strongly in a regularly recurring pattern. The intercontractile intervals lengthen in duration, and the intensity of the contractions progressively diminishes.

D. Resolution Phase

Along with the onset of the resolution phase, retrogressive changes develop first in the outer one-third of the vaginal lumen. The localized vasocongestion is dispersed rapidly, leading to an increase in the diameter of the central lumen of the outer one-third of the vagina.

The previously expanded inner two-thirds of the vaginal lumen also gradually shrinks back to the original collapsed, unstimulated state (Figure 3). This shrinking process is an irregular, zonal-type relaxation of the lateral and posterior walls. The anterior wall and the cervix of the anteriorly positioned uterus descend rapidly toward the vaginal floor, leading to a quick resolution of the tenting effect created earlier during the excitement phase.

IV. BIOPHARMACEUTICS OF INTRAVAGINAL CONTROLLED DRUG ADMINISTRATION

A. Theoretical Model of Intravaginal Controlled Drug Release

Vaginal absorption of a therapeutic agent from a controlled-release drug delivery system, such as a medicated vaginal ring, should be visualized as a consecutive process of several definable steps. For the drug-dispersing vaginal ring, the steps consist of the dissolution of the finely ground, well-dispersed drug particles into the surrounding polymer structure, diffusion through the polymer matrix to the device surface, partitioning into and then diffusion across the vaginal secretion fluid (which is sandwiched between the device surface and vaginal walls), uptake by and then penetration through the vaginal mucosa, and, finally, transport and distribution of the absorbed drug molecules by circulating blood and/or lymph to a target tissue (Figure 5).

The important features of the physical model proposed in Figure 5 are the receding interface of the drug dispersion zone/drug depletion zone in the polymer matrix as drug is released with time; an aqueous hydrodynamic diffusion layer (which represents the vaginal secretion fluid) sandwiched between the vaginal drug delivery device and the vaginal epithelium; and the vaginal wall, composed of a lipid continuum with interspered "pores" (or an aqueous shunt) pathway. The rate-controlling step in the whole process of intravaginal absorption of a drug varies depending upon the duration of intravaginal residence of the medicated vaginal ring (23).

Assume that the finely ground drug particles are homogeneously dispersed throughout the polymer matrix structure such that dissolution of drug in the polymer

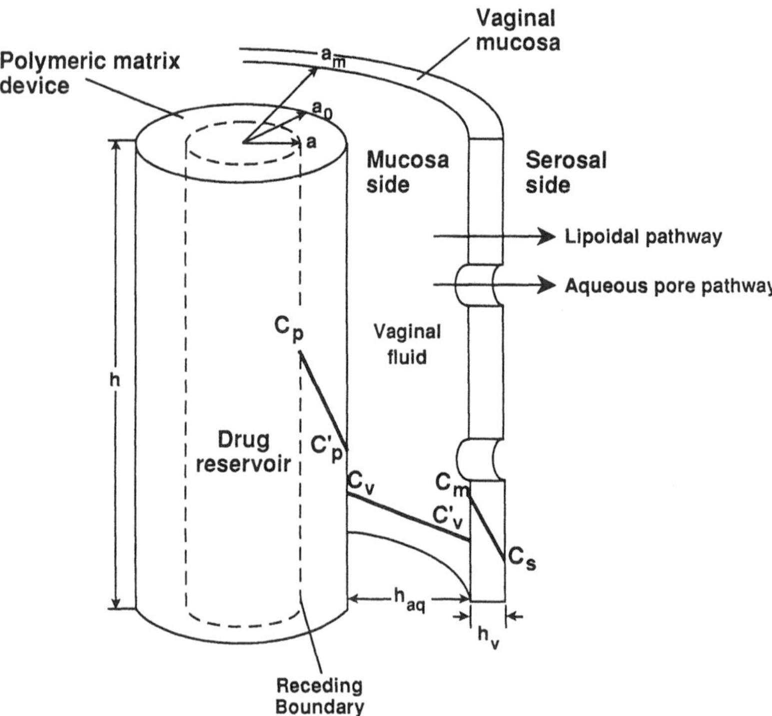

Figure 5 Various kinetic steps involved in the release of a drug from a drug-dispersing polymeric device and its transport across the vaginal fluid layer and then parallel diffusion through the lipoidal and aqueous pore pathways of the vaginal membrane. The cylindrical region (enclosed by the dashed line) in the polymer matrix, where drug particles disperse, diminishes inward uniformly as drug is released. (Modified from Flynn et al., in: ACS Symposium Series 33 on Controlled Release Polymeric Formulations, American Chemical Society, 1976.)

phase is not a rate-limiting step and a sharp interface is maintained between the drug dispersion zone and the drug depletion zone, which recedes continuously into the core of the device with time. It is also assumed that the drug has a finite solubility C_p in the polymer phase and that the total drug content per unit volume A, including the undissolved solid particles, is much greater than C_p. The drug molecules are visualized to reach the device surface by diffusion through the polymer matrix with a negligible end diffusion (for simplicity of mathematical treatment). Under these conditions a drug concentration gradient is established that extends from the receding interface of the drug depletion zone/drug dispersion zone to the outer reaches of the vaginal microcirculation, where the drug molecules are taken up and transported to the target tissues. This gradient is depicted in Figure 5 as a series of discontinuous concentration gradients with slopes depending on the physicochemical properties of the drug species and the polymer matrix.

Under these conditions the rate of intravaginal release of a drug species from the matrix-type medicated vaginal ring is thus defined as

$$\frac{dQ}{dt} = \frac{D_p C_p A}{\{[D_p K_s(1/P_{aq} + 1/P_v)]^2 + 4D_p C_p A t\}^{1/2}} \tag{1}$$

where:

D_p = diffusivity of the drug in the polymer matrix
C_p = solubility of the drug in the polymer phase
A = initial drug loading per unit volume of the polymer matrix
K_s = interfacial partition coefficient (C_p'/C_s), where C_p' is the drug concentration in polymer phase at the device surface and C_s is the solubility of drug in the aqueous diffusion layer at the device surface
P_{aq} = permeability coefficient across the aqueous diffusion layer with thickness h_{aq}
P_v = permeability coefficient across the vaginal mucosa (D_v/h_v), where D_v is the diffusivity in the vaginal mucosa with thickness h_v

At the very beginning of intravaginal insertion the thickness of the drug depletion zone ($a_0 - a$) is very narrow and the rate-controlling step of vaginal drug absorption resides in the permeation of drug through either the hydrodynamic diffusion layer or the vaginal mucosa. Under such circumstances Equation (1) can be simplified to

$$\frac{dQ}{dt} = \frac{C_p P_{aq} P_v}{K_s(P_{aq} + P_v)} \tag{2}$$

and a zero-order drug release profile is attained.

As the intravaginal residence time of the medicated ring is prolonged, the receding interface of the drug depletion zone/drug dispersion zone progresses to the extent that the matrix-controlled process becomes the predominant step in determining the intravaginal release of drug molecule. Under this new condition Equation (1) is simplified to

$$\frac{dQ}{dt} = \frac{1}{2}\left(\frac{D_p C_p A}{t}\right)^{1/2} \tag{3}$$

or

$$\frac{Q}{t^{1/2}} = (2AC_pD_p)^{1/2} \tag{4}$$

and a Q versus $t^{1/2}$ drug release profile is followed (Figure 6). The value of $Q/t^{1/2}$ can be calculated from the slope of Q versus $t^{1/2}$ plots, and its magnitude should be directly proportional to $(2A)^{1/2}$, the square root of the initial drug loading in a unit volume of the vaginal ring (Figure 7). Mechanistic analysis of this polymer matrix diffusion-controlled drug release process is detailed in Chapter 2.

B. Vaginal Pharmacokinetics

If the absorption, distribution, and elimination of a drug molecule after its release from a vaginal drug delivery device in the vaginal lumen follow the pharmacokinetic sequences

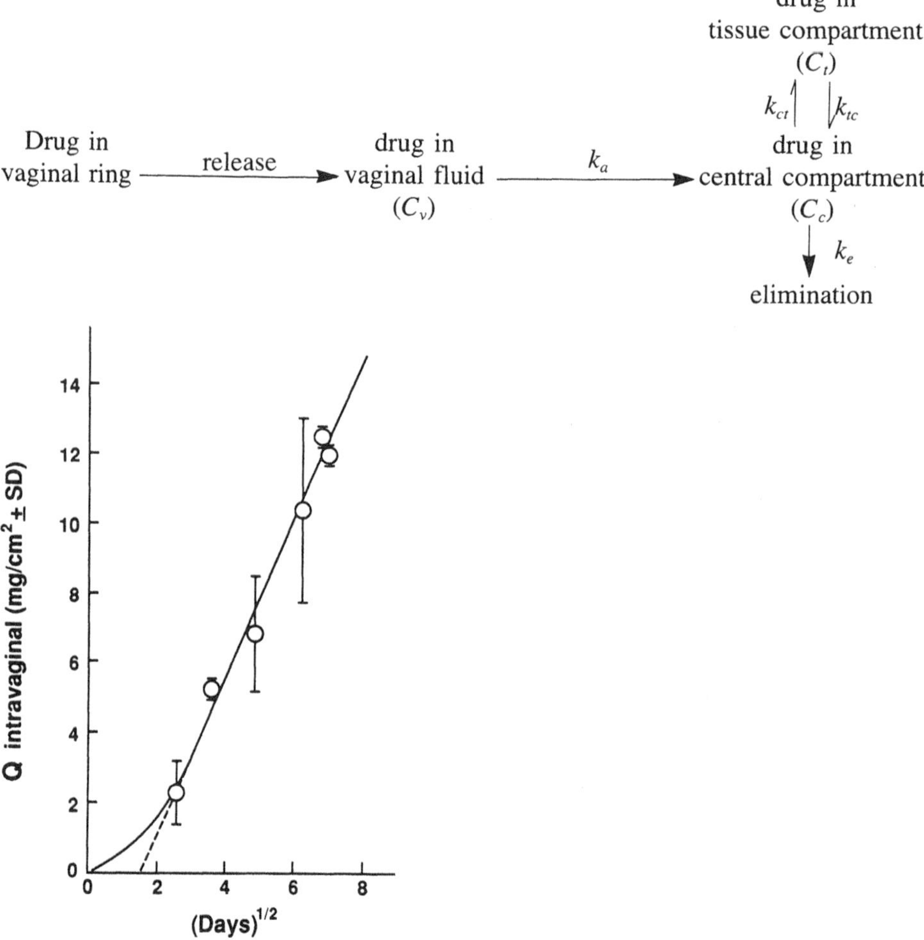

Figure 6 Linear relationship between the cumulative amount Q of ethynodiol diacetate released from a matrix-type silicone vaginal device in the rabbit and the square root of time ($t^{1/2}$). [Replotted from the data by Chien et al., J. Pharm. Sci., 64:1776 (1975).]

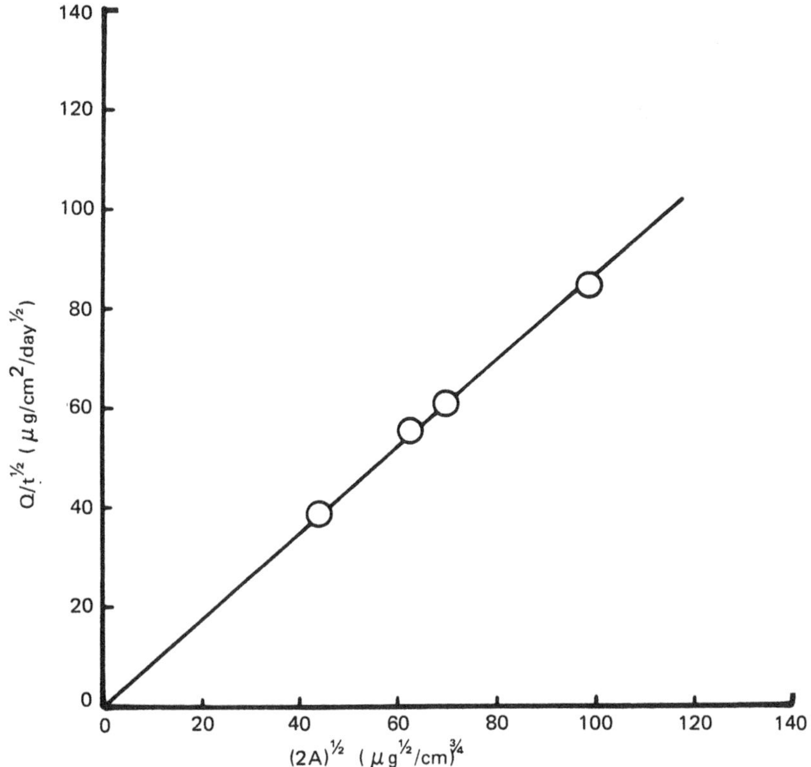

Figure 7 Dose dependence of the intravaginal release flux ($Q/t^{1/2}$) of levonorgestrel from vaginal rings in women. [Plotted from the data by Mishell et al., Contraception, 12:253 (1975) and Victor et al., Contraception, 12:261 (1975).]

then the instantaneous rate of change in drug concentration in the central compartment can be expressed by

$$\frac{d(C_c)}{dt} = k_a C_v + k_{tc} C_t - (k_{ct} + k_e) C_c \tag{5}$$

where k_a, k_e, k_{ct}, and k_{tc} are the rate constants for absorption, elimination, and central compartment/tissue compartment exchange, respectively, and C_v, C_c, and C_t are the drug concentrations in the tissue fluid surrounding the vaginal ring, in the central compartment, and in the tissue compartment, respectively.

The vaginal absorption of drug following its release from vaginal drug delivery devices can be described by a simplified one-compartment open model with first-order drug absorption (Figure 8) (34). Following this simplified model, Equation (5) is reduced to

$$\frac{d(C_B)}{dt} = k_a C_v - k_e C_B \tag{6}$$

where C_v and C_B are the drug concentrations in the vagina and in the body (including blood, tissues, and related compartments with fast drug exchange rates), respectively.

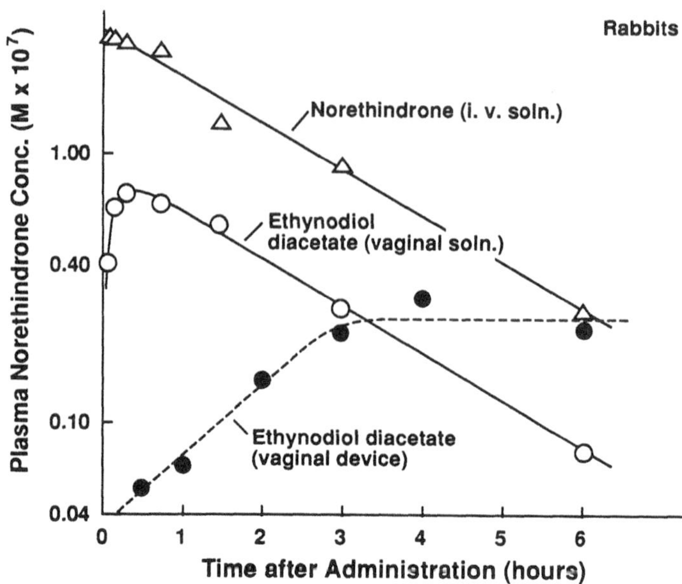

OAc CH3 C≡CH AcO Ethynodiol diacetate

OH CH3 C≡CH O Norethindrone

Figure 8 Plasma concentration profiles of norethindrone following the intravenous administration of a single solution dose (1 mg in 0.2 ml) of norethindrone as well as the intravaginal absorption of ethynodiol diacetate from a solution dose or delivered by a vaginal device in the rabbit. [Replotted from the data by Chien et al., J. Pharm., Sci., 64:776 (1975).]

At steady state the change in the body concentration of the drug is relatively small (Figure 9), that is, $d(C_B)/dt \simeq 0$, and the body concentration of norethindrone (C_B), the major metabolite of ethynodiol diacetate, is then related to the amount of ethynodiol diacetate (Q) released at time t as follows:

$$C_B = \frac{k_a \Sigma R_v}{2k_e} \left(\frac{Q}{t} \right)_v \tag{7}$$

where ΣR_v is the total diffusional resistance across the vaginal wall. Its definition and physical meaning are discussed in Chapter 2.

Equation (7) suggests that the norethindrone concentration (C_B) in the body of each test animal should be directly proportional to the amount of ethynodiol diacetate released from the vaginal device $(Q/t)_v$ for a given duration of intravaginal residence (Figure 10).

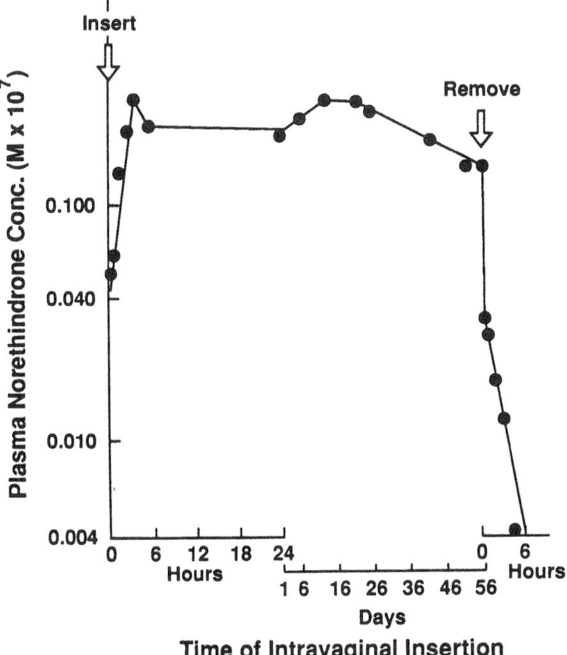

Figure 9 Plasma profile of norethindrone following the intravaginal insertion of ethynodiol diacetate-releasing vaginal devices in rabbits for 56 days and after device removal. [Replotted from the data by Chien et al., J. Pharm. Sci., 64:776 (1975).]

From the slope of the C_B versus $(Q/t)_v$ plots and the values of k_a and k_e from the vaginal absorption studies of the same drug from a solution, the magnitude of ΣR_v, the total diffusional resistance for the vaginal permeation of the drug, can be estimated; for example, ΣR_v is 0.134 sec./cm for the vaginal absorption of ethynodiol diacetate in rabbits (34).

As expected from Equation (7), the mean plasma drug levels in women wearing medicated vaginal rings are correlated fairly well with the intravaginal drug release rate profiles throughout the course of the cyclic 3-week treatment (Figure 11).

C. Vaginal Mucosa Permeability

A series of vaginal permeation studies were conducted in rabbits as the animal model to learn the fundamentals of the vaginal absorption of drugs (26,35).

The female rabbit does not exhibit an estrus cycle, so its vaginal tissues show a constancy in the histological, biochemical, and physiological properties not ordinarily seen with most other mammals (36). The lack of a sexual cycle is therefore expected to produce a minimal variability in the permeability of the vaginal membrane (26,35,37). Hence, the use of the doe rabbit as the animal model has made measurements of vaginal drug permeation more controllable and accessible without the complication of estrus cycles.

The vaginal mucosal permeability of the doe rabbit has been examined by continuous perfusion of straight-chain alkanols and alkanoic acids (26,35). Similar to

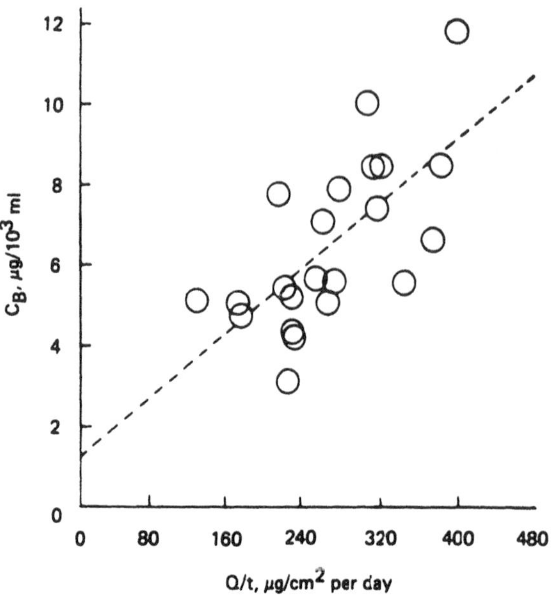

Figure 10 Correlation between the plasma concentration C_B of norethindrone, the major metabolite of ethynodiol diacetate, in each rabbit and the amount of ethynodiol diacetate (Q/t) released from per unit surface area of the vaginal devices at a given duration of vaginal insertion ($p < 0.01$). [Reproduced with permission from Chien et al., J. Pharm. Sci., 64:1776 (1975).]

the vaginal absorption of ethynodiol diacetate (Figure 8), the vaginal uptake of both alkanols and alkanoic acids also follows a first-order rate process and is dependent upon the drug concentration in the vaginal fluid. The results agree well with a physical model having a hydrodynamic diffusion layer in series with the mucosal membrane, which consists of two parallel pathways: a lipoidal pathway and an aqueous "pore" pathway (Chapter 4, Figure 4). Immediately beyind the mucosa (serosal side) a perfect sink is maintained by hemoperfusion.

The apparent permeability coefficient P_{app} for vaginal membrane permeation is defined by

$$P_{app} = \frac{1}{1/P_{aq} + 1/P_v} \tag{8a}$$

or

$$P_{app} = \frac{1}{1/P_{aq} + 1/(P_p + P_1)} \tag{8b}$$

Since

$$P_v = P_p + P_l \tag{9}$$

where P_{aq}, P_v, P_p, and P_l are the permeability coefficients of the aqueous diffusion

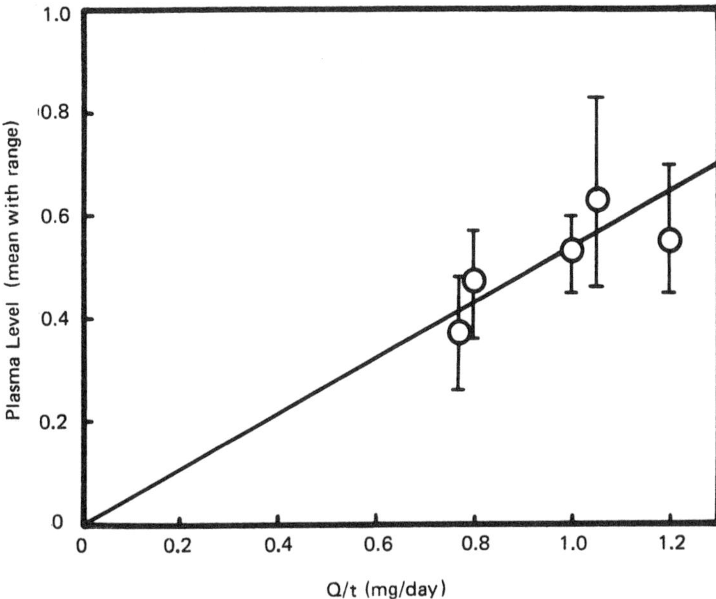

Figure 11 Correlation between the plasma levels of medroxyprogesterone acetate in five women wearing vaginal rings and their corresponding intravaginal drug release rate during a 3-week insertion. [Plotted from the data by Victor and Johansson, Contraception, 14:319 (1976).]

layer, the vaginal membrane, the aqueous pore pathway, and the lipoidal pathway, respectively.

The vaginal permeation kinetics of a series of straight-chain alkanols was investigated (35). Using methanol as a reference permeant a normalized permeability coefficient $P_{app}(alc/\text{MeOH})$ was determined for each of the alkanols. The normalized permeability coefficient was observed to increase in value as the alkyl chain length of the alkanols increased (Table 1). This increased permeability can be attributed to the increase in the permeability coefficient for the lipoidal pathway P_l [Equation

Table 1 Effect of Alkyl Chain Length on the Normalized Permeability Coefficients of Straight-Chain Alkanols

Alkanol	$CH_3(CH_2)_nOH$	P_{app} (Alc/MeOH)[a]
Methanol	$n = 0$	1.00
Propanol	$n = 2$	1.11
Butanol	$n = 3$	1.13
Pentanol	$n = 4$	1.20
Hexanol	$n = 5$	1.48
Heptanol	$n = 6$	1.91
Octanol	$n = 7$	2.15

[a]Normalized permeability coefficient = P_{app}(alcohol)/P_{app}(methanol). Mean value from three rabbits at pH 6.0 and 37°C.

Source: Compiled from the data by Hwang et al. (1976).

(8b)]. It is estimated that for straight-chain aliphatic alcohols the P_l value increases by 2.5 for the addition of each methylene (CH_2) group (35). On the other hand, the P_l value increases by 3.5 for the series of straight-chain alkanoic acids (26).

For the vaginal absorption of ionizable compounds, such as the homologous series of n-alkanoic acids, the apparent permeability coefficient P_{app} becomes pH dependent and is defined by

$$P_{app}(n, pH) = \frac{1}{P_{aq}(n)} + \frac{1}{\frac{[H]}{K_a + [H]} P_l^0 10^n \pi + P_p} \tag{10}$$

where n is the number of methylene (CH_2) groups in the alkyl chain; [H] is the concentration of protons; K_a is the dissociation constant of the acid; P_l^0 is the permeability coefficient of the lipoidal pathway for the hypothetical acid with zero carbon atom ($n = 0$); and π is the methylene group incremental constant, that is, 3.5 for straight-chain alkanoic acids.

As illustrated earlier in the homologous series of n-alkanols, the normalized permeability coefficient of n-alkanoic acids also shows a dependence on alkyl chain length (Table 2). In addition, the straight-chain alkanoic acids demonstrate a pH dependence in their normalized permeability coefficients (26). It should be pointed out that the rabbit's vaginal secretion has an effective pK_a value of 6.3 ± 0.1. However, the rate of vaginal secretion is relatively small, which leads to a surface pH of around 2.0–2.1 (38). This acid surface pH affects the extent of dissociation of n-alkanoic acids and thus the magnitudes of P_l and P_{app} [Equation (10)].

The vaginal uptake of steroids was also studied and found to follow a first-order rate process as well (23). The normalized permeability coefficient of steroids appears to be dependent upon steroidal structure (Table 3). The permeability coefficient across the vaginal membrane (P_v) also shows the same trend of structure dependence. However, the P_{aq} values, the permeability coefficient across the hydrodynamic diffusion layer (Chapter 4, Figure 4), are very much the same among the four steroids (Table 3).

For drugs with high P_v values, such as progesterone and estrone (6.10–7.6 × 10^{-4} cm/sec), vaginal absorption is mainly controlled by their permeability across

Table 2 Effect of Alkyl Chain Length and pH on the Normalized Permeability Coefficients of Straight-Chain Alkanoic Acids

	P_{app}(acid/MeOH)[a]		
Acids	pH 3	pH 6	pH 8
Acetic	1.22	0.73	0.25
Butyric	1.62	1.94	0.34
Hexanoic	1.89	2.06	0.81
Octanoic	1.74	2.49	1.24
Decanoic	—	—	1.26

[a]Normalized permeability coefficient = P_{app}(acid)/P_{app}(methanol). Mean value of three experiments involving different rabbits at pH 3, 6, and 8.

Source: Compiled from the data by Hwang et al. (1977).

Table 3 Vaginal Permeation Parameters of Representative Steroids

Steroids	P_{app} [a]	$P_v \times 10^4$ (cm/sec)	$P_{aq} \times 10^4$ (cm/sec)
Estrone	1.00	7.60	2.81
Progesterone	0.93	6.10	2.80
Testosterone	0.29	0.75	2.76
Hydrocortisone	0.23	0.58	2.79

[a]Normalized permeability coefficient = P_{app}(steroid)/P_{app}(methanol).
Source: Compiled from the data by Flynn et al. (1976).

the hydrodnamic diffusion layer on the surface of the vaginal mucosa ($P_v > P_{aq}$). On the other hand, for drugs with low P_v values, such as testosterone and hydrocortisone (5.8–7.5 × 10^{-5} cm/sec), vaginal uptake is determined predominantly by their molecular permeation through the vaginal membrane ($P_v \ll P_{aq}$) (39).

The apparent permeability coefficient P_{app} is related to the first-order rate constant for the disappearance of drug from vaginal lumen (k_v) as follows:

$$P_{app} = k_v \frac{V_v}{S_v} \tag{11}$$

where V_v is the volume of vaginal fluid and S_v is the geometric surface area of vaginal lumen.

D. Cyclic Variation in Vaginal Mucosa Permeability

As pointed out earlier, the female rabbit does not exhibit an estrus cycle and its vaginal tissues show a constancy in their histological, biochemical, and physiological properties not seen with most other mammals (36). The rabbit thus appears to be an ideal animal model for studying vaginal mucosa permeation kinetics.

Unfortunately, the doe rabbit may not be a suitable animal model for studying the long-term vaginal absorption of contraceptive agents from various controlled-release drug delivery systems, such as medicated vaginal rings, because it lacks the typical cyclic variations observed in the human vaginal tract associated with the rhythmic pattern of hormones during a sexual cycle.

In the human female the cyclic secretion of estrogenic hormones in the ovarian cycle induces variation in the histology, biochemistry, and physiology of vaginal tissues; it is therefore reasonable to expect that the vaginal mucosa may also undergo a corresponding cyclic variation in its drug permeation behavior.

The macaque rhesus monkey has an ovarian cycle of approximately 28 days, as does the human female, and it also exhibits an estrus pattern very similar to the menstrual pattern of the human female. It is widely believed by researchers in the fertility field that rhesus monkeys and humans have comparable anatomy and physiology as well as similar reproductive functions (40). The female rhesus monkey therefore could be an excellent animal model for studying the vaginal absorption of various drugs from controlled-release drug delivery systems designed for use in human females.

The effect of the estrus cycle on the permeability of the vaginal mucosa has been demonstrated in the vaginal absorption of a small molecule, like methanol, which has a vaginal membrane-controlled permeation, and a larger molecule, such as *n*-octanol, whose vaginal permeation is controlled by the hydrodynamic diffusion layer (Figure 12). Further studies in intact and ovariectomized monkeys could not establish any systematic relationship between the menstrual cycle and vaginal membrane permeability (41). Conflicting observations were also reported in the vaginal absorption of penicillin in humans (42,43) as well as in rats (44,45).

Overall, the vaginal permeability of cycling monkey during the period immediately following menstruation is lower than that of noncyclic rabbit (Figure 13). The difference in vaginal permeability between rhesus monkey and rabbit is greater for hydrophilic molecules, such as the short-chain alkanols (e.g., methanol), whose vaginal permeability is controlled by vaginal membrane permeation. This difference

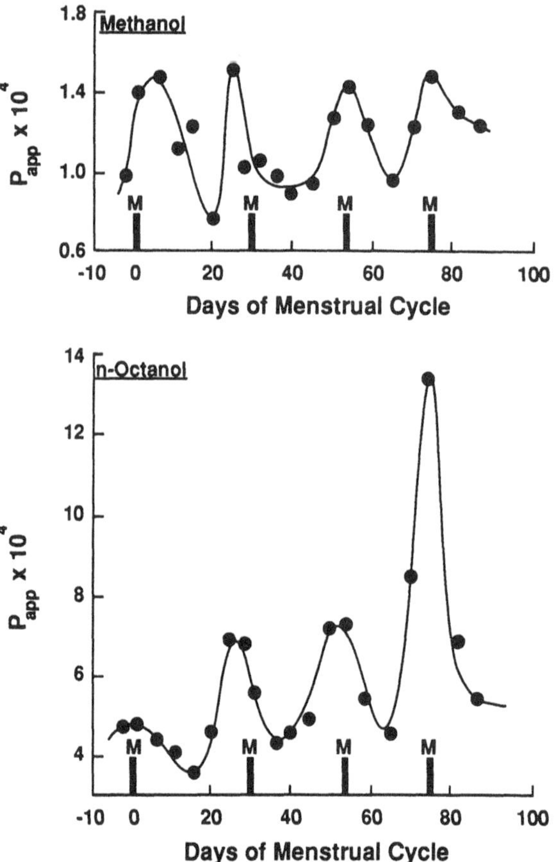

Figure 12 Cyclic variation in the apparent permeability coefficient P_{app} of methanol (top) and *n*-octanol (bottom) in the female rhesus monkey in response to its estrus cycle. The bars indicate the time of observed menstruation. (Replotted from the data by Flynn et al., in ACS Symposium Series 33 on Controlled Release Polymeric Formulations, American Chemical Society, 1976.)

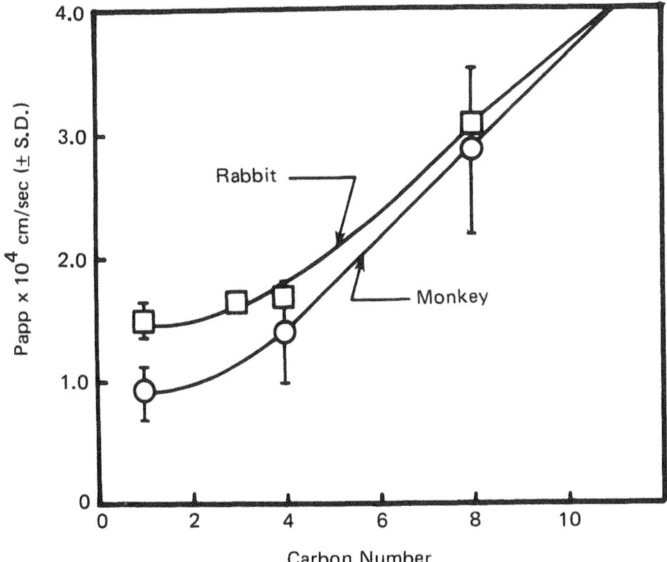

Figure 13 Comparison of apparent permeability coefficients P_{app} for the vaginal absorption of straight-chain alkanols in noncyclic rabbits and cyclic rhesus monkeys. [Plotted from the data by Hwang et al., J. Pharm. Sci., 65:1574 (1976) and Owada et al., J. Pharm. Sci., 66:216 (1977).]

lessens as the alkyl chain length of alkanols increases, which leads to an increase in molecular lipophilicity at the expense of hydrophilicity. At ovulation the monkey's vaginal permeability is several folds lower than that of the noncyclic rabbit (23).

The cyclic variation in vaginal drug permeability observed in rhesus monkeys in association with the rhythmic pattern of the sexual cycle suggests that the vaginal absorption data generated in the rhesus monkeys may be more reflective of what will occur in humans and can be directly extrapolated to humans. The use of rhesus monkeys as the animal model should benefit the research and development of an intravaginal contraceptive device.

E. Comparative Bioavailability of Vaginal Contraceptive Devices

The serum levels of medroxyprogesterone acetate in rhesus monkeys after oral administration of a solution dose and intravaginal administration from a vaginal device are compared in Figure 14. Following oral administration the peak serum level (13.8 ng/ml) was achieved within 2 hr. The serum level then declined gradually, and no drug could be detected 48 hr after treatment. On the other hand, monkeys receiving medroxyprogesterone acetate intravaginally from a silicone-based vaginal device achieved a serum level of approximately 3.5 ng/ml by 4 hr after insertion. This drug concentration increased continuously and reached a level of approximately 9 ng/ml by the second day of treatment. This serum level was then maintained throughout the remainder of the 3-week treatment. After removal of the vaginal device on Day 21, the serum concentrations dropped markedly during the next 24 hr. It was reported that within 3–7 days after termination of the treatment all monkeys

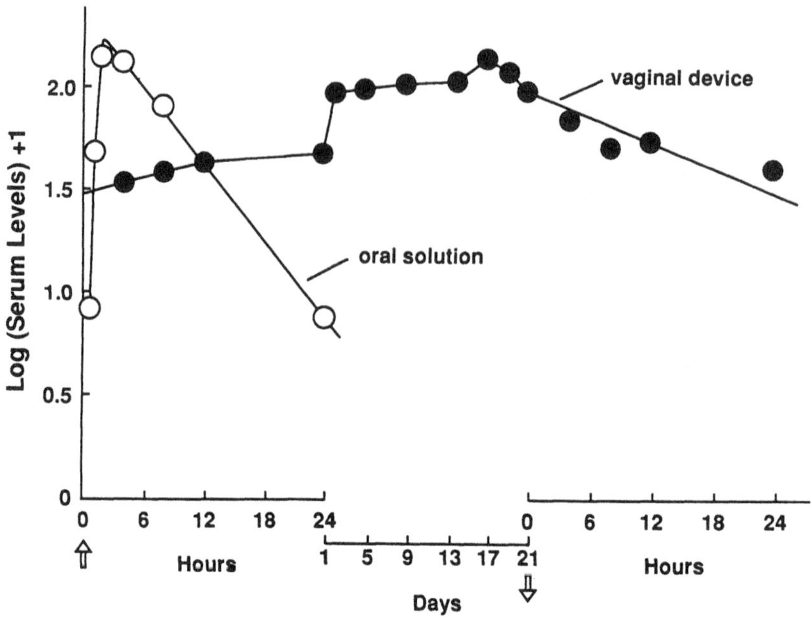

Figure 14 Comparative serum levels of medroxyprogesterone acetate in monkeys following the oral administration of drug (1 mg) in 10 ml solution (○) and the intravaginal delivery of drug (270 mg) by a vaginal ring (●). [Plotted from the data by Cornette et al., J. Clin. Endocrinol. Metab., 33:459 (1971).]

began to menstruate (46). Similar serum profiles were also observed in humans (Figure 15).

In Figure 15, serum drug levels were monitored in healthy women who took 10 mg medroxyprogesterone acetate in tablets orally following the same dosing schedule as that for oral contraceptive pills for 5 consecutive days (47). Circulating medroxyprogesterone acetate was detectable at 30 min after oral administration of the first dose and reached peak levels, in the range of 3.4–4.4 ng/ml, within 1–4 hr. Serum drug levels were noted to decrease rapidly thereafter and fell to a level of 0.3–0.6 ng/ml within 24 hr after oral intake. After each consecutive dose serum medroxyprogesterone acetate levels rose from a range of 0.7–0.9 ng/ml to a peak level within 2 hr, which is similar to that attained earlier by the initial dose. However, serum levels showed a fluctuating pattern in response to the consecutive daily intake of one oral dose every 24 hr.

On the other hand, following the insertion of a medroxyprogesterone acetate-releasing vaginal ring, a rapid elevation in serum drug levels was observed, indicating rapid drug uptake by the vaginal mucosa. Plateau drug concentrations were reached within 2–3 days after intravaginal insertion of the controlled-release vaginal device, which were maintained throughout the course of treatment with minimal fluctuation.

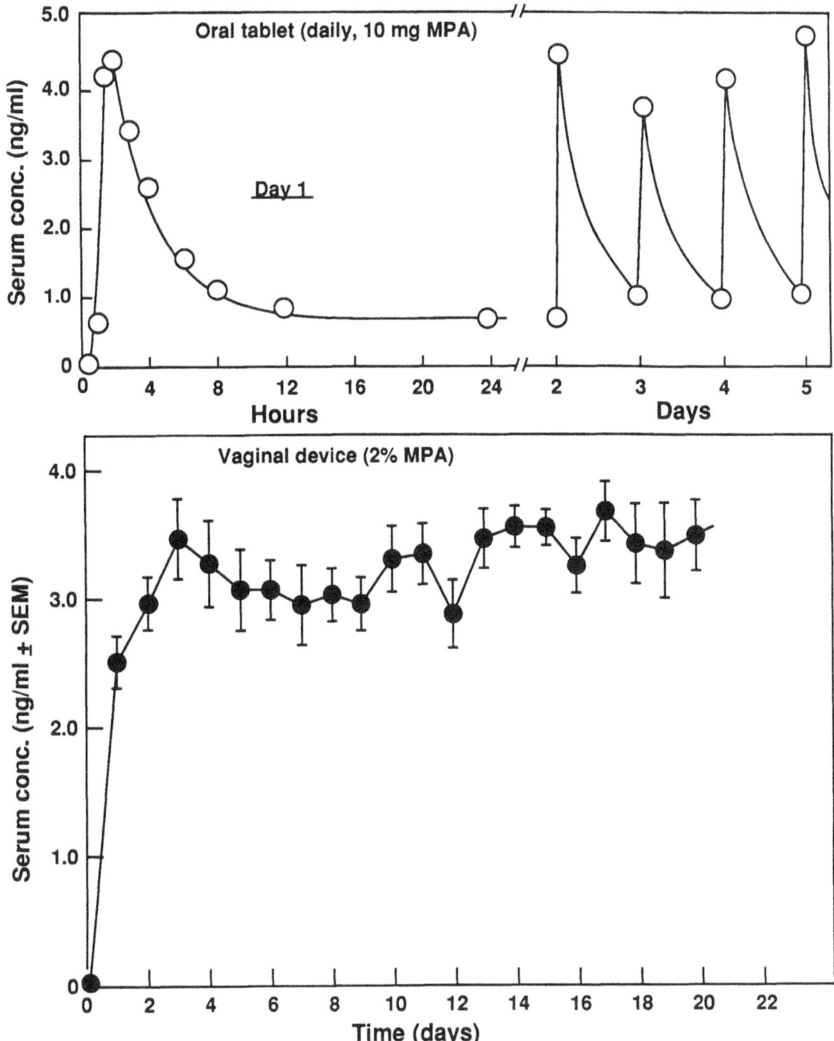

Figure 15 (Top) Serum concentrations of medroxyprogesterone acetate after a daily oral administration of a medroxyprogesterone acetate (10 mg) tablet taken by a healthy woman before breakfast for 5 consecutive days. [Plotted from the data by Horoi et al., Steroids, 26:373 (1975).] (Bottom) Daily serum concentrations of medroxyprogesterone acetate in women wearing medicated silicone vaginal rings (containing 2% medroxyprogesterone acetate) for 20 days.

It was also noted that the peak serum concentrations of drug resulting from the dissolution of oral tablets in the gastrointestinal tract and subsequent absorption of drug show a linear dependence on drug loading doses in the tablet formulation (Figure 16); the peak plasma concentrations of drug from the continuous intravaginal delivery of drug from a controlled-release matrix-type vaginal delivery system show

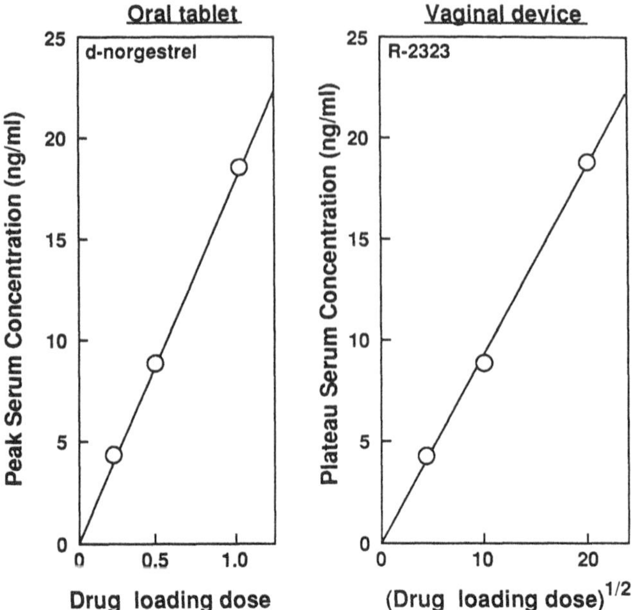

Figure 16 Linear dependence of peak and plateau serum drug concentrations on drug loading dose in the formulations: (left) *d*-norgestrel (levonorgestrel) in oral pills [Victor and Johansson, Contraception, 16:115 (1977)]; (right) R2323 (gestrinone) in vaginal rings [Viinikka et al., Contraception, 12:309 (1975)].

a dependence on the square root of the drug loading doses in the device, as expected from the polymer matrix diffusion-controlled process (Chapter 2, Section II.B).

V. BIOMEDICAL APPLICATIONS OF INTRAVAGINAL CONTROLLED DRUG ADMINISTRATION

A. Intravaginal Drug Delivery for Human Applications

1. *Medroxyprogesterone Acetate-Releasing Vaginal Contraceptive Rings*

Clinical Pharmacokinetics. After intravaginal insertion of a vaginal contraceptive ring, a ring-shaped silicone elastomer-based device (with a thickness of 9 mm and an outer diameter of 55 mm, Figure 1) that contained either 100 or 200 mg medroxyprogesterone acetate on Day 5 of the menstrual cycle, serum drug concentrations were observed to rise rapidly and reached the contraceptive plateau levels within 2–3 days (Figure 17). These plateau drug concentrations were maintained at relatively steady levels throughout the 3-week treatment period. Following removal of the ring the residual drug in the body diminished very rapidly (48).

The results in Figure 17 demonstrate that a significantly higher serum level of medroxyprogesterone acetate was achieved by the vaginal ring containing 200 mg medroxyprogesterone acetate than by the ring containing a 100-mg dose (48). This dependence is essentially a reflection of the effect of drug loading dose on the intravaginal release flux of drug (Figure 7), as expected from Equation (4). As shown

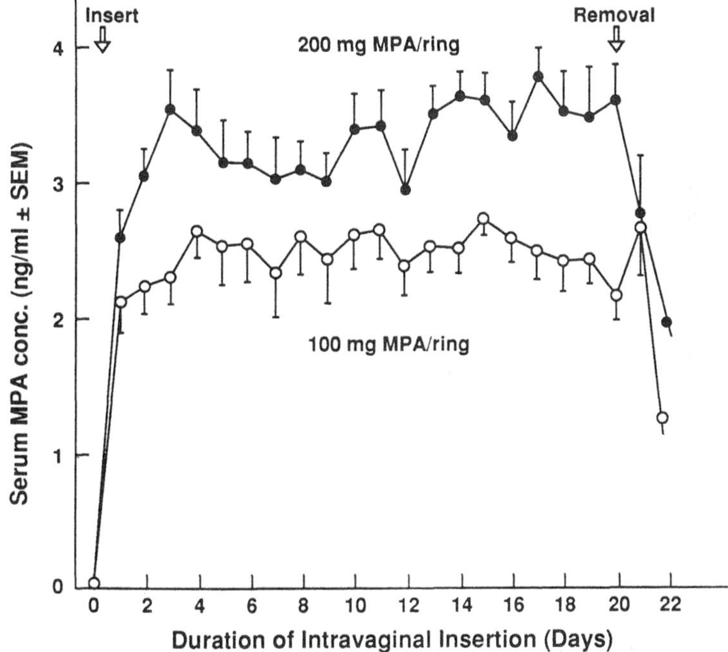

Figure 17 Daily serum concentration profiles of medroxyprogesterone acetate (MPA) in women wearing a medicated silicone vaginal ring for 21 days. Each vaginal ring contains either 100 or 200 mg medroxyprogesterone acetate. [Replotted from the data by Thier et al., Contraception, 13:605 (1976).]

in Figure 16, a linear relationship exists between the plateau serum levels of drug and the square root of drug loading doses.

During the 21-day intravaginal application a total of 33 ± 7 and 20.6 ± 6 mg, respectively, of medroxyprogesterone acetate was delivered (48). These doses of medroxyprogesterone acetate delivered intravaginally are substantially lower than the total dose (210 mg) expected from a daily dose of 10 mg taken orally in a tablet formulation for the same 21-day treatment duration. It becomes obvious that the vaginal absorption of medroxyprogesterone acetate delivered continuously at a controlled rate from the medicated vaginal ring provides a steady plateau of effective contraceptive serum concentrations with a total dose that is only one-tenth to one-sixth of the oral dose required. Apparently, a more efficient systemic bioavailability was achieved with the controlled-release vaginal drug delivery systems.

Clinical Pharmacodynamics. The pharamcodynamic effect of the intravaginal controlled administration of medroxyprogesterone acetate from a controlled-release vaginal ring is demonstrated by the suppression of rhythmic serum hormonal patterns in Figure 18. Before the insertion of a medroxyprogesterone acetate-releasing vaginal ring, the hormonal pattern in the pretreatment cycle shows a midcycle surge of luteinizing hormone (LH) and follicle stimulating hormone (FSH). This midcycle surge of gonadotropins is followed by a gradual increase in serum estradiol concentration which reaches a peak level on the day or 1 day after the gonadotropins surge. The serum concentration profile of estradiol shows another peak level, which is lower in

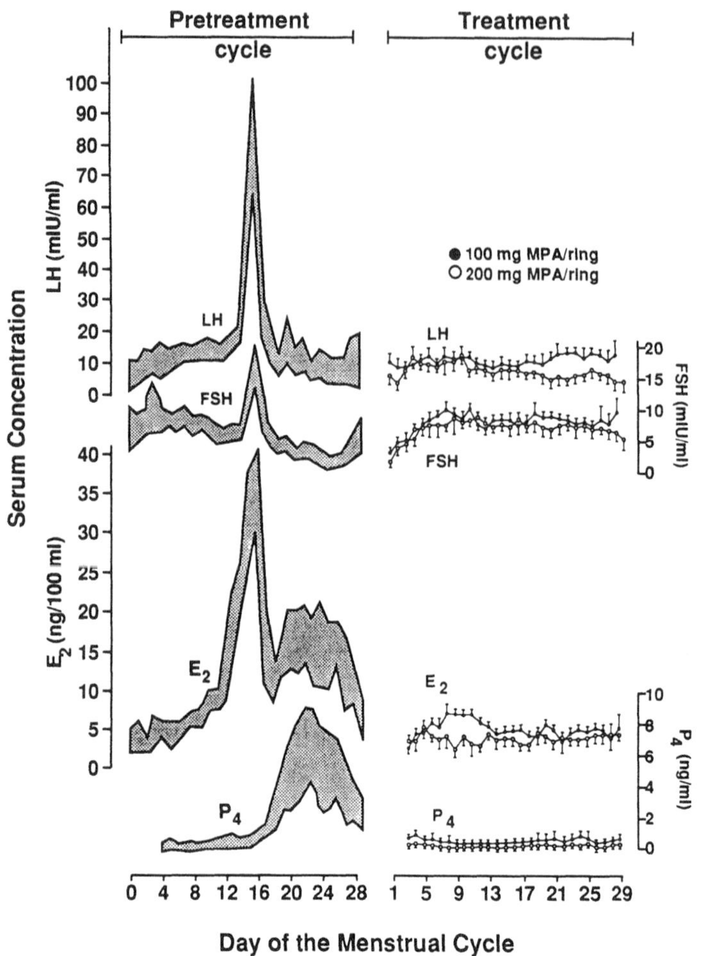

Figure 18 Daily serum concentration profiles of LH (luteinizing hormone), FSH (follicle stimulating hormone), E_2 (estradiol), and P_4 (progesterone) in 14 women wearing vaginal contraceptive rings, each containing either 100 or 200 mg medroxyprogesterone acetate. The shaded areas represent the 95% confidence limits of mean hormonal values during the control cycle. Vertical bars represent 1 SEM of the mean values during the treatment cycle. [Replotted from the data by Thier et al., Contraception, 13:605 (1976).]

magnitude but broader in duration, coinciding with the postovulatory rise in progesterone levels (48).

Following insertion of a vaginal ring containing either 100 or 200 mg medroxyprogesterone acetate, the serum levels of LH fall to the normal range of postovulatory values of the pretreatment cycle, whereas serum FSH concentrations remain within the preovulatory range of the pretreatment cycle. No significant rise in the serum levels of estradiol or progesterone occurs during the 21-day treatment period. The mean serum concentrations of estradiol and progesterone are maintained at a level similar to those of the early follicular phase (48,49).

Interestingly, there is no significant difference in the hormonal levels between these two dosage groups even though the plateau serum levels of medroxyprogesterone acetate shows dose dependence (Figure 17). The results apparently suggest that the vaginal ring that delivers 20.6 ± 6 mg medroxyprogesterone acetate during the 21-day treatment could be sufficient in achieving contraceptive efficacy. It was also reported that ovulation was effectively inhibited, and androgen secretion by the ovaries was also substantially reduced (49).

The effect of the medroxyprogesterone acetate-releasing vaginal ring on serum profiles of gonadotropins is further demonstrated in Figure 19. After treatment with a nonmedicated (placebo) vaginal ring for 28 days there was no significant alteration in the midcycle ovulatory peaks of LH and FSH. In addition, the luteal elevation of basal body temperature and the appearance of late secretory endometrium in the biopsies were not affected. On the other hand, treatment with a medicated (medroxyprogesterone acetate-releasing) vaginal ring resulted in abolition of the midcycle LH peak but no significant variation in the serum levels of FSH. The serum

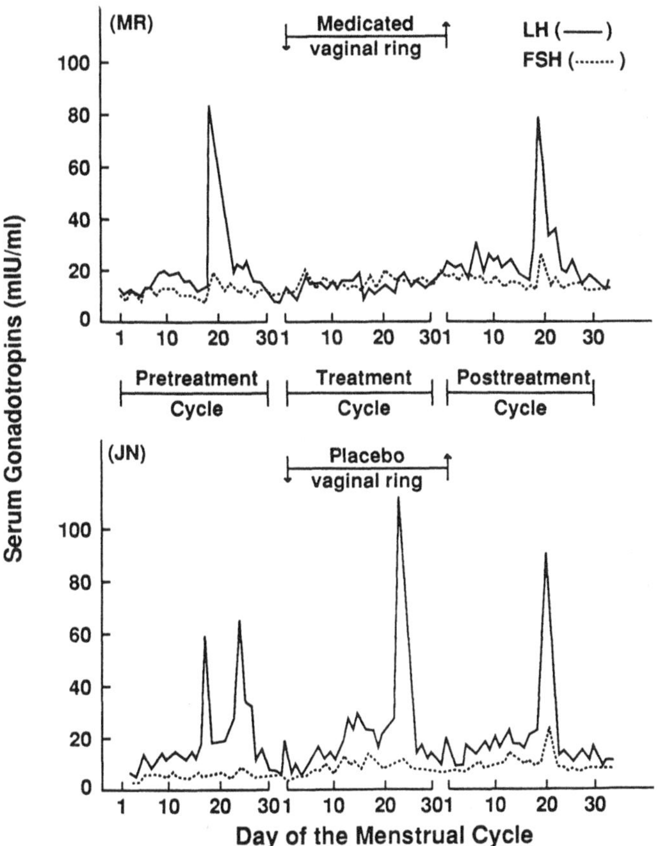

Figure 19 Daily serum concentration profiles of luteinizing hormone (LH, solid line) and follicle stimulating hormone (FSH, dotted line) during three consecutive menstrual cycles in two females. Throughout the second cycle a nonmedicated (placebo) or a medicated (medroxyprogesterone acetate-releasing) vaginal ring was inserted in the vagina for 28 days. [Replotted from the data by Mishell et al., Am. J. Obstet. Gynecol., 107:100 (1970).]

LH concentrations were similar to the levels detected in the luteal phase of the pre- and posttreatment cycles. The basal body temperature rose promptly within the first 2 days of insertion of the medicated vaginal ring, and this elevation persisted as long as the vaginal ring was in place (4).

Cyclic Administration of Vaginal Contraceptive Rings. The effect of multicyclic intravaginal administration of medroxyprogesterone acetate-releasing vaginal rings was investigated in women for six cycles. The resultant serum drug concentration profiles are illustrated in Figure 20. In each cycle a new unit of vaginal ring (each contains 100 mg medroxyprogesterone acetate) was inserted on either day 5 or day 10, but both were removed on day 26 of the cycle to allow withdrawal bleeding to occur (46).

In the pretreatment cycles the serum levels of endogenous progesterone showed the normal midcycle elevation around day 22. With the intravaginal administration

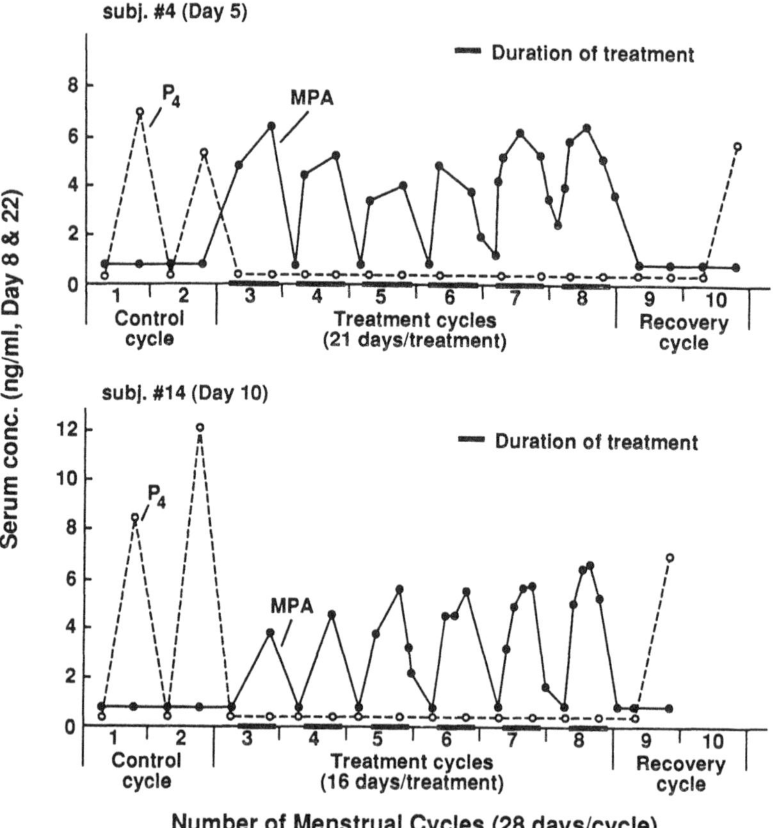

Figure 20 Serum concentration (days 8 and 22) profiles of medroxyprogesterone acetate and endogenous progesterone before, during, and after the cyclic intravaginal administration of medroxyprogesterone acetate-releasing vaginal rings. A new vaginal ring is inserted on either day 5 (subject 4) or day 10 (subject 14) and removed on day 26 of the cycle. Blood samples were collected and analyzed on days 8 and 22 of each cycle. [Replotted from the data by Cornette et al., J. Clin. Endocrinol. Metab., 33:459 (1971).]

of a medroxyprogesterone acetate-releasing vaginal ring this midcycle elevation did not occur, indicating that ovulation was effectively suppressed by the treatment. This suppression was observed to continue while the serum concentrations of progesterone remained low or nondetectable throughout medication with the cyclic intravaginal insertion of medroxyprogesterone acetate-releasing vaginal rings. In the meantime, serum levels of contraceptive medroxyprogesterone acetate measured on day 22 of each treatment cycle were maintained at a relatively constant level of approximately 6 ng/ml in all the wearers.

For those women who had the vaginal ring inserted on day 5 of the cycle, a serum level of medroxyprogesterone acetate, in the range of 4–6 ng/ml, was achieved on day 8. However, there was no detectable level of medroxyprogesterone acetate in those women receiving the medicated vaginal ring on day 10 of the cycle. However, very much the same serum drug concentration profile was achieved in both groups of women with cyclic insertion of the medroxyprogesterone acetate-releasing vaginal rings on either day 5 or day 10 of the cycle.

It is interesting to note that following completion of the last treatment cycle the vaginal ring is removed and the concentration of medroxyprogesterone acetate in the systemic circulation falls rapidly to an undetectable level. On the other hand, the regular midcycle elevation in progesterone levels reappears on day 22 of the first cycle following the termination. The observations demonstrate the ready reversibility of intravaginal contraception (6,46).

The intravaginal release profiles of medroxyprogesterone acetate from vaginal rings were found to be quite reproducible from one cycle to another (Table 4), with 18.2 ± 1.3 mg (or 1.14 mg/day) delivered in the 16-day treatment group and 27.3 ± 1.2 mg (or 1.30 mg/day) delivered in the 21-day treatment group. These results indicate a high degree of reproducibility in the uptake of medroxyprogesterone acetate by the vaginal mucosa (6).

Breakthrough bleeding was reportedly to be at a minimum. Some women (~20%)

Table 4 Intravaginal Release of Medroxyprogesterone Acetate from Silicone-Based Vaginal Rings[a]

	Total amount of drug released (mg ± SD)[b]	
Treatment cycles	16 Day treatment	21 Day treatment
1	20.3 (± 3.4)	29.0 (± 3.6)
2	18.7 (± 3.1)	27.8 (± 3.7)
3	17.7 (± 2.0)	27.5 (± 2.6)
4	18.0 (± 3.6)	25.9 (± 2.8)
5	18.1 (± 2.5)	27.5 (± 3.5)
6	16.2 (± 5.7)	26.1 (± 1.7)
Mean (± SD)[c]	18.2 (± 1.3)	27.3 (± 1.2)

[a]Each ring contains 100 mg medroxyprogesterone acetate and measures 65 mm in outside diameter and 9 mm in thickness.
[b]Mean (± standard deviation) of determinations conducted in six women wearing a vaginal ring from day 10 through day 26 (16-day treatment) or from day 5 through day 26 (21-day treatment) for six consecutive cycles.
[c]Mean (± standard deviation) of six treatment cycles.
Source: Compiled from the data by Mishell et al. (1972).

experienced a small degree of vaginal erosion that healed spontaneously following removal of the vaginal rings. Also, no discomfort to either partner was experienced during coitus (6).

2. Norgestrel-Releasing Vaginal Contraceptive Rings

Silicone-based vaginal rings have also been applied to the cyclic intravaginal administration of *d,l*-norgestrel in women for either single (11) or multiple cycles (50). Following the insertion of norgestrel-releasing vaginal rings, each containing 50 or 100 mg *d,l*-norgestrel, serum concentrations of *d*-norgestrel were observed to rise rapidly, reaching a peak level of 5–8 ng/ml within 1–2 days and then declining gradually during the remainder of the 21-day treatment to a level which was 30–40% of the peak concentration (Figure 21). In all cases the serum drug levels were found to fall off rapidly following removal of the vaginal ring.

Serum levels of *d*-norgestrel become stabilized at around 3 ng/ml with reinsertion of the same vaginal ring in the second and third cycles. During treatment serum concentrations of endogenous progesterone and estradiol were found suppressed to the levels detected in the early follicular phase. This observation suggests that ovulation was consistently inhibited (11).

B. Intravaginal Drug Delivery for Veterinary Applications

One of the well-demonstrated practical applications of intravaginal drug delivery in veterinary medicine is for estrus synchronization in domestic animals. Estrus syn-

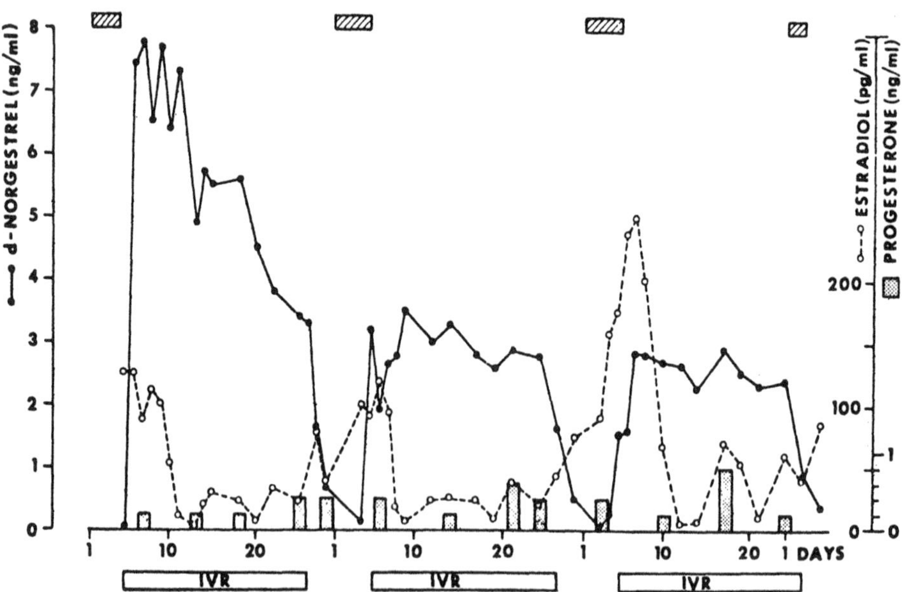

Figure 21 Serum concentrations of *d*-norgestrel and of endogenous progesterone and estradiol in a healthy woman wearing a silicone-based vaginal ring (IVR) that contains 100 mg *d,l*-norgestrel for three 21-day periods separated by 7-day intervals. Menstrual bleeding (including spotting) is indicated by hatched bars. [Reproduced with permission from Stanczyk et al., Contraception, 12:279 (1975).]

chronization is a treatment used to establish a stage at which one can induce fertile estrus as desired and also make female animals fertilizable in a desired season.

The administration of progesterone by daily injections throughout the duration of a cycle was reported to suppress estrus and ovulation in cows (51). The majority of the cows treated were observed to regain estrus within 2–5 days following termination of daily progesterone injection (52). Unfortunately, subfertility was often encountered if artificial insemination was carried out during this period. However, the normal fertility level returns at the subsequent estrus. Furthermore, estrus detection has been recognized as a time-consuming task, and inaccurate timing of insemination due to the incorrect detection of estrus has often led to a reduction in the rate of conception (53–55). Apparently, any means to relieve the burden of estrus detection would be beneficial. This can be accomplished by the development of an estrus synchronization treatment that produces a sexual response so precise that artificial insemination can be carried out on a predetermined time schedule following termination of the treatment rather than depending upon estrus detected in individual animals. Such a treatment was made possible with the development of a progestin-releasing vaginal pessary (Figure 22). This was first applied to sheep (56) and then successfully used in cyclic and noncyclic ewes (57,58).

1. Fluorogestone Acetate-Releasing Vaginal Pessaries

The progestin-releasing vaginal pessary is fabricated by dispersing a progestin, such as fluorogestone acetate, in a pessary made of porous polyurethane sponge. The pessary can be readily inserted intravaginally and removed according to a predetermined schedule.

Evaluation of seven synthetic progestins in spayed ewes identified fluorogestone acetate as the most active estrus-synchronizing agent. It has an activity profile iden-

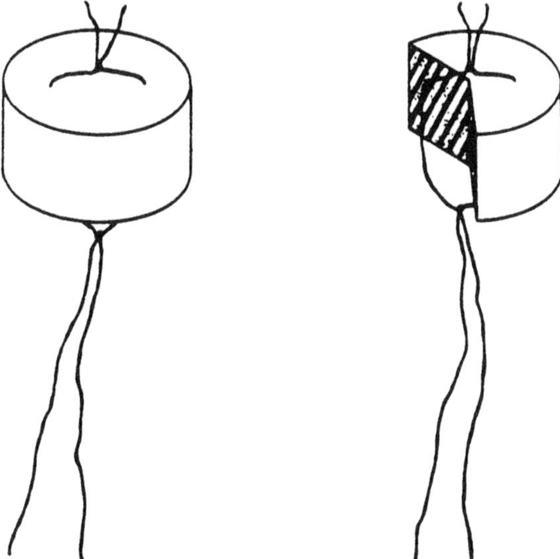

Figure 22 Fluorogestone acetate-releasing vaginal sponges (Syncro-Mate pessary, G. D. Searle & Co.).

tical to that of progesterone but is 20–25 times more potent (59). Further testing in the cyclic ewe confirmed that fluorogestone acetate is an effective ovulation inhibitor at a daily dose of 0.4 mg. The onset of estrus and ovulation following termination of the treatment is precise and dependable, with a timing comparable to that achieved by daily intravenous progesterone injections (60).

The effectiveness of fluorogestone acetate and other synthetic progestins was evaluated in merino ewes for the control of the timing of estrus and ovulation and of artificial insemination by delivering these progestins intravaginally from the vaginal pessary for 15 days. These progestin-releasing vaginal pessaries, which contain fluorogestone acetate (10–40 mg), medroxyprogesterone acetate (20–80 mg), or SC-9022 (80 mg or more), were found effective in blocking estrus and ovulation during insertion. In only 2–4 days after withdrawal of the pessaries 85% of the treated ewes were in estrus. With the first artificial insemination a lambing rate of 43% was achieved based on the number of ewes in estrus. This rate was increased to 65.4%, based on the total number of ewes treated, in the second cycle. The timing for the onset of estrus following treatment with fluorogestone acetate and SC-9022 was found to occur earlier and appeared to be more predictable than with medroxyprogesterone acetate (60).

The clinical efficacy of estrus synchronization using vaginal pessaries impregnated with either 500 mg progesterone or 30 mg fluorogestone acetate was compared in three field trials involving a total of 3388 Merino ewes. The results demonstrated that fluorogestone acetate-releasing vaginal pessaries were superior to those releasing progesterone. More effectively controlled estrus, a higher conception rate, and a lower sponge expulsion rate were achieved with fluorogestone acetate-releasing vaginal pessaries. A maximum rate of conception was obtained when artificial insemination was conducted on the second day following the withdrawal of sponge after a 15–18 day insertion. It was concluded that the fluorogestone acetate-releasing vaginal pessary offered a simple, effective method for estrus synchronization in Merino ewes. The lambing rate achieved with pooled semen was found comparable to that expected in untreated ewes, but with the benefits of inseminating all the animals on a scheduled basis (without requiring time-consuming estrus detection) and of eliminating the use of teaser rams (61).

Further studies in the anestrus Merino X Border Leicester ewes also reached the same conclusion that fluorogestone acetate-releasing vaginal pessaries are as effective as progesterone by daily injection in conditioning the anestrus crossbred ewe. The estrus response with ovulation was achieved following a single injection of pregnant mare serum gonadotropin (PMSG). Reasonable fertility was produced if the PMSG injection was administered within 24 hr after progestin treatment (62). On the other hand, PMSG injection was not required to advance the breeding season of the British-bred ewes for estrus synchronization by fluorogestone acetate-releasing vaginal pessaries (63).

The concept of using a progestin-releasing vaginal pessary as a simple, effective method for estrus synchronization was then patented by Robinson (64) and marketed by G. D. Searle & Co. under the trade name Syncro-Mate.

The intravaginal release of fluorogestone acetate from the Syncro-Mate pessary was observed to follow a linear Q versus $t^{1/2}$ relationship after a burst release in the initial phase (Figure 23). The release flux ($Q/t^{1/2}$) was found to be dependent upon the loading dose of fluorogestone acetate in the polyurethane sponge, as expected

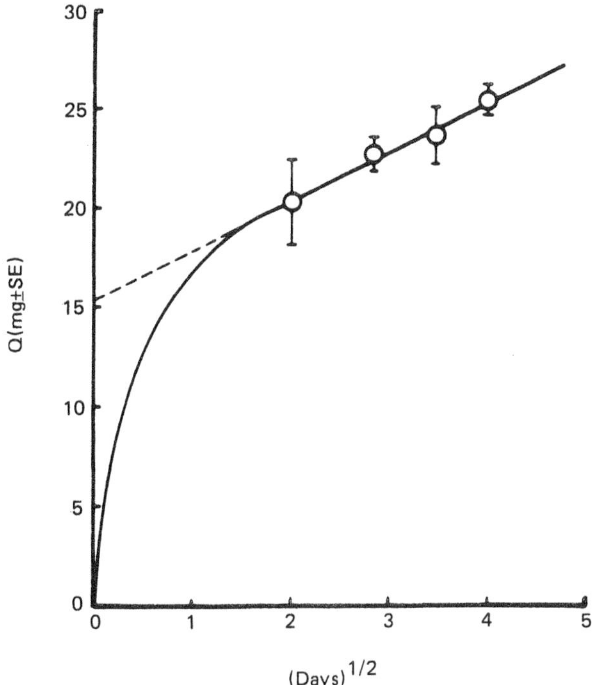

Figure 23 Intravaginal release of fluorogestone acetate from vaginal sponges in 40 lactating anestrus ewes. A release flux ($Q/t^{1/2}$) of 54.8 μg/cm^2 per day$^{1/2}$ was achieved.

from the polymer matrix diffusion process, but independent of the drying conditions used in its manufacture (Figure 24).

After insertion of the fluorogestone acetate-releasing vaginal pessary, the plasma concentration of fluorogestone acetate rose rapidly and then stayed within the range of 2–6 μg/ml for the 20-day application period (Figure 25). After removal of the vaginal pessary the concentrations of fluorogestone acetate in the blood and in the milk fell slowly as the compound was excreted through the urine and feces. A total of 12.3% of the drug loading in the sponge was absorbed during the 20-day treatment, and 98.2% of the absorbed dose was recovered in the excreta.

The time relationship between ovulation and estrus following treatment with the fluorogestone acetate-releasing vaginal pessary was examined in cyclic Merino ewes (65). The results suggested that the timing of ovulation is fairly reliable, and there is no evidence of any deviation from the normal estrus-ovulation time relationship. However, a highly significant relationship was detected between the doses of fluorogestone acetate and the timing of estrus and ovulation as well as the rate of conception (Table 5). The higher the loading dose of fluorogestone acetate in the vaginal sponges, the greater was the rate of conception. However, both estrus and ovulation tend to occur at a later time for higher doses.

Fluorogestone acetate-releasing vaginal pessaries were found not to affect the transport of the ovum and the capacity for fertilization. However, a significant effect ($p < 0.01$) on fertility was noted in the cycle immediately following completion of the synchronization treatment (58.1% in the first post-synchronization cycle com-

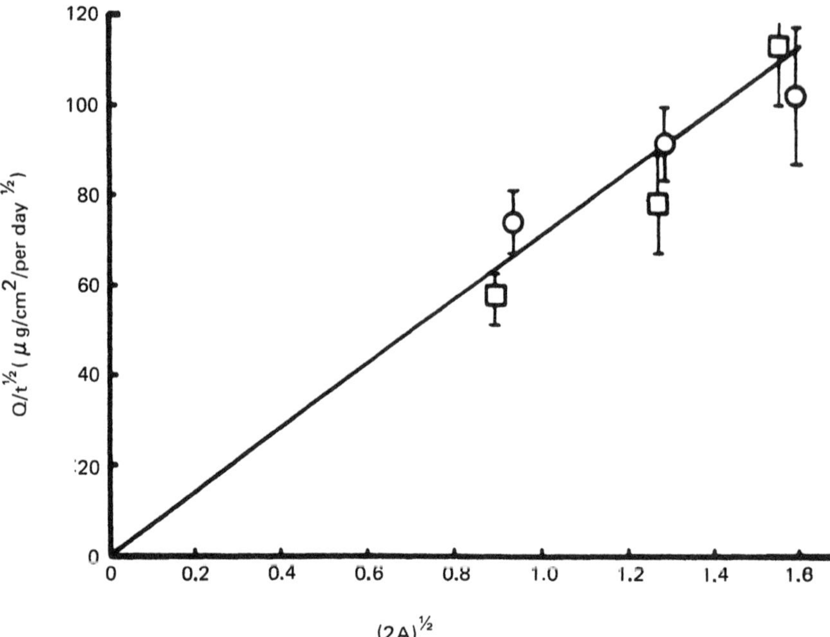

Figure 24 Dependence of the intravaginal release flux of fluorogestone acetate in Australian ewes on drug loading dose $(2A)^{1/2}$ in vaginal sponges: (○) drying at 50°C for 4 hr and (□) drying at room temperature for 12 hr.

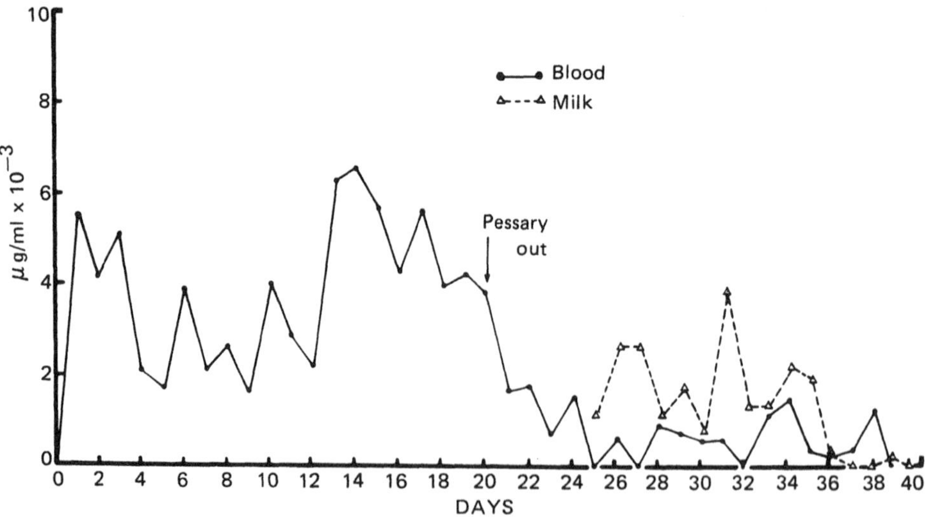

Figure 25 Concentration profiles of fluorogestone acetate in blood and milk during and after the 20-day insertion of a vaginal pessary in ewes.

Table 5 Effect of Drug Loading Dose in Vaginal Sponge on Times of Estrus and Ovulation and Fertility Rate

Fluorogestone acetate (mg per sponge)	Time[a] (hr) Estrus	Ovulation	Fertility rate (%)
10	29	48	18.4
20	34–48	60	31.7
40	48–53	72	41.7

[a]After the withdrawal of vaginal sponges.
Source: Tabulated from the data of Robinson and Smith (1967).

pared to 79.5% in the second cycle). Furthermore, it was reported that the first cycle following the treatment is more apt to be infertile. The percentage of fertilized ova recovered in the first cycle was maximized by artificial insemination with undiluted semen (66). On the other hand, in the second cycle no difference could be detected among the three methods of insemination (Table 6). These observations suggest a possible change in the environment of the reproductive tract in the first synchronized cycle, which influences the requirements of spermatozoa for maximum fertilization (67).

These studies were extended to investigate the possible effect of the fluorogestone acetate-releasing vaginal pessary on sperm transport. Results from investigations in ewes demonstrated that the number of spermatozoa recovered in the fallopian tube 24 hr after artificial insemination with 80, 200, or 500 million spermatozoa is substantially different between the treated and the untreated ewes. The number of spermatozoa recovered showed a trimodal distribution in the untreated ewes and a bimodal distribution in the treated animals. The ova recovered at 60 hr after artificial insemination were 85.7% fertilized in the untreated ewes but only 41.1% in the synchronized ewes. However, no effect could be attributed to the initial number of spermatozoa used for insemination, and no significant difference could be detected at 4, 12, and 36 hr. Thus the major cause for the reduction in fertilization in the synchronized ewes was attributed to the impairment of sperm transport (67).

The potential of fluorogestone acetate-releasing vaginal pessary as an estrus synchronizer was further evaluated in a large-scale field trial involving 9552 Merino ewes (68). The results confirmed the earlier observations that estrus is effectively synchronized, but the fertility from artificial insemination conducted at the first synchronized estrus is exceedingly variable. On the other hand, the timing of estrus in

Table 6 Cycle Dependence of the Fertility Rate Following Estrus Synchronization with Fluorogestone Acetate-Releasing Vaginal Pessaries

	Percentage of fertilized ova recovered		
	Artificial insemination with		
Cycle after synchronization	Diluted semen	Undiluted semen	Natural service
First cycle	39.1	80.8	52.0
Second cycle	82.1	77.8	78.6

Source: From Moore et al. (1967).

the second cycle is less variable and insemination is able to achieve a higher rate of fertility. It was thus recommended that for a large-scale artificial insemination with diluted semen the second estrus cycle is more desirable.

2. *Progesterone-Releasing Vaginal Spirals*

It was reported earlier that the intravaginal administration of progesterone failed to synchronize estrus in ewes. The three field trials conducted in Merino ewes also suggested that progesterone is less effective than fluorogestone acetate for estrus synchronization in ewes when both are administered through drug-impregnated vaginal sponges (61).

A new treatment was later developed for estrus control in cattle, which consists of a 21-day continuous intravaginal administration of progesterone from a controlled-release drug delivery device plus an intramuscular injection of gonadotropin-releasing hormone. This treatment permits the inseminators to schedule the time of artificial insemination without the need for behavioral detection of estrus (69).

This progesterone-releasing vaginal device was fabricated by coating a progesterone-dispersed silicone polymer matrix around a stainless steel spiral. It was designed to be inserted into the vagina as close to the cervix as possible by winding it around a rod. The spiral was left in the vagina for a duration of 18–21 days and then removed by pulling the nylon cord attached to the lower end. The spiral produced spontaneous luteal regression. Intramuscular administration of gonadotropin-releasing hormone (GRH) in saline solution 28–30 hr after removal of the spiral was found to induce the release of luteinizing hormone and ovulation at a predictable time.

A constant daily dose of progesterone was delivered to each cow from this progesterone-releasing vaginal spiral (Figure 26). This pseudo-zero-order drug release profile was attributed to the effect of the difference in time lags between the inner and outer surfaces (70). Following vaginal absorption a rapid elevation in the plasma levels of progesterone was observed, which reached the luteal phase concentration of 5–8 ng/ml within 60–90 min after insertion (Figure 27). The plasma progesterone level declined slowly to 1–3 ng/ml by Day 14 and then remained stable until Day 21. After removal of the vaginal spiral plasma progesterone concentrations dropped to 0.5 ng/ml within 24 hr (69). The vaginal absorption of progesterone from the spiral was reportedly independent of the cycle stage of the cattle. In both the follicular and luteal stages progesterone released from the vaginal spiral was rapidly absorbed through the vaginal mucosa and the peak plasma level was reached within 60–90 min.

The interval from the removal of the vaginal spiral to the return of estrus was prolonged, and a high incidence of silent estrus was noted in cows treated with the progesterone-releasing vaginal spiral. However, the rate of conception in cows inseminated at a predetermined time schedule, that is, 18–24 hr after the injection of GRH, showed no difference between the cows exhibiting behavioral estrus and those having silent estrus. Apparently the silent estrus was due to the fact that LH was released and ovulation occurred at a time earlier than what would normally occur. This resulted in elimination of the normal rise in the estrogen level and behavioral estrus. Nevertheless, the induced ovulation was equally fertile.

The plasma levels of LH near the onset of estrus in those cows treated with

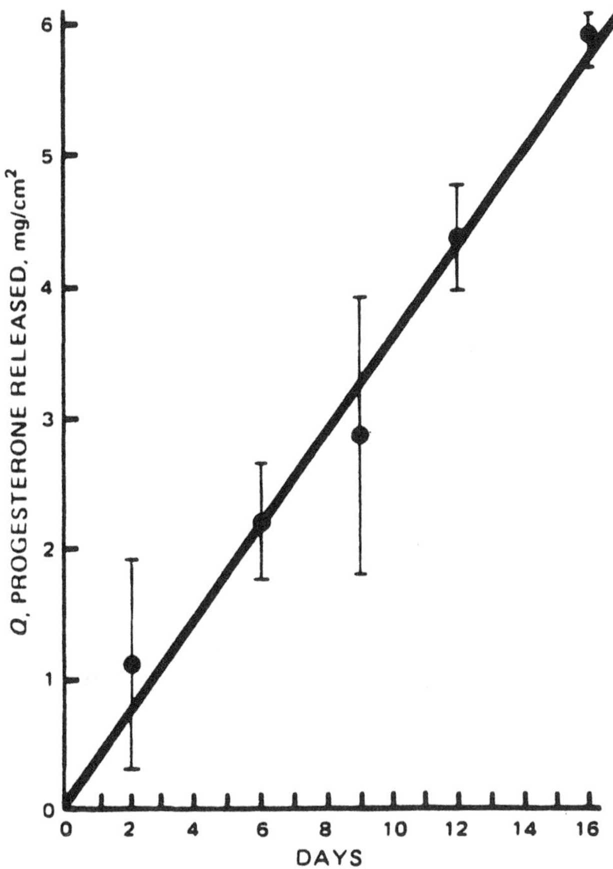

Figure 26 Cumulative intravaginal release of progesterone from the total (outside and inside) surface of a vaginal spiral in cows. A dose of 80.2 mg/day of progesterone was released from a surface area of 224 cm^2. [Replotted from the data by Winkler et al., J. Pharm. Sci., 66:816 (1977).]

progesterone-releasing vaginal spiral were similar to those of untreated controls, but a shorter duration of estrus was observed in cows treated with progesterone-releasing vaginal spirals plus GRH injection, which lasts 4–6 hr compared to a duration of 6–10 hr observed in untreated animals. Furthermore, it was also reported that the time and height of the LH peaks induced by GRH depend upon the route of administration as well as the dose of GRH and the time of dosing relative to the time of removal for the vaginal spiral. The intramuscular injection of 100 μg GRH at 28–30 hr after removal of the vaginal spirals was determined as the optimal condition for achieving the maximum LH levels. It is known that the LH peak is closely associated with the onset of behavioral estrus (69).

Furthermore, it was also found that the proportion of the cows that exhibit estrus and the time of estrus are both influenced by the plasma level of progesterone at the time of vaginal spiral removal. From a study of 44 cows receiving different doses of progesterone (but no GRH) it was observed that as the circulating level of pro-

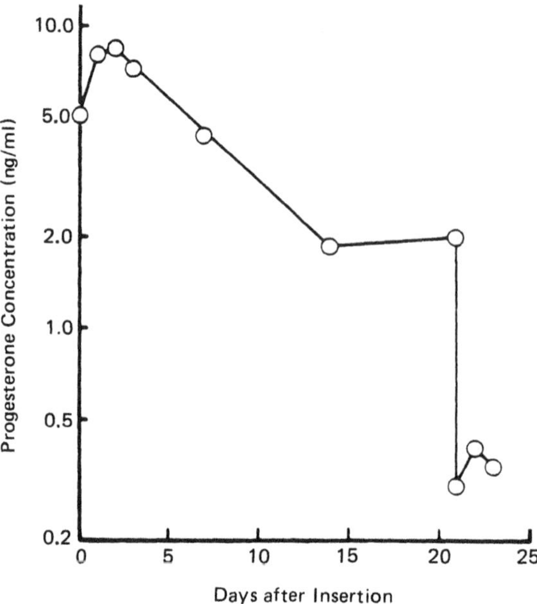

Figure 27 Plasma progesterone levels in 10 cows treated with progesterone-releasing vaginal spirals for 21 days. [Plotted from the data by Mauer et al., Ann. Biol. Anim. Biochem. Biophys., 15:291 (1975).]

gesterone is relatively high ($\geqslant$3 ng/ml) at the end of the 21-day treatment, the interval from the removal of the vaginal spirals to the appearance of estrus is extended and more variable (70 $\pm$ 30 hr). On the other hand, if the progesterone level is low ($\approx$1 ng/ml), some of the treated animals come into estrus during the treatment but the others come into estrus within the first 24 hr after removal of the vaginal spiral (with an interval of 30 $\pm$ 7 hr). Animals with a plasma progesterone level of 1–3 ng/ml at the time of vaginal spiral removal showed a synchronized estrus at 48 $\pm$ 16 hr after removal. The cows that did not exhibit estrus after the treatment returned to estrus within 21 days. This incidence of silent estrus was reportedly increased with the injection of GRH (69).

The proportion of silent estrus associated with progesterone-releasing vaginal spiral treatment was reportedly reduced by a modified vaginal spiral in which the progesterone-dispersed silicone polymer matrix was affixed to a nylon sleeve. Intravaginal administration of this new vaginal spiral yielded a lower but similar plasma profile of progesterone and a lower proportion of cows in estrus, as well as a shorter interval from the removal of the vaginal spiral to the return of estrus (Table 7). However, no significant difference was noticed in their rates of retention and conception.

The effect of the treatment duration with progesterone-releasing vaginal spirals on the rate of conception was studied in heifers (71). The spirals were inserted and left in the vagina for 9–21 days. At the time of spiral removal a supplementary injection of progesterone and estradiol benzoate (50 and 5 mg, respectively) was administered. All heifers in estrus were then inseminated with frozen semen from

Table 7 Relative Effectiveness of Progesterone-Releasing Vaginal Devices in Cows

	Proportion of cows in estrus[a] (%)				Efficacy[b] (%)	
Vaginal device	26–30 hr[c]	31–38 hr	39–46 hr	Silent	Retention	Conception
Nylon sleeve	32	41	12	15	96	49
Steel coil	6	26	32	35	99	50
Control	—	—	—	—	—	58

[a]Nylon sleeve treatment, 34 cows; steel spiral treatment, 31 cows.

[b]Nylon sleeve studies, 342 nonlactating cows and heifer; steel spiral studies, 54 nonlactating cows and heifer; control, 420 cows.

[c]Hours between the removal of vaginal device and the appearance of estrus.

Source: Compiled from the data by Mauer et al. (1975).

the same bull. The results showed that the fertility of the heifers is influenced by the length of intravaginal progesterone treatment. The proportion of heifers in estrus was observed to increase with the duration of vaginal spiral treatment, but the percentage of synchronized heifers that become pregnant decreased in proportion (Table 8).

The effect of the timing of artificial insemination on the rate of conception in cows treated with progesterone-releasing vaginal spirals was also investigated in 1496 cows (72). It appeared that single or double inseminations at 56 and 74 hr after the removal of the vaginal spirals improve the pregnancy rate from 62.8% to 71.0 and 68.9%, respectively (73).

VI. POTENTIAL DEVELOPMENTS OF INTEREST

A. Sandwich-type Vaginal Contraceptive Rings for Multicycle Administration

The intravaginal contraception treatments discussed earlier (Section V.A) require that a vaginal contraceptive ring be inserted continuously for a period of 21 days, followed by removal of the ring for 7 days between treatments to permit withdrawal bleeding.

A new treatment schedule was recently explored to minimize bleeding problems and to reduce the risk of ovulation during treatment. This new treatment schedule requires the user to insert the medicated vaginal ring into the vagina herself im-

Table 8 Effect of Treatment Duration of Progesterone-Releasing Vaginal Spirals on Biologic Efficacy

		Duration of treatment (days)			
Proportion	Control	9	12	18	21
Heifers in estrus (%)	96	70	82	100	92
Synchronized heifers become pregnant (%)	68	54	65	38	37

Source: Compiled from the data by Roche (1974).

mediately after the ceasing of menstrual bleeding and to leave it in place until menstruation-like bleeding (or spotting) occurs for 5 consecutive days. The vaginal ring is removed for 5 days and then it is reinserted again. In this way, breakthrough bleeding is transformed into withdrawal bleeding (14).

For this multicycle intravaginal contraception a new generation of sandwich-type vaginal rings (Figure 28) was developed, in which the drug-dispersed silicone polymer matrix is coated by a nonmedicated silicone polymeric membrane. It was designed to reduce the initial drug concentration spike frequently observed in the first treatment cycle, like that shown for *d*-norgestrel in Figure 21.

1. Norgestrel-Releasing Multicycle Vaginal Rings

This sandwich-type vaginal contraceptive ring was first utilized in the development of a norgestrel-releasing multicycle vaginal contraceptive ring to overcome the burst release of *d*-norgestrel in the initial phase of intravaginal release. The effect of the overcoat on the release rate profile of *d*-norgestrel is demonstrated in Figure 29, which shows that the addition of overcoat minimizes or eliminates the burst release of drug and shifts the non-zero-order drug release profile to the constant zero-order release rate profile. The greater the thickness h_m of the overcoat, the lower is the magnitude of the release rate.

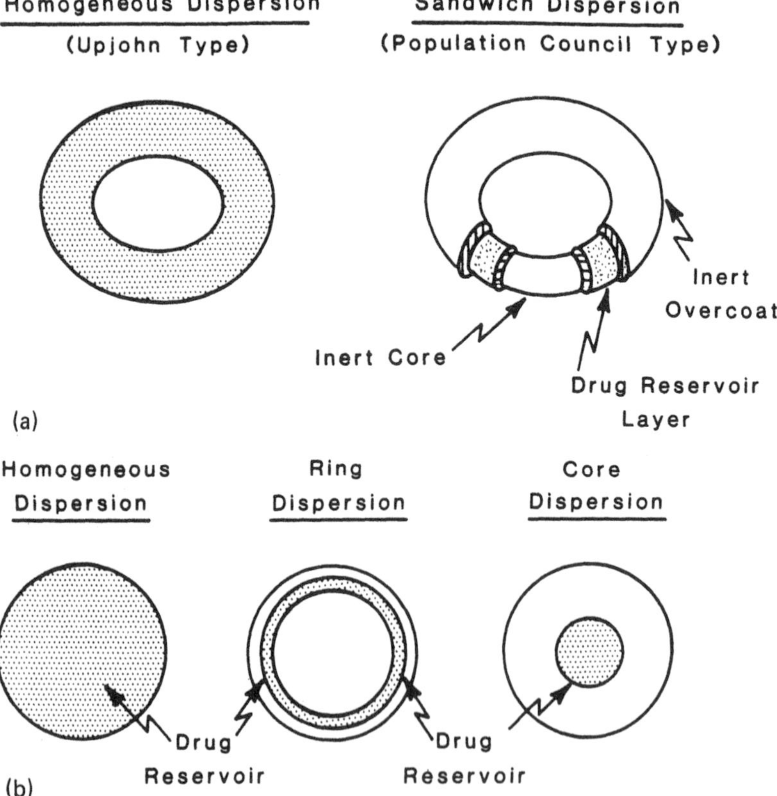

Figure 28 Various designs of intravaginal contraceptive delivery devices: (a) doughnut-shaped; and (b) disk-shaped devices (WHO).

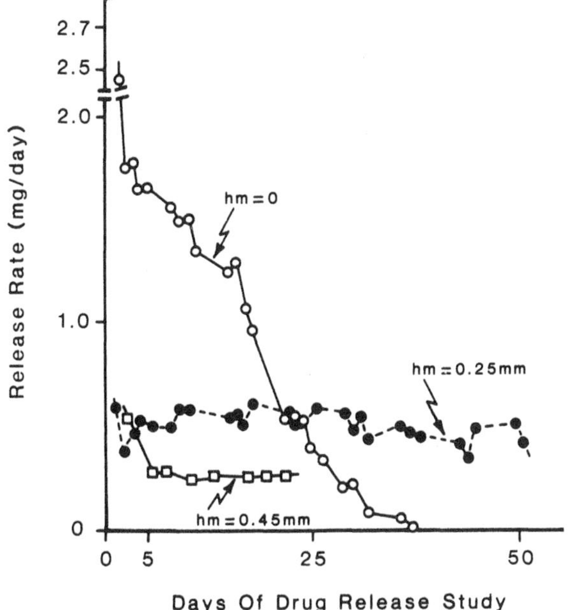

Figure 29 Comparative in vitro release rate profile of levonorgestrel from the Upjohn-type vaginal ring, which contains a homogeneous dispersion of drug in a silicone-based polymer matrix (○) and from the population council-type vaginal ring, which has an inert overcoat covering the drug reservoir layer (●, □). Also shown is the effect of overcoat thickness h_m on the release rate of levonorgestrel.

The total amounts of *d*-norgestrel released after either the single or the multiple intravaginal insertions for varying durations (different numbers of cycles) are compared in Figure 30. A fairly linear Q versus t relationship is achieved with this sandwich-type vaginal ring when one considers the inherent physiological variability among the subjects. Apparently, the number of insertions has no obvious effect on the intravaginal release profile of *d*-norgestrel from the sandwich-type vaginal contraceptive ring.

The pharmacokinetic profile of *d*-norgestrel following the multicycle intravaginal administration of this sandwich-type norgestrel-releasing vaginal ring was investigated in women continuously for up to a period of seven menstrual cycles (i.e., 194 days). The results in Figure 31 indicate that the plasma levels of *d*-norgestrel show an initial spike with concentrations varying in the range of 1.6–8.7 ng/ml, which decrease gradually to a rather stable level within 3–4 weeks of insertion. In the steady-state phase the variation in plasma drug concentrations in each subject is small (with the largest standard deviation only 0.4 ng/ml). Also, intersubject variation in the mean values of the plasma drug concentration has a range of 0.83–2.17 ng/ml. During multicycle treatment the midcycle peak of serum progesterone was suppressed, indicating the effective inhibition of ovulation.

The most advanced development of the contraceptive vaginal ring is a 90-day progestogen-releasing vaginal ring that releases levonorgestrel at an in vitro rate of 20 ± 3.5 μg/day (74). This vaginal ring is doughnut-shaped and is fabricated from

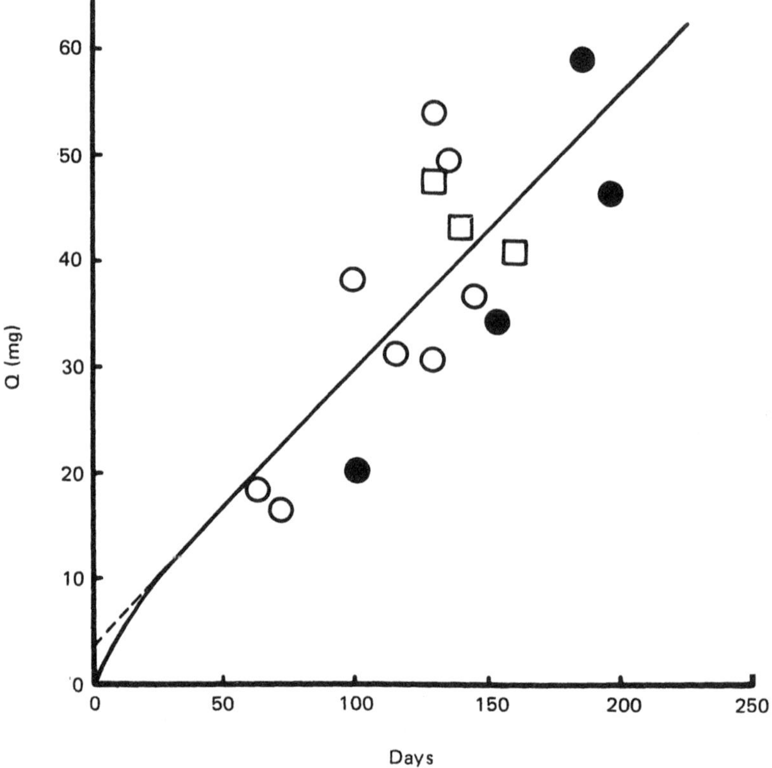

Figure 30 Intravaginal release of *d*-norgestrel from sandwich-type vaginal rings in 15 women during multicycle treatment: (●) single insertion, (○) two or three insertions, (□) four or five insertions. [Plotted from the data by Victor and Johansson, Contraception, 16:137 (1979).]

silicone elastomer by a three-stage molding process (75) to contain a drug-loaded silicone polymer matrix core containing 5 mg levonorgestrel and a drug-free silicone polymer overcoat. The rings are sterilized by ethylene oxide sterilization.

Following intravaginal insertion of the levonorgestrel-releasing contraceptive ring, plasma concentrations comparable to the final plateau levels were attained within approximately half an hour. Peak concentrations, which are approximately 200% of the plateau levels, were achieved within 1–8 hr after the insertion. A more or less constant plateau level of levonorgestrel was observed on Day 8. The in vivo release rate of levonorgestrel was noted to decline gradually after insertion for a few days, resulting in a reduction of 23–26% over the 90-day treatment period. Extension of the duration of insertion time beyond the 3-month period revealed a high variability in plasma levels, which varied in the range 36–86% after 6 months and 11–73% of the initial plasma levels after 12 months of continued use (76).

The contraceptive action of the vaginal ring was reported to be mainly based on its effect on the cervical mucus and the endometrium (77), even though the ring prevents ovulation for at least half of the cycles. The continuous intravaginal administration of levonorgestrel from the vaginal ring was observed to produce an ac-

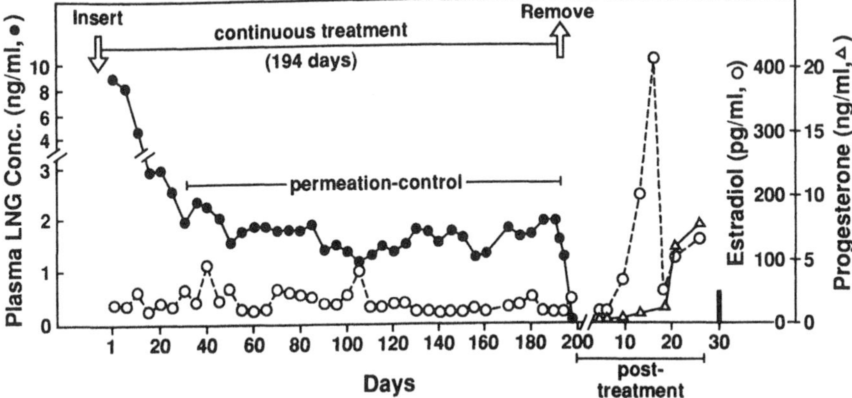

Figure 31 Plasma concentrations of *d*-norgestrel (●) and estradiol (○) during treatment and of progesterone (△) and estradiol (○) during recovery in a woman wearing a *d*-norgestrel-releasing sandwich-type vaginal ring for 194 days. Each data point in the treatment curve represents the mean for a 5-day period. [Replotted from the data by Victor and Johansson, Contraception, 16:137 (1977).]

ceptable bleeding pattern, especially in those cycles in which normal ovarian activities were not interfered by the levonorgestrel dose delivered. On the other hand, it was found that the continuous delivery of levonorgestrel at a dosage rate of 25 μg/day yielded an unacceptably high incidence of bleeding irregularities (78).

2. *(Progestin-Estrogen)–Releasing Vaginal Contraceptive Rings*

To improve the contraceptive efficacy and to overcome the irregular bleeding problem during treatment with progestogen-only vaginal rings, the sandwich-type vaginal ring (Figure 28) was later improved to provide the simultaneous release of a combination of progestin, such as levonorgestrel, and estrogen, such as estradiol, just like oral combined contraceptive pills (15). This doughnut-shaped vaginal ring developed by the population council contains levonorgestrel (39–77 mg) and estradiol (20–66 mg) in the middle silicone layer, which is sandwiched between a steroid-free central silicone polymer core and an outer silicone polymer coating, and was reported to deliver a combination of levonorgestrel and estradiol simultaneously at respective dosage rates of 230–290 and 150–180 μg/day (79).

A pilot study conducted in 10 healthy women for a period of up to six cycles has shown good clinical acceptance. Ovulation was inhibited in all treatment cycles and resumed within 1 month after completion of the trial. Bleeding control was excellent, with regular withdrawal bleeding (i.e., no episodes of failure of withdrawal bleeding), and only 3 days of breakthrough spotting were noted.

Serum levonorgestrel levels were relatively constant in each subject except for the first half of the first treatment cycle, which had slightly higher levels (Figure 32). On the other hand, serum estradiol concentrations rose rapidly to a peak of 100–300 pg/ml within 1–2 hr after insertion of the vaginal ring and then gradually declined over the next few days to a level that was generally less than 50 pg/ml.

After completion of the treatment no significant elevation in the mean levels of the binding capacity of corticosteroid-binding globulin was seen. On the other hand,

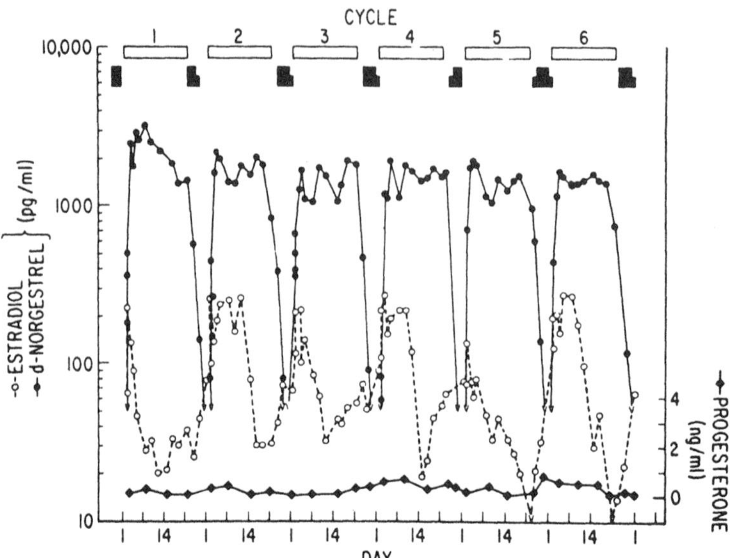

Figure 32 Serum profiles of *d*-norgestrel (levonorgestrel) and estradiol (on a logarithmic scale) and progesterone levels during six treatment cycles (vaginal rings were inserted on Day 1 and removed on Day 21 during each cycle). Horizontal bars represent a 3-week treatment period with rings in place. Solid bars represent bleeding days (full height for bleeding and half-height for spotting). [Reproduced with permission from Mishell et al., Am. J. Obstet. Gynecol., 130:55 (1978).]

the serum triglyceride concentration was found to decline from 59 mg per 100 ml (before treatment) to 49.8 ± 6.0 mg per 100 ml (after 6 months of treatment). These observations are in contrast to the increase in both concentrations usually noted in women taking oral contraceptive pills. They suggest that cyclic intravaginal contraception with the continuous administration of an estrogen-progestin combination via a controlled-release drug delivery system, such as the levonorgestrel-estradiol–releasing vaginal contraceptive rings reported here, has the advantages of inhibiting ovulation and controlling uterine bleeding without the disadvantages of adverse effects on systemic metabolic pathways and hepatic protein synthesis (15).

The sandwich-type contraceptive vaginal ring was also applied to the cyclic intravaginal administration of gestrinone, a synthetic progestin from Roussel-Uclaf. Addition of a thick coat of nonmedicated silicone elastomer over the progestin-containing core polymer matrix improved the release profile and gave a fairly constant plasma level (80).

The concept of intravaginal dual administration of progestin and estrogen in combination was recently extended to the development of a combined contraceptive vaginal ring to deliver simultaneously a combination of 3-keto-desogestrel, an active metabolite of desogestrel (which is the progestogenic constituent in the oral contraceptive Marvelon/Organon), and ethinyl estradiol, a synthetic estrogen. This new design of contraceptive vaginal ring (Figure 33) is constructed from two drug reservoir compartments; the major compartment consists of a 3-keto-desogestrel-loaded

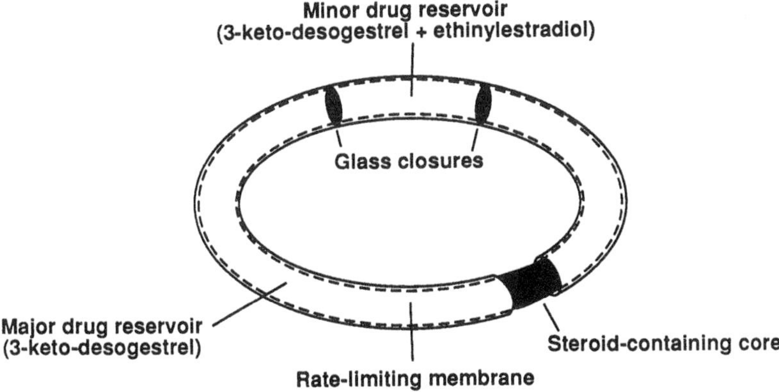

Figure 33 The Organon-type contraceptive vaginal ring showing various structural components. (Modified from A. P. Sam, 1991.)

core, and the other, minor compartment consists of a core loaded with a combination of 3-keto-desogestrel and ethinylestradiol. These drug reservoir compartments are separated by two steroid-impermeable glass closures, as the partitions, and release the steroids at a fixed ratio through a rate-limiting silicone membrane (81).

Comparative clinical evaluations of this contraceptive vaginal ring, with oral contraceptive Marvelon tablets as the reference product, were conducted in female volunteers (81). The ring was inserted on day 1 of a normal menstruation period and continuously used for a period of 21 days. The ring was then removed on day 21 to permit withdrawal bleeding to occur. The used ring was cleaned by washing with tap water and then reinserted 1 week later. This process was repeated for another cycle. After use for three consecutive cycles, a new vaginal ring was applied. The results in Figure 34 indicate that the Organon contraceptive vaginal ring achieved a rapid vaginal absorption of both progestin and estrogen, which reach plateau serum levels within 24 hr, and maintained steady-state serum levels throughout the course of the 21-day treatment. On the other hand, the daily oral administration of both steroids from Marvelon tablets took a much longer period of time to reach the same steady-state levels (82). The serum profiles in Figure 34 suggest that the vaginal ring appears to achieve a plateau serum level of 3-keto-desogestrel which is more than 10 times higher than that of ethinylestradrol, while the loading dose of 3-keto-desogestrel is 5- to 10-fold greater than that of ethinyl estradiol. The data in Figure 34 also demonstrate that the plateau serum level of 3-keto-desogestrel can be increased by increasing the loading level of 3-keto-desogestrel in the vaginal ring.

Instead of directly dispersing the drug particles in the polymer matrix, a microreservoir partition-controlled drug delivery technology (Chapter 1, Section III.C) was applied to the development of contraceptive vaginal rings (83). This type of vaginal ring contains a combination of progestin and estrogen in numerous microscopic drug reservoirs and uses a dissolution/partition process to achieve a constant delivery of drugs at a fixed daily dosage rate ratio (Chapter 2, Section II.D). A typical set of release profiles for the simultaneous release of progestin and estrogen is shown in Figure 35.

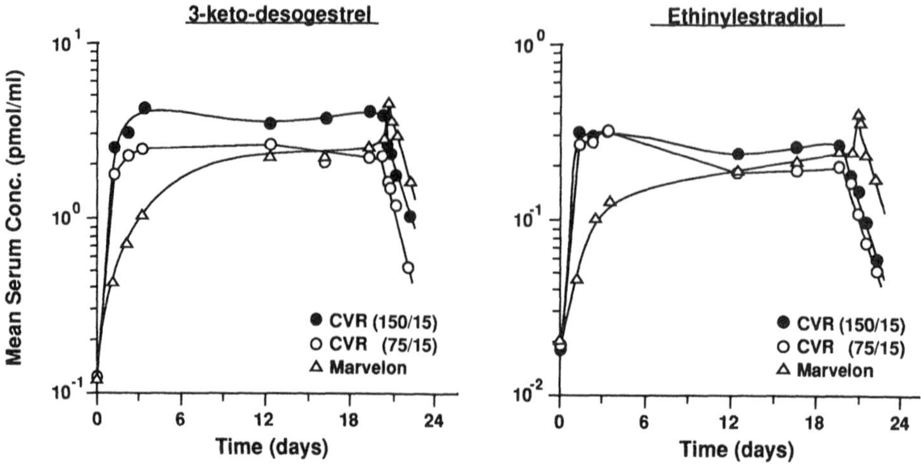

Figure 34 Comparative mean serum profiles of 3-keto-desogestrel (3-KD) and ethinyl estradiol (EE) resulted from the 21-day continuous intravaginal delivery of 3-KD and EE from two prototype combined contraceptive vaginal rings (CVR) as well as the trough serum levels of 3-KD and EE from the oral daily administration of Marvelon tablets. (Replotted from A. P. Sam, 1991.)

B. Prostaglandin-Releasing Vaginal Devices for Pregnancy Termination

Prostaglandins, notably PGE_2 and PGF_2, can promote normal phasic contractions of the uterine smooth muscle in the same way as oxytocin secreted by the posterior pituitary. However, the oxytocic action of prostaglandins is not as dependent on the estrogen-progesterone balance as that of oxytocin, so that prostaglandins can induce labor considerably in advance of term and hence can be used to terminate pregnancy. Certain prostaglandins have been used successfully as abortifacients (84).

The concept of intravaginal drug delivery has also been applied for some years to the vagina-targeted delivery of natural or synthetic prostaglandins using vaginal delivery systems, such as vaginal suppositories (85). Even though clinical efficacy for menses induction, second-trimester abortion, and cervical dilation has been demonstrated when using these prostaglandin-containing vaginal suppositories, several problems, such as the physical stability of suppository formulations, which are mostly made of lipid-based vehicles, and the chemical stability of certain prostaglandins in the formulation vehicles, have been reported with some of the vaginal suppositories (86). Furthermore, substantial variation in the intravaginal release and vaginal absorption of drug delivered by the suppository dosage form occurs as the suppository base melts and the surface area varies as the base changes in physical form from a solid to an oil state (87).

It is thus critically important to develop a controlled-release vaginal delivery system that would improve the intravaginal administration of very potent drugs, such as prostaglandins, by (1) controlling the release rate of prostaglandin in a predictable and reproducible fashion, (2) requiring only a single administration, which avoids the multiple dosing regimens normally required for vaginal suppositories, and (3) allowing self-administration and the termination of drug administration by the facile

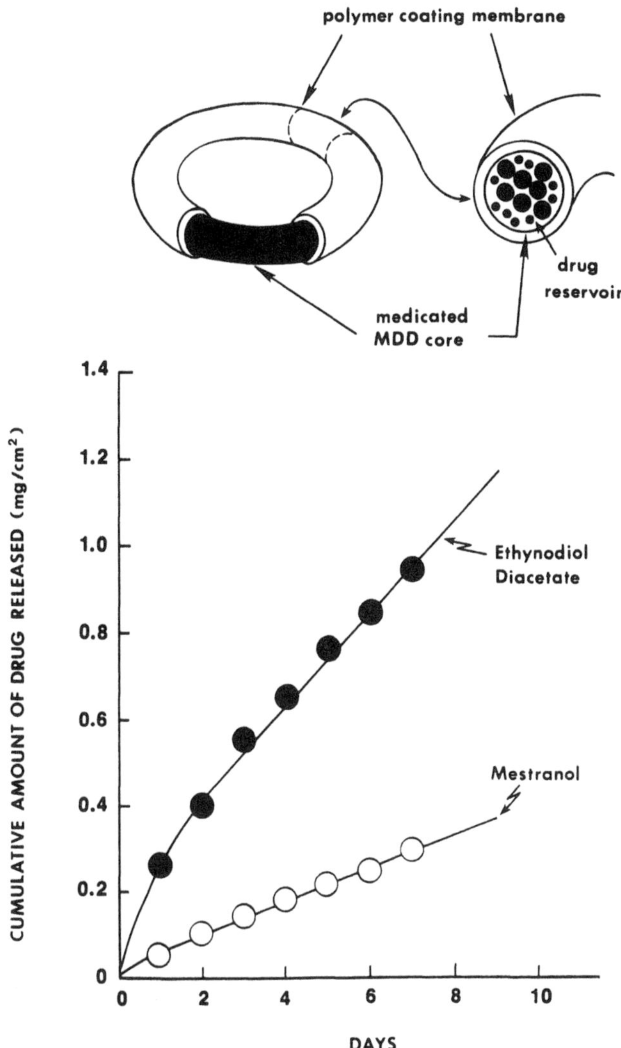

Figure 35 The Searle-type contraceptive vaginal ring showing various structural components and in vitro release profiles of ethynodiol diacetate, a synthetic progestin, and mestranol, a synthetic estrogen, at a daily dosage rate ratio similar to that of oral contraceptive pills.

removal of the system. Several monolithic silicone-based prostaglandin-releasing vaginal delivery systems have been developed (19,20,85,88,89).

A typical example is the development of a prostaglandin-releasing vaginal device, such as the silicone-based vaginal insert that releases the methyl ester of 15(*S*)-15-methylprostaglandin $F_{2\alpha}$ for the induction of abortion. The clinical efficacy of the vaginal device that releases this PGF_2 analog (0.5 or 1.0%) for the induction of abortion was studied in 90 healthy women in either the first or second trimester of gestation (19,20).

The peripheral plasma levels of this synthetic analog of prostaglandin $F_{2\alpha}$ in the

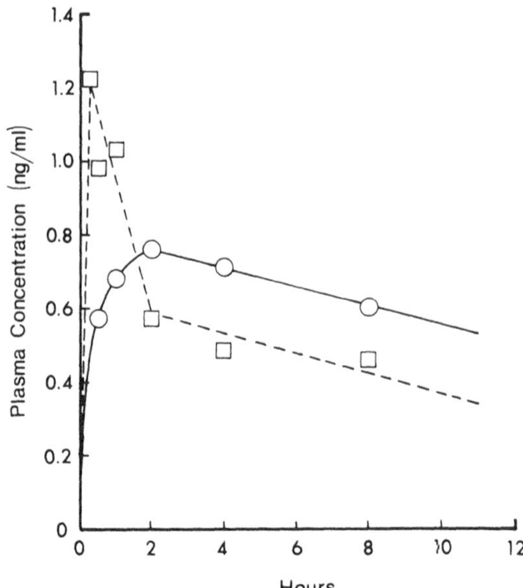

Figure 36 Plasma levels of 15(*S*)-15-methylprostaglandin $F_{2\alpha}$ methyl ester (15-Me-$PGF_{2\alpha}$ methyl ester) in 23 patients after the intravaginal administration of a silicone device containing either (○) 0.5% or (□) 1.0% 15-Me-$PGF_{2\alpha}$ methyl ester. [Plotted from the data by Lauersen and Wilson, Prostaglandins, 12(Suppl.):63 (1976).]

patients wearing vaginal inserts containing a loading dose of 0.5% reached peak concentrations within 2 hr after treatment; vaginal inserts with 1.0% drug loading achieved a higher peak concentration within a shorter period (15 min) of insertion followed by a rapid decline to a much more stable plasma level (Figure 36).

The abortifacient activity of the prostaglandin-releasing vaginal insert is illustrated in Table 9. It appears that the continuous intravaginal administration of prostaglandin from a controlled-release drug delivery device induces a high percentage (75%) of women in either the first or second trimester to successfully abort with a

Table 9 Abortifacient Effects of Prostaglandin-Releasing Vaginal Devices

Drug content (%)	Efficacy of abortion[a] (%)		Mean abortion time (hr)
	First trimester	Second trimester	
0.5	75[b] (n = 20)	—	11.3[e]
	—	74.3[c] (n = 35)	15.2[e]
1.0	—	74.3[d] (n = 35)	15.6[f]

[a]n is the number of patients treated.
[b]Improved to 100% by applying suction aspiration.
[c]Improved to 88.6% by concomitant oxytocin infusion.
[d]Improved to 91.4% by concomitant oxytocin infusion.
[e]Multiparous patients aborted significantly faster than the nulliparous patients.
[f]No significant difference between multiparous and nulliparous patients.
Source: Compiled from the data by Lauersen and Wilson (1976).

mean abortion time of 11.3 hr for the first-trimester patients and 15.2–15.6 hr for the second-trimester patients. This abortifacient effect can be further improved with a concomitant oxytocin infusion or suction aspiration.

The effect of the variation in prostaglandin dosages on the efficiency of induced abortion and the mean abortion time is not significantly different. However, with the 0.5% device multiparous women in both the first and the second trimesters aborted in an abortion time significantly faster than nulliparous women. On the other hand, with the 1.0% device there is not significant difference in the mean abortion time between nulliparous and multiparous patients.

If the loading dose of prostaglandin in the vaginal insert was reduced to 0.25%, all the fetuses at 15–16 weeks of pregnancy were aborted before 22 hr, with a mean time to fetal abortion of 12.7 $\pm$ 4.2 hr (84).

The prostaglandin-releasing vaginal insert was also found effective in inducing abortion in pregnant women who are no more than 50 days beyond the first day of their last normal menstrual period. Following intravaginal insertion the serum levels of human chorionic gondotropin fall rapidly, reaching about one-fourth the baseline level within 24 hr. The fall in plasma progesterone and estradiol levels was less pronounced; plasma 17α-hydroxyprogesterone levels increase during the first 2 hr but fall to an average value that is 27.4% of the baseline level by 24 hr (90).

Diarrhea was the most frequently observed side effect. The severity of the gastrointestinal disturbance appears to be dose related and can be substantially eliminated by premedication with antiemetic and antidiarrheal drugs. Except for this side effect, this abortion induction technique could offer a valid alternative to the surgical termination of pregnancy.

Recently, a new generation of prostaglandin-releasing vaginal device, which is a reservoir-type membrane permeation-controlled drug delivery system, was developed to achieve the intravaginal controlled delivery of 16,16-dimethylprostaglandin E_2 *p*-hydroxybenzaldehyde semicarbazone ester (87). The clinical efficacy of this PGE_2 analog in a vaginal suppository has been documented (91,92).

This prostaglandin-releasing vaginal device is a vaginal tampon-shaped drug delivery system fabricated by coating a cotton tampon with a composite of an inner prostaglandin-containing reservoir layer and an outer rate-controlling membrane (Figure 37) by a double dip-coating technique. The rate of prostaglandin release was altered by varying the thickness of the rate-controlling membrane (Chapter 2, Section II.A). Two vaginal devices that release this PGE_2 analog at dosage rate of either 20 or 40 μg/hr were developed.

The in vivo efficacy of these prostaglandin-releasing vaginal devices was evaluated in four pregnant (second-trimester) rhesus monkeys. The results indicated that these devices, with different dosage rates of prostaglandin, caused a slightly different effect on the uterine motility of these pregnant monkeys. The frequency and amplitude of uterine contractions were both observed to begin with different timings of onset; at approximately 5 min for the 40 μg/hr device and at 15–30 min for the 20 μg/hr device following vaginal insertion. Also, the uterine activity was observed to be significantly higher for the 40 μg/hr device than for the 20 μg/hr device. The vaginal devices were left in place for 8 hrs; the increased uterine activity was observed to persist throughout the duration of 4.5 hr recording session (Figure 38). The two pregnant monkeys treated with the 40 μg/hr device were aborted within 7 days after the treatment, and the other two treated with the 20 μg/hr device were

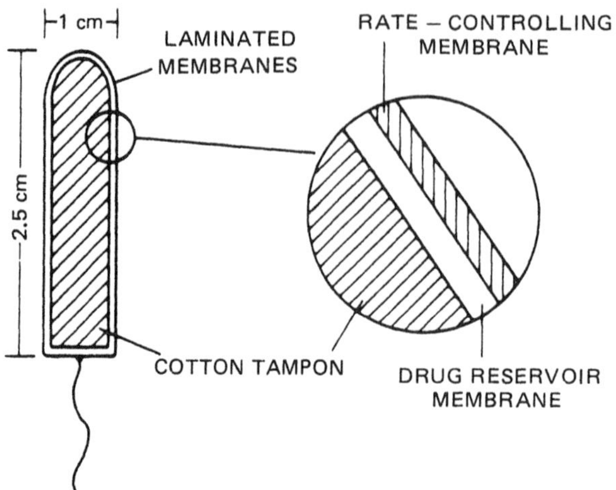

Figure 37 A reservoir-type membrane permeation-controlled prostaglandin-releasing vaginal device. (Reproduced from Reference 87.)

each aborted within 48 hr and 8 days, respectively (87). The results led the investigators to conclude that the vaginal devices developed are capable of controlling the intravaginal delivery of this prostaglandin analog and of promoting continuous uterine stimulation throughout the duration of vaginal insertion at a level sufficient to terminate pregnancy in monkeys.

C. Progestin-Releasing Vaginal Devices for Estrus Synchronization

As reported in Section V.B.1, a progestin-releasing vaginal pessary was developed for the intravaginal controlled delivery of fluorogestone acetate, a synthetic progestin

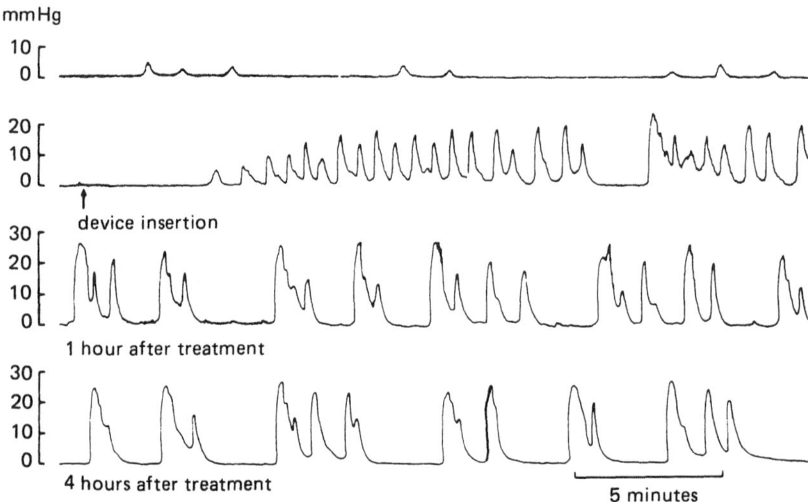

Figure 38 Uterine motility in a pregnant monkey treated with a prostaglandin-releasing vaginal device (shown in Figure 37) with a release rate of 40 μg/hr. (Reproduced from Reference 87.)

that is 20–25 times more potent than progesterone (59), for estrus synchronization in domestic animals, such as ewes. Unfortunately, the intravaginal delivery profile of fluorogestone acetate was found not at a constant rate but followed a matrix diffusion process. A Q versus $t^{1/2}$ profile was obtained, and a loading dose of fluorogestone acetate as high as 40 mg was incorporated into each vaginal pessary (Chronogest or Syncro-mate pessaries, Searle), even though a total of only 12.3% of the dose were actually delivered during the 20-day treatment. This has raised some regulatory concerns, especially by the U.S. Food and Drug Administration.

Extensive developmental efforts were recently devoted to the redesign of this progestin-releasing vaginal pessary, which aimed to minimize the loading dose, to overcome the Q versus $t^{1/2}$ intravaginal release and absorption profiles, as well as to improve the systemic bioavailability (16,17). These efforts have resulted in the development of two new generations of progestin-releasing vaginal pessaries (Figure 39). Both types have made use of the polyurethane sponge in the vaginal pessary as the mechanical support for vaginal insertion and retention, but the drug reservoir was relocated from the porous sponge matrix to a sheet-type rate-controlled silicone device that covers the circumferential surface of the sponge. The type I rate-controlled silicone device consists of a homogeneous dispersion of drug in a silicone polymer matrix; in type II the drug-dispersing polymer matrix is sandwiched between two sheets of silicone polymer membrane to form a three-layered laminate.

The in vitro vaginal permeation profiles of fluorogestone acetate delivered by

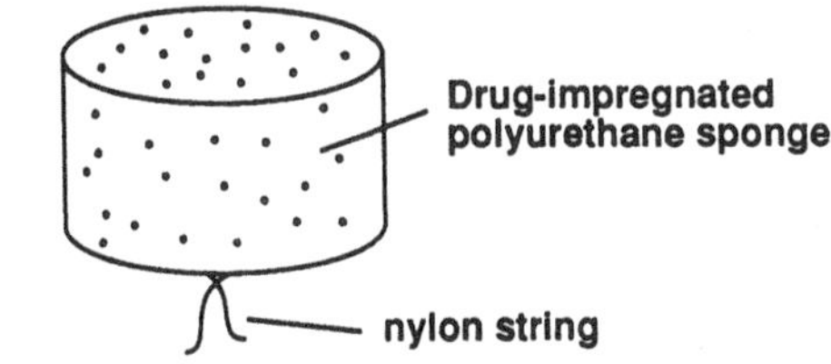

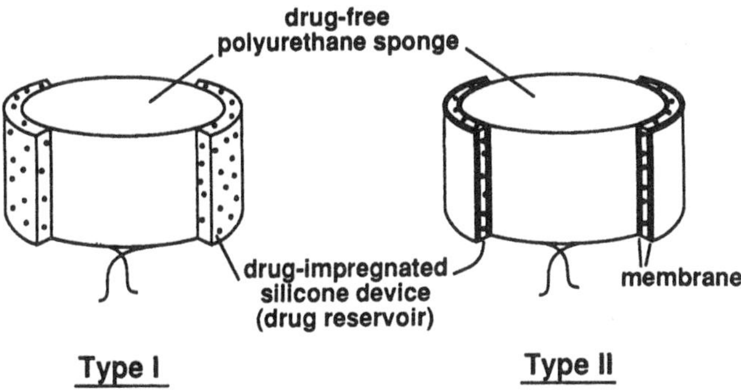

Figure 39 Comparison of the new generation of progestin-releasing vaginal pessaries (types I and II) with the old (Syncro-Mate) vaginal pessary.

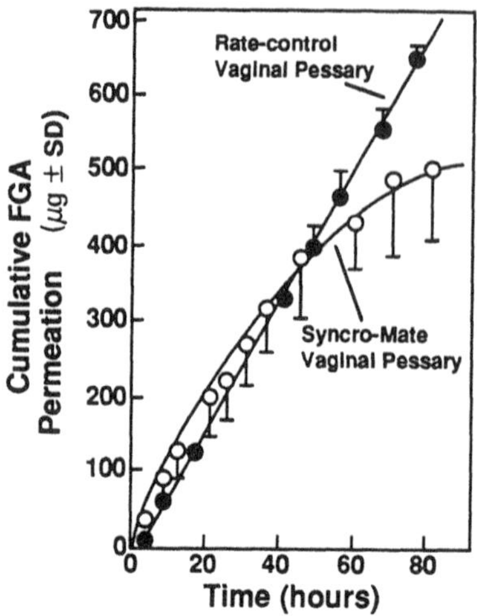

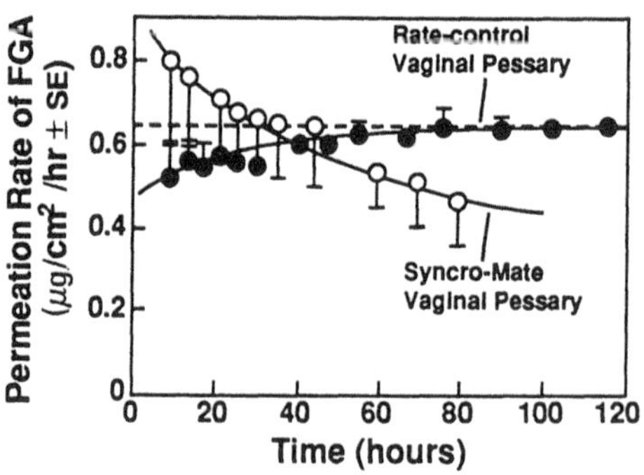

Figure 40 (Top) Comparison of the vaginal permeation profiles of fluorogestone acetate (FGA) from the rate-controlled vaginal pessary (type I) and from the Syncro-Mate vaginal pessary (both containing 20 mg FGA). The rate-controlled vaginal pessary (●) shows a constant $Q - t$ relationship with a Q/t value of 275 ± 5 $\mu g/cm^2$ per day; the Syncro-Mate pessary (○) shows a nonlinear $Q - t^{1/2}$ relationship with a $Q/t^{1/2}$ value of 397 ± 60 $\mu g/cm^2$ per day$^{1/2}$. Each data point represents the mean value (± SD) of three determinations. (Bottom) Comparison of the time course for the vaginal permeation rates of FGA from the rate-controlled (type I) and Syncro-Mate vaginal pessaries. The permeation rate profile increases and reaches a plateau and constant value of 0.64 ± 0.01 $\mu g/cm^2/hr$ (or 15.28 ± 0.28 $\mu g/cm^2$ per day) at around 40 hr for rate-controlled vaginal pessaries (●); the permeation rate for the Syncro-Mate pessary (○) continues to decrease in a curvilinear fashion ($Q/t^{1/2} = 27.95 \pm 4.22$ $\mu g/cm^2$ per day$^{1/2}$).

the type I rate-controlled vaginal pessary and the Syncro-Mate vaginal pessary, both containing the same drug loading dose of fluorogestone acetate (20 mg per pessary), were first evaluated simultaneously using excised sheep vaginal mucosa as permeation barrier (16). The results indicated that fluorogestone acetate permetes through the vaginal mucosa at a constant rate; a Q versus t vaginal permeation profile is attained when delivered from the rate-controlled pessaries, but a nonlinear (Q versus

$t^{1/2}$) permeation profile is observed with Syncro-Mate pessaries (Figure 40). The rate of vaginal permeation of fluorogestone acetate from the type I rate-controlled vaginal pessary increased with time initially and then reached steady state at around 40 hr and remained at the steady-state rate of $0.64 \pm 0.01\ \mu g/cm^2$ per hr throughout the course of the studies; the rate from the Syncro-Mate pessary was not constant and continued to decrease with time (17). The in vivo intravaginal release of fluorogestone acetate from the type I rate-controlled vaginal pessaries was also studied in sheep, and the results confirmed the in vitro observations that fluorogestone acetate is delivered at a constant rate from the rate-controlled vaginal pessaries and the rate of intravaginal release is relatively independent of drug loading dose in the polymeric device and its location on the surface of the polyurethane sponge (18).

Further in vivo studies indicated that the cumulative amount of fluorogestone acetate delivered intravaginally from the rate-controlled vaginal pessary in the sheep is linearly dependent upon the surface area of the drug-releasing device covering the circumference of the polyurethane sponge (Figure 41). Correspondingly, the plasma levels of fluorogestone acetate in the sheep also increase in proportion with the surface area of the drug-releasing polymeric device. This proportionality permits one to determine the proper surface area of the drug-releasing polymeric device needed

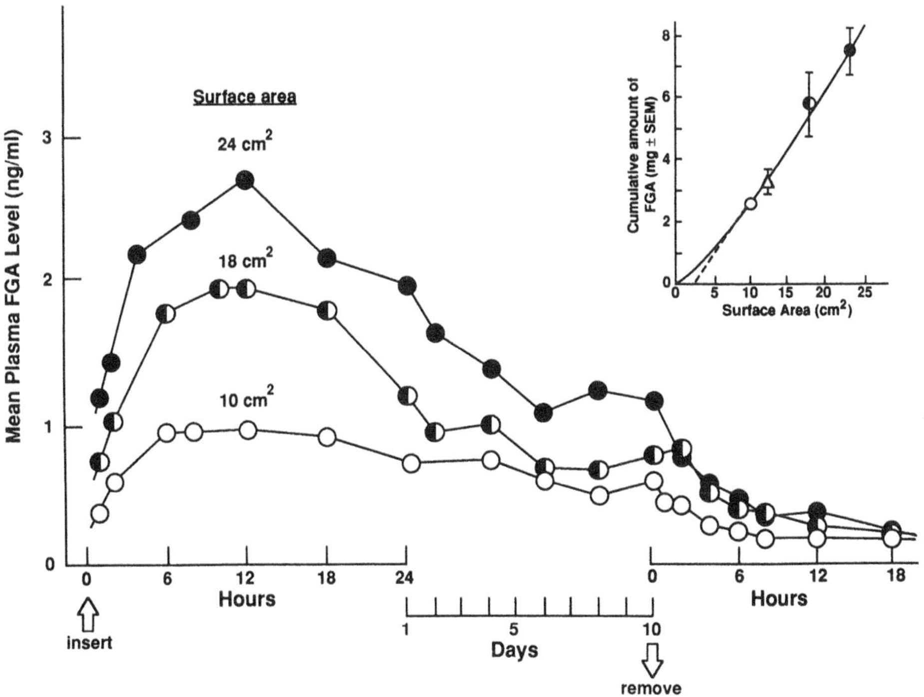

Figure 41 Mean plasma concentration profiles of fluorogestone acetate (FGA) as a function of the surface area of a FGA-releasing device on type I rate-controlled vaginal pessaries. (Insert) Linear relationship between the cumulative amount of FGA delivered and the surface area of the FGA-releasing device located on the circumferential surface of the rate-controlled vaginal pessary (type I), which was inserted in the sheep's vagina for 10 days. Each data point represents the mean value (± 1 SEM) of four to five animals.

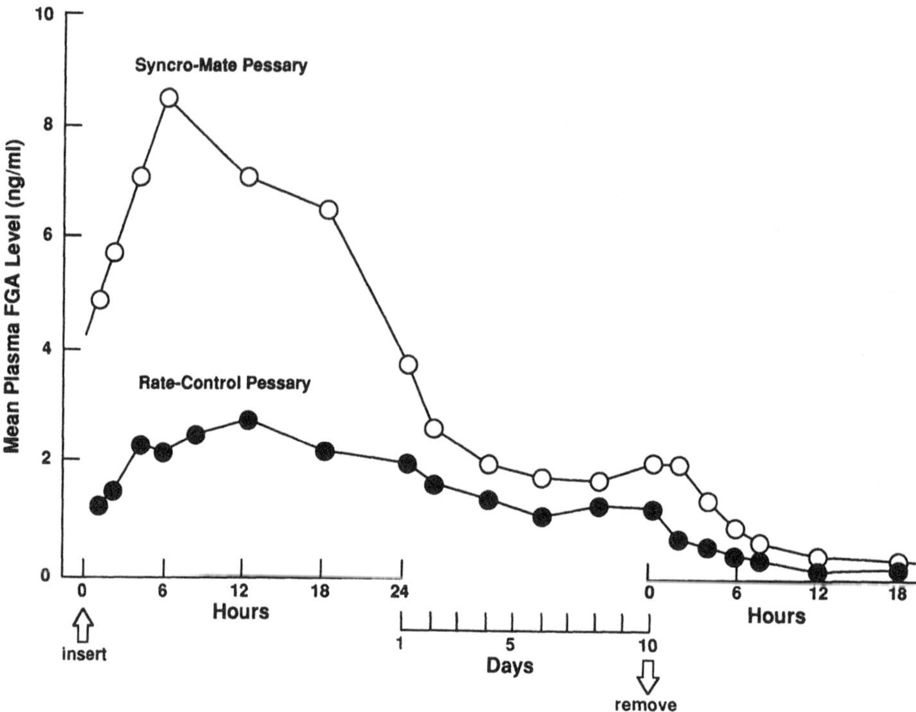

Figure 42 Comparative plasma concentration profiles of fluorogestone acetate delivered intravaginally in sheep from rate-controlled vaginal pessaries ($n = 6$, with a drug-releasing surface of 24 cm^2) and from Syncro-Mate vaginal pessaries ($n = 6$) for 10 days.

to achieve a therapeutically effective level. It is interesting to note that rate-controlled vaginal pessaries achieve a lower but more steady plateau plasma concentration profile of fluorogestone acetate than Syncro-Mate vaginal pessaries (Figure 42).

The clinical efficacy of type I rate-controlled vaginal pessaries was evaluated in 60 sheep for 19 days. These vaginal pessaries were found to be easily introduced into the sheep vagina (18). Moreover, only 1 of the 60 pessaries evaluated was lost during the entire clinical trial. Most importantly, rate-controlled pessaries with the

Table 10 Comparative Clinical Efficacy in Ewes[a]

	Biological effectivess (%)		
Treatment	Cycle blocking	Estrus synchronization[b]	Ovulation
Chronogest Pessaries (40 mg FGA)[c]	100	62.5	100
Rate-controlled vaginal pessaries (20 mg FGA)[c]	100	63.6	100

[a]Adult, cyclic Ile de France ewes, 12-day treatment.
[b]Detected within 24–56 hr following termination of treatment.
[c]Loading dose of fluorogestone acetate (FGA) in each pessary.

18 cm^2 drug-releasing polymeric device on the circumferential surface were observed to be clinically effective in blocking the cycle during the 19-day treatment and then synchronizing estrus following termination of the treatment. All the ewes treated were detected to ovulate normally by endoscopic examination of the ovaries a few days later. Artificial insemination could then be carried out during this period to achieve a greater rate of conception. The results in Table 10 demonstrated that the biological effectiveness of this rate-controlled vaginal pessary (containing 20 mg fluorogestone acetate) is comparable to that achieved by the Chronogest pessaries (which contain 40 mg fluorogestone acetate each) on the market.

ACKNOWLEDGMENTS

The author expresses his appreciation to Dr. J. C. Bouffault of Synkron Corporation for supporting development and animal studies of the rate-controlled vaginal pessaries and the Dr. A. P. Sam of Organon for sharing technical data on the Organon-type contraceptive vaginal ring.

REFERENCES

1. J. R. Rock, H. Baker and W. Bacon; Science, 105:13 (1947).
2. D. R. Mishell, Jr.; Contraception, 12:249 (1975).
3. G. W. Duncan; U.S. Patent 3,545,439 (December 8, 1970).
4. D. R. Mishell, Jr., M. Talas, A. F. Parlow and D. L. Moyer; Am. J. Obstet. Gynecol., 107:100 (1970).
5. D. R. Mishell, Jr. and M. Lumkin; Fertil. Steril., 21:99 (1970).
6. D. R. Mishell, Jr., M. Lumkin and S. Stone; Am. J. Obstet. Gynecol., 113:927 (1972).
7. M. R. Henzl, D. R. Mishell, Jr., J. G. Velasquez and W. E. Leitch; Am. J. Obstet. Gynecol., 117:101 (1973).
8. D. R. Mishell, Jr., M. Lumkin and T. Jackanicz; Contraception, 12:253 (1975).
9. E. D. B. Johansson, T. Luukkainen, E. Vartianinen and A. Victor; Contraception, 12:299 (1975).
10. O. Akinla, P. Lahteenmaki and T. M. Jackanicz; Contraception, 14:671 (1976).
11. A. Victor, L. Edquist, P. Lindberg, K. Elamsson and E. D. B. Johansson; Contraception, 12:261 (1975).
12. A. Victor and E. D. B. Johansson; Contraception, 14:215 (1976).
13. A. Victor, H. A. Nash, T. M. Jackanicz and E. D. B. Johansson; Contraception, 16:125 (1977).
14. A. Victor and E. D. B. Johansson; Contraception, 16:137 (1977).
15. D. R. Mishell, Jr., D. E. Moore, S. Roy, P. F. Brenner and M. A. Page; Am. J. Obstet. Gynecol., 130:55 (1978).
16. M. B. Kabadi and Y. W. Chien; Intravaginal controlled administration of fluorogestone acetate. II. Development of in vitro system for intravaginal release and permeation of fluorogestone acetate. J. Pharm. Sci., 73:1464 (1984).
17. M. B. Kabadi and Y. W. Chien; Intravaginal controlled administration of fluorogestone acetate. III. Development of rate-control vaginal devices. Drug Develop. & Ind. Pharm., 11(6 & 7):1271–1312 (1985).
18. M. B. Kabadi and Y. W. Chien; Intravaginal controlled administration of fluorogestone acetate. IV. In vitro-in vivo correlation for intravaginal drug delivery from rate-control vaginal pessary. Drug Develop. & Ind. Pharm., 11(6 & 7):1313–1361 (1985).
19. N. H. Lauersen and K. H. Wilson; Prostaglandins, 12(Suppl.):63 (1976).

20. N. H. Lauersen and K. H. Wilson; Contraception, 13:697 (1976).
21. E. S. Nuwayser and D. L. Williams; Adv. Exp. Med. Biol., 47:145 (1974).
22. Y. W. Chien, J. Opperman, B. Nicolova and H. J. Lambert; Medicated tampons—intravaginal sustained administration of metronidazole and in vitro/in vivo relationships. J. Pharm. Sci., 71:767 (1982).
23. G. L. Flynn, N. F. H. Ho, S. Hwang, E. Owada, A. Molokhia, C. R. Behl, W. I. Higuchi, T. Yotsuyanagi, Y. Shah and J. Park; in: *Controlled Release Polymeric Formulations* (D. R. Paul and F. W. Harris, Eds.), Am. Chem. Society, Washington, D.C., 1976, p. 87.
24. S. W. Jacob and C. A. Francone; *Structure and Function in Man*, 2nd ed., W. B. Saunders Co., Philadelphia, 1970, Chapter 17.
25. W. H. Masters and V. E. Johnson; *Human Sexual Response*, Little, Brown, Boston, 1966, Chapter 6.
26. S. Hwang, E. Owada, L. Suhardja, N. F. H. Ho, G. L. Flynn and W. I. Higuchi; J. Pharm. Sci., 66:781 (1977).
27. J. R. Udry, N. M. Morris and L. Waller; Arch. Sex. Behav., 2:205 (1973).
28. R. P. Michael and D. Zumpe; J. Reprod. Fertil., 21:199 (1970).
29. D. W. Bullock, C. A. Paris and R. W. Goy; J. Reprod. Fertil., 31:225 (1972).
30. G. S. Saayman; Folia Primatol. (Basel), 12:81 (1970).
31. M. J. Baum, B. J. Everitt, J. Herbert, E. B. Keverne and W. J. DeGreef; Nature, 263:606 (1976).
32. E. B. Keverne; in: *Advances in the Study of Behavior* (J. S. Rosenblatt, Ed.), Academic Press, New York, 1976, Vol. 7.
33. R. A. Sparks, B. G. A. Purrier, P. J. Watt and M. Elstein; Br. J. Obstet. Gynecol., 84:701 (1977).
34. Y. W. Chien, S. E. Mares, J. Berg, S. Huber, H. J. Lambert and K. F. King; J. Pharm. Sci., 64:1776 (1975).
35. S. Hwang, E. Owada, T. Yatsuyanagi, L. Suhardja, N. F. H. Ho, G. L. Flynn and W. I. Higuchi; J. Pharm. Sci., 65:1574 (1976).
36. L. P. Bengtsson; Acta Endocrinol. (Copenh.), 13(Suppl.):11 (1953).
37. T. Yotsuyanagi, A. Molokhia, S. Hwang, N. F. H. Ho, G. L. Flynn and W. I. Higuchi; J. Pharm. Sci., 64:71 (1975).
38. S. Hwang, E. Owada, L. Suhardja, N. F. H. Ho, G. L. Flynn and W. I. Higuchi; J. Pharm. Sci., 66:778 (1977).
39. N. F. H. Ho, L. Suhardja, S. Hwang, E. Owada, A. Molokhia, G. L. Flynn, W. I. Higuchi and J. Y. Park; J. Pharm. Sci., 65:1578 (1976).
40. C. G. Hartman; Endocrinology, 25:670 (1939).
41. C. R. Behl; Ph.D. dissertation, University of Michigan, Ann Arbor, 1979.
42. J. Rock, R. H. Barker and W. B. Bacon; Science, 105:13 (1947).
43. M. Shudmak and H. C. Hesseltine; Am. J. Obstet. Gynecol., 62:669 (1951).
44. D. D. Baker; Anat. Rec., 39:339 (1928).
45. E. P. Laug and F. M. Kunze; J. Pharm. Exp. Ther., 95:460 (1949).
46. J. C. Cornette, K. T. Kirton and G. W. Duncan; J. Clin. Endocrinol. Metab., 33:459 (1971).
47. M. Hiroi, F. Z. Stanczyk, U. Goebelsmann, P. F. Brenner, M. E. Lumkin and D. R. Mishell, Jr.; Steroids, 26:373 (1975).
48. M. Thiery, D. Vandekerckhove, M. Dhondt, A. Vermeulen and J. M. Decoster; Contraception, 13:605 (1976).
49. A. Vermeulen, M. Dhondt, M. Thiery and D. Vandekerckhove; Fertil. Steril., 27:773 (1976).
50. F. Z. Stanczyk, M. Hiroi, U. Goebelsmann, P. F. Brenner, M. E. Lumkin and D. R. Mishell, Jr.; Contraception, 12:279 (1975).

51. R. E. Christian and L. E. Casida; J. Anim. Sci., 7:540 (1948).
52. D. R. Lamond; Anim. Breed. (Abstr.), 32:269 (1964).
53. J. P. Frappell; Vet. Rec., 84:381 (1969).
54. J. A. Laing; *Fertility and Infertility in Domestic Animals*, Bailliere, Tindall & Cassell, London, 1970, p. 355.
55. D. W. Deas; Vet. Rec., 86:450 (1970).
56. T. J. Robinson; *The Control of the Ovarian Cycle in the Sheep*, Sydney University Press, Sydney, Australia, 1967.
57. T. H. McClelland and J. F. Quirke; J. Anim. Prod., 13:323 (1971).
58. R. J. Cooper, D. N. Wallace, D. F. Wishart and B. D. Hoskin; Vet. Rec., 88:381 (1971).
59. J. N. Shelton, T. J. Robinson and P. J. Holst; in: *The Control of the Ovarian Cycle in the Sheep* (T. J. Robinson, Ed.), Sydney University Press, Sydney, Australia, 1967, p. 14.
60. J. N. Shelton and T. J. Robinson; in: *The Control of the Ovarian Cycle in the Sheep* (T. J. Robinson, Ed.), Sydney University Press, Sydney, Australia, 1967, p. 39.
61. T. J. Robinson and N. W. Moore; in: *The Control of the Ovarian Cycle in the Sheep* (T. J. Robinson, Ed.), Sydney University Press, Sydney, Australia, 1967, p. 116.
62. N. W. Moore and P. J. Holst; in: *The Control of the Ovarian Cycle in the Sheep* (T. J. Robinson, Ed.), Sydney University Press, Sydney, Australia, 1967, p. 133.
63. T. J. Robinson and J. F. Smith; in: *The Control of the Ovarian Cycle in the Sheep* (T. J. Robinson, Ed.), Sydney University Press, Sydney, Australia, 1967, p. 144.
64. T. J. Robinson; U.S. Patent 3,916,898 (November 4, 1975).
65. T. J. Robinson and J. F. Smith; in: *The Control of the Ovarian Cycle in the Sheep* (T. J. Robinson, Ed.), Sydney University Press, Sydney, Australia, 1967, p. 158.
66. N. W. Moore, T. D. Quinlivan, T. J. Robinson and J. F. Smith; in: *The Control of the Ovarian Cycle in the Sheep* (T. J. Robinson, Ed.), Sydney University Press, Sydney, Australia, 1967, p. 169.
67. T. D. Quinlivan and T. J. Robinson; in: *The Control of the Ovarian Cycle in the Sheep* (T. J. Robinson, Ed.), Sydney University Press, Sydney, Australia, 1967, p. 177.
68. T. J. Robinson, S. Salamon, N. W. Moore and J. F. Smith; in: *The Control of the Ovarian Cycle in the Sheep* (T. J. Robinson, Ed.), Sydney University Press, Sydney, Australia, 1967, p. 208.
69. R. E. Mauer, S. K. Webel and M. D. Brown; Ann. Biol. Anim. Biochem. Biophys., 15:291 (1975).
70. V. W. Winkler, S. Borodkin, S. K. Webel and J. T. Mannebach; J. Pharm. Sci., 66:816 (1977).
71. J. F. Roche; J. Reprod. Fertil., 40:433 (1974).
72. J. F. Roche; World Rev. Anim. Prod., XII(2):79 (1976).
73. A Victor and E. D. B. Johansson; Contraception, 16:115 (1977).
74. WHO Special Programme of Research Development and Research Training in Human Reproduction, Task Force on Long-acting Systemic Agents for Fertility Regulation; Microdose intravaginal levonorgestrel contraception: A multicentered clinical trial I. Efficacy and side effects. Contraception, 17:105–124 (1990).
75. F. G. Burton, W. E. Skiens, N. R. Gordon, J. T. Veal, D. R. Kalkwarf and G. W. Duncan; Contraception, 17:221 (1978).
76. B. M. Laudgren, A. R. Aedo, S. Z. Cekan and E. Diczfalusy, Contraception, 35:473 (1986).
77. B. M. Laudgren, E. Johannisson, B. Masironi and E. Diczfalusy; Contraception, 26:567 (1982).
78. B. M. Laudgren, E. Johannisson, S. Xing, A. R. Aedo and E. Diczfalusy; Contraception, 32:581 (1985).

79. H. A. Nash; in: *Medical Applications of Controlled Release* (R. S. Langer and D. L. Wise, Eds.), CRC Press, Boca Raton, Florida, 1984, pp. 35–64.
80. L. Viinikka, A Victor, O. Janne and J. P. Raynaud; Contraception, 12:309 (1975).
81. A. P. Sam; in: *Minutes of 5th International Pharmaceuticl Technology Symposium on New Approaches to the Controlled Drug Delivery* (A. A. Hincal et al., Eds.), Editions de Santé, Paris, 1991, pp. 271–284.
82. C. J. Timmer, D. Apter and G. Voortman; Pharmacokineetics of 3-keto-desogestrel and ethinylestradiol released from different types of contraceptive vaginal rings, Contraception (in press).
83. Y. W. Chien; Methods Enzymol., 112:461–470 (1985).
84. P. G. Stubblefield; Contraception, 15:75 (1977).
85. T. J. Roseman and C. H. Spilman; in: *Controlled Release Systems* (S. K. Chandrasekaran, Ed.), AICHE Symp. Serice 206, Vol. 77, New York, 1981.
86. L. J. Cohen and N. G. Lordi; J. Pharm. Sci., 69:955 (1980).
87. C. H. Spilman, T. J. Roseman, R. W. Baker, M. E. Tuttle and H. K. Lonsdale; in: *Controlled Release Delivery Systems* (T. J. Roseman and S. Z. Mansdorf, Eds.), Dekker, New York, 1983, pp. 133–140.
88. C. H. Spilman and T. J. Roseman; Contraception, 11:409 (1975).
89. C. H. Spilman, D. C. Beuving, A. D. Forbes, T. J. Roseman and L. J. Larion; Prostaglandins, 12(Suppl.):1 (1976).
90. J. H. Duenhoelter, R. S. Ramos, L. Milewich and P. C. MacDonald; Contraception, 17:51 (1978).
91. S. M. M. Karim and S. S. Ratnam; Br. J. Obstet. Gynecol., 84:135 (1977).
92. S. M. M. Karim, S. S. Ratnam, R. N. V. Prasad and Y. M. Wong; Br. J. Obstet. Gynecol., 84:269 (1977).

10 Intrauterine Drug Delivery and Delivery Systems

I. INTRODUCTION

A survey conducted in 1975 revealed approximately 27 million couples of child-bearing age in the United States alone (1). Of those couples surveyed, 76.3% expressed a desire to prevent conception, either temporarily or permanently, and a willingness to try various methods of contraception (Table 1). The statistics suggested that oral contraceptives are the most widely used reversible method of fertility regulation.

The benefits and risks associated with the use of the three most popular methods of temporary contraception, oral contraceptive pills, condoms or diaphragms, and intrauterine devices, were compared (2). The statistics indicated that if a population of 100,000 fertile women were exposed without the use of any contraception method for a whole year, these women were expected to have 60,000 pregnancies; this pregnancy rate would produce 50,000 births and approximately 12 pregnancy-related deaths at the maternal mortality levels than in the United States (Table 2). On the other hand, if the same population used a perfectly safe method of fertility regulation, like a condom or a diaphragm, the pregnancy rate would be reduced by approximately fivefold to 13,000 pregnancies and would be expected to yield 10,833 births and 2.5 pregnancy-related deaths.

Oral contraceptive pills and intrauterine contraceptive devices (IUDs) are the most effective, temporary methods of fertility regulation. The administration of oral contraceptive pills provides the maximum level of contraception with only 100 pregnancies, 83 births, and 3 method-related deaths. On the other hand, with the use of either nonmedicated or medicated IUDs, the same population is expected to have 2190 pregnancies, 1825 births (in the absence of induced abortions), and 0.44 pregnancy-related deaths as well as 0.3 IUD-related deaths.

An ideal method of contraception should be the one that has a minimum number of births with the lowest number of method-related deaths. The comparison has demonstrated that all three methods of contraception outlined in Table 2 are more desirable than non-use, since the expected number of births as well as the expected number of deaths are substantially reduced. In comparison to barrier methods, like

Table 1 Contraception Practices of U.S. Couples of Child-Bearing Age

Method of contraception	Percentage of those surveyed[a]
Oral contraceptive pills	26.3
Condom or diaphragm	10.0[b]
Intrauterine devices	6.4
Foam	2.6
Rhythm	2.2
Other	28.8[c]

[a]Based on a sampling of 3403 women nationwide by the Office of Population Research, Princeton University.
[b]The sum of condom users (7.5%) and diaphragm users (2.5%).
[c]Includes 25.1% for men and women sterilized surgically for contraceptive reasons.
Source: Compiled from data in Reference 1.

the condom or diaphragm, the IUD and the pill are more desirable because the expected number of births as well as the expected number of deaths are significantly lower.

However, the choice between pills and IUDs is difficult because pills have a lower rate of birth but a higher probability of death than IUDs. This difficulty can be resolved by comparing the potential benefits (in terms of births averted) and the absolute risks (in terms of deaths associated with the use of a given method). A mortality-benefit ratio (MBR), which is the ratio of deaths per 1000 births averted, can be established. Analysis indicated that pills are associated with a higher benefit as well as a greater risk than IUDs. Contraception using oral contraceptive pills has attained an MBR value four times greater than that with IUDs (Table 2).

Although oral contraceptive pills are associated with systemic contraceptive activity and meet many of the criteria for an ideal method of fertility regulation, an IUD exerts its contraceptive action locally in the uterine cavity and continuously for a prolonged period without requiring sustained patient motivation. Thus, contraception with an IUD is long-acting and poses less of a problem with patient noncompliance. A recent survey indicated that IUDs, which are used by approximately 90

Table 2 Relative Benefits and Risks of the Most Popular Methods of Temporary Contraception

Methods of contraception	Pregnancies	Births	Deaths[a] P	Deaths[a] M	Deaths[a] Total	MBR[b]
None	60,000	50,000	12.0	0.0	12.0	—
Condom or diaphragm	13,000	10,833	2.5	0.0	2.5	0.064
Oral pills	100	83	0.0	3.0	3.0	0.060
IUDs	2,190	1,825	0.44	0.3	0.74	0.015

[a]Deaths per 1000 births averted as related to pregnancy P or method M.
[b]MBR = Mortality/benefit ratio.
Source: Compiled from data in Reference 3.

million women worldwide, have been found to be a very effective means of preventing pregnancy (3).

There are two types of IUDs: one is nonmedicated, and the other is medicated. The nonmedicated IUD exerts its contraceptive action by producing a sterile inflammatory response in the endometrium by its mechanical interaction. Medicated IUDs are those IUDs that are capable of delivering pharmacologically-active antifertility agents. Two classes of medicated IUDs are available: the copper-bearing IUD (e.g., CU-7, G. D. Searle & Co.) or the progesterone-releasing IUD (e.g., Progestasert, Alza Corporation). The nonmedicated IUDs currently available for intrauterine contraception consist of ring-shaped IUDs made of stainless steel, which have been used by 50 million women in China, and plastic IUDs fabricated from polyethylene or polypropylene, which are sold in Asia, South America, and Africa, after discontinuation of the Dalkon shield, the Lippes loop, and the Saf-T-Coil (3). Nonmedicated IUDs are no longer commercially available in Europe. In the United States, only Lippes loop IUD (Ortho) is still available commercially. Some of these IUDs simply vanished from the U.S. market in the litigious 1980s, although most were never shown to be unsafe (*U.S. News & World Report*, 12/24/90). A report in 1989 estimated that approximately 90 million women worldwide currently use IUDs as the method of choice for contraception (3), compared to 15 million users surveyed in 1977 (4).

II. HISTORICAL DEVELOPMENTS

The development of intrauterine contraceptive devices began in the 1920s with the first generation of IUDs constructed from silkworm gut and flexible metal wire, such as the Grafenberg star and the Ota ring. These early types of IUDs soon fell into disrepute because of the difficulty of insertion, the need for frequent removal as a result of pain and bleeding, and other serious complications.

Subsequently, several plastic-based IUDs of varying shapes and sizes were constructed from various inert, biocompatible polymeric materials, including polyethylene, polypropylene, ethylene-vinyl acetate copolymer, and silicone elastomer, are made available. A review of patent literature disclosed that in the period 1968–1974, the U.S. Patent Office alone granted at least 77 patents.

The development of a plastic spiral by Margulies and a plastic loop by Lippes opened up the modern era of IUD development (4). The contraceptive efficacy of these nonmedicated plastic IUDs was reported to be proportional to the surface area of the device in direct contact with the endometrium: that is, larger IUDs are more effective in preventing pregnancy than small IUDs. Unfortunately, large devices also cause more endometrial compression and myometrial distension, leading to increased uterine cramps, bleeding, and expulsion of the IUD (5).

A T-shaped polyethylene device was later developed by Tatum to conform better to the contours of the uterine cavity (6). This new design was reported to significantly reduce the IUD-related side effects, including pain, bleeding, and expulsion. Because of its small surface area, however, it was also linked with a pregnancy rate as high as 18%. Recently, its good uterine tolerance has been recognized and has made this device the carrier of choice for the intrauterine delivery of contraceptive agents, such as copper and progestins. This development initiated a new era of research and development for long-term intrauterine contraception, leading to the generation of a new generation of IUDs, the medicated intrauterine contraceptive devices. Copper-bearing IUDs, such as Cu-7, and progesterone-releasing IUDs, such as Progestasert, thus evolved (Figure 1).

III. UTERUS

A. Anatomy

The uterus is a pear-shaped, thick-walled, muscular organ suspended in the anterior part of the pelvic cavity (Figure 2). In its normal state, it measures about 3 inches long and 2 inches wide (7). Fallopian tubes enter its upper portion, one on each side, and the lower portion of the uterus projects into the vagina. The uterus moves freely; consequently, its position varies with the state of distension of the bladder and rectum. The uterine cavity is normally triangular in shape and flattened anteroposteriorly, making the cavity appear a mere slit when it is observed from the side.

The wall of the uterus consists of three layers:

1. The *endometrium* is the inner coat of the uterine wall and is a mucous membrane. It consists of an epithelial lining and connective tissue. Two types of arteries supply blood to the endometrium: the straight arteries supply the deeper layer; the coiled arteries supply the superficial layer. The coiled arteries are important in menstruation when the superficial portion of the endometrium is sloughed off.

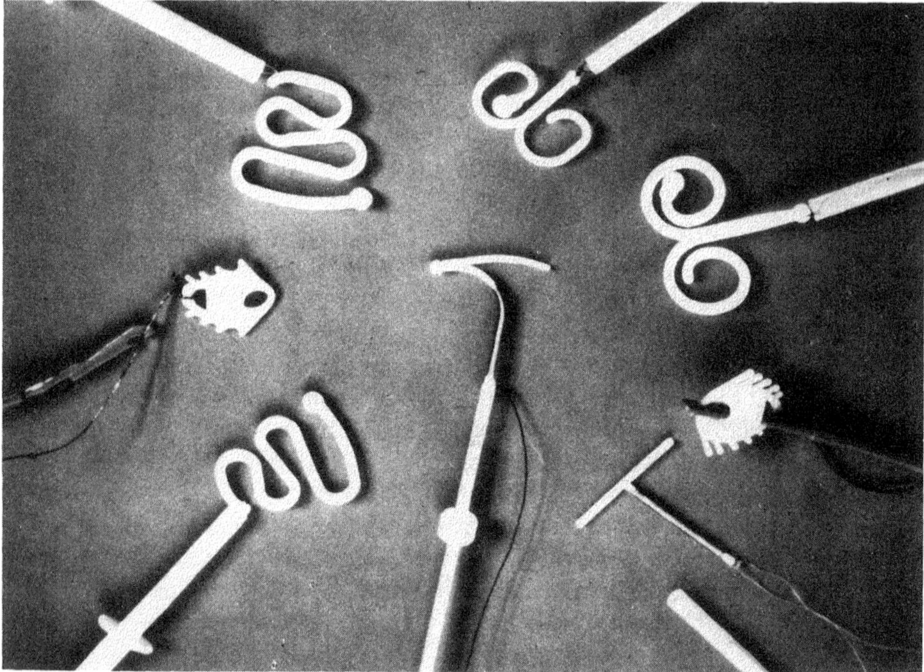

Figure 1 Representative intrauterine contraceptive devices. (Courtesy of Dr. W. C. Steward, Searle Laboratories.)

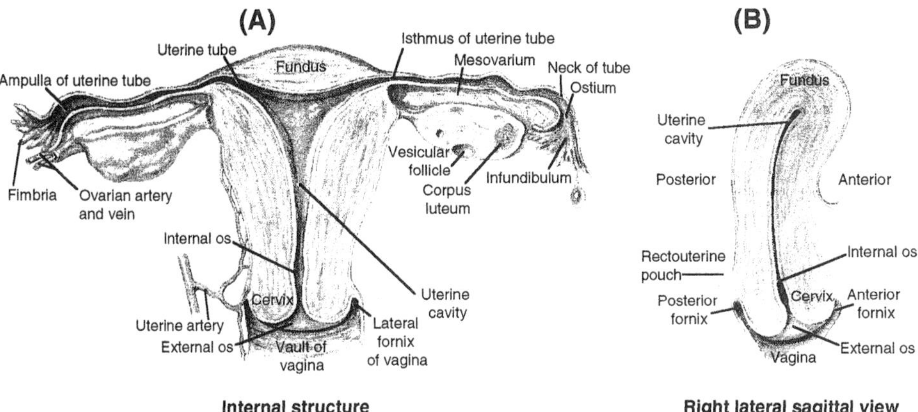

Figure 2 Female reproductive system with uterus sectioned to show its internal structure and various adjacent organs (*A*) and the right lateral sagittal view of the uterus (*B*). (Modified from Jacob and Francone, 1970.)

2. The *myometrium* is the thick, muscular middle layer made up of bundles of interlaced, smooth muscle fibers embedded in connective tissue. The myometrium is subdivided into three ill-defined, intertwining muscular layers, the middle of which contains many large blood vessels of the uterine wall. It is this intertwining arrangement of muscle that presses against the blood vessels and stops bleeding after delivery of a baby.
3. The *peritoneum* covers the external surface of the uterus, which is then attached to both sides of the pelvic cavity by broad ligaments through which the uterine arteries cross.

B. Ovarian Hormones and Control of the Menstrual Cycle

From the time of puberty the endometrium undergoes cyclical changes under the control of ovarian hormones. Menstrual bleeding from the uterus, which is called menstruation, is observed to recur approximately every 28 days and to last for 4–6 days (8).

Histologically, the menstrual cycle can be divided into two phases: the first 14 days of the cycle (starts from the first day on which bleeding occurs) is called the follicular phase and the second 14 days the luteal phase. During the follicular phase, estradiol is secreted from the ovary and an ovarian follicle enlarges and moves to the surface of the ovary. Ovulation occurs normally on Day 14 (Figure 3). In the subsequent luteal phase, the corpus luteum is very active initially and then degenerates just before the next menstruation.

In both the follicular and luteal phases of the cycle the blood circulation continuously supplies the endometrium through the coiled arteries. When the time of menstruation approaches, there is a reduction in the blood circulation in the coiled arteries supplying the superficial layers of the endometrium, followed by prolonged closure of these vessels that results in a blanching of the endometrium. After a period of time these arteries open again and blood escapes through the walls with the for-

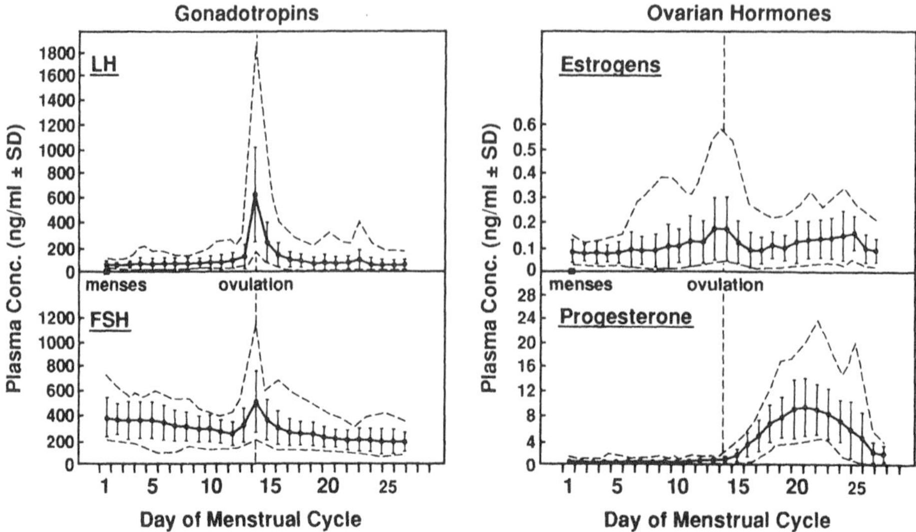

Figure 3 Plasma concentration profiles of gonadotropins—LH (luteinizing hormone) and FSH (follicle stimulating hormone)—and ovarian hormones—estrogens and progesterone as a function of the menstrual cycle. (Replotted from Vande Wiele and Dyrenfurth, 1973.)

mation of subepithelial hematomata. The prolonged ischemia leads to degeneration of the superficial layers of the endometrium and hemorrhage from capillaries, venules, and arterioles. These superficial layers are cast off, leaving the endometrium about 1 mm in thickness. The cause for the vasoconstriction is not clear but may be a consequence of the fall in the concentrations of ovarian hormones in the circulation at the end of a menstrual cycle.

Menstrual flow consists of blood mixed with mucus and also contains numerous leukocytes at the beginning. It does not clot. The fluidity of menstrual blood may be due to the anticoagulant substances manufactured by the endometrium or to liquefaction of the blood by uterine enzymes after clotting in the uterine cavity. The volume of menstrual blood varies greatly in different women, but it is usually in the range of 50–250 ml. Menstrual flow contains slowly acting stimulants of smooth muscle, which are probably long-chain fatty acids.

At the end of a menstrual cycle, the endometrium is regenerated from the deeper regions of the glands. During the period from day 5 to day 14, the endometrial glands progressively increase in length and tortuosity and the cells of the stroma become hypertrophy, so that at the end of this proliferative stage the endometrium is thickened to some 3–4 mm. All these uterine changes are brought about by estradiol. In the subsequent 14 days (luteal phase), the progesterone produced by the follicular lutein cells, in conjunction with estrogenic hormones in a suitable proportion, builds up the endometrium for the reception of a fertilized ovum. The glands continue to grow, becoming more irregular in outline and very convoluted and showing secretory activity. Approximately at the end of this secretory stage, the glands are dilated and contain mucus and glycogen; their walls are folded, and the free edges of the cells are indistinct. The superficial cells of the stroma enlarge and are tightly packed together. The cells in the deep layers are also enlarged, but the stroma has a more open and edematous appearance. On the other hand, the basal layer of the endometrium (next to the myometrium) does not show edema or hypertrophy. The vascularity of the endometrium increases, and the coiled arteries grow toward the surface of the endometrium. When these premenstrual variations are completed, the endometrium has a thickness of approximately 6–7 mm (9).

If the ovum is not fertilized and implantation does not occur, small collections of blood appear in the endometrium and the process of degeneration begins again.

Ovulation may take place any time between day 6 and day 20 of the menstrual cycle. At the time of ovulation, the cervix secretes a watery mucus rich in glucose that presumably facilitates the penetration of spermatozoa.

As discussed earlier, bleeding normally takes place in the secretory stage, but it may also occur even if the secretory stage is not reached and the endometrium is histologically still in the proliferative stage. In this case, ovulation has not happened and the corpus luteum has not developed. This is called anovular menstruation.

Like the smooth muscle of the intestine, uterine muscle shows rhythmic contractions and relaxations, which although present at all phases of the menstrual cycle, are usually more marked in the luteal phase. Contraction of the myometrium is stimulated by vasopressor, not by oxytocic action.

Menstruation shows a considerable range of variation among individuals. Even in an individual woman, successive cycles may vary in length by 1–2 days. At the onset of menstruation and also when the menopause is near, the cycle may become irregular and anovulatory. The process is absent during pregnancy and ceases finally at the menopause, usually between 42 and 52 years of age, when cyclical uterine and ovarian changes cease.

Both estradiol and progesterone are inactivated or metabolized to varying extents in the liver. Estradiol is biotransformed or metabolized to estrone, estriol, and other closely related metabolites and then excreted in the urine. On the other hand, Progesterone is produced at the rate of approximately 15 mg/day. This amount of progesterone is first metabolized to pregnanediol and then excreted in the urine as a biologically inactive glucuronide. The daily excretion of pregnanediol is 1–2 mg in the proliferative phase and around 4 mg in the luteal phase. The output of urinary pregnanediol usually increases about when the basal temperature rises and the output begins to fall before the onset of menstruation. On the other hand, estrogens are excreted throughout the cycle, with two peaks of urinary excretion at about the time of ovulation and during the luteal phase.

A recent in vitro estrogen pretreatment study conducted in ovariectomized rats demonstrated that estrogen causes an increase in protein content in uterine tissue, accompanied by an increase in the binding sites for progesterone binding components and a reduction in the metabolism of progesterone in uterine tissue; after saturation of these binding components, the metabolic rate of progesterone starts to increase. This investigation suggested that estradiol plays a key role in the enzymatic conversion of progesterone in the uterus by controlling the production of progesterone binding components (10).

C. Uterine Proteins

The amount of protein in the uterine secretion depends on the stage of the menstrual cycle, varying from 1 to 21 mg (11). In the secretory phase, less than 5 mg uterine proteins could be recovered, but the protein content increased just before menstruation and in the early stage of the proliferative phase. The quantity of uterine proteins recovered from postmenopausal women is high, ranging from 8 to 14 mg per uterus.

Uterine proteins consist mainly of serum proteins, such as albumins, gamma-globulins, and transferrin, plus small fractions of uterine-specific proteins, such as posttransferrin. Posttransferrin was found to vary in intensity with the menstrual cycle, and its concentration in uterine fluid was reported to decrease in the postovulatory phase (12). It was also detected in genital tract fluids from primates (13,14) and women (15,16). While one uterine protein was found only in postovulatory uterine collections, with a maximum frequency of occurrence during the midsecretory phase, another could be detected predominantly in preovulatory uterine fluids. The remaining uterine-specific proteins appeared randomly throughout the menstrual cycle.

The progesterone binding activity of uterine flushings was found greater than that illustrated by serum but indistinguishable from that exhibited by serum proteins for steroid specificity and chromatographic behavior. No significant differences in progesterone binding activity were detected between the preovulatory and the postovulatory phases of the menstrual cycle.

Uterine proteins originate primarily from the selective filtration of serum proteins through the endometrium. The contribution of uterine-specific proteins is of only minor significance in humans, compared with rabbits, which have a specific uteroglobin.

D. Cyclic Viscoelasticity of Cervical Mucus

Human cervical mucus shows a cyclical fluctuation. The viscoelasticity and nondialyzable solids in cervical mucus are at a minimum at or near the midcycle. Viscoelasticity increases significantly in association with the ovulatory phase of the cycle. The cyclical fluctuation in mucous viscoelasticity is related to the variation in mucin concentration (17–19).

E. Microbiology of Uterus and Cervix

A group of 50 women who had a hysterectomy were examined using a multiple-biopsy technique to determine the existence of normal microbial flora in the uterine cavity and cervical canal (20). The uterine cavity of all subjects was found to be sterile, and no bacteria could be isolated; however, bacteria were isolated from 54% of the cervical canal biopsy samples. These isolated microorganisms, predominantly lactobacilli and species of Bacteroides, were found primarily within the lower half of the canal. From the sites of bacterial colonization, a diminishing gradient of concentration of bacteria was observed as the canal was ascended. These bacteria were also detected in the ectocervix and vagina.

F. Dynamics of Uterus

The anteriorly positioned or midpositioned uterus was observed to respond to sexual stimulation as a composite organ (11). As the excitement-phase levels of sexual tension progress toward the plateau state, the entire uterine body elevates from the true into the false pelvis. Concurrently, the cervix slowly retracts from its resting position, which is in direct contact with the posterior vaginal floor and superior plane, as the vaginal walls expand under the influence of sexual stimulation, creating a tenting effect in the transcervical depth of the vaginal lumen (21). It has been theorized that the elevation of uterus is due to increased negative pressure in the abdominal cavity in response to the partial elevation and fixing of the diaphragm or the passive vasocongestion of broad ligaments in the pelvis.

Conversely, the posteriorly-positioned uterus, either retroverted or retroflexed, does not elevate from the true into the false pelvis in response to sexual stimulation; on the other hand, expansion of vaginal wall occurs in the transcervical depth of the anterior, posterior, and lateral planes.

With the onset of the resolution phase, the elevated uterus begins to descend rapidly to its unstimulated resting position in the true pelvis. This return drops the cervix into the anatomically-contrived seminal basin in the transcervical depth of the vagina.

The cervix does not secrete lubricating fluids during any phase of the sexual response cycle. The only frequently observed cervical response to sexual stimulation is a minimum dilation of the external cervical os, occurring immediately after an orgasm. When dilation of the cervical os develops, the resolution phase (20–30 min in duration) must intervene before a constrictive effect closes the external cervical os.

Normally the external cervical os is slightly patulous with a larger entrance to the cervical canal in parous than in nulliparous women. This is obviously a result of the obstetric trauma of delivery.

Uterine dynamics has been extensively studied using intrauterine and abdominal electrode placements and physiological recording techniques. An identifiable recurrent pattern of uterine muscle contraction was observed, which oriented specifically to the orgasmic phase of sexual excitement. Typical corpus contractions start in the fundus, progress through the midzone, and expire in the lower uterine region. The contractile pattern is similar to those developed by the uterine musculature during the first stage of labor; however, orgasmic-phase contractions showed a reduction in excursion and an increase in contractile frequency.

A significant increase in uterine size was observed in parous women in pre- and post-menstruation weeks. This increase can range from 50 to 100% over uterine size measured immediately before the onset of sexual stimulation. On the other hand, the majority of nulliparous women do not develop an increase in uterine size of clinical magnitude in response to sexual tension.

There is no longer any question about the increase in the size of the uterus during sexual response, particularly when the excitement and plateau phases are extended in time sequence and the responding woman is parous. Under the influence of sexual stimulation, the uterus responds specifically with a marked vasocongestive increase in the volume of uterine fluid and, consequently, in the size of the uterus.

IV. DEVELOPMENT OF MEDICATED INTRAUTERINE DEVICES

In the 1960s, intrauterine devices (IUDs) were available as an alternative method of fertility control to women who experienced adverse reactions when taking oral contraceptive pills. The effectiveness of conventional (nonmedicated) IUDs in preventing pregnancy depends primarily on their mechanical effect on the endometrial surface. Their reliability in contraception is greater when they are designed to achieve a greater area of contact with the endometrial surface; therefore, the larger the size (or the surface area) of the IUD, the more effective it is expected to be (22). Unfortunately, the endometrium is more likely to develop an irritation to larger IUDs and thereby provoke bleeding, cramping, and expulsion. Hence, some women develop undesirable responses after wearing IUDs.

Numerous IUDs have been developed in the last 20 years that aim to eliminate these side effects. Unfortunately, variations in the shape and size of IUDs have failed to minimize the side effects and, concurrently, to improve the contraceptive efficacy as much as was hoped. Therefore, some biomedical researchers have shifted their efforts to the development of medicated IUDs. The controlled release of antifertility agents from medicated IUDs was conceived independently in 1968 by Zipper et al. (22), who added contraceptive metals, such as copper, onto the IUD frame, and Doyle and Clewe (23), who developed a progestin-releasing IUD. The objectives were to add (i) antifertility agents to the more easily tolerated, smaller devices, such as the T-shaped device, to enhance their contraceptive effectiveness; or (ii) antifibrinolytic agents, such as ϵ-aminocaproic acid and tranexamic acid, to the larger, more effective IUDs to minimize the incidence of bleeding and pain (24).

A. Logical Considerations

A logical consideration in the development of an IUD is that the device designed should conform to the contours of the uterine cavity, rather than the endometrial contour and myometrium being obliged to conform to the device. To accomplish this, certain anatomic factors and functional characteristics of the human uterus should be considered.

As discussed in Section III.A, when the uterine cavity is empty, the endometrial surfaces are separated from one another only by a thin layer of mucus and other secretion fluids of the endometrial glands and tubal epithelium. The volume and shape of the uterine cavity depend on the contractile state of the myometrium, normally reflecting the summation of myometrial forces. The average dimensions of the endometrial cavity at several levels have been computed and are summarized in Figure 4 (25).

As myometrial fibers contract, the uterine wall thickens and shortens, and in response, the endometrial cavity becomes smaller in all dimensions. As the contraction increases, the lateral walls of the cavity approximate one another and the cavity assumes the shape of a letter T. The sequential changes in the uterine walls and endometrial cavity during the progress of myometrial contraction are illustrated in Figure 5. Apparently, an intrauterine device in the shape of a T can easily conform to the shape as well as the size of the endometrial cavity, causing minimal myometrial distension and endometrial compression.

These theoretical considerations were tested in 1967 by Tatum and Zipper (25). Preliminary clinical data showed that the T-shaped IUD results in approximately one-fifth the incidence of pain and bleeding as the popular Lippes loop D IUD. In addition, the expulsion rate was found to be only half that of the conventional Lippes loop. This low incidence for the T-shaped IUD was rationalized as due to two distinct features of the T configuration: (i) the T shape conforms well to the uterine cavity, minimizing the distortion of myometrium; and (ii) displacement and rotation of the T-shaped IUD inside the endometrial cavity are resisted by the three points of contact between the device and the endometrial walls (26).

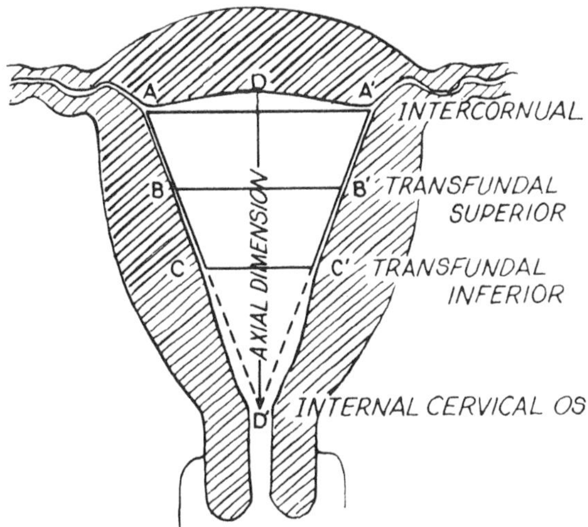

Figure 4 Endometrial dimensions (mean ± SD): *A-A'* = 30.3 ± 4.3 mm; *B-B'* = 22.5 ± 5.2 mm; *C-C'* = 14.3 ± 4.3 mm; *D-D'* = 38.5 ± 5.3 mm; *AA'-BB'* = 10 mm; and *BB'-CC'* = 10 mm. [Reproduced from H. J. Tatum; Am. J. Obstet. Gynecol., 112:1000 (1972).]

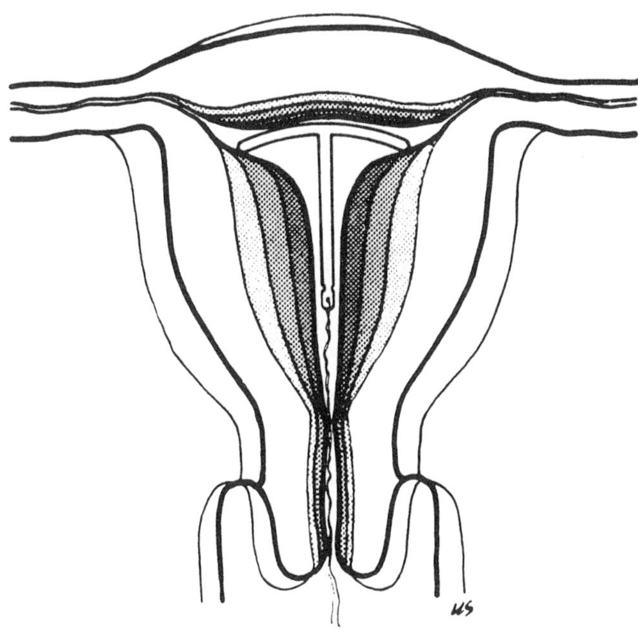

Figure 5 Sequential change in the endometrial cavity during a myometrial contraction. [Reproduced from H. J. Tatum; Am. J. Obstet. Gynecol., 112:1000 (1972).]

Unfortunately, this low incidence of side effects was also accompanied by a reduction in the level of contraceptive efficacy. It was reported that at the end of 1 year of clinical evaluation, the pregnancy rate was approximately 18.3%. Apparently, this low contraceptive effectiveness could be related to the relatively small surface area of the T-shaped IUD (~3.15 cm^2). As discussed earlier, the antifertility activity of a nonmedicated IUD is closely related to its surface area of contact with the endometrial walls (27,28).

Although it is moderately ineffective in preventing conception, the T-shaped IUD possesses some useful features as a potential carrier for the long-term intrauterine administration of antifertility agents. The first major testing of this idea was initiated by Zipper et al. (28–30) by winding the contraceptive copper wire around the dependent (or vertical) arms of the T-shaped plastic frame. Further development of the concept of intrauterine contraception by metallic copper has resulted in the commercialization of a copper-bearing IUD (Cu-7, G. D. Searle & Co.), which contains 89 mg copper wire winding around the vertical limb of a 7-shaped polypropylene plastic device developed by Abramson (31) to give an effective surface area of approximately 200 mm^2 (Figure 6). The Cu-7 was approved by the U.S. Food and Drug Administration in 1974, first for up to 24 months of use and later revised for 3 years of intrauterine contraceptive treatment (32,33). Later, T-shaped copper-bearing IUDs also received regulatory approval. One representative product, Cu-T 380A, is still available on contraceptive market.

The T-shaped IUD was also used by Scommegna et al. (34, 35) for the intrauterine delivery of natural progesterone to achieve a localized contraceptive activity. Further development of this silicone-based progesterone-releasing IUD by the Alza Corporation resulted in the marketing of Progestasert (Figure 7), which contains and releases progesterone at controlled rate from a suspension of progesterone N.F. (38 mg) in a silicone medical fluid encapsulated in the vertical limb of the T-shaped device. Progestasert was approved in 1976 by the U.S. Food and Drug Administration for 1 year of intrauterine contraceptive use.

B. Copper-Bearing Intrauterine Devices

Zipper and his associates (28,29,36) reported that copper (and other metals, such as zinc) attached to an intrauterine device has markedly enhanced the contraceptive effectiveness of an IUD in both experimental animals and in humans.

The first clinical studies were conducted by Zipper et al. (28,29) in 1969 using a T-shaped polyethylene plastic device wound with 30 mm^2 copper wire (Cu-T-30). The pregnancy rate was found to be reduced to 5% from the 18% attained by a nonmedicated (placebo) T-shaped IUD. This result encouraged the initiation of additional clinical evaluations using IUDs with larger surface areas of copper wire. The Cu-T-200, which contains 200 mm^2 copper wire, was then identified as the IUD with the maximum contraceptive efficacy (Figure 8) (37). A further increase in the surface area of copper wire from 200 to 300 mm^2 did not produce any significant improvement in contraceptive effectiveness (38,39).

1. *Fundamentals of the Antifertility Action of Copper*

Copper is known to be cytotoxic if present in sufficiently high concentrations. Metallic copper was reported to interfere with the implanted rat fetus (40) and enhance significantly the spermatocidal and spermatodepressive action of an IUD (41–44).

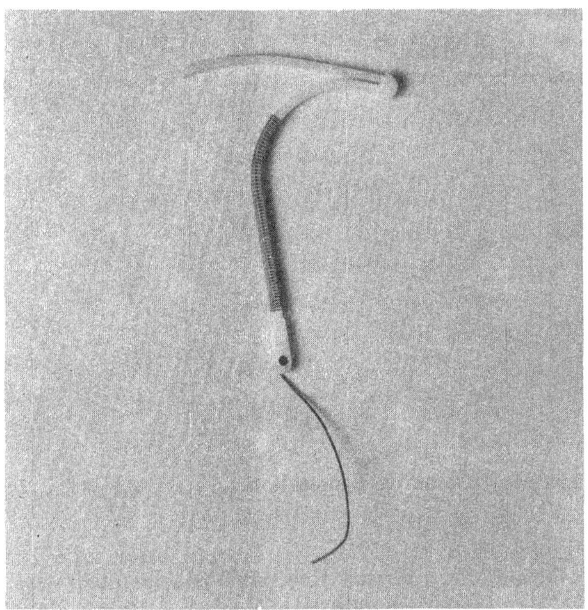

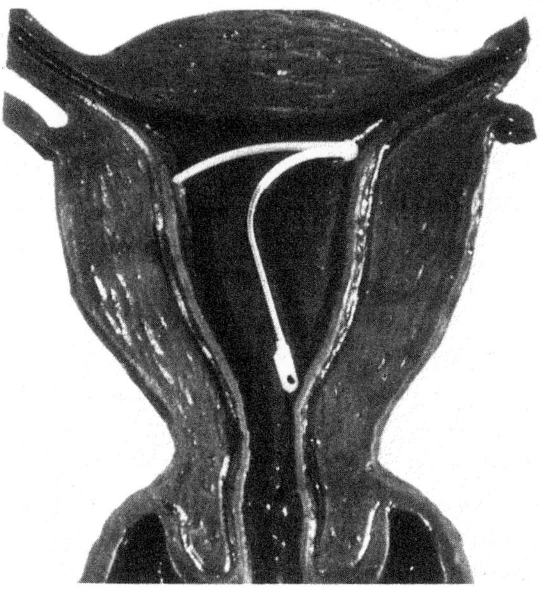

Figure 6 Cu-7, a 7-shaped copper-bearing intrauterine contraceptive device (top) and a uterine model showing the position of the Cu-7 in the uterine cavity following intrauterine insertion (bottom).

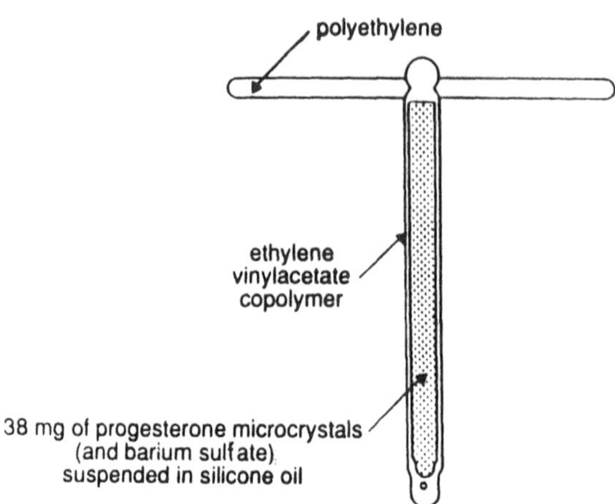

Figure 7 Cross section of the Progestasert IUD and its composition.

The effect of copper ions on the binding of steroid hormones to receptors was investigated in rabbit uterus (45). The results revealed that cupric ion (Cu2+) is a competitive inhibitor of steroid-receptor interaction by acting on the receptor sites via the dissociation and aggregation of receptor macromolecules. The progesterone receptor was found to be more susceptible to cupric ions than the estrogen receptor.

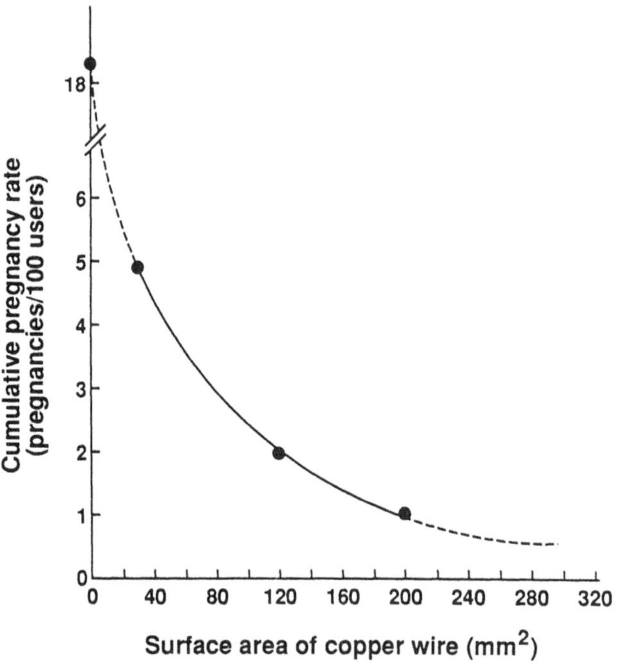

Figure 8 Relationship between the surface area of copper wire on the T-shaped IUD and the cumulative pregnancy rate after a year of use (first insertion). [Replotted from the data by H. J. Tatum; Am. J. Obstet. Gynecol., 112:1000 (1972).]

The influence of cupric and other metallic ions on the binding of 17β-estradiol to human endometrial cytosol was studied (46). Cupric ion was observed to be the most potent inhibitor, followed by Cd, Zn, and Pb ions. The possible site of action was projected to be the sulfhydryl (SH) group(s) of the binding protein.

Copper was found to be taken up by endometrial epithelium and superficial stromata. The copper concentration in the cytoplasm of uterus bearing the copper-releasing IUD was reportedly increased to approximately 1.4×10^{-6} M, at which the inhibitory activity of cupric ion on the binding of steroids to their specific receptors became apparent. Morphologically, progestational proliferation was noted to be severely inhibited and estrogenic action also seemed to be interrupted (45).

Pregnancy in rabbits was observed to occur unilaterally and was reversibly inhibited when copper wire was placed in a uterine horn (29). Copper wire was also found to be blastocystocidal, and the death rate of blastocyst was observed in proportion to the surface area of copper wire. This blastocystocidal activity may be attributed to its interference with the active transport of ions (47). It was further reported that the lysosomes of the blastocyst incorporate the copper, thus releasing lysosomal enzymes and causing cellular autolysis and death (48).

In a number of species, such as rats, hamsters, and rabbits, if copper wire is inserted into the uterine cavity after implantation takes place, gestation proceeds normally with no gross malformations (49,50). In addition, no abnormalities were observed in the 26 human pregnancies conceived with a Cu-T IUD in place and allowed to progress to term (49).

On the other hand, the copper-bearing IUD was reported to have little effect on sperm mobility (44) and no interference with fertilization (51).

The presence of large amounts of copper, as either metal or salts, in the uterine cavity with coexistent exogenous estrogen was noted to increase the electrical activity of the rabbit uterus.

A copper-bearing IUD was reported to produce a significant increase in the alkaline phosphatase activity in the uterine fluid and endometrium. The acid phosphatase in the tissue was observed to increase gradually but was unchanged in the uterine fluid; β-glucuronidase activity, on the other hand, was slightly decreased in the endometrium but significantly increased in the fluid (52,53). Metallic copper and cupric ion have also been shown to reduce the alkaline phosphatase activity in homogenates of human endometria (54).

2. *Biopharmaceutics of Intrauterine Controlled Copper Administration*

Kinetics of Intrauterine Copper Release. Copper appears to be released continuously from a copper-bearing IUD by a combination of ionization and chelation processes. The mean diameter of the copper wire was observed to reduce with time, a process that is apparently accompanied by corrosion and flaking of the metal (49).

Analysis of 284 Cu-7 IUDs inserted intrauterinally in women for up to 40 months indicated that the dosage of copper delivered, after statistical analysis, can be expressed (55) by

$$\text{Dosage (mg)} = 0.30 \text{ month} + 3.79 \tag{1}$$

which is equivalent to the intrauterine release of copper at a daily dose of 9.87 μg/day. The linear relationship between the cumulative amount of copper released intrauterinally with the duration of in utero residence is illustrated in Figure 9. The results appear to suggest that a constant release (Q versus t) profile was achieved for the intrauterine controlled delivery of copper from the Cu-7.

The antifertility activity of a copper-bearing IUD, as reported earlier, was primarily dependent upon the surface area of the contraceptive copper wire (Figure 8); thus, the effectiveness of its intrauterine contraception should be associated with the dosage of copper released daily and should also depend on the concentration of copper in uterine secretion fluids.

The copper concentration in uterine secretion fluids was monitored by atomic absorption spectroscopy in 63 Cu-7 wearers for up to 4 years (56). The results showed that it increases from a mean value of 0.05 $\mu g/g$ of uterine secretion fluids before IUD insertion to 0.67 $\mu g/g$ within 1–6 months after intrauterine insertion and continues to increase gradually thereafter as the duration of intrauterine residence is prolonged (Figure 10), indicating that the mean uterine copper concentration is increased linearly in proportion to the duration of in utero residence of the Cu-7 (r = 0.93). This linearity closely resembles the earlier observation on the intrauterine release profile of copper from the Cu-7 [Figure 9 and Equation (1)]. The copper concentration in the uterine secretion fluid was calculated to increase at a rate of 30.9 $\mu g/g$/month.

Uptake of Copper. The copper released from copper-bearing IUDs was found to be partly taken up by the endometrium. The copper concentration increased in both the proliferative and the secretory endometria during the first 6 months, whereas only the secretory endometrium showed an increased level of copper following 1 year of use (56–58), even though the copper levels in uterine secretion fluid were observed to increase continuously throughout the cycle (Figure 10). Endometrial copper levels were noted to rapidly return to normal values when the devices were removed (59).

Studies performed in rats using radioactive copper demonstrated that the metal

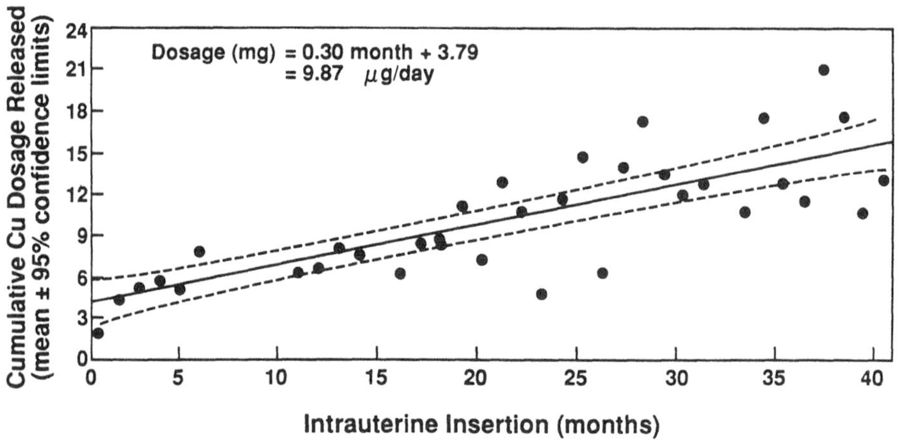

Figure 9 Intrauterine release of copper from 284 Cu-7 devices in women for up to 40 months. The data points are the mean amounts (mg) released cumulatively in utero as defined by Equation (1), and the broken lines are the 95% confidence curves. (Replotted from the data by Dr. F. B. O'Brien, Searle Laboratories.)

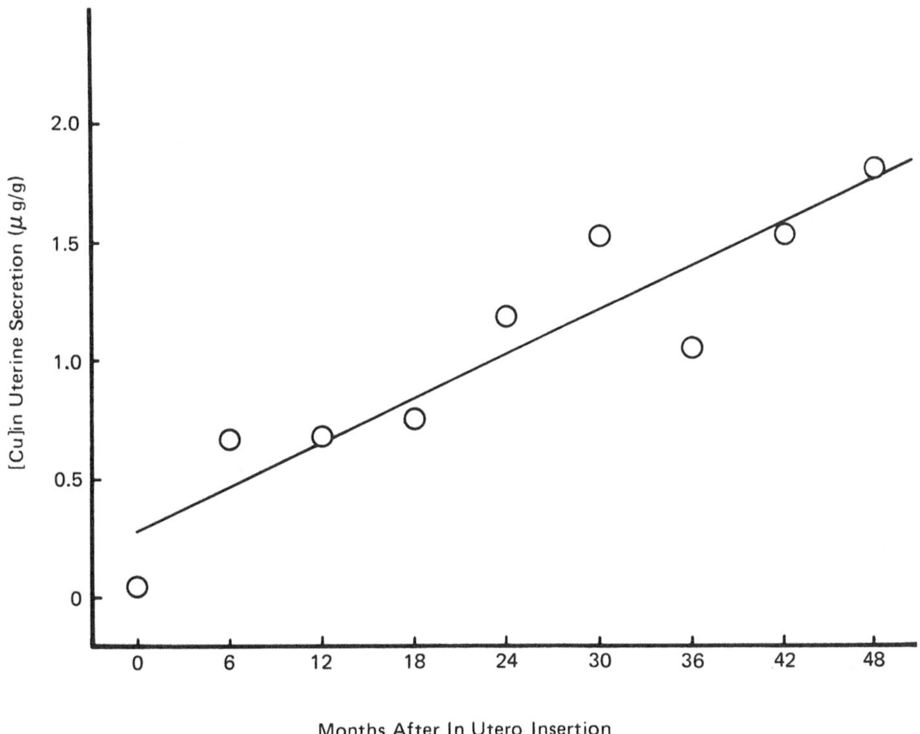

Figure 10 Copper concentration profile in the uterine secretion of 63 women wearing a Cu-7 for up to 48 months. [Plotted from the data by B. Larsson and L. Hamberger; Fertil. Steril., 28:624 (1977).]

released from copper-bearing IUDs is also absorbed systemically and deposited, in detectable amounts, in such organs as liver and kidney (60). On the other hand, when a copper wire was inserted into the uterine cavity of pregnant rabbits after implantation, no changes were detected in copper levels of any maternal tissues other than the uterus. However, an elevated level of copper was detected in placentas and fetal livers (61).

Effect of Calcium Deposition and Protein Coating. Electron microscopic examination of Cu-7 IUDs left in utero for various lengths of time has shown the development of a composite of a "corrosion layer" and an "encrustation layer" that cover the surface of the copper wire. The corrosion layer was reported to contain protein and hypothesized to play an important role in the intrauterine delivery of cupric ion. The encrustation layer, on the other hand, was found to develop over the corrosion layer and contained predominantly the deposition of calcium. This layer may be attributed to a reduction in the release rate of copper (62). A large intersubject variability was noted in the formation of these two layers. Formation of the encrustation layer could be prevented or reduced by coating the copper wire with polymeric membrane, such as collodion, permitting the release of copper ions for antifertility activity.

The chemical analysis of clinically tested copper-bearing IUDs revealed that the corrosion layer consists primarily of cuprous (Cu^{+}) and cupric (Cu^{2+}) ions with small amounts of calcium, iron, and nonmetallic elements. In the 23 clinical samples assayed, an average of 46% cuprous oxide (Cu_2O) was measured. The cupric ion was found to form a precipitate with proteins. A large intersubject variation was noted in the composition of the corrosion layer (63).

Studying the corrosion of copper in the aqueous solution of human female serum and of oxidized glutathione suggested that the formation of cuprous and cupric ions is a result of the oxidation of copper by oxygen (62).

Extensive calcium carbonate ($CaCO_3$) deposits were detected on copper-bearing IUDs removed from women because of pregnancy. The $CaCO_3$ deposits apparently resulted in the formation of a compact, impermeable layer, which may impair the intrauterine delivery of contraceptive copper species, causing the failure of achieving active contraception in these women. The calcification observed may begin as early as within 6 months of insertion and increased with prolongation of the in utero residence time (64,65). The clinical significance of this calcification has been further substantiated by the observation of an increased pregnancy rate in those women wearing a copper-bearing IUD for longer than 24 months (66). The pregnancy rate can be reduced if a new IUD is inserted every 2 years.

The release of copper from copper-bearing IUDs with varying surface areas of copper wire was recently examined in a simulated uterine fluid (67). The daily rate of release was observed to decrease with time. This time-dependent decrease in the in vitro release rate of copper could be a consequence of the continuous deposition of a protein layer on the surface of the copper wire. Mechanistic analysis of the copper release profiles, in light of any possible barrier effects from protein deposition, suggested that the release of copper from protein-coated IUDs follows a linear Q versus $t^{1/2}$ relationship (Figure 11). The cumulative amount of copper released for a given time period was found to be dependent upon the surface area of copper wire winding around the IUD. This result supported the earlier observation of the dependence of the antifertility effectiveness of a copper-bearing IUD on the surface area of the copper wire (Figure 8). In fact, however, the magnitude of the $Q/t^{1/2}$ values, the release flux of copper from a unit surface area of copper wire, was not found to be totally dependent on the copper surface area (Table 3).

3. *Clinical Contraceptive Effectiveness of Copper-Bearing IUDs*

The net annual rates of pregnancy and of termination for 3-year uses of Cu-T-200 IUDs in 16,345 first-insertion women are shown in Table 4. The annual pregnancy rates are approximately the same for the first 2 years and show a slight but significant increase during the third year of use. On the other hand, the annual rates of expulsion and of removal due to bleeding, pain, or other medical reasons have declined. However, the removal rates for planned pregnancy and other personal reasons have increased sharply with the length of insertion. Consequently, the Cu-T-200 yields approximately the same annual rate of continuation over the 3 years of use (2).

In addition, the age of the copper-bearing IUD users at the time of insertion was noted to strongly influence the effectiveness of contraception and the rate of termination (Table 5). The probability of becoming pregnant is reduced substantially with the increase in age at insertion. Similarly, the rate of termination decreases with the increase in the age of the users as a result of a reduction in the number of ex-

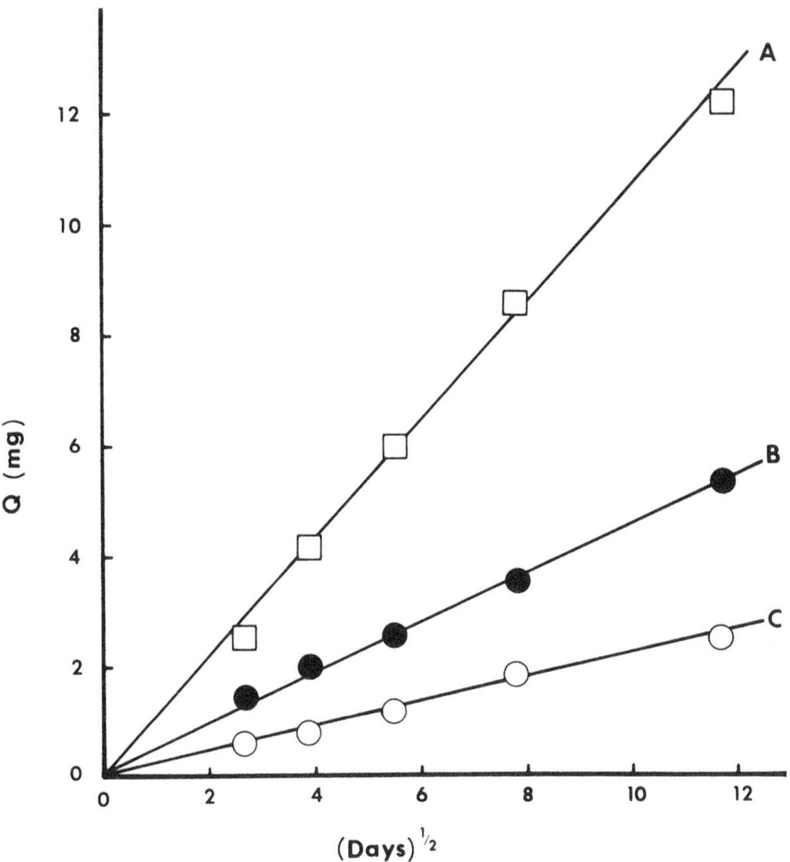

Figure 11 In vitro release profile of copper from IUDs in simulated uterine fluid. A = 1, 112 mm^2; B = 471 mm^2; and C = 198 mm^2. [Plotted from the data by E. Chantler et al.; Brit. Med. J., 2:288 (1977).]

pulsions and termination. Thus, the overall rate of continuation is greater for older users (2).

As discussed in Section IV.A, the Cu-7 (G. D. Searle & Co.) was the first copper-bearing IUD approved by the U.S. Food and Drug Administration for 3-year intrauterine contraceptive treatment (33,34). Each unit of the Cu-7 is a polypropylene plastic device shaped like the number 7, with 89 mg copper wire winding around the vertical limb to give an effective surface area of approximately 200 mm^2 copper (Figures 6 and 8). Contraceptive copper species is released at a mean daily dose of 9.87 μg/day continuously in the uterine cavity for up to 40 months (Figure 9). The Cu-7 appears to be substantially smaller than other commercially available intrauterine devices (Figure 1), and thus it can be easily inserted because of its special 7 configuration in addition to its small size (with a volume of only 0.09 cm^3), even in nulliparous women. Insertion can usually be accomplished without the need for cervical dilation, and furthermore, removal is generally painless. Life table analysis

Table 3 Independence of $Q/^{1/2}$ Values on the Surface Area of Copper Wire in Copper-Bearing IUDs

Copper surface area (cm^2)	$Q/t^{1/2}$ ($\mu g/cm^2 \cdot day^{1/2}$)	γ[a]
1.98	112.66	0.993
2.52	94.18	0.996
3.12	82.60	0.990
3.91	78.91	0.978
4.71	91.89	0.999
5.55	74.92	0.996
7.34	58.13	0.999
8.68	72.76	0.976
9.40	70.72	0.985
11.12	97.39	0.998

[a]Correlation coefficient γ for $Q - t^{1/2}$ relation.
Source: Data were recalculated from Reference 64.

of the 3-year clinical effectiveness of Cu-7 for both parous and nulliparous women is summarized in Table 6.

In collaboration with the U.S. Population Council, G. D. Searle & Co. also manufactured another T-shaped copper-bearing IUD for distribution to nonprofit family-planning clinics at cost (67). The Cu-T IUD was developed by Dr. H. Tatum of the Population Council; it also contains 200 mm^2 contraceptive copper wire winding around the vertical leg of a T-shaped polyethylene device in the manner as in Cu-7 device. The Cu-T IUDs were first introduced outside the United States in 1972 and have been made available since then in more than 34 countries. Its 3-year clinical effectiveness is illustrated in Table 4.

This T-shaped device has a volume of 0.16 cm^3, which is almost twice as large

Table 4 Clinical Effectiveness for 3-Year Use of Cu-T-200

	Clinical effectiveness[a]		
Events	Year 1	Year 2	Year 3
Pregnancies	2.6	2.5	3.6
Expulsions	7.3	2.3	2.5
Removals			
Bleeding or pain	8.7	7.3	5.2
Other medical reasons	2.7	2.4	2.0
Planning pregnancy	1.9	4.4	7.0
Other personal reasons	2.3	3.4	7.9
Termination rates	25.5	22.3	28.3
Continuation rates	74.5	77.7	71.7
Women-months of use[b]	116,155	38,862	4,821

[a]Events per 100 women.
[b]Number of first insertions, 16,345.
Source: Compiled from data in Reference 2.

Table 5 Effect of Wearer's Ages on the Contraceptive Effectiveness and Termination Rates of Cu-T-200

Events	Age of women[a]				
	15–19	20–24	25–29	30–34	35–49
Pregnancies	5.3	5.7	3.9	2.8	0.5
Expulsions	14.2	8.9	8.1	6.0	2.6
Removals					
Bleeding or pain	16.1	14.8	13.8	9.6	13.3
Other medical reasons	5.2	4.2	4.5	4.3	3.4
Planning pregnancy	4.2	6.5	5.6	4.5	0.3
Other personal reasons	4.4	4.5	4.4	4.9	6.6
Termination rates	50.4	44.6	40.3	32.1	26.7
Continuation rates	49.6	55.4	59.7	67.9	73.3
No. of first insertion	3,112	6,776	3,915	1,569	950
Woman-months of use	27,588	61,704	37,883	16,363	11,250

[a]Cumulated data at 24 months of use per 100 women.
Source: Compiled from data from Reference 2.

as the 7-shaped device (0.09 cm^3). Comparative studies of the pregnancy rates for the two copper-bearing IUDs were performed in nulliparous women. The results, outlined in Figure 12, demonstrate that the Cu-7 has a lower cumulative rate of pregnancy than the Cu-T (68).

A copper-bearing IUD produced only a minimal effect on the endometria of women and rhesus monkeys (58,69). Some changes, which include leukocytic infiltration, were observed but they were no more, and probably less, marked than with conventional, nonmedicated IUDs. Moreover, there were no changes observed in cytology for up to 5 years of use (49,70). An avalanche of product liability lawsuits in the mid-1980s involving oral contraceptives and IUDs drove product liability premiums sky-high, which unfortunately made G. D. Searle & Co., the manufacturer, to discontinue the production of Cu-7 IUDs (*U.S. News & World Report*, 12/24/90). This decision left the Cu-T-380A (paragard T380A, Grynophanua), which is discussed later, the only copper-bearing IUDs in the U.S. marketplace. On the other hand, more copper-bearing IUDs are available commercially outside the United

Table 6 Life Table Analysis of the Clinical Effectiveness of Cu-7[a]

Events	Clinical effectiveness[b]					
	12 Months		24 Months		36 Months	
	Parous	Nulliparous	Parous	Nulliparous	Parous	Nulliparous
Pregnancies	1.9	1.6	2.9	2.5	3.4	3.3
Expulsions	5.7	8.0	6.8	9.2	7.3	9.6
Medical removals	10.7	13.7	18.1	20.8	24.4	28.4
Continuation rate	73.9	67.5	51.4	48.3	25.8	19.5

[a]Compiled from the package insert by Searle Laboratories. This encompasses 399,054 woman-months of use, including 12 months for 11,852 women, 24 months for 8309, and 36 months for 3885.
[b]Expressed as cumulative events per 100 women through 12, 24, and 36 months of use.

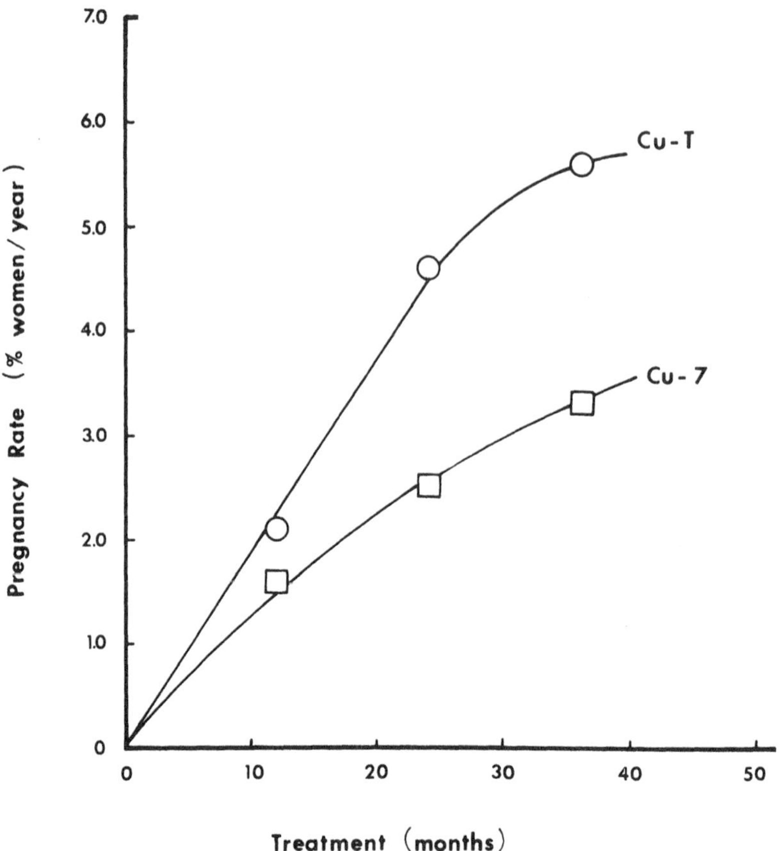

Figure 12 Comparative pregnancy rates in nulliparous women for Cu-T (with 192,562 women-months of use) and Cu-7 (with 403,427 women-months of use). (Plotted from the data in Washington Drug & Device Letter, 10/24/77.)

States. Examples are the polyethylene Nova T (manufactured by Leiras and Outokumpu Oy, Finland) and the polyethylene multiload (manufactured by Multilan, Switzerland). These commercial copper-bearing IUDs vary in wire thickness (0.2–0.4 mm), in the surface area of exposed copper (200–380 mm^2), and in shape.

4. *New Developments*

Both copper-bearing IUDs approved for U.S. commercialization contain 200 mm^2 contraceptive copper wire winding around the depending leg of the 7- or T-shaped plastic device. However, it was reported that the copper-bearing IUD is most efficacious when the copper wire is located on the transverse arm of the device, which is in close contact with the fundus (71). This finding led to the development of a new generation of copper-bearing T-shaped IUDs, the Cu-T-380A, with two collars of copper positioned on the transverse arms of the letter T (also developed by the Population Council, United States). Each collar provides an additional surface area of 30 mm^2. The results of clinical trials demonstrated that the addition of this relatively small surface area of copper (2 × 30 mm^2) in close contact with the upper portion of the endometrial cavity significantly enhanced the antifertility efficacy of copper-bearing IUDs. Thus, the contraceptive effectiveness of metallic copper very much depends on the location as well as the surface area of copper exposed to the endometrial cavity (38,39). Cu-T-380A received regulatory approval from the U.S. FDA for marketing in the 1980s and is now the only copper-bearing IUD available in the United States.

Further development has resulted in the evolution of the Cu-T-220C, with seven copper sleeves (~30 mm^2 each) enveloping both the transverse arms and the vertical stem. Clinical studies on the effectiveness of Cu-T-220C indicate that it has an antifertility effectiveness similar to that of Cu-T-380A IUD but considerably greater than the commercially available Cu-T-200 device (38,39,71,72). From analyses of copper content lost from the sleeves after 2 years of in utero use, the Cu-T-220C device is projected to retain its physical integrity for 15–20 years of continuous intrauterine administration of contraceptive copper. This type of long-acting contraceptive device will be particularly beneficial to populations in which medical care is not readily available (73).

Another new version of copper-bearing IUD, the combined multiload copper IUD (or MLCu-250), was also introduced. It resembles a compromise between the copper-T IUD and the Dalkon shield IUD, but without the central plastic membrane, and has a surface area of 250 mm^2 copper wire. The blunt apex of the device fits into the vault of the uterine cavity without penetrating the endometrial walls, and its two teeth-studded side arms fully adapt to the contours of the uterine cavity. Preliminary clinical testing for 1 year yielded a pregnancy rate of only 0.3 pregnancies per 100 users with an extremely low rate of expulsion (1%). This expulsion rate was substantially lower than that for other IUDs, such as the Lippes D and the Cu-T-200. The low expulsion rate observed with multiload copper IUDs was rationalized as follows. During the uterine contraction, the fundus presses against the upper edge of the multiload IUD; this results in bending of the two side arms, leading to enhanced resistance to expulsion. Various types of multiload copper-bearing IUDs are now commercially available (manufactured by Multilan, Switzerland), with variations in copper surface area (MLCu 250 and MLCu 325) or in size (MLCu 250 standard, MLCu 250 mini, MLCu 250 short, and MLCu 375SL), which makes adaption of the IUD to the size of the individual uterine cavity possible (74). The intrauterine corrosion of copper wire is linearly dependent on the duration of insertion, resulting in a copper release of approximately 0.1 $\mu g/day/mm^2$ of copper surface area for the MLCu 250 IUD, for example. After 5 years of use of this type of IUD, more than half of the cross section was uncorroded (75,76). As discussed earlier, all copper-bearing IUDs act by producing a sterile inflammatory response in the endometrium and by reducing sperm penetration and its ascent upward into the uterus and tubes through modification of the cervical mucus by the copper ions released. Copper was found not to be absorbed to a measurable extent into the systemic circulation from the uterus (77).

The tissue compatibility of copper-bearing IUDs was reportedly improved by coating the IUDs with hydrogel, leading to a reduction in inflammatory responses and local morphological effects (73,74).

C. Hormone-Releasing Intrauterine Devices

The use of hormone-releasing IUDs was first initiated by Doyle and Clewe (23) with the objective to enhance the intrauterine retention of a silicone device in animals. They showed that the slowly releasing steroids, such as melengestrol acetate, have improved the retention time of the IUD and also produced a typical secretory endometrium in estrogen-primed, castrated monkeys.

Croxatto et al. (78) showed 2 years later that a progestin, such as megestrol acetate, released at a controlled rate from a silicone capsule inserted in rabbit uterine cavity, is able to prevent implantation in the experimental horn, whereas normal implantation proceeds in the control (contralateral) horn. This investigation clearly demonstrated localization of the antifertility activity of a progestin within the uterine cavity.

The concept of long-term, continuous intrauterine administration of contraceptive steroids was carried one step further to human testing using a conventional IUD as a carrier by Scommegna and coworkers (34) in 1970. Preliminary investigations (35) were conducted by affixing progesterone-containing silicone capsules to a modified Lippes loop. The results of these short-term studies demonstrated that the intrauterine administration of progesterone, which is inactive when taken orally, induced histological changes in the endometrium. This localized activity could presumably be a result of the interference of the normal reproductive process. Several prototypes of hormone-releasing IUDs were subsequently developed and evaluated by attachment of the steroid-containing silicone capsules to various carriers of different shapes or sizes. Scommegna was later granted a U.S. patent for his development of progestin-releasing IUDs (79). However, these early models of medicated IUDs often had high expulsion rates or side effects, such as bleeding.

Further development has resulted in the evolution of a T-shaped progesterone-releasing IUD in which a drug-containing silicone capsule forms an integral part of the vertical limb of the polyethylene T device (36). The clinicl efficacy was found encouraging, but the relatively rapid release of progesterone (~300 μg/day) from the silicone capsule made this mode of steroid delivery impractical (39). This problem was overcome by encapsulating the progesterone in a polymeric device with a lower progesterone permeability (80).

Working independently, Pharris et al. (80) developed a new version of the progesterone-releasing IUD fabricated from encapsulating a medicated core, which is a suspension of progesterone microcrystals in a silicone medical fluid, in a rate-limiting barrier prepared from ethylene-vinyl acetate copolymer (EVA). As in the Scommegna IUD, the progesterone-releasing compartment also forms an integral part of the vertical limb of the T-shaped IUD. By varying the permeation characteristics of the copolymer, it became possible to deliver progesterone at constant rates that are substantially lower than from the Scommegna progesterone-releasing silicone-based IUD. The progesterone-releasing EVA-based IUD, with a release rate of 65 μg/day, was later selected as the final design of the IUD, called Progestasert (Figure 7), for clinical trials. Progestasert received regulatory approval in 1975 from the U.S. Food and Drug Administration as a medicated IUD for 12-month intrauterine contraception (39).

1. Fundamentals of the Antifertility Action of Progesterone

As the human blastocyst attaches itself to the endometrium, a unique transformation, called the endometrial decidual reaction, takes place between days 5 and 6 of age in the connective tissue of the endometrium. The stromal cells enlarge and grow as they gradually differentiate into large epitheliallike, polyhedral cells rich in glycogen and lipids. In the presence of the implanted blastocyst, this decidual reaction persists. On the other hand, if there is no embryo present, the endometrium is sloughed in the process of menstruation. Once the decidual reaction has occurred, however, implantation of the blastocyst cannot take place. The principal function of the decidual reaction is to limit the invasiveness of the trophoblast (82,83).

Implantation of the blastocyst takes place on the secretory endometrium. The development of secretory endometrium is hormonally controlled, and optimal amounts of estrogen and progesterone are required for a properly developed secretory endometrium (84). An excess or deficiency of either hormone results in abnormal endometrial development. Endometrial hypermaturation was reported to be unfavorable for the implantation of a blastocyst (85). The maturation of endometrium in humans is associated with decidual formation, which has been reportedly induced by progesterone (86,87).

The effects of the estrogen-progesterone system (Figure 3) on the implantation of the fertilized ovum in most mammals, including humans, were recently related to the presence of a membrane eletrical potential that favors or inhibits the ovum-endometrium contact before the occurrence of implantation (88).

These observations were substantiated by reports (89–92) that demonstrated that the suppression of ovulation is not essential for the hormonal prevention of pregnancy.

Progesterone, the naturally-occurring progestin (Figure 3), is rapidly inactivated by hepatic first-pass metabolism when taken orally. Thus, a large dose of natural progesterone or a more potent synthetic analog, such as levonorgestrel (Chapters 8 and 9), which resists hepatic elimination, or a prodrug of a potent synthetic analog, such as ethynodiol diacetate, which is subject to hepatic first-pass metabolism to regenerate the active progestogen, such as norethindrone, is required to achieve fertility control via oral administration.

Oral or systemic administration of potent synthetic progestins or large doses of natural progesterone affect the target organs, like the pituitary and uterus, as well as nontarget organs simultaneously and indiscriminantly. Thus, for contraception, most of these effects are not necessary clinically and are considered undesirable (93). These problems can be overcome by delivering the natural progesterone directly to the target organ (uterus) at low doses to induce the decidual reaction and thereby prevent blastocyst implantation. By localizing the antifertility action to the target tissue site, contraception can be achieved even without inducing undesired hypothalamic-pituitary inhibition or other systemic effects, such as interruption of ovulation.

This localization of drug administration and pharmacological action yields a practically negligible concentration of progesterone in the bloodstream or systemic activity in other nontarget tissues. This absence of a systemic effect is attributed to the ability of endometrium to metabolize progesterone as it traverses various endometrial layers. Therefore, the deep layers of endometrium would not be subjected to the progestational action of the exogenous progesterone administered intrauterinally, and in the meantime, they would continue to respond to the hormonal actions of the endogenously-secreted ovarian hormones (94,95). The cyclical changes and normal desquamation are thus maintained.

2. *Biopharmaceutics of Intrauterine Progesterone Administration*

The uterine bioavailability and tissue distribution of progesterone following intrauterine administration were compared with that attained by oral delivery and subcutaneous injection in mature female rats (96). Following the gastric intubation of $[H^3]$-progesterone, most radioactivity was detected in the liver, a nontarget organ, throughout the study (as much as 40% of the radioactivity administered orally was found 4 hr after administration; Figure 13). However, the target organs, such as uterus and vagina, and other nontarget organs, including blood and brain, were observed to retain only an extremely small fraction of the administered dose (for instance, a maximum of only 0.15% of the total radioactivity administered was measured in the uterine tissue 4 hr after the ingestion). On a concentration basis, the progesterone-specific tissues, uterus and vagina, retained high concentrations of radioactivity per gram tissue for a prolonged period, but the nontarget liver still had the highest concentration of radioactivity, which was approximately 20 times more than that of the uterus for up to 8 hr after oral administration. A similar pattern and level of radioactivity distribution in various tissues were also observed when the same dose of radioactive progesterone was administered by subcutaneous injection. Most of the radioactivity was again detected in the nontarget liver.

When the radioactive progesterone was directly instilled, via laparotomy, into the luminal cavity of the rat uterine horns, large amounts of radioactivity were first taken up by the uterine tissue soon after the instillation. The uterine radioactivity was observed to decrease in a biphasic exponential pattern: an initial (α) phase of rapid distribution, which lasted for about 60 min, followed by a second (β) phase with a much slower rate of elimination (Figure 14). This decrease in uterine radioactivity coincided with the increases in radioactivity in liver and blood (96).

With intrauterine administration, uterine tissue retained a high concentration of radioactivity, as predicted, for a rather long period of time. Only a small fraction of the progesterone was observed to reach various nontarget tissues (brain, lung, diaphragm, and muscles). The bioavailability of progesterone in the uterus, as estimated from the area under the radioactivity-time curve (Figure 14), was approximately 45 times higher than that obtained by either gastric intubation or subcutaneous injection. The uptake of progesterone by the endometrium was found to be extremely fast. Essentially, all radioactive progesterone (95%) was absorbed by the uterine tissue 5 min after intraluminal instillation (α phase). Similar observations were also reported in spayed mice (97). Apparently, the endometrium is an extremely effective tissue for progesterone absorption (96).

Following rapid endometrial absorption, the radioactivity in uterine tissue was observed to be eliminated in a kinetic pattern closely resembling a biphasic exponential curve (Figure 15). A two-compartment open model was proposed to describe this phenomenon (scheme I: Figure 15), where $[A]$ and $[B]$ represent the progesterone concentrations in compartments A and B of the uterus, respectively; $[C]$ is the amount of progesterone excreted from uterine tissues; k_{12} and k_{21} are the rate constants for the reversible exchange between compartment A, representing easily accessible tis-

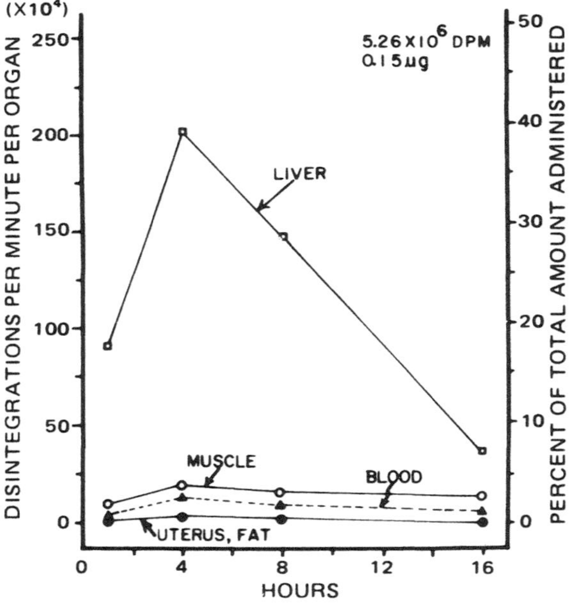

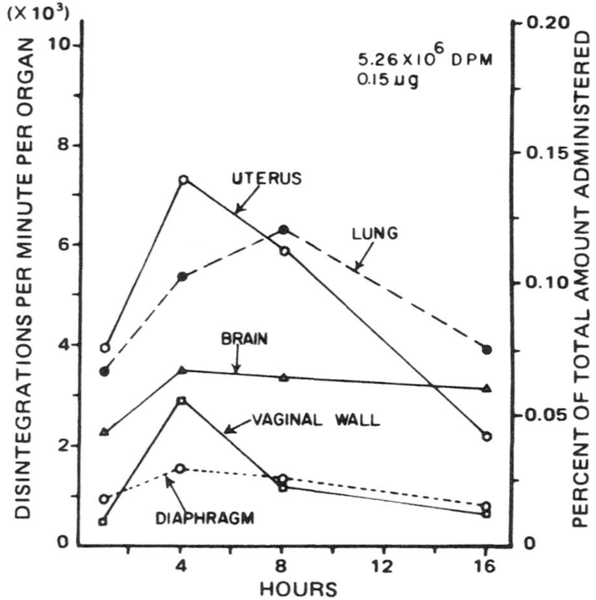

Figure 13 Comparative distribution and retention profiles of radioactivity by various nontarget tissues, such as liver, and target tissues, such as uterus, following the oral administration of [^{3}H]progesterone via gastric intubation. [Reproduced from Fang et al; J. Pharm. Sci., 66:1744 (1977).]

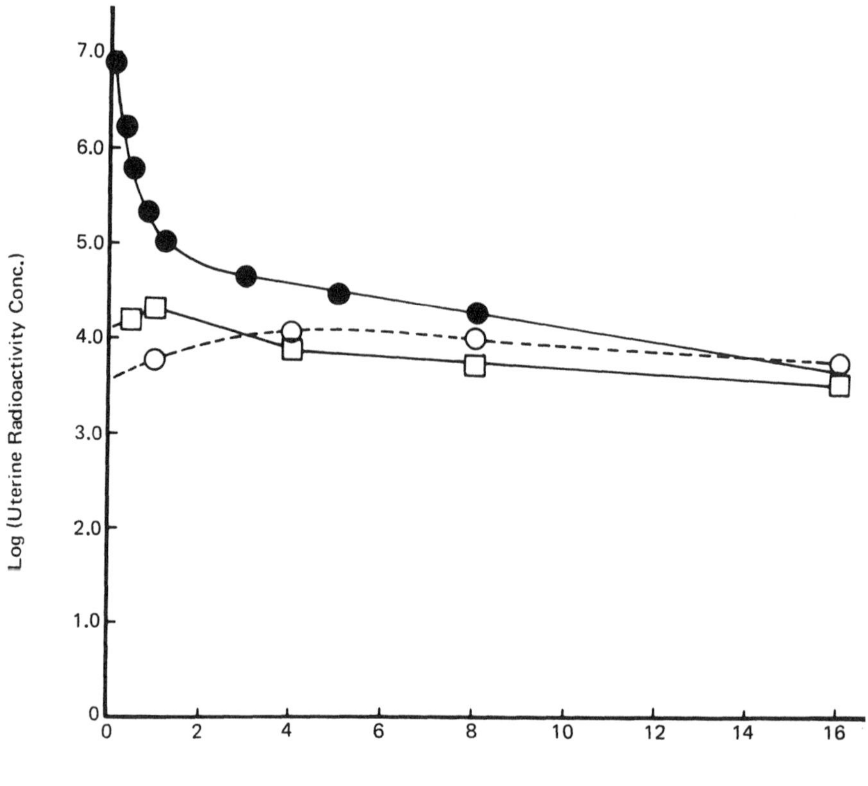

Figure 14 Relative bioavailability of [^{3}H]-progesterone to rat uterus following three routes of administration: (●) uterine intraluminal instillation, (○) gastric intubation, and (□) subcutaneous injection. [Plotted from the data by Fang et al; J. Pharm. Sci., 66:1744 (1977).]

sues, and compartment B, deep tissues in the uterus; and k_e is the rate constant for the elimination of progesterone to other parts of the body, such as the systemic circulation and liver. Mathematical analyses suggested that the total concentration of radioactivity in the uterus, C_u, following the intrauterine administration of [H^3]-progesterone is described by

$$C_u = M_1 e^{-\alpha t} + M_2 e^{-\beta t} \tag{2}$$

where α, and β are the slopes of the distribution and elimination phases, respectively (Figure 15), and M_1 and M_2 the y-axis intercepts are determined by

$$M_1 = \frac{D_u(k_e - \beta)}{V_d(\alpha - \beta)} \tag{3}$$

$$M_2 = \frac{D_u(\alpha - k_e)}{V_d(\alpha - \beta)} \tag{4}$$

where D_u and V_d are the total amount of progesterone administered intrauterinally

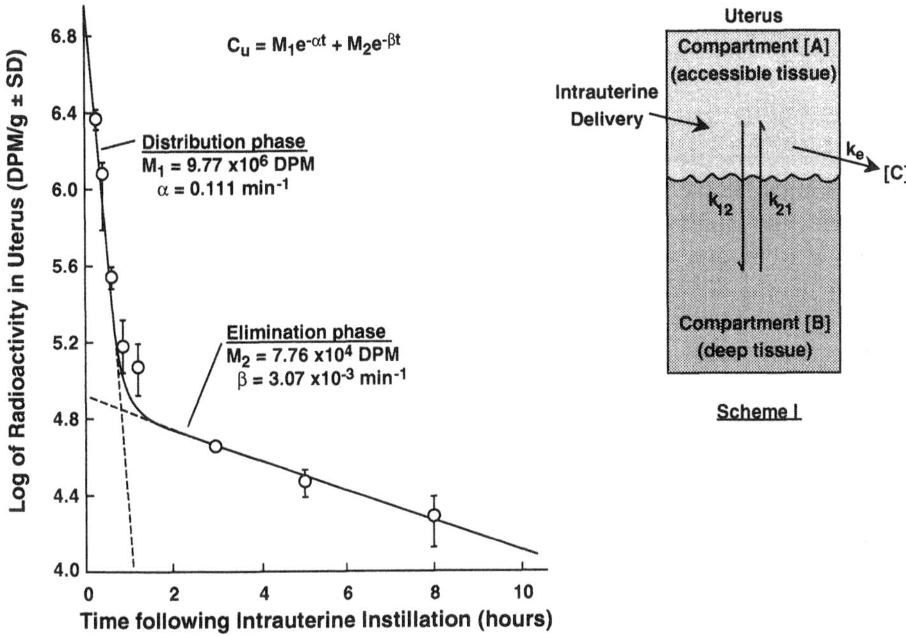

Figure 15 (Left) Pharmacokinetic analysis of the uterine retention of radioactivity after the uterine intraluminal instillation of [^{3}H]-progesterone. (Right) Two-compartment open model (scheme I) developed to describe the distribution and elimination of [^{3}H]progesteronoe in the uterine tissues following intrauterine delivery. [Replotted from the data by Fang et al; J. Pharm. Sci., 66:1744 (1977).]

and the apparent volume of distribution, respectively. The first-order rate constants k_{12}, k_{21}, and K_e can be calculated from the relationships

$$k_e = \frac{\alpha M_1 + \beta M_2}{M_1 + M_2} \tag{5}$$

$$k_{21} = \frac{\alpha\beta}{k_e} \tag{6}$$

$$k_{12} = \alpha + \beta - k_{21} - k_e \tag{7}$$

where M_1, M_2, α, and β can be estimated from the log C_u versus t plots (Figure 15).

Biopharmaceutical analysis of this intrauterine progesterone uptake study demonstrated that progesterone by intrauterine delivery is first taken up by the endometrium and then rapidly distributed, upon reaching uterine tissues (represented collectively by compartment A), into other parts of the body, particularly the liver, in a first-order kinetic pattern with a rate constant k_e of 0.110 min^{-1}. Simultaneously, a diffusion process also takes place within the uterine tissues, where progesterone molecules diffuse to a deep compartment (represented by compartment B) within the uterine cells (scheme I). This intrauterine distribution is a reversible process ($A \rightleftharpoons B$), with $k_{12} = 8.27 \times 10^{-4}$ min^{-1} and $k_{21} = 3.09 \times 10^{-3}$ min^{-1}.

The sequence for the controlled release of progesterone from a progesterone-releasing IUD and its subsequent uptake by the endometrium, exhibition of pharmacological activity, metabolism, and elimination is illustrated in Figure 16.

3. *Clinical Effectiveness of Progesterone-Releasing IUDs*

The clinical contraceptive efficacy of progesterone-releasing IUDs was reportedly dependent upon the daily dose of progesterone released from the devices (93). In 1 year studies performed with a placebo T-shaped device containing no progesterone, the rate of pregnancy was 22%, which was close to the result of 18.3% reported previously (Figure 8). The pregnancy rate was considerably reduced to 5.2%, when the IUD was medicated to release a daily dose of 10 μg progesterone. As the dose of progesterone released was increased to 25 μg/day, the rate decreased to 2.7%. With progesterone released at 65 μg/day, the pregnancy rate was further reduced to 1.1%. However, a further increase in the release rate of progesterone to 120 μg/day was noted to yield only a slight reduction in the pregnancy rate to 0.6% (Figure 17). The IUD that releases progesterone at an in utero daily dose of 65 μg/day was thus selected as the final design of Progestasert^{R}. The results of extensive clinical trials for longer than 400 days are illustrated in Figure 18.

A life table analysis of the 1-year clinical effectiveness of Progestasert for both parous and nulliparous women is presented in Table 7. The results are compared

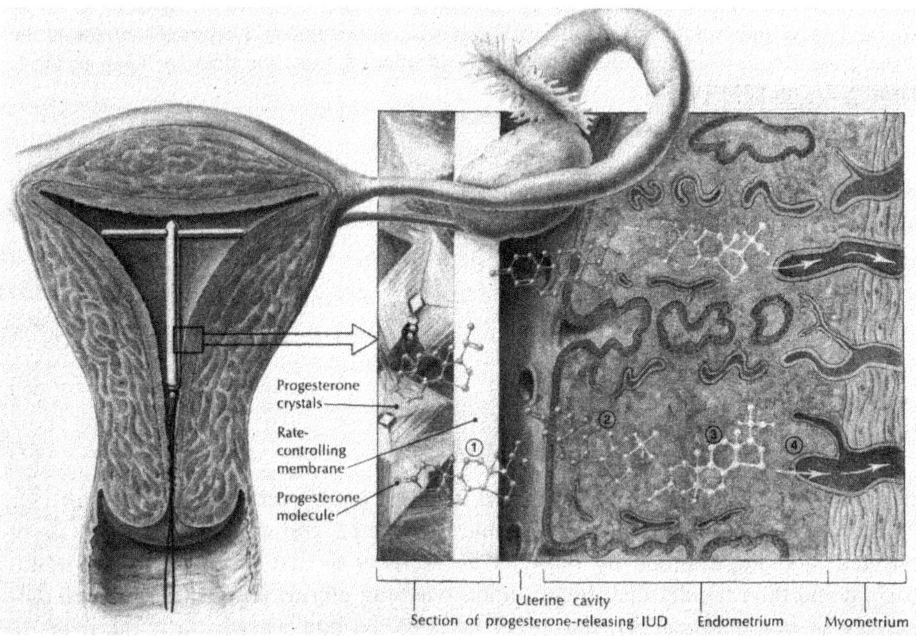

Figure 16 Local effects of progesterone-releasing IUDs. (Insert) Progesterone molecules: (1) diffusing through the rate-controlling membrane in the vertical bar of a T-shaped IUD; (2) penetrating the endometrium, where they exert their pharmacological effects; (3) being rapidly metabolized; and (4) leaving the endometrium. [Reproduced from Rao and Scommegna; Am. Fam. Phys., 16:177 (1977).]

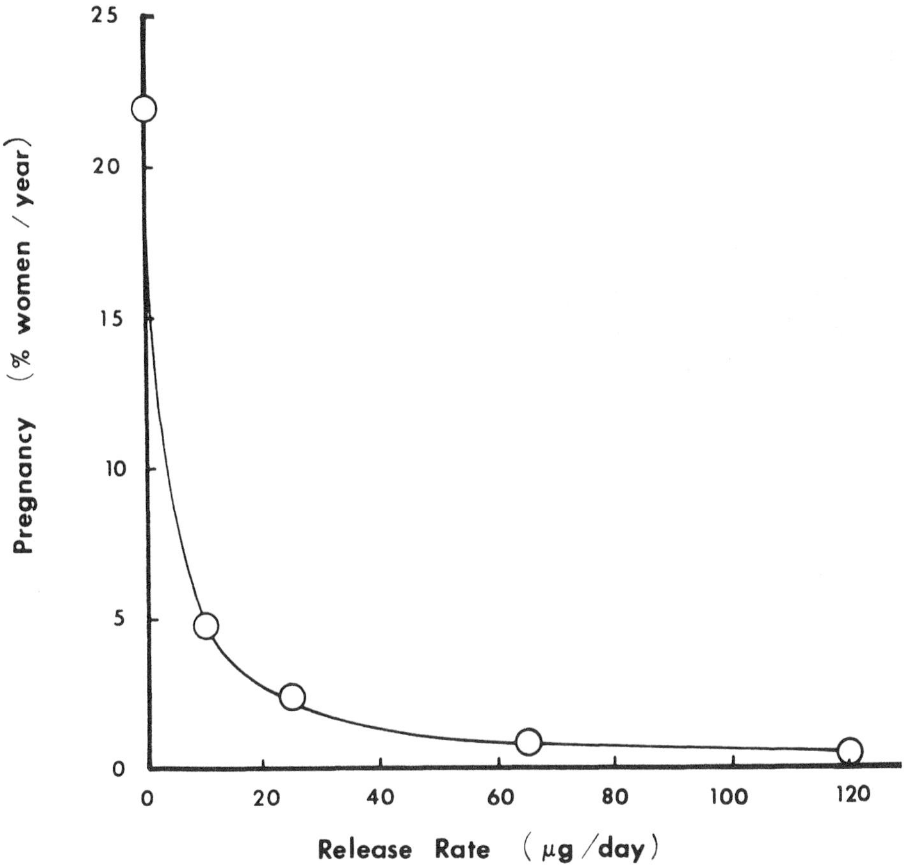

Figure 17 Dependence of the 1-year clinical efficacy (as indicated by the rate of pregnancy) on the daily dose of progesterone released from medicated IUDs in utero. [Plotted from the data by J. Martinez-Manautou; J. Steroid Biochem., 6:889 (1975).]

favorably with those of nonmedicated IUDs, such as Lippes loop, especially the rate of expulsion (Table 8).

Effects on Endometrial Physiology. The time course for histological changes produced by a progesterone-releasing IUD was examined (94). The device induced a mild perifocal "arrested secretion" of the endometrium in its upper layers after the first month of in utero use. After the IUD was in place for 3 months, the histological changes became more pronounced, with decidually transformed stromal cells and atrophic glands. Following 12 months of use, these changes had partially progressed to fibrous atrophy. The endometrial tissue underlying the perifocal arrested secretion showed either proliferative or secretory changes that closely resembled those of a normal menstrual cycle. The histological changes were apparently localized to the upper layers of the endometrium that are important to the implantation of blastocysts.

The observed perifocal arrested secretion is very different from the generalized arrested secretion of endometrium induced by steroids following oral or parenteral administration, and also from the decidualization produced by the mechanical effect

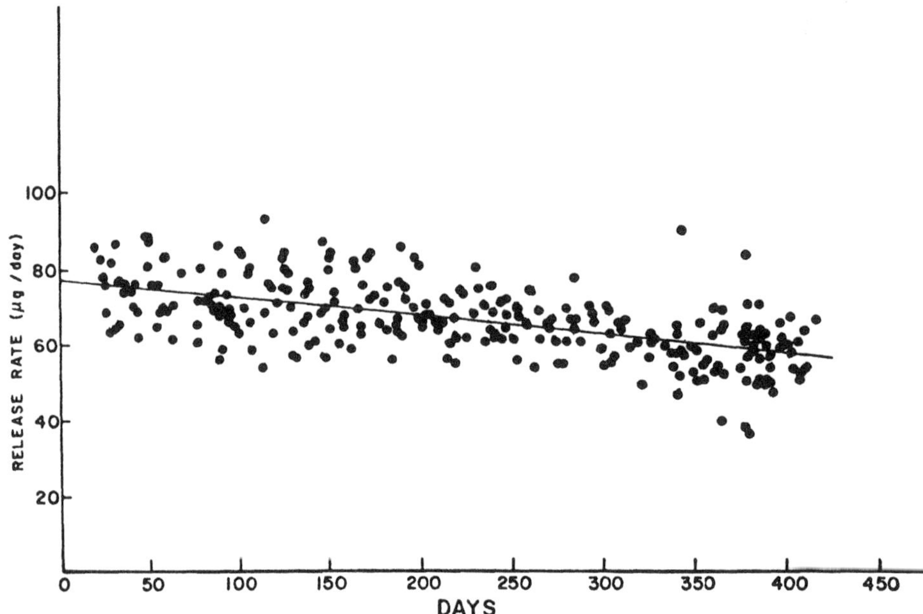

Figure 18 Intrauterine release rate profile of progesterone from Progestasert in women for more than 400 days. [Reproduced from J. Martinez-Manautou; J. Steroid Biochem., 6:889 (1975).]

of a nonmedicated IUD. Compared with a nonmedicated IUD, the glandular atrophy of the arrested secretion enhances the contraceptive effectiveness of a progesterone-releasing IUD.

Extensive histological exmination of the endometria of 400 users of Progestasert (which releases progesterone at the rate of 65 $\mu g/day$) demonstrated a progesterone-induced decidual response. This response was more intense around the devices initially and later spread to the distal endometrium. Proliferative and secretory endome-

Table 7 Life Table Analysis of the Clinical Effectiveness of Progestasert[a]

Events	Clinical effectiveness[b]	
	Parous	Nulliparous
Pregnancy	1.9	2.5
Expulsion	3.1	7.5
Medical removals	12.3	16.4
Continuation rate	79.1	70.9

[a]Compiled from the package insert by Alza Corporation (January 1976). This statistic is based on 45,848 woman-months of use with 5495 women, of whom about 25% were nulliparous; 2157 women completed 12 months of use.

[b]Expressed as events per 100 women through 12 months of use.

Table 8 Life Table Analysis of the Clinical Effectiveness of Lippes Loop Intrauterine Double-S Devices[a]

	Clinical effectiveness[b]			
Events	Loop A	Loop B	Loop C	Loop D
Pregnancy	5.3	3.4	3.0	2.7
Expulsion	23.9	18.9	19.1	12.7
Medical Removals	12.2	15.1	14.3	15.2
Continuation rate	75.2	74.6	76.5	77.4

[a]Compiled from the data in Physicians' Desk Reference (45th edition, 1991). This statistic is based on 121,489 woman-months of use, about 7.2% for Loop A, 7.9% for Loop B, 25.5% for Loop C, and 59.3% for Loop D.

[b]Expressed as events per 100 women through 12 months of use.

tria became histologically indistinguishable. Menstruation continued normally, which suggests overall control by ovarian hormones (95).

The effect of daily doses of progesterone released from IUDs on endometrial morphology was also studied (96). With variations in the release rate of progesterone from 10 to 120 μg/day, the general pattern of endometrium showed a trend of morphologicl variation from a normal secretory to a suppressed endometrium. Additionally, a significant inflammatory infiltration and a frequent predecidual reaction were observed. These morphological changes appeared to be dose dependent and were not observed in women wearing a placebo device.

These endometrial changes were observed to occur evenly over all the superficial regions of the endometrium and created an image of pseudopregnancy very similar to that seen in women taking oral progesterone-estrogen combinations.

The effect of progesterone administered during the preovulatory phase on embryonic mortality was recently investigated in rabbits (97). The results indicated that fertilization proceeds normally, but embryonic death due to the reduced ability of implantation occurs by day 4 following mating. This embryonic mortality was related to the delay in the arrival of embryos into the uterus coupled with the earlier secretion of uteroglobin. The asynchrony of approximately 1 day between embryo arrival and uteroglobin secretion resulted in death of the embryo. On the other hand, no alteration in the physiology of cervical mucus was observed.

Effects on Endometrial Biochemistry. The progesterone-releasing IUD was reported to induce a statistically significant reduction in the concentrations of fucose and sialic acid as well as in the fucose to sialic acid ratio but an increase in the concentrations of Na and K ions in both the proliferative and secretory phases (96). On the other hand, the concentrations of Zn and Ca ions showed a reduction in the proliferative phase but did not vary in the secretory phase, but the concentration of Mg ion was unchanged in the proliferative phase and was enhanced in the secretory phase. The variation in the concentrations of Zn and Mg ions resulted in abolition of the physiological difference between the proliferative and the secretory phases (98). These changes showed no dependence on the daily doses of progesterone released from medicated IUDs (96).

There was a significant decrease in alkaline phosphatase and β-glucuronidase activities, but acid phosphatase increased and the total lactic dehydrogenase showed a biphasic variation with an increase in the proliferative and a decrease in the secretory phase (98). Progeterone-releasing IUDs also lessened the production of lactate and the incorporation of glucose to proteins and lipids during the proliferative phase. No significant changes were noted in the secretory phase, however (96). No statistically significant modifications were detected in proteins, DNA, RNA, or total hexoses (96).

Uterine washings from progesterone-releasing IUD wearers were found to produce a significant reduction in oxygen uptake and glucose utilization, an inhibition of benzoyl-*d,l*-arginine β-naphthylamide hydrolytic activity, and a change in the tetracycline binding and release processes in human spermatozoa (96). This experimental evidence suggested a direct inhibiting effect of progesterone-containing uterine secretions on the capacitation of human spermatozoa.

Systemic Effects of Progesterone-Releasing IUDs. The 65 μg/day of progesterone released from the Progestasert produced a 5- to 10-fold increase in uterine progesterone levels but a 50–75% decrease in the estradiol level (95). These variations in intrauterine steroid levels did not cause any concomitant change in the plasma levels of these and other hormones (95,98).

Plasma profiles of progesterone, estradiol, luteinizing hormone (LH), and follicle stimulating hormone (FSH) showed no differences between women wearing a progesterone-releasing IUD and those using only a placebo device (82,99). Pituitary hormonal signals and ovarian hormonal responses occurred as expected in a normal ovulatory menstrual cycle. The results suggest that when administered intrauterinally, progesterone does not have systemic endocrine activity. Its mechanism of action is not associated with the inhibition of ovulation, since the dosage (65 μg/day) of progesterone delivered by a Progestasert is considerably less than the average production of endogenous progesterone by the corpus luteum (4 mg/day) before ovulation (99). The progesterone released intrauterinally acts directly on progesterone receptors present in the uterine wall to produce an endometrium in its exhausted "secretory" phase, which is not conducive to nidation. By localizing the antifertility action at the target tissue site, contraception is achieved without inhibition of LH and FSH production and therefore without interruption of ovulation.

4. *Potential Developments*

Review of the ongoing research activities on development of medicated IUDs has revealed that several intrauterine drug delivery systems are under development which may be categorized into the following groups.

Membrane-Controlled Reservoir-type Drug Delivery Devices. These drug delivery devices consist of a polymeric membrane that both encapsulates and controls the release of contraceptive agent(s) (Chapter 2, Section II.A). These can be divided into two subgroups.

Single-component system. In this system, the contraceptive agent(s) is encapsulated in its pure solid form in a capsule fabricated from biocompatible polymeric materials, notably silicone elastomers and polyethylene polymer. Scommegna's silicone-based IUD used in the early years of intrauterine controlled progesterone delivery is a typical example. The release of progesterone follows essentially zero-order kinetics, as expected from a polymer membrane permeation-controlled drug delivery mechanism (Chapter 2, Section II.A).

Silicone elastomers have been widely utilized in the fabrication of membrane-controlled reservoir-type drug delivery devices (Chapters 7–9) because of their high permeability to many drugs and excellent tissue biocompatibility. However, they do not possess a tensile strength or elastic modulus high enough to meet most structural requirements of IUDs. These drawbacks can be overcome by forming copolymers of poly(dimethylsiloxane) with polycarbonate or polyurethane. Silicone elastomers give higher release rates than glassy or semicrystalline polymers, such as polyethylene, because of their high internal free volume (100).

Multicomponent system. A constant drug release profile can also be maintained by encapsulating a liquid medium saturated with excess drug particles in a rate-controlling polymeric membrane. Progestasert is a typical example. Progesterone is released at a constant rate of 65 μg/day in utero from a rate-controlling membrane of ethylene-vinyl acetate copolymer. The correlation between the in vivo and the in vitro release rate profiles of progesterone from the Progestasert as shown in Figure 19 is satisfactory.

In theory, a zero-order drug release profile is maintained until the encapsulated drug solution becomes unsaturated. The controlled release of drugs at various rates of delivery can be tailored by modifying the physicochemical properties, such as permeability of the rate-controlling polymeric membrane. By varying the volume of the drug-saturated solution and the surface area of solution/polymer interface, the duration of medication can be adjusted by incorporating an appropriate amount of drug into the solution system (100).

One of the drawbacks of the Progestasert discussed in Section IV.C is its useful life. Because during the first year of use (Figure 19), as much as 60% of the progesterone loading dose (38 mg) in the reservoir compartment has been depleted during the first year of use, the release rate declines approximately 20% and the useful life of the Progestasert is thus limited to 1 year. A 3-year version of a progesterone-releasing IUD has been developed and is currently undergoing clinical evaluation by the World Health Organization (74).

Polymer Matrix Diffusion-Controlled Drug Delivery Devices. This drug delivery device is prepared by homogeneously dispersing drug particles in a cross-linked po-

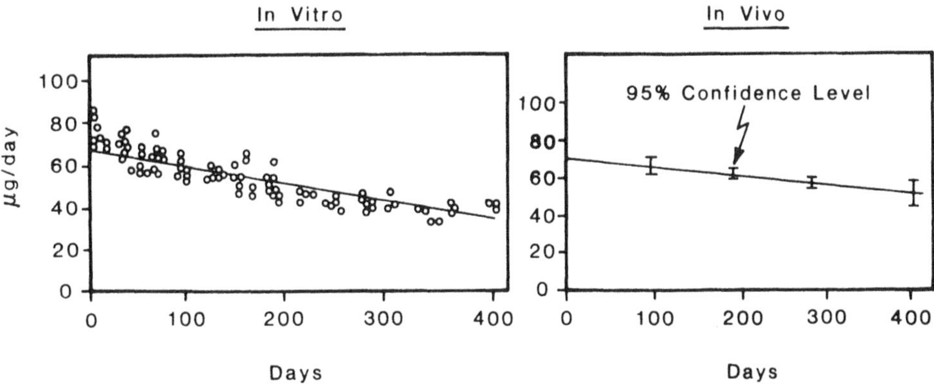

Figure 19 Comparison between the in vitro and in vivo release rate profiles of progesterone from Progestasert.

lymeric matrix (Chapter 2, Section II.B). There ar two main subgroups of this type of drug delivery system (101).

Retrievable matrix device. This drug delivery system is designed for feasibility of retrieval at termination of treatment. It can be easily fabricated from silicone elstomers by premixing a drug in powder form with a semisolid silicone elastomer before vulcanization at room or low temperature (102), or from polyethylene by dry mixing the drug powder with low-density polyethylene particles before melt extrusion (103). It is exemplified by the progesterone-releasing IUDs developed and tested by Doyle and Clewe (23). The rate of drug release is not constant but time dependent. The amount of drug released is linearly proportional to the square root of time (102). A Q versus $t^{1/2}$ drug release profile was obtained, as expected from the matrix diffusion process (Chapter 2, Section II.B).

Biodegradable matrix device. This drug delivery system is designed to eliminate the need for retrieving at the end of treatment. It can be prepared by dissolving both drug and biodegradable polymers, such as poly(lactic acid), in a common organic solvent and then melt pressing at an elevated temperature, after flashing off the solvent to dryness, to produce drug-dispersing biodegradable matrix devices of varying shapes and sizes (104). The rate of drug release from this type of drug delivery system is a combination of polymer hydrolysis and drug diffusion (105,106).

Sandwich-type Drug Delivery Device. This type of drug delivery device is a hybrid of a polymer membrane permeation-controlled drug delivery system with a polymer matrix diffusion-controlled drug delivery system. Several examples can be found in the U.S. patent literature (107).

The surface area of the uterine cavity has physiological limits, but the surface area and configuration of IUDs are basically unconstrained design variables (101). The release rate of contraceptive agent(s) and the surface area of IUDs can be either increased or decreased, and the needed mechanical integrity of an IUD can be maintained by using a thin, rate-controlling membrane to encapsulate a highly permeable drug-dispersing matrix of either dense or porous polymeric materials. High release rates can be achieved for most drugs using this technique, especially when a highly drug permeable membrane, such as silicone elastomer, is used to coat a porous support.

This hybrid-type drug delivery technology was applied to the development of medicated IUDs to release norethindrone (108) or levonorgestrel (109) at a controlled rate. Preliminary clinical testing in small groups of women has shown some promising results (108–110) and resulted in the registration of the levonorgestrel-releasing Nova-T IUD (Figure 20) in Finland (Leiras Pharmaceuticals). It closely resembles the Nova-T copper IUD and is fabricated by replacing the copper wire on the vertical shaft of the T-shaped polyethylene support by a sandwich-type silicone-based levonorgestrel reservoir. As discussed earlier for Norplant (Chapter 8, Section III.C.1), levonorgestrel is a much more potent synthetic progestogen than the natural progesterone and contraceptive effectiveness can be achieved with a daily release as low as 20 μg levonorgestrel (compared to a daily release of 65 μg progesterone for the Progestasert). In vivo studies indicated that the initial release is 24 μg/day during the first month, which is then decreased to an average of 16 μg/day during the following 23-month period (111). After 5 years of use, approximately 60% of the

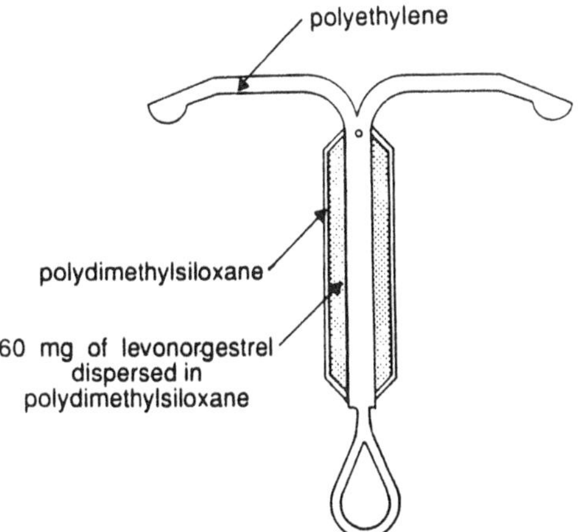

Figure 20 Cross section of a Nova-T IUD and its composition.

levonorgestrel loading dose was released (112). Based on the in vitro release rate, the effective lifetime of this new IUD is estimated to exceed 5 years.

Plasma concentrations obtained in the intrauterine levonorgestrel group were found comparable to those in the oral levonorgestrel group, suggesting that levonorgestrel is not metabolized in the uterine cavity or during endometrial absorption. However, the tissue concentration of levonorgestrel in the endometrium was much higher for the IUD group, which created a profoundly suppressed endometrium even though the total amount of levonorgestrel in the endometrial tissue was only a few percent of that released daily (113). Menstruation was also found to be extremely scanty or disappeared completely. Effects on ovulatory function and partial inhibition of ovulation were also observed (112).

Estriol-Releasing IUDs. Implantation of the fertilized ovum onto the endometrium is known to depend on a series of precisely timed endocrine events in which the secretion of both estrogen and progesterone plays a key role (Section IV.C.1). The synthesis of estradiol-dependent uterine RNA has been found essential for the success of implantation in many species. Estriol is capable of binding with the uterine cytoplasmic receptor in the same way as estradiol but is incapable of inducing true uterine growth, and it may thus interfere with the synthesis of estradiol-induced uterine RNA (114). As shown in animal testing, estriol by systemic administration indeed interferes with conception (115,116); however, this antifertility effect can be reversed by concomitant administration of estradiol (117). On the other hand, the uterotropic effect of estradiol in castrated female rats was reportedly diminished by the coadministration of estriol (118). Estriol was also found to be a potent inhibitor of estradiol binding to several target tissues (114,119). This inhibition is competitive in nature, and the extent of inhibition is proportional to the dose of estriol administered (120).

The antiestradiol properties of estriol were soon recognized, and the experience accumulated over the years with the intrauterine administration of progesterone was thus extended to the development of an estriol-releasing IUD (Figure 21). Preliminary testing in female rabbits indicated that estriol released at a steady rate of 1.25

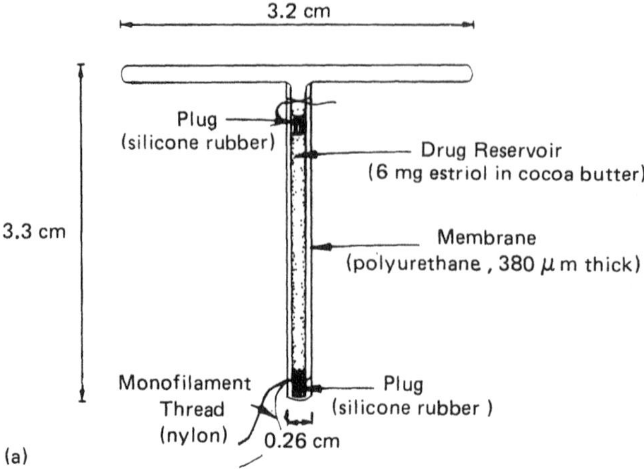

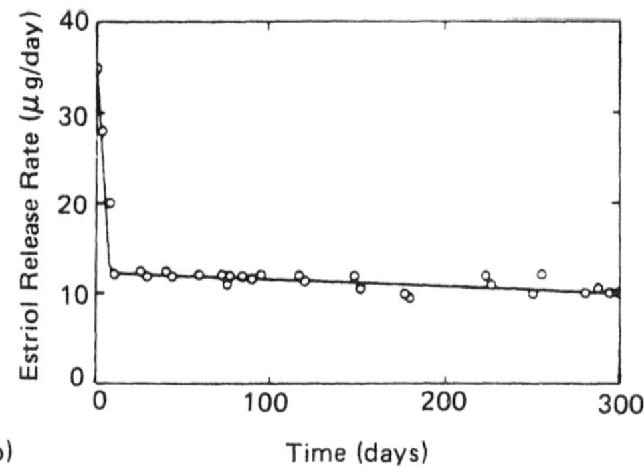

Figure 21 (Top) Cross section of an estriol-releasing IUD and its composition. (Bottom) In vitro release profile of estriol from the IUD.

μg/day effectively inhibits the development and implantation of blastocyst (121). This contraceptive activity was found to be clearly local, since the estriol delivered intrauterinally appeared to have no effect on spontaneous ovulation or ovum fertilization. Further investigations in baboons demonstrated that an estriol-releasing T-shaped IUD showed no appreciable effect on the endometrium. No detectable increase was observed in the proliferative activity of the endometrial glands or stroma. It exhibited little or no estrogenic activity in castrated animals but displayed a significant antiestrogenic action. In addition, intrauterinally delivered estriol showed no appreciable effect on the reproductive cycle or the plasma concentration of estriol.

Antifibrinolytic IUDs. IUD-associated pain and bleeding were reportedly reduced by medicating intrauterine contraceptive devices with antifibrinolytic agents, such as aminocaproic acid. Preliminary studies in monkeys demonstrated a 50% reduction in menstrual bleeding with a dose that was only 1/500 of the oral dose (122).

Table 9 Comparative Clinical Effectiveness of Cu-7 and Progestasert

	Clinical effectiveness[a]	
Events	Cu-7[b]	Progestasert[c]
Pregnancy	1.5	1.4
Expulsion	5.8	4.2
Medical removals	3.0	2.8
Continuation rate	84.6	87.2

[a]Expressed as events per 100 women-years.
[b]G.D. Searle & Co.
[c]Alza Corporation.

D. Comparative Efficacy of Medicated and Nonmedicated IUDs

The progesterone-releasing IUD (Progestasert, Alza Corporation) and copper-bearing IUD (Cu-7, G. D. Searle & Co.) were compared, under the sponsorship of the World Health Organization, in a small randomized clinical trial in Chile (123). This investigation demonstrated that the clinical performance of these two medicated IUDs is very similar (Table 9). The pregnancy rates were found to be comparable (1.4 versus 1.5 per 100 women-years) between the Progestasert and the Cu-7. All medical removals in the Cu-7 group were due to the problem of excessive bleeding. Irregular bleeding was higher in the Cu-7 group (13.4%) than in the Progestasert group (7.5%). The menstrual blood loss in Progestasert users was reduced compared to preinsertion blood loss. However, the Progestasert has a limited life span of 1 year, which is disadvantageous as compared to the 3-year user's life of the Cu-7.

It is known that endometrial lysosomes play a distinct role in destroying their own cytoplasm just before or during implantation (124–127). This has triggered an investigation of the effects of progesterone-releasing, copper-bearing, and nonmedicated IUDs, all T-shaped, on the activities and subcellular distribution of three lysosomal enzymes in the endometrium of IUD wearers (128). The changes in the activities and subcellular distribution of lysosomal enzymes induced by a nonmedicated placebo IUD were found to be quantitatively small and of limited biological significance. The insertion of a copper-bearing IUD yielded some significant variations, primarily in the secretory phase of the endometrium, with the total enzyme activities increased at least twice. A remarkable increase was also detected in the activities of glucoasminidase, acid phosphatase, and β-glucoronidase, which were more than three-, four-, and sixfold higher, respectively, than the activities observed in women wearing a placebo IUD. On the other hand, the progesterone-releasing IUD induced no (or only a small) change in the activity of lysosomal enzymes and increased the stability of lysosomal membranes during the secretory phase.

The effect of copper, progesterone, and mestranol, using the spring coil IUD as the carrier, on the histology of endometrium was compared in 400 parous women (129). With the plain and copper-bearing spring coil IUDs, the cyclical pattern of the endometrium was preserved. The only significant change was found in the leukocytic infiltration of the endometrium, which was more intense with the copper-

Table 10 Comparison of Contraceptive Efficacy Between Intrauterine Devices and Other Major Contraceptive Methods Available in the United States

Method	User[a] (%)	Efficacy (%) Effectiveness[b]	Pregnancy[c]	Failure rate[d]
Intrauterine Device	1.2	94	—	—
Cu-T380A	—		0.8	6
Progestasert	—		2.0	6
Oral contraceptive	18.5	94–97	—	—
Combo	—		0.1	3–6
Minipill	—		0.5	3–9
Barrier method				
Condom	8.8	86	2	14
Diaphragm	3.5	84	6	16
Cervical cap	1.0	73–92	6	8–27
Sponge	0.7	72–82	6–9	18–28
Spermicide	0.6	79	3	21
Subdermal implant:				
Norplant	—[e]	99.8	0.2	0.2
Rhythm	1.8	80	1–9	20
Sterilization				
Male	7.0	99.8	0.1	0.15
Female	16.6	99.6	0.2	0.4

[a]Percentage of couples who use a particular method of contraception.
[b]Percentage of users who avoid pregnancy by using a specific method in a given year.
[c]Percentage of women who expect to conceive in a year assuming they always use the method correctly.
[d]Percentage of women who actually conceive in the first year of using a particular method.
[e]Clinical studies were conducted in some 55,000 women for 20 years. Roughly 500,000 women in 15 countries have used Norplant.
Source: S. Findlay (U.S. News & World Report, 12/24/90).

bearing device. Although progesterone-releasing IUDs produced histological changes that made the endometrium unsuitable for implantation, mestranol-releasing devices produced advanced proliferative or even hyperplastic changes in both glandular and stromal cells from week 1 to week 5 with the prevention of secretory changes in the endometrium, which became unreceptive to ova. From week 6 to 12 of use, regression of these changes occurred with a tendency toward atrophic changes of the endometrium.

An increase in menstrual blood loss is one of the main causes for discontinuing the use of intrauterine devices (130). IUD-associated uterine bleeding was noted not to be related to the mechanically induced surface erosion of the endometrium but could be better explained by the observed formation of vascular gaps from the degeneration of endothelial cells. These gaps may provide a potential route for the escape of blood from endometrium exposed to the pressure generated by an IUD (131–133).

Recently it was reported that the insertion of inert or copper-bearing IUDs has resulted in an increase in menstrual blood loss (by 21.85 and 39.12 ml, respectively) and a decrease in hemoglobin (0.41 and 0.52 g per 100 ml, respectively) compared to the preinsertion cycle. On the other hand, the insertion of progesterone- or progestin-releasing IUDs yielded either no change or a reduction in the menstrual blood loss by as much as 25.7 ml and no significant variations in hemoglobin concentration (110).

V. PROSPECTS FOR INTRAUTERINE CONTRACEPTION

A report was recently published to compare the contraceptive efficacy of IUDs with that of other major contraceptive methods currently available in the United States (*U.S. News & World Report,* 12/24/90). Compared to the statistics collected in the 1975 survey, the data in Table 10 suggest that IUD users in the United States have declined substantially, from 6.4% in 1975 to 1.2% in 1990. The same trend is also followed by oral contraceptive pills, but the use of barrier methods, such as the condom and diaphragm, has increased from 10% in 1975 to 12.3% in 1990. This change could be attributed to the avalanche of product liability lawsuits in the mid-1980s involving oral contraceptives and IUDs, such as the Dalkon shield. This has driven liability premiums sky-high and eight U.S. pharmaceutical firms to cancel contraceptive research. One of the outcomes is that five types of IUDs simply vanished in the litigious 1980s, even though most of them were never shown to be unsafe, and now only two types of IUDs, the Cu-T-380A and Progestasert, are available for intrauterine contraception.

A recent survey conducted in 1988 by the Centers for Disease Control in women indicated that sexual activity among young American women has soared in the last two decades to the point that 51.5% of those aged 15–19 reported premarital sex, nearly double the 28.6% for 1970 (*The Star-Ledger,* 1/5/91). The rates of teenage pregnancy and abortion in the United States, which are the highest among industrialized nations, argue for the development of safer and more efficacious contraceptive methods for long-term continuous fertility regulation, including intrauterine contraceptive devices.

REFERENCES

1. C. Westoff and E. Jones; Washington Drug & Device Letter, (August 1, 1977).
2. A. K. Jain; Contraception, 11:243 (1975).
3. T. O. M. Dieben; in: *State of the Art of the IUD* (Van der Pas et al., Eds.), Kluwer, Dordrecht, 1989, pp. 185–187.
4. R. P. Rao and A. Scommegna; Am. Fam. Phys., 16:177 (1977).
5. C. Tietze and S. Lewit; *Intrauterine Contraception Devices*, International Congress Series, No. 54, Excerpta Medica, Amsterdam, 1962.
6. H. J. Tatum; U.S. Patent, 3,533,406, October 13, 1979.
7. S. W. Jacob and C. A. Francone; *Structure and Function in Man*, 2nd Ed., W. B. Saunders, Philadelphia, 1970, Chapter 17.
8. G. H. Bell, J. N. Davidson and H. Scarborough; *Textbook of Physiology and Biochemistry*, 5th Ed., E. & S. Livingston, Edinburgh, 1963, Chatper 37.
9. D. Egert; Steroids, 31:269 (1978).
10. G. P. Roberts, J. M. Parker and S. R. Henderson; J. Reprod. Fertil., 48:153 (1976).
11. D. P. Wolf and L. Mastroianni; Fertil. Steril., 26:240 (1975).
12. L. Mastroianni, M. Urzua and R. Stambough; Fertil. Steril., 21:817 (1970).

13. V. Peplow, W. G. Breed, C. M. J. Jones and P. Eskstein; Am. J. Obstet. Gynecol., 116:771 (1973).
14. K. S. Moghissi; Fertil. Steril. 21:821 (1970).
15. H. M. Beier and K. Beier-Hellwig; in: *6gh Symposium on Protein Synthesis in Reproductive Tissue*, Karolinska Symposia on Research Methods in Reproductive Endocrinology, 1973, p. 404.
16. D. P. Wolf, L. Blasco, M. A. Khan and M. Litt; Fertil. Steril., 28:41 (1977).
17. D. P. Wolf, L. Blasco, M. A. Khan and M. Litt; Fertil. Steril., 28:47 (1977).
18. D. P. Wolf, J. Sokoloski, M. A. Khan and M. Litt; Fertil. Steril., 28:53 (1977).
19. R. A. Sparks, B. G. A. Purrier, P. J. Watt and M. Elstein; Brit. J. Obstet. Gynecol., 84:701 (1977).
20. W. H. Masters and V. E. Johnson; *Human Sexual Response,* Little, Brown, Boston, 1966, Chapter 8.
21. R. G. Wheeler, G. W. Duncan and J. J. Speidel; *Intrauterine Devices: Development, Evaluation, and Program Implementation*, Academic Press, New York, 1974.
22. J. A. Zipper, M. Medel and R. Prager; in *Abstracts of the 6th World Congress on Fertility and Sterility*, Tel Aviv, Israel, May 20–27, 1968, p. 154.
23. L. L. Doyle and T. Clewe; Am. J. Obstet. Gynecol., 101:564 (1968).
24. H. J. David and R. Israel; *Intrauterine Contraception*, International Congress Series No. 86, Excerpta Medica, Amsterdam, 1965, p. 135.
25. H. J. Tatum and J. A. Zipper; *Proc. VI Northest Obstetrics-Gynecology Congress*, Bahia, Brazil, Oct. 4–9, 1968, p. 78.
26. J. A. Zipper, H. J. Tatum, L. Pastene, M. Medel and M. Rivera; Am. J. Obstet. Gynecol., 105:1274 (1969).
27. L. Andolsek; *The Ljubljana IUD Experience: 10 Years*, Program Abstracts of the Third International Conference of Intrauterine Contraception, Cairo, Egypt, 1974, p. 17.
28. J. A. Zipper, M. Medel and R. Prager; Am. J. Obstet. Gynecol., 105:529 (1969).
29. J. A. Zipper, M. Medel, L. Pastene, M. Rivera and H. J. Tatum; in: *Control of Human Fertility* (E. Diczfalusy and U. Borell, Eds.), John Wiley, New York, 1971, pp. 199–218.
30. J. A. Zipper; U.S. Patent 3,563,235, Feb. 16, 1971.
31. H. J. Abramson; U.S. Patent 3,777,748, Dec. 11, 1973.
32. S. Tejuja, S. D. Choudhury, U. Malhotra and N. C. Saxena; Contraception, 10:351 (1974).
33. J. A. Zipper, M. Medel, L. Pastene, M. Rivera, I. Torres, A. Osorio and C. Toscanini; Contraception, 13:7 (1976).
34. A. Scommegna, G. N. Pandya, M. Christ, A. W. Lee and M. R. Cohen, Fertil. Steril., 21:201 (1970).
35. A. Scommegna, T. Avila, M. Luna and W. P. Dmowski; Obstet. Gynecol., 43:769 (1974).
36. J. A. Zipper, H. J. Tatum, M. Medel, L. Pastene and M. Rivera; Am. J. Obstet. Gynecol., 109:771 (1971).
37. H. J. Tatum; in: *Analysis of Intrauterine Contraception* (F. Hefnawi and S. Segal, Eds.), North Holland, Amsterdam, 1975, p. 155.
38. H. J. Tatum; Fertil. Steril., 28:3 (1977).
39. C. C. Chang and H. J. Tatum; Contraception, 11:79 (1975).
40. F. Hefnawi, O. Kandil, A. Askalani, G. I. Serour, D. Zaki, F. Nasr and M. Mousa; in: *Analysis of Intrauterine Contraception* (F. Hefnawi and S. Segal, Eds.), North Holland, Amsterdam, 1975, p. 459.
41. K. Loewit; Contraception, 3:219 (1971).
42. G. Ullmann and J. Hammerstein; Contraception, 6:71 (1972).
43. E. W. Jecht and G. S. Bernstein; Contraception, 7:381 (1973).

44. T. Tamaya, Y. Nakata, Y. Ohno, S. Nioka, N. Furuta and H. Okada; Fertil. Steril., 27:767 (1976).
45. P. C. M. Young, R. E. Cleary and W. D. Ragan; Fertil. Steril., 28:459 (1977).
46. M. H. Cross; J. Reprod. Fertil., 320485 (1973).
47. R. Abraham, R. Mankes, J. Fulfs, L. Goldberg and F. Coulston; J. Reprod. Fertil., 36:59 (1974).
48. H. J. Tatum; Am. J. Obstet. Gynecol., 117:602 (1973).
49. H. W. Hawk, B. S. Cooper and H. H. Conley; Am. J. Obstet. Gynecol., 118:480 (1974).
50. F. T. Webb; J. Reprod. Fertil., 320429 (1973).
51. K. Hagenfeldt; Contraception, 6:191 (1972).
52. K. Hagenfeldt; Contraception, 6:219 (1972).
53. E. W. Wilson; J. Obstet. Gynecol. Br. Commonw., 80:648 (1973).
54. F. B. O'Brien; Searle Laboratories, Personal Communication (1981).
55. B. Larsson and L. Hamberger, Fertil. Steril., 28:624 (1977).
56. K. Hagenfeldt; Acta Endocrinal. (Copenh.) 1971(Suppl.):169 (1972).
57. K. Hagenfeldt; Contraception, 6:37 (1972).
58. E. Johannisson; Contraception, 8:99 (1973).
59. T. Okereke, I. Sternlieb, A. G. Morell and I. B. Scheinberg; Science, 177:358 (1972).
60. A. Moo-Young and H. J. Tatum; Contraception, 9:487 (1974).
61. J. McEwan; Seminar on "Cu-7 Devices" to Searle Laboratory's Sexual Disorder & Reproduction Committee, June 14, 1977.
62. K. M. Lewis; Dissertation Abstr. Intern. B, 38:185 (1977).
63. C. Gosden, A. Ross and N. B. Louden; Brit. Med. J., 1:202 (1977).
64. A. B. Johnson, R. F. Maness and R. G. Wheeler; Contraception, 14:507 (1976).
65. J. Newton, R. Illingworth, J. Elias and J. McEwan; Brit. Med. J., 1:197 (1977).
66. E. Chantler, F. Critoph and M. Elstein; Brit. Med. J., 2:288 (1977).
67. Washington Drug & Device Letter, October 24, 1977.
68. A. J. Moo-Young, H. J. Tatum, A. O. Brinson and W. Hood; Fertil. Steril., 24:843 (1973).
69. J. Zipper, M. Model, R. Segura and L. Torris; in: *Clinical Proceedings of International Planned Parenthood Federation, South-East Asia and Oceania Congress, 1972*, IPPF, 1973, pp. 193–204.
70. D. R. Mishell; in: *Analysis of Intrauterine Contraception* (F. Hefnawi and S. Segal, Eds.), North Holland, Amsterdam, 1975, p. 27.
71. D. L. Cooper, A. K. Millen and D. R. Mishell; Am. J. Obstet. Gynecol., 124:121 (1976).
72. H. Scott, P. L. Kronick, R. C. May, R. H. Davis and J. Balin; Biomat. Med. Dev. 1:681 (1973).
73. R. H. Davis, J. Scott, G. S. Kyriazis and H. Balin; Proceedings of the Society for Experimental Biology and Medicine, 147:407 (1974).
74. A. P. Sam; Controlled-Release Contraceptive Devices, in: *Minutes of 5th Int'l. Pharm. Techn. Symp.* on New Approaches to the Controlled Drug Delivery (Hincal et al., Eds.), Editions de Sante, Paris 1991, pp. 271–284.
75. R. H. Drost, M. Thiery and R. A. A. Maes; Long-term release of copper from two multiload IUD models: MLCu 250 and MLCu 350, Adv. Contracept., 3:315–318 (1987).
76. U. J. Koch, W. Stichel and E. Stange; Copper corrosion and life span of IUD's (MLCu 250), in: *Advances in Fertility and Sterility Series*, Vol. 6, *Contraception* (Ratnam et al., Eds.), 1987, Chapter 12.
77. H. J. Tatum and E. B. Connell; Intrauterine contraceptive devices, in: *Contraception, Science and Practice* (Filshie and Guilebaud, Eds.), Butterworths, London, 1989, pp. 144–171.

78. H. B. Croxatto, R. Vera and M. A. Parga; in: *Proceedings of IV Annual Meeting of ALIRH*, Ixtapan, Mexico, April 5–9, 1970, p. 77.
79. A. Scommegna; U.S. Patent 3,911,911, October 14, 1975.
80. B. B. Pharriss, R. Erickson, J. Bashaw, S. Hoff, V. A. Place and A. Zaffaroni; Fertil. Steril., 25:915 (1974).
81. E. E. Baulieu; in: *Hormones and Breast Cancer* (M. Namer and C. M. Lalanne, Eds.), 1975.
82. A. Scommegna; in: *Regulation of Human Fertility* (K. S. Moghissi and T. N. Evans, Eds.), Wayne State University Press, Detroit 1976, Chapter 12.
83. R. G. Good and D. L. Moyer; Fertil. Steril., 19:37 (1968).
84. R. W. Noyes, Z. Dickmann, L. L. Doyle and A. J. Gates; in: *Delayed Implantation* (A. C. Enders, Ed.), University of Chicago Press, Chicago, 1963.
85. R. K. Meyer and R. L. Cochrane; J. Reprod. Fertil., 4:67 (1962).
86. V. R. Mallikarjuneswara; Indian J. Pharm., 39:161 (1977).
87. J. Martinez-Manautou; Excerpta Medica Int'l. Cong. Ser., 188:999 (1967).
88. H. W. Rudel; Excerpta Medica Int'l. Cong. Ser., 188:994 (1967).
89. J. Zanarta, M. Pupkin, D. Rosemberg, R. Guerrero, R. Rodriguez-Bravo, M. Garcia-Huidobro and J. A. Puga; Brit. Med. J., 2:266 (1968).
90. F. Murad; Drug Therap., May:119 (1977).
91. S. M. Fang, C. S. Lin and V. Lyon; J. Pharm. Sci., 66:1744 (1977).
92. B. F. Clark; J. Endocrinol., 58:555 (1973).
93. J. Martinez-Manautou; J. Steroid Biochem., 6:889 (1975).
94. D. Dallenbach-Hellweg and S. Sievers; Virchows Arch. [A] Pathol. Anat. & Histol., 368:289 (1975).
95. B. B. Pharriss; Ann. NY Acad. Sci., 286:226 (1977).
96. J. Martinez-Manautou, J. Aznar, A. Rosadao and M. Maqueo; in: *Analysis of Intrauterine Contraception* (F. Hefnawi and S. J. Segal, Eds.), North Holland, Amsterdam, 1975, p. 173.
97. S. M. McCarthy, R. H. Foote and R. R. Maurer; Fertil. Steril., 28:101 (1977).
98. K. Hagenfeldt and B. M. Landgren; J. Steroid Biochem., 6:895 (1975).
99. S. A. Tillson, M. Marian, R. Hudson, P. Wong, B. Pharriss, R. Aznar and J. Martinez-Manautou; Contraception, 11:179 (1975).
100. R. W. Baker and H. K. Lonsdale; in: *Controlled Release of Biologically Active Agents* (A. C. Tanquary and R. E. Lacey, Eds.), Plenum Press, New York, 1974, pp. 15–71.
101. H. J. Davis; *Intrauterine Devices for Contraception: The IUD*, Williams and Wilkins, Baltimore, 1971.
102. Y. W. Chien, H. J. Lambert and L. F. Rozek; in: *Controlled Release Polymeric Formualtions* (D. R. Paul and F. W. Harris, Eds.), ACS Symposium Series 33, American Chemical Society, Washington, D.C., 1976, p. 72.
103. D. R. Kalkwarf, M. R. Sikov, L. Smith and R. Gordon; Contraception, 6:423 (1972).
104. S. Yolles; U.S. Patent 3,887,699, June 3, 1975.
105. C. G. Nilsson, E. D. B. Johansson, U. M. Jackanicz and T. Luukkaineni, Am. J. Obstet. Gynecol., 122:90 (1975).
106. T. M. Jackanicz, H. A. Nash, D. L. Wise and J. B. Gregory; Contraception, 8:227 (1973).
107. A. Zaffaroni; U.S. Patent 3,854,480, Dec. 17, 1974.
108. D. G. Nilsson, T. Luukkainen and P. Lahteenmaki; Contraception, 17:115 (1978).
109. C. G. Nilsson and T. Luukkainen; Contraception, 15:295 (1977).
110. A. J. Gallegos, R. Aznar, G. Merino and E. Guizer; Contraception, 17:153 (1978).
111. C. G. Nilsson, P. Lahteenmaki, D. N. Robertson, T. Luukkainen; Plasma concentra-

tions of levonorgestrel as a function of the release rate of levonorgestrel from medicated intrauterine devices, Acta Endocrinol. (Coposh.), 93:380–384 (1980).

112. K. Ratsula, J. Toivonen, P. Lahteenmaki, T. Luukkainen; Plasma levonorgestrel levels and ovarian function during the use of a levonorgestrel-releasing intracervical contraceptive device, Contraception, 39:195–204 (1989).
113. T. Luukkainen, C. G. Nilsson, H. Allonen, M. Haukkamaa and J. Toivonen, Intrauterine release of levonorgestrel, in: *Long-Acting Contraceptive Delivery Systems* (Zatuchni et al., Eds.), Harper & Row, Philadelphia, 1984, pp. 1–19.
114. P. I. Brecher and H. H. Wotiz; Steroids, 9:431 (1967).
115. A. Scublinsky and H. H. Wotiz; J. Reprod. Fertil., 26:139 (1971).
116. A. Scublinsky and H. H. Wotiz; J. Reprod. Fertil., 26:365 (1971).
117. H. H. Wotiz, S. Smith, B. Shapiro and A. Scublinsky; J. Reprod. Fertil., 26:237 (1971).
118. F. L. Hisaw, J. T. Velardo and C. M. Goolsby; J. Clin. Endocrinol. Metab., 14:1134 (1954).
119. A. J. Eisenfeld and J. Axelrod; Endocrinology, 79:38 (1966).
120. A. E. Wicks and S. J. Segal; Proc. Soc. Exp. Biiol. Med., 93:270 (1956).
121. W. P. Dmowski, A. Shih, J. Wilhelm, F. Auletta and A. Scommegna; Fertil. Steril., 28:262 (1977).
122. S. T. Shaw, D. L. Moyer, D. E. Aaronson, J. Underwood and R. V. Forino; Contraception, 11:395 (1975).
123. E. Pizarro, C. Comez-Rogers, P. J. Rowe and S. Lucero; Contraception, 16:313 (1977).
124. R. Abraham, R. Hendy, W. J. Doubherty, J. C. Fulfs and L. Goldberg; Exp. Mol. Pathol., 13:329 (1970).
125. E. Lindford and J. M. Iosson; J. Reprod. Fertil., 44:249 (1975).
126. J. C. Wood; Adv. Reprod. Physiol., 6:221 (1973).
127. A. Rosado, E. Mercado, A. J. Gallegos, M. A. Wens and R. Aznar; Contraception, 16:287 (1977).
128. E. Mercado, R. Aznar, A. J. Gallegos, R. Dominguez and A. Rosado; Contraception, 16:299 (1977).
129. M. I. Ragab and I. A. Senna; in: *Analyses of Intrauterine Contraception* (F. Hefnawi and S. J. Segal, Eds.), North Holland, Amsterdam, 1975, p. 325.
130. C. Tietze; in: *Human Fertility: Conception and Contraception* (E. S. E. Hafez and T. N. Evans, Eds.), Harper & Row, New York, 1972.
131. W. R. Hohman, S. T. Shaw, Jr., L. Macaulay, D. L. Moyer; Contraception, 16:507 (1977).
132. A. Gonzalez-Angulo, R. Aznar-Ramos and A. Feria-Velasco; J. Reprod. Med., 10:44 (1973).
133. M. B. Sammour, S. G. Iskander and S. F. Rifai; Am. J. Obstet. Gynecol., 98:946 (1967).

11
Systemic Delivery of Peptide-Based Pharmaceuticals

I. INTRODUCTION

Management of illness through medication is entering a new era in which a growing number of biotechnology-produced peptide-based pharmaceuticals are available for therapeutic use. This has become a reality as a result of the increased understanding of the role of peptides and proteins in physiology and therapy as well as advances in biotechnology. The advances made in recent years in biotechnology have led to the development of processes and the establishment of facilities for producing a large quantity of peptide-based pharmaceuticals on an economical scale, which has made the therapeutic use of many peptide and protein pharmaceuticals feasible and practical. Ailments that can be treated more effectively by this new class of therapeutic agents include cancers, autoimmune diseases, memory impairment, mental disorders, hypertension, and certain cardiovascular and metabolic diseases (Table 1) (1,2).

The recent development of tissue plasminogen activator (t-PA) is a typical example. This is a macromolecule consisting of 567 amino acid residues and it has a molecular weight of 59,050 daltons. t-PA is a naturally-occurring thrombolytic protein that clears debris from the bloodstream and is capable of causing blood clots to dissolve. It has been discovered that therapeutically, t-PA can be used effectively for the treatment of certain heart attacks and strokes. This life-saving therapeutic application of t-PA has become feasible as the result of the successful development of a biotechnology process to produce t-PA commercially. This biotechnology-produced t-PA has only recently received U.S. Food and Drug Administration (FDA) approval for marketing (Activase, Genentech) (3). The results of recent clinical trials have demonstrated that t-PA is efficacious if it is administered intravenously within a few hours of the onset of a heart attack. Superoxide dismutase, which is an enzyme, is another protein-based pharmaceutical that has also been found beneficial in the treatment of heart attacks.

A large family of hormonelike growth-regulating agents, which are called polypeptide growth factors (PGFs), are now being characterized (4). They are important physiologically and have been reported as capable of stimulating the growth as well as maintaining the viability of a wide range of cell types. Epidermal growth factor

Table 1 Representative Peptide-Based Pharmaceuticals and Potential Functions and Biomedical Applications

Peptide and protein drugs	Function and applications
Cardiovascular-active peptides	
Angiotension II antagonist	Lowers blood pressure
Antriopeptins	Regulate cardiovascular function as well as electrolyte and fluid balance
Bradykinin	Improves peripheral circulation
Calcitonin gene-related factor	Vasodilation
Captopril	Management of heart failure
Tissue plasminogen activator	Dissolution of blood clots
CNS-active peptides	
Cholecystokinin (CCK-8 or CCK-32)	Suppress appetite
Delta sleep-inducing peptide (DSIP)	Improves disturbed sleep
β-Endorphin	Relieves pain
Melanocyte inhibiting factor I	Improves mood of depressed patients
Melanocyte stimulating hormone	Improves attention span
Neuropeptide Y	Controls feeding and drinking behavior
Nerve growth factor	Stimulates nerve growth and repair
Gastrointestinal-active peptides	
Gastrin antagonist	Reduces secretion of gastric acid
Neurotension	Inhibits secretion of gastric juice
Pancreatic enzymes	Digestive supplement
Somatostatin	Reduces bleeding of gastric ulcers
Immunomodulating peptides	
Bursin	Selective B cell differentiating hormone
Colony stimulating factor	Stimulates granulocyte differentiation
Cyclosporine	Inhibits functions of T lymphocyte
Enkephalins	Stimulate lymphocyte blastogenesis
Interferons	Enhances activity of killer cells
Muramyl dipeptide	Stimulates nonspecific resistance to bacterial infections
Thymopoietin	Selective T cell differentiating hormone
Tumor necrosis factor	Controls polymorphonuclear functions
Metabolism-modulating peptides	
Human growth hormone	Treats hypopituitary dwarfism
Gonadotropins	Induce ovulation, spermatogenesis, and cryptorchidism
Insulin	Treats diabetes mellitus
Luteinizing hormone-releasing hormone (LHRH)	Induce ovulation in women with hypothalamic amenorrhea
Oxytocin	Maintains labor
Thyrotropin-releasing hormone (TRH)	Prolongs infertility and lactation in women who are breast feeding
Vasopressins	Treat diabetes insipidus

Source: Expanded from Chien (1987) and Lee (1987).

(EGF), a PGF that has been characterized in detail, is a polypeptide known to affect the growth and/or differentiation of a wide variety of tissues (5). There has been growing interest in the therapeutic potential of EGF, since it has been shown that EGF may be useful for the medication of burns and wounds, as well as for cataract surgery and other ophthalmic applications. EGF and its receptor may also play a role in carcinogenesis (6). Interleukin-2 (IL-2), another well-characterized PGF, is a lymphokine that affects a number of immune systems and their immunological responses. IL-2 has recently been evaluated for the treatment of cancer, either alone or in combination with other lymphokines and chemotherapies.

Erythropoietin, which is also a PGF, has been produced by recombinant DNA, but with only partial characterization. It is a circulating glycoprotein type of hormone produced naturally by the kidney and has also been reported to stimulate the production of red blood cells. In a recent study (7), the human erythropoietin produced by recombinant DNA technology was administered to anemic patients with end-stage renal disease who were undergoing hemodialysis, and the results suggested that it is effective in eliminating the need for blood transfusion as well as in restoring the hematocrit to normal levels. Epoetin-α, a genetically-engineered form of erythropoietin that was originally approved in 1989 for use in patients with chronic renal failure, has recently received FDA approval for the treatment of anemia in acquired immunodeficiency syndrome (AIDS) patients on azidothymidine (AZT) therapy. It will be comarketed by Amgen (Epogen) and Ortho (Procrit) (*The Star-Ledger*, 1/3/91).

Atrial natriuretic factor (ANF), which is a peptide hormone secreted naturally by the heart (8), is a natural diuretic and has hypotensive activity. ANF is a potential therapeutic agent for the treatment of hypertension, and a number of biotechnology companies have been active recently to develop technologies for its biosynthesis. Other cardiovascularly active peptide-based pharmaceuticals that have received substantial interest recently are bradykinin and related kinins (9). They are useful therapeutically for the improvement of peripheral circulation.

These therapeutic peptides and proteins are only a few examples of the peptide-based pharmaceuticals currently under active development for possible production by genetic engineering and biotechnology process.

All this progress can be attributed to the advances made in biotechnology, which has played a key role in the development of therapeutic peptides and proteins (10–15). A new series of peptide- and protein-based pharmaceuticals has now arrived with the advent of recombinant DNA and hybridoma techniques and also with the recent progress made in large-scale fermentation and purification processes (16). A recent survey conducted by the Pharmaceutical Manufacturers Association (PMA) indicated that 104 biotechnology-produced pharmaceuticals are currently in clinical trials or under FDA regulatory review (*Pharmacy Today*, 5/25/90). These genetically engineering pharmaceuticals include 59 for cancer therapy and 15 for AIDS or human immunodeficiency virus (HIV)-related conditions. This latest PMA survey also indicates that monoclonal antibodies make up the largest therapeutic category, with 37 drugs; vaccines are the second largest category with 10, and other therapies include 8 interferons, 7 colony stimulating factors, 7 growth factors, 4 human growth hormones, 4 recombinant soluble CD4s, 3 tumor necrosis factors, 2 each for dismutases, erythropoietins, and factor VIIIs, as well as 1 each for cloned peptide, tissue plasminogen activator, ribosome, and vesicle, and 1 to treat children's ailments.

Product-oriented recombinant DNA research was initiated by a number of small biotechnology research and development firms in the late 1970s, and many of the multinational pharmaceutical companies and other industries followed this lead either by establishing their own in-house biotechnology research programs or by entering into joint ventures with fledgling biotechnology firms (17). Economic forecasts predict a bright future for this new industry (Table 2). The United States and Japan are expected to be the major competitors with the greatest potential for success, but a number of European nations have also recently mounted considerable efforts to commercialize biotechnology (12). The PMA report published in 1990 indicated that U.S. corporations remain the leading source of patents on biotechnology products, with patents accounting for 47% of therapies using genetically-engineered pharmaceuticals. However, this lead has decreased from 84% in 1986 and 78% in 1989.

With the advancements made in biotechnology, several therapeutically-useful proteins have been produced successfully through recombinant DNA technology, such as human growth hormone, human insulin, interferons, hepatitis B subunit vaccine, t-PA, and factor VIII. These biotechnology products received FDA regulatory approval in 1982–88 for marketing and have been projected to possibly contribute more than one-quarter of the 1992 biotechnology market in the United States alone (Table 2). The market could expand as a result of the further discovery of new treatments for these biotechnology products. For example, Epoetin-α was recently approved for the treatment of anemia in AIDS patients on AZT therapy, and γ-interferon has also recently received FDA approval for the therapy of chronic granulomatous disease. Furthermore, two kinds of colony stimulating factors, granulocyte colony stimulating factor (Amgen) and granulocyte-macrophage colony stimulating factor (Immunex) recently received FDA approval for cancer treatment. They prevent deadly infections in patients receiving myelosuppressive chemotherapy during cancer treatment (*USA Today*, 3/15/91 and 3/29/91).

Table 2 Economic Forecast for Biotechnological Products

	Economic growth ($ milllion)			
	Harford (1985)[a]		Frost and Sullivan (1989)[b]	
Market sectors	1980	1990	1988	1992
Pharmaceuticals	28	>8,000	725	5,000
Diagnostics	8	>3,000	—	—
Vaccines	—	1,600	—	—
Total market sales	36	>12,600	—	6,000[c]

[a]International health care market.
[b]U.S. health care market.
[c]A total of $1.6 billion is expected to be generated by the seven biotechnology products approved by the FDA in 1982–88: human insulin, human growth hormone, interferons, hepatitis B subunit vaccine, tissue plasminogen activator, factor VIII, and a monoclonal antibody to suppress kidney transplant rejection.

These new biotechnology-produced peptides and proteins are the exact biochemical replicas of the natural products. Although they are extremely potent and highly specific in their therapeutic activity, the majority are difficult to administer clinically. They are therapeutically effective only by parenteral administration. However, they have additional shortcomings: they are extremely short acting intrinsically, and repeated injections are often required to maintain therapeutic efficacy. Therefore, therapeutic applications and commercialization of these therapeutic peptides and proteins rely on the successful development of viable delivery systems to improve their biochemical and biophysical stability and systemic bioavailability.

Several of these peptide- and protein-based pharmaceuticals, notably growth hormone, luteinizing hormone-releasing hormone, interferons, cyclosporins, and tissue plasminogen activator, are unfortunately therapeutically useful only by following a therapeutic regimen that requires daily multiple injections. This therapeutic regimen is highly risky and can be administered only under close medical supervision; hospitalization is often required. Thus, the commercial success of therapeutic peptides and proteins for medication will greatly depend on the development of nonparenteral routes of administration, such as nasal, pulmonary, ocular, buccal, vaginal, rectal, and transdermal (2), or, alternatively, on the successful development of other novel technologies to improve the capability and programmability of parenteral administration in controlling the systemic delivery of drugs, such as the design of implantable, programmable, and/or self-regulating delivery system, while overcoming the drawbacks associated with parenteral routes of administration. However, each of the nonparenteral routes of administration just mentioned has imposed additional biological barriers to systemic delivery, especially for peptide and protein molecules, because of their unique biophysical and biochemical characteristics. The barrier properties inherent to these routes can be very different in terms of tissue permeability, protease activity, and metabolic processes (18).

II. STRUCTURAL COMPLEXITY OF THE PEPTIDE AND PROTEIN MOLECULE

Peptides and proteins are the most abundant components of biological cells. These naturally-occurring macromolecules exist not only as the structural components of cells but also as functioning moieties (e.g., enzymes, antibodies, hormones, and transport mediators) (19). Even though all peptide and protein molecules are constructed from the same 20 amino acids, they are functionally the most diverse of all biological substances (20). Basically, they are macromolecules with molecular weights ranging from approximately a few hundred to several million. Each peptide or protein molecule is a polymer chain with α-amino acids linked together in a sequential manner by peptide bonds, which are peptide linkages formed by interaction between the α-carboxyl and α-amino groups of the adjacent amino acids. The resulting polymers are generally called peptides. The term "polypeptide" refers to the peptide molecule that contains about eight or more units of amino acids, whereas an "oligopeptide" is a peptide molecule with the peptide chain of fewer than eight amino acids. An amino acid unit in a polypeptide is called a residue. A polypeptide that contains from about 50 to as many as 2500 units of amino acids in a peptide chain is called a protein. A protein with two or more peptide chains linked together via disulfide bond(s) is known as an "oligomeric protein"; each of its component chains is called a subunit or a protomer.

The peptide chains in a protein molecule are often folded into a specific three-dimensional structure to fulfill a certain biological function. A typical example is illustrated in Figure 1 for the globular three-dimensional structure of insulin monomer, which consists of residues for linkage with C-peptide in proinsulin and a residue for binding to receptor sites (20a). The functional groups on each of the amino acids and the sequence of the amino acids in the peptide chain determine the specific three-dimensional folding of the protein molecule (19), which is referred to as the "conformation" of the protein. Hydrophobic residues, such as amino acids with aliphatic hydrocarbon groups (e.g., alanine and leucine) or with aromatic rings (e.g., phenylalanine and tryptophan), tend to inhabit the interior of the protein molecule. On the other hand, hydrophilic residues, such as amino acids with charged and/or polar groups (e.g., aspartic acid and lysine), tend to reside on the surface of the protein molecule and are in contact with the surrounding water.

Based on their conformation, proteins can be classified into two major classes: fibrous and globular proteins. Fibrous protein consists of polypeptide chains arranged

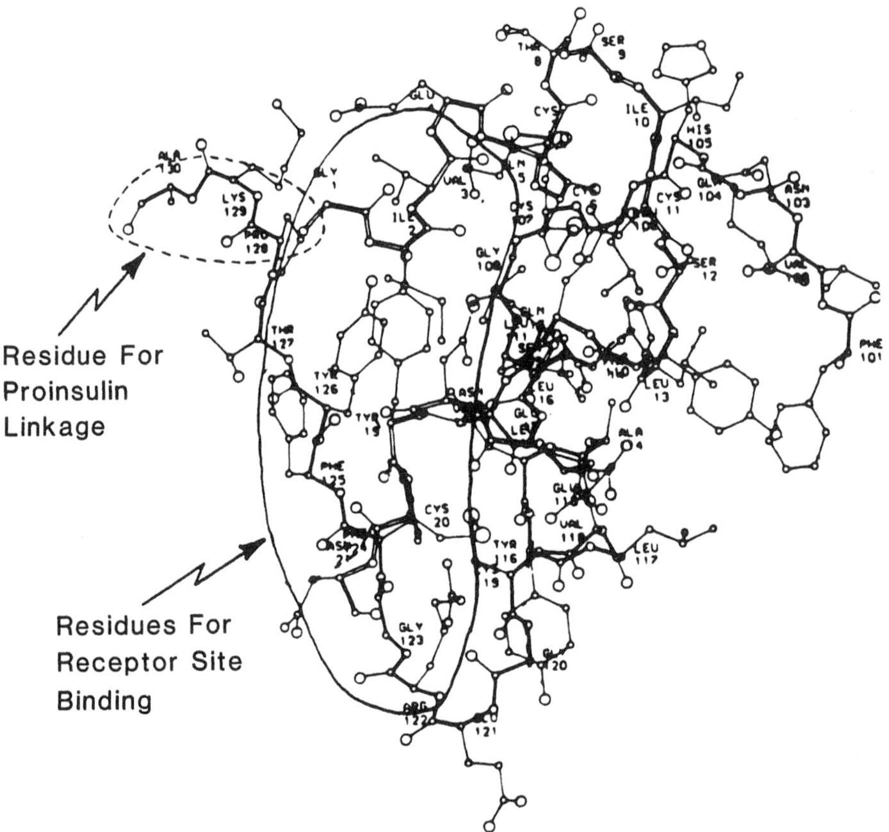

Figure 1 The molecular structure and conformation of insulin molecule. (Modified from Blundell and Wood, 1975.)

in parallel along a single axis to yield long fibers or sheets. These are insoluble in water and form the basic structural elements of connective tissue, like collagen, keratin, and elastin (21). In a globular protein, the polypeptide chains are tightly folded into compact spherical (or globular) shapes. Most globular proteins are soluble in aqueous media. The protein-based pharmaceuticals discussed in this chapter are primarily the globular proteins.

A typical globular protein is pig insulin, whose monomer has the globular tertiary structure shown in Figure 1. It consists of an hydrophobic core and two predominantly hydrophobic surfaces. These surfaces are also responsible for the association of insulin molecules, first as a dimer, and then, in the presence of zinc ions, to form a hexamer (20a). The hexamer has the shape of a torus with two zinc atoms bound in the central, cylindrical channel, and its surface contains most of the hydrophilic amino acid residues.

In fact, protein molecules have several levels of structure, and the general term "conformation" refers only to these structures in combination (21). A protein molecule consists of a primary structure, the covalent backbone of the polypeptide chain constructed from a sequence of amino acid residues; a secondary structure, a regular, recurring arrangement in the polypeptide chain along one dimension; a tertiary structure, the three-dimensional structure in which the polypeptide chain is bent or folded to form a compact, tightly folded structure of globular proteins; and a quaternary structure, the arrangement and the relationship among various polypeptide chains of a protein molecule having two or more polypeptide chains.

III. DELIVERY OF PEPTIDE-BASED PHARMACEUTICALS FOR SYSTEMIC MEDICATION

Because of their susceptibility to the strongly acid environment and the proteolytic enzymes in the gastrointestinal tract, the systemic bioavailability of most peptide- and protein-based pharmaceuticals by oral administration is extremely low. Also, peptides and proteins are high-molecular-weight macromolecules and thus do not easily permeate the intestinal mucosa. Even after successful gastrointestinal absorption, they are further subjected to first-pass elimination in the liver (Figure 2). Thus, the oral bioavailability of therapeutic peptides and proteins is often too low to be therapeutically useful for effective systemic medication.

A. Parenteral Systemic Delivery

For the systemic delivery of therapeutic peptides and proteins, parenteral administration is currently believed to be the most efficient route and also the delivery method of choice to achieve therapeutic activity. However, many peptide- and protein-based pharmaceuticals cannot accomplish their full range of therapeutic benefit when delivered by the parenteral route because they are limited by the extremely short duration of their biological functions. Thus, research programs have been initiated over the years in several institutions to develop a viable drug delivery technique in the hope of retaining the efficiency of systemic delivery of peptide and protein pharmaceuticals via parenteral administration while reducing its drawbacks. One viable approach is the development of controlled-release drug delivery systems to modulate the parenteral delivery of peptide and protein pharmaceuticals, which produces pro-

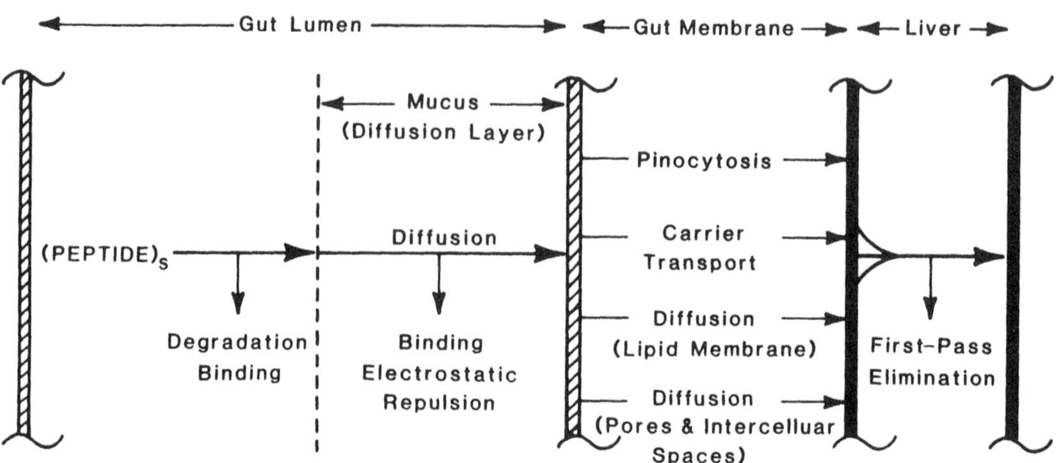

Figure 2 Various oral absorption pathways as well as diffusional and metabolic barriers for the systemic delivery of peptides and proteins by oral administration. (Modified from Humphrey, 1986.)

longation of biological activity. Coupled with an implantability and/or biofeedback mechanism, these parenteral controlled-release drug delivery systems have already shown some potential biomedical applications in the systemic controlled administration of organic-based as well as peptide-based pharmaceuticals (Chapter 8).

1. General Considerations

Parenteral systemic delivery consists of three major routes: intravenous (IV), intramuscular (IM), and subcutaneous (SC). A judicious choice of these routes must be made. For instance, the optimal systemic levels of antibody that are needed for the treatment of primary immune deficiency disea ses cannot be achieved with the IM administration of gammaglobulin, a proteineous molecule, because of limitations in the muscle mass and volume (22). By IV administration, on the other hand, gammaglobulin can achieve the optimal blood levels of antibody needed for treating patients with a broad spectrum of antibody deficiencies. In contrast, the intramuscular administration of gammaglobulin has been found beneficial to provide long-term protection from hepatitis infection.

As discussed earlier, intravenous administration is currently the method of choice for the systemic delivery of therapeutic peptides and proteins that are either excessively metabolized by and/or bound to tissues at the site of intramuscular or subcutaneous injection.

Besides the usual complications often resulting from intravascular administration, such as thrombophlebitis and tissue necrosis, additional complications associated with the therapeutic use of peptide- and protein-based pharmaceuticals are related to their potential immunogenicity. Amino acids and small peptides are not themselves immunogenic, but macromolecular proteins are often recognized as "foreign" by the body, which may respond with the production of a specific antibody. The most potent immunogens are the proteins with a molecular weight in the range of 100,000 daltons or greater (23). In addition to molecular size, biochemical complexity is also an important attribute; for example, the aromatic amino acid components in a protein molecule contribute more to its immunogenicity than the non-aromatic amino acid residues. Impurities, such as contaminating bacterial protein, represent another source of immunogenic substances. For example, insulin preparations prepared before 1973 may contain antigenic impurities, like proinsulin, which is the biosynthetic precursor of insulin, and also other pancreatic hormones as well as its incompletely converted products (24). Current insulin preparations should contain no more than 10 parts per million (ppm) proinsulin. In fact, "purified" porcine insulin has been considered the least immunogenic of the nonhuman insulin preparations available commercially.

Insulin (25), interferons (26), and gammaglobulins (27) have been reportedly metabolized and/or bound to tissues at injection sites following IM administration, and as a result, the systemic bioavailability of these protein drugs following IM administration is often less than that obtained by IV injection. On the other hand, with the use of proper adjuvants and/or the application of electrical current, the systemic bioavailability of peptides and proteins from IM injection sites can be improved. For example, the bioavailability of tissue plasminogen activator following IM administration was reportedly facilitated by the coadministration of hydroxylamine or by electrical stimulation of muscle, which results in prompt attainment of therapeutic blood levels and achievement of coronary thrombolysis (28). Currently, however, t-PA is administered either intravenously or intraarterially directly into the coronary arteries. Lysis of coronary clots can be achieved with blood t-PA levels as low as 77 ng/ml.

For SC administration, insulin represents the best example of a therapeutic protein that is administerd subcutaneously on a long-term basis for the treatment of diabetes. Several long-acting injectable insulin preparations are commercially available, and their development is discussed later (Section III.A.3).

2. *Approaches to Subcutaneous Controlled Delivery*

The controlled delivery of peptide- and protein-based pharmaceuticals from subcutaneously implanted polymeric devices was first reported by Davis (29,30). He used a gel formulation of cross-linked polyacrylamide-polyvinylpyrrolidone to achieve the prolonged release of immunoglobulin, luteinizing hormone, bovine serum albumin, insulin, and prostaglandin. Subsequent studies by Langer and Folkman (31,32) using hydroxyethyl methacrylate (Hydron) polymer and ethylene-vinyl acetate copolymer showed that macromolecules with molecular weights of up to 2 million daltons can be released steadily from the implanted polymeric device over periods of longer than 3 months. However, the reproducibility of release kinetics was poor and was later improved by using a low-temperature solvent casting method (33,34). Since this procedure was time consuming and required the use of organic solvent, an alternative sintering technique was then developed to prepare polymer matrix to control the release of macromolecules (35,36). It has also been reported that the delivery rate of macromolecules from polymeric devices can be modulated with the application of magnetism (37–40).

The biocompatibility of subdermal implants is an important concern since these implants may reside in the body for an extended period of time (41). In general, the biocompatibility of a polymeric material is expressed by the degree of acute and chronic inflammatory responses occurring locally at the site of implantation and the extent of fibrous capsule formation around the implanted polymeric material (42).

In addition to inflammatory responses, the biocompatibility of an implant can also be measured in terms of sensitivity reactions and infections. These consist of an initial adhesion of macrophages at the implant/tissue interface and then phagocytosis of the polymer by macrophages and giant cells.

From the biocompatibility point of view, biodegradable polymers are attractive, especially when their degradation products are known to be innocuous or biocompatible, since they need not be surgically removed at the end of a treatment. Bioerodible polymers that have been investigated for the controlled delivery of peptide- and protein-based pharmaceuticals include copolymers of d, l-lactide-glycolide (PLGA), cross-linked serum albumin, and a homopolymer of poly(lactic acid) (PLA). PLA and PLGA are linear polyesters that hydrolyze by an acid- or base-catalyzed reaction to form innocuous breakdown products like d-lactic, l-lactic, and glycolic acids. Lactic and glycolic acids are reportedly metabolized by the Krebs cycle, and d-lactic acid is excreted intact, so these final breakdown products are physiologically innocuous. PLGA has been utilized in the development of subdermal implants for the monthly subcutaneous controlled delivery of goserelin (Zoladex implant, ICI), a potent synthetic analog of LHRH (Figure 3), and used clinically to suppress the secretion of luteinizing hormone (LH) and of testosterone in patients with prostate carcinoma (Chapter 8, Section III.C.1).

Unfortunately, PLGA and PLA often undergo bulk rather than surface erosion, and thus they produce a delivery rate that changes with time, rendering the drug release profiles neither constant nor predictable (41). In view of this drawback, hydrophobic polymers that undergo surface erosion, such as poly(ortho ester) and polyanhydride, are being investigated as potential alternatives (41).

Ethylene-vinyl acetate copolymer (EVAc), a biocompatible but nonbiodegradable polymer, has been studied extensively as a polymer matrix to control the delivery of polypeptides (33–40). EVAc has been used successfully in some nonparenteral controlled-release drug delivery systems that have already been approved by the FDA for therapeutic use in humans (39). An example is the Ocusert system for the ocular controlled delivery of pilocarpine for glaucome treatment (Chapter 6). The release rate of polypeptide from EVAc-based polymer matrix can be controlled by varying the particle size of polypeptide and its loading dose. It has been suggested (39) that incorporation of the powdered polypeptide during casting of the polymer matrix creates a series of interconnecting channels through which water can diffuse into the matrix to dissolve the macromolecule for release through the channels. At higher loadings, polypeptide particles are more likely to be in contact with each other and thus create an efficient or less tortuous channel network that would facilitate the movement of peptide or protein molecules out of the matrix. At low loadings, on the other hand, most polypeptide particles disperse discretely in the polymer matrix and are completely surrounded by the polymer phase. Thus, only the particles on the surface of the polymer matrix could be released.

In addition to being delivered in a controlled manner from an EVAc-based polymer matrix (43,44), insulin has also been reportedly released from a porous poly(ϵ-caprolactone) matrix (45) as well as from poly(lactic acid) microbeads and pellets (46). The feasibility of using the biocompatible but lipophilic silicone elastomers to replace EVAc for controlling the delivery of proteinaceous macromolecules has also been demonstrated (47).

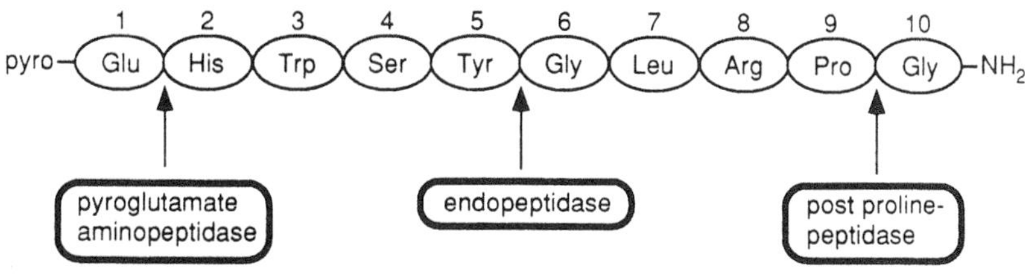

Synthetic LHRH	Sites of Modification 6th	9th	10th	Biological Activities Potency	$t_{1/2}$ (min)	Binding
Natural LHRH	Gly	Pro	Gly	1	8	?
Leuprorelin	D-Leu	Pro–N-Et	0	5-15	-	+
Buserelin	D-Ser–t-Bu	Pro–N-Et	0	25	80	+
Goserelin	D-Ser–t-Bu	Pro	Aza–Gly-NH_2	75	-	+
Tryptorelin	Trp–N-Me-Leu	Pro	Gly	100	30	?
Histrelin	D-His–Bzl	Pro–N-Et	0	100	-	+
Nafarelin	D-2-napht.-Alanine	Pro	Gly	200	144	?

Figure 3 Amino acid sequence of luteinizing hormone-releasing hormone (LHRH) and the enzymes reportedly responsible for its degradation. Also illustrated are various synthetic modifications of LHRH and their effect on the potency and the duration of the biological activities of LHRH.

3. *Biomedical Applications*

The systemic controlled delivery of peptide- and protein-based pharmaceuticals via various parenteral routes of administration is exemplified by biomedical applications of the following therapeutic peptides and proteins.

LHRH and Analogs. Luteinizing hormone-releasing hormone is a naturally occurring decapeptide hormone (Figure 3) with a molecular weight of 1182 daltons. LHRH has been reportedly subjected to enzymatic degradation by pyroglutamate aminopeptidase, endopeptidase, and/or postprolin peptidase, resulting in a substantial reduction in its biological activity with a half-life of only 8 min. Several synthetic analogs have been prepared to minimize the enzymatic degradation and thus enhance the potency of LHRH and/or prolong the duration of its biological activity (Figure 3). For example, buserelin, in which the glycine at position 6 and the proline at position 9 of the native LHRH molecule have been replaced by t-butyl-D-serine and N-ethylproline, respectively, to reduce the enzymatic degradation of the tyrosine-glycine bond by endopeptidase and of the proline-glycine bond by the postproline peptidase. It is a potent LHRH analog with an increase in potency by 25-fold and prolongation of biological half-life by approximately 10 times. Furthermore, the use of benzyl-D-histidine to replace 6-glycine has improved the potency even more by as much as 100-fold (Histrelin). On the other hand, the replacement of 6-glycine by D-leucine to form leuprorelin has enhanced the potency by only 5–15 times. It is interesting to note that nafarelin, in which only the 6-glycine has been substituted by D-naphthylalanine, is a superior agonist with a 200-fold increase in potency and 18-fold prolongation in the duration of biological activity (48).

LHRH is a neuropeptide of hypothalamic origin and has the function of stimulating the synthesis and release of gonadotropins from gonadotropes in the anterior pituitary. However, the biological activity of LHRH depends greatly on the mode of its delivery. Administration of LHRH in a pulsatile manner, which mimics the natural secretory pattern (Figure 4), causes a sustained secretion of gonadotropins (49,50) as well as the synthesis and release of testosterone in the male and of estradiol in the female in a biological rhythmic pattern. On the other hand, the long-acting synthetic analogs of LHRH have been shown to desensitize the pituitary gland

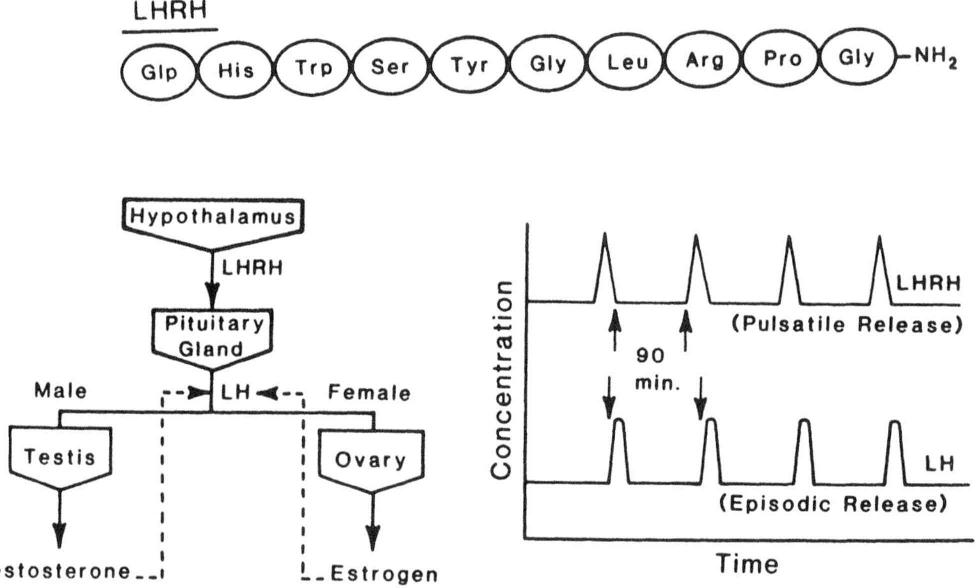

Figure 4 The release of LHRH, in pulsatile manner, from hypothalamus, which activates the release of LH in an episodic pattern from pituitary gland. In turn, LH triggers the secretion of testosterone from the testis or estradiol from the ovary. Continuous delivery of LHRH and its synthetic analogs stops the secretion of testosterone in males and of estradiol in females by downregulation.

and thus inhibit the release of gonadotropins and testicular or ovarian steroids when administered as a single daily subcutaneous injection (51–54). This is now being exploited clinically for various biomedical applications, including fertility regulation and the treatment of prostate carcinoma.

However, LHRH and its synthetic analogs are poorly absorbed when taken orally. Thus, the native decapeptide hormone and its synthetic analogs require parenteral administration, and research has been initiated to develop long-acting parenteral delivery system (55–57). One approach has been the use of an injectable gel system (55) to extend the duration of the biological activities of LHRH agonist. The gel formulation was prepared by suspending the LHRH agonist in sesame oil and then gelling it with aluminum monostearate to form a thixotropic gel, which could be injected subcutaneously using a standard syringe. This by itself is not a very long-acting drug delivery system, but when used together with a mixed zinc tannate salt of a potent LHRH analog, the gel can exert an action that is very prolonged in duration. This approach is very similar to the development of long-acting antinarcotic preparations (Chapter 8, Section II.B.8) with the formation of a naltrexone-Zn-tannate complex (Figure 20 in Chapter 8). Using an implantable cholesterol pellet, the subcutaneous controlled delivery of LHRH analog has also been achieved (55,56). On the other hand, LHRH analogs are known to have a very low polymer solubility and diffusivity through silicone elastomer, presumably because of their hydrophilic nature and macromolecular size, and therefore a silicone device, which has been used successfully for the controlled delivery of organic-based pharmaceuticals, is not a suitable delivery system to achieve the controlled delivery of LHRH. Since LHRH agonists are required to be administered only at very low doses, they could be delivered by biodegradable polymer-based delivery systems (57–61). One such example, which has been successfully used clinically, is the development of a biodegradable subdermal implant for the subcutaneous controlled delivery of goserelin (Zoladex), a potent synthetic analog of LHRH (Figure 3) for the monthly treatment of prostate carcinoma (Chapter 8, Section III.C.1). Another example is the subcutaneous controlled delivery of nafarelin, also a potent LHRH agonist, in acetate form either by dispersing it in microspheres (59) or encapsulating it in microcapsules (60); both are injectable parenteral delivery systems prepared from PLGA copolymer (61). PLGA has also been utilized for the controlled delivery of other peptide-based pharmaceuticals through the parenteral route of administration, and some clinical trials have even been carried out (62–64).

Insulin. Insulin is a pancreatic hormone that, in healthy humans, is secreted by the beta cells in the islets of Langerhans to control the utilization of glucose in practically all the organs (Figure 5), and maintain the glucose concentration at the normoglycemic level. In normal subjects, the pancreatic secretion of insulin varies in response to the variation in blood glucose levels (Figure 6). In diabetic patients, this response to blood glucose variation is slower and inefficient (Figure 7).

Insulin is the drug of choice for the treatment of diabetes mellitus, a disease that affects approximately 14 million Americans (The Star-ledger, 6/25/91). Insulin, a protein molecule with a molecular weight of approximately 6000 daltons, consists of 51 amino acid residues in two polypeptide chains linked together by disulfide bridges (Figure 8). It is easily inactivated by the gastric acid and intestinal enzymes when taken orally and is thus generally given by parenteral administration via the

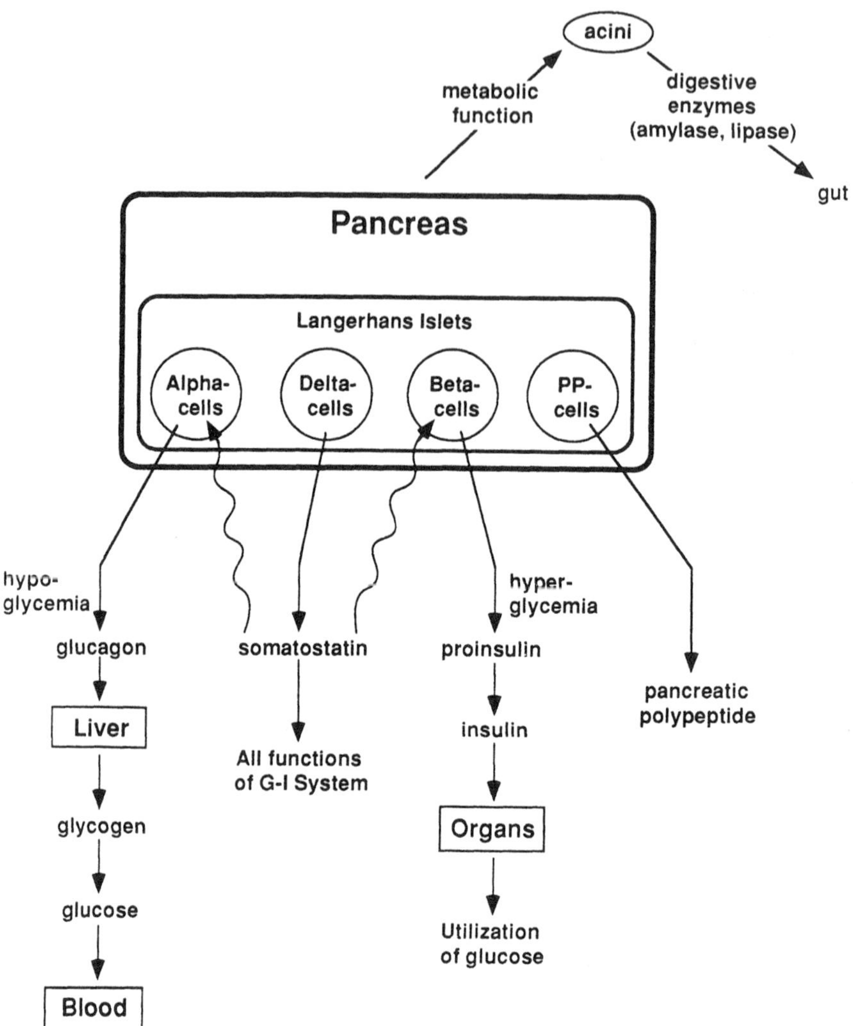

Figure 5 Various cell types in the Langerhans islets and the secretion of various peptide and protein hormones from these pancreatic cells. The metabolic functions of these pancreatic hormones and their interrelationship are also illustrated.

subcutaneous route. Several parenteral formulations with varying durations of normoglycemic activity are commercially available (Table 3). On the other hand, extensive research has been initiated recently to evaluate other nonparenteral routes, such as nasal, ocular, buccal, rectal, and transdermal. These nonparenteral routes of insulin delivery are discussed later (Section III.B).

An extensive review of the chemistry and biochemistry of insulin was published by Klostermeyer and Humbel (65). An interesting historical discussion of the development of insulin, in collaboration with the Eli Lilly Company, was recently completed by Swann (66). Human insulin (Humulin, Genentech-Eli Lilly), which has been produced by genetic engineering from a special laboratory strain of *Esch-*

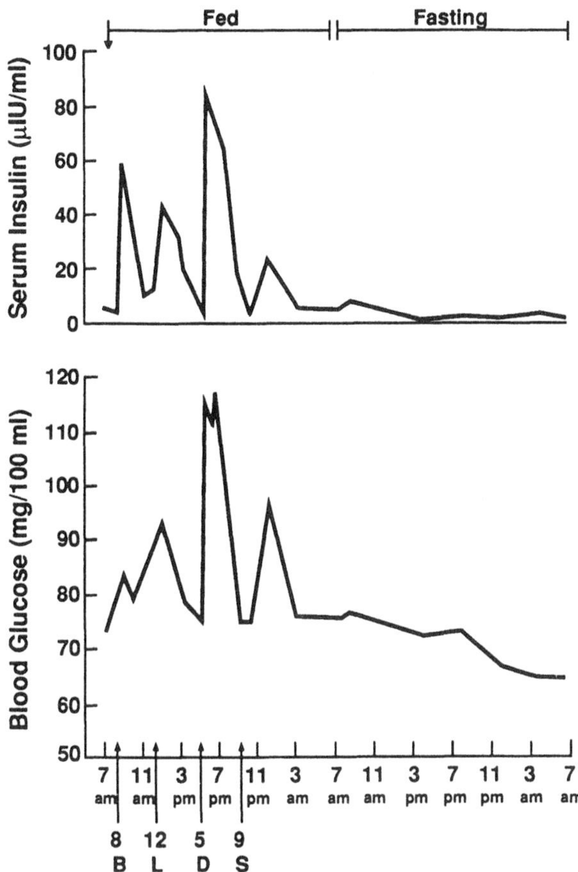

Figure 6 The mean daily response profiles of insulin and glucose in six normal men fed with a 30 cal/kg meal at breakfast (B), lunch (L), dinner (D), and snack (S). The response profiles in the fasting condition are also shown for comparison.

erichia coli that was genetically altered by the addition of the human gene (67), was the first biotechnology-produced peptide-based pharmaceutical approved by the FDA for marketing, in 1982 (68). Another type of human insulin (Novolin, Novo-Nordisk) is a product produced from porcine insulin by an enzymatic transpeptidation process, which uses a naturally occurring enzyme to selectively substitute the alanine in the porcine insulin by threonine to form human insulin. The Novolin series human insulin products have also received regulatory approval for the treatment of insulin-dependent diabetic patients (Table 3). The details of production and comparison with other insulins have been reported (67,69–71). Depending upon the response types of diabetic patients, the efficiency of Ultralente insulin, a long-acting insulin, in controlling blood glucose levels varies from one patient to another (Figure 9). The optimum technique for the systemic delivery of insulin has been the subject of much debate (72,73).

Several delivery techniques and systems have been developed for the parenteral controlled and/or sustained delivery of insulin. They are outlined as follows.

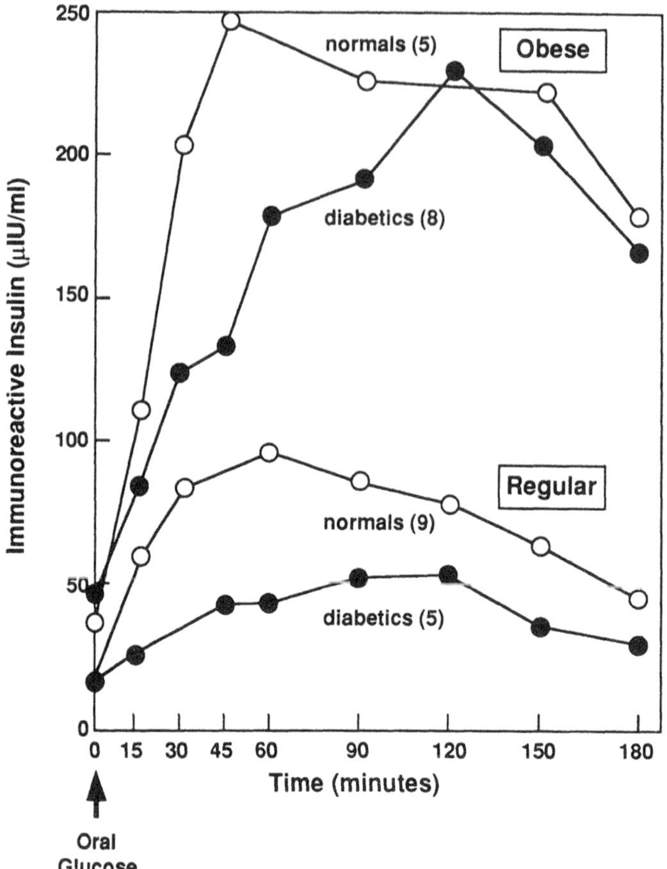

Figure 7 The mean serum response profiles of immunoreactive insulin in normal subjects and diabetic patients in response to the oral intake of glucose as well as the effect of body size on the response profiles.

Long-acting insulin injectables. An early approach to the development of long-acting insulin preparations included the complexation of insulin with zinc salt and basic proteins; for example, the formation of protamine-Zn-insulin suspension (Chapter 8, Section II.B.2) from protamine, a water-soluble basic protein isolated from the sperm of the mature testes of fish. Further investigations have applied the controlled crystallization of zinc-insulin complex, without the use of proteins (like protamine, isophane, or globin). This approach used a buffer medium with controlled pH in which insulin can be precipitated by zinc ion to yield a water-insoluble Zn-insulin complex in either a crystalline or an amorphous form. The crystallinity of the complex can be controlled by varying the pH of the buffer medium (74). The crystalline Zn-insulin complex formed is an extremely long-acting insulin, whereas the amorphous Zn-insulin complex is a moderately long-acting insulin.

Three long-acting Zn-insulin preparations, Semilente, Lente, and Ultralente insulin, have been formulated and are still widely prescribed today for the treatment of diabetes. As compared in Table 3, these preparations differ in their onset, inten-

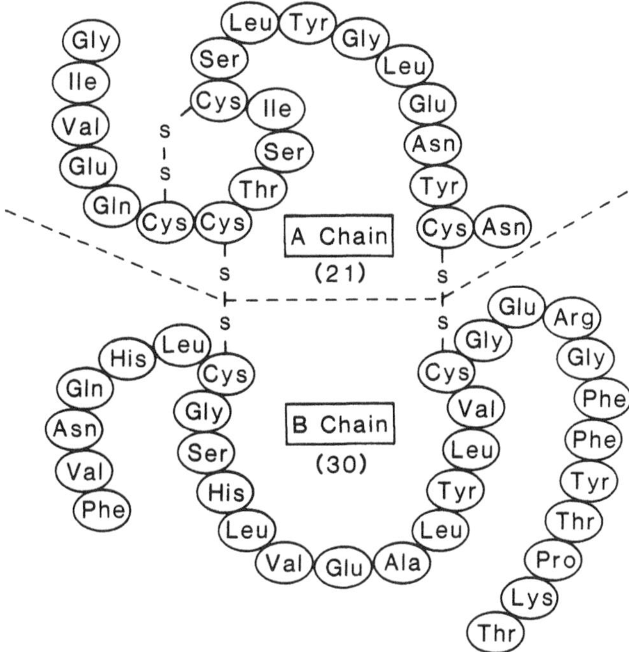

Figure 8 Molecular structure of insulin, a protein hormone for diabetes mellitus, and its amino acid sequence. It consists of two peptide chains linked together through two disulfide linkages. The A chain contains 21 amino acid residues; the B chain is composed of 30 amino acid residues.

sity, and duration of action following subcutaneous administration. The amorphous Zn-insulin complex (Semilente insulin), which forms at pH 6–8, has a rapid onset (0.5–1.0 hr) and a moderately long duration of action (12–16 hr) compared to regular insulin (6 hr). On the other hand, the cloudy suspension of crystalline Zn-insulin complex (Ultralente insulin), which is prepared at pH 5–6, has a much slower onset (4–8 hr) but a rather long duration of action (>36 hr). A mixture of crystalline (seven parts) and amorphous (three parts) insulins forms a preparation called Lente insulin, which is intermediate in both the onset (1.0–1.5 hr) and duration of action (24 hr) (75,76). The normoglycemic activity of these long-acting Zn-insulin complex preparations was compared with that of regular insulin and was reported earlier (Chapter 8, Figure 7).

Alternatively, attempts have also been made to prolong the normoglycemic action of insulin by entrapping it in a liposome-collagen gel matrix for subcutaneous administration (77).

New injection devices for insulin delivery. A new jet-type injector (Precijet-50) was recently developed (78,79). The features of this device, in addition to being small in size, are simple in design with the capability of mixing two types of insulin before injection (Figure 10); these features were not available in the earlier model of jet injection technology (80–82). On the other hand, a sprinkler type of needle for insulin injection has also been recently designed (83). Another later development was the design of an insulin injection "pen" called the Novolin Pen (Chapter 8,

Table 3 Normoglycemic Activity and Duration of Some Commercial Insulin Products

Insulin preparations	Normoglycemic activity[a] Onset (hr)	Peak (hr)	Duration (hr)
Standard or purified insulin			
Regular[b]	~0.5	2.5–5	~8
Semilente[c]	~1.5	5–10	~16
Lente[d]	~2.5	7–15	~24
Ultralente[e]	~4.0	10–30	~36
NPH insulin[f]	~1.5	4–12	~24
Human insulin[g]			
Novolin R/Humulin BR or R	~0.5	2.5–5	~8
Novolin L/Humulin L	~2.5	7–15	~22
Novolin N/Humulin N	~1.5	4–12	<24
Novolin 70/30	~0.5	2–12	<24

[a]The time at which activity is evident.
[b]Solution of pork insulin.
[c]Suspension of amorphous beef insulin-Zn complex.
[d]Suspension of 70% crystalline and 30% amorphous beef insulin-Zn complex.
[e]Suspension of crystalline beef insulin-Zn complex.
[f]Suspension of protamine-Zn-beef insulin complex.
[g]Novolin series human insulin is a product produced by enzymatic transpeptidation process, which uses a naturally occurring enzyme to selectively substitute the alanine in pork insulin by threonine to form human insulin. Humulin series human insulin is a product produced by recombinant DNA, which is synthesized in a special laboratory strain of *E. coli* been genetically altered by the addition of the human gene for insulin production.
Source: Compiled from *Physicians' Desk Reference*, 43rd ed. Medical Economics Co., Oradell, NJ, 1989.

Figure 9), which is a pocket-sized apparatus that resembles a fountain pen. When fitted with a disposable needle and a unit-dose ampule of insulin, it becomes a portable, self-contained insulin syringe (84).

Infusion pumps for insulin delivery. A continuous, subcutaneous insulin infusion (CSII) device (or pump) has been in use for almost a decade now (85–87). However, a possible increase in mortality (88), morbidity due to mechanical failure (89), and lack of data on the safety of its long-term use (90) suggest that further long-term assessment of the CSII device is needed. The feasibility of intraperitoneal delivery of insulin using an implantable micropump has also been investigated in dogs (91,92). An implantable controlled-release micropump (Figure 11) was recently developed for the intraperitoneal modulated delivery of insulin. It is an open-loop control system characterized by operation at two levels: basal delivery for the between-meal period and augmented delivery for short periods following the ingestion of meals. The rate of delivery is adjusted to meet the insulin requirement of the respective meal. With an adequate supply of insulin to the pump, the concentration gradient from the reservoir produces the delivery of insulin at the basal rate (when no external power source is applied). Augmented delivery is achieved by repeated compression of the foam membrane above the steel piston, which applies a current

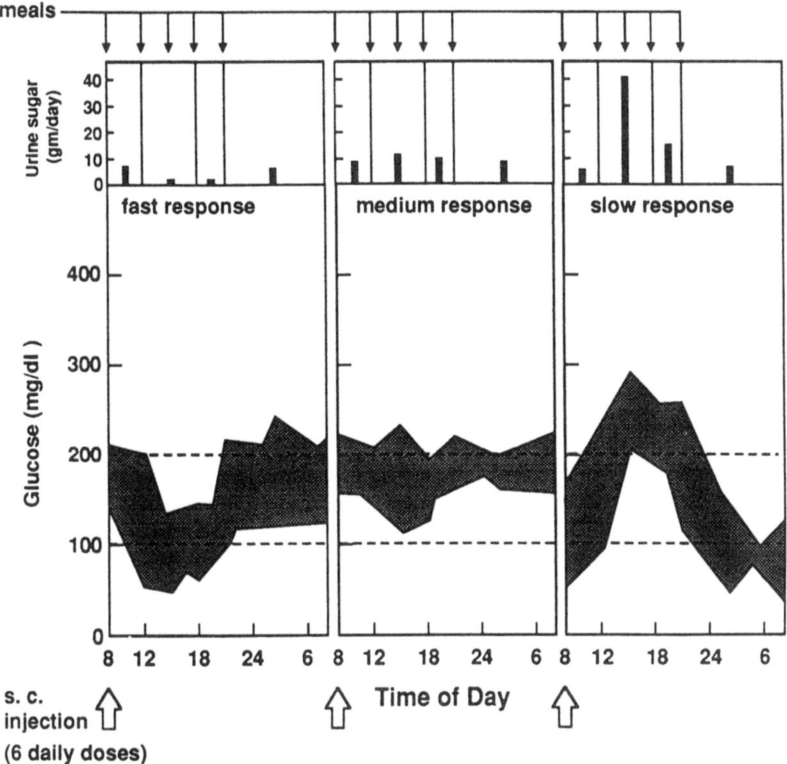

Figure 9 Blood glucose profiles in three responsive types of diabetic patients following daily subcutaneous injections of Ultralente insulin, a long-acting Zn insulin preparation, at 8 A.M. for 6 days and the cumulative amount of sugar collected in the urine throughout the period. The patients were on a regular meal course. (Modified from Hallas-Moller et al., 1952.)

to the solenoid cell. Further investigations have resulted in the development of a piezoelectrically-controlled micropump (P-CRM) for the programmed delivery of insulin (93).

Self-regulating delivery systems. An interesting approach was recently reported with the development of an artificial beta cell that consists of a glucose-sensitive hydrogel membrane for the feedback-controlled delivery of insulin (94,95). The glucose-sensitive membrane is fabricated by entrapping a glucose oxidase in a hydrogel polymer with pendant amine groups (Figure 12). As glucose diffuses into the polymer, glucose oxidase catalyzes its conversion to gluconic acid (pKa = 3.6), thereby lowering the microenvironmental pH in the membrane. The reduced pH results in increased ionization of the pendant amine groups. Electrostatic repulsion between the ionized amine groups increases the swelling and thus the permeability of the hydrogel membrane to insulin contained in the reservoir. Ultimately, the membrane permeability to insulin is thus a function of the glucose concentration surrounding the membrane, and the release of insulin is accelerated by the increase in the glucose level (96). Another potential technique to achieve the self-regulating de-

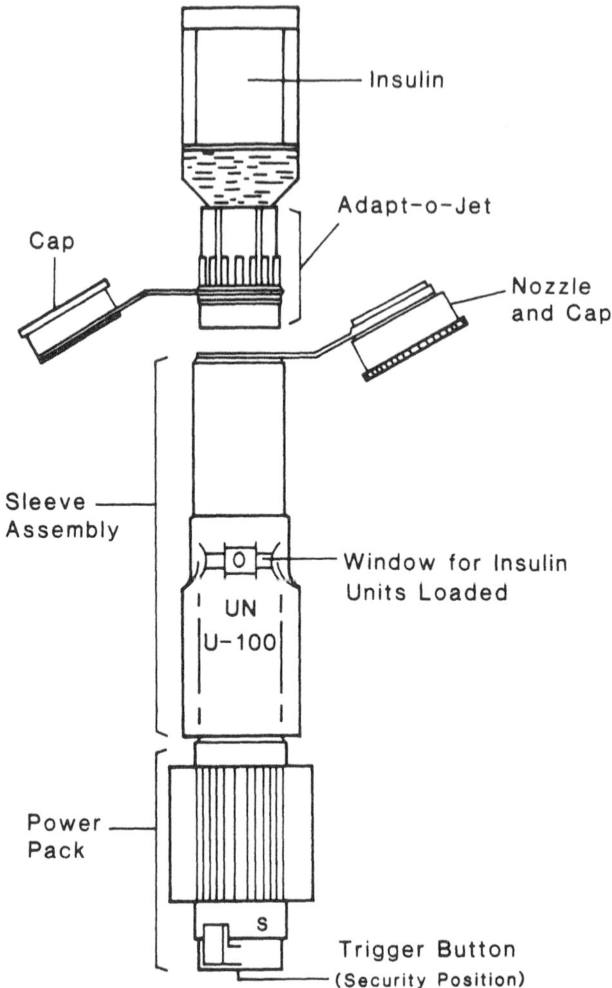

Figure 10 The main components of the Preci-Jet 50 and the Adapt-o-Jet (vial holder). For filling with insulin, the nozzle is unscrewed and replaced with the Adapt-o-Jet and an insulin vial. The injector is filled with insulin by turning the power rack counterclockwise until the number on the small window indicates the number of insulin units required. The nozzle is then screwed back into position, and pressure is generated by turning power pack clockwise until a click is heard. (Modified from Lindmayer et al., 1986.)

livery of insulin is a biochemical approach based on the principle of the competitive and complementary binding behavior of concanavalin A (ConA) with glucose and glycosylated insulin (G-insulin) (97–101). The ConA-G-insulin complex is encapsulated inside a device with a polymeric membrane that is permeable to glucose and G-insulin but not to ConA or its complex (Figure 13). As the glucose level increases, the influx of glucose to the device increases, resulting in the displacement of G-insulin from the ConA complex and its efflux to the body. The results in Figure 14 indicate that the blood glucose levels of pancreatectomized dog can be controlled by the G-insulin delivery device even with the daily ingestion of food.

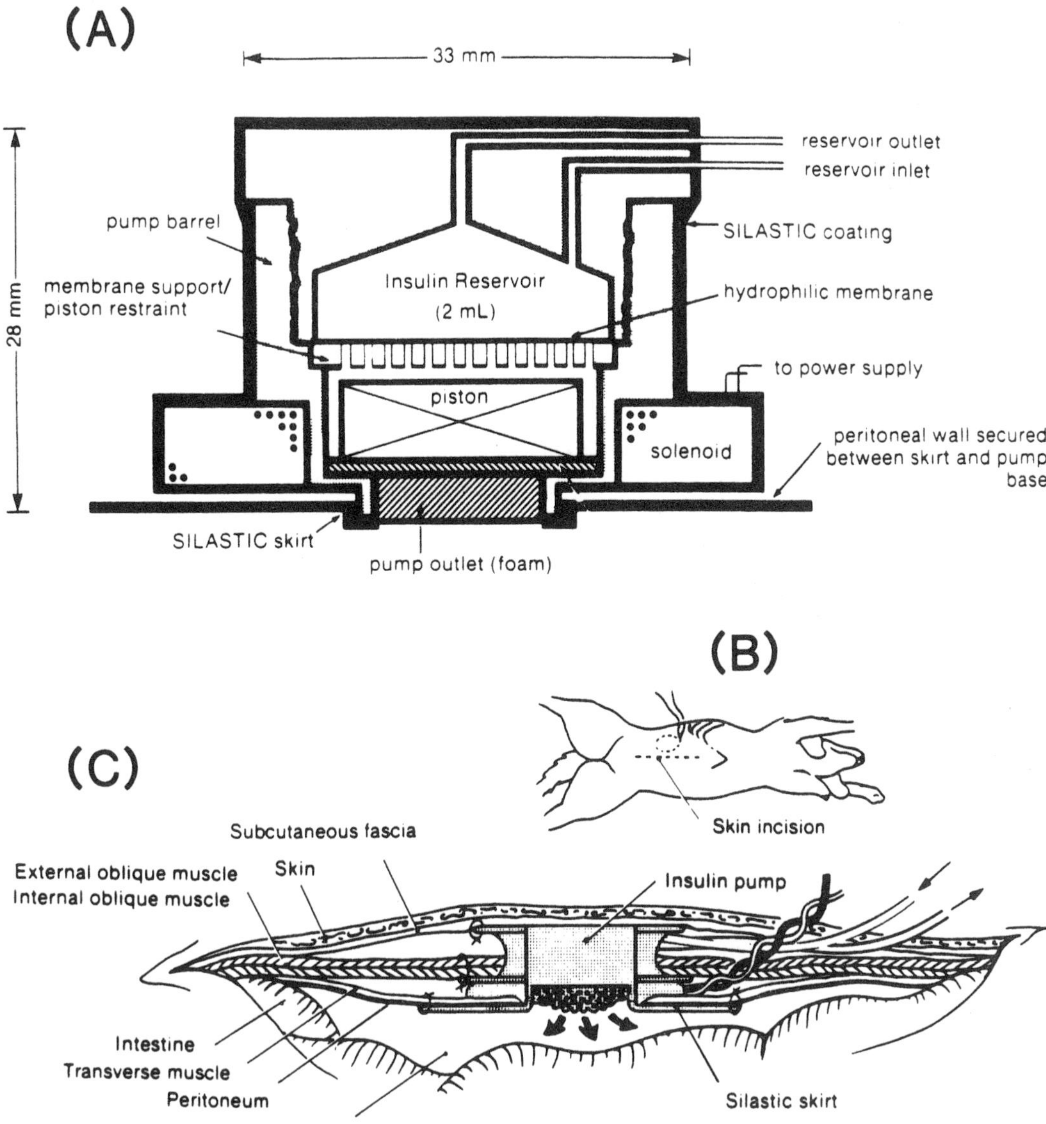

Figure 11 The controlled release micropump (A) and intraperitoneal implantation in pancreatectomized dogs (B and C). (Replotted from Sefton et al., 1984.)

Vasopressin. Vasopressin, a nonapeptide (Figure 15) with a molecular weight of 1084 daltons, is an antidiuretic hormone (ADH). As secreted by the posterior pituitary glands, its physiological role is to maintain serum osmolality within a narrow range. It acts on the renal cells responsible for the reabsorption of free water from the glomerular filtrate. A deficiency of ADH makes a patient incapable of producing a concentrated urine, a condition called diabetes insipidus.

Using a device prepared by covering a section of microporous polypropylene (Accurel) tubing with collodion, a long-lasting and constant in vitro release of va-

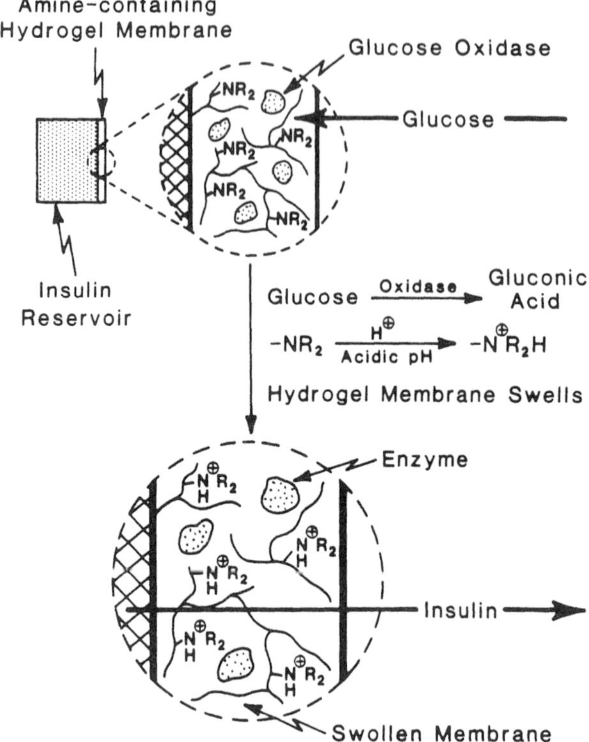

Figure 12 Cross-sectional view of the bioresponsive insulin delivery system, which uses a glucose-sensitive membrane to control the delivery of insulin in response to the influx of glucose. (Based on Horbett et al., 1983.)

sopressin was achieved for periods of up to 50 days (102). The in vivo potency of this device was also demonstrated in adult vasopressin-deficient Brattleboro rats by subcutaneous implantation. Normal levels of urine production and osmolality were achieved and maintained for at least 1 month. This successful result led to further investigations of the Accurel-collodion device (103,104), and it was found that the device appears to be biocompatible and can be used as an implant in adult rats.

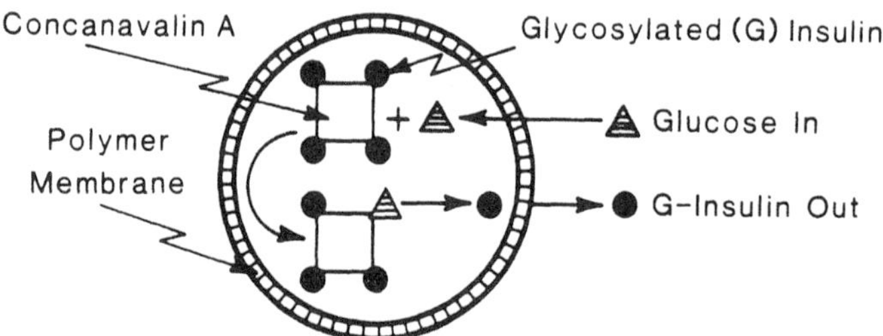

Figure 13 A self-regulating insulin delivery system that releases glycosylated insulin in response to the influx of glucose. (Based on Jeong et al., 1984.)

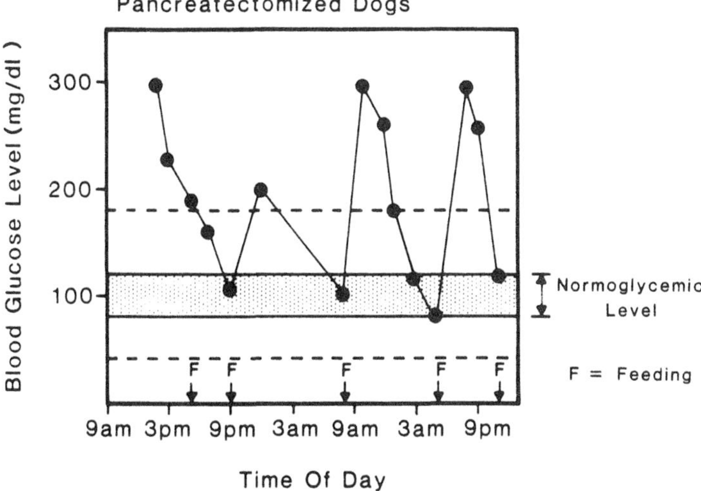

Figure 14 Blood glucose profiles in pancreatectomized dogs implanted with a self-regulating insulin delivery system in response to a daily feeding schedule. (Replotted from Jeong et al., 1984.)

Utilizing an osmotic pressure-activated minipump, called the Alzet pump (Chapter 8, Section III.B.5), the subcutaneous controlled delivery of vasopressin was also investigated in Brattleboro rats (105). The results demonstrated that the volume of daily urinary excretion was substantially reduced and maintained at a normal level and the osmolality of the urine was remarkably increased and maintained at a constant elevated level (Figure 16). Both responses were maintained at constant levels throughout the course of the 7-day implantation and were linearly dependent upon the doses of vasopressin delivered subcutaneously (Figure 17).

B. Nonparenteral Systemic Delivery

As discussed previously, the systemic delivery of most peptide- and protein-based pharmaceuticals, including endogenous hormones and biological response modifiers (Table 1), as well as their potent synthetic analogs, has been mostly accomplished by parenteral administration. However, congenital and acquired metabolic disorders usually require long-term therapy by these peptide- and protein-based pharmaceuticals. Moreover, the rapid disappearance of most peptides and proteins from the body, owing to their degradation by proteolytic enzymes, mandates implementation of a therapeutic regimen that requires multiple daily injections to maintain therapeutic efficacy. Many attempts and research efforts have therefore been devoted to search for an easily accessible route for nonparenteral administration and to develop a reliable method for systemic delivery with the convenience of self-administration.

The potential nonparenteral routes of administration for the systemic delivery of peptide-based pharmaceuticals include the ocular, nasal, pulmonary, buccal, oral, rectal, vaginal, and transdermal routes (Figure 18). Without coadministration of an absorption-promoting adjuvant, these routes are generally known to achieve a much

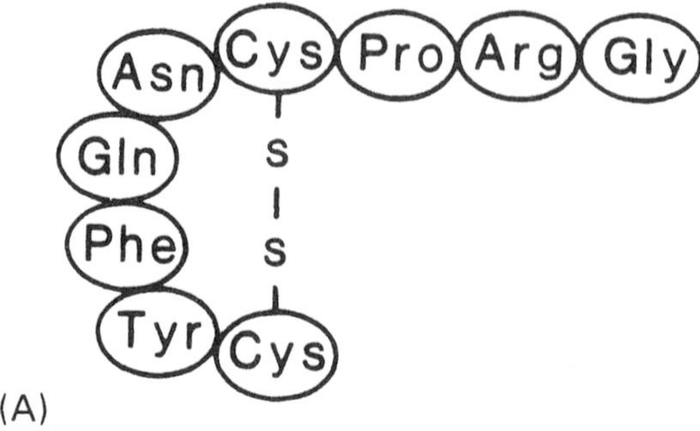

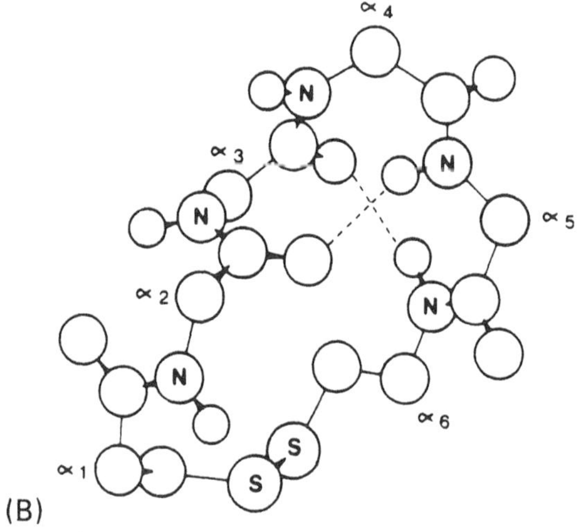

Figure 15 (A) Amino acid sequence of vasopressin, an antidiuretic peptide hormone for diabetes insipidus, and (B) its molecular conformation in solution.

lower bioavailability than parenteral administration (which could be a result of an incomplete absorption of peptide or protein drugs). Incomplete absorption is probably due to a combination of poor mucosal permeability and extensive metabolism at the absorption site. A recent study conducted in Lewis rats (106), which compared the hypoglycemic activity of insulin delivered through several mucosal routes of administration, indicated that the rectal delivery of insulin is more efficacious than the nasal, buccal, or sublingual route, when insulin is administered without an absorption promotor (Table 4). However, the hypoglycemic activity of insulin delivered by all mucosal routes was substantially lower than that by intramuscular administration. Sodium glycocholate, a commonly-used absorption-promoting bile salt for transmucosal permeation, was found to greatly improve the hypoglycemic activity of insulin delivered by all mucosal routes with the rank order: nasal > rectal > buccal > sublingual. The nasal and rectal delivery of insulin achieved a therapeutic efficacy almost half that of intramuscular insulin administration. It has been reported (42) that protease activities in homogenates of the nasal, buccal, rectal, and vaginal mu-

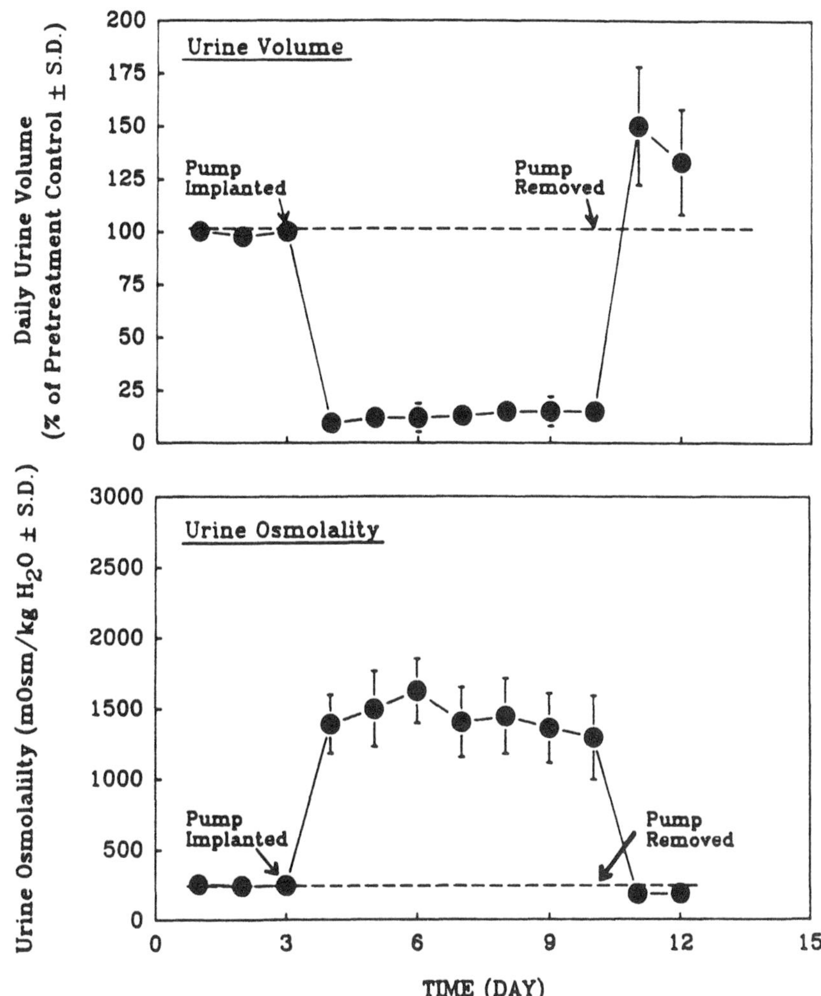

Figure 16 Effect of subcutaneous controlled administration of vasopressin by an osmotic pressure-activated minipump (10 IU/kg per week) on the volume of daily urinary excretion (top) and the osmolality of the urine (bottom) in vasopressin-deficient Brattleboro rats. (Plotted from data by P. Lelawongs, 1990.)

cosae of the albino rabbit are substantial and comparable to those in the ileal homogenate. On the other hand, a review of the literature has suggested that dermal tissues lack proteolytic enzymes, which are known to be responsible for the enzymatic degradation of peptides and proteins in various tissues (105,107). The enzymatic barrier involved in these nonparenteral routes of administration is discussed in more detail in Section III.C.

The efficiency of absorption and thus the systemic bioavailability of peptide- and protein-based pharmaceuticals can be considerably improved when administered through the nonoral routes (108). A typical example is shown in Table 5, in which the absorption of leuprorelin (leuprolide, Takera) a synthetic LHRH analog (Figure

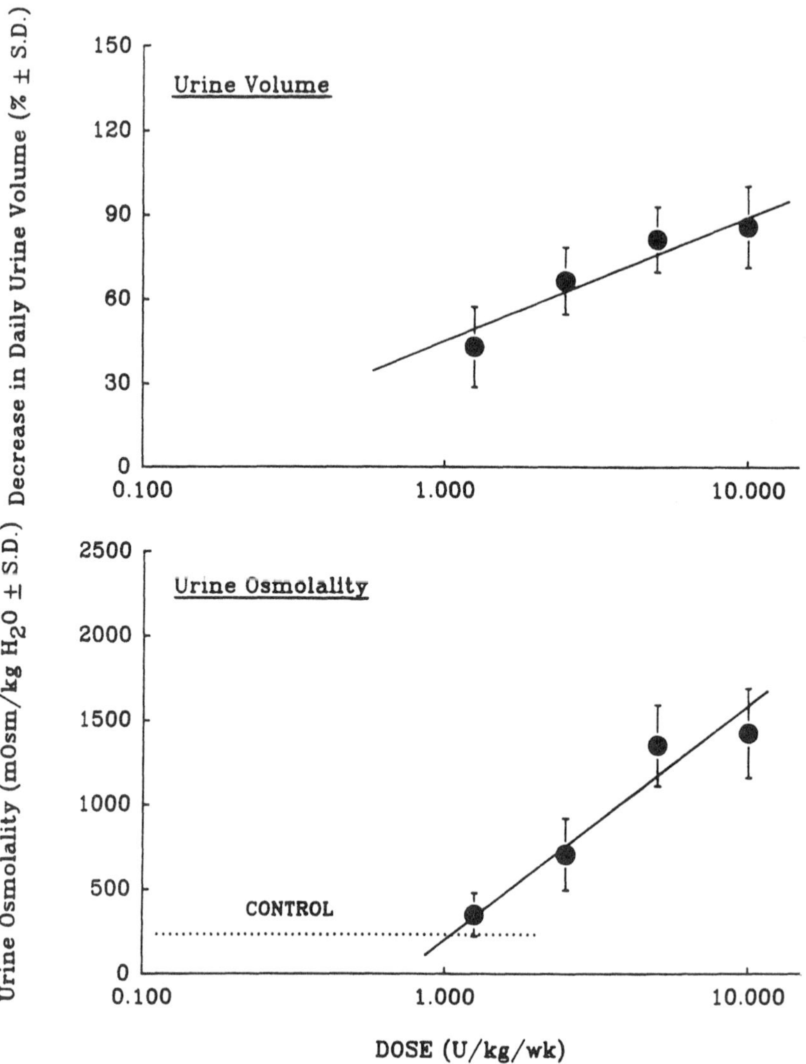

Figure 17 Linear dependence of the volume of daily urinary excretion (top) and the osmolality of the urine (bottom) with the administered dose of vasopressin delivered subcutaneously by an osmotic pressure-activated minipump to vasopressin-deficient Brattleboro rats. (Plotted from data by P. Lelawongs, 1990.)

3), through such nonoral routes as the nasal, rectal, and vaginal mucosae was substantially improved over absorption by oral administration. However, the systemic bioavailability of this synthetic LHRH analog through nonoral routes is still far from that afforded by parenteral administration. As demonstrated earlier for insulin (Table 4), a protein, coadministration of adjuvants to overcome permeation and enzymatic barriers should also be considered to improve the systemic bioavailability of leuprorelin, a peptide, as well as other peptide- and protein-based pharmaceuticals through these mucosal membranes. The feasibility of systemic delivery of therapeutic pep-

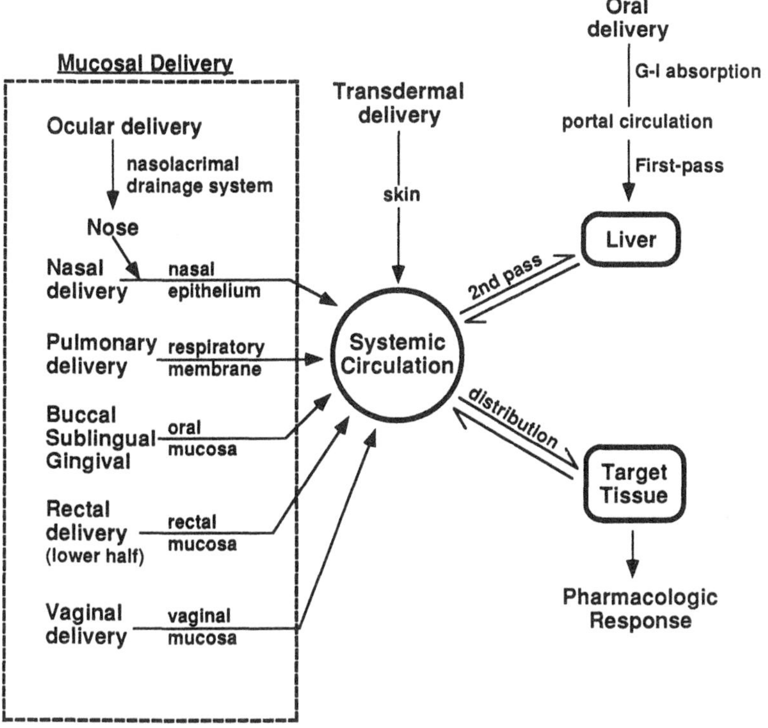

Figure 18 Various pathways for the systemic delivery of peptide and protein pharmaceuticals that are labile to hepatogastrointestinal first-pass elimination when taken orally.

tides and proteins through the various nonparenteral routes is discussed individually in the following sections.

1. *Nasal Delivery*

While the nasal route has been commonly used for the delivery of topically active drugs to alleviate histaminic symptoms in the nasal cavity that result from local in-

Table 4 Relative Hypoglycemic Activity of Insulin[a] Delivered Through Various Mucosal Routes

	Relative efficacy[c] (%)	
Route[b]	No adjuvant	Na glycocholate[d]
Nasal	0.4 ± 0.2	46.8 ± 4.7
Sublingual	0.3 ± 0.2	11.9 ± 7.6
Buccal	3.6 ± 2.8	25.5 ± 7.5
Rectal	17.0 ± 5.6	40.1 ± 8.2

[a]Bovine insulin solution (pH 7.4, 10 IU/kg).
[b]Group of six male Lewis rats.
[c]Relative to IM administration.
[d]Absorption promotor (5%).
Source: Modified from Aungst et al. (1988).

Table 5 Effect of Administration Routes on the Absorption of Leuprorelin

Route	Absolute bioavailability[a] (%)
Nonparenteral	
Oral[b]	0.05
Nasal[c]	0.11
Rectal	1.2
Vaginal[d]	3.8
Parenteral	
Subcutaneous	65.0
Intravenous	100.0

[a]Calculated from the ovulation-inducing activity versus leuprorelin dose relationship in diestrous rats.
[b]Administered in an absorption-promoting formulation containing a mixed micellar solution of monoolein, sodium taurocholate, and sodium glycocholate.
[c]Drainage of the nasal solution occurred. It increases to 1.8–3% with coadministration of an absorption-promoting adjuvant like 1% sodium glycocholate.
[d]Vaginal absorption increases fivefold with coadministration of an absorption-promoting adjuvant like 10% citric acid.
Source: From Okada et al. (1982).

fection and/or inflammatory responses, there is growing interest in the possibility of using this route for the systemic delivery of systemically active drugs, such as insulin for the long-term treatment of hyperglycemia in diabetic patients. A host of peptide-based systemically active pharmaceuticals have been evaluated for the feasibility of nasal delivery (Table 6). Within the next few years, many drugs are likely to be formulated in nasal preparations and made available in the prescription and/or over-the-counter market (109).

As discussed earlier in Chapters 4 and 5, there exists, underneath the nasal mucosa, an extensive network of microcirculation that enables the effective nasal absorption of orally inactive drugs, which are highly susceptible to gastrointestinal degradation when administered orally, for systemic delivery. Furthermore, it is also known that drug molecules absorbed nasally can directly enter the systemic circulation before passing through the hepatic circulation (Figure 18). Thus, nasal delivery should potentially benefit the systemic delivery of peptides and proteins that are otherwise subject to extensive hepatogastrointestinal first-pass elimination. Extensive reviews in this field have recently been completed (110–112).

The nose is primarily an olfactory organ, but it also plays a role in the clearance of foreign substances, including dust, allergens, and bacteria. The nasal mucosa consists of ciliated, columnar epithelium; the cilia help to remove foreign substances by transporting them posteriorly toward the nasopharynx, where they are swallowed into the stomach. The cilia also transport some materials anteriorly for removal by blowing or wiping. Therefore, any therapeutic peptides and proteins and their formulations targeted for nasal delivery must be critically evaluated for their potential effect

Table 6 Biopharmaceutics of Nasal Delivery of Peptide-Based Pharmaceuticals

Peptide and protein drugs	Number of amino acid residues	Time to plasma peak (min)	Relative bioavailability[a] (%)	Testing model
Thyrotropin-releasing hormone	3	5–15	10–20	Rats, humans
Enkephalin analogs	5	5–10	70–90	Rats, humans
Oxytocin	9	5–10	30–40	Humans
Vasopressin analogs	9	10–20	6–12	Humans
LHRH agonist and antagonists	9–10	10–30	2–5	Monkeys, humans
Glucagon	29	5–10	70–90	Humans
Growth hormone-releasing factor	40–44	20–40	2–20	Rats, dogs, humans
Insulin	51	5–10	10–30	Rats, dogs, humans

[a]Relative to the IV, SC, or IM dose. Based on the data obtained in different studies conducted in various testing models.

Source: Modified from Su (1986).

on nasal mucociliary functions (113). Any potential physical damage to the nasal epithelium must also be investigated. For instance, histopathological examination of rat nasal mucosa following the nasal deliver of enkephalins has shown mild mucosal necrosis, but the degree of irritation was considered slight in view of the long (5-hr) contact time (114).

The nasal mucosa is bathed by secretions that contain proteolytic enzymes. This enzymatic barrier can significantly reduce the systemic bioavailability of peptide and protein pharmaceuticals. The aminopeptidases present in the nasal mucosa were evaluated for their enzymatic activity and degradative effect on peptides and proteins (115). The enzymatic degradation of peptide-based pharmaceuticals is discussed later in Section III.C.

Bile salts have been used extensively to enhance the nasal absorption of peptide-based pharmaceuticals. It has been proposed that the effect of sodium glycocholate in enhancing the nasal absorption of insulin is due to its inhibition of proteolytic activity (116,117). The investigation by Lee and Kashi (115) concluded, however, that the inhibition of aminopeptidase by bile salts is not necessarily a suitable predictor of their enhancement effect.

Investigation of the antidiuretic activities of posterior pituitary gland extracts has led to commercialization of synthetic lysine-8-vasopressin (lypressin; Diapid, Sandoz) and 1-desamino-8-D-arginine vasopressin (desmopressin; DDAVP, Ferring AB) in nasal dosage forms for the treatment of diabetes insipidus (118). Further studies of the nasal delivery of a number of peptide-based pharmaceuticals have demonstrated that systemic bioavailability can be significantly improved by the nasal route (Table 6) (114,116–119). Several investigations have been carried out to evaluate the feasibility of using the nasal route as a simple and practical way to achieve the systemic delivery of LHRH analogs and also to determine their efficacy as contraceptive agents (120–124). Buserelin, for example, is a synthetic nonapeptide with a biological activity 25 times more potent than that of the natural LHRH (Figure 3). Although a much higher dose has been found necessary for systemic medication via nasal delivery than parenteral administration, this synthetic LHRH analog was observed to be therapeutically effective when administered intranasally (123). In addition to lypressin and desmopressin, two more peptides have been successfully developed and marketed as nasal pharmaceutical products in the United States: oxytocin (Syntocinon, Sandoz) and nafarelin acetate (Synarel, Syntex).

The nasal absorption of peptides was found to achieve a lower bioavailability than that of its structural component, amino acids. However, this low transnasal bioavailability was shown to be unrelated to the size of the peptide molecule but rather was attributed to one or more of the following factors: polarity of the molecule, absorption via passive diffusion or a specific carrier-mediated mechanism, and susceptibility to hydrolytic degradation in the nasal cavity and/or mucosal membrane (125).

There is a perception that the molecular size of peptides and proteins, which are one to three orders of magnitude larger than organic-based drug molecules, is a major factor affecting their permeation across biological membranes. The effect of molecular weight on the permeation of biological membranes like the nasal and oral mucosae was compared and discussed in Chapter 5. The relationship between the transmucosal bioavailability and molecular weight of 25 organic- and peptide-based penetrants with molecular weights (MW) in the range of 160–34,000 was analyzed

(126). This analysis included the following peptide and protein molecules: thyrotropin-releasing hormone (MW 362), met-enkephamide (MW 661), a somastostatin analog (MW 806), oxytocin (MW 1007), lysine vasopressin (MW 1056), desmopressin (MW 1069), LHRH (MW 1182), buserelin (MW 1297), nafarelin (MW 1337), alsactide (MW 2100), secretin (MW 3052), growth hormone-releasing factor (MW 4800), and horseradish peroxidase (MW 34,000). The analysis revealed that molecules with a molecular weight less than 1000 daltons are adequately absorbed with an average bioavailability of 70 ± 26%; for instance, thyrotropin-releasing hormone (MW 362) achieved a transnasal bioavailability of 40 ± 1.5%. For molecules with a molecular weight greater than 1000 daltons, the percentage of dose absorbed nasally was found to decline as the molecular weight increased (Figure 19) according to the relationship

$$\% \text{ Absorption} = \frac{100}{1 + a(\text{MW})^{-b}} \tag{1}$$

where $a = 0.001$ and $b = 1.35$ for human nasal mucosa; $a = 0.003$ and $b = 1.3$ for rat nasal mucosa. Deviation of absorption from the values predicted by this model was not found to be correlated with such factors as the charge, hydrophobicity, and susceptibility of the peptide or protein molecule to aminopeptidases. This observation probably reflects the relative homogeneity of charge and the hydrophobicity of the molecules examined and their resistance to aminopeptidase action (127). As discussed in Chapter 5, nasal absorption is about three times less dependent upon the molecular weight of penetrant molecules than oral absorption.

Nasal Delivery of Oligopeptides. *Dipeptides:* The nasal absorption of the dipeptides, l-tyrosyl-l-tyrosine, and its methylester, l-glycyl-l-tyrosine, and l-glycyl-l-ty-

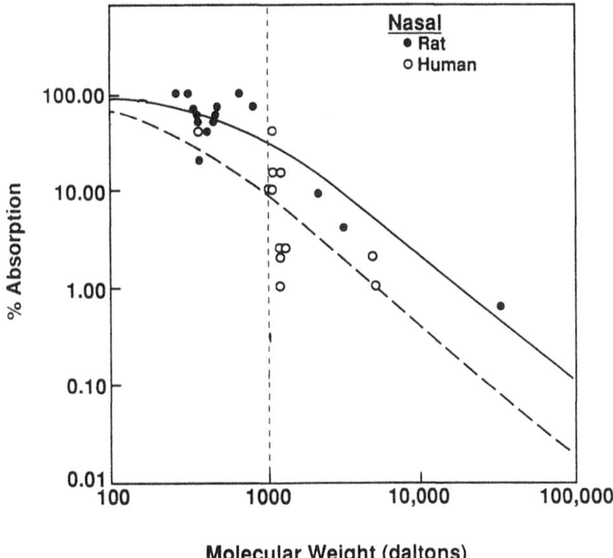

Figure 19 Dependence of nasal absorption on the molecular weight of various penetrants. (Modified from McMartin et al., 1987.)

rosinamide was investigated by in situ nasal perfusion studies (125). The results indicated that the transnasal bioavailability of dipeptides is poor as a result of their hydrolytic degradation by peptidase in the nasal cavity.

Tripeptides: The nasal absorption of thyrotropin-releasing hormone (TRH), a tripeptide neurotransmitter frequently used in the diagnosis of thyroid function, was studied in rats by nasal instillation (128). The results showed a rapid rise in the serum concentration of thyroid stimulating hormone (TSH), with its peak level reached within 15 min, and a nasal bioavailability of TRH estimated at around 20%. It was reported that the use of TRH nasal spray for the clinical diagnosis of thyroid function can reduce the side effects, such as flushing and hypotensive, associated with the intravenous TRH test.

Pentapeptides: The feasibility of the nasal delivery of enkephalins, a group of naturally occurring pentapeptides, and their synthetic analogs, such as DADLE and met-enkephamide, was investigated. Using the in situ nasal perfusion technique, the nasal absorption of leucine-enkephalin (Tyr-Gly-Gly-Phe-Leu) was studied in anesthetized rats (129). The extent of nasal absorption was calculated as less than 10%. The results indicated that when administered at low concentrations, leucine-enkephalin undergoes extensive hydrolysis. However, the extent of hydrolysis is considerably reduced by the addition of excess dipeptides, such as l-tyrosyl-l-tyrosine. This investigation demonstrated that the nasal bioavailability of therapeutic peptides can be improved by coadministration of a competing pharmacologically inactive peptide.

The nasal absorption of DADLE (Tyr-D-Ala-Gly-L-Phe-D-Leu-OH), the synthetic analog of leucine-enkephalin, was also examined in rats (130). A relative bioavailability of 59% (compared to SC administration) was achieved, which was increased to 94% with the incorporation of 1% sodium glycocholate into the formulation.

On the other hand, the intranasal administration of met-enkaphamide, a stable analog of enkephalins, was observed to produce an area under the curve (AUC) value approximately equivalent to that by IV administration and linearly dependent upon the dose administered nasally. Its nasal absorption was found to be insignificantly influenced by surfactant, which could be attributed to the fact that met-enkephamide is less susceptible to enzymatic degradation (131).

Hexapeptides: The nasal absorption of SS-6 [cyclo(Pro-Phe-D-Trp-Lys-Thr-Phe)], a hexapeptide, was recently evaluated in male Wistar rats (132). A nasal bioavailability of 73% was achieved.

Nasal Delivery of Polypeptides. *Nonapeptides:* The vasopressin isolated from human and most mammalian animals is usually arginine vasopressin [Cys-Tyr-Phe-Gln-Asn-Cys-Pro-L-*Arg*-Gly]; that from the pig is lysine vasopressin [Cys-Tyr-Phe-Gln-Asn-Cys-Pro-*Lys*-Gly]. Both are nonapeptides and are called antidiuretic hormone which decreases urine flow by increasing the reabsorption of water in the kidney and facilitates the excretion of sodium and chloride. In patients who are deficient in the hypothalamicopituitary secretion of arginine vasopressin, diabetes insipidus develops with polyuria and polydipsia as the primary symptoms. The long-term treatment of diabetes insipidus has been the IM injection of aqueous vasopressin or vasopressin tannate in peanut oil, which is painful and unpleasant, or the nasal insufflation of posterior pituitary extract powder, which was found as effective as by parenteral administration but required frequent administration (three times or more a day).

A nasal spray preparation has been developed for the intranasal administration of synthetic lysine vasopressin, which has been proven effective in treating diabetes insipidus patients (133–141) and successfully commercialized (Diapid, Sandoz). It has a short duration of action, so patients need repeated dosings several times a day at 2–4 hr intervals.

The feasibility of nasal delivery has also been investigated for other synthetic vasopressin analogs, such as phenylalanyl-lysine vasopressin (PLV-2) and 1-desamino-8-D-arginine vasopressin (desmopressin). PLV-2 has been reported to be 5-fold stronger in hypotensive activity but 10–40 times weaker in diuretic action than lysine vasopressin. It has been formulated in a nasal spray dosage form for topical vasoconstriction, which is as effective as 4% cocaine in achieving nasal decongestion (142) and has shown a greater degree of safety than epinephrine in controlling hypertension.

On the other hand, desmopressin has a greater antidiuretic potency than either natural or any known synthetic vasopressins and also produces prolonged antidiuresis (143). Its intranasal administration has been evaluated for the treatment of various forms of vasopressin-sensitive diabetes insipidus (144–151) and found especially advantageous in the management of central diabetes insipidus. It has been formulated in a metered-dose nasal spray formulation and commercialized (DDAVP, Ferring AB). The metered-dose nasal sprayer was found to deliver a well-controlled dose to the nasal cavity, with the spray droplets deposited mainly in the atrium, and the viscosity, particle size, and nasal clearance were observed to be the important parameters in the design of nasal delivery systems.

Oxytocin, another naturally-occurring nonapeptide with a molecular structure (Cys-Tyr-*Ilu*-Gln-Asn-Cys-Pro-*Leu*-Gly) closely similar to that of vasopressins, is known to stimulate the contraction of smooth muscle in the uterus as well as to induce the contraction of the myoepithelial cells around the breast and alveoli to squeee milk into the large ducts and increase milk flow through the nipple.

Synthetic oxytocin (Syntocinon, Sandoz and Partocon, Ferring AB) is the drug of choice for the induction and enhancement of labor. The intranasal administration of synthetic oxytocin has been investigated for the stimulation of uterine contraction for the induction and enhancement of labor (152–161) and stimulation of lactation in lactating women (162–165). Recently it has also been evaluated for its potential in evoking the secretion of insulin and glucagon (166).

Decapeptides. Luteinizing hormone-releasing hormone LHRH, a decapeptide, is known to be secreted in a pulsatile manner to stimulate the gonadotropes of the anterior pituitary to trigger and to maintain the release of luteinizing hormone (LH) and follicle stimulating hormone (FSH) in an episodical fashion (Figure 4). Administration of LHRH in a continuous manner has reportedly produced downregulation of the homologous pituitary receptors and interruption of the normal release patterns of these gonadotropins (167–169), resulting in progressive desensitization of gonadotropic secretion responses as well as downregulation and suppression of testicular or ovarian steroid secretion (170). LHRH and its analogs (Figure 3) have a relatively short circulatory half-life and require repeated daily administration to achieve the desired effect. LHRH agonists have distinct clinical advantages as potent therapeutic agents that can be self-administered by nasal spray. For example, buserelin [D-Ser(TBU)^{6}-LHRH-EA^{10}, Hoechst], a synthetic LHRH agonist, was observed to produce a prolonged effect on the release of LH and FSH, which lasted for 8–10 hr, by parenteral injections. On the other hand, intranasal administration yielded a rise in the serum concentration of gonadotropins that was smaller, however, than that by IV injection. The investigations conducted to date have indicated that the nasal route is a safer, relatively rapid, and more convenient means for the systemic delivery of buserelin and is an attractive alternative for the long-term therapy of mal- or undescended testes and uni- or bilateral cryptorchidism (171). The intranasal administration of buserelin has been evaluated clinically for the treatment of precocious and delayed puberty, induction or inhibition of ovulation, contraception, luteolysis, endometriosis, hypogonadotropic hypogonadism, secondary amenorrhea, mammary or prostate carcinoma, and uterine leiomyomas.

Nafarelin acetate [6-D-(2-naphthylalanine)-LHRH, Syntex] is another potent synthetic LHRH agonist (Figure 3) that has been studied extensively for the feasibility of nasal delivery. A dose-dependent inhibition of ovulation was observed following the daily intranasal administration of nafarelin acetate from a metered-dose pump (172). Nasal absorption of nafarelin acetate in rhesus monkeys was found to be rapid and reproducible, achieving a peak level within 15 min and a bioavailability of 2%. On the other hand, the intranasal administration of nafarelin acetate to women was observed to attain a bioavailability of 5% (173) and produce an inhibition of normal ovulation and corpus luteum function (174). With the administration of a higher dose of nafarelin acetate, there were significantly fewer presumed ovulatory cycles, fewer bleeding episodes, fewer bleeding days, and longer cycles (175).

A 6-month double-blind study was conducted in 213 patients with laparoscopically confirmed endometriosis to compare the clinical efficacy of nasal nafarelin spray and oral danazol (176). The results concluded that the nasal delivery of nafarelin is effective for the treatment of endometriosis and has fewer side effects. The nafarelin acetate nasal sprayer (Synarel, Syntex) has recently received FDA approval for marketing.

Undecapeptides: Substance P (Arg-Pro-Lys-Pro-Gln-Gln-Phe-Phe-Gly-Leu-Met-NH_2), an undecapeptide in the brain, the spinal ganglia, and the intestines that acts as a vasodilator as well as a depressant, was evaluated for the possibility of nasal absorption in rats (177). The results demonstrated that 18 of the 22 rats studied showed a rapid increase in plasma levels of immunoreactive substance P, with peak concentrations reached within 2–5 min and a return to baseline levels within 20 min. Nasal delivery was shown to achieve a more sustained systemic level than IV administration.

Glucagon: This polypeptide consists of 29 amino acid residues with a molecular weight of 3483 daltons. This hyperglycemic factor secreted by the alpha cells in the pancreatic islets of Langerhans plays an important role in the physiological regulation of blood sugar (Figure 5), and its IM or SC administration has been shown to be an effective treatment of hypoglycemic shock emergencies commonly observed in the daily management of insulin-dependent diabetes. The feasibility of nasal delivery of glucagon was investigated in seven normal subjects (178), and the results demonstrated that with the coadministration of sodium glycocholate, glucagon is well absorbed through the nasal mucosa to increase blood glucose levels; however, the efficacy of nasal glucagon is lower than that of intramuscular glucagon, with a potency ratio of 1:2. The glucagon delivered intranasally produced a sharp increase in the plasma concentration of immunoreactive glucagon as rapidly as the IM injection, but it also activated the release of insulin from beta cells in normal subjects (Figure 20) as a result of its normal metabolic effect.

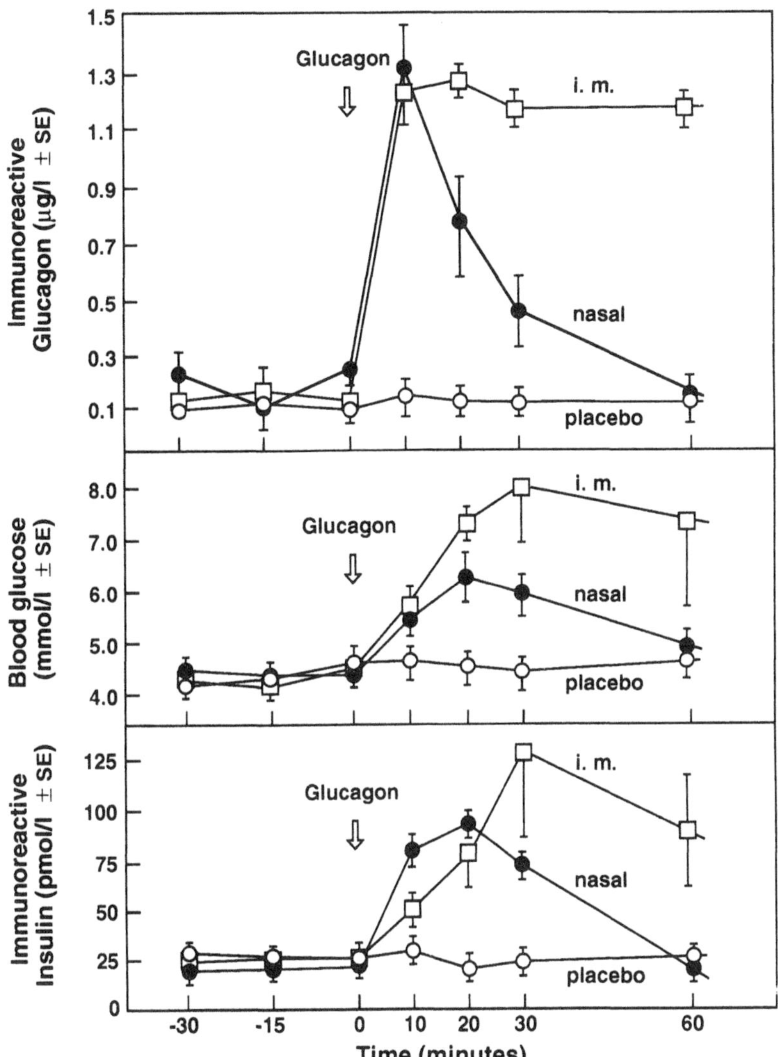

Figure 20 Effect of glucagon administered intranasally (●) or intramuscularly (□) on the plasma profiles of glucagon and insulin as well as the blood profiles of glucose. The profiles resulting from the intranasal administration of diluent (○) are also shown for comparison. (Replotted from Pontiroli et al., 1983.)

Calcitonin: This polypeptide consists of 32 amino acid residues with a molecular weight of 3418 daltons. This calcium-regulating hormone secreted from the thyroid gland plays an important role in the physiological regulation of calcium concentration in plasma. The feasibility of nasal delivery of calcitonin was recently studied using a nose drop formulation (Cibacalcin, Ciba-Geigy) in six normal subjects (179). The results indicated that nasal calcitonin achieved a similar effect in reducing the plasma calcium level as IV injection, even though the plasma calcitonin concentration attained by transnasal absorption is lower than that by IV administration. These data also demonstrated that although the plasma calcitonin concentrations achieved are dependent upon the calcitonin doses administered intranasally, the reduction in plasma calcium levels show no dose dependence. The aqueous polyacrylic acid gel was found to improve the nasal absorption of [$Asu^{1,7}$]-eel calcitonin in rats, with the maximal hypocalcemic effect achieved within 30–60 min (180). The hypocalcemic effect of calcitonin delivered nasally in the polyacrylic acid gel showed a dependence on the doses administered.

Adrenal corticotropic hormone (ACTH): This polypeptide consists of 39 amino acid residues and has a molecular weight of around 4500 daltons. It is useful in the symptomatic treatment of rheumatoid arthritis and asthma by IV or IM administration. The feasibility of nasal delivery of natural ACTH was investigated in six normal subjects and 112 rheumatoid patients by nasal nebulization of ACTH in solution (181,182). There was no significant drop in circulating eosinophil levels in the normal subjects, but a reduction in circulating eosinophil levels was achieved in the rheumatoid patients by the daily intranasal administration of ACTH, which induced a remission of the rheumatoid arthritis.

In addition to natural ACTH, the nasal delivery of a long-acting ACTH that is a synthetic ACTH with only 18 amino acids (D-Ser^1-,$Lys^{17,18}$-α^{1-18}-ACTH) was also studied (183). The results indicated that the α^{1-18}-ACTH is well absorbed and produces a pharmacokinetic profile similar to that achieved by IM and SC injections. Further studies in 14 normal subjects by intranasal administration of an oily suspension of α^{1-18}-ACTH in an aerosol formulation, via an automatic dose-controlled nebulizer, demonstrated that nasal delivery of α^{1-18}-ACTH may provide a simple and effective way to stimulate the secretion of endogenous corticosteroids.

Nasal delivery of Alsactide (β-Ala^1,Lys^{17}-$ACTH^{1-17}$ heptadecapeptide-4-aminobutylamide), a potent ACTH derivative with a potency five to eight times greater than the natural ACTH, was also investigated. A bioavailability of 12% was achieved as estimated from its effect on corticosterone release and adrenal ascorbic acid depletion (184). Significant prolongation in the duration of cortisol release (longer than 4 hr) was observed in humans.

Nasal Delivery of Proteins. *Insulin:* This protein hormone has been studied extensively for systemic delivery by intranasal administration (180,185–188). The transnasal permeability and nasal absorption of insulin were found to be enhanced by the coadministration of absorption promotors, such as bile salts (e.g., glyco- and deoxycholate), naturally occurring surfactants, or synthetic surfactants. By using these adjuvants, the therapeutically effective plasma levels of insulin required for its normoglycemic effects have been achieved (116,189–191). Gordon et al. (191) reported that in the case of sodium deoxycholate, a minimum concentration of 2.4 mM is required to enhance the transnasal permeation of insulin. By varying the concentration of sodium deoxycholate coadministered, therapeutically useful amounts of insulin were absorbed nasally in healthy human volunteers (Figure 21). The nasal absorption of insulin was found to correlate positively with the hydrophilicity of the bile salt with the rank order: deoxycholate > chenodeoxycholate > cholate > ursodeoxycholate (191). Recently, sodium taurodihydrofusidate, a detergentlike adjuvant, was also shown to be an excellent absorption promotor for enhancing the transnasal permeation of insulin (192), and the results of clinical trials demonstrated

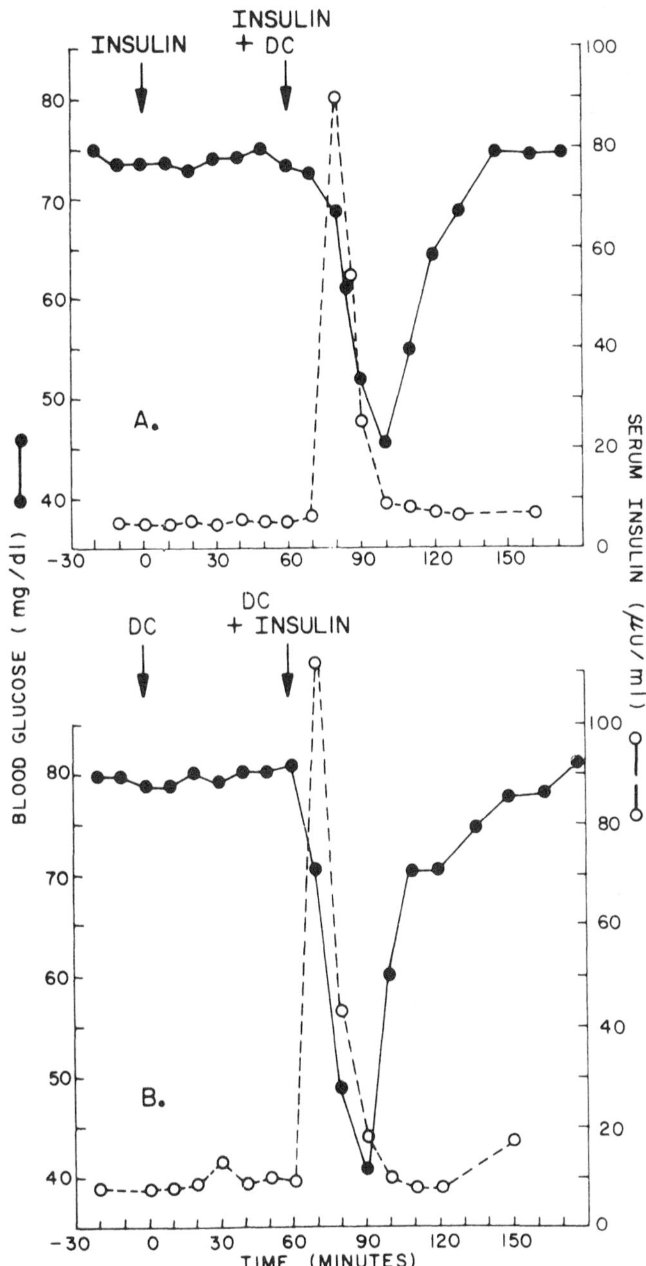

Figure 21 Effect of the intranasal administration of insulin (0.5 IU/kg) in an aerosol formulation with and without coadministration of 1% (w/v) deoxycholate (DC) on the serum insulin profile (○) and blood glucose profile (●) in a normal subject (A). The effect of nasal delivery of deoxycholate (DC) with and without insulin is also shown (B) for comparison. (Based on Moses et al., 1983.)

that insulin is absorbed readily into the systemic circulation when delivered intranasally in formulations containing this promotor (193).

Interferon: This is a protein type of macromolecule released by biological cells following exposure to a virus that enables other cells to resist viral infections. It is now recognized that one cell type can produce several types of interferon (194): α-, β-, and γ-interferons. The advance in recombinant DNA technology has led to the isolation of human interferon genes and their cloning in bacteria, as well as the production in commercially-viable quantities of recombinant human interferon by fermentation and purification using monoclonal antibodies (195). Extensive evaluation of interferon efficacy against a variety of viral diseases and cancers has been underway. A recent PMA survey indicated that eight interferons are in human clinical trials or with the FDA for review (*Pharmacy Today*, 5/25/90). γ-Interferon has recently received FDA approval for the therapy of chronic granulomatous disease, a hereditary immune disorder that strikes young men and boys, leaving them susceptible to infection (*Pharmacy Today*, 1/18/91).

The feasibility of administering interferons intranasally has been investigated. Merigan et al. (196) studied the inhibition of respiratory virus infection by intranasal administration of human leukocyte (α) interferon. By applying a high daily dose of interferon intranasally using a treatment schedule of 1 day of prophylaxis and 3 days of treatment, they observed a statistically significant reduction in the severity of symptoms as well as the frequency of virus shedding when challenged with rhinovirus 4. Other studies have suggested that nasal epithelial cells can be made antiviral, under in vivo conditions, by intranasal administration of human α-interferon, depending upon the method used to deliver interferon to the nasal mucosa and the extent of interferon contact with nasal epithelial cells. Scott et al. (197) reported that repeated nasal sprays of a purified human α-interferon reduced the incidence and the severity of colds in volunteers challenged with human rhinovirus 9. The tolerance and histopathological effects of long-term intranasal administration of α_2-interferon was evaluated in humans (198).

Horseradish peroxidase: This is a protein molecule with a molecular weight of around 34,000 daltons. Its nasal absorption was recently investigated in the mouse, rat, and squirrel monkey (199), and the results indicated that horseradish peroxidase passes freely through the intercellular junctions of olfactory epithelium and reaches the olfactory bulbs in the central nervous system (CNS) within 45–90 min. A systemic bioavailability of only 0.6% was achieved in male Wistar rats, compared to 73% for an octapeptide with a molecular weight of around 800 daltons (132).

For intranasal administration of drugs, nasal sprays or inhalers are the delivery devices of choice because they achieve a better absorption as compared to nasal drops, probably because they reach the nasal mucosa in a more diffuse form and the drugs delivered are distributed to a greater area of mucosa surface. Furthermore, these delivery devices were observed to produce far less pathologic changes than those resulting from the use of nasal drops (113,199). Some new generations of nasal delivery systems, such as the metered-dose nebulizer, which is operated by mechanical actuation, and the metered-dose aerosol, which is operated by pressurized actuation (110,200), have recently been introduced into the marketplace. Proper selection of nasal delivery systems will depend on the physicochemical properties of the peptide-based pharmaceuticals to be delivered.

Conclusion. It appears that the nasal mucosa offers a practical route of administration for the systemic delivery of peptide- and protein-based pharmaceuticals. Compared to parenteral administration, nasal systemic delivery has several benefits, including rapid absorption, which leads to fast onset of action, ease of administration, which should improve patient compliance, and good local tolerance, which should permit long-term application for chronic medication. The biopharmaceutics of some peptide and protein drugs following transnasal delivery is summarized in Table 6. For certain peptide and protein pharmaceuticals, such as insulin, the reported biopharmaceutical responses can be achieved only with the assistance of absorption promotors.

2. *Pulmonary Delivery*

Delivery of medication to the respiratory tract for the localized therapy of respiratory diseases, such as the use of inhalation aerosols for the pulmonary delivery of bronchodilators to achieve the relief of asthma, has been practiced for several decades.

The lungs are an attractive site for the systemic delivery of therapeutic peptides and proteins in view of the enormous surface area (70 m^2) that they offer compared to other absorptive mucosae, including the nasal mucosa just discussed. Approximately 90% of the absorptive surface area offered by the lungs is attributed to the alveoli. The epithelium of the alveoli consists of a heterogeneous population of epithelial cells, approximately 97% type I (or squamous) cells and 3% type II (or granular cuboidal, surfactant-secreting) cells. A very few type III (brush) cells and free alveolar macrophages are present in the alveolar spaces (201,202). Type I cells and macrophages have been reportedly involved in the uptake of horseradish peroxidase, a protein with a molecular weight of around 34,000 daltons, by endocytosis (203). These epithelial cells are generally in intimate contact with the underlying vasculature, where the air-blood separation has a thickness of much less than 1 μm (204). Such an anatomic architecture permits the free exchange of gases, such as O_2 and CO_2, but it is a major barrier to the permeation of much larger organic- and peptide-based drug molecules because the cells in the alveolar epithelium are tightly interlaced (205). The alveolar epithelium was shown to have a permeability approximately 1000 times less permeable to sucrose, a water-soluble neutral molecule, than the microvascular endothelium (206). This may explain the observation that horseradish peroxidase deposited on the air-side surface of alveolar epithelium hardly reaches the interstitium, even though it has reached the interstitium from the blood side (207,208). The alveolar epithelium has been estimated to have a pore size of only 6–10 Å, which is much smaller than the 40–58 Å calculated for the pulmonary capillary membrane (207–209). Furthermore, lung surfactants, which coat the air/water interface on the surface of the alveoli to maintain appropriate surface tension and thereby preserve airway patency, could also play an important role in regulating the permeability of alveolar epithelium. Any peptides or proteins that are delivered by way of the lungs could vary in their systemic delivery if they affect these lung surfactants.

The absorption of drugs in the lungs is similar to that in the intestine, where both simple diffusion and carrier-mediated transport operate. Simple diffusion was reportedly involved in the pulmonary absorption of a number of drugs with molecular weights ranging from 60 to 75,000 following intratracheal administration in the anesthetized rat. Absorption rates were found to be dependent upon molecular size (210–

215). For example, the half-time of absorption from the alveolar region is 26.5 min for mannitol, 9.2 hr for heparin, and 28 hr for dextran. On the other hand, carrier-mediated transport appeared to exist in several animal species and was reportedly involved in the pulmonary absorption of cycloleucine (216,217), α-methyl-D-glucose pyranoside (218), and sodium cromoglycate (219). However, the involvement of carriers in the pulmonary absorption of peptides and proteins has yet to be determined.

Very few reports have been published in the literature on the pulmonary absorption of peptide and protein pharmaceuticals. Leuprolide, insulin, and albumin are three of the few peptides and proteins whose absorption from the lungs has been examined. The pulmonary absorption of leuprolide (leuprorelin, Figure 3) was investigated in human volunteers (220) and found to be absorbed to the extent of 18%. On the other hand, it was reported that insulin, when administered in an aerosol formulation and delivered by a nebulizer, achieved an absorption of 7–16% in healthy subjects and diabetic patients compared to about 40% in anesthetized rabbits (221,222). Complete absorption of insulin in rats was achieved by coadministration of absorption promotors, such as azone, fusidic acid, or glycerol. Furthermore, albumin was observed to be largely absorbed in guinea pigs and dogs, probably via a pinocytotic process, within 48 hr following instillation into the lungs (207,223).

One of the major challenges in the pulmonary delivery of drugs is to achieve reproducibility in the deposition site of the applied dose. The rate of pulmonary absorption for a drug is expected to vary from one level of the respiratory tree to another, owing to the variation in the thickness of epithelial lining cells as well as other anatomic and physiological variables. To deliver drugs efficiently to the alveolar region would require overcoming the intense power of the upper airways, which filter and remove particles by mucociliary action. This in turn would require the generation of particles (or droplets) with sizes in a more monodisperse distribution. For example, a monodisperse aerosol with a mass-median aerodynamic diameter (MMAD) of 3 μm was reported to achieve an alveolar deposition of 50% or higher, compared to the deposition of 30% or less for a polydisperse aerosol with an equivalent MMAD (224,225).

Three devices are currently available for the pulmonary delivery of drugs: metered-dose inhaler, nebulizer, and powder inhaler (or insufflator). Proper use of the delivery device plus an appropriate coordination between breathing and dispensing of the aerosol are essential to the attainment of a reproducible dosing and deposition (226). Other sources of variability include the absorption and accumulation of moisture (in the highly humid lungs) on the drug particles, evaporation of the propellant from the aerosol formulation after spraying, and the patient's manner of inspiration. None of these devices has been reported as capable of delivering more than 10% of the applied dose to the lungs, since a major fraction of the dose administered has been found to be deposited outside the respiratory tree (226). Of the dose successfully delivered into the lungs, only 30–40% can actually reach the alveolar region when administered by metered-dose inhalers and nebulizers, and the efficiency is even less when administered by dry powder inhalers; the remainder is left behind in the conducting airways (226–229). Deposition in the alveolar region has been reportedly improved by breath holding if the aerosol droplets have an aerodynamic diameter of 1.5–3 μm (227,229). However, such maneuvers have been noted to have only a minimal effect on dose deposition in the ciliated airways. The time it takes to deplete 99% of the administered dose in the ciliated airways also depends on the rate of dissolution, in addition to the size of the droplets and the mode of inhalation. The entrapment of drugs in liposomes (230,231), the use of sparingly soluble drugs (225), or the slowly releasing coprecipitates of drugs (232,233) have been proposed to achieve a prolongation of drug action following pulmonary delivery.

Several safety issues are associated with the pulmonary delivery of peptide and protein pharmaceuticals: one is the potential elicitation of an immune response, which could produce changes in epithelial permeability. For example, in the monkey, the bronchial permeability to macromolecules was found to be increased upon antigen challenge, which may have resulted from alteration in the tight junctions (234), and both immunologic and inflammatory reactions were found to control the pulmonary absorption of the soluble proteins that were inhaled. In the rabbit, a chronic hypersensitivity reaction was observed to alter the pulmonary absorption of inhaled proteins, whereas acute hypersensitivity was not. On the other hand, the permeability to an immunologically unrelated protein was found to be increased.

Another safety issue is related to destabilization of the surfactant film coating around the alveolar surface by the peptides and proteins delivered, as briefly discussed earlier. For example, ACTH, porcine β-lipoprotein, α-endorphin, and human fibrinopeptide A have been demonstrated to interact with the monolayers of dimyristoyl-phosphatidylcholine, which yields changes in the surface tension of the monolayers, and the effect was noted to be inversely proportional to the lipophilicity of peptides and proteins (235). The nature of the polar amino acids and their electrostatic charges could also play a role.

Apparently, the lungs could be an attractive site for the systemic delivery of therapeutic peptides and proteins, but their full potential in the systemic delivery of peptide and protein pharmaceuticals will not be realized until the biochemical and biophysical aspects of pulmonary absorption have been properly characterized and the reproducibility of the dose deposition site has been satisfactorily achieved (236).

3. *Ocular Delivery*

As in the nasal delivery discussed earlier, the ocular route may also be utilized for the systemic delivery of peptide- and protein-based pharmaceuticals (2,237,238). As early as 1931, Christie and Hanzal (239) reported the observation of a dose-dependent reduction in blood glucose levels following the ocular administration of insulin to the rabbit.

Several peptide and protein pharmaceuticals, such as enkephalins, thyrotropin releasing hormone, LHRH, glucagon, and insulin (240–246), were reportedly absorbed to some extent into the bloodstream of the albino rabbit. A typical example is the pharmacodynamic responses illustrated by insulin and glucagon delivered in an eye drop formulation to anesthetized rabbits (Figure 22) (241–243). On the other hand, Stratford et al. (245) and Yamamoto et al. (246) studied the systemic delivery of topically applied [D-Ala^2]-met-enkephalinamide and insulin, respectively, in fully awake rabbits and reported the attainment of peak plasma concentrations within 15–20 min following a solution instillation, with a bioavailability of around 36% achieved for the enkephalin analog and less than 1% for insulin. The systemic bioavailability of insulin was improved to 4–13% with the incorporation of absorption promotors, such as bile salts (the sodium salt of glycocholic, taurocholic, and deoxycholic acid) and nonionic surfactant (polyoxyethylene-9-lauryl ether).

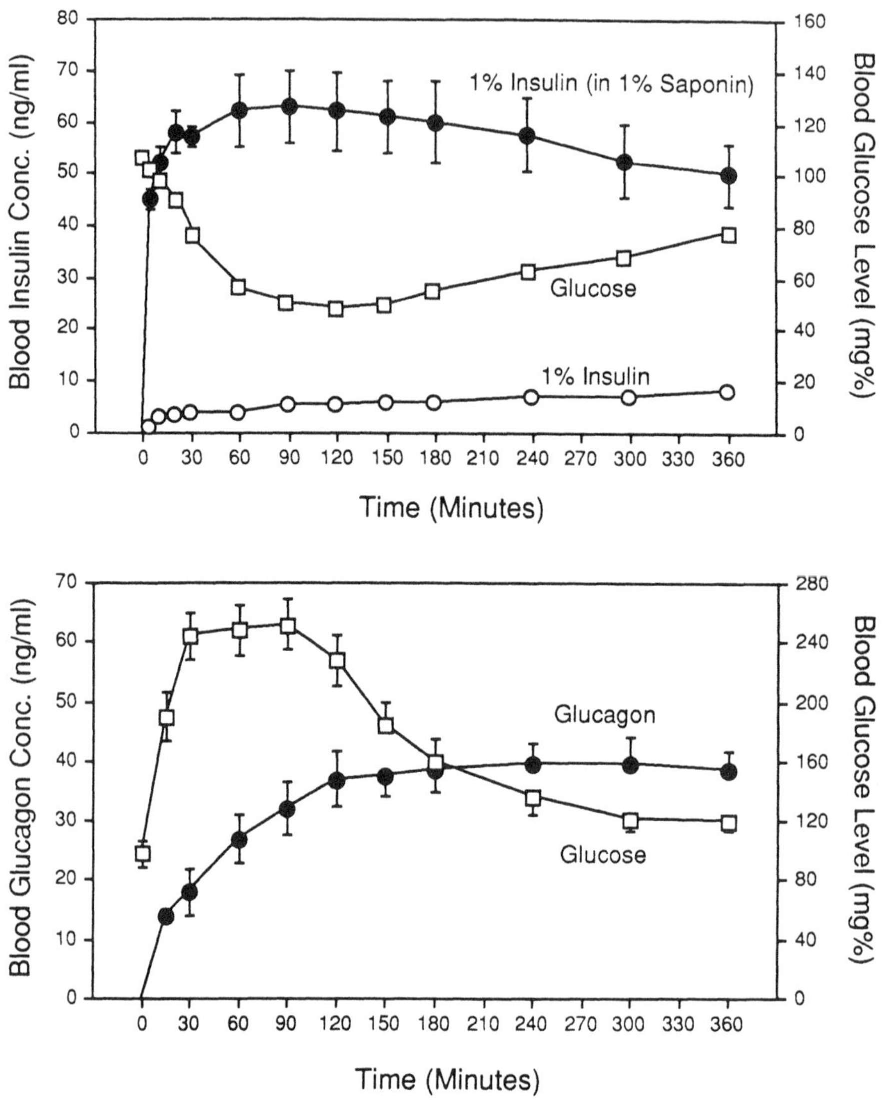

Figure 22 (Top) Blood concentration profile of insulin following the ocular delivery of insulin (1%) and enhancement by saponin (1%). Following the saponin-enhanced ocular absorption of insulin, the blood glucose level was observed to reduce. (Replotted from Chiou et al., 1989.) (Bottom) Blood concentration profile of glucagon following the ocular delivery of glucagon (1%) and the increase in blood glucose level. (Replotted from Chiou and Chuang, 1988.)

Drugs instilled into the precorneal cavity can reach the systemic circulation via the blood vessels underlying the conjunctival mucosa or via overflow of drug solution into the nasolacrimal drainage system followed by absorption through the nasal mucosa (Figure 23). The relative contribution of these two routes to the overall systemic delivery is drug dependent. The nasal mucosa has reportedly contributed approximately four times more to the systemic absorption of insulin than the conjunctival mucosa (246), but nasal and conjuctival mucosae contributed equally to the systemic absorption of [D-Ala^2]-met-enkephalinamide (245). By incorporating this enkephalin analog in a viscous vehicle, like 5% poly(vinyl alcohol), its systemic bioavailability was improved from 36 to 51%.

On the other hand, the ocular route is the site of choice for the localized delivery of ophthalmologically-active peptides and proteins for the treatment of ocular diseases that affect the anterior segment tissues of the eye (Table 7).

In permeation through the corneal epithelium, the molecules of peptides and proteins encounter several diffusional barriers arising from their unfavorable hydrophilicity and large molecular size as well as their susceptibility to enzymatic barriers in various ocular tissues. Furthermore, some physiological and toxicological factors as well as the formulation used could also affect the ocular absorption of peptides

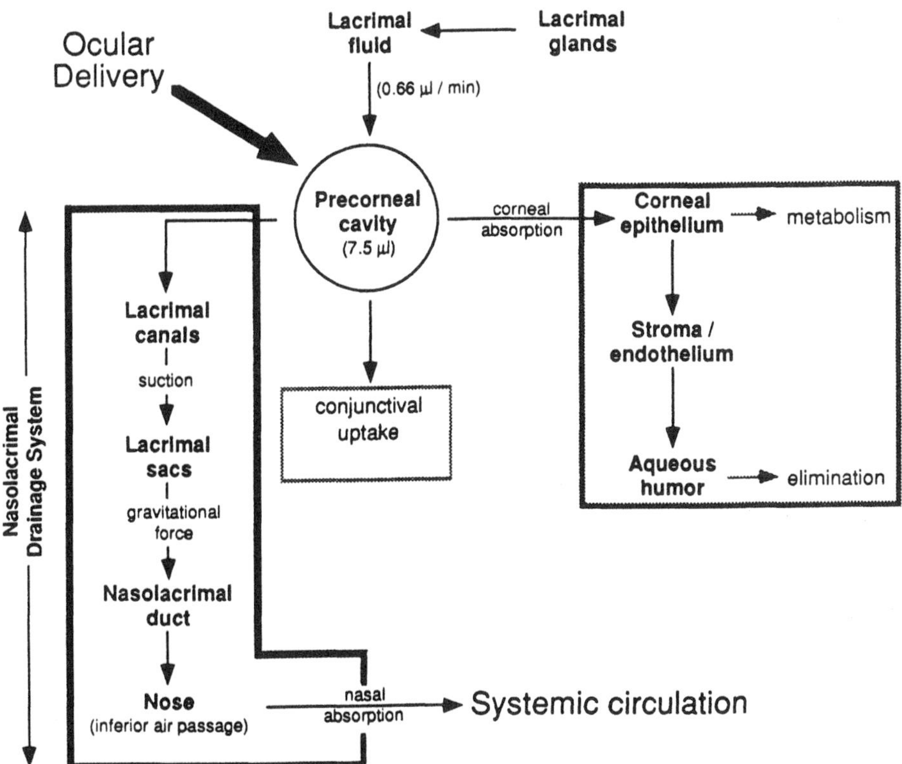

Figure 23 Various pathways for the localized and systemic delivery of drugs administered to the precorneal cavity.

Table 7 Peptides and Proteins with Ophthalmological Activities

Affect aqueous humor dynamics
Atrial natriuretic factor
Calcitonin gene-related factor
Luteinizing hormone-releasing hormone
Neurotensin
Vasoactive intestinal peptide
Vasopressin
Immunomodulating activities
Cyclosporine
Interferons
Act on inflammation
Substance P
Enkephalins
Affect wound healing
Epidermal growth factor
Eye-derived growth factor
Fibronectin
Insulinlike growth factor
Mesodermal growth factor

Source: Modified from Lee (1987).

and proteins. Some of the general approaches that have been found useful in enhancing the ocular absorption of organic-based pharmaceuticals, such as the use of nanoparticles, liposomes, gels, ocular inserts, bioadhesives, or surfactants (247,248), may also be utilized to improve the ocular delivery of peptide-based pharmaceuticals. In addition to the effect of the permeation barrier, the enzymatic barrier created by the existence of peptidases in the ocular tissues is very important in affecting the systemic bioavailability of therapeutic peptides and proteins when delivered by the ocular route (249–252). For example, enkephalins have been found to be readily hydrolyzed in the albino rabbit's eyes (253–255). One solution to overcome this enzymatic barrier is to administer analogs of an active peptide that are resistant to metabolism by the peptidase while maintaining its intrinsic pharmacological activities; for example, [D-ala^2]-met-enkephalinamide is resistant to the metabolic action of aminopeptidase (249). Recent work by Lee et al. (256) demonstrated that less than 0.4% inulin (with molecular weight of 5000 daltons), administered in an eye drop formulation is absorbed into the anterior segment tissues of the rabbit eye. The polypeptide antibiotics (e.g., cyclosporine, tyrothricin, gramicidin, tryocidine, bacitracin, and polymyxins) have often been considered potential candidates for achieving local pharmacological actions in the eye (257).

Further work needs to be done in the ocular delivery of peptide and protein pharmaceuticals, since reports in this field of biomedical research have been rather scarce. However, this route of administration is not likely to become popular for the systemic delivery of peptide-based pharmaceuticals since the systemic bioavailability achieved is extremely low as a result of extensive precorneal clearance and the limited volume of dosage that can be accommodated by the precorneal cavity, that is,

7.5 μl (Chapter 6). Furthermore, ocular tissues are extremely sensitive to the presence of foreign substances and patient acceptance could be rather low.

4. *Buccal Delivery*

Buccal delivery has been practiced for many years as evidenced by the development of pharmaceutical dosage forms, like tablets and lozenges, that have been introduced for several decades and are commonly used for delivering the conventional organic-based pharmaceuticals to the oral mucosa for either local or systemic medication. By buccal delivery, drugs are absorbed rapidly into the reticulated vein, which lies under the oral mucosa, and enter the systemic circulation directly, bypassing the liver (Figure 18).

The penetration of macromolecules through oral epithelia has been studied by several investigators (258–260), and some results are outlined in Table 8. Merkle et al. (261) developed a self-adhesive buccal patch and reported that it is feasible to deliver peptide-based pharmaceuticals, such as protirelin (a thyrotropin-releasing hormone) and buserelin (a synthetic LHRH analog), through the buccal mucosa.

At approximately the same time, a mucosal adhesive delivery system was developed independently by Nagai with Machida (262–265) for the buccal delivery of insulin (a hypoglycemic protein). It was found that the systemic delivery of insulin through the buccal mucosa is significantly affected by the formulation composition used. Insulin could not be effectively absorbed when using a simple disk-shaped dosage form prepared by the direct compression of insulin in a mixture of hydroxypropylcellulose (HPC) and Carbopol 934 (CM). Buccal absorption was achieved by using a dome-shaped two-phase mucosal adhesive device (264) prepared by dis-

Table 8 Investigations of the Buccal or Sublingual Delivery of Therapeutic Peptides and Proteins

Peptide and protein	No. amino acids	References
Thyrotropin-releasing hormone	3	Anders et al. (1983)
		Schurr et al. (1985)
		Merkle et al. (1986)
Tripeptide (model peptide)	3	Veillard et al. (1987)
Oxytocin	9	Wespi and Rehsteiner (1966)
		Bergsjo and Jenssen (1969)
		Sjostedt (1969)
Vasopressin	9	Laczi et al. (1980)
LHRH analogs	10	Merkle et al. (1986)
Calcitonin	32	Anders et al. (1983)
		Nakada et al. (1988)
Insulin	51	Earle (1972)
		Ishida et al. (1981)
		Nagai and Machida (1985)
		Lee et al. (1987)
		Aungst et al. (1988)
		Oh and Ritschel (1988)
		Ritschel et al. (1988)
		Yamamoto et al. (1988)
Interferon-α_2	165	Paulesu et al. (1988)

persing insulin crystals with sodium glycocholate, an absorption promotor, in an oleaginous core and then overlaying the medicated core by an adhesive dome (Figure 24). The dome-shaped adhesive peripheral layer was fabricated from a blend of HPC and CM. This mucosal adhesive device was observed to adhere tightly to dog oral mucosa when it became gelated, as activated by saliva, and its shape was maintained for longer than 6 hr. A systemic bioavailability of only 0.5% was achieved when no absorption promotor was used, which was improved by coadministration of bile salts, such as sodium glycocholate, as absorption promotors. Even though the systemic bioavailability was still very low, therapeutically effective plasma concentrations of insulin were achieved and blood glucose levels were substantially reduced (Figure 25).

Because of anatomic and physiological differences among various oral mucosal tissues, the location for buccal delivery should be carefully selected. A methodology was recently developed to compare the permeabilities of different oral mucosal sites (266).

5. *Oral Delivery*

As discussed earlier, the problems that make the oral route unsuitable for the systemic delivery of therapeutic peptides and proteins are the potential degradation by the strongly acid environment in the stomach and by proteolytic enzymes in the intestinal tract, as well as presystemic elimination in the liver (Figure 2). Furthermore, it is known that macromolecular peptides and proteins have a very low permeability across the gastrointestinal mucosa. For an orally administered peptide to reach its site of action, it must be able to resist any chemical and enzymatic degradation in the gastrointestinal tract and then, after penetration of the mucosal membrane, to escape first-pass metabolism and clearance by the gut mucosa and liver (267). It has been reported that only a very small fraction of an oral insulin dose

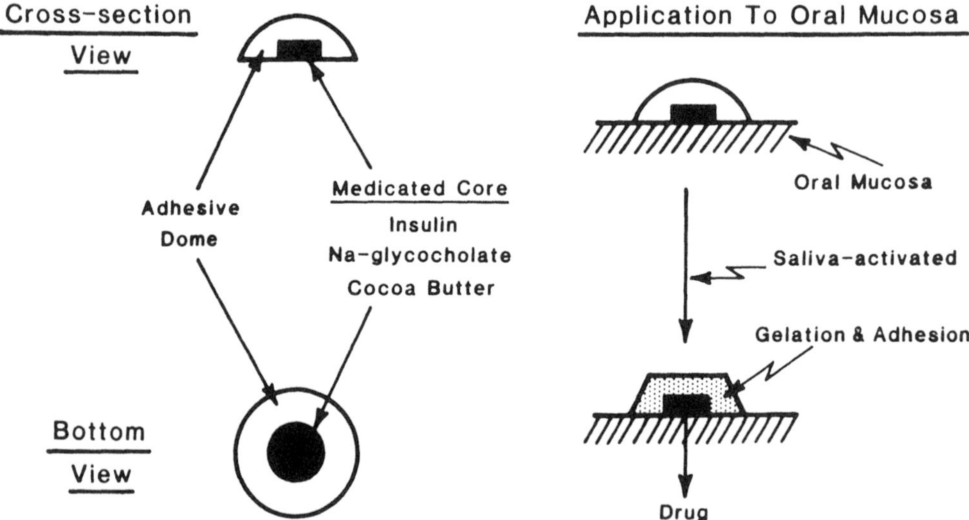

Figure 24 The dome-shaped mucosal adhesive device and its application to oral mucosa. (Based on Nagai and Machida, 1985.)

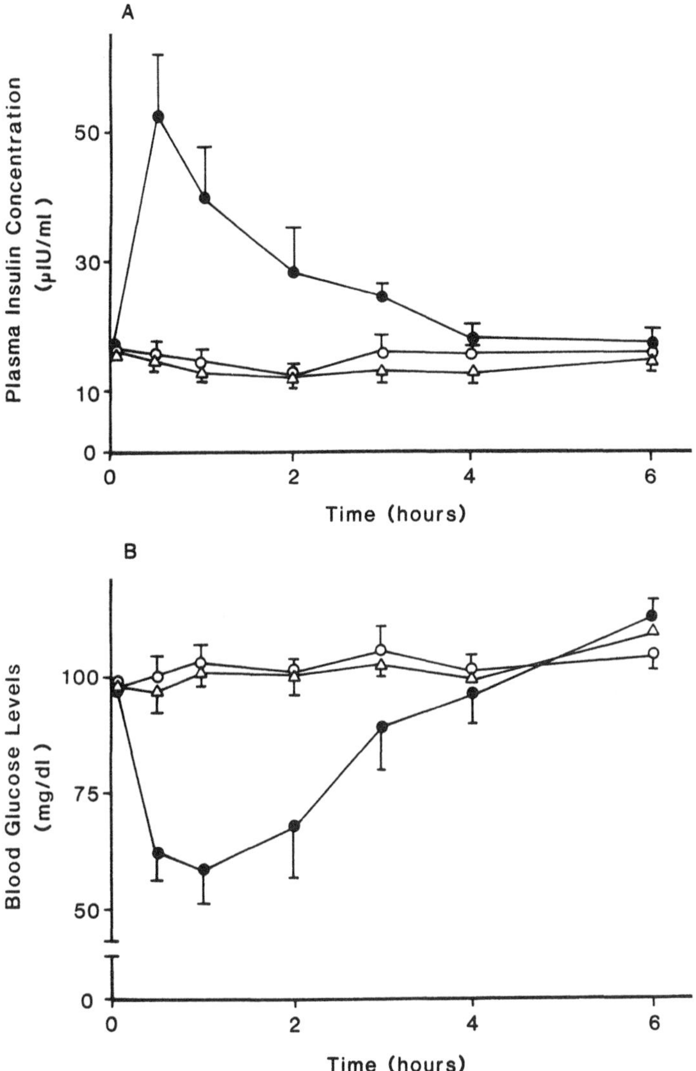

Figure 25 Time course for the increase in plasma insulin levels (A) and the reduction in blood glucose levels (B) in beagle dogs following the buccal delivery of insulin from the mucosal adhesive device shown in Figure 24: (○) control device; (△) mucosal adhesive device containing insulin (10 mg); (●) mucosal adhesive device containing insulin (10 mg) and sodium glycocholate. (Replotted from Nagai and Mahida, 1985.)

becomes available for absorption through the gastrointestinal membrane (24,268). The systemic bioavailability of peptide and protein pharmaceuticals by oral delivery is thus very low, generally less than 2%.

The absorption of insulin from the intestine was shown to be feasible if it is delivered directly into the ascending colon with sodium deoxycholate (269) and achieved a 50% reduction in blood glucose levels. On the other hand, no lowering of blood glucose levels was observed when insulin was delivered directly into the ileum, unless it was coadministered with a naturally occurring trypsin inhibitor, such as aprotinin. Without the naturally occurring trypsin inhibitor, insulin was rapidly digested by the proteolytic enzymes present in the ileum. However, oral delivery is feasible for certain peptides, such as arginine and lysine vasopressin (AVP and LVP), and a synthetic analog, 1-deamino-8-D-arginine vasopressin (DDAVP). These could be administered orally to rats, and a rapid antidiuretic response was achieved (270). The antidiuretic activities of AVP and LVP following oral delivery was substantially enhanced by the simultaneous administration of aprotinin (Figure 26); however, the effect of aprotinin on the oral antidiuretic activity of DDAVP was reported to be inconsistent.

The following approaches have been evaluated as potential means to enhance the oral delivery of peptide-based pharmaceuticals.

Entrapment in Liposomes. The feasibility of using liposomes as a potential oral delivery system for the systemic delivery of therapeutic peptides and proteins has been extensively studied (271). Stefanov et al. (272) investigated the feasibility of delivering insulin systemically by the oral route using a liposome prepared from phosphatidylcholine (PC) and cholesterol (CH) as the delivery system and reported that no change in blood glucose levels is noted in normal animals, but a significant reduction is obtained with diabetic rats, with the maximum effect observed within 3 hr. On the other hand, Moufti et al. (273) were able to produce a 50% reduction in blood glucose levels in normal rats by an insulin-containing liposome. Patel et al. (274) published the results of an extensive study on the delivery of insulin-entrapped liposomes to dogs, in the duodenal region, via a catheter. Arrieta-Molero et al. (275) demonstrated that the oral delivery of an insulin-entrapped liposome is effective in reducing the blood glucose level of diabetic animals, but its stability and effectiveness are rather unpredictable. Dobre et al. (276) illustrated a lowering of blood glucose levels in normal rats following the oral administration of insulin entrapped in

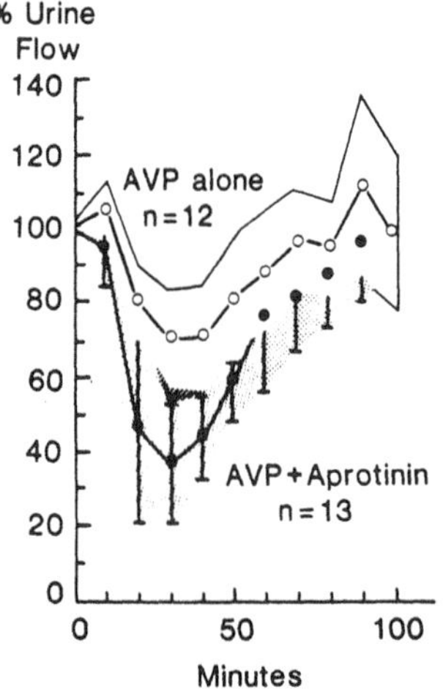

Figure 26 Antidiuretic response in rats following the oral delivery of arginine vasopressin (○; 1 nM) and the enhancing effect of aprotinin (●; 1000 units), a natural inhibitor of trypsin. (Reproduced from Saffran et al., 1988.)

PC/CH liposomes. However, negative results (277,278) were later reported. Thus, the feasibility of delivering insulin by oral liposome formulations needs further work with standardization of liposome composition and improvement of physicochemical stability.

Encapsulation in Azopolymer Coating. An interesting approach was recently developed and applied to the oral delivery of peptide-based pharmaceuticals that involved the coating of peptides with polymers with azoaromatic groups and the cross-linking of azopolymers to form an impervious film to protect the orally administered peptide or protein molecules from degradation and metabolism in the stomach and small intestine (279). When the azopolymer-protected peptide or protein reaches the large intestine, where the microflora reduces the azo bonds to break the cross-linkings in the polymer film and hence releases the peptide or protein molecules in the colonic region for systemic absorption or for local action. The ability of the azopolymer coating to protect the orally administered peptide or protein drugs for systemic delivery was demonstrated with peptides, such as vasopressin, and proteins, such as insulin, in rats.

Other Approaches. Shichiri et al. (280) used a water/oil/water type of multiple-emulsion formulation to deliver insulin orally to rabbits and diabetic rats via an indwelling catheter in the jejunum and observed a reduction in the urinary glucose level in the diabetic rats. Bird et al. (281) carried out an in vivo study to show that incorporation of enzyme inhibitors during the encapsulation of insulin in erythrocytes may reduce its degradation in the human body. Erythrocytes have been investigated as a drug carrier for the intravenous sustained delivery of drugs, but their potential as carriers for the oral delivery of peptide-based pharmaceuticals remains to be seen. It has also been suggested that the oral absorption of insulin may be enhanced by the simultaneous administration of inhibitors for proteolytic enzymes. Another approach for the targeted enteral delivery of insulin was also developed by encapsulating insulin in a small, soft gelatin capsule coated with polyacrylic polymer (Eudragit) with pH-dependent permeability properties (282). On the other hand, nanocapsules fabricated from biodegradable poly(isobutylcyanoacrylate) were also evaluated as the oral delivery system for insulin to enhance its systemic bioavailability and therapeutic efficacy (283). The results indicated that when administered intragastrically, the biodegradable nanocapsules were observed to achieve normalization of hyperglycemia in diabetic rats.

6. *Rectal Delivery*

The use of the rectum for the systemic delivery of organic- and peptide-based pharmaceuticals is a relatively recent idea, even though the rectal delivery of drugs, in a suppository dosage form, for localized medication is a very old practice. In contrast to the oral delivery, the rectal route may provide, in addition to the possibility of bypassing hepatic first-pass elimination (Chapter 3, Section IV.C.2), the advantage of reduced proteolytic degradation for peptides and proteins and thus improve systemic bioavailability (Figure 18), especially when they are coadministered with absorption-promoting adjuvants. The coadministration of an absorption-promoting adjuvant, such as sodium glycocholate, has been reported to enhance the rectal absorption of insulin (Table 4). In the rectum, the upper venous drainage system (superior hemorrhoidal vein) is connected to the portal system, whereas the lower venous drainage system (inferior and middle hemorrhoidal veins) is connected directly to the systemic circulation by the iliac veins and the vena cava (Chapter 3, Figure 45). Thus, there exists an opportunity to reduce the extent of presystemic elimination by rectal delivery, especially when the drug is administered in the low region of the rectum (Chapter 3, Section IV.C.2). The rectum is also comprised of a large number of lymphatic vessels that may offer an opportunity to target drug delivery to the lymphatic circulation through rectal absorption.

Extensive studies have been conducted to investigate the rectal delivery of peptide-based pharmaceuticals, such as insulin (284–290), and on the feasibility of enhancing the rectal mucosal permeability of insulin by using adjuvants (291–303). Some results are summarized in Table 9. Without the use of an absorption-promoting adjuvant, the absorption of peptides and proteins from the rectum is less and has a low systemic bioavailability (304). The rectal absorption of insulin from a microenema was reported to be significantly promoted when coadministered with sodium 5-methoxysalicylate (SMS), whereas sodium salicylate was found to be less effective as an absorption promotor (296). The addition of 4% gelatin was observed to have a synergistic effect on the enhancement of rectal absorption of insulin by SMS (Figure 27). A systemic bioavailability of approximately 25% was achieved. Phenylglycine enamines of various β-diketones, such as ethylacetoacetate, have also been found effective in promoting the rectal absorption of insulin (294,299). Touitou et al. (291) achieved hypoglycemia in rats by administering insulin, via rectal (and also vaginal) routes, in a dosage form that contained polyethylene glycols and a surfactant. The influence of different suppository bases on the systemic bioavailability of insulin was also reported (305).

It was recently reported that a solid dispersion of insulin with sodium salicylate (or mannitol) can produce a rapid release of insulin from the suppositories and achieve a significant reduction in plasma glucose levels in normal dogs, even at doses as low as 0.5 IU/kg (306). On the other hand, the addition of lecithin to the suppository base was observed to prolong the hypoglycemic effect of insulin because of the slowed release of sodium salicylate (306). Bile salts, such as the sodium salts of cholic, deoxycholic, and glycocholic acids, have also been shown to enhance the rectal absorption of insulin in rats (106,307) and in human volunteers (308).

Although not as extensive as the studies conducted on insulin, there are also several reports in the literature investigating the rectal delivery of other peptide-based pharmaceuticals, such as vasopressin and its synthetic analogs (270), pentagastrin and gastrin (309), calcitonin analogs (310,311), and human albumin (312). Some results are summarized in Table 10. For instance, calcitonin, a peptide drug with 32 amino acid residues, has been reportedly useful in the clinical treatment of diseases involving hypercalcemia and osteoporosis. The rectal absorption of its synthetic analog, $[Asu^{1,7}]$-eel calcitonin, was studied in rats and incorporation of polyoxyethylene-9-lauryl ether, a nonionic surfactant, in a polyacrylic acid gel base was observed to enhance the rectal absorption of this calcitonin analog (311).

7. *Vaginal Delivery*

As discussed earlier, the vaginal route has already been utilized for several decades for the administration of a number of therapeutic agents, especially those designated specifically for women or targeted for vaginal tissues (Chapter 9). The vaginal preparations currently on the market contain topically effective therapeutic agents: (i) antimicrobial agents, such as metronidazole for the treatment of vaginal infections;

Table 9 Enhancement of Rectal Delivery of Insulin by Adjuvants and Formulation

Adjuvant or formulation	Testing model	Bioavailability[a] or response	Reference
Polyacrylic acid gel	Rats		Morimoto et al. (1983)
Alone		5%	
With 1% oleic acid		10%	
Sodium phenylalanine enamine of ethyl acetoacetate	Beagle dogs (depancreatized)		Nishihata et al. (1985)
Insulin-enamine suppository		19.4%	
Insulin-enamine suppository followed by enamine suppository		38.2%	
Sodium salicylate	Human volunteers	Better than enamine	Nishihata et al. (1986)
Microenema with 4% gelatin	Beagle dogs	Better than enamine	Kim et al. (1983)
	Diabetic rabbits		Nishihata et al. (1983).
Glyceryl esters of acetoacetic acid	Rabbits	Effective and rapid response	Nishihata et al. (1983)
Solid dispersion of sodium salicylate-mannitol	Beagle dogs	Effective and rapid response; high in vitro dissolution rates	Nishihata et al. (1987)
Sodium deoxycholate-sodium cholate	Rats	50% Reduction in blood glucose levels within 1 hr	Ziv et al. (1981)
Sodium cholate	Human volunteers	50% reduction in blood glucose levels within 30 min	Raz et al. (1984)

[a]Compared to IM administration.

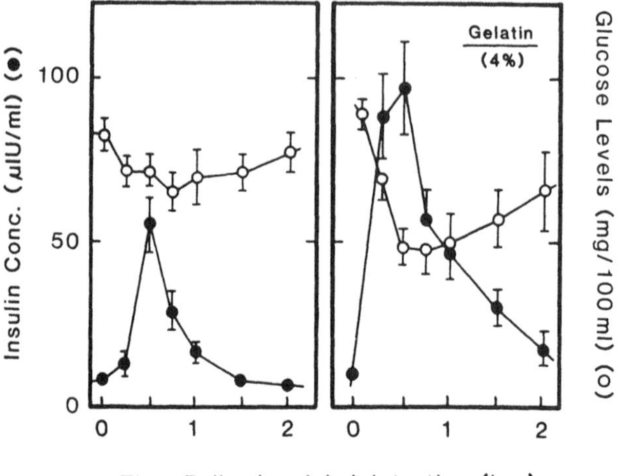

Figure 27 Time course for the increase in plasma insulin levels (●) and the reduction in blood glucose levels (○) in dogs following the rectal delivery of insulin (20 IU) via a microenema containing 150 mg sodium 5 methoxysalicylate and the effect of 4% gelatin. (Replotted from Nishihata et al., 1982.)

(ii) estrogenic steroids, such as estradiol for restoring the normal physiology of vaginal mucosa; (iii) prostaglandins, such as 15-Me-PGF$_{2\alpha}$-methyl ester, which induce labor and therapeutic abortion; and (iv) spermicidal agents, such as nonoxynol-9, and progestins, such as levonorgestrel, which achieve local or systemic contraception. With the exception of prostaglandins and estrogenic and progestational steroids, most of these agents exert their therapeutic activities mainly locally and their systemic absorption has been considered only from the standpoint of toxicity (313).

For long-term systemic medication, the vaginal route offers several advantages: (i) the feasibility of self-administration, (ii) the possibility of prolonged retention of delivery system (Chapter 9), (iii) the potential avoidance of hepatogastrointestinal first-pass elimination (Figure 18), and (iv) minimization of proteolytic degradation. These advantages are particularly beneficial to the chronic administration of peptide and protein pharmaceuticals for systemic medication. Unfortunately, to date only a few systematic investigations have been conducted to evaluate the potential application of the vaginal route for the systemic delivery of therapeutic peptides and proteins, even though the feasibility of vaginal absorption of peptide-based macromolecules, including insulin, was examined as early as in 1923 (314).

Fisher (314) reported the observation of a rapid and temporary reduction in blood glucose levels in depancreatized dogs when insulin was administered intravaginally in either single or consecutive doses. The maximal hypoglycemic effect, at which the blood glucose concentrations were reduced to one-sixth pretreatment levels, was achieved within 3 hr of administration. This effect was found to be greater than that achieved by a larger dose of insulin delivered through an intestinal fistula. Similar results were also observed later in anesthetized cats (315). These pioneering studies provide the first experimental evidence of the systemic delivery of peptides and proteins through the vaginal mucosa.

Table 10 Rectal Delivery of Peptide- and Protein-Based Pharmaceuticals

Peptide and protein	Number of amino acid residues	Testing model	Formulation	Bioavailability (%)	Reference
Pentagastrin	5	Rats	Rectal enema	2–10[a]	Yoshioka et al. (1982)
			Rectal enema with sodium 5-methoxysalicylate	23–33[a]	
Gastrin	17	Rats	Rectal enema	11–25[a]	Yoshioka et al. (1982)
			Rectal enema with sodium 5-methoxysalicylate		
Calcitonin	32	Rats	Saline solution	0	Morimoto et al. (1984)
			Polyacrylic acid gel base	2–3[b]	Morimoto et al. (1985)
			Gel base with POE-9-lauryl	30–50[b]	
Albumins	Variable	Humans	Radiolabeled albumin in 0.85% saline solution instilled into rectum	Radioactivity appeared in plasma	Dalmark (1968)

[a]Compared to IV administration.
[b]Compared to IM administration.

Survey of the literature has revealed that few studies have been conducted to evaluate the vaginal absorption of antigens and antibodies (316). Using an immunological method, for instance, it was shown that in humans, the absorption of peanut protein from the vagina is slower than that from the cervix (317). The 3-day vaginal insertion of a tampon impregnated with a bacterial antigen, such as *Candida albicans*, in human females was observed to cause the local secretion of antibodies and to exert strong local immunological responses (318). This phenomenon has demonstrated the fundamentally important role that the vaginal absorption of bacterial antigens plays in the local production of antibodies to prevent bacterial infection of the female genital organs. Furthermore, the absorption of surface antigens on spermatozoa from the vagina and cervix was reported to stimulate the production of spermatozoal antibodies in the serum as well as in cervicovaginal secretions of women (319–323), which may be involved in infertility (322). Antisperm antibodies against the sperm head, which may prevent sperm from penetrating the cervical mucus to cause immunological infertility, were recently elucidated (323). Intercellular channels in the vaginal epithelium play an essential role in the passage of various immunogens from the lumen to the lamina propria and, finally, to the blood or lymphatic vessels (324). In the lamina propria of the human vagina, there exist several immunoglobulin-producing cells that, depending on the phase of the ovarian cycle, mature to produce and secrete immunoglobulins (IgA and IgG). At midcycle, the concentration of immunoglobulins in cervicovaginal secretions decreases to an extremely low level (323). This decrease is probably due to the preovulatory rise in endogenous estrogens that produces a reduction in the dimension of intercellular channels (325) and the production of cervical mucus (326).

Repeated vaginal application of a plastic sponge impregnated with two model antigens was reported to induce a humoral immune response in rhesus monkey (327), which is weaker than that achieved by systemic immunization. Although there is no specifically developed lymphoid tissue in the vagina like the local immune system found in the gastrointestinal tract, one should recognize the possibility that antibodies induced by chronic administration of a peptide or a protein may neutralize this peptide or protein or even prevent its vaginal absorption (313).

Intravaginal administration of insulin in a formulation containing a nonionic surfactant (polyethylene glycol 1000 monoacetyl ether) and polyethylene glycol 400 to streptozotocin-induced diabetic rats was found to produce a reduction in the blood glucose level by 66.3–48.9% over 1–4 hr (291). However, this hypoglycemic effect was comparatively lower than that attained by the rectal delivery of insulin in the same formulation. Using a formulation containing insulin suspension in a polyacrylic acid aqueous gel base, on the other hand, the vaginal absorption of insulin was observed to produce a rapid and pronounced hypoglycemic effect in alloxan-induced rats and rabbits (328). A dose-dependent change in the plasma levels of both insulin and glucose was observed. Apparently, the difference in pharmacokinetic profiles and pharmacodynamic responses is related to the composition of vaginal preparations: the release of insulin from an aqueous gel formulation was found to be rapid, whereas it was slow and sustained from an oleaginous suppository base.

As reported in Chapter 9 (Section IV.D and Figure 12), the vaginal absorption of insulin in mature female rats was also observed to be affected by the estrous cycle (329). Following the intravaginal administration of insulin in an oleaginous suppository formulation at different stages of the cycle, a slight decrease in the glucose level was observed at the early period of the proestrus stage, whereas a distinct decrease was observed during the estrus stage and a more remarkable decrease during the metestrus and diestrus stages (Figure 28). Similarly, the vaginal absorption of phenol red, a water-soluble marker, was also similarly affected by the cycle. The effect of estrous cycle can be explained by cyclical variations in the structure of vaginal epithelium, including the porelike intercellular pathway. The apparent porosity at the metestrus and diestrus is presumably more than 10 times higher than that during the proestrus and estrus stages (329).

The intravaginal administration of several synthetic LHRH analogs to obtain systemic efficacy was recently investigated in estrogen- or progesterone-blocked ovariectomized rats (330), which have a thin vaginal epithelium. A sustained effect on LH secretion was elicited. Using the same animal model as well as intact immature rats, the vaginal delivery of LHRH and its synthetic analogs, (D-Ala^6,des-Gly^{10})- and (D-Leu^6,des-Gly^{10})-LHRH ethylamide (Figure 3), was observed to induce a greater elevation of serum LH and FSH levels than oral administration. A bioavailability of approximately 1–2% was estimated for vaginal delivery, compared to 0.1% for oral administration (331,332). Further studies on the vaginal delivery of (D-Leu^6,des-Gly^{10})-LHRH ethylamide (leuprorelin) were performed in 10 women by delivering it in a tablet formulation inserted into the posterior fornix of the vagina during the early or midfollicular phase of the cycle (333). Elevation of the plasma levels of gonadotropins, which peaked within 4–6 hr, and of estrogen were obtained. A bioavailability of around 0.6% was achieved. The vaginal delivery of this synthetic LHRH analog was observed to produce a more prolonged response than that induced by SC injection. The results suggested that the vaginal route is useful for the administration of LHRH and its synthetic analogs, especially when a low but long-lasting

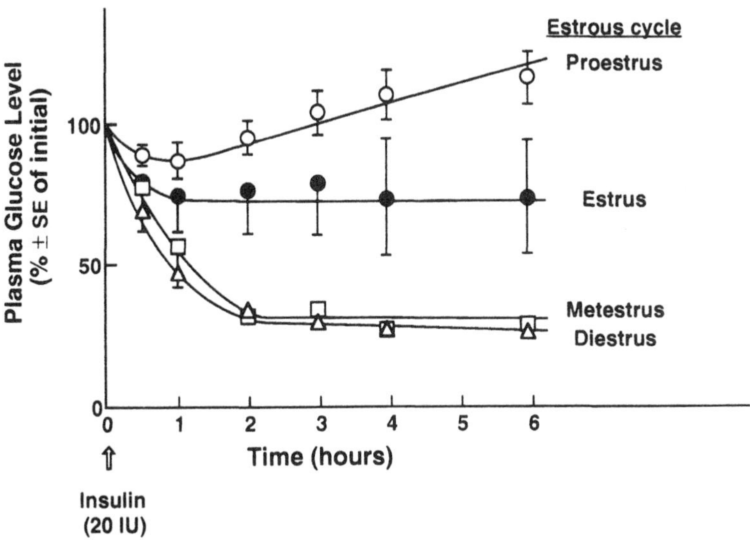

Figure 28 Effect of estrous cycle on the time course of the hypoglycemic effect of insulin (20 IU) in an oleaginous suppository (which contains 10% citric acid) delivered intravaginally in rats (n = 5). The initial plasma glucose level was set as 100%. (Replotted from the data by Okada et al., 1983.)

release of gonadotropins is required, as is the case for the stimulation of follicular maturation and estrogen secretion.

Vaginal delivery of leuprorelin was also investigated for the treatment of LHRH-dependent tumors, endometriosis, and other conditions (329,333–341). Using the dose-response relationship for the ovulation-inducing activity in diestrus rats, it was calculated that the intravaginal administration of leuprorelin achieved an absolute bioavailability of 3.8% from both an oleaginous suppository formulation and an aqueous gel base (with no absorption promotor) which compares favorably with the 0.05% by oral (in an absorption-promoting formulation), 0.11% by nasal, and 1.2% by rectal (Table 5). On the other hand, the pregnancy-terminating efficacy of leuprorelin achieved by intravaginal administration in rats was found to be almost the same as that demonstrated by SC injection (342–344). The observation is probably due to the attainment of a persistent blood level of leuprorelin by vaginal delivery, which could promote artificial abortion. However, the systemic bioavailability of leuprorelin by vaginal delivery may be too low and too variable to be useful clinically.

Using ovulation-inducing activity in diestrous rats, the effect of various additives on the vaginal absorption of leuprorelin from an oleaginous suppository base was investigated (336). The results suggested that the ovulation-inducing activity of leuprorelin is markedly enhanced by the addition of polybasic carboxylic acids (such as glycocholic acid, succinic acid, and citric acid), with the absolute bioavailability increased from 3.8% to around 20% and slightly improved by hydroxycarboxylic acids and acidic amino acids. However, vaginal absorption was found to be poorly enhanced by the addition of fatty acids and their salts, such as oleic acid and sodium oleate, and surfactants, such as polyoxyethylene-9-lauryl ether, even though they demonstrated their absorption-promoting activity in the nasal and rectal absorption of hydrophilic molecules. The observed difference could be due to the difference in biochemical and biophysical properties among various mucosal membranes (Chapter 4). It was surprising to note that the absorption-promoting efficacy of polybasic carboxylic acids was adversely affected when forming sodium salt; for instance, the efficacy was reduced by 5-fold for sodium citrate and 3-fold for sodium glycocholate. On the other hand, the potassium salt of EDTA is more effective than hydroxycarboxylic acids or acidic amino acids. Both citric and succinic acids showed a similar dose-dependent absorption-promoting effect, with the maximal effect achieved at 10% in an oleaginous base (336) and an aqueous solution (337). It was found that the polymer composition in the vaginal jelly formulation also affects the absorption-promoting activity of citric acid. Incorporation of 10% citric acid into an oleaginous suppository formulation was observed to enhance the ovulation-inducing activity of leuprorelin by as much as 30-fold, with an absolute bioavailability of 18% estimated for intravaginal administration. Recently, α-cyclodextrin was discovered to be capable of facilitating the vaginal absorption of leuprorelin in rats by approximately 6-fold, just as it does to the nasal absorption of leuprorelin and insulin in rats and dogs (345).

The intravaginal administration of leuprorelin from a cotton ball presoaked with a 5% citric acid solution (pH 3.5) was observed to produce a high and long-lasting serum level of leuprorelin (Figure 29), which achieved an absolute bioavailability of 25.8% at 6 hr. The chronic administration of a large leuprorelin dose via the vaginal route was noted paradoxically to cause the downregulation of receptors in the pituitary, inhibition of the gonadotropin-releasing response, and functional gon-

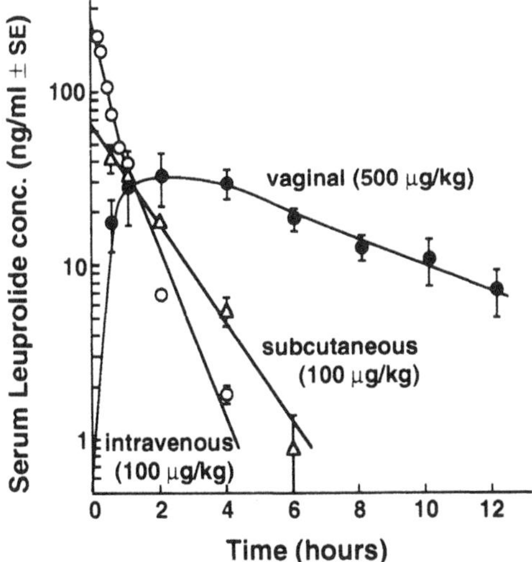

Figure 29 Comparative serum concentration profiles of immunoreactive leuprolide (leuprorelin) following intravaginal (●), subcutaneous (△), and intravenous (○) administration of leuprolide to rats. Leuprolide was delivered intravaginally using a cotton ball presoaked with a solution formulation (pH 3.5) containing 5% citric acid. The biological half-life was 8.4 min for the α phase and 33.2 min for the β phase following intravenous injection. (Replotted from Okada et al., 1984.)

adotrophy (313). As a result, leuprorelin, a potent synthetic LHRH agonist, exerts its therapeutic effects against LHRH-dependent tumors (346), endometriosis (347), and precocious puberty (348). It could also be useful in fertility regulation (349).

The intravaginal continuous administration of leuprorelin was found to induce an elevation of serum LH levels at the beginning of treatment, which is similar to that following subcutaneous injection, and then drastically suppress serum LH levels for a prolonged period (339). This phenomenon is very similar to the results observed earlier in the monthly subcutaneous controlled delivery of goserelin, also a synthetic LHRH analog, from a biodegradable subdermal implant (Zoladex implant, ICI) in the treatment of prostate carcinoma (Chapter 8, Section III.C.1). The investigation was extended to evaluate the effect of the consecutive daily intravaginal administration of leuprorelin (500 μg/kg) in a 5% citric acid jelly (pH 3.5) on LHRH-dependent DMBA-induced mammary tumors in rats (388). A highly significant regression of the tumors was obtained (Table 11).

As demonstrated earlier for insulin (Figure 28), the vaginal absorption of leuprorelin was also reportedly affected by the estrous cycle (Figure 30). The effect of a cyclic variation of the reproductive system apparently caused a substantial fluctuation in the absorption of leuprorelin during the course of continuous therapy. This therapeutic dilemma was overcome by 10-day pretreatment with daily subcutaneous injections of leuprorelin, which was found not only to enhance the vaginal absorption of leuprorelin by more than twofold but also to reduce the effect of cyclic variation observed earlier (Figure 30). The results demonstrate that in the chronic administration of peptide and protein pharmaceuticals for long-term therapy, the fluctuation in

Table 11 Effect of Consecutive Daily Vaginal Delivery of Leuprorelin on LHRH-Dependent Mammary Tumors in Rats

Treatment	Dose (μg/kg/day)	No. initial Tumor[a]	Antitumor activity[b]				No. new tumor
			Growing	Static	Regressing	%[c]	
Control	0	14	71.4	0	28.6	21.4	13
Intraperitoneal	500	15	0	0	100.0	86.7	2
Vaginal[d]	500	19	15.8	0	84.2	52.6	2
	1000	20	10.0	5	85.0	50.0	1
	2500	17	11.8	17.6	70.6	58.8	0
	5000	18	5.6	5.6	88.9	50.0	2

[a]Induced by DMBA (dimethylbenzanthracene).
[b]Percentage of the initial tumor number: growing (increased by at least 10% in mean diameter), static (decreased by less than 10% in mean diameter), regressing (decreased by at least 10% in mean diameter).
[c]Percentage of initial tumor that disappeared.
[d]Administered in aqueous jelly formulation (added with 5% citric acid, pH 3.5).
Source: Modified from Okada (1990).

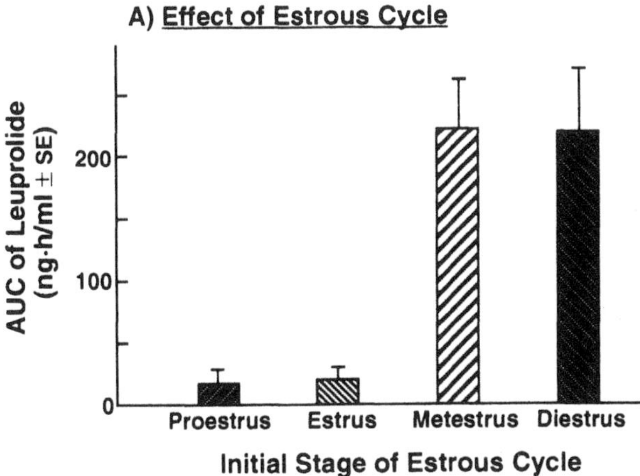

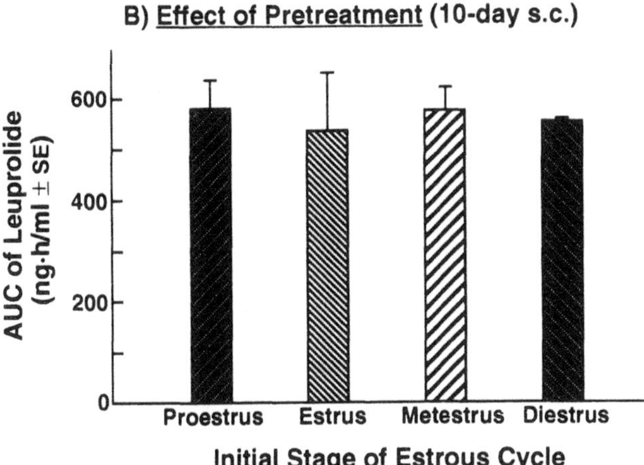

Figure 30 (Top) Effect of estrous cycle on the systemic bioavailability of leuprolide (leuprorelin) in five rats over a 6-hr period following intravaginal administration of leuprolide (500 μg/kg) in a solution formulation (pH 3.5; 0.2 ml) containing 5% citric acid. (Bottom) Effect of a 10-day pretreatment with daily subcutaneous injections of leuprolide on the systemic bioavailability of leuprolide at various stages of the estrous cycle of five rats over a 6-hr period following intravaginal administration. (Replotted from the data by Okada et al., 1984.)

vaginal absorption resulting from cyclic variation should be taken into consideration. The vaginal permeability to peptides and proteins is thus expected to be strongly influenced by variation in the serum estrogen level throughout the life cycle of a woman from birth to menarche and to menopause, as well as during a menstrual cycle. Thus, a strategy must be developed to overcome or minimize the impact on long-term therapy and incorporated into the development of vaginal peptide (or protein) delivery systems (313).

8. *Transdermal Delivery*

Because of the hydrophilicity and large molecular size of peptide and protein molecules as well as the lipophilic barrier properties of the stratum corneum, the idea of delivering therapeutic peptides and proteins through the intact skin has received very limited acceptance because of skepticism about its feasibility. This is probably the reason that not much work has been done in this area of biomedical research. However, a few reports exist in the literature. As early as 1966, Tregear investigated the feasibility of administering proteins and polymers through skin excised from human and animals (350). More recently, Menasche et al. (351) studied the percutaneous absorption of elastin peptides through rat skin and its subsequent distribution in the body. It was found that elastin penetrates the dermis, and 30–40% of the administered dose can be detected in the skin even 48 hr later. Resorption from the skin appeared to be a slow process, with very little or no radioactivity detectable at any time in the blood. Some water-soluble proteins and protein hydrolysates find topical application in cosmetics or toiletries (352–354). For example, collagen has been shown to be an effective moisturizing agent and has been formulated in some skin preparations (353).

One advantage of using the transdermal route for the systemic delivery of therapeutic peptides and proteins is that the skin has a very low proteolytic activity (105,107). Most of the other nonparenteral routes discussed in this chapter have a significant proteolytic enzyme barrier that could drastically reduce the systemic bioavailability of peptides and proteins. However, the skin is known to be much more impermeable than all the absorptive mucosal membranes discussed earlier, particularly to a hydrophilic macromolecules like peptides and proteins. This is indicated by the experience with small peptides, such as thyrotropin-releasing hormone (TRH) (355), a tripeptide molecule with a molecular weight (362 daltons) closely resembling that of the skin-permeable organic-based pharmaceuticals, such as nitroglycerin (Chapter 7), and vasopressin (356–358), a nonapeptide. Both were noted to have great difficulty in permeating the skin barrier. It was found that stripping the stratum corneum, the permeation barrier, from the skin is required to permit the transdermal permeation of vasopressin; treatment of the skin with sodium lauryl sulfate has been found to be ineffective even at a concentration as high as 20% (356). On the other hand, the transdermal permeation rate of vasopressin across intact hairless rat skin was observed to be markedly enhanced by application of a pulsed current delivered from an iontophoretic device (358). The permeation of TRH across nude mouse skin was also facilitated by using dc-iontophoresis (355). Thus, the skin is a potential route for the systemic delivery of peptide- and protein-based pharmaceuticals provided that the penetration of hydrophilic macromolecular peptide or protein through the permeation barrier, the stratum corneum, can be facilitated by a skin permeation-enhancing technique.

Based on experimental results with TRH and vasopressin (355,358), iontophoresis, which facilitates the permeation of ions and/or charged molecules into the body under a stream of electrical current, appears to be a potential noninvasive skin permeation-enhancing technique for achieving the systemic delivery of peptide or protein molecules. A comprehensive review of the principles and technique of iontophoretic drug delivery has been completed very recently (359). In addition to TRH and vasopressin, several studies were also conducted to investigate the feasibility of applying iontophoresis to facilitate the transdermal delivery of protein, like insulin (Figure 8). Contradictory results were obtained in diabetic animals (360,361). A series of systematic investigations were recently carried out to evaluate the systemic delivery of several peptide-based pharmaceuticals, including vasopressin and insulin, by iontophoresis-facilitated transdermal delivery (357,360,362–373).

Review of the scientific literature indicated that the potential of applying iontophoresis as a noninvasive drug delivery technique to facilitate the transdermal systemic delivery of peptide-based pharmaceuticals has been recognized only recently (359), even though it was proposed by Pivati as early as in 1747, as a means to facilitate the transport of ionic species (374). Essentially, two types of iontophoresis have been developed and evaluated: the early type used a simple direct current; while the recent type utilizes a pulsed direct current (or pulse current) as the current source (Table 12). Their application in facilitating the transdermal delivery of therapeutic peptides and proteins is discussed in the following sections.

DC Iontophoresis. In addition to in vitro studies on the iontophoresis-facilitated skin permeation of small peptides, like TRH (355), the feasibility of applying a dc-generating iontophoretic delivery device to facilitate the transdermal systemic delivery of a protein molecule like insulin was recently investigated in live animals by Stephen et al. (360). The results indicated that a highly ionized monomeric form of insulin can be delivered through pig skin by dc iontophoresis with attainment of some systemic effects. However, these investigators could not reproduce the same effects in human subjects. Karl (361) also observed that glucose levels in diabetic rabbits can be effectively controlled by insulin delivered transdermally by a dc-producing iontophoretic delivery device. They found that for an effective delivery of insulin, however, the permeation barrier, the stratum corneum, must first be removed by stripping. On the other hand, Siddiqui et al. (362) used the Phoresor system, a recently marketed dc iontophoretic device, as the power source for the direct current and were able to deliver insulin transdermally to diabetic hairless rats, with attainment of a reduction in hyperglycemia even without stripping of the stratum corneum (Figure 31). The extent of reduction in the blood glucose level from the hypergly-

Table 12 Representative Iontophoretic Delivery Systems Applied to the Transdermal Delivery of Peptide and Protein Drugs

Iontophoretic TDD system	Drug delivery mode
Direct current mode	
Phoresor (Motion Control)	Continuous drug delivery under constant intensity of direct current
Powerpatch (DDS)	
Pulse current mode	
ADIS-4030 (Advance depolarizing pulse iontophoresis system, Advance Co.)	Continuous drug delivery under constant intensity of pulse current
TPIS (transdermal periodic iontotherapeutic system, Rutgers University)	Programmed (continuous or pulsatile) drug delivery under periodic applications of constant pulse current

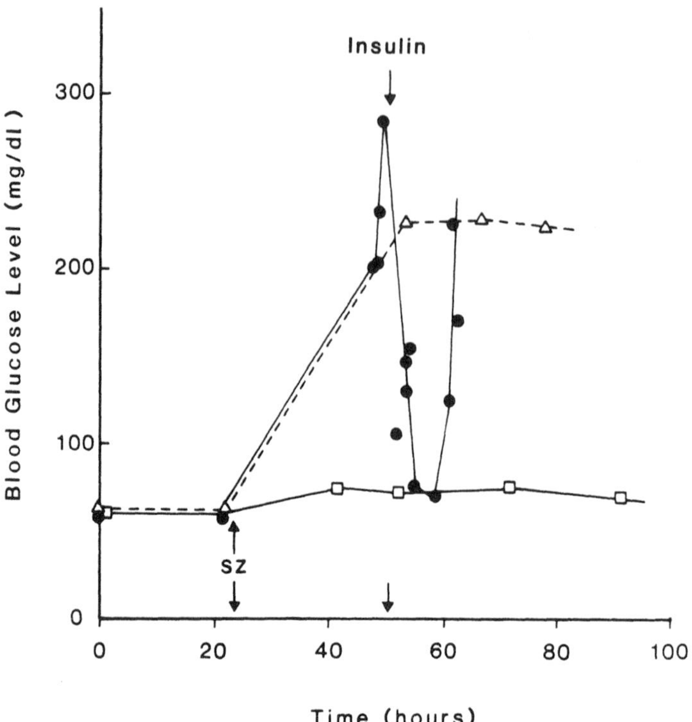

Figure 31 Time course for the reduction in the blood glucose level in diabetic hairless rats from the hyperglycemic state to the nomoglycemic state as a result of the iontophoresis-facilitated transdermal delivery of insulin (at pH 3.68) by an 80-min application of phoresor at a current intensity of 4 mA (equivalent to a current density of 0.67 mA/cm^2). (Reproduced from Chien et al., 1989.)

cemic state was observed to depend upon the solution pH in the insulin reservoir electrode as well as the duration of iontophoresis treatment (Table 13). These investigators also reported the observation of no significant difference between the regular and hairless rats in the extent of reduction in blood glucose levels by insulin (at pH 7.1) delivered transdermally by iontophoresis. In vitro permeation studies using freshly-excised skin from hairless rats demonstrated that the skin permeation of insulin is substantially facilitated by iontophoresis (Figure 32) and the degree of enhancement in the skin permeation rate is pH dependent, with enhanced permeation observed at a solution pH that is lower and higher than the isoelectric point of the insulin molecule.

Recently, a body-wearable dc iontophoresis delivery device, called the Powerpatch applicator, was developed (370) and applied to diabetic rabbits (371,372). Using a direct current of 0.4 mA applied continuously for 14 hr, insulin was successfully delivered transdermally through a negative reservoir electrode. A therapeutically useful serum insulin level was attained, with a peak concentration reached within 4 hr, and a hypoglycemic effect was accomplished at 4 hr and beyond (371). The system was successfully applied to achieve the systemic delivery of human insulin and the control of hyperglycemia in albino rabbits with alloxan-induced acute

Table 13 Effect of Iontophoresis Treatment Duration on Hypoglycemic Activity of Insulin Delivered Transdermally

Duration[a] (min)	% Reduction in Blood Glucose Levels[b]			
	pH 3.7	pH 5.1	pH 7.1	pH 8.0
20	1.02	—	—	—
40	—	—	16.67	—
60	21.43	—	—	—
80	71.15	8.00	54.55	37.50

[a]Application of Phoresor system with applied current of 4 mA (current density = 0.67 mA/cm^2).
[b]Percentage reduction in blood glucose level = (initial blood glucose level − final blood glucose level)/initial blood glucose level × 100%.
Source: Calculated from the data by Siddiqui et al. (1987).

diabetes mellitus (372). The system was also recently used clinically to deliver leuprorelin, a synthetic LHRH agonist (Figure 3), in 13 normal subjects (373). Substantial elevation of serum LH concentrations was produced, with a plateau level attained at 4 hr after continuous treatment with a direct current of 0.2 mA (applied over a patch area of 70 cm^2); however, serum testosterone levels were not significantly altered (Figure 33).

Pulsed dc Iontophoresis. The skin is known to produce a large diffusional resistance to the transport of charged molecules driven by an applied electrical field. The electrical properties of the skin are also reportedly dominated by the least conductive stratum corneum layer. As discussed earlier, the stratum corneum layer is constructed

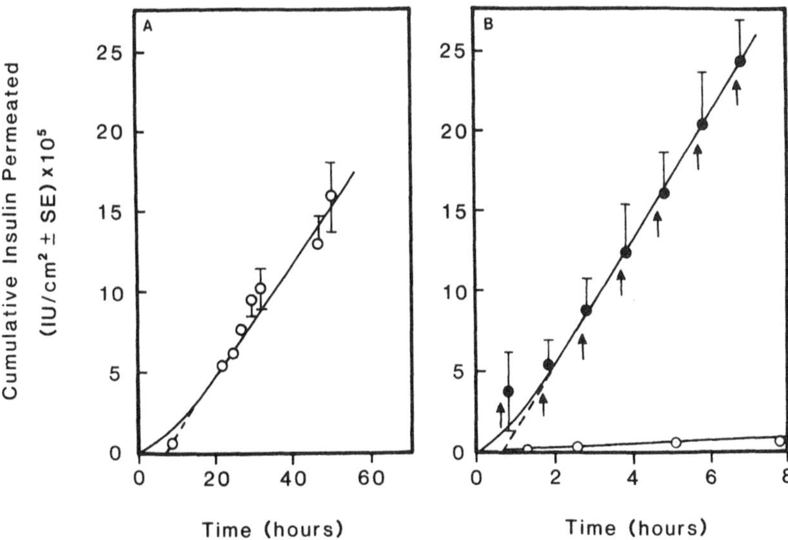

Figure 32 In vitro permeation profile of insulin across hairless rat skin by passive diffusion alone (○) and the enhancement by iontophoresis treatment with a pulse current delivered by a transdermal periodic iontotherapeutic system (●). The current was applied once per hour for 5 min (↑) and repeated periodically for 7 hr.

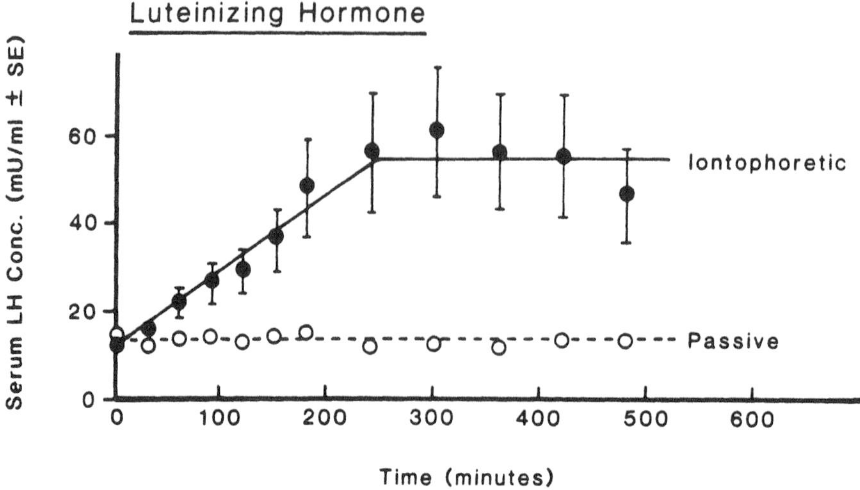

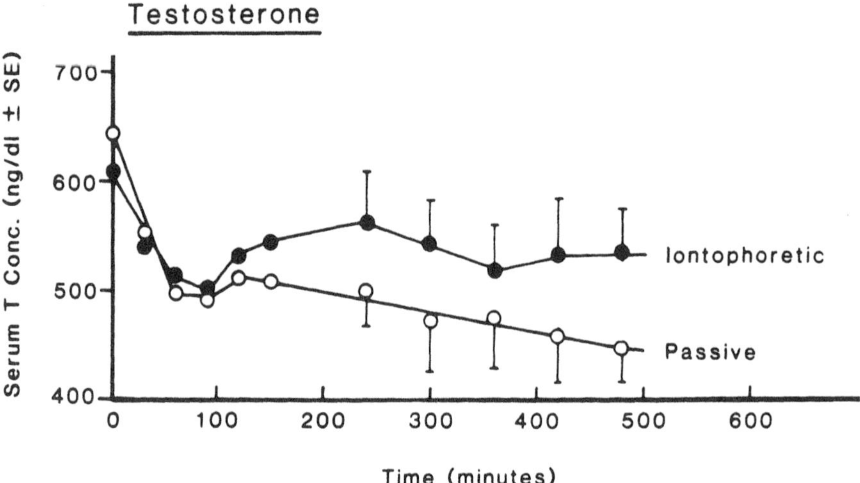

Figure 33 Comparative serum concentration profiles of luteinizing hormone (LH) and testosterone in 13 normal subjects following the transdermal delivery of leuprorelin, a synthetic LHRH analog, by iontophoresis treatment with the direct current (0.2 mA over 70 cm^2) generated by the Powerpatch applicator (●) and by passive diffusion alone (○). (Replotted from the data by Meyer et al. 1988.)

from multilayers of horny cells that are breached by hair follicles and sweat ducts. These skin appendages could act as the pathway for shunt diffusion across the skin. This "shunt" pathway may be significant, especially for charged penetrants that show extremely poor skin permeation via the transcellular route (375). Under the applied electrical field, charged penetrants are driven through the skin, possibly via the shunt pathway and/or the "artificial shunt" pathway in the stratum corneum created by perturbation of the intercellular lipid matrix during iontophoresis treatment, which may disrupt the organization of intercellular lipids (Figure 34).

The stratum corneum has two important electrical features. First, it tends to

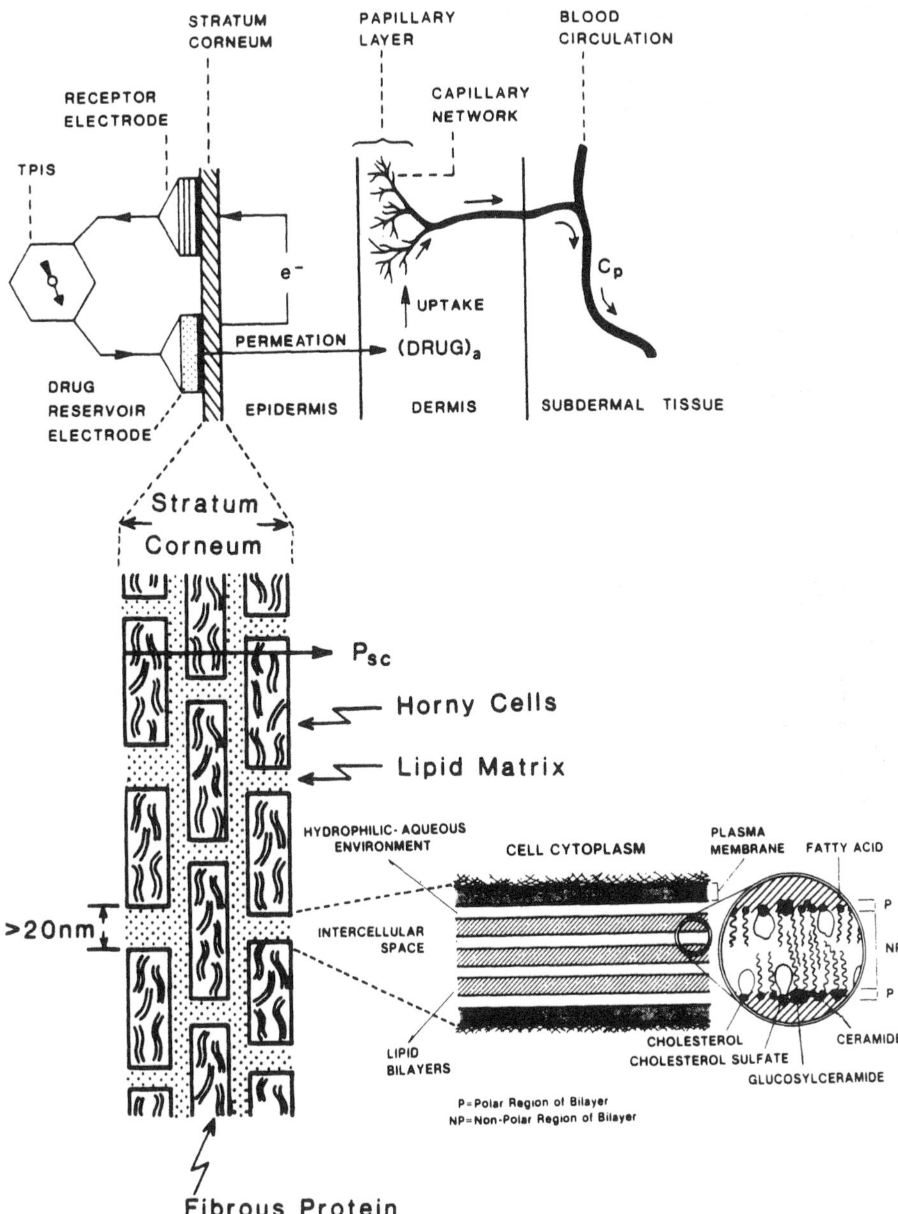

Figure 34 The iontophoresis-facilitated transdermal delivery of charged molecules or ions and in situ application of an iontophoretic delivery device, such as the transdermal periodic iontotherapeutic system (TPIS). An expanded view of the stratum corneum shows its microstructure, in which keratinized horny cells are dispersed as multilayers of bricklike organization in the lipid matrix. (Modified from Chien et al., 1987 and Chien, 1988.)

become polarized as an electrical field is continuously applied. Second, its impedance changes with the frequency of the applied electrical field. Therefore, as an electrical field with direct current is applied in a continuous manner to the stratum corneum to facilitate the transdermal permeation of charged molecules, an electrochemical polarization may occur in the skin. This polarization often operates against the applied electrical field and greatly reduces the magnitude of effective current across the skin (Figure 35). Thus, the current gradient across the skin decays exponentially. Consequently, the efficiency of iontophoresis-facilitated transdermal delivery is reduced as a function of the duration of dc iontophoresis treatment.

To avoid counterproductive polarization, the electrical current should be applied in a pulsatile (or periodic) manner, which is called a pulse current. With the pulse current, the electrical field is switched on and off, alternately, in a periodic pattern. As the electrical field is switched on, charged molecules are delivered iontophoret-

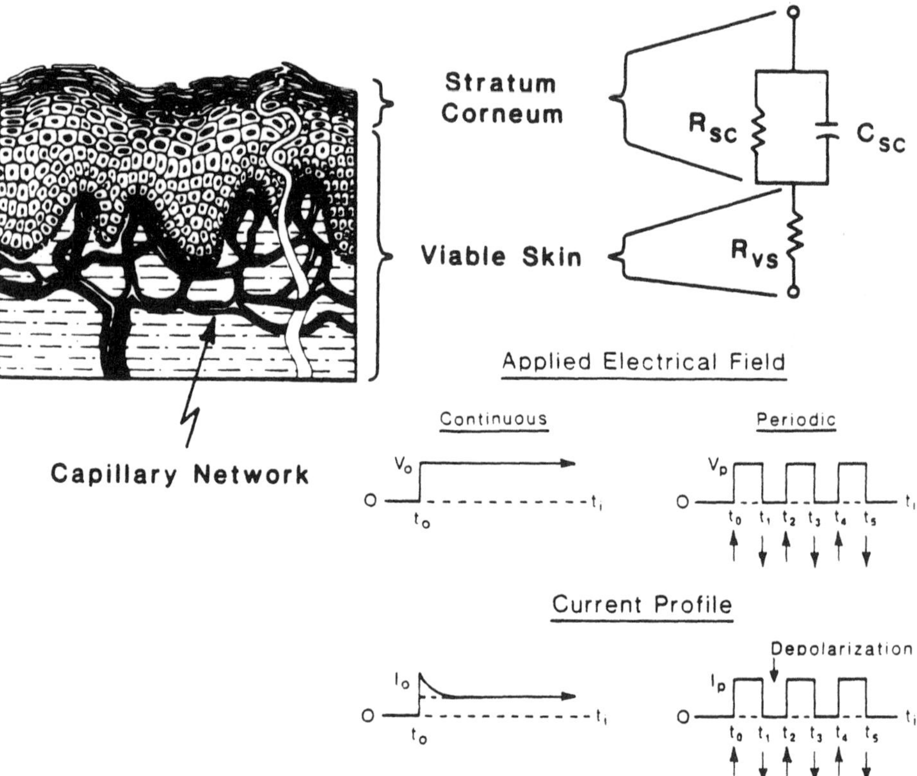

Figure 35 (Top) Analogous equivalent circuit of skin impedance in which R_{vs}, the resistance generated from the deep tissues of the skin, forms a series with a parallel combination of the resistance R_{sc} and the capacitance C_{sc} from the stratum corneum. (Bottom) Current profiles across the skin barrier as a function of the electrical field applied: Periodic versus continuous application. Use of a pulse current allows the skin to depolarize during the off state so that polarization is avoided or minimized and the intensity of the effective current across the skin does not decay exponentially as it does in the continuous application of a direct current. (Replotted from the data by Chien et al., 1987.)

ically into the skin. Before the skin becomes polarized, the electrical field is switched off to relieve the skin and give the skin a chance to depolarize, a process equivalent to discharging the electrical current from the skin (365). Recently, a pulsed dc-generating iontophoretic delivery system, called the advance depolarizing pulse iontophoresis system (Table 12), was introduced. By delivering a pulse current with a 20% duty cycle (4 μsec), followed by an 80% depolarizing period (16 μsec), a β-blocker was successfully delivered systemically to five human subjects without polarization-induced skin irritation (376). With the pulsed dc mode, every new cycle started with no residual polarization remained in the skin from the previous cycle, if the proper frequency was selected.

With the theoretical foundation just outlined, a transdermal periodic iontotherapeutic system (TPIS) was designed (365). It is capable of delivering the pulsed direct current with variable combinations of waveform, frequency, on/off ratio, and current intensity for a specific duration of treatment (Figure 36). The TPIS has been extensively evaluated for potential applications in facilitating the transdermal systemic delivery of several therapeutically-important peptides and proteins. The results of some representative investigations are summarized as follows.

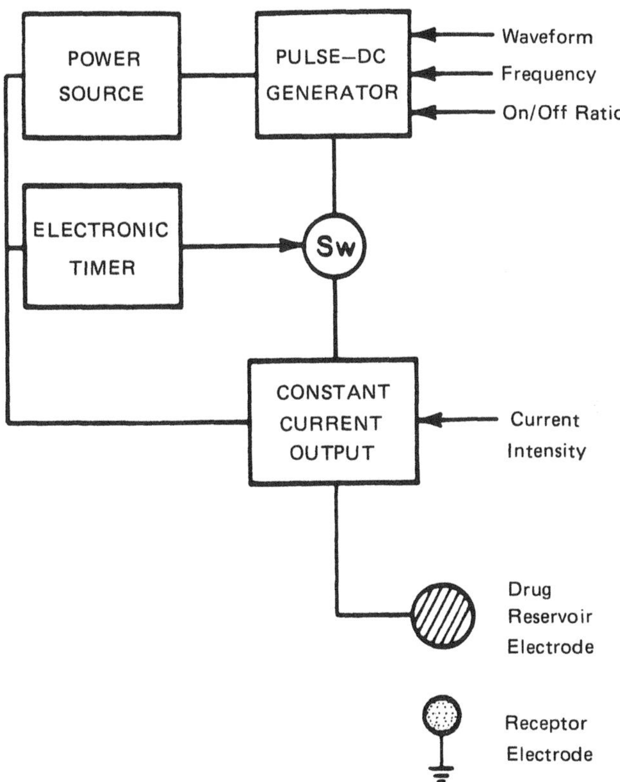

Figure 36 The transdermal periodic iontotherapeutic system (TPIS). The pulsed direct current generated with a preset combination of waveform, frequency, on/off ratio, and intensity is delivered via the drug reservoir electrode to the site of application. (Adapted from Chien et al., 1989.)

Transdermal delivery of therapeutic peptides: The feasibility of applying the pulse current generated by TPIS to facilitate the transdermal delivery of therapeutic peptides was investigated using vasopressin (Figure 15), an antidiuretic peptide, as the model therapeutic peptide.

To gain a better understanding of the mechanisms involved in the pulsed dc iontophoresis-facilitated transdermal delivery of peptides, a series of systematic in vitro skin permeation studies was conducted. The skin permeation profiles shown in Figure 37 demonstrate that under passive diffusion (without TPIS treatment), the skin permeation rate of the positively-charged vasopressin molecule is extremely low (0.94 ± 0.62 ng/cm$^2 \cdot$ hr), with a lag time of 9.12 ± 1.06 hr. Following the periodic application of TPIS, the lag time is reduced to less than 0.5 hr and the skin permeation rate of vasopressin ions is facilitated substantially as much as 190 times (178.0 ± 25.0 ng/cm$^2 \cdot$ hr). The pulsed dc iontophoresis-facilitated skin permeation profile appears to consist of two phases: (i) the activation phase, in which a pulse current with a physiologically-acceptable intensity is applied periodically from the TPIS, and (ii) the postactivation phase, in which no TPIS treatment is applied (358). The skin permeation profiles in Figure 37 appear to suggest that after termination of the TPIS treatment, the barrier properties of the skin appear to be recovered as shown by the gradual return of the skin permeation profile to that under passive diffusion alone. The enhancement of the skin permeation rate and the reversibility of skin

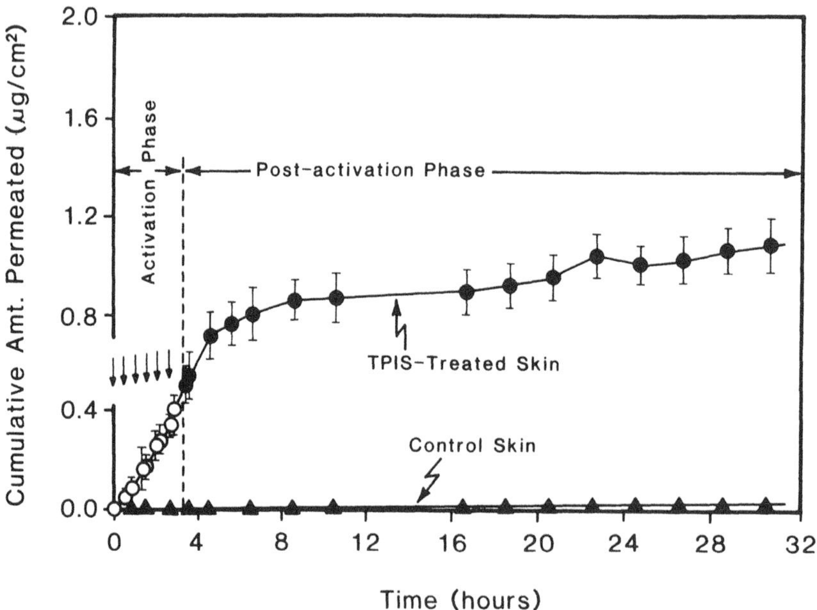

Figure 37 In vitro permeation profiles of vasopressin, from donor solution at pH 5, across the abdominal skin of hairless rats under passive diffusion (▲) and under iontophoresis-facilitated diffusion (●) by the pulse current (1 mA) generated by TPIS, which was switch on for 10 min (↓) and then off for 30 min, cyclically, six times for a total treatment time of 1 hr over an Activation phase of 4 hr (0). (Adapted from Chien et al., 1989.)

permeability are dependent upon the intensity and duration of the pulse current applied (Table 14).

The pulse current generated by TPIS can also be applied in a continuous manner to the skin surface for a specific duration of application (Figure 38). Again, the cumulative amount of vasopressin permeating the skin increases with time during the period of TPIS treatment and gradually returns to the skin permeation profile of passive diffusion after termination of the treatment. Also shown in Figure 38 is the rate profile of vasopressin permeation. Analysis of the skin permeation rate profile suggests that the pulsed dc iontophoresis-facilitated transdermal transport of peptide molecules can be characterized by four phases: (i) the facilitated-absorption phase, in which the skin permeation of peptide molecules is facilitated by iontophoresis treatment and the skin permeation rate linearly increases with the duration of treatment; (ii) the equilibrium phase, in which the skin permeation rate has reached the plateau level even though iontophoresis treatment is still applied without interruption; (iii) the desorption phase, in which the skin permeation rate decreases linearly with time immediately after termination of iontophoresis treatment (posttreatment period), in which the peptide molecules that have already permeated into the skin tissues are gradually desorbed into the receptor solution; and (iv) the passive diffusion phase, in which the skin permeation rate returns to the baseline level as defined by passive diffusion. The skin permeation rate of vasopressin as facilitated by TPIS treatment was observed to increase linearly with the density of pulsed current applied, which also produces a linear increase in the cumulative amount of vasopressin permeating the skin (Figure 39). The data in Figure 40 indicate that although the cumulative amount of vasopressin permeating the skin increases linearly as the duration of TPIS application time is prolonged, the skin permeation rate of vasopressin increases proportionally, but not in a linear manner, with the duration of the application time (357).

To evaluate the possibility of applying TPIS many times to facilitate the transdermal delivery of peptide drugs, studies were carried out to investigate the effect of multiple TPIS treatments on the skin permeation profiles of vasopressin (368). The results in Figure 41 demonstrate that the cumulative permeation profiles of va-

Table 14 Effect of TPIS Treatment on Skin Permeation Profiles of Vasopressin

Conditions		Skin permeation profile ($\bar{x} \pm$ SD)		
			Permeation rate[a] (ng/cm^2-hr)	
Treatment	Current density (mA/cm^2)	Lag time (hr)	A	P
Control	0.0	9.12 ± 1.06	0.94 ± 0.62	
TPIS-treated[b]	0.78	<0.5	116.2 ± 10.7	0.7 ± 0.4
	1.56	<0.5	178.0 ± 25.0	5.3 ± 0.5

[a]In vitro permeation across freshly-excised hairless rat skin mounted in the modified V-C skin permeation cell. Activation phase (A): treatment with TPIS for a duration of 10 min followed by a 30-min no-application period. Treatment was repeated cyclically six times. Postactivation phase (P): after termination of TPIS treatment.

[b]Pulse current (frequency, 2 kHz; on/off ratio, 1:1).

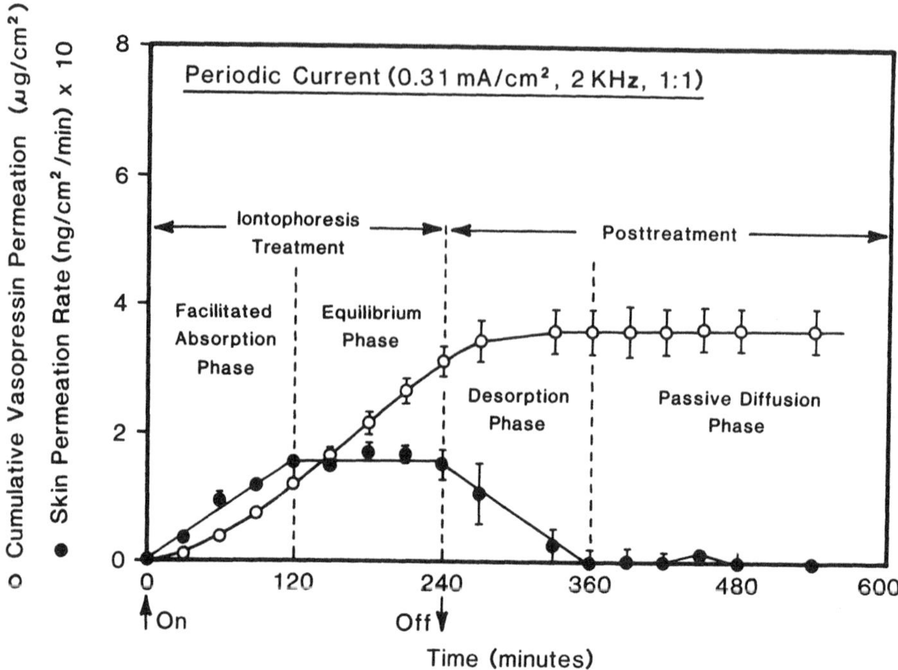

Figure 38 In vitro skin permeation profiles of vasopressin under iontophoresis-facilitated diffusion (○) by the pulse current (0.31 mA/cm^2) generated by TPIS, which was applied continuously for 240 min. The skin permeation rate profile (●) is also shown to illustrate various phases of iontophoretic transdermal delivery.

sopressin and the corresponding skin permeation rate profiles remain very much the same in both magnitude and time course among the multiple TPIS treatments. Further analysis of the skin permeation data indicates that for all three treatments, the skin permeation rate of vasopressin increases as the duration of the application time is extended, which reaches the plateau level at 40 min (Figure 42); beyond the critical period of 40 min, the skin permeation rate levels off for the first two treatments but begins to drop for the third treatment. The results apparently suggest that 40 min is the optimum duration for multiple TPIS applications without affecting the reversibility of the skin's barrier properties.

The iontophoresis-facilitated transdermal delivery of vasopressin was also conducted in rabbits to compare the relative efficiency of its antidiuretic activity between pulsed direct current and simple direct current (357). The results in Figure 43 demonstrate that 40 min of iontophoresis treatment with the pulsed direct current (delivered by TPIS) produced an antidiuretic activity approximately twofold greater than that achieved by the simple direct current (delivered by Phoresor system) when the same current density (0.22 mA/cm^2) is applied.

Transdermal delivery of therapeutic proteins: The feasibility of applying the pulse current generated by TPIS to facilitate the transdermal delivery of therapeutic proteins was investigated using insulin, a hypoglycemic protein (Figures 1 and 8), as the model therapeutic protein.

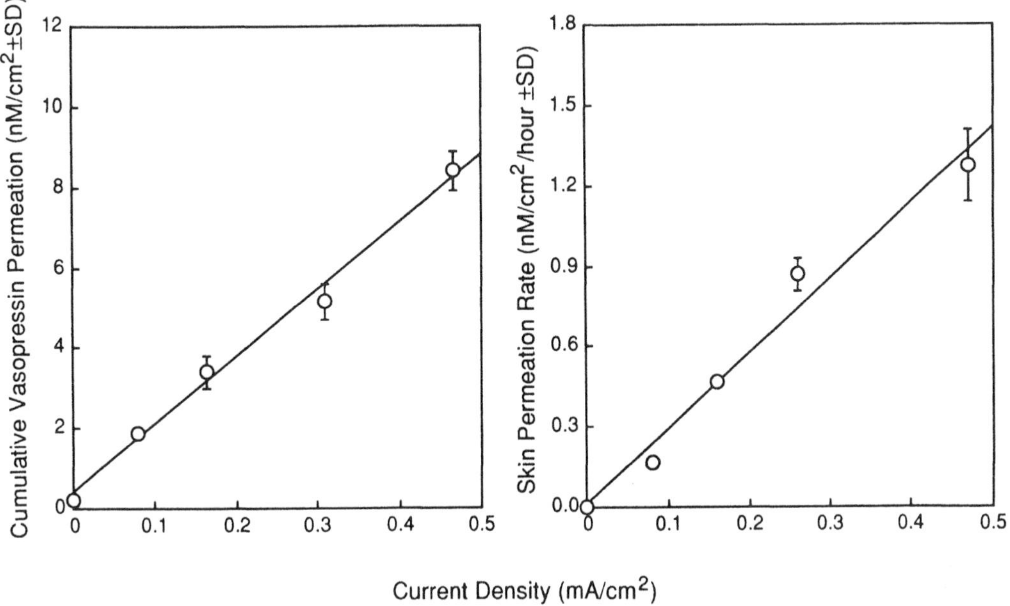

Figure 39 Linear dependence of the cumulative amount of permeation and the skin permeation rate of vasopressin on the current density of pulse current applied to hairless rat skin.

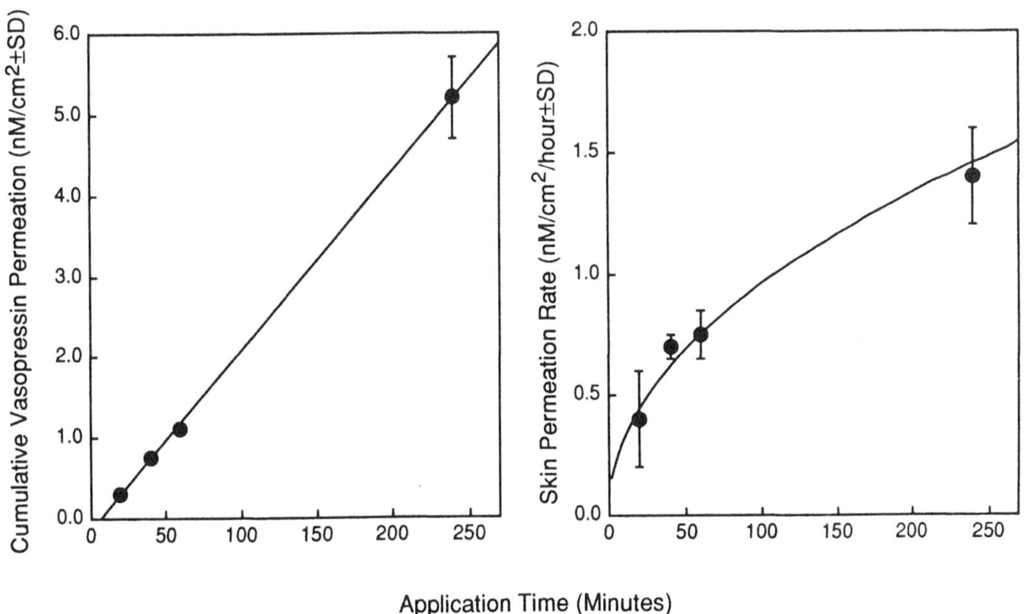

Figure 40 Effect of TPIS application time on the cumulative amount of permeation and the skin permeation rate of vasopressin.

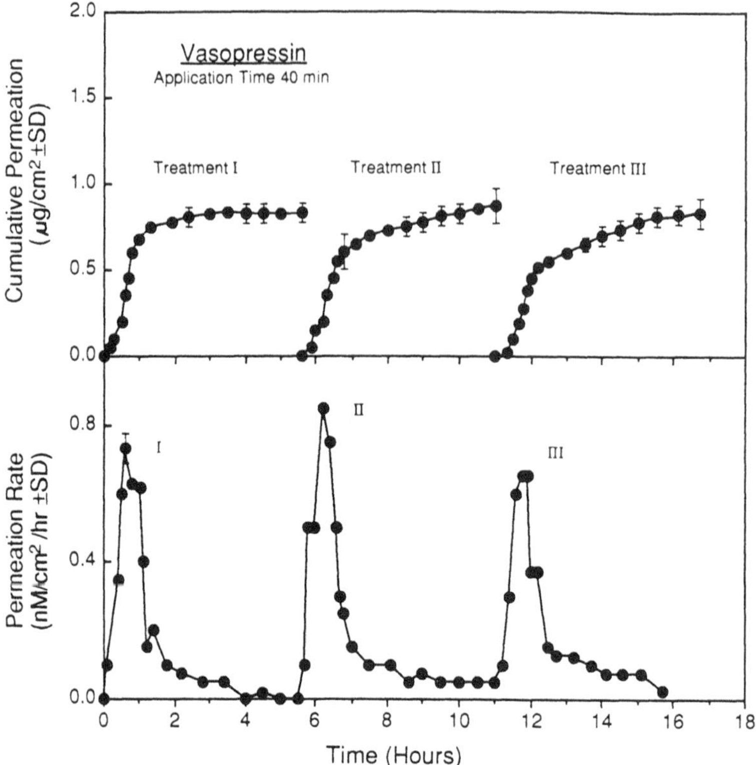

Figure 41 Cumulative permeation profiles (top) and skin permeation rate profiles (bottom) of vasopressin following the consecutive triplicate application of TPIS treatment with a pulse current of 0.3 mA for 40 min each.

To investigate the efficiency of iontophoresis-facilitate transdermal permeation in enhancing the systemic delivery of therapeutic proteins, in vivo studies should be performed using an appropriate animal model. In the case of insulin, a diabetic animal model should be used. By parenteral administration of streptozotocin, a diabetic model was successfully developed in both hairless rats and New Zealand white rabbits (364). Streptozotocin was found to produce extensive damage of the beta cells and necrosis of the Langerhans islet in the pancreas (Figure 5), leading to elevation of blood glucose levels from the normoglycemic state of around 80 mg/dl to the hyperglycemic state of higher than 200 mg/dl (Figure 44). This hyperglycemic state was found to remain fairly stable for around 1 week in diabetic hairless rats but for only 1 day in diabetic rabbits. It was unaffected by fasting, anesthesia, or TPIS treatment with placebo formulation having no insulin in the reservoir electrode (Figure 45).

The pulse current generated by TPIS is characterized by four features: waveform, frequency, on/off ratio, and intensity (Figure 36). All these system parameters must be characterized and optimized to maximize the efficiency of TPIS in facilitating the transdermal delivery of peptides or proteins. Current intensity controls the amount

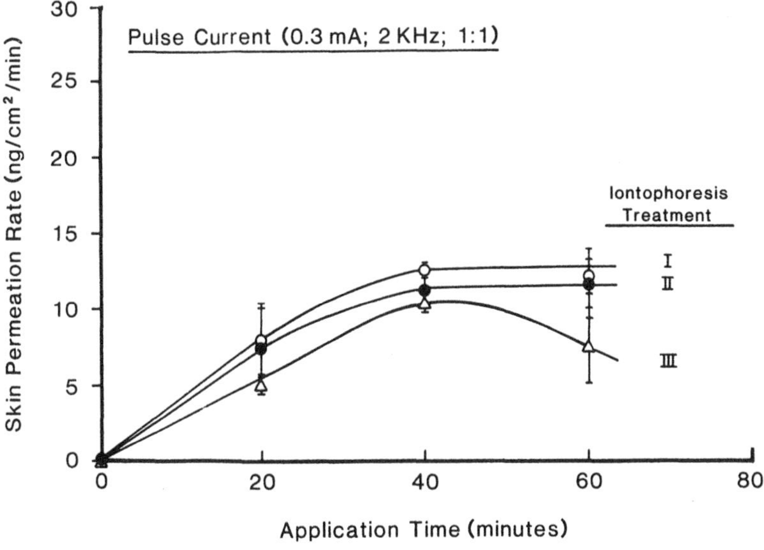

Figure 42 Relationship between the skin permeation rate of vasopressin and application time as well as the sequence of TPIS treatment.

and rate of charged macromolecules of peptide or protein penetrating the skin (Faraday's law), and the waveform, frequency, and on/off ratio of the pulse current determine its effectiveness in overcoming the impedance of the least conductive stratum corneum. A series of systemic in vivo studies carried out in diabetic hairless rats demonstrated that the time course for the reduction of hyperglycemic levels is dependent upon the waveform (Figure 46) frequency (Figure 47), and on/off ratio (Figure 48) of the pulse current applied as well as the duration of TPIS application (364). The data in Figures 46 through 48 indicate that the onset, extent, and duration

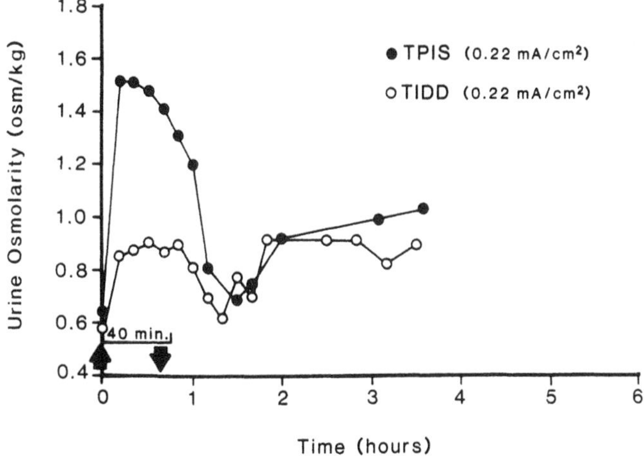

Figure 43 Urine osmolarity profiles in rabbits following the iontophoresis-facilitated transdermal delivery of vasopressin using the pulse current (●) from TPIS and the direct current (○) from the Phoresor system.

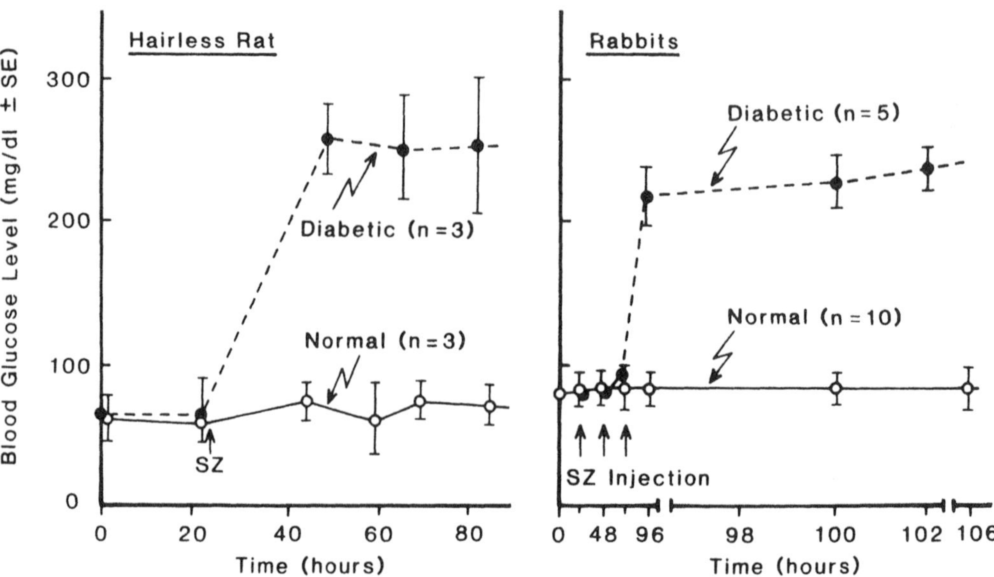

Figure 44 Successful induction of the hyperglycemic state in hairless rats and rabbits, as animal models for diabetic treatment studies, by the parenteral administration of streptozotocin (SZ). (Adapted from Chien et al., 1987.)

of hypoglycemic effect, an indicator of the efficiency of the systemic delivery of insulin by TPIS-facilitated transdermal transport, are varied. They can be tailored to meet treatment needs by selecting a specific combination of waveform, frequency, and on/off ratio of the pulse current.

To provide a direct assessment of the systemic delivery of insulin following TPIS-facilitated transdermal permeation and to study its relationship with pharmacodynamic responses, in vivo studies were also conducted in diabetic rabbits. Pharmacokinetic and pharmacodynamic profiles were established by taking blood samples from the marginal ear vein at regular time intervals and then simultaneously assaying glucose levels in the blood by glucose analyzer and the insulin concentration in the plasma by radioimmunoassay (358).

The results in Figure 49 illustrate that during the 40-min TPIS treatment, insulin is rapidly delivered transdermally, with the peak plasma concentration (0.72 mIU/ml) of immunoreactive insulin attained within 30 min. On the other hand, subcutaneous insulin takes almost 2 hr to reach the same peak level. The subcutaneous dose maintains the elevated plasma insulin level for a sustained period, but the plasma insulin profile resulted from TPIS-facilitated transdermal delivery is rather short-acting in duration. In response to the difference in the pharmacokinetic profiles of insulin between transdermal and subcutaneous delivery, pharmacodynamic responses also show a characteristic difference in hypoglycemic effect (Figure 50). The blood glucose levels drop rapidly in both cases, but the glucose concentration continues to decline following the conventional subcutaneous insulin injection, even after the normoglycemic level has been reached, while it stays slightly above the normoglycemic level after transdermal insulin delivery. The observed hypoglycemic profiles are in good agreement with the pharmacokinetic profiles (366).

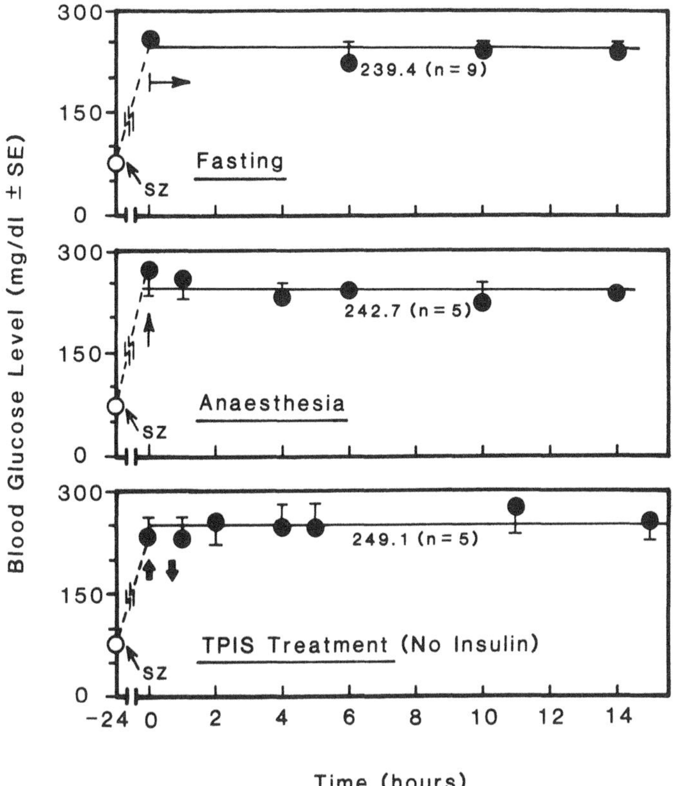

Figure 45 Stable hyperglycemic state in diabetic hairless rats, which was unaffected by fasting, anesthesia, or treatment with the transdermal periodic iontotherapeutic system (TPIS, with a reservoir electrode containing an insulin-free placebo formulation).

The efficiency of the pulse current in facilitating the systemic delivery of insulin is also compared with the simple direct current by delivering both current modes at the same intensity (e.g., 1 mA) from the same iontophoretic delivery device (357). The results in Figure 51 indicate that the systemic delivery of insulin is rapidly achieved by the pulse current, with the peak plasma concentration of immunoreactive insulin attained within 10 min; a much lower plasma insulin level is yielded by the simple direct current. The pharmacodynamic profiles also show that the higher systemic delivery of insulin by pulse current achieved a greater reduction in blood glucose levels and sustained hypoglycemic effect than that by direct current.

The effectiveness of TPIS in facilitating the systemic delivery of insulin via the transdermal route was also compared with the Phoresor system (358). The results (Figure 52) generated in diabetic rabbits demonstrate that by applying the pulse current from the TPIS at 1 mA for 40 min, the transdermal delivery of insulin was rapidly achieved, with the peak plasma level reached within 30 min compared to the 1–2 hr required for the Phoresor system (which applies simple direct current at a current intensity of 4 mA for a treatment duration of 80 min). In comparison, pharmacodynamic profiles show that the rapid systemic delivery of insulin by the TPIS

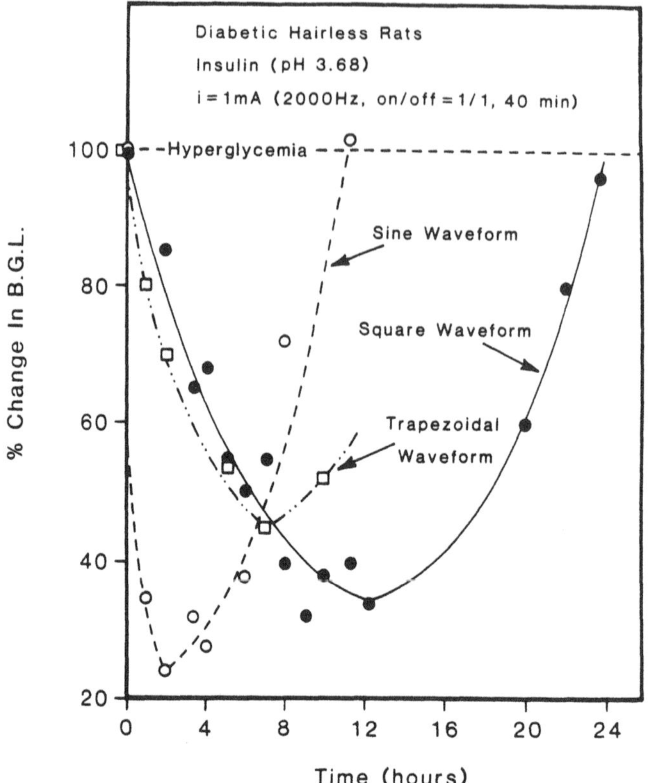

Figure 46 Effect of various waveforms of pulse current on the onset, extent, and duration of the hypoglycemic effect of insulin delivered transdermally by TPIS. (Modified from the data by Chien et al., 1987.)

produced a more prompt reduction in the blood glucose level, but for shorter duration, than the Phoresor system.

In summary, macromolecular drugs like insulin (a protein molecule) and vasopressin (a peptide molecule) can be successfully delivered through the intact skin by applying iontophoresis as a noninvasive drug delivery technique. Using proper animal models, such as diabetic hairless rats and rabbits, the systemic bioavailability of therapeutic peptides and proteins by iontophoresis-facilitated transdermal delivery can be quantitatively assessed by simultaneously monitoring pharmacokinetic profiles and pharmacodynamic responses. Comparative studies have also demonstrated that therapeutically effective systemic levels of peptide and protein pharmaceuticals can be achieved by transdermal delivery with application of a iontophoretic delivery device. The results also confirm that the iontophoresis-facilitated transdermal delivery of peptide or protein molecules can be more efficiently accomplished by pulsed dc iontophoresis than by conventional dc iontophoresis.

A survey of the literature has suggested that an enzyme-responsive transdermal delivery system, which could be viewed as an ideal insulin delivery device, has been conceptualized and envisioned for noninvasive insulin delivery (377). In concept, it consists of an insulin reservoir device, which is attached to the skin, that would

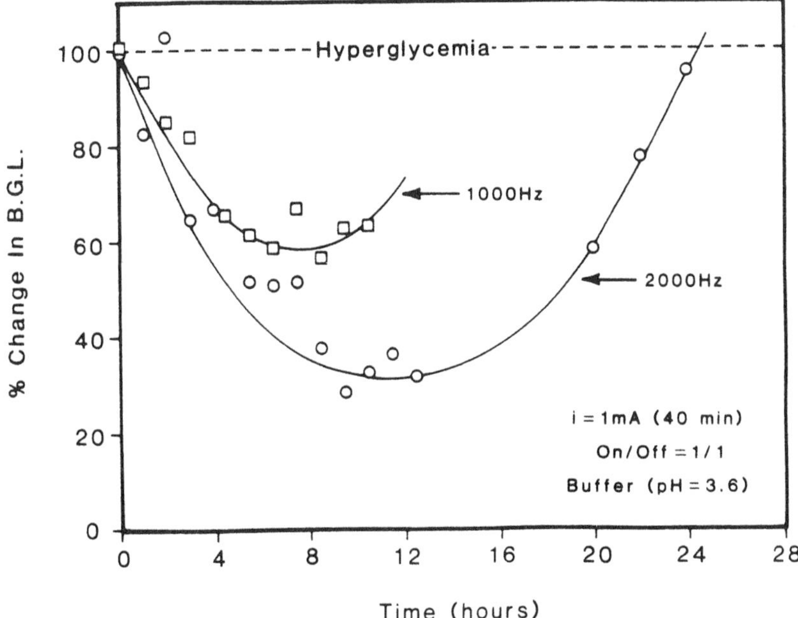

Figure 47 Effect of the frequency of pulse current on the onset, extent, and duration of the hypoglycemic effect of insulin delivered transdermally by TPIS. (Modified from the data by Chien et al., 1987.)

generate a minute pulse of electricity to temporarily open the skin pores, and in the meantime, the device would sample the blood and process it through a glucose oxidizing enzyme, by which the device would monitor physiological indicators, such as glucose level, and adjust the release of insulin accordingly. A second brief electrical pulse would be applied to open the skin pores again for delivering a therapeutic dose of insulin into the body.

C. Systemic Delivery and Enzymatic Barriers

A major challenge in the systemic delivery of peptide- and protein-based pharmaceuticals is to overcome the enzymatic barrier that limits the amount of therapeutic peptide (or protein) molecules from reaching their target tissues. Degradation usually begins at the site of administration and can be extensive. Even by parenteral (IM or SC) administration, less than complete bioavailability can be achieved. For instance, the SC and IM administration of thyrotropin-releasing hormone, a simple and small tripeptide molecule, in mice has not achieved a total systemic bioavailability but 67.5 and 31.1%, respectively (378). On the other hand, the subcutaneous administration of leuprorelin, a synthetic stable analog of LHRH (Figure 3), and of insulin, a structurally more complex protein molecule (Figure 8), has attained a bioavailability of 65 and 80%, respectively (379,380). Furthermore, proinsulin, which is converted to insulin in the beta cell by the action of two proteases (Figure 5) during the maturation of secretory vesicles (381,382), was found to degrade extensively in subcutaneous tissues to four metabolites; none of them was insulin (383).

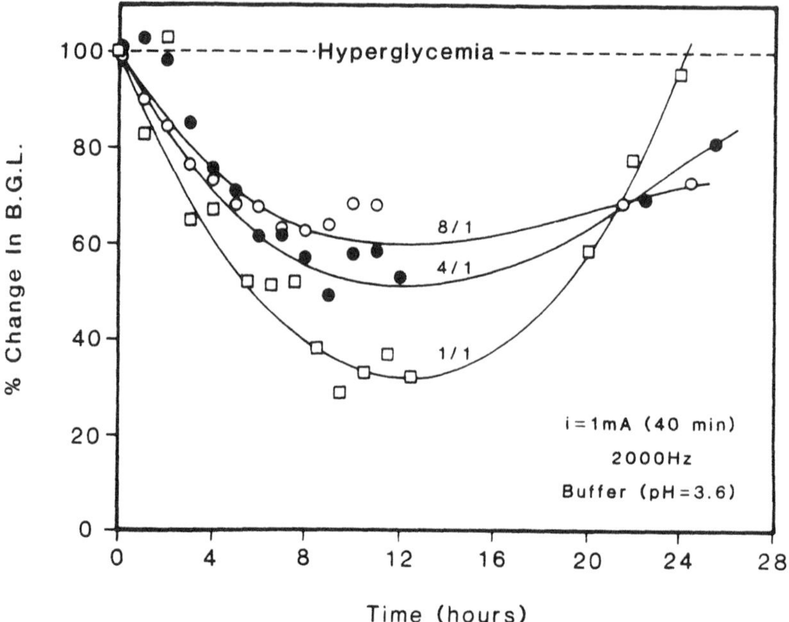

Figure 48 Effect of the on/off ratio of pulse current on the onset, extent, and duration of the hypoglycemic effect of insulin delivered transdermally by TPIS. (Modified from the data by Chien et al., 1987.)

Peptide- and protein-based pharmaceuticals are subjected to degradation by numerous enzymes (or enzyme systems) throughout the body. This degradation can occur in either one of the following two ways: (i) proteolysis, which is the hydrolytic cleavage of peptide bonds by proteases, such as enkephalinases and insulin-degrading enzyme, and (ii) biochemical modification of the peptide (or protein) molecule, which causes the molecule to denature by aggregation or fragmentation, such as oxidation by glucose oxidase and phosphorylation by kinases. However, proteolysis is by far the more common. There is usually a rate-limiting event in degradation. Beyond this initial event, peptide or protein molecules are rapidly degraded to small peptides and then further metabolized to amino acids.

The enzymatic barrier for the transmucosal permeation of peptides and proteins has been investigated. For example, enkephalins are naturally-occurring pentapeptides that act as neurotransmitters or neuromodulators in pain transmission (384,385) and thus have analgesic properties. However, their analgesic activity is rather short in duration, with a half-life as short as less than 1 min, and remains purely transient even by intracerebroventricular administration. This rapid loss in activity has reportedly resulted from their rapid inactivation by enzymes present in plasma (386,387), brain membrane and brain homogenate (388,389), calf and bovine brain (390,391), rat and mouse brain extracts (392), and other tissues (393–395). The results accumulated to date have suggested that several enzymes are responsible for the degradation of enkephalins, and their relative contribution to the total amount of enkephalins metabolized varies from one enzyme to another. Differences also exist from

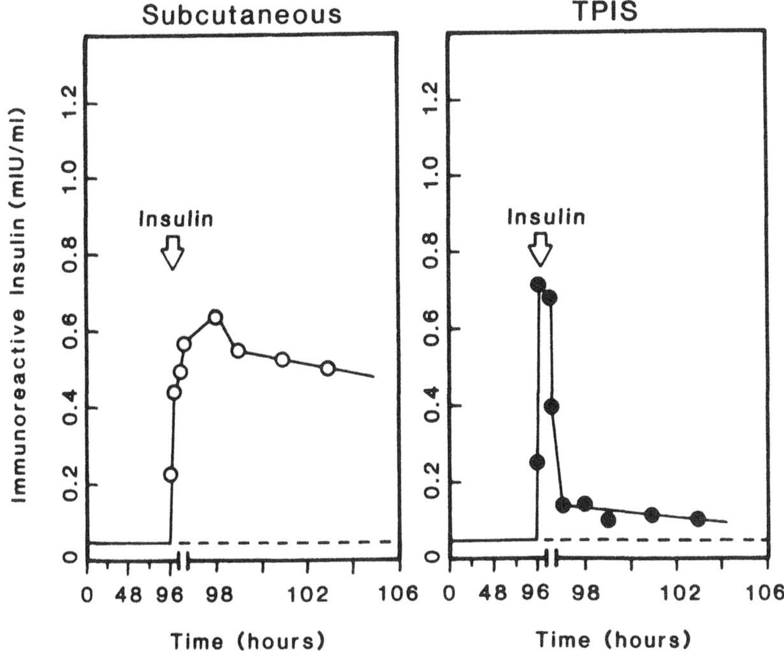

Figure 49 Comparative plasma concentration profiles of immunoreactive insulin following the subcutaneous administration and TPIS-facilitated transdermal delivery of insulin.

one species to another and even among various tissues or preparations within a given species.

The hydrolysis of enkephalins was studied in homogenates of various rabbit mucosal tissues, and the results indicated that enkephalins are most rapidly hydrolyzed in buccal and rectal homogenates, followed by nasal and then vaginal homogenates, but the difference in the rate of hydrolysis is small (396). Aminopeptidases were found to be the major enzymes responsible for the hydrolysis of both leucine- and methionine-enkepthalins, to which dipeptidylpeptidase and dipeptidylcarboxypeptidase contributed to a much lesser extent. In anterior segment tissues homogenates from the albino rabbit eye, these enkephalins were found to be equally susceptible to hydrolysis. Peptidases were also involved in their hydrolysis in these ocular tissues in a similar manner as in mucosal homogenates (397). These results led the investigators to conclude that a similarity exists in the magnitude of enzyme activities among various mucosal routes in terms of the rate constant for the hydrolysis of methionine-enkephalin in homogenates of various mucosal tissues (398). In contrast, dipeptidylcarboxypeptidase was found to be the primary enzyme responsible for the hydrolysis of $[\text{D-ala}]^2$-Met-enkephalinamide, a synthetic analog of methionine-enkephalin, in homogenates of these nonoral mucosae; it was designed to be more resistant to hydrolysis by aminopeptidase (396,397). It has also been reported that in the rat, endopeptidase activity toward $[\text{D-ala}]^2$-Met-enkephalinamide is significantly lower in buccal than in intestinal homogenates; in the hamster there is no significant difference in activity between buccal and intestinal homogenates (399).

The enzymatic barrier for the permeation of peptides and proteins across various

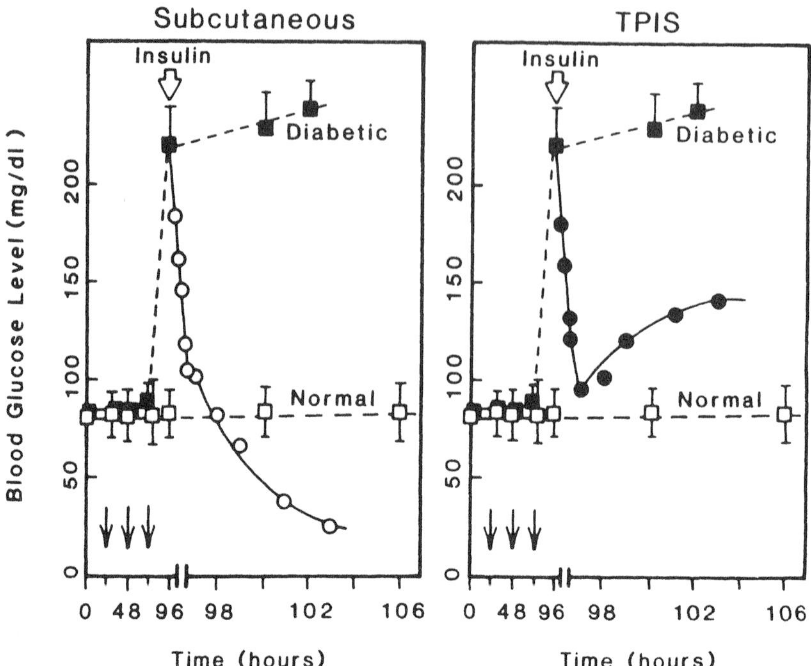

Figure 50 Comparative blood concentration profiles of glucose following the subcutaneous administration and TPIS-facilitated transdermal delivery of insulin.

absorptive mucosae was recently characterized using mucosal extracts, not homogenates, to better simulate conditions in vivo (400). The extracts were prepared by separately exposing the mucosal and serosal surfaces of various freshly excised mucosae from albino rabbits to the buffered solution at physiological pH for 24 hr. Leucine- or methionine-enkephalin was then added to these mucosal and serosal extracts and incubated at 37°C for 6 hr. Samples were taken and assayed by high-performance liquid chromatography for intact enkepthalin and its degradation products. The enzymatic degradation profiles of methionine-enkepthalin are shown in Figures 53 and 54.

The results in Figure 53 susggest that the enzymatic degradation profiles of methionine-enkephalin can be described by first-order kinetics process. The rate of degradation varies from one mucosa to another, with a rank order: rectal > vaginal > nasal mucosa. The apparent rate constants for the enzymatic degradation of methionine-enkephalin are compared in Table 15, which indicates that the rate constants for the enzymatic degradation of enkephalin in the mucosal and serosal extracts are not substantially different. On the other hand, hydrolytic degradation, without the involvement of extracted enzymes, is insignificant at these pH levels, with a rate constant of $3.84–5.87 \times 10^{-7}$ min^{-1}.

By monitoring the formation of various degradation products, one may characterize the enzymatic activity in various absorptive mucosae. The results in Figure 54 suggest that aminopeptidase, which is known to be the enzyme responsible for the breakdown of enkephalins by hydrolyzing the Tyr-Gly bond with the formation

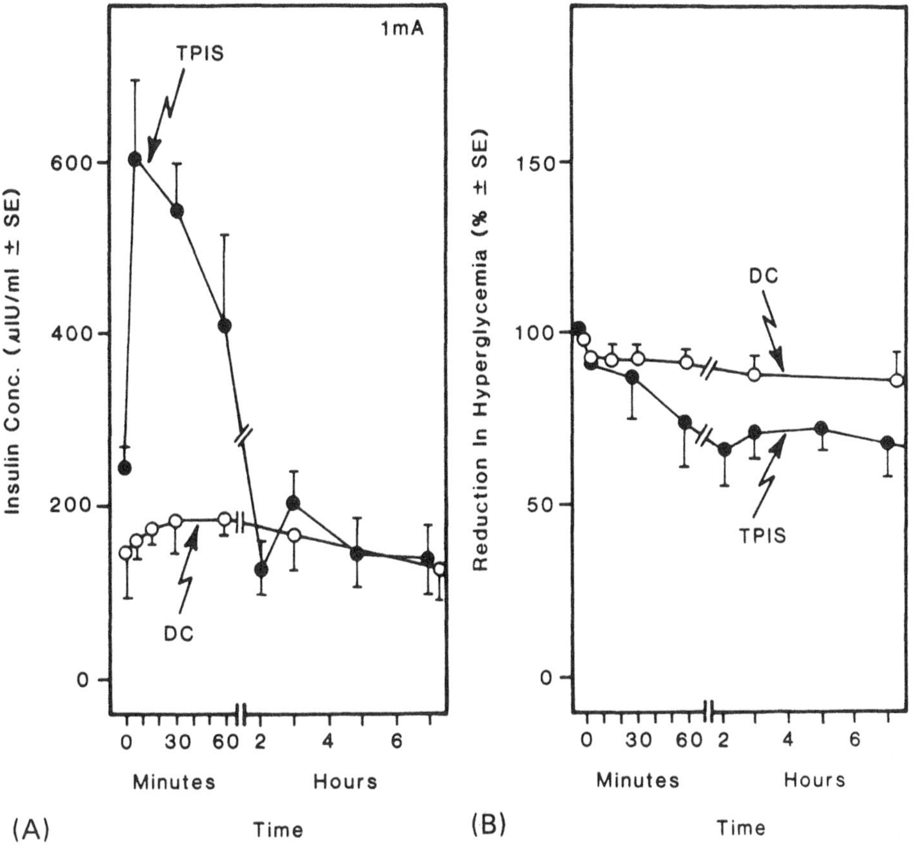

Figure 51 Comparative plasma insulin concentration profiles (A) and blood glucose concentration profiles (B) following the iontophoresis-facilitated transdermal delivery of insulin by pulse current and direct current, both at 1 mA.

of tyrosine, has shown greater activity in rectal mucosa than in vaginal and nasal mucosae; dipeptidase, whose action is responsible for the formation of the Tyr-Gly fragment by its attack on the Gly-Gly bond, has an activity that is substantially greater in the nasal mucosa than in rectal and vaginal mucosae; enkephalinase, which has been attributed to the formation of a Phe-Met fragment by its attack on the Gly-Phe bond, shows an activity pattern similar to that of the dipeptidase. On the other hand, the formation profile of phenylalanine resulted from the further degradation of Phe-Met fragment by the action of carboxypeptidase suggests that carboxypeptidase activity is greater in the rectal mucosa than in nasal and vaginal mucosae.

Studies were also conducted to investigate the enzymatic degradation kinetics of leucine-enkephalin, another naturally occurring brain analgesic pentapeptide with the methionine at position 5 in the methionine-enkephalin replaced by leucine. The results indicated that leucine-enkephalin is also rapidly degraded in extracts of various absorptive mucosae and also follows a first-order kinetic process (Figure 55) (401), with the rate constants lower than those for methionine-enkephalin, especially in the rectal mucosal extract (Table 16). The first-order rate constant for the enzymatic

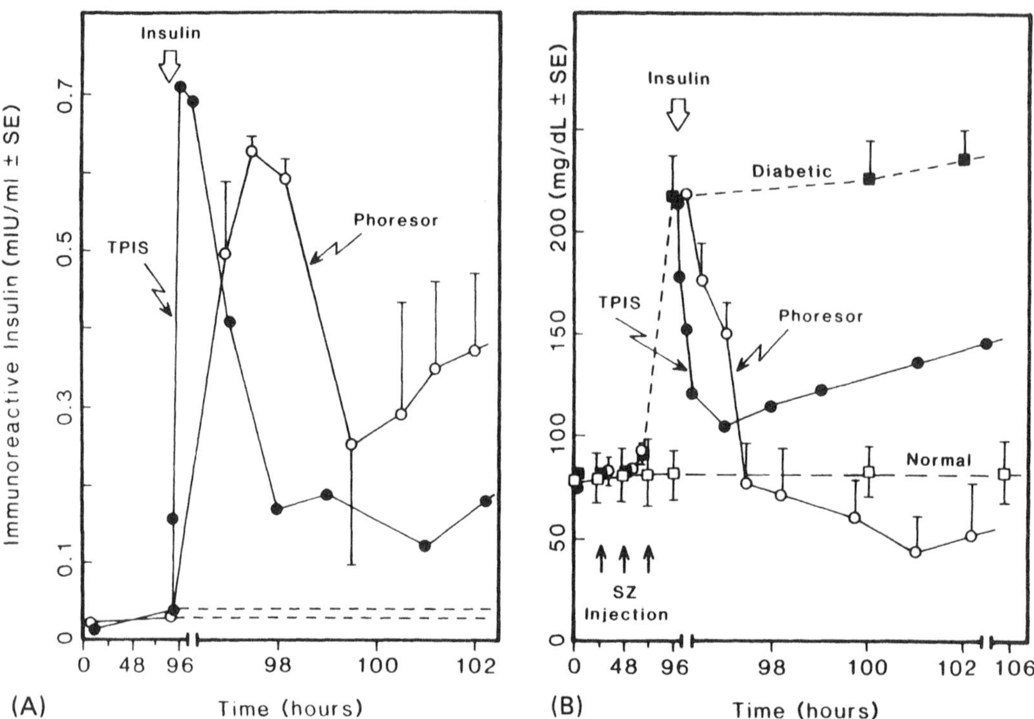

Figure 52 Comparative plasma insulin concentration profiles (A) and blood glucose concentration profiles (B) following the iontophoresis-facilitated transdermal delivery of insulin by the pulse current (1 mA, 40 min) generated by TPIS and the direct current (4 mA, 80 min) generated by the Phoresor system.

degradation of leucine-enkephalin in rectal mucosal extracts was found to be twofold slower than that for methionine-enkephalin.

Efforts were extended to investigate means to stabilize enkephalins in extracts of various absorptive mucosae by adding an antibacterial agent, such as thimerosal (Merthiolate, Lilly), a proteolytic enzyme inhibitor, such as amastatin (402), and a chelating agent, such as EDTA (401). The degradation kinetic profiles outlined in Figure 56 demonstrate that the enzymatic degradation of leucine-enkephalin is progressively retarded with the addition of thimerosal, amastatin, and EDTA. After incorporation of 0.01% thimerosal, the rate constant for enkephalin degradation was reduced by as much as 4-fold in extracts of nasal and rectal mucosae and 9-fold in vaginal mucosa extract (Table 17). After the addition of 0.1 mM amastatin, the stabilizing effect of thimerosal is further enhanced another 2–4 fold. After further addition of 10 mM EDTA, the enzymatic degradation of leucine-enkephalin is substantially minimized to a rate constant of approximately 10^{-4} min^{-1} in all the extracts. The overall stability of leucine-enkaphalin was improved by 35-fold in the nasal extract, 82-fold in the vaginal extract, and 107-fold in the rectal extract. Similar results were also obtained for methionine-enkephalin (403).

The cutaneous metabolism of leucine-enkephalin was recently studied using homogenates of human epidermis and cultured human keratinocytes (404). The results

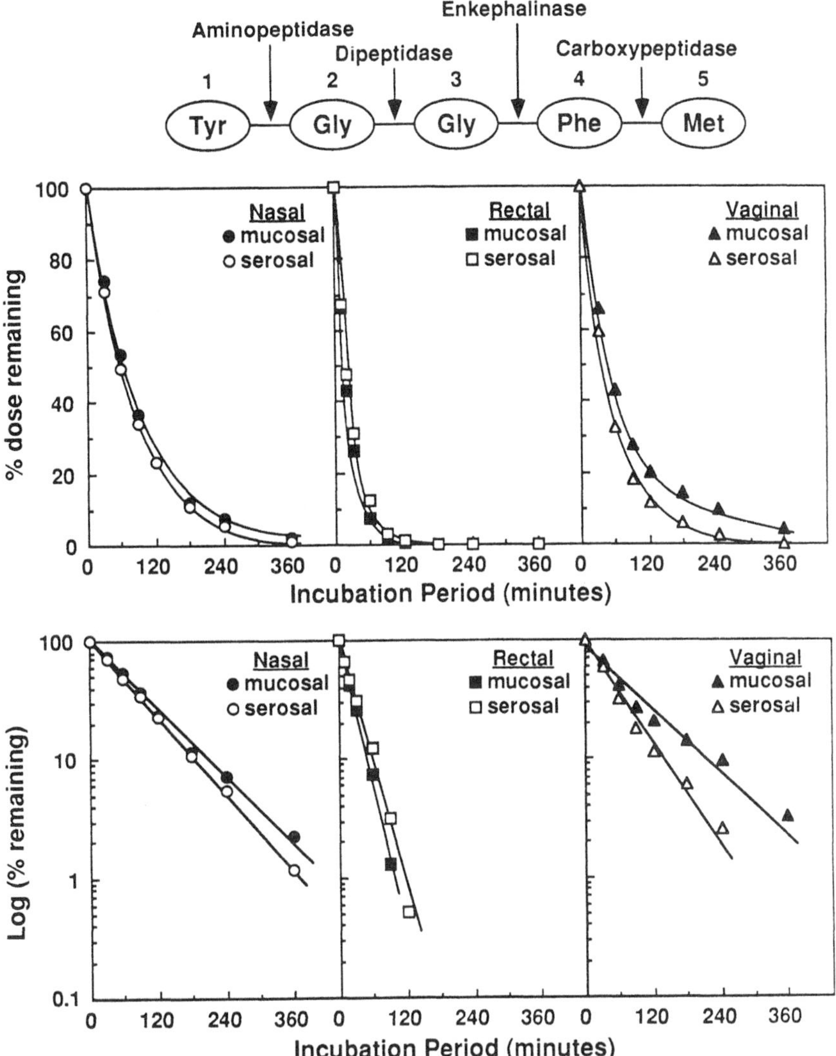

Figure 53 (Top) Comparative degradation profiles of methionine-enkephalin incubated in mucosal and serosal extracts of various rabbit absorptive mucosae at physiological pH. (Bottom) First-order kinetic plots for the enzymatic degradation profiles of methionine-enkephalin.

indicated that both homogenates produce similar Km values, apparent first-order rate constants, and metabolite profiles, and homogenates of foreskin, breast skin, and cloned cells yield comparable aminopeptidase activities. On the other hand, it was found that at least two types of aminopeptidase activity were responsible for metabolizing the N-terminal amino acid successively to the complete degradation of leucine-enkephalin in skin homogenates (405). Endopeptidase activity in skin homogenates was found to be negligibly small. Amastatin and puromycin were observed to produce the highest inhibitory effect on the cutaneous metabolism of leucine-enkephalin in skin homogenates but only slightly effective in stablizing leucine-en-

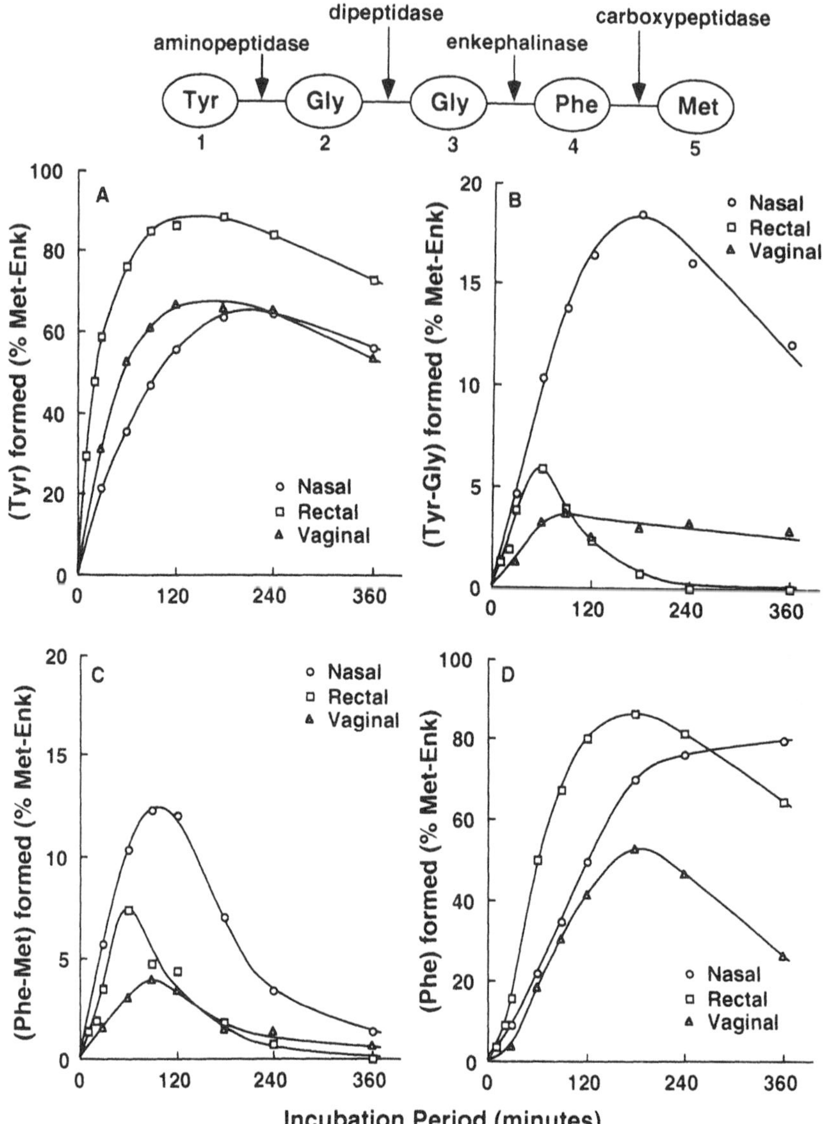

Figure 54 Time course for the appearance of some fragments from the enzymatic degradation of methionine-enkephalin incubated in extracts of various rabbit absorptive mucosae at physiological pH.

kephalin during skin permeation studies. The results suggested that complex proteolytic enzyme activities during the course of skin permeation are different from those data generated from skin homogenates (405).

Using the results obtained in enzyme inhibition studies as summarized in Table 17, a permeation medium was developed in which the enkephalins have the maximal biochemical and biophysical stabilities for studying the transmucosal permeation kinetics of leucine- and methionine-enkephalins (401,406). The results Table 18 in-

Table 15 Apparent First-order Rate Constants for the Enzymatic Degradation of Methionine-Enkephalins in Extracts of Various Absorptive Mucosae

Extracts		Physiologic pH[c]	Rate constant ($min^{-1} \times 10^3$)
Mucosa[a]	Surface[b]		
Nasal	Mucosal	6.0	4.60
	Serosal	6.0	5.23
Rectal	Mucosal	7.2	19.49
	Serosal	7.2	16.20
Vaginal	Mucosal	8.0	4.69
	Serosal	8.0	7.01

[a]Freshly excised from six New Zealand white rabbits just before extraction by buffered solution at physiological pH.
[b]Exposed mucosal and serosal surfaces separately in V-C permeation cell to extracting solution buffered to physiological pH.
[c]Isotonic phosphate buffer adjusted to the physiological pH of each specific mucosa for rabbit.

dicated that nasal and vaginal mucosae are approximately three and two times more permeable to leucine-enkephalin than to methionine-enkephalin. On the other hand, no permeation across the rectal mucosa was achieved for leucine-enkephalin, whereas some permeation (at a low permeation rate) was attained for methionine-enkephalin. For both enkephalins, the rate of transmucosal permeation showed the rank order: nasal >> vaginal >> rectal mucosa. The observation is in good agreement with the transmucosal permeation rate profile of mannitol, a hydrophilic marker, reported earlier (Chapter 4, Figure 6).

IV. DEVELOPMENT OF DELIVERY SYSTEMS FOR PEPTIDE-BASED PHARMACEUTICALS

A. Formulation Considerations

The increased biochemical and structural complexity of peptide- and protein-based pharmaceuticals compared to that of conventional organic-based pharmaceuticals makes formulation design for the systemic delivery of therapeutic peptides and proteins a very challenging and difficult task (407). The development of delivery systems for therapeutic peptides and proteins and their evaluation depend on the biophysical, biochemical, and physiological characteristics of the peptide or protein molecules, including their molecular size, biological half-life, immunogenicity, conformational stability, dose requirement, site and rate of administration, and pharmacokinetics and pharmacodynamics (408). Biological, pharmacological, and toxicological issues can also play a critically important role in the rational design of peptide and protein delivery systems. The immunogenic potential of any impurities or the peptide (or protein) molecule itself must also be taken into consideration. Because the majority of the therapeutic peptides and proteins are extremely potent in their therapeutic activity, the delivery systems developed must be extremely precise in the rate of delivery. Furthermore, the delivery pattern must be properly designed to suit the

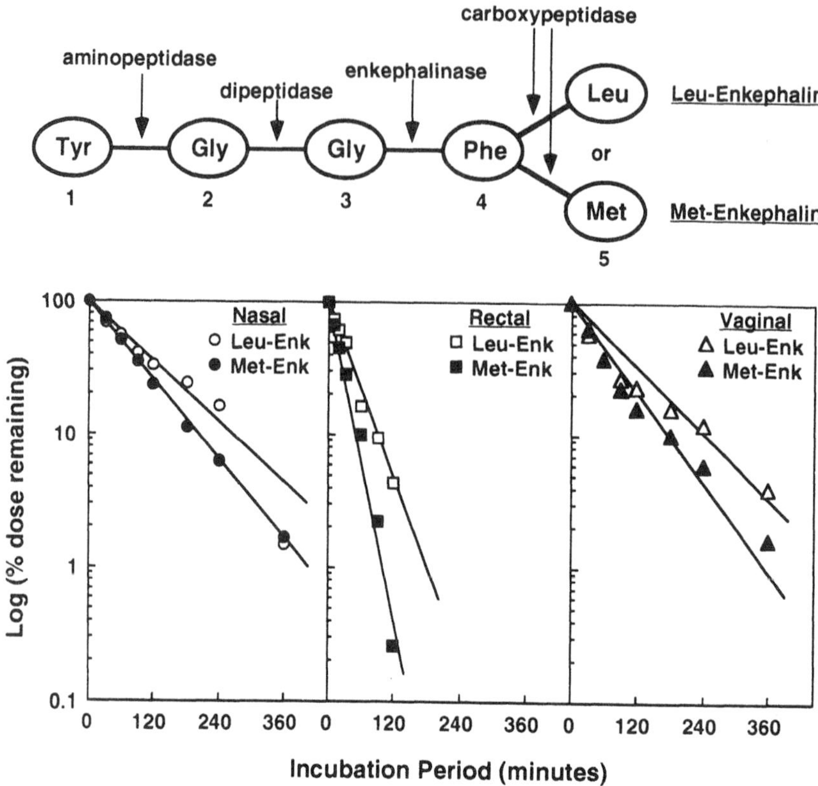

Figure 55 Comparison of amino acid sequence between leucine- and methionine-enkephalin and their first-order degradation kinetic profiles in extracts of various rabbit absorptive mucosae at physiological pH.

Table 16 Comparison of Rate Constants for Enzymatic Degradation of Enkephalins in Extracts of Various Absorptive Mucosae

	Rate constant[b] ($min^{-1} \times 10^3$)	
Extracts[a]	Met-Enk	Leu-Enk
Nasal mucosa	4.81	3.69
Rectal mucosa	20.64	10.77
Vaginal mucosa	5.42	4.02

[a]Mucosal and serosal extracts of freshly excised mucosae from New Zealand white rabbits (n = 6 for Met-Enk and n = 3 for Leu-Enk).

[b]Mean values of first-order degradation rate constant [n = 12 for methionine-enkephalin (Met-Enk) and n = 6 for leucine-enkephalin (Leu-Enk)].

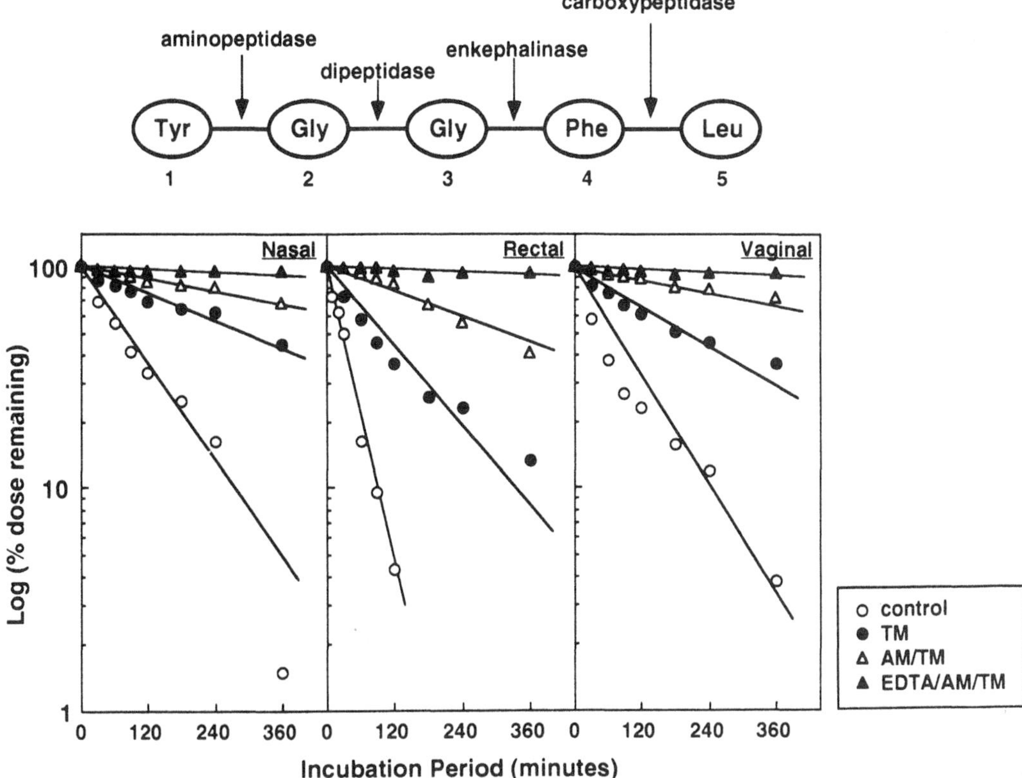

Figure 56 Effect of thimerosal (TM) with amastatin (AM) and ethylenediaminetetraacetate (EDTA) on the first-order degradation kinetic profiles of leucine-enkephalin in extracts of various rabbit absorptive mucosae at physiological pH.

Table 17 Stabilizing Effect of Some Additives on the Enzymatic Degradation of Leucine-Enkephalin in Extracts of Various Absorptive Mucosae

	Rate constant[a] ($min^{-1} \times 10^4$)		
Additive	Nasal mucosa	Rectal mucosa	Vaginal mucosa
No inhibitor	30.98	123.36	94.12
With inhibitor			
Thimerosal, 0.01%	7.08	26.18	10.14
Amastatin, 0.1 mM			
Thimerosal, 0.01%	4.73	8.40	2.69
EDTA, 10 mM			
Amastatin, 0.1 mM			
Thimerosal, 0.01%	0.89	1.15	1.15

[a]First-order rate constant for enzymatic degradation. Each data point is the mean value of the data generated in six extracts (three each for mucosal and serosal extracts) of each mucosa freshly excised from three New Zealand white rabbits.

Table 18 Steady-State Rate of Transmucosal Permeation of Enkephalins

Rabbit mucosa	Transmucosal permeation rate[a] ($\mu g/cm^2$ per hr ± SEM) Leu-Enkephalin	Met-Enkephalin
Nasal	24.13 ± 0.14	8.28 ± 1.88
Rectal	0.00 ± 0.00	1.43 ± 0.01
Vaginal	4.00 ± 0.80	2.88 ± 1.47

Both donor and receptor solution contained an enzyme inhibitor combination: Thimerosal (0.01%), Amastatin (0.1 mM), and EDTA (10 mM).

pattern and mechanism of the pharmacological action of the therapeutic peptides and proteins to be delivered. Pulsatile release, rather than constant release, may be required for peptides and proteins with regulatory functions, such as LHRH (Figure 4) and insulin (Figure 6). Therefore, in carrying out the formulation development task for peptide and protein pharmaceuticals, issues associated with the route and the pattern of delivery and the biochemical and biophysical properties, as well as the enzymatic and physicochemical stability of the peptide (or protein) molecule to be delivered, must all be considered. These are outlined and discussed in the following sections

1. *Preformulation Studies of Therapeutic Peptides and Proteins*

Preformulation data must be generated to serve as the basis for the formulation development of dosage forms or for the design of delivery systems to achieve optimum physicochemical stability and maximum systemic bioavailability. Data on isoelectric point and the pH profiles of stability and solubility, as well as sensitivity to light, heat, moisture, pH, and so on, must be generated. The effect of formulation excipients and delivery system composition on the physical and chemical stability of peptide and protein pharmaceuticals should be investigated. A typical example of the effect of environmental temperature and solution pH on the physicochemical stability of therapeutic peptides and proteins is illustrated by the temperature-dependent degradation profiles of methionine-enkephalin (Figure 57) and its stability-pH profile (Figure 58).

Many peptides do not exhibit a strong tendency to crystallize and are often isolated as an amorphous powder. The lack of crystallinity of many peptides and proteins may thus result in a higher water uptake, and the moisutre content may also vary from one batch to another, creating additional problems. Peptide and protein molecules, being amphoteric in nature, usually have a complex solubility- and stability-pH profile. Generally, the solubility is minimal at the isoelectric point at which the peptide or protein molecule is neutral or has no net charge. Intermolecular interactions can lead to precipitation or gelling of peptides and proteins, which results in minimal solubility . A typical example is illustrated by the preformulation studies of t-PA (409).

2. *Surface Adsorption Behavior of Peptide and Protein Molecules*

Protein and peptide molecules have a tendency to be adsorbed to a variety of surfaces, including glass and plastic (410–415). The losses due to surface adsorption

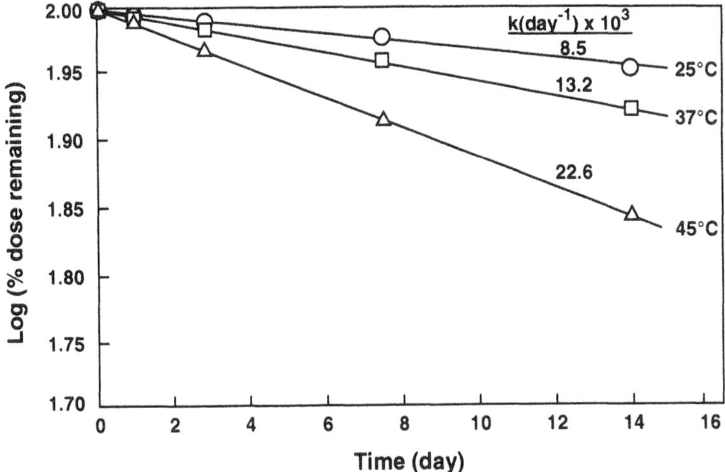

Figure 57 Effect of environmental temperature on the first-order degradation kinetic profiles of methionine-enkephalin in borate buffer (pH 9.8; μ = 0.12).

are particularly significant with peptide (or protein) at low concentrations, and this can result in insufficient dosage for treatment (416,417). If the adsorption is due to the ionic interaction of the peptide molecules with the silanol groups on the glass surface, this can be prevented by silylation of the glass surface. Other approaches that have been applied successfully include the use of a carrier protein, such as albumin or gelatin (416), surfactants, such as sodium lauryl sulfate, amino acids (418), or sodium chloride (419). A quantitative assessment of adsorption under various conditions (411) and the effect of adsorption on the biological activity of therapeutic peptides and proteins have been reported in the literature (412).

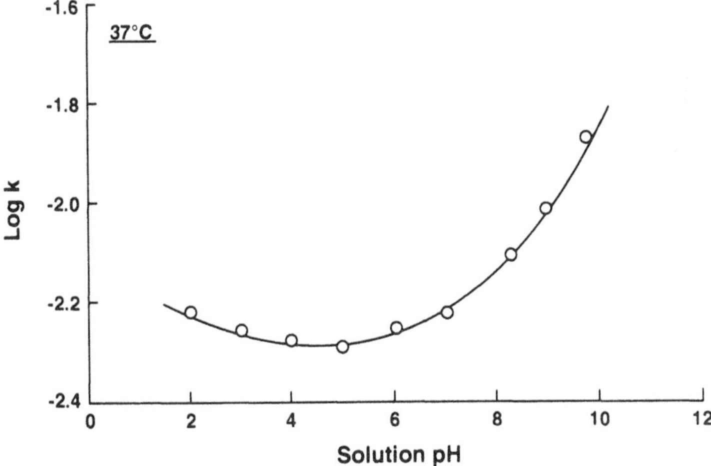

Figure 58 Effect of solution pH on the first-order degradation rate constant (K) of methionine-enkephalin in various buffer solutions at 37°C.

3. Aggregation Behavior of Peptide and Protein Molecules

Another potential problem is the self-aggregation of peptide and protein molecules, such as insulin (420–422). This has been reportedly minimized or prevented by the incorporation of additives, for example, urea (423), dicarboxylic amino acids, such as aspartic acid and glutamic acid (424), or other reagents, such as glycerol (425), EDTA, lysine, Tris or bicarbonate buffer (421). The results of an extensive study involving 60 additives and 1125 formulations (422) suggested that nonionic surfactants like Pluronic F68 (Poloxamer 188), a polyoxyethylene and polyoxypropylene glycol-based surfactant, appear to be promising stabilizers. The study also concluded that (i) human insulin molecules tend to aggregate more readily than insulin molecules from pork or beef; (ii) ionic ingredients and phenolic preservatives tend to accelerate the aggregation of insulin; and (iii) zinc insulin is more stable than zinc-free insulin. Also, many protein drugs adopt several different conformations in solution, and this could be a problem in preserving their pharmacologically-active conformation during processing, formulation development, and sterilization.

4. Other Stability Considerations

The large molecular size and structural complexity of protein and peptide molecules makes them amenable to multiple inactivation pathways. These may be chemical (e.g., fragmentation, enzymatic "clipping," deamidation, covalent dimerization, disulfide scrambling, and oxidation) or physical (e.g., unfolding, noncovalent aggregation, and surface adsorption) in nature or a combination of the two (16). A comprehensive treatment of protein stability has been reported (426). The physical and chemical stability of insulin formulations has been extensively studied (427–435).

B. Pharmacokinetic Considerations

The pharmacokinetics of therapeutic peptides and proteins can be handled in the same way as other conventional organic-based pharmaceuticals, but the manner of collecting experimental data should be more critical, largely because the half-life of most peptides and proteins tends to be very short, in the order of a few minutes or shorter, as a result of the complex patterns of metabolism. Metabolic degradation by peptidases and proteinases can occur in the vascular endothelium, liver, kidney, and/or other nontarget tissues and even at the site of administration. Since the doses administered are often very small and the metabolites and degradation products may be closely related to the parent drug itself, this can give rise to analytical errors (436). Furthermore, if the metabolites are also biologically active, then the pharmacokinetics of the parent compound alone may not be adequate to determine the dosing requirements.

For instance, the pharmacokinetics of insulin has been extensively studied, but estimates of biological half-life and other pharmacokinetic parameters vary significantly, depending upon how the disappearance of the hormone, after a bolus injection, was followed. The variation could be attributed to the time interval chosen for sampling and analysis and also to the choice of radiolabeled or unlabeled hormone. One study (437), however, approached the problem by infusion not to the steady-state plasma concentrations of insulin but to the steady-state blood levels of glucose. A three-compartment model was found adequate to fit all the data from diverse experimental protocols. Compartment III, which could not be sampled but in which the concentration of insulin could be calculated, was the compartment pertinent to glucose utilization. This compartment probably consisted of the interstitial fluid in the muscle and adipose tissues and was in slow equilibrium with plasma insulin levels. As reported earlier, the pancreatic secretion of insulin responds differently to the oral intake of glucose between normal humans and diabetic patients, which is further affected by the body size of the subject (Figure 7).

The relationship between pharmacokinetics and pharmacodynamics for peptide and protein pharmaceuticals is interesting and coordinating, illustrated by the similarity in the daily insulin and glucose response profiles following the intake of meals (Figure 6). It can also be as complex as in the case of LHRH and its analog, in which the systemic administration in a pulsatile manner or a steady-state pattern has produced totally opposite pharmacodynamic responses (Figure 4). Also, for many other regulatory agents, like vasopressin (438), pulsatile delivery may be required for therapeutic effect. This is due to the possibility that the phenomenon of down-regulation or tolerance can result from continuous administration; that is, the continual presence of the agent at a receptor site can lead to a reduction in activity. Thus, continuous dosing could lead to desensitization of receptor; pulsatile dosing mimics the normal physiological rhythm or the circadian pattern.

Basal insulin secretion in healthy subjects also shows a circadian rhythm with a peak time at 15:00 hr. It has been suggested that a larger amount of insulin is needed in the afternoon as well as at night. This may be achieved by delivering insulin using a pump programmable in time that can mimic the physiological circadian baseline of insulin (439). Another recent study (440) using normal and diabetic rodents also reached similar conclusions. Normal mice showed a circadian fluctuation in basal blood glucose levels, which yielded a peak concentration at 14:30 hr, and the greatest sensitivity to insulin also occurred at 14:30, which produced a 60% reduction in blood glucose. From 18:30 to 10:30 hr, on the other hand, the insulin produced only a 38% reduction in blood glucose level. Diabetic mice showed a circadian variation with phases like those of normal mice, with basal glucose levels peaking somewhere between 10:30 and 14:30 hr (440).

It appears that the time of administration can also affect the amount of therapeutic peptide or protein absorbed. For instance, following nasal delivery, the increase in the serum level of salmon calcitonin (sCT) was found to be dependent upon the time of sCT administration. The sCT concentration at 10 min after dosing was considerably greater at midnight than at other times of the day or night (441).

C. Analytical Considerations

For some peptides and proteins, bioassay has been the only method available for detection and potency determination until today. Bioassays are known to be very time consuming, labor intensive, and highly variable. They are not suitable for automation and thus cannot be used on a routine basis. Multiple replication and the use of standards, blanks, and controls are required for statistical interpretation of the results, since variability in the biological response itself can be as high as 50%. In view of the disadvantage of bioassays, there has been a constant effort to develop sensitive and specific analytical methods, both physical and chemical, for routine use, such as spectroscopy, chromatography, and electrophoretic methods and conformationally dependent immunoassays (442).

The more commonly used analytical methodologies include high-performance liquid chromatography (HPLC) (443–446) and radioimmunoassay (RIA) (447–452). HPLC has a very high resolution power and is being increasingly used as an analytical tool to determine the structure and purity of peptides and proteins. However, the low sensitivity of HPLC has limited the direct measurement of peptides in certain biological samples, for example, in tissue extracts. Sensitivity can be improved by using derivatizing reagents to enable detection at the picomole level (453). HPLC techniques with the specificity of detecting proteins that differ by a single amino acid have been developed (454). For insulin, HPLC has been suggested as a more precise measurement of potency than the rabbit assay (67) or the mouse blood glucose assay (455). HPLC was found to be capable of differentiating the insulins from cow, pig, and human and to be both reproducible and stability indicating. For insulin and insulin injections under accelerate stability tests, the HPLC method can detect the decomposition that cannot be detected by either the mouse blood glucose assay or the immunochemical assay. It has been suggested that the U.S. Pharmacopeia should replace the animal response assay for insulin with HPLC assays (455).

Radioimmunoassay has the advantage of specificity and sensitivity, but it lacks the resolving power of HPLC. RIA techniques exploit the specific and tight association of antibody with a peptide or protein molecule, as the antigen, to determine peptides and proteins at very low concentrations in a variety of complex matrices. The ability of a protein to interact with its corresponding antibody is a structurally and conformationally specific interaction. Thus, if the protein molecule or the antibody is conformationally altered, that is, denatured, a less than optimal protein-antibody interaction occurs. However, a drop in the RIA readings does not necessarily imply a reduction in biological activity since the decomposition products or the conformationally altered protein may still be bioactive. On the other hand, decomposition of the protein molecule may not be reflected in the immunochemical assay as long as the antigenic determinant fractions of the protein molecule are intact and capable of reacting with the antibody.

Since HPLC assays are becoming popular for the analysis of peptides and proteins, it is important to determine and to assure that the solvents and/or buffer systems used as the mobile phase do not alter the native conformation of the proteins. One such study for insulin has been reported (456) that evaluated the effect of temperature, buffer composition, pH, ionic strength, and solvents on insulin binding. Optimum insulin-antibody binding occurred at 22°C and pH 6 with a buffer strength of 0.1 M or lower. For all solvents tested, it was found that as the volume (or fraction) of solvent increases, the amount of insulin binding decreases. The results indicated that ethylene glycol and methanol are the least denaturant, while *n*-propanol and acetonitrile are among the most denaturant (456).

Fast atom bombardment mass spectrometry (457) is also very useful in peptide and protein analysis. A radioreceptor assay has also been described for insulin (458,459). Enzyme assays also have a very high specificity and sensitivity. A method for determining the particle size of zinc insulin and its distribution in suspension formulations, based on the measurement of absorbance in the high ultraviolet-visible region, has been reported (460).

In view of the complex structure of proteins and the deficiency of various analytical methods in one respect or another, it is important to utilize several analytical techniques to gain greater confidence in the data generated. In particular, the results obtained by radiolabeled peptide or protein must be verified by some other technique to prove that the label is associated with the intact, bioactive protein, not with a metabolic fragment or denatured protein.

Protein purification is also a very important aspect in the handling of peptide and protein pharmaceuticals. In the early days, fractional or differential precipitation was widely used. However, the techniques commonly used today include electrophoresis, isoelectric focusing, and liquid chromatography (461–466). An electrophoretic method has also been widely used for protein analysis (467,468).

D. Regulatory Considerations

Unlike the conventional organic-based pharmaceuticals, peptide- and protein-based pharmaceuticals have primary, secondary, and even tertiary structures, all of which must be taken into account to gain complete control of the identity, strength, quality, and potency of these biotechnology products (469). As a result, establishing specific standards for the identity, purity, potency, and stability of peptide and protein pharmaceuticals is a complex procedure. Furthermore, the use of recombinant DNA techniques or hybridoma manufacturing processes for the production of peptides and proteins has introduced additional complexity (470).

Biotechnology products are regulated under the statutory authority of four federal agencies: the Food and Drug Administration, the Environmental Protection Agency (EPA), the Occupational Safety and Health Administration (OSHA), and U.S. Department of Agriculture (USDA) (68). The FDA has developed a series of publications to provide guidelines to prospective manufacturers of drugs and biological products by recombinant DNA and hybridoma technology. The U.S. Pharmacopoeial Convention (USPC) must develop additional tests and assays for these pharmaceuticals and products on a case-by-case basis until extensive feedback becomes available in the field to allow generalizations. In the field of biotechnology, the interpretation and enforcement of patent law can be difficult because we may be dealing with patent applications for new organisms. A recent decision of the U.S. Supreme Court seems to suggest that new forms of life can be patented, but the legal language of the ruling may still be ambiguous. Several other reports have also appeared in the literature to address the regulatory issues of peptide- and protein-based pharmaceuticals (471–474).

V. CONCLUSION

Peptide- and protein-based pharmaceuticals are rapidly becoming a very important class of therapeutic agents and are likely to replace many existing organic-based pharmaceuticals in the very near future. The field of biotechnology and genetic engineering is rapidly developing, and an increasing number of such peptide- and protein-based pharmaceuticals will be produced on a large scale by biotechnology processes and will become available commercially for therapeutic use. This poses an urgent challenge to the pharmaceutical industry to develop viable delivery systems for the efficient delivery of these complex therapeutic agents in biologically-active forms. Much work needs to be done on the development of viable delivery systems for nonparenteral administration to make peptide and protein pharmaceuticals commercially viable and therapeutically useful.

REFERENCES

1. A. K. Banga and Y. W. Chien; Systemic delivery of therapeutic peptides and proteins. Int. J. Pharmaceutics, 48:15–50 (1988).
2. V. H. L. Lee; Ophthalmic delivery of peptides and proteins. Pharm. Technol., 11:26–38 (April 1987).
3. F-D-C Reports, 49:3 (46) (1987).
4. R. James and R. A. Bradshaw; Polypeptide growth factors. Annu. Rev. Biochem., 53:259–292 (1984).
5. C. M. Stoscheck and L. E. King; Functional and structural characteristics of EGF and its receptor and their relationship to transforming proteins. J. Cell. Biochem., 31:135–152 (1986).
6. C. M. Stoscheck and L. E. King; Role of epidermal growth factor in carcinogenesis. Cancer Res., 46:1030 (1986).
7. J. W. Eschbach, J. C. Egrie, M. R. Downing, J. K. Browne and J. W. Adamson; Correction of anemia of end-stage renal disease with recombinant human erythropoietin, N. Engl. J. Med., 316:73–78 (1987).
8. S. A. Atlas; Atrial natriuretic factor: A new hormone of cardiac origin. Recent Prog. Horm. Res., 42:207–249 (1986).
9. D. Regoli and J. Barabe; Pharmacology of bradykinin and related kinins. Phamacol. Rev., 32:1–46 (1980).
10. R. P. Elander; Biotechnology: Present and future roles in the pharmaceutical industry. Drug Dev. Ind. Pharm., 11:965–999 (1985).
11. S. Harford; Genetic engineering and the phamaceutical industry. Can. Pharm. J., 118:467–469 (1985).
12. M. D. Dibner; Biotechnology in Europe. Science, 232:1367–1372 (1986).
13. W. Sadee; Protein drugs: A revolution in therapy? Pharm. Res., 3:3–6 (1986).
14. R. M. Baum; Biotech industry moving pharmaceutical products to market. Chem. Eng. News, 65:11–32 (July 1987).
15. T. Itch; Biotech trends in the Japanese pharmaceutical industry. Biotechology, 5:794–799 (1987).
16. R. E. Jones; Challenges in formulation design and delivery of protein drugs. Proc. Int. Symp. Control. Rel. Bio. Mat., 14:96 (1987).
17. L. R. Beck and V. Z. Pope; Controlled release delivery systems for hormones: A review of their properties and current therapeutic use. Drugs, 27:528–547 (1984).
18. J. R. Robinson; Constraints of nonparenteral routes of administration. Pharm. Technol., 11:34 (August 1987).
19. G. Zubay; *Biochemistry*, Addison-Wesley, Reading, MA, 1983.
20. E. L. Smith, R. L. Hill, I. R. Lehman, R. J. Lefkowitz, P. Handler and A. White; *Principles of Biochemistry: General Aspects*, McGraw-Hill, New York, 1983.
20a. T. L. Blundell and S. P. Wood; Is the evolution of insulin Darwinian or due to selectively neutral mutation? Nature, 257:197–203 (1975).
21. A. L. Lehninger; *Biochemistry*, Worth Publishers, 1975.
22. R. H. Buckley; Long-term use of intravenous immune globulin in patients with primary immunodeficiency diseases: Inadequacy of current dosage practices and approaches to the problem. J. Clin. Immunol., 2:155 (1982).
23. L. Martis; Use of specific routes of administration for peptides and proteins: Parenteral, in: *28th Annual National Industrial Pharmaceutical Research Conference*, Madison, Wisconsin, 1986.
24. A. G. Gilman, L. S. Goodman and A. Gilman; *The Phamacological Basis of Therapeutics*, Macmillan, New York, 1980, Chapter 64.
25. G. F. Maberly, G. A. Wait, J. A. Kilpatrick, E. G. Loten, K. R. Gain, R. D. H.

Stewart and C. J. Eastman; Evidence for insulin degradation by muscle and fat tissue in an insulin-resistant diabetic patient. Diabetologia, 23:333–336 (1982).

26. R. J. Wills, S. Dennis, H. E. Spiegel, D. M. Gibson and P. I. Nadler; Interferon kinetics and adverse reactions after intravenous, intramuscular and subcutaneous injection. Clin. Phamacol. Ther., 35:722–727 (1984).
27. R. A. Good; Intravenous gamma globulin therapy. J. Clin. Immunol., 2:55 (1982).
28. B. E. Sobel, L. E. Fields, A. K. Robison, K. A. A. Fox and S. J. Sarnoff; Coronary thrombolysis with facilitated absorption of intramuscularly injected tissue-type plasminogen activator. Proc. Natl. Acad. Sci. USA, 82:4258–4262 (1985).
29. B. K. Davis; Control of diabetes with polyacrylamide implants containing insulin. Experientia, 28:348 (1972).
30. B. K. Davis; Diffusion in polymer gel implants. Proc. Natl. Acad. Sci. USA, 71:3120–3123 (1974).
31. R. Langer and J. Folkman; Polymers for the sustained release of proteins and other macromolecules. Nature, 263:797–800 (1976).
32. R. Langer and J. Folkman; Sustained release of macromolecules from polymers, in: *Polymeric Delivery Systems* (R. J. Kostlenik, Ed.), Gordon and Breach, New York, 1978 pp. 175–196.
33. W. D. Rhine, D. S. T. Hsieh and R. Langer; Polymers for sustained macromolecule release: Procedure to fabricate reproducible delivery systems and control release kinetics. J. Pharm. Sci., 69:265–270 (1980).
34. D. S. T. Hsieh, W. D. Rhine and R. Langer; Zero-order controlled-release polymer matrices for micro- and macromolecules. J. Pharm. Sci., 72:17–22 (1983).
35. J. Cohen, R. A. Siegel and R. Langer; Sintering technique for the preparation of polymer matrices for the controlled release of macromolecules. J. Pharm. Sci., 73:1034–1037 (1984).
36. R. A. Siegel, J. M. Cohen, L. Brown and R. Langer; Sintered polymers for sustained macromolecular drug release, in: *Recent Advances in Drug Delivery Systems*, (J. M. Anderson and S. W. Kim, Eds.), Plenum Press, New York, 1984, pp. 315–320.
37. D. S. T. Hsieh, R. Langer and J. Folkman; Magnetic modulation of macromolecules from polymers. Proc. Natl. Acad. Sci. USA, 78:1863 (1981).
38. D. S. T. Hsieh and R. Langer; Zero-order drug delivery systems with magnetic control, in: *Controlled Release Delivery Systems* (T. J. Roseman and S. Z. Mansdorf, Eds.), Dekker, New York, 1983, Chapter 7.
39. R. Langer, R. Siegel, L. Brown, K. Leong, J. Kost and E. Edelman; Controlled release and magnetically-modulated systems for macromolecular drugs, in: *Macromolecules as Drugs and as Carriers for Biologically-Active Materials* (D. A. Tirrell, L. G. Donaruma and A. B. Turek, Eds.), Ann. N.Y. Acad. Sci., 446:1–13 (1985).
40. J. Kost and R. Langer; Magnetically modulated drug delivery systems. Pharm. Int., 7:60–63 (1986).
41. J. Heller; Biodegradable polymers in controlled drug delivery. CRC Crit. Rev. Ther. Drug. Carrier Systems, 1:39–90 (1984).
42. V. H. L. Lee; Peptide and protein drug delivery systems. Biopharm. Manufact., 1:24–31 (March 1988).
43. L. Brown, L. Siemer, C. Munoz and R. Langer; Controlled release of insulin from polymer matrices: In vitro kinetics. Diabetes, 35:684–691 (1986).
44. L. Brown, C. Munoz, L. Siemer, E. Edelman and R. Langer; Controlled release of insulin from polymer matrices: Control of diabetes in rats. Diabetes, 35:692–697 (1986).
45. S. Bechard, N. Yamaguchi and J. N. McMullen; In vitro and in vivo release of insulin from porous polymeric implants. Proc. Int. Symp. Control. Rel. Bio. Mat., 14:57–58 (1987).
46. A. F. Kwong, S. Chou, A. M. Sun, M. V. Sefton and M. F. A. Goosen; In vitro

and in vivo release of insulin from poly(lactic acid) microbeads and pellets. J. Control. Rel., 4:47–62 (1986).
47. D. S. T. Hsieh, C. C. Chiang and D. S. Desai; Controlled release of macromolecules from silicone elastomer. Pharm. Technol., 39–49 (June 1985).
48. S. T. Anik; LHRH analogs—a case history: Pharmaceutical considerations in the use of peptides and proteins, *28th Annual National Industrial Pharmaceutical Research Conference*, Madison, Wisconsin, June 1986.
49. G. Leyendecker, T. Struve and E. J. Plotz; Induction of ovulation with chronic intermittent (pulsatile) administration of LHRH in women with hypothalmic and hyperprolactinemic amenorrhea. Arch. Gynecol., 229:177–190 (1980).
50. W. F. Crowley and J. W. McArthur; Stimulation of the normal menstrual cycle in Kallman's syndrome by pulsatile administration of luteinizing hormone-releasing hormone (LHRH). J. Clin. Endocrinol. Metab., 51:173–175 (1980).
51. P. E. Belchetz, T. M. Plant, Y. Nakai, E. J. Keogh and E. Knobil; Hypophysical responses to continuous and intermittent delivery of hypothalamic gonadotropin-releasing hormone. Science, 202:631–633 (1978).
52. A. R. Hoffman and W. F. Crowley; Induction of puberty in men by long term pulsatile administration of low-dose gonadotropin releasing hormone. N. Engl. J. Med., 307:1237–1242 (1982).
53. G. Skarin, S. J. Nillius, L. Wibell and L. Wida; Chronic pulsatile low dose GnRH therapy for induction of testosterone production and spermatogenesis in a man with secondary hypogonadotrophic hypogonadism. J. Clin. Endocrin. Metab., 55:723–726 (1982).
54. G. B. Cutler, A. R. Hoffman, R. S. Swerdloff, R. J. Santen, D. R. Meldrum and F. Comiti; Therapeutic applications of luteinizing hormone-releasing hormone and its analogs. Ann. Intern. Med., 102:643–657 (1985).
55. B. H. Vickery, G. I. McRae, L. M. Sanders, J. S. Kent and J. J. Nestor; In vivo assessment of long-acting formulations of luteinizing hormone-releasing hormone analogs, in: *Long-Acting Contraceptive Delivery Systems*, (G. I. Zatuchni, A. Goldsmith, J. D. Shelton and J. J. Sciarra, Eds.), Harper and Row, Philadelphia, 1984.
56. J. S. Kent, B. H. Vickery and G. I. McRae; The use of cholesterol matrix pellets implants for early studies on the prolonged release in animals of agonist analogues of luteinizing hormone-releasing hormone. Proc. Int. Symp. Control. Rel. Bio. Mat., 7:67–e176 (1980).
57. J. S. Kent, L. M. Sanders, G. I. McRae, B. H. Vickery, T. R. Tice and D. H. Lewis; Feasibility studies on the controlled release of an LHRH analogue from subcutaneously injected polymeric microspheres: Early in vivo studies. Controlled Delivery Systems, 3:60 (1982).
58. L. M. Sanders, J. S. Kent, G. I. McRae, B. H. Vickery, T. R. Tice and D. H. Lewis; Feasibility studies on the controlled release of and LHRH analogue from subcutaneously injected polymeric microspheres: Formulation characteristics. Controlled Delivery Systems, 3:80 (1982).
59. L. M. Sanders, J. S. Kent, G. I. McRae, B. H. Vickery, T. R. Tice and D. H. Lewis; Controlled release of luteinizing hormone-releasing hormone analogue from poly(d, l-lactide-co-glycolide) microspheres. J. Pharm. Sci., 73:1294–1297 (1984).
60. J. S. Kent, L. M. Sanders, T. R. Tice and D. H. Lewis; Microencapsulation of the peptide Nafarelin acetate for controlled release, in: *Long-Acting Contraceptive Delivery Systems* (G. I. Zatuchni, A. Goldsmith, J. D. Shelton and J. J. Sciarra, Eds.), Harper and Row, Philadelphia, 1984, pp. 169–178.
61. L. M. Sanders, R. Bruns, K. Vitale, G. McRae and P. Hoffman; Clinical performance

of nafarelin controlled release injectable: Influence of formulation parameters on release kinetics and duration of efficacy, in: *Abstracts of 15th Int. Symp. Control. Rel. Bioact. Mat.*, Controlled Release Society, Lincolnshire, IL, 1988, p. 73.
62. L. R. Beck, R. A. Ramos, C. E. Flowers, G. Z. Lopez, D. H. Lewis and D. R. Cowsar; Clinical evaluation of injectable biodegradable contraceptive systems. Am. J. Obstet. Gynecol., 140:799–806 (1981).
63. D. H. Lewis, T. R. Tice, W. E. Meyers, D. R. Cowsar and L. R. Beck; Biodegradable microcapsules for contraceptive steroids. Contracept. Delivery Systems, 3:55 (1982).
64. F. G. Hutchinson and B. J. A. Furr; Biodegradable polymer systems for the sustained release of polypeptides. J. Control. Rel., 13:279 (1990).
65. H. Klostermeyer and R. E. Humbel; The chemistry and biochemistry of insulin. Angew. Chem. Int. Ed., 5:807–822 (1966).
66. J. P. Swann; Insulin: A case study in the emergence of collaborative pharmacomedical research. Pharm. Hist., 28:3–13 (1986).
67. I. S. Johnson; Human insulin from recombinant DNA technology. Science, 219:632–637 (1983).
68. M. D. Giddins, R. Dabbah, L. T. Grady and C. T. Rhodes; Scientific and regulatory aspects of macromolecular drugs and devices. Drug. Dev. Ind. Pharm., 13:873–968 (1987).
69. J. L. Gueriguian; *Insulins, Growth Hormone and Recombinant DNA Technology*, Raven, New York, 1981.
70. A. P. Bollon; *Recombinant DNA Products: Insulin, Interferon and Growth Hormone*, CRC Press, Boca Raton, FL, 1984.
71. J. L. Selam, H. Eichner, L. Woertz, D. Turner, A. Lauritano and M. A. Charles; The effectiveness and comparison of human and bovine longacting insulins. Clin. Res., 35:194A (1987).
72. D. S. Schade and R. P. Eaton; Insulin delivery: How, when, where. N. Engl. J. Med., 312:1120–1121 (1985).
73. P. J. Thomas; Optimum administration of insulin: A complex story. Aust. J. Pharm., 67:745–746 (1986).
74. J. Haleblian and W. McCrone; Pharmaceutical applications of polymorphism. J. Pharm. Sci., 58:911–929 (1969).
75. K. Hallas-Moller, M. Jersild, K. Petersen and J. Schlichtkrull; Zinc insulin preparations for single daily injection. J. Am. Med. Assoc., 150:1667–1671 (1952).
76. K. Petersen, J. Schlichtkrull and K. Hallas-Moller; Injectable insulin preparation with protracted effect. U.S. Patent 2,882,203, April 14, 1959.
77. A. L. Weiner, S. S. G. Carpenter, E. C. Soehngen, R. P. Lenk and M. C. Popescu; Liposome-collagen gel matrix: A novel sustained drug delivery system. J. Pharm. Sci., 74:922–925 (1985).
78. I. Lindmayer, K. Menassa, J. Lambert, A. Legendre, C. Legault, M. Letendre and J. P. Halle; Development of a new jet injection for insulin therapy. Diabetes Care, 9:294–297 (1986).
79. J. P. Halle, I. Lindymayer, K. Manessa, J. Lambert, C. Legault and G. Lalumiere; Clinical trials of a new insulin jet injector allowing the mixing of two types of insulin. Diabetes, 35:144A (1986).
80. R. A. Hingson and J. G. Hughes; Clinical studies with jet injection: New method of drug administration. Anesth. Analg., 26:221–230 (1947).
81. O. Weller and M. Linder; Jet injection of insulin versus the syringe and needle method. J. Am. Med. Assoc., 195:844–847 (1966).
82. M. L. Cohen, R. A. Chez, R. A. Hingson, A. E. Szulman and M. Trimmer; Use of jet insulin injector in diabetes mellitus therapy. Diabetes, 21:39–44 (1972).

83. C. Kuhl, B. Edsberg, P. Hildebrandt and D. Herly; Insulin bolus given by sprinkler needle: Improved absorption and glycaemic control. Diabetologia, 29:561A (1986).
84. P. Haycock; Experience with and acceptance of multiple daily insulin injections using novopen, a new insulin delivery device. Diabetes, 35:145A (1986).
85. J. C. Pickup, H. Keen, J. A. Parsons, and K. G. M. M. Alberta; Continuous subcutaneous insulin infusion: An approach to acheiving normoglycaemia. Br. Med. J., 1:204–207 (1978).
86. W. V. Tamborlane, R. S. Sherwin, M. Genel and P. Felig; Normalization of plasma glucose in juvenile diabetics by subcutaneous administration of insulin with a portable infusion pump. N. Engl. J. Med., 300:573–578 (1979).
87. D. Nathan; Successful treatment of extremely brittle, insulin dependent diabetes with a novel subcutaneous insulin pump regimen. Diabetes Care, 5:105–110 (1982).
88. S. M. Teutsch, W. H. Herman, D. M. Dwyer and J. M. Lane; Mortality among diabetic patients using continuous subcutaneous insulin infusion pumps. N. Engl. J. Med., 310:361–368 (1984).
89. A. E. Kitabchi, J. N. Fisher, R. Matteri and M. B. Murphy; The use of continuous insulin delivery systems in treatment of diabetes mellitus. Adv. Intern. Med., 28:449–490 (1983).
90. S. B. Leichter, M. E. Schreiner, L. R. Reynolds and T. Bolick; Long-term followup of diabetic patients using insulin infusion pumps: Considerations for future clinical application. Arch. Intern. Med., 145:1409–1412 (1985).
91. M. V. Sefton, H. M. Lusher, S. R. Firth and M. U. Waher; Controlled release micropump for insulin administration. Ann. Biomed. Eng., 7:329–343 (1979).
92. M. V. Sefton, D. G. Allen, V. Horvath and W. Zingg; Insulin delivery at variable rates from a controlled release micropump, in: *Recent Advances in Drug Delivery Systems* (J. M. Anderson and S. W. Kim, Eds.), Plenum Press, New York, 1984, pp. 349–365.
93. P. K. Watler and M. V. Sefton; A piezoelectric micropump for insulin delivery. Proc. Int. Symp. Control. Rel. Bio. Mat., 14:231 (1987).
94. T. A. Horbett, T. Kost and B. D. Ratner; Swelling behavior of glucose sensitive membranes. Am. Chem. Soc. Div. Polym. Chem. Prepr., 24:34–35 (1983).
95. T. A. Horbett, B. D. Ratner, J. Kost and M. Singh; A bioresponsive membrane for insulin delivery, in: *Recent Advances in Drug Delivery System* (J. M. Anderson and S. W. Kim, Eds.), Plenum Press, New York, 1984, pp. 209–220.
96. G. Albin, T. A. Horbett and B. D. Ratner; Glucose sensitive membranes for controlled delivery of insulin: Insulin transport studies, in: *Advances in Drug Delivery Systems* (J. M. Anderson and S. W. Kim, Eds.), Elsevier, Amsterdam, 1986, pp. 153–163.
97. S. Y. Jeong, S. W. Kim, M. J. D. Eenink and J. Feijen; Self-regulating insulin delivery systems. I. Synthesis and characterization of glycosylated insulin. J. Control. Rel., 1:57–66 (1984).
98. S. Y. Jeong, S. W. Kim, D. L. Homberg and J. C. McRea; Self-regulating insulin delivery systems. III. In vivo studies, in: *Advances in Drug Delivery Systems* (J. M. Anderson and S. W. Kim, Eds.), Elsevier, Amsterdam, 1986, pp. 143–152.
99. S. W. Kim, S. Y. Jeong, S. Sato, J. C. McRea and J. Feijen; Self-regulating insulin delivery system—a chemical approach, in: *Recent Advances in Drug Delivery Systems* (J. M. Anderson and S. W. Kim, Eds.), Plenum Press, New York, 1984, p. 123.
100. S. Sato, S. Y. Jeong, J. C McRae and S. W. Kim; Self-regulating insulin delivery systems. II. In vitro studies. J. Control. Rel., 1:67–77 (1984).
101. L. A. Seminoff and S. W. Kim; In vitro characterization of a self-regulating insulin delivery system. Pharm. Res., 3:40S (1986).

102. G. J. Boer, J. Kruisbrink and H. V. Pelt-Heerschap; Long-term and constant release of vasopressin from accurel tubing: Implantation in the Brattleboro rat. J. Endocrinol., 98:147–152 (1983).
103. J. Kruisbrink and G. J. Boer; Controlled long-term release of small peptide hormones using a new microporous polypropylene polymer: Its application for vasopressin in the Brattleboro rat and potential perinatal use. J. Pharm. Sci., 73:1713–1718 (1984).
104. J. Kruisbrink and G. J. Boer; The use of [^{3}H]-vasopressin for in vivo studies of controlled delivery from an Accurel/collodion device in the Brattleboro rat. J. Pharm. Pharmacol., 38:893–897 (1986).
105. P. Lelawongs; Ph.D. dissertation, Rutgers University, New Brunswick, NJ, 1989.
106. B. J. Aungst, N. J. Rogers and E. Shafter; Comparison of nasal, rectal, buccal, sublingual and intramuscular insulin efficacy and the effects of a bile salt absorption promotor. J. Pharmacol. Exp. Ther., 224:23–27 (1988).
107. A. Pannatier, P. Jenner, B. Testa and J. C. Etter; The skin as a drug metabolizing organ. Drug Metab. Rev., 8:319–343 (1978).
108. H. Okada, I. Yamazaki, Y. Ogawa, S. Hirai, T. Yashiki and H. Mima; J. Pharm. Sci., 71:1367–71 (1982).
109. I. Nudelman; Nasal delivery: A revolution in drug administration, in: *1987 Conference Proceedings on the Latest Developments in Drug Delivery Systems*, Pharm. Technol., Aster Publishing, 1987.
110. Y. W. Chien; *Transnasal Systemic Medications*, Elsevier, Amsterdam, 1985.
111. Y. W. Chien and S. F. Chang; Intranasal drug delivery for systemic medication. Crit. Rev. Ther. Drug Carrier Syst., 4:67–194 (1987).
112. Y. W. Chien, K. S. E. Su and S. F. Chang; *Nasal Systemic Drug Delivery*, Dekker, New York, 1989.
113. K. S. E. Su; Intranasal delivery of peptides and proteins, Pharm. Int., 7:8–11 (1986).
114. K. S. E. Su, K. M. Campanale, L. G. Mandelsohn, G. A. Kerchner and C. L. Gries; Nasal delivery of polypeptides. I. Nasal absorption of enkephalins in rats. J. Pharm. Sci., 74:394–398 (1985).
115. V. H. L. Lee and S. D. Kashi; Nasal peptide and protein absorption promoters: Aminopeptidase inhibition as a predictor of absorption enhancement potency of bile salts. Proc. Int. Symp. Control. Rel. Bio. Mat., 14:53–54 (1987).
116. S. Hirai, T. Yashida and H. Mima; Mechanisms for the enhancement of the nasal absorption of insulin by surfactants, Int. J. Pharm. 9:173–184 (1981).
117. S. Hirai, T. Ikenaga and T. Matsuzawa; Nasal absorption of insulin in dogs. Diabetes, 27:296–299 (1978).
118. H. G. Solbach and W. Wiegelmann; Intranasal application of luteinizing hormone-releasing hormone. Lancet, 1:1259 (1973).
119. W. S. Evans, J. L. C. Borges, D. L. Kaiser, M. L. Vance, R. P. Sellers, R. M. McLeod, W. Vale, J. Rivier and M. O. Thorner; Intranasal administration of human pancreatic tumor GH-releasing factor-40 stimulates GH-release in normal men. J. Clin. Endocrinol. Metab., 57:1081–1083 (1983).
120. G. I. Zatuchni, J. D. Shelton and J. J. Sciarra; *LHRH Peptides as Female and Male Contraceptives*, Harper & Row, Philadelphia, 1976.
121. R. Illig, H. Bucher and A. Prader; Success relapse and failure after intranasal LHRH treatment of cryptorchidism in 55 prepubertal boys. Eur. J. Pediatr., 133:147–150 (1980).
122. R. Illig, T. Torresani, H. Bucher, M. Zachmann and A. Prader; Effect of intranasal LHRH therapy on plasma, LH, FSH and testosterone, and relation of clinical results in prepubertal boys with cryptorchidism. Clin. Endocrinol. (Oxf.), 12:91–97 (1980).

123. H. Koch; Buserelin: Contraception through nasal spray. Pharm. Int., 72:99 (1981).
124. S. T. Anik, G. McRea, C. Neerenberg, A. Worden, J. Foreman, J. Y. Hwang, S. Kushinky, R. E. Jones and B. Vickery; Nasal absorption of nafarelin acetate, the decapeptide LHRH, in rhesus monkeys. J. Pharm. Sci., 73:684–685 (1984).
125. A. A. Hussain, R. Bawarshi-Nassar and C. H. Huang; Physicochemical condiderations in intranasal drug administrations in: *Transnasal System Medications* (Y. W. Chien, Ed.), Elsevier, Amsterdam, 1985, pp. 121–137.
126. C. McMartin, L. E. F. Hutchinson, R. Hyde and G. E. Peters; Analysis of structural requirements for the absorption of drugs and macromolecules from the nasal cavity. J. Pharm. Sci., 76:535–540 (1987).
127. V. H. L. Lee; *Peptide and Protein Drug Delivery*, Dekker, New York, 1991, Chapter 1.
128. J. Sandow and W. Petri; Intranasal administration of peptides: Biological activity and therapeutic efficacy, in: *Transnasal Systemic Medication* (Y. W. Chien, Ed.), Elsevier, Amsterdam, 1985, pp. 183–199.
129. A. Hussain, J. Faraj, Y. Aramaki and J. E. Truelove; Hydrolysis of leucine enkephalin in the nasal cavity of the rat—a possible factor in the low bioavailability of nasally administered peptides. Biochem. Biophys. Res. Commun., 133:923 (1985).
130. K. S. E. Su, K. M. Campanale, L. G. Mendelsohn, G. A. Kerchner and C. L. Gries; Nasal delivery of polypeptides. I. Nasal absorption of enkephalins in rats. J. Pharm. Sci., 74:394 (1985).
131. P. D. Gesselchen, C. J. Parli and R. C. A. Frederickson; Peptides: Synthesis-structure-function, in: *Proceedings of the 7th American Peptide Symposium* (D. H. Rich and E. Gross, Eds.), Pierce Chemical Co., Rockford, IL, 1986, p. 637.
132. B. J. Balin, R. D. Broadwell, M. Salcman and M. El-Kalliny; Avenues for entry of peripherally administered protein to the central nervous system in mouse, rat and squirrel monkey. J. Comp Neurol., 251:260 (1986).
133. A. R. Spiegelman; Treatment of diabetes with synthetic vasopressin. J. Am. Med. Assoc., 184:657 (1963).
134. H. Sjoberg and R. Luft; Nasal spray of synthetic vasopressin for the treatment of diabetes insipidus. Lancet, 1:1159 (1963).
135. N. D. Fabricant; Nasal spray treatment of diabetes insipidus. Eye Ear Nose Throat Monthly, 42:64 (1967).
136. A. M. Moses; Synthetic lysine vasopressin nasal spray in the treatment of diabetes insipidus. Clin. Pharmacol. Ther., 5:422 (1964).
137. A. M. Dashe, C. R. Kleeman, J. W. Czaczkes, H. Rubinoff and I. Spears; Synthetic vasopressin nasal spray in the treatment of diabetes insipidus. J. Am. Med. Assoc., 190:1069 (1964).
138. J. F. Dingman and J. H. Hauger-Klevene; Treatment of diabetes insipidus: Synthetic lysine vasopressin nasal solution. J. Clin. Encocrinol., 24:550 (1964).
139. R. L. Fogel; Treatment of diabetes insipidus with lysine vasopressin spray. J. Med. Soc. N.J., 63:203 (1966).
140. W. Hung; Treatment of diabetes insipidus in children with synthetic lysine-8-vasopressin nasal spray. Med. Ann. Dis. Col., 36:400 (1967).
141. N. Mimica, L. C. Wegienka and P. H. Forsham; Lypressin nasal spray: Usefulness in patients who manifest allergies to other antidiuretic hormone preparation, J. Am. Med. Assoc., 203:802 (1968).
142. H. D. Green and J. B. Blumberg; The use of a synthetic analogue of post-hypophysial vasopressin (PVL-2) for local hemostatis. Surgery, 58:524 (1965).
143. I. Vavra, A. Machova, V. Holecek, J. Cort, H. Zaoral and F. Sorm; Effect of a synthetic analogue of vasopressin in animals and in patients with diabetes insipidus. Lancet, 1:948 (1968).

144. F. Ziai, R. Walter and I. M. Rosenthal; Treatment of central diabetes insipidus in adults and children with desmopressin. Arch. Intern. Med., 138:1382 (1978).
145. W. N. Lee, B. M. Lippe, S. H. Franchi and S. A. Kaplan; Vasopressin analog DDAVP in the treatment of diabetes insipidus. Am. J. Dis. Child., 130:166 (1976).
146. W. E. Cobb, S. Spare and S. Reichlin; Neurogenic diabetes insipidus: Management with DDAVP (1-desamino-8-arginine vasopressin). J. Intern. Med., 88:183 (1978).
147. D. R. Brown and D. L. Ulden; the use of 1-desamino-8-arginine vasopressin (DDAVP) in the management of neurogenic diabetes insipidus. Minn. Med., 62:427 (1979).
148. A. Grossman, A. Fabbri, P. L. Goldberg and G. M. Nesser; Two new modes of desmopressin (DDAVP) administration. Br. Med. J., 280:1215 (1980).
149. J. P. Rado, J. Marosi, L. Szende, L. Borbely, J. Tako and J. Fischer; The antidiuretic action of 1-desamino-8-arginine vasopressin (DDAVP) in man. Int. J. Clin. Pharmacol., 13:196 (1976).
150. A. G. Robinson, S. M. Seif, F. F. Ciarochi, T. V. Zenser and B. B. Davis, DDAVP (1-desamino-8-D-arginine vasopressin)—kinetics of prolonged antidiuresis in central diabetes insipidus. Clin. Res., 24:277A (1976).
151. A. S. Harris, E. Svensson, Z. G. Wagner, S. Lethagen and I. M. Nilsson; Effect of viscosity on particle size, deposition and clearance of nasal delivery systems containing desmopressin. J. Pharm. Sci., 77:405 (1988).
152. J. E. Clement, V. C. Harwell and J. R. McCain; Use of intranasal oxytocin for induction and/or stimulation of labor. Am. J. Obstet. Gynecol., 83:778 (1962).
153. F. C. Hinde; The value of intranasal oxytocin spray in obstetrics. Med. J. Aust., 1:268 (1963).
154. C. Jones; Clinical experience with intranasal oxytocin spray. Med. J. Aust., 2:1099 (1966).
155. I. V. Gent, T. Eskes and J. C. Seelen; Changes in intrauterine pressure due to intranasal administration of oxytocin (Partocon). Acta Obstet. Gynecol. Scand., 46:340 (1967).
156. K. Muller and M. Osler; Induction of labor: A comparison of intravenous, intranasal and transbuccal oxytocin. Acta Obstet. Gynecol. Scand., 46:59 (1967).
157. K. Devoe, W. C. Rigsby and C. McDaniels; The effect of intranasal oxytocin on the pregnant uterus. Am. J. Obstet. Gynecol., 97:208 (1967).
158. R. T. Hoover; Intranasal oxytocin in eighteen hundred patients: A study on its safety as used in a community hospital. Am. J. Obstet. Gynecol., 110:788 (1971).
159. J. Laine; Experience of the use of intranasal, buccal and intravenous oxytocin as methods of inducing labor. Acta Obstet. Gynecol. Scand., 49:149 (1970).
160. P. Bergsjo and H. Jenssen; Nasal and buccal oxytocin for the induction of labor: A clinical trial. J. Obstet. Gynecol., Cwlth., 76:131 (1969).
161. S. Sjostedt; Induction of labour: A comparison of intranasal and transbuccal administration of oxytocin. Acta Obstet. Gynecol. Scand. (Suppl. 7), 48:1 (1969).
162. M. Newton and G. E. Elgi; The effect of intranasal administration of oxytocin on the let-down of milk in lactating women. Acta Obstet. Gynecol. Scand., 76:103 (1958).
163. P. J. Huntingord; Intranasal use of synthetic oxytocin in the treatment of breast feeding. Br. Med. J., 1:709 (1961).
164. E. A. Friedman and M. R. Sachtleben; Oxytocin in lactation: Clinical applications. Am. J. Obstet. Gynecol., 82:846 (1961).
165. L. E. Sandholm; The effect of intravenous and intranasal oxytocin on intramammary pressure during early lactation. Acta Obstet. Gynecol. Scand., 47:145 (1968).
166. N. Altszuler and J. Hampshire; Intranasal instillation of oxytocin increases insulin and glucagon secretion. Proc. Soc. Exp. Biol. Med., 168:123 (1981).
167. A. V. Schally, A. Arimura, A. J. Kastin, H. Matsuo, Y. Baba, T. W. Redding, R. M. G. Nair and J. Debeljuk; Gonadotropin-releasing hormone: One polypeptide-reg-

ulated secretion of luteinizing and follicle-stimulating hormones. Science, 173:1036 (1971).

168. C. Bergquist, S. J. Nillius, G. Skarin and L. Wide; Inhibitory effects on gonadotropin secretion and gonadal function in men during chronic treatment with a potent stimulatory luteinizing hormone-releasing hormone analog. Acta Endocrinol. (Copenh.), 91:601 (1979).
169. J. DeKonig, J. A. M. J. Van Dieten and G. P. Van Rees; Refractoriness of the pituitary gland after continuous exposure to luteinizing hormone-releasing hormone. J. Endocrinol., 79:311 (1979).
170. J. Sandow; Inhibition of pituitary and testicular function by LHRH analogues, in: *Progress Towards a Male Contraceptive* (S. L. Jeffcoate and M. Sandler, Eds.), Wiley, London, 1982, pp. 19–39.
171. Y. W. Chien, K. S. E. Su and S. F. Chang; *Nasal Systemic Drug Delivery*, Dekker, New York, 1989, Chapter 4.
172. S. T. Anik, G. McRae, C. Nerenberg, A. Worden, J. Foreman, J. Y. Hwang, S. Kushinsky, R. E. Jones and B. Vickery; Nasal absorption of Nafarelin acetate, the decapeptide [E-NaI $(2)^6$]-LHRH in rhesus monkeys. I. J. Pharm. Sci., 73:684 (1984).
173. B. H. Vikery, S. Anik, M. Chaplin and M. Henzl; Intranasal administration of nafarelin acetate: Concentration and therapeutic applications, in: *Transnasal Systemic Medications* (Y. W. Chien, Ed.), Elsevier, Amsterdam, 1985, pp. 201–215.
174. S. J. Nillius, C. Bergquist, J. Gudmendsson and L. Wide; Superagonists of LHRH for contraception in women. J. Steroid Biochem., 20:1373 (1984).
175. P. F. Brenner, D. Shoupe and D. R. Mishell; Ovulation inhibition with nafarelin acetate nasal administration for six months. Contraception, 32:531 (1985).
176. M. R. Henzl, S. L. Carson, K. Moghissi, V. C. Buttram, C. Bergquist and J. Jacobson; Administration of nasal nafarelin as compared with oral danazol for endometriosis: A multicenter double-blind comparative clinical trial. N. Engl. J. Med., 318:485 (1988).
177. E. Karasek, R. Rathsack, K. Fechner and M. Grafenberg; Nasal absorption of substance P in rats. Pharmazie, 41:289 (1986).
178. A. E. Pontiroli, M. Alberetto and G. Pozza; Intranasal glucagon raises blood glucose concentrations in healthy volunteers. Br. Med. J., 287:462 (1983).
179. A. Pontiroli, M. Alberetto and G. Pozza; Intranasal calcitonin and plasma calcium concentrations in normal subjects. Br. Med. J., 290:1390 (1985).
180. K. Morimoto, K. Morisaka and A. Kamada; Enhancement of nasal absorption of insulin and calcitonin using polyacrylic acid gel. J. Pharm. Pharmacol., 37:134–136 (1985).
181. M. L. Glefand, and M. A. Shearn; Absence of eosinopenic response to ACTH administration by the aerosol route in normal subjects. Proc. Soc. Exp. Biol. Med., 80:134 (1952).
182. F. Paulson and E. Nordstrom; Pharmacologic and clinical results with ACTH in 112 cases. Swed. Med. J., 49:2998 (1952).
183. J. Keenan, J. B. Thompson, M. A. Chamberlain and G. M. Besser; Prolonged corticotropic action of a synthetic substituted $^{1-18}$ACTH. Br. Med. J., 3:742 (1971).
184. G. Baumann, A. Walser, P. A. Desaulles, F. J. A. Paesi and L. Geller; Corticotropic action of an intranasally applied synthetic ACTH derivative. J. Clin. Endocrinol. Metab., 42:60 (1976).
185. A. E. Pontiroli, M. Alberetto, A. Secchi, G. Dossi, I. Bosi and G. Pozza; Insulin given intranasally induced hypoglycemia in normal and diabetic subjects. Br. Med. J., 284:303–306 (1982).
186. A. C. Moses, J. S. Flier, G. S. Gordon and R. S. Silver; Intranasal insulin delivery: Structure-function studies of absorption enhancing adjuvants. Clin. Res., 32:245A (1984).

187. J. S. Flier, A. C. Moses, G. S. Gordon and R. S. Silver; Intranasal admixture of insulin efficacy and mechanism, in: *Transnasal Systemic Medication* (Y. W. Chien, Ed.), Elsevier, Amsterdam, 1985, Chapter 9.
188. R. Salzman, J. E. Manson, G. T. Griffing, R. Kimmerle, N. Ruderman, A. McCall, E. I. Stolz, J. Armstrong and J. C. Melby; Intranasal aerosolized insulin: Mixed meal studies and long-term use in type I diabetes. N. Engl. J. Med., 312:1078–1084 (1985).
189. S. Hirai, T. Yashidi and H. Mima; Effect of surfactants on the nasal absorption of insulin in rats. Int. J. Pharmaceutics, 9:165–172 (1981).
190. A. C. Moses, G. S. Gordon, M. C. Carey and J. S. Flier; Insulin administered intranasally as an insulin bile salt aerosol: Effectiveness and reproducibility in normal and diabetic subjects. Diabetes, 32:1040–1047 (1983).
191. G. S. Gordon, A. C. Moses, R. D. Silver, J. S. Flier and M. C. Carey; Nasal absorption of insulin: Enhancement by hydrophobic bile salts. Proc. Natl. Acad. Sci. USA, 82:7419–7423 (1985).
192. J. P. Longnecker, A. C. Moses, J. S. Flier, R. D. Silver, M. C. Carey and E. J. Dubovi; Effects of sodium taurodihydrofusidate on nasal absorption of insulin in sheep. J. Pharm. Sci., 76:351–355 (1987).
193. L. C. Foster and W. A. Lee; The pathway for insulin absorption across the nasal mucosal membrane in the presence of the absorption enhancer sodium taurodihydrofusidate. Intl. Symp. Control. Rel. Bio. Mat., 15:87–88 (1988).
194. J. Hiscott, K. Cantell and C. Weissmann; Differential expression of human interferon genes. Nucleic Acids Res., 12:3727–3746 (1984).
195. S. Pestka; The purification and manufacture of human interferons. Sci. Am., 249:36–43 (1983).
196. T. C. Merigan, S. E. Reed, T. S. Hall and D. A. J. Tyrell; Inhibition of respiratory virus infection by locally applied interferon. Lancet, 1:563–567 (1973).
197. G. M. Scott, R. J. Phillpotts, J. Wallace, D. S. Secher, K. Cantrell and D. A. J. Tyrell; Purified interferon as protection against rhinovirus infection. Br. Med. J., 284:1822–1825 (1982).
198. F. G. Hayden, S. E. Mills and M. E. Johns; Human tolerance and histologic effects of long term administration of intranasal interferon α 2. J. Infect. Dis., 148:914–921 (1983).
199. J. G. Hardy, S. W. Lee and C. G. Wilson; Intranasal drug delivery by spray and drops. J. Pharm. Pharmacol., 37:294–297 (1985).
200. W. Petri, R. Schmiedel and J. Sandow; Development of a metered-dose nebulizer for intranasal peptide administration, in: *Transdermal Systemic Medications* (Y. W. Chien, Ed.), Elsevier, Amsterdam, 1985, Chapter 6.
201. J. Gil; Comparative morphology and ultrastructure of the airways, in: *Mechanisms in Respiratory Toxicology*, Vol. 2 (H. Witschi and P. Netlesheim, Eds.), CRC Press, Boca Raton, FL, 1982, pp. 3–25.
202. R. Breezer and M. Turk; Environ. Health Perspect., 55:3–24 (1984).
203. K. G. Bensch and E. A. M. Dominguez; Yale J. Biol. Med., 43:236–241 (1971).
204. B. Corrin; Cellular constituents of the lung, in: *Scientific Foundations of Respiratory Medicine* (J. G. Scadding and G. Cummings, Eds.), W. B. Saunders, Philadelphia, 1981, pp. 78–90.
205. M. Simeonescu; Ciba Found. Symp., 78:11–36 (1980).
206. O. D. Wagensteen, L. E. Wittmers, Jr., and J. A. Johnson, *Am. J. Physiol.*, 216:719–727 (1969).
207. E. E. Schneeburger; Fed. Proc., 37:2471–2478 (1978).
208. A. E. Taylor and K. A. Garr, Jr.; J. Am. Physiol., 218:1133–1139 (1970).
209. J. T. Gatzy; Exp. Lung Res., 3:147–161 (1982).

210. R. C. Lanman, R. M. Gillilan and L. S. Schanker; J. Pharmacol. Exp. Ther., 187:105–111 (1973).
211. J. Burton, T. H. Gardiner and L. S. Schanker; Arch. Environ. Health, 29:31-33 (1974).
212. L. S. Schanker, S. J. Ena and J. A. Burton; Xenobiotica, 7:521–528 (1976).
213. A. R. Clark and P. R. Byron; J. Pharm. Sci., 74:939–942 (1985).
214. S. G. Woolfrey, G. Taylor, I. W. Kellaway and A. Smith; J. Pharm. Pharmacol., 38(Suppl.):34P (1986).
215. R. A. Brown and L. S. Schanker; Drug Metab. Dispos., 11:355–360 (1983).
216. Y. J. Lin and L. S. Schanker; Biochem. Pharmacol., 30:2937–2943 (1981).
217. Y. J. Lin and L. S. Schanker; Am. J. Physiol., 240:C215–221 (1981).
218. J. Kerr, A. B. Fisher, and A. Kleinzeller; Am. J. Physiol., 241:E191–195 (1981).
219. Y. J. Lin and L. S. Schanker; Drug Metab. Dispos., 11:75–76 (1983).
220. M. Vadnere, A. Adjei, R. Doyle and E. Johnson; Evaluation of alternate routes for delivery of leuprolide, in: *Second International Symposium on Disposition and Delivery of Peptide Drugs*, Leiden, 1989, Abstract P22.
221. H. Yoshida, K. Okumura, R. Hori, T. Anmo and H. Yamaguchi; J. Pharm. Sci., 68:170–171 (1979).
222. F. M. Wigley, J. H. Londono, S. H. Wood, J. C. Shipp and R. H. Waldman; Diabetes, 20:552–556 (1971).
223. E. A. M. Dominguez, A. A. Liebow and K. G. Bensch; Lab. Invest., 16:905–911 (1967).
224. P. R. Byron and A. J. Hickey; J. Pharm. Sci., 76:60–64 (1987).
225. I. Gonda, A. F. A. E. Khalik and A. Z. Britten; Int. J. Pharm., 27:255–265 (1985).
226. S. P. Newman, F. Moren, D. Pavia, F. Little and S. W. Clark; Am. Respir. Dis., 124:317–320 (1981).
227. S. P. Newman, D. Pavia, N. Garland and S. W. Clarke; Eur. J. Respir. Dis., 63(Suppl. 119):57–65 (1982).
228. S. P. Newman; Chest, 88:152S–160S (1985).
229. P. R. Byron; J. Pharm. Sci., 75:433–438 (1986).
230. H. N. McCullough and R. L. Juliano; J. Natl. Cancer Inst., 63:727–731 (1979).
231. R. L. Juliano and H. N. McCullough; J. Pharmacol. Exp. Ther., 214:381–387 (1980).
232. A. J. Hickey and P. R. Byron; J. Pharm. Sci., 75:756–759 (1986).
233. R. W. Niven and P. R. Byron; Pharm. Res., 5:574–579 (1988).
234. J. C. Hogg, P. D. Pare and R. C. Boucher; Fed. Proc., 38:97–201 (1979).
235. N. A. Williams and N. D. Weiner; Int. J. Pharm., 50:261–266 (1989).
236. P. S. Banerjee and W. A. Ritschel; Int. J. Pharm., 49:199–204 (1989).
237. V.H. L. Lee; Topical ocular drug delivery: Recent advances and future perspectives. Pharm. Int., 6:135–138 (1985).
238. V. H. L. Lee and J. R. Robinson; Topical drug delivery: recent developments and future challenges. J. Ocular Pharmacol., 2:67 (1986).
239. C. D. Christie and R. F. Hanzal; J. Clin. Invest., 10:787–793 (1931).
240. G. C. Y. Chiou and C. Y. Chuang; Systemic delivery of polypeptides with molecular weights of between 300 and 3500 through the eyes. J. Ocular Pharmacol., 4:165–178 (1988).
241. G. C. Y. Chiou, C. Y. Chuang and M. S. Chang; Reduction of blood glucose concentration with insulin eye drops without using needles and syringes. Diabetes Care, 11:750–751 (1988).
242. G. C. Y. Chiou, C. Y. Chuang and M. S. Chang; Systemic delivery of insulin through eyes to lower the blood glucose concentration. J. Ocular Pharmacol., 5:81–91 (1989).
243. G. C. Y. Chiou and C. Y. Chuang; Treatment of hypoglycemia with glucagon eye drops. J. Ocular Pharmacol., 4:179–186 (1988).

244. G. C. Y. Chiou, C. Y. Chuang and M. S. Chang; Systemic delivery of enkephalin peptide through eyes. Life Sci., 43:509–514 (1988).
245. R. E. Stratford, Jr., L. W. Carson, S. Dodda-Kashi and V. H. L. Lee; J. Pharm. Sci., 77:838–842 (1988).
246. A. Yamamoto, A. M. Luo, S. Dodda-Kashi and V. H. L. Lee; J. Pharmacol. Exp. Ther., 249:249–255 (1989).
247. V. H. L. Lee, P. T. Urrea, R. E. Smith and D. T. Schanzlin; Ocular drug bioavailability from topically applied liposomes. Surv. Ophthalmol., 29:335 (1985).
248. H. W. Hui and J. R. Robinson; Ocular drug delivery of progesterone using a bioadhesive polymer. Int. J. Pharm., 26:203–213 (1985).
249. C. B. Pert, A. Pert, J. K. Chang and B. T. W. Fong; [D-Ala^2]-Met-Enkephalinamide: A potent long-lasting synthetic pentapeptide analgesic. Science, 194:330–332 (1976).
250. R. E. Stratford and V. H. L. Lee; Aminopeptidase activity in the albino rabbit extraocular tissues relative to the small intestine. J. Pharm. Sci., 74:731–734 (1985).
251. R. E. Stratford and V. H. L. Lee; Ocular aminopeptidase activity and distribution in the albino rabbit. Curr. Eye Res., 4:995 (1985).
252. K. Inagaki and V. H. L. Lee; Ocular peptide delivery: Proteolytic activities in ocular tissues of the albino rabbit. Pharm. Res., 4:S-39 (1987).
253. V. H. L. Lee, L. W. Carson, S. D. Kashi and R. E. Stratford; Metabolic and permeation barriers to the ocular absorption of topically applied enkephalins in albino rabbits. J. Ocular Pharmacol., 2:345 (1986).
254. S. D. Kashi and V. H. L. Lee; Enkephalin hydrolysis in homogenates of various mucosae of the albino rabbit: Similarities in rates and involvement of aminopeptidases. Life Sci., 38:2019–2028 (1986).
255. S. D. Kashi and V. H. L. Lee; Hydrolysis of enkephalins in anterior segment tissue homogenates of the albino rabbit eye. Invest. Ophthalmol. Vis. Sci., 27:1300 (1986).
256. V. H. L. Lee, L. W. Carson and K. A. Takemoto; Macromolecular drug absorption in the albino rabbit eye. Int. J. Pharm., 29:43–51 (1986).
257. O. Siddiqui and Y. W. Chien; Nonparenteral administration of peptides and protein drugs. CRC Crit. Rev. Ther. Drug Carrier Syst., 3:195–208 (1987).
258. J. Wieriks; Resorption of alpha amylase upon buccal application. Arch. Int. Pharmacodyn. Ther., 151:126–135 (1964).
259. K. J. Tolo; A study of permeability of gingival pocket epithelium to albumin in guinea pigs and Norwegian pigs. Arch. Oral Biol., 16:881–888 (1971).
260. K. Tolo and J. Jonsen; In vitro penetration of tritiated dextrans through rabbit oral mucosa. Arch. Oral Biol., 20:419–422 (1975).
261. H. P. Merkle, R. Anders, J. Sandow and W. Schurr; Self-adhesive patches for buccal delivery of peptides. Proc. Int. Symp. Control. Rel. Bio. Mat., 12:85 (1985).
262. M. Ishida, Y. Machida, N. Nambu and T. Nagai; New mucosal dosage forms of insulin. Chem. Pharm. Bull., 29:810–816 (1981).
263. T. Nagai; Adhesive topical drug delivery systems. J. Control. Rel., 2:121–134 (1985).
264. T. Nagai and Y. Machida; Mucosal adhesive dosage forms. Pharm. Int., 6:196–200 (1985).
265. T. Nagai; Topical mucosal adhesive dosage forms. Med. Res. Rev., 6:227–242 (1986).
266. M. M. Veillard, M. A. Longer, I. G. Tucker and J. R. Robinson; Buccal controlled delivery of peptides. Proc. Int. Symp. Control. Rel. Bio. Mat., 14:22–23 (1987).
267. M. J. Humphrey; The oral bioavailability of peptides and related drugs, in: *Delivery Systems for Peptide Drugs* (S. S. Davis, L. Illum and E. Tomlinson, Eds.), Plenum Press, London, 1986.
268. C. W. Crane, M. C. Perth and G. R. W. H. Luntz; Absorption of insulin from the human small intestine. Diabetes, 17:625–627 (1968).

269. M. Kidron, H. Bar-On, E. M. Berry and E. Ziv; The absorption of insulin from various regions of the rat intestine. Life Sci., 31:2837–2841 (1982).
270. M. Saffran, C. Bedra, G. S. Kumar and D. C. Neckers; Vasopressin: A model for the study of effects of additives on the oral and rectal administration of peptide drugs. J. Pharm. Sci., 77:33–38 (1988).
271. J. F. Woodley; Liposomes for oral administration of drugs. CRC Crit. Rev. Ther. Drug Carrier Sys., 2(1):1–18 (1986).
272. A. V. Stefanov, N. I. Kononenko, V. K. Lishko and A. V. Shevchenko; Effect of liposomally entrapped insulin administered per os on the blood sugar level in normal and experimentally diabetic rats. Ukr. Biokhim. Zh., 52:497 (1980).
273. A. Moufti, C. Weingarten, F. Puisieux, T. T. Luong and G. Durnad; Hypoglycemia after liposomized insulin in the rat. Pediatr. Res., 14:174 (1980).
274. H. M. Patel, R. W. Stevenson, J. A. Parons and B. E. Ryman; Use of liposomes to aid intestinal absorption of entrapped insulin in normal and diabetic dogs. Biochim. Biophys. Acta, 716:188–193 (1982).
275. J. E. Arrieta-Molero, K. Aleck, M. K. Sinha, C. M. Brownscheidle, L. J. Shapiro and M. A. Sperling; Orally administered liposome entrapped insulin in diabetic animals. Horm. Res., 16:249–256 (1982).
276. V. Dobre, D. Georgescu, L. Simionescu, E. Aman, U. Stroescu and C. Motas; The entrapment of biologically active substances into liposomes. I. Effects of oral administration of liposomally entrapped insulin in normal rats. Rev. Roum. Biochim., 20:15 (1983).
277. J. Kawada, N. Tamaka and Y. Nozaki; No reduction of blood glucose in diabetic rats after oral administration of insulin liposomes prepared under acidic conditions. Endocrinol. Jpn., 28:235 (1981).
278. C. Weingarten, A. Moufti, J. P. Desjeux, T. T. Luong, G. Durand, J. P. Deivssaguet and F. Puisieux; Oral ingestion of insulin liposomes: Effects of the administration route. Life Sci., 28:2747–2752 (1981).
279. M. Saffran, G. S. Kumar, C. Savariar, J. C. Burnham, F. Williams and D. C. Neckers; A new approach to the oral administration of insulin and other peptide drugs. Science, 233:1081–1084 (1986).
280. M. Shichiri, R. Kawamori, M. Yoshida, N. Etani, M. Hoshi, K. Izumi, Y. Shigeta and H. Abe; Short-term treatment of alloxan diabetic rats with intrajejunal administration of water-in-oil-in-water insulin emulsions. Diabetes, 24:971–976 (1975).
281. J. Bird, D. A. Best and D. A. Lewis; The encapsulation of insulin in erythrocytes. J. Pharm. Pharmacol., 35:246–247 (1983).
282. E. Toiutou and A. Rubinstein; Targeted enteral delivery of insulin to rats. Int. J. Pharm. 30:95–99 (1986).
283. C. Damge, C. Michel, M. Aprahamian and P. Couvreur; Advantage of a new colloidal drug delivery system in insulin treatment of streptozotocin-induced diabetic rats. Diabetologia, 29:531A (1986).
284. K. Ichikawa, I. Ohata, M. Mitomi, S. Kawamura, H. Maeno and H. Kawata; Rectal absorption of insulin suppositories in rabbits. J. Pharm. Pharmacol., 32:314–318 (1980).
285. Y. Yamasaki, R. Shichiri, M. Kawamori, M. Kikuchi, T. Yaki, S. Arai, R. Tohdo, N. Hakui, N. Oji and H. Abe; The effectiveness of rectal administration of insulin suppository in normal and diabetic subjects. Diabetes Care, 4:454–458 (1981).
286. Y. Yamasaki, M. Shichiri, R. Kawamori, T. Morishima, N. Hakui, T. Yagi and H. Abe; The effect of rectal administration of insulin on short term treatment of alloxan diabetic dogs. Can. J. Physiol. Pharmacol., 59:1–6 (1981).
287. M. Mesiha, S. Lobel, D. P. Salo, L. D. Khaleeva and N. Y. Zekova; Biopharmaceutical study of insulin suppositories. Pharmazie, 36:29–32 (1981).
288. G. G. Liversidge, T. Nishihata, K. K. Engle and T. Higuchi; Effect of rectal sup-

pository formulation on the release of insulin and on the glucose plasma levels in dogs. Int. J. Pharm., 23:87–95 (1985).
289. G. G. Liversidge, T. Nishihata, K. K. Engle and T. Higuchi; Effect of suppository shape on the systemic availability of rectally administered insulin and sodium salicylate. Int. J. Pharm., 30:247–250 (1986).
290. T. Nishihata, Y. Okamura, H. Inagaki, M. Sudho and A. Kamada; Trials of rectal insulin suppositories in healthy humans. Int. J. Pharm., 34:157–161 (1986).
291. E. Touitou, M. Donbrow and E. Azaz; New hydrophilic vehicle enabling rectal and vaginal absorption of insulin, heparin, phenol red and gentamicin. J. Pharm. Pharmacol., 30:662–663 (1978).
292. E. Touitou, M. Donbrow and A. Rubinstein; Effective intestinal absorption of insulin in diabetic rats using a new formulation approach. J. Pharm. Pharmacol., 32:108–110 (1980).
293. K. Morimoto, I. Hama, Y. Nakamoto, T. Takeeds, E. Hirano and K. Morisaka; Pharmaceutical studies of polyacrylic acid aqueous gel bases: Absorption of insulin from polyacrylic acid aqueous gel bases following rectal administration in alloxan diabetic rats and rabbits. J. Pharmacobiodyn., 3:24 (1980).
294. A. Kamada, T. Nishihata, S. Kim, M. Yamamoto and N. Yata; Study of enamine derivatives of phenylglycine as adjuvants for the rectal absorption of insulin. Chem. Pharm. Bull., 29:2012–2019 (1981).
295. T. Nishihata, J. H. Rutting, T. Higuchi and L. Caldwell; Enhanced rectal absorption of insulin and heparin in rats in the presence of non-surfactant adjuvants. J. Pharm. Pharmacol., 33:334–335 (1981).
296. T. Nishihata, J. H. Rutting, A. Kamada, T. Higuchi, M. Routh and L. Cladwell; Enhancement of rectal absorption of insulin using salicylates in dogs. J. Pharm. Pharmacol., 35:148–151 (1983).
297. T. Nishihata, G. G. Liversidge and T. Higuchi; Effect of aprotinin on the rectal delivery of insulin. J. Pharm. Pharmacol., 35:616–617 (1983).
298. T. Nishihata, S. Kim, S. Morishita, A. Kamada, N. Yata and T. Higuchi; Adjuvant effects of glyceryl esters of acetoacetic acid on rectal absorption of insulin and inulin of rabbits. J. Pharm. Sci., 72:280–285 (1983).
299. S. Kim, A. Kamada, T. Higuchi and T. Nishihata; Effect of enamine derivatives on the rectal absorption of insulin in dogs and rabbits. J. Pharm. Pharmacol., 35:100–103 (1983).
300. T. Yagi, N. Hakui, Y. Yamasahi, R. Kawamori, M. Shichiri, H. Abe, S. Kim, M. Miyake, K. Kamikawa, T. Nishihata and A. Kamada; Insulin suppository: Enhanced rectal absorption of insulin using an enamine derivative as a new promoter. J. Pharm. Pharmacol., 35:177–178 (1983).
301. K. Morimoto, E. Kamiya, T. Takeeds, Y. Nakamoto and K. Morisaka; Enhancement of rectal absorption of insulin in polyacrylic acid aqueous gel bases containing long chain fatty acids in rats. Int. J. Pharm., 14:149–157 (1983).
302. E. Touitou and M. Donbrow; Promoted rectal absorption of insulin: Formulation parameters involved in the absorption from hydrophilic bases. Int. J. Pharm., 15:13–24 (1983).
303. T. Nishihata, Y. Okamura, A. Kamada, T. Higuchi, T. Yagi, R. Kawamori and M. Shichiri; Enhanced bioavailability of insulin after rectal administration with enamine as adjuvant in depancreatized dogs. J. Pharm. Pharmacol., 37:22–26 (1985).
304. L. Caldwell, T. Nishinata, J. Fix, S. Salk, R. Cargill, C. R. Gardner and T. Higuchi; *Rectal Therapy*, Prous Publishers, Barcelona, 1984, pp. 57–61.
305. A. G. Sitnik, T. V. Degtyarova, I. M. Pertsev and L. D. Khaleeva; Biological investigation of suppositories with insulin and lidase. Farm. Zh., 64:66 (1986).
306. T. Nishihata, M. Sudoh, H. Inagaki, A. Kamada, T. Yagi, R. Kawamori and M.

Shichiri; An effective formulation for an insulin suppository; Examination in normal dogs. Int. J. Pharm., 38:83–90 (1987).
307. E. Ziv, M. Kidron, E. M. Berry and H. Bar-On; Bile salts promote the absorption of insulin from the rat colon. Life Sci., 29:803–809 (1981).
308. I. Raz, M. Kidron, H. Bar-On and E. Ziv; Rectal administration of insulin. Isr. J. Med. Sci., 20:173–175 (1984).
309. S. Yoshioka, C. Caldwell and T. Higuchi; Enhanced rectal bioavailability of polypeptides using sodium 5-methoxysalicylate as an absorption promoter. J. Pharm. Sci., 71:593–594 (1982).
310. K. Morimoto, H. Akatsuchi, R. Aikawa, M. Morishita and K. Morisaka; Enhanced rectal absorption of [Asul^{1,7}]-eel calcitonin in rats using polyacrylic acid aqueous gel base. J. Pharm. Sci., 73:1366–1368 (1984).
311. K. Morimoto, H. Akatsuchi, K. Morisaka and A. Kamada; Effect of nonionic surfactants in a polyacrylic acid gel base on the rectal absorption of [Asul^{1,7}]-eel calcitonin in rats. J. Pharm. Pharmacol., 37:759–760 (1985).
312. M. Dalmark; Plasma radioactivity after rectal instillation of radioiodine-labelled human albumin in normal subjects and in patients with ulcerative colitis. Scand. J. Gastroenterol., 3:490–496 (1968).
313. H. Okada; Vaginal route of peptide and protein drug delivery, in: *Peptide and Protein Drug Delivery* (V. H. L. Lee, Ed.), Dekker, New York, 1991, Chapter 14.
314. N. F. Fisher; Am. J. Physiol., 67:65 (1923).
315. G. D. Robinson; J. Pharmacol. Exp. Ther., 32:81 (1927).
316. D. P. Benziger and J. Edelson; Drug Metab. Rev., 14:137 (1983).
317. C. Aron and R. Aron-Brunetiere; Ann. Endocrinol., 14:1039 (1953).
318. J. Govers and J. P. Girard; Gynecol. Invest., 3:184 (1972).
319. R. R. Franklin and C. D. Dukes; Am. J. Obstet. Gynecol., 89:6 (1964).
320. D. M. Israelstam; Fertil. Steril., 20:275 (1969).
321. R. H. Glass and R. A. Vaidya; Fertil. Steril, 21:657 (1970).
322. R. H. Waldman, J. M. Cruz and D. S. Rowe; Clin. Exp. Immunol., 12:49 (1972).
323. C. Wang, H. W. G. Baker, M. G. Jennings, H. G. Burger and P. Lutjen; Fertil. Steril., 44:484 (1985).
324. G. F. B. Schumacher, M. H. Kim, A. H. Hosseiman and C. Dupon; Am. J. Obstet. Gynecol., 129:629 (1977).
325. C. D. Roig de Vargas-Linares; in: *The Human Vagina* (E. S. E. Hafez and T. N. Evans, Eds.), Elsevier/North-Holland Biomedical Press, Amsterdam, 1978, p. 193.
326. J. A. Holt, G. F. B. Schumacher, H. I. Jacobson and D. P. Schwartz, Fertil. Steril., 32:170 (1979).
327. S. -L. Yang and G. F. B. Schumacher; Fertil. Steril., 32:588 (1979).
328. K. Morimoto, T. Takeeda, Y. Nakamoto and K. Morisaka; Int. J. Pharm., 12:107 (1982).
329. H. Okada, T. Yashiki and H. Mima; J. Pharm. Sci., 72:173 (1983).
330. R. R. Humphrey, W. C. Dermody, H. O. Brink, F. G. Bousley, N. H. Schottin, R. Sakowski, J. W. Vaitkus, H. T. Veloso and J. R. Reel; Endocrinology, 92:1515 (1973).
331. N. Nishi, A. Arimura, D. H. Coy, J. A. Vilchez-Martinez and A. V. Schally; Proc. Soc. Exp. Biol. Med., 148:1009 (1975).
332. A. De La Cruz, K. G. De La Cruz, A. Arimura, D. H. Coy, J. A. Vilchez-Martinez, E. J. Coy and A. V. Schally; Fertil. Steril., 26:894 (1975).
333. M. Saito, T. Kumasaki, Y. Yaoi, N. Nishi, A. Arimura, D. H. Coy and A. V. Schally; Fertil. Steril., 28:240 (1977).
334. I. Yamazaki and H. Okada; Endocrinol. Jpn., 27:593 (1980).
335. H. Okada, I. Yamazaki and T. Yashiki; J. Pharmacobio. Dyn., 4:S-17 (1981).

336. H. Okada, I. Yamazaki, Y. Ogawa, S. Hirai, T. Yashiki and H. Mima; J. Pharm. Sci., 71:1367 (1982).
337. H. Okada, I. Yamazaki, T. Yashiki and H. Mima; J. Pharm. Sci., 72:75 (1983).
338. H. Okada, Y. Sakura, H. Kawaji, T. Yashiki and H. Mima; Cancer Res., 43:1869 (1983).
339. H. Okada, I. Yamazaki, Y. Sakura, T. Yashiki, T. Shimamoto and H. Mima; J. Pharmacobio. Dyn., 6:512 (1983).
340. H. Okada; J. Takeda Res. Lab., 42:150 (1983).
341. H. Okada, I. Yamazaki, T. Yashiki, T. Shimamoto and H. Mima; J. Pharm. Sci., 73:298 (1984).
342. I. Yamazaki; Endocrinol. Jpn., 29:197 (1982).
343. I. Yamazaki; Endocrinol. Jpn., 29:415 (1982).
344. I. Yamazaki; J. Reprod. Fertil., 72:129 (1984).
345. S. Hirai, H. Okada, T. Yashiki and T. Shimamoto; *Abstracts of 105th Ann. Meet. Pharm. Soc. Jpn.*, Kanazawa, Japan, 1985, p. 797.
346. E. S. Johnson, J. H. Seely, W. H. White and E. R. DeSombre; Science, 194:329 (1976).
347. A. Lemay, R. Maheux, N. Faure, C. Jean and A. T. A. Fazekas; Fertil. Steril., 41:863 (1984).
348. P. A. Boepple, M. J. Mansfield, M. E. Wierman, C. R. Rudlin, H. H. Bode, J. F. Crigler, Jr., J. D. Crawford and W. F. Crowley, Jr.; Endocrinol. Rev., 7:24 (1986).
349. K. L. Sheeham, R. F. Casper and S. S. C. Yen; Science, 215:170 (1982).
350. R. T. Tregear; The permeability of skin to albumin, dextrans and polyvinylpyrrolidone. J. Invest. Dermatol., 46:24–27 (1966).
351. M. Menasche, M. P. Jacob, G. Godeau, A. M. Robert and L. Robert; Pharmacological studies on elastin peptides (kappa-elastin). Blood clearance, percutaneous penetration and tissue distribution. Pathol. Biol., 29:548–554 (1981).
352. G. Schuster and L. A. Domsch; Protein chemistry as related to cosmetics and toiletries. Cosmet. Toilet., 99:63–74 (1984).
353. P. Alexander; Protein derivatives: Part I. Manuf. Chem., 57:47–51 (1986).
354. P. Alexander; Protein derivatives: Part II. Manuf. Chem., 57:63–67 (1986).
355. B. R. Brunette and D. Marrero; Comparison between the iontophoretic and passive transport of thyrotropin-releasing hormone across excised nude mouse skin. J. Pharm. Sci., 75:738–743 (1986).
356. P. S. Banerjee and W. A. Ritschel; Int. J. Pharm. 49:189–197 (1989).
357. P. Lelawongs, J. C. Liu, O. Siddiqui and Y. W. Chien; Transdermal iontophoretic delivery of vasopressin. I. Physicochemical consideration. Int. J. Pharm., 56:13–22 (1989).
358. Y. W. Chien, P. Lelawongs, O. Siddiqui, Y. Sun and W. M. Shi; Facilitated transdermal delivery of therapeutic peptides/proteins by iontophoretic delivery devices. J. Control. Rel., 13:263–278 (1990).
359. A. K. Banga and Y. W. Chien; Iontophoretic delivery of drugs: Fundamentals, developments and biomedical applications. J. Control. Rel., 7:1–24 (1988).
360. R. L. Stephen, Petelenz and S. C. Jacobsen; Potential novel methods for insulin administration: Iontophoresis. Biomed. Biochim. Acta, 43:553–558 (1984).
361. B. Karl; Control of blood glucose levels in alloxan-diabetic rabbits by iontophoresis of insulin. Diabetes, 35:217–221 (1986).
362. O. Siddiqui, Y. Sun, J. C. Liu and Y. W. Chien; Facilitated transdermal transport of insulin. J. Pharm. Sci., 76:341–345 (1987).
363. Y. Sun, O. Siddiqui, J. C. Liu and Y. W. Chien; Transdermal modulated delivery of polypeptides: Effect of DC pulse wave form on enhancement. Proc. Int. Symp. Control. Rel. Bio. Mat., 13:175–176 (1986).

364. O. Siddiqui, W. Shi and Y. W. Chien; Transdermal iontophoretic delivery of insulin for blood glucose control in diabetic rabbits. Proc. Int. Symp. Control. Rel. Bio. Mat., 14:174–175 (1987).
365. Y. W. Chien, O. Siddiqui, Y. Sun, W. M. Shi and J. C. Liu; Transdermal iontophoretic delivery of therapeutic peptides/proteins. I. Insulin. Ann. N.Y. Acad. Sci., 507:32–51 (1988).
366. J. C. Liu, Y. Sun, O. Siddiqui and Y. W. Chien; Blood glucose control in diabetic rats by transdermal iontophoretic delivery of insulin. Int. J. Pharm., 44:197–204 (1988).
367. Y. W. Chien, O. Siddiqui, W. M. Shi, P. Lelawongs and J. C. Liu; Pulse dc-iontophoretic transdermal delivery of peptide and protein drugs. J. Pharm. Sci., 78:376–383 (1989).
368. P. Lelawongs, J. C. Liu and Y. W. Chien; Transdermal iontophoretic delivery of arginine-vasopressin. II. Evaluation of electrical and operational factors. Int. J. Pharm., 61:179–188 (1990).
369. D. Sibalis; Transdermal drug application and electrodes thereof. U.S. Patent 4,640,689, February 3, 1987.
370. D. Sibalis; Transdermal drug applicator. U.S. Patent 4,708,716, November 24, 1987.
371. B. R. Meyer; Electro-osmotic transdermal drug delivery, in: *1987 Conference Proceedings on the Latest Developments in Drug Delivery Systems*, Aster Publishing, Eugene, Oregon, 1987, p. 40.
372. B. R. Meyer, H. L. Katzeff, J. C. Eschbach, J. Trimmer, S. B. Zacharias, S. Rosen and D. Sibalis; Transdermal delivery of human insulin to albino rabbits using electrical current. Am. J. Med. Sci., 297:321–325 (1989).
373. B. R. Meyer, W. Kreis, J. Eschbach, V. O'Mare, S. Rosen and D. Sibalis; Successful transdermal administration of therapeutic doses of a polypeptide to normal human volunteers. Clin. Pharmacol. Ther., 44:607–612 (1988).
374. K. Stillwell; *Therapeutic Electricity and Ultraviolet Radiation*, 3rd ed., Williams and Wilkins, Baltimore, 1983, p. 33.
375. Y. W. Chien; *Novel Drug Delivery Systems*, Dekker, New York, 1982, Chapter 5.
376. K. Okabe, H. Yamaguchi and Y. Kawai; New iontophoretic transdermal administration of the beta blocker metoprolol. J. Control. Rel., 4:79–85 (1986).
377. W. A. Check; New drugs and drug-delivery systems in the year 2000. Am. J. Hosp. Pharm., 41:1536–1547 (1984).
378. T. W. Redding and A. V. Schally; Neuroendocrinology, 9:250–256 (1972).
379. H. Okada, I. Yamazaki, Y. Ogawa, S. Hirai, T. Yashiki and H. Mima; J. Pharm. Sci., 71:1367–1371 (1982).
380. M. J. Hageman; Drug Dev. Ind. Pharm., 14:2047–2070 (1988).
381. K. Docherty, R. Carroll and D. F. Steiner; Proc. Natl. Acad. Sci. USA, 79:4613–4617 (1982).
382. S. O. Emdin, G. G. Dodson, J. M. Cutfield and S. M. Cutfield; Diabetologia, 19:174–182 (1974).
383. B. Given, R. Cohen, B. Brank, A. Rubenstein and H. Tager; Clin. Res., 33:569A (1985).
384. M. J. Brownstein; Peptidergic pathways in the central nervous system. Proc. R. Soc. (Lond. B), 210:79–90 (1980).
385. R. A. Nicholl, B. E. Alger and C. E. Jahr; Peptides as putative excitatory neurotransmitters: Carnosine, Enkephalin, Substance P and TRH, *Proc. Royal Soc. Lond.* [*Biol.*], 210:133–149 (1980).
386. S. B. Weinberger and J. L. Martinez, Jr.; Characterization of hydrolysis of [Leu] enkephalin and D-ala^2-[1-leu] enkephalin in rat plasma. J. Pharmacol. Exp. Ther., 247:129–135 (1988).

387. Z. Vogel, T. Miron, M. Altstein and M. Wilchek; Spontaneous inactivation of enkephalin. Biochem. Biophys. Res. Commun., 85:226–233 (1978).
388. J. M. Hambrook, B. A. Morgan, M. J. Rance and L. F. C. Smith; Mode of deactivation of the enkephalins by rat and human plasma and rat brain homogenates. Nature, 262:782–783 (1976).
389. L. Graf, A. Nagy and A. Lajtha; Enkephalin-hydrolyzing peptidases of rat brain membrane: Are they topographically/functionally coupled to opiate receptors? Life Sci., 31:1861–1865 (1982).
390. J. G. C. van Amsterdam; K. J. H. van Buuren, A. M. de Jong and W. Soudijn; Inhibitors of calf-brain enkephalinase A and B. Life Sci., 33(suppl. I):109–112 (1983).
391. L. B. Hersh and J. F. McKelvy; An aminopeptidase from bovine brain which catalyzes the hydrolysis of enkephalin. J. Neurochem., 36:171–178 (1981).
392. N. Marks, A. Grynbaum and A. Neidle; On the degradation of enkephalins and endorphins by rat and mouse brain extracts. Biochem. Biophys. Res. Commun., 74:1552–1559 (1977).
393. M. N. Gillespie, J. W. Krechniak, P. A. Crooks, R. J. Altiere and J. W. Olson; Pulmonary metabolism of exogeneous enkephalins in isolated perfused rat lung. J. Pharmacol. Exp. Ther., 232:675–681 (1985).
394. M. L. Cohen, L. E. Geary and K. S. Wiley; Enkephalin degradation in the guinea pig ileum: Effect of aminopeptidase inhibitors, puromycin and bestatin. J. Pharmacol. Exp. Ther., 224:379–385 (1983).
395. H. K. Choi, G. L. Flynn and G. L. Amidon; Transdermal delivery of bioactive paptides: The effect of n-decylmethyl sulfoxide and pH on enkephalin transport. Pharm. Res., 6:S-148 (1989).
396. S. Dodda Kashi and V. H. L. Lee; Enkephalin hydrolysis of various absorptive mucosae of the albino rabbit: Similarities in rates and involvement of aminopeptidases. Life Sci., 38:2019–2028 (1986).
397. S. Dodda Kashi and V. H. L. Lee; Hydrolysis of enkephalins in homogenates of anterior segment tissues of the albino rabbit eye. Invest. Ophthalmol. Vis. Sci., 27:1300–1303 (1986).
398. V. H. L. Lee; Ophthalmic delivery of peptides and proteins. Pharm. Technol., April:26–38 (1987).
399. K. W. Garen and A. J. Repta; Buccal drug absorption. I. Comparative levels of esterase and peptidase activities in rat and hamster buccal and intestinal homogenates. Int. J. Pharm., 48:189–194 (1988).
400. I. K. Chun and Y. W. Chien; Methionine enkephalin. I. Kinetics of degradation in buffered solution and metabolism in various mucosa extracts. Pharm. Res., 7:S-48 (1990).
401. A. P. Sayani, I. K. Chun and Y. W. Chien; Transmucosal delivery of leucine enkephalin: Stabilization in and permeation through rabbit membranes. American Association of Pharmaceutical Scientists' 6th annual meeting, Washington, D.C., Nov. 17–21, 1991.
402. D. H. Rich, B. J. Moon and S. Harbeson; Inhibition of aminopeptidases by amastatin and bestatin derivatives. Effect of inhibitor structure on slow-binding processes. J. Med. Chem., 27:417–422 (1984).
403. I. K. Chun and Y. W. Chien; Transmucosal delivery of methionine enkephalin (Met-Enk) II. Stabilization of Met-Enk in various rabbit mucosa extracts by combined enzyme inhibitors. International Congress of New Drug Development, Seoul, Korea, August 18–24, 1991.
404. P. K. Shah and R. T. Borchardt; Cultured human keratinocytes as an in vitro model to study peptide metabolism in human skin. Pharm. Res., 6:S-148 (1989).
405. H. K. Choi, G. L. Flynn and G. L. Amidon; Transdermal delivery of bioactive pep-

tides: The effect of n-decylmethyl sulfoxide, pH, and inhibitors on enkephalin metabolism and transport. Pharm. Res., 7:1099–1106 (1990).

406. Y. W. Chien; Systemic delivery of therapeutic peptides: Overcome of enzymatic barrier and enhancement of permeation. Second Jerusalem Conference on Pharmaceutical Sciences and Clinical Pharmacology, Jerusalem, Israel, May 24–29, 1992.
407. S. J. Shire; Formulation design of protein pharmaceuticals. Pharm. Technol., 11:34 (August 1987).
408. V. H. L. Lee; Peptide and protein drug delivery: Opportunities and challenges. Pharm. Int., 7:208–212 (1986).
409. T. H. Nguyen and R. E. Jones; Preformulation studies on recombinant tissue plasminogen activator. Pharm. Res., 4:S-35 (1987).
410. J. B. Hill; Adsorption of insulin to glass. Proc. Soc. Exp. Biol. Med., 102:75–77 (1959).
411. T. Mizutani and A. Mizutani; Estimation of adsorption of drugs and proteins on glass surfaces with controlled pore glass as a reference. J. Pharm. Sci., 67:1102–1105 (1978).
412. T. Mizutani; Decreased activity of proteins adsorbed onto glass surfaces with porous glass as a reference. J. Pharm. Sci., 69:279–281 (1980).
413. S. T. Anik and J. Y. Hwang; Adsorption of D-Nal $(2)^6$ LHRH, a decapeptide, onto glass and other surfaces. Int. J. Pharm., 16:181–190 (1983).
414. Z. J. Twardowski, K. D. Nolph, T. J. McGary and H. L. Moore; Nature of insulin binding to plastic bags. Am. J. Hosp. Pharm., 40:579–582 (1983).
415. J. C. McElnay, D. S. Elliott and P. F. D'Arcy; Binding of human insulin to burette administration sets. Int. J. Pharm., 36:199–203 (1987).
416. C. Petty and N. L. Cunningham; Insulin adsorption by glass infusion bottles, polyvinyl chloride infusion containers and intravenous tubing. Anesthesiology, 40:400–404 (1974).
417. J. L. Hirsch, M. J. Fratkin, J. H. Wood and R. B. Thomas; Clinical significance of insulin adsorption to polyvinyl chloride infusion systems. Am. J. Hosp. Pharm., 34:583–588 (1977).
418. T. Mizutani and A. Mizutani; Prevention of adsorption of protein on controlled-pore glass with amino acid buffer. J. Chromatogr., 111:214–216 (1975).
419. H. Furberg, A. K. Jenson and B. Salbu; Effect of pretreatment with 0.9% sodium chloride or insulin solutions on the delivery of insulin from an infusion system. Am. J. Hosp. Pharm., 43:2209–2213 (1986).
420. A. H. Pekar and B. H. Frank; Conformation of proinsulin. A comparison of insulin and proinsulin self-association at neutral pH. Biochemistry, 11:4013–4016 (1972).
421. R. Quinn and J. D. Andrale; Minimizing the aggregation of neutral insulin solutions. J. Pharm. Sci., 72:1472–1473 (1983).
422. E. H. Massey and T. A. Sheliga; Development of aggregation resistant insulin formulations. Pharm. Res., 3:26S (1986).
423. S. Sato, C. D. Ebert and S. W. Kim; Prevention of insulin self-association and surface adsorption. J. Pharm. Sci., 72:228–232 (1983).
424. J. Bringer, A. Heldt and G. M. Grodsky; Prevention of insulin aggregation by dicarboxylic amino acids during prolonged infusion. Diabetes, 30:83–85 (1981).
425. B. Wigness, F. Dorman, T. Rhode, K. Kernstine, E. Chute and H. Buchwald; Improved glycerol-insulin solutions for use in implantable pumps. Diabetes, 35:140A (1986).
426. P. L. Privalov; Stability of proteins: Small globular proteins. Adv. Protein Chem., 33:167–241 (1979).
427. N. R. Stephenson and R. G. Romans; Thermal stability of insulin made from zinc insulin crystals. J. Pharm. Pharmacol., 12:372–376 (1960).
428. W. O. Storvick and H. J. Henry; Effect of storage temperature on stability of commercial insulin preparations. Diabetes, 17:499–502 (1968).

429. M. Bingel and A. Volund; Stability of insulin preparations. Diabetes, 21:805–813 (1972).
430. G. Federici, S. Dupre, E. Barboni, A. Fiori and M. Costa; Insulin dissociation at alkalline pH. FEBS Lett., 32:27–29 (1973).
431. P. D. Jeffrey, B. K. Milthrope and L. W. Nichol; Polymerization pattern of insulin at pH 7.0. Biochemistry, 15:4660–4665 (1976).
432. B. V. Fisher and P. B. Porter; Stability of bovine insulin. J. Pharm. Pharmacol., 33:203–206 (1981).
433. W. D. Longheed, A. M. Albisser, H. M. Martindale, J. C. Chow and J. R. Clement; Physical stability of insulin formulations. Diabetes, 32:424–432 (1983).
434. U. Grau; Chemical stability of insulin in a delivery system environment. Diabetologia, 28:458–463 (1985).
435. P. S. Adams, R. F. Haines-Nutt and R. Town; Stability of insulin mixtures in disposable plastic insulin syringes. J. Pharm. Pharmacol., 39:158–163 (1987).
436. M. Hickers; Peptides and proteins: Pharmacokinetics and analytical considerations, in: *28th National Industrial Pharmaceutical Research Conference*, Madison, Wisconsin, 1986.
437. R. S. Sherwin, K. J. Kramer, J. D. Tobin, P. A. Insel, J. E. Liljenuist, M. Berman and R. Andres; A model of the kinetics of insulin in man. J. Clin. Invest., 53:1481–1492 (1974).
438. B. Koch and B. Lutz-Bucher; Specific receptors for vasopressin in the pituitary gland: Evidence for down-regulation and desensitisation to adrenocorticotropin-releasing factors. Endocrinology, 116:671–676 (1985).
439. A. Reinberg, P. Drouin, M. Kolopp, L. Mejean, F. Levi, G. Debry, M. Mechkouri, G. DiCostanzo and A. Bicakora; Pump delivered insulin and home monitored blood glucose in a diabetic patient: Retrospective and chronophysiologic evaluation of 3-year time series, in: *Proc. 3rd Int. Conf. Chronopharm.*, March 1988, Nice, France.
440. J. Hunter, J. McGee, J. Saldivar, T. Tsai, R. Feuers and L. E. Scheving; Circadian variations in insulin and alloxan sensitivity noted in blood glucose alterations in normal and diabetic mice, in: *Proc. 3rd Int. Conf. Chronopharma.*, March 1988, Nice, France.
441. B. Tarquini, V. Cavallini, A. Cariddi, M. Checchi, V. Sorice and M. Cecchettin; Prominent circadian absorption of intranasal salmon calcitonin (SCT) in healthy subjects, *Chronobiol. Int.*, 4:199–202 (1987).
442. T. R. Malefyt; Analytical aspects in the development of peptide and protein drugs. Pharm. Technol., 11:34–43 (August 1987).
443. M. Ohta, H. Tokunaga, T. Kimura, H. Satoh and J. Kawamura; Analysis of insulins by high-performance liquid chromatography. III. Determination of insulins in various preparations. Chem. Pharm. Bull., 32:4641–4644 (1984).
444. P. S. Adams and R. F. Haines-Nutt; Analysis of bovine, porcine and human insulins in pharmaceutical dosage forms and drug delivery systems. J. Chromatogr., 351:574–579 (1986).
445. M. C. Sammons, B. R. Demark and M. S. McCracken; Determination of total and encapsulated insulin in a vesicle formulation. J. Pharm. Sci., 75:838–841 (1986).
446. G. N. SubbaRao, J. W. Sutherland and G. N. Menon; High performance liquid chromatographic assay for thyrotropin releasing hormone and benzyl alcohol in injectable formulation. Pharm. Res., 4:38–41 (1987).
447. C. R. Morgan and A. Lazarow; Immunoassay of insulin: Two antibody system. Diabetes, 12:115–126 (1963).
448. R. C. Talamo, E. Haber and K. F. Austen; A radioimmunoassay for bradykinin in plasma and synovial fluid. J. Lab. Clin. Med., 74:816–827 (1969).
449. M. Ceska, F. Grossmuller and U. Lundkvist; Solid-phase radioimmunoassay of insulin. Acta Endocrinol. (Copenh.), 64:111–125 (1970).

450. C. G. Beardwell; Radioimmunoassay of arginine vasopressin in human plasma. J. Clin. Endocrinol. Metab., 33:254–260 (1971).
451. R. S. Yalow; Radioimmunoassay methodology: Application to problems of heterogeneity of peptide hormones. Pharmacol. Rev., 25:161–178 (1973).
452. R. C. Talamo and T. L. Goodfriend; Bradykinin radioimmunoassays. Handb. Exp. Pharmacol., 25:301–309 (1979).
453. J. L. Meek; Derivatizing reagents for high-performance liquid chromatography detection of peptides at the picomole level. J. Chromatog., 266:401–408 (1983).
454. R. E. Chance, E. P. Kroeff, J. A. Hoffmann and B. H. Frank; Chemical, physical and biological properties of biosynthetic human insulin. Diabetes Care, 4:147–154 (1981).
455. B. V. Fisher and D. Smith; HPLC as a replacement for the animal response assays for insulin. J. Pharm. Biomed. Anal., 4:377–387 (1986).
456. S. P. Musial, M. P. Duran and R. V. Smith; Analysis of solvent-mediated conformational changes of insulin by radioimmunoassay (RIA) techniques. J. Pharm. Biomed. Anal., 4:589–600 (1986).
457. M. E. Hemling; Fast atom bombardment mass spectrometry and its application to the analysis of some peptides and proteins. Pharm. Res., 4:5–15 (1987).
458. D. Baxter; A radioreceptor assay for pharmaceutical preparations of insulin. J. Biol. Stand., 14:319–330 (1986).
459. L. Sjodin and E. Viitanen; Radioreceptor assay for insulin formulations. Pharm. Res., 4:189–194 (1987).
460. M. Z. Beigel, N. N. Soboleva and V. I. Trofimov; Determination of the dimensions of particles in suspensions of zinc-insulin by a turbidity spectrum method. Khim. Farm. Zh., 19:1395–1400 (1985).
461. B. J. Davis; Disc electrophoresis. II. Method and application to human serum proteins. Ann. N.Y. Acad. Sci., 121:404–427 (1964).
462. A. Stockell Hartree; Separation and partial purification of the protein hormone from human pituitary glands. Biochem. J., 100:754–761 (1966).
463. L. E. Reichert; Electrophoretic properties of pituitary gonadotropins as studied by electrofocussing. Endocrinology, 88:1029–1044 (1971).
464. P. O. Farrell; High resolution of two-dimensional electrophoresis of proteins. J. Biol. Chem., 250:4007–4021 (1975).
465. A. J. Parcells; Protein purification: Past, present and future. Pharm. Technol., 8:78–88 (September 1984).
466. A. Stockell Hartree, J. B. Lester and R. C. Shownkeen; Studies of the heterogeneity of human pituitary LH by fast protein liquid chromatography. J. Endocrinol., 105:405–413 (1985).
467. I. Kerese; *Methods of Protein Analysis* (translated by R. A. Chalmers), Halsted Press, 1984.
468. J. M. Walker; *Methods in Molecular Biology*, Vol. 1, Proteins, Humana Press, 1984.
469. F. M. Bogdansky; Considerations for the quality control of biotechnology products. Pharm. Technol., 11:72–74 (September 1987).
470. M. D. Giddins and C. T. Rhodes; Developing functionally relevant standards for protein and polypeptide drugs produced using biotechnological methods, in: *8th Annual Meeting, Northeastern Regional Pharmaceutical Association*, New Haven, Connecticut, 1987.
471. E. L. Korwek; FDA, OSHA and EPA regulation of the recombinant DNA technology. J. Parent. Sci. Technol., 36:251–255 (1982).
472. H. I. Miller; Recombinant DNA as a paradigm of new technology: Its impact on reg-

ulation by the Food and Drug Administration. J. Parent. Sci. Technol., 36:248–250 (1982).

473. J. C. Petricciani; Regulatory considerations for products derived from the new biotechnology. Pharm. Manuf., January:31–34 (1985).
474. W. Szkrybalo; Emerging trends in biotechnology: A perspective from the pharmaceutical industry. Pharm. Res., 4:361–363 (1987).

12
Regulatory Considerations in Controlled Drug Delivery

I. INTRODUCTION

The pharmaceutical industry in the United States is a relatively small, but highly research-intensive industry, and is one of the most important U.S. high-technology industries. In 1989, the Pharmaceutical Manufacturers Association (PMA) member firms had a total worldwide sales of $51.2 billion but reinvested 16.7% of their revenues ($7.3 billion) in biomedical research and pharmaceutical development (1).

The essential characteristics of this research-based pharmaceutical industry is its continued commitment to the discovery, development, and marketing of new medicines. However, innovations in new drug therapy have become increasingly more complex, time consuming, and costly over the years (Table 1). The cost of developing a new drug is high and has increased more than four times over a time span of 14 years (Table 2). It has been estimated that, on average, it cost $231 million in 1990 to bring a new drug through discovery, preclinical research and development (R&D) and clinical testing for Food and Drug Administration (FDA) approval before marketing (2). More than 12 years is necessary to complete the R&D activities required for regulatory approval of a new drug entity (Table 3). The regulatory review process for a new drug application (NDA) is long, an average of 31.6 months in 1981–1990 (Table 4). Over the last 10 years, the FDA has approved an average of 23 new drugs per year but its work load for NDA review has remained high with a backlog of 67 NDAs per year. It has been found that of the 135 new drug entities approved by the FDA between 1984 and 1989, 106 have already been approved in other countries (Table 5): 70% were first approved in Germany, 55.7% in Switzerland, 54.7% in the United Kingdom, 48.1% in France, 43.4% in Italy, and 34.9% in Japan (1). A trend that has become increasingly common in recent years is that pharmaceutical companies often seek regulatory approval for marketing of their pharmaceutical products first in foreign nations.

The lack of FDA-approved drugs, the high cost of new drug development, and the increasing number of patent expiration for existing drugs (Tables 6 and 7) means that a growing number of pharmaceutical companies have been faced with a decreasing number of patent-protected drugs from which they may generate revenue.

Table 1 Research Expenditure and Development Productivity in the U.S. Pharmaceutical Industry

	1960	1965	1975	1990
Research Expenditure (U.S. $ millions)	<100	365	1062	8229[a]
Development productivity (no. new drug entity receives NDA approval for maketing)	50	25	15	23
R&D efficiency				
R&D cost ($ millions/new drug entity)	<2	15	71	358[b]
R&D time (years/new drug entity)	2	—	10–15	>12

[a]R&D expenditures (projected) for ethical pharmaceuticals by PMA member firms in 1990.
[b]$231 million (estimated by J. DiMasi, Tufts University) or $284 million (when the actual R&D expenditures for 1988 is used for calculation).
Source: PMA *Annual Survey Report* (1988–90 edition) and *Facts at a Glance* (1990 edition).

A recent survey conducted by Grabowski and Vernon (3) indicated that only 3 of every 10 drugs introduced in the period 1970–1979 subsequently recovered R&D costs.

This dilemma, which has been faced by most pharmaceutical companies in recent years, could be resolved by developing a new R&D strategy. Instead of a constant search for new drugs in the traditional random, hit-or-miss approach, new R&D programs should be developed to focus on making clinically established drugs do their therapeutic best. The development of novel and patentable technologies and/or delivery systems, by applying the concepts and techniques of *rate-controlled drug delivery*, may not only extend the patent protection of the existing drugs but also minimize the scope and expenditure of testing required for FDA approval. The re-

Table 2 Developmental Cost of a New Drug in the United States

Year of estimate	Developmental cost (direct and indirect)	Reference
1976	$54 million	R. Hansen (University of Rochester)
1982	$87 million[a]	R. Hansen (University of Rochester)
1987	$125 million	S. N. Wiggins (Texas A&M University)
1990	$231 million[b]	J. DiMasi (Tufts University)

[a]Adjusted by PMA for inflation.
[b]Based on FDA approval success rate of 23% (DiMasi, 1990).
Source: PMA *Facts at a Glance* (1990 edition).

Table 3 Steps and Time Toward Regulatory Approval of a New Drug Entity for Marketing in United States

Step	R&D activities for regulatory approval		Time
0	Discovery of a new drug entity		Varies
1	Preclinical R&D activities		3.5 years
2	Filing of IND application		Varies
3	Clinical studies		
	Phase I	Safety and bioavailability study	1.0 year
	Phase II	Evaluation of therapeutic efficacy	2.0 years
	Phase III	Extensive clinical trials	3.0 years
4	Filing of NDA with FDA		Varies
5	Review of NDA documentation		2.5 years
6	FDA approval		—
		Total	>12 years

Source: Compiled from PMA *Facts at a Glance* (1990 edition) and analysis of J. A. DiMasi's data (Tufts University).

quirements for the regulatory approval of controlled-release drug delivery systems is elaborated in Section IV.

II. HISTORIC DEVELOPMENTS

The regulatory requirements related to controlled-release dosage forms first appeared in a regulation that was really a statement of policy and was published 30 years ago by the FDA. It defined the conditions under which drugs delivered to patients in a controlled-release formulation over a prolonged period would be regarded as new

Table 4 Approval and Backlog of New Drug Application at FDA (10-Year Statistics)

	New drug application/year		Mean approval time (months)	
Year	Approved	Pending	All	Fast track
1981	27	48	31	23.5
1982	28	65	28.5	14.5
1983	14	74	28	19
1984	22	83	39	25.5
1985	30	64	31.5	28.5
1986	20	66	34	32
1987	21	63	32.4	23
1988	20	72	31.3	—
1989	23	69	32.5	22.5
1990	23	—	27.7	—
Mean ± SD	22.8 ± 4.6	67.1 ± 9.5	31.6 ± 3.3	23.6 ± 5.4

Source: U.S. Food and Drug Administration.

Table 5 Regulatory Approval: United States versus Foreign Countries

	Number of new drug entity approved		
		First Approval	
Year	Total	United States	Foreign
1984–89	135	29	106
1989	23	5	18

Source: Compiled from PMA *Facts at a Glance* (1990 edition).

drugs within the meaning of the Federal Food, Drug and Cosmetic Act, Section 201(p). Since then, there has been a proliferation of controlled-release dosage forms for pharmaceutical products that may have little rationale and provide no advantage over the same drugs in conventional dosage forms. There has also been an increase in the use of controlled drug release labeling claims for drugs, and some of them may not be warranted.

Controlled-release dosage forms have been referred to by a number of terminologies, such as delayed-action, extended-action, gradual-release, prolonged-release, protracted-release, depot, repeated-action, slow-release, sustained-release, retard, and timed-release dosage forms. Any of these terms, or similar terms that impart the same idea, may have been used in the labeling for designation of controlled-release dosage forms.

Table 6 Pharmaceuticals with Patent Already Expired in Last 10 Years

Year of patent expiration		Pharmaceuticals[a]	
1981	Dyazide (250)	Tolinase (70)	Afrin (29)
	Depakene (23)	Polycilin (14)	
1982	Vibramycin (64)	Minocin (52)	Reglan (46)
	Keflin (44)		
1983	Mellaril (76)	Apresazide (15)	Atromid S (14)
1984	Inderal (335)	Indocin (106)	Ativan (88)
	Dalmane (56)	Norpace (47)	Navane (31)
	Onvocin (28)	Serax (24)	Restoril (23)
	Symmetral,(14)	Dymelor (12)	
1985	Motrin (209)	Aldomet (204)	Microdantin (50)
1986	Haldol (76)	Sinequan (44)	Flexeril (40)
	Carafate (25)	Tinactin (19)	
1987	Keflex (224)	Ancef (86)	Omnipen (31)
	Duricef (26)	Ser-ap-es (26)	Cleocin (12)
1988	Naflon (46)		
1989	Diabinese (151)	Clinoril (142)	
Total (1983 sales) = $2872 million			

[a]Only those pharmaceuticals with 1983 sales of at least $12 million U.S. The figures in parentheses are 1983 sales in $ millions U.S.

Table 7 Best-Selling Pharmaceuticals with Patent Expiring Within 5 Years

Year of patent expiration	Pharmaceutical product	Annual sales[a] ($ millions)
1991	Procardia	365
1992	Cardizem	510
	Calcor	330
	Feldene	310
1993	Xanax	340
	Naprosyn	320
	Calan	270
1994	Tagamet	500
1995	Zantac	1050
	Capoten	426
Total (1988) sales = $4421 million		

[a]1988 sales.
Source: D. Saks (*Pharmacy Today*, 1/26/90).

III. RECENT DEVELOPMENTS AND PERSPECTIVES

A milestone for the evolution of controlled-release dosage forms was recently made with the development of several new and innovative controlled-release drug delivery systems, primarily the invention of polymer-mediated controlled-release drug delivery devices for the long-term medication of chronic ailments. These are exemplified by the development of the nifedipine-releasing gastrointestinal therapeutic system (Procardia XL, Pfizer) for the once-a-day management of vasopastic and chronic stable angina or the treatment of hypertension (Chapter 3), pilocarpine-releasing ocular inserts (Ocusert system, Ciba) for the weekly management of glaucoma (Chapter 6), nitroglycerin-releasing transdermal therapeutic systems (Deponit, Schwarz; Minitran, 3M; Nitrodisc, Searle; Nitro-Dur, Key; Transdermal-NTG, Warner Chilcott; Transderm-Nitro, Summit) for the 12–24 hr prevention and treatment of angina pectoris (Chapter 7), levonorgestrel-releasing subdermal implants (Norplant, Wyeth-Ayerst) for 5-year fertility regulation in females (Chapter 8), a progesterone-releasing intrauterine device (Progestasert IUD, Alza) for 1-year intrauterine contraception (Chapter 10), and the goserelin acetate-releasing biodegradable implant (Zoladex, ICI) for the once-a-month subcutaneous palliative treatment of advanced carcinoma of the prostate (Chapters 8 and 11). Regulatory approval of these controlled-release drug delivery systems for marketing has been granted only in the past 10 years.

IV. REGULATORY REQUIREMENTS

As with new drugs in conventional dosage forms, the regulatory approval of a controlled-release pharmaceutical product (or drug delivery system) requires submission of scientific documents from pharmaceutical firms to substantiate the clinical safety and efficacy of the controlled-release drug delivery system and to demonstrate its controlled drug release characteristics. The requirements for regulatory approval may be outlined as follows (4).

A. Requirements to Demonstrate Safety and Efficacy

To demonstrate the safety and efficacy of a controlled-release formulation (or drug delivery system), the firm needs to meet, depending upon the classification of the drug, the following requirements (5):

I. Drugs that have been published in the *Federal Register* as safe and effective in conventional dosage forms (classification A)
   A). Controlled clinical studies may be required to demonstrate the safety and efficacy of drugs in the controlled-release formulations.
   B). Bioavailability data of drugs delivered in controlled-release formulations are also required and may be acceptable in lieu of clinical trials.

II. Drugs that have been published in the *Federal Register* as safe and effective in controlled-release dosage forms (classification B)
   A). Bioavailability data are required and acceptable when comparable to an approved controlled-release pharmaceutical product of the same drug.
   B). The labeling must be identical to the reference standard with regard to effectiveness and side effects. Without appropriate clinical studies the labeling cannot be modified to make any different claims of clinical effectiveness or side effects.
   C). Bioavailability studies performed under steady-state conditions to demonstrated comparability to an approved immediate-release drug product are acceptable for supporting labeling for dosage administration.

The bioavailability data are required by law to be included in the submission of a new drug application in accordance with the "Bioavailability Requirements for Controlled Release Formulations" as specified in the *Federal Register*, CFR320.25(f).

FDA bioavailability regulations call for firms to provide the following information for their controlled-release pharmaceutical product:

1. The product meets the controlled-release claims made for it.
2. The bioavailability profile established for the product rules out the occurrence of dose dumping.
3. The product's steady-state performance is equivalent to that of currently marketed noncontrolled-release or controlled-release pharmaceutical products that contain the same active drug ingredient (or therapeutic moiety) and is subject to an approved, complete new drug application.
4. The product's formulation provides consistent pharmacokinetic performance between individual dosage units.

For comparative studies of controlled-release pharmaceutical products, the following reference standards may be used:

1. A solution or suspension of the same active drug ingredient (or therapeutic moiety)
2. A currently marketed FDA-approved noncontrolled-release pharamceutical product containing the same active drug ingredient (or therapeutic moiety)
3. A currently marketed controlled-release pharmaceutical product subject to an approved, complete new drug application containing the same active drug ingredient (or therapeutic moiety)

The bioavailability data, which consist of blood levels and/or urinary excretion rate profiles performed under steady-state conditions, may be acceptable in lieu of clinical trails if it can be demonstrated that the blood levels and/or urinary excretion rate profiles are comparable to those achieved by the administration of multiple doses of the same drug in appropriate conventional dosage forms (for drugs in classification A) or to an equivalent dose of the same drug in appropriate controlled-release dosage forms (for drugs in classification B). In this case the labeling must clearly state the recommended dosing regimen, and the claims of effectiveness and side effects must be identical to those for reference dosage forms. At times, a multiple-dose steady-state study design in normal subjects or patients may be required for drugs in classification A to establish comparability of dosage forms and/or to support drug labeling.

Any labeling claims of clinical advantage, such as greater effectiveness and/or reduced incidence of side effects, must be substantiated by appropriate well-controlled clinical studies.

Clinical studies are also required if there is substantial evidence that indicates that the drug's effectiveness is related to a rate of change in drug level, that is, pseudo-steady state, rather than to the absolute blood level achieved at steady state.

B. Requirements to Demonstrate Controlled Release Characteristics

To demonstrate the controlled-release characteristics of drug(s) delivered from a controlled-release pharmaceutical product (or drug delivery system), the manufacturer must submit the following information:

I. In vitro drug release data. Drug release profiles should be generated by a well-designed, reproducible in vitro testing method, such as the dissolution test for solid dosage forms. This test should be sensitive enough to discriminate any change in formulation parameters and lot-to-lot variations. A meaningful in vitro–in vivo correlation should have been established. The key elements are as follows:
   A). Reproducibility of the method
   B). Proper choice of medium
   C). Maintenance of perfect sink conditions
   D). Good control of solution hydrodynamics

II. In vivo bioavailability data

A). Pharmacokinetic profiles.
B). Bioavailability data, either comparable to the reference dosage form to support the same labeling as to indications and side effects or not equivalent to the reference dosage form but with demonstration of safety and efficacy to support different labeling
C). Reproducibility of in vivo performance

It should be feasible to demonstrate the controlled-release characteristics of a drug delivered by a controlled-release dosage form (or drug delivery system) through both in vitro and in vivo evaluations. The manufacturers of controlled-release pharmaceutical products are required to develop reproducible and sensitive in vitro testing methods to characterize the release mechanisms of the controlled-release pharmaceutical products they have developed and intend to market. The in vitro tests developed can be utilized to predict the bioavailability of the controlled-release dosage forms and can also be relied upon to assure lot-to-lot performance.

The controlled-release dosage form (or drug delivery system) developed should aim to accomplish two important objectives: (i) it should allow a maximum possible percentage of the dose in the formulation form to be absorbed; and (ii) it should be capable of minimizing patient-to-patient variability. For the development of controlled-release dosage form (or drug delivery system), one commonly-used approach is to modify the rate of release of a drug from the formulation (or delivery system) by pharmaceutical manipulation. This may yield an alteration in the absorption rate and thus the plasma level of the drug to be delivered. Therefore, in so doing, one must assure with scientific evidence that the absorption efficiency of the drug is not appreciably impaired and the variability is not adversely increased.

The selection of a drug candidate for the development of a controlled-release dosage form (or drug delivery system) is also of vital importance. Drugs that are known or suspected of undergoing extensive hepatic first-pass metabolism should not be formulated in a controlled-release dosage form by slowing their rate of oral delivery, unless this is justified by the fact that an active metabolite is generated in the process.

On the other hand, for drugs with a long biological half-life, that is, greater than 12 hr, there is little medical rationale for the development of a controlled-release dosage form (or drug delivery system), since these drugs are long-acting themselves, unless the development of a controlled-release pharmaceutical product will provide some advantages or offer convenience to patients, which could yield better patient compliance and occasionally may avoid some unwanted side effects resulting from peaking in the plasma level, if the drugs are administered as a single daily dose. These are demonstrated from the standpoint of bioavailability in the following section.

V. PHARMACOKINETIC CONSIDERATIONS

Using a simple, one-compartment, open pharmacokinetic model (Scheme 1), the effect of the release rate of a drug from a controlled-release dosage form (or drug delivery system) on plasma levels can be illustrated:

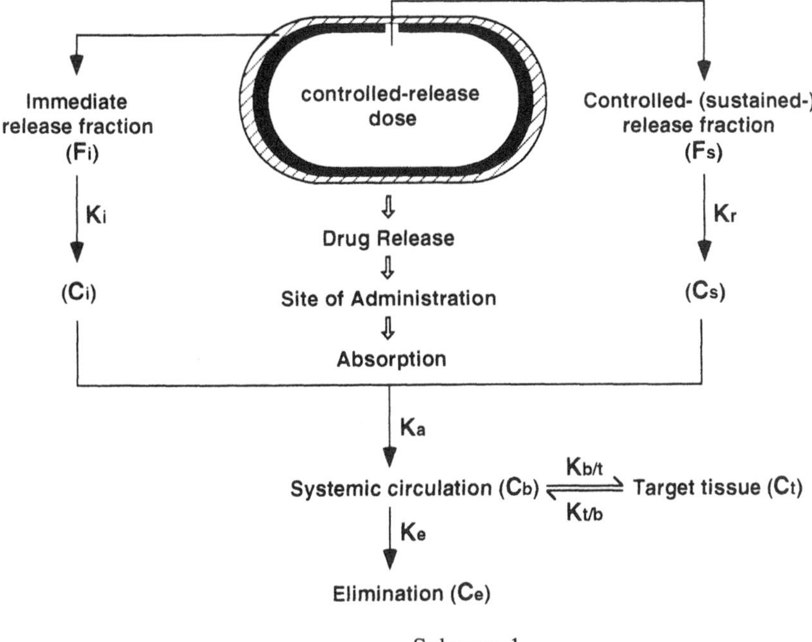

Scheme 1

where f_i and f_s are the fractions of the dose in the controlled-release formulation that provide immediate release and controlled (or sustained) release of the drug, respectively; k_i and k_r are the rate constants of drug release for the immediate-release fraction and for the controlled- (or sustained-)release fraction, respectively, where k_i is much greater than k_r; and k_a and k_e are the rate constants for the absorption and elimination of the drug, respectively; C_i and C_s are the concentrations of drug released to the site of administration from the immediate-release and controlled- (or sustained-)release fractions of the controlled-release formulation, respectively; and C_b, C_t, and C_e are the concentrations of drug absorbed into the systemic circulation, distributed to the target tissues, and eliminated from the body, respectively.

A. Controlled-Release Formulation with Single Dose

The effect of the release rate of drugs from controlled-release formulations on the plasma level of drugs with either a long biological half-life ($t_{1/2}$ = 14 hr) or a short biological half-life ($t_{1/2}$ = 1.7 hr) is illustrated in Figures 1 and 2, respectively. Both drugs are absorbed at a relatively fast rate (with k_a = 3 hr^{-1}). The influence of the rate constant of drug release is best seen in the initial rate of appearance of the drug in the plasma.

The data in Figure 1 indicate that the most significant change produced by the reduction in the release rate is the reduction in the peak level with a flattening of the plasma drug concentration profile. It should be noted that the bioavailability of the drug delivered by preparations 4 and 5 is significantly reduced in the initial 24 hr but is expected to result in similar steady-state plasma levels at a later time. It should also be noted that with the exception of preparation 5, the plasma levels beyond 18 hr do not appear to differ significantly; that is, tailing is similar. It appears

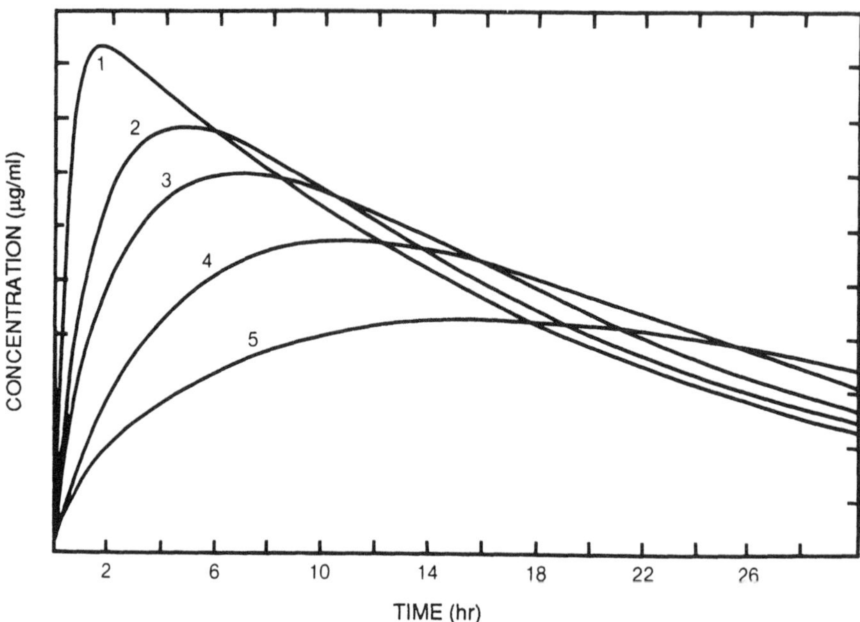

Figure 1 The simulated plasma concentration profiles of a long-acting drug ($t_{1/2}$ = 14 hr) based on a one-compartment open kinetic model. Curve 1 represents the plasma drug concentration resulting from the administration of a conventional dosage form with rapid drug release. The remaining curves describe four controlled-release formulations with an identical drug dose released at different release rates k_r. Curves 2 and 3 represent the controlled-release formulations that immediately release 25% of the loading dose followed by a controlled- (sustained-)release of the remaining drug dose at a rate of 0.60 and 0.30, respectively. Curves 4 and 5 represent the controlled-release formulations that immediately release 10% of the loading dose followed by a controlled- (sustained-)release of the remaining drug dose at a rate of 0.15 and 0.07, respectively. (Replotted from Taylor and Wiegand, 1962.)

that controlled-release formulations for drugs with a long biological half-life are best evaluated by multiple doses at pseudo-steady state.

Similarly, controlled-release formulations of a short-acting drug also result in a flattening of the plasma peak (Figure 2). Additionally, a significant difference in the tail end of the plasma drug concentration curves is observed as the release rate is reduced. The use of any of the controlled-release formulations in this case would be justified by clinical studies that demonstrate efficacy and the incidence of side effects. Demonstration that steady-state drug levels are required for therapeutic activity, rather than the initial peak plasma concentration, may warrant the use of such a controlled-release pharmaceutical product.

B. Controlled-Release Formulation with Multiple Doses

1. *Controlled Release of Short-Acting Drugs*

The plasma drug concentration profile resulting from a controlled-release formulation containing multiple doses of a short-acting drug ($t_{1/2}$ = 1 hr) is shown in Figure 3 in comparison with three consecutive dosings of the same drug in a conventional immediate-release formulation. The use of the controlled-release formulation has

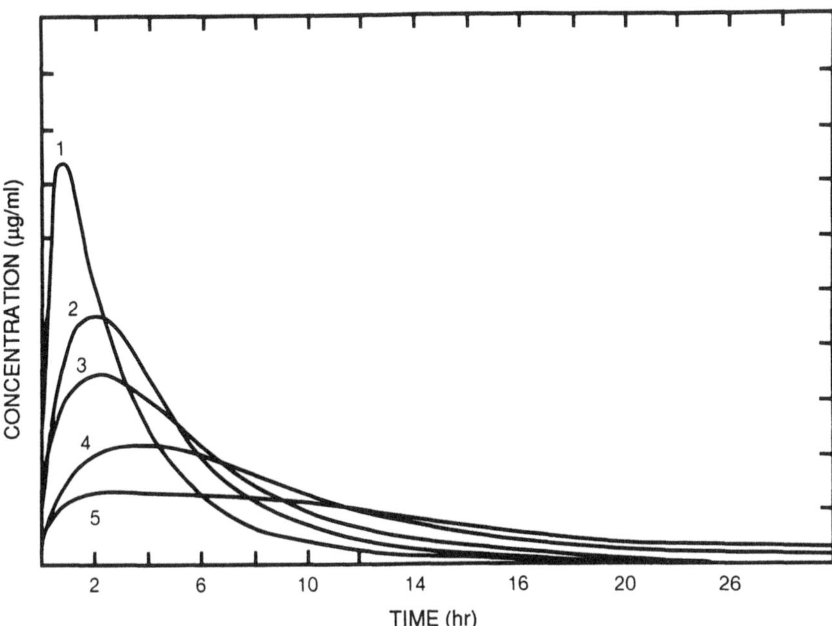

Figure 2 Simulated plasma concentration profiles of a short-acting drug ($t_{1/2}$ = 1.7 hr) based on a one-compartment open kinetic model. The experimental conditions for curves 1–5 are the same as in Figure 1. (Replotted from Taylor and Wiegand, 1962.)

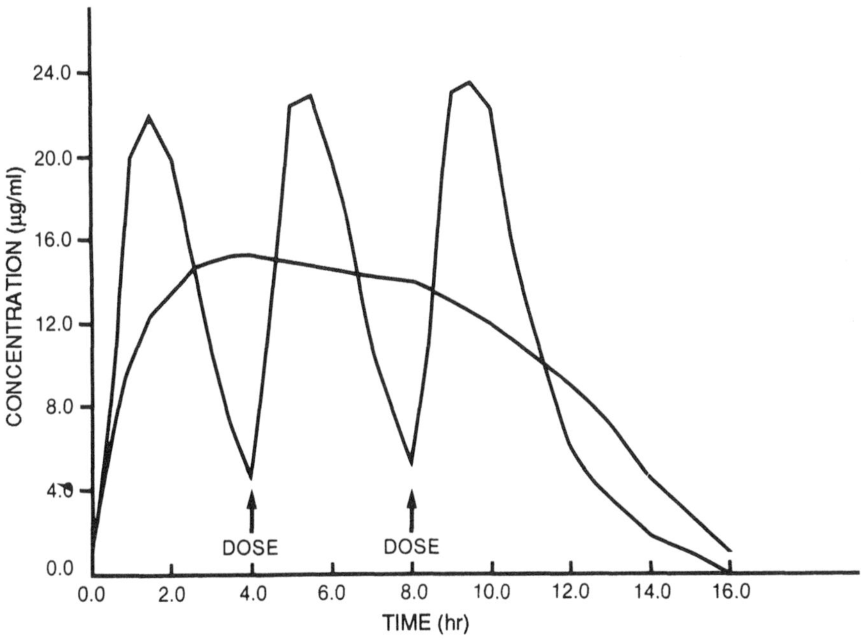

Figure 3 Comparison of the simulated plasma concentration profiles for a short-acting drug that has a biological half-life of 1 hr delivered by a single dose of a controlled-release formulation or by multiple doses of an immediate-release dosage form. (Replotted from Cabana and Chien, 1982.)

achieved an equivalent bioavailability [area under the curve (AUC) = 165 μg-hr/ml] while producing a sustained plateau plasma level with a substantial reduction in peak-to-valley fluctuations in the drug concentration profile. The peak concentrations were reduced by 30%.

Regulatory approval of such a controlled-release pharmaceutical product would require demonstration that the sustained plateau plasma level achieved by the controlled-release formulation is in the therapeutic effective range specific for the drug. Alternatively, demonstration of the clinical effectiveness of such a long-acting dosage form by appropriate clinical evaluations would be needed. Claim of a lower incidence of side effects for such a controlled-release pharmaceutical product would also require clinical verification.

The comparative plasma drug concentration profiles shown in Figure 3 also demonstrate the bioavailability and controlled drug delivery characteristics of the controlled-release formulation. However, it is also necessary to demonstrate the clinical efficacy of this controlled-release pharmaceutical product, especially for drugs whose onset of pharamcological activity is thought to be associated with peaking of their plasma concentrations.

2. *Controlled Release of Intermediate-Acting Drugs*

The bioavailability data of a drug with an intermediate biological half-life ($t_{1/2}$ = 3 hr) delivered by controlled-release and immediate-release formulations are compared in Figure 4. Administration of a standard dose every 4 hr in an immediate-release dosage form yields a peak plasma concentration approximately 1 hr after each dos-

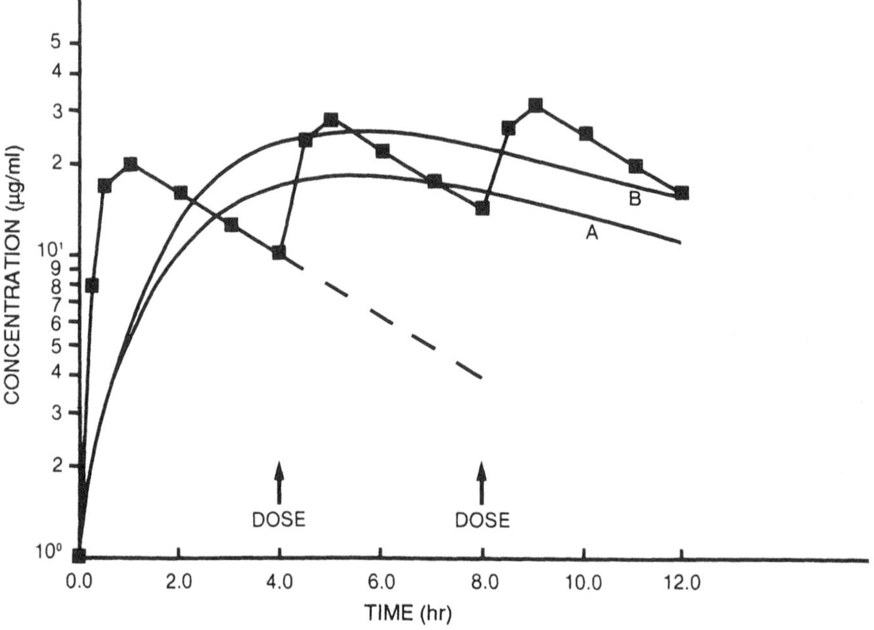

Figure 4 Comparison of the simulated plasma concentration profiles for an intermediate-acting drug that has a biological half-life of 3 hr delivered by a single dose of a controlled-release formulation (formation A or B) or by multiple doses of an immediate-release dosage form, one dose every 4 hr.

ing. The C_{max} and C_{min} values at pseudo-steady state are about 32 and 16 μg/ml, respectively, with an AUC of 228 μg-hr/ml. On the other hand, formulation A of a controlled-release pharmaceutical product reaches the C_{max} value of 20 μg/ml at 4 hr and has a C_{min} value of about 12.5 μg/ml at 12 hr after administration and an AUC of 185 μg-hr/ml; formulation B peaks at a C_{max} of approximately 25 μg/ml and has a C_{min} value of 16 μg/ml and an AUC of 221 μg-hr/ml.

Basically, both formulations A and B are comparable in bioavailability, but regulatory approval of formulation A would require the performance of a multiple-dose steady-state study to establish the comparativeness of its steady-state plasma drug levels with those of the conventional immediate-release formulation or would otherwise require clinical studies to demonstrate its clinical efficacy and/or to support any claim for a lower incidence of toxicity. On the other hand, regulatory approval of formulation B would require only bioavailability data, which would be limited to the same labeling, however, in terms of efficacy and side effects, as the conventional immediate-release dosage form.

3. *Controlled Release of Long-Acting Drugs*

The bioavailability of a long-acting drug (with $t_{1/2}$ = 12 hr) from a controlled-release formulation compared with that from an immediate-release dosage form is illustrated in Figure 5. The data appear to suggest that the bioavailabilities of both dosage forms are comparable. By using a multiple-dose regimen, it is possible to demonstrate that comparable pseudo–steady-state plasma drug levels are achieved with both preparations. Very likely clinical trials would not be necessary for the regulatory approval

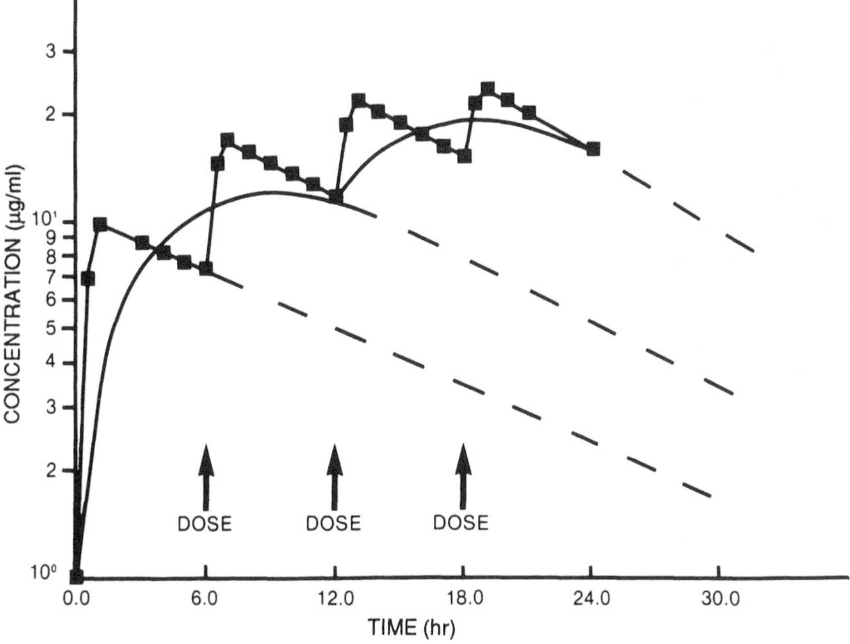

Figure 5 Comparison of the simulated plasma concentration profiles for a long-acting drug that has a biological half-life of 12 hr delivered by a controlled-release formulation, one dose every 12 hr, or by multiple doses of an immediate-release formulation, one dose every 6 hr.

of this controlled-release formulation other than to demonstrate its reproducibility in terms of bioavailability.

C. Needs of Multiple-Dose Steady-State Study

Often it is not possible for a controlled-release pharmaceutical product to be properly evaluated by single-dose administration. At times, there may be the need to determine whether a controlled-release formulation can achieve a well-defined therapeutic plasma level or a plasma level equivalent to that attained by an immediate-release dosage form when administered to the same group of healthy volunteers or patients as labeled. This can be demonstrated by multiple-dose steady-state study.

It should be emphasized that bioequivalence employing the *superimposition principle* does not need to be established. By definition, the rate of absorption of a drug delivered from a controlled-release formulation differs appreciably from that delivered from immediate-release dosage forms or other controlled-release formulations. What needs to be established is the generation of satisfactory bioavailability data to support drug labeling. The following criteria should be met:

1. Satisfactory steady-state plasma levels should be attained with both the test and reference formulations in a sufficient number of patients (or volunteers) to warrant comparison.
2. Establishment of an adequate steady state should be assured by comparing the C_{min} (trough) values obtained in 3 or more consecutive days.
3. Failure to achieve a satisfactory steady state in a large percentage of subjects studied indicate either a lack of patient compliance in these subjects or failure in the performance of a controlled-release formulation.
4. Comparison of pharmacokinetic parameters, such as C_{min} and AUC values, should be limited only to those subjects who have achieved steady-state conditions.
5. Comparison of the AUC during a dosing interval is only proper if both the plasma levels of the test and the reference formulations are at steady state. Otherwise, the comparison is theoretically invalid.

In considering the regulatory requirements for controlled-release pharmaceutical products, one needs to take into account each of the following factors in the development and evaluation of any controlled-release formulation:

I. Pharmacokinetic properties of the drug
    A). Hepatic first-pass metabolism
    B). Michaelis-Menten kinetics
    C). Biological half-life
    D). Rate of absorption
II. Pharmacological properties of the drug
    A). Minimum therapeutic effective concentration
    B). Influence of peaking (C_{max}) versus steady-state plasma concentration
    C). Desirability of steady-state plasma levels
    D). Conventional dosing regimen
III. Toxicological properties of the drug
    A). Minimum toxic level
    B). Frequency and type of toxicity encountered

VI. DEVELOPMENT OF IN VITRO–IN VIVO CORRELATION AND BIOAVAILABILITY ASSURANCE

A. pH-Dependent Sustained-Release Formulations

In the evaluation of controlled-release dosage forms, investigators should be cognizant of formulations with pH-dependent release characteristics. In recent years, the FDA has been gathering in vivo and in vitro data on drugs like papaverine and theophylline that strongly suggest that the pH specificity of the drug or the formulation may often affect the controlled-release profile of the drug. A formulation with a pH-dependent in vitro dissolution profile may behave poorly under in vivo conditions because the bioavailability of the drug it delivers is subjected to pH variation in the gastrointestinal tract and is also at the mercy of gastrointestinal motility, in particular gastric emptying (Chapter 3). Furthermore, there is a possibility that the prevailing gastric physiology, for example achlorhydria, may also impact adversely on such dosage forms. For instance, a formulation that rapidly dissolves at pH 6–7 can be expected to dump the entire drug dose in the upper intestine, and therefore the plasma levels of drug are a function of gastric emptying. Similarly, one should be aware of poorly-soluble drugs that undergo rapid metabolic or renal clearance. Although they are often considered, for therapeutic reasons, as potential drug candidates for controlled delivery, such drugs are extremely difficult to formulate and to evaluate pharmacokinetically. Such drugs often require clinical studies to establish their safety, efficacy, and labeling.

The use of different media in dissolution testing often allows anomalies to surface. This is best illustrated by the findings on papaverine using the NF XIV rotating bottle method, which calls for variation in pH conditions over a 7-hr study period (Figure 6).

Regardless of the total fraction of the dose released by individual products, the bulk of the dissolution occurs generally within the first 2 hr, with a small amount in the next 1.5 hr, and no further release occurs beyond 3.5 hr. These findings correlate with observations that no further dissolution occurs at pH above 4.5 and only a small amount of papaverine dissolves at pH 7.0 and higher. Therefore, because of rapid gastrointestinal motility under fasting conditions and the prevailing pH conditions of the upper and lower intestine, a large difference in the bioavailability profile can be anticipated from these sustained-release papaverine products.

In vivo studies showed a significant difference in the bioavailability of papaverine hydrochloride from nine sustained-release pharmaceutical products compared to a conventional tablet and an elixer (6). Of particular interest was the finding of failure to substantiate the controlled-release characteristics of such products. The bioavailability of these products ranged from 19 to 64% in comparison with that attained by papaverine HCl in oral solution.

Attempts to correlate the bioavailability data with the dissolution profiles indicated that the best correlation is obtained in relating the relative maximum plasma concentrations C_{max} and the extent of drug absorption, such as, relative bioavailability (in terms of the AUC in each subject), as a function of dissolution profiles at pH 1.2 (Figures 7 and 8). The correlation demonstrates that the relative bioavailabilty of papaverine hydrochloride from these sustained-release pharmaceutical products is

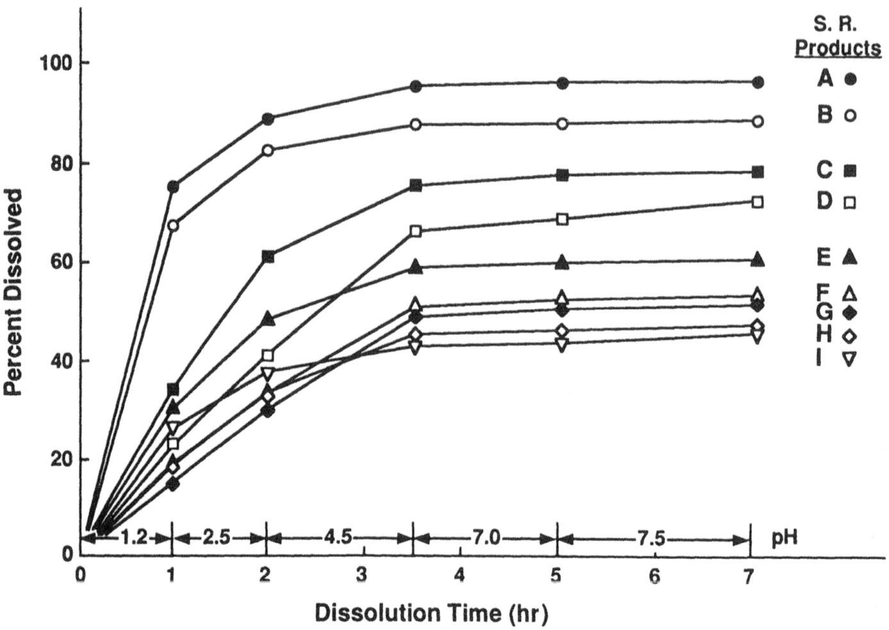

Figure 6 Dissolution profiles of papaverine hydrochloride (160 mg) from various sustained-release capsule and tablet formulations employing the rotating bottle method with continuous pH changes over a 7-hr study period.

directly related to the dissolution obtained during the first hour at gastric pH; the correlation progressively worsens as the pH of the dissolution medium increases (6).

B. Enteric-Coated Sustained-Release Formulations

Aminophylline- and theophylline-containing enteric-coated sustained-release products are perhaps the best examples of areas in which the Food and Drug Administration has attempted to customize requirements to a specific need to support claims in labeling and to assure lot-to-lot performance through the development of in vitro–in vivo correlations.

The in vitro release profiles of aminophylline from one plain product and three enteric-coated products by dissolution studies are compared in Figure 9 for illustration. The results demonstrate that although the plain (uncoated) aminophylline tablet formulation is readily dissolved in simulated gastric fluid (SGF; pH 1.2), all three enteric-coated tablet formulations require simulated intestinal fluid (SIF; pH 7.2) for dissolution. One of the enteric-coated products (product C) even requires 5 hr under simulated intestinal conditions to achieve total dissolution.

Bioavailability studies of these enteric-coated products demonstrated a good in vitro–in vivo correlation (7). The plasma profiles in Figure 10 indicate that the two enteric-coated products (products A and B), which show a rapid dissolution profile of aminophylline in simulated intestinal fluid (Figure 9), achieved peak plasma levels of aminophylline equivalent to that of the uncoated aminophylline product, but the value of t_{max}, the time to reach the peak plasma concentration, was found to be

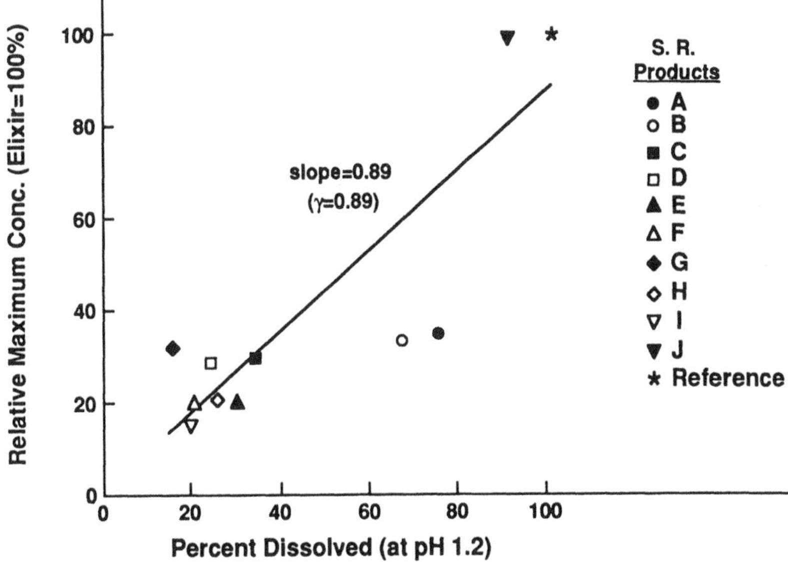

Figure 7 In vitro–in vivo correlation between the relative maximum serum concentration of papaverine attained by various sustained-release papaverine HCl products [relative to that by elixir (Cooper) formulation] and the percentage of a papaverine dose dissolved at pH 1.2 during the first hour of the dissolution study.

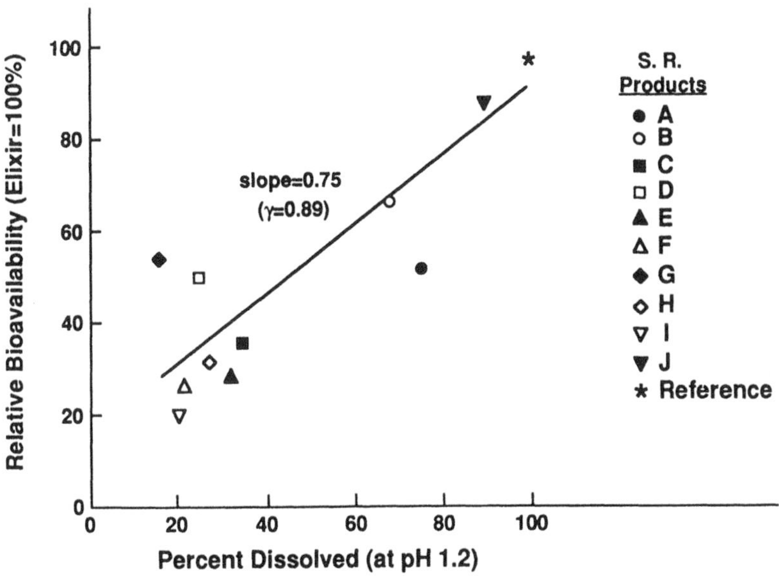

Figure 8 In vitro–in vivo correlation between the relative bioavailability of papaverine achieved by various sustained-release papaverine HCl products [relative to that by elixir (Cooper) formulation] and the percentage of a papaverine dose dissolved at pH 1.2 during the first hour of the dissolution study.

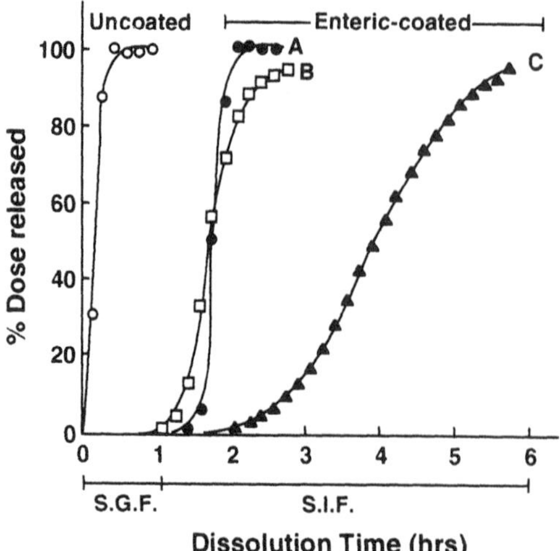

Figure 9 Comparative dissolution profiles of aminophylline from plain (uncoated) and enteric-coated aminophylline products studied by a dissolution rate test developed at the National Center for Drug Analysis: (○) uncoated tablets (Searle); (●) enteric-coated tablet A (Tablecaps, Robinson); (□) enteric-coated tablet B (Richlyn, Columbia); and (△) enteric-coated tablet C (Tablicaps, Wales). (Replotted from Cabana and Chien, 1982.)

delayed by 1–2 hr (Figure 10). On the other hand, the slowly-dissolving enteric-coated product (product C) yielded a substantially lower plasma level at later time (5–8 hr after oral administration) (8). On the basis of these data, it is possible to establish both in vitro and in vivo requirements for enteric-coated sustained-release products.

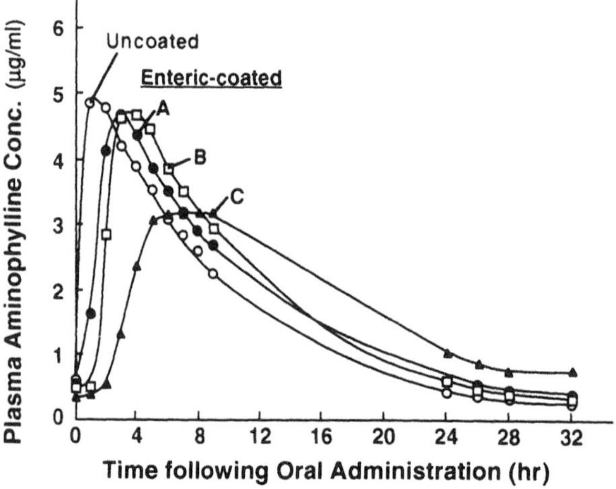

Figure 10 Comparative plasma concentration profiles of aminophylline delivered from the same uncoated and enteric-coated aminophylline products as investigated in Figure 9. (Replotted from the data by Dr. S. Riegelman, University of California.)

C. Nonenteric-Coated Sustained-Release Formulations

The result obtained with a nonenteric-coated sustained-release theophylline formulation (Theodur extended-release tablets, Key) is another classical example.

Using both the USP rotating basket and FDA paddle methods and under conditions that simulate the gastrointestinal tract, it was revealed that this sustained-release theophylline product releases theophylline at a fairly constant rate up to 90% of the label dose (Figure 11). A sustained plasma profile of theophylline was obtained (Figure 12) as predicted from the in vitro dissolution profiles (Figure 11). A plateau plasma level of 7–8 μg/ml of theophylline was achieved within 5–6 hr after oral administration and then maintained for 5–6 hr. In contrast, a higher peak level of theophylline was achieved, but maintained for a shorter duration, with the solution formulation.

Development of a controlled-release pharmaceutical product with desirable controlled drug release characteristics requires extensive effort in formulation design. The in vitro–in vivo correlation can be used in the formulation development process to assess the rate-limiting formulation parameters, to screen various formulation designs, and also to assure lot-to-lot performance of the production batches. As demonstrated in Figure 13, the percentage of the labeled dose absorbed from commercial lots of Theodur tablets (300 mg), particularly lots #6806 and #6126, correlate extremely well with their dissolution profiles over the 12-hr period (9).

In light of the narrow therapeutic index of theophylline and the variation in its rate of clearance in various patient populations, such as children versus adults and smokers versus nonsmokers, there is additional need to assure that therapeutic levels

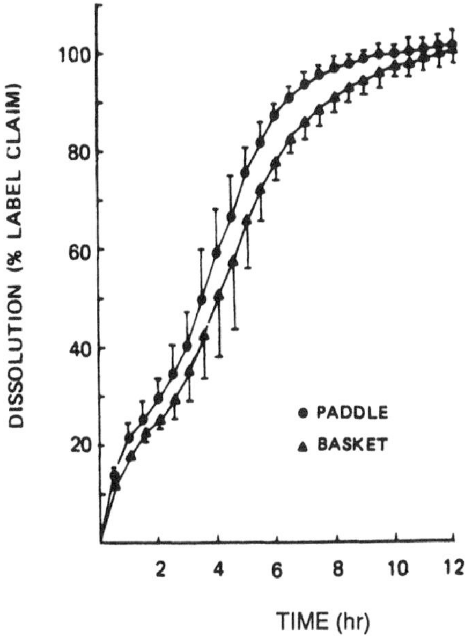

Figure 11 Dissolution profiles of theophylline from Theodur, a theophylline sustained-release formulation under two dissolution conditions. (Courtesy of Dr. D. Cohen, Key Laboratories.)

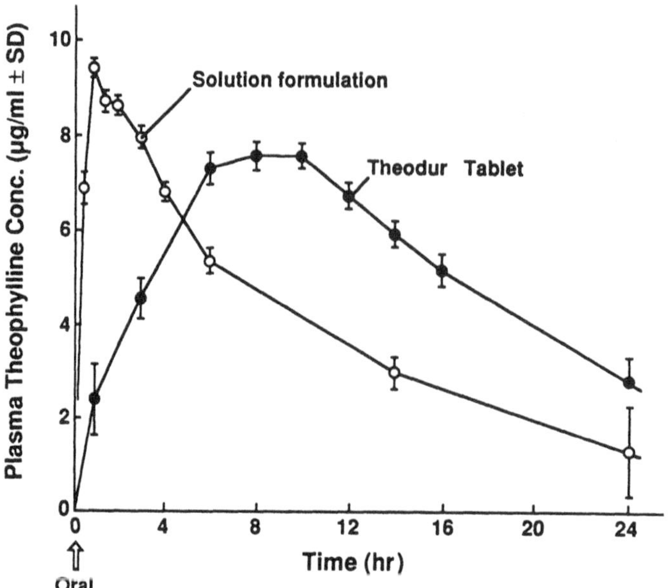

Figure 12 Plasma profile of theophylline from Theodur, a theophylline sustained-release formulation, compared to that from a solution formulation. (Courtesy of Dr. D. Cohen, Key Laboratories.)

are maintained when doses are taken in accordance with labeling instruction (i.e., every 8 hr or every 12 hr). The FDA has imposed a specific bioavailability requirement, which consists of both single- and multiple-dosing steady-state studies, as a basis for NDA approval of theophylline-containing sustained-release pharmaceutical products. A typical multiple-dosing steady-state study conducted in patients over a 5-day period is illustrated in Figure 14. The results clearly demonstrate that therapeutic levels are maintained throughout the dosing interval without any observation of dose dumping which, if occurred, could have yielded adverse side effects. The controlled-release behavior of Theodur observed in in vitro dissolution studies (Figure 11) appears to give a reasonably steady plasma level of theophylline in in vivo bioavailability studies (Figure 14).

D. Controlled-Release Drug Delivery Systems

Over the last 10–15 years, several controlled-release drug delivery systems have been developed to deliver drugs at predetermined rates over a prolonged period of time (Chapter 1). Among these are such therapeutic systems as the pilocarpine-releasing ocular inserts (Ocusert System, Ciba), progesterone-releasing intrauterine device (Progestasert IUD, Alza), scopolamine-releasing transdermal therapeutic system (Transderm-Scop, Ciba), clonidine-releasing transdermal therapeutic system (Catapres-TTS, Boehringer–Ingelheim), estradiol-releasing transdermal therapeutic system (Estraderm, Ciba) and several nitroglycerin-releasing transdermal therapeutic systems, levonorgestrel-releasing subdermal implants (Norplant, Wyeth-Ayerst), nifedipine-releasing gastrointestinal therapeutic system (Procardia XL, Pfizer), and

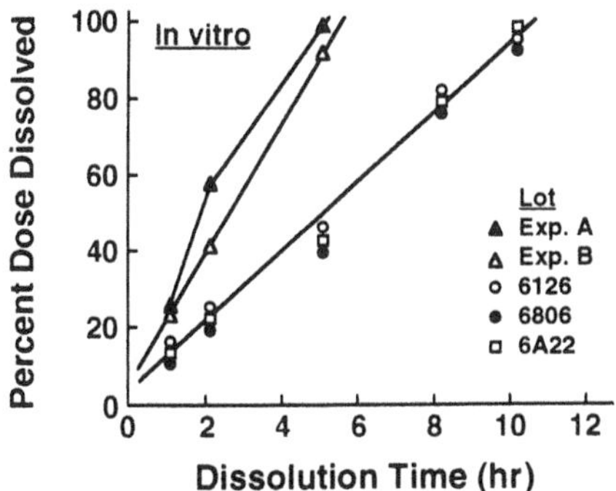

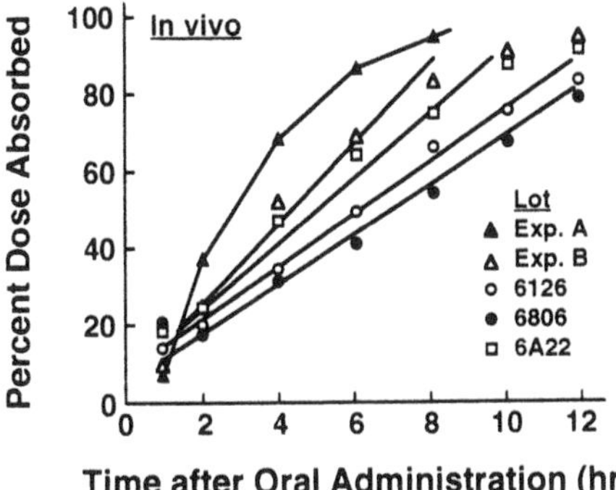

Figure 13 In vitro–in vivo correlation between the bioavailability of theophylline from experimental formulations and commercial lots of 300 mg Theodur tablets. (Replotted from the data by Dr. D. Cohen, Key Laboratories.)

goserelin acetate-releasing biodegradable implant (Zoladex, ICI). All these new drug delivery systems are considered from the regulatory point of view new drugs, and they require full new drug applications as a basis of regulatory approval for marketing.

In addition to safety and efficacy considerations of such new drug delivery systems, biopharmaceutics and pharmacokinetic issues need to be addressed by the manufacturers. The key elements that need to be established are as follows:

1. Reproducibility of the drug release kinetics
2. A defined bioavailability profile that rules out the possibility of dose dumping
3. Demonstration of reasonably good absorption relative to an appropriate standard
4. A well-defined pharmacokinetic profile that supports the drug labeling

These key elements are illustrated in the following representative approved drug delivery systems.

1. *Transdermal Scopolamine Delivery Systems*

The scopolamine TTS system (the scopolamine-releasing transdermal therapeutic system; Transderm-Scop, Ciba) exemplifies the FDA biopharmaceutics requirements

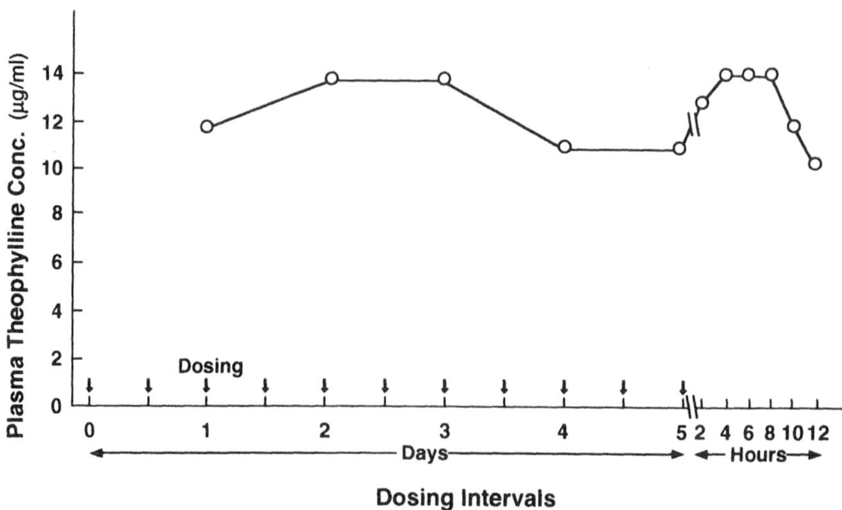

Figure 14 Mean plasma theophylline levels achieved in 12 subjects taking Theodur tablets (8 mg/kg) every 12 hr for 5 days. (Replotted from the data by Dr. D. Cohen, Key Laboratories.)

for the regulatory approval of controlled-release drug delivery systems. The drug, scopolamine, is delivered to the postauricular area and then absorbed through the intact skin. The scopolamine TTS system was developed with the objective of providing effective control of motion-induced nausea associated with minimal adverse parasympatholytic effects (10).

The biopharmaceutics basis for FDA approval of scopolamine TTS system consisted of a comparison in healthy adult volunteers of two lots of scopolamine TTS systems programmed to deliver a 0.5 mg dose over a 72-hr period in reference to a continuous 72-hr intravenous infusion of scopolamine. By means of gas-liquid chromatographic–mass spectrometric analysis of urinary excretion rates of scopolamine, the manufacturer was able to demonstrate comparable bioavailability for both lots of the scopolamine TTS system (Figure 15). In addition to the demonstration of comparability and reproducibility of the scopolamine TTS systems developed, the manufacturer also established that the scopolamine is delivered transdermally mostly at a zero-order rate over the 24–84 hr period (Figure 16). The urinary excretion of scopolamine from the scopolamine TTS systems differed from that by continuous intravenous (IV) infusion only in its excretion rate during the initial 12–36 hr period and following the termination of transdermal application (72–120 hr). The scopolamine TTS systems produced a significantly greater rate of urinary excretion in the initial 12–36 hr but thereafter demonstrated equivalent steady-state systemic delivery characteristics over the remaining treatment period. Such data permit defining the pharmacokinetic profile of the scopolamine TTS systems relative to an absolute reference standard to support drug labeling. Of great importance clinically is that the scopolamine TTS systems delivered scopolamine levels far below toxic levels (Figure 17) (10).

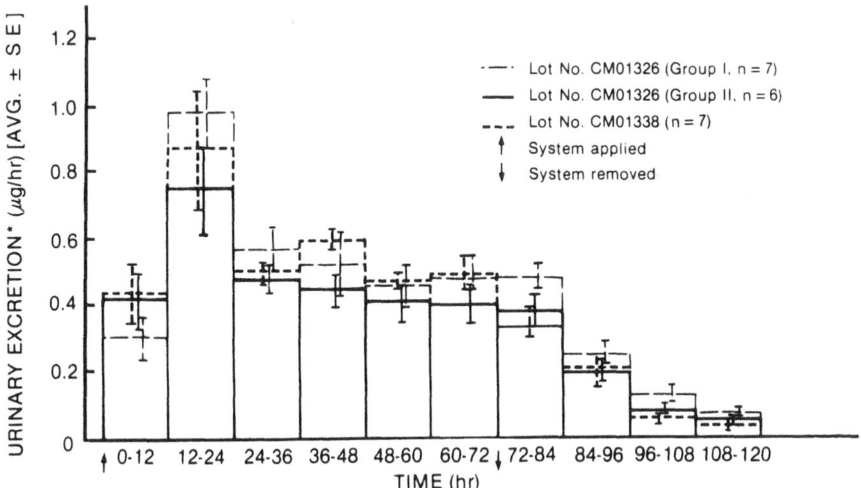

Figure 15 Urinary excretion of free scopolamine following the transdermal administration of two different lots of scopolamine TTS systems. (Courtesy of Dr. J. Shaw, Alza Corporation.)

2. *Transdermal Nitroglycerin Delivery Systems*

The toxicological and clinical requirements for systemically-active drugs delivered dermally and transdermally should parallel those for other systemic drugs of the same pharamcological class. From the boipharmaceutics point of view, there is a need to define the bioavailability and pharmacokinetics of these drugs relative to, ideally, an intravenous dose (or an oral dose if IV dosing is not feasible) (11). As well as defining the pharmacokinetic profile of drugs delivered by a new drug delivery system, it is necessary to demonstrate the reproducibility of such a system. This involves demonstration of reproducibility in manufacturing as well as intra- and intersubject variabilities. Manufacturing reproducibility can be illustrated by comparing the pharmacokinetics and bioavailability of different production batches of a new drug delivery system in the same group of patients (Figure 15).

Demonstration of bioavailability and bioequivalence of transdermal nitroglycerin delivery system is illustrated by a comparison made with a sublingual nitroglycerin tablet. Studies performed and assayed by GC-MS clearly demonstrated that the cutaneous absorption of nitroglycerin from the transdermal nitroglycerin delivery system produced constant plasma levels of nitroglycerin throughout the 24-hr application (Figure 18) in contrast to the short duration (16-min) maintained by the sublingual administration (Figure 19).

The basis for the approval of such a controlled-release drug delivery system from a biopharmaceutics standpoint consists of the following data:

1. Reproducibility of plasma drug levels
2. Defining pharmacokinetic parameters to support drug labeling.
3. Demonstration that the steady-state plasma concentrations of nitroglycerin achieved

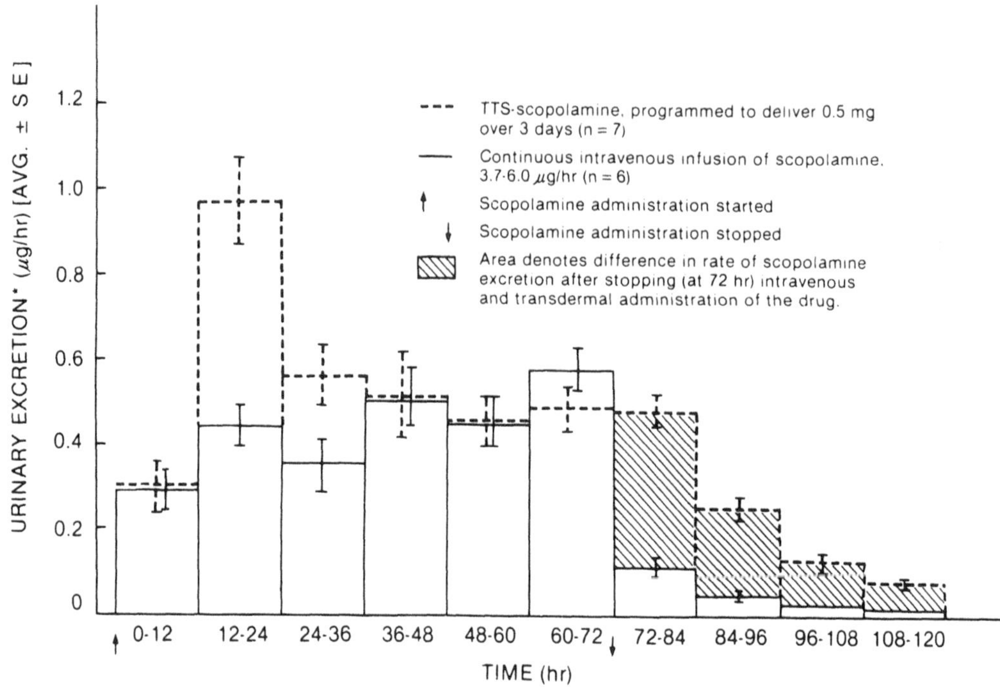

Figure 16 Comparison of urinary excretion rates of free scopolamine following the transdermal delivery and intravenous infusion of scopolamine. (Courtesy of Dr. J. Shaw, Alza Corporation.)

by a transdermal drug delivery system are within the reasonable therapeutic range attained by the sublingual dosage form

It should be pointed out that clinical evaluations may be needed to support the dosage regimen, that is, every 12 hr and, in particular, a 24-hr "activity claim" (or every 24 hr). Such clinical studies may not only support maintenance dosage recommendations but may also document the reliability of systemic drug delivery.

For extensive reviews of the regulatory requirements for the marketing approval of transdermal therapeutic systems, interested readers should refer to *Transdermal Controlled Systemic Medications* (12).

VII. SUMMARY

In summary, justification for the regulatory approval of controlled-release formulations (or drug delivery systems) of established and new drug entities should be based solely on scientific documentation for the drug in terms of safety and efficacy. Regulatory approval of a controlled-release pharmaceutical product in terms of bioavailability requirements requires demonstration of the following:

1. Bioavailability
2. Controlled release characteristics
3. Reproducibility of in vivo performance

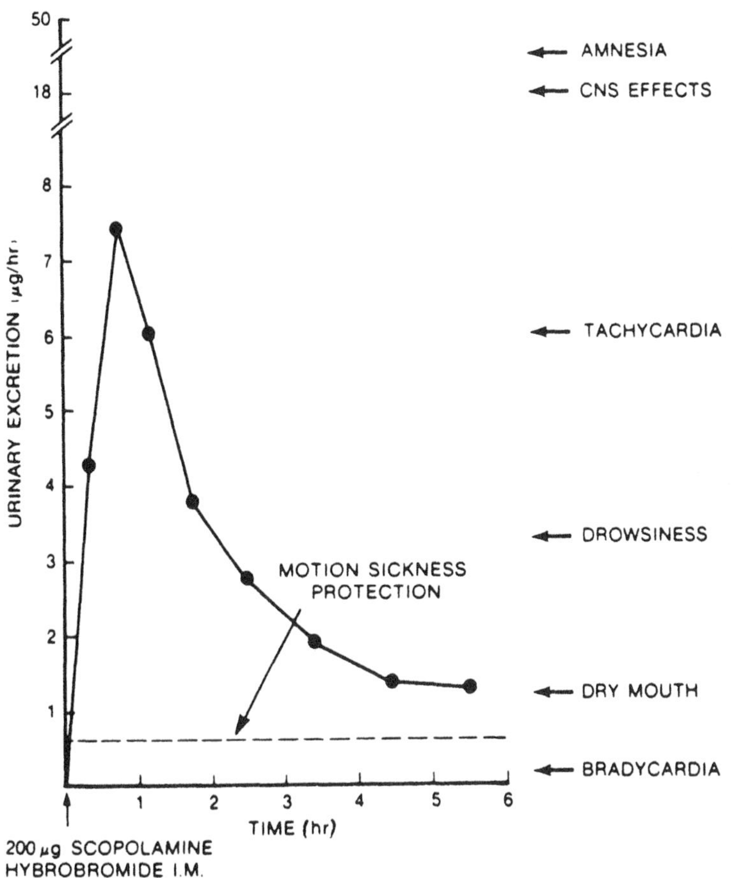

Figure 17 The pharmacological and toxicological effects of scopolamine as a function of its urinary excretion rate following an intrasmucular administration of scopolamine (200 μg).

4. Evidence to support clinical safety and efficacy, as well as the rationale as reflected in the labeling

The same standard of requirements is applied to both human and veterinary pharmaceutical products.

VIII. FUTURE PROSPECTS

The following three areas of development may have potential impacts on regulatory requirements for controlled- (sustained-) release drug delivery systems.

A. Pharmacodynamic Considerations

In a recent interview (13), Dr. C. Peck, Director of the Center for Drug Evaluation and Research, discussed the role of pharmacodynamics in the FDA philosophy in the regulatory review of new drug applications. Pharmacodynamics, in a more restricted definition, has been viewed as involving the relationship between drug con-

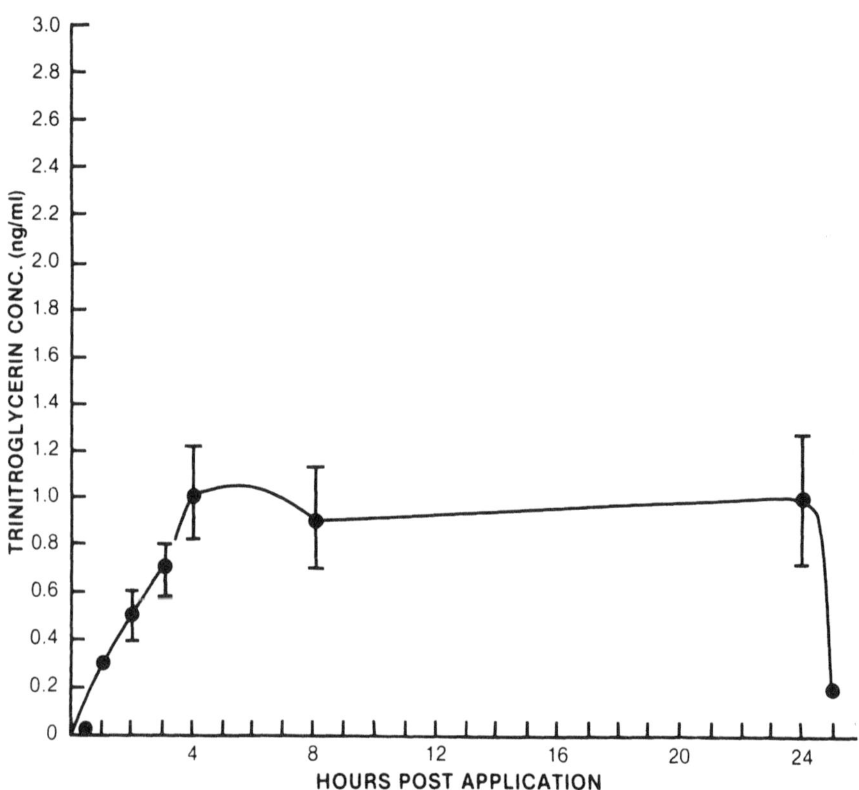

Figure 18 Plasma concentration profile of trinitroglycerin in five human subjects following a 24-hr topical application of a transdermal nitroglycerin delivery system. (Courtesy of Dr. D. Cohen, Key Laboratories.)

centrations and therapeutic effects; pharmacokinetics, because of the disparity between kinetics and effects—the time delay in the onset of therapeutic responses from drug administration and inter-subject variation—is considered a useful basis for regulatory considerations only when plasma (or tissue) levels have been linked to some pharmacologically or toxicologically meaningful effects or clinical responses. Once the relationship between pharmacokinetics and pharmacodynamics has been established, plasma levels can then be used as a substitute for therapeutic effects. Philosophically, pharmacodynamics is considered the intellectual and conceptual backdrop that justifies pharmacokinetic bioequivalence standards. Hence, the role of pharmacodynamics is not to establish a new standard but to provide a scientific means for verification of the existing standard on bioequivalence. This philosophy is applicable to NDA and ANDA for both conventional immediate-release dosage forms and nonconventional controlled-release drug delivery systems.

Dr. Peck also revealed that the Center for Drug Evaluation and Research is committed to working with the Center for Biologics and the Center for Devices to develop an overall strategy for the regulatory review of pharmaceutical products involving biotechnology-produced drugs and drug-device combinations.

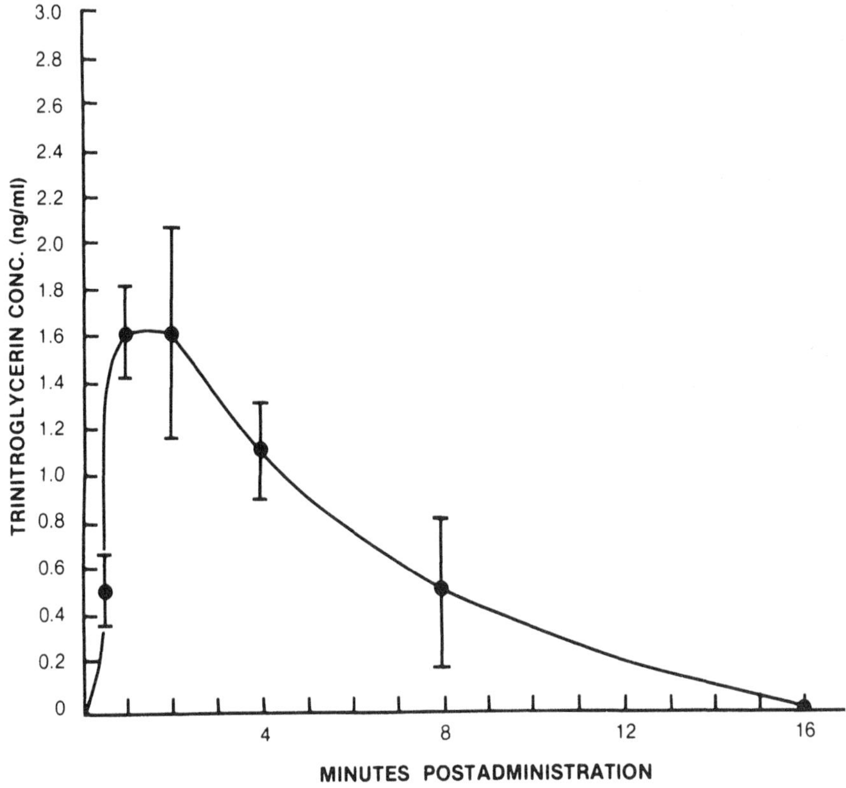

Figure 19 Plasma concentration profile of trinitroglycerin following the sublingual administration of a sublingual nitroglycerin tablet. (Courtesy of Dr. D. Cohen, Key Laboratories.)

B. Biotechnology Medicines

Human recombinant insulin was the first biotechnology product to receive regulatory approval for routine clinical use (14). Since then, eight more biotechnology-produced pharmaceuticals and vaccines have been approved by the FDA (Table 8), and numerous others are either pending FDA approval or are in various stages of clinical development (Table 9).

A recent PMA publication (1) disclosed a growing new challenge for the FDA, the regulatory review of biotechnology medicines. Although a few biotechnology products have only recently received regulatory approval for marketing, there are mounting numbers of genetically-engineered pharmaceuticals and vaccines that have either already been submitted to the FDA for review or are currently in clinical trials (Table 10). Furthermore, advances in biotechnology have opened new avenues for drug innovation. In 1989, for example, the U.S. Patent Office issued 1948 biotechnology-related patents, of which more than 900 were pharmaceutical or health related. This performance in the development of patentable biotechnologies is a significant growth over that of 1988, when 1391 biotechnology-related U.S. patents were issued and more than 800 patents were in pharmaceutical- or health-related fields. It is reasonable to project that in the foreseeable future, the FDA is expected

Table 8 FDA-Approved Biotechnology-Produced Pharmaceuticals and Vaccines

Human insulin
Interferon α_{-2a}
Interferon α_{2b}
Human growth hormone (γ-hGH)
Hepatitis B vaccine
Tissue plasminogen activator (t-PA)
Factor VIII_c
Epoetin-α (recombinant human erythropoietin; rHuEPO)
Granulocyte colony stimulating factor (G-CSF)
Granulocyte-macrophage colony stimulating factor (GM-CSF)

Source: Compiled from *Pharmaceutical Biotechnology Monitor* (February 1991).

Table 9 Biotechnology-Produced Pharmaceutical and Veterinary Products Under Development

Atrial natriuretic factor (ANF)
Bovine growth hormone (bGH)
Epidermal growth factor (EGF)
Interferon β (IFN-β)
Interferon-γ (IFN-γ)
Interleukins 1, 2, 3, 4, 6
Lymphokine-activated killer (LAK) cells
Macrophage colony stimulating factor (M-CSF)
Monoclonal antibodies
Platelet-derived growth factor
Porcine growth hormone
Stem cell growth factor
Superoxide dismutase
Tumor-infiltrating lymphocytes
Tumor necrosis factor

Source: Compiled from *Pharmaceutical Biotechnology Monitor* (February 1991)

Table 10 Biotechnology Medicines in Clinical Development

	Number of biotechnology medicines		
Developmental state	1988	1989	1990
Regulatory review	14	10	18
Clinical trials	67	70	86

Source: PMA *Facts at a Glance* (1990 edition).

to receive a growing number of submissions on biotechnology-produced pharmaceuticals and vaccines for regulatory review.

Many of these biotechnology products are peptide- and protein-based pharmaceuticals, and most of them are known to be therapeutically active only by parenteral administration. A classic example is the systemic delivery of genetically-engineered human insulin by subcutaneous administration (Chapter 8). To optimize the therapeutic regimen of these peptide- and protein-based pharmaceuticals, extensive research activity has recently been devoted to the development of implantable delivery systems with controlled drug release characteristics to overcome the disadvantages of parenteral drug administration, such as the recently FDA-approved goserelin acetate-releasing biodegradable subdermal implant (Zoladex, ICI) (Chapter 8), or the search for noninvasive (nonparenteral) routes for the systemic delivery of therapeutic peptides and proteins and the development of topically-applied drug delivery systems for facilitated transdermal or mucosal delivery, such as the recently FDA-approved nafarelin acetate nasal spray (Synarel, Syntax) (Chapter 11). Because of their biochemical and biophysical complexities, the systemic delivery of this new generation of macromolecular therapeutic agents and the development of viable delivery systems have presented great scientific challenges with a scope far beyond our experience with traditional organic-based pharmaceuticals. The same situation will be faced by the FDA when these nonconventional, innovative drug delivery systems that evolve from these R&D activities are submitted for regulatory approval.

C. Geriatric Medicines

Pharmaceutical R&D activities in the development of geriatric medicines have grown substantially in recent years. In 1990 alone, PMA member firms were projected to spend approximately half the industry's $8.2 billion R&D investment on the development of medicines for treating diseases that commonly affect the elderly population (1). Among the 259 new geriatric medicines they are now testing, 92 are for cancer therapy, 91 for cardiovascular diseases, and 76 for Alzheimer's, arthritis, diabetes, osteoporosis, and other diseases. All these ailments require long-term medication, and for achieving such an objective, controlled-release drug delivery systems are known to have a significant therapeutic advantage over conventional immediate-release dosage forms. Several controlled-release drug delivery systems designed for elderly patients have recently received regulatory approval for marketing, such as the nifedipine-releasing gastrointestinal therapeutic system for the once-a-day management of angina or treatment of hypertension (Chapter 3), the estradiol-releasing transdermal system for postmenopausal syndromes (Chapter 7), the human insulin-zinc suspension and delivery device for the control of diabetes (Chapter 8), and the goserelin acetate-releasing biodegradable implant for the once-a-month subcutaneous palliative treatment of advanced prostate carcinoma (Chapter 11). In view of the intensity of research activity in this area, a growing number of controlled-release drug delivery systems are expected to be developed to provide a more rational therapy of chronic ailment affecting the elderly with improved patient compliance. Regulatory review of these novel pharmaceutical products is expected to grow.

ACKNOWLEDGMENTS

Portions of this chapter were presented at the First International Symposium on Dermal and Transdermal Absorption of Pharmaceutical Substances, Munich, Germany,

January 11–14, 1981; at the 39th International Congress of Pharmaceutical Sciences, Brighton, England, September 3–7, 1979; and at the APhA/APS Midwest Regional Meeting, May 21, 1979, Chicago, by Dr. B. E. Cabana, who also contributed a chapter to the first edition of this book (1).

REFERENCES

1. Pharmaceutical Manufacturers Association, *Facts at a Glance*, Washington, D.C., 1990.
2. J. DiMasi (Tufts University), personal communication, 1991.
3. H. Grabowski and J. Vernon, Duke University, 1991.
4. B. E. Cabana and Y. W. Chien; Regulatory considerations in controlled-release medication, in: *Novel Drug Delivery Systems* (by Y. W. Chien), Dekker, New York, 1982, Chapter 10.
5. B. E. Cabana and C. S. Kunkumian; A view of controlled release dosage forms, in: *Proceedings of Industrial Bioavailability and Pharmacokinetics,* University of Texas, Austin, TX, February 1974.
6. W. H. Pitlick, B. E. Cabana and M. C. Meyer; Absorption kinetics and bioavailability of nine sustained-release papaverine products, in: *Abstract book of APhA Academy of Pharmaceutical Sciences 26th National Meeting,* Kansas City, MO, November 1979.
7. R. A. Upton, J. F. Thiercelin, J. R. Powell, Jr., L. Sansom, T. W. Geuntert, P. E. Coates, P. Ravencroft, V. P. Shah, B. E. Cabana and S. Riegelman; Absorption from some commercial sustained- and enteric-release preparations of theophylline, in: *Abstract book of APhA Academy of Pharmaceutical Sciences 24th National Meeting*, Montreal, Canada, November 1978.
8. R. A. Upton, J. R. Powell, Jr., T. W. Guentert, J. F. Thiercelin, L. Sansom, P. E. Coates and S. Riegelman; Evaluation of the absorption from some commercial enteric release theophylline products. J. Pharmacokinet Biopharm., 8:151 (1980).
9. D. L. Spangler, D. D. Kalof, F. L. Bloom and H. G. Wittig; Theophylline bioavailability following oral administration of six sustained-release preparations. Ann. Allergy, 40:6 (1978).
10. J. Shaw and J. Urquhart; Programmed, systemic drug delivery by the transdermal route. Trends Pharmacol. Sci., pp. 208–211 (April 1980).
11. B. E. Cabana and J. G. Harter; Biopharmaceutic, clinical and toxicological consideration in evaluation of dermal and transdermal drug products, presented at the *1st International Symposium on Dermal and Transdermal Absorption,* sponsored by the International Association for Pharmaceutical Technology. Munich, Germany, January 12–14, 1981.
12. Y. W. Chein; *Transdermal Controlled Systemic Medications,* Dekker, New York, 1987, Chapters 16–18.
13. D. M. Cohen and A. L. Golub; F.D.A. perspectives on drug delivery: An exclusive interview with C. Peck, M.D. Controlled Release Newsletter, pp. 1–6, (February 1991).
14. Pharmaceutical Biotechnology Monitor, 1(1):1 (1991).

Index

Milton Keynes UK
Ingram Content Group UK Ltd.
UKHW050308111024
449327UK00044B/2521